Commercial Ornamental Crops
Traditional and Loose Flowers

The Editors

Dr. R.L. Misra served the cause of floricultural science for more than 16 years in temperate regions of Himachal Pradesh and about 24 years in sub-tropical conditions of New Delhi. After his joining at IARI, New Delhi in 1984, he was associated with all the 7-8 floricultural courses being offered to the M.Sc. and Ph.D. students, apart from looking his mandated research programmes. Out of his whole official career, as a chairman, he guided 16 Ph.D. and 6 M.Sc. students apart from being members of the advisory committees of many more students, and developed 30 gladiolus varieties which have high demand countrywide. Now all these students are serving various institutions as Professors and Associate Professors and in other departments holding a very high position. His contributions are acclaimed worldwide. As a teacher as well as a researcher, his contribution is well known as he has in all some 500 publications out of which some 170 are research publications, 80 symposium papers, 80 book chapters, a few review articles including 6 books. He is the founder member of Indian Society of Ornamental Horticulture from 1993 to 2004 (12 years continuously) he edited its regular publication Journal of Ornamental Horticulture. He is Fellow of Indian Society of Genetics and Plant Breeding since 80's, and Indian Society of Ornamental Horticulture since 2010. For his most significant contribution in the field of floriculture, he was honoured with 'Gold Medal' of Horticultural Society of India in 2013. He is life member in 11 professional societies. Once he was Officer-in-Charge, Indo-Israel Project of Research and Development on Farms, IARI, New Delhi which confines to the protected cultivation of vegetables and flowers; and joined as a regular Project Coordinator of All India Coordinated Research Project on Floriculture of ICAR at Division of Floriculture and Landscaping, IARI, New Delhi where he coordinated some 21 centres across 17 states. He served IARI for 40 years and retired as Project Coordinator *cum* Principal Scientist in May 2008. Only because of his highly significant contributions in the field of teaching and research, after his retirement he was inducted as adjunct Faculty of IARI in Floriculture.

Dr. Sanyat Misra is a graduate from UAS, Bangalore; postgraduate from GBP University of Agriculture and Technology, Pantnagar; and Ph.D. from Dr. YSP University of Horticulture and Forestry, Solan (H.P.), and then after completing his Ph.D. degree served for 3 years as Research Associate at NBPGR, New Delhi and Shimla, and in 2004 joined as regular Scientist *cum* Assistant Professor (Floriculture) in the Department of Horticulture, Birsa Agricultural University, Kanke, Ranchi-834006 (Jharkhand). So far he has guided three M.Sc. students for their thesis work. He is a very active member of the horticulture faculty looking after research, teaching and extension. He has to his credit about 20 publications, two bulletins and three books of international repute.

Commercial Ornamental Crops
Traditional and Loose Flowers

– Editors –

R. L. Misra

Sanyat Misra

2023

Daya Publishing House®
A Division of
Astral International Pvt. Ltd.
New Delhi – 110002

ISBN: 9789359191027 (HB)

Published by : **Daya Publishing House®**
A Division of
Astral International Pvt. Ltd.
– ISO 9001:2015 Certified Company –
4736/23, Ansari Road, Darya Ganj
New Delhi-110 002
Ph. 011-43549197, 23278134
E-mail: info@astralint.com
Website: www.astralint.com

Trust for Advancement of Agricultural Sciences

Avenue-II, Indian Agricultural Research Institute, New Delhi - 110 012
Phone: 011-65437870 Telefax: 011-25843243
E-mail: taasiari@gmail.com Website: www.taas.in

taas
Progress Through Science

Dr. R.S. Paroda
Founder Chairman

Foreword

Since the times immemorial, the plants and flowers have remained an integral part of Indian rituals and source of ecofriendly pigments, besides being a source of chemicals of industrial and medicinal use. India is bestowed with wide range of agroclimatic conditions that support rich plant diversity.

In the recent years horticulture in India has fast developed on commercial lines and today we are among the leading producers of fruits, vegetables, flowers and spices. In the present scenario floriculture holds potential for agricultural diversification and improving farmer's income in the otherwise predominant cereal based farming. Though India is among the leading countries in flower cultivation but our share in global flower trade is less than one percent. In India, floriculture as a subject was introduced at UG and PG level in early 1950s. The establishment of All India Coordinated Research Project on Floriculture, brought the research efforts under one umbrella. In recognition of increasing importance of floriculture in India the AICRP later on was upgraded as into Directorate of Floriculture Research at Pune.

The authors, Drs. R.L. Misra and Sanyat Misra, have already brought out two books: (i) **Commercial Ornamental Crops: Cut Flowers** and (ii) **Commonly Used Ornamental Plants**, both published from Kruger Brentt, England, and now current publication **Commercial Ornamental Crops: Traditional and Loose Flowers** is the third book in the series to be published from the same publisher.

The book **"Commercial Ornamental Crops: Traditional and Loose Flowers"** deals exclusively on production technology of 29 most commonly cultivated ornamental plants and related aspects of post harvest management for various commercial ventures.

It is expected that the book will cater to the needs of students, teachers, researchers, horticultural training centres and development officers engaged in the field of horticulture, and over all to the growers to generate more income, and wish the team all success in their all the future endeavours.

(R. S. Paroda)
Chairman, TAAS
Former **Director General**, ICAR and
Secretary, DARE, Govt. off India

Preface

Since the time immemorial, the flowers in Asian system are being used traditionally irrespective of the caste, creed and religions. There are many plants and trees which are therefore attached with our day-to-day life and many a plants have been associated with gods and goddesses. All over India and elsewhere, people from different walks of life, irrespective of their sex, faith, creed and caste grow some ornamentals near their homes to be used in needs. Also in the villages, mostly there used to be one patch where local flowers grow on their own from where villagers used to pluck them for ceremonial occasions and for offering to gods and goddesses. The trees or shrubs emitting pleasant odours especially by the sunset used to be planted in villages. Then there was no flower market, no flower business or any other thing alike and the people used to collect them unhesitantly from anywhere, irrespective of the fact that who is growing these. For making the look of the house and its surrounding beautiful, people were and are using ornamentals by planting around. Even in the orchard one can see such plantings and there had been no restrictions that who will use them. These are *Amaranthus, Amaryllis, Barleria, Bougainvillea, Calotropis, Canna, Catharanthus, Celosia, Clerodendron, Codiaeum, Crossandra, Dahlia, Dendranthema, Eranthemum, Ervatamia, Gaillardia, Helianthus, Heliconia, Hibiscus, Impatiens, Ixora, Jasminum,Narcissus, Nerium, Pandanus, Polianthes, Rosa, Tagetes, Thevetia* and so on, which people used to grow and enjoy, and out of which a few were and are being used even for extraction of essential oils. These flowers were being plucked or harvested to be used for certain arrangements, decorations, for making garlands or for offering at the religious ceremonies or meant for social celebrations. In Thailand no marriage is solemnized without exchange of floral garlands of *Calotropis* between brides and bridegrooms. Loose flowers are, in fact, not really the loose ones as the name spells, they may be quite compact but the people pluck them from the parent plant without stalks for net instant use. Since there was no concept in Asian countries for enjoying them indoors by arranging them in flower vases so such flowers were without a part of the stalk as in the cut flowers but now cut flowers have also achieved an important place in the country. However, Asians residing in western countries, for their traditional use on certain occasions, import loose flowers from India. This country has several recognized markets in different cities. The cut flower concept we borrowed from western cultures. Here in this book, we have put all our efforts to include almost all the important ornamentals which not only in India are being grown but world over. We have tried to furnish here their 'A to Z' proper cultivation practices in a scientific manner. Moreover, there are various other chapters dealing with all branches of flowers to make this book truely comprehensive.

Indian Council of Agricultural Research, New Delhi has formulated an uniform course programme in Floriculture for Bachelor, Master's and Doctoral programmes, therefore, this book has been written to avoid any confusion that what are important **Traditional and Loose Flowers**, the second publication towards **Commercial Flower Series**. It is being followed by the third publication, **Ornamental Plants in most Common Use**., and fourth **Soilless Culture.** Mostly, even the teachers confuse that apart from barleria, crossandra, jasmine, marigold, rose and tuberose, is there any other ornamental to be used as traditional and loose flower or what is the difference between loose and cut flowers. Yes there is, hence this book as from those formulated in the course, we have added many other regional flowers to substitute a few as suggested in the course contents. We assure that not only to the students and their professors, this book will be useful but will also serve an useful guide to the scientific community of floriculture, the landcapers and horticulturists, the growers and other garden enthusiasts. As All the crops are of universal appeal, hence this will certainly be appreciated worldwide. The book is first of its kind exclusively written on **Traditional and Loose Flowers**.

Though a few of the contributed chapters were found satisfactory, but certain chapters were so casual that we had to improve to a tune of some 90 per cent, therefore, hence forth the contributors are requested to be more careful.

R. L. Misra

Sanyat Misra

Contents

Part II: General Aspects

List of Contributors

Ajit Kumar

(*Amaranthus,* Flower Seed Production, *Gaillardia, Gomphrena, Helianthus,* Insect-Pests of Ornamentals and their Management, *Tagetes*)

Department of Horticulture, G.B. Pant University of Agriculture & Technology, Pantnagar-263145, Distt. Udham Singh Nagar (Uttarakhand).

Alka Singh

(*Hymenocallis,* Flower Arranging and Designing)

ASPEE College of Horticulture and Forestry, Navsari Agricultural University, Navsari-396450.

Anirban Roy

(Diseases of Ornamentals and their Management)

Advanced Centre for Plant Virology, Division of Plant Pathology, IARI, New Delhi-110 012.

Atul Batra

(*Asclepias* & *Calotropis, Dahlia, Ervatamia divaricata, Nerium, Thevetia*)

Floriculture section, National Botanical Research Institute, Rana Pratap Marg, Lucknow-226001.

Babita Singh

(*Nyctanthes arbor-tristis*)

Department of Horticulture, G.B. Pant University of Agriculture & Technology, Pantnagar-263145, Distt. Udham Singh Nagar (Uttarakhand).

Banerji, B.K.

(*Asclepias* & *Calotropis, Dahlia, Ervatamia divaricata, Nerium, Thevetia*)

Floriculture Section, National Botanical Research Institute, Rana Pratap Marg, Lucknow-226001. E-mail: banerjibk@yahoo.co.in

Bhavya Bhargava

(Bio-colours from Ornamentals)

Institute of Himalayan Bioresource Technology, CSIR, Palampur-176061 (H.P.).

Chandrashekar, S.Y.

(*Eranthemum*)

Department of Floriculture and Landscape Architecture, College of Horticulture, Mudigere-577132, Chikmangalur dist. (Karnataka).

Dubey, R.K.

(*Lawsonia inermis*; Post Harvest Management of Loose Flowers, Cut Flowers and Cut Foliage)

Department of Floriculture and Landscaping, P.A.U., Ludhiana-141004 (Punjab).

Dwivedi, A.K.

(*Asclepias* & *Calotropis, Dahlia, Ervatamia divaricata, Thevetia*)

Institute of Himalayan Bioresource Technology, CSIR, Palampur-176061 (H.P.).

Ganga, M.

(*Barleria, Crossandra, Jasminum*)
Horticultural College and Research Institute, T.N.A.U., Coimbatore-641003 (T.N.).

Grewal, H.S.

(Genetically-Engineered Ornamentals, Light in Regulating Growth and Flowering)
Department of Floriculture and Landscaping, Punjab Agricultural University, Ludhiana-141 004.

Gupta, Y.C.

(Bio-colours from Ornamentals)
Department of Floriculture and Landscaping, Dr Y S Parmar University of Horticulture and Forestry, Nauni, Distt.- Solan (H.P.).

Hada, B.S.

[Nutraceutical Pigments from Ornamental Crops (Scope, Potential and *In Vitro* Expression)]
Division of Floriculture and Landscaping, Indian Agricultural Research Institute, New Delhi-110012.

Himanshu Trivedi

(*Helianthus*)
Department of Horticulture, G.B. Pant University of Agriculture & Technology, Pantnagar-263145, Distt. Udham Singh Nagar (Uttarakhand).

Indra Mani

(Mechanization of Microbial Composting Technology for Commercial Productin to Address Depleting Soil Health)
Division of Agricultural Engineering, I.A.R.I, New Delhi-110012

Jagadev, P.N.

(*Pandanus*)
Department of Plant Breeding and Genetics, O.U.A.T., Bhubaneswar-751003 (Odisha).

Janakiram, T.

[Nutraceutical Pigments from Ornamental Crops (Scope, Potential and *In Vitro* Expression), Nutrient Management in Floral Crops]
Division of Floriculture and Landscaping, Indian Agricultural Research Institute, New Delhi-110012.

Jayoti Majumder

(Edible Flowers, *Hibiscus, Ixora, Nyctanthes arbor-tristis, Paeonia,Rudbeckia*)
Faculty of Horticulture, Department of Floriculture and Landscaping, B.C.K.V., Mohanpur-741252, Nadia (W.B.).

Jegadeeswari, V.

(*Barleria, Crossandra, Jasminum*)
Horticultural College and Research Institute, T.N.A.U., Coimbatore-641003 (T.N.).

Jyoti Bajeli

(Flower Seed Production, *Helianthus*)
Department of Horticulture, G.B. Pant University of Agriculture & Technology, Pantnagar-263145, Distt. Udham Singh Nagar (Uttarakhand).

Kushal Singh

(Post Harvest Management of Loose Flowers, Cut Flowers and Cut Foliage)
Department of Floriculture and Landscaping, P.A.U., Ludhiana-141004 (Punjab).

Legha, M.R.

[Nutraceutical Pigments from Ornamental Crops (Scope, Potential and *In Vitro* Expression)]
Division of Floriculture and Landscaping, Indian Agricultural Research Institute, New Delhi-110012.

Livleen Shukla

(Mechanization of Microbial Composting Technology for Commercial Productin to Address Depleting Soil Health)
Division of Microbiology, I.A.R.I, New Delhi- 110012

Mamta Bohra

(Flower Seed Production)
Department of Horticulture, College of Agriculture, G.B. Pant University of Agriculture & Technology, Pantnagar-263145, Distt. Udham Singh Nagar (Uttarakhand).

Manas Kumar Bag

(Diseases of Ornamentals and their Management)
Germplasm Evaluation Division, NBPGR, Pusa Campus, New Delhi-110 012.

Manish Kapoor

(Flower Seed Production, *Gaillardia, Gomphrena, Helianthus,Tagetes*)

Department of Horticulture, G.B. Pant University of Agriculture & Technology, Pantnagar-263145, Distt. Udham Singh Nagar (Uttarakhand).

Manju Rana

(*Gaillardia, Gomphrena, Tagetes*)

Department of Horticulture, G.B. Pant University of Agriculture & Technology, Pantnagar-263145, Distt. Udham Singh Nagar (Uttarakhand).

Mekala, P.

(*Barleria, Crossandra, Jasminum*)

Horticultural College and Research Institute, T.N.A.U., Coimbatore-641003 (T.N.).

Misra, R.L.

(*Amaranthus,* Annual *Crysanthemum, Asclepias* & *Calotropis, Celosia argentea, Crossandra, Dahlia, Dianthus,* Diseases of Ornamentals and their Management, Flower Drying, *Hibiscus,* Insect-Pests of Ornamentals and their Management, *Ixora, Jasminum,* Light in Regulating Growth and Flowering, *Nelumbo, Nerium, Nyctanthes arbor-tristis, Nymphaea, Paeonia, Polianthes tuberosa, Rosa, Rudbeckia, Tagetes, Thevetia*)

(Ex-Project Coordinator, AICRP on Floriculture, ICAR; & Former Officer-in-Charge, Indo-Israel Project on R&D, IARI, New Delhi-110 012) C-4, Brahma Apartments, Plot 7, Sector 7, Dwarka, New Delhi-110 075 (Mob.: 9968287841).

Namita

(Nutrient Managemenat in Flower Crops)

Division of Floriculture and Landscaping, I.A.R.I., New Delhi-110 012.

Nataraj, S.K.

(*Eranthemum*)

Department of Floriculture and Landscape Architecture, College of Horticulture, Mudigere-577132, Chikmangalur dist. (Karnataka).

Padmadevi, K.

(*Barleria, Crossandra, Jasminum*)

Horticultural College and Research Institute, T.N.A.U., Coimbatore-641003 (T.N.).

Pardeep Kaur

(Post Harvest Management of Loose Flowers, Cut Flowers and Cut Foliage)

Department of Floriculture and Landscaping, P.A.U., Ludhiana-141004 (Punjab).

Pragya Ranjan

(*Amaranthus, Celosia argentea,* Diseases of Ornamentals and their Management, Insect-Pests of Ornamentals and their Management)

Indian Institute of Vegetable Research, P.O. Jakhani, Varanasi-221 305.

Prasad, K.V.

[Annual *Chrysanthemum*, Nutraceutical Pigments from Ornamental Crops (Scope, Potential and *In Vitro* Expression)]

Directorate of Floricultural Research, M.P.K.V. Campus, Pune (Maharashtra).

Prativa Lakhotia

[Nutraceutical Pigments from Ornamental Crops (Scope, Potential and *In Vitro* Expression)]

Division of Floriculture and Landscaping, I.A.R.I., New Delhi-110012.

Priyanka Sharma

(Bio-colours from Ornamentals)

Department of Floriculture and Landscaping, Dr Y S Parmar University of Horticulture and Forestry, Nauni, Distt.- Solan (H.P.).

Pushpalatha, G.

(Biotechnological Interventions in Floriculture)

Department of Bioitechnology, Centurion University, Paralakhemudi, Gajapathi district. Odisha-761211

Raghava, S.P.S.

(*Celosia argentea*)

(Ex-Project Coordinator, AICRP on Floriculture; and Former Head , Division of Floriculture and Landscaping, IARI, New Delhi-110 012) 3/80, Sector 2, Rajinder Nagar (Sahibabad), Ghaziabad-201 005.

Ramesh Kumar

(*Amaranthus*)

Directorate of Floricultural Research, M.P.K.V. Campus, Pune (Maharashtra).

Ranjan, J.K.

(*Celosia argentea*)

Indian Institute of Vegetable Research, P.O. Jakhani, Varanasi-221 305.

Ranjan Srivastava

(Flower Seed Production, *Gomphrena*)

Department of Horticulture, G.B. Pant University of Agriculture & Technology, Pantnagar-263145, Distt. Udham Singh Nagar (Uttarakhand).

Ranjit Singh

(Post Harvest Management of Loose Flowers, Cut Flowers and Cut Foliage)

Department of Floriculture and Landscaping, P.A.U., Ludhiana-141004 (Punjab).

Rao, V.K.

(*Tagetes*)

Department of Horticulture, G.B. Pant University of Agriculture & Technology, Pantnagar-263145, Distt. Udham Singh Nagar (Uttarakhand).

Ritu Jain

(Nutrient Managemenat in Flower Crops)

Division of Floriculture and Landscaping, I.A.R.I., New Delhi-110 012.

Saha, T.N.

(*Hibiscus, Paeonia*)

Directorate of Floricultural Research, M.P.K.V. Campus, Pune (Maharashtra).

Sanjeev Kumar Sharma

(*Gaillardia, Gomphrena*)

Department of Horticulture, G.B. Pant University of Agriculture & Technology, Pantnagar-263145, Distt. Udham Singh Nagar (Uttarakhand).

Sanyat Misra

(*Amaranthus,Asclepias* & *Calotropis, Barleria, Celosia, Crossandra, Dahlia, Dianthus,* Diseases of Ornamentals and their Management, Edible Flowers, *Ervatamia divaricata,* Flower seed Production, *Gaillardia, Gomphrena, Helianthus, Hibiscus, Hymenocallis,* Insect-Pests of Ornamentals and their Management, *Jasminum, Lawsonia inermis,* Light in Regulating Growth and Flowering, *Nelumbo, Nerium, Nyctanthes arbor-tristis, Nymphaea, Paeonia, Polianthes tuberosa, Rosa, Rudbeckia, Tagetes, Thevetia*)

Department of Horticulture, Birsa Agricultural University, Kanke, Ranchi-834 006 (Jharkhand).

Sapna Panwar

(Nutrient Managemenat in Flower Crops)

Division of Floriculture and Landscaping, I.A.R.I., New Delhi-110 012.

Sashikala Beura

(*Pandanus*)

Biotechnology-cum-Tissue Culture Centre, O.U.A.T., Bhubaneswar-751003 (Odisha).

Sathyanarayana Reddy, B.

(*Eranthemum*)

Department of Floriculture and Landscape Architecture, College of Horticulture, Mudigere-577132, Chikmangalur dist. (Karnataka). Dr.reddybs@gmsil.com

Satish Chand

(*Gaillardia*)

Department of Horticulture, College of Agriculture, G.B. Pant University of Agriculture & Technology, Pantnagar-263145, Distt. Udham Singh Nagar (Uttarakhand).

Satish Lande

(Mechanization of Microbial Composting Technology for Commercial Production to Address Depleting Soil Health)

Division of Agriculture Engineering, I.A.R.I., New Delhi-110012

Sellam Perinban

(Edible Flowers, *Ixora, Paeonia, Rudbeckia*)

Division of Agricultural Engineering, I.A.R.I., New Delhi-110 012.

Singh M.K.

(Polianthes tuberosa)

Division of Floriculture and Landscaping, I.A.R.I, New Delhi-110012

Sunder Pal

(Flower Seed Production,)

Department of Horticulture, College of Agriculture, G.B. Pant University of Agriculture & Technology, Pantnagar-263145, Distt. Udham Singh Nagar (Uttarakhand).

Surendra Kumar

[Nutraceutical Pigments from Ornamental Crops (Scope, Potential and *In Vitro* Expression)]

Division of Floriculture and Landscaping, Indian Agricultural Research Institute, New Delhi-110012.

Tanya Thakur

(Genetically-Engineered Ornamentals, Light in Regulating Growth and Flowering)

Department of Floriculture and Landscaping, Punjab Agricultural University, Ludhiana-141 004.

Part I

Crops

1

Amaranthus

Ramesh Kumar, Sanyat Misra, Ajit Kumar, Pragya Ranjan and R.L. Misra

Family: **Amaranthaceae**

Common names: Amaranthus, Fountain plant (*A. salicifolius*), Joseph's coat (*A. tricolor*), Cats's tail/Love-lies bleeding/Queue de Renard (Fox-tail in French) (*A. caudatus*), Prince's feather (*A. hypochondriacus*)

Introduction

In Greek '*a*' means without and '*marain*' means to wither, also *amarantos/amaranthos* meaning unfading as the flower clusters of some species retain their colour like-everlastings. It is native to temperate, sub-tropical and tropical regions of the world. Amaranths have been popular garden plants since the 16th century. Though there are some 60 species of annuals but only 3-4 are of ornamental importance for growing during summer, some for their brightly coloured tail-like clusters of densely packed tiny flowers, especially the red-flowered ones while certain others for their beautifully coloured foliage. The leaves and tender stem parts even of the ornamental species are edible. Those cultivated in gardens are mostly hybrids. These are excellent for summer borders or as pot plants. These also last longer in ordinary water as cut flower, especially *A. caudatus* and *A. hypochondriacus*, and the green-flowered *A. caudatus* 'Viridis' is very suitable for flower arrangers and for drying, and its carefully dried cut flowers retain their colour for a considerable time. Usually, the life of cut blooms is for a week.

Botany

The genus *Amaranthus* consists of some 60 species from tropical and temperate regions (Bailey, 1942; Hay and Beckett, 1971; Pizzetti and Cocker, 1975; Beckett, 1983; Anthony Huxley *et al.*, 1992) of weedy annuals, monoecious or dioecious herbs, erect, spreading or prostrate, often glandular-pubescent; leaves alternate, often pilose on veins beneath, entire or sinuate, lanceolate to ovate and sometimes brightly coloured; petioles long; flowers with 3-5 persistent sepals but without petals, unisexual and tiny where each flower is subtended by 3 chaffy bracts, bright red or green in terminal or axillary, dense and spike-like clusters, and catkin-like cymes; bracts 3, at base of perianth; perianth segmernts 3-5, free or fused below, persistent, often rudimentary in female flowers; stamens 3-5, free to base, numbers as perianth segments; ovary ovoid to lenticular, ovule 1; stigmas 2-3; styles 1-3; fruit a small, indehiscent utricle; seeds smooth to minutely verrucose. For chromosome numbers, *Amaranthus cruentus* has been recorded with 2n = 32 while *A. tricolor* with 2n = 34.

Propagation

Seeds of amaranths are sown in seed compost in February-March at a temperature of 15-24 ^{0}C where these germinate within 10 days, and the seedlings are planted out in April-May. These may successfully be sown directly at the site of flowering in March-April and when have attained some 5 cm height these are thinned out to required spacing. A humid atmosphere and regular watering will enhance growth of the plants for seed production. After about one month of full flowering and when plants are turning off-colour, the plants are harvested for seed extraction. These are dried and then threshed for seed extraction. After thorough seed drying, the seeds are packed in moisture-proof envelops and cold-stored. Uptake of oxygen by seeds of *Amaranthus hypochondriacus* which is grown as summer vegetables and as a flowering ornamental plant in Korea showed a marked peak between 9 and 24 h after the beginning of imbibition in darkness in red light with 1000 micro l/litre ethephon in the light or with ethephon in red light but did not show this peak in light alone, however, carbohydrate and protein contents decreased less in light than with other treatments from 6 to 9 h after the beginning of imbibition (Kang and Kwack, 1990). They further observed that total sugar content decreased less rapidly, though reducing sugar and amino acid contents remained more or less the same, in the light, but both were found increased in other treatments. The respiratory enzymes (hexokinase, glucose-6-phosphate dehydrogenase and 6-phosphogluconate dehydrogenase), *vis-à-vis* alpha-amylase and catalase activities increased during germination in almost all the treatments except after imbibition in the light without ethephon. They did not observe any difference in the bands produced by electrophoresis of soluble proteins but crossed immunoelectrophoresis showed greater protein degradation in germinating seeds.

Species and Varieties

The **species** of ornamental importance are ***A. caudatus*** (syn. *A. paniculatus, A. sanguineus*), a most popular species native to India and other tropical regions, growing to 90-120 cm high, leaves entire, ovate and pale-green, and it is valued for its more than 45 cm or much longer drooping racemes of clustered crimson flowers with catkin-like tails which sometimes touch the ground, appearing from June to October. During autumn the stems also turn crimson. Its vars 'Green Balls' bears ball-like clusters while 'Viridis' (flowers pale-green) is excellent in flower arrangement. ***A. cruentus*** cv. 'Kerala Red' with intense red colouring was discovered in a home garden in Ames (Iowa) in well adapted form which was being grown there through seeds originally introduced from Kerala (India). It is edible as vegetable and as salad being unusually mildly flavoured but is most favoured as an attractive bedding plant growing some two metres tall (Brenner and Makus, 1997). ***A. hybridus*** (syn. *A. hypochondriacus*), a native of America is a hybrid species growing to 90-150 cm high. Though it is similar to *A. caudatus* but bears coppery flushed foliage and erect plumes of blood-red flowers. Sometimes, ***A. hypochondriacus*** is referred to as a different species than *A. hybridus* though both are similar as many genuine publications refer these two as synonyms with erect flowering tails. ***A. salicifolius***, a native to Japan but introduced to Europe in 1871, and this grows some 90 cm in height bearing long pendulous willow-like leaves that change from bronze-green to orange-red when maturing. ***A. tricolor*** (syn. *A. gangeticus*), a native of India grows to 90 cm tall, is strong growing and has many hybrid forms being valued for their attractively coloured, 20 cm or more long and ovate foliage. The leaves are pointed, ovate, scarlet or crimson overlaid with yellow, bronze and green colouration. Floral racemes are dense, erect and brush-like, more than 15 cm long, often fasciated and deep red in colour but hidden, appearing from June to September. The species and the forms of this species make striking bedding and pot plants. Its var. 'Aurora' bears red, yellow and dark green leaves, var. 'Joseph's Coat' bears leaves patterned bronze, red, golden-yellow and deep green, var. *marmoratuus* bears green and yellow leaves mottled pink, var. *melancholicus ruber* (syn. *A. bicolor*) from Japan is a compact growing crimson-leaved form, var. *tricolor* has curiously coloured leaves with the base being crimson-scarlet and gradually merging to bright-yellow in the centre and green at the point, often confused with 'Joseph's Coat', var. 'Molten Fire' has deep copper-crimson leaves with upper leaves scarlet and 'Salicifolius' bears willow-shaped drooping and lanceolate leaves of orange, carmine and bronze colours.

After studying the RAPD profiles of various species and accessions of *Amaranthus* using 65 arbitrary sequence decamer primers, Ranade *et al.* (1997) drew an unweighted pair group method with arithmetic means (UPGMA) dendograms illustrating the inter- and intra-specific relationships based on the RAPD data and concluded that the putative hybrid origin of *A. dubius* from *A. hybridus* and *A. spinosus* can not be supported.

Cultural Practices

Amaranths excell in beddings and in pots as specimen plants. Being of tropical interests, these dislike cold. For their cultivation, these require sunny situation. Mattson and Erwin (2005) grew *Amaranthus hybridus* at 8 h photoperiod, ambient irradiance averaging 12.0-13.2 and 6.4-8.3 mol m^{-2} day^{-1} with or without supplemental high-pressure sodium lighting (+50, 100 or 150 micro mol m^{-2} s^{-1}), or under long days using natural daylengths and irradiance with night interruption lighting (2200-0200 h at 2 micro mol m^{-2} s^{-1} from incandescent lamps) or under ambient daylight + supplemental irradiance during day and as a day extension to 18 h (0800-0200 h) with supplemental high pressure sodium lighting (+50, 100 or 150 micro mol m^{-2} s^{-1}) to know the impact of photoperiod and irradiance on flowering, and obseved least mean dry weight gain per day (MDWG), i. e. 0.01 g day^{-1} across photoperiod and irradiance levels. For its successxful cultivation the soil should be sandy-loam with a pH range of 5.5 to 7.5. Soil should be rich in organic matter and quite deep, not less than 30 cm. Some 150 quintals/ha of farmyard manure should be mixed in the soil at the time of final ploughing. Before planting or direct sowing the soil should be well pulverized and should have sufficient moisture so that it may require watering only after germination. The beds of convenient sizes are made with ample provision of bunds and channels for tillage practices and for watering. The beds should be properly levelled. The size is normally kept 1.0-1.6 metres in a length of convenient size. A humid atmosphere and regular watering will deepen the leaf colour of the plants. In an experiment with *Amaranthus cruentus* taking three NPK fertilization rates, Skwarylo-Bednarz *et al.* (2011) in Poland found 90 kg N, 60 kg P and 60 kg K/ha giving highest yield of herbage and seeds irrespective of spacings and as compared to its higher level application, i.e. 130 kg N, 70 kg P and 70 kg/ha K. They further noted higher Cu, Zn and Fe contents in leaves and stems, and higher Cu and Mn contents in seeds due to NPK fertlization. For pot culture, its seeds may be sown in seed compost into 12-15 cm pots in February-March at a temperature of 13-15 ^{0}C. When half of the spikelets have started flowering these may be harvested for vase decoration, and slightly later for drying. For drying these are normally hanged in inverted postion. When after flowering the plants start showing the dull colours and yellowing of leaves these should be harvested for seed extraction. Their flowers can be kept preferably at 2-5 ^{0}C.

Insect-Pests and Diseases

The plants are generally trouble-free, however, sometimes aphids are found infesting the crop which may be killed through jet water spraying.

References

Anthony Huxley, Mark Griffiths and Margot Levy, 1992. The New Royal Horticultural Society Dictionary of Gardening (A-C), pp. 145. The Macmillan Press Ltd., London.

Bailey, L.H. 1942. The Standard Cyclopedia of Horticulture, pp. 269-270. The Macmillan Co., New York.

Beckett, K.A. 1983. The Concise Encyclopedia of Garden plants, pp. 21. Orbis Publishing Ltd., London, Great Britain.

Brenner, D.M. and D.J. Makus, 1997. Kerala Red ~ ornamental amaranth. *HortSci.*, **32**(4): 749-750.

Hay, R. and K.A. Beckett, 1971. Reader's Digest Encyclopaedia of Garden Plants and Flowers, pp. 32-35. The Reader's Digest Association Ltd, London.

Kang, H. and B.H. Kwack, 1990. The effects of light and ethephon on the germination of light-inhibited *Amaranthus hypochondriacus* seeds in relation to respiration, stored nutrients and certain enzymatic activities (Korean). *J. Korean Soc. hort. Sci.*, **31**(2): 150-161.

Mattson, N.S. and J.E. Erwin, 2005. The impact of photoperiod and irradiance on flowering of several herbaceous ornamentals. *Sci. Hort.*, **104**(3): 275-292.

Pizzetti, I. and H. Cocker, 1975. Flowers ~ A Guide for Your Garden, pp. 51-52.Harry N. Abrams Inc., Publishers, New York.

Ranade, S.A., Anil Kumar, Mamta Goswami, Nuzhat Farooqui and P.V. Sane, 1997. Genome analysis of amaranths : determination of inter- and intra-species variations. *J. Biosci.*, **22**(4): 457-464.

Skwarylo-Bednarz, B., M.S. Brodowska and R. Brodowski, 2011. Rvaluating the influence of varied NPK fertilization on yielding and microelements contents at amaranth (*Amaranthus cruentus* L.) depending on its cultivar and plant spacing. *Scientiarum Polonorum -Hortorum Cultus*, **10**(4): 245-261.

2

Annual Chrysanthemum

K.V. Prasad and R.L. Misra

Family: **Asteraceae**

Common names: Alecoste (*Chrysanthemum majus*), Corn daisy/Corn marigold (*C. segetum*), Chop suey green/Chrysanthemum greens/Crown daisy/Edible chrysanthemum/Garland chrysanthemum/Japanese green (*C. coronarium*, syn. *Glebionis coronarium, G. coronaria*), Daisy flower/Golden flower (*Chrysanthemum*), Feverfew (*C. parthenium*, syn. *Matricaria eximia*), Paris daisy/Marguerite (*C. frutescens*), Shasta daisy (*C. maximum*), German flag/ Painted daisy/Tricolor chrysanthemum or daisy (*C. carinatum*, syn. *C. atrococcineum, C. tricolor*), Yellow daisy (*C. multicaule*), *etc.*

Introduction

Chrysanthemum genus name owes to J.P. de Tournefort. *Chrysanthemum* is a Greek name made of two words, *chrysos* meaning 'gold' and *anthos* meaning 'flower', i.e. 'golden flower' which is now its common name, and the genus was finally codified by Linnaeus in 1735. This may have been named so, probably owing to common yellow daisy or corn marigold (*Chrysanathemum segetum*). The members of this genus are confined to Eurasia and the Mediterranean area but it has been construed by some in the past to include another 200 species which extend into Africa and America (Everard and Morley, 1970). Annual chrysanthemum is a winter annual that can be grown in herbaceous or annual borders, for bedding, and for loose flowers for various ornamental purposes. The genus *Chrysanthemum* constitutes of more than 200 species belonging to the family Asteraceae (with some 900 genera) which is one of the largest families of flowering plants, but now perennial ones have been put under the genus *Dendranthema* though annual ones being only a few are retained as *Chrysanthemum*. It is native of Asia and Southeastern Europe. Chrysanthemums were first cultivated in China about 3,500 years ago, and at the time of Confucius, about 500 B.C., the chrysanthemums were widely cultivated and very popular, and a thousand years later it became national flower of Japan where it was initially introduced from Korea to Japan in 8th century, i.e. 73rd year of the emperor Nintoku who ascended the throne in A.D. 313, as per Japanese history. In Japan, every year on November 9, a chrysanthemum festival is held in the Imperial gardens. In Europe, the chrysanthemum (*Chrysanthemum coronarium*) was introduced in 17th century (probably 1629). The first chrysanthemums cultivated in European gardens were imported into France from China in 1789, from where it went to England and then to Italy. Annual chrysanthemum produces daisy-like flowers of many different colours, *viz.* yellow, red, white and pink, often the petals having two shades of differing colours. White and yellow blooming cultivars are used for bedding and for making garland, *veni* and home decorations.

Some of the annual or annual-like chrysanthemums and their uses

Scientific name	Common name	Uses
C.carinatum (syn. *C. atrococcineum, C. tricolor*)	Tricolor daisy	Cultivated as bedding plants, with edible leaves
C. coronarium (syn. *Glebionis coronarium, G. coronaria*)	Crown daisy	With black pepper used for treating syphilis and gonorrhea
C. majus	Alecoste	Leaves to flavour the soup and petals for making conserves
C. maximum	Shasta daisy	Evergreen with attractive foliage, ideal as ground cover
C. multicaule	Yellow daisy	Suitable for small pot-planting
C.parthenium	Feverfew	A popular plant though used for dubious headache remedy
C. segetum	Corn marigold	An ideal meadow plant, and good for naturalizing
C. viscosum (*syn. C. viscidi-hirtum*)	Viscous chrysanthemum	Grown for bedding and for naturalizing at sunny location
C. morifolium	Perennial florist's chrysanthemum	Ideal for growing in beds and pots

Note: Here, *C.* × *morifolium* is well known perennial florist's chrysanthemum which has now been placed as *Dendranthema* × *grandiflorum* Kitam. (Anderson, 1987) which shall be described separately in another book though here it has been mentioned just for comparison. *C. parthenium* is also though a perennial species and has also been renamed as *Tanacetum parthenium* (L.) Schultz-Bib. but it is being cultivated as an annual chrysanthemum, hence is being described here. Likewise, *C. frutescens* is another perennial species, renamed as *Argyranthemum frutescens* (L.) Schultz-Bib. but is being cultivated as an annual species, therefore has been included here. Other *Chrysanthemum* species whose names have been revised are *C. balsamita* as *Tanacetum balsamita* L., *C. ptarmiciflorum* as *T. ptarmiciflorum* (Webb and Berlin.) Schultz-Bieb., *C. serotinum* as *Leucanthemella serotinum* (L.) Tzveleu., *C. leucanthemum* as *Leucanthemum vulgare* Lam., *C. cinerariifolium* as *Tanacetum cinerariifolium* (Trev.) Schultz-Bib., *C.* × *superbum* as *Leucanthemum superbum* (J. Ingram) Bergmans ex Kert., *C. vulgare* as *Tanacetum vulgare* L., and *C. macrophyllum* as *Tanacetum macrophyllum* (Wldst. and Kit.) Schultz-Bib. (Anderson, 1987).

Other than Aesthetic Uses

Organic amendments and green manure are potential alternatives to the harmful chemical control means currently used against plant-parasitic nematodes. *Chrysanthemum coronarium* was applied to the soil as a green manure to control the root-knot nematodes (*Meloidogyne incognita, M. javanica*) which significantly reduced nematode infestation of tomato roots and improved plant top fresh weight, both in the greenhouse and in microplots (Meira *et al.*, 2006). Insect antifeedant compounds and antijuvenile hormones from the aerial parts have also been isolated. *C. coronarium* has insecticidal activity against confused flour beetle (*Tribolium confusum*) and many other insects as its flowerheads contain sesquiterpene lactones that are toxic for insects. Floral methanolic extracts are more toxic than those of leaf extracts. Methanolic extracts of all the *Chrysanthemum* species except *C. segetum* were toxic by ingestion or by topical application (flower methanolic extracts of *C. coronarium, C. fuscatum, C. grandiflorum, C. macrotum, C. paludosum* and *C. segetum*; and leaf methanolic extracts of *C. coronarium*) (Haouas *et al.*, 2008). Essential oils contain terpenoids, *viz.* monoterpenes (alpha and beta pinene, camphor) and sesquiterpenes (germacrene D, alpha bisabolene) that have insecticidal properties against stored product-pests (Perez and Pascual-Villalobos, 1999). Pizzetti and Cocker (1975) stated that leaves and petals of *Chrysanthemum vulgare* (*Tanacetum vulgare*) when in full bloom are gathered and dried in shade, produces a bitter substance tanacetone, various acids and an oil. Tanacetin is used as a vermifuge, in cases of kidney debilities and for dyspepsia. The plant itself is quite poisonous and was being used in the folk medicines for abortion. Its poisoning occurs in the form of vomiting, diarrhea, abdominal cramps and nervous system paralysis. The oil extracted from the costmary (*C. balsamita*) grown for herb or condiments, is used in cases of chronic gall bladder problems, catarrh, coughs, agitation and insomnia. *C. parthenium* is used in case of bruises and dermal ulcers. *C. cinerariifolium* (pyrethrum) flowers contain pyrethrin insecticide which is sold in dry powder form and is used as insecticide, though is not injurious to human beings and animals. Other *Chrysanthemum* species such as *C. carneum, C. caucasicum, C. coccineum, C. parthenium* and *C. roseum* are also used for preparation of insecticides (Pizzetti and Cocker, 1975).

Alpha-cadinol, murolol, beta-elemene and beta-myrcene of the flower heads have antibacterial activity against gram positive and gram negative bacteria and also have a significant antifungal activity against *Candida albicans*. The extracts also have antimicrobial and/or anti-HSV-1 activities. Some of the species are used medicinally to cure influenza, liver and menstrual disorders and have anti-inflammatory and antispasmodic effects.

Botany and Biotechnology

The flower is an inflorescence with compact head of small flowers (florets), underneath surrounds by whorls of bracts called an involucre which function as calyx, and in many cases the florets are of two types, first in the outer whorls usually strap-shaped and petal-like (ray floret), and the second less prominent and often tubular (disc floret) and here ovaries and stamens are located, or sometimes all the florets may be tubular or all strap-shaped so they may be hermaphrodite or only female. The anthers in the centre of the florets are fused into a tube around the style which normally has a brush-type or ring of hairs which help sweeping the pollen out of the tube formed by the anthers so that these may be picked up by any insect crawling over the flowerhead. Annual chrysanthemum is though self-fertile but is generally cross-pollinated, pollinated by bees, flies, beetles, butterflies or moths. Interspecific and intergeneric hybridization has efficiently been used with the aid of embryo-rescue techniques such as embryo and ovule culture to breed novel cultivars. Intergeneric crossing between marguerite (*Argyranthemum frutescens*) as a seed parent and annual chrysanthemum (*C. carinatum*) or crown daisy (*C. coronaria*) as the pollen parent yielded novel plants in flower shape and colour by applying ovule culture method. Some also showed novel flowering habits such as perpetual flowering (Ohtsuka and Inaba, 2008).

Efficient shoot regeneration and selection are essential for ***Agrobacterium*-mediated genetic transformation** though success with *Chrysanthemum coronarium* has been limited. Transgenic plants were obtained by first co-culturing of leaf discs with *A.tumefaciens* strain EHA105, which harbours plasmid pRCVII containing the hygromycin resistance *(hpt)* and ~ glucuronidase *(GUS)* genes. The optimal concentrations of NAA and BA were 0.2 mg/l and 0.5 mg/l, respectively. Hygromycin at 5 mg/l proved adequate as a selectable marker, whereas 500 mg/l carbenicillin was effective in eliminating excessive *Agrobacterium* after co-cultivation (Chung and Park, 2005).

Species and Cultivars

Chrysanthemum *carinatum* (*C. atrococcineum, C. tricolor*) is a non-hardy evergreen herbaceous annual native to North Africa, especially Morocco, attaining 60 cm height, sometimes growing even up to one metre tall with long, erect, rigid and multi-branched stems. It was introduced into European gardens in 1796. The plants of the species have monopodially branched stem bearing somewhat fleshy, bipinnatifid, deeply cut, graceful, stiff, bright green and 5-8 cm or more long leaves. The plant flowers freely carrying numerous hemispherical terminal flowerheads (capitula) some 5.0-6.25 cm across which are attached to a long peduncle. The flowers are mostly single though in some cases even double and excellent for cutting. The floral disc is purple and ray florets white with a yellow base, banded with different colours. The flowering season is from April to June in the sub-tropical parts whereas from June to September in temperate parts and the fruits are cypsela. This species has numerous beautiful varieties such as 'Court Jesters' (red, yellow or white flowers zoned with crimson, bronze and gold), 'Pole Star' (large white flowers zoned with primrose-yellow), 'Rainbow Mixed' (combinations of yellow, scarlet, red, bronze, lavender, white and so on), 'Torch' (flowers bronze and red zoned with crimson and purple), 'Merry Mixed' (various colours, large single flowers), 'Monarch Court Jesters' (flowers in various colours, large and single), 'Northern Star' (single, large, pure white with yellow zones), 'Dunnett's Mixed' (profusion of double flowers in various colours), *etc.*

C. coronarium (*Glebionis coronaria*, syn. *G. coronaria*) is a diploid herbaceous annual native to Mediterranean regions and East Asia with 9 as basic chromosome number (Razaq *et al.*,1988). It is hardier, vigorous, erect-growing with much-branching habit, taller (up to 1.2 m), leaves very attractive, pale-green, deeply cut and much-divided, and the flowers are 3.75- 5.0 cm across in various shades of yellow, golden-yellow, cream and white with single, semi-double or double forms, which appear very freely from July to August. They are hermaphrodite in nature. The plant is self-fertile and seed propagated. Seeds are rich in chlorogenic acid (a type of hydroxycinnamic acid), carotene, flavonoids, vitamins and potassium, and can offer a multitude of health benefits. Some of the beneficial effects associated with eating its leaves include weight loss, antioxidant protection, a reduced risk of lung cancer, as well as protection against cardiovascular problems. It produces many viable seeds that germinate readily. In India, the crop has been naturalized and is locally called 'Bijli' in Nagpur (Meshram *et al.,* 2008); 'Baboona' in Haryana (Misra *et al.,* 2002), western U.P. (also 'Gendi') and Delhi (also 'Market'); and 'Guldhak' in Punjab. It has numerous beautiful varieties such as 'Flore-Plenum' (flowers golden-yellow or white, semi-double or double, and grows up to 75 cm tall), 'Golden Crown' (height up to 1.2 m and erect, foliage silver-green, flowers single and deep yellow), 'Golden Gem' (plants compact, height 15-30 cm hence good for pot planting, flowers button-like, double and lemon-yellow), 'Golden Glory' (height 60-90 cm, flowers single and deep canary-yellow and best for cutting), *C. nanum compactum* is quite dwarf and suitable for pot cultivation, *etc.*

C. frutescens (*Argyranthemum frutescens*), a non-hardy and semi-shrubby evergreen perennial species though grown as an annual pot plant, is native to Canary Islands with height and spread of 30-45 cm having rounded habit, but as perennial it grows up to 90 cm. Its leaves are pale- to glaucous-green, very graceful, 4-8 cm long, usually bipinnatisect, much divided and ferny. The flowers are single, some 4.0-5.0 cm across, white or pale to sulphur-yellow but rays white and are borne from May to October or even longer. White form bears slightly larger blooms. Its flowers are suitable for cutting. It is propagated through seeds and cuttings.

C. multicaule, a hardy herbaceous annual species originating from Africa (Algeria), and has been introduced to many countries; and in China it is a quite popular ornamental; and it grows up to 30 cm in height with base-branching, forming a dense mass of short branches and spathulate, trifid or pinnatifid, 5-10 cm long, glaucous and somewhat fleshy leaves, and it flowers from mid-spring to early summer bearing single, 2.5-5.0 cm across and golden-yellow flowers during July-August. Its leaves are blue-grey, somewhat fleshy and coarsely dentate. It is an ideal subject for covering the ground, in borders, for edging and in rock gardens, for window-boxes, for bedding and as a pot plant. The species has been shown to be highly resistant to a spectrum of diseases and pests. *C. multicaule* is therefore regarded not only as a useful ornamental plant in itself, but also as a potential reservoir of useful genetic variation for chrysanthemum improvement via wide hybridization (Deng *et al.*, 2010). The species is highly fertile.

C. parthenium (*Matricaria eximia, Tanacetum parthenium*) is a hardy, bushy and short-lived perennial species native to Great Britain, growing 20-45 cm tall and is grown as an annual as from seeds it flowers the same season. Its light green leaves are aromatic with pungent aroma, flowers are white, some 2.0 cm across and are borne freely from July to September. Its plants are ideal for bedding, pots and window-boxes. Its all the varieties are seed-grown, flower the same year and bear long-lasting flowers. Its varieties are 'Aureum' (45 cm tall, flowers single white and foliage golden-green), 'Ball's Double White' (growing 45 cm high, bearing plumes of long-stemmed small double white flowers suitable for cutting), 'Golden Ball' (25 cm tall and compact, ball-shaped golden-yellow flowers appear in clusters), 'Gold Star' (grows 20 cm tall and the yellow centre is surrounded by a row of white ray petals), 'Snowball' (plants 30 cm tall, flowers are pompon-like and ivory-white), 'Snow Puffs' (grows up to 30 cm high and an ideal pot plant, bearing profusion of small button-like white flowers that last longer), 'White Bonnet' (plants 75 cm tall, flowers double and pure white), *etc.*

C. segetum (*Glebionis segetum*) is native to Great Britain and also probably to the eastern Mediterranean region, and grows some 60-80 cm tall with rigid stems. It is an herbaceous species bearing stalkless, oblong & spirally arranged, dark green and deeply lobed leaves 5-20 cm long. It is suitable as border plant and for cutting. The flowers which appear from July to September are single, solitary, 3.5-6.5 cm in diameter, bright yellow and are produced in capitulae (flowerheads) with a ring of ray florets and a centre of disc florets. It is fast-growing and widely naturalised outside of its native range, colonising western and central Europe with early human agriculture. It can be an invasive weed in some of the areas. Its most of the varieties grow some 45 cm tall. 'Eastern Star' (flowers primrose-yellow with brown disc), 'Eldorado' (canary-yellow with black centre), 'Evening Star' (bright golden-yellow), 'Golden Glow' (double, golden-yellow), 'Morning Star' (yellow, shading to primrose), *etc.*

C. viscosum (*C. viscidi-hirtum*) is a vigorous bush-like annual species native to North Africa and Spain. This grows up to 45 cm in height with stems bearing abundance of sticky hairs. It flowers from July to September with yellow flowers. It is good for naturalizing in poor soil but at sunny position.

C.* × *spectabile is a hybrid between *C. carinatum* and *C. coronarium* which grows up to 1.2 m high and is ideal for growing in the borders and for cutting. Its foliage is grey-green and deeply dentate. Variously coloured single flowers some 7.5-10.0 cm across appear from June to September. 'Cecilia' (a tetraploid form growing to 75-90 cm tall with long and strong-stemmed flowers which are 10 cm across, white banded yellow with bright yellow disc, ideal for cutting). 'Sunray' (growing vigorous up to 1 m tall, bearing single large canary-yellow flowers).

Out of all these, though *C. frutescens* and *C. parthenium* are perennial species but are grown just as other annual species from seeds easily.

Propagation

Annual chrysanthemums can be propagated by seeds. Sowing time depends on climatic conditions and it varies in different parts of country. In the northern plains, the seeds of annual chrysanthemums are sown for winter season cropping during September–October in nursery beds, earthen pots, seed pans or wooden seed trays. The seed compost should consist of one part each of garden soil, coarse sand, farmyard manure (FYM) and leaf-mould, at a sunny location. For preparing the nursery beds, the soil should be dug up thoroughly and sufficient FYM should be mixed in the soil. Raised nursery beds of convenient size (normally 60cm wide and 15cm high) should be prepared. If soil is heavy, some quantity of sand may be added. It is better if the soil of nursery bed or earthen pots is sterilized with 2% formalin. For this, soil is drenched with formalin solution and is covered with polythene sheet for 45h. Afterwards the polythene is removed and soil is dried before sowing the seeds.

Before sowing, the seeds should be treated with Cerasan (0.2%) or Captaf (0.2%) to prevent the seedlings from damping off disease. The seeds should be sown thinly and evenly as thick sowing causes poor growth and damping off of the seedlings. In nursery beds, the seeds are sown in rows spaced 6.0 cm apart and are covered with finely sieved leaf-mould. Watering is to be done with a watering can having a fine rose both in beds and pots. In beds, when germination is over, water is given for proper moisture. When the seedlings attain 10-12 cm height, these are gently removed from the nursery beds after moistening the medium and planted in the prepared beds.

Chrysanthemum frutescens is also raised through cuttings similar to those of the seeds apart from the seeds. The plants are lifted in September and are cut to their half after planting in a proper medium and the plants are kept to grow throughout the winter at a temperature of 7-10 ^{0}C. Half cutting of the plants promotes lateral growths for taking cuttings in February.

Cultural Practices

Annual chrysanthemums can be grown in a variety of soils having pH range of 5.6-7.5 but those having light, fertile and well-drained soils are more preferable. However, site should always be sunny. The seedlings when have attained 4-leaf stage, i.e. 10-12 cm height or 25-30 days after seed sowing can be **transplanted** when sun is not bright in the afternoon or on a cloudy day. Transplanting should follow watering immediately to minimize the shock to the roots from lifting. Transplanting in evening is good as the night cool temperature is beneficial for the establishment. Light watering every day in early morning or late in the afternoon is required for 3-4 days for proper establishment of the seedlings. Since annual chrysanthemum is a tall plant with branching habit, it should be planted 40-45 cm apart, though there are reports that these have been planted 30 × 30 cm to 60 × 60 cm **spacings**. Plant spacing of 30 × 30 cm is optimum for both *kharif* and *rabi* season crops. The number of flowers per plant, to some extent, increases with increase in the spacing, highest being recorded at 60 × 60 cm. Maximum values in respect of leaf area index, crop growth rate and net assimilation rate were recorded by closer spacing level of 30 × 30 cm. Plants spaced closely at 30 × 30 cm recorded more plant height, however, significantly lesser leaf area and aboveground dry matter accumulation per plant compared to widely spaced plants at 60 × 60 cm (Dorajeerao and Mokashi, 2013). These show indeterminate growth habit. **Pinching** is the act of cutting or nipping off the growing tips in order to maintain proper plant architecture and to force more branching so that loose flower yield is maximised. Pinching at 20 days after planting, and again after 10-15 days is effective to encourage more lateral branching for maximum production of flowers. The superiority of early pinching treatments in garland chrysanthemum can be attributed to the efficient photosynthetic area, better assimilation into reproductive parts and putting up optimum vegetative growth without interrupting floral bud initiation (Dorajeerao and Mokashi, 2012). Friedrich and Sackmann (1985) through spraying of CCC on *Chrysanthemum coccineum* plants obtained reduction in the stem growth but neither flowering time was altered nor there was any change in the flower quality or vase life, however, the marginal leaf chlorosis and necrosis, and occasionally shorter flower pedicels were obtained, so they suggested that watering CCC on the soil may give better results as it is then less affected by weather conditions. When the purpose is to harvest the cut flowers, the pinching of the plant is carried out only once but first floral buds are removed to encourage longer stemmed side-shoots. At initial stages it requires to be kept **weed-free** though after two months of growing the crop becomes so rampant that weeds are suppressed. Lamont and O'Connell (1986) found Oxyfluorfen (2.0 l/ha) weedicide quite safe only on transplanted *Chrysanthemum morifolium* crop.

Feeding and Irrigation

Farmyard manure or **compost** at 3kg/m^2 is mixed in the soil at second ploughing. Chemical **fertilizers** at 60-120g superphosphate and 30-60g muriate of potash/m^2 should also be added at the time of bed preparation, i.e. on the day of planting for better results but should be applied only after soil analysis. Our Indian soils are normally not deficient in potash. Urea 2% at 20g/m^2 can be sprayed when the plants are well established, i.e. after one month of planting. Spraying plants with 2% urea twice or thrice is beneficial for good growth and flowering. Fertilizers should never come in the direct contact of the foliage since they cause scorching. Fertilizers should never be applied in the pot-grown annual flowers. However, some readymade pot-mixtures can be used. The pot-mixture should consist of 2 parts of garden soil and one part each of coarse sand and FYM. Instead of fertilizers, it is better if pot-grown plants are given liquid feeding. The liquid manure is prepared by fermenting 1-2kg each of fresh cow dung and oil cake in 10 litres of water in a drum for one week. It is diluted to tea colour and sieved with the help of a muslin cloth. It is applied at 500-1,000ml/pot at 10-15 days intervals.

Little water is needed everyday up to 3-4 days after transplantation. When the seedlings start new growth, profuse **watering** once in a week is required during summer and once in 10-12 days during winters in the beds. Later, frequency and quantity of watering depend upon soil condition and weather condition. In lighter soils, more frequent irrigation is needed than that in heavy soils. The season of planting also determines the frequency of irrigation. Irrigation during rainy season depends upon prevailing weather conditions. Potted plants need daily in summer or at alternate days during winter.

Postharvest Management

Most of the annual flowers are grown for garden display in various ways, and in pots for instant landscaping. However, annual chrysanthemums are grown commercially for the purpose of either cut flowers or loose flowers, mostly for loose flowers. Their flowers are harvested when they are fully open and are sold in the local markets. The flowers, in general, are cut either late in the afternoon or very early in the morning. After harvesting, cut flowers should be put in a bucket of water filled up to one-fourth of the volume as it helps in their recovery from the shock of being cut away from the plant. As far as possible, the freshly opened flowers should be cut as freshness enhances their shelf-life. If annual chrysanthemums are used as cut flowers, proper postharvest management is necessary for prolonging their vase-life. The flowers are graded according to stem length, flower size, flower shape, flower colour and freshness. The cut flowers are marketed in local markets. If the flowers are not to be sold the day they are harvested, these are cold stored dry after wrapping in cellophane sleeves or in polythene bags at 4-5 ^{0}C for up to 4-5 days. For loose flowers, these are harvested by the evening when the flowers are not wet with dew or rain water, kept in cold store overnight and then in the morning are taken out, packed in bamboo baskets, hessian cloth or in the gunny bags and taken to the market for sale or auction.

The technique of drying flowers can play a vital role as dehydrated flowers can be stored for several years and also occupy a large share in floriculture trade. Best quality of dried flowers is obtained by embedding them in silica gel and keeping at 50 ^{0}C for 48 hours in the oven as per cent decrease in weight and moisture content is maximum and per cent decrease in carotenoid content is less, *vis-à-vis* best visual score (Dahiya *et al.*, 2003).

Insect-Pests and Diseases

Several insect-pests and diseases attack chrysanthemum plants. Insects that attack annual chrysanthemum are aphids, four-lined plant bug, tarnished plant bug, spittle bugs, thrips, certain beetles and caterpillars, stem borers and spider mites,. **Four-lined plant bug** adults are greenish-yellow in colour with 4 black lines running down their backs. They are generally flat and slender and about 6 mm long. The insect, as a result of its feeding, causes brownish spots on the upper leaf surface. These could be easily mistaken for fungal leaf spots. **Tarnished plant bugs** are oval and flattened, measuring approximately 6 mm in length. The adults are mottled brown with irregular yellow-brown, red-brown or black markings, and with very prominent eyes. Distinctive markings along the side of the posterior third of the body are triangular and yellow and tipped with a small, dense black triangle. Injury is similar to that of the four-lined plant bug. Likewise, spittle bugs have also been found infesting on this crop. Certrain beetles and caterpillars have been found feeding on its leaves, flowers and the buds by boring them, making the crop totally unfit for marketing. A lepidopterous borer feeds on the stem, bores these and makes complete loss. Aphids of various kinds and thrips suck plant sap and hamper plant growth. During warm weather their population increases dramatically and then these are also found in flowers which wither with their attack. Spider mites web the leaves, flowers and floral buds and by feeding them deteriorate their quality. These become serious when weather is quite dry and dusty or when the planting has been done in the side of vehicular road. Proper field sanitation is a best way to manage these pests. Spraying of the crop with 0.2% methyl parathion will control all these pests. Whenever, the infestation of any pest is noticed, immediately insecticide spraying should be followed.

Chrysanthemum **foliar nematode** (*Aphelenchoides ritzema-boi*) is a widespread problem on chrysanthemum. Dark spots on areas on the undersides of the leaves characterize first symptoms. About 5 days after infestation the veins become discoloured and stand out sharply on the upper leaf surface. The diseased leaves turn brown or black, starting in wedge-shaped areas between the veins. The leaves eventually dry, wither, and hang down along the stems. By keeping the foliage dry, avoiding of overhead watering if possible, use of mulches to prevent splashing, avoiding crown divisions and instead making tip cuttings in case where plants are to be multiplied through cuttings are the measures which may keep the crop free from nematode infestation. Root-knot nematode (*Meloidogyne incognita*) also infests by galling its roots which can be controlled by applying carbofuran in the soil followed by irrigation (Misra *et al.*, 2002).

Leaf spot (*Septoria chrysanthemella, S. obese*) appear initially as small and yellowish but later on turning to brown and finally black spots on leaves and even on stems which later on coalesce and make larger spots. Sometimes the spots coalesce into blotches, minute black fruiting bodies are faintly visible in these spots. Leaves may turn yellow and drop prematurely or dry and hang down along the stems. Spores are splashed from plant to plant through rains or when watering. No field operation should be carried out when the leaves are wet. This problem can be controlled by spraying with maneb or zineb. **Powdery mildew** (*Erysiphe cichoracearum*) appears in the form of white fungal growth primarily on the upper surface of the leaves. Usually this disease is not considered serious since it forms late in summer. To help manage this disease, planting should be done to space the plants properly so that there may be free and proper air circulation in the plantings. Avoid frequent overhead watering, especially at night. Remove all infected plant parts in autumn and destroy them. Use of any sulphur fungicide will control this problem. **Vascular wilt diseases** caused by *Fusarium* spp. and *Verticillium alboatrum* are symptomised as of water stress, partly because they attack the vascular system of the plant. This is partly because the vascular system is physically clogged and partly because the fungus may release harmful toxins. Plants may show stunted growth, yellowing or scorched leaves, and ultimate wilt and death. Affected plants should be removed and destroyed. Bavistin spraying at 0.1% followed by Captafol 0.2% spraying will manage these diseases. **Rusts** (*Puccinia chrysanthemi, P. horiana*) also attack chrysanthemum. *P. chrysanthemi* is usually found in greenhouses and *P. horiana* is common outdoors. Removal of infected plants and/or plant parts and destroying them will prevent their further spread. Proper plant spacing to permit proper air circulation and sunlight, and watering early in the day will also minimize the incidence of the rusts. Species that appear to be resistant to white rust include *C. carinatum, C. coronarium*, and *Tanacetum coccineum* (*Chrysanthemum coccineum*).

References

Anderson, N.O. 1987. Reclassification of the genus *Chrysanthemum* L. *HortSci.*, **22**: 313.

Chung, K.M and Y.D. Park, 2005. Development of an *Agrobacterium*-mediated transformation system for regenerating garland chrysanthemum (*Chrysanthemum coronarium* L.). *J. Pl. Biol.*, **48**(1): 136-141.

Dahiya, D.S., D. Unnikrishnan, A.K. Gupta, S.K. Sehrawat and S. Siddiqui, 2003. Dehydration of annual chrysanthemum (*Chrysanthmum coronarium*). *Acta Hort.*, No. 624, pp. 385-388.

Deng, Y., S. Chen, N. Teng, F. Chen, F. Li, A. Song and Z. Guan, 2010. Flower morphologic anatomy and embryological characteristics in *Chrysanthemum multicaule* (Asteraceae). *Scientia Horti.*, **124**: 500-505.

Dorajeerao, A. V. D. and A.N. Mokashi, 2012. Branching pattern as influenced by pinching time in garland chrysanthemum (*Chrysanthemum coronarium* L.). *Indian J. Pl. Sci.*, **1**(1): 9-15.

Dorajeerao, A.V.D. and A.N. Mokashi, 2013. Growth analysis as influenced by planting geometry in garland chrysanthemum (*Chrysanthemum coronarium* L.). *Global J. Biosci. & Techn.*, **2** (1): 21-26.

Everard, B. and Brian D. Morley, 1970. Wild Flowers of the World, plate 39. Peerage Books, London.

Friedrich, H. and G. Sackmann, 1985. Cycocel: not suitable for use on perennials for cut flowers. Outdoor spray treatments did not improve quality (German). *Gb+Gw.*, **85**(20): 751-753.

Haouas, D., M. Ben Halima-Kamel and M.H. Ben Hamouda, 2008. Insecticidal activity of flower and leaf extracts from *Chrysanthemum* species against *Tribolium confusum*. *Tunisian J. Pl. Prot.*, **3**: 87-93.

Lamont, G.P. and M.A. O'Connell, 1986. An evaluation of pre-emergent herbicides in field-grown cut flowers. *Plant Prot. Quart.*, **1**(3): 95-100.

Meira Bar-Eyal, Edna Sharon and Yitzhak Spiegel, 2006. Nematicidal activity of *Chrysanthemum coronarium*. *European J. Pl. Path.*, **114**: 427-433.

Meshram, N., S. Badge, S.A. Bhongle and S.D. Khiratkar, 2008. Effect of bio-inoculants with graded doses of NPK on flowering, yield attributes and economics of chrysanthemum. *J. Soil & Crops*, **18**(1): 217-220.

Misra, R.L., S.D. Mishra and S. Misra, 2002. Annual chrysanthemum - a good host of root-knot nematode (*Meloidogyne* spp.). *J. Ornam. Hort.*, **5**(2): 65.

Ohtsuka, H. and Z. Inaba, 2008. Intergeneric hybridization of marguerite (*Argyranthemum frutescens*) with annual chrysanthemum (*Glebionis carinatum*) and crown daisy (*G. coronaria*) using ovule culture. *Plant Biotech.*, **25**: 535–539.

Perez, M.P. and M.J. Pascual-Villalobos, 1999. Efectos del aceite esencial de inflorescencias de *Chrysanthemum coronarium* L. en mosca blanca y plagas de almace´n. *Investigacio´n Agraria, Produccio´n y Proteccio´n Vegetal*, **14**(1/2): 249-258.

Pizzetti, I. and H. Cocker, 1975. *Flowers A Guide for Your Gardens*, pp. 225-244. Harry N. Abrams, Inc., Publishers, New York.

Razaq, Z.A., S. Khatoon and S.I. Ali, 1988. A contribution to the chromosome numbers of Compositae from pakistan. *Pakistan J. Bot.*, **20**(2): 177-189.

3

Asclepias & Calotropis

R.L. Misra, B.K.Banerji, Sanyat Misra, Atul Batra and A.K.Dwivedi

Family: **Asclepiadaceae,** formerly **Apocynaceae**

Common names: Blood flower (*Asclepias curassavica*), Butterfly milkweed/Butterfly weed/Indian paintbrush/Pleurisy root/Swallow-wort/Tuber root (*A. tuberosa*), Calotrope/Calotropis /Giant milkweed (*Calotropis gigantea*), Dead sea-fruit/Desert wick/ Milkweed/Mudar fibre/Rubber bush/rubber tree/Silkweed/Sodom apple (*C. procera*), Silp plant (*A. syriaca*), Swamp milkweed (*A. incarnata*), *alarka in Sanskrit, aak in Ayurveda,* etc.)

Introduction and Origin

Asclepias is ancient Greek name in Latinized form deriving from *Asklepios* or *Arsculapius* (son of Apollo and Coronis), the Greek god of medicine, due to the medicinal properties of some species, and *Calotropis* is Greek for beauty of the flowers. These are found growing up to 1,300 m elevation in the areas experiencing a mean rainfall of 300-400 mm. In most part of the world including India, *Calotropis procera* (syn. *Asclepias procera*) is considered a weed though *C. gigantea* (syn. *Asclepias gigantea*) is grown by gardeners, and also in India especially in the northern zone it is considered highly auspicious when planted in front of the house. Its flowers are a favourite as offering to lords Shiva and Hanuman, and is also used in preparing the garlands. The flowers are fragrant and are often used in making floral tassels in some mainland SE Asian cultures. The inner bark of *C. procera* yields *mudar, mader* or *yercum* fibre which makes strong rope but is difficult to extract. In saline and alkaline soils of tropical India, as well as in Rajasthan deserts, and in the wastelands of Delhi, Punjab, Haryana, Gujarat, Maharashtra, U.P., Bihar, Jharkhand, M.P., Chhattisgarh, Odisha, Andhra Pradesh, Tamil Nadu, Karnataka, Goa, etc., *C. procera* grows rampantly as weed with abundance of blooms from spring to autumn. *Calotropis procera* is a suitable indicator of exhausted soil and adapts nicely to the meagre water supply of its arid environment, a source of green manure,

Plate 1: (a) Calotropis plant, (b) Mature Stem, (c) Young stem with foliage, (d) Buds, (e) Flowers, (f) Fruit

improves soil water condition, acts as a soil binder, and monitors atmospheric SO_2. In Thailand, *C. gigantea* is considered so auspicious that its garlands are exchanged during each and every wedding. *C. gigantea* is native to India and probably W. Indies while *C. procera* is native to India. *Asclepias* are native to North and South America, *A. curassavica* to Tropical America, and *A.fruticosa, A. incarnata* & *A. tuberosa* to N. America and *A. fruticosa* also to Africa. Many other countries are reported to be native haunts for many of the *Asclepias* and *Calotropis* and these countries are Afghanistan, Algeria, Burkina Faso, Cameroon, Chad, Cote d'Ivoire, Democratic Republic of Congo, Egypt, Eritrea, Ethiopia, Gambia, Ghana, Guinea-Bissau, Iran, Iraq, Israel, Kenya, Kuwait, Lebanon, Libyan Arab Jamahiriya, Mali, Mauritania, Morocco, Mozambique, Myanmar, Nepal, Niger, Nigeria, Oman, Pakistan, Saudi Arabia, Senegal, Sierra Leone, Somalia, Sudan, Syrian Arab Republic, Tanzania, Thailand, Uganda, United Arab Emirates, Vietnam, Yemen, and Republic of, Zimbabwe (Eggeling, 1940; Aronson, 1984; von Maydell, 1986; Abbiw, 1990; Bekele-Tasemma *et al.*, 1993; Beentje, 1994). Exotic cultivars of *Calatropis* are widely distributed in different countries throughout the world, *viz.* Antigua and Barbuda, Argentina, Australia, Bahamas, Barbados, Bolivia, Brazil, Chile, Colombia, Cuba, Dominica, Dominican Republic, Ecuador, French Guiana, Grenada, Guadeloupe, Guatemala, Guyana, Haiti, Honduras, Jamaica, Martinique, Mexico, Montserrat, Netherlands Antilles, Nicaragua, Panama, Paraguay, Peru, Puerto Rico, St Kitts and Nevis, St Lucia, St Vincent and the Grenadines, Surinam, Trinidad and Tobago, Uruguay, Venezuela and Virgin Islands (USA). The flowers are fragrant and are often used in making floral tassels in some mainland SE Asian cultures. Fibers of these plants are called *mudar* or *mader*.

Asclepias are ideal for wild gardens and sheltered borders though in the wild these are usually found in abandoned farmland. *A. gigantea*, the tree-type woody species is cultivated for its small white or purplish flowers for use in garland making and as loose flowers for offering to gods and goddesses, especially Lords Shiva and Hanuman. *A. procera* which is seldom cultivated for loose flowers except in South America and in Caribbean Islands for production of fibres yielding 5 quintals/ha annually. Its stem fibre is cylindrical, white to silky, strong, flexible and durable which are used in making ropes, gunny bags, bow strings, fishing nets, paper, pulp and duplicating stencils (Arora, 1982). Its large inflated ovoid fruits are packed with a mass of hairs (floss) attached to the seeds and which provide inferior substitute for kapok. Floss is white-silky, though not so strong and about 2.0-3.5 cm long cottony mass suitable for stuffing pillows and mattresses and for weaving into strong cloth, *vis-à-vis* a substitute to surgical cotton wool. Strong inner bark fibres produce a binding material suitable in fabrics. Stems of *C. procera* are used for building huts and roofing as these are termite-proof. Stem pith makes a good timber and the whole stems good charcoal, producing an effective and sustained smoky fire most suitable for drying fish. Charcoal has also been used as gun powder in India. A macerated bark extract is used for dehairing-hides, as dyes and for tanning. The bark and latex are used in brewing and to curdle milk. Liquid latex is suitable source of hydrocarbons and an intermediate energy source, and also contains 11-23% of rubber. *A. curassavica, A. fruticosa*, *A. incarnata* and *A. tuberosa* are used as cut flowers.

Medicinal Uses

Cattle often stay away from the plants because of their unpleasant taste and due to presence of cardiac glycosides in their sap even then a little quantity is sometimes swallowed by some of the livestocks, however, goats and sheeps normally feed regularly on its flowers, young fruits and senescing leaves albeit in little quantity. The flower is digestive, *vis-a-vis* a tonic used for treating asthma and catarrh-like diseases. The twigs are applied against asthma and for preparation of diuretics, stomach tonic and anti-diarrhoetics. It is also used in abortion, as an anthelmintic, for colic, cough, whooping cough, dysentery, headache, lice treatment, jaundice, sore gums and mouth, toothache, sterility, swellings and ulcers. The latex of *C. procera* is used for treatment of skin diseases (Kamal *et al.*, 2009), ring worms (Ajaib *et al.*, 2010) and the crushed plant parts mixed with new sugar is effective against dog bite (Kamal *et al.*, 2009). Its latex is laxative and is also used for treating guineaworm blisters, scorpion stings, venereal sores and ophthalmic disorders. In India, it is used in the treatment of skin diseases which has caused severe bullous dermatitis leading sometimes to hypertrophic scars. The local effect of the latex on the conjunctiva is congestion, epiphora and local anaesthesia. The milky exudation from the plant is a corrosive poison (Ajaib *et. al.*, 2010), and its latex as well as bark are widely used as arrow and spear poisons. The latex from the stem when mixed with salt removes the hairs from the hides. Roots are extremely poisonous and are applied against snakebite. In Ayurveda, it is used in cases of cutaneous diseases, intestinal worms, cough, ascites, asthma, bronchitis, dyspepsia, paralysis, swellings, intermittent fevers, anorexia, inflammations and tumours (Puri, 2003). In large doses, its juice is known to act as a purgative and an emetic. Root bark has digitalin-like effect on heart though earlier it was being used as a substitute of ipecacuanha. The plant contains calotropin (similar in structure to two cardiac glycosides; Kupchan *et al.*, 1964) in the latex which is more toxic than strychnine. Though its leaves are used in various types of folk medication but sometimes these are fried in oil and then used as medicine. The latex is said to have mercury-like effects on the human body, and is sometimes referred to as vegetable mercury and used in place of mercury in aphrodisiacs. The milky sap is used as a rubefacient and is also strongly purgative and caustic. The proteolytic enzyme, calotropin which is cardio-toxic is found in its latex. The leafy branches are said to deter ants. Latex of *C. procera* is 80% effective in inhibiting the activity of the tobacco mosaic virus. Certain principles found in the plant have bactericidal and vermicidal properties. Compounds derived from the plant have been found to have emeto-cathartic and digitalic properties due to active principles being asclepin and mudarin. An infusion of fresh bark powder is used in the treatment and cure of leprosy and elephantiasis. The root bark is an emetic. Bark and wood stimulate lactation in cattle.

Botany

Asclepias are reported with 120 to 150 species, all being evergreen in nature. Though *Asclepias gigantea* is woody growing up to 5.0 metres, other cultivated species are herbaceous and sub-shrubby perennials of vgorous habit and yield white or clear latex, however, *A. procera* is a weed growing some 3.0 metres. Other *Asclepias* grow from 4.0 cm to 2.0 metres in height with simple or divaricate, erect or decumbent stems. Leaves in opposite pairs or in whorls in so-called *Calotropis* species are ovate, though in so-called *Asclepias* these are narrow, lanceolate and spirally attached to the stem (Bailey and Bailey, 1976). The inflorescence is an umbel, flowers small, numerous, in terminal or axillary flat-topped cymes and gamopetalous, each flower consisting of 5 sepals and corolla rotate with 5 lobes, often reflexed at anthesis, corolla conspicuous, composed of 5 cucullii, i.e. hooded lobes originating as an outgrowth of the stamina column, stamens 5, the stamens fused to the stigma to form complex stamina column, one filament developing into a horn and the second into a hood which is wrapped around the horn and acts as a block to the nector (Lyons, 1986; Dole and Wilkins, 1999), pollinia solitary in each cell and hanging from the apex, the globose, narrow ellipsoid to filiform paired fruits with erect pedicels are pod-like follicles or bilocular containing large but flat seeds crowned with long silky hairs to aid in dispersal through wind. The plant is highly cross-pollinated. The somatic chromosome number in *Calotropis procera* is 2n=22 (Raghavan, 1957; Eisikowitch, 1986).

Species and Varieties

Under *Asclepias* there are some 120 species. *Calotropis gigantea* is a slow-growing perennial woody shrub growing to a height of up to 5 m with branches and sub-branches. Leaves sub-sessile, opposite, simple and ovate but cordate at base. Flowers in umbellate lateral cymes, non-fragrant, pure white, purplish or greenish-white, and beautiful. Fruits fleshy follicles filled with black or brown seeds attached with silky threads. *C. procera*, a very common weed in the neglected soils, grows to a height of 2.4 to 3.0 m and throws out densely white tomentose branches from the ground but rarely with aerial branching. The decussate opposite and white tomentose leaves are similar in shape to *C. gigantea* but larger, and the pedicellate and pentamerous flowers though bigger but not so attractive and compact as *C. gigantea*, size being about 3.8 to 5.1 cm, with umbellate lateral cymes and are coloured white to purplish-pink and fragrant. The fruit is a fleshy, inflated, subglobose to obliquely ovoid follicle to 10 cm or more in length and 5 cm or more in diameter. The seeds are numerous, compressed, broadly ovoid, with a tufted micropylar coma of long silky hairs of 3 cm long or more. Extracts from the flowers of *Calotropis procera* have shown strong cytotoxic activity in the patients of colorectal cancer, *vis-a-vis* harmful to the eyes. *Asclepias curassavica* is though perennial but can be grown as an annual in warmer areas, growing up to 90 cm in height, plant glabrous and is woody-based, semi-shrubby with an ererct habit, slender-pointed mid-green leaves in opposite or alternate pairs, oblong-lanceolate, 5-15 cm long and short-petioled, and the flowers in the umbels of 5-10, borne at the summit of a stalk about 10 cm long, 2 cm wide, orange-red with a yellow staminal crown, and fruits glabrous and erect. Its var. *aurea* is orange-yellow. *A. fruticosa* is a sparsely branched perennial sub-shrub with slender stem, growing to 1m, leaves 12.5 cm long, lanceolate to oblanceolate, short petioled and opposite, inflorecence axillary, 1.2 cm long corolla lobes with infolded hoods but lacking horns, colour cream to white, fruits 7.5 cm long and ovoid, inflated and softly spiny. *A. hallii* is a hardy, tuberous and upright perennial growing up to 90 cm in height with 60 cm stem length, leaves 13 cm long and oblong, umbels of small, 5-horned and dark pink flowers, and fruits narrowly ovoid, 15 cm long and tightly packed with silky mass attached with seeds. A. *incarnata* was first recorded in Canada in 1710 growing along the banks of rivers, grows up to 1.2 m in height, stems glabrous or nearly so, slightly reddish and branched towards the top, plants slightly pubescent, leaves opposite, oblong-lanceolate, glabrous, mostly acuminate at the apex and mid-green, umbels 5 cm wide, the flowers 6 mm wide, crown-like with oblong lobes, flesh-pink (rose-purple), 6 mm across and with a mild vanilla-like fragrance, and the pods erect, oval and glabrous. Its var. *pulchra* is hairier and has larger leaves. Its roots in powdered form is emetic and cathartic and is used in case of rheumatism, asthma, in catarrhal ailments and as a vermifuge. *A. speciosa* (syn. *A. douglasii*) grows up to 90 cm in height, suitable as cut flower, stems covered with grey hairs, leaves large, ovate-elongated but base cordate with short petioles, inflorescence numerous, umbrella-shaped, terminal or axillary, and flowers greenish-red and externally hairy. *A. syriaca* (syn. *A. cornutii*), growing up to 1.8 m tall, is rhizomatous throwing underground runners and suckers so may become invasive, stems greyish and hairy, undersurface of leaves woolly, opposite and oval-obtuse, inflorescence shorter than leaves, axillary on shorter stems, flowers more or less nodding and with mild honey fragrance, calyx 5-lobed, corolla with 5 segments, pale-pink and waxy, the 5 stamens forming a 5-pointed miniature crown, fruit looking-like bird's body. *A. tuberosa* is found in the meadows and due to atttraction of the butterflies in abundance, this is commonly known as butterfly weed, rhizomatous rootstock is tuberous and can be grown as an annual, height up to 90 cm, stems hairy, erect and branched at the tip, leaves mid-green, opposite to alternate (more or less alternate), oval- to linear lanceolate and narrower at both the extremes, inflorescences many, short stalked, corymbose and umbrella-shaped, flowers 6 mm wide, orange- to reddish-yellow and changing to reddish-scarlet with age, and the fruits hairy, oblong and narrower at both the ends. It has no known variety. Its powdered roots are effective against smallpox, and are also used as stomachic, diuretic, and laxative. Other important species are *A. amplexicaulis* (green-purple), *A. eriocarpa* (dull white), *A. hallii* (flowers greenish-white and purple), *A. mexicana* (greenish-white or purplish), *A. phytolaccoides* (greenish), *A. quadrifolia* (pink to white), *A. variegata* (white and pink), *A. verticillata* (greenish-white), etc.

Propagation

All the *Asclepias* produce seeds freely whose dispersal is through wind, and it keeps on multiplying in nature, growing more comfortably at neglected nooks and corners. When growing the desired species, especially non-calotropis asclepias or their varieties through seeds, the freshly harvested seeds are immediately sown in the prepared beds in February at 15-21 ^{0}C which germinate within 30-90 days though older seeds require a chilling temperature for germination. For 1,000 seedlings some 14-21 g of seeds is required (Dole and Wilkins, 1999). When grown sufficiently to handle, these seedlings are transplanted first in 5 cm pots, after some time in 10 cm pots and again after some time finally in 15-20 cm pots. This whole process is complete within five months and then these may be planted at their permanent place spaced suitably as per the requirement of the species or varieties. Commercially, *A. tuberosa* is multiplied only through seeds (Borland, 1987).

Vegetative propagation through half stumps assumes a special importance as compared with the entire stumps because they help in faster multiplication of the parent genotype with added characters, as each plant gives rise to two half stumps. Vegetative propagation through stem and root cuttings is very useful in large scale multiplication of the superior genotypes and for getting uniform plants (Ecker and Barzilay, 1993). *Asclepias incarnata* roots and crowns measuring 5.0-7.5 cm in length are taken and planted when new growths are to start in February-March in the soil containing leaf mould or peat. Tap roots are sectioned into many bits, each bit having sufficient length and planted vertically at 21 ^{0}C and the medium is kept continuously moist where these root within 3-4 weeks (Bhowmilk and Bandeen, 1978). After root and shoot formation, these young plants perform well and flower for long term if given a cold treatment. After their proper establishment in the beds, the adventitious buds start forming the flowering shoots at the crown providing flowers for a long time (Wyatt, 1980, 1981; Lucansky and Clough, 1986). Stem cuttings some 10 cm in length are taken prior to flowering and planted as to that of root cuttings and these also provide optimum success. The response of cuttings taken after flowering is poor.

Cultural Practices

The plant is drought-resistant, and salt-tolerant to a relatively high degree. All the species quickly become established as a weed along degraded roadsides, lagoon edges and in over-grazed native pastures (Abbas *et al.*, 1992). Population of the asclepias in natural habitat can be seen dominant in areas of abandoned cultivation, especially sandy soils in areas of low rainfall, assumed to be an indicator of over-cultivation. The plant prefers disturbed sandy soil. It requires dry conditions during winter and fairly hard cutting back in early spring to encourage multibranched vigorous growth of the plants. Though **soil** for its cultivation may not be rich but it should be porous, well-drained and at a fairly sunny situation though it is only *Asclepias incarnata* which grows usually in moist areas. However, for pot-growing, the medium should consist of good garden soil:sphagnum peat:perlite into 1:1:1 plus dolomite limestone at 3 kg/m^3 (Albrecht and Lehmann, 1991). *Asclepias curassavica* is a good long term house plant. *A tuberosa* requires deep, sandy and peaty soil with little or no lime. *A. curassavica* does not tolerate a persisting winter temperature of 7 ^{0}C and below. During winters the aerial parts may die down but while forking the soil during winters care should be taken not to damage them. Albrecht and Lehmann (1991) stated that *Asclepias* prefer high light levels because floral buds abort in continuous cloudy weathers, i.e. SD though the plants continue blooming under LD, however, they recommended a forcing temperature of 17-25 ^{0}C and stated that forcing temperatures below 16 ^{0}C or above 30 ^{0}C cause abortion of flowers. When they stored the glasshouse-grown dormant crowns at 10 ^{0}C for 12 weeks, these failed to form shoots though crowns stored at 4 ^{0}C for 12, 14 or 16 weeks emerged in 16, 16 and 9 days, respectively.

Asclepias gigantea seedlings are planted 2 metres apart, *A. procera* 1.2 × 1.0 m apart, and other asclepias 45 × 45 cm apart with minor adjustments as per the height and spread of the individual species. Once planted these do not like to be disturbed at least for three years and at the most for five years. However, seedlings remain vegetative for one year (Lyons and Booze, 1983; Lyons, 1986; Shannon and Wyatt, 1986). Initially, these require staking but when fully grown these grow up and support themselves. However, the field should be kept weed-free. Pot-planted ones are given periodic applications of liquid fertilizers and for this Ecker and Barzilay (1993) recommended 100 ppm N from a 20-9-16 fertilizer at each irrigation. Ecker and Barzilay (1993) when used watering daily to *Asclepias tuberosa,* no any adverse effect was observed. *A. curassavica, A. incarnata* and *A. tuberosa* are normally grown as annuals.

Growth and Flowering

Asclepias when were vernalized at 4 ^{0}C for 12, 14 and 16 weeks and then grown at a 25 ^{0}C day and 17 ^{0}C night temperatures under long days, these flowered in 10, 10 and 8 weeks, respectively, and those vernalized outside and were then forced at 26 ^{0}C day and 17 ^{0}C night temperatures under 13 or 17 h photoperiods, the plants were found flowering in 71 and 61 days, respectively (Albrecht and Lehmann, 1991). SD is not conducive to plant growth as shoot elongation, *vis-à-vis* leaf formation stops, and thus increases root-to-shoot dry weight ratio though LD inhibits basal branching (Lyons and Booze, 1983). Armitage (1996) recommended 13 h or longer photoperiods for floral induction as well as for further development though 9 h and below photoperiod causes flower bud abortion, *vis-à-vis* development of blind shoots, however, photoperiods above 13 h do not make any impact. Certainly, there had been advantage of increased number of flowering shoots at 13 or 15 h photoperiod if the plants were first vernalized for 12 or 15 weeks at 4 ^{0}C, and the vernalization at 4 ^{0}C for 16 weeks and then 17 h lighting doubled the number of flowers per inflorescence though on the cost of number of flowering stems per plant (Abrecht and Lehmann, 1991). If the plants have protection from heat and are given with LD, the plants continue flowering continuously. Abrecht and Lehmann (1991) reported that after flowering when crowns become dormant under natural autumn or SD conditions, further growth can be initiated by cold treatment. For controlling the height of the plants in *A. curassavica,* ancymidol, chlormequat and daminozide can be used.

Insect-Pests and Diseases

Susceptibility to pests is relatively high. Aphids, grasshoppers and certain other insect-pests feed on leaves despite the toxic latex within. The caterpillar of the plain tiger butterfly (*Danaus chrysippus*), feeds on *C. procera.* These pests can be controlled by spraying with 0.2% Metacid-50. Sometimes stunted plants with yellow-green mottled deformed leaves are observed due to infection of cucumber mosaic virus. In such cases the plants should be uprooted and burnt. Leaf spot (*Cercospora asclepiadorae, C. clavata, C. vernturioides, Phyllosticta tuberosa,* etc.), rusts (*Puccinia bartholomaei, Uromyces asclepiadis*), stem blight (*Phoma asclepiadea*) and root rot (*Phymatotrichum omnivorum*) diseases have been reported on asclepias by Horst (1990). Root rot can be controlled through soil drenching with 0.3% Thiride or Captan, leaf spots and stem blight by spraying with Dithane Z-78 at 0.2%, and other pathogens by use of Bavistin 0.1% spraying alternate with Captan at fortnightly intervals. Low light intensities or when the plants are shifted from LD to SD, floral abortion occurs (Abrecht and Lehmann, 1991).

Post Harvest

Inflorescences of *Asclepias* for cut flowers are harvested when ½ to ⅔rd of the florets are open and such inflorescences last for 7 to 8 days (Wyatt, 1980), however, immediately after harvesting such flowers should be placed in STS and stored at 4-7 ^{0}C temperature (Armitage, 1993).

References

Abbas, B. , A.E. El-Tayeb and Y.R. Sulleiman, 1992. *Calotropis procera*: feed potential for arid zones. *Veterinary Record,* **131**(6):132.

Abbiw, D. 1990. Useful plants of Ghana. Intermediate Technology Publications and the Royal Botanical Gardens, Kew.

Ajaib, M. , Z. Khan, N. Khan and M. Wahab, 2010. Ethnobotanical studies on useful shrubs of district Kotli, Azad Jammu and Kashmir, Pakistan. *Pakistan J. Bot.,* **42**(31) :1407-1415.

Albrecht, M.L. and J.T. Lehmann, 1991. Day length, cold storage, and plant production method influence growth and flowering of *Asclepias tuberosa. HortSci.*, **26**: 120-121.

Armitage, A.M. 1993. *Asclepias tuberosa.* In: *Specialty Cut Flowers*, pp. 159-163. Varsity Press/Timber Press, Portland, Oregon, USA.

Armitage, A.M. 1996. Forcing perennials in the greenhouse. *GrowerTalks*, **60**(3): 86, 88, 93, 94, 96, 97.

Aronson, J.A. 1984. Energy plants for desert agriculture, current state of knowledge: an Israeli perspective. *Israel Ministries of Agriculture and Energy*, File No. 74, p. 12.

Arora, S.S. 1982. *Calotropis procera* (Ait.) R. Br. - Ak: A new and free source of fibre and renewable hydrocarbons. *Agricultural Mechanization in Asia, Africa and Latin America*, **13**(3): 71-75.

Bailey, L.H. and E.Z. Bailey, 1976. *Asclepias.*In: *Hortus Third ~ A Concise Dictionary of Plants Cultivated in the United States and Canada*, pp. 116-117. Macmillan Publishing, New York, USA.

Beentje, H.J. 1994. Kenya trees, shrubs and lianas. National Museums of Kenya.

Bekele-Tesemma, A., A. Birnie and B. Tengnas, 1993. Useful trees and shrubs for Ethiopia. Regional Soil Conservation Unit (RSCU), Swedish International Development Authority (SIDA).

Bhowmilk, P.C. and J.D. Bandeen, 1978. The biology of Canadian weeds, *Asclepias syriaca* L. *Canadian J. Pl. Sci.*, **56**: 579-589.

Borland, J. 1987. Clues to butterfly milkweed germination from a literature search. *Amer. Nurseryman*, No. 165, pp. 91-92, 94-96.

Dole, J.M. and H.F. Wilkins, 1999. *Asclepias*. In: *Floriculture Principles and Species*, pp. 206-209. Prentice Hall, New Jersey, USA.

Eggeling, 1940. Indigenous trees of Uganda. Govt. of Uganda (FAO, 1986). Some medicinal plants of Africa and Latin America. FAO Forestry Paper 67. Rome.

Eisikowitch, D. 1986. Morpho-ecological aspects on the pollination of *Calotropis procera* (Asclepiadaceae) in Israel. *Plant Systematics and Evolution*, **152** (3-4) :185-194.

Ecker, R. and A. Brazilay, 1993. Propagation of *Asclepias tuberosa* from short root segments. *Scientia Hort.*, **56**: 171-174.

Horst, R.K. 1990. Butterfly weed. In: *Westcott's Plant Disease Handbook* (5th ed.), p. 568. Van Nostrand Reinhold, New York, USA.

Kamal, M., S.N. Wazir, N. Hasan, M.S.S.U. Khan, A. Mohammad and S. Taj, 2009. Ethnobotanically important plants of district Bannu, Pakistan. *Pakistan J.Pl. Sci.*, **15**(2): 87-93.

Kupchan, S.M., R.J. Nox, J.E. Kelsey and J.A.S. Renauld, 1964.Calotropin, a cytotoxic principle isolated from *Asclepias curassavica* L. *Science*, 25 (December 1964).

Lucansky, T.W. and K.T. Clough, 1986. Comparative anatomy and morphology of *Asclepias perennis* and *Asclepias tuberosa* subspecies *rolfsii. Bot. Gaz.*, **147**(3): 290-301.

Lyons, R.E. 1986. *Asclepias tuberosa.* In: *Handbook of Flowering*, vol. V (ed. Halevy, A.H.), pp. 22-28. CRC Press, Boca Raton, Florida, USA.

Lyons, R.E. and J.N. Booze, 1983. Effect of photoperiod on first year growth of two *Asclepias* species (abstr.). *HortSci.*, **18**(4): 575.

Puri, H.S. 2003. Rasayana: Ayurvedic herbs for longevity and rejuvenation. Taylor &Francis, London.

Raghavan, R.S. 1957. Chromosome numbers in Indian medicinal plants. *Proceedings Plant Sciences*, **45** (6): 296-298.

Shannon, T.R. and R. Wyatt, 1986. Reproductive biology of *Asclepias exaltata. Amer. J. Bot.*, **73**(1): 11-20.

von Maydell, H.J. 1986. Trees and shrubs of the Sahel - their characteristics and uses. GTZ 6MBH, Eschborn.

Wyatt, R. 1980. The reproductive biology of *Asclepias tuberosa.* I. Flower number, arrangement and fruit-set. *New Physiol.*, **85**: 119-131.

Wyatt, R. 1981. The reproductive biology of *Asclepias tuberosa.* II. Factors determining fruit-set. *New Physiol.*, **88**: 375-385.

4

Barleria

M. Ganga, K. Padmadevi, V. Jegadeeswari, P. Mekala and Sanyat Misra

Family: **Acanthaceae**

Common names: Barleria, Bluebell barleria, Crested Philippine violet, Philippine violet, Hindi-*Jhate*, Kannada-*Spatika*, *Mullujaji* or *Gorate* and Tamil-*December poo'it*

Introduction and Botany

The genus *Barleria* was erected after J. Barrelier, a French botanist who died in 1673. Barleria is a small evergreen shrubby ornamental which grows 60-150 cm high with many branches on the main stem. The flowers are violet, pink or white and bell shaped. It is native to India and Africa and is grown in many tropical and subtropical countries, out of which the majority are developed in Africa though firmly naturalized in Hawaii. *Barleria* is a pantropical but predominantly an Old World genus with about 100 cultivated species mostly confined to the Old World tropics (Bailey, 1957) though a total of some 150 species occur worldwide (Bailey, 1942), with its greatest centre of species diversity in tropical East Africa, followed by South Africa and Asia, especially India, China and Myanmar where it is cultivated in rural home gardens for loose flowers (Balkwill & Balkwill, 1998). Jayanthi (2006) stated that the genus contains some 230 species out of which 19 species have been reported to occur in South India. Balkwill & Balkwill (1997) and Mabberley (2008) stated that it is the third largest genus in the family~ Acanthaceae with 300 species. Balkwill & Balkwill (1997) stated that India ia home for 32 species. Whereas Karthikeyan *et al.* (2009) described 29 species, one subspecies and six varieties. Shendage and Yadav (2010) recognised 26 species, one subspecies and one variety in India. In India it is grown on commercial scale and there are 21 species indigenous to India that are spread over the hills of Bengal, the Deccan Plateau, Tamil Nadu, Karnataka, Maharashtra, Madhya Pradesh and the north-eastern Himalayan region (Bhattacharjee, 1980, Jayanthi, 2006). Flowers of *Barleria* are used as offerings to Gods and Goddesses, for making garlands, for ceremonial

Barleria cristata

offerings and for hair decorations by ladies. Plants are used in mixed borders, ornamental hedges and also for fencing due to presence of spines in certain species.

Barleria can easily be distinguished from other genera of Acanthaceae by a combination of three characters; calyx 4-partite with 2 large outer segments and 2 smaller inner ones, spheroidal, honey-combed pollen grains and the predominance of double cystoliths (calcium oxalate crystals) in the epidermal cells. It is an evergreen shrub growing from 0.6 to 1.5 metres. Young shoots are dark green in colour and turn to brown as they mature. The nodes are swollen, leaves are dark green, stalked, simple, opposite, oval-shaped with pointed tips, margins not serrated, about 7 cm in length and 4-5 cm in width having long spines at the axils of the leaves. Flowers axillary or terminal and are borne either solitary or in spike inflorescence. The thorny calyx is present at the base of the flowers and has 4 sepals with 2 outer larger. Corolla tubes long and are swollen at the tip at bud stage and, when open, it is funnel shaped with 5 rounded ovate lobes, and wider at the top and split. Flowers are delicate which may be white, yellow, blue, purple or pink in colour. Sepals four in twins, fruits are capsules with 2-4 seeds, which are hairy and compressed. When the seeds are mature, the capsules split open and seeds are thrown out like fire crackers in the sunlight.

The chromosome numbers that have been determined in the family are almost exclusively confined to a few species. Narayanan (1951) gave the chromosome numbers of 28 species belonging to 18 genera. Raman and Kesavan (1964) studied meiosis in three wild species of *Barleria, viz. Barleria buxifolia, Barleria prionitis* and *Barleria noctiflora* and also four cultivated varieties of *Barleria cristata* and stated that Narayanan's (1951) study in respect of *Barleria buxifolia, Barleria prionitis* and two varieties of *Barleria cristata* does not conform to the counts made in this investigation. Narayanan (1951) determined the somatic chromosome number of *Barleria prionitis* and *Barleria buxifolia* as 2n=30 and 38, respectively, besides the diploid number of four flower-colour varieties of the cultivated species, *Barleria cristata, viz.* white, pink, striped and blue were also reported to be 34, 36, 38 and 40, respectively. The haploid chromosome number for three wild species of *Barleria, viz. B. buxifolia, B. noctiflora* and *B. prionitis* and four cultivated varieties of *B. cristata* has been determined to be n = 20. Meiosis has been studied and aberrations such as heterozygosity for one translocation, degeneration of P.M. cells, non-synchronisation of bivalents, chromosome mosaic cells, lagging chromosomes and sporads with a varying number of spores have been recorded (Raman and Kesavan, 1964).

Species and Varieties and Their Propagation

Barleria cristata L is the most common species grown widely. It is an erect, branched, leafy, hairy shrub, originally from India, growing below 900 metres a.s.l. It is about one metre tall and during the flowering period is completely covered with flowers. The twigs are four-sided and elliptical and the leaves decussately placed. Corolla white, purple or purplish, red in colour, or purple with white stripes or spots, 4 to 5 cm long. Corolla tube obliquely, funnel-shaped, widening towards the top, and containing a great deal of honey, the limb of the flower is 5-lobed and spreading out, one of the lobes being almost spherical and the other four obliquely oval. These little shrubs are commonly used as hedges. It is propagated by cutting and seeds. The important varieties are: 'Aliporens', 'Carnoceana', 'Candida', 'CO 1 Barleria' developed in 1984 at TNAU and bears pink flowers after 7 months of planting, producing 2.11 kg flowers per year per plant, 'Dichotoma', 'Lanchestry', 'Lilacyna', 'Pulcherrima', 'Punitata', 'Selestris', 'Striata', etc. A new variant of the indigenous plant, *Barleria cristata,* with attractive leaf variegation, derived through selection, is reported as a new addition and is named as

'Variegated Philippine Violet' (Rajvikraman *et al.*, 2000). 'Philippine Violet' which arose as a spontaneous mutant in *Barleria cristata* is a soft-wooded shrub with striped-violet and white flowers, and is submitted for registration of plant variety right in Australia, which has been named as 'Jestreak' (Hilder's Nursery, 2000). Chezhiyan *et al.* (1985) evaluated six clones of *B. cristata* and recorded significant differences among them with regard to branch and spike number, plant spread, yield, 100-flower weight and flower colour and suggested that as there had been great variation in these characters among different clones so there is further scope for genetic improvement under this species.

Barleria involucrata Nees (bush type) originates in India, growing below 800 metres a.s.l. and grows 90 to 180 cm tall, with green branches and decussately placed, pointed leaves, 16 cm long and 6.5 cm wide. The flowers are very large, mostly 1-3 together in the leaf axils, calyx green with lobes, two of which are very large and enclose the other two, 5 cm long. Corolla 5-lobed, pretty blue-lilac in colour, 8.5 cm wide at the top, and the tube is 5 cm long and dark purple in the throat. This species, and especially its var. 'Elata Clarke' is most suitable in the lowlands for planting in beds and for forming groups. The plants need some shade and should be pruned each year to keep them in shape. These are propagated by cutting. The other species like *B. cerodia, B. nitida* and *B. strigosa* can also be grouped as bushy type (Jayanthi, 2006). *B. cortalica* grows up to 2 metres height bearing white, yellow or deep blue flowers. Smaller types are *B. acanthoides* which bears ash-coloured flowers and grows 30-45 cm tall; *B. buxifolia*, a prickly bushy shrub with small ovate leaves and white flowers; *B. cuspidata*, another bushy species producing large yellow flowers; *B. highlose*, also a bushy shrub which bears flowers of different colours; *B. mysorensis*, also a bushy shrub producing blue flowers; and *B. tomentosa*, a South Indian species bearing rose-coloured flowers.

Hedge (fencing) types are *Barleria prionitis*, a medicinal plant, which is very thorny and much branched, originates in South Africa, growing 45 cm to 2 metres tall with green twigs and in the axils there are one or many little prickles, leaves narrow and long, and it can be grown up to 800 metres a.s.l. The flowers are sessile, yellow, strongly resembling those of *Barleria lupulina*, but somewhat larger. This like the previous species, is suitable both for hedges and planting in beds, especially in arid regions. It is propagated by seeds and cuttings. It prefers full sunlight and well drained soil. *Barleria gibsoni* originates in India which is a glabrous and branching shrub with pale-bluish-purple flowers.

Herbaceous ones are *Barleria montana* and *B. grandiflora* where plants are soft and succulent and the flowers white, rose or ash-coloured. **Thornless** species are *B. flava* which originates in Arabia. It grows 90 cm tall and bears yellow flowers and *B. strigosa*, a native of India which bears large blue flowers. **Indoor** types are *B. lupulina*, a spiny shrub from Madagascar with erect and strongly branched stem which grows from 1-2 metres high with dark purplish-brown branches, a decorative shrub excellent for hedges, for beds and for group planting. The leaves are shining dark green, entire-margined, linear-lanceolate, 8 to 10 cm long, 2 cm broad, with red margins and middle vein. The flowers are golden-yellow, with round bracts, dark brown or green at the base, all united into spicate, with 10 cm long inflorescence. These are propagated by seeds or cuttings. *B. macconoii* bears attractive blue flowers and grows to a height of about 1 metre. **Night-flowering** type is *B. noctiflora* with oblong and acute leaves, the spines with leafy margins and the flowers which open after the sunset are dull white having a very long corolla tube. **Climbing** type is *Barleria repens* with attractive pink flowers.

Propagation

Barleria is propagated by both sexual (seeds) and vegetative means (stem cuttings). Commercial method of propagation is seeds. The seeds are sown on raised beds or in plastic bags with proper ventilation. Water soaked seeds are sown in straight lines at 10 cm distance in the seed bed to a depth of 2-3 cm. Seeds germinate in 7-8 days and within 1-2 months plants will be ready for transplanting. For vegetative propagation, 4-5 nodes are selected for making cuttings and planted in a rooting medium. Before planting, the cuttings are treated with IBA and placed in the mist chamber for rooting (Jayanthi, 2006).

Cultural Practices

All types of soils that are well-drained are suited for cultivation of barleria, the porous sandy-loam soil with a pH range of 6-7 which is rich in organic matter is most suitable for its cultivation, and the red sandy soils that are found in parts of Maharashtra, Karnataka and other parts of South india are also quite suitable for its cultivation. However, mild climate is best suited for its commercial cultivation as loose flowers. The period and intensity of flowering is much less in more humid areas.

The land selected for cultivation is ploughed 2-3 times and brought to a fine tilth. Pits of 0.75 m x 0.75 m size are dug at 1.5 metre distances. The dug up soil is mixed thoroughly with well rotten farmyard manure at the rate of 4.5 kg per pit. Well decomposed leaf mould is also preferred. After planting, the soil around the plant should be raised to a height of 15-20 cm to prevent water-logging. Application of 15: 15: 15 of NPK per plant is recommended. In a trial at Solan (H.P.) when Priyanka Sharma *et al.* (2013) gave 0, 15 and 30 g per square metre of nitrogen, phosphorus and potassium, they recorded maximum plant height, plant spread, primary and secondary branch numbers, and leaf number per plant under 30 g/m^2 each of NPK though flowering characters such as earlier bud formation, flowering, flowers per cluster, flower clusters per plant and duration of flowering were observed maximum under 15 g/m^2 of N and 30 g/m^2 each of P and K. Around 4,500 plants can be planted in one hectare area. The spacing recommended is 1.5 x 1.5 metres. February to March is the best planting season. Initially, it requires plenty of water, in summer twice a week but in winter once in 10-12 days is enough. Weeding is carried out whenever the need be but hoeing is necessary to loosen the soil. Pruning is done after flowers are harvested. It helps in sprouting of new shoots and to get more flowers per plant. Pruning is done every year. Dried and criss-cross branches are also pruned.

Growth and Flowering

A pot culture experiment was carried out to study the effects of growth promoters (maleic hydrazide, TIBA, IAA or GA at 100 or 1,000 ppm) and growth retardants 4,000 ppm CCC (chlormequat) or 5,000 ppm B-Nine (daminozide) which were applied as foliar sprays to *Barleria cristata* plants. Growth retardants were applied three days after application of growth promoters or growth promoters were applied 3 or 15 days after application of growth retardants. GA resulted in increased growth compared with untreated plants when applied either before or after growth retardants (Samanta and Das, 1991). The growth retarding chemicals, CCC and B-Nine have been found to be very effective in dwarfing shoot length of *Barleria cristata*. Pronounced inhibition in shoot length was also recorded with foliar sprays of MH and TIBA. GA_3 treated plants promoted shoot length markedly, irrespective of all treatments. Synergistic promotion was observed in GA_3 interactions with growth retardants (Bhattacharjee, 1976).

Two *Barleria cristata* cultivars, one with violet and other with white flowers, were treated with a single foliar spray of Cycocel (chlormequat) at 0, 250, 500, 750 or 1,000 mg/l in trials to examine if they would make good pot plants. Both cultivars showed a significant linear reduction in canopy height, more pronounced being with increased application rates as there had been drastic reduction in internodal length. Canopy diameter was reduced only in the white cultivar at rates 500 mg/l. Final plant quality as assessed by compactness and foliage colour, improved only in white cultivar in response to plant growth regulator treatment (Henny, 1995).

Plants grown in 10-cm pots were given a single foliar spray of B-9 (daminozide, 2,500 rng/litre), Bonzi (paclobutrazol, 25 mg/litre), BA (500 mg/l) or Cycocel (chlormequat, 1,000 mg/litre). BA was used in an attempt to stimulate lateral branching while the other growth regulators were applied to reduce plant height. At 14 weeks after treatment, Cycocel treated plants were significantly shorter than untreated plants, the other treatments had no significant effect on growth. Leaves of Cycocel treated plants of the white cultivar showed a significant increase in width compared with control plants (Henny *et al.,* 1994).

Post Harvest

Harvesting is done daily or on alternate days manually. Yield of flowers is higher during the months of June, July and also in February. The plants can be maintained for up to five years. The average yield of flowers is about 5 t/ha and during summer it is 8-10 t/ha. The harvested flowers are placed in bamboo baskets, covered and then transported to the local market. The flowers are sold as early as possible due to their low keeping quality.

Insect-Pests and Diseases

Leaf eating caterpillars make snaky channels and blister-like excavations on the leaf surface. They bore holes in the growing flower buds. Such buds either do not open or the opened flowers are badly deformed. They attack the interior tissues of leaves and are controlled by spraying malathion at 0.2 per cent or Fenitrothion at 0.05

per cent. Preventive spray of Sevin (0.15 to 0.2 per cent) or 5.0 per cent wettable powder is effective or Metasystox at the rate of 0.1 per cent can also be used. **Mealy bugs** are a very common pest in *Barleria* which suck the sap. They are controlled by spraying any systemic insecticide at 20 day's intervals.

Necrotic leaf spots can be controlled through fortnightly spraying of Blitox or any other copper fungicide. **Rusting** or reddish brown symptoms appear on the leaves and branches which is controlled by spraying Dithane Z-78 or Dithane M-45 at 0.2 per cent. Prophylactic sprays with Bordeaux mixture is also recommended.

References

Bailey, L.H. 1942. The Standard Cyclopedia of Horticulture (vol.I). The Macmillan Co., N. Y.

Bailey, L. H. 1957. *Manual of Cultivated Plants.* The Macmillan Co., New York, U.S.A.

Balkwill, M.J. and K. Balkwill, 1997. Delimitation and intra-generic classification of *Barleria* (Acanthaceae). *Kew Bull.*, **52**(3): 535-573.

Balkwill, M.J. and K. Balkwill, 1998. A preliminary analysis of distribution pattern in a large, pantropical genus, *Barleria* L. (Acanthaceae). *J. Biogeography*, **25**: 95-110.

Bhattacharjee, S. K. 1976. Interaction of growth promoting and inhibiting chemicals on shoot length of *Barleria cristata. Bull. Bot. Soc. Bengal*, **30** *(*1-2): 143-146.

Bhattacharjee, S.K. 1980. Grow barleria for profitable flower production. *Indian Farmers' Digest*, **13**(5-6): 43-44.

Chezhiyan, N., V. Ponnuswami, S. Thamburaj and J.B.M. Abdul khader, 1985. Clonal evaluation in *Barleria cristata* L. (commercial flower crop). *South Indian Hort.*, **33**(1): 38-41.

Henny, R.J. 1995. Effects of Cycocel on growth of two cultivars of *Barleria cristata* L. (Philippine violet). *Plant Gr. Reg. Soc. America Quart.,* **23**(4): 214-217.

Henny, R.J., B. Beall, N. Freeman and D. Schmaltz, 1994. Effect of BA, B-nine, Bonzi and Cycocel on growth of Philippine violet (*Barleria cristata* L.). *Proc. Fla St. hort. Soc.*, **107**: 177-178.

Hilder's Nursery, Upperstone via Ingham, QLD, Australia, 2000. Variety: Jetstreak, application No. 2000/055. *Plant Varieties Journal* **13**(2): 16.

Jayanthi, R. 2006. Barleria. Advances in Ornamental Horticulture, Vol.II (ed. Bhattacharjee, S.K.). Pointer Publishers, Jaipur.

Karthikeyan, S., M. Sanjappa and S. Moorthy, 2009. Acanthaceae. In: *Flowering Plants of India-Dicotyledons* (vol. I) *(Acanthaceae - Avicenniaceae).* Botanical Survey of India, Kolkata, pp. 1-62.

Mabberley, D.J. 2008. *Mabberley's Plant-Book. A Portable Dictionary of Plants, their Classification and Uses* (3rd ed.). Cambridge University Press, Cambridge.

Narayanan, C. R. 1951. Somatic chromosomes in Acanthaceae. *J. Madras Univ.*, **21B**: 220-231.

Priyanka Sharma, Priyanka Thakur, Y.C. Gupta and S.R. Dhiman, 2013. Effect of nitrogen, phosphorus and potassium on growth and flowering of *Barleria cristata* Linn. *Indian J. Hort.*, **70**(3): 442-447.

Rajvikraman, R., J. Thomas and A.G. Pandurangan, 2000. Variegated Philippine Violet (*Barleria cristata* L.) - a new ornamental plant. *J. Eco. Tax. Bot.*, **24**(3): 723-726.

Raman, V. S. and P. C. Kesavan, 1964. Cytological investigation on some cultivated and wild species of *Barleria. J. Indian bot. Soc.*, **43**: 363-368.

Samanta, P.K.S. and P. Das, 1991. Interaction of growth retardants and growth promoting substances in *Barleria cristata* L. *Orissa J. Hort.*, **19**(1-2): 22-26.

Shendage, D.J. and D.S. Yadav, 2010. Revision of the genus *Barleria* (Acanthaceae) in India. *Rheedea*, **20**(2): 81-130.

5

Celosia argentea

S.P.S. Raghava, J.K. Ranjan, Pragya Ranjan, Sanyat Misra and R.L. Misra

Family: **Amaranthaceae**

Common names: Cockscomb (*Celosia argentea, C. a. cristata*), Prince of Wales feathers (*C. a. pyramidalis,* syn. *C. a. plumosa*), Red fox (*C. argentea*), Woolflower

Introduction and Botany

It takes its name from the Greek *keleos* meaning burned due to the burned look of the flowers in some species. Various literature sources mention some 60 species under the genus *Celosia*, all native to Tropical Asia. The dwarf ones which grow up to 30 cm in height are most suitasble for growing in pots while the taller ones can be grown in beds and borders, for cutting and for drying. In certain countries it is used as vegetables. Scientifically proved antidiarrhoeal effect has been observed through use of whole plant extract (Verma *et al.*, 2011). The colour range in the varieties are white, yellow, orange, pink, scarlet, red, maroon, purple, violet, crimson and amaranth.

These are tender and very popular summer flowering annuals, grown for their brightly coloured, showy and agglomerated flower-heads and in some cases even for its coloured foliage, most suitable being in beddings and borders as well as pot plants. The inflorescences can be cut and used as cut flower, *vis-a-vis* dried for indoor decoration. These bear mainly two types of flowers ~ plume-like sprays such as in *Celosia argentea, C. a. pyramidalis* and *C. a. plumosa,* and curiously crested and flattened into a fan-like or cockscomb-shape clusters as in *C. a. cristata.* Plants are herbaceous annual with simple, entire or lobed, alternate and petiolate leaves; cymes dense or lax, bracteate, plumose or cristate in terminal and axillary spikes; stem inflorescences in cultivated types being mostly fascicled; flowers hermaphrodite, small, shiny, white or coloured; perianth quite small, dry, 5-parted, free to base, scarious; stamens 5 and fused below into a cup with free filaments being united at base, filiform above and deltoid below and some species with rudimentary pseudostaminodes; style elongate to obsolete; stigmas 2-3, filiform; and fruit a circumscissile corky utricle bearing 2 to many black, shiny, finely to strongly reticulate or grooved seeds. The chromosome numbers in the celosias have been recorded from diploid to decaploid, i.e. 2n=4x=36 (*Celosia cristata*) to 2n=8x=72 (*C. argentea*)

Species and Varieties

Various sources mention *Celosia* having 35 to 60 species (Anthony Huxley *et al.*, 1992 mentions 50 species) but the fact is otherwise as various species mentioned by various literature sources are, in fact, various ecotypes of one species, *Celosia argentea*, a native of India. Many catalogues mention *Celosia argentea, C. childsii, C. chrysanthaeflora, C. cristata, C. globosa, C. magnifica, C. plumosa, C. pyramidalis, C. spicata, C. thompsonii*, etc., but after all, the botanists have agreed that either all these are forms or synonyms of one species, *Celosia argentea* (Pizzetti and Cocker, 1975). Beckett (1987) also mentions only one species as *Celosia argentea* and its various forms such as *C. argentea cristata, C. a. plumosa, C. a. pyramidalis*, etc. Bailey (1960) mentions some 35 species under the genus and describes *C. argentea* producing plumed inflorescence in almost pyramidal shape [almost all the forms, including *C. cristata* (with crested or cockscomb-type inflorescence) seem to have been derived from this species], *C. cristata, C. huttoni* (a hybrid species valued for its foliage which was developed in Java), *C. pyramidalis* (again a form of *C. argentea*), *C. spicata* (again a hybrid species, different from the originally described *C. spicata*, perhaps a form of *C. cristata*) and *C. thompsonii magnifica* (not a true species but a trade name). Hay and Beckett (1971) mentions *C. argentea* (a native of Tropical Asia, especially India), *C. a. cristata* (cockscomb) and *C. a. pyramidalis* syn. *C.a. plumosa*. Once these most popular flowers started declining their fame but due to introduction of many new hybrids, especially by Japanese seedsmen in various delicate colours, these have regained their former position once again. *Celosia* produces a showy heads of plumed or crested flowers, predominantly red or yellow though original species which is rarely cultivated has white flowers, however, secondary colours available are purple, violet, crimson, orange and amaranth. The best form in *Celosia* is that which produces largest possible crest or plume on smallest plant. The colour variation being more under *C. argentea* than in *C. a. cristata*. The varieties are available in both the types, i.e. tall and dwarf ones under both the kinds, *viz.* plumed or crested ones. Best kind being the ones having 22 to 30 cm plant height and 7.5 to 15.0 cm panicle width but always with minimum of leaf-bearing part of stem. The well-marked colour forms are 8 to 9 in crested type though even more under plumed type. *A. argentea* grows up to 90 cm tall bearing pale-green linear-lanceolate to ovate leaves up to 7 cm or more long with shorter petioles, flowers and their bracts silvery-white being borne in panicles which may be erect or plume- or spike-like, or somewhat drooping sometimes, long peduncled, pyramidal or cylindrical. This species has many forms including *C. cristata*. Its forms are available in shades of red, orange and yellow. *C. a. cristata* is real cockscomb, a form of *C. argentea* with crested flowerheads, predominantly in red, orange and yellow colours. *C. cristata* Childsii group bears globose terminal bosses of flowers, 'Big Chef Mixed' growing to one metre tall and the cauliflower-shaped flowers are large to 15 cm in diameter, 'Jewel Box Mixed' has crested flowers having 12.5 cm diameter and the colours being deep red, gold, yellow, pink and salmon, 'Jewel Box' is available in wide colour range (white, yellow, orange, pink and red), 'Jewel Box Red Dark Leaf' bears leaves of deep bronze colour and the flowers of 10 cm diameter of dark red colour, 'Jewel Box Improved' is dwarf and bears velvety-orange, rose or red flowers having 13 cm diameter, 'Dwarf Fairy Fountain' grows very compact and is branching type bearing long-lasting red, pink and yellow flowers, 'Torcador' bears red flowers of some 30 cm diameter, and the var. 'Fire Chief' is scarlet. *C. a. pyramidalis* (syn. *C. plumosa*) also grows up to 90 cm tall and bears mid-green and ovate leaves, floral plumes are feathery, dwarf types, especially the 'Liliput' growing to 15 cm tall are grown in pots and the tall ones for cutting. The varieties under Plumosa group are 'Apricot Brandy' bears up to 30 cm diameter of flowers in red, yellow, rose and cream colours, 'Fairy Fountain Mix' is free-branching and bears spike of 10-12.5 cm with red or orange flowers, 'Forest Fire' bears maroon leaves and 15 cm tall spikes bearing bright scarlet flowers, Geisha series is very dwarf with heavy spikes and the flower colours are orange, salmon, yellow and red,Kewpee series is extremely dwarf with flower colours of yellow, orange, rose and red, 'New Look' bears deep bronze leaves and feathery flowers of intense scarlet colouring, 'Pink Castle' bears striking rose-pink flowers and Spicata group bears metallic pink or yellow flowers in slender heads. Cv. 'Pampas Plumes' is quite popular worldwide. 'Argentea Plumosa', 'Fairy Fountains', 'Feather', 'Fiery Feather', 'Golden Feather', 'Liliput Mixed', etc. are the dwarf types for growing in pots, while some 60-90 cm tall strains are 'Thompsonii Magnifica' with red flower colour and 'Thompson's Magnifica Mixed' in wide range of colours which can be used for cutting as well as for drying.

Propagation

Celosia is multiplied only through seeds. Daniel *et al.* (2012) stored the seeds of *Celosia* at 28-90% RH and under 45 ^{0}C temperature and reported that drying seeds down to 5% moisture content before storage can improve their longevity. Celosia seeds with full cover against light germinate within 5-10 days. Seeds may be sown in late winter when temperature is around 18-24 ^{0}C. However, after germination, these make luxurious growth at 18-20 ^{0}C. When the seedlings are large enough to handle these are pricked out into a porous, sandy and high-fertility loam-based medium. After pricking out, these can be grown in direct sunlight at 15±5-6 ^{0}C in a humid and buoyant atmosphere. In advanced countries, the plug-growing is most common and these plugs can be stored in light at 10 ^{0}C up to one week.

Cultural Practices and Post Harvest

Mostly the celosias being tropical in nature, they dislike the cold though at a sheltered position these survive the rigours of the winter when temperature goes down and persists even as minimum as 4 ^{0}C, especially the crested types, as in Delhi and other sub-tropical conditions in the plains of north and central India. However, the atmospheric temperature of 40 ^{0}C or above is not injurious to the plants. Under direct sun the colour of the inflorescence is full of shine when summer is mild but in shade that is dull and dark in colour. Short days have been recorded promoting growth while long days increasing the growth and plant height. Winter protection makes them perennial in habit and then these continue for many years. Its shiny-black seeds may be sown in February-March in a frame or greenhouse and planted out in April-May or directly by second week of March and planted out by last week of April in a light and fertile but well-drained soil at a sunny position at a distance of 20-30 cm across, dwarfer ones closely while taller ones distantly. It can be grown effectively in the soil having a pH range of 5.5 to 8.5. Before planting the soil should be dug up at least thrice for pulverization and at the time of final ploughing it should be incorporated at least with 150 quintals of farmyard manure and thoroughly mixed. However, in greenhouses, these can be grown at a temperature of 16 ^{0}C with proper ventilation and regular watering and when once roots fill the pots these may be applied with liquid fertilizer every fortnight. Carter and Grieve (2006) investigated into the salinity tolerance of cut flowers of *Celosia argentea* var. *cristata* 'Chief Rose' and 'Chief Gold' at the U. S. Salinity Laboratory to determine marketability based on stem length, where plants were exposed to differing water ionic compositions and salinity levels which though caused reduction in plant height with increase in the levels but the flowers were of marketable quality at moderate salinity levels. While growing, mulching with straw, dry leaves, organic manures or with black polythene keeps away the weeds, otherwise weeds will have to be taken out manually when and where these are observed. **Disbudding** once in *Celosia* cvs 'Carmine' and 'Chief Gold' in Ghana increased the leaf area, relative growth rate, harvest index, floral head size and flower weight (Adjei-Frimpong *et al.*, 2011).

In a hydroponically grown *Celosia argentea* cv. TLV 8 under greenhouse conditions in Nigeria with an average temperature of 27.7 ^{0}C and light at 1.32 × 10^3 μEinstein M-2 S-1 (PAR) and sulphur at 0, 13.5, 27.0, 40.5 and 54.0 μg g^{-1} of sand gave good growth and development (Ojo and Oputa, 1999). Carter *et al.* (2005) tried saline waste-water for **irrigating** *Celosia argentea* cvs 'Chief Gold' and 'Chief Rose' and recorded these doing well even in between 8-12 dS m^{-1} EC, the former cultivar doing well at 12 dS while latter at 10 dS m^{-1}. The **manure** of water hyacinth (*Eichhornia crassipes*) applied at the rate of 30 g/plant and 60 g/plant to *Celosia argentea* and the assessment of plants after three months of application revealed that 60 g/plant application yielded best quality plants with better floral attributes (Sanni and Adesina, 2012). When nitrogen was applied in the form of urea at 0, 30, 45 and 60 kg/ha to four varieties of *Celosia*, Olaniyi and Ojetayo (2012) in Nigeria observed excellent plant height, number of leaves and fresh & dry shoot yields at 45 kg/ha treatment. Out of three levels of nitrogen in two split doses (basal and top-dressing), and sprayed with CCC (500 or 1000 ppm), PBZ (100 or 150 ppm) or daminozide (1000 or 1500 ppm) at 20 and 40 days after transplanting, Easwaran *et al.* (1999) recorded nitrogen at 150 kg/ha, CCC 1000 ppm and PBZ at 100 ppm increasing leaf N, P and K contents. For dwarfing, Bonzi (paclobutrazol) and Sumagic (uniconazole) have been found very effective while A-Rest (ancymidol), Cycocel (CCC, chlormequat chloride) and B-Nine (daminozide) quite effective (Dole and Wilkins, 1999).

For cut flower, only plume-types are used and when a few plumes in the inflorescence have started flowering, the whole inflorescence is detached and kept in vase water having hydroxyquinoline and citric acid. Usually fully developed inflorescences are harvested but prior to significant seed formation. Optimum transport and storage temperatures for celosia flowers are 10-13 ^{0}C. For drying the inflorescences are removed before start of the seed setting as may be observed by whitening in the basal part of the inflorescence. These are dried through inverted hanging.

Insect-Pests and Diseases

Celosia crop is badly infested with root-knot **nematode** (*Meloidogyne incognita*). Tobih *et al.* (2012) mulched the plantings of *Celosia* with leaves of *Chromolaena odorata, Thevetia peruviana* and *Ocimum viride* to find out if these have any effect on the root-knot nematode infestation and recorded enhanced growth and yield components of the crop in treated plots though there was more root-galling, *vis-à-vis* more crop damage under control. However, they concluded that *C. odorata* mulching was rather more effective as compared to the two others. It is also attacked by red spider mites so plants require to be syringed often and more so as the plant's aerial parts like more moisture.

Sometimes it is observed to be infested by aphids, beetle and lepidopterous larvae, leafhoppers, mites, thrips and whiteflies where aphids, leafhoppers and whiteflies act as vectors for spreading viral diseases. These all can be controlled through 0.2 per cent Rogor.

Foot and **root rot** of celosia is associated with one or more fungi by which root and/or collar portion of the plants rot which eventually causes collapse or wilt of the plants. Elm disease caused by *Ceratocystis ulmi* may cause extensive damage to this crop. These can be controlled through drenching with 0.2% Thiride or Captan. In Israel, the phytoplasma and/or spiroplasma disease was identified in *Celosia* spp. where it caused general leaf yellowing, stunting, witch-broom of axillary shoots and floral phyllody (Gera *et al.*, 2006).

References

Adjei-Frimpong, P., J. Ofosu-Anim and J.C. Norman, 2011. Effects of disbudding on growth analysis of two *Celosia* cultivars. *Adv. Hort. Sci.*, **25**(4): 223-231.

Anthony Huxley, Mark Griffiths and Margot Levy, 1992. *The New Royal Horticultural Society Dictionary of Gardening (A-C)*, pp. 560. The Macmillan Press Ltd., London.

Bailey, L.H. 1960. *The Standard Cyclopedia of Horticulture* (vol I), pp. 709. The Macmillan Co., New York, USA.

Beckett, K.A. 1987. *The RHS Encyclopedia of House Plants Including Greenhouse Plants*, pp. 138-139. Salem House Publishers, 462 Boston St., Topsfield, Massachustts 01983, USA.

Carter, C.T., C.M. Grieve, J.A. Poss and D.L. Suarez, 2005. Production and ion uptake of *Celosia argentea* irrigated with saline waste waters. *Scientia Hort.*, **106**(3): 381-394.

Carter, C.T. and C.M. Grieve, 2006. Salt tolerance of floriculture crops. In: *Ecophysiology of High Salinity Tolerant Plants*, pp. 279-287 (eds Almal Khan, M. and D.J. Weber). Springer Science +Business Media, Dordrecht, Netherlands.

Daniel, I.O, M. Kruse and A. Borner, 2012. Predicting longevity of *Celosia argentea* L. seeds during storage. *Acta Hort.*, No. 953, pp. 319-324.

Dole, J.M. and H.F. Wilkins, 1999. Plant Growth Regulation. In: *Floriculture ~ Principles and Species*, pp. 90-104. Prentice Hall, Upper Saddle River, New Jersey 07458.

Easwaran, S., K. Balakrishnan, C.M. Pappiah and A. doraipandian, 1999. Influence of nitrogen and growth retardants on foliar nutrient content in *Celosia cristata. South Indian Hort.*, **47**(1-6): 345-347.

Gera, A., L. Maslenin, A. Rosner, M. Zeidan and P.G. Weintraub, 2006. Phytoplasma diseases in ornamental crops in Israel. *Acta Hort.* (*Proceedings of the XIth Int. Symp. on Virus Dis. of Ornam. Plants*, held at Taichung, Taiwan, on March 9-14, 2004; ed. Chang, C.A.), No. 722, pp. 155-161.

Hay, R. and K.A. Beckett, 1971. *Reader's Digest Encyclopaedia of Garden Plants and Flowers*, pp. 127-128. Reader's Digest Association, London.

Ojo, D.O. and O.C. Oputa, 1999. Influence of sulphur on growth, yield and quality of *Celosia.J. Pl. Nutr.*, **22**(9): 1379-1388.

Olaniyi, J.O. and A.E. Ojetayo, 2012. Effect of N on growth, yield, nutrient uptake and quality of *Celosia* varieties. *J. Agric. Biol. Res.*, **3**(1): 227-231.

Pizzetti, I. and H. Cocker, 1975. *Flowers ~ A Guide for Your Garden*, pp. 203-204. Harry N. Abrams Inc., Pulishers, New York, USA.

Sanni, K.O. and J.M. Adesina, 2012. Response of water hyacinth manure on growth attributes and yield of *Celosia argentea* L (Lagos spinach). *J. agric. Tech.*, **8**(3): 1109-1118.

Tobih, F.O., A.A. Adegbite and C.C. Ononuju, 2011. Evaluation of some plant materials as organic mulch for the control of root-knot nematode (*Meloidogyne* spp.) and its impact on growth and yield of *Celosia argentea* L. *J. agric. Sci.*, **1**(4): 195-202.

Verma Harjinder, Sanjeev Mittal, Sandeep Goyal and Arun Mittal, 2011. Anti-diarrhoeal activity of ethaholic extract of *Celosia argentea* (Linn.). *Intern. J. pharma. Innovations*, **1**(2): unpaginated.

6

Crossandra

M. Ganga, Sanyat Misra, R.L. Misra, P. Mekala, V. Jegadeeswari and K. Padmadevi

Family: **Acanthaceae**

Common names: Crossandra, Firecracker flower/Firecracker plant/Flame flower, Hindi-*karsaraiya*, Telugu-*kanakambaram*)

Introduction

Crossandra is an important loose flower crop, especially in South and East India. It is commonly known as 'Firecracker plant' because of the cracking sound produced during splitting of seed-pod. The word crossandra is derived from Greek words *krossoi* for 'fringe' and *andros* (*aner*) (genitive of 'aner') for 'male', referring to its fringed anthers. All the species of crossandra are native of the Arabian Peninsula, Tropical Africa, Madagascar, India and Sri Lanka, and in South India where it is a commercial crop, its cultivation is estimated to be more than 4,000 hectares. Bhattacharjee (1980) stated its description, commercial utility and its scope of cultivation in South India. This is also a valuable pot flower in Denmark as well as in Sweden and Hungary. In Denmak, it is only cv. 'Deep Orange' (a triploid) which is being grown commercially. It is a small evergreen shrub that is quite hardy and has the ability to produce beautiful flowers with a remarkable range of colours (yellow, pink, salmon-pink, orange and red) almost round the year. Crossandra ranks fourth next to rose, jasmine and tuberose in loose flower production in India.

Flowers are commonly used for hair adornment. Though not fragrant, but array of colours and the brightness, the flower shape, being light in weight, good keeping quality and its floriferousness has made this loose flower very popular in India. The flowers are used for making garlands either alone or in combination with jasmine flowers. Crossandra flowers in combination with jasmine flowers are known for the fragrance and contrasting colours. Crossandra can be grown in home gardens, rockeries and in herbaceous borders. The varieties with medium and small stature having proper architectural management can be very well grown as potted plants.

Crossandra infundibuliformis (syn. *C. undulaefolia*) var. 'Danica' is a valuable flowering potted plant that has been introduced to the international floriculture market. In Sri Lanka also it is very popular plant for pot-planting and for landscape use. This is in great demand in many of the European and Asian countries. Sri Lanka alone exports more than six lakh rooted cuttings of crossandra per annum. The flower is a valuable pot-plant in Sweden, Denmark and Hungary.

Origin and Botany

Crossandra Salisb. is a genus of about 50 species which are shrubs or herbs from Tropical Africa, Madagascar and Arabia, where these are grown as perennial shrubs. *Crossandra undulaefolia* Salisb. (Syn *C. infundibuliformis* Nees) is a widely cultivated species, native to East Indies. However, it is sometimes considered to be native of South India and Sri Lanka. *C. guineensis* is native of East Africa while *C. nilotica* has its origin in East Africa and Mozambigue. *C. flava* is a native of tropical West Africa. The genus is also native to India, Sri Lanka and Arabia. Crossandra plants are distributed in tropical as well as subtropical regions of the world such as South Asia, South America, South Africa and Madagascar. In India, it is commercially cultivated in the southern states of Karnataka, Tamil Nadu and Andhra Pradesh.

Crossandra is an evergreen sub-shrub which may sometimes grow up to 90 cm tall. The leaves are upright, glossy gardenia-like, opposite or verticillate, hairy, 5.0-12.5 cm long and moderately toothed. Hairy inflorescences appear in dense erect raceme-like sessile spikes, yellow to red shades, flowers overlapping one another with prominent bracts. Individual flower is a cylindrical corolla, somewhat enlarged at the throat, more or less curved, composing of a slender tube forming five long-lasting and rounded lobes, with a flat or spreading oblique limb having four stamens in pairs. The common orange type has handsome four-sided spikes of scarlet-orange flowers. The perianth has five segments, the two upper ones being smaller. The fruit a capsule is oblong, acute and 4-seeded.

Its basic chromosome number is n=10. Chezhian (1990) treated seeds of crossandra (orange type) with gamma rays and EMS and obtained good results with plant vigour and floral yield in M_1, M_2 and recurrent M_2 generations, frequency of chlorophyll mutants being higher under gamma rays at 10 kR while the high frequencies of viable mutants were higher under EMS, and this way he isolated 120 useful mutants. Lalitha (1991) continued with M_2 mutants of Chezhian (1990) taking 47 families and continued in M_4 by selecting 46 families from M_3 and recorded high magnitude of phenotypic and genotypic variances for number of spikes, *vis-à-vis* flowers per plant both in M_3 and M_4 generations, high co-heritability estimate between number of flowers per plant and yield, and finally isolated eight M_4 progenies from five M_4 families as superior plants.

Crossandra comes to flowering 2-3 months after planting and continues to bear flowers throughout the year, with a drop in production during the rainy season. The flowers open from the base of the spike. Two flowers which are diagonally opposite in the spike, open at the same time. It takes two days for complete opening. Therefore, picking of flowers has to be done on alternate days. Depending on the length of the spike, it takes nearly 15-25 days to complete flowering on a spike. After flowering, the spent spikes should be removed. Fully open flowers remain fresh on the plant for about three days but when picked they fade away in 36-48 hours.

Species and Varieties

There are about 50 species but only a few, *viz. Crossandra infundibuliformis, C. guineensis, C. mucronata, C. nilotica* and C. *subacaulis* are cultivated (Sambandamurthy, 1989).

C. flava Hook., a native to tropical Africa, this is an unbranched shrub, about 15-20 cm tall with erect and green stems. Leaves are dark green, glabrous, opposite, obovate to lanceolate, base tapering, lower leaves stalked and upper ones sessile, spike densely packed, 4-sided with spiny yellow green bracts and bright yellow flowers.

C. guineensis Nees, a native to forest floors of western tropical Africa, this grows under sun as well as in partial shade to about 30-60 cm high with terminal, narrow, solitary and free-flowering spike bearing pale-lilac or white bracts which bear no spines. Its elliptic leaves are very attractive, and bear golden reticulated nerves.

C. guineensis

C. nilotica

C. flava

C. pungens

C.p. cv. 'Orange'

'Delhi'

'Lutea Yellow'

'Sebaculis Red'

C. undulaefolia

'Arka Ambara'

'Arka Kanaka'

C. mucronata Lindau, a native to Ethiopia, Somalia, Uganda, Kenya, Tanzania, Zimbabwe and South Africa, it is an erect, evergreen and clump-forming sub-shrub growing to a height of 30 cm. Its broadly elliptic leaves appear in whorls of four, flowering spike is erect and up to 12 cm long and bright orange to dark red flowers appear in densely hairy bracts.

C. nilotica Oliver, a native to Kenya and Mozambique, it is a semi-woody free-flowering species suited for partial shade, *vis-à-vis* moist conditions, growing to about 60 cm in height, bearing dark green, glossy and elliptic leaves in whorls of four and some 5-7 cm long. Bracts are hairy, green, veined and about 2 cm long. Yellow-orange to brick-red flowers are borne in dense spikes and each flower bears a long slender corolla tube at the end of which there are 5 petal-like lobes where 3 lower petals form a lip.

C. pungens Lindav, a native to East Africa, it is a dense sub-shrub growing up to 60 cm tall. Oblong leaves are dull green with conspicuously coloured veins, and the flowers are orange or yellow. The orange type is the only variety of *Crossandra undulaefolia* being grown commercially in South India but it is highly prone to wilt and nematodes (root-knot and lesion; Sivagami Vadivelu and Muthukrishnan, 1980).

C. undulaefolia Salisb (*C. infundibuliformis* Nees) is a native of India, Pakistan and Sri Lanka. It is a short branching, evergreen perennial shrub producing flowers freely in dense sessile spikes. It is the only commercially cultivated species growing to some 30-90 cm in height. Leaves are dark green, lustrous, opposite, narrowly oval, pointed, about 5-12 cm long and 2-5 cm wide with wavy margins and are crowded near the inflorescence. Spikes are terminal and axillary, large, densely packed, 4-sided and 7.5-15.0 cm long. Calyx is 5-lobed, irregular and bright orange. Salmon to scarlet showy flowers have prominent spiny bracts, individual flowers consisting of long slender tube some 2.5 cm long which terminate in a 5 lobed limb, and each flower consists of normally 4 stamens. The four cultivars of the species are (i) 'Orange' with bright orange flowers, a tetraploid with chromosome numbers 2n = 40 which bears seeds profusely and breeds true, (ii) 'Delhi' is a triploid (2n=30) and produces attractive deep orange flowers, (iii) 'Lutea Yellow', a tetraploid (2n=40) with orange-yellow flowers, most suitable for hanging baskets and for planting in the front rows of shrubbery, and (iv) 'Sebaculis Red', a tetraploid (2n=40) and hardy cultivar with red flowers, possessing high degree of tolerance to nematodes.

The three forms like 'Yellow', 'Deep Orange' and 'Sebaculis Red' are grown to a small extent. The deep orange type does not set seed. It is a triploid, similar to 'Mona Wallhead', a cultivar which is grown in Denmark. Variability available in crossandra is very low as compared to other flowers even then there is a remarkable range of colours, varying from orange, pink, red, yellow to red via pink to orange, and double coloured blue types with white throat. Recently white, light green and violet types have become popular. Important varieties of crossandra developed by various agencies are 'Mona Wallhead', an introduction from Sweden, growing to 50 cm height, has compact growth with shiny black-green foliage, and deep salmon to pink irregular flowers having prominent bracts; 'Fortuna', a selection which has extended plant life, quite resistant to temperature fluctuations due to greatly improved root system and can tolerate as low as 10 ^{0}C, foliage quite glossy, flowering lavish with longer duration and flowers quite bright; 'Diane', a new and robust form of the species *C. infundibuliformis* having compact growth bearing large and wide leaves, flowers dark salmon-orange, and petals much larger and rounded than the commonly cultivated variety; 'Savindi', a mutant of pink-flowered 'Danica' with orange flowers, developed through *in vitro* gamma irradiation; 'Lakshmi', a high yielding (up to 75 kg flowers per day per hectare from second year of planting) gamma irradiated mutant developed through cv. 'Delhi' bears large orange flowers; 'Maruvur Arasi', a high yielding bright deep-red variety with longer floral stalks which bears over 75 flowers per stalk was developed through cv. 'Delhi' by gamma irradiation; 'Dr. A.P.J. Abdul Kalam', a deep red variety developed through cv. 'Delhi' by gamma irradiation which flowers all the year round, and is also resistant to major pests; 'Kanakadhara', a mutant of cv. 'Delhi' developed through chemical mutagenesis (sodium azide), is a vegetatively propagated crossandra which produces brilliant orange flowers yielding 25% more than the mother cultivar; 'Vijaya Kanakambaram', a tall mutant of free-branching and profusely flowering type having stout stems bearing long inflorescences with large light reddish-orange flowers, developed through seed treatment of a local cultivar with colchicine, which yields 75 kg flowers per day per hectare from 90 days after planting; 'Raj Kanakambaram', a leaf-variegated mutant of cv. 'Delhi' with deep orange flowers bearing 15 cm long inflorescences and has potentiality of producing up to 75 kg flowers daily per hectare through proper management practices; 'Subasu', a mutant of cv. 'Delhi' with long inflorescences bearing light orange flowers of medium size; 'Neelambari', a mutant of local yellow cultivar, bearing dense green foliage and normal size of flowers on long inflorescences; 'Arka Ambara', a cross of 'Local' × 'Lakshmi' with large flowers, the flower weight being double of the cv. 'Local'; 'Arka Kanaka', a cross of 'Local Yellow' × 'Delhi' with large and showy orange flowers which weigh roughly double of the mother parent; etc. TNAU, Coimbatore developed two hybrids, one highly tolerant to nematodes and with large flowers having yellowish-orange petals, by crossing two local types collected from Palani and Madhuramalai hills, and the other with dark pink flowers by crossing an orange cultivar with 'Sebaculis Red'. 'Tropic Flame', an orange-flowered cultivar evolved in USA multiplies through seeds. 'Florida series' which is more cold tolerant than others is vegetatively propagated and has four colours, *viz.* yellow, orange, red-orange and red (Dole and Wilkins, 1999). Apart from these, 'Aboli', a variety of *C. undulaefolia* is being grown in Andaman & Nicobar Islands.

Propagation

Propagation is by seeds, stem cuttings and layering. Triploid varieties since do not set seeds, hence are propagated only through cuttings.

Approximate seed count in crossandra is 140-218 per gramme and for planting one hectare of land it requires some 2.0-2.5 kg seeds. A spacing of 50 cm x 50 cm has been reported as ideal for seed production. Seeds attain physiological maturity 55-60 days after flowering and can be stored for six months after treating with Captan @ 2g/kg of seeds otherwise these will remain viable only for three months. However, fresh extracted **seeds** are the best for nursery raising. Seeds are sown in June-July on 10-cm raised beds having one metre width or in polythene bags, lightly covered with 3 mm of medium, at the rate of 2.25 kg per hectare under cover at 26 ^{0}C day and 21 ^{0}C night temperatures (Nau, 1993) where these germinate but sporadically within 3-4 weeks (Harthun, 1991) though Langefeld (1991) recommends 29/21 ^{0}C day/night temperatures for optimum germination within 14 days and after germination these grow comfortably at 18-20 ^{0}C temperature. Carbaryl 15% dusting of the seed-sown beds will protect the seedlings against cut worm infestation in the nursery beds.

Triploid crossandras can be propagated by **ground layering** as these do not set seeds. Well conditioned layers can be planted from June to July in pits of 15 cm x 15 cm x 15 cm at a spacing of 90 x 90 cm. About 12,500 layers are needed to cover one hectare area. Plants will start flowering within a month after planting. **Softwood tip cuttings** of 5-8 cm length are taken during March-June when day/night temperatures are 29/21 ^{0}C and light intensity 1500-1800 fc, treated with some rooting hormone and planted in 6.5-cm pots or directly under mist, root in 3-4 weeks time which are now called as liners (Dole and Wilkins, 1999). During summer white plastic tents while during winters clear plastic tents should be used (Harthun, 1991; Langefeld, 1991). Triploids are also commercially multiplied only through cuttings which are transplanted when sufficient roots have developed. For better results, rooting hormone IBA at 3000 ppm can be used. Propagation by cutting has lower rates of multiplication and hence **micropropagation** is resorted to. Multiple shoots from shoot tips have been regenerated on full strength MS medium fortified with BAP 1 mg/l. Callus induction from leaf bits has been reported on full MS medium fortified with 2,4-D 1 mg+BAP 0.5mg/1.

Cultural Practices

Crossandra can be grown in almost all types of **soils**. However, well drained loamy soils rich in organic matter with a pH of 6.0 -7.5 are well suited. The liners when after rooting are provided with high nutrition, warm temperatures and long days of summer, the flowering is encouraged (Dole and Wilkins, 1999). In Danish greenhouses, natural flowering occurs from March to September (Wilkins, 1985) though under open conditions in India it is for more extended period. The crop should not be cultivated in nematode infested soils. In an experiment, Velu (1988) reported that increasing levels of calcium carbonate adversely affected plant growth and proportionally reduced the availability of Fe, regardless of the from in which it was supplied. Alkaline and saline soils are not suitable as plant develops deficiency symptoms such as chlorosis leading to improper growth of plant and poor flower production (Sambandamurthy, 1989). Clay or heavy soil is not suitable since poor drainage will induce root rot.

Crossandra is a crop suited to tropical areas and it cannot tolerate **temperature** below 13 ^{0}C and dies when frost occurs. It performs well in warm humid areas with temperature ranging from 20 to 32 ^{0}C. Luxuriant growth occurs at 30 ^{0}C. A temperature of 25-28 ^{0}C is most ideal for seed germination, while for cotyledon emergence and true leaf expansion a temperature of 21-22 ^{0}C is required. The optimum minimum night temperature is 18-24 ^{0}C, while minimum day temperature requirement is 21-28 ^{0}C. Plants should be held at 21 to 22 ^{0}C till transplanting and lower temperatures turn the foliage purple due to unavailability of phosphorus. Till emergence of cotyledon, 95% RH is maintained, while after germination excess moisture is harmful and can lead to damping off and then 70-80% RH is maintained for first six weeks of its growth and then reduced to 30-50% (Dole and Wilkins, 1999). Plants can be grown under sunlight as well as partial shade. Light is not required for germination but after germination it needs light levels between 1000 and 1500 foot candles. **Photoperiod** does not affect flower induction. However, recommended light intensity during production is 3000-3500 fc as below 1800 fc the growth is slow (Dole and Wilkins, 1999). *C. infundibuliformis* cv. 'Mona Wallhead' with light durations of >12 h or light intensities of >20μmol s^{-1} m^{-2} produces good quality plants with enhanced flowering duration. Plants raised through seeds require 31 weeks to attain marketable size though from liners it requires 12-14 weeks, albeit it takes additional 6-7 weeks for producing liners from rooted cuttings.

The land is thoroughly **ploughed** 2-3 times and about 25 tonnes FYM/ha is incorporated in the soil at the time of first ploughing. Seedlings are first raised in the nursery beds in April-May. The seed requirement per hectare of land is 2.0-2.5 kg. Seedlings 30-days old, i.e. 4-5 leaf stage are transplanted. Proper spacing between plants is a pre-requisite for better development of plant and higher flower yield. Seedlings are transplanted at 30 x 50 or 60 × 40 cm **spacing**. At 30 x 50 cm spacing, 66,666 seedlings per hectare can be accommodated, however, at 60 × 40 cm spacing as is being practiced in Andhra Pradesh for commercial cultivation, it can accommodate some 41,667 seedlings per hectare of land (Girwani *et al.*, 2011). A spacing of 45 x 30 cm is recommended for rooted cuttings of cv. 'Delhi'. In yellow crossandra the spacing followed is 90 x 90 cm. For pot culture as is being done in Denmark, seedlings and cuttings are transplanted into 6.25-cm pots singly, and the liners which are transplanted into 10-cm, 11 or 15-cm pots with final spacings of 16.5 × 18.0, 18.0 × 18.0 or 23.0 × 23.0 cm, respectively, are kept pot-to-pot commercially until shipped.

Weeding, application of manure, fertilizers and earthing-up are combined for easy **upkeep** of crop and for labour saving. Application of Metachlor 1.25 kg per hectare can control weed population. In a chemical weedicide trial taking 1.0 and 1.25 kg a.i./ha metolachlor, alachlor and pendimethalin, Sharadamma *et al.* (2002) recorded good control of weeds with high yield in crossandra crop with 1.25 kg/ha of metolachlor and alachlor though pendimethalin was even injurious to the plants. Mulching can be done to check weeds and to save water.

Pinching is done to encourage bushy growth and axillary branches. In Denmark under greenhouse conditions, pinching of *Crossandra infundibuliformis* cv. 'Mona Wallhead', four weeks after planting leaving three pairs of leaves, induces more lateral shoots and increases number of inflorescences. The pinching should be carried out three weeks after final potting so that 5-6 leaves remain on the plants. However, disbudding is not required in crossandra crop. **Pruning** of crossandra is done to improve plant vigour and flower quality. Light pruning is followed in late winter after flowering.

Once the seeds are formed and reach the advanced stage of development, growth and flowering decline. When the flowering is over, the spent spikes are removed regularly.

Feeding and Irrigation

For increased growth and flowering in crossandra, application of **farmyard manure** 25 t/ha, N, P, K and zinc sulphate is recommended. Application of 25 kg N, 20 kg P_2O_5 and 75 kg K_2O per hectare per year is recommended. Entire P and K are applied in the field before planting. After the cuttings have rootings, a high dose of nutrient is required for good vegetative growth. A complete liquid fertilizer with 200 ppm N should be regularly provided (Harthun, 1991; Langefeld, 1991). In Andhra Pradesh, they recommend FYM 25 t/ha + 5 t/ha *neem* cake and 38 kg N, 62kg of P_2O_5, and 62 kg of K_2O/ha (Girwani *et* al., 2011). Nitrogen should be applied in two split doses 60 days and 120 days after planting. Fe, Mn and Mg being critical nutrients should be certainly applied. Foliar application of $ZnSO_4$ at the rate of 0.5% once in 60 days is recommended for better quality flowers. Fe deficiency is a common problem in crossandra. Subramanian (1982) with Fe application obtained 46-89% flower yield increase in calcareous soil while 39.4% in black soil. In fact, Fe fertilization interferes with P nutrition while it does not affect K and Ca nutrition. Foliar application of 1% Fe_2SO_4 + 2% urea is recommended once in every 30 days. Application of 100% NPK + *Azospirillum* + phosphobacteria increases flower yield (41.72 g/plant) with maximum returns per rupee invested (1:3.5). Twenty five per cent reduced rate of application of inorganic nutrients is sufficient if applied along with *Azospirillum* to get yield on par with blanket recommendation of inorganic fertilizers in crossandra. Venkatesh and Nalawadi (1991) when applied 150 kg N/ha, obtained maximum flower yield, i.e 10.07 quintals per hectare in crossandra. Swaminathan (1995) reported improved vegetative parameters (plant height, number of branches, number of leaves, fresh and dry weight of shoot) by application of nitrogen at 120 kg per hectare. Nitrogen @ 180 kg/ ha along with foliar spray of $ZnSO_4$ @ 0.5 % and Fe_2SO_4 @ 1% recorded higher flower yield (Aruna, 1999). Swaminathan and Sambandamurthy (2000) reported improved vegetative (plant height, number of branches, number of leaves, leaf area per plant) and flowering parameters (number of spikes per plant, spike length, number of flowers per spike and flower yield per plant) with N 120 kg + K_2O 70 kg/ha + *Azospirillum* 2 kg/ha + FYM 30 t/ha. Bhavanisanker and Vanangamudi (1999) found that the combined application of 100% recommended N as urea + *Azospirillum* and 75 % reccommended P as superphosphate + phosphobacteria gave higher flower yield (1.106 and 1.474 kg/plant) at 3 and 17 months after planting.

Crossandra needs adequate moisture in soil immediately after planting. **Irrigation** is repeated the third day after planting to facilitate easy rooting and better establishment. It requires irrigation once in 4-5 days. Crossandra is drought tolerant but requires free watering for better yield and at the time when the plants are flowering. Overwatering is detrimental due to root being susceptible to *Rhizoctonia* and starved watering burns the leaves.

Growth and Flowering

Flower yield can be increased by applying GA_3 +1% urea 1 month after planting. Rooted cuttings of *Crossandra infundibuliformis* cv. 'Delhi' treated with gibberellic acid and/or BA (100 or 200 ppm) with or without 1% urea, one month after planting and subsequently at bimonthly intervals, results in maximum leaf area and number of branches, yield per hectare, number of spikes per month and spike length. Total floret number per spike is highest with 200 ppm GA_3. Veluswamy *et al.* (1974) reported routine sprayings of 1500-5000 ppm B-Nine (more pronounced being 2500 ppm after 2 weeks of pinching or when new axillary shoots are up to 5 cm long) on crossandra plants for reducing internodal length and for inducing more branching without delaying the flowering, however, Adriansen (1979) recommended A-Rest in water culture in recirculating ebb-and-flow irrigation solution. Sreekala (1997) when tried GA_3 at 100 and 200 ppm alone or in combination of 1% urea, recorded increased height especially under 200 ppm GA_3 though branching , spreading habit and certain other vegetative characters were found significantly increased under BA 200 ppm + urea 1%, followed by 200 ppm GA_3. She recorded earliest flower emergence under CCC 500 ppm, followed by paclobutrazol 1,000 ppm + urea 1% though paclobutrazol suppressed the growth and flowering significantly. BA 200 ppm + urea 1%, followed by 200 ppm GA_3 though were effective in increasing number of flowers per truss and trusses per plant as well as stalk length but CCC 500 ppm was found more consistent for these characters from October to January under Thrissur conditions

Sandved (1974) and Christensen (1975) reported that when growing at 7-13 ^{0}C when light intensity in Denmark is low during winter, after six weeks the plants die, though at 26-28 ^{0}C day and 21 ^{0}C night temperatures for six weeks after propagation and then 21/18 ^{0}C day/night temperatures encourage comfortable growth in USA (Harthun, 1991; Langefeld, 1991), while at 24 ^{0}C there had been rapid flowering (Pedersen, 1975; Adriansen, 1976). Moes (1976) recommended 22 ^{0}C medium air temperature as ideal for its growing. Pedersen *et al.* (1973), Pedersen and Moes (1973) and Moes (1976) stated that when high plant quality is desired during low light intensities, low forcing temperature is always better. Harthun (1991) stated that whether photoperiod is 8 or 16 hours long, the flowering time is the same, i.e. 100-105 days after planting in the final pot sizes.

Post Harvest

Crossandra commences flowering 70-75 days after planting and continues to flower throughout the year with a drop in production during rainy season. Flower opening in sequence starts from the base of the spike. Two diagonally opposite flowers open together. It takes about two days for complete opening of the flower. The flowers are harvested on alternate days in the early morning by pulling corolla out of the calyx. Flowers are ethylene sensitive. After flowering is over, dried spikes are removed. For local market, flowers are packed in cloth or polythene bags. Its potted flowering plants can be transported under ethylene-free atmosphere at 10-13 ^{0}C temperature, as temperatures higher than 18 ^{0}C are damaging and lower than 6 ^{0}C blacken the leaves. If 1000 fc light is being provided, a flowering crossandra plant will last up to one month though ethylene is damaging causing buds to abscise (Nowak and Rudnicki, 1990; Nell, 1993) so 0.4-0.5 mM STS can be used to negate its effect and to prolong the flower longevity (Moe and Fjeld, 1987).

Yield of flowers depends on fertility status of soil, cultural operations, variety and climate of the area. There is a drop in production during rainy season. With a plant population of 40,000 per hectare with spacing of 50 x 50 cm, an yield of 4.5 tonnes per hectare can be obtained. 'Delhi' crossandra yields 6-7 tonnes per hectare. However, the high yielding cultivars may yield up to 7-9 tonnes per hectare. Crossandra flowers are very light and on an average about 15,000 flowers make one kilogram. In Andhra Pradesh the yield per hectare is reported as 2.5-3.75 t under 60 × 40 cm spacing with an income of Rs. 1.0-1.5 lac (Girwani *et al.*, 2011). The crop can be retained in the field for about three years after which it has to be removed as it becomes uneconomical.

While fully opened flowers may remain fresh for about three days on the plants, the harvested flowers will fade away in 36-48 hours. Freshly harvested flowers should be packed in cloth or polythene bags for local market. Shelf life of flowers is 2-3 days and soaking flowers in 0.5% boric acid or 1% aluminium sulphate can extend the shelf life by three hours. Sreekala (1997) reported that pulsing of crossandra flowers with 15% sucrose + 500 ppm 8-HQ and refrigerated storage in 200-300 gauge polythene sleeves extends the postharvest life of the stalked trusses, however, without storage, the vase life was recorded maximum from the plants treated with 100 ppm BA or those kept in a vase solution of 3% sucrose + 100 ppm $CoCl_2$.

Insect-Pests and Diseases

Colonies of nymphs and mature females of **scales** [(brown scale, *Saissetia nigra*; white scale, *Orthezia insignis*, Ranjith, 2013); *Saissalia viridis*] infest leaves, stems and flowers. They feed on sap and excrete honeydew, which makes plants sticky and encourages growth of sooty moulds. Persistent infestations weaken the growth. Chemical treatments are only effective when applied to kill young crawlers before they have settled and started to form protective scales. Application of malathion at 0.3 - 0.6 kg a.i./ha kills the scale insects. Adults and eggs of **white fly** (*Lipaleyrodes* sp.) are seen on underside of leaves and the upper surfaces are often fouled with sticky honeydew and sooty moulds, *vis-à-vis* leaf chlorosis. Infestations reduce plant vigour and sooty moulds make infested plants look unsightly. Yellow spots and other discolourations develop on leaves. Serious damage can be prevented by examining plants regularly. If adult or immature whiteflies are seen, spraying should be done thoroughly with appropriate contact insecticides such as Malathion @ 0.5 kg/ha or systemic insecticides such as dimethoate @ 0.4 kg/ha at weekly intervals.

Fusarium and Phytophthora wilt in crossandra

Sorghum midge (*Contarinia maculipennis*, Ranjith, 2013), a small fly lays eggs within the floral buds. Orange crawling maggots after hatching start feeding developing buds and ovaries causing failure to bud opening, their rotting in severe cases and pupating therein. The nymphs and adults of **pentomid bug** (*Gynencia affinis*) infest the shoots and flowers during winter causing brown lesions on the infested parts (Ranjith, 2013). Control measures for both the pests, *viz.* sorghum midge and pentomid bug are the same as for white fly. Spider mites and aphids can also be a problem in crossandra cultivation, especially under greenhouse.

Nematodes commonly infesting on crossandra are *Meloidogyne incognita, M. arenaria, M. javanica* (root-knot nematode), *Pratylenchus delattrei* (lesion nematode) and *Longidorus africanus* (needle nematode). The roots exhibit brown to black lesions and prominent galls. The needle nematode causes forking and clubbing of the root tips. It causes chlorotic leaves which ultimately turn white at the advanced stage of infestation. In case of severe attack, plants become stunted without any side shoots and remain defoliated. Application of Phorate @1g/plant weekly after planting checks the nematodes. Mites have also been found infesting the crossandras in Andhra Pradesh.

Wilt in crossandra is caused by *Phytophthora* and *Fusarium solani* and due to their attack the leaves turn pale, margins start showing pinkish-brown discolouration and these finally droop. Discolouration may extend up to the midrib within 7-10 days. Stems also shrivel and the dark lesions can be seen on the roots extending to the collar region, and at this time the invasion by *Pratylenchus* nematodes will predispose the plants to the fungus, causing severe wilting. To get rid of this problem, the crossandras should not be grown in nematode infested soils, however, nematodes can be controlled by soil application of Phorate @ 1 g/plant 10 days after transplanting. Crop rotation with marigold will also reduce nematode population in the field. Carbendazim 0.1 per cent or copper oxychloride 0.25 per cent at 30 days interval soil drenching will control the disease. Pulling and destroying the infected plants will also minimize the disease. Application of *Trichoderma viride* and *T. harzianum* cultured in FYM or on neem cake before transplanting effectively reduce the incidence of *Fusarium oxysporum*. **Collar rot** (*Sclerotium rolfsii*) saprophytic fungal pathogen is present in the soil. Its infection causes sunken and brown-coloured spots on bark on the collar region which later on extend to the roots and its brown to black discolouration is seen after peeling such barks. The leaves of infected plants show pink discolouration and fall off. It invades the plants mostly at the stage of transplanting. Seedling root dip with 0.1% thiram and in a mixture of *Pseudomonas* sp. + *Trichoderma viride* will control this problem. **Stem rot** occurs in the form of pre-emergence damping off through rotting of the stems and roots. Brown to black lesions develop on stem just above soil level and girdle the stem. Lesions extend to the upper part of the stem and result in collapse of seedlings. Fortnightly spray of benomyl at 0.2% and Captan at 0.25% at early stage of infection controls the disease. **Alternaria leaf spot** is caused by *Alternaria amaranthi* var. *crossandrae* where small, circular or irregular yellow spots on the upper surface appear which soon enlarge, turn brown and develop into dark brown concentric rings. Infected leaves become yellow and drop off prematurely. Spraying with benomyl 0.1% or mancozeb 0.2% or carbendazim 0.1% can control the disease. **Root and crown rot** is caused by *Phytophthora nicotinae*, the soil borne pathogen, which enters the plants through damaged tissues. It causes violet discolouration of leaves and the basal regions of plants rot. It is better to destroy the affected plants. *Rhizoctonia* and *Thielaviopsis* (**black root rot**) may be a serious problem during and after propagation (Daughtrey and Chase, 1992) so deep planting should be avoided to avoid their incidence including that of even *Pythium*. **Flower rot** is caused by *Fusarium pallidoroseum* where leaves turn purplish, most prominently on the veins, internodes are shortened and flowers decay with cottony colonies of the fungus. These symptoms begin on older leaves and extend upward into younger leaves towards the apical branches.

The branchlets also turn purple. **Leaf blight** is caused by *Collectorichum crossandrae* and in its infection the leaf margins are slightly rolled up with brownish to reddish and depressed necrotic areas, more prominently on lower leaves and that too near the margins. Such leaves shrivel and drop off leaving a whorl of young leaves only at the top. Proper sanitation, removal of affected parts, sulphur spraying at 7-10 days interval at its first appearance and spraying with 0.1% benomyl or carbendazim, or 0.2% mancozeb will control this malady effectively.

References

Adriansen, E. 1976. Effect of SADH and ethephon on *Crossandra infundibuliformis 'Mona Wallhead' Tidsskrift for Planteavl*, **80**:31-40.

Adriansen, E. 1979. Growth retardation of potted plants cultivated in water culture with ancymidol applied to the nutrient solution. *Acta Hort.*, No. 91, pp. 387-394.

Aruna, 1999. *Yield and Physiological Responses to Nitrogen Levels and Micronutrients in Crossandra* [*Crossandra infundibuliformis* (L) Nees]. Ph.D (Hort) Thesis, Tamil Nadu Agricultural University, Coimbatore.

Bhattacharjee, S.K. 1980. Crossandra- a boon to flower growers. *Indian Farmer's Digest*, **13**(2): 29-30.

Bhavanisanker, K. and K. Vanangamudi, 1999. Integrated nutrient management in crossandra (*Crossandra infundibuliformis* L.). *South Indian Hort.*, **47**(1/6): 125-129.

Chezhiyan, N. 1990. *Studies on Induced Mutations in Crossandra* [*Crossandra infundibuliformis* (L.) Nees]. Ph. D. Thesis, TNAU, Coimbatore.

Christensen, O.V. 1975. The influence of low temperature on flowering of *Beloperone, Crossandra, Jacobinia* and *Mackaya. Tidsskrift for Planteavl*, **79**: 459-462.

Daughtrey, M. and A.R. Chase, 1992. *Crossandra.* In: *Ball Field Guide to Diseases of Greenhouse Ornamentals*, p. 49. Ball Publishing, Geneva, Illinois.

Dole, J.M. and H.F. Wilkins, 1999. Plant Growth Regulation, pp. 90-104; and Crossandra, pp. 279-282. In: *Floriculture~ Principles and Species.* Prentice Hall, Upper Saddle River, New Jersey.

Girwani, A., Lalitha Kameswari, A.L.N. Prasad, K. Purushotham, Ramesh Kumar, P. Naveen Kumar, Gunjeet Kumar and Tarak Nath Saha, 2011. *Prospects of Floriculture in Andhra Pradesh*, pp. 19. Directorate of Floricultural Research (ICAR), Pusa Campus, New Delhi.

Harthun, E. 1991. *Crossandra.* In: *Ball Red Book* (5th edn, ed. Ball, V.), pp. 474-475. Ball Publishing, West Chicago, Illinois.

Lalitha, S. 1991. *Evaluation of M_3 and M_4 Progenies of Crossandra [Crossandra infundibuliformis*(L.) Nees] *for Economic Characters.* M.Sc. (Hort.) Thesis, TNAU, Coimbatore.

Langfeld, J. 1991. *Crossandra.* In: *GrowerTalks' on Crop Culture* (ed. Hameick, D.J.), pp. 63-65. Ball Publishing, Geneva, Illinois.

Moe, R. and T. Fjeld, 1987. Keeping quality of potted plants as influenced by ethylene. *Gartner Tidende*, **101**: 1580-1583.

Moes, E. 1976. Temperature for *Crossandra. Gartner Tidende*, **11**: 163-164.

Nau, J. 1993. *Crossandra.* In: *Ball Culture Guide ~ The Encylopedia of Seed Germination* (2nd edn), p. 28. Ball Publishing, Batavia, Illinois.

Nell, T.A. 1993. *Crossandra.* In: *Flowering Potted Plants ~ Prolonging Shelf Performances*, pp. 31-32. Ball Publishing, Batavia, Illinois.

Nowak, J. and R.M. Rudnicki, 1990. *Postharvest Handling and Storage of Cut Flowers, Florist Greens and potted Plants.* Timber Press, Portland, Oregon.

Pedersen, A.M. 1975. Standard growing of *Crossandra infundibuliformis* (L.) Nees 'Mona Wallhed'. *Tidsskrift Planteavl*, **79**: 463-473.

Pedersen, A.M., E. Adriansen and E. moes, 1973. Trials at Sørbus with *Crossandra. Gartner* Tidende, **89**: 307-310.

Pedersen, A.M. and E. Moes, 1973. Trials with *Crossandra. Gartneryrket*, **63**: 606-610.

Ranjith, A.M. 2013. Crossandra. In: *Identification and Management of Horticultural Pests*, pp. 137-138. New India Publishing Agency, Pitam Pura, New Delhi-110 088.

Sambandamurthy, S. 1989. Crossandra. In: *Commercial Flowers* (eds Bose, T.K. and L.P. Yadav), pp. 856-858. Naya Prokash, Calcutta.

Sandved, G. 1974. Spring flowering of *Crossandra* cultivar 'Mona Wallhead' after overwintering at different temperatures. *Gartneryrket*, **64**: 607-609.

Sharadamma, B.C., J.V. N. Gowda and T.V. Ramachandra Prasad, 2002. Chemical weed management in crossandra. In: *Floriculture Research Trend in India* (eds Misra, R.L. and Sanyat Misra), pp. 212-215. ISOH, Div. of Floriculture and Landscaping, IARI, New Delhi.

Sivagami Vadivelu and T.S. Muthukrishnan, 1980. *Nat. Sem. Prod. Tech. Comm. Flower Crops.* T.N.A.U., Coimbatore.

Sreekala, C. 1997. Effect of Bioregulators on Growth, Flowering and Postharvest life of Crossandra {*Crossandra infundibuliformis* (L.) Nees.}. M.Sc. Thesis. Kerala Agricultural University, Vellanikkara, Thrissur (Kerala).

Subramanian, T.L. 1982. *Studies on the Iron Nutrition of Crossandra using Tracer Technique.* Ph. D. Thesis, TNAU, Coimbatore.

Swaminathan, V. 1995. *Nutritional Studies in Triploid Crossandra.* Ph.D. (Hort.) Thesis, Tamil Nadu Agricultural University, Coimbatore.

Swaminathan, V. and S. Sambandamurthy, 2000. Studies on integrated nutrient management on growth and yield of triploid crossandra [*Crossandra infundibuliformis* (L.) Nees] cv. Delhi. *Crop Research* (Hisar), **20**(2): 334-337.

Velu, G. 1988. *Research and Development Report* (TNAU), **5**(1-2): 89-90.

Veluswamy, P., T. Thangaraj and S. Muthuswamy, 1974. A study of complete morphology of four cultivars of *Crossandra* (*Crossandra undulaefolia*, Salisb.). *South Indian Hort.*, **22**: 81-83.

Venkatesh, N.H. and U.G. Nalawadi, 1991. Influence of intra-row spacing with different levels of nitrogen and phosphorus on growth and flower yield in crossandra (*Crossandra undulaefolia*). *Karnataka J. agric. Sci.*, **4**(1-2): 117.

Wilkins, H.F. 1985. *Crossandra infundibuliformis.* In: *Handbook of Flowering* (ed. Halevy, A.H.), pp. 356-359. CRC Press, Boca Raton, Fla, USA.

7

Dahlia

Sanyat Misra, R.L. Misra, B.K.Banerji, Atul Batra and A.K.Dwivedi

Family: **Asteraceae**

Common names: Cactus dahlia (*Dahlia juarezii*), Collarette dahlia, Dahlia, Decorative dahlia, Old-fashioned dahlia, Fancy dahlia/Pompon dahlia/Show dahlia/Single dahlia (*D. rosea* syn. *D. barkeriae*/*D. coronata*/*D. crocata*/*D. nana*/*D. purpurea*/*D. royleana*/*D. superflua*/*D. variabilis*), Paeony-flowered dahlia, Tom Thumb dahlia, Tree dahlia (*D. imperialis, D. excelsa* syn. *D. arborea*), etc.

Introduction and Origin

Dahlia is named so after 18th century Swedish botanist Anders Dahl (1751-1789) who had been author of '**Observationes Botanicae**' and who was also pupil of Linnaeus. It is native to Mexico, Central America, and Colombia. Garden dahlia (*Dahlia variabilis*) was already domesticated in Mexico before 1575. There are at least 27 species of dahlia such as *D. imperialis* (syn. *D. arborea, D. lehmannii, D. maximillainan* and *D. maxonii*), a native of southern Mexico and Central America though distributed in the rocky slopes and fields extending from Chiapas in southern Mexico to Guatemala, El Salvador, Costa Rica and northern Colombia at elevations ranging from 750 to 2,700 metres, is a herbaceous perennial possessing similar morphology and growth habits as to those of usual dahlias except that it grows faster, attains a tree-size and flowers only in the temperate climatic conditions during autumn. This has two variants, white and purple, and further bearing single or double flowers, growing up to 8.0 metres in height. Francisco Hernández visited Mexico in 1615 and noticed two spectacular varieties of dahlia,

Single-flowered

Anemone

Star shaped

Decorative

Collarette

Peony

Semi-double

Pompon

Cactus

Plate-1: Close up view of flower heads of different types of dahlias

which he mentioned in his account of medicinal plants of New Spain. The French botanist Nicolas Joseph Thiéry de Menonville sent the dahlias to Mexico to steal the 'Cochineal Insect' valued for its scarlet dye. He noted in his official report the strangely beautiful flowers he had seen there, and this was published in 1787. Seeds sent to Madrid from botanical garden of Mexico flowered for the first time in the botanical garden in October 1789, which was named *Dahlia coccinea* by Antonio José Cavanilles, the head of the Madrid Botanical Garden, in his *Icones Plantarum* published in 1791. A few seeds were secured by Lord Bute and sent to England, where these flowered but were lost afterwards. Introduction of dahlia to the Netherlands florists from mexico occurred when a box of its root-tubers was sent from Mexico there, where only one plant survived the trip which produced spectacular red flowers with pointed petals and this was named as *Dahlia juarezii*. European nurserymen used this in crosssing with certain other dahlias already present there, as parents and these became the earliest progenitors of all modern dahlia hybrids. The *Jardin des Plantes* in Paris received *Dahlia variabilis* in 1802 from Madrid which was grown in 1804 by the gardener at Holland House, Kensington. It was known to Aztecs long before as they gathered and cultivated dahlias for food, ceremonies, as well as for social occasions and there the long woody stem of one variety was being used for small pipes. An early breeder of dahlia was Comte Léon Charles LeLieur de Ville-sur-Arce who had four varieties to work with and by 1806 he had produced three double-flowered dahlias. Since 1813, many amateur breeders produced thousands of cultivars, usually chosen for their stunning and brightly coloured waxy flowers. Dahlia was named the national flower of Mexico in 1963. In Japan it is known as *Tenjikubotan* which literally means 'paeony of India'. According to the Japanese language of flowers, it means 'good taste'. In Germany, the dahlia was known during most of the 19th century as Georgia being named after the naturalist Johann Gottlieb Georgi of St. Petersburg, Russia. It has diverse flower colours ranging from white, cream, yellow, orange, pink, to scarlet, red, purple, magenta, bicolours, multicolours and variegated though lacks in true blues, greens and blacks, and flowerhead size starting from some 2 cm to 30 cm in diameter, with enormous shapes and so the height of the plant which ranges as low as 30 cm to as high as 2.4 metres. It is used as pot plant for exhibition, instant landscape use, for bedding, in the borders, as loose flowers especially the bedding types grown through seeds, as dry flowers, for cut flowers especially the pompon types where their flowers last for up to 10 days, and in the use for various other landscape works. Apart from its various ornamental uses, in southern mountainous Mexico the plant has culinary properties though petals are not that tasty, however, the petals are used in a foot-soak in tired feet (Robert, 2000). Its petals and thinly sliced tubers were used by Aztecs and later by the Mexicans as excellent skin treatment. Its roots are valued for containing inulin inside though certain antibiotic compounds are found in the tuber skins. Dahlia tubers contain inulin and fructose and small quantities of medicinally active compounds such as phytin and benzoic acid (Whitley, 1985). Flavonoid **glycosides** such as cyanidin, apigenin and luteolin have been extracted from *D. variabilis* cv. 'Dandy' (Nordström and Swain, 1953), malonylated chalcone glucoside, i.e. 4'- malonylsophoroside and 4'-malonyglucoside of butein (the yellow pigment) in *D. variabilis* (*D. pinnata*) cv. 'Cocquencourt' and *D. coccinea* petals (Harborne *et al.*, 1990) while flavanone 3-hydroxylase, catalysing the conversion of flavanones to dehydroflavonols was detected in flower extracts of cyanic strains of dahlia but not in the white-flowering mutants (Forkmann and Slotz, 1984).

Botany and Cytology

It is a tuberous-rooted herbaceous perennial, the tuber having growing points only in the crown where it is attached with the aerial stem. Stems are hollow, mostly erect and branched, glabrous or scabrous, leaves opposite, pinnatifid, heads long-peduncled consisting of several hundred individual florets in a cyme, small to large (10 to 30 cm in diameter), open to ball shape, disc yellow and fertile, ray florets neutral or pistillate and spreading, and fruits oblong or obovate with many flat seeds. Capitulum inflorescence, consisting of ray and disc florets both, blooms in succession where beauty lies in the ratio of ray versus disc florets and their arrangement on the head where ray florets (the only female flowers) open first from the outermost side towards the centre. In singles these are a few or only one while in doubles these are many. The disc florets are complete flowers having stigma in the tip of style while the ray florets are only the female flowers with stigma contained in the inner base of the petals. Due to its capitulum-head being quite large, compact and composite, complete emasculation in dahlia heads is not an easy task (Swami Vinayanand, 1991).

Cytological investigations of 24 double and four single horticultural varieties of *D. variabilis* and *D. inperialis* have been carried out by Singh and Roy (1973). *D. variabilis* is octaploid and *D. inperials* tetraploid. Gupta *et al.* (1972) reported two different base numbers in dahlia, *viz.* x=8 and x=18, the latter being derived from x=9. Mehra and Ramanandan (1974) determined 2n=64 chromosomes (Octaploid) in *D. coccinea* and *D. rosea* with base number of x=8. The great variety results from dahlias being octoploids (they have eight sets of homologous chromosomes, whereas most plants have only two). They have haploid **chromosome** numbers of 16, 17, 18 and 32 with most of the garden cultivars including *Dahlia variabilis* being tetraploids having 2n = 64. Lawrence (1942) stated that since Humboldt sent home the seeds of dahlia in 1804, it took 12 years for evolution of every colour which we find today. The reasons for the profusion of form and colour are ~ (i) that dahlias being self-incompatible can never breed true, (ii) the factors producing flower colour are in duplicate, 2 for flavone (ivory and yellow) and 2 for anthocyanin (pale and deep pigmentation), and (iii) each of these factors is represented 4 times. It was therefore evidently suggested that the plant arose as a hybrid between polyploid members of 2 colour groups and that doubling of the chromosome number accompanied this hybridization.

Classification, Species and Varieties

Dahlia is a genus of some 27 to 36 perennial, bushy and tuberous species with more than 20,000 hybrids and varieties which had been developed through hybridization and selection, *vis-à-vis* mutation breeding. Present day dahlia grown in the gardens is mostly cultivars and hybrids. The *Dahlia* species are *D. apiculata, D. australis, D. barkeriae, D. brevis, D. campanulata, D. cardiophylla*, *D. coccinea* (syn. *D. bidentifolia, D. cervantesii, D. crocea*), *D. cuspidata, D. dissecta, D. excelsa* (syn. *D. arborea*), *D. foeniculifolia, D. hjertingii, D. hortensis* (referring to all the garden cultivars), *D. juarezii, D. linearis, D. macdougallii, D. merckii* (syn. *D. glabrata*), *D. mollis, D. neglecta, D. parvibracteata, D. pinnata, D. pteropoda, D. rudis, D. rupicola, D. scapigera, D. scapigeroides, D. sherffii, D. sorensenii, D. spectabilis, D. subliginosa, D. tenuicaulis, D. tenuis, D. tubulata* and *D. variabilis*. Dahlias like cool hilly climate where flower colours can be seen with a larger size of flowerheads. These bloom during winter and springs in the Indian plains and during summer to autumn in the temperate regions. Modern garden dahlias are derived from three Mexican species, ***Dahlia pinnata*** (low growing up to 90 cm height, leaves roundish, leaflets up to 25 cm long, flowers large and double of purple colour or red with bluish tinge), ***D. coccinea*** (*syn. D. bidentifolia, D. cervantesii, D. crocea*) having pinnate or bipinnate leaves, flowers on the stem appear at 90 cm height, as single red or scarlet, and sometimes with orange or yellow centres, and ***D. rosea*** (leaves typically 1-pinnate, seldom bipinnate, flowers single rose, and is a parent of old fashioned dahlias, *viz.*, Single, Pompon, Show and Fancy types, sometimes double forms are also available) which may themselves all originate from one species, *D. variabilis* (sometimes semi-double type of *D. rosea* is called *D variabilis*). ***D. imperialis***, an erect species up to 8-9 meters tall with 2-3-pinnately parted leaves, flowers 10-18 cm across, white or purple and single or double during summer and autumn, grows as escape on Indian hills throughout the country and is known as tree dahlia. ***D. scapigera*** is a dwarf species, some 50 cm high, leaves pinnate, segments 3-7, capitula few, usually solitary, slightly drooping, ray florets white to pale purple, base sometimes yellow, disc florets yellow, apex purple, and flowering during summer and autumn. ***D. excelsa*** (syn. *D. arborea*) grows up to 6 metres or more, stems become woody in mild climates, leaves up to 75 cm long and 60 cm wide with as many as 25 leaflets, flowers purple or crimson-pink and up to 12 cm across, ***D. juarezii*** grows to 90-150 cm, is cactus type with florets of irregular length and overlapping, flowers bright scarlet, and flowering in mid- to late-summer, and ***D. merckii*** (syn. *D. glabrata*) has bipinnate leaves, flowers lilac, rays pistillate and is much valued for its more attractive foliage. Broadly, dahlias are divided into two groups ~ A) **border dahlias** most suitable for mixed borders, and ~ B) **bedding dahlias** which are grown as annuals from seeds.

A. **Border dahlias** flower freely in India for 4-5 months, on the hills from July onward until cut down by the frost though in the plains from December to April when the flowers dry up due to scorching sun. Border dahlias are classified into 10 groups.

   Single-Flowered. Height 45-75 cm, planting distance 45-60 cm, flower diameter up to 10 cm, each flower has only one outer ring of florets which may overlap and a central disc, and flowering by July end on the hills and in January in the plains. This can further be divided into ***show singles*** (flower heads not more than 7.5 cm across, only 8 rays, smooth, slightly recurved at tips and overlapping in a perfect round form), ***singles*** (rays not so overlapping as in *show singles*, and tips separated) and ***mignon*** dahlias (similar to *singles* but plants not more than 45 cm high). It is ideal for pot culture or patio plantings. The varieties are 'Bambino' (white), 'Chessy' (bronze-yellow), 'Coltness Gem' (a dwarf bedder, seed-sown various colours), 'G.F. Hemerik' (soft orange), 'Inflammation' ('Red Liliput,' red), 'Irene van der Zwet' (soft yellow), 'Murillo' (lilac-pink with deeper centre), 'Nellie Geerlings' (scarlet), 'Orangeade' (flame), 'Princess Marie Jose' (lilac-pink), 'Reddy' (red), 'Sion' (bronze), etc.

Anemone-Flowered. Height 60-105 cm, planting distance 60 cm, flower diameter up to 10 cm with dome-shaped central disc of tubular florets, double flowers having flat outer florets surrounding a densely packed group of shorter tubular florets, and flowering from July end on the hills and in January in the plains. Ideal for cutting and for floral arrangements. Varieties are 'Asahi Chohje' (red and white outer petals and white centre), 'Bridesmaid' (white outer petals, lemon centre), 'Comet' (maroon), 'Evita' (lilac with orange centre), 'Floorimoor' (pink-ornage), 'Guinea' (yellow), 'Honey' (apricot-pink), 'Lucy' (outer petals purple, centre yellow), 'Purple Puff' (purple), 'Scarlet Comet' (a lighter red sport of Comet), 'Siemen Doorenbosch' (light magenta), 'Slow Flox' (pink-yellow), etc.

Collerette. Height 75-120 cm, planting distance 60-75 cm, flower diameter 10 cm, flowerheads with one or more series of flat ray florets looking as single as in group I but second layer of small ray florets (the collar) being roughly half the length of the outer florets that surround the central disc, ring of florets (the collerette) only half the length of rays and usually of contrasting colour to the main rays. This may further be divided into (a) ***Collerette Singles*** (heads with single series of true rays and one collerette with yellow disc), (b) ***Collerette Paeony-flowered*** (flower heads with 2-3 series of rays and collars with yellow disc), and (c) ***Collerette Decorative*** [as to that of b) but fully double]. These start flowering from late July in the hills. The varieties are 'Ace of Hearts' (pink), 'Alpen Cherub' (white), 'Can Can' (pink petals, yellow collar), 'Cascade Ken' (deep red), 'Christmas Star' (red and white), 'Claire de Lune' (yellow, paler-yellow collar), 'La Cierva' (purple, tipped white on the outer petals, collar white), 'La Giaconda' (scarlet with golden yellow collar), 'Komeet' (red and buff), 'Little Snowoff' (purple), 'Mars' (deep red), 'Mrs. H. Brown' (orange and flame petals, yellow collar), 'Nonsense' (cream white petals, orange collar), 'Ruwenzori' (scarlet petals, yellow collar), 'Thais' (purple and white), etc.

Waterlily or *Nymphaea*-Flowered. Flowers fully double with broad and generally sparse flat or slightly with recurved or incurved margins of petals, giving overall effect of a flat or shallow bloom, height 90-120 cm, flower diameter 10-12 cm and blooming in early August on the hills. The varieties are 'Christopher Taylor' (red), 'Kari Blue' (bluish-purple), 'Glebane' (yellow), 'Golden Cloud' (light golden), 'Island Dawn' (pink blended yellow), 'Peace Pact' (white), 'Pearl of Heemstede' (dark pink), 'Porcelain' (white), 'Red Velvete' (red), 'Snowhill Rose' (pink), 'Taratahi Sunrise' (light pinkish-yellow), 'Yellow Petite Glory' (yellow), etc.

Decorative. Blooms fully double, showing no central disc, with broad, flat or slightly involute ray florets, seldom twisted and usually bluntly pointed. This may further be classed into two: ***Formal Decorative*** (fully double with no visible disc and arrangement of florets regular with slightly incurved margins, tips flattened and pointed or rounded) and ***Informal Decorative*** (as *Formal Decorative* but rays irregularly arranged). As per size of the blooms, these are further divided into **Giant** with flower size >10 cm [varieties include 'Burgess Ray' (yellow), 'Crossfield Festival' (bicoloured, light red and white), 'Hamari Girl' (pink), 'Holland Festival' (rich orange with white-tipped petals), 'Jocondo' (purple), 'Lavender Perfection' (lavender-pink), 'Liberator' (crimson), etc.], **Large** with flower size 20-25 cm [varieties include 'Bunratty' (purple), 'Dutch Triumph' (pink, yellow centre),

'Enfield Salmon' (pink), 'Robert Damp' (yellow), etc.], **Medium** with flower diameter of 15-20 cm [varieties include 'Betty Russell', syn. 'Yellow Terpo' (yellow), 'Breckland Joy' (bronze), 'Deuil du Roi Albert' (violet-purple tipped white), 'Edinburgh' (maroon and white), 'First Lady' (yellow), 'Golden Turban' (shades of yellow, orange), 'Majuba' (blood-red), 'Peter' (purple-pink), 'Peter's Glory' (lilac-rose, white centre), 'Red & White' (red & white), 'Rosella' (lilac-pink), 'Snow Country' (white), 'Sterling Silver' (white), 'Terpo' (crimson), etc.], **Small** with flower diameter of 10-15 cm [varieties include 'Arabian Night' (maroon, appearing quite black), 'Chinese Lantern' (yellow splashed orange-red), 'Dedham' (shades of pink and mauve), 'Gerrie Hoek' (deep pink), 'Glorie Van Heemstede' (yellow), 'House of Orange' (amber-orange), 'Millbank Inferno' (shades of red orange), 'Snow Queen' (white), 'That's It' (flame), 'Towneley Class' (white, flushed lavender), etc.], and **Miniature** with up to 10 cm flower diameter [varieties include 'David Howard' (orange), 'Doris Duke' (pink), 'Jo's Choice' (scarlet), 'Lilianne Ballego' (bronze), 'Newby' (peach), etc.]. Certain other popular varieties are 'Barbara Marshall' (red), 'Bhikkus Mother' (bicoloured), 'B.J. Beauty' (white), 'Eastwin' (purplish-red), 'Hillcrest Kismet' (deeper pink), 'Islander' (deeper pink), 'Klondike' (white), 'Lady Linda' (yellow striped lavender), 'Mascot Maya' (yellow), 'Queen Elizabeth' (lemon-yellow), etc.

Ball. Fully double ball-shaped blooms, seldom flattened at top. The sides of the ray florets curving inwards for more than half its length, tips blunt or round and spirally arranged. This is divided into **Ball** with 90-120 cm height, planting distance 75 cm and flower diameter 10-15 cm and the varieties are 'Bernard Colwyn Hayes' (deep pink), 'Doreen Hayes' (scarlet), 'Esmonde' (yellow), 'Gloire de Lyon' (white), etc., and **Miniature Ball** with 90-120 cm height, planting distance 75 cm and flower diameter up to 10 cm, and the varieties are 'Attlami Cherry' (red), 'Camano Candy' (pink) 'Cornel' (red), 'Crusader' (red), 'Dr. John Grainge' (orange), 'Florence Vernon' (lilac), 'Nelly Birch' (scarlet), 'Snowfall' (white), 'Risca Miner' (purple), 'Sulphurea' (yellow), 'Swiss Miss' (shades of crimson-purple and dark pink), 'White Polventon' (white), etc.

Pompon. Height 90-120 cm, planting distance 75 cm, flowerheads double, diameter from 5 cm to 10 cm (large from 7.5-10.0 cm, medium from 5.0-7.5 cm and small up to 5 cm), similar to ball dahlias but much smaller and more globular, florets curving inwards along their whole length and flowering late July. The varieties are 'Albino' (white), 'Andrew Lockwood' (lilac), 'Crossfield Ebony' (deep red), 'Deepest Yellow' (deep yellow), 'Eden Barbarossa' (red), 'Elffin' (purple), 'Hallmark' (pink), 'Little Conn' (dark red), 'Lydia' (red), 'Miss Kennedy' (orange), 'Moor Place' (purple), 'Nero' (dark red with purple sheen), 'Pom of Poms' (scarlet), 'Potgieter' (primrose yellow), 'Rhondda' (lilac, centre white), 'Small World' (white), 'Stolze von Berlin' (pink), 'Willo Violet' (violet), 'Zonnegoud' (canary yellow), etc.

Cactus. Cactus and semi-cactus groups are largely derived by crossing *Dahlia hortensis* with other species. Both groups are further divided on the basis of flower size. The cactus types have fully double blooms with invisible central disc and with long, narrower and more pointed ray florets than the decorative types, finely quilled (margins strongly revolute but sometimes incurving or straight) for more than half its length. They are further divided into **Giant** with 120-150 cm plant height, 120 cm planting distance, flower diameter >25 cm and flowering by last August on the hills [varieties are 'Danny' (pink), 'Gladys M. Reynolds' (bronze), 'Polar Sight' (cream white), etc.], **Large** with 120-150 cm height, planting distance 120 cm, flower diameter 20-25 cm, and flowering in August [varieties are 'Cum Laude' (deep pink), 'Drakenburg' (purple), 'Irish Visit' (crimson), 'Royal Highness' (pink), 'Soest Vooruit' (orange), etc.), **Medium** with105-135 cm plant height, planting distance 90 cm, flower diameter 20-25 cm and flowering in early August [varieties are 'Authority' (shades of bronze and orange), 'Priscilla' (flame), 'Sure Thing' (red), 'Topaz' (yellow), 'Tornado' (orange-red), etc.), **Small** with 105-120 cm height, planting distance, 75 cm, flower diameter 10-15 cm and flowering in late July [varieties are 'Doris Day' (crimson),

'Klankstad Kerkrade' (yellow), 'Paul Chester' (orange with yellow centre), 'Worton Sally Ann' (lilac), etc.], and **Miniature** with 90-120 plant height, planting distance 75 cm, flower diameter up to 10 cm and flowering by late July [varieties are 'Charmer' (purple), 'Chasamay' (pink), 'Hazel' (lilac), 'Pirouette' (yellow), 'Poppet' (blended salmon and yellow), etc.]. Some other popular varieties are 'Alvas Doris' (red), 'Athalie' (deeper pink), 'Camano Classic' (orange), 'Done Huston' (orange), 'Doris Daj' (red), 'Eastwood' (pink), 'Jyotsna' (bicoloured), 'Reputation' (purple-pink), 'Snow' (white), 'Yellow Galator' (yellow), etc.

Semi-Cactus. Blooms are fully double with pointed ray florets, broader and quilled for less or half their length. Similar to the **CACTUS** group, this is also divided in five sub-groups with the same characters as to their counterparts in **CACTUS** group. However the varieties are 'Cocorico' (scarlet), 'Frontispiece' (cream white), 'Hamari Boy' (yellow), 'Respectable' (yellow), etc. in **Giant** sub-group; 'General Mourer' (purple), 'Nantenan' (yellow); 'Royal Sceptre' (orange to deep yellow), 'Royal Wedding' (flame with yellow centre), etc. under **Large** sub-group; 'Apache' (bright red), 'Autumn Fire' (apricot -yellow tinged flame at tip), 'Countess Dunraven' (pink), 'Hamari Bride' (white), 'Rotterdam' (crimson), 'Sweet Secret' (mixed deep pink and yellows), etc. under **Medium** sub-group; 'Hoek's Yellow' (yellow), 'Marilyn' (pink), 'Rothesay Red' (crimson), 'White Swallow' (white), etc. under **Small** sub-group; and 'Happy Mood' (yellow), 'Little Ann' (pink), 'Lynne Bartholomew' (cherry-purple), 'Snip' (bronze tinged orange), 'Yellow Mood' (yellow), etc. under **Miniature** sub-group. Certain other popular varieties are 'Davenport Sunlight' (yellow), 'Frigoulet' (bicoloured), 'Gay Triumph' (bronze), 'Mandarine Napolean' (orange), 'Pink Jupiter' (pink), 'Pianella' (reddish), 'Preference' (light yellow-pink), 'Super' (red), 'Weston Miss' (yellow), etc.

Miscellaneous. This group contains small, dissimilar classes not covered above, *viz.* **Orchid** cultivars which are similar in shape to the Single dahlias except that their ray florets are revolute for at least two-thirds of their length as var. 'Giraffe', banded yellow and bronze; the **fimbriated** dahlias where flowers are double having fimbriated petals starting from the tips to a considerable length and this also has some five groups such as giant, large, medium, small and miniature with the varieties such as 'Ana Mari' (centre white and tip pink), 'Cabaret' (bicoloured), 'Frilly Dilly' (pink), 'Frontispiece' (white), 'Kaiwera Gold' (golden-yellow), 'Show-n-Tell' (red),'Sultan' (pink), 'Terry' (lavender), 'Tioga Dawn' (pinkish-white), 'Willow Night' (deep red), etc.; the **Star** dahlias with small incurving flowers, consisting of 2-3 rows of scarcely overlapping pointed rays encircling a central disc in the form of a cup; the **Chrysanthemum** types looking incurved exhibition chrysanthemum as var. 'Akita'; and the tiny **Liliput** series with 2.5 cm blooms. Others which are included in this group are **Paeony-flowered** (originally group 4 of the **British National Dahlia Society** classification) which grows 90 cm, planting distance 60 cm, flower diameter 10 cm, the flowers consist of two or more rings of flat ray florets and a central disc , i.e. semi-double type, flowering from late July and the varieties are 'Bishop of Llandoff' (scarlet), 'Fascination' (purple), 'Grenadier' (scarlet), 'Oranje Flora' ('Orange Flora', orange), etc. Some other popular varieties are 'Honka' (yellow), 'Golden Star' (yellow), 'Jescot Julie' (orange-purple), 'Pink Giraffe' (pink), 'Tiny Tot' (pink), 'White Orchid' (white), etc.

B. **Bedding dahlias** are grown from seeds. These grow 30-50 cm in height, planted at 30-50 cm across, flowers of almost all the hues and colours and may be single, semi-double or double. Flowers measure 5.0-7.5 cm across and appear from June end and continue until the occurrence of first frost on the hills, but in the plains these continue blooming from December to April or May when flowers burn due to scorching sun. This group has many seed strains and mixtures out of which 'Coltness Hybrids' is the best known group, seeds with mixtures and separate colours being available in this group. 'Pink Radiance' has single rose-pink flowers with a bold crimson zone. 'Unwin's Dwarf Hybrids' produces semi-double flowers in a wide range of bright colours having yellow, orange, pink and crimson. 'Early

Bird' is a selected dwarf form (growing hardly to 30 cm high) with flowers in the similar colour range. 'Fascination' is a pink or purple, and paeony-flowered dwarf variety. 'Princess Mary Jose' produces single lilac-pink flowers. 'Redskin' (mixed colours, double and semi-double) and 'Rigoletto' (mixed colours: white, yellow, orange, pink and red, with double flowers) are two other outstanding varieties being grown from seeds.

Senaratna (1947) developed two new paeony-flowered dahlia hybrids which he named 'Fr. Le Goc' and 'Ceylon Sunrise'. Protich (1988) from Bulgaria also described 5 new decorative-flowered cultivars, *viz.* 'Topajo', 'Sourie de Cozon', 'Esmeralda', 'Faluna' and 'Rosela', and 2 cactus-flowered, *viz.* 'Véritable' and 'Top Choice', all flowering for two months during July, August and September and are suitable for cut flower production. Broertjes and Ballego (1968) developed five cultivars through irradiation of the cv. 'Arthur Godfrey'. Singh (1970) isolated one spontaneous mutant of rosy-magenta flowers with more attractive shape than its parent cultivar 'White Pearl', and suggested that an inhibitor gene was responsible for the new type. Dohare *et al.* (1974) released one cv. 'Manali' of pure white colour obtained through spontaneous mutation. Das *et al.* (1978) through ^{60}Co gamma irradiation of 14 dahlia cultivars developed 19 flower colour mutants under 2-3 krad, out of which 11 have been released as varieties. In dahlia, there is progressive degeneration in terms of branching, the flower numbers and head structures, and tuber size the cause for which was attributed by Lavrichenko (1975) as the ontogenic character associated with age changes in the shoot and root structure whether grown from cuttings, tubers or seeds.

In a varietal trial at Lucknow with 22 dahlia cultivars, Sharga (1975) recommended 'Axford', 'Triumph', 'Black-Out', 'Black-Out Sport', 'Cheroky Beauty', 'Ginga', 'Kelvin Rose', 'Mrs. John Ralph' and 'Nirmal Chandra' as most promising. Through a trial of new and old cultivars of dahlia in Germany, Mehlis (1976, 1978a,b) found 'Bodenseeperle', 'Kurdirektor Diekmann' and 'Rote Spinne' as outstanding in the first study while out of 27 and 25 cultivars in the second and third studies, respectively, 'Hanna Hofbauer' (yellow, semi-cactus), 'Pipers Pink' (semi-cactus), 'Stadt Wiehl' (wine-red, cactus) and 'Wolstad' (lilac, decorative) were found outstanding. Mishra *et al.* (1990) with seven cultivars reported that cvs 'Kenya' and 'Vigour' were outstanding.

Propagation

Dahlias are propagated through **seeds**, by tuberous rooted divisions, stem cuttings and through *in vitro* methods. Small flowered and bedding dahlias are propagated through seeds. In dahlia one gramme may contain some 98 seeds and seeds germinate within a week while these require 6-8 weeks from sowing to attain planting size. At planting, these should have attained at least two pairs of leaves. Safranova (1969) obtained fully viable seeds from open pollination of double dahlias and obtained best germination by sowing the seeds directly in the open ground, and the plants were healthier, flowers were better and the tubers were sturdier than their parents. For seedling sale these are raised in the plugs, just covered for 3-4 days and kept at 26-27^0 C, then they are provided with 20-21^{0}C temperature for 7 days and fed with 50 ppm N per week, then for 7 days at 18-21^0 C with 100 ppm N once or twice per week, and finally they are kept at 15-17^0 C for 7 days and fed with liquid fertilizer as per requirement (Ball, 1991). This way the plugs are ready for sale within 3-4 weeks. For usual seed growing, the bedding dahlias are grown through seeds. Seeds are sown 3 cm apart in boxes or pans filled with seed compost. For early growing in the temperate regions these may be started inside the glasshouse in February otherwise in the open in March though in the subtropical and tropical regions of the country these may be sown in September. After sowing they are covered with a very thin layer ($^1/_3$ cm) of soil and watered with a fine rose can where within 15 days all the seeds will germinate at 13-16^0 C temperature. As soon as these grow enough for handling, these are planted outside individually in the standard potting mix at around 10-13^0 C night temperature for further growth and hardening. These require full sun but not piercing one at seedling stage. The seedlings are planted in the field when have attained 5 cm height in the nursery. Now these are treated as usual as other plants where these will flower the same season like to those grown through tubers, *vis-à-vis* will form even tubers. Aoba *et al.* (1960) when planted dahlia seedlings in first week of June in Japan and at every 10 days until November a batch was being dug out to observe their root development. It was found that adventitious roots started appearing from early June and continued till early August and these roots developed into tubers by October. Early foliage removal (not late) and shade treatment reduced root diameter but SD (8 h for 20 or 40 days) caused reduction in the number of roots though increased root diameters. Germinating dahlia seeds on moist filter paper and when their roots are 0.2 to 0.5 mm long the halving of young plants longitudinally and then the two halves are induced to grow independently into normal plants (Rudionenko and Zaar, 1951), and this way one can get two **identical genetic plants** from one seedling.

Propagation through **stem cuttings** is the most commercial method. For this, healthy mother tubers are selected and planted in March (temperate areas) or in October (plains) under full sunlight with their collars just peeping out and the night temperature should not be less than 13^0 C. The soil at planting should have sufficient moisture so that up to sprouting there may not be any need for watering. After a week young shoots may emerge. These are allowed to develop until these attain 8-15 cm height. Now these are removed with a sharp knife, cutting close to the tuber but without removing any part of it. Preferably the cuttings should be 7.5 cm long. This way each mother tuber shall be producing 20-30 cuttings, depending on the cultivar. However, in case of tree dahlia the size of cuttings taken from the new growths should be 25-30 cm having at least 3 nodes. After dipping them in rooting hormone, these are planted by inserting its half lower part by removing the leaves if any, in pure white sand in flats or pots at 18-24 ^{0}C temperature where it may take 10-18 days for rooting, depending on the temperature range provided. After planting, the pots should be watered and covered with a plastic tent quite above the seedlings with the help of some frames. After 3-4 days these may be opened through side flaps for ventilation and after 8-10 days these tents are completely removed. After rooting, these may be potted in individual pots containing 7 parts top soil, 3 parts peat and 2 parts sharp sand and then taken to a frame until planting outdoors in beds in May on the hills or in November-December in the plains. In the frame, a night temperature of 13^0 C is provided. Within 8 weeks, the cuttings will produce a tuber having 2.5-4.0 cm in diameter, along with three or more tubers. It takes some six weeks from the time they are first taken until of planting size. The most suitable soil for its growing is sandy loam at a sunny situation having pH from 5.5 to 7.5. Marston and Evans (1972) took nodal cuttings of dahlia cvs 'Autumn Fire' and 'Hamari Bride', inserted in a peat or perlite compost under mist without bottom heat in last week of June, and after rooting these were potted in JIP compost No. 1 in 9-cm plastic pots on a capillary bench in a cool greenhouse. Fibrous roots grew through the pot holes into the sand and anchored the pots, and without feeding these plants were allowed to flower. After flowering, in mid-October the pots were removed and the plants dried off gradually before the tops were cut to leave about 7.5 cm of stem, and then average fresh weight per pot-tuber was recorded as 48 g for 'Autumn Fire' and 50.5 g for 'Hamari Bride'. These tubers stored well and were later used to produce cuttings.

After planting the mother **tuber**, it sends out one or many aerial shoots which afterwards form many enlarged tuberous roots as the primary storage organ. The development of these roots is facilitated by the short days. When plants have matured, the aerial part is cut down and the bunch of these tuberous roots are gently pulled out from the soil with the help of fork or shovel, cleaned and stored in a well ventilated room in saw dust or sand from June to September in the plains and from December to March on the hills. At planting time they are separated gently with the help of a sharp knife or just by pulling them gently from one another in a manner that crown (where it is attached to the aerial shoot) is intact on the tuberous root, each crown having at least one eye (adventitious bud). Sometimes there may be only one-eyed two-tuberous roots jointed together, and in such cases both the tubers should be taken as one entity while separating from the aerial shoot. On planting, these root tubers will produce new plants which in turn will produce some 6-15 root tubers per plant by the time they are lifted. If minimum size has been attained there is insiginificant difference between various sizes of root tubers, and the performance of tubers weighing 80 grammes goes at par to those tubers which weigh 120 grammes. Reduction in number and/or size of the flowers increases the tuber weight (Wasscher, 1955; Newton, 1960). Newton and Tullis (1962) stated that flower bud removal, especially in late flowering cultivars, increases tuber weight, and the presence of flowers even for a brief period reduces the weight of tubers. Dahlia cuttings taken in April, May or June, grown in peat/perlite compost in pots containing JIP-2 or peat/perlite in greenhouse under natural daylength and subjected to 10-h SD from mid-June formed good tubers by mid-October (Lloyd, 1968). Read (1968) through foliar sprayings of Alar at 2500 ppm and CCC at 1500 ppm found increased size (up to 200%) and number of tuberous roots in dahlia grown under short day conditions conducive to tuberous root formation, and the development of tuberous roots on cuttings grown under LD conditions was also enhanced by these treatments whereas control produced only fibrous roots. Moser and Hess (1968) found critical day length of 11-12 hours and 15.6 or 21.0 ^{0}C night temperatures for maximum tuberization in cv. 'Sneezy' though 10.0 or 26.7 ^{0}C night temperatures inhibited it, and temperature could not alter the SD requirement for tuberization. Spraying with ABA or ethephon promotes tuberization in dahlia plants in long days though less effective than at short days, SADH and ethephon treatments of budless leaf-cuttings inhibit tuberization whereas ABA slightly enhances it, and after the start of SD treatments endogenous ethylene evolution reaches a peak between second and third week, i.e. one week before the onset of tuberization and then decreases to the low level under LDs (Biran *et al.*, 1972). Read *et al.* (1972) when sprayed SADH at 2500 ppm and chlormequat at 1500 ppm found increased number and size (sometimes even three times) of tuberous roots grown under SD in plants of *Dahlia pinnata* cultivars irrespective of whether the cultivars were quick tuberous-root forming or those forming poor quality roots, and these plants were also excellent in growth. Even under LD conditions these treatments caused formation of tuberous roots but untreated cuttings produced only fibrous roots. Čaplygin and Šahova (1961) suggested supply of cheaper planting material by taking of **cuttings** in June instead of usual March or April when heating is also necessary and these cuttings root to a tune of 87% within 10 days. They also advocated that keeping of rooted cuttings in soil in pots at 2-4 ^{0}C overwinter better than those stored in the usual way in sand. Contrary to this, Mustafajeva (1985) advocated taking of cuttings in March-April in a perlite + sand mixture for up to 100% rooting, and then planting 50 cm apart both ways. The cuttings trimmed to a node were when inserted in the soil or medium though produced more but smaller tubers than untrimmed cuttings which produced slightly less but heavier and longer tubers and these tubers produced more cuttings (Newton, 1962). Biran and Halevy (1973) found an inverse relationship between rooting percentage of the cuttings and growth rate of the buds in the cuttings which sprouted during the rooting period, reproductive buds suppressing rooting more than the vegetative ones, and removing the growing point together with a few terminal leaves there was an increase in the rooting percentage, particularly in largest cuttings which root poorly (33%) than the smaller ones (85%). Maiko and Yashchenko (1980) found early tuber formation with improved subsequent tuber growth through pre-plant soaking of dahlia cuttings in Alar at 20 mg/litre though foliar spraying with ethrel at 100 mg/litre or Alar at 200 mg/litre was even more effective than dipping, and the flowering in these treatments was also stimulated but only in dwarf cultivars. Cayeux (1936) and Safranova (1972) advocated dahlia propagation through heel cuttings (single leaf cutting with a small portion of the main stem attached) by taking two cuttings from each node (leaves are opposite on the stem) in April-May and the new growth strikes from the axil of the leaf, irrespective of whether at the time of taking the cutting the buds were visible or not, if planted in turf soil and sand in a 1:2 ratio in a cold frame. These cuttings should be planted out at 50 x 30 cm spacings, followed by saw dust mulching immediately.

This may be cultured *in vitro* through **meristem tip culture** but it is not a commercial practice (Morel and Martin, 1952). Through this practice one can get rid of 'dahlia mosaic viruses' (Mullin and Schlegel, 1978) and 'tomato spotted wilt viruses'. Gavinlertvatana *et al.* (1979, 1982) when grew *in vitro* the leaf segments of stock plants grown under 8 h SD or 16 h LD, on a modified MS agar medium containing various levels of NAA and/or kinetin, larger callus and higher levels of ethylene and CO_2 were detected when stock plants grown under SD were further incubated under SD (SD to SD) than SD to LD, and tissues under LD to LD or LD to SD formed similar amounts of callus to SD-LD cultures but with higher ethylene and CO_2 levels. Daminozide spray on SD stock plants one day before taking the explants promoted callus formation, *vi-a-vis* ethylene and Co_2 production in tissues incubated under either SD or LD though spray on LD stock plants had no effect on callus or ethylene but CO_2 level was found increased under SD or LD incubation. They further stated that the ethylene accumulation in the flask atmospheres was dependent mainly on NAA and to a lesser extent on kinetin levels. The usual propagation methods in dahlia provide ample opportunity in multiplying the elite germplasm lines (varieties) rapidly therefore tissue culture remains only of academic interest and not a commercial proposition.

Cultural Practices

Dahlias planted too early or late, mature and produce poor flowers. For healthy growth of dahlia one should be very careful about soil types, the climate, spacing, mulching, feeding, watering and various other practices. Plant enjoys **full sun** and well at soil which is rich in humus. These grow well at 13-25 ^{0}C **temperatures**. They can't grow under cooler climate with plenty of rainfall. Dahlias aren't fussy about soil, though sandy-loam and well-drained soil which is deep, porous and has a **pH** range of 5.5-7.5, preferably 6.5-7.0, is most suitable for its growing. Heavy soils are improved through addition of organic matter and coarse sand. It responds well to feeding, therefore, the soil for its cultivation should have sufficient humus, at least 3 per cent of organic matter. Compost or farmyard manure should be added into soil every year at the rate of 50 tonnes per hectare at the time of field preparation. Ploughing of the soil starts about two months before planting. While ploughing, all the foreign materials, i.e. rootstocks of the weeds, wood pieces, crocks and grits should be taken out and the soil should be planked. Now the soil is watered and again cultivated deeply, followed by planking and taking out of foreign materials and, if need be, some green manuring crop should be taken and when this becomes four weeks old it is knocked down in the soil to decompose. Third ploughing is carried out when soil is having sufficient moisture, about 4-7 days before planting, soil is again planked and levelled, and the beds are prepared of the convenient sizes, preferably 2 metres in width and to any convenient length, keeping at least 50 cm space between two beds for intercultural operations and these spaces may act as irrigation channels also. **Planting** on the hills is carried out during March-April while in the plains during October. **Planting distances** kept are 45 × 45 cm for smaller plants and bedding dahlias, dwarfs and the pompons though for exhibition dahlias it should be 75 × 75 cm. While planting the tubers, the crowns should be just at the soil level or slightly buried. In light soils, the large tubers are planted a little deeper compared to small tubers and the crown is either barely covered or only with little soil over it. Root tubers require 8-12 weeks to flower from planting, and the seeds 10-12 weeks, provided seeds are already grown and shifted to 10-cm pot which requires 1-3 weeks further. Bose and Mukhopadhaya (1967) recorded that in rooted cuttings of dahlia, tuber formation starts within 30 days of planting, being most active during the period of maximum vegetative growth and flower initiation, contents of total and non-reducing sugars in the stems increases markedly during flowering, followed by an abrupt decline at the peak flowering, leaf contents of sugars decline sharply when the rooted cuttings are planted and remains low until the final sampling when flowers start fading, the contents of soluble N in leaves and stems decline appreciably when buds become visible. Since after planting in the process of starting the new growth, the old planted tuber disintegrates and daughter tubers start forming so to protect these from physical damage and the feeder roots being formed close to surface, the intercultural operations are stopped when plants attain 30 cm of height.

In **pots**, only the ones of shorter stature and the bedding types are grown. The medium in pots should be well-drained and should contain sandy garden loam, sharp sand and well rotten farmyard manure, all in equal volume with a little bone meal (15-20 g per pot). Winter (1977) used four types of media with or without a basic slow-acting inorganic fertilizer, and drenching with four rates (0, 2, 4 or 8 ml/15-cm pot) of ancymidol (Reducymol) to dahlia cv. 'Miramar', and recorded best growth and shoot production with advanced flowering in compost medium with 6 kg/m^3 of 14:14:14 Osmocote but 8 ml/pot of ancymidol delayed flowering only by 2-3 days though the same strength was quite effective in a mixture of black soil, sphagnum moss and FYM to overecome strong vegetative growth. He found 4 ml adequate for a 1:1:1:1 mixture of black soil, sphagnum peat, clay and perlite mixture or a similar mixture + sand, whereas only 2 ml was required for a sand medium. Except certain bedding dahlias which grow up to 50 cm in height, others require **staking**. Stakes of roughly 1 m length are used for dahlias and these are inserted to three sides of the plants slightly away from the plant base as a safeguard from tuber damage, at a time when plants have attained some 30-40 cm length, however, the stakes are threaded with strings only when the plants require it. Also the 3-4 tier strings around the beds as well as between the rows at 30 cm, 45 cm, 60 cm and 75 cm will also save the plants from breaking.

Barrett and Hertogh (1978) stated that **pinched** plants produce more flowers though the flowers remain small and flowering is also delayed. Pinching the terminal at node four was found optimum, and the pinching at the more distal side encourages formation of more laterals. Pinching is practiced when, as per the requirement of the variety the shoots have not appeared or only one shoot is developing per tuberous root though more shoots are required when profusion of blooms are required or when flowering is to be delayed. Vigorous dominant shoots should also be pinched on multistem plants. **Disbudding** the terminal flower should only be done when flowering is ahead of schedule for shows or for marketing, and this practice will force the axillary flowers to develop, though for cut flower growing reverse is practiced, axillary flower buds are removed to increase size of the terminal flower (Dole and Wilkins, 1999). For precocious flowering, the main shoot is pinched when it has attained 30 cm of height.

In dahlia plantings, **mulching** is carried out with grass clippings, old hay, saw dust, straw, dried leaves, compost, etc. which helps in conserving moisture, maintains an even soil temperature, keeps down the weeds and in the end decomposes and provides organic matter to the field. Black polythene does not allow weeds to grow, conserves the moisture and increases soil temperature. Simazine, lenacil and linuron are effective **weedicides** for controlling weeds in dahlia plantings. Clay and Ivens (1964) recommended simazine application under moist soil conditions in sandy loam soil at 1.15 kg/ha 11-14 days after planting and chlorpropham at 2.25 kg/ha + diuron at 0.9 kg/ha as quite safe though there may be varietal differences. Clay and Davison (1968) recorded lenacil at 2.25 kg and 4.5 kg/ha spraying as the most promising in dahlia field as compared to chloramben, linuron, prometryne, simazine and trifluralin.

Irrigation and Nutrient Management

Dahlias require copious **watering** throughout the growing period. In the plains they are grown during winter where these require watering at 10-days interval from November to February and at 5-7 days interval in October, March and April but on the hills they are watered every week as it is grown during summer, however, during the rains it is adjusted accordingly. These are also watered immediately after application of fertilizer. The plants in pots require watering every fourth day. Cavalchini (1968) recommended the soil moisture tension in a plastic house to dahlia plants at 0.25 and 0.1 atm during summer and 0.25 atm during late summer and early autumn.

For successful dahlia cultivation, balanced **nutrition** levels are required at right time. Nitrogen at 100 kg, phosphorus at 125 kg and potassium at 100 kg per hectare of land should be applied for getting proper flower yield with good quality. Though phosphorus and potassium are applied as basal fertilizers but nitrogen can be given in three split doses, $^1/_3{}^{rd}$ as basal dose, and the other $^1/_3{}^{rd}$ at the time when these are at the youth of their vegetative growth and the remaining $^1/_3{}^{rd}$ when the plants have started producing reproductive buds. Potassium fertilizers may be applied only when soil is lacking in this element as Indian soils, by and large, are not deficient in K_2O. However, the use of fertilizers should be based on soil and leaf analysis. Sang (1974) applied $(NH_4)_2SO_4$, NH_4NO_3 and urea to dahlia plants grown in sandy-loam soil, in sandy-clay-loam soil and in a loam soil in Korea and found ammonium sulphate promoting most growth, followed by ammonium nitrate but urea least effective, however, tuber yields were found highest with ammonium nitrate and lowest with urea where flowering was also poor. Regarding the media, ammonium nitrate was more effective as the clay content increased. Through four applications (mid-July to mid-October) of ammonium nitrate at 20, 40 or 60 g N/plant in Italy, only flower numbers were found significantly increased with the lowest rate (Sasso, 1961). El-Gamassy and Moustafa (1963a, b) had grown dahlias in Cairo with high, medium and low levels of N, P, K, Ca and Mg solutions in various combinations and recorded best growth, flowering and tuber yield from highest dose of complete treatments, N and P being the most important elements but where K and Ca were omitted the result was slightly poor. They stated that though with age N, P and K contents in plant parts decreased in dry weight basis but total amounts were found increased though Ca increased with age, and the omission of N reduced the N, P, K and Ca contents of the plants and omission of P, Ca or K also reduced N content of the plants, however, absence of K resulted in a slight increase in the percentage of Ca. El-Gamassy and Moustafa (1964a,b) applied various amounts of calcium nitrate, calcium superphosphate and potassium sulphate to dahlias in pot and field trials and recorded best growth, flowering and tuber yield with a fertilizer mixture of 46:60:10 g/plant of calcium nitrate:calcium superphosphate:potassium sulphate, followed by one of 40:40:20, those receiving only P and K were poor than those also receiving N. No superphosphate or only 20 g/plant reduced the number of flowers, and increase in K levels in presence of N and P increased the number of flowers but high rates of P or 30 g of potassium sulphate reduced the flower length and weights. Calcium nitrate 40 g, calcium superphosphate 60 g and potassium sulphate 20 g to the potted dahlias with nitrate being added as 3-monthly dressings and other compounds either at or after planting, they obtained best results in terms of growth, flowering and tuber yield with P and K applied in 3-monthly doses starting one month after planting. Rahman and Mitra (1974) applied N at 0, 28, 56 or 84 kg/ha and P and K at 28 kg/ha each, and obtained maximum flower size and number at 56 kg N though 84 kg N yielded highest tuber yield and earliest flowering, though plants receiving no N were poor in growth. Boertje (1979) in Holland found 1.5-2.0 g/l of 17:6:18 NPK fertilizer when applied four weeks after pricking out in 160 ml potting pots filled with standard potting compost. Bhattacharjee and Mukherjee (1983) when applied P_2O_5 at 30, 40 or 50 kg/ha along with a basal dose of 40 kg/ha K_2O, and first half of the 20, 30 or 40 kg N/ha at planting and the other half 30 days after planting to dahlia cv. 'Blackout', they obtained best growth and flower production with the highest NP rates.

Potted dahlias may be given with slurry every fortnight. Slurry is prepared by diluting the fresh cow dung to its four times and then kept in an earthen pot under sun with covered mouth for decomposition for 10-15 days by stirring daily and then again diluting it to its four times, so one kg of cow dung will become 16 litre in which 320 g of urea is added at the final stage, and this is the slurry which is very effective in case of potted as well as to those plants growing in the field.

Growth, Development, Flowering, Dormancy and Height Control

Freshly harvested root tubers of dahlia do not sprout immediately because of **dormancy**. Summer grown ones as in the hills where these are grown during summer, i.e. from April to October, these have deeper dormancy compared to those grown in winters as in the subtropical and tropical regions of India. These sprout only when these have attained physiological maturity. At 15^{0}C temperature and at 13-14 h photoperiods, many dahlia cultivars will grow continuously. Konishi and Inaba (1967a, b) obtained reducing percentage of sprouting in dahlia tubers planted between the end of October and early December and those sprouted also the shoots did not elongate normally, but those planted from mid-December onwards the sprouting as well as those planted at the end of December even shoots also grew normally, and digging of the tubers after November 20 and stored for 40 days at 0 ^{0}C sprouted well with normal growth of shoots though there had been varietal differences. Adventitious or axillary shoots from newly planted tubers do not elongate if photoperiod is less than 12 hours (Konishi and Inaba, 1967b). They further stated that with tubers from plants propagated from cuttings taken during or after June the dormancy was weak whereas the dormancy of tubers from cuttings taken in or before April was similar to that of the mother plants. A **photoperiod** of 13 h and 10 ^{0}C night temperature enhances the flower quality as compared to 5 or 15 ^{0}C. Photoperiods also determine number of total florets in the inflorescence, 14 h photoperiod producing more number of flowers though 8 h period drastically reducing it. Short days, apart from reducing the number of florets, increase the proportion of disc florets as well as induce abortion. Low light intensities also behave like photoperiods, these delay flowering and reduce the percentage of flowering plants. Tuberization is induced by short days (11-12 h) for 5 days at 16-21 ^{0}C temperature where maximum fresh weight of the tuberous root is obtained, though the root tuber growth is inhibited at 10 and 27 ^{0}C. Potted dahlia plants can be started in March on the hills if the tubers are stored at 7-9 ^{0}C for 2-3 months or in October in the plains when stored for 4 months. For glasshouse forcing, in the northern hemisphere, the tubers are planted from January to April (De Hertogh, 1996). Photoperiods less than 12 h promote tuberous root formation and check shoot growth. Brøndum and Heins (1993) reported optimum production conditions for dahlia plants of cv. 'Royal Dahliette Yellow' grown in 10-cm pot at 12-14 h photoperiod and 20 ^{0}C temperature; with decrease in photoperiod from 16-h to 10-h, the number and length of axillary shoots decreased; and photoperiods shorter than 12-h inhibited shoot growth though promoted tuberous root formation. Maatsch and Rünger (1955) reported that in dahlia lateral shoots are formed only in the upper leaf axils in 10- to 12-hour days and with increase in the day-length up to 12 h tuber growth also increases but tuber growth under long day conditions in cuttings is inhibited though all the leaf axils produce shoots under long day conditions (14 h days), however, a short day treatment from beginning of August for 20-30 days promotes tuber formation. Short days treatment of 20-30 days, especially for 4-weeks, is adequate between July end to beginning of September or from latter half of July to early August to increase tuber size 3-5 times in varieties which are slow in tuber formation (Rünger, 1955; Wasscher, 1955) though flower number becomes considerably less while long days stimulate aerial growth (Rünger, 1955; Yasuda and Yokoyama, 1959; Konishi and

Inaba, 1964). Yasuda and Yokoyama (1959) obtained 446 flowers under 13-h photoperiod, 235 under 10-h and only 158 under 7-h photoperiod. Increase in day length (up to 13 or in some varieties 14 hours) with supplementary lighting with 20-36 lux increased the total number of florets and the number of ray florets but decreased the number of disc florets. Konishi and Inaba (1966a) obtained floral primordia initiation under 8 to 16 h day lengths but their further development was found inhibited by LDs, optimum day length for flower initiation being 10 hours or less but later stages of flower development required 12-13 hours of LD. They further recorded that growing the plants for 40 days at optimum day length after pinching and then shifting to less than 12-h day length many buds become blind. Exposure of dahlia cuttings to 12-h SD and 14-h LD with and without interruptions of SD and LD treatment during early period of lateral growth, floral primordia were formed under both day lengths but the longer the SD treatment the earlier was the flowering with reduced shoot length and weight, floret number, *vis-à-vis* ray florets and double flowers but increased the disc floret number when applied only for five days, though LD for less than 20 days did not affect the time of flowering, however, beyond this the percentage of ray florets and double flowers increased. Konishi and Inaba (1966b) recorded a day length of 13-h with a minimum night temperature of 10 ^{0}C optimum for flowering, a high night temperature of 15 ^{0}C extending the flowering period while 5 ^{0}C night temperature retarded the flowering and shortened the flowering duration. Reduced light intensity conditions was compensated by increasing the critical day length for flowering with same amount of blooms but with delayed flowering. Konishi and Inaba (1966c) reported that photoperiodic conditions prior to stopping of shoot had no effect when after stopping the plants were grown in 13-h days but when plants were grown at SD below the critical day length of 12-h prior to stopping, the flowering was considerably retarded. Early photoperiod treatment did not influence the number of florets developed. Mastalerz (1976) found that in some of the dahlia cultivars the flower initiation and development could not be prevented by continuous LDs, though in certain other cultivars plants remained vegetative in 16-h or longer days and initiated flowers in days shorter than 14-h and when these second type of plants were given 9-h days until the buds opened it became too late as stems were very short with reduced floret number but this problem was overcome by subjecting the vegetative plants to 14 SDs of 9-h followed by 18- to 24-h long days until the buds opened. Post (1949) stated that a great profusion of blooms in dahlia is usually in September and October as plants respond well to short days for bud formation, and high quality blooms are produced in regions of cool nights but shorter nights during autumn promote tuber formation though frosty nights during autumn terminate the growth cycle.

Temperature affects storage of root tubers, after interacting with photoperiod it controls tuberization of the roots, affects the growth rate of the vegetative shoot, after interacting with photoperiod it affects flowering of the shoot, and it also affects termination of shoot growth. For vegetative shoot development the optimal temperature is 13-25 ^{0}C, and for flower initiation and development it is 10-15 ^{0}C which is optimum. After tuber harvesting, they exhibit dormancy to save them from climatic adversities because in the plains it is very hot from May to August and on the hills from November to March it is very cold while their lifting time in the plains and on the hills is May and October, respectively. There is no significant difference in flower production due to size of the root tubers planted if this has attained minimum required size and the tubers having 80 g weight will behave similar to those having 120 g weight. After lifting, if the root tubers are stored at 0 ^{0}C for 40 days, the dormancy is released and the tubers sprout but then the climatic conditions are not congenial for its growing until in the glasshouses the required temperatures and photoperiods are provided artificially. If such tubers are exchanged from the plains to the hills and *vice versa,* there too the crop fails as the tubers have not attained physiological maturity though dahlias grown during winter season (as in the plains) have little dormancy than those grown during summers (as on the hills) which have maximum dormancy. Okada and Harada (1955) found small to very large disc to ray floret ratios due to seasonal variations (some due to temperature, some due to photoperiod while others due to both the factors) in various dahlia varieties, and reported that total number of florets per head was affected due to temperature and not the photoperiod. Tubers of five dahlia cultivars planted in late February at a minimum temperature of 13 ^{0}C with half the plants being exposed to 2-h night break through fluorescent lighting until mid-May, started flowering between 5 to 19 April, and their floret count revealed that night break influenced the plants favourably (Hall, 1969). He noted semi-daisy-eyed blooms in cv. 'David Howard' until April 26 in lighted plants and until May 31 in unlighted plants; in 'Horn of Plenty' the normal blooms were produced five weeks ahead in lighted plants than the unlighted ones in 'Larkford' the blooms became normal by April 26, i.e. three weeks ahead of unlighted plants; in cv. 'Minley Boy' such blooms appeared until May 3 in lighted plants though further up to two weeks after in unlighted plants; and in 'Brandaris' such blooms appeared up to April 26 under lighted plants and the blooms were normal four weeks later on the unlighted plants.

When root tubers of dahlia are planted, it is not so that immediately these start **flower initiation**. In fact, in glasshouses, flower initiation occurs after the formation of five leaf-pairs while in the field after seven leaf-pairs. At 17 ^{0}C night temperature in the greenhouse tubers start floral initiation after 15-25 days of planting. Various **growth regulators** have been found altering growth and flowering in dahlia. Moser and Hess (1968) stated that GA inhibits tuberization either under long or under short days but Alar promotes it under LDs though simultaneous spray of GA again suppresses the effect of Alar. Pappiah and Muthuswamy (1974) found increased number of laterals (nearly double of control) through ethrel at 1000 ppm but its 500 ppm followed by 1000 ppm shortened the time for flowering, with 90% increase in flower yield under 1000 ppm, and where one concentration was effective other was also found effective but slightly less. Biran *et al.* (1974) tried ABA and GA to budless dahlia leaf cuttings under LD, and found GA inhibiting the growth of tubers and roots and enhancing thickening of the petiole base though ABA promoted growth of the tuberous roots. SADH and ethephon promote tuberization but not in budless cuttings, ABA also enhances tuberization but GA inhibits it, and SD conditions which promote tuberization increases the endogenous level of ABA (Halevy and Biran, 1975). Mittal (1967) found height increase with fewer internodes and more lower branches, *vis-à-vis* advanced flowering when dahlia plants were sprayed twice with GA, especially with 100 to 500 ppm. Mather and Sharma (1969) stated that dahlias behave as short day plants with a critical day length of 16 hours, with SDs flowering occurs significantly earlier but with LDs or in continuous light no flowering occurs, and such plants flower immediately on being transferred from non-inductive to inductive day lengths. They stated that GA application accelerates flowering under inductive day lengths but the critical day length is not affected. Bhattacharjee *et al.* (1976, 1978, 1984) when used IAA and GA both at 100 and 1000 ppm, found marked promotion of shoot growth of dahlia with increased number of flowers under GA treatment, and IAA at higher concentration and GA at both the dilutions counteracted the retarding effect of B9 while IAA at higher rate and GA at lower rate suppressed the effect of floral initiation induced due to B9 or CCC treatments. Durso and Hertogh (1977) forced dahlia cvs 'Kolchelsee' and 'Park Princess' from January to June where plant heights were controlled by applying 0.5 mg/pot ancymidol two weeks after planting, and obtained best height control at 25 ^{0}C day and 16 ^{0}C night temperatures, accelerated flowering at 28/17 ^{0}C and 29/20 ^{0}C day/night temperatures but with poor quality of flowers, and delayed flowering at 12/12 ^{0}C day/night temperatures. They found natural spring photoperiod increase from 10 to 14 h as optimal but with high light intensities for forcing, long days given either as a 16-h photoperiod or as a 4-h night break delayed the flowering slightly, and those grown at 8-h photoperiod though flowered earliest but only scarcely, and under 50% shading the plants were too tall to control even with ancymidol so were unfit for pot growing. Kumar *et al.* (1978) had grown the dahlia plants in 8-h SD or 24-h LD and applied 10 or 100 mg/litre of GA to the plant apices on alternate days, which resulted into much taller plants especially under LD with more leaves and bud initiation in several plants after 62 and 85 days at the two concentrations, respectively, though with SD all the plants flowered and with GA number of days for floral initiation was found reduced from 56 to 40 days.

Height control is required only for potted plants and not for cut flower production. Internode elongation is controlled with ancymidol (A-Rest) drench at the rate of 0.5-2.0 mg a.i. per 15-cm pot in moist medium and at the time when the shoots are 0.6 cm long or after two weeks of planting. Ancymidol is consistently effective in reducing total plant height at flowering without affecting the flowering date, flower size or number of shoots per tuber. Paclobutrazol (Bonzi) at 2-4 mg and uniconazole (Sumagic) at 0.25-0.5 a.i. are applied per 15-cm pot (Whipker and Hammer, 1997). De Hertogh (1996) advocated application of drenches to moist media and when the shoots are at least 0.5 cm long. In rooted cuttings, B-Nine (daminozide) also reduces the total plant height. Daminozide or chlormequat increases the number and size of the tuberous roots produced under short days. Pappiah and Muthuswamy (1974) through twice sprayings at fortnightly intervals of MH (1000, 2000 ppm), chlormequat (1,000, 2,000 ppm) and ethrel (ethephon 39.56% at 500 and 1,000 ppm) to single-flowering dahlia plants 45 days after planting, recorded highly reduced height under MH but only little through chlormequat though its both the concentrations greatly increased the number of tubers per plant. To the cuttings of 10 dahlia varieties when B9 was applied at 5,000 and 10,000 ppm as a spray, Bhattacharjee *et al.* (1971) obtained height reduction by 27% and 49% though CCC as drench or spray did not inhibit growth and in some cases this even promoted it. Bhattacharjee *et al.* (1974) obtained retardation of 17.4-64.7% plant height in 15 cultivars at 10,000 ppm B-Nine application though flower size was larger; when pinched and non-pinched plants of cv. 'Avalanche' were treated with B-Nine, its 5,000 ppm caused more retardation in the pinched plants; and chlormequat as drench or foliar spray at 1,000 or 2,000 ppm promoted plant height by 0.6-12.0% but its 4,000 and 8,000 ppm concentrations retarded the growth by 1.5-12.6%. De Hertogh *et al.* (1976) used ancymidol, chlormequat and daminozide at various rates as soil drenches or foliar applications to potted plants of various cultivars 2 or 4 weeks after planting and then the plants were forced under glass in natural days at 16/17 ^{0} night and 18/20 ^{0}C day temperatures, and found that only ancymidol consistently reduced plant height at flowering, its 2 mg/plant soil drench being more effective than its 0.5 mg or the spraying, though noted that for early planting on February 8 the time of drenching was not important but for late planting on March 8 the treatment two weeks after planting was best. Durso and Hertogh (1977) controlled the plant heights in dahlias through ancymidol application at 0.5 mg/pot two weeks after planting at 25 ^{0}C/16 ^{0}C day/night temperatures and obtained best height control. Bhattacharjee *et al.* (1978) found highly suppressed plant height in dahlias through B-nine 500 ppm application and in *Dahlia conzattii* by CCC at 500 ppm. Bhattacharjee (1984) recorded markedly reduced plant height through 2,000 ppm TIBA or 500 ppm MH when applied at 15-18 cm tall plants of dahlia cv. 'Kelvin Rose'. Suma (1993) stated that though CCC was not as effective as to that of Alar which retarded the plant height, increased the branching, delayed the flowering by 3.9-5.9 days, improved the floral size, caused the flowers remaining fresh for longer period on stalk & in vase and increased the weight of the root-tubers with maximum tuber production under 4,000 ppm but CCC increased the thickness of the shoots and its branches and induced 8-10 days earlier flowering.

Post Harvest

The dahlia flowers for traditional use are plucked when almost open though for vase decoration these are cut when half-open. After cutting, they are immediately placed in water which is 50 ^{0}C hot, to prevent wilting and to encourage water uptake (Post, 1949; Nowak and Rudnicki, 1990) though Arnosky and Arnosky (1997) recommend 71 ^{0}C water and a floral preservative. Holding solutions should be adjusted at a pH 3.5 to 4.0. They are cut with as much stem as possible. Too immature buds if cut will fail to open. Transportation or storage for up to 5 days at 4-5 ^{0}C is possible (Nowak and Rudnicki, 1990). Bar-Sella (1959) stated that cut dahlias have short vase life as two days after picking, uptake of water is impeded by plugs originating in the xylem vessels, and antibiotics increase the storage life only by about one day, therefore, reduction of the respiration rate is important and dipping the stems in MH at 50 ppm may solve this problem to some extent. Staden (1976) recommended the product Anger Chrysal (Carnation) as better than Tulpen (Tulip) Chrysal by which a vase life of at least one week was achieved. Lukaszewska (1983a,b) tried semi-open dahlia cv. 'Purple Gem' cut blooms with 30 cm leafless stalks by placing them for 24 h in glucose at 10% + $AgNO_3$ 0.2 mM or with glucose 10% + $AgNO_3$ at 0.2% + 8-HQS at 200 mg/litre, and obtained vase life of 8.7 days in the former case while 8.2 days in the latter though cut flowers lasted only for 4.3 days in control and the life in sugar was also poor. Further, he sampled the flowers at harvest, 24 h later and again at wilting and found that when flowers were kept continuously in glucose or sucrose solutions, it resulted into accumulation of reducing sugars and sucrose in the flowers, the amount of endogenous carbohydrates doubled in the inner rays and receptacles after 24 h treatment, at wilting this amount was higher in flowers held in water but lower than in flowers held continuously in sugar solutions, and the presence of sugars in the holding solution prevented an undesirable accumulation of free amino acids in the flowers. Further, he reported that changes in the level of free amino acids in the outer rows were negligible but in the inner rays a significant increase occurred at wilting but least of all in flowers held continuously in the sugar solutions. Swart (1989) advocated picking of dahlia flowers at a relatively immature stage of flower opening and their immediate placing in Chrysal AKC solution at 24 g/litre or Anjer-Chrysal at 13 g/litre for improving the vase life, and low temperature storage at 2 ^{0}C for two days also improves subsequent vase life at 20 ^{0}C. Dahlias respond well to glucose, $AgNO_3$ and 8-hydroxyquinoline sulphate provided the water pH is adjusted to 3.5-4.0. Glucose 10% + $AgNO_3$ 0.2 mM or 10% glucose + 0.2mM $AgNO_3$ + 200 mg 8-HQS/litre are most favourable for prolonging the vase life in dahlia cut flowers. The decreasing effect of Ethephon 50 ppm is negated when glucose 10% + $AgNO_3$ 50 ppm treatment is given to dahlia cut flowers, and this treatment also increases the vase life.

The pot dahlias are ready for marketing when first flowers start opening. When in flower the pot dahlias are not recommended for cold storage, however, in case of too early flowering, the flowers are disbudded and the laterals are allowed to develop which may take 5-7 days further. In case flowering is ahead of the schedule, the greenhouse temperature may be reduced to 10-13 ^{0}C which will slow down the development of the plants. For post-greenhouse handling, a plastic sleeve is used.

Lifting of Tubers and Storage

On the hills, the plants appear as burnt by the first frost in October-November when these should be lifted leaving some 15 cm aerial stems on the plants. Tubers are lifted gently with the help of fork, shovel, spade or *khurpi* just by loosening the soil and pulling the whole clump intact by catching the left out portion of the aerial stem. A frost of -2 ^{0}C before lifting in October kills all the dahlia leaves but only partially injures the flowers and the osmotic values of frozen cells are considerably higher than normal cells (Pardatscher, 1951). Dogonadze (1957) also observed that the seedlings of a dahlia grafted on a yellow chrysanthemum acquired an improved resistance to light autumn frosts as its resistance became similar to that of chrysanthemum. Instead of manual removal of the top growth at lifting which may disseminate viral and bacterial infection, Berthier (1987) found dinoseb at 12 kg/ha quite effective when applied in last week of September at 17-20 ^{0}C temperature and 80% RH, then tubers were lifted in first week of November and stored in moist peat at 12-15 ^{0}C in mid-February, these all the tubers sprouted cent per cent. He recommended diquat and dinoseb quite promising and safe but suggested to apply dinoseb at a minimum temperature of 15 ^{0}C. Root tubers after lifting and cleaning are shade-dried for 5-7 days, then stored at 5-10 ^{0}C in peat with 50% water content to prevent moisture loss, as in storage the tubers should not be allowed to dry out, alternatively the root tubers can also be dipped in paraffin but in any case the freezing or lethal temperature for tubers is -2 ^{0}C which should be avoided (Allen, 1937; De Hertogh and Le Nard, 1993). On the hills, either these are not lifted from the field until planting time or the clumps of root tubers intact with a portion of aerial shoots are stored in a room with open windows for ventilation and these start throwing out the sprouts in March-April and that is the proper planting time on the hills. If the lifted clumps as such are buried elsewhere in the field, these rot, and the situation is aggravated when divided tubers are buried because of the injury at cut portion from where cold or infection enters the tubers. Room storage may not pose this problem. In the plains, in May when flowers start burning and drying, these are lifted the same way as on the hills, cleaned and stored at 7-9 ^{0}C temperature with peat having 50% moisture. These are taken out from cold storage in September for planting in October. Steffen (1958) advocated the storage of tubers at 10 ^{0}C in boxes containing dry sand/peat mixture or in plastic bags after drying the tubers. Ailincai (1972) stated that dahlia tubers stored at 6 or 9 ^{0}C had lower respiration rates, *vis-à-vis* lower peroxidase, catalase and ascorbinoxidase activities and a higher polyphenoloxidase activity than those held at 20 ^{0}C.

Insect-Pest, Diseases and Physiological Disorder

Aphids (*Aphis fabae, Aulacorthum circumflexum, A. solani, Brachycaudus helichrysi* and *Myzus persicae*) suck plant sap from tender parts and deform the plants. These are found clustering on newly emerging leaves and buds and on the tips. These produce honey dews which attract sooty mould. These are also carrier of various viruses which transmit diseases from one to other plants. Thrips (*Frankliniella occidentalis, F. tritici, Heliothrips haemorrhoidalis* and *Thrips tabaci*), **whiteflies** (*Trialeurodes vaporariorum*), **capsid bugs**, **tarnished plant bugs**, flea hoppers (*Halticus bractatus*) and leaf hoppers (*Empoasca fabae,* and *E. solana)* also feed on tender leaves causing their discolouration along the margins, affected areas becoming brown and brittle and the infected plants showing deformed and poor growth. **Thrips** rasp the surface of the petals and feed on the exuding juice and the lower surface of the petals turns whitish and petals wither. **Flea hoppers**, **leaf hoppers** and **grasshoppers** feed on the aerial parts of the plants and wireworms on roots and tubers so while controlling beetles these insects will automatically be controlled. Neiswander (1955) found 0.065 g of 25% Malathion powder + 0.048 g of 50% DDT powder per litre of water when sprayed at every 10 days starting when plants were 30 cm, found excellent control of potato leafhopper, tarnished plant bug, onion thrips, green peach aphid and common red spider mite (two-spotted spider mite). Widmer (1965) found effective insect control when applied 10% granular Thimet (phorate) in the well prepared soil at 21 g and 35 g/plant and 10% granular Di-Syston (disulfoton) at the same rates as Thimet and also at 12 g/plant, followed by immediate irrigation, and found effective control with both the chemicals at 35 g per plant (12 sq. ft.). At 35 g of Dimetilan and Thimet, with Metasystox-R for an area of $0.6 \times 1.8\ m^2$ containing 2 plants were also recorded very effective. These all the pests can also be controlled through use of tobacco decoction or fortnightly spraying with 0.2% Malathion or Metacid-50. Olisevič (1959) found good control of aphids (*Myzodes dianthi* and *Neomyzus circumflexus*) through spraying with 0.1-0.2% Mercaptophos and octamethyl (Schradan), first remaining toxic to aphids up to or beyond three months while the second for two months. While controlling the aphid, this will control the thrips also. **European corn borer** (*Ostrinia nubilalis* syn. *Pyrausta nubilalis)* caterpillars feed on the young shoot tips, bud ends and flowers and bore the stems causing wilt of the affected parts. Measures applied for aphids will control this pest also. **Earwigs** (*Euborellia annulipes*) feed on the petals by making holes in the buds. Its activities are manifested during summer. These hide within flowers during day time but become active in the nights. These are controlled through dry straw traps by covering whole plant during day time or spraying some contact insecticide after darkness of the night. Japanese beetles (*Popillia japonica*), the carrot beetle (*Ligyrus gibbosus*), mordellid beetle (*Mordellistena ghanii),* the flea beetle (*Nodonta clypealis),* grasshoppers and **wireworms** also feed on dahlia plants and cause considerable loss. Adult **Japanese beetle** feeds on the leaves, flowers and stems but grubs damage the roots and tubers. Commercial preparations of *Bacillus popiliae* (milky spore disease) in soil infects the grubs and kill them. **Mordellid beetles** bore into the stems and cause extensive damage. **Flea beetle** adults and larvae feed on leaves, flowers and seed pods. The **carrot beetle** larvae feed on the root and tuberous roots of dahlia. Soil application of Temik 10-G in the soil and spraying of Nuvacron at 0.2% on the aerial parts will control all sorts of beetles. Red spider (*Tetranychus urticae*) and certain other mites are also serious pests of dahlia. Both adult and larval stages of **broad mite** (*Hemitarsonemus latus*) causing curling and distortion of foliage can be controlled by dusting with finely ground sulphur or spraying with water-soluble sulphur at 6 g/litre of water (Anon., 1951) and **red spider mites** (*Tetranychus urticae*) which feed on the underside of the leaves and cause white specklings, make the webs, and transmit dahlia mosaic virus (Korneeva *et al.*, 1970) are controlled with Plictran (cyhexatin) within 11 days and through anise oil giving acceptable control (Wickizer and Meisch, 1977). Fluvanilate tuber dip or foliar spray will also control the mites. Harris (1971), because of the resistance being developed of organophophorus compounds to red spider mite, advocated introduction of the predacious mite (*Phytoseiulus persimilis*) which gave effective control even in the absence of chemical sprays. Spray of Propargite (0.1%) is also very effective against mites. **Nematodes** like *Meloidogyne hapla* causes root galling, *Trichordoridorus pachydermus* found in the soil has been recorded from Michigan causing poor development of root system with only few feeder roots and leaf notching in *Dahlia pinnata* (Bird and Knobloch, 1976), *Aphelenchoides ritzemabosi* attacks leaves and floral buds, and *Rotylenchus robustus* infests the aerial parts, especially the leaves. All these nematodes can be controlled by use of Furadan @ 4-5 kg when mixed in the soil at the time of preparation of the soil and by spraying the infested crop with Dimecron or Nuvacron @ 0.1%. Pearman (1959) suggested crop rotation and burning of all the infected plants, *vis-à-vis* soil fumigation with ethylene dibromide, methyl bromide, DD or Namagon. Gill (1981) recommended HWT of infected cuttings or shoots at 50^0C for 5 minutes or at 44.4^0C for 30 minutes, and phorate or aldicarb at 1.5 kg/ha applied as basal dressing at the time of transplanting. Slugs are particularly active in humid and dark areas, especially during nights and batter the entire plant. This damage will coincide with characteristic slime trails. These can be controlled through formaldehyde traps.

Beaumont (1950) stated that in dahlia **damping off** is caused by *Phytophthora cryptogea, Rhizoctonia solani* and *Botrytis cinerea*; **wilting** by *Verticillium dahliae* and *Sclerotinia sclerotiorum*; **leaf spots** by *Entyloma dahliae* and *Ascochyta dahliaecola*; **crown gall** by *Bacterium tumefaciens*; **leafy gall** by *Corynebacterium dasciens*; and **dahlia mosaic** such as tomato-spotted wilt and cucumber mosaic through viruses. **Powdery mildew** (*Erysiphe polygoni* and *E. cichoracearum)* causes white to grey powdery mass on the upper surface of the leaves which can be controlled by spraying 0.1% Bavistin or Karathane. Pearman (1959) recommended sulphur preparations for control of powdery mildew (*Oidium* sp.), and Roy (1965) against the infection of *O. erysiphoides* recommended spraying with Thiovit, barium polysulphide or lime-sulphur at the first appearance after removal of infected leaves and again a fortnight later. **Dahlia smut** (*Entyloma calendulae* f. sp. *dahliae*) though is not a serious disease of dahlia but it causes black spots and shot holes on the foliage. Dithane M-45 or Dithane Z-78 at 0.2% sprayings will control this disease. Pearman (1959) recommended Thiram and Captan for its control. Copper spraying in summer also controls this disease (Pape, 1953). Grouet (1962) recommended spraying the crop with 0.25% Captan. Joshi (1967) when sprayed infected dahlia plants with Bordeaux mixture at 4 gal/100 plants thrice, i.e. July, August and September and found good control of smut. Misra (1978) sprayed several fungicides to the smut infected dahlia plants at 15-days intervals from June 20 to October 20 but obtained good control only with Dithane Z-78 (zineb) and Dithane M-45 (mancozeb), each at 0.2%. Upadhyay and Bhandari (1985) found good control of dahlia smut through Dithane Z-78, Blitox (copper oxychloride) and benomyl. **Rots** (foot, crown, root or brown core rot) are caused by *Pythium* spp., *Rhizoctonia solani* and *Sclerotinia sclerotiorum* and the symptoms produced by these fungi are almost similar so visual identification becomes almost impossible. Their action and treatment are also almost the same. Earliest signs of their infection are indicated by production of smaller leaves which turn yellow before wilting, flowering is reduced and the plant loses its turgidity due to its decaying root system. **Stem rot** (*Sclerotinia sclerotiorum*) in the form of white mould encircles the main stem and sister branches near the base of the plant which causes them to wilt and die. To avoid this disease, the soil should be quite porous with good drainage facilities, and the spacing among the plants should be sufficient to permit proper aeration. Soil sterilization with chemicals or steam will prevent this problem. Pearman (1959) suggested burning of all the infected plants and crop rotation for *S. sclerotiorum* as well as *Sclerotium rolfsii*. For dahlia tubers infected with white rot, Šumilenko and Šaranova (1963) treated the tubers with 50% TMTD at 8 g per kg of tubers four months before planting which produced vigorous plants with large number of flowers and when tubers were lifted, there appeared no sign of white rot. Sometimes, its infection is found reduced due to proper management of the drainage system, nutrient supply and soil amendments for changing the pH levels as there is no any effective chemical cure for this problem. **Root** and **stem rots** occur due to attack of *Verticillium alboatrum, Fusarium, Rhizoctonia solani, Pythium* and *Phytophthora* fungi and some of these fungi attack the tubers even in storage and rot them. Only healthy tubers should be used for planting after treating them with Captan 0.2%. Regular fortnightly spraying of benomyl 0.2% alternate with Captan 0.2% throughout the growing period will save the crop from all the fungal diseases. Jain and Nema (1952) recorded **blossom blight** of dahlia (*Choanephora infundibulifera*) which can be controlled through use of Dithane M-45. **Leaf blight/grey mould** (*Botrytis cinerea*) and **downy mildew** cause burning or rotting of plants and its parts, especially leaves during poor ventilation, cloudy, cold, windy and excessive humid weathers and unfortunately they are not host-specific. This grey mould infects the young shoots and the flower buds, and afterwards causes death of entire plant. This causes soft rot and in chilly and humid weathers the white mould envelops whole plant. It also infects the floral buds and flowers causing the buds to rot or shrivel. Dithane Z-78 (Zineb), Dithane M-45 (mancozeb) or Captan 0.2% sprayings at weekly intervals will keep these diseases under check. Pearman (1959) recommended Thiram and Captan for controlling grey mould. Sulphur dusting will control the downy mildew. Shrotri *et al.* (1984) stated that out of 21 **fungi associated with dahlia seeds**, *Alternaria alternata, Curvularia pollescens, Phome putaminum, Aspergillus flavus, A. niger, A. repens, Cochliobolus spicifer, Drechslera hawaiiensis* syn. *Cochliobolus hawaiiensis* and *Fusarium* spp. caused varying reductions in seed germination in pathogenicity tests, and for complete control of seed-borne fungi they recommended Ceresan, Dithane M-45 (mancozeb) and Aureofungin.

Crown gall disease (*Agrobacterium tumefaciens*) causes large tumours of abnormal tissues at the base of the plants and on roots. Infected plants become stunted and deformed. The disease can be prevented by dipping the roots in streptomycin solution before planting and after lifting. Before planting, soil should also be disinfected. Clark and Paton (1956) in Scotland observed a **wet and black rot** in dahlia, on young cuttings in propagation beds and sometimes even of mature plants due to infection of *Pseudomonas marginalis*, which to some extent could be controlled by pre-planting dip or spraying the plants in field after planting with streptomycin. Pearman (1959) suggested burning of all the infected plants and crop rotation against this bacterium. Plants affected with *Pseudomonas solanacearum* droop and wilt suddenly from the base. Control measures are the same as described under crown gall disease. *Erwinia carotovora, E. chrysanthemi* and *E. cytolytice* cause **soft**

rot and browning of stems and soft rot of root tubers. The infection is aggravated during cold weathers. These can also be controlled as described under crown gall disease. Saaltink (1969) suggested careful selection of planting material which can eliminate the bacterial wilt of dahlia (*E. chrysanthemi*) almost completely. Kamerman and Saaltink (1968) recommended formalin or steaming treatment of infected soil, use of clean water for irrigation and avoidance of any dubious planting material. The most common bacterial die back or canker is fire blight caused by *Erwinia amylovora*. The bacterium enters the plant parts through wounds or dead tissues and causes death of tissues on dahlia stems. **Bacterial wilt** is caused by *Ralstonia solanacearum* where symptoms in the plant appear in the form of stem drooping, and in severe cases even simultaneous wilting of the plants. Infected stems have a wet soft rot near the soil. The first step to control this disease is destroying of infected plants. Crop rotation is also recommended to control this disease.

Brierley (1933a, b) stated that dahlias are infected by mosaic (stunt as dwarf, rugose mosaic and veinal mosaic) virus transmitted by aphid *Myzus persicae*, ring-spot, yellow ring-spot and oakleaf viruses. *Dahlia* plants are also attacked by cucumber mosaic virus, dahlia mosaic virus, tobacco stripe virus, tomato fleck virus and tomato spotted wilt virus (bronzing). Marini (1951) from Italy reported dahlia virus 1, the spotted wilt lycopersicon virus 3 and the cucumber virus 1. Lawson (1966) isolated from dahlia cv. 'Helen Anita' plants a strain of CMV producing oakleaf patterns. Brunt (1968) isolated tobacco streak virus for the first time from dahlia plants. Sansdrap (1969) described the symptoms of dahlia mosaic, bronzing (tomato spotted wilt virus), potato virus Y and cucumber virus 1. Holmes (1955) suggested elimination of spotted wilt virus by growing them through their small stem-tip cuttings as this virus does not move freely towards the growing tips of dahlia plants. Brierley (1933b) stated that mosaic symptoms appear as yellowish or pale-green banding along the midribs and larger branch veins of affected leaves, and the dwarfing is a second type of symptoms evident in mosaic plants of all but the most tolerant varieties and includes shortening of the roots. For its control he suggested fumigation of glasshouses with nicotine once a week, periodic rogueing and control of *Myzus persicae* with insecticides. Limasset (1946) reported from France that tomato spotted wilt virus may be transmitted by *Thrips* spp. Martin (1954) stated that biochemical activity of the enzymes cytochrome-oxidase, peroxidase and tyrosinase were greater in dahlia mosaic, cucumber mosaic and spotted wilt infected than in healthy dahlia plants. Martin (1959) also stated that mosaic-affected leaves of dahlia had higher contents of phenols. Korneeva *et al.* (1970) suggested destruction of mosaic virus infected plants, control of aphids and mites, and weeds harbouring the mites, taking the cuttings only from the healthy plants and sanitation measures. Korneeva (1971) stated that through healthy shoot grafting of dahlia on mosaic infected plants, symptoms appeared after three weeks and grafting of mosaic infected plants on healthy plants the symptoms were found developing only in the third year. Breman (1990) stated that though dahlia mosaic caulimovirus (DMV) is present in USA since 1927 and as there is no any effective control so the plants should be started from seeds as it does not transmit through seeds. Pearman (1959) stated that thrips being the vector for spotted wilt and aphids for stunt which should be controlled by spraying the plants with 0.05% Malathion every 10 days. Asjes (1968) suggested placing of 12 mm wide strips of reflective aluminium foil horizontally just above dahlia plants in the field. In any case the infected plants should be rogued out and burnt immediately. Aphids, hoppers, mites and thrips which spread the disease should also be controlled effectively by use of insecticides. The field should also be kept clean of debris.

Due to imbalanced temperature and/or photoperiod, flower bud abortion and terminal shoot blindness are caused while forcing the potted dahlias. Those cultivars which are amenable to the standard treatment of temperature and photoperiod should only be forced while others may be forced only when proved through experiments. Blindness usually occurs due to improper propagation practices through stem cuttings when a piece of crown was not taken while removing the cuttings from the crown of the root tuber.

References

Ailincai, N. 1972. Cîteva procese biochimice caracteristice rădăcinilor tuberizate de dahlia, depozitate în conditii diferite de temperatură. *Lucrări Stiintifice, Institutul agronomic* "Ion Ionescu de la Brad", **I**: 325-331.

Allen, R.C. 1937. Temperature and humidity requirements for the storage of dahlia roots. *Proc. Amer. Soc. hort. Sci.*, **35**:770-773.

Anon. 1951. The broad mite (*Hemitarsonemus latus*) as a pest of dahlias. *Agric. Gaz. N.S.W.*, **62**: 252-253.

Aoba, T., S. Watanabe and C. Saito, 1960. Studies on tuberous root formation in dahlia. I. Periods of tuberous root formation in dahlia (Japanese). *J. hort. Ass. Japan*, **29**: 247-252.

Arnosky, P. and F. Arnosky, 1997. Dahlias are difficult but worth the effort. *Growing for Market*, **6**(10): 11-13.

Asjes, C.J. 1968. Effect of aluminium strips on the spreading of two aphid-borne viruses in dahlias. *Meded. Rijksfac. LandbWetensch., Gent*, **13**: 1171-1177.

Ball, V. 1991. Plugs – The way of the 1990s. In: *Ball Redbook* (15th edition), pp. 137-153. Geo. J. Ball Publishing, W. Chicago, Illinois.

Barrett, J.E. and A.A.De Hertogh, 1978. Pinching forced tuberous-rooted dahlias. *J. Amer. Soc. hort. Sci.*, **103**(6): 775-778.

Bar-Sella, P. 1959. Causes of rapid wilting of cut dahlias and means to improve their keeping qualities. *Bull. Res. Counc. Israel, Sect. D*, **7D**: 35-38.

Beaumont, A. 1950. Diseases of dahlias. *Gdnrs' Chron.*, **128**: 247.

Berthier, G. 1987. Etude de défoliants appliqués sur dahlia. *P.H.M.-Revue Horticole*, No. 278, pp. 39-42.

Bhattacharjee, S.K. 1984. Effect of growth regulating chemicals on growth, flowering and tuberous-root formation of *Dahlia variabilis* Desf. *Punjab hort. J.*, **24**(1/4): 138-144.

Bhattacharjee, S.K. and T. Mukherjee, 1983. Influence of nitrogen and phosphorus on growth and flowering in dahlia. *Pb. Hort. J.*, **23**(1/2): 111-115.

Bhattacharjee, S.K., K.D. Gupta and T.K. Bose, 1971. B-Nine, and effective growth retardant on *Dahlia;* in effectiveness of CCC. *Phyton, Argentina*, **28**(1): 61-65.

Bhattacharjee, S.K., T. Mukhopadhyay and T.K. Bose, 1974. Experiments with growth retardants on *Dahlia. Indian J. Hort.*, **31**(1): 86-90.

Bhattacharjee, S.K., T. Mukhopdhyay and T.K. Bose, 1976. Interaction of auxin and gibberellin with growth retardants on growth and flowering of *Dahlia variablils* and *Malvaviscus conzattii. Indian Agricult.*, **20**(3): 193-199.

Bhattacharjee, S.K., T. Mukhopadhyay and T.K. Bose, 1978. Changes in nitrogen and carbohydrate fraction in *Dahlia variabilis* cv. "Avalanche" and *Malvaviscus conzattii* treated with CCC, B-nine and GA. *Indian Agricult.*, **22**(1): 41-48.

Biran, I., I. Gur and A.H. Halevy, 1972. The relationship between exogenous growth inhibitors and endogenous levels of ethylene, and tuberization of dahlias. *Physiologia Plantarum*, **27**(2): 226-230.

Biran, I. and A.H. Halevy, 1973. The relationship between rooting of dahlia cuttings and the presence and type of bud. *Physiologia Plant.*, **28**(2): 244-247.

Biran, I., B. Leshem, I. Gur and A.H. Halevy, 1974. Further studies on the relationship between growth regulators and tuberization of dahlias. *Physiologia Plantarum*, **31**(1): 23-28.

Bird, G.W. and N. Knobloch, 1976. Occurrence of *Trichodorus pachydermus* in Michigan. *Pl. Dis. Reptr*, **60**(1): 76-77.

Boertje, G.A. 1979. Bijmesten perkplanten in setjes : hoe vaak en hoeveel? *Vakblad Bloemist.*, **34**(12): 26-29.

Bose, T.K. and A. Mukhopadhaya, 1967. Changes in the carbohydrate and nitrogen during growth, flowering and tuber formation in dahlia. *Indian J. Hort.*, **24**: 191-196.

Breman, L.L. 1990. Dahlia mosaic virus. *Plant Pathology Circ.* (Gainesville), No. 331, 2pp.

Brierley, P. 1933a. Studies on mosaic and related diseases of dahlia. *Contrib. Boyce Thompson Inst.*, **5**: 235-288.

Brierley, P. 1333b. Dahlia mosaic and its relation to stunt. *Boyce Thompson Inst.*, **1**(25) 240-246.

Broertjes, C. and J.M. Ballego, 1968. Addendum to mutation breeding of *Dahlia variabilis. Euphytica*, **1**: 507.

Brøndum, J.J. and R.D. Heins, 1993. Modeling temperature and photoperiod effects on growth and development of dahlia. *J. Amer. Soc. hort. Sci.*, **118**: 36-42.

Brunt, A.A. 1968. Tobacco streak virus in dahlias. *Plant Path.*, **17**: 119-122.

Čaplygin, B.K. and G.I. Šahova, 1961. Accelerated propagation of dahlias (Russian). *Priroda*, **50**(9): 116-117.

Cavalchini, A. 1968. Studies on dahlia irrigation (Italian). *Ann. Fac. Sci. agrar. Torino*, **4**: 345-358.

Cayeux, H. 1936. Un mode nouveau de bouturage du dahlia. *Rev. hort.. Suisse,***9**: 9-12.

Clark, M.R.M. and A.M. Paton, 1956. A new bacterial disease of dahlias caused by *Pseudomonas marginalis. Plant Path.*, **5**: 32-35.

Clay, D.V. and J.G. Davison, 1968. The response of dahlias to soil-applied herbicides. *Proc. 9th Brit. Weed Control Conf.*, pp. 965-970.

Clay, D.V. and G.W. Ivens, 1964. Herbicide trials on dahlias. *Proc. 7th Brit. Weed Control Conf.*, pp. 241-247.

Das, P.K., S. Dubey and A.K. Dey, 1978. New dahlias through irradiation. *Indian Hort.*, **23**(1): 19, 21.

De Hertogh, A.A. 1996. Dahlia - Potted Plants. In: *Holland Bulb Forcer's Guide* (5th edition). International Flower Bulb Centre, Hillegom, The Netherlands.

De Hertogh, A.A., N. Blakely and W. Szlachetka, 1976. The influence of ancymidol, chlormequat and daminozide on the growth and development of forced *Dahlia variabilis* Willd. *Scientia Hort.*, **4**(2): 123-130.

De Hertogh, A.A. and M. Le Nard, 1993. *Dahlia*. In: *The Physiology of Flower Bulbs*, pp. 273-283. Elsevier, Amsterdam.

Dogonadze, D.S. 1957. Vegetative hybrids of dahlia and chrysanthemum (Russian). *Agrobiologija*, No. 1, pp. 127-128.

Dohare, S.R., H.S. Gill and R.S. Arora, 1974. 'Manali' a new dahlia. *Indian Hort.*, **19**(2): 21.

Dole, J.M. and H.F. Wilkins, 1999. Dahlia. In: *Floriculture Principles and Species*, pp. 287-291. Prentice Hall, New Jersey.

Durso, M. and A.A. De Hertogh, 1977. The influence of greenhouse environmental factors on forcing *Dahlia* Willd. *J. Amer. Soc. hort. Sci.*, **102**(3): 314-317.

El-Gamassy, A. and M.B. Moustafa, 1963a. Nutritional requirements of *Dahlia pinnata*. 1. Effect of nutrient solutions on the growth, flower quality and yield of tuberous roots of dahlia plants in sand culture. *Ann. Agric.Sci., Cairo*, **8**(2): 357-381.

El-Gamassy, A. and M.B. Moustafa, 1963b. Nutritional requirements of *Dahlia pinnata*. 2. Effect of nutrient solutions on the chemical composition of dahlia plant parts in sand culture. *Ann. agric.Sci., Cairo*, **8**(2): 383-416.

El-Gamassy, A. and M.B. Moustafa, 1964a. Nutritional requirements of dahlia plants. 3. Effect of fertilizer rates on the growth rate, flower quality and yield of tuberous roots of dahlia plants grown in pot and field. *Ann. agric.Sci., Cairo*, **9**(1): 395-421.

El-Gamassy, A. and M.B. Moustafa, 1964b. Nutritional requirements of *Dahlia pinnata*. 4. Effect of appliction time of phosphatic and potassic fertilizers on the growth, flower quality and yield of tuberous roots of dahlia plants. *Ann. agric.Sci., Cairo*, **9**(1): 423-431.

Forkmann, G. and G. Stotz, 1984. Selection and characterization of flavanone 3-hydroxylase mutants of *Dahlia, Streptocarpus, Verbena* and *Zinnia. Planta*, **161**(3): 261-265.

Gavinlertvatana, P., P.E. Read, H.F. Wilkins and R. Heins, 1979. Influence of photoperiod and daminozide on stock plant pretreatments, on ethylene and CO_2 levels and callus formation from dahlia leaf segment cultures. *J. Amer. Soc. hort. Sci.*, **104**(6): 849-852.

Gavinlertvatana, P., P.E. Read, H.F. Wilkins and R. Heins, 1982. Ethylene levels in flask atmospheres of *Dahlia pinnata* Cav. Leaf segments and callus cultured *in vitro. J. Amer. Soc. hort. Sci.*, **107**(1): 3-6.

Gill, J.S. 1981. Fighting the foliar nematode of chrysanthemum. *Indian Hort.*, **25**(4): 21-22.

Grouet, D. 1962. A contribution to the study of dahlia *Entyloma* (French). *Ann. Épiphyt.*, **13**: 5-22.

Gupta, P.K., D.K. Agarwal and A.K. Srivastava, 1972. Further cytological investigations in Indian Compositae. *Cytologia*, **37**: 581-593.

Halevy, A.H. and I. Biran, 1975. Hormonal regulation of tuberization in dahlia. *Acta Hort.*, No. 47, pp. 319-329.

Hall, O.G. 1969. Dahlias for early cut flower production. *Shinfield Progr.*, **14**: 36-37.

Harborne, J.B., J. Greenham and J. Eagles, 1990. Malonylated chalcone glycosides in *Dahlia*. *Phytochem.*, **29**(9): 2899-2900.

Harris, K.M. 1971. A new approach to pest control in the dahlia trial at Wisley. *J. Roy. hort. Soc.*, **96**: 200-206.

Holmes, F.O. 1955. Elimination of spotted wilt from dahlias by propagation of tip cuttings. *Phytopath.*, **45**: 224-226.

Jain, A.C. abd K.G. Nema, 1952. Blossom blight of dahlia. *Curr. Sci.*, **21**: 71-72.

Joshi, N.C. 1967. Varietal resistance of dahlia to smut disease (*Entyloma dahliae*) and efficacy of bordeaux mixture against it. *Sci. Cult.*, **33**: 452-453.

Kamerman, W. and G.J. Saaltink, 1968. Bacterial wilt in dahlias. *Praktijmeded. Lab. BloembollOnderz., Lisse*, No. 28, pp. 10.

Konishi, K. and K. Inaba, 1964. Studies on flowering control of dahlia. I. On optimum day length (Japanese). *J. Jap. Soc. hort. Sci.*, **33**: 171-180.

Konishi, K. and K. Inaba, 1967a. Studies on flowering control in dahlias (Japanese, summary English). VII. On dormancy of crown-tubers. *J. Jap. Soc. hort. Sci.*, **36**: 131-140.

Konishi, K. and K. Inaba, 1967b. Studies on flowering control in dahlias. VIII. Effect of day-length on dormancy in axillary bud (Japanese, summary English). *J. Jap. Soc. hort. Sci.*, **36**: 243-249.

Konishi, K. and K. Inaba, 1966a. Studies on flowering control of dahlias. III. The effect of day length on the initiation and development of flower buds (Japanese). *J. Jap. Soc. hort. Sci.*, **35**: 73-79.

Konishi, K. and K. Inaba, 1966b. Studies on flowering control of dahlias. IV. The effect of day length at the early stage of shoot growth upon the flowering date and the quality of cut flowers (Japanese). *J. Jap. Soc. hort. Sci.*, **35**: 195-202.

Konishi, K. and K. Inaba, 1966c. Studies on flowering control of dahlias. V. The effects of night temperature and light intensity and duration on flowering (Japanese). *J. Jap. Soc. hort. Sci.*, **35**: 317-324.

Korneeva, I. T. 1971. The development of mosaic virus symptoms in dahlias. *Byulleten' Glavnogo Botanicheskogo Sada*, No. 82, pp. 120-123.

Korneeva, I.T., L.A. Zinov'eva and N.L. Rostova, 1970. The role of the mite *Tetranychus urticae* in the spread of mosaic virus on dahlias (Russian). *Bjull. Glav. Bot. Sada*, No. 76, pp. 88-92.

Kumar, S., R. Sharma, K.S. Datta and K.K. Nanda, 1978. Gibberellic acid causes induction of flowering in *Dahlia palmata* under non-inductive photoperiods. *Indian J. Pl. Physiol.*, **21**(3): 261-264.

Lavrichenko, E.V. 1975. Morphogenesis of the vegetative parts and causes of degeneration of *Dahlia cultorum*. *Botanicheskiĭ Zhurnal*, **60**(3): 412-424.

Lawrence, W.J.C. 1942. The origin of the garden dahlia. *Publ. John Innes hort. Inst.*, 8 pp.

Lawson, R.H. 1966. Oakleaf chlorosis symptomatic of cucumber mosaic virus infection in dahlia. *Phytopath.*, **56**: 343-344.

Limasset, P. 1946. Les maladies à virus du dahlia. *Rev. hort. Paris*, **118**: 11-14.

Lloyd, F.G. 1968. Dahlias - production of healthy tubers under glass. 9th *Progr. Rep. exp. Husb. Fms exp. Hort. Stats*, pp. 66-67.

Lukaszewska, A.J. 1983a. The effect of continuous and 24 hour sugar feeding on the keeping quality of cut dahlias. *Prace Instytutu Sadownictwa i Kwiaciarstwa w Skierniewicach, B* (*Rośliny Ozdobne*), **8**: 199-205.

Lukaszewska, A.J. 1983b. The effect of continuous and 24 hour sugar feeding on carbohydrates Free aminoacids in the inflorescences of cut dahlias. *Prace Instytutu Sadownictwa i Kwiaciarstwa w Skierniewicach, B (Rośliny Ozdobne)*, **8**: 207-214.

Maatsch, R. and W. Rünger, 1955. Über die photoperiodische Reaktion einiger Sorten von *Dahlia variabilis* Desf. *Gartenbauwiss.* (n.s.), **1**: 366-390.

Maiko, T.K. and N.P. Yashchenko, 1980. Effect of retardants on tuber formation and flowering in dahlias. *Byulleten' Glavnogo Botanicheskogo Sada*, No. 118, pp. 61-64.

Marini, E. 1951. Segnalazione di virosi della dalia nelle culture di produzione in Italia. *Not. Mal. Piante*, No. 17, pp. 13-16.

Marston, M.E. and G. Evans, 1972. Capillary bench technique could prove invaluable in commercial production of dahlia pot tubers. *Grower*, **77**(17) 950.

Martin, C. 1954. Recherches sur les maladies à virus du dahlia. *Ann. Epiphyt.*, **5**: 63-78.

Martin, C. 1959. Study of some metabolic abnormalities in plants affected by virus diseases (French). *Ann. Physiol. Vég.*, **1**: 59-82.

Mastalerz, J.W. 1976. Dahlias have potential as new cut flower crop. *Science in Agric.*, **24**(1): 11.

Mather, S.N. and R.A. Sharma, 1969. Influence of gibberellic acid on flowering and critical daylength of dahlia. *Abstracts 11th int. bot. Congr., Seattle*, p. 142.

Mehlis, E. 1976. Alte und neue Sorten gezeigt. *Deutsche Gärtnerbörse*, **76**(49): 1070-1071.

Mehlis, E. 1978a. Dahlien - Neuheiten 1978 (2. Prüfung). *Gb + Gw*, **78**(44): 1055.

Mehlis, E. 1978b. Dahlien - Neuheitenprüfung 1978 (zweite Prüfung). *Deutscher Gartenbau*, **32**(44): 1833.

Mehra, P.N. and P. Ramanandan, 1974 Cytological investigations on the Indian compositae, II Asteraceae, Heliantheae, Hebriaceae and Anthenudeas, *Caryologia*, **27**:255-284.

Mishra, H.P., K.P. Singh, G.M. Mishra and B. Prasad, 1990. Performance of some dahlia (*Dahlia variabilis*) varieties under late planted condition in calcareous soil of plains. *Haryana J. hort. Sci.*, **19**(3-4): 284-290.

Misra, R.L. 1978. Combating dahlia smut. *Progressive Hort.*, **10**(1): 43-46.

Mittal, S.P. 1967. studies on the effect of gibberellin on growth and flowering of dahlia. *Madras agric. J.*, **54**: 103-107.

Morel, G. and C. Martin, 1952. Guérrison de dahlias atteints d'une maladie á virus. *Comptes Rendus academie des Sciences*, **235**: 1324-1325.

Moser, B.C. and C.E. Hess, 1968. The physiology of tuberous root development in dahlia. *Proc. Amer. Soc. hort. Sci.*, **93**: 595-603.

Mullin, R.H. and D.E. Schlegel, 1978. Meristem-tip culture of dahlia infected with dahlia mosaic virus. *Plant Dis. Reptr*, **62**(7): 565-567.

Mustafajeva, A.K. 1985. Propagation of dahlias by cuttings in Apsheron (Russian). In: *Introduktsiya i Akklimatizatsiya Rastenii, Baku, Azerbaidzhan SSR*, pp. 89-92.

Neiswander, R.B. 1955. Control of insects results in more beautiful dahlias. *Ohio Fm Home Res.*, **40**: 80-81.

Newton, P. 1960. Dahlia researches. *Gdnrs' Chron.*, **148**: 70.

Newton, P. 1962. Investigating tuber production by dahlias. *Gdnrs' Chron.*, **152**: 140-141.

Newton, P. and D. Tullis, 1962. Investigating tuber production by dahlias. *Gdnrs, Chron.*, **152**: 176-177.

Nordström, C.G. and T. Swain, 1953. The flavonoid glycosides of *Dahlia variabilis.* Part I. General introduction. Cyanidin, apigenin, and luteolin glycosides from the variety "Dandy". *J. Chem. Soc. London*, pp. 2764-2773.

Nowak, J. and R.M. Rudnicki, 1990. *Postharvest Handling and Storage of Cut Flowers, Florist Greens, and Potted Plants.* Timber Press, Portland, Oregon.

Okada, M. and H. Harada, 1955. Effects of daylength and temperature on the ratio of ray-flowers to disk-flowers in dahlia flower heads (Japanese). *J. hort. Ass. Japan*, **23**: 259-263.

Olisevič, G.P. 1959. The use of chemicals with systemic activity for the control of aphids under glass. *Bjull. Glav. Bot. sada,* No. 35, pp. 91-95.

Pape, H. 1953. Eine häufige Blattfleckenkrankheit der Dahlie. *Gartenwelt,* **53**: 255-256.

Pearman, J.A. 1959. Diseases of dahlias. *Agric. Gaz. N.S.W.,* **70**: 369-375.

Pappiah, C.M. and S. Muthuswamy, 1974. Effect of some plant growth regulants on growth and flowering of dahlia (*Dahlia variabilis*). *South Indian Hort.,* **22**(3/4): 77-80.

Pardatscher, G. 1951. Protoplasmatische Studien an Blütenzellen von Dahlia. *Portugal Acta biol., Ser. A,* **3**: 171-186.

Post, K. 1949. *Dahlia pinnata.* In: *Florist Crop Production and Marketing,* pp. 439-444. Orange Judd Publishing, New York.

Protich, N. 1988. New dahlia cultivars. *Rasteniev"dni Nauki,* **225**(4): 52-55.

Rahman, M.M. and S.N. Mitra, 1974. Effects of nitrogen on growth and flowering of *Dahlia variabilis* Willd. *Bangladesh Hort.,* **2**(2): 15-18.

Read, P.E. 1968. Effects of succinic acid 2, 2-dimethyl hydrazide and (2-chloroethyl) trimethylammonium chloride on tomatoes (*Lycopersicon esculentum* Mill.) and on tuberous root formation in *Dahlia pinnata* Cav. *Diss. Abstr., sect. B,* **29**: 838.

Read, P.E., C.W. Dunham and D.J. Fieldhouse, 1972. Increasing tuberous root production in *Dahlia pinnata* Cav. with SADH and chlormequat. *HortSci.,* **7**(1): 62-63.

Robert, M.J. 2000. Edible and Medicinal Flowers. New Africa Book. Published by Spearhead, an Imprint of New Africa Books Pvt. Ltd., 99 Garfield Road, Claremont 7708.

Roy, A.K. 1965. Occurrence of powdery mildew caused by *Oidium erysiphoides. FAO Plant Prot. Bull.,* **13**: 42.

Rudionenko, G.I. and E.I. Zaar, 1951. Obtaining dahlia twin plants (Russian). *Priroda,* **40**(5): 62-63.

Rünger, W. 1955. Die photoperiodische reaction von dahlien. *Gartenwelt,* **55**: 223-224.

Saaltink, G.J. 1969. Bacterial infections of flower bulbs (Dutch). *Tuinbouwmededelingen,* **32**: 274-285.

Safranova, A.I. 1969. Seed propagation and selection of double dahlias in Crimean conditions. *Bjull. gos. nikitsk. Bot. Sada,* No. 1(8), pp. 14-17.

Safranova, A.I. 1972. Characteristics of dahlia (*Dahlia* x *cultorum*) cultivation in the Crimea (Russian). *Trudy Gosudarstvennogo Nikitskogo Bonicheskogo Sada,* **59**: 27-43.

Sang, C.G. 1974. Studies on propagating dahlias from cuttings. II. The effect of nitrogen forms on plant growth, flowering and tuberous root formation (Korean).*J. Korean Soc. hort. Sci.,* **15**(2): 178-186.

Sansdrap, A. 1969. Virus diseases of dahlias (French). *Pépiniéristes Horticulteurs, Maraîchers,* No. 96, pp. 5657-5663.

Sasso, G. 1961. An experiment on mineral nitrogen fertilizing of dahlias grown for cut flowers (Italian). *Riv. Ortoflorofruttic. ital.,* **45**: 119-126.

Senaratna, J.E. 1947. Two new dahlias bred in Ceylon. *Ceylon J. Sci.,* **12**: 217-219.

Sharga, A.N. 1975. Horticultural description and performance studies of some dahlia cultivars. *Progressive Hort.,* **7**(3): 29-38.

Shrotri, S.C., J.S. Gupta and R.N. Srivastava, 1984. Mycoflora associated with dahlia seeds, its pathogenicity and control. *Indian J. Mycol. Pl. Path.,* **14**(3): 266-269.

Singh, J.P. 1970. a spontaneous mutant for flower colour and shape in a white flowering dahlia. *Euphytica,* **19**: 261-262.

Singh, V.K. and S.K. Roy, 1973. Cytological studies in *Dahlia. The Nucleus,* **16**(3):180-184.

Staden, O.L. 1976. Houdbaarheidsmiddelen bijzonder waardevol. *Vakblad Bloemist.,***31**(34):44-45.

Steffen, L. 1958. Winter storage of dahlias. *GartenweltI,* **58**: 398.

Suma, B. 1993. Effect of Growth Retardants on Growth, Flowering, Vase-life and Tuber Fornation of Dahlia (*Dahlia variabilis* Desf.) Propagated through Cuttings (M.Sc. Thesis). College of Agriculture, Kerala Agricultural University, Vellayani (Kerala).

Šumilenko, E.P. and A.J. Šaranova, 1963. The distinction of planting material of flower crops. *Zasč. Rast. Vred. Bolez.*, **8**(4): 38.

Swami Vinayanand, 1991. Assumption on dahlia breeding. *Dahlia of Today*, pp. 62-63. 991, Puget Sound Dahlia Asscn., U.S.A.

Swart, A. 1989. Dahlia als snijbloem. Korte houdbaarheid blijft problem voor afzet. *Bloembollencultuur*, **100**(5): 24-25.

Upadhyay, J. and T.P.S. Bhandari, 1985. Fungicidal control of leaf smut of dahlia caused by *Entyloma dahliae. Indian Phytopath.*, **38**(2): 338-339.

Wasscher, J. 1955. De invloed van een kortedagbehandeling op de knolvorming bij dahlia's, knolbegonia's en gesneria's. *Meded. Dir. Tuinb.*, **18**: 342-352.

Whipker, B.E. and P.A. Hammer, 1997. Efficacy of ancymidol, paclobutrazol, and uniconazole on growth of tuberous-rooted dahlias. *HortTechn.*, **7**: 269-273.

Whitley, G.R. 1985. The medicinal and nutritional properties of *Dahlia* spp. *J. Ethnopharmacol.*, **14** (1): 75–82.

Wickizer, S.L. and M.V. Meisch, 1977. Effects of non-commercial miticides against mites on dahlias and soybeans. *Arknsas Farm Res.*, **26**(6): 13.

Widmer, R.E. 1965. Systemic insect control for dahlias. *Minn. Hort.*, **93**: 26-27.

Winter, J.A.T. De, 1977. De teelt van dahlia's als pot plant. *Bloembollencultuur*, **87**(30): 602-603.

Yasuda, I. and N. Yokoyama, 1959. Effect of day length on growth and root formation of dahlia. I. Short-day treatment in summer (Japanese). *Sci. Reps Fac. Agric. Okayama Univ.*, No. 13, pp. 57-62.

8

Dianthus

R.L. Misra and Sanyat Misra

Family:	**Caryophyllaceae,** the Pink family
Common names for carnation:	Carnation/Clove pink/Cut carnation/Divine flower/Florist carnation/Gillyflower/Grenadine/Picotee (*Dianthus caryophyllus*), Miniature carnation/Dwarf carnation/Pot carnation/Carthusian pink/Clusterhead pink (*D. carthusianorum*, syn. *D. hybriden* though latter name does not have any botanical standing), *Giroflées* (French), *Garofolo* (Italian), etc.
Common names for Pinks:	Alpine pink (*D. alpinus*), Common grass/Chinese pinks/Dianthus/Indian pinks/Garden pinks/Pheasant's eye pink/Scotch pink/Pinks (*D. chinensis* or *D. sinensis*), Cottage pinks (*D. plumarius*), Amoor pink (*D. chinensis* var. *macrosepalus,* syn. *D. dentosus*), Cheddar pink (*C. deltoides* var. *gratinopolitanus,* syn. *D. caesius*), Glacier pinks (*D. deltoides* var. *glacialis*), Fringed pinks (*D. superbus*), Maiden pink (*D. deltoides*), Modern pinks (Old-fashioned pinks × Perpetual carnation cultivars), Old-fashioned pinks (the old varieties of pinks), etc.
Common name for Sweet Williams:	Sweet William (*D. barbatus*)

crimson, and (iii) the transition group with salmon-yellow, orange and maroon (Mehlquist, 1940). Crossing yellow with white where red character is latent results into white as white is dominant over yellow or red. Mehlquist (1940) describes that pink, red and crimson are dominant over salmon, red is recessive to crimson and deep pink is dominant to light pink. In some yellow cultivars, the presence of red is associated with the presence of perfect sexual organs (Bhatt, 1989). Expression of self-colour in carnation depends on six independent factors, *viz.* *Y* controls the production of yellow anthoxanthin, and *I* the ivory anthoxanthin which is epistatic to *Y*. *A* is the basic anthocyanin factor which is effective only in presence of *Y* and *I*. *S* controls the production of scarlet-red anthocyanin and its recessive allele *s* dilutes red or salmon, *R* controls crimson-red anthocyanin while its recessive allele *r* is responsible only for production of scarlet-red anthocyanin. *M* allele modifies the red series colour to the corresponding magenta series (Bhatt, 1989). Gimelli *et al.* (1982) when analysed the segregation for flower colour in selfed progenies of six carnation cultivars, reported that white, variegation and clear colours were controlled by recessive genes.

Due to single recessive factor, the green, yellow (*yel*) and albino (*alb*) occur, albinos occurring even due to double recessive factors, and in all these cases the yellow and albino seedlings die at cotyledon stage. Lutescent seedlings (*lut* monogenic character, single recessive factor) having greenish-yellow colouration and weak growth also die mostly at cotyledon stage and some even survive up to 8-leaf pairs. Virescent (chlorophyll deficiency in longitudinal stripes on the leaves and stems) is also due to single recessive factor being designated as *vir* and such plants grow and flower. Lazy virescent (chlorosis occurring at the leaf edges but stems nearly white and internodes greatly elongated) is also monogenic and is represented as *laz*. Dwarfness (short internodes) occurs due to single recessive factor with *dw* gene.

The objective of breeding in the cultivars of standard carnations is (i) the branching only at basal nodes and not the top ones as such shoots produce flowers of poor quality with small peduncles, (ii) production of more than 15 shoots from the basal nodes, (iii) tolerance or resistance to calyx splitting so should have broader calyx with less number of petals, (iv) the introduction of fragrance in nonfragrant but otherwise good cultivars, *vis-à-vis* improved flower size, colour, vase life, and yield in Sim type of cultivars, (v) resistance to major diseases such as *Fusarium dianthi* (wilt), *Alternaria dianthi* (leaf blight), *Oidium dianthi* (mildew), *Uromyces dianthi* (rust), *Rhizoctonia solani* (wet stem rot) and *Pseudomonas caryophyllus* (bacterial wilt), and (vi) development of male sterility in the seed parents for production of F_1 hybrids.

For improving the spray carnations in terms of floriferousness and flower size, Thia Segers (1987) suggested use of *D. knappii* and *D. chinensis*. Guba (1952) developed 'Waltham Pink', 'Regal Pink', 'Spicy Rose' and 'E.F. Guba' which are resistant to *Fusarium* wilt, rust and leaf blight. Gupta *et al.* (1995) tested 12 cultivars and found 'Vermillion Protruding' and 'William Sim' less susceptible to stem rot. Ben-Yephet *et al.* (1997) stated that cultivars 'Arbel' and 'Scarlette' are resistant to *Fusarium* wilt. Onozaki *et al.* (1998) crossed spray type carnation cv. 'Super Gold' with *Dianthus capitatus* and developed a line 91BO4-2 resistant to bacterial wilt. In the Sims, the cvs 'Arthur Sim' and 'Rose' are highly resistant to rust. The cultivars 'Angella Solone', 'Naveda', 'Revada' and 'Vanya' are also resistant to *Fusarium* wilt. Yoders Nursery Catalogue of California (1984) mentions various degrees of resistance to *Fusarium* wilt in various carnation (Standard and Miniature types) varieties as (i) most tolerant, *viz.* 'Big Red', 'Carbasio', 'Melody' and 'White Melody' under Standard type, and 'Barbi', 'Candy Maj', 'Etna', 'Siri-I' and 'Teresa' under Miniature type; (ii) good tolerant, *viz.* 'Chianti Chianti', 'Crowley Pink Sim', 'Dusty', 'Felicia', 'Lucy Carrier', 'Orchid Beauty', 'Peterson's Red Sim' and 'Pink Ice' under Standard type, and 'Besty', 'Capello', 'Danilo', 'Goldstar','Improved Lilac', 'Shiro' and 'Siri-2' under Miniature type; (iii) average tolerant, *viz.* 'Atlantis', 'Barsemi Yasmina', 'Colorado Red Sim', 'Improved White', 'Portrait' and 'Yellow Dusty' under Standard type, and 'Georgia Anne', 'Peachy', 'Sam's Pride', 'Sweetheart', 'Toni', and 'Twinkle' under Miniature type; (iv) poor tolerant, *viz.* 'Corrida', 'Flamingo Sim', 'Lolita', 'Parade' and 'Sandra' under Standard type, and 'Jack', 'Kirishima', 'Moonlight' and 'Winsome' under Miniature type; (iv) least tolerant, *viz.* 'Tangerne Sim' under Standard type, and 'Corona', 'Ocean Spray' and 'Orange Corona' under Miniature type.

Physical and chemical mutagens for induction of mutations and colchicines for inducing polyploidy have been used for improving carnation. The first radiation induced flower colour and single flowered mutants were reported by Richter and Singleton (1955). The works on commercial mutation breeding of carnation by Saparnaay and Demmink (1971), Sparnaay *et al.* (1974) and Sparnaay (1974) are noteworthy. Samata *et al.* (1979) obtained 3.3 to 30 per cent colour variation in carnation due to Co^{60} gamma irradiation. Ferrero *et al.* (1987) stated that French cv. 'Londorga' produced some 35 new cultivars in terms of colour mutations. Singh *et al.* (2000) through *in vitro* EMS treatment to axillary bud explants of cv. 'Espana' induced flower colour changes. Sarafi and Mahadavian (1979) obtained tetraploid seedling after treating the diploid seeds of carnation with colchicines, and in this case the diameter of flower, the branch number, petal size, leaf size and pistil size were found increased compared to its diploid parent. Datta (2002) has stated about the possibility of transgenic carnations with increased axillary bud breaks, more stem cuttings, better rooting ability and increased flowering through expression of *rol C* gene. Novak (1991) stated that through *in vitro* mutagenesis, variability at genetic level can be obtained more rapidly and at higher pace. Roest and Bokelmann (1981) produced stable mutants in carnation varieties by splitting the cuttings lengthwise in two, irradiating them and culturing in a medium inducing axillary bud proliferation. Simard *et al.* (1992) induced variations in cv. 'Niky' through organogenesis of 40 Gy irradiated petals when treated on the fourth day of culturing. Cassells *et al.* (1993) when tried X-ray irradiations of nodes *in vitro* in cv. 'Mystere', obtained vigorous and stable colour mutants. Jerzy and Zalewsky (2000) when got irradiated the internodal segments of cv. 'Mini Pinky' with ^{60}Co gamma and X-rays (5-30 Gy) and cultured them on regeneration medium, obtained certain mutants.

To save labour cost for pinching, disbudding, staking, tieing, etc., people are opting for miniature carnations as these are high yielding and also quite attractive with compact growth most suitable for bouquets. Moreover, the recent trend in the development of modern carnation is also changing. Demmink in 1982 in Holland raised a new group of glasshouse carnations with compact growth and small flowers, involving *Dianthus chinensis, D. pancicii* and *D. pontederae*; and van Staaveren in Holland developed a race of micro-carnation which produces button-like flowers in great profusion, some 1000 per square metre. Institute of Horticultural Plant Breeding at Wageningen (Netherlands) developed cv. 'Pico' micro-carnation with spray type flowers having prominent dark centre. Gold Smith in Glory (California) developed F_1 'Knight' series with compact growth, producing 15-20 flowers per stem with a stem length of 25-30 cm. Burpee Seed Co., California developed 'Luminette' and 'Pixie Delight' (Bhatt, 1989).

Classification, Species and Varieties

French system of classification then had been putting carnation into three classes: (i) ***Grenadins*** which included those with strong perfumes, flowers of medium size and with frnged petals whether single or double but of one colour, (ii) ***Flamands*** with large, round and double flowers having entire petals of one colour or striped with two or more colours but rising in the centre to form a convex surface, and (iii) ***Fancies*** which may or may not be toothed but the colours are arranged in bands on light grounds; however, English classification has four categories, *viz.* (i) ***Selfs*** which have only one colour of the petals with no stripes or spots, (ii) ***Flakes*** having a pure ground of white or yellow and flaked or striped with one colour, e.g. scarlet, purple or rose, (iii) ***Bizarres*** having a pure ground as in the *Flakes* but marked with two or three colours such as crimson, scarlet, pink or purple, and (iv) ***Picotees*** which have a pure ground of white or yellow and each petal bordered with a band of colour at the margin (Bailey, 1942).

There are 'the Border Carnations', 'Picotees', 'Flakes' and 'Bizarres'; 'Chabaud (Annual) or Marguerite Carnations'; 'Perpetual Flowering Carnations', 'the Cottage Carnations', 'the Malmaisons', and all kinds of hybrids, *viz.* magnificent types to the baby carnations.

Border carnations are hardy perennials often with fragrant flowers, plants being more condensed and bushy but treated as annuals and are propagated through seeds, layering and through cuttings. These grow to 60-90 cm high with a planting distance of 30-45 cm apart. Flowers comprise of smooth-edged petals some 5 cm across. Varieties of Border carnations are though numerous but the representative selections are 'Alice Forbes Improved' (white marked dusty rose), 'Beauty of Cambridge' (yellow), 'Border Fancy' (flowers large, colour white flecked red lightly), 'Catherine Glover' (yellow flecked scarlet), 'Consul' (apricot), 'Crystal Clove' (white, clove scent), 'Dainty' (primrose yellow, splashed and edged pink), 'Down's Sunset' (blooms large, apricot marked rose-pink), 'Down's Unique' (cream marked scarlet), 'Ebor' (dwarfer, chocolate flaked red), 'Eva Humphries' (white edged beetroot purple), 'Fiery Cross' (scarlet), 'Harmony' (grey flecked cerise), 'Imperial Clove' (violet-carmine, fragrant), 'Lavender Clove' (lavender, fragrant), 'Leslie Rennison' (purple overlaid rose, fragrant), 'Madonna' (flowers camellia-like, white), 'Majestic Clove' (crimson-purple, strongly fragrant), 'Merlin Clove' (white marked crimson, fragrant), 'Mrs. Edmund Charrington' (flowers large, white striped lilac), 'Oakfield Clove' (glowing crimson, fragrant), 'Old Crimson Clove' (crimson, strongly fragrant), 'Perfect Clove' (crimson, fragrant), 'Robin Thain' (white marked crimson, fragrant), 'Santa Claus' (yellow edged purple), 'Snow Clove' (flowers large, white, clove-scent), 'Teviotdale' (ruby rose splashed apricot), 'Thomas Lee' (pale-yellow, flecked and edged scarlet), 'W.B. Cranfield' (geranium scarlet), 'Zebra' (yellow marked crimson), etc. These have various divisions such as (i) **Selfs** only with one colour, (ii) **Fancies** comprising more than one colour, and (iii) **Picotees** comprising of white or yellow ground colour contrasting with a narrow and darker edge of petals, petal colour free from spots or specks but the edge of the flowers should be regular bright contrasting the base colour. Picotees are well known in England though not so popular in America. These divisions may further be sub-divided for exhibition purposes. These all the classes flower only once in the season, blooming more freely the second year. The plants being short-lived are discarded and replaced every two years. Plants require firm staking. This does not require to be disbudded as this looks spectacular in sprays of 3-4 blooms per stem. Picotees are most suitable for planting in the borders 30 cm apart for production of cut blooms. Their cultural requirements are the same as the border carnations. The most popular Picotee varieties are 'Eva Humphries' (white edged with beetroot purple), 'Fair Maiden' (white edged light scarlet), 'Firefly' (yellow edged heavily with crimson), 'Ganymede' (white edged crimson), 'Mrs. J.J. Keen' (yellow edged rose-pink), 'Perfection' (white edged deep scarlet), 'Santa Claus' (yellow edged purple), etc. The representative varieties of Perpetual-flowering Picotee carnations are 'Dainty Lady' (sulphur-yellow edged rose-scarlet), 'Dainty Lass' (cream edged rose-pink), etc. **Flakes** or **Flake Border carnations** have combination of two distinct colours such as 'Cherry Flake' (flowers cherry flaked maroon), 'Scarlet Flakes' (orange-scarlet flaked deep chestnut), etc., though **Bizarres** or **Bizarre Border carnations** have combination of more than two colours such as 'Apricot Bizarre' (apricot splashed carmine and rose), 'Rose Bizarre' (shell pink splashed rose and purple); and these both the types are allowed to grow naturally even without stopping, however, the large blooms can be obtained through disbudding.

Chabaud (**Marguerite**) or **Annual carnations** are described further in detail under the sub-heading 'Cultural Practices' of this crop (*Dianthus*).

'**Perpetual-flowering carnations'** are less hardy, grown as greenhouse perennials, the most familiar ones among all types of carnations, and are generally called as 'Standard carnations'. These have grey-green leaves, long and pointed buds, one large flower on an individual stem, flowers double, often fragrant and are grown for cut flowers. These are the carnations available with the florist's shops, hence, is also known as 'florist carnation'. These also consist of two groups – 'the Sims' and the 'Mediterranean varieties'. These were though developed in France in 1840 where these grew until their introduction into America in 1856, and since then these were regarded as 'American carnation' as the Sim cultivar group in the United States as in 1858 through hybridization the first variety 'Mrs. Degraw' was developed there. The other variety of white flower colour in USA was developed in 1864 with the name of 'Flatbush'. Afterwards, so many amateur breeders in France and America developed many varieties including 'Astoria' which is now conceded as the ancestor of all the yellow varieties grown today. These carnations flower profusely at 10 ^{0}C temperature, though at 7 ^{0}C it produces a fair number of good winter blooms, at 5 ^{0}C the plants grow but only a few blooms are produced while at 1 ^{0}C no growth takes place though plants survive (Hay and Beckett, 1971). This 'perpetual flowering group' consists of thousands of cultivars, and the representative ones are 'Aleck Sparkes' (crimson), 'Aledo' (variegated white), 'Allwoods Cream' (cream, fragrant), 'Allwood's Heather Pink' (free flowering, mauve), 'Allwood's Market Pink' (an excellent pot plant, deep flesh-pink), 'Allwood's Pink' (pink, heaviest cropping), 'Allwood's Yellow' (free flowering, yellow), 'Amstel' (orange), 'Arthur Sim' (white marked red), 'Atletico' (variegated white), 'Bailey's Amarillo' (fancies), 'Bailey's Masterpiece' (crimson), 'Bailey's Splendour' (pink), 'Bailey's Supreme' (salmon-pink), 'Baltico' (white), 'Barbara Bingham' (apricot and rose-pink), 'Beauty of Cambridge' (yellow), 'Black Baccara' (dark maroon, looking almost black), 'Bogota' (white), 'Bologna' (pink), 'Brigadoon' (yellow), 'Britannia' (scarlet), 'Calibra' (yellow), 'Candy Sim' (fancy), 'Canup's Pride' (cerise), 'Cherry Solar' (cherry), 'Consul' (bronze),

'Crowley Sim' (pink), 'Dairymaid' (white flaked pink), 'Dakar' (red), 'Dazzler' (fancy), 'Decio' (red), 'Diplomat' (crimson), 'Domingo' (red), 'Dona Brecas' (cherry), 'Dover' (white), 'Dumas' (red), 'Dusty Rose' (fancy), 'Dusty Sim' (fancy), 'Emily Allwood' (fragrant, white marked lavender), 'Epoca' (pink), 'Eroico' (yellow), 'Esty' (yellow), 'Exotica' (yellow), 'Exquisite' (lavender), 'Fiery Cross' (pink), 'Forever' (striped white), 'Fragrant Ann' (white), 'Galatee' (bronze), 'George Allwood' (fragrant, fringed white), 'Giamaica' (variegated white), 'Golden Rain' (yellow), 'Harvest Moon' (apricot), 'Helios' (yellow), 'Icecap' (white), 'Impala' (red), 'Imperial Clove' (purple), 'Ivonne Orange' (variegated), 'Joker' (crimson), 'Jumbo Sim' (scarlet), 'Killer' (cherry), 'Kiro' (yellow), 'Laddie Sim' (salmon), 'Laurella' (cherry), 'Lavender Lace' (purple), 'Lelie Rennison' (purple), 'Linda' (pink), 'La Royale' (mauve), 'Machioness of Headfort' (excellent pot plant, salmon-red edged creamy-white), 'Madonna' (white), 'Majestic' (scarlet), 'Manabi' (yellow), 'Mandarin Sim' (apricot), 'Margaret' (mauve-purple), 'Margaret Sparta' (purple), 'Marian Allwood' (cardinal pink), 'Marie Chabaud' (yellow), 'Mary' (salmon-pink), 'Master' (red), 'Michelle' (variegated yellow), 'Minerva' (variegated red), 'Monte Lisa' (red), 'Nelson' (red), 'Nicole' (variegated red), 'Nordika' (white), 'Northwood' (white), 'Novona' (pink), 'Olympus' (striped white), 'Orange Isac' (orange), 'Orange Pinto' (orange), 'Orange Prestige' (variegated orange), 'Orange Triumph' (fancy), 'Oriana' (pink), 'Paola' (pink), 'Paris' (pink), 'Perfect Clove' (red), 'Peter Fisher' (free-flowering, silver-pink), 'Petra' (pink), 'Pink Diamante' (pink), 'Pink Dover' (pink), 'Pink Helena' (pink), 'Pink Sim' (pink), 'Pinto' (yellow), 'Pirandello' (cherry), 'Pontiac' (variegated yellow), 'Portrait' (salmon-pink), 'Princess of Wales' (salmon-pink), 'Record' (cherry), 'Red-edged Skyline' (fancy), 'Red Pimpernel' (excellent for pot, bright scarlet), 'Rimini' (red), 'Rivera' (white), 'Roderic' (variegated red), 'Rose Splendour' (cerise), 'Royal Crimson' (large, fragrant, crimson), 'Royal Harlequin' (flesh-pink speckled and flaked carmine), 'Royal Mail' (scarlet), 'Sangria' (red), 'Saugus White' or 'Saygue White' (white), 'Scania' (scarlet), 'Shocking Pink' (pink), 'Simstripe' (fancy), 'Snow Clove' (white), 'Solar' (orange), 'Solat Chiaro' (orange), 'Sonsara' (white), 'Star' (red), 'Storm' (lavender), 'Super Star' (variegated yellow),'Tahiti' (yellow), 'Tanga' (red), 'Tangerin Sim' (apricot), 'Tangette' (fancy), 'Tetra' (salmon), 'Tropea' (variegated white), 'Vanessa' (purple), 'Vapna' (pink), 'Viking' (cerise), 'Virginie' (white), 'White Giant' (white), 'White Perfection' (white), 'White Sim' (scarlet), 'White William Sim' (white sport of 'William Sim'), 'William Sim' (scarlet), 'Wivelsfield' (white), 'Yellow Sim' (yellow), 'Zuni' (cerise flecked crimson-maroon), etc. 'William Sim' and 'Sydney Little Field' are the two varieties (strains) through which most of the cultivars presently under cultivation have arisen as bud sports. In India, the cultivars being cultivated are 'All Yellow', 'Aristo', 'Arthur Sim', 'Candy', 'Cream Valencia', 'Dusty', 'Espana', 'Famba', 'Flair Laspalms', 'Light Pink Candy', 'Murga', 'Pamir', 'Red Corso', 'Sam's Pride', 'Scania', 'Scarlet Elegance', 'Star Light', 'Stylo', 'White Candy', 'William Sim', etc. Around 1996, a company used genetic manipulation through certain genes extracted from petunia and snapdragon and produced a blue-mauve carnation which was commercialized as 'Moondust', subsequently in 1998, a violet carnation named as 'Moonshadow' was commercialized, and by 2004, three additional blue-violet/purple varieties have been commercialized.

Cottage carnations, the true perennial are sturdier and hardy, standing out to any frost or rains and continue blooming for a very long period. These are large to button-type carnations. These are perpetual in nature and are as beautiful as Perpetual-flowering glasshouse types. It was the variety 'Allwood's Rifleman', a scented and lovely brick-red which gave us the variety 'Cottage Scarlet' (brighter in colour) in the early part of twentieth century. Their cultural practices are almost the same as for Border carnations except that these are planted 20 to 25 cm apart in September-October to flower during summer, i.e. June in temperate regions where these will continue flowering until October. However, these can also be planted during February-March. These are most suitable for planting as the edging plant. These prefer non-acidic light loamy soils at an exposed situation. These may require staking and disbudding of the side buds. Propagation of these carnations are through layering. A few of the promising varieties are 'Cottage Gem' (large blush white flecked crimson), 'Cottage Mauve' (mauve), 'Cottage Orange' (apricot, very hardy), 'Cottage Pink' (large flesh pink), 'Cottage Primrose' (yellow, dwarf, erect and hardy), 'Cottage Vivid' (brilliant scarlet, excellent for bedding), 'Cottage White' (white, highly fragrant), etc.

Malmaison or **Souvenir de la Malmaison** strain though was leading carnation in England prior to advent of the 'Perpetual Flowering strain' but did not command the same status in USA, albeit its flowers are large and had been used in breeding programme in USA. However, in

Carnation) with *D. caryophyllus* (old Perennial carnation). *D. chinensis* is tetraploid with 2n = 60 while *D. caryophyllus* is a diploid with 2n = 30. Chabaud carnation is therefore tetraploid, *vis-à-vis* fertile. Shewell-Cooper (1953) has mentioned divisions of Chabaud carnations into: I. **Compact Dwarf Chabauds** which hardly grow taller than 38 cm, bear strongly scented full size double blooms, are available in mixed colours and are excellent for pot culture and window boxes; II. **Giant Chabauds** have wide variety of colours with fragrant flowers which are borne on strong stems, the varieties being numerous but the choicest ones are 'Avranchin' (striped red on yellow ground), 'China' (white with mauve centre), 'Enticelant' (blood-red), 'Giroflee' (violet-pink), 'La France' (pale-pink), 'La Perle' (lilac), 'Legion d'Honneur' (geranium-red), 'Marie Chabaud' (sulphur-yellow), 'Mikado' (nice blue), 'Nero' (royal purple), etc.; III. **Enfant de Nice** which have been developed out of Giant Chabauds but their colour richness and the fragrance of the flowers mask the beauty of the latter, are free flowering and hardy and possess sturdy stems and attractive grey-green foliage, though the group does not have definite varieties but seeds are available in various colours such as white, yellow, pink, crimson, mauve, purple and stripes; IV. **Fleur de Camelia** is also a specific strain of Giant Chabauds where flowers strongly resemble the camelia, and though there may be varieties but the seed packets are available only of mixed colours; V. **Margarita** grows as tall as the Compact Dwarf Chabauds, blooming within 5 months of sowing, though there is no any known variety but these can be purchased only in mixed colours; VI. **Iron Stem** strain which bears flowers much larger than the normal Margarita types on the strong stems and has a range of exceptionally bright colours. Other noteworthy spray cultivars are (i) **White**: 'Amalia', 'Close Up', 'Excel', 'Hermon', 'Manatovani' (variegated), 'Opale', 'Regis' (striped), 'Saffora' (striped), 'Virgo', 'West Crystal', 'White Prestige', etc.; (ii) **Yellow**: 'Annelies', 'Castillo', 'Eureka', 'Garfield' (striped), 'Guernsey Yellow', 'Jacobine' (striped), 'Koreno', 'Limon', 'Lior', 'Naomi' (striped), 'Picaro' (striped), 'Prestige', etc.; (iii) **Orange**: 'Amaya', 'Challenger' (striped), 'Fuego' (striped), 'Macarena', 'Marisol', 'Niky', 'Sintonia' (striped), 'Target', 'Tip Top', etc.; (iv) **Cherry/cyclamen**: 'Cherry Bag', 'Ciska', 'Claudia', 'Kortina Cherry', 'Marisma', 'Ondelia', 'Veleta', etc.; (v) **Pink**: 'Donora', 'Fantasia', 'Happiness', 'Krystie', 'Medea', 'Medley', 'Noelia', 'Nuria', 'Opale', 'Rossini', etc.; (vi) **Red**: 'Alister', 'Darling', 'Diablo', 'Electra', 'Etna', 'New Elsy', 'Red Arrow', 'Rony Arony', etc.; and (vii) **Purple**: 'Boreal', 'Duraongo', 'Purple Chopin', 'Roxette', etc.

For its cultivation, the seeds may be sown in September-October in the plains or during February in greenhouses in the temperate regions of the country and grown at 13-15 °C in John Innes Seed Compost (loam 2 parts + loose peat 1 part + coarse sand 1 part, and to one litre of this mixture some 1.2 g of lime superphosphate + 0.6 g of finely ground chalk or limestone) sieved through 1.12 mesh. Ten to thirty days after germination, these are pricked out in boxes at 4 × 4 cm apart or 5 × 5 cm apart (in case when planting is to be delayed) of John Innes Potting Compost No. I, i.e. loose loam 7 parts, peat 3 parts coarse sand 2 parts, and to each litre of this mixture some 3 grammes of a mixture consisting of bone meal 2 parts + lime superphosphate 2 parts + potassium sulphate 1 part, and some 0.6 g of finely ground chalk are added. For about a month of pricking, the plants are allowed to remain at 13-15 °C temperature and then temperature should be lowered to 10 °C and then hardened off in a cold frame and planted outdoors at their permanent position in the beds or in pots in May with a fair ball of soil to their roots. Pricking out (transplanting of the seedlings when these are very young and their basal leaves are resting on the soil) is normally done in the boxes from 10 to 30 days of germination so that baby roots go down their full length. While pricking out, the compost should be firmed up in the boxes until it is within 6 mm of the top. After sometime the dose of fertilizers and chalk or limestone are doubled when the seedlings are changed to more than 10 cm pot or when planted to their permanent position. Their final planting is effected 30 × 30 cm apart in a well prepared field where 3-5 kg of compost/m^2 has been mixed and the soil has been made to a fine tilth. These carnations do not require to be stopped, however, sometimes side flower buds are disbudded to facilitate the end bloom to develop properly.

Another method of raising the plants through seeds is that these carnations should be grown in greenhouses in January-February and when the seedlings attain 2.5-3.0 cm height these are potted up in 7.5 cm pot where these are grown for 6-8 weeks and then potted finally in 15 cm pot and this way in the temperate regions one can enjoy the blooms from mid-June to full December, however, in the plains under open conditions it is not possible to obtain the blooms after May until these are given greenhouse conditions. In the temperate regions of the country these may be scheduled to bloom throughout the year provided neither there occurs hard frost nor the hot temperatures (above 21 °C). Wherever these are grown, place should have plenty of light and air.

'Crowley Sim' (pink), 'Dairymaid' (white flaked pink), 'Dakar' (red), 'Dazzler' (fancy), 'Decio' (red), 'Diplomat' (crimson), 'Domingo' (red), 'Dona Brecas' (cherry), 'Dover' (white), 'Dumas' (red), 'Dusty Rose' (fancy), 'Dusty Sim' (fancy), 'Emily Allwood' (fragrant, white marked lavender), 'Epoca' (pink), 'Eroico' (yellow), 'Esty' (yellow), 'Exotica' (yellow), 'Exquisite' (lavender), 'Fiery Cross' (pink), 'Forever' (striped white), 'Fragrant Ann' (white), 'Galatee' (bronze), 'George Allwood' (fragrant, fringed white), 'Giamaica' (variegated white), 'Golden Rain' (yellow), 'Harvest Moon' (apricot), 'Helios' (yellow), 'Icecap' (white), 'Impala' (red), 'Imperial Clove' (purple), 'Ivonne Orange' (variegated), 'Joker' (crimson), 'Jumbo Sim' (scarlet), 'Killer' (cherry), 'Kiro' (yellow), 'Laddie Sim' (salmon), 'Laurella' (cherry), 'Lavender Lace' (purple), 'Lelie Rennison' (purple), 'Linda' (pink), 'La Royale' (mauve), 'Machioness of Headfort' (excellent pot plant, salmon-red edged creamy-white), 'Madonna' (white), 'Majestic' (scarlet), 'Manabi' (yellow), 'Mandarin Sim' (apricot), 'Margaret' (mauve-purple), 'Margaret Sparta' (purple), 'Marian Allwood' (cardinal pink), 'Marie Chabaud' (yellow), 'Mary' (salmon-pink), 'Master' (red), 'Michelle' (variegated yellow), 'Minerva' (variegated red), 'Monte Lisa' (red), 'Nelson' (red), 'Nicole' (variegated red), 'Nordika' (white), 'Northwood' (white), 'Novona' (pink), 'Olympus' (striped white), 'Orange Isac' (orange), 'Orange Pinto' (orange), 'Orange Prestige' (variegated orange), 'Orange Triumph' (fancy), 'Oriana' (pink), 'Paola' (pink), 'Paris' (pink), 'Perfect Clove' (red), 'Peter Fisher' (free-flowering, silver-pink), 'Petra' (pink), 'Pink Diamante' (pink), 'Pink Dover' (pink), 'Pink Helena' (pink), 'Pink Sim' (pink), 'Pinto' (yellow), 'Pirandello' (cherry), 'Pontiac' (variegated yellow), 'Portrait' (salmon-pink), 'Princess of Wales' (salmon-pink), 'Record' (cherry), 'Red-edged Skyline' (fancy), 'Red Pimpernel' (excellent for pot, bright scarlet), 'Rimini' (red), 'Rivera' (white), 'Roderic' (variegated red), 'Rose Splendour' (cerise), 'Royal Crimson' (large, fragrant, crimson), 'Royal Harlequin' (flesh-pink speckled and flaked carmine), 'Royal Mail' (scarlet), 'Sangria' (red), 'Saugus White' or 'Saygue White' (white), 'Scania' (scarlet), 'Shocking Pink' (pink), 'Simstripe' (fancy), 'Snow Clove' (white), 'Solar' (orange), 'Solat Chiaro' (orange), 'Sonsara' (white), 'Star' (red), 'Storm' (lavender), 'Super Star' (variegated yellow),'Tahiti' (yellow), 'Tanga' (red), 'Tangerin Sim' (apricot), 'Tangette' (fancy), 'Tetra' (salmon), 'Tropea' (variegated white), 'Vanessa' (purple), 'Vapna' (pink), 'Viking' (cerise), 'Virginie' (white), 'White Giant' (white), 'White Perfection' (white), 'White Sim' (scarlet), 'White William Sim' (white sport of 'William Sim'), 'William Sim' (scarlet), 'Wivelsfield' (white), 'Yellow Sim' (yellow), 'Zuni' (cerise flecked crimson-maroon), etc. 'William Sim' and 'Sydney Little Field' are the two varieties (strains) through which most of the cultivars presently under cultivation have arisen as bud sports. In India, the cultivars being cultivated are 'All Yellow', 'Aristo', 'Arthur Sim', 'Candy', 'Cream Valencia', 'Dusty', 'Espana', 'Famba', 'Flair Laspalms', 'Light Pink Candy', 'Murga', 'Pamir', 'Red Corso', 'Sam's Pride', 'Scania', 'Scarlet Elegance', 'Star Light', 'Stylo', 'White Candy', 'William Sim', etc. Around 1996, a company used genetic manipulation through certain genes extracted from petunia and snapdragon and produced a blue-mauve carnation which was commercialized as 'Moondust', subsequently in 1998, a violet carnation named as 'Moonshadow' was commercialized, and by 2004, three additional blue-violet/purple varieties have been commercialized.

Cottage carnations, the true perennial are sturdier and hardy, standing out to any frost or rains and continue blooming for a very long period. These are large to button-type carnations. These are perpetual in nature and are as beautiful as Perpetual-flowering glasshouse types. It was the variety 'Allwood's Rifleman', a scented and lovely brick-red which gave us the variety 'Cottage Scarlet' (brighter in colour) in the early part of twentieth century. Their cultural practices are almost the same as for Border carnations except that these are planted 20 to 25 cm apart in September-October to flower during summer, i.e. June in temperate regions where these will continue flowering until October. However, these can also be planted during February-March. These are most suitable for planting as the edging plant. These prefer non-acidic light loamy soils at an exposed situation. These may require staking and disbudding of the side buds. Propagation of these carnations are through layering. A few of the promising varieties are 'Cottage Gem' (large blush white flecked crimson), 'Cottage Mauve' (mauve), 'Cottage Orange' (apricot, very hardy), 'Cottage Pink' (large flesh pink), 'Cottage Primrose' (yellow, dwarf, erect and hardy), 'Cottage Vivid' (brilliant scarlet, excellent for bedding), 'Cottage White' (white, highly fragrant), etc.

Malmaison or **Souvenir de la Malmaison** strain though was leading carnation in England prior to advent of the 'Perpetual Flowering strain' but did not command the same status in USA, albeit its flowers are large and had been used in breeding programme in USA. However, in

England, this group is still popular though not to the extent as it had been initially. Neither these are suitable for growing outdoors nor for the cold greenhouse. The old-fashioned Malmaisons are quite distinct from the modern Perpetual Malmaison carnations whose cultivation requirements are similar to Perpetual-flowering carnations except that their temperature requirement during night is more, i.e. 12-16 ^{0}C. Malmaisons are stiffer than Perpetual-flowering, the wide foliage, round and chubby buds, large double fragrant flowers with a well-filled centre, and mostly of pink colours but with short calyx so more sensitive to splitting. These are propagated through layers. Varieties under the old-fashioned Malmaisons are the three main kinds, *viz.* 'Old Blush' (pink with pleasing fragrance), 'Princess of Wales' (salmon pink with pleasing fragrance), 'Rothschild's Variety' (deep salmon pink, probably a sport of 'Princess of Wales'); and under Perpetual Malmaisons are 'Adriatic' (pure white), 'Homeric' (crimson), 'Jessie Allwood' (yellow), 'Mrs. C.F. Raphael' (cherry-red), 'Olympic' (crimson), etc.

Propagation

Carnations are propagated **sexually**, especially in marguerite (spray) and border types and for the purpose of hybridization, whereas asexually through cuttings, layering, grafting and micropropagation. Perpetual-flowering carnations are usually propagated through cuttings but for evolution of the new varieties also through seeds during January-February at 15 ^{0}C. **Seeds** in the tremperate regions are sown from April to June whereas in the plains in September as for the species and the garden pinks. Seeds are sown thinly in pans or boxes which may be maintained at 15 ^{0}C until germination. Nau (1993) described that seeds of garden cultivars of carnations are successfully germinated at 18-21 ^{0}C. The medium for seed sowing should consist of loam 2 parts + loose peat 1 part + coarse sand 1 part, and to one litre of this mixture some 1.2 g of lime superphosphate + 0.6 g of finely ground chalk or limestone, and this mixture is sieved through 1.12 mesh. After 10-30 days of germination, these are pricked out in boxes at 4 × 4 cm, or 5 × 5 cm (in case of delayed planting) apart in John Innes Potting Compost No. I, i.e. loose loam 7 parts, peat 3 parts, coarse sand 2 parts, and to each litre of this mixture some 3 grammes of a mixture consisting of bone meal 2 parts + lime superphosphate 2 parts + potassium sulphate 1 part, and some 0.6 g of finely ground chalk are added. For about a month of pricking, the plants are allowed to remain at 13-15 ^{0}C temperature and then temperature should be lowered to 10 ^{0}C and then hardened off in a cold frame and when seedlings attain 3 cm of height these should be planted outdoors at their permanent position in the beds or in pots in October in the plains or from March to May in temperate regions of the country with a fair ball of soil to their roots. During germination there should be sufficient moisture in the medium, but after germination these are moderately watered, however, during frosty weathers no watering should be done. The seeds of perpetual-flowering carnations should be sown in February in the temperate regions in the greenhouse as for Chabauds to get the blooms in majority of the cases by September, and by all means by October when inferior seedlings should be discarded. Similar to the perpetual-flowering carnations, the seeds of perpetual border carnations are also grown. In both these cases the temperature requirement for seed germination is 15.5 ^{0}C. Seedlings appear normally within 10 days, and then any time after 10 days of germination the seedlings should be pricked out and planted either directly in 5.0-7.5 cm pots or some 5 cm apart in other boxes containing JIPC No. I. No pinching is required in case of perpetual-flowering and perpetual border carnations if raised through seeds so that immediately after flowering the singles and inferior ones are discarded and good ones retained.

Commercially, the carnations are propagated only through **top cuttings** obtained from healthy mother stock plants which are in vegetative phase. Such stock plants are established in a cool and well illuminated area and regularly stopped to ensure regular supply of cuttings. SD forces the plants to increase axillary bud activity which in turn encourages constant vegetative cutting production (Pokony and Kamp, 1960, 1965). However, these cuttings can be stored from 0.0 to 0.25 ^{0}C for up to six months (Salinger, 1985) and can be replanted when required in the prepared beds. The cuttings are taken when dry and kept in polythene lined boxes, polythene lightly closed over the cuttings so that it may limit gaseous exchange. For small scale plantings, the cuttings may be stored for 2-3 weeks in cold storage at 2-4 ^{0}C packed in polythene lined boxes to limit desiccation (Salinger, 1985).

Border carnations are also propagated through **layers** apart from seeds, especially during late-July to August in the temperate regions as at this time the plants are sturdy and can easily be potted up and overwintered in a cold frame. Layering before this time is not appropriate as the plants are weak and leggy though layering later is not proper due to improper weather conditions and rooting may be poor. Cottage carnations are also propagated through layering. Best layers are provided by the young plants. While propagating through layers only the healthy plants are selected and layers should be prepared only through the basal shoots. The soil around the plant in a circle is little loosened, removed and substituted with some good compost or sifted loam mixed with equal proportions of sedge peat and sharp sand and the mixture is firmed up in a little mound shape. The lower leaves of the shoots to be layered are stripped off by upward pull but these shoots should have some five pairs of leaves above. A cut with a sharp knife in the centre of the stem just above a node but quite low down, and this cut is extended upward up to just below the next node in a tongue shape. This tongue is also sheared off slightly for giving it sufficient room for callusing and further rooting. Now the stem is downed in a fashion that incised part is quite open. This incised part is pressed firmly in the compost some 5 cm deep with the help of a bend wire, pegged and the compost is firmed up. The number of layers taken from a plant depends upon the vigour of the plant. Some 6 to 12 layers can be prepared from one healthy plant. The compost around the layers is kept moist regularly but water should be sprinkled through a fine can. The plants may root properly from 4 to 6 weeks depending upon the cultivars, and then these are severed from the mother plant close to the layers. Within further 3-4 weeks these plantlets (layers) are lifted and planted in the prepared beds outside where these will take growth and establish themselves during autumn so that these are fully ready to overwinter. In case the soil of the bed is clayey, the layers should be planted in the pots which should be kept in the cold greenhouse until the following March with proper ventilation otherwise the plants may be infected with mildew and rust. Through **grafting** also the carnations are propagated successfully. 'Gus Royalette' and 'Scania' act as the good cultivars not succumbing in soil inoculated with *Fusarium oxysporum* f. sp. *dianthi*. Other wilt resistant cultivars as rootstocks are 'Arancio 25 D', 'Exquisite', 'Fulvio Rose', 'Heidi' and 'May Britt'.

For mass multiplication, it is **micropropagation** through which one can produce disease-free carnation plants in bulk for self-use and for sale to others. Manisha Thakur *et al.* (2002) through 0.2-0.5 cm internodal segments of carnation cvs 'Cabaret' and 'Feyenoord' established the culture on MS medium containing NAA 0.5-2.0 ppm+KIN 1.0-2.0 ppm+2,4-D 0.5-2.0 ppm and without NAA, and obtained shoot initiation only after culturing the calli on MS basal medium containing 0.3% activated charcoal, and maximum number of shoots was obtained when the calli were cultured on MS medium containing 2 ppm KIN and 0.5 ppm NAA. Raja Ram and Zaidi (1991) cultured meristem tips to obtain virus-free plants. Miroshnichenko *et al.* (1998) regenerated adventitious shoots from leaf explant. The leaves larger than 5 mm are removed by hand while the smaller leaves, leaf primordia, axillary buds and the large leaf remainings are removed by sterilized dissecting needle to ensure the explant consisting of apical dome, and these explants are placed with apex facing up in an autoclaved nutrient medium (pH 5.7 before autoclaving) in petridish, sealed with paraffin and maintained at 25 ^{0}C for developing multiple shoots from the explants. After development of multiple shoots these along with the explant are transferred to the fresh medium for further proliferation. These plantlets can be grown at 20 ^{0}C in growth chamber with 16-h light for proper establishment. The MS modified medium containing 100 mg/l myo-inosital, 0.4 mg/l thiamine HCl, 30 g/l sucrose, 6 g/l agar, 0.5 mg/l kinetin and 0.1 mg/l α-NAA is most suitable (Bhatt, 1989). Apical or axillary meristems ranging from 0.6-1 mm in length are when cultured in modified MS medium give good results (Bhatt, 1989). Anther culture on MS medium supplemented with 1 mg IAA/l also provides satisfactory results (Bhatt, 1989).

Cultural Practices

1. Chabaud/marguerite/miniature/ spray carnation (*Dianthus chinensis* × *D. caryophyllus*)

These carnations should, in fact, be called only as Marguerite carnations instead of Chabaud, and these are though half-hardy perennials but treated as annuals with large single or double flowers having fringed petals and coming quite true from seeds though their flowers do not stand in the vase water as long as the Perpetual Flowering types but are very freely produced over a long period. These carnations have several shorter branches with smaller flowers on each branch. Since these were initially developed in 1870 by Monsieur Chabaud of Toulons (France) so these famous annual carnations are named as Chabaud carnations. Its culture is comparatively much easier than Perpetuals. These carnations were developed by crossing *Dianthus chinensis* (the Annual Wood

Carnation) with *D. caryophyllus* (old Perennial carnation). *D. chinensis* is tetraploid with 2n = 60 while *D. caryophyllus* is a diploid with 2n = 30. Chabaud carnation is therefore tetraploid, *vis-à-vis* fertile. Shewell-Cooper (1953) has mentioned divisions of Chabaud carnations into: I. **Compact Dwarf Chabauds** which hardly grow taller than 38 cm, bear strongly scented full size double blooms, are available in mixed colours and are excellent for pot culture and window boxes; II. **Giant Chabauds** have wide variety of colours with fragrant flowers which are borne on strong stems, the varieties being numerous but the choicest ones are 'Avranchin' (striped red on yellow ground), 'China' (white with mauve centre), 'Enticelant' (blood-red), 'Giroflee' (violet-pink), 'La France' (pale-pink), 'La Perle' (lilac), 'Legion d'Honneur' (geranium-red), 'Marie Chabaud' (sulphur-yellow), 'Mikado' (nice blue), 'Nero' (royal purple), etc.; III. **Enfant de Nice** which have been developed out of Giant Chabauds but their colour richness and the fragrance of the flowers mask the beauty of the latter, are free flowering and hardy and possess sturdy stems and attractive grey-green foliage, though the group does not have definite varieties but seeds are available in various colours such as white, yellow, pink, crimson, mauve, purple and stripes; IV. **Fleur de Camelia** is also a specific strain of Giant Chabauds where flowers strongly resemble the camelia, and though there may be varieties but the seed packets are available only of mixed colours; V. **Margarita** grows as tall as the Compact Dwarf Chabauds**,** blooming within 5 months of sowing, though there is no any known variety but these can be purchased only in mixed colours; VI. **Iron Stem** strain which bears flowers much larger than the normal Margarita types on the strong stems and has a range of exceptionally bright colours. Other noteworthy spray cultivars are (i) **White**: 'Amalia', 'Close Up', 'Excel', 'Hermon', 'Manatovani' (variegated), 'Opale', 'Regis' (striped), 'Saffora' (striped), 'Virgo', 'West Crystal', 'White Prestige', etc.; (ii) **Yellow**: 'Annelies', 'Castillo', 'Eureka', 'Garfield' (striped), 'Guernsey Yellow', 'Jacobine' (striped), 'Koreno', 'Limon', 'Lior', 'Naomi' (striped), 'Picaro' (striped), 'Prestige', etc.; (iii) **Orange**: 'Amaya', 'Challenger' (striped), 'Fuego' (striped), 'Macarena', 'Marisol', 'Niky', 'Sintonia' (striped), 'Target', 'Tip Top', etc.; (iv) **Cherry/cyclamen**: 'Cherry Bag', 'Ciska', 'Claudia', 'Kortina Cherry', 'Marisma', 'Ondelia', 'Veleta', etc.; (v) **Pink**: 'Donora', 'Fantasia', 'Happiness', 'Krystie', 'Medea', 'Medley', 'Noelia', 'Nuria', 'Opale', 'Rossini', etc.; (vi) **Red**: 'Alister', 'Darling', 'Diablo', 'Electra', 'Etna', 'New Elsy', 'Red Arrow', 'Rony Arony', etc.; and (vii) **Purple**: 'Boreal', 'Duraongo', 'Purple Chopin', 'Roxette', etc.

For its cultivation, the seeds may be sown in September-October in the plains or during February in greenhouses in the temperate regions of the country and grown at 13-15 ^{0}C in John Innes Seed Compost (loam 2 parts + loose peat 1 part + coarse sand 1 part, and to one litre of this mixture some 1.2 g of lime superphosphate + 0.6 g of finely ground chalk or limestone) sieved through 1.12 mesh. Ten to thirty days after germination, these are pricked out in boxes at 4 × 4 cm apart or 5 × 5 cm apart (in case when planting is to be delayed) of John Innes Potting Compost No. I, i.e. loose loam 7 parts, peat 3 parts coarse sand 2 parts, and to each litre of this mixture some 3 grammes of a mixture consisting of bone meal 2 parts + lime superphosphate 2 parts + potassium sulphate 1 part, and some 0.6 g of finely ground chalk are added. For about a month of pricking, the plants are allowed to remain at 13-15 ^{0}C temperature and then temperature should be lowered to 10 ^{0}C and then hardened off in a cold frame and planted outdoors at their permanent position in the beds or in pots in May with a fair ball of soil to their roots. Pricking out (transplanting of the seedlings when these are very young and their basal leaves are resting on the soil) is normally done in the boxes from 10 to 30 days of germination so that baby roots go down their full length. While pricking out, the compost should be firmed up in the boxes until it is within 6 mm of the top. After sometime the dose of fertilizers and chalk or limestone are doubled when the seedlings are changed to more than 10 cm pot or when planted to their permanent position. Their final planting is effected 30 × 30 cm apart in a well prepared field where 3-5 kg of compost/m^2 has been mixed and the soil has been made to a fine tilth. These carnations do not require to be stopped, however, sometimes side flower buds are disbudded to facilitate the end bloom to develop properly.

Another method of raising the plants through seeds is that these carnations should be grown in greenhouses in January-February and when the seedlings attain 2.5-3.0 cm height these are potted up in 7.5 cm pot where these are grown for 6-8 weeks and then potted finally in 15 cm pot and this way in the temperate regions one can enjoy the blooms from mid-June to full December, however, in the plains under open conditions it is not possible to obtain the blooms after May until these are given greenhouse conditions. In the temperate regions of the country these may be scheduled to bloom throughout the year provided neither there occurs hard frost nor the hot temperatures (above 21 ^{0}C). Wherever these are grown, place should have plenty of light and air.

2. Pinks or garden pinks (*Dianthus chinensis* and *D. plumarius*)

Pinks or Garden pinks [*Dianthus chinensis* (*D. sinensis*) and *D. plumarius*] are probably derived from Cottage pink (*D. plumarius*) and certain other species through hybridization. *D. chinensis* is a tetraploid species with 2n = 60, basic number being 15. *D. plumarius*, a native of Austria to Siberia, is a hardy low tufty plant growing to 30 to 45 cm in height, stems simple or forked, leaves elongate-linear, spreading or recurved, and 1-nerved, flowers medium (5.0 cm wide), highly fragrant, rose-coloured but varying in cultivated forms and varieties with pink, purple, red, white and variegated having petal-blade fringed and calyx cylindrical with mucronate bracts. It is much used in the old-fashioned gardens and is most suitable for edging. It has double forms, *vis-à-vis* var. *semperflorens* is perpetual type with many varieties. This has four defined groups: (i) ***Selfs***, only of one colour throughout, (ii) ***Bicolours***, largely of one colour but with a basal blotch of contrasting tint, (iii) ***Laced***, where the basal blotch extends along the entire petal margins, and (iv) ***Fancies***, where marking is quite irregular. These colour markings are found in both, i.e. Old-fashioned which are slower than the Modern pinks and need propagation every 4-5 years and are normally not pinched, as well as the Modern pinks which are more vigorous and free-flowering with all the parts larger than Old-fashioned ones because of these being derived by crossing the Old-fashioned ones with the Perpetual carnations though this requires regular propagation every 2nd or 3rd year due to their rapid deterioration though through seeds many colours appear. The noteworthy Old-fashioned pinks are 'Blairside Crimson' (deep pink with a crimson centre, single type with 5 cm diameter, being borne on a 15 cm stalk), 'Charles Musgrave' (syn. Musgrave Pink, flowers single, slightly fringed, white with a green eye and some 3.75 cm in diameter, peduncle length some 30 cm), 'Cockenzie' (lovely double-flowered pink, pink in colour, peduncle length some 30 cm), 'Dad's Favourite' (base white, laced purple), 'Delmondham Fairy' (though a pink but looks more as carnation, flowers shrimp pink with a scarlet tinge in the centre), 'Elsenham Seedling' (flowers large, pink and single with darker central zone, peduncle length 30 cm), 'Emil Paré' (deep pink with long flowering span), 'Ernest Ballard' (flowers double, deep crimson and clove-scented), 'Excelsior' (syn. Pink Mrs. Sinkins, carmine, eye darker), 'F.W. Millard' (flowers double and dull crimson, peduncle some 30 cm, it seeds abundantly and majority of seedlings appear true to colour), 'Holmsted' (flowers rose-pink and 5 cm in diameter, propagated through layering or cuttings), 'Inchmery' (pale-pink), 'Ipswich Crimson' (deep crimson), 'Isolde' (palest-pink with irregular purple lacing), 'Little Jock' (semi-double, pink in colour with darker eye, flower diameter 3.75 cm and peduncle length 120 cm), 'Mars' (a dwarf with double and vivid crimson flowers, sometimes with a trace of blue, can be propagated through cuttings or layers though these easily divide regularly), 'Miss Sinkins' (flowers double white and fragrant but peduncle length is hardly 10 cm), 'Mrs. Sinkins' (white), 'Ruby Castle' (semi-double red flowers, petals striped black, flowers highly scented, it looks more like a carnation than a pink), 'Whiteladies' (vigorous white), etc.; and the noteworthy Modern pinks are 'Cherryripe' (bright cherry-pink), 'Daphne' (pale-pink with a crimson eye, single), 'Diane' (salmon-red), 'Doris' (palest salmon-pink), 'Freckles' (silver-pink flecked red), 'London Poppet' (palest-pink with a red eye and lacing), 'Prudence' (palest-pink with a crimson eye and lacing), 'Show Beauty' (deep rose-pink with a maroon eye), 'Show Pearl' (white), 'Show Portrait' (crimson), etc. The showy miniature hybrids most suitable for planting in the rock gardens and growing up to 15 cm are 'Baby Doll' (a dwarf strain growing only up to 15 cm high, and has mixed colours), 'Bombardier' (red), 'Fay' (mauve, single), 'Little Jock' (pink), 'Mercury' (pink with a maroon eye), 'Pluto' (white, eye purple), 'Spencer Bickham' (pink, single), 'Wisp' (white, eye purple, single), etc. Through cuttings or layerings these are propagated to maintain its colour. Indian pinks or Chinese pinks (*Dianthus chinensis*), a native of East Asia, particularly China and Japan, and west of Portugal, is a tufted biennial or short-lived perennial, growing to 25-45 cm height and requires 20-30 cm across planting distance. The plants are glabrous, stems forking, angled, more or less grooved, and branching at top, leaves lanceolate to linear and almost flat or slightly trough-shaped and 3-5 nerved, and pale to mid-green, flowers large (>4 cm wide), solitary or more or less clustered, basically lilac to pink or red to white often with a darker central zone but the forms and varieties possess various others and intermediate colours, petals in the original species are hairy towards the base, and calyx-bracts 4, and the modern varieties have single and double-flowered strains.

The European extension of the species now referred to *D. chinensis* are *D. caucasicus* Sims, *D. collinus* Waldst. & Kit., *D. ibericus* Willd., *D. montanus* Bieb., and

D. ruthenicus Roem. Its one of the forms, the European pink is from Portugal. Amoor pink (*D. dentosus*) is, in fact, *D. chinensis* var. *macrosepalus* growing some 30 cm high with bright-red flowers having a spot at the base of each petal, *D. chinensis* var. *asper* (syn. *D. seguieri*) produces flowers in panicles, *D. semperflorens* Hort. (*D. c. semperflorens)*, a hardy perennial form grows to 50 cm high with silvery foliage and red-eyed deep pink fragrant flowers, and *D. chinensis* var. *heddewigii* (*D. heddewigii* Hort.) is also a beautiful race of garden pinks originating in Japan, which is though annual and is grown from seeds but can be perennialized, the flowers are single or double of many colours with beautiful markings and sometimes fragrant and has been parents of most annual garden pinks. *D. chinensis* var. *laciniatus* (*D. laciniatus* Hort.) bears slashed and cut petals, *D. imperialis* Hort. is a strain with strong growing habit having mostly double flowers, *D. cincinnatus* is a red form with shredded petals, *D. hybridus* Hort. is another race though it is sometimes regarded to have been derived from *D. dentosus* and some other species, as *D. hybridus* is similar to *D. dentosus* in many aspects. Var. *stellaris* (*D. stellaris* Hort.) bears narrow star-like petals. *D. diadematus* Hort. is its another garden strain. The established varieties of *D. chinensis* are 'Bravo' (bright red), 'Fireball' (double scarlet), 'Gaiety' (mixed), 'Gaiety Double' (mixed colours), 'Heddewigii' (single or double mixed), 'Snowball' (large double white), etc. Indian pinks thrive best at a sunny location in a well-drained soil having 6.5 to 8.0 pH. It likes dressing of bonemeal at 120 g/m^2 and in the following spring some 60 g/m^2 potassium sulphate. Flowering stems are pinched to build up bushy plants. Dead flowers should be removed regularly to ensure a long flowering span. However, its cultivation practices are just the same as the Sweet Williams. Organic mulches are not preferable as these promote stem rot. For cutting, the flowers are removed early in the morning and held to their necks in water for four hours before being arranged.

Maiden pink (*Dianthus deltoides*), a prettiest border pinks and a native from Europe (Scotland to Norway) to Japan, is densely tufted and creeping but stems ascending, growing to 15 to 25 cm and the solitary flowers are fragrant and rose to deep red with a crimson eye, petals are toothed with an inverted V-shaped patch at the base hence the name *deltoides*; its var. *albus* is white while 'Brilliant' crimson; var. *glacialis* (glacier pink) is about 10 cm tall and is native to Europe and bears pink and white flowers; and Cheddar pink [var. *gratianopolitanus* (*syn. caesius*)], a native of Great Britain and which is available in tetraploid and hexaploid forms with basic number being n = 15, grows to 15 cm and bears fragrant rose flowers. *D. knappii*, a native of Europe grows to 20 to 45 cm and bears yellow flowers; *D. neglectus*, a native of Pyrenees grows only 5 cm high and bears rose-coloured flowers; Fringed pink (*D. superbus*), a native of Europe and Asia grows to 20 to 45 cm and bears the flowers of lilac colour; *D. allwoodii*, a hybrid between *D. plumarius* and *D. caryophyllus* cultivars which grows to 15 to 40 cm in height bears white, pink, salmon, scarlet and crimson flowers; etc.

3. Sweet William (*Dianthus barbatus*)

Sweet William, one of the oldest garden flowers, a native to Eurasia (East Europe to East Asia, *viz.* Russian Carpathian and down to the Balkan Peninsula to China and south to the temperate mountains of the Pyrenees) and well naturalized in areas of North America and China; is though hardy herbaceous biennial plant but in the temperate as well as in tropical regions it can be grown as a perennial provided each year the flowers are immediately cut or removed. The basic chromosome number in *Dianthus* is 15 and most of the species are diploid with 2n = 30. Originally, the species is a short-lived perennial but some of the garden forms are very large, showy and produce the blooms over a long season. Its single or double flowers are 7.5 to 13 cm across, in various colours, from white to red or marked with other colours, even colouring being in concentric zones as in annulatus or auricular-eyed Sweet William (Hay and Beckett, 1971). It sets copious seeds. When perennialized, every 2-3 years the plants should be divided (Armitage, 1989, 1993). It grows 30-90 cm high with glabrous, 4-angled and erect stems but branched at the top with 7.5 to 15 cm long leaves which are broad, flat and 5-nerved and requires 20 to 25 cm planting distance. Inflorescence is a flat or round-headed dense cyme composed of several to many flowers, bracts 4, subtending the calyx and long-pointed, petals bearded, toothed and the garden forms are available in red, rose, purple and violets, pink or white, bicolours and multicolours (Cockshull, 1985; Dole and Wilkins, 1999). Its form *D. laucheanus* Bolle is a hybrid between *D. barbatus* and *D. deltoides* (Bailey, 1942), and *D. barbatus* when crossed with *D.* x *allwoodii*, produced hybrid strain 'Sweet Wivelsfield' which has loose but large flowerheads up to 20 cm across with greater range of flower colours than Sweet William. An hybrid *D. barbatus magnificus* is a double form of Sweet William with deep crimson flowers being borne on a 22-25 cm stalk. Seed companies quote various other cultivars of Sweet William. Though their flowers can be used as cut flowers but these plants are most suitable for herbaceous borders and in garden beds. Sweet Williams can be multiplied through division of stocks or by vegetative cuttings (Post, 1949). Usually, it is multiplied through seeds as it produces copious seeds, though through continuous intercrossing the appearance of colour may be different. Seeds are sown thinly in open nursery beds in India in September and transplanted at its permanent place in October-November or after 8-10 weeks of germination, normally to flower during spring and summer the following year. Within 1-2 weeks when the seeds germinate at 21 ^{0}C temperature these may be grown at 13-16 ^{0}C, and then planted in the field to overwinter (Dole and Wilkins, 1999). Seedlings can also be prepared in plugs or in flats in a well-drained porous medium, i.e. 1:1:1 soil/vermiculite: peat: perlite (Holcomb and Messinger, 1986) having pH range of 6.0 to 7.0. Even directly the seeds can be sown in the prepared beds which should be thinned out at 15 x 20 cm after germination. Seeds may also be sown in temperate climate of India when winter is coming to an end in March inside greenhouses at 13 ^{0}C (Hay and Beckett, 1971), the seedlings should be pricked out in the glasshouses 7.5 cm apart in boxes filled with good potting mixture at a temperature of 10 ^{0}C (Hay and Beckett, 1971), and in May these after hardening should be planted in open beds to get flowers after a month. The beds should have sufficient organic matter, and after planting the beds are immediately watered. Afterwards also these are cared well through proper watering, if need be feeding, regular weeding and timely control of diseases and other pests if any. Watering of Sweet William should be on demand and it is watered when the soil has become slightly dry. Higher temperatures promote flowering but several weeks after end of the cold treatment at 4 to 7 ^{0}C (Waterschoot, 1957; Cockshull, 1985) to grown up plants, i.e. 12-15 weeks old though 35 ^{0}C temperature immediately after cold treatment negates the effect of cold treatment (Dole and Wilkins, 1999). However, there is greater variation due to varieties, though photoperiod is of little response to flowering regardless of cooling but in case when there is filtered light during growing, HID (high density discharge) lighting can accelerate growth at any stage. It becomes biennial in temperate regions though perennial in mild climates and also when flowers are immediately harvested. When 10-20% flowers on the cymes are open, the flower stems are cut and held in water for 6-10 days at 1-2 ^{0}C or dry at 5-7 ^{0}C for 3-4 days. Nowak and Rudnicki (1990) stated that the preservatives having STS prolong the longevity of cut flowers. Normally the cut flower life in Sweet William is 7-10 days. After two years of production, the plants usually die after seed formation. Spider mites, *Fusarium oxysporum* and *Rhizoctonia solani* are the serious problems in Sweet William. Horst (1990) reported aster yellows and certain leaf spot diseases which attack Sweet William in open beds.

4. Pot/baby/dwarf/miniature carnation (*Dianthus carthusianòrum*)

Dianthus carthusianòrum (syn. *D. atrórubens*) has origin in SWC Europe, i.e.Denmark to Portugal, and Egypt. It grows from 30 to 90 cm and has limited flower colours ranging from white to purple via pink, mostly deep pink. It is hardy, evergreen, glabrous, stems stiffly, lanky and angled, leaves green, grassy, slender, linear, pointed and some 7 cm long & 5 cm wide, flowerheads tight with 5-30 densely arranged terminal flowers surrounded by purple-brown bracts, flowers non- or slightly fragrant, 2 cm wide and petals toothed. It is a popular potted flowering plant in Europe and to a lesser extent in North America (Dole and Wilkins, 1999) but is most suitable as border plant. Sometimes these are also grown in hanging baskets and other outdoor containers. *Dianthus carthusianòrum multiflorus* bears up to 70 flowers per head. Seed lines are normally true to colour though may not be uniform in height and days to flower. Seeds as well as the cuttings only from the healthy plants should be propagated at 18 to 20 ^{0}C (Goldsberry, 1979; Wandas, 1989). Its various cultivars may be propagated through cuttings or *in vitro*. The cultivars may have single or double flowers, sometimes scented, in single or multicolours including yellow, red, maroon and lavender apart from the original species colours, *viz.* white, pink and purple (Anonymous, 1983; Dole and Wilkins, 1999; Goldsberry, 1979; Kuack, 1986). The noteworthy cultivars of pot carnation are 'Cerratop' (fancy), 'Charmtop' (red), 'Davinci' (pink), 'Maldeves' (pink), 'Pinky' (red), 'White Sunny' (white), etc. Its most popular cultivar is 'Napoleon III' which grows some 20-25 cm high producing masses of cerise scarlet flowers but is difficult to propagate.

The medium for growing *Dianthus carthusianorum* should be well-drained, having a pH range of 5.5 to 7.0 though some of the forms of this can tolerate soils containing heavy metals such as lead and zinc and are therefore useful for revegetating polluted areas. The planting may be done in 10 cm pots spaced 15 × 15 cm with only one seedling or rooted cutting per pot. These can also be planted in 13 cm pots (two plants per pot and 22 pots per square metre) or 15 cm pots (three plants per pot and 16 pots per square metre) (Goldsberry, 1997; Dole and Wilkins, 1999). While planting, the crown should remain above the medium to avoid infection by pathogens. From seeds it requires 16-18 weeks for flowering from seed sowing though through cuttings it requires 12-15 weeks from rooting though this period is influenced by the cultivars used and prevailing weather conditions (Goldsberry, 1979; Kuack, 1986; Wandas, 1989). Irrespective of the provenance of the material (seed, cutting or through *in* vitro) these are pinched only once to force development of 4-5 breaks. Flowering plants are severely pruned in the spring to force them coming into bloom again in three weeks (Goldsberry, 1979). The plants being propagated through vegetative means should be pruned within a few days of repotting (Wandas, 1989) in a way that second pinching may be avoided as second pinching will though increase the number of flowers but will delay the flowering, *vis-à-vis* marketing (Goldsberry, 1997). Though it does not require support but in highly fertile soils with copious of watering the growth may be rampant, especially in case of seed-grown plants, height of which can be controlled by spraying with 3000 ppm chlormequat 50 days after germination (Anonymous, 1983), however, in case of plants grown from cuttings the height may be controlled through drenching of chlormequat at 5000 ppm at the stage when the shoots have attained 8-10 cm of height (Dole and Wilkins, 1999). Goldsberry (1997) recommended uniconazole or paclobutrazol at 15 ppm for getting reduced height, however, his preference was for uniconazole as it increased the number of axillary shoot formation. Mostly the vegetatively propagated ones grow only up to 15 cm in height where no treatment is required to stunt the growth. Adriansen (1985) reported even ancymidol checking its growth. Moe (1983) recommended negative DIF (monitoring of difference between the day and the night temperatures) for getting reduced stem elongation.

Dianthus carthusianorum plants (young plants, cuttings and the stock plants) are required to be kept at short photoperiods, and at 12-15 ^{0}C (for plants raised through cuttings) for the first five weeks for good vegetative growth, followed by 14 h or at longer photoperiods and at 15-18 ^{0}C for accerating floral initiation (Moe, 1983; Wandas, 1989). However, in case of seed-grown plants the night temperature is required from 11-16 ^{0}C (Goldsberry, 1979) and increase in temperature up to 21 ^{0}C accelerates flowering though beyond this it has no response. These require full light, from 18-20 hours of 290-380 fc supplemental light for quality blooms (Wandas, 1989) though for protection against excess temperature or for promoting colouration these are shaded (Goldsberry, 1979). Short photoperiods coupled with low temperatures accelerate vegetative growth and axillary shoot development but delays flowering though the long days of 14 h and above (supplementary lighting through incandescent bulbs) with low temperatures hasten floral initiation with increased internodal length due to incandescent lighting. Photoperiod is not responsible for the development of the floral buds, it is only tempearature from 18-21 ^{0}C which controls the flowering (Moe, 1983). CO_2 concentration up to 1500 ppm in the greenhouse has been found to increase net photosynthesis, thereby increasing the production of flowers and increases flower weight, and CO_2 in combination with high photoperiods accelerates floral induction (Enoch and Hurd, 1977; Bunt *et al.*, 1981).

Flowers are harvested when apical flowers begin to open. At this stage if the plants are kept at a sunny and cool location, such plants continue flowering for up to four months as one by one axillary buds will continue to flower, and for delayed sales the terminal buds which has opened should be removed so that in succession axillary buds may continue coming up. Plants can be transported at 2-5 ^{0}C and these will tolerate darkness even up to six days. Spraying with STS at 150 ppm the plants can last for three weeks (Dole and Wilkins, 1999).

Aphids, caterpillars, leaf miners, spider mites, thrips, and slugs and snails feed on the crop which may be kept under check through proper sanitation and the use of appropriate insecticides and poison baits. *Alternaria, Botrytis, Heterosporium, Rhizoctonia, Fusarium, Corynebacterium, Pseudomonas,* etc. attack this crop which can be kept under check by taking the cuttings only from the healthy plants (culture-indexed cuttings, especially against *Fusarium, Corynebacterium* and *Pseudomonas*) and through maintaining proper crop spacing, by keeping the foliage dry, use of sterilized media and the use of fungicides and bactericides, as the case may be. Virus-infected plants should be rogued out and destroyed.

No plant can survive without nutrition or watering and so is the case even with *D. carthusianorum* but this does not like moist medium continuosly otherwise the plants will grow taller and weak and flowering will also be delayed. Overhead watering is not preferred as it encourages spread of diseases. To avoid ammonium toxicity, especially during winters when light is also low, it is only nitrate form of nitrogen which is applied. The requirement of various nutrients, *viz.* N, P, K, Ca, Mg and B though depends on the nutrient status of the soil, the season and the location, therefore, it is necessary to analyse the medium before any application (Langhans, 1961; Anonymous, 1995) as medium should contain 25-40 ppm each of the N and K_2O, 5-10 ppm P_2O_5, 150-200 ppm Ca, 30-40 ppm Mg and 30-35 ppm boron. A constant liquid fertilizer of 200 ppm N and K should be applied fortnightly (Dole and Wilkins, 1988).

5. Florist or perpetual carnation (*Dianthus caryophyllus*)

Perpetual-flowering carnation is usually grown only in the glasshouses. Roots of carnation being highly susceptible to water-logging, hence, wherever it is grown the **soil** should be well-drained, aerated and should be rich sandy-loam. Heavy soils are not suitable as these restrict the growth of roots and plants so any such soil should be amended by incorporating sufficient compost or any other organic matter. In case the crop is to be perennialized in the same field, the soil of the field requires to be fertilized adequately, however, in the same soil the crop should not be taken for more than three years. It does not like soil with acidic reaction, so the pH of the soil should be around 7.0 (6.0-7.0). Acidic soils should be amended by incorporating sufficient quantity of calcium. High light intensity during winter is quite helpful in obtaining quality blooms, however, during summer there should be provision for bringing down the temperatures. It grows well and produces quality blooms at **temperature** range of 10 (night)-16 ^{0}C (day) (Misra, *et al.*, 2002). However, these tolerate the summer temperature of 23 ^{0}C. Gill and Aulakh (1973) stated that in southern India, the places like Kodaikanal, Ooty and Coonoor are quite suitable for year round carnation production. Direct sunlight may burn the leaves and plants, and hot dry winds are also injurious. The mean minimum temperature at main carnation producing sites in the world is from 6-21 ^{0}C. Increased temperature increases yield, the number of leaf pairs with lessening of the leaf curling, *vis-à-vis* calyx splitting though decreases flower size, the stem length, petal numbers per flower and the keeping quality of the flowers, however, calyx splitting increases when there is substantial difference between maximum and minimum temperatures. Plants grown under **short days** produce more cuttings though root well under long days. If the plants are kept in vegetative phase during summer, it can tolerate the maximum temperature of 33 ^{0}C. In case the crop is taken in the pots or in some medium under glasshouse, care should be taken that pH of the medium is roughly neutral, is rich in nutrients, free of pathogens, nematodes and insects (Langhans, 1961). To keep the soil pest-frree, pre-plant soil sterilization through 2 per cent formalin and then covering the soil with black polythene for 5-7 days and then uncovering for next one week will be quite effective.

The **glasshouses** may be constructed through polythene sheets with provision of vents on the top and exhaust fans below to exit the hot air during summer, and pad and fan cooling system to keep the temperature under control during summer. The beds inside the polyhouse may be 1.0-1.2 metres wide with 60 cm clearance in between the beds for moving and cultural operations (Sehgal and Pathania, 2001) with a covering of black polythene sheets to check the weed germination and to generate heat during winters (Arora *et al.*, 2002). In polyhouses, the carnation is planted in the pots, raised benches or ground beds. While preparing the soil or medium, the soil should be cultivated thrice to make it to a fine tilth, all the rootstocks of the weeds are removed and the soil is incorporated with five kilogrammes per square metre of well-rotted organic manure along with basal doses of NPK fertilizers (full dose of K_2O and P_2O_5 and half dose of N). Apart from these, the ground limestone should be incorporated in non-calcareous soil at the rate of 400 grammes per square metre. The 7.5 cm pots where rooted cuttings had been planted are to be replaced by 17.5-20.0 cm pots filled with JIPC No. 2 and in such pots these will flower.

While **planting** in the beds, on raised benches or in pots, the top of the ball of the soil attached with the cuttings lifted from 7.5 cm pots should be within 1.25 cm of the top of the new pot though in the raised benches or in the beds these should be planted akin to the level of their 7.5 cm pot planting from where these have been taken out and then pressed firmly all around (Shewell-Cooper, 1953; Besemer, 1980). Deep planting makes it more prone for attack of *Rhizoctonia solani* which causes stem rot. To obtain adequate number of flowers from mid-May to mid-November from the hills and from December to April in the plains, the carnations in India are planted in October-February in the foothills with no snowfall, February-April in temperate areas and October-November in the plains. However, mid-hills of Himachal Pradesh, whole of the NEH regions, Nilgiri hills, Darjeeling area of West Bengal, Koraput region of Orissa, whole of Gangtok, Nagaland, etc. are most suitable for growing of carnation throughout the year. Verma and Sharma (2002) when used carnation var. 'Impala' in Solan (H.P.), recorded October planting as the best for maximum plant height, stem length, flowers per plant and minimum disease incidence. In another trial at Solan with cv. 'Impala', Verma *et al.* (2002) planted the rooted cuttings on Feb. 15, June 15 and Oct. 15 at 20 × 20 sm spacings and recorded October planting as the best for quality and quantity blooms, however, February planting induced earlier flowering. Early planting increases the production while late planting hastens flowering (Ramesh Kumar *et al.*, 2002). **Distances** apart for perpetual flowering carnations is normally 20 × 20 cm while for spray type (Chabaud) it is 30 × 30 cm. Wider spacing is always better for perennial cropping. Sawwan (1998) recorded highest production with a plant density of 32 plants/m^2. Langhans (1961) advocated 10 × 15 cm for 1-year production, and 15 × 15 cm & 15 × 20 cm for 2-year production.

Nutritional regimes and practices vary with season and location, however, the correct doses should be worked out through soil and tissue analysis. Peterson (1982) and Dole and Wilkins (1988) stated that for quality carnation plants, the tissue nutrient levels should be nitrogen 3.2-5.2%, phosphorus 0.2-0.3%, potassium 2.5-6.0%, calcium 1.0-2.0%, magnesium 0.2-0.5%, iron 100-300 ppm, zinc 25-75 ppm, copper 10-30 ppm and boron 30-100 ppm. Dole and Wilkins (1999) stated that optimum levels of nutrients in the medium should be 25-40 ppm nitrate, 5-10 ppm phosphorus, 25-40 ppm potassium, 150-200 ppm calcium, 30-40 ppm magnesium, and 30-35 ppm boron. Nitrogen in the nitrate form is preferred over ammonium form in low light areas to avoid soft winter growth, *vis-à-vis* ammonium toxicity due to cold though in excess nitrate form also causes weak stems (Dole and Wilkins, 1999). Bhatt (1989) stated that nitrogen deficiency limits the erect growth of the plants, causes thin stems with stiff leaves curving at the tip, yellowing of mature and lower leaves, precocious leaf formation at the base so flowering is delayed and encourages calyx splitting though its excess weakens the stems, delays flowering and reduces the post harvest life. Phosphorus increases flower yield, its diameter, weight, petal number and pigmentation intensity but its deficiency restricts the growth of the plants with narrow leaves and there appears death of the tips of lower leaves coupled with yellowing, especially during the second year. Verma and Sharma (2002) when used fortnightly foliar spray of urea at 500, 1000 and 1500 ppm starting after 15 days of transplanting to bud formation stage, and observed maximum number of flowers, *vis-à-vis* flower yield. Potassium is responsible for the nitrate changes in the plant, invigorates the plants and brings resistance to certain diseases. Its deficiency causes leaf blade scorching, development of white spots on the foliage close to flowers, calyx yellowing, loss of petal colour and floral malformation. Calcium deficiency causes thin and weak stems, foliage scorching, calyx splitting and small, *vis-à-vis* improper opening of flowers. Magnesium deficiency causes weak stems with small flowers. Boron deficiency causes internodal length reduction, pale-yellowing zone turning to purple afterwards along the midrib of leaves, calyx splitting and failure of flower buds to reach anthesis while its excess causes leaf tip burn. A generally recommended dose for carnation is FYM 5-10 kg/m^2, N 30 g/m^2, P_2O_5 20 g/m^2 and K_2O 10 g/m^2 for standards while for sprays it is the same except that N is 40 g/m^2 instead of 30 g/m^2. Nitrogen should be provided in two split doses, half at planting and the remaining half one month after planting. Urea and KNO_3 at 0.1 per cent and boron at 1 ppm should be sprayed at interval of 10 days till flowering (Sehgal and Pathania, 2001). Apart from the nutrients mentioned above, carnations also require manganese and molybdenum in very small amount. Every fertilizer or micronutrient application in the soil should be followed by irrigation. The planting should also be followed by immediate **watering**. Under mist chamber for rooting of cuttings, the atmosphere and medium are regularly wet but when planted outside or in polyhouse, immediately after planting the pot or field should be watered, especially when rooted cuttings are planted otherwise the roots may desiccate. Even in case of unrooted cuttings, these require watering. If there is provision of sprinkler system of watering, it should be watered a few times daily for proper establishment of cuttings or plants. Overhead irrigation system is stopped and replaced by surface irrigation when floral buds start initiating. However, drip system is the best alternative. Frequency and quantum of watering depend upon prevailing weather conditions, air temperature, wind velocity, soil type and its temperature, light intensity and relative humidity. Though it requires thorough watering to keep the soil moist but it does not like wet feet hence in no case the water should stagnate in the field. During summer every four days while during winters it may require watering at 10-12 days intervals. However, before irrigation it would be better to determine moisture content of the soil through tensiometer and irrigation applied at tensiometer reading of 30-50 is taken as the best. Irrigation water should regularly be checked for its EC as carnations are highly sensitive to high EC (Langhans, 1961; Laurie *et al.*, 1969).

A quality young carnation mother plant is that which has been grown from healthy stock plants in full light, has a thick central stem with short internodes, bright coloured broad leaves on the shoot, lower leaf apices rolling under, and the upper leaves growing upright and terminating into a vigorous growing point. To regulate flowering and for quality blooms, stopping is practiced when stem lengthens for flowering, leaves get wider apart and the floral buds start initiating, and at the same time secondary growths are initiated in the axils of the leaves which also in their turn flower afterwards, though this stage may vary from variety to variety. When main floral stem is about 15 cm high with six pairs of leaves which occurs normally 4-5 weeks after planting when lateral shoots from lower leaf pairs are about 5 cm long, the tip with one pair of leaves (at or below the sixth node) is nipped off and this is known as **stopping** (**pinching** or **disbudding**) and this practice forces the plant to initiate 4-6 shoots lower down if yet not initiated which are stopped later on, better if stopped one by one each at 3-day's intervals when these also have attained six pairs of leaves. Stopping encourages the number of flowering stems though delays flowering (Sawwan and Samawi, 2000). In India, there are three types of pinching methods: (i) single pinch where for uniform and early flowering the main shoot is nipped off below the sixth node to force 5-6 flowering laterals, (ii) pinch-and-a-half where in 'single pinchs' when lasterals have developed 5-6 leaf pairs, half of these laterals are pinched as to that of single pinch to reduce the load of first crop and to induce early flowering in half of the non-pinched laterals, and (iii) double pinch where all the laterals emerging after single pinch are repinched at 2-3 node about 5-6 weeks after single pinch to produce more number of flowers at later time but this produces flowers with weak stems (Besemer, 1980, Sawwan and Samawi, 2000). Verma *et al.* (2002) with cv. 'Impala' reported that 'single pinch method' produced flowers of better quality but total yield of flowers is obtained better in 'pinch and a half method'. However, in case of standards, within seven days of emergence the axillary buds are removed (disbudding) when the apical bud is about 15 mm in diameter, to produce best quality terminal flowers though in sprays the terminal buds are removed. The plants grown through cuttings are stopped only when these have fully established in their 7.5 cm pots. To obtain flowers throughout autumn and winter, no stopping should be followed after July in the temperate regions of India. **Staking** (support) with 4-5 tier nylon rope or galvanized metal wire having squares of 7.5 × 7.5 cm (the first one) and fixed to a frame is placed 12 cm above the ground or medium, while subsequent layers having squares of 15 × 15 cm are placed one above the other at 15 cm distance. Laurie *et al.* (1969) suggested usually four layers of plastic mesh with 15 × 20 cm opening, first level of support being 15 cm above the medium and subsequent levels being at 20 cm, 25 cm and 30 cm apart, respectively above the first layer.

Weed control saves the nutrients to be consumed by the weeds for their main crop, keeps off the hinderance for sunlight competition and minimizes the diseases and pests. Lenacil at 2.2-4.4. kg/ha applied one week after planting *Dianthus barbatus* in autumn, was safe and controlled autumn-germinating annual broad-leaved weeds (Anon., 1972).Tenoran 7 kg/ha, Ramrod (propachlor 65%) 7 kg, Amiben-Granulat (chloramben), Kerb 50W 3 kg, Legurame flussig [(carbetamide (70%)] 10 litres and Ronstar (oxidiazon [75%]) 8 litres were tested for their selectivity in *Dianthus chinensis* where Tenoran was observed giving best results; Kerb 50W and Legurame flussig giving satisfactory control; Amiben-Granulat, Kerb 50W and Legurame flussig inhibiting the growth but were well tolerable; Ramrod lacked adequate herbicidal activity; and Ronstar caused severe necrosis though the plants recovered fully after 3-4 weeks (Anon., 1974). Alachlor (1.125 or 2.25 l/ha) was safe on direct sowings of *Dianthus barbatus* and controlled at least 70% of broadleaved weeds and 90% of grasses in both summer and winter, Chloroxuron (2.25 or 4.5 kg/ha) was found quite safe only on its transplants controlling about 50% weeds, Oryzalin (2.25 or 4.5 kg/ha) was though found safe in case of transplanted crop controlling 60% of broad-leaved weeds and 70% grasses, but not in direct-sown one, though Oxadiazon (1, 2 or 4 kg/ha) and Oxyfluorfen (2 l/ha) were safe on transplanted *Dianthus caryophyllus* (Lamont and O'Connell, 1986). Spraying 20-30 litres per hectare of sodium 2, 4-dichlorophenoxy acetate before flower bud formation, Ronstar (25% oxidiazon) at 16 l/ha, chlorpropham (Nexoal) at 6 l/ha or Alipur (cycluron) at 4 l/ha either 10 days before planting or immediately after transplanting results into good control of weeds (Bhatt, 1989). Oxidiazon and napropamide at 2.24 and 4.48 g/m^2 show very good results to glasshouse carnations (Bhatt, 1989).

Growth and Flowering

Carnation produces flowers through termination of each set of lateral buds so lateral buds extend to produce flowers. The number of leaves being formed before flower initiation differs in axillary vegetative shoots and the shoots from nodes just below the flowers form few leaves before flower induction though basipetal shoots form numerous shoots before flowering, and this fact should not be ignored in selecting cuttings, pinching plants and harvesting of flowers. The basal nodes are better for producing long stems (Hanzel *et al.*, 1955; Harris and Bradbeer, 1966; Powell, 1979; Bunt and Cockshull, 1985; Dole and Wilkins, 1999). Flushing time depends on time of planting, the spacing provided, the stopping stage, photoperiod and light intensity, the temperature range and the use of plant growth regulators. Floral induction in carnation is determined by **light intensity** as under low light intensity the induction is slow with production of more leaves though it does not influence the rate of flower development (Bunt *et al.*, 1981). However, initial vegetative growth is directly proportional to light intensity. It is a facultative long day plant as floral induction is rapid under long day than short day conditions (Misra *et al.*, 2002). Misra *et al.* (2006) also stated that carnations respond effectively to light intensity (21.5 klx, i.e. 2000 fc) and day length though associated with growing temperatures and further stated that the modern cultivars are quantitatively long day plants where flowering as such is not inhibited by short days but is definitely hastened by long days. Misra *et al.* (2006) stated that more flowers are produced during long days with high light intensity at low temperature. Shoots become receptive to photoperiod at 6-8 leaf pair stage (Kumar *et al.*, 1999). Photoperiod should be provided from dusk to dawn for four weeks at an intensity of 110 lux or higher, and the shoots with 5-7 leaf pairs will respond by flowering about six weeks later though the next flush is delayed (Byrne, 1982; Salinger, 1985) and this way the flowering can be advanced or regulated at the required dates. Under normal environmental conditions, 15-18 pair leaves are formed prior to floral initiation. Short days produce more leaves while long days fewer with more flowers but continuous long days cause an excessive elongation of internodes. Long days through cyclic lighting is as effective as continuous night break lighting though day extension lighting is comparatively less effective (Harris and Ashford, 1966; Bunt and Cockshull, 1985; Dole and Wilkins, 1988; Neelofar Lolapuri and Arora, 1995). For commercial production, alternating 21 LD followed by 21 days of natural day length (if shorter than 12 hours) will time the flower flushes in carnation (Blake and Whitehead, 1961; Pokony and Kamp, 1960, 1965; Harris, 1968, 1972; Harris and Harris, 1977; Heins and Wilkins, 1977; Heins *et al.*, 1979; Bunt *et al.*, 1981; Bunt and Cockshull, 1985; Dole and Wilkins, 1988). Heins and Wilkins (1977) and Heins *et al.* (1979) stated that the LD cycle should not be over 10 days in length if longer axillary buds are induced to flower. Koon (1974) suggested that before removing the old plants for replanting, for maximum harvest the LD should be applied continuously to induce flowering on every possible shoot for maximizing flower production. In warmer weathers, always the flower yield is higher than the colder weathers even if the plants are grown at optimum growing temperatures because of the better light levels and the photoperiod on warmer days (Salinger, 1985).

Temperature is also most important factor influencing rate of growth, flower production, quality and post harvest life in carnation. The day/night temperature of 16/10 ^{0}C is though most suitable for Sim carnation growing where calyx splitting and bullhead formation are also reduced but high temperature accelerates flower anthesis. Abrupt high fluctuation in day and night temperatures encourages calyx splitrting. A 4 ^{0}C temperature induces flowering though 10 ^{0}C delays the same only in standard and not in the spray carnations as becomes evident by counting the leaf pairs below the flower (Bunt and Cockshull, 1985; Dole and Wilkins, 1999). Once the flower induction is initiated, further development of the floral bud becomes temperature dependent. Bunt (1973) stated that it took some 100 days from floral initiation to anthesis at 10 ^{0}C in carnation though only 33 days at 20 ^{0}C. However, low temperature with long photoperiods advances floral initiation (Bunt and Cockshull, 1985). Temperature above 16 ^{0}C deteriorates the flower quality. Spray carnations require a slightly higher temperature for its growing, up to 6 ^{0}C higher in summer and up to 4 ^{0}C in winter (Linas and Schypula, 1982).

Similar to other ornamentals, in carnation also **growth regulators** have played a significant role. GA_3 application at 100 ppm, one at first pinch and the second at 8-10 cm stage of axillary shoots accelerates flowering with long stems (Gupta and Bhattacharjee, 2002; Ramesh Kumar *et al.*, 2002). GA_3 application at 200 and 300 ppm was found improving only shoot number and not the other characters (Neelofar Lolapuri and Arora, 1995). Rugge *et al.* (2010) conducted an experiment in the hi-tech park of JAU, Junagadh taking four carnation varieties, *viz.* 'Begumande', 'Firato', 'Gaudina' and 'Viking' and GA_3 sprayings (0, 50, 100 and 150 ppm) at 60 days after transplanting and again after 15 days of the first treatment, and recorded maximum plant height, highest number of axillary branches, early bud initiation and maximum number of flowers per plant under 150 ppm though flower diameter was found highest under 50 ppm, however, 'Gaudina' provided highest plant height while 'Firato' produced maximum number of axillary branches, largest flower diameter and maximum number of cut flowers per plant. Gupta and Bhattacharjee (2002) recorded increase in shoot number and flower yield with 1.1 or 2.2 ppm morphactin, and reduced calyx splitting, *vis-à-vis* healthy stems with 30 ppm mepiquat chloride (Chamatkar) in carnation cvs 'Keefers Cherism' and 'William Sim'. Chlormequat at 10-100 ppm or 0.25 per cent and SADH at 0.4 per cent increased flower production and advanced flowering while MH delayed it (Bhatt, 1989). Sonvir Sooch *et al.* (2002) with carnation var. 'Corella' at Ludhiana tried NAA and kinetin, each at 25, 50 and 100 ppm and their various combinations and recorded that kinetin promoted lateral bud growth and 100 ppm of NAA or kinetin stimulated precocious branching thus increased number of flowers per plant while interactions between NAA and kinetin at 100 ppm increased carnation vase life.

The plug-grown *Dianthus chinensis* 'Telstar Mix' was transplanted into jumbo six packs and when the seedlings had attained 5 cm in height or width, Kuehny *et al.* (2001) sprayed them with 1000/800, 1250/1250 or 1500/5000 mg l^{-1} of chlormequat/daminozide solutions which reduced the plant height significantly.

Enoch and Hurd (1977) and Bunt *et al.* (1981) reported that **CO_2** enrichment at 1500 ppm and high light intensity increase the photosynthetic rates, and thereby increases meristematic leaf making activity rate and floral induction. Besemer (1980) reported that addition of CO_2 under favourable conditions increases flower production by 10-20 per cent.

Post Harvest

The proper stage for flower harvesting in carnations is when the petals above the calyx are expanding and the flower is rectangular in shape (15-20 mm diameter size), however, paint brush stage (20-25 mm diameter) is not very preferable. Carnations for long distance transportation can also be harvested at bud opening stage (15 mm diameter when the petals are 0.5-1.5 cm above the calyx with clear visibility of the colour) where post harvest life is always better. Tight bud stage with little or without colour showing (the stage just prior to colour showing) can be harvested but should be pulsed with silver thiosulphate which then can be stored dry up to 24 weeks (though it is being done commercially for 4-5 weeks) in an illuminated chamber and at 0-1 ^{0}C, open flowers for 2-4 weeks at 0 ^{0}C and 90 per cent relative humidity though such flowers may be attacked with *Botrytis* for which maneb or zineb should be used before storage (Dole and Wilkins, 1999). STS treated flowers held in a preservative solution last for up to 30 days as compared to 6-9 days in untreated ones. Only sucrose-based preservative solution causes lasting of cut flowers for 12-16 days (Dole and Wilkins, 1999). Even a very low level of ethylene to the floral buds causes failure to opening of the flowers, and the petals of the open flowers turn inward and upward, lose turgidity and become sleepy (Nowak and Rudnicki, 1990; Sacalis, 1993). In case of spray type carnations, these are harvested when 2-3 flowers have openend on each stem and the remaining buds start showing colour (Besemer, 1980).

Where stems are dehydrated, 38-43 ^{0}C warm water should be used (Dole and Wilkins, 1999). Flower harvesting is done with sharp scissors either to the top vegetative shoot or at the seventh node. Flowers are harvested in the morning when turgid, kept with their cut ends dipped in a bucket of water containing some preservative solution, graded and placed at 1-6 ^{0}C refrigerated conditions to reduce field heat (Dole and Wilkins, 1999). The flower yield depends upon the season of growing, the cultivar used, planting time, plant density, pinching and other management operations. Besemer (1980) reported that in Israel, the average flower yield in spray carnation is about 12,00000 stems per hectare. Standard carnations yield some 100-200 stems per square metre. These buds can be pulsed in 10-20 per cent sugar + 8-HQC or Physan 200 ppm + 25 ppm silver nitrate + 75 ppm aluminium sulphate or 75 ppm citric acid instead of aluminium sulphate and then packed but after thorough rinsing of the pulsing syrup to avoid development of any mould. Aluminium sulphate causes fading of the red floral colour, hence, on its place citric acid can be used in red cultivars. Dole and Wilkins (1999) suggested that carnation stems and flowers should be pulsed at pH 3.5 for 20 h at 20 ^{0}C with STS, sucrose 10% and 8-HQC (a microorganism inhibitor), however, repeated use of STS causes phytotoxicity so it should not be used for more than once though sucrose may be used at all the times when the cut ends are recut and placed in water every time.

Carnation flowers are highly sensitive to ethylene (even 1 part per billion) and flowers become incurled and sleepy in its environemnt. Through adequate ventilation of the store, keeping flowers away from the fruit stores and restriction in the production of ethylene are the measures to get rid of this problem to some extent. Silver nitrate and acetate are ethylene inhibitors which also kill micro-organisms. The stems stood in the STS solution are cool stored at 2-4 ^{0}C in a nutritive solution for 12-14 hours. Another anti-ethylene compound 1-MCP (1-methyl cyclopropene) is being encouraged as it is easy to dispose off being in the gaseous state. These are then graded and bunched in one dozen or in 20 or 25 stems by putting rubber bands at the base and at the necks of the flowers. These are then packed in corrugated cardboard boxes (30 cm height × 50 cm width × 122 cm length) and transported in a refrigerated van at 2-4 ^{0}C temperature. The box of this size accommodates some 800 flowers.

The grades and standards in the United States were developed jointly by Society of American Florists and the United States Department of Agriculture while European standards were devised by the United Nations Economic Commission for Europe (Anon. 1982) as are being given below:

Grades	**United States**			**UNEC for Europe**	
	Description	**Flower dia (cm)**	**Minimum stem length (cm)**	**Code**	**Stem length (cm)**
Fancy (blue)	Tight	5.0	55	0	Stemless
	Fairly tight	6.2	55	5	5-10
	Open	7.5	55	10	10-15
Standard (red)	Tight	4.4	43	15	15-20
	Fairly tight	5.6	43	20	20-30
	Open	6.9	43	30	30-40
Short (green)	Tight		30	40	40-50
	Fairly tight		30	50	50-60
	Open		30	60	60-80
				80	80-100
				100	100-120
				120	Over 120

Class	**Description**
Extra	Flowers of best quality, free of extraneous matter, properly developed with strong stems and conforming to a given species or cultivar. Only 3 % flowers with slight faults allowed.
I-First	As above but with good quality of flowers and stem rigidity. Only 5% flowers with slight faults allowed.
II-Second	Not so good as above but satisfying minimal requirements and useful for decoration. Only 10% flowers with slight faults allowed.

After grading, the flowers are bunched normally in units of 25 stems, stem ends are slightly trimmed and placed in a preservative solution with a pH of 3.5-4.5 at 37 ⁰C, then shifted to room temperature of 21 ⁰C for 2-4 hours, and then finally in a cold room at 0-2 ⁰C for 12-14 hours. Besemer (1980) also advocated flower pulsing in 10% sugar solution with a biocide for 12-18 hours before marketing. The best vase life is found when flowers are stored at 10 ⁰C and 98 per cent relative humidity (Bhatt, 1989). Flowers can be stored for 6-8 weeks at 0 ⁰C and 90 per cent relative humidity though commercially these are never stored for more than two weeks (Bhatt, 1989). Kushal Singh *et al.* (2002) harvested 60 cm long stems of carnation var. 'Tasman' at paint-brush stage, kept in water for 30 minutes by removing 10 cm base of stems under water and leaves from 1/3rd of the lower part of stem and then pre-cooled for 6 h at 4±0.5 ⁰C by putting cut ends in water, and then a set was weighed and sealed in 10 cm wide polythene sleeves and another set after weighing was kept in plastic bucket with 6-7 cm lower stem dipped in water, and finally both the sets were stored in cool chamber at 4±0.5 ⁰C and 80-85 per cent R.H. for 3, 6, 9, 12, 15 and 18 days. After the storage for stipulated period the cut blooms were taken out and put in vases containing distilled water and placed in A.C. lab having 23±2 ⁰C temperature and 16 h illumination with 1000 lux intensity. Vase life was recorded always better under dry storage up to six days though afterwards and under all the durations of wet storage, it was only control which proved superior.

Senescence in carnation is evident by the inrolling (sleepiness) of corolla and further withering, growth of gynoecium and finally wilting of the flowers. The major factor which contributes to its withering is build up of ethylene. The enlargement of gynoecium involves drawing of food from other parts of the flower which leads to its eventual death. Physiological maturity, temperature and relative proportion of O_2 and CO_2 contribute towards sensitivity of cut flowers to ethylene. During normal metabolism of the flowers, due to stress, the infection and wounding of the flowers, its stem and the leaves cause production of ethylene. The genetical make up of the cultivar, and proper supply of nutrients, soil moisture and relative humidity, light, temperature and CO_2 during growing season, the harvesting time and method, and treatment to the flowers after cutting are the factors which contribute towards vase life of the cut flowers. So far the nutrient is concerned, potassium deficiency during growth drastically reduces the vase life. Vase life is further threatened due to high temperature and low humidity when enjoying the cut flowers indoors and that is why the cut flowers received in Delhi during summer from temperate regions such as Srinagar, Kullu and Solan show less than half of its usual vase life as compared to winters, *vis-à-vis* the flowers grown at higher temperatures show poor vase life even if displayed indoors in air conditioned rooms. Dusk to dawn cyclic or continuous lighting during growth period also reduces vase life. The amount of cell wall polysaccharides, especially pectins also influence the vase life positively. More so, the higher level of ABA and lower levels of cytokinins and gibberellins also cause early senescence. BA and sucrose enhance the vase life. A good preservative solution is that which maintains water absorption and turgidity in floral tissues. Soaking of stems in 1200 ppm silver nitrate for 5-20 minutes enhances the life by about 50%. However, microbial growth should be checked through addition of 8-HQC at 200-1000 ppm or sodium benzoate 100 ppm to check stem plugging by ubiquitous microbial growth. Sucrose 3% + 8-HQC 200 ppm or sucrose 10% + silver nitrate 2 ppm are quite beneficial as pulsing solutions. Verma *et al.* (2002) subjected carnation cv. 'Impala' rooted cuttings to different planting dates (Feb. 15, June 15 and Oct. 15), urea sprayings (nitrogen 500, 1000 and 1500 ppm) fortnightly starting after 15 days of planting to bud formation stage, and the cut blooms treated with 200 ppm 8-HQC+2% sucrose, 0.5 mM STS+2% sucrose and 200 ppm 8-HQC+0.5 mM STS+2% sucrose, and recorded maximum vase life in 200 ppm 8-HQC+0.5 mM STS+2% sucrose, in June planting and under 1500 ppm nitrogen (urea). Life in distilled water was recorded 6.78 days as compared to 17.23 days under June planted with 1500 nitrogen treatment and under 200 ppm 8-HQC+0.5 mM STS+2% sucrose.

Insect-Pests, Diseases and Physiological Disorders

Fusarium wilt (*Fusarium oxysporum* f. *dianthi*) causes serious damage to carnation and the situation aggravates when the temperature is high and soil drainage is poor. The disease is soil-borne and enters the plants through roots and damages the vascular system which results into wilting and complete death of the plant. Initially this may cause abnormal growth and stunting of the young shoots with yellowing of the leaves. Besemer (1980) recommends soil sterilization or soil fumigation with methyl bromide (0.5 kg/m^2), mylone (135 kg/ha), chloropicrin (180 kg/

ha) and vapam, use of pathogen-free plants and general sanitation of the polyhouse. Soil solarization with quite transparent polyethylene film (0.1 mm thick) for 30 days is also quite satisfactory. Elena and Tjamos (1997) obtained satisfactory crop through soil solarization for seven weeks. Spraying the plants with 0.1 per cent Bavistin alternate with Captan 0.2 per cent throughout the growing period but at fortnightly intervals is quite effective even in infected soil. Manaka and Fruzynska Jozwiak (1996) recommended use of saprophytic forest fungi (*Trichoderma viride, T. harzianum, Penicillium adametzii, P. funiculosum* and *P. nigricans*) for effective control of *Fusarium* wilt. One can get rid of this problem by grafting the susceptible cultivars on to resistant rootstocks such as 'Arancio 25D', 'Exquisite', 'Gus Royalette', 'Heidi', 'May Britt' and 'Scania'. Most susceptible cultivars are 'Alice', 'Jony', 'Fulvio Rosa', etc. However, planting of resistant cultivars is the best option. **Phialophora wilt** is caused by *Phialophora cinerescens* which is a soil-borne fungus. Its incidence is favoured by cool soil temperatures. Its infection causes dull colouration of the foliage, uniform light discolouration of the basal wood, loss of cell turgidity and ultimate death of the plant. Use of healthy seeds or cuttings, proper watering and sanitation, shallow planting and proper aeration in the polyhouse throughout the growing period will keep this disease under check. Fumigation or steam sterilization of soil is the best method of controlling this disease. Its control measures are the same as for *Fusarium* wilt. However, in polyhouses, fortnightly spraying of methyl thiophanate at 1.4 or 2.8 g/m^2 or with benomyl at 0.5 g/m^2 gives complete control of this disease. **Fusarium stem rot** caused by *Fusarium roseum* f. sp. *cerealis* is also a soil-borne disease entering the stem at soil line, first infecting the surface and afterwards affecting the inner part of the stem and causes plant to topple over. Its attack also causes longitudinal cracks at the bases of the shoots and grooves in the flowering stems coupled with twisted internodes. Same control measures as stated with *Fusarium* wilt will control stem rot also. Deep planting should also be avoided. Dipping of infected cuttings in the suspensions of *Bacillus subtilis* or mixing the suspensions with infected perlite prior to inserting healthy cuttings controls the rot significantly. **Fusarium cutting and stub rot** is caused by *Fusarium culmorum* which causes basal rot of unrooted cuttings in poorly managed fields though it is not a problem in well managed crops. However, this fungus under high humid conditions with poor aeration infects the stubs after pinching and harvesting, entering through the wounds and sometimes creating an havoc. To get rid of this problem one will have to lower the humidity level in the growing chamber and the crop will have to be sprayed with thiophanate methyl. Lesions may girdle the stems which may result into wilting. **Stem rot** is also caused by *Rhizoctonia solani* which infects the plants at soil level and plant dies within 10 days of its infection. Due to its infection brown lesions appear at the soil level apart from chlorosis, and finally the plant wilts. Proper soil drainage, shallow planting and low to medium fertility level reduce its occurrence. While controlling *Fusarium* stem rot, this pathogen will also be controlled. Pencycuron and flutolanil at 2 and 1 g/m^2, respectively as soil dusting before transplanting and 0.1 g/l as foliar spraying will keep this problem under check. Copper oxychloride at 0.4 per cent is also effective against this disease. *Trichoderma harzianum* is also very effective in controlling this pathogen. **Sclerotinia rot** (*Sclerotinia sclerotiorum*) occurs normally under poor management. Its infection causes poor growth and a white web-formation occurs around the stem which contains white sclerotia later on. This afterwards turns black. Unuseful humid conditions in the environment should be avoided, diseased plants should be rogued out and destroyed, and the fungicides like Iprodione, thiophanate methyl, Thiram or Vinchlozolin should be used. Before planting, soil should be disinfected.

Fusarium **bud rot** caused by *F. tricinctum* f. *poae* is transmitted to the carnation plants through a mite (*Siteroptes gramineum*) which takes the fungus deep into the floral centre, hence some effective acaricide such as Kelthane should be applied to kill larvae and adults after the autumn rains when these mites hide themselves or when even slightest infestation is observed. **Bud rot** (*Sporotrichum anthophilum*) is observed on petals and other floral organs enclosed by the calyx. Its carrier is also one of the mites so to check its spread Kelthane should be used. **Calyx rot** is caused by *Pleospora herbarum* which spreads rapidly under high humidity and this has no any effective control as yet in the open though proper aeration in the polyhouses helps in preventing this problem to some extent. **Floral bud rot** is caused by *Alternaria dianthicola* which is kept under control by application of zineb and maneb. **Gray mould** (flower blight) is caused by *Botrytis cinerea* in the field as well as in store. In its infection the petal edges become water-soaked initially but afterwards whole flower is affected. Situation aggravates when humidity is high. Such flowers have no market value. In the open there is no any effective remedy though proper aeration in the polyhouses prevents this problem to some extent. Thiram application also controls this disease.

Fairy ring spot (*Heterosporium echinulatum*) causes fairy ring-like spots which coalesce afterwards and cause foliage death. Spraying 0.1 per cent Bavistin or 0.2 per cent Dithane M-45 regularly at 10-day intervals will control this problem effectively. Either in combination with both protective (probineb) and systemic (triforine) fungicides or their simultaneous use has been recommended as the best strategy for control of fairy ring spot (Penaranda *et al.*, 1992). **Branch rot** or black mould (*Alternaria* leaf spot) is most common foliage disease of carnation which is caused by *Alternaria dianthi*. This causes spots on leaves and stems and therefore leaves wilt and die prematurely. To prevent the incidence of this pathogen, the cuttings should be collected from healthy plants, and dipped in 1000 ppm benomyl or Dithane M-45. Cv. 'Smokey' is also resistant to this pathogen. **Greasy blotch** (*Zygophiala jamaicensis*) causes greasy blotches on carnation leaves during humid weathers. Proper sanitation and aeration in the greenhouse will prevent this disease. *Heteropatella valtellinensis, Phytophthora nicotiana, P. parasitica, Pythium vexans* and *Sclerotium rolfsii* also infect carnation plants but are of little significance.

Rust (*Uromyces caryophyllinus*) exhibits chocolate brown patches on leaves and stems which causes stunting and weakening of the plants, leaf curling and early defoliation. The incidence of the disease is favoured by high moisture conditions. Plantvax (oxycarboxin) at 0.1% + Dithane M-45 (mancozeb) at 0.2% provides almost full control of the incidence when applied at every 21 days. The cv. 'Minirosa' is resistant to this pathogen. **Powdery mildew** (*Oidium* spp.) is a problem only on spray carnations and is characterized as flour-like patches on leaves, buds and flowers. This is controlled through sulphur dusting.

Bacterial wilt (*Pseudomonas caryophylli*) is encountered only at high temperatures, i.e. 13.8 ^{0}C and above and if night temperature is low this disease is rarely found. However, temperature above 22 ^{0}C is most favourable for the incidence of this disease. Due to its infection first the leaves turn grayish-green, then yellow and die. More or less the symptoms are similar to *Fusarium* wilt but in bacterial wilt stems and roots ooze out a sticky substance. Control measures are the same as to those of *Fusarium* wilt. Treating the planting material with 0.01% streptocycline minimizes this disease. **Bacterial leaf spot** (*Pseudomonas woodsii*) is symptomised as the elongated brown spots on the leaves. Plants are killed only when the incidence is severe. Use of healthy planting material, control of insect-pests and diseases with copper compounds and burning of infected material minimizes the incidence of this disease. **Fasciation** is caused due to infection of *Corynebacterium fascians* where plants stunt and leaves become distorted. Soil surface treatment with mercury bichloride 1 : 1000 kills the bacterium (Pirone *et al.*, 1960). *Erwinia tunefaciens, E. chrysanthemi, E. parthenii* var. *dianthicola* and *Xanthomonas oryzae* var. *dianthi* also infect carnation crops causing various diseases.

Carnation is infected with many of the **viruses** such as 'carnation etched ring virus', 'carnation latent virus', 'carnation yellow fleck virus', 'mosaic virus', 'mottle virus', 'ring spot virus', 'streak virus' and 'vein mottle virus'. 'Vein mottle virus', 'mottle virus' (also through tools and handling), 'carnation yellow fleck virus', 'carnation etched ring virus' and 'carnation latent virus' are transmitted by aphids, vector for 'streak virus' is not known though it does not spread by tools or handling, and 'ring spot virus' is spread by cutting tools and handling. 'Mottle virus' is the most common virus in carnation fields where it produces leaf mottling and colour breaking in flowers. *Saponaria vaccaria* cv. 'Pink Beauty' is the sensitive indicator of this virus where symptoms develop within a fortnight of its inoculation. Viruses reduce the flower quality and yield. Shoot tip culture will provide virus-free plants. However, infected plants should be rogued out immediately to check their spread.

Bulhead formation and **calyx splitting** are associated with high temperature fluctuations during the periods of day and night so reduction in such variations will help out checking the calyx splitting. In fact, calyx develops under rapid growth conditions such as high temperatures followed by optimum petal initiation and development at cool temperatures and so the mechanical pressure generated in the process causes calyx to split to give way to the petals coming out. Since the level of calyx splitting varies from variety to variety, hence this abnormality is thought to be governed by genes. Increase in potassium levels causes splitting though nitrogen after a certain level reduces it. Higher plant density also causes more splitting than lower density. However, giving nitrogen in nitrate form than in ammonical form and application of boron will lessen calyx splitting. Moreover, the non-splitting type cultivars should be selected, and rubber or plastic band ring around each calyx should be placed in splitting type cultivars. **Curly tip** where growing tips curl and become distorted, usually appears under low light intensity and a very low temperature (<4 ^{0}C) conditions, as well as nitrogen deficiency may also be one of the causes (Bailey, 1971).

Some 21 species of **nematodes** have been recorded attacking carnation but the most serious ones are *Criconemoides curvatum* and *C. xeroplex*. *C. curvatum* feeds on the roots as endo- as well as ectoparasites which, to some extent, can be controlled through soil steaming. *C. xeroplex* reduces root growth, flower yield and stunt the plants. Nemafos granules (10%) whose active principle is diethyl pyrazinyl phosphorothioate, when applied at 6 kg/ha before transplanting and 2 kg/ha $9^1/_2$ weeks later, provides better economic results.

Red spider mite (*Tetranychus urticae*) is the most serious pest which feeds on the underside of the leaves making dusty coating and fine webs causing leaf yellowing, *vis-à-vis* plant stunting. In its serious attack, flowers are also invaded. Proper ventilation and watering may prevent this pest. Weeding and sanitation also reduce its population. Propargite is very effective in controlling adults, eggs and larvae.

Carnation fly maggots (*Hylemyia briennescens*) are tiny larvae of a fly which bore the leaves and tunnel down the growing tips of young plants, especially when grown outdoor, and plants die. Metasystox spraying at 0.2 per cent will control this pest. **Carnation caterpillar** or **bud borer** [*Helicoverpa armigera, Cacoecimorpha pronub* (**leafrollers**), and certain others] being polyphagous feeds on every part of the plant, and flower yield becomes nil. All such pests are controlled through regular fortnightly sprayings of Ripcord 0.1% or Polytrin. Larvae of **carnation *Tortrix* moth** (*Tortrix pronubata*) immediately after hatching fasten a few of the leaves together with silken threads and feed hungrily beneath this shelter. Damage is observed in the floral buds in which these bore and rest inside and around the growing points of the shoots and the leaves. The pest is observed during summers and autumns. Spraying of 0.5% at 10 day's intervals of 25% DDT + 10% parathion or Sevin + Kelthane controls this pest. **Thrips** [*Thrips tabaci, Haplothrips* sp. (Ranjith, 2013) , *Frankliniella tritici* and *F. occidentalis*] suck the sap from the leaves and flowers causing leaves becoming yellow, patchy (often with black specks) and crinkled and the flowers are streaked. In its severe attack, the plant growth is adversely affected, even otherwise the pink and dark red cultivars suffer most. Regular BHC fumigation or dusting the lower part of the plants with 5% BHC will control this pest. **Aphids** (*Myzus persicae*) distort the young plants by sucking the cell sap of the leaves and by leaving a sticky deposit on leaves and floral buds. These act as vectors and transmit carnation ring spot and carnation mosaic virus. Rogor 0.35% or demeton methyl 0.25% drenching once on young plants will save the plants up to 45 days, however, on older plants the spraying of both these chemicals are effective only for 1-2 weeks but drenching is not effective. Primor (primicarb) 50 w.p. sprayings thrice at fortnightly intervals using 0.125 or 0.24 g a.i./l is very effective. The maggots of **leaf miner** (*Liriomyza* sp.) and **white maggots** (*Hylemyia* sp.) bore into the leaves and growing tips of young plants destroying the flower quality (Ranjith, 2013). These are controlled through use of systemic insecticides.

References

Adriansen, E. 1985. Kemisk vaekstregulering. In: *Potteplanter I-Produktion, Metoder, Midler* (eds Christensen, O.V., A. Klougart, I.S. Pedersen and K. Wikesjö), pp. 142-162. GartnerINFO, København, Denmark.

Anonymous, 1972. Weed control in fall-planted annual flowers. Research Branch Report 1971, Canada Department of Agriculture, p. 337.

Anonymous, 1974. *Ann. Rep. German Pl. Prot. Serv.* (German), No. 21 Jg, pp. 221-223.

Anonymous, 1982. Cut flowers document AGRF/Wp. 127/ Review 1. United Nations Economic Commission for Europe. Geneva, Switzerland.

Anonymous, 1983. 'Knight' series mini-carnation popular pot plant. *Greenhouse Management and Production*, **2**(1): 14.

Anonymous, 1995. *Interpretation of Soil Analysis.* Greenhouse Soil Testing Service, Univ. of Minnesota, St. Paul, Minnesota.

Armitage, A.M. 1989. *Dianthus.* In: *Herbaceous Perennial Plants*, pp. 171-176. Varsity Press, Athens, Georgia.

Armitage, A.M. 1993. *Dianthus barbatus*. In: *Specialty Cut Flowers*, pp. 194-199. Varsity press/Timber Press, Portland, Oregon.

Arora, J.S., Amanpreet Kaur and G.S. Sidhu, 2002. Performance of carnation in polyhouse. *J. Ornam. Hort.* (N.S.), **5**(2): 58.

Bailey, L.H. 1942. *The Standard Cyclopedia of Horticulture* (vol. I), pp. 665-672. The Macmillan Company, New York.

Bailey, L.H. and E.Z. Bailey, 1976. *Dianthus*. In: *Hortus Third: A Concise Dictionary of Plants Cultivated in the United States and Canada*, pp. 376-380. Macmillan Publishing, New York.

Bailey, S. 1971. Perpetual Flowering Carnations. Faber and Faber, London.

Ben-Yephet, Y., M. Reuben, and D. Shtienberg, 1997. Complete resistance by carnation cultivars to *Fusarium* wilt induced by *Fusarium oxysporum* f. sp. *dianthi* race 2. *Plant Dis.*, **81**(7): 777-780.

Besemer, S.T. 1980. Carnation. In: *Introduction to Floriculture* (ed. Larson, R.A.), pp. 49-79. Academic Press, New York, USA.

Bhatt, N.R. 1989. Carnation. In: *Commercial Flowers* (eds Bose, T.K. and L.P. Yadav), pp. 355-416. Naya Prokash, Kolkata.

Blake, J. and T. Whitehead, 1961. The influence of extended day-length on flower production of carnations. *Exp. Hort.*, **5**: 1-8.

Bunt, A. 1973. Effect of season on carnation (*Dianthus caryophyllus*). II. Flower production. *J. Pl. Sci.*, **48**: 315-325.

Bunt, A.C. and K.E. Cockshull, 1985. *Dianthus caryophyllus*. In: *Handbook of Flowering*, **Vol II** (ed. Halevy, A.H.), pp. 433-440. CRC Press, Boca Raton, Florida, U.S.A.

Bunt, A.C., M.C. Powell and D.O. Chanter, 1981. Effects of shoot size, number of continuous light cycles and solar radiation on flower initiation in carnations. *Sci. Hort.*, **15**: 267-276.

Byrne, T.G. 1982. Lighting carnations. *Flower and Nursery Report* (Summer), pp. 3-4.

Cassells, A.C., C. Walsh and C Periappuram, 1993. Diplontic selection as a positive factor in determining the fitness of mutants of *Dianthus* 'Mystere' derived from X-ray irradiation of nodes in *in vitro* culture. *Euphytica*, **70**(3): 167-174.

Cockshull, K.E. 1985. *Dianthus*. In: *Handbook of Flowering*, vol. II (ed. Halevy, A.H.), pp. 430-432. CRC Press, Boca Raton, Florida.

Datta, S.K. 2002. Biotechnological approach of breeding ornamental crops. In: *Souvenir* based on *Nat. Symp. on Indian Floriculture in the New Millenium*, org. by ISOH, New Delhi, in Bangalore, on Feb. 25-26, pp. 51-56.

Dole, J.M. and H.F. Wilkins, 1988. University of Minnesota. Tissue analysis standards. *Minnesota St. Florist Bull.*, **37**(6): 10-13.

Dole, J.M. and H.F.Wilkins, 1999. Dianthus, Cut, pp.304-310; Dianthus, Pot, pp. 310-314; Dianthus, Sweet William, pp. 314-316. In: *Floriculture Principles and Species*. Prentice Hall, New Jersey.

Elena, K. and E.C. Tjamos, 1997. Soil solarization for the control of *Fusarium* wilt of greenhouse carnation. *Phytopathologica-Mediterranea*, **36**(2): 87-93.

Enoch, H.Z. and R.G. Hurd, 1977. Effects of light intensity, carbon dioxide concentration, and leaf temperature on gas exchange of spray carnation plants. *J. exp. Bot.*, **28**: 84-95.

Ferrero, F., A. Silvy, M. Jay and P. Ledeme, 1987. Experimental evidence proving the mutational origin of carnation cultivars obtained from the "Londorga" genotype. *Acta Hort.*, No. 216, pp. 205-214.

Gill, A.P.S. and M.S. Aulakh, 1993. Programming carnation production for export from India. *J. Ornam. Hort.*, **1**(1): 87-93.

Gimelli, F., G. Ginatta, A. Panizzi, R. Venturo and M. Buiatti, 1982. Genetic analysis of the flower colour in Mediterranean carnation cultivars. *Rivista-della-Ortoflorofrutticoltura-Italiana*, **66**(2): 97-102.

Goldsberry, K.L. 1979. *Calceolaria* and *Dianthus* for mini pot program. *Colorado Flower Grs Assocn*, No. 348, pp. 1-3.

Goldsberry, K.L. 1997. Dwarf carnation. In: *Tips on Growing Specialty Potted Crops* (eds Gaston, M.L., S.A. Carver, C.A. Irwin and R.A. Larson), pp. 47-53. Ohio Florists' Asscn, Columbus, Ohio.

Guba, E.F. 1952. Breeding carnations for resistance to *Fusarium* wilt. *Phytopath.*, **42**: 112-113.

Gupta, Y.C. and S.K. Bhattacharjee, 2002. Carnation. Tech Bull. No. 15. AICRP on Floriculture, ICAR, New Delhi.

Gupta, Deepak; Madhu Mita and N.S. Pathania, 1995. Evaluation of carnation germplasm for resistance against *Rhizoctonia* stem rot. *J. Ornam. Hort.*, **3**(1-2): 1-3.

Hanzel, R.J., K.S. Nelson and D.C. Kiplinger, 1955. Floral initiation and development in the carnation var. Northland. *Proc. Amer. hort. Soc.*, **65**: 455-462.

Harris, G.P. 1968. Photoperiodism in the glasshouse carnation. The effectiveness of different light sources in promoting flower initiation. *Ann. Bot.*, **32**: 187-197.

Harris, G.P. 1972. Intermittent illumination and the photoperiod control of flowering in carnation. *Ann. Bot.*, **36**: 345-352.

Harris, G.P. and M. Ashford, 1966. Promotion of flower initiation in the glasshouse carnation by continuous light. *J. hort.* Sci., **41**:397-406.

Harris, G.P. and A.P. Bradbeer, 1966. The flowering behavior of carnation cuttings produced in long and short days. *J. hort. Sci.*, **41**: 73-83.

Harris, G.P. and J.P. Harris, 1977. Effects of environment on flower initiation in carnation. *J. hort. Sci.*, **37**: 219-234.

Hay, R. and K.A. Beckett, 1971. *Dianthus*. In: *Reader's Digest Encyclopaedia of Garden Plants and Flowers*, pp. 222-228. The Reader's Digest Association Ltd., Berkeley Square, London.

Heins, R.D. and H.F. Wilkins, 1977. Influence of photoperiod on "Improved White Sim' carnation (*Dianthus caryophyllus* L.) branching and flowering. *Acta Hort.*, No. 71, pp. 69-74.

Heins, R.D., H.F. Wilkins and W.E. Healy, 1979. The effect of photoperiod on lateral shoot development in *Dianthus caryophyllus* L. cv. 'Improved White Sim'. *J. Amer. Soc. hort. Sci.*, **104**: 314-319.

Holcomb, E.J. and N.L. Messinger, 1986. Producing *Dianthus*. *Pennsylvania Flower Grs Asscn Bull.*, No. 365, pp. 5-6.

Horst, R.K. 1990. Sweet William. In: *Westcott's Plant Disease Handbook* (5th ed.), p. 840. Van Nostrand Reinhold, New York.

Jerzy, M. and M. Zalewsky, 2000. Effect of X- and gamma rays on *in vitro* adventitious bud production of pot carnation (*Dianthus gratianpolotanus* Vill.). *Revista Chapingo Serie Horticultura*, **6**(1): 49-52.

Koon, G. 1974. Timing carnations in Colorado with lights. *Colorado Fl. Gr. Assocn Bull.*, No. 284, pp. 3-4.

Kuack, W.L. 1986. Mini carnations go to pots in Colorado. *Greenhouse Gr.*, **4**(11): 90-92.

Kuehny, J.S., A. Painter and P.C. Branch, 2001. Plug source and growth retardants affect finish size of bedding plants. *HortSci.*, **35**(2): 321-323.

Kumar, P. Naveen; B. Singh; S.R. Voleti; and S.S. Sindhu, 1999. Influence of continuous illumination during vegetative growth on flowering response of carnation (*Dianthus caryophyllus* L.). *Indian J. Pl. Physiol.* (N.S.), **4**(3): 185-188.

Kushal Singh, P.J. Singh, J.S. Arora and R.P.S. Mann, 2002. Studies on refrigerated storage of carnation flowers. In: *Floriculture Research Trend in India* (eds Misra, R.L. and Sanyat Misra), pp. 303-304. ISOH, Division of Floriculture and Landscaping, I.A.R.I., New Delhi-110012.

Lamont, G.P. and M.A. O'Connell, 1986. An evaluation of pre-emergent herbicides in field-grown cut flowers. *Plant Prot. Quart.*, **1**(3): 95-100.

Langhans, R.W. 1961. *A Manual of the Culture, Insects and Diseases and Economics of Carnations.* New York St. and County agric. Ext. Services, New York St. Fl. Grw's Assocn, Ithaca, New York.

Laurie, A., D.C. Kiplinger aand K.S. Nelson, 1969. Carnation. In: *Commercial Flower Forcing*, pp. 262-282. McGraw- Hill, New York.

Linas, G. and D. Schypula, 1982. Minicarnations - a revolutionary flower for the American market. *Florist Review*, **171**(4436) 32-34.

Manaka, M. and D. Fruzynska Jozwiak, 1996. Biocontrol of greenhouse carnation *Fusarium* wilt with saprophytic forest soil fungi. *Folia Horticulturae*, **8**(1): 93-105.

Manisha Thakur, D.R. Sharma and N.S. Pathania, 2002. *In vitro* callus initiation and organogenesis from internodal segments in perpetual carnation. In: *Floriculture Research Trend in India* (eds Misra, R.L. and Sanyat Misra), pp. 164-168. ISOH, Division of Floriculture and Landscaping, I.A.R.I., New Delhi-110012.

Mehlquist, G.A.L. 1940. Flower colour inheritance in carnation. *Proc. Am.Soc. hort. Sci.*, **37**: 1019-1021.

Misra, R.L., Vinod Kumar and Sanyat Misra, 2002. Greenhouse Management of Ornamental Plants - An Overview. In : *Floriculture Research Trend in India*, pp. 13-18. Indian Society of Ornamental Horticulture, Division of Floriculture and Landscaping, I.A.R.I., New Delhi-110012.

Misra, Sanyat; Pragya; and R.L. Misra, 2006. Carnation. In: *Advances in Ornamental Horticulture*, **Vol. 2**: *Herbaceous Perennials and Shade loving Foliage Plants* (ed. Bhattacharjee, S.K.), pp. 67-80. Pointer Publishers, Jaipur.

Moe, R. 1983. Temperature and daylight responses in *Dianthus carthusianorum* cv. Napoleon III. *Acta Hort.*, No. 141, pp. 165-171.

Nakayama, M., M. Koshioka, H. Oshida, Y. Kan, F. Fukui, A. Koike and M. Yamaguchi, 2000. Cyclic malyl anthocyanins in *Dianthus caryophyllus*. *Phytochem.*, **55**(8): 937-939.

Nau, J. 1993. *Ball Culture Guide, The Encyclopedia of Seed Germination* (2nd ed.). Ball Publishing, Batavia, Illinois.

Neelofar Lolapuri and J.S. Arora, 1995. Response of Sim carnations (*Dianthus caryophyllus* L.) to photoperiodism (extended day length) and gibberellic acid (GA_3). *J. Ornam. Hort.*, **3**(1-2): 14-22.

Novak, F.J. 1991. *In vitro* mutation system for crop improvement. *Plant Mutation Breeding for Crop Improvement*, **2**: 327-342.

Nowak, J. and N.M. Rudnicki, 1990. *Postharvest Handling and Storage of Cut Flowers, Florist Greens and Potted Plants*. Timber Press, Portland, Oregon.

Onozaki, T., H. Ikeda, T. Yamaguchi, M. Himeno, J.A. Considine and H. Gibbs, 1998. Introduction of bacterial wilt (*Pseudomonas caryophylli*) resistance in *Dianthus* wild species to carnation. In: *Proc. Third int. Symp. on New Floricultural Crops* (ed. Considine, J.A.), held on Oct. 1-4, 1996, at Perth, Australia. *Acta Hort.*, No. 454, pp. 127-132.

Penaranda, L.A., H.J. Torres and G. Arbelaez, 1992. Chemical control of fairy ring spot of miniature carnation caused by *Heterosporium echinulatum*. *Acta Hort.*, No. 207, pp. 447-450.

Peterson, J.C. 1982. Monitoring and managing nutrition. Part IV- Foliar analysis. *Ohio Florists' Assocn Bull.*, No. 632.

Pirone, P.P., B.O. Dodge and H.W. Rickett, 1960. Diseases and Pests of Ornamental Plants. Constable and Company Limited, London.

Pokony, F.A. and J.R. Kamp, 1960. Photoperiodic control of growth and flowering of carnations. *Illinois St. Fl. Asscn. Florist Bull.*, No. 202, pp. 6-8.

Pokony, F.A. and J.R. Kamp, 1965. Influence of photoperiod on the rooting response of cuttings of carnations. *Proc. Amer. Soc. hort. Sci.*, **86**: 626-630.

Post, K. 1949. *Dianthus barbatus.* In: *Florists Crops Production and Marketing*, pp. 483-484. Orange Judd Publishing, New York.

Powell, M.C. 1979. Observations on the growth of carnation (*Dianthus caryophyllus* L.) in natural and long days. *Ann. Bot.*, **43**: 579-591.

Ramesh Kumar, Kartar Singh, B.S. Reddy and K. Singh, 2002. Effect of planting time, photoperiod, GA_3 and pinching on carnation. *J. Ornam. Hort.* (N.S.), **5**(2): 20-23.

Ranjith, A.M. 2013. Carnation. In: *Identification & Management of Horticultural Pests*, pp. 129-130. New India Publishing Agency, Pitam Pura, New Delhi-110 088.

Richter, A. and W.R. Singleton, 1955. The effect of chronic gamma radiation on the production of somatic mutation in carnations. *Proc. Natl. Acad. Sci.,* USA, **45**(5): 295-300.

Roest, S. and G.S. Bokelmann, 1981. Vegetative propagation of carnation through *in vitro* shoot development. *Scintia Hort.*, **14**(4): 357-356.

Rugge, V.C., D.V. Delvadia, P.B. Bhor and V. Singh, 2010. Effect of foliar application of gibberellic acid on carnation (*Dianthus caryophyllus* Linn.) cultivars growing under polyhouse in south Saurashtra condition. *Haryana J. hort. Sci.*, **39**(3-4): 279.

Sacalis, J.M. 1993. *Dianthus caryophyllus*. In: *Cut Flowers, Prolonging Freshness* (2nd ed., ed. J.L. Seals), pp. 52-54. Ball Publishing, Batavia, Illinois, USA.

Salinger, J.P. 1985. Carnation. In: *Commercial Flower Growing* , pp. 151-161. Butterworths Horticultural Books, Wellington, New Zealand.

Samata, Y., Y. Tsuyuki, K. Inazu and K. Iizuka, 1979. Induction of flower colour changes in carnation (*Dianthus caryophyllus*) by gamma ray irradiation. *Bull. Fac. Agric. Tamagawa Univ.*, Japan, No. 19, pp. 29-41.

Sarafi, A. and M. Mahadavian, 1979. Comparison between diploid and tetraploid plants of an Iranian carnation variety. In*:* Eucarpia Meet. on Genetics and Plant Breeding of Carnation and Gerbera (April 1978), Italy.

Sawwan, J. 1998. Carnation production and quality of three cultivars as affected by plant density. *Divasat Agric. Sciences*, **25**(3): 375-378.

Sawwan, J. and M. Samawi, 2000. Effect of pinching on plastic house grown spray-type carnation yield and yield distribution. *Divasat Agric. Sciences,* **27**(1): 106-111.

Sehgal, O.P. and N.S. Pathania, 2001. Carnation. In:*Handbook of Horticulture* (ed. Chadha, K.L.), pp. 548-554. ICAR, New Delhi.

Shewell-Cooper, W.E. 1953. *Shewell-Cooper's Guide to Carnations*. English Universities Press Limited, London.

Simard, M.H., N.M. Ferriere and A. Silvy, 1992. Variants of carnation (*Dianathus caryophyllus* L.) obtained by organogenesis from irradiated petals. *Plant Cell Tissue Organ Cult.*, **29**: 37-42.

Singh, K.P., B. Singh, S.P.S. Raghava and C.S. Kalia, 2000. Induced flower colour mutations in carnation through *in vitro* application of chemical mutagen. *Indian J. Gen. Pl. Br.*, **60**(4): 535- 539.

Sonvir Sooch, Ramesh Kumar and Manreech Sooch, 2002. Effect of NAA and kinetin on plant growth and flower production in carnation. In: *Floriculture Research Trend in India* (eds Misra, R.L. and Sanyat Misra), pp. 295-297. ISOH, Division of Floriculture and Landscaping, I.A.R.I., New Delhi-110012.

Sparnaay, L.D. 1974. Progress report on clonal selection in carnation after irradiation. In: *Meeting of the Mutation Breeding Contact Group* (ed. C. Broertjes). External Report No. 23, Assoc. Euratom-ITAL, Wageningen, pp. 28-31.

Sparnaay, L.D. and J.F. Demmink, 1971. Mutation breeding as a means of improving productivity in commercial carnation cultivars. In: *Eucarpia Meeting on Ornamentals*. Inst. Hort. and Plant Breed., Wageningen, pp. 38-44.

Sparnaay, L. D., J.F. Demmink and F. Gaeretsen, 1974. Clonal selection in carnation. In: *Eucarpia Meeting on Ornamentals.* Frejus Inst. Hort. and Plant Breed, Wageningen, pp. 39-50.

Thia Segers, 1987. The development of inter-specific carnation hybrids. *Acta Hort.*, No. 216, pp. 373-375.

Verma, V.K. and Y.D. Sharma, 2002. Effect of planting dates and nitrogen levels on carnation under protected conditions. In: *Floriculture Research Trend in India* (eds Misra, R.L. and Sanyat Misra), pp. 154-158. ISOH, Division of Floriculture and Landscaping, I.A.R.I., New Delhi-110012.

Verma, V.K., Y.D. Sharma and Y.C. Gupta, 2002. Effect of planting dates and pinching methods on carnation under protected conditions. In: *Floriculture Research*

Trend in India (eds Misra, R.L. and Sanyat Misra), pp. 159-163. ISOH, Division of Floriculture and Landscaping, I.A.R.I., New Delhi.

Verma, V.K., Y.D. Sharma and Y.C. Gupta, 2002. Effect of holding solutions under different planting dates and nitrogen levels on vase life of cut carnation. In: *Floriculture Research Trend in India* (eds Misra, R.L. and Sanyat Misra), pp. 169-172. ISOH, Division of Floriculture and Landscaping, I.A.R.I., New Delhi-110012.

Wandas, F. 1989. *Dianthus carthusianorum.* In: *Utplanteringsväxtert Trädgrad,* No. 314, pp. 30-32. Swedish Univers. of agric. Sci., Alnarp, Sweden.

Waterschoot, H.E. 1957. Effects of temperature and day length on flowering in *Dianthus barbatus* L. *Verhandelinger Koninklijke Akademie van Wetenschappen Series C,* **60**: 318-323.

9

Eranthemum

B. Sathyanarayana Reddy, S.Y. Chandrashekar and S.K. Nataraj

Family: **Acanthaceae**

Common names: Blue eranthemum/Blue sage (*Eranthemum pulchellum*), Blue eranthemum/Rosy eranthemum (*E. roseum*), Eranthemum, Eldorado/ False eranthemum/Golden eldorado/ Golden net bush/ Golden pseuderanthemum/Speckled eranthemum/Yellow vein eranthemum (*Pseudoeranthemum reticulatum*)

Introduction and Origin

A group of 30 species of shrubs or sub-shrubs which grow in shady and semi-shady places where one of them has long been grown for its pure blue winter flowers. *Eranthemum* is an ancient Greek name used for this genus by Linnaeus. *Eranthemum* consists of two words *errano's* meaning 'lovely' and *anthemon* meaning 'flower'. These shrubs are grown in tropical or sub-tropical climates mainly for their colourful foliage except *Eranthemum laxiflorum* which bears conspicuous flowers. *Pseudoeranthemum repandum* was first described in 1786 under the name *Justitia repanda*. It has since been known by other names and ascribed to different genera (*Eranthemum* and *Anthacanthus*). *Pseuderanthemum* has several species listed as cultivated one (Anderson *et al.*, 1998). *Eranthemum* is used for making hedges, as screens or as pot plants, *vis-à-vis* as specimen plant, in flower beds and in borders. It comes up very well both in open sunlight under moderate climate as well in partial shade under dry regions. This genus is native to India, Polynesia (Hawaii and Somoa) and Melanesia (Solomon Islands, Fiji and New Caledonia).

Ethano-botanically, root of *Eranthemum roseum* (Vahl) R.Br. boiled in milk is popular remedy for leucorrhoea. Roots are also given to pregnant cattle to promote growth of foetus. Antimicrobial activity of the roots of *Eranthemum roseum* were tested against different bacteria and fungi by cup-plate diffusion method and results obtained showed strong activity of the petroleum ether extract of the roots against bacteria and fungi as

The plants should be moderately **fertilized** and the feeding is completely discontinued when active growth stops in winter. Use of a balanced fertilizer such as 15-15-15 of NPK or a blooming fertilizer such as 7-9-5 NPK at the rate of 25 grammes per square metre every fortnight is quite suitable. Before **watering** the soil should be brought to a state of visual dryness. While watering, the soil is fully saturated until a little water runs out of the bottom of the pot. Growing in a clay pot will help maintain a healthy root system. The foliage can show staining on the leaves if watered with ice cold water.

Growth and Development

The plant growth must fit in standard pot sizes and produce harmonious and uniform plants in a reasonably short time. The growth retardation is often necessary in case of *Pseuderanthemum repandum.* Wide range of compounds are available such as flurprimidol and uniconazol which are more effective than paclobutrazol. Another compound, daminozide is not efficient since it does not reduce desired plant height even at a concentration of 1%. Internodal lengths and number of leaves were found reduced significantly in all treatments with retardants. This is perhaps not desirable but has also been reported with *Epipremnum* (Anderson *et al.,* 1998). *Pseuderanthemum repandum* does not have a certain day length requirement, rather than the irradiance must exceed a certain level for flowering to occur. Probably an interaction occurs between day length and irradiance where irradiance plays the major role. The 'ripeness to flower' stage caused no problems since flowering was observed in small as well as large plants (Andersen *et al.,* 1998).

Post Harvest

For successful commercial production a plant must possess good keeping quality, i.e. should be able to withstand stresses in transport, trade and home. Many of these stresses result in increased endogenous ethylene production followed by ethylene induced leaf, flower or bud abscission . *Pseuderanthemum repandum* keeps well under simulated interior conditions. Single flowers have a life span of 4 to 5 days and buds are produced in great numbers. When these buds open they may develop into smaller and paler flowers than those developed in the greenhouse. Flowers and to a lesser extent the buds though not the leaves are sensitive to prolonged ethylene exposure. Under these conditions bud drop can be prevented by STS pretreatment, which also may prolong the display life in ethylene-free storage (Andersen *et al.,* 1998).

Insect-pests and Diseases

Among the insect-pests, white flies prefer underside of leaves to feed and breed. Whitefly infested plants are when disturbed a cloud of fleeing insects are seen. They also produce a sweet substance called honeydew which leads to an unattractive black surface fungal growth called sooty mould. This pest can be controlled by using yellow sticky cards and also by spraying systemic insecticides. The root systems are strong and all almost problem-free. They are prone to exudates (the excretion of excess fluids) through the surface of the leaves, producing tiny, raised beads along stems and leaf surfaces. This can be mistaken for mites. It is unsightly, but not harmful to the plant. The spider mites also damage the plants. Among the diseases, fungal leaf spots occur though not common. Any systemic fungicide such as Carbendazim @ 1 g/ 1 of water will control the disease.

References

Andersen, A.S., U. Brekman and L.S. Moller, 1998. *Pseuderanthemum repandum*: potential as a new flowering potted plant. *Proc. III int. Symp. New Flor. Crops* (*Acta Hort.*), Perth, Australia, p. 454.

Ashish S. Jain, Sanjay J. Surana, S.B. Gokhale, A.U. Tatiya and R. C. Bothara, 2006. Antimicrobial properties of *Eranthemum roseum* (Vahl) R.Br. *Iranian. Phar. Res.,* **6** (2): 131-133.

Patil, P.H. and Surana S.J, 2010. Toxicological Evaluations of *Eranthemum roseum* R. Br Linn Root extracts, *Interl. J. Pharm Tech Res.,* **2**(1):753-759.

Surainpong and Te-Chato, 2010. *In vitro* propagation of a woody ornamental *Eranthemum nervosum* (Vahl) R. Br. *J. agr. Tech.,* **6**(3): 579-588.

10

Ervatamia divaricata

B.K.Banerji, A.K.Dwivedi, Atul Batra and Sanyat Misra

Formerly: ***Tabernaemontana coronaria* and *T. divaricata***

Family: **Apocynaceae**

Common names: Adam's apple, Banana bush (*T. pandacaqui*), Carnation of India, Common milkweed (*T. citrifolia*), Coffee rose/Crepe jasmine/Crepe gardenia/Nero's crown/Paper gardenia/Pinwheel flower/Spiral gardenia (*T. coronaria*, syn. *T. divaricata, Ervatamia divaricata*), East Indian roseway, Forest toad tree (*T. ventricosa*), Samoan gardenia (*T. africana*, syn. *T. grandiflora, Conopharyngia chippii, C. longiflora*), Toad tree (*T. elegans*), White milkwood (*T. alba*); *Chandni* in Hindi, *Nandiar vattai* in Tamil, *Sagar* in Gujarati, *Ananta Tagar* in Marathi and *Nanta Phool* in Konkani

Introduction

Though derivation of *Ervatamia* is not known but *Tabernaemontana* is named so due to Jakob Theodor von Bergzaben (*d.* 1590), physician and botanist at Heidelberg who Latinized his name to *J.T. Tabernaemontanus*. Under *Tabernaemontanus ventricosa* (*ventricosa* meaning to having a swelling on one side, being pot-bellied and probably referring to the shape of the fruits of the plant) under which pseudonym he wrote the classic herbal *Neuw Kreuterbuch*. The species name *divaricata* means at an obtuse angle due to its branches growing almost horizontal. It has a pan-tropical distribution. It is native to North India, other parts of Asia, especially Yunnan, throughout tropical America and Tropical and subtropical Africa including Madagascar, and is widely grown for its evergreen shrubs having shining leaves and flowering almost perpetually with pure white (central region yellowish) fragrant flowers (single and double) in frost-free areas around the world. Its double forms appear as *Gardenia jasminoides*. Due to fragrance of its flowers being most intense at evenings when it flowers, it is also known as 'Crepe jasmine'. It is a very fast growing shrub most suitable for landscape use, especially at filtered locations, for planting near the porch or patio, as a specimen plant in the corner of a lawn, as a hedge plant, for mass planting and for loose flower production in bulk.

Plate-1
Cultivars of tabernaemontana

a. Tree
b. Shrub
c. Variegated shrub
d. Dwarf variegated
e. Shrub with multipetalled flower

Plate-2
Close up view of flowers of different cultivars of tabernaemontana

a. Tree
b. Shrub
c. Variegated shrub
d. Dwarf variegated
e. Shrub with multipetalled flower

T. stapfiana contains plant metabolites such as alkaloids, flavonoids, coumarins and saponins well known for antimicrobial activity and Keiyo community in Rift Valley of Kenya use its roots and stem barks in case of some abdominal ailments, dermatitis, sexually transmitted infections and upper respiratory tract infections. Coumarins and flavonoids, *in vitro*, inhibit certain fungal and other microbial pathogens. Likewise, alkaloids and saponins also inhibit certain fungal infections. Apart from *T. stapfiana, T. cymosa* and *T. pachysiphon* also show significant antifungal activities. Ruttoh *et al.* (2009) found stem and root bark extracts inhibiting certain bacterial growth such as *Staphytococcus aureus*. *T. sananho* preparations have been found useful against eye injuries while *T. heterophylla* is used to treat dementia (Pande and Rao, 1990). Ayahuasca Jonathan (1995) mentions *T. undulata* containing ibogaine and *T. africana* containing voacangine which are used in psychedelic drinks, even by Krahô Indians in Brazil (Rodrigues and Carlini, 2006).

Botany

Different species of *Tabernaemontana*, from time to time, has been reported from various parts of the world. Crepe jasmine (*T. divaricata*) is a glabrous bushy and spreading (1.0-2.4 m) shrub dichotomously branched, exuding milky latex (hence its common name milkwood) and grows 1.0-3.0 m tall. The leaves are thin-textured or coriaceous, mostly shiny and waxy, evergreen, opposite, entire, petiolate (though not prominent and seldom sessile), 3-15 cm long having acute apex and with milky sap. It is a multi-branched plant, branches being 3 to 10 depending upon the species grown or on the type and stage of pruning, and where branches tend to grow almost parallel to the ground. Crepe jasmine blooms in spring but flowers may appear sporadically whole the year, with profusion of blooms appearing during summer and rainy seasons, however, in winter in the subtropical climate it blooms shyly. The waxy salver-form flowers appear in a forked cymose inflorescence and in small pinwheel clusters of 27 to 35, opening one by one on the stem tips, and are 1-5 cm in diameter, pedicellate (about 1.7 cm), calyx usually small and 5-parted, corolla salver-form, throats lacking scales, lobes contorted, white, 5-petalled and nocturnally fragrant. Petal margin at left slightly fringed, basal end gamopetalous and forms a small tube and the petal aestivation is imbricate. Anthers are sagittate or cordate, epipetalous, lobes very small and yellow, partly sterile and present at the distal end of the floral tube, and filaments very short. Fruits bicarpellary having globose, oblong, ovoid or recurved-reniform follicles, smooth or 3-ribbed, free or partly to completely fused, dehiscent or not, and fleshy with thick wall. Seeds are obliquely ellipsoid and deeply grooved to one side. Chromosome association and male sterlity in a single flowering *T. divaricata* due to seasonal variation has been found as 2n=22 chromosomes. However, the comparative studies of diploid, triploid, tetraploid and hexaploid forms of *T. divaricata* with special reference to morphology at meiosis shows laggards, unequal distribution, bivalents, bridges and polyspory.

Species and Varieties

Though some 100-150 species are reported under *Tabernaemontana* but only a few are of some significance horticulturally. The important ones are *T. africana* (Africa), i.e. Tropical West Africa and Senegal to Ghana), *T. alba* (C. America), *T. amygdalifolia* (S. Mexico, CS America), *T. bovina* (China to Thailand), *T. bufalina* (China to Thailand), *T. calcarea* (Madagascar), *T. capuronii* (Madagascar), *T. catharinensis* (S. America), *T. cerifera* (New Caledonia) *T. ciliata* (Madagascar), *T. citrifolia* (Caribbean), *T. coffeooides* (Madagascar), *T. crassa* (Madagascar), *T. crassifolia* (Madagascar), *T. debrayi* (Madagascar), *T. cymosa* (South Africa), *T. divaricata* (India to Thailand), *T. elegans* (South Africa to Somalia), *T. eusepala* (Madagascar), *T eusepaloides* (Madagascar), *T. humboldtii* (Madagascar), *T. mocquerysii* (Madagascar), *T. pachysiphon* (syn. *T. holstii*; Tropical East Africa), *T. pandacaqui* (SE Asia, Australasia, Western Pacific Ocean Islands), *T. phymata* (Madagascar), *T. retusa* (Madagascar), *T. rostrata* (SE Asia), *T. sambiranensis* (Madagascar), *T. sessilifolia* (Madagascar), *T. stapfiana* (Tropical East Africa), *T. stellata* (Madagascar), *T. ventricosa* (Cameroon, South Africa), etc. Various species such as *T. apoda, T. antheonycta, T. cordata, T. coronaria* (syn. *T. divaricata*), *T. corymbosa, T. cumata, T. panamensis, T. fuchsiaefolia, T. gamblei, T. heterophylla, T. heyneana, T. longipes, T. muricata, T. ochroleuca, T. oppositifolia, T. orientalis, T. ovalifolia, T. persicariaefolia, T. polyneura, T. remota, T. stenosiphon, T. thurstonii, T. undulata, T. wullschlaegelii*, etc. are of unknown origin.

Five cultivars, *viz.* one tree and other four shrubby ones were studied at NBRI, Lucknow and it was recorded that shrubby types consisted of one usual type of shrub, one variegated type of shrub, one dwarf compact type which is sometimes mentioned as *Ervatamia divaricata* 'Coronaria' and this also has a variegated type, and fourth usual shrub type with multi-petalled flowers and all these are being maintained at NBRI, Lucknow. The species *Ervatamia divaricata* is a five-petalled almost sterile species having pinwheel type white petals. Variegated type (*E. divaricata* 'Variegata') originated somewhere in South Asia which has a variously shaded mixed colouration of white, yellow and green on the leaves but these are shy bloomer. Under variegated types there are varieties such as 'Gold', 'Silver Rain', etc. *Ervatamia divaricata* 'Flore Pleno' or 'Plena' has double flowers with ruffled wavy petals and larger leaves. Another multi-petal variety is also described as var. 'Grandifolia'. The second species is *Tabernaemontana elegans* where leaves are though straight but curving up near the margin and flowers appear in lax corymbs which are fragrant. *T. africana* has elliptic dark green and long leaves (up to 15 cm) and pinwheel-shaped 10-cm across sweet-scented flowers emitting its fragrance all the 24 hours, and its var. 'Compacta' from Hawaii is quite dwarf and compact. *T. stapfiana* is a plant of high medicinal values.

Propagation and Cultural Practices

It can bear as low as 2 ^{0}C **temperature** only for a few days but it prefers tropical and sub-tropical climatic conditions as prevailing in whole of the southern, eastern, western and subtropical northern India though sometimes during winters of December and January temperatures fall down below 0 ^{0}C which is also tolerated provided it does not persist for long. Lower temperature causes yellowing and subsequent falling of the leaves and the frosts burn the leaves and tender parts. Though it is a versatile plant surviving rigours of the climates, i.e. dry or humid conditions but prolonged dry periods cause them requiring watering for survival and to maintain continuous flowering. These come up well in filtered as well as **sunny** situations with compact growth but complete shade is neither good for their health nor for flowering. Excessive shade may cause the plants becoming leggy with a few leaves and that too pale in colour. **Soil** should be well-drained but water-retentive and fertile sandy- to clay-loam with a pH range of 5.5 to 7.5. Planting is either carried out during monsoon or at the onset of spring at a distance of 1.2 metres in case of taller plants and 90 cm in case of dwarf ones. *Ervatamia divaricata* is almost sterile so it is **propagated** only through hardwood and semi-hardwood stem cuttings. Cuttings with more than five nodes should be taken for propagation, inserting two nodes in the propagating medium and at least three nodes above. Before inserting into the medium these are dipped in IBA rooting hormone and kept under high humidity conditions or in mist chamber for better rooting and sprouting where these root from 2 to 6 weeks. It would be always better to multiply these through cuttings when monsoon has started as at this time natural humidity is quite high. Though there is no any strict schedule for its **watering** and **fertilization** for better flowering since it is a most common plant in India giving flowers perpetually for years without any care but watering during dry periods and manure or fertilizer application before spring will boost the plants for better growth and flower production. Immediately after planting it should be drenched properly to settle the roots and then after every 3 to 4 days until the plants are fully established. Once established properly, it may remain rainfed for years but the pot-planted ones are watered regularly otherwise the plants may die during dry periods. Young plants require extra phosphorus to encourage good root development. In case of flowering plants, fertilizers will improve flowering and flower yield. At the time of planting if sufficient quantity of well decomposed cow dung manure or farmyard manure (10 kg) has been added to each pit, there is no any further need of fertilizer for whole of the year. Per pit each of N, P and K at the rate of 40, 40 and 20 g, respectively, should be used at the time of soil filling after mixing these thoroughly with the soil if no manure has been used. In case of potted plants, at the beginning of spring these should be repotted after fortifying the soil with 5 kg of FYM and 40 g P_2O_5, 20 g N and 10 g K_2O. Fertilizers that are high in N will promote excessive vegetative growth but on the cost of flowering. It is best to avoid fertilizing late in the growing season.

For the topiaries and in case of hedges, the plants are **sheared** to conform to their specific shape. This is normally done only once in a year. **Pruning** is generally carried out to maintain proper shape of the plants to conform the matching look with the surrounding and for production of more flowers of better quality. Over-grown or old neglected plants are rejuvenated through pruning by removing old branches or for reduction of the size of the shrub for looking more beautiful. In any case, not more than one-third of a plant is removed at a time. Congestion of twigs should be avoided by removing straggling and precocious branchings whether growing inside or outside. Cutting back of canes at various heights will provide ample space for development of more floral buds and the landscape beauty will also be enhanced. **Thinning** is carried out to permit more light and air circulation inside the plant and here the branches are removed either from the main trunk or a few sub-branches in each main branch. This ensures proper health of the plant. **Pinching** is removal of the stem tips of a young plant to promote enormous branching for more flower production and to give a balanced and attractive shape to the plant.

Insect-Pests and Diseases

The nymphs and adult **aphids** belonging to various genera are small, soft-bodied and slow-moving creatures that suck plant sap from tender parts and can be found clustering around the shoot tips. They are green to brown to black and may have wings, and multiply quickly through parthenogenesis, especially during springs to autumn though during summer their population is quite less. They attack a wide range of plant species causing stunting and deformed leaves & buds. They also transmit viruses from one plant to other while feeding on them. However, aphids do produce a sweet substance called honeydew which attracts ants and which also leads to an unattractive black surface growth, sooty mould. Field sanitation, jet water spraying on infested plants and use of lady bird beetles & lacewings will keep the aphids under check. Spraying of nicotine concoction will also control these pests. **Scales** (*Coccus hesperidum*) are insects related to mealy bugs that can be a problem on a wide variety of indoor and outdoor ornamentals. Young scales crawl until they find a good feeding site. The adult females remain on a spot protected by its hard shell layer. They appear as bumps, often on the lower side of leaves and remain there sucking cell sap which causes yellowing of the foliage and their subsequent drop, *vis-à-vis* exude honeydew which attracts ants and causes sooty mould. Once established they are hard to control. Isolate infested plants away from those that are not infested. Diseased twig should be immediately pruned to eradicate the huge population from the growing plants. Confidor 0.01% fortnightly spraying will control scales as well as mealy bugs.

Leaf spots of *E. divaricata* is caused by *Phytophthora citrophthora as* reported by Sushma and Sharma (1999) from Jabalpur which spreads through garden tools, rains and insects. Brown to black spots and patches may be either ragged or circular with a water-soaked or yellow-edged appearance. *Pseudocercospora* spp. causes brown leaf spots as reported by Yen and Lim (1980). **Crepe jasmine rust** (*Uredo manilensis*) was first reported by Perez *et al.* (2008) in the Western Hemisphere and Key West and Miami (USA) are worst affected areas in the U.S. Both the young and old leaves become severely affected due to formation of subepidermal chlorotic flecks which afterwords turn into necrotic spots with orange to reddish-brown colouration. Removal of infected leaves, *vis-à-vis* pruning of infected branches followed by burning these along with the plant debris will keep these diseases under check. Overhead irrigation should be avoided. Spraying with maneb or zineb will control the disease. Benomyl, copper-based fungicides such as copper oxychloride, chlorothalonil and mancozeb orthio phanate methyl may be used when disease has been noticed first. **Wilt** is caused by *Fusarium oxysporum* f. sp. *tabernaemontanae* as reported by Pande and Rao (1990) from Pune. Other hosts of this pathogen are *Catharanthus roseus* and *Thevetia neriifolia.* In checking the incidence of this disease only healthy plants should be planted and as it is soil-borne, soil should also be treated when planting in infected or suspected soil. **Virus** infection on the plants of *Tabernaemontana divaricata* was recorded first in Israel by Cohen *et al.* (2006). The symptoms varied widely and included chlorotic ring spots and banding, oak-leaf pattern and mosaic. At the end of the winter, large yellow spots were found appearing which later on became necrotic as are clearly seen as shot holes on fully expanded leaves. Preliminary analysis suggested that the disease was associated with a tobacco mosaic virus. The disease spreads through cuttings taken from such plants, therefore, such plants should be uprooted and burnt.

References

Cohen, J., A. Rosner, S. Kagan, M. Lampel, L. Maslenin, M. Zeidan and A. Gera, 2006. A new viral disease in *Tabernaemontana divaricata* associated with tobacco mild green mosaic virus (TMGMV). *Ann. Appl. Biol.*, **138** (2): 153-159.

Jonathan, O. 1995. *Ayahuasca Additive Plants.* In: *Ayahuasca Analogues,* Pangaean Entheogens.

Pande, Alka and V.G. Rao, 1990. Wilt disease of *Tabernaemontana coronaria* (T. *divaricata*) caused by *Fusarium oxysporum* f. sp. *tabernaemontanae* reported from Pune, Maharashtra (other hosts: *Catharanthus roseus* and *Thevetia neriifolia*). *Biovigyanam,* **16** (1): 58-61.

Perez, J.M., A.J. Palmateer, R.C. Ploetz, R.A. Cating and J.M. Lynn, 2008. First report of *Uredo manilensis* in the Western Hemisphere. *Plant Dis.*, **92**(12): 1711.

Rodrigues, Eliana and E.A. Carlini, 2006. Plants with possible psychoactive effects used by the Krahô Indians, Brazil. *Revista Brasileira de Psiquiatria,* **28**(4): 277-282.

Ruttoh, E., K. Paul, K. Tarus, C. Christine, Alex Bii, K. Machocho, K. Lucas, K. Karimie and Okemo Paul, 2009. Antibacterial activity of *Tabernaemontana stapfiana* Britten (Apocynaceae). *Extracts Afr. J.* (Tradit. Complement. Altern Med.), **6**(2): 186-194.

Sushma, Nema and N.D. Sharma, 1999. A new leaf blight disease of *Tabernaemontana coronaria. J. mycopath. Res.*, **37**(1): 45-46.

Yen, J.M. and G. Lim, 1980. *Cercospora* and allied genera of Singapore and the Malay Peninsula. *Garden's Bulletin Singapore,* No. 33, pp. 151-263.

11

Gaillardia

Ajit Kumar, Manju Rana, Manish Kapoor, Sanyat Misra*, Sanjeev Kumar Sharma and Satish Chand

Family: **Asteraceae**

Common names: Amber wheels (*G. aristata* Pursh.), Blanket flower/Firewheel/Indian blanket/Sundance (*G. pulchella*)

Introduction and Origin

The generic name *Gaillardia* was proposed in honour of M. Gaillard de Charentonneau, a French patron of botany cum magistrate there. With bright and colourful daisy-like flowers which are usually multi-coloured with yellow, red and orange being most common, it is no wonder that this flower has earned a spot on the heart of gardeners everywhere. Though it is a beautiful summer-flowering annual (sometimes perennial) but can be grown all the year round even in the winter months. Plants are bushy, free-flowering and bloom continuously for a long time under wide range of climatic conditions. Commonly known as blanket flower, gaillardias are hardy, sun-loving and easy to grow and maintain, they deserve a place in every garden.

Gaillardias are conspicuous for profusion and duration of flowers. The plants are grown as annuals, biennials and perennials. A constant succession is produced all summer until very late into the autumn. Besides their use as border or bedding plants, they are good as cut flowers, as they last well in water (Bailey and Bailey, 1976). The flowers are used as loose flowers as well as for flower arrangements, both as excellent cut flowers and dried with the outer petals removed. *Gaillardia aristata* is suitable for dry land and requires low maintenance. Apart from its uses as landscape plant, *G. pulchella* is also useful as an edging plant along the walks or driveways, as an accent in perennial garden or in front of a shrubbery border and suitable for growing along the beaches right on the dunes.

Gaillardia spp. is as effective as *Tagetes* in nematicidal properties, when grown as catch crop or green manure

(Gommers, 1972). Duschatzky *et al.* (2003) identified 33 components of *G. megapotamica* var. *radiata* essential oil of which beta-pinene (24.2%), beta-ocimene (16.5%), alpha-pinene (7.7%), limonene (7.5%) and beta-caryophyllene (6.7%) constitute the major portion. *G. pulchella* could be used to manage the nematode *Meloidogyne incognita* (causing root gall) as a rotation crop, a co-planted crop or as a soil amendment for control of root-knot nematode (Tsay *et al.*, 2004).

Gaillardia is native of North and South America. Of the total species of *Gaillardia*, 12 are native to North America, though rest originate in South America mainly Argentina. *Gaillardia* x *grandiflora*, short lived perennial plant is a tetraploid hybrid resulting from a cross between perennial gaillardia (*G. aristata*) and annual gaillardia (*G. pulchella*). It arguably inherited its perennial habit from the former and its long flowering habit and rapid growth rate from the latter. Gaillardia is sometimes commonly called blanket flower in probable reference to the resemblance of its rich and warm flower colours and patterns to blankets woven by native Americans. However, some authorities suggest that the name blanket flower was originally in reference to the habit of wild species plants to form colonies which blanketed the ground.

Botany, Genetics and Breeding

It consists of about 30 species of annual, biennial and perennial herbs, out of which *G. pulchella*, *G. aristata* and *G.* x *grandiflora* (*G. aristata* x *G. pulchella*) are of horticultural importance.Gaillardias, the bushy herbs are tufted but grow erect with spreading branches and bear lanceolate to spathulate hairy foliage, leaves are basal as well as on the stems, alternate, toothed and spotted. Flower heads are daisy-like, solitary, terminal, very showy, up to 10 cm (usually 5-7 cm) in diameter, ray florets broad, 3-lobed, yellow and red, and always neutral, certain varieties bearing even double flowers with tubular-shaped petals that hide the central disc completely, disc florets purple and fertile, involucres broad and the hairy bracts are formed in 2-3 series. The flowers last for several weeks. The gametophytic chromosome number is 36, i.e. 2n = 36 in *Gaillardia aristata* as reported by Atwood (1937). *G. pulchella* is diploid (2n = 36) and *G. aristata* is both diploid and tetraploid 2n=36, 72 (Pal, 2003). Two new species of gaillardia, *viz. G. gypsophila* and *G. powellii* have been described by Turner (1972). *G. gypsophila* is a diploid with 2n = 17 and is most closely related to the Mexican species, *G. comosa* which is tetraploid with 2n = 34. *G. powellii* appears to be most closely related to *G. multiceps.* The latter is a tetraploid or hexaploid (2n = 34 and 51) while *G. powellii* is diploid with 2n = 17. The latter two species are clearly related to the diploid *G. pinnatifida* (2n = 17).

Genetic factors strongly influence growth and flowering in gaillardia. High heritability values were reported for all the characteristics in *G. pulchella* (Hegde and Gopinath, 2003a). It was observed for number of leaves (99.8%), height of plants (99.7%) and dry weight of flowers (99.5%). However, low heritability was found for yield of seeds (41.1%). Hegde and Gopinath (2003b) observed significant positive correlation between yield of flowers per plant and number of leaves and branches per plant, mainly due to their direct effects. Other traits like length of stalks, dry weight per plant and height of plants had non-significant correlation with the yield of flowers per plant.

Species and Varieties

G. pulchella Foug. is a short lived hairy annual. The branching stem of this plant is hairy and upright, growing to 60 cm tall. The leaves are alternate, mostly basal, 4-8 cm long, with edges smooth to coarsely toothed or lobed. Flowers resemble daisies and are single, semi-double or double and held on long stems above the foliage. Appearing throughout the summer, 5-7 cm wide flowers are available in yellow, orange, red or bicolours and make excellent long lasting cut flowers. The central disc florets of the flower head tend to be more red-violet, with the outer ray florets being yellow. The brilliant blossoms are quite attractive to butterflies and these annual flowers will normally reseed themselves quite readily. This is largely represented by cultivation of variety 'Lorenziana' which has numerous and enlarged funnel shaped disc florets of red, yellow or bicoloured type. Other varieties include 'Yellow Sun' (bright yellow), 'Red Plume' (vibrant dark red) , 'Gaity Double Mixed' and 'Double Tetra Fiesta' (tetraploid).

G. aristata Pursh. is a hairy perennial. Leaves oblanceolate to lanceolate, entire to toothed or lobed, sessile or shortly petioled and densely hairy. It also has large daisy-like flowers but they have bright yellow petals with orange-red centres. One cultivar recently introduced is 'Fan Fare'. The flowers of this cultivar are 7.5 cm wide and have red tubular petals with flared open yellow tips that resemble tiny trumpets. It is a truely stunning cultivar to add to a container as a focal point or plant in a mass for dramatic effects.

G. × grandiflora hort. ex Van Houtte (*G. aristata × G. pulchella*) is like *G. aristata* but larger and more vigorous than *G. aristata.* It is a short-lived tetraploid perennial hybrid. It has small green leaves. Flowers are single, showy orange and red. Small cultivars include 20-25 cm 'Arizona Sun' (syn. 'All America Winner'), plants compact and bushy, highly tolerant to heat and drought conditions, early with abundance of bright-red blooms 5-7 cm across, with yellow tips; 'Summer Fire' 35 cm tall yellow, red-zoned; 'Red Plume' 40 cm tall brick-red; and 'Kobold' (syn. 'Goblin') red and yellow bicoloured; and those with 60 cm plant height include the 'Chloe' (yellow), 'Sunset' (orange), 'Aurea' (gold), 'Gold Kobold', 'Yellow Queen' and 'Burgundy' (wine-red). Another gaillardia cv. 'Oranges and Lemons' produces orange tipped, yellow flowers in great numbers. There can be over 70 flowers on an established plant making this one of the most prolific gaillardias available. Several cultivars have yellow ray florets with broad thick red zones like 'Bremen', 'Baby Cole' and 'Attraction'. Tall cultivars with height of 70 cm are 'Monarch Strain Mixed' with bright range of colours, *viz.* 'Tokajer' (vivid dark orange), 'Dazzler' (golden yellow with maroon centre) and 'Torchlight' (very large flowers having dark red petals with yellow borders). Hybrids include 'Gaiety Mixed', 'Stokes Royal Monarch' and 'Summer Kiss'.

G. pulchella var. Picta (Sweet) A. Gray includes cultivars like 'Sunshine Mixed' (50 cm tall with double globe-like large flowers on strong and long stems) having shades of orange, yellow, red and rose and a combination of these in varying degrees; 'Sundance Bicolour', a dwarf (25 cm tall) cultivar with outstanding heat and drought resistant, uniquely bicoloured with 5 cm globe-shaped in red and yellow colours; 'Picta', a robust compact and rounded bush, much-branched with secondary branches widely spreading; *G. picta* (Plume Series) growing 30-45 cm tall with uniform growth habit ideal for mass planting in full sun, flowers double and 3-5 cm wide, and the cultivars in this series are 'Red', 'Yellow' and 'Mixture'.

Other species are *G. amblyodon* (an annual); *G. comosa* (flowers yellow-pink, purple veined); *G. lanceolata* (red or yellow); *G. multiceps*; *G. parryi* (yellow), *G. pinnatifida* (yellow, tinged pink or red at base, purple or red veined); and *G. suavis* (yellow or red brown). Gaillardia varieties fall under two categories: annual kinds and perennial kinds. Perennial types are the original wild types or species. *G. aristata* has been superseded by large flowered varieties of brilliant colours commonly classed as *G.* × *grandiflora*, though some of them have been given separate names. They are more commonly sold as English hybrids. Annual kinds include *G. amblyodon*, *G. lanceolata* and *G. pulchella*. Other important varieties of gaillardia are 'Indain Chief', 'Lollypop Yellow', 'Lollypop Orange', 'Sunshine Strain', 'Regalis' and 'Sanguinea'.

Propagation

The annual gaillardia is commonly propagated through seeds. Perennial kinds can also be propagated either through division, cuttings or seeds. Division and cutting method of propagation is followed to get true to type of plants as it is a cross-pollinated crop. Seeds can be sown in pots or raised nursery beds from January to October to obtain almost year round flowering.

Micropropagation is also used to propagate gaillardias. Young leaf and internodal stem segments of *G. pulchella* were used for callus culture on MS medium supplemented with NAA (2 mg/l) and BA (0.4 mg/l). Flowers were produced by regenerated shoots maintained on MS medium, but no development of roots was observed until the plantlets were transferred to soil (Bourque *et al.*, 1989). Pillai *et al.* (1992) reported callus differentiated somatic embryos and subsequent plant regeneration from isolated leaf epidermis on B5 basal medium supplemented with either NAA + BAP or IBA + BAP and *in vitro* root formation on medium supplemented with IBA + kinetin in *Gaillardia pulchella* cv. 'Picta'.

Cultural Practices

Being a easy-going plant, gaillardia can be grown in a wide range of **climatic conditions**. It does well in both temperate and tropical climates. It has a high drought tolerance and does best with a dry and hot climate in full sun. Drought conditions are tolerated once the blanket flower plant is established.

Its cultivation is very easy and for **soil** types it is not selective, though prefers light sandy loam which is very well-drained. The pH range of soil should be 5.5-8.0. Tjia and Rose (1987) reported gaillardia as salt tolerant bedding plant. However, Sharma *et al.* (1989a) reported an increase in plant nitrogen and phosphorus content per cent from 1.26 and 0.049% under control to 1.57 and 0.065%, respectively with increase in salinity and decrease in plant potassium content per cent from 1.48% under control to 1.05% with increasing level of salinity. Different salinity levels had also marked influence on leaf relative water content per cent and leaf water potential (Sharma *et al.*, 1989b). They observed that leaf relative water content per cent decreased from 82.34% under control to 80.46% at higher levels of soil salinity (9 mmhos/cm). However, leaf water potential (-bar) increased significantly from 3.67 to 6.83 from control to salinity levels of 9 mmhos/cm. Khimani and Patil (1994) conducted an experiment to study the response of *G. pulchella* to varying levels of naturally saline soil and found that total plant, stem, flower and root dry weights as well as flower number and yield per plant decreased significantly with increase in salinity level from 1.20 to 2.80 dS/m. However, non-significant reduction was recorded in plant height, number of branches and leaves, leaf area and leaf dry weight by increasing salinity levels. Leaf proline content increased with increasing salinity levels from 21.92 mg/g DW at 1.20 dS/m to 128.42 mg/g DW at 8.70 dS/m.

Gaillardia can be grown throughout the year and the seeds are sown in the three growing seasons and seedlings are transplanted accordingly. Seeds are sown in the pots, boxes and on raised nursery beds depending upon the quantity of seeds to be sown. Twelve grammes of seeds are required to obtain 1,000 plants in gaillardia. Seeds germinate in 2-3 weeks and are ready for transplanting in 4-8 weeks. Ramakrishnan *et al.* (1970) reported that gaillardia seeds need a 4-month after-ripening period before germination and remain viable for storage period of 31 months. Patil *et al.* (2005) studied the influence of planting time on growth and productivity of *G. pulchella*. Flower yield was recorded highest (33.33 t/ha) from January planting. They attributed the increase in flower yield due to improved growth parameters, *viz.* plant height, number of branches and leaves per plant and yield components like number of flowers per plant, flower size and yield per plant. Mishra (1998) recommended 30 x 30 cm spacing as the best for all the vegetative and floral parameters except flower size. In a study on interaction effects of nitrogen and plant spacing in *G. pulchella* cv. 'Picta', Hugar and Nalawadi (1999) obtained highest flower yield with a spacing of 30 x 20 cm and 75 kg N/ha. However, seed yield was highest with 100 kg N/ha at a spacing of 30 x 20 cm. Generally, hoeing is done 3-4 times during the crop period as hoeing makes the soil porous and also controls weed population in the field.

Perennials need to be cared like any other garden plant. One thing that distinguishes perennials is that they tend to be active growers. As perennials establish, it is important to prune them back and **thin** them out occasionally. This will prevent them from completely taking over an area to the exclusion of other plants and will also increase air circulation, thereby reducing the incidence of diseases like botrytis and powdery mildew.

Weeds should be removed periodically as they compete with main crop for light, water and nutrients. The presence of weeds also reduces the aesthetic value of the crop. Weeds can harbour pests and diseases and can be kept to a minimum only if proper weed control practices are carried out regularly. Weeds are removed through manual weeding in the form of hoeing. **Mulching** is also very effective to reduce weed population. Besides manual weeding, weedicides are also used to control weeds. Dithiopyr, pendimethalin and prodiamine as pre-emergence herbicides provided excellent control of spotted spurge (*Euphorbia maculata* L.) and yellow woodsorrel (*Oxalis stricta* L.) with little injury or growth reductions to the gaillardia plants (Derr, 1994).

Irrigation and Nutrition

Seedlings should be watered immediately after transplanting. The plants are watered once in a week or in a fortnight depending upon weather conditions. Water-logging conditions are avoided in the field as it is highly sensitive to water-logging similar to most of the flower crops.

Proper **nutrition** is very essential for obtaining optimum plant growth and higher yield of good quality flowers. The requirements for manures and fertilizers for gaillardia vary with climatic conditions and soil types. The soil should be well prepared by incorporating well rotten farmyard manure or cattle manure at the rate of 3-5 kg/sq m. Singatkar *et al.* (1995) reported that optimum rates for growth and flower yield in gaillardia are 173 kg N + 103 kg P + 63 kg K/ha. The growth, days to flowering, total flowering period and flower production per plant increases significantly with increasing NPK application. Mishra (1998) studied the effect of nitrogen and planting density on growth and flowering of gaillardia. The height and spread of the plant, days to flowering, size of flowers and flower weight were greatest at 30 g followed by 20 g N/m^2 along with application of 5 kg farmyard manure, 15 g P_2O_5 and 10 g K_2O/m^2. Shafi *et al.* (2002) studied the response of *Gaillardia pulchella* cv. 'Picta' to different levels of nitrogen with constant doses of phosphorus and potassium. It was found that maximum days to flowering (117.7) were recorded in N + P + K (40 + 10 + 10 g/m^2), while minimum days for flowering (88.3) were recorded in control (N=0). Maximum number of flowers per plant (209.7), flowering period (185.3 days) and root weight (8.2 g/plant) were recorded in treatment having N + P + K (20 + 10 + 10 g/m^2). Minimum values for these parameters were 90, 145.3 days and 5.23 g, respectively in control. Plants treated with N + P + K (25 + 10 + 10 g/m^2) showed maximum plant survival (97.22%).

Integrated nutrient management (INM) studies were conducted by Rathod *et al.* (2003). They observed that application of 75% of the recommended NPK rate in combination with *Azospirillum* + phosphate solubilizing bacteria (PSB) resulted in the highest number of branches (41), leaves (218.67), number of flowers per plant (70.23), flower diameter (5.03 cm), stalk length (7.37 cm), weight of single flowers (2.03 g), yield per plant (142.57 g) and total yield (89.11 q/ha). Similarly, use of INM practices in gaillardia was also suggested by Gadagi *et al.* (2004) in *G. pulchella*. The combinations of *Azospirillum* strain OAD-2 and 150 kg N/ha showed the highest nitrogen uptake at 120 days after transplanting. This strain inoculation significantly increased plant height, number of leaves and branches per plant and total dry mass accumulation in the plants. Plant growth and yield parameters were also increased when inoculated with non-diazotrophic, but efficient IAA producing *Azospirillum* strain OAD-57. Patil *et al.* (2004) recommended application of nitrogen @ 150 kg/ha in two split doses, i.e. 50% as basal dose and 50% at initiation of flowering. They obtained highest flower yield

(39.04 and 33.16 t/ha) on red and black soils, respectively by applying nitrogen @ 150 kg/ha in two split doses.

Growth and Flowering

Environmental factors have great influence on the growth and flowering of gaillardia. Yuan *et al.* (1998) studied the effect of temperature and light on flowering in *Gaillardia* x *grandiflora* cv. 'Goblin'. The plants were exposed to 5 °C for 10 weeks, and were then grown in greenhouse sections set at 15, 18, 21, 24 or 27 °C under 4-h night interruption lighting until anthesis. Days to visible bud, days to anthesis and days from visible bud to anthesis decreased as temperature increased. The rate of progress towards days to anthesis increased linearly with temperature. Cathey (1976) studied the effect of day length on plant growth and flowering in gaillardia. He stated that long days induce flowering in gaillardia and no flowering occurs under short days. His findings were corroborated by Cameron *et al.* (1996) who studied vernalization and photoperiodic response in *Gaillardia* x *grandiflora* cv. 'Goblin'. They told that gaillardia is a obligate long day plant and vernalization was required for flowering in *Gaillardia* x *grandiflora* cv. 'Goblin'.

Vige *et al.* (2003) reported that *G.* x *grandiflora* is heat tolerant, however, it is affected by temperature over 35 °C. In another study, Zollinger *et al.* (2006) studied the drought responses of *G. aristata* and found that the plants displayed drought avoidance mechanisms, dying back when water was limiting and showing new growth when the plants were watered again. Dhutraj *et al.* (2003) studied the effect of various bioenzymes, *viz.* Plantozyme, Multizyme, Novazyme and Shaktazyme @ 2-3 ml/l on the flowering and yield of gaillardia. It was found that all treatments significantly enhanced flowering and reduced number of days to 50% flowering. The lowest number of days to 50% flowering (62.0) was recorded for 3 ml Novazyme/l. The length of flowering spikes was greatest (21.77 cm) with Shaktazyme @ 3 ml/l. The highest yield per plant (412.11 g) and per hectare (9.25 tons) was obtained with Novazyme at 3 ml/l.

Vegetative and floral characters can be regulated by application of **plant growth regulators**. Mukherjee and Bose (1972) reported that Daminozide (B-Nine) @ 2,500-10,000 ppm improved number of flowers in gaillardia. Evans and Lyons (1988) studied the gibberellin-induced growth and flowering responses of *Gaillardia* x *grandiflora* in cvs 'Dazzler' and 'Goblin' (both long day cultivars). They found that application of GA_{4+7} substituted for long days and promoted flowering under short days in the same amount of time required by untreated plants, although GA effects were cultivar and photoperiod dependent. GA also increased scape length and reduced flower head diameter and number of ray florets. Leaf production was accelerated by several applications of GA at weekly intervals (Table 1 and Table 2).

Table 1. *Gaillardia* × *grandiflora* flowering response 3 months after treatment with 100 ppm GA_{4+7} under long and short photoperiods*

No. of GA applications	Per cent plants flowered^z		No. of GA applications	Per cent plants flowered^z	
Long days	**Dazzler**	**Goblin**	**Short days**	**Dazzler**	**Goblin**
0	76	44	0	0	4
1	100	80	1	60	50
5	100	100	5	80	100
10	100	100	10	100	90
				$\chi2=71.6$**	60.1**

^{z}Results presented as percentages; actual data analyzed via χ^2 test;

**null hypothesis = non-association of treatments.

* Evans and Lyons (1988)

Table 2. Flower head diameter, ray floret number, scape length and caliper at first flowering after weekly treatment with 100 ppm GA_{4+7} under long and short photoperiods*

No. of GA Applications	[y]Leaf number after 6 weeks	Inflorescence diameter (mm)		Ray floret No.		Scape length (mm)	
	Dazzler	Dazzler	Goblin	Dazzler	Goblin	Dazzler	Goblin
Long days							
0	58	81	98	21.1	19.0	286	106
1	56	95	88	18.4	20.3	253	95
5	62	79	85	19.7	18.4	273	109
10	82	72	72	19.3	17.6	311	128
Short days							
0	58	---	---	---	---	---	---
1	56	88	90	13.0	18.0	153	63
5	65	77	69	15.0	21.5	95	99
10	72	61	76	16.4	17.5	263	228
Standard Error		±1.3	±1.2	±0.25	±0.29	±8.1	±4.0

[y]Recorded for 'Dazzler' only; by this time a maximum of six weekly GA applications had been made

* Evans and Lyons (1988).

Koning (1983) stated that auxin controlled filament elongation in *G. grandiflora*. Application of ethylene slightly promoted filament elongation *in vitro*, and ethylene production of the flowers slightly increased prior to filament growth. Thus it was interpreted that auxin in controlling filament elongation in gaillardia may involve in stimulation of ethylene biosynthesis and acid growth. Khimani *et al.* (1994) found that chlormequat and Alar sprayed on gaillardia plants produced significantly more flowers than control and maleic hydrazide sprayed plants. Application of Alar @ 500 ppm produced highest number of flowers per plant, seed yield per plant and yield per hectare as 160.6, 165.2 g and 27.43 tonnes, respectively. According to Pal (2003), TIBA at 100, 200 and 300 ppm, CCC at 750 ppm and kinetin at 50 and 75 ppm produced significantly higher flower yield in *G. pulchella* cv. 'Picta'.

Post Harvest

It takes about 3 months for the appearance of first bloom in gaillardia plants. The flower heads should be harvested when they attain full size. Harvesting of flowers should be done in cool hours of the day. For table decoration as cut flowers, fully opened flowers with long stems are harvested which remain fresh for 5-6 days in vase. It is advisable to remove spent flowers on the plant to prolong the flowering period of gaillardia. Hugar and Nalawadi (1999) recorded highest flower yield of 19.6 and 18.0 t/ha and seed yield of 7.0 and 6.8 q/ha in *rabi* and summer seasons, respectively. Norcini *et al.* (2002) reported that mechanical harvesting of flowers was more efficient as compared to manual harvesting but manual harvesting yielded substantially more yield of clean seeds as compared to former method. They also observed that July harvesting of seeds resulted in higher clean seed yields (67%), 45% higher clean seed yield rate (a measure of harvesting efficiency), 9% greater seed mass and 95% greater germination rate than the October harvest.

Certain chemicals affect post-harvest life of gaillardia flowers. Harode *et al.* (1993) reported that 1% sucrose increased vase life of *G. pulchella* cv. 'Picta' over other treatments comprising of NaCl, GA_3 and charcoal. The flowers had also better colour and freshness after pulsing with sucrose. Effects of tap or distilled water, sucrose, cobalt nitrate, sodium benzoate, citric acid, IAA, kinetin and gibberellic acid on loose flowers of gaillardia were studied by Parikh and Mankad (2003). The weight of gaillardia flowers after one day was highest with tap water treatment, while at 2, 3 and 4 days the weight of gaillardia as well as their total water balance were highest with gibberellic acid 15 ppm.

McGregor (1989) reported that the gaillardia flowers should be transported in a cool atmosphere (4.5°C) and at a high relative humidity (90-95%) for higher longevity of flowers. Koning (1984) stated that application of gibberellin promoted opening in several flowers including *Gaillardia grandiflora*.

Insect-Pests and Diseases

Slugs and **snails** favour moist climates. They can be voracious feeders, eating just about anything that is not woody or highly scented. They batter the leaves by feeding in between the veins, strip entire stems, or completely devour seedlings and tender transplants, leaving behind tell-tale silvery slimy trails. Ground cover in shady places and heavy mulches provide favourite hiding places. Sanitation of the area is the best option, eliminating hiding places such as leaf debris and over-turned pots. In the spring, watching of the clusters of small translucent spheres of eggs and destroying them, and at dusk and dawn watching the adults and destroying them will keep its infestation at the minimum. The adults of **leafhopper** (*Macrosteles fascifrons*) are grayish and are about 3 mm long, *vis-a-vis* spread 'aster yellows' infection. Application of carbaryl is useful when leafhoppers are troublesome. Imidacloprid applied as a systemic to be taken up by the roots will also provide season-long control. **Four-lined plant bug** (*Poecilocapsus lineatus*) lays eggs on the soft stems. They hatch about middle of May and young bugs suck the sap from tender leaves. They moult five times and when mature, about the middle of June, they have wings and are nearly 1 cm long. The insect body is yellow, marked lengthwise on the wings with four black stripes alternating with three green stripes. The injury to the leaves consists of sunken areas around the punctures. These areas later appear as circular transparent spots and finally as circular holes. This insect injures the new leaves of many different kinds of annual and perennial plants, and shrubs. There is one generation each year. The nymphs can be managed by spraying with ultra-fine horticultural oil, insecticidal soap or Malathion. Imidacloprid applied as a soil drench will provide season-long systemic control for this pest. Occasionally, gaillardia plants are also attacked by aphids, thrips and spider mites. **Aphids** and **thrips** can be controlled by spraying Malathion and **spider mites** by spraying Kelthane @ 2%. Pallavi *et al.* (2002) conducted a survey to record the pests of *G. pulchella* at different stages of crop growth, and for the first time, recorded *Helicoverpa armigera, Porthesia scintilans, Plusia orichalsia, Spodoptera litura, Nesidiocornis tenuis, Empoasca* sp., *Empoascanra* sp., *Austroagallia* sp. and *Eusarcocoris montivagus* infesting gaillardia in Maharashtra.

Root and **stem rots** are caused by *Pythium* sp. (damping off), *Rhizoctonia, Sclerotinia sclerotiorum* (foot rot) and *Thielaviopsis*, symptoms appearing first on stem surface with distinct colour change and then lower part turning tan to dark brown afterwards. This problem is kept under control by avoiding mulching around the plant and aerial watering. Glawe *et al.* (2006) reported **powdery mildew** (*Erysiphe cichorearum*) on *G.* x *grandiflora* cv. 'Baby Cole'. *Sphaerotheca* also infects this crop. White or gray powder is usually found on the upper surface of leaves. Leaves will often turn yellow or brown, curl up, and drop off. New foliage emerges crinkled and distorted. Powdery mildew is usually found on plants that do not have enough air circulation or adequate light. Problems are worse where nights are cool and days are warm and humid. Plants should be spaced properly so that they receive adequate light and air circulation. Planting resistant varieties is recommended. Sanitation of the field and removal of the spent plant parts are necessary. Powdery mildew is controlled by using karathane, myclobutanil, sulphur and horticultural oil. **Downy mildew**, a fluffy white fungal growth that develops on the underside of leaves is most common during cool and humid conditions. Foliage often discolours and is stunted. Use of disease-free plants and proper spacing for proper air circulation can keep this problem under check. Discarding infected leaves or even entire diseased plants can save other plants from its infection. It can be controlled by spraying Bordeaux mixture (0.1%) or Dithane M-45 (0.2%) at fortnightly intervals. **Rust** (*Puccinia* sp.) over-winters on leaves, stems and spent flower debris. Rust often appears as small, bright orange, yellow or brown pustules on the underside of leaves. It is spread by splashing water or rain, and during moist weather its attack is very serious. The field should be properly cleaned of plant debris and the provision will have to be made for proper air circulation. Resistant varieties should be planted out. Only day-watering should be carried out so that water on aerial parts dries up by the sunset. Application of mancozeb is quite effective. **White smut** (*Entyloma polysporum*) first causes light green spots on the leaves, some with tan centres but spores in the spots are white. Leaf spots at later stage become dark brown. Dithane Z-78 at 0.2 per cent can control this disease. **Leaf spot** is caused by fungi *Cercospora gaillardiae* and *Septoria gaillardiae*. The ragged or circular brown to black spots and patches are seen, with a water-soaked or yellow-edged appearance and it spreads through insects, rain water, dirty garden tools, or through people movement. Infected leaves should immediately be removed when the plant is dry. Overhead irrigation should be avoided. Leaves that collect around the base of the plant should be gathered and destroyed. Fungal leaf spots can be controlled by spraying copper or dithiocarbamate fungicides. *Pseudomonas* bacterium causes discolouration between veins of leaves which turn dark brown in severe infection, and plants also die. Sprinkler irrigation should be avoided and plants showing such symtoms should be uprooted and destroyed.

Yellows are caused by phytoplasmas. Symptoms consist of yellowing or clearing of the veins in newly infected leaves, shortening of the internodes of the main stem, and production of long axillary breaks which create a yellowish 'witches broom'. If plants are infected young, they remain stunted, and if flowers are produced, they are small, abnormal, and exhibit phyllody, a condition in which flower parts revert to green tissues. Symptoms do not always appear on the entire plant, but may show on only one stalk, depending on the time of infection. This phytoplasma is transmitted by leafhoppers. Since the pathogen infects many plants like daisies, chrysanthemum, rudbeckia, etc., gaillardia in the vicinity may also be infected. It is difficult to control this malady though roguing of infected plants as well as control of the leafhopper vector will keep this problem under check. 'Impatiens necrotic spot' and 'tomato spotted wilt' spread through thrips, while 'cucumber mosaic virus' through aphids so these vectors also require to be controlled.

References

Atwood, S. 1937. The last premeiotic mitosis and its relation to meiosis in gaillardia. *Proc. Natl. Acad. Sci.*, USA, **23:** 1-5.

Bailey, L. H. and E. Z. Bailey, 1976. *Hortus Third: A Concise Dictionary of Plants Cultivated in the United States and Canada*, 1290 p. Macmillan Press, New York.

Bourque, J. E., S. Tanner and T. J. Mabry, 1989. *In vitro* regeneration of *Gaillardia pulchella* Foug. *Plant Cell Tissue Organ Cult.*, **16** (1): 67-72.

Cameron, A., H. Royal and C. Will, 1996. Forcing perennials 101. *Greenhouse Grower*, **14** (3): 19-20.

Cathey, H. M. 1976. Growth regulators in bedding plants: A Pennsylvania State Manual (ed. Mastalerz, 2nd Edition), pp. 177-189. *Pennsylvania Flower Growers*, Univ. Park, PA.

Derr, J.F. 1994. Weed control in container grown herbaceous perennials. *HortSci.*, **29** (2): 95-97.

Dhutraj, S.V., P.R. Narwadkar, V.V. Bhagat, N.G. Rathod and B.M. Kalalbandi, 2003. Effect of various bioenzymes on flowering and yield of gaillardia. *Annals Plant Physiol.*, **17**(2): 161-163.

Duschatzky, C., N. Almeida, S. Colombres and M.P. de Lampasona, 2003. Essential oil of *Gaillardia megapotamica* (Spreng.) Backer var. Radiata (Gris.) Backer from San Luis, Argentina. *J. Essential Oil Res.*, **15** (2):112-113.

Evans, M.R. and R.E. Lyons, 1988. Photoperiodic and gibberellin induced growth and flowering responses in *Gaillardia* x *grandiflora*. *HortSci.*, **23** (3): 584-586.

Gadagi, R.S., P.U. Krishnaraj, J.H. Kulkarni and T. Sa, 2004. The effect of combined *Azospirillum* inoculation and nitrogen fertilizer on plant growth promotion and yield response of the blanket flower (*Gaillardia pulchella*). *Sci. Hort.*, **100** (1-4): 323-332.

Glawe, D.A., G.G. Grove and M. Nelson, 2006. First report of powdery mildew of *Gaillardia* × *grandiflora* (Blanket flower) caused by *Leveillula taurica* in North America. *Plant Health Progress*, **1**: 1-3.

Gommers, F.J. 1972. Nematicidal principles from roots of some Compositae. *Acta Botanica Neerlandica*, **21** (1): 111-112.

Harode, S.M., V.G. Kotake and J.P. Zend, 1993. Effects of selected treatments on longevity and freshness of selected flowers in home decoration. *J. Maharashtra agric. Univ.*, **18**(3): 448-450.

Hegde, P.S. and G. Gopinath, 2003a. Genetic variability, heritability and genetic advance in *Gaillardia pulchella*. *J. Ornam. Hort.* (*New Series*), **6** (3): 277-279.

Hegde, P.S. and G. Gopinath, 2003b. Correlation and path coefficient analysis studies in *Gaillardia pulchella*. *J. Ornam. Hort.* (*New Series*), **6** (4): 385-9.

Hugar, A.H. and U.G. Nalawadi, 1999. Studies on influence of spacings and nitrogen levels on growth and yield parameters in gaillardia (*Gaillardia pulchella* var. 'Picta' Fouger). *Karnataka J. Agric. Sci.*, **12** (1-4): 137-141.

Khimani, R.A. and A.A. Patil, 1994. Response of gaillardia (*Gaillardia pulchella* Fouger) to varying levels of naturally saline soil. *Recent Hort.*, **1**(1): 97-101.

Khimani, R.A., A.A. Patil and M.B. Chetty, 1994. Influence of CCC, Alar and MH on yield and physiological growth parameters of gaillardia during *rabi* season. *Recent Hort.*, **1**(1): 91-96.

Koning, R.E. 1983. The roles of auxin, ethylene and acid growth in filament elongation in *Gaillardia grandiflora* (Asteraceae). *American J. Bot.*, **70**(4): 602-610.

Koning, R.E. 1984. The role of plant hormones in the growth of the corolla of *Gaillardia grandiflora* (Asteraceae) ray florets. *American J. Bot.*, **71**: 1-8.

McGregor, B.M. 1989. Flowers and Florist Greens. *Tropical Products Transport Handbook*, USDA Office of Transportation, 668 pp.

Mishra, H.P. 1998. Effect of nitrogen and planting density on growth and flowering of gaillardia. *J. Ornam. Hort.* (*New Series*), **1**(2): 41-7.

Mukherjee, D. and T.K. Bose, 1972. Comparative effects of B-Nine and CCC on some species of plants under Compositae. *Plant Sci.*, **4**: 95-100.

Norcini, J.G., J.H. Aldrich and F.G. Martin, 2002. Effect of harvest method on seed yield of *Coreopsis lanceolata* L. and *Gaillardia pulchella* Foug. *J. environ. Hort.*, **20**(1): 20-23.

Pal, P. 2003. Gaillardia. In: Commercial Flowers (eds Bose, T. K., L. P. Yadav, P. Pal, P. Das and V. A. Parthasarathy), pp. 637-640. Naya Prokash, Kolkatta, India.

Pallavi, T., A.E. Rane, S.S. Ali and P.S. Burange, 2002. Occurrence of insect pests of gaillardia (*Gaillardia pulchella*). *J. Soil & Crops.* (*New Series*), **12**(1): 101-104.

Parikh, S. and A. Mankad, 2003. Effect of certain chemicals on *Gaillardia* and *Dianthus* loose flowers. *J. Ornam. Hort.* (*New Series*), **6**(2): 151-152.

Patil, B.C., B.S. Kulkarni and B.H. Naik, 2005. Growth and productivity of blanket flower (*Gaillardia pulchella* Foug.) as influenced by planting time. *J. Ornam. Hort.* (*New Series*), **8**(2): 100-103.

Patil, B.C., H. Naik, B.S. Kulkarni, S.L. Jagadeesh and U.G. Nalawadi, 2004. Comparison of effect of split application of nitrogen at different stages on growth and flower yield of gaillardia on red and black soils. *J. Ornam. Hort.* (*New Series*), **7**(3-4): 138-143.

Pillai, K.G., I.U. Rao, I.V.R. Rao and H.Y.M. Ram, 1992. Induction of division and differentiation of somatic embryos in the leaf epidermis of *Gaillardia picta*. *Plant Cell Rep.*, **10**(12): 599-603.

Ramakrishnan, V., W.M.A. Khan and A.M.M.A. Khan, 1970. Studies on the germination of seeds of a few ornamental flowering annuals. *South Indian Hort.*, **18**(3-4): 93-95.

Rathod, N.G., P.R. Narwadkar, A.K. Sajindranath and T. Prabu, 2003. Effect of integrated nutrient management on growth and yield of gaillardia. *J. Maharashtra agric. Univ.*, **27**(3): 318-319.

Shafi, M., M. Ishtiaq and N. Rehman, 2002. Response of *Gaillardia pulchella* (cv. 'Picta') to different levels of nitrogen with constant doses of phosphorous and potassium. *Sarhad J. Agri.*, **18**(2): 189-191.

Sharma, S.S., R. Yamdagni, and R.K. Arora, 1989a. Effect of different salinity levels on mineral composition of garden annuals. *South Indian Hort.*, **37**(5): 291-294.

Sharma, S.S., R. Yamdagni and R.K. Arora, 1989b. Effect of salinity levels on physiological aspects of garden annuals. *South Indian Hort.*, **37**(6): 337-340.

Singatkar, S.S., R.B. Sawant, S.A. Ranpise and K.N. Wavhal, 1995. Effects of different levels of N, P and K on growth and flower production of gaillardia. *J. Maharashtra agric. Univ.*, **20**(3): 392-394.

Tjia, B. and S.A. Rose, 1987. Salt tolerant bedding plants. *Proc. Florida State Hort. Soc.*, USA, **100**: 181-182.

Tsay, T.T., S.T. Wu and Y.Y. Lin, 2004. Evaluation of Asteraceae plants for control of *Meloidogyne incognita*. *J. Nemat.*, **36**(1): 36-41.

Turner, B. 1972. Two new gypsophilous species of *Gaillardia* (Compositae) from North-Central Mexico. *The Southwestern Naturalist*, **17**(2): 181-190.

Vige, M.K.D., J.S. Kuehny, J.E. Board and G. McClure, 2003. Effects of supra-optimal temperature on coreopsis and gaillardia. *Acta Hort.*, No. 618, pp. 169-176.

Yuan, M., W.H. Carlson, R.D. Heins and A.C. Cameron, 1998. Effect of forcing temperature on time to flower of *Coreopsis grandiflora*, *Gaillardia* × *grandiflora*, *Leucanthemum* × *superbum* and *Rudbeckia fulgida* . *HortSci.*, **33**(4): 663-667.

Zollinger, N., R. Kjelgren, T. Cerny-Koenig, K. Kopp and R. Koenig, 2006. Drought responses of six ornamental herbaceous perennials. *Sci. Hort.*, **109**(3): 267-274.

12

Gomphrena

Ajit Kumar, Sanyat Misra, Manish Kapoor, Ranjan Srivastava, Sanjeev Kumar Sharma and Manju Rana

Family: **Amaranthaceae**
Common names: Bachelor's button, Globe amaranth

Introduction and Origin

Gomphrena is native to tropical and sub-tropical countries like Panama and Guatemala in Central and South America, SE Asia (mainly India) and Australia. From C. America, it gradually spread to rest of the world but with a slower pace. It was cultivated in 1714 A.D. by Dutchess of Beaufort, but was not common in the English gardens till 1725 A.D. It was first raised in Holland about 1670 A.D. The name *Gromphrena* (*Gomphrena globosa* L.) was originally used by Pliny the Elder for a kind of amaranth (Bailey and Bailey, 1976) which in Latin means 'a type of everlasting amaranth' on account of retention of colour and shape of the flower for a very long period after drying, and *globosa* means spherical, due to spherical shape of flower heads. Its flower heads appear as a small woolly ball. It is a common everlasting flower of easy culture. Though *Gomphrena* consists of some 100 species but only one species (*G. globosa*) and its varieties are in general cultivation.

When there are only a few flowers available for enjoyment in the tropical and subtropical climate of India, gomphrena ornaments the beds and borders (including edgings) in the gardens or elsewhere. It is grown for its 'clover-like' flowers in strong colours like rose, red and purple range as in *Gomphrena globosa* Linnaeus, and in vivid orange as in *G. haageana* Klotzsch. It may also

have white, red, carmine and different shades of pinks. *Gomphrena* is commonly grown in beds, borders and in pots. It is also useful in edging and rock gardens. By using contrasting colour scheme of planting, splashing mass planting effect is created through gomphrenas. The cone-like flower heads are beautiful in dried flower arrangements and will hold their shape and colour indefinitely when dried. Pot-grown gomphrenas are used for indoor decoration and also in hanging baskets. Both fresh and dried flowers are used for flower arrangement indoors. Once started, gomphrenas are very easy to grow and tolerant to extremes of heat, drought or humidity. *Gomphrena globosa* acts as a potential indicator of the phytotoxicity and mutagenicity of pollutants (Bessonova *et al.*, 1997). It is also a potential summer cut flower as is available in vivid colours, for long blooming period and for its good stand in vases (Cocozza-Talia, 1993). *G. officinalis* has medicinal properties also (Mercier *et al.*, 1992). Occurrence of fructose-containing carbohydrates by leaf and node callus of *G. macrocephala* was also recorded by Candida *et al.* (1995). Though in India, it is being cultivated everywhere but commercially it is being grown in southern districts of Tamil Nadu for use as cut flowers as well as for loose flowers for making garlands (Rajasekharan *et al.*, 1983).

Botany and Breeding

Gomphrena is an annual bedding plant that grows 30-60 cm tall with a spread of about 30 cm. The plants are herbaceous in nature, may be erect or prostrate in habit, and simple to much branched from base. The branched stems are erect and stiff and the plants have a bushy appearance. Leaves are opposite, sessile or petiolate, and blade ovate to obovate. The true flowers are insignificant and tiny and white to yellow trumpets that are only visible when seen closely. It is the bright magenta bracts arranged in globose, papery-textured and, clover-like flower heads that provide the real show in summer and its dried flowers even during winter. The inflorescence is dense, terminal or axillary, capitate or glomerate and 2.5-3.8 cm long. Flowers are small, bisexual, deltoid to ovate, five tepals, and filaments connate basally into tube. Stamens are 5, style one (short or long), stigmas 2 and ovary unilocular. Chromosome number for gomphrena as described by Darlington and Wylie (1955) is n = 9 and 10.

Badr *et al.* (2004) treated the seeds of *G. globosa* with ^{60}Co gamma rays (0-30 kr) and recorded a lot of variations in terms of plant height (dwarf plants particularly under 25 kR) with more secondary branching, *vis-à-vis* various leaf abnormalities (in almost all the doses), and inflorescence yield and size in M_1, while in M_2 though most of such variations disappeared but even some dwarf plants with more primary branching were recorded.

Species and Varieties

Gomphrena consists of some 100 species of annuals, biennials and occasionally perennial, but only one species *Gomphrena globosa*, a native of tropical Asia, mainly India, with its cultivars is in general cultivation. Linnaeus described eleven species of *Gomphrena* on the basis of inflorescence characteristics rather than to stamen and stigma number. The first edition of *Species Plantarum* (1753) and the 5th edition of *Genera Plantarum* (1754) included nine species. The published concept of *Gomphrena* dates from Linnaeus' publication on the Clifford Herbarium (1737, 1738) in which only two species were included, one eventually named *G. globosa* L. and the other *G. flava* L. *Gomphrena globosa* is consistent with the genus characters noted by Linnaeus in 1754. Of the nine species of *Gomphrena* in *Genera Plantarum* (1754), only two species, *viz. G. perennis* L. and *G. serrata* L. are now considered co-generic with *G. globosa* (Mears, 1980).

In the recent past in Australia, as per the taxonomical revision of the genus some 33 species were recognized where 12 were new ones such as *G. arida, G. atrorubra, G. connata, G. cucullata, G. eichleri, G. floribunda, G. humifusa, G. kanisii, G. lacinulata, G. magentitepala, G. occulata* and *G. rosula* (Palmer, 1998). Though several other species such as *G. caespitosa* Torrey, *G. martiana* Gillies ex Moquin, *G. nealleyi* J.M. Coulter & Fisher, *G. nitida* Rothrock, *G. sonorae* Torrey and *G. viridis* Wooton & Standley are documented but *G. globosa* is the only available species with one of its dwarf strains (15 cm tall). *G. globosa* is an half-hardy annual to 60 cm with tall, erect and branched stature; leaves elliptic to oblong or ovobate, acute, narrowing towards the base and 5 cm long; spikes are solitary, axillary and terminal with ovoid heads, 2.5-4.0 cm long and everlasting with tiny flowers having a 5-lobed perianth almost hidden by chaffy dark violet flowers though now in the hybrids the bract's colours available are white, yellow, orange, pink and purple. The varieties under *Gomphrena haageana* are (i) 'Aurea Superba' (syn. *G. haageana*) with orange-yellow flowers having red-tinged bracts, (ii) 'Strawberry Fields' with 60-75 cm tall plants, large inflorescence, blooms strawberry-red and most suitable as cut flower or for dry arrangements, (iii) 'Mixture' having mixture of all of the colours (white, pink, rose, red-purple, etc.) in the series of *haageana* types, (iv) 'QIS Series', being a strain (varieties 'Carmine', 'Orange', 'Red', etc.) with strong stems requiring 18 weeks to flower, best for fresh flower production, plants 70-75 cm tall and flowers large (4-5 cm); under *G. disperva* is 'Pink Pinheads', a very dwarf (10 cm height) variety most suitable for ground cover and hanging baskets, with beautiful pink flowers covering the compact creeping mounted plants; under *G. globosa* are (i) 'Alba' with white flowers, (ii) 'Lavender Lady', plant height to 70 cm with lavender-liliac flowers, (iii) 'All Round Purple', a strong care-free variety ideal for cut flower and dry flower arrangements, growing to 60 cm with good basal branching and most suitable for growing in areas with hot and humid summers, where leaves are dark green and 2-3 cm, and the flowers clover-like and rich deep purple, (iv) 'Bicolor Rose', ideal for beds, borders or container planting and for cut and dry flower arrangements, plant height some 50-60 cm, an attractive bicolour variety with lavender rose blooms and a white centre, (v) 'Dwarf White' with clear white everlasting flowers, (vi) 'Nana' of dark red colour, (vii) 'Nana Compacta Rubra' with purple flowers and plant height 15 cm, (viii) 'Rosea' (pink), (ix) 'Rubra' (purple-red with edible leaves), (x) 'Buddy Series', quite weather-tolerant, ball-shaped, compact and dwarf (20 cm) plants, excellent for bedding and pot-planting and the varieties are 'Purple', 'Rose', 'White' and 'Mixture; '**Gnome Series**', with dwarf (15 cm), ball-shaped, compact, heat-resistant plants flowering in all the seasons, flowers beautiful and 2 cm in diameter, and the varieties are 'Pink', 'Purple', 'White' and 'Mixture', and '**QIS Series**', with tall (50-60 cm) and properly branched plants having strong stems, and most suitable for cut flower and dry arrangements, flowers 4-5 cm which appear in 18 weeks of planting and the varieties are 'Lilac', 'Pink', 'Purple', 'Rose', 'White' and 'Mixture'.

Propagation

Gomphrena is commercially propagated by seeds. Seeds are sown in seed compost on raised nursery beds or pans during January-March for blooming during summer and rainy seasons, i.e. from April to September. If January-March period at a place remains chilly, the seeds may be sown in 8-cm pots under polyhouses with a temperature range of 15-18 ^{0}C. These pots may be replaced with 12.5-cm pots after the plants have attained sufficient growth. It is very tolerant to high temperature and humidity of summer and rainy seasons, so it has very long lasting flowering period from different sowing periods (Hong and Jeong, 1988). Germination of seeds is adversely affected by immersion of seeds in water for 24 hours (Sena-Filho *et al.*, 2003). The seeds can also be sown during May-June to grow the plants during rainy season. In temperate regions, the seeds are sown during March-April. Gomphrena may also be multiplied through leaf cuttings. Bhargavi *et al.* (1977) reported that rooting in leaf cuttings of *G. globosa* was encouraged by plant growth regulators in the order daminozide > kinetin > $CaCl_2$ > abscisic acid (Table 1).

Gomphrena can also be propagated through micropropagation. Using hypocotyl tissues of gomphrena as explants, Ruiz and Valadez (1985) obtained callus in MS medium supplemented with 3, 5 or 7% sucrose and NAA, BA and 2,4-D in various concentrations. Media supplemented with 10^{-5} M NAA + 10^{-8} M BA induced shoot growth with a few or no roots. However, sucrose at 5 and 7% was required for induction of shoots and roots. Mercier *et al.* (1992) obtained callus from leaf and stem segments of *G. officinalis*. Callus was produced after 10-15 days of culture. Shoot regeneration was best in BA (5.0 or 10.0 mg/l) + NAA (0.1 mg/l), while optimum root formation was found in 10.0 mg/l IBA.

Table 1: Effect of some growth regulators and Ca^{2+} on rooting and senescence of detached leaves of *G. globosa* in light*

Treatments (mg/ litre)	Rooting/leaf			Chlorophyll (mg/g fresh weight)	Protein (mg/g fresh weight)	Dry weight (mg/g fresh weight)
	Days for root initiation	Av. root number	Av. Root length (mm)			
After 10 days						
				9.82**	0.35**	155**
Control	---	---	---	3.38	0.13	93.0
Kinetin, 10	6	3	20	7.48	0.27	130.0
ABA, 5	---	---	---	5.22	0.14	125.0
CaCl2, 200	---	---	---	6.66	0.15	140.0
B-Nine, 15	6	3	23	7.8	0.47	140.0
After 20 days						
Control	13	3	13	5.07	0.2	63.0
Kinetin	6	6	58	8.29	0.6	129.0
ABA	12	4	37	6.68	0.15	135.0
CaCl2	12	5	25	9.01	0.3	142.0
B-Nine	6	6	68	9.58	0.7	140.0

**Initial level at the beginning of experiment

* Bhargavi *et al.* (1977)

Cultural Practices

Gomphrena is a quite easy-growing annual for the warm regions. It can be grown in a wide range of **soils**, soil pH ranging from 5.5 to 8.0 or even more. It can be grown successfully at a sunny situation in modertately dry, warm (being heat-tolerant) and not-too-rich garden soil but it should be light and well-drained. Kang and Iersel (2002) reported that gomphrena growth was found best at relatively high EC of the growing medium (2.3-6.5 ds/m). This suggests that gomphrena can tolerate relatively high levels of root zone salinity. Different salinity levels had also marked influence on leaf relative water content per cent leaf water potential (Sharma *et al.*, 1989a). Similarly, different salinity levels also affect mineral composition of gomphrena (Sharma *et al.*, 1989b). At all stages of growth, it needs full sun, however, it can tolerate the slight shade from the strongest sunshine during summer (Griffiths *et al.*,1992). It does well in both tropical and sub-tropical climates. Under mild climatic conditions, gomphrena will flower throughout the year. Puczel and Wazbinska (2004) compared the effects of soil or non-heated foil tunnel cultivation on the yield and morphology of *G. globosa* and found that cultivation of *G. globosa* in non-heated foil tunnel results in prolonged period of inflorescence harvest, increased flower yield (22-90%) and wider flower diameter.

Seedlings attain the size suitable for transplanting in one month. At this stage seedlings have 3-4 true leaves and are 3-4 cm in height. Seedlings are transplanted in the prepared beds at spacing of 20-30 cm apart, the dwarfer ones requiring lowest in the range while tallest ones highest. Planting should follow gentle watering to settle the roots and to recover the plants from lifting shock. For bed preparation, it requires 2-3 ploughings some 20 cm deep, followed by levelling and planking each time. At the time of first ploughing, some 100-150 quintals of farmyard manure per hectare should be incorporated in the soil so that before second or final ploughing all the grass seeds germinate which may be knocked down at the time of second ploughing and all the rootstocks of the perennial weeds, previous year's woody plant debris and other foreign materials which may hinder its cultivation are removed manually or with forks and then the beds of convenient sizes are made and levelled after making the water channels and bunds.

Pinching, i.e removal of terminal growing portion has been found effective in regulating and maximizing flower production through precocious branching. Effect of pinching on the yield and quality of flowers in gomphrena is given in Table 2. Rajasekharan *et al.* (1983) reported that the plants pinched 15 days after transplanting (the optimum stage) registered higher number of flowers per plant (602.5) with higher flower yield (3165 g), and larger size of flower diameter (2.85 cm) when compared to non-pinched plants which gave poor number of flowers per plant (518.5) with lower flower yield (2455.5 g/plant), and smaller size of flower diameter (1.5 cm). However, stalk length was found reduced in pinched plants.

Maintaining weed-free growing condition is necessary to produce high quality flowers. Insects and diseases can be kept to a minimum only if proper weed control practices are carried out regularly, along with appropriate control measures. Weeds are also a primary source of insects such as aphids, whiteflies, thrips, mites and slugs. Weeds pose

Table 2: Effect of pinching on the yield and quality of flowers in gomphrena*

Treatments	No. of flowers/ plant	Yield (g/plant)	Calculated yield (t/ha)	Stalk length (cm)	Diameter of flower (cm)
15 days a. t.	602.5	3165.0	23.48	5.0	2.85
30 days a. t.	560.2	2916.0	21.63	6.0	2.30
45 days a. t.	553.0	2845.0	21.19	11.0	2.55
Control	518.5	2455.5	18.22	12.5	1.50
CD (P = 0.01)	72.7	164.2	---	2.94	1.14

* Rajasekharan *et al.* (1983)

a. t. = after transplanting

a serious threat to gomphrena cultivation at an early stage of plant growth as these rob the main crop of its nutrients and suppress plant growth. Weeding should be carried out manually.

Irrigation and Feeding

Since gomphrena is a summer crop and has shallow root system, therefore, shallow but frequent watering should be carried out. Once the seedlings should be watered immediately after transplanting. The plants require adequate watering during active vegetative growth, therefore, continuous supply of moisture should be ensured in the field. Irrigation is done through flooding once in a week during hot and dry spells.

It is not a heavy feeder, hence, can be grown successfully with a small amount of fertilizers. However, Kang and Iersel (2002) found that gomphrena cv. 'Gnome White' prefers high fertilizer concentrations (1.0-2.0 X) of Hoagland nutrient solution which results in maximum dry mass production and more number of flowers per plant. Application of well rotten FYM @ 2-3 kg/m^2 at the time of bed preparation gives quality flower production with high yield. Parthiban *et al.* (2001) reported that number and yield of flowers were highest when supplied with FYM 200 kg + *Azospirillum* per hectare. In general, NPK fertilizers should be applied at the rate of 15-20 g/m^2 to improve growth and flowering.

Growth and Flowering

Environmental factors have great influence on the growth and flowering of gomphrena. High temperature and humidity affect flowering of gomphrena plants (Hong and Jeong, 1988). However, gomphrena is an irradiance indifferent plant. Erwin *et al.* (2003) found that when light levels are at or above 6.4 moles per day, gomphrenas do not flower earlier even when supplemental lighting is provided.

Growth and flowering can be regulated by application of plant growth regulators. Shanks (1981) used paclobutrazol (PP_{333}) granules @ 2% in the soil or its single spray at 10-80 ppm on gomphrena foliage and recorded dwarfed plants with reduced shoot length. Bailey and Whipker (1998) also recorded retarded growth when sprayed gomphrena plants with B-nine (2,500-5,000 ppm) or cycocel (400-3,000 ppm). *Pseudomonas fluorescens*, an antibiotic-producing bacterium, is assumed to be important in protecting plants from soilborne diseases. Nakata *et al.* (2000) when applied *Pseudomonas fluorescens* S272 culture-broth (0.01-1% v/v) and H_{12} polysaccharide, it resulted in enhancement of stem length of gomphrena by 160%. The results suggest that combination of polysaccharide and *P. fluorescens* S272 promotes plant growth as these increase antibiotic effect in the plants.

Growth and development behaviour of gomphrena is also regulated either by a single or by an interaction of several endogenous growth hormones like abscisic acid, kinetin, daminozide, B-Nine and cycocel. Bhargavi *et al.* (1977) found daminozide quite effective in delaying leaf senescence (Table 1).

Post Harvest

Gomphrena starts flowering after 75-80 days of transplanting. Plants pinched after 15 days yield some 20-25 tonnes of flowers per hectare as compared to only 18 tonnes from non-pinched plants (Rajasekharan *et al.*, 1983). Harvesting of flowers should be done either early in the morning or in the evening in cool hours of the day. Gomphrena flower heads should be harvested when they attain full size. For table decoration or for dry flower arrangements, flowers with stems are harvested. For drying, the flowers are harvested when the flower ball is fully expanded and then these are hanged in bunches upside-down in a cool and airy but at shaded place. To prolong the life of cut flowers, a few leaves are retained on the flower stalks.

Loose flowers are packed in bamboo baskets and are transported to the nearby wholesale market where they are sold by weight. However, in case of cut flowers the stems are graded as per their lengths, quality of individual flower and overall appearance. Short stems with faded and distorted flowers should be discarded in grading. The graded flowers are then bunched into round bundles. The stem portion of the bundles is wrapped in wet newspaper sheets. Though flower heads remain fresh for longer period, use of floral preservatives like SADH, sucrose and 8-HQC in appropriate proportion are found useful for enhancing the vase life of flowers.

Insect-Pests and Diseases

Gomphrena is comparatively free from insect-pests and diseases. Sometimes it is observed to be infested by aphids, beetle and lepidopterous larvae, leafhoppers, mites, thrips and whiteflies (*Bemisia tabaci*) where aphids, leafhoppers and whiteflies act as vectors for spreading viral diseases. These all can be controlled through 0.2 per cent Rogor. *Alternaria gomphrenae* causes significant losses in commercial planting of *G. globosa* cultivars 'Alba', 'Rubra', 'Aureo' and 'Soft Pink'. Das *et al.* (1996) reported a new leaf spot disease from Bhubaneswar caused by *Phaeotrichoconis crotalarie.* Benomyl, Iprodione and Mancozeb provide significant control of these diseases. However, *G. haageanaa* cv. 'Strawberry' was not affected by this disease (Rane and Wick, 1990). The effect of seed borne pathogens on gomphrena was studied by Wu *et al.* (2001). Seeds infected with *Nimbya gomphrenae* resulted in abnormal seedlings or un-emergent seeds of gomphrena. Treating the fungi-contaminated gomphrena seeds with *Bacillus megaterium* [1.3 x 10^9 colony forming units/ml] or 1% NaOCl significantly increases seed emergence. The disease could also be controlled through spraying of 10 ppm pyrifenox. Leaf spot disease caused by *Nimbya gomphrenae* on *G. globosa* cvs 'Buddy' and 'Globosa' was first reported from Taiwan (Chou and Wu, 2003). The fungus could grow in the temperature range of 16-32°C with maximum growth at 32°C. This disease can be controlled by spraying of 10 ppm pyrifenox.

Sutarya (1994) reported that *G. globosa* is somewhat resistant to mosaic diseases. However, a few viral diseases are found to infect gomphrena plants. The virus affected plants should be removed immediately as soon as they appear in the field. The symptoms of 'aster yellows' on gomphrena may appear as failure of flower formation, distorted leaves and flowers, and/or foliar yellowing. The best method for disease management is to control leafhoppers as soon as they appear. Leafhoppers can be controlled by spraying Carbaryl (Sevin 80 WSP) or Dimethoate (4 EC). Subbarayudu and Wilcoxson (1967) reported that the infection of 'red clover vein mosaic virus' of gomphrena can be controlled by spraying mannose. Gomphrena plants are very susceptible to 'tomato bushy stunt virus' infection. Severe mottle symptoms are observed followed by formation of necrotic lesions. The mottled tissue consists of small patches of necrotic, chlorotic and green tissues closely intermingled as reported by Appiano *et al.* (1978). Martinez-Soriano *et al.* (1996) reported difference in replication and symptoms of three viroid, *viz.* potato spindle tuber viroid (PSTVd), Mexican papita viroid (MPVd) and tomato plant macho viroid (TPMVd) on *G. globosa.* Symptoms consists of epinasty, stunting and some veinal necrosis. Such plants should be rogued out and destroyed.

References

Appiano, A., S. Pennazio and P. Redolfi, 1978. Cytological alternations in tissues of *Gomphrena globosa* plants systemically infected with tomato bushy stunt virus. *J. Gen. Virol.*, **40**: 277-286.

Badr, M., B.A. Abdel-Maksood, and S.S. Omar, 2004. Growth, flowering and induced variability in *Gomphrena globosa* L. plant grown from dry and water-soaked seeds treated with gamma rays. *Alexandra J. agric. Res.*, **49** (1): 49-70.

Bailey, D. and B. Whipker, 1998. Height control of commercial greenhouse flowers. *Horticulture Information Leaflet*, USA, 28 pp.

Bailey, L. H. and E. Z. Bailey, 1976. *Hortus Third: A Concise Dictionary of Plants Cultivated in the United States and Canada.* Macmillan Press, New York, 1290 pp.

Bessonova, V.P., L.M. Fendyur and T.N. Peresypkina, 1997. The impact of environmental pollution on male fertility in ornamental flowering plants. *Botanicheskii-Zhurnal*, **82**(5): 38-45.

Bhargavi, K., I.M. Rao and P.M. Swamy, 1977. Promotion of rooting and delay of senescence in detached leaves of *Gomphrena globosa* by growth regulators and Ca^{2+}. *Indian J. exp. Biol.*, **15** (11): 1069-1070.

Candida, C.J. V., R.B. Marcia and C.L.F.R. Rita, 1995. Fructans in callus of *Gomphrena macrocephala* St. Hil. *Plant Cell Tissue Organ Cult.*, **42**(3): 233-238.

Chou, H.H. and W.S. Wu, 2003. The etiology and control of leaf spot on globe amaranth. *J. Plant Path.*, **19**: 23-28.

Cocozza-Talia, M.A. 1993. Research on *Gomphrena globosa* L. and *Cirsium japonicum* DC as cut flower (Apulia). *Colture Protette* (Italy), **22**(3): 105-106.

Darlington, C.D. and A.P. Wylie, 1955. *Chromosome Atlas of Flowering Plants*, pp. 77-78. G. Allen and Unwin Ltd., London.

Das, S.R., Minati Kar and M. Kar, 1996. A new leaf spot disease of bachelor's button. *Plant Dis. Res.*, **11**(2): 167-168.

Erwin, J., N. Mattson and R. Warner, 2003. Fundamentals of Flowering in Plants : Supplemental Lighting and Earliness of Flowering. *Minnesota Commercial Flower Growers Bull.*, 5 p.

Griffiths, M., A. Huxley and M. Levy, 1992. *Dictionary of Gardening. The New Royal Horticultural Society. Macmillan Press,* New York, **4,** pp. 888.

Hong, Y.P. and J.H. Jeong, 1988. Effect of seeding time on the flowering of several garden annuals. 2. Studies on the 10 annuals including dahlia, gomphrena, aster etc. *The Research Rep. Rural Development Administration Hort.* (Korea), **30**(2): 104-110.

Kang, J.G. and M.W. van Iersel, 2002. Nutrient solution concentration affects growth of subirrigated bedding plants. *J. Plant Nutrition,* **25**(2): 387-403.

Martinez-Soriano, J.P., J. Galindo-Alonso, C.J.M. Maroon, I. Yucel, D.R. Smith, and T.O. Diener, 1996. Mexican papita viroid: Putative ancestor of crop viroid. *Proc. Natl. Acad. Sci.*, USA, **93**: 9397-9401.

Mears, J. A. 1980. The Linnaean Species of *Gomphrena* L. *Taxon,* **29**(1): 85-95.

Mercier, H., C.C.J. Vieira, and R.C.L. Figueiredo-Ribeiro, 1992. Tissue culture and plant propagation of *Gomphrena officinalis* - a Brazilian medicinal plant. *Plant Cell Tissue Organ Cult.*, **28** (3): 249-254.

Nakata, K., N. Harada, K. Sumitomo and K. Yoneda, 2000. Enhancement of plant cell growth by flocculation of the antibiotic producing bacterium, *Pseudomonas fluorescens* S 272, on the roots. *Biosci. Biotechnol. Biochem.*, **64**(3): 459-65.

Palmer, J. 1998. A taxonomic revision of *Gomphrena* (Amaranthaceae) in Australia. *Australian Systematic Bot.*, **11**(1):73-161.

Parthiban, S., T.E. Vadivelu and C.M. Pappiah, 2001. Integrated nutrient management in globe amaranth (*Gomphrena globosa* L.*). South Indian Hort.*, **49**: 187-190.

Puczel, U. and J. Wazbinska, 2004. Yield of globe amaranth (*Gomphrena globosa* L.) cultivated in soil and a non-heated foil tunnel. *Folia Universitatis Agriculturae Stetinensis, Agricultura,* **94**: 169-173.

Rajasekharan, L.P., K.G. Shanmugavelu and K.M.P. Nambisan, 1983. Studies of pinching in gomphrena (*Gomphrena globosa* L.). *South Indian Hort.*, **31**: 248-249.

Rane, K.K. and R.L. Wick, 1990. Occurrence of *Alternaria gomphrenae* in North America. *Canadian J. Plant Path.*, **12**(4): 442-444.

Ruiz, A.Q. and J.M. Valadez, 1985. Effect of different growth factors on *Gomphrena globosa* callus tissue. *J. Natural Products,* **48**(6): 976-980.

Sena-Filho, A., B.M. Ferreira, V.J. Frorez-Roncancio and M.E.S.P. Dematte, 2003. Effect of light and immersion in water on germination of globe amaranth seeds produced in the state of Sao Paulo, Brazil. *Revista Brasileira de Horticulture Ornam.*, **9**(2): 167-170.

Shanks, J. B. 1981. Chemical dwarfing of several ornamental greenhouse crops with PP_{333}. *Proc. Seventh Annual Meeting, Plant Growth Regulator Working Group,* Longmost, Colorado, USA, Great Western Sugar Co., pp. 46-51.

Sharma, S.S., R. Yamdagni and R.K. Arora, 1989a. Effect of salinity levels on physiological aspects of garden annuals. *South Indian Hort.*, **37**(6): 337-340.

Sharma, S.S., R. Yamdagni, and R.K. Arora, 1989b. Effect of different salinity levels on mineral composition of garden annuals. *South Indian Hort.*, **37**(5): 291-294.

Subbarayudu, S. and R.D. Wilcoxson, 1967. Effect of mannose on infection of *Gomphrena globosa* by red clover vein mosaic virus. *Phytopath.*, **57**: 1292-1295.

Sutarya, R. 1994. Study of mosaic disease of celery. *Buletin Penelitian Hortikultura* (Indonesia), **26**(3): 104-111.

Wu, W.S., H.H. Chou, S.M.Lin and H.C. Wu, 2001. The effect of seed borne pathogens on emergence of globe amaranth, calendula and *Tagetes* and the methods of control. *J. Phytopath.*, **149**(2): 91-96.

13

Helianthus

Ajit Kumar, Jyoi Bajeli, R.L. Misra, Manish Kapoor, Himanshu Trivedi, V.K. Rao and Sanyat Misra

Family: **Asteraceae**

Common names: Japanese sunflower/Miniature sunflower (*Helianthus cucumerifolius,* syn. *H. debilis*), Jerusalem artichoke/Sun choke/Girasole/Canada potato (*H. tuberosus*), Prairie sunflower (*H. maximillianii*), River sunflower (*H. decapetalus*), Silver leaf sunflower (*Helianthus argyrophyllus*), Sunflower (*H. annus*, syn. *H. lenticularis*), swamp sunflower (*H. angustifolius*), Hindi – *súrajmukhi*)

Introduction and Origin

Helianthus derives from the Greek *helios* meaning 'sun' and *anthos* meaning 'flower', as it was thought that the flowers follow the course of the sun though it is incorrect. Its native haunts are North and South America, sunflower being from Mexico and Peru, from where it was introduced to Spain and grown in the Royal Gardens in Madrid in 1562, and by the end of the 16th century, it is thought to have been cultivated in Florence (Italy) where that time it was known by its Portuguese name meaning 'Indian Sun' or 'Sunflower'. From Spain, it was intoduced into England. Initially these were being valued for the ornamental qualities of their yellow flowerheads but now they are also considered as a major cooking oil yielding plant since the start of its cultivation as oilseed crop in the 19th century in USSR, and so it was introduced into India in 1969 with many of the improved varieties having higher oil yield.

Sunflower (*Helianthus annus*) is an important summer season annual, which attracts attention in the garden due to its attractive flowers. It can also be used for its large, bright flowers in floral and wedding displays. Sunflower is also used for screening and as a background plant. Even before introduction of this crop in India from Russia for production of edible oil, the crop was being grown in British India but only for its aesthetic values as cut or loose flowers, but the specimens were only of primitive types and not the large-flowered improved ones which yield good cooking oil. Whether these are small

flowered or large ones but in all the cases the flowers are highly ornamental. A bed or field full of sunflower when is in flowering with its majestic yellow and nodding flowers, it looks quite spectacular. *H. debilis* has showy flowers which attract butterflies and the plants make wonderful drought tolerant ground cover in landscaping. Most are tall enough for planting in clumps at the back of wide borders. In the language of flowers, sunflower is a symbol of adoration, pure and lofty thoughts and love and respect. It also serves as an ideal catch crop when land is left fallow.

The most common sunflower (*Helianthus annus*) produces good quality of seeds which after roasting and grinding serve as a beverage similar to coffee, seeds as such yield edible oil used for cooking, and they also serve as food for birds, especially parrots. The roasted seeds of sunflower are eaten as they contain many nutritional constituents such as vitamin D, B, niacin and protein. Vitamin A, D and certain valuable minerals such as iron and calcium are found in the sunflower oil which make it more effective for many other therapeutic uses. Seeds make the flour more nutritious when ground by mixing with wheat. A bitter liquor is prepared against fever in Europe by infusing its petals in alcohol. Its petals also yield an yellow dye. The roots of sunflower were used by the native American Indians to treat snake bites. Its oil extracted from flowerhead was used as a hair conditioner. Sunflower has been reported as an important resource of natural oil and lipid-rich nutrients for centuries. Its oil contains therapeutic properties as it contains 40-44% high quality protein and used for curing skin diseases, haemorrhoids and leg ulcers. Bronchitis and rheumatism-like diseases were treated by flowerheads and leaves of sunflower by Russians.Certain primitive tribes eat raw tubers of *H. tuberosus* with pepper and salt or after boiling in water similar to Jerusalem artichoke or by roasting in mild fire. The flour prepared from the tubers of *H. tuberosus* are also used in the preparation of dietic and health foods. Some American Indian tribes relish the tubers of *H. maximilianii*, and use the seeds of *H. giganteus* to make a flour mixed with maize flour for making bread (Pizzetti and Cocker, 1968).

The worldwide cultivation of sunflower is more than 35 metric tonnes as in the year 2010 (The Statistical Division of the Economic and Social Department, FAO Report, 2011). India is at 12th rank in sunflower production with the yield of 1.06 metric tonnes (FAO Report, 2010). In India, sunflower is cultivated in around 18 lakh hectares (10% of the total global sunflower growing areas) and production is around 12.52 lakh tonnes (4% of the total global production). The major sunflower producing states are Karnataka (54.86%), Andhra Pradesh (20.83%) and Maharashtra (14.58%) (Anon., 2009). Since the people of this country are now fully aware about its potential for oil production, so commercial production of the crop for ornamental purpose is limited only for local use.

Botany and Breeding

These are hardy, erect, unbranced and fast-growing annuals and herbaceous perennials having height of 0.7-3.5 m. The perennial kinds of sunflower are tall and vigorous, fibrous-rooted, tuberous or rhizomatous, forming clumps or colonies of erect, thick and branched stems bearing alternate leaves (lower leaves opposite and cordate) which are rough on both the surfaces, ovate to lanceolate, irregularly toothed on the margins, and acute-acuminate with 3-veined, 10-30 cm long and 5-20 cm wide lamina. The flowering head is daisy-like on the terminal position on the main stem with 10-40 cm in diameter rotating to face the sun and something drooping in nature. This head consists of sterile outer ray florets and inner brown or black numerous disc florets spirally arranged. A definite pattern of florets in the head of a sunflower was given by H. Vogel in 1979 in polar coordinates expressed as $r = c\sqrt{n}$, and $\theta = n \times 137.5°$, where θ is the angle, *r* is the radius or distance from the centre, *n* is the index number of the floret and *c* is a constant scaling factor. It is a form of Fermat's spiral. The angle 137.5^0 is related to the golden ratio and gives a close packing of florets. This model has been used to produce computer graphics representations of sunflowers. A single head has capability of producing 350 to 2,000 greyish-green seeds covered by an outer shell (pericarp) that appears black, white or striated. The length of the basal seed may be up to 1.5 cm long. Flowering in sunflower takes place in late summer and autumn bearing large, golden-yellow blooms. Ovary found in *Helianthus* is inferior with single basal ovule, achenes obovoid, compressed, slightly 4-angled, variable in size, usually from 1.0-1.5 cm long with fully coloured or striped. Sunflower is protandrous in nature, and lagging time between maturation of stamens and pistils is around 18-24 hours, favouring cross pollination.

The first step towards improvement in sunflower is through **introduction** of high-yielding cultivars worldwise. In India, sunflower was introduced in 1969 from USSR. Some varieties released through introduction were EC68414, EC68415 and EC69874. After its introduction in the country, it is being grown in Karnataka, Maharashtra, Tamil Nadu and Andhra Pradesh and is spreading fast in other states, especially Punjab and Haryana. Second step is genetic improvement by combining good characters of various parents through **hybridization** and this way the emphasis on developing the hybrids in India reached to sound footing after 1980, and now around 80% of sunflower growing area in the country is under the hybrids. Sunflower is a highly cross-pollinated crop and is mainly pollinated through insects and to some extent by wind. Hand pollination increases seed setting up to 30% if followed in morning hours. Firstly, flower head opens which is preceded by the centre of the head blooms. This whole procedure takes place in 5 to 10 days which depends upon size of flower and season. Occurrence of anthers usually takes place between 5 to 8 a.m. with viable pollen whose viability remains for 12 hours, and stigma remains receptive for 2 to 3 days. Emasculation in sunflower is done by two methods, i.e. by hand and by chemical means (application of GA_3 @ 100 ppm after bud initiation for atleast 3 days). The emasculated heads are bagged to ensure that no foreign pollen reaches over the emasculated heads afterwards, and when stigma becomes receptive hand pollination with the help of cotton or brush is effected and again bagged. For hybrid seed production, generally **cytoplasmic male sterile lines** (CMS) are used. Leclercq of France discovered the first usable source of cytoplasmic male sterility in a cross *H. petiolaris* Nutt. × *H. annus* L. and Kinman of the United States developed fertility restorer lines RHA 265 and RHA 266 which allowed practical development of sunflower hybrids. Likewise, BSH-1 (CMS 234 × RHA- 274) was the first hybrid of sunflower which was developed by private sector by using CMS line used in hybrid seed production and was released commercially in 1980. Since then the hybrid programme was boosted up and 29 productive hybrids were developed, 11 by private sector and 18 in the public sectors. Non-functional male gamete (pollen), i.e. male-sterility, but with functional female gamete in the flowers of a variety offers the best opportunity for crossing with desired males without going for emasculation. Tripathi and Singh (2013) studied the effect of various concentrations of benzotriazole ($C_6H_5N_3$), an hybridizing agent, on plant morphology, days taken for first flowering, pollen fertility and yield in *H. annus* var. MSFH-17 through male sterility, and recorded that 1-2 sprays at 1% caused reduction in plant height, pollen fertility (almost complete pollen sterility) and yield components, *vis-à-vis* delayed flowering. Plants treated once and twice with 1.0% benzotriazole showed 23.01 and 22.22 g total seed yield/plant, respectively, in comparison to 23.68 g total seed yield/plant in control. **Heterosis** may be defined as the superiority of an F_1 hybrid over both its parents in terms of yield or some other character. But in some cases, the hybrid may be inferior to the weaker parent. Ahmad *et al.* (2005) studied the effect of heterosis and inbreeding depression in sunflower (*H. annus*) with an objective of recording heterotic effect in F_1 hybrids, and inbreeding depression in F_2 population. The yield and leaf area showed highly significant heterosis in F_1 hybrids ranging from 102 to 309% and 46.3 to 163.9%, respectively, while inbreeding depression in the F_2 population ranged from 17-71% and -9.7 to 4.3%, respectively. However, leaves per plant showed low level of heterosis in F_1 hybrids (-0.9 to 39.7%) whereas the effect of inbreeding depression in the F_2 population was comparatively high (1.1 to 22.2%) for this character. The parent RHA-882 proved itself to be a good general combiner by making higher contribution towards heterosis both in F_1 hybrids and in F_2 populations. A breeding method based on repeated **backcrossing** of the F_1 and the subsequent generations to one of the parents is known as a method of further improvement. Sujata *et al.* (2008) studied a set of 250 distinct, stable and uniform backcross-derived inbred lines which were developed in sunflower through 5 interspecific cross combinations involving four wild diploid annual species (*H. argophyllus, H. petiolaris, H. annuus* and *H. debilis*). The presence of wild-species genome in these inbred lines was confirmed through higher chromosome associations (tri- and quadrivalents) at diakinesis. Maximum structural rearrangements of chromosomes were observed in lines derived from *H. petiolaris*. There were 46 unique alleles and the number of unique alleles was highest in the lines derived from the cross involving *H. petiolaris*. The results indicate that the sunflower gene pool could benefit from introgression of novel alleles from the latent genetic diversity present in the wild species and particularly through exploitation of the diploid annual *H. petiolaris*.

Species and Varieties

Genus *Helianthus* comprises of some 80 species out of which 38 are perennial. The commonly cultivated species include *Helianthus angustifolius, H. annus, H. argophyllus, H. atrorubens, H. cucumerifolius, H. decapetalus, H. laetiflorus, H. maximillianii, H. mollis, H. salicifolius,* and *H. tuberosus*. The two species, *Helianthus annus* and *H. tuberosus* are more commonly grown for their exquisite blooms and *H. annus* for oil content. All the annual species described below such as *H. annus, H. argophyllus* and *H. cucumerifolius* produce exquisite blooms, though remaining ones here are perennials producing beautiful flowers but are least cultivated.

Helianthus angustifolius: A rhizomatous herbaceous perennial from United States (New York to Florida and west to Texas) is always found growing in wet, boggy or swamp-like positions to a height of 2.1 m with flowering from August to October. This tends to spread by rhizomes to form clumps. Its all the parts are hairy, leaves are lanceolate and 15-18 cm long, flowers are solitary (rarely 3), terminal and 7.5 cm across, the petals (ligules) are yellow and the disc florets are purple.

Helianthus annus (syn. *H. lenticularis*): An annual from United States (Minnesota to Washington and California), growing from 1.8 to 3.0 m in height and flowering from July to September producing huge black-centred flowers. Apart from being ornamental, this is the major oil-seed crop with bright yellow giant flowers (up to 45 cm across flowerheads) grown in every part of India, especially in all the five states of South India with many of the improved varieties. Much-branched smaller flowered bushy types are in plenty in the country which are mostly used for aesthetic taste. The dwarf type much-branched forms are mostly used as 'cut-and-come-again'. A dwarf vigorous form but with short fibrous roots requires support to save the plants from falling off due to lodging and winds. The leaves are stalked, alternate, heart-shaped or elongated-oval, irregularly toothed, 35-40 cm long and are hairy on both the surfaces. The terminal and solitary flowerheads have numerous orange-yellow and oval-lanceolate ligules and radiate round the velvet-black central disc, however, later in the season, in the axils of the leaves, relatively insignificant smaller blooms are borne. The disc bears several small pale-yellow tubular florets and when young these dominate the central disc with their colour.The oil content in seed ranges from 22 to 36% while the kernel consists of 45-55% oil. The colour of oil is pale-yellow with mild and pleasant flavour. In 1910, Mrs. T.D.A. Cockerell from Colorado collected a form suffused with a beautiful chestnut-red tint instead of pure golden-yellow and named it var. *coronatus* which she crossed with *H. annus* and many other varieties, and obtained one with reddish petals tipped yellow which she named var. *bicolor*, one with a red circle which she named var. *zonatus*, and one more where all the petals were completely chestnut-red which she named var. *ruberrimus*, and when in 2014 she crossed var. *ruberrimus* with var. *primulinus* with primrose-yellow flowers, she obtained a wine-red form which she named var. *vinosus*. Apart from these she also developed var. *vinosissimus* (petals completely dark wine-red) and var. *niger* (black petals suffused red at the tips). Apart from the Cockerell varieties, there are numerous other cultivars such as 'Abendsonne' (head diameter 15 cm, petals purplish changing to chestnut-red or yellow and centre maroon), '*californicus*' (semi-double with bright yellow disc), '*chrysanthaeflorus*' (large double flowers with cut petals), '*citrinus*' (primrose-yellow), 'Gelber Knirps' (dwarf with only 1.35 cm height, double flowered), 'Globe of Gold' (dwarf with up to 1.35 m height, double flowered), '*globosus fustulosus*' (enormous globose flowers), 'Goldener Neger' (much-branched, colour golden with dark centre), 'Indianerin' (much-branched, disc small, golden with dark centre), '*intermedius*' syn. *Helianthus hybridus* (as the original species but with smaller flowers), '*nanus flore pleno*' (height up to 1.25 m, double-flowered), '*purpureus*' (flowers suffused with pink to purple), 'Red' (red), 'Russian Giant' (very tall oil-deed crop), 'Sonnengold' syn. 'Sungold' (90 cm high, flowers quite double), '*variegatus*' (leaves variegated), *etc.* (Pizzetti and Cocker, 1968). Statewise recommended varieties for oilseed crop in India are 'Badshah', 'Divyamukhi', 'Haryana Surajmukhi-1' and 'Peredovik' for Punjab and Rajasthan; former three varieties and 'Jawalamukhi' for Haryana; 'APSH-11', 'Badshah', 'Divyamukhi', 'Laxmi', 'Mega-363', 'MSHF-8', 'MSHF-17', 'MSHF-30', 'Peredovik' and 'Sunrise Selection' for U.P.; 'Badshah', 'Divyamukhi', 'Peredovik' and 'Vinimik' for M.P. and Gujarat including 'MSHF-30' only for the latter state; 'APSH-11', 'CO-1', 'CO-2', 'Garden', 'MSHF-8', 'MSHF-17', 'MSHF-30', 'RSH-1', 'SS-56' and 'Surya' for Maharashtra; 'KBSH-1', 'MSHF-17', 'MSHF-30', 'Peredovik' and 'RSH-1' for Karnataka; and 'APSH-11', 'Armaverts', 'Armavirskij', 'Peredovik' and 'RSH-1' for Andhra Pradesh. ***H. annus*- Sunny Smile:** It is best known dwarf cultivar of *H. annus* which grows up to the height of 30 to 37.5 cm. It is perfect for growing inside the houses as no pollen is found. Growing this cultivar in the pot yields quality blooms. Some of the cultivars recently developed are 'Banilla Ice' (stems multi-branched, blooms nearly white and 12.5 cm across, suitable for cutting), 'Moonshadow' (height 1.2 m, male-sterile near white blooms with 10 cm across blooms, good for cutting and drying), 'Velvet Queen' (blooms orange and yellow but centre black, 20 cm across), 'Claret Hybrid' (1.5-1.8 m tall with side-branching, blooms male-sterile, deep red and 15 cm across), *etc.*

Helianthus argophyllus: An annual from North America growing up to 1.8 m and bears silky and soft hairy-clothed leaves giving silvery appearance. It generally produces 1.2-1.5 m long solitary stems which are usually soft grey and silky-hairy, especially the upper part, otherwise it is similar to *H. annus* with yellow flowers.

Helianthus atrorubens (syn. *H. sparsifolius*): A herbaceous perennial from United States (Virginia to Florida and westwards as far as southern Illinois, Missouri, Arkansas and Louisiana) found wild from light woodlands in dry soil, growing up to 2.1 m in height and flowering from August to October bearing large golden-yellow blooms. Its roots are wide-spreading, stems hairy and cylindrical, leaves mostly opposite, hairy on both the sides, oval-acuminate to lanceolate, dentate and prominently 3-veined, flowers several on naked stems with up to 15 cm long and 1.25 cm wide ligules of yellow colour, and the florets in the central disc purplish-brown. The species was one of the parents in evolving the most elegant and graceful autumn-flowering var. 'Monarch', a quite hardy bushy plant with height up to 2.0 m, flowerhead some 15 cm across with semidouble yellow flowers, which as cut flower lasts for a couple of weeks.

Helianthus cucumerifolius (syn. *H. debilis*): A much-branched annual from southern United States (Florida to Texas) is found growing up to 1.2 m with cucumber-like leaves, and free-blooming from July to October, bearing numerous small yellow, golden-yellow or yellow with dark eye flowers on long stalks. Its stems are generally reddish-blue spotted-yellow, leaves are cordate to triangular, dentate and some 10 cm long, disc florets purple-brown but other flowers yellow, solitary, terminal and axillary, 7.5 cm across, and the petals are narrow-lanceolate and spaced. They are excellent for cutting. Some of the most popular varieties are 'Dazzler' (orange, spotted-marron), 'Diadem' (large lemon-yellow but disc almost black), 'Herbstschönheit' (yellow-bronze, central disc ringed maroon), 'Orion' (golden-yellow with a little twisted petals), 'Perkeo' (free-flowering, blooms small and in various shades of yellow), '*plumosus*' (central disc changing from pale to intense yellow), '*purpureus*' (petals changing from pale-pink to intense purple), 'Rotstern' (flowers suffused with different shades of reddish-brown), 'Stella' (flowers intense golden-yellow with curled petals, disc dark violet-maroon), *etc.*

Helianthus decapetalus: A herbaceous perennial from North America (Maine and Quebec to Wisconsin, and as far as South Carolina and Missouri) is found in woodlands and in the vicinity of rivers. It is thought to have been introduced into cultivation in 1688. It grows up to 1.5 m with flowering from August to October. In mature plants the roots swell and become tuber-like. The stems are cylindrical and smooth but top slightly hairy, leaves thin and opposite, 10-20 cm long and 5.0-7.5 cm wide, oval to lanceolate, dentate, and little wrinkled above and smooth below, and the flowers are terminal, and yellow with some 2.5 cm long ligules. Some of the popular varieties are 'Flore Pleno' syn. 'Duplex'(fully double), 'Grandiflorus' (fully double), 'Major' (flowers large), 'Maximus' (flowers large and simple but ligules pointed), 'Simplex' (flowers like cactus dahlia), 'Soleil d'Or' (flowers as cactus dahlia), *etc.*

Helianthus laetiflorus (syn. *H. rigidus, H. scaberrimus*): A herbaceous perennial with spreading rhizomatous roots from North America (Minnesota, Indiana and Missouri, and westwards as far as Montana and New Mexico) has been found growing up to 2.4 m from August to September in dry prairieland. Stems downy-silky, leaves short-stalked, opposite towards base but alternate upward on the stem, oval-lanceolate, dentate and wrinkled, while flowers are short-stalked and several together, the ligules are yellow and 15-20 cm long, and the disc florets are purplish but sometimes yellow.

Helianthus maximillianii: A herbaceous perennial from North America growing up to 3 m in height with 7.5 cm diameter of flowerheads, and flowers during July to October. It has a tendency of spreading quickly. Leaves are trough-shaped, flower peduncles short and densely hairy, and the flowers are deep yellow with about 15-30 rays which are some 3.5 cm or more long.

Helianthus mollis: A frost-tender herbaceous perennial from United States (Florida to Midwest) which grows up to 3 m and flowers from September to October. It grows vigorously having been clothed throughout with soft white hairs. Leaves are 12.5 cm long, broadly lanceolate and dentate. Flowers some 7.5 cm across appear either solitary or in small groups in yellow colour but disc being darker. It is most suitable for planting at dry locations.

Helianthus salicifolius (syn. *H. orgyalis*): A herbaceous perennial from North America (Missouri and Kansas to Oklahoma, to as far east as Ohio), growing up to 3 m with flowering during September-October. It is found growing in dry prairielands in calcareous soils. Stems are reddish-brown, sylindrical, smooth and thickly covered with shortly-stalked, 15-40 cm long and some 8 mm wide, attractive, alternate, thin, lanceolate, willow-like and acuminate, drooping, slightly rough, leaves up to the crotch of the stem from where it has started producing pedunculate flowers though quite less on the portion of the raceme, i.e. stem bearing flowers. Flowers are numerous together on some 1.2 m long and branched spikes, 5 cm across, first greenish and then yellow, and the petals are lemon-yellow. This species does very well in poor, dry and sandy soils.

***Helianthus* tuberosus**: A tuberous-rooted hardy herbaceous perennial from eastern United States extending to western rocky mountains, and is the state flower of Kansas. It was introduced into England possibly in 1616. It grows up to 3.6 m in height with 10-40 cm leaf length, and flowering during September-October. Apart from its ornamental value, its tubers are also edible having texture similar to potato, rich in insulin and carbohydrates, and can easily be digested by a diabetic patient. Mostly used for growing around the corners of a garden. The stems are erect, cylindrical, leafy and hairy. Leaves are stalked, opposite or alternate, 15-20 cm long and 5-10 cm wide, oval-acuminate, roughly wrinkled and coarsely dentate. Flowers are long-stalked, terminal, solitary, 6.0-7.5 cm across and yellow along with ligules. Its tubers are edible by human beings and are also used as fodder for cattle and pigs.

Some of the other promising cultivars are 'American Giant' (height up to 5 m, blooms golden-yellow and 25 cm across, suitable as background crop), 'Mammoth Russian' (very vigorous, orange-yellow), 'Prado Red' (height 1.2-1.8 m, floral colours various, vase life poor), 'Zohar' (improved version of 'Jerusalem Gold', male-sterile, single-flowered, blooms 11.5 cm across, petals deep orange and centre dark brown, good for cutting), *etc.* All these are giant types which mature late with 1.8-4.2 m height bearing 30-50 cm across blooms. Semi-dwarf types are early maturing with 1.3-1.8 m plant height, flowerheads 17-23 cm across such as 'Jupiter' and 'Pole Star'. Dwarf types are also early maturing with 0.6-1.4 m height and 14-16 cm across flowerhead such as 'Advance' and 'Sunset'. Some other cultivars suitable for oil yielding as well as for flowers are 'PKVSH-27' developed by Krishi Vidyapeeth, Pune; and 'CO-2', 'CO-4' (TNAUSUF-7), and 'COSFV-5' by TNAU, Coimbatore. Some of the newly introduced F_1 hybrids suitable for home gardening are 'Full Sun' (large yellow head), 'Moonwalker' (attractively branched with pale heads), 'Sunbeam' (attractive golden-yellow with green eyes), and 'Valentine' (creamy-yellow with black disc). Krishi Vidyapeeth, Pune has developed three open pollinated varieties in sunflower namely 'SF9', 'Surya PKV', and 'TAS 82'. 'Sunny Smile' is excellent as pot plant.

Propagation

Propagation of sunflower is done through sexual (seeds) means in case of all the annual types, as well as asexual (cuttings) through division or by soft cuttings in spring of annual as well as of perennial types. Usually seed propagation is preferred over cuttings as it is more convenient and the plants raised from seeds are tall, vigorous and heavy yielder. The **seeds** of *H. annus* and its various cultivars, *vis-à-vis* other species and their varieties, in most of the cases, are easily available and germinate quickly at temperature ranging from 15 to 25 ^{0}C, preferably at 16 ^{0}C. The seeds can be sown during spring, i.e. in mid-March in sub-tropical conditions, in April under sub-temperate and temperate conditions whereas in February under tropical conditions, at their permanent positions, setting 2-3 seeds at a place followed by thinning later on, in pots, boxes or on raised beds. Since the plants grow very fast so they do not relish seedling-transplanting as shock by these plants is unbearable. For winter season crop, sowing of seed is done during mid-September to mid-October. Similarly, for summer and rainy season crop, sowing of seed is done in January-February and mid-June, respectively. With utmost care and to take the early crop, the crop may be grown through seeds in the nursery beds, in polybags or in plugs and gently transplanted one month after sowing. However, seeds of perennial sunflowers should be sown in nursery beds, then pricking them off 15 cm apart and then these are transplanted to their permanent positions or in the beds one month after sowing. The seeds of hardy types are sown in September and transplanted in October though tender ones from February to April and transplanted from March to May.

Vegetative propagation is used to derive uniform plants for various vegetative and floral attributes of a particular variety. Tip-cuttings of perennials from shoots of about 5-10 cm length from apical portion of the purely vegetative phase are taken which give out roots adventitiously. Cuttings root best when planted in sand or vermiculite in the plugs under mist chamber or in a highly humid atmosphere. Root promoting hormones such as IBA and IAA can also be used. The rooted cuttings are successfully transplanted in beds or pots for flowering. Apart from this, perennial plants can also be multiplied by division of rootstock before or when have initiated sprouting from October to April and replanted direct in permanent positions.

Cultural Practices

Sunflower can be grown in a wide range of well-drained **soils** with a pH range of 6.0 to 7.5. Drought-prone soil can also be easily accepted by sunflower as it is a drought tolerant crop. Waterlogged and more acidic soils should always be avoided. However, there are certain species and varieties which can easily be grown in alkaline (calcareous) soils, and certain others even in the wet or boggy soil. In any case, it should be at an exposed site. Commercial production of sunflower is mainly seen in the warm temperate regions, but due to breeding approaches, its varieties are adapted to the wide range of environments. Sunflower is grown 2,500 m above sea level and maximum oil yield/ha is recorded below 1,500 m a.s.l.. Usually, it is a **day-neutral** plant and its development is seen more rapid at 12 h daylight. Flowering is affected by **daylength** duration. High temperature drastically reduces its time of maturity. Sunflower holds great potential as it is short duration, photo-insensitive, widely adapted and drought-tolerant. Since it is a photo-insensitive crop, it can be grown in any season, i.e. *kharif, rabi* and spring-summer. All stages of growth including immature seed are likely to get damaged by frost. A frost-free period of 120 days is recommended for a commercial crop. The young plants of 4-6-leaf stage can withstand temperature of 5 to 6 °C for short period. Sunflower grows well within range of temperature of 20-25 °C. Temperature above 25 °C adversely affects the flowering, seed yield and the oil content though for ornamental purposes there is no problem of temperature as it grows luxuriantly even up to 40 °C temperature.

Sunflower crop raised by transplanting method gives higher yield and early maturity than that of seeded crop. Therefore, the crop should be raised by **transplanting** the nursery. About 100 m^2 of nursery is sufficient for transplanting one hectare area. Nursery for raising seedlings should preferably be made on raised beds due to more aeration and better drainage for healthy growth of seedlings for round the year cultivation of sunflower. Seed rate for raising nursery in one hectare area is 4.0 kg. Nursery should be raised by sowing seed 30 days before transplanting in well prepared seed-bed after mixing 1.5 kg urea and 4.5 kg single superphosphate per 100 m^2 nursery bed area. Seeds should be spread thinly in the nursery beds and covered with fine soil/sand or FYM. Seeds should be sown in lines spaced at 6-8 cm and at a depth of 5-6 cm. Thick sowing of seed should be avoided as it results in week seedlings which take longer time to regain growth. After seed sowing, beds should be covered with mulch materials of grasses or dried leaves for better germination. The beds should be watered with a fine nozzle can and should be kept moist. When the seed starts germinating, mulch should be removed in the evening hours otherwise etiolation of seedlings will take place. To minimize mortality of plants in the nursery, the soil should be drenched with 0.2% Captan. Land should be properly manured before transplanting of the seedlings. About one month old seedlings at 4-leaf stage should be transplanted at a distance of 30 cm between plant to plant and in rows 60 cm apart. Transplanting should be done during cooler hours of the day followed by irrigation. Uniformity in spacing ensures uniform growth and good yield. Ali *et al.* (2011) studied the performance of sunflower hybrids ('Hysun-38' and 'FH- 331') at 17.5, 20.0, 22.5 and 25.0 cm spacings and recorded maximum achene's yield (1920 kg/ha) in spacing of 22.5 cm, whereas 17.5 cm plant spacing resulted in minimum yield.

For **direct sowing of the crop**, pre-treatment with Apron® XL and Maxim® 4FS fungicides available in Cruiser Maxx at the rate of 6 g per kg of seed offers protection against a broad-spectrum of fungal diseases such as 'damping off', 'seedling blight' and 'downy mildew' usually caused by fungi such as *Pythium, Rhizoctonia* and *Fusarium.* Seed treated with Cruiser Maxx demonstrates better stand and establishment, more vigour, faster early-season growth and higher yield potential. Pre-treatment of sunflower seeds with 10 mg IAA and 10 mg NAA/litre of water for 6 hours gives high yield also. Sowing period for *kharif, rabi* and *zaid* (spring season) crops is first fortnight of July, Second fortnight of October and first fortnight of March, respectively. The seed should be sown at a depth of 4 cm in lines 60 cm apart. It takes about 10-12 days for germination. The crop should be thinned to maintain a plant to plant spacing of 20 cm. Normal recommended seed rate of sunflower is 8-10 kg per hectare. Seed rates of sunflower varieties are 7 kg/ha under rainfed conditions and 6 kg/ha under irrigated, whereas of the hybrids are 5 kg/ha under rainfed conditions and 4 kg/ha under irrigated conditions. **Earthing up** is an important operation in sunflower cultivation. It should be carried out when the crop attains a knee high stage to prevent the crop from lodging which generally occurs at heading stage in high speed windblown areas.

This crop responds well to the application of manures and fertilizers. Application of 25-30 tonnes per hectare of well rotten farmyard manure at the time of field preparation should be incorporated into the field. Generally, sunflower requires 75-80 kg N/ha under irrigated conditions and 50-60 kg N/ha for rainfed crop. Excessive application of nitrogen may reduce the oil content of sunflower and make it more prone towards diseases and lodging. Recommended rate of phosphorus and potash are 60-80 kg P_2O_5/ha and 50-150 kg K_2O/

ha which should be provided only after soil analysis. Application of 50 per cent nitrogen along with full dose of phosphorus and potash should be given as basal dose. Remaining 50 per cent of the nitrogen should be given in two split doses, i.e. 25 per cent at buttoning and 25 per cent at flowering stage. Beneficial effects of sulphur application have been observed in sunflower. Recommended dose of 37.0 kg/ha is beneficial for qualitative and quantitative oil production in sunflower. As per recommendations of Indian Fertilizer Association in 1992, the crop producing 3.5 tonnes of seeds per hectare requires 131 kg N, 87 kg P_2O_5, 385 kg K_2O and 70 kg MgO per hectare and micronutrients like Fe 732 g, Zn 348 g, Mn 412 g, Cu 59 g and B 396 g/hectare. However, for ornamental crops the doses may be lowered to its half. Sunflower is a good water drinker though it varies from season to season. *Kharif* crop hardy needs any irrigation but if rainfall is uneven, atleast one irrigation per week should be given. For proper germination and good crop stand, pre-sowing irrigation is must for both *rabi* and *zaid* crops. *Rabi* season crop should be irrigated after around 40, 75, 110 days of sowing which will coincide with four to five leaf stage, flowering stage and grain filling stage, respectively. It should be kept in mind that sunflower is highly sensitive to water stress during flowering and grain filling stage, so one irrigation is necessary during this period. Carter and Grieve (2006) investigated into the salinity tolerance of cut flowers of *Helianthus annuus* 'Moonbright' and 'Sunbeam' at the U. S. Salinity Laboratory to determine marketability based on stem length, where plants were exposed to differing water ionic compositions and salinity levels which though caused reduction in plant height with increase in the levels but the flowers were of marketable quality at moderate salinity levels. Abdou *et al.* (2011) studied the response of sunflower yield with irrigation schedule and found highest seed yield of 2,825 kg/ha by applying irrigation at 1.2 cumulative pan evaporation (C.P.E.), while the lowest seed yield of 1,490 kg/ha was obtained from 0.8 C.P.E.

Good **weed control** should be effected as sunflowers are poor competitors, especially in first four weeks of establishment. Application of Fluchloralin (Basalin) at 1.0-2.0 kg/ha should be applied as pre-plant incorporation (PPI) into soil or sprayed into soil before sowing of a crop followed by slight stirring of surface soil for incorporation into soil to reduce vaporization. Uses of Pendimethalin @ 1.5 kg a.i./ha 1-2 days after sowing (pre-emergence sprays) also show positive results. Weeds can also be effectively controlled by applying Oxadiazon @ 0.5 - 1.0 kg/ha or Alachlor @ 1.0-2.0 kg/ha as pre-emergence sprays. In case no pre-emergence weedicide is used, two hand weedings at monthly intervals should be practised to ensure weed-free conditions up to 60 days after sowing which results in higher yield. **Hoeing** should also be practiced as it will not only kill the weeds but also aerate the soil.

Growth and Flowering

Plant growth regulators play an important role on growth and yield attributes of sunflower. Koutroubas *et al.* (2004) studied the effect of PGRs on plant height at various stages of growth. Chlormequat chloride at 3.8 kg/ha reduced plant height at all the three stages under observation. Conversely, mepiquat chloride at 0.025 kg/ha had a significant effect at both flowering (9.2% reduction compared to untreated control) and maturity (11.7%) stages. Split applications at 0.0375+0.0375 kg/ha and 0.05+0.05 kg/ha revealed the shortening effect at maturity (9.5%) and 10 days after transplanting (6.5%), respectively. Paclobutrazol at 0.0125 kg/ha had only a slight reduction effect at maturity (11.4%), while no effect at the other stages was observed. Tahsin and Kolev (2006) investigated the response of sunflower hybrid 'Super Star' to various plant growth regulators, *viz.* H-40 (alfanaftilvinegar acid), M-2 (amide of dicarbonovy acids of benzolovid number), 31 (derivate of naphthalene with auxin efficacy), XP (flalaminova acid) and Agat 25 EK- 2.5 g/da application during pinhead and flowering stage. The results revealed that there was no significant difference between the seed yield (kg/da) and oil content (l/da) in pinhead period by the use of plant growth regulator 31 (1 cm^3/l water) and Agat 25 EK (2.5g/da). During the flowering period, the seed yield was increased by 15.3% and oil content by 16.4% with Agat 25 EK but it was decreased by 18.6% by application of 31.

Post Harvest of the Blooms

They are harvested early in the morning when outer florets in the heads have opened. If humidity on the plants is more at the time of cutting, better these should be harvested in the evening. Immediately after harvesting the cut-ends of the cut flowers should be put in distilled or mineral water. As a cut flower these last from one to two weeks depending upon the cultivars, growing conditions and prevailing weather conditions. Quality sunflowers are of uniform maturity and free from blemishes and any other defect. They should have straight stems and good quality foliage. Smaller flowered cultivars should be bunched in 10's or 12's but large flowered are packed individually. Pre-treatment of sunflower includes treating with detergent like Tween-20, Trilon X-100, dishwashing detergent in clean water for 15 to 30 minutes which avoids the premature wilting of sunflower in the vase. After that sunflower can be safely stored at 0-1^0 C and packed in standard horizontal flower boxes. *Helianthus* is also gravity-sensitive as horizontal packaging may permanently bend the flower heads upwards due to accumulation of auxins in the lower side of the stem which decreases the beauty of the flower stem. It is important to maintain cool temperature during transportation and storage of cut flowers. As the loose flowers also this is an important crop, and the ones only with smaller heads are used for the purpose. After harvesting these are filled in the baskets and taken to the narket. Packing in gunny bags damages the flowers and these degenerate soon. If need be these may be cold-stored at 5-7 ^{0}C for 3-4 days. These as cut flowers are used for worshipping and for making garland.

Seed Harvesting, Curing and Value addition

When black flowerheads of sunflower become yellow, it is said to be physiologically mature. Maturation of crop takes about four months after sowing. The seeds are ready to **harvest** when colour of head turns brown on the back or when involucral bracts turn yellow and loosening of seeds takes place. High moisture content at time of harvesting reduces seed shattering and saves the plants from lodging. Moisture in the sunflower seeds should not exceed 20 per cent at the time of harvesting. All the heads are not ready for harvesting at a time, therefore, harvesting is carried out twice to thrice to avoid seed shattering. The harvested heads should be spread on the floor and **sun-dried**, followed by their threshing with a small stick. In commercial production, threshers can also be used after speed adjustment in the thresher. Cold air treatment can be used for reducing the moisture up to 15% and can be said as the best method of drying. However, the moisture percentage of seeds should be brought down to 10 per cent before storage or oil extraction. If seed has to be used for long term storage then reduction of moisture to 9 % is necessary. On an average, one can get seed yield up to 20 q/ha by following improved package of practices. In case when the crop is harvested for fodder use or for making silage, it is done when the crop is in flowering.

Its flower petals are also used for **value addition**. Lutein extracted from the flower petals of sunflower is the best source of carotene for colouring food stuffs. Essential oil is also extracted from sunflower (*H. annus*) which is a potential source of protein in poultry diets. After extraction of oil from sunflower seeds, they are passed through a 1.4 mm screen which produces low-fibre sunflower residues and results in a feed that improves poultry performance (Mupeta *et al.*, 2001). North Dakota State University has done research on purple hulled sunflower which gave red food colorant extract which can be used for colouring the food stuff or as a dye. Since the petals of the sunflower are rich in coloured source compound, so an attempt has also been made to study its ability to quench reactive oxygen species. The petals from their flowers add bright colour and a spicy tang to tossed salads. The petals are chopped to make a tangy garnish for boiled eggs, steamed vegetables or fish dishes. Sunflowers show high tolerance to heavy metals, therefore, are used in phytoremidation studies (Pilon-Smits, 2005). It is also having the potential of removing heavy metals such as Zn or Cu as well as several radio nuclides from contaminated environment (Nehnovajova *et al.*, 2005).

Insect-Pests and Diseases

Sunflower crop is susceptible to various insect-pests and diseases which cause decrease in yield qualitatively and quantitatively. Various insect-pests that attack the crops are elaborated herewith. **Head borer** (*Helicoverpa armigera*) is a polyphagous insect and a serious pest of sunflower responsible for causing 20-25 per cent loss in yield under normal conditions. However, sometimes the damage is very severe and loss goes up to 40-70 per cent. Newly hatched and grown up larvae feed on leaves, buds and flowers for a short period of time and after making a hole in the disc may enter in it to feed the developing seed. The full grown larvae are greenish in colour and about 3.5 cm long. Deep ploughing of the field is helpful to kill the hibernating larvae. Spray of nuclear polyhedrosis virus (HaNPV) @ 350 LE (larval equivalent)/ ha for the control of 1st and 2nd instar larvae and release of egg parasitoid, *Trichogramma* sp @ 50,000 adults/ ha at weekly interval will keep the head borer at bay. **Hairy caterpillars** (*Spilosoma obliqua*) are polyphagous pests found throughout the year. Among various hairy caterpillars, Bihar hairy caterpillar is one of the most devastating causing severe damage to the sunflower crop. The female lays eggs in cluster on the lower surface of leaves. After hatching, the tiny larvae up to second instar feed gregariously on the chlorophyll content of the leaves. The infested leaves look like a dirty paper, which can be recognized from a distance. The eggs which are laid in cluster should be collected and destroyed manually and application of *Bacillus thuringiensis* (Bt) @ 1.0 kg/ha has been found effective in controlling hairy caterpillars. They can also be controlled by spraying of Quinalpos 25 EC @ 1.5ml/l or Chlorpyriphos 20 EC 1.0ml/l. The nymphs and adults of **jassids** (*Amrasca biguttula*) are important pests of sunflower as they suck cell sap. They are small, greenish-yellow and very active and can be seen in clusters on the lower surface of the leaves. The symptom is characterized by typical yellowish-white spots on the leaves, shrivelling of seeds and the drastic reduction in oil content. Balanced dose of fertilizers should be applied as excess nitrogen makes the plants more sappy and susceptible. Spraying of 5% *neem* **seed kernel extract** protects the crop from insect damage and application of oxydemeton methyl 25 EC or Dimethoate 30 EC @ 1.0 ml/l water controls almost all the pests effectively. **Cut worm** (*Agrotis ipsilon*) is also a polyphagous pest causing serious problem with sunflower cultivation. Greasy brown coloured larvae feed on the young roots and basal portion of the plant below the ground and kill the plant by cutting at the base. Deep ploughing of the field should be done after harvesting of the crop in order to expose the pupal stage of pest so that these are destroyed being exposed from their safe places to open environments immaturely where these may be eaten by various insects and birds. Treating the seeds with Chlorpyriphos 20 EC @ 12 ml/kg seed will keep these worms away from being swallowed by them after sowing. **Ants** also feed on the seeds by taking these away to their burrows so the burrows should be dusted with pure and fresh turmeric powder or their access to the field should be checked by drawing a line of powdered turmeric as a barrier around the field.

The **birds** like parrot, house sparrow and doves eat away seeds which are not fully covered at the time of sowing. They also cause damage to the crop at the time of seed formation. Seed treatment with Chlorpyriphos 20 EC @ 12 ml/kg seed protects the sown seeds from bird damage. Use of bird scarer is very important from the starting of seed development stage to harvest, *vis-à-vis* at the time of sowing.

Various diseases which infect sunflowers are elaborated here under. **Alternaria blight** (*Alternaria helianthi*) causes more than 80 per cent of yield losses under severe epiphytic conditions (Shankergoud *et al.*, 2006). Symptoms of alternaria blight first appear on lower leaves which later spread to middle and upper leaves. Appearance of dark brown irregular leaf spots with very dark border and grey centre are its major symptoms. This disease affects the quality of seeds by reducing the germination percentage and infection of the disease causes losses to an extent of 30 per cent. Till now, no blight-resistant variety is available. Therefore, it becomes inevitable to go for the use of fungicides for the management of this disease. Mesta *et al.* (2011) used Propiconazole and Hexaconazol @ 0.1% sprayings fortnightly from seed germination stage and recorded maximum yield of 10.06 q/ha through Propiconazole, followed by 9.42 q/ha through Hexaconazol. The highest benefit cost ratio was obtained in Hexaconazol, though net returns were highest in Propiconazole.

Botrytis head rot or grey mould (*Botrytis cinerea*) is favoured by root-zone temperature of 15-25 ^{0}C and wet conditions. It usually develops on seedlings, stems, leaves and causes major destruction when it infects the head. Symptoms first appear on the back of the head as sunken brown spots. If infection is found in the head late in the season, i.e. during harvesting period, maximum seed losses take place but both yield and quality get reduced during early infections. As there is no chemical options for controlling *Botrytis* for this crop, application of a desiccant in the crop should be applied when seed moisture reaches about 30 per cent. Zineb 0.2 per cent spraying when its first appearance is visible and then repeated 2-3 more times will control this problem. **Sclerotinia wilt and rot** (*Sclerotinia sclerotiorum*) causes stem and head rot in sunflower. Stem rot is caused by the fungal threads known as mycelium. This leads to wilting and finally collapsing of the plant. An air-borne spore causes head rot which infects the plants during the growing season and finally results in total destruction of the head and plant. The attack of this disease is related to the number of crops grown in the rotation. If the interval between susceptible crops gets reduced, the risk of infection decreases. Till today there is no chemical control available in sunflower against this problem. However, proper sanitation and crop rotation are only the alternatives left. When the seeds of sunflowers were inoculated with *Pseudomonas fluorescens* strain E6, Yuen and Schroth (1986) recorded increased growth and fresh weight (18-41% greater) after 3-4 weeks of treatment. Among the **viral diseases** affecting the crop, the sunflower necrosis viral disease (SND) caused by 'tobacco streak virus' (TSV) belonging to the *Ilarvirus* class, and bud necrosis caused by 'tospovirus' (Lavanya *et al.*, 2005) are most important. Ravi *et al.* (2001) also reported that SND in India is caused by *Ilarvirus* related to TSV and in their study they observed that the symptoms of this disease comprised of chlorotic & necrotic ringspots, and leaf distortion. They also observed that in electron microscopy *Ilarvirus*-like particles were detected in crude sap of SND-affected sunflower inoculated with leaf extracts prepared from SND-affected sunflower plants. Back transmission to healthy sunflower seedlings with leaf extracts of systemically infected indicator plants resulted in identical symptoms of SND, hence confirming the *Ilar*-like virus as the causative agent of SND. In DAC-ELISA, SNV was detected up to 0.370µg/ml in purified virus preparation and up to 10^{-4} dilution in leaf sap extracted from virus infected sunflower leaves. In Dot-ELISA, SNV was detected up to 0.0370µg/ml in purified virus preparation and up to 10^{-5} dilution in leaf sap from infected sunflower leaves. Whereas in DAS-ELISA, the virus was detected up to 370 µg/ml and 10^{-3} dilution in purified virus preparation and in leaf sap extracted from infected sunflower leaves, respectively. Hence, it is concluded that DAC-ELISA and Dot-ELISA were found to be the suitable tests for the detection of sunflower necrosis virus (Kotakadi *et al.*, 2012).

References

Abdou, S.M.M., K.E. Abad El-Latif, R.M.F. Farrag and K.M.R. Yousef, 2011. Response of sunflower yield and water relations to sowing dates and irrigation scheduling under middle Egypt conditions. *Advances in Applied Science Res.*, **2**(3):141-150.

Ahmad, S., M.S. Khan, M.S. Swati, G.S. Shah and I.H. Khalil, 2005. A study on heterosis and inbreeding depression in sunflower (*Helianthus annus* L.). *Songklanakarin. J. Sci. Technol.*, **27**(1): 1-8.

Ali, A., M. Afzal, I. Rasool, S. Hussain and M. Ahmad, 2011. Sunflower (*Helianthus annus* L.) hybrids performance at different plant spacing under agro-ecological conditions of Sargodha Pakistan. *International Conference on Food Engineering and Biotechnology IPCBEE* vol. 9 (2011) @ (2011) IACSIT Press, Singapore.

Anonymous, 2009. Sunflower. Annual Report 2009-10, p. 225. All India Coordinated Research Project. Directorate of Oil Seed Research, Rajendranagar, Hyderabad.

Carter, C.T. and C.M. Grieve, 2006. Salt tolerance of floriculture crops. In: *Ecophysiology of High Salinity Tolerant Plants*, pp. 279-287 (eds Almal Khan, M. and D.J. Weber). Springer Science+Business Media, Dordrecht, Netherlands.

Kotakadi, V.S., S.A. Gaddam and D.V.R.S. Gopal, 2012. Serological tests for detection of sunflower necrosis Tospo Virus causing necrosis disease of sunflower (*Helianthus annus* L.). *The Bioscan.*, **7**(3): 543-545.

Koutroubas, S.D., G. Vassiliou, S. Fotiadis and C. Alexoudis, 2004. Response of sunflower to plant growth regulators "New directions for a diverse plant", *Proceedings of the 4th International Crop Science Congress*, Brisbane, Australia, 26 September-1 October, 2004.

Lavanya, N., R.M. Ramiah, A. Sankaralingam, P. Renukadevi and R. Velazhahan, 2005. Identification of hosts for ilarvirus associated with sunflower necrosis disease. *Acta Phytopathol. Entomol. Hungarica*, **40**: 31-34.

Mesta, R.K., V.I. Benagi, S. Kulkarni and M.P. Basavarajappa, 2011. Management of alternaria blight of sunflower through fungicides. *Karnataka. J. Agric. Sci.*, **24**(2):149-152.

Mupeta, B., J. Wood, F. Mandonga and J. Mhlanga, 2001. The performance of indigenous under improved feed management compared with the hybrid chickens in Zimbabwe. DFID project R7524. NR International, Aylesford, Kent, U.K.

Nehnovajova, E., R. Herziq, G. Fedcrer, K.H. Erismann and J.P. Schwitzguebel, 2005. Screening of sunflower cultivars for metal phyto extraction in a contaminated field prior to mutagenesis, *International Journal of Phytoremediation*, **7**: 337-349.

Pilon-Smith, E. 2005. Phytoremediation. *Ann. Rev. Plant. Biol.*, **56**, 15-39. Online at: arjournals. Annual reviews.org.

Pizzetti, I. and H. Cocker, 1968. *Helianthus*. In: *Flowers ~ A Guide for Your Garden*, pp. 557-565. Harry N. Abrams, Inc., Publishers, New York, USA.

Ravi, K.S., A. Buttgereitt, A.S. Kitkaru, S. Deshmukh, D.E. Lesemann and S. Winter, 2001. Sunflower necrosis disease from India is caused by an ilarvirus related to *tobacco streak virus*. *New Disease Reports*, **3**: 10.

Shankergoud, I., K.G. Parameshwarappa, H.T. Chandranath, P. Katti and R.K. Mesta, 2006. Sunflower and Castor Research in Karnataka. An Overview, *Univ. Agric. Sci.*, Dharwad (India), p.1-41.

Sujatha, M., A.J. Prabakanan, S.L. Diwedi and S. Chandra, 2008. Cytomorphological and molecular diversity in backcross-derived inbred lines of sunflower (*Helianthus annus* L.). *Genome*, **51**: 282-293.

Tahsin, N. and T. Kolev, 2006. Investigation of the effect of some plant growth regulator on sunflower (*Helianthus annus* .L). *Journal of Tekirdag Agricultural Faculty*, **3**(2):229-232.

Tripathi, S. and K.P. Singh, 2013. Hybrid seed production in *H. annus* L. through male sterility induced by Benzotriazole. **2**: 645.

Yuen, G.Y. and M.N. Schroth, 1986. Interactions of *Pseudomonas fluorescens* strain E6 with ornamental plants and its effect on the composition of root-colonizing microflora. *Phytopath.*, **76**(2): 176-180.

14

Hibiscus

Jayoti Majumder, Tarak Nath Saha, Sanyat Misra and R.L. Misra

Family: **Malvaceae**

Common names: African rose mallow (*Hibiscus acetosella*), Arizona rosemallow (*H. biseptus*), Blacking plant/China rose/Hawaiian hibiscus/Hibiscus/Rose of China/Shoe black plant/Shoe flower/Shrubby althea (*Hibiscus rosa-sinensis*), Bladder Ketmia/Flower-of-an-hour/Trailing hollyhock (*H. trionum*, syn. *H. vesicarius*), Brackenridge's rosemallow (*H. brackenridgei, H. brackenridgei* ssp. *brackenridgei*), Brazilian rosemallow (*H. phoeniceus*), Brown Indian hemp (*H. cannabinus*), Comfortroot (*H. aculeatus*), Congo mahoe (*H. clypeatus*), Coral hibiscus/Fringed rose mallow/Japanese lantern/Skeleton hibiscus (*Hibiscus schizopetalus*), Cotton rose/Confederate rose/Dixie rosemallow/Rose of Sharon, (*H. mutabilis*), Crimson-eye rosemallow (*H. moscheutos, H. oculiroseus*), Desert rosemallow (*H. coulteri*), Fork-bracted rosemallow (*H. bifurcatus*), Halberdleaf rosemallow (*H. laevis*), Heartleaf rosemallow (*H. martianus*), Hollyhock tree (*H. splendens*), Largeleaf rosemallow (*H. macrophyllus*), Lemon-yellow rosemallow (*H. calyphyllus*), Lindenleaf rosemallow (*H. furcellatus, H. furcellatus* var. *youngianus*), Mahoe/Mountain mahoe (*H. elatus*), Mokulei rosemallow (*H. brackenridgei* ssp. *mokuleianus*), Monarch rosemallow (*H. radiatus*), Neches river rosemallow (*H. dasycalyx*), Paleface (*H. denudatus*), Poeppig's rosemallow (*H. poeppigii*), Punaluu rosemallow (*H. arnottianus* ssp. *punaluuensis*), Queen of the summer hibiscus (*H. manihot*), Red Kauai rosemallow (*H. clayi*), Red rosemallow (*H. kokio, H. kokio* ssp. *kokio*), Roselle (*H. sabdariffa*), Rosemallow (*H. lasiocarpos*), Rose of Sharon/Tree mallow (*H. syriacus*), Scarlet rosemallow (*H. coccineus*), Seaside mahoe (*H. pernambucensis*), St. John's rosemallow (*H. kokio* ssp. *saintjohnianus*), Striped rosemallow (*H. striatus, H. striatus* ssp. *lambertianus*), Sunset hibiscus/Swamp rose-mallow (*H. grandiflorus, H. moscheùtos*), Three-lobed rosemallow (*H. trilobus*), Tropical rosemallow (*H. vitifolius*), White Kauai rosemallow (*H. waimeae, H. waimeae* ssp. *hannerae*s), White rosemallow (*H. arnottianus, H. arnottianus* ssp. *arnottianus, H. arnottianus* ssp. *immaculatus*), English-Hibiscus, Mallow, Rose mallow, Hindi- Gudhal, Japa, Javakusum, Bengali-Jaba, Malayalam-Bunga raya, Tamil-Sembaruthi, Telugu-Dasananu.

Introduction

The family Malvaceae comprises 88 genera with 23,000 species which are distributed in tropical, subtropical and temperate regions (Willis, 1973; Mabberley, 1987), and under this family, *Hibiscus* is a polymorphic genus of some 250–300 species of trees, shrubs, perennial herbs and annuals, widely distributed in mainly tropical and subtropical regions (Bailey, 1950; Bates, 1965; Beckett, 1987), such as Tropical Africa, Old World tropics, Tropical Asia, India, Java, Mautitius Islands, Seychelles Islands, China, Japan, Mediterranean, Australasia, Pacific Islands, United States (Pennsylvania to Minnesota, Massachussets to Florida, Michigan, New Jersey, etc.), Brazil, West Indies, etc. Father Ferrari in his book *'De Florum Cultura'* mentioned *Ketmia* or 'Rose of China', i.e. *Hibiscus syriacus* as was then called, 'It is such a pretty plant, with such charming flowers, that when the Indians wish to stress the beauty of a man they describe him as more beautiful than the Chinese Rose'. *Hibiscus*, a name given by Linnaeus which in Greek means 'mallow' or it is an old Latin name which includes even *Abelmoschus* and *Paritium*, and is allied to *Abutilon, Althaea, Gossypium* and *Malva*. The name probably derives from *ibis* due to a bird living in such shrubs. When Linnaeus named *H. syriacus* he may not have been aware of its native being India and China but surely it is not native to Syria as it is named so. *H. syriacus* before being named so, it was named as *Altheaea frutex* (Pizzetti and Cocker, 1968). Ornamental hibiscus for loose flowers (*Hibiscus rosa-sinensis* and *H. schizopetalus*) is very popular in India and its flowers are used for religious offerings, for making garlands, as pot plant for use in porches, balconies, patios and elsewhere, for hedging, for use in landscaping in the garden and for growing in the backyard for instant collection of its beautiful flowers whenever the need be. It is widely grown as an ornamental plant throughout the tropics and subtropics. Its flower colours range from white through yellow and orange to scarlet and shades of pink, with both single and double sets of petals and bicolours. The species *Hibiscus rosa-sinensis*, the most beautiful of all the shrubby hibiscus has numerous hybrids, and the species was introduced into Europe from East Indies in 1731. *H. rosa-sinensis* is the national flower of Korea (Van de Laar, 1997), Malaysia (known as *Bunga Raya* in Malay and the Republic of Haiti (Lawton, 2004). The hibiscus blooms throughout the year, most freely in summer, after a period of rain when weather is relatively cool. *H. syriacus* is known in England since 1597 when its seeds were first sown by John Gerard. John Parkinson was growing it in 1629 and its first varieties started appearing in 1659, in 1730 one variety was evolved with flowers bearing red and white stripes, and in 1838 John Claudius London mentioned that the double-flowered were most common. In China, *H. syriacus* is in cultivation since the time of antiquity and at one time the leaves were used as a substitute for tea while the flowers were eaten as a delicacy. *H. mutabilis* was introduced into European gardens in 1690. The common name of *H. mutabilis* given in Chinese language means 'public official' due to its flowers opening white and then turning dark red.

They are also a source of food, fibre and medicines. Many of the species yield bast fibres, *viz. H. cannabinus* (Anderson and Pharis, 2003), *H. sabdariffa, H. surattensis, H. planifolius, H. flococcus* and *H. macrophyllus* which are locally used for fibre in Asia. In West Africa, throughout the Pacific, in Malaysia and in the Sundarbans and Andaman Islands of India, *H. honarifolius, H. rostellatas, H. squamosus* and *H. tiliaceus* (a small tree of tropical shores throughout the world) yield bast fibres for making ropes and mats (Cobley, 1976). *H. cannabinus* and *H. sabdariffa* leaves are consumed as green vegetables while the swollen calyx of *H. sabdariffa* are eaten raw in salads or used in making marmalades, jams, jellies and preserves. Its calyces have medicinal value due to possession of saccharose, malic acid and oxalic acid as active principles. The dried calyx of *H. sabdariffa* is used for making an infusion beverage called 'zobo' in Nigeria. Leaves of *H. acetosella* and *H. starattensis* are used as medicines to cure certain stomach ailments (Mohamed *et al.*, 2012). Seeds of *H. moscheutos* have been reported to be used medicinally (Winters, 1970). Various *Hibiscus* species are used in Ayurveda as a remedy for many ailments and diseases such as fever, menorrhagia, gonorrhoea, blood vomiting, stomachic troubles, irritable geneto-urinary tract conditions, swellings, boils, ulcers and for hair strengthening as well.

Botany, Genetics and Breeding

Hibiscus rosa-sinensis has long been cultivated in China, India, Japan and the Pacific islands where this has been recorded growing even long before the European discovery era, and is now one of the most widely grown ornamental shrub throughout the tropics and sub-tropics for its showy flowers (Kimbrough, 1997) which are epheral in nature ~ opening in the morning and senescing by the evening. This is so versatile that it can be used even as house plant. It is a very important potted plant in Europe, USA and other temperate countries but only in the regions where minimum persisting temperature does not fall below 12 ^{0}C (Hughes and Hughes, 2003). It is a upright growing (1.5-6.0 m) evergreen woody plant, often described as being either a small tree or large shrub, which at maturity becomes almost round to oval in shape, the mature tropical hibiscus can grow with a densely branched, rounded silhouette or with a low, mounded but spreading shape having fewer branches though dependent upon varieties and with hundreds of extant cultivars (Kumar and Singh, 2012). The bark is grayish-brown to gray and becomes almost corky on large and old plants. The evergreen and glossy green leaves on tropical hibiscus plants are ovate, 7.5-15.0 cm long and 6.25-8.75 cm wide, often coarsely but irregularly toothed and pointed though sometimes slightly lobed. The leaf blade is wide, tapering and oval with a pointed tip. The base of the leaf where it attaches to the petiole is either rounded or wedge-shaped. Leaf edges lack any imperfections, or more commonly sparse, short teeth. Tropical hibiscus leaves never display lobes. If temperatures are warm and soil moisture is adequate, tropical hibiscus blooms year-round (Lee, 2007). Flower buds arise from the branch tips anywhere on the plant. Several flowers may be produced on each branch tip, but only one at a time. Each bud opens into five green sepals enclosing five papery petals. Each petal is oval, and depending on the cultivar, overlaps onto the next or just touches it. In the centre of the wide-opening, funnel-shaped bloom, there is a long staminate column about 5.0-10.0 cm long and 10-15 cm across. At the outer reaches of the column are numerous horizontal anthers that shed pollen. At the very tip is a pentafid style (Kholkute *et al.*, 1972). Pollination of the style's five branches leads to seed production in a rounded green capsule fruit. The flower petals wither and fall away but the green sepals remain and partially hide the swelling capsule, sometimes just referred to as a seed pod. When ripe, the pod is 2.5 cm long and on drying becomes light brown. The fruit naturally splits open to release up to 60 seeds (usually 10 to 20), that are dark brown and roughly the size of apple seeds (Pal *et al.*, 1985). Borssum Waalkes (1966) suggested a key for the separation of the genus in the tribe Hibisceae. The ovary of the genus contains more than one ovule, *vis-à-vis* more than one seed per cell. Calyx not or rarely splitting on one side during anthesis, 5- lobed to 5-parted, not adnate to the corolla and persisting after flowering. The five antisepalous stamens are represented by teeth at the top of the pollen tube (Willis, 1988). The flowers are actinomorphic and bisexual while the calyx, corolla and carpel are pentamerous, and the stamens monadelphous. East African Coast, Mauritius and Rodriguez Islands are the homes of ***H. schizopetatus*** which is even more beautiful than *H. rosa-sinensis* and has produced several hybrids with it. It is also evergreen and glabrous shrub growing 2-3 m tall with slender and somewhat arching branches, leaves ovate-elliptic, toothed and 2-13 cm long, flowers solitary, pink to orange-red to red pendent some 6.0 cm across are borne on long jointed peduncles in summer with deeply fringed reflexed petals and a long stamen column, stamens long exserted and fruits long bearing smooth seeds.

Hibiscus species are largely self-compatible (Allard, 1960). The large solitary flowers are bisexual and perfect. Outer whorls consist of 6-10 epicalyx-segments and a fused 5-parted calyx at the top and completely surrounds the inner whorls in the bud. The corolla consists of five petals which are relatively large and brightly coloured in shades of white to pink, to orange, to red, depending on the variety (Kimbrough, 1997). The stamens are formed into a stamina-tube which encloses the style and is five-pointed at the top; a mass of short filaments arises towards the top third of the tube bearing the monadelphous anthers. From the top of the stamina-tube, emerges the stigma which consists of five very short style branches bearing a patch of stigma at the tip (Roe Sister, 1961; Borssum-Waalkes, 1966). The petals, the stamina column and stigma may be variously modified, especially in double flowered varieties.

Hibiscus variety 'Butterfly' is reported to have resulted from a cross between *H. schizopetalus* and *H. liliiflorus* at the Singapore Botanic Garden (Singh and Khoshoo, 1989). JuanJuan and DongHong (2001) investigated the root-tip metaphase chromosome numbers of 6 *Hibiscus* taxa by wall degradation hypotonic staining method and recorded 2n=42 for *H. schizopetalus*, 2n=92 for *H. mutabilis,* 2n=84 for *H rosa-sinensis*, 2n=105 for *H.rosa-sinensis* cv. 'Double Rainbow', 2n=138 for *H. rosa-sinensis* cv. 'Flavo Plenus' and 2n=147 for *H. rosa-sinensis* cv. 'Carminatus'. It is shown that the ploidy relationship exists among them except cv. 'Flavo Plenus', the basic number being x=21. From the observation on chromosome numbers and morphological characters, *H. rosa-sinensis* is thought to be a hybrid which arose from cv. 'Carminatus' and *H. schizopetalus*.

On the day before anthesis, the growth of the corolla suddenly spurts and it breaks through the calyx. The exposed portion of the corolla may be from a half to two-thirds of the flower bud; the petals are less tightly packed but nevertheless do not open. It is at this stage that emasculation of the flower bud may be done (Akpan, 1988). The anthers are then carefully picked-off from the stamina column. *Hibiscus* flowers are very sensitive to damage, and when all petals and anthers are removed the flowers become unfit for taking part in successful crossing (Akpan 1988). Because of this reason only the anthers are removed and not the other parts (Akpan, 1991). Early in the morning on the day of anthesis, the pollen grains from the just open flowers of the desired male parent are placed over the stigma of the emasculated female parent. When capsule of such buds grows to maturity, the hybrid seeds are harvested and sown. Hybrid *Hibiscus* is very easy to detect at the seedling stage because they tend to blend the genotypic and phenotypic characteristics of both parents thereby becoming easily distinguishable, provided the parents differ (Akpan and Hossain 1996). *H. rosa-sinensis* is reported to have a fairly high pollen fertility of over 60% (Singh and Khoshoo, 1989). There are no reports on the viability of the ova but it is probable that these are viable because female gametogenesis is not as easily upset by chromosomal and genic disharmonies as male gametogenesis (Allard 1960). Seed set in *H. rosa-sinensis* is very rare (Sharma and Sharma, 1962). The causes of this occurrence in the species has not been reported though in the related genus *Gossypium*, Brown and Menzel (1952) found that hybrids between some wild species aborted after meiosis had taken place. Silow (1941) attributed this behaviour to general genic disharmony in the hybrids. Other studies of interspecific hybrids in cotton revealed that hybrid inviability was controlled by simple genetic mechanisms (Stephen 1950; Gerstel, 1973). However, in the winter hardy *Hibiscus syriacus*, Korean workers succeeded on interspecific hybridisation between *H. syriacus*, *H. sinosyriacus* and *H. paramutabilis* (Kyung and Kim 2001a,b; Kyung *et al.*, 2001a,b). Tachibana (1958) and Kuwada (1964) made successful crosses between *H. mutabilis* and *H. moscheutos*, but the F_1 seedlings were sterile. Embryo rescue in *H. syriacus* has already succesfully been performed following intraspecific tetraploid × octoploid crosses (Kim *et al.*, 1996a,b). Also ovule culture was followed in *Hibiscus* (Paek *et al.*, 1989a,b). Attempts to create interspecific hybrids between *H. syriacus* and *H. rosa-sinensis* by somatic fusion, especially to introduce new flower colours and forms were not successful (Yu *et al.*, 1976; Paek *et al.*, 1989c). With *H. syriacus*, breeding work resulted in the development of polyploid cultivars, which in general have larger flowers and a longer flowering period (Egolf, 1971, 1981; Shim *et al.*, 1993).

Mutation breeding in *H rosa-sinensis* was reported by Das *et al.* (1974) who obtained several flower form and colour mutants (from double to semi-double and single) such as from carmine-red to deep red and from deep mauve to white. The optimum dose was somewhere from 10-20 kR at the rate of 30-50 R/h ^{60}Co gamma rays. An acute dose of approximately 5 krad X-rays at the rate of 75 rad/minute is the optimum treatment for rooted cuttings. By mutagenesis, a new cultivar 'Shirasagi-no-yume' was developed in Japan.

Classification, Species and Varieties

A few species, including *Hibiscus syriacus* L., *Hibiscus sinosyriacus* Bailey and *Hibiscus paramutabilis* Bailey are natural to temperate regions of the world. *H. sinosyriacus* is allied to *H. syriacus* but has broader leaves, and the involucral bracts outside the calyx are as long as the calyx or even longer (Bates, 1965; Bean, 1973). Out of these three species, *H. syriacus* of the section Bombicella, a deciduous shrub with more or less distinctly trilobate leaves, is the most popular one. According to Bates (1965), these three species are native to China and the similarity in their natural distribution pattern is an indication of a similar tolerance to environmental factors. At the moment about 40 different cultivars are commonly in cultivation in Europe though Korea has its largest collection. *H. paramutabilis* has large round leaves (Krüssmann, 1962; Bates, 1965). Both the species are also more vigorous compared to *H. syriacus*. Observations of calyx, capsule and seeds suggest that *H. paramutabilis* can be classified close to *H. syriacus*, while *H. sinosyriacus* is intermediate between both the species. However, interrelationships among these Chinese species are not very well understood (Bates, 1965). Close relationships with a few other native Chinese species (*H. mutabilis* and *H. indicus*) are also suggested. The ornamental *Hibiscus* species are *H. aculeatus*, *H. arnottianus*, *H. calycinus*, *H. cisplatinus*, *H. coccineus*, *H. denisonii*, *H. diversifolius*, *H. elatus*, *H. grandiflorus*, *H. hamabo*, *H. heterophyllus*, *H. incanus*, *H. lasiocarpos*, *H. liliiflorus*, *H. l.* var. *californicus*, *H. militaris*, *H. moscheutos*,

H. mutabilis, H. oculiroseus, H. radiatus, H. rosa-sinensis, H. schizopetalus, H. syriacus, H. tiliaceus, H. trionum and *H. tasiacarpus*. *H. moscheutos* is a deciduous hardy perennial growing 0.6-1.8 m, most suitable for bedding, and normally grown through hybrid seed lines which take 4-5 months (Bailey and Bailey, 1976) for flowering from seeds producing 23-28 cm across flowers which last only for a day but new ones opening each morning (Dole and Wilkins, 1999). Apart from these, there are many others such as *H. apeneùrus* (a shrub from Tropical East Africa, 1.2 m tall with solitary and bright scarlet flowers 2.5 cm across), *H. cámeronii* (probably a form of *H. rosa-sinensis*, a shrub from Madagascar with solitary, large and red-veined cream flowers), *H. cùpreus* (from Sri Lanka, a shrub allied to *H. rosa-sinensis* with coppery-purple branches and 5.0 cm across purple flowers with carmine eye), *H. kòkio* (a tree from Hawaii bearing axillary, solitary and red flowers with 6.25 cm long petals), *H. acóttii* (a large shrub or small tree from Island of Socotra, bearing 8.75 cm across bright golden-yellow flowers with carmine eye), *H. spléndens* (a 3.6-6.0 m tall prickly shrub from Australsia bearing very large rose-red flowers), *H. suratténsis* (a spiny trailer from India with yellow flowers), *H. venùstus* (similar to *H. mutabilis*, from Java with broad involucre bracts), *H. waìmeae* (a 6.0-9.0 m tall tree from Hawaii bearing axillary, large and white or pink tinged flowers), and *H. archeri* (*H. rosa-sinensis* × *H. schizopetalus*, much like *H. rosa-sinensis* with red flowers).

According to the USDA classification, the *Hibiscus* is having 40 species and 50 acceptable taxa overall and they are *H.* acetosella Welw. Ex Hiem, *H.* aculeatus Walter, *H.* archeri W. Watson, *H. arnottianus* A. Gray, *H. arnottianus* A. Gray ssp. *arnottianus*, *H. arnottianus* A. Gray ssp. *immaculatus* (Roe) D.M. Bates, *H. arnottianus* A. Gray ssp. *punaluuensis* (Skottsb.) D.M. Bates, *H. bifurcatus* Cav., *H. biseptus* S. Watson, *H. brackenridgei* A. Gray, *H. brackenridgei* A. Gray ssp. *brackenridgei*, *H. brackenridgei* A. Gray ssp. *mokuleianus* (M.J. Roe) D.M. Bates), *H. calyphyllus* Cav., *H. cannabinus* L., *H. clayi* O. Deg. & I. Deg., *H. clypeatus* L., *H. coccineus* Walter, *H. coulteri* Harv. ex A. Gray, *H. denudatus* Benth., *H. dasycalyx* S.F. Blake & Shiller, *H. elatus* Sw., *H. furcellatus* Lam., *H. furcellatus* Lam. var. *youngianus* (Gaudich. ex Hook. & Arn.) Hochr., *H. grandiflorus* Michx., *H. kokio* Hillebr. ex Wawra, *H. kokio* Hillebr. ex Wawra ssp. *kokio*, *H. kokio* Hillebr. ex Wawra ssp. *saintjohnianus* (Roe) D.M. Bates, *H. laevis* All., *H. lasiocarpos* Cav., *H. lunariifolius* Willd., *H. macrophyllus* Roxb. ex Hornem., *H. martianus* Zucc., *H. moscheutos* L., *H. mutabilis* L., *H. pernambucensis* Arruda, *H. phoeniceus* Jacq., *H. poeppigii* Spreng., *H. radiatus* Cav., *H. rosa-sinensis* L., *H. sabdariffa* L., *H. schizopetalus* (Dyer) Hook. f., *H. striatus* Cav., *H. striatus* Cav. ssp. *lambertianus* (Kunth) Blanch. ex Proct., *H. syriacus* L., *H. tiliaceus* L., *H. trilobus* Aubl., *H. trionum* L., *H. vitifolius* L., *H. waimeae* A. Heller, *H. waimeae* A. Heller ssp. *hannerae* (O. Deg. & I. Deg.) D.M. Bates), and *H. waimeae* A. Heller ssp. *waimeae*. *H. acetosella* is widely planted in Nigeria and other West African countries. *H.mutabilis* has long been cultivated in China and here the double flowers open white in the morning and turn red by evening (Wilson and Menzel, 1964; Purseglove, 1974).

Under *H. rosa-sinensis* there are hundreds of overseas varieties, the most famous ones being from Hawaii but in India also many varieties have been bred such as 'Aikta' (single, red), 'Anuradha' (single, golden buff with reddish border), 'Arunodaya' (nasturtium orange), 'Ashirwad' (single, orange), 'Basant' (sulphur yellow), 'Benazeer' (bright yellow), 'Bharat Sundari' (single, deep rose), 'Chitralekha' (single, frilled rose), 'Dilruba' (single, golden buff), 'Geetanjali' (single, golden), 'Jogan' (azalea pink), 'Nartaki' (marigold orange), 'Nazneen' (tangerine orange), 'Neelofer' (magenta rose), 'Pakeezah' (Carmine red), 'Priya' (Bengal rose), 'Queen of Hessaraghatta' (orange), 'Ratna' (single, yellow with white centre), 'Red Gold' (Dutch vermilion), 'Red Saturn' (double, ruffled-red), 'Phulkari' (single, rose), 'Shanti' (primrose yellow), 'Smt. Indira Gandhi' (single, yellow with orange-red margin), 'Smt. Kamla Nehru' (Bengal rose), 'Tribal Queen' (single, red), etc. Under *H. syriacus* also there are several popular varieties such as 'Alba Plena' (double white), 'Ardens' (double, rose tinted violet), 'Blue Bird' (medium blue with red centre), 'Boule de Feu' (vivid red), 'Coeleste' (deep blue), 'Coeruleus Plenus' (double violet), 'Compte de Flandre' (double red), 'Dorothy Crane' (white with crimson centre), 'Elegantissimus' (double, white with maroon centre), 'Hamabo' (blue-white with crimson centre), 'Jeanne d'Arc' (double white), 'Leopoldi Plena' (double white and red), 'Lutea Plena' (double cream-white), 'Mauve Queen' (mauve with maroon centre), 'Meehani' (flowers blue and foliage variegated), 'Monstrosus' (white with maroon centre), 'Purpurea Variegata' (flowers red and foliage variegated), 'Rubris' (red), 'Rubra grandiflora' (red), 'Spectabilis' (flesh pink), 'Totus Albus' (white), 'Violacea atropurpurea' (purple-violet), 'Violaceous Plenus' (double, wine-red), 'Woodbridge' (deep rose-pink, deeper at centre), 'W.R. Smith' (white with crinkled petals), etc.

Propagation

For evolution of the new varieties or the annual growing hibiscus are raised from **seeds** (Kudoh and Whigham, 2001). *H. moscheutos* seeds (92 seeds/g) germinate in 21 days at 21 °C temperature (Dole and Wilkins, 1999). *H. trionum* seeds to a tune of 90% possessed physical dormancy which after immersing for 0.5 h in concentrated sulphuric acid germinated though still 29% exhibited primary embryo dormancy, for which Chachalis *et al.* (2008) stated that pre-chilling exposure of seeds even for shorter duration (15 days) induces secondary dormancy with only 10% germination. 8-Nonynoic, 9-decynoic acids and their methyl esters as germination inhibitors were isolated for the first time as natural products from *H. rosa-sinensis*. Seeds can be planted in a commercial seed starting medium or a mix which can be made out of some combination of sand, perlite, vermiculite and peat. The seeds are sown 6-13 mm deep in the mixture which should be moist though not soggy and at about 28 °C temperature where these may take from a week to a month for germination. For placing seeds, styrofoam coffee cups with drainage holes poked in the bottom may be used with 100% perlite to start with. Many seeds can be sown in 10 cm (preferably 5) pots in a tray kept in light to make a head start (Popenoe, 2008). After the seedlings have developed several sets of true leaves and the stems have begun to harden and woody, they may be moved to a larger pot. A water soluble 20-20-20 NPK fertilizer can be used per week with irrigation water. When the seedlings attain 10-12 cm height, these are transplanted to the larger pots or directly in the prepared field. These small seedlings should be exposed to sun gradually over several days by avoiding sudden fierce sun.

Commercially, *Hibiscus rosa-sinensis*, *H. schizopetalus* and others are propagated through **cuttings**, especially softwood (tip), though these can be propagated even by semi-hardwood (medium wood) and hardwood cuttings (Beers and Howie, 1990). The **tip cuttings** from stock plants of *H. rosa-sinensis* which have received HID (high intensiaty discharge) lighting less than 4,500 fc root more rapidly as compared to those receiving higher irradiances (Johnson and Hamilton, 1977; Kelty, 1984) having at least two eyes, of the size of 9-15 cm cut at 45° angle with a basal eye or node, and two mature basal leaves intact as has been recorded in 'Purity' (Ryan *et al.*, 1958), are usually taken in summer when vigorous new growths are appearing and when medium temperature is in between 26-30 °C. Basal leaf retention is also necessary while assessing the rooting capacity of various cultivars (Ryan *et al.*, 1958). This temperature range has been reported as optimum for *H. rosa-sinensis* either for easy-to-root cultivars or for difficult-to-root (Carpenter, 1989) though in case of *H. syriacus* 20 °C was reported the best (Whalley, 1974). Proper handling of these cuttings is of prime importance in successfully developing new plants (Evans and Blazitch, 1999). *H. rosa-sinensis* tip cuttings are dipped in water and wrapped in thick and wet cloth or in a plastic bag with some water sprinkled inside the bag and are kept at aerated and shaded place but not under direct sun (Beers and Howie, 1992). Before planting, their cut ends are dipped in IBA from 1,250-10,000 ppm aqueous solution for the durations of 0.1, 0.5, 1.0, 5.0 or 10.0 minutes depending upon the cultivar to be propagated, and then these are placed in the rooting medium. With increase in the medium temperature from 18 to 34 °C, Carpenter and Cornell (1992) recorded decrease in the optimum concentration of IBA. Cuttings of certain clones of *H. rosa-sinensis* develop roots rapidly during propagation while certain others requiring considerable time (Kelty, 1984). Van Overbeek *et al.* (1946) propagated difficult-to-root 'Purity' by grafting onto stem cuttings of the easily rooted 'Brilliant' which responded to auxin application. Warm propagation medium temperatures hasten root initial formation in most of the plant species (Halma, 1931; Stoutmeyer *et al.*, 1935), with temperatures below 21 °C delaying rooting and above 27 °C inhibiting it (Hartmann and Kester, 1983). Golby (1967) stated that as per American Hibiscus Society, the temperature of propagation medium (coarse river sand, perlite and peat, vermiculite and peat, or a mixture of perlite, sand and peat) should be 20-25 °C. Glasshouses with misting devices are preferred for tip cuttings. For the home gardener a small frame covered with clear plastic will work provided the plants are sprinkled several times throughout the day to prevent excessive transpiration. Wikesjö (1981) suggested hibiscus cutting propagation during spring and summer under white plastic tents and during autumn under clear plastics with 100% relative humidity. **Semi-hardwood cuttings** are usually taken by the nurserymen in autumn as the softwoods begin to mature, mainly to build up stock of a particular variety. With home gardeners such cuttings in case of *H. rosa-sinensis* are not popular as these do not establish properly since cooler months are ahead and also as flowering wood is removed which hampers flowering. But in this case, firm wood slightly thinner than pencil is selected, all but the three top leaves are removed, then these are reduced by half and then a diagonal cut through an eye about 7.5 cm below the bottom leaf is made on the stem and planted in individual propagating tubes containing the same type of propagating medium as in case of tip cutting after dipping in aqueous solution of the required concentration of IBA akin to its type or in Seradix No. 2 powder. They should be placed under glass or polythene cover. The plants should not be repotted until spring. Two methods for *H. sabdariffa* are proposed as a technique of ensuring maximum genetic stability. Softwood and semi hardwood cuttings from two-month-old plants are rooted by dipping in IBA or NAA at 0-1.0 g/l and then planted in a medium containing soil, compost and rocks. Rooting is significantly affected by the type of cuttings taken and the concentration of auxin used (Carpenter, 1989) as softwood cuttings responded more positively to lower auxin level (0.5 g/l), and the response in auxin-treated cuttings was more pronounced (Soulange *et al.*, 2009). The easiest way of propagating hibiscus is by **hardwood cuttings** taken by the end of winter or in early spring (Beers and Howie, 1990), usually in the normal pruning time. However, such cuttings may require extra warmth to encourage proper root growth. (Hamilton and Midcap, 2009). In this case the straight and strong woods of pencil thickness or a little thicker should be selected, all the leaves removed by making a cut at about 45 degrees through an eye at the base of the cutting by using a sharp knife in one attempt (Larsen, 1914). The cutting is then trimmed to a length of 12.5 - 15.0 cm and placed either in individual propagating tubes or in 15 cm community pots which may hold some 30 cuttings. Hardwood cuttings seem to prefer coarse river sand with a little peat mixed through it as a rooting medium, however various mixtures of perlite and peat, or perlite, sand and peat have all been successful (Hartmann *et al.,* 1996). Once the cuttings are struck they may be potted on into larger pots using a mixture of two parts sandy loam, one part coarse sand, one part peat, one part old cow manure (fine), and one part leaf mould. Rooted cuttings from the community pots are gently removed without injuring their roots and are kept in a bucket of clear water, patiently separated, dipped in IBA or in Seradix No. 2 or 3, planted 2.5 to 3.75 cm deep (Hartman *et al.*, 1996) and watered.

Those rooting with difficulty are propagated through **grafting** (Hartman *et al.*,1996) on the easy to root understock having matured twigs as soft and succulent green twigs are more susceptible to wilting or injury by the heated wax (Larsen, 1914). **Side-wedge grafting** is a convenient and most commonly used method (Nakasone, 1980) though a few propagators also use the cleft-grafting (Beers and Howie, 1990). For grafting, the stocks should be made ready. Tip, semi-hardwood or hardwood cuttings 15 to 20 cm long are inserted in a well-prepared rooting medium and kept constantly moist in the greenhouse or elsewhere under filtered location. Hibiscus cuttings will root in rooting medium within 2-3 months and then these are transplanted individually into containers and when there these are well established after a few weeks, these can be used for grafting. The cleft method consists of cutting off the stock completely of easy-to-root varieties just a few distance above the soil surface leaving a stub upon which a slit in the centre is made where the wedge-shaped sharpened scion base is inserted by bending slightly the stock away from the cut so that the scion is tightened in the position of the stock and then further sealed with just melting paraffin or grafting wax so that scion remains tightly positioned, stock growth retards, disease infection through the wound is halted, evaporation rate is reduced and the reserved food and moisture are diverted towards growth of the scion (Fullaway and Krauss, 1945), however, the wax should not have high temperature to injure the scion tissues. Any growth on the scion after grafting is not allowed to occur and when the buds on the scion wood have sprouted and grown 10-15 cm, entire stock is removed and planted elsewhere or at its permanent place. Grafted hibiscus comes into flowering in about four months. 'Peachblow' (common pink hedge hibiscus bearing double pink flowers) and *H. rosa-sinensis* (single red) are excellent understocks (Gilman, 1999) as these root easily. *H. schizopetalus* and its hybrids though root easily but are not recommended because of their whip-like quick growth (Ingram and Rabinowitz, 2004) with poor girth and so the main trunk of the grafted scion becomes more thickened than the stock on which grafted upon, *vis-à-vis* causes enormous graft-union bulge. Cuttings for rootstocks are rooted in the usual manner.

Topworking is a rejuvenation of a hibiscus shrub where entire top is cut back and then grafted with some desirable varieties on the stubs of the main branches which should be well distributed on the plant so that complete balance of the shrub is achieved, however, other unimportant branches should be removed. Bark grafting is the usual method used (Nakasone, 1980) in top working. The bark is slit and loosened about 2.5 cm at the end of the stub on both the sides.The diagonally cut base of the scion is inserted between bark and wood (Neal,1948) and then all the cut surfaces are properly waxed. Those scions showing weaker growth should be removed.

Air layering is a popular and a labour saving method where a branch of more than a pencil thickness is girdled for 1.25 cm or so, or a strip of 1.25-3.0 cm wide bark is removed and girdled or stripped part with a bit of adjoining areas of the branch is covered with a ball of moist sphagnum moss some 7.0-10.0 cm in diameter and then securely wrapped with plastic film where rooting may occur within 1-3 months depending on the variety and the season and then from below the girdled area the layer is separated and planted at its permanent place. Nakasone and Rauch (1973) tried air layers of difficult-to-root through cuttings of hibiscus, *viz.* 'Apple Blossom', 'Betty Yellow', 'Fluffy Ruffles', 'Mrs. John Walker', 'Glorious' and 'Perfection' where former four showed no response even after five months of insertion though latter two vigorous growing cultivars rooted.

Production of disease-free plants is a major concern due to viruses present in *H. rosa-sinensis* as reported by various workers (Adkins *et al.*, 2003; Huang *et al.*, 2004; Rajeshwari *et al.*, 2005). When an embargo was placed upon Mainland Export of Plants, due to occurrence of a virus disease, the commercial propagation and sale of hibiscus were restricted to the local trade on September 12, 1947 by the U.S. Bureau of Entomology and Plant Quarantine. Methods are available to eradicate virus from plants (Kasseanis, 1957; Baker and Phillips, 1962), and if these methods are employed to obtain virus-free material to be used in the initial culture, certified virus-free plants of *H. rosa-sinensis* can be produced in large numbers using ***in vitro*** methods. *In vitro* methods have been reported for some related species such as *H. cannabinus* (Zapata *et al.*, 1999; Srivatanakul *et al.*, 2000; Herath *et al.*, 2004), *H. moscheutos* (West and Preece, 2004), *H. acetosella* (Reynolds and Blackmon, 1983) and *H. syriacus* (Yang *et al.*, 1995; Jenderek and Olney, 2001). In these studies, somatic embryos, callus and explants from seedlings were used for regeneration and genetic transformation. Initial contamination resulting from mite infestation was also reported to be a problem (West and Preece, 2006). Moreover, regeneration through somatic embryogenesis and callus appears complicated and it involves long processes. Although, multiple shoot regeneration was reported to be efficient using explants from seedlings of *H. cannabinus* and *H. syriacus* (Yang *et al.*, 1995; Srivatanakul *et al.*, 2000; Jenderek and Olney, 2001; Herath *et al.*, 2004) and not *H. rosa-sinensis* as its seedlings may not come true due to hybrid nature of *H. rosa-sinensis* itself.

Cultural Practices

Different kinds of hibiscus vary in their growing habits, their need for sun, and the time of year they are in their optimum blooms. None of them like persisting wet feet while many others need some protection from wind if they are to keep in good condition. For proper shaping, the top of the plant should be free to develop properly and the plant should have proper spacing for good root development, plant growth and flowering. *Hibiscus* species are primarily tropical in origin (Ryan *et al.*, 1958), though there are many found in temperate and warm-temperate regions. Commercially grown *Hibiscus* are both, day-neutral such as *H. rosa-sinensis* (Dole and Wilkins, 1999) and long-day plants such as *H. syriacus* (Thomas and Vince-Prue, 1997) and *H. moscheutos* (Runkle *et al.*, 1996). *H. coccineus* flowers naturally in July (Bates, 1965) suggesting that it may also be a long-day plant. *H. hiernianus* Exell & Mend., *H. mastersianus* Hiern., and *H. mechowii* Garcke are short-day plants (Menzel *et al.*, 1986). However, little information is available on the floral induction requirements of *Hibiscus* species not commercially cultivated. **Temperature** and **lighting** treatments interact to affect leaf number below the first flower and/or flower diameter on some species. *H. acetosella*, *H. cisplatinus* St.-Hil., *H. radiatus* and *H. trionum* were selected as potential new commercially significant ornamental species (Ryan *et al.*, 2001). Outdoor *Hibiscus* prefers good morning sun with a little afternoon shade. This condition will provide more vivid and larger blooms. As it needs full sun, requires protection in the winter so may be kept in home. Light intensities reported for proper growth of *Hibiscus rosa-sinensis* are from 1,750 fc to 5, 000 fc (Osborne and Chase, 1990; Steinberg *et al.*, 1991; Wang, 1991). Tropical hibiscus needs warm and steady temperatures to thrive though temperature fluctuations, especially when extreme can shock and kill the plant. Tropical hibiscus has the best bud growth between 18-21 ^{0}C night and 24-29 ^{0}C day temperatures and high light (Dole and Wilkins, 1999) though a temperature even of 39 ^{0}C is not injurious to this shrub (Woodson and Raiford, 1986; Neumaier *et al.*, 1987; Osborne and Chase, 1990). Buds stop growing around 12 ^{0}C, and at persisting 7-10 ^{0}C temperatures the plants may suffer chilling injury, more pronounced being at 5 ^{0}C (Karlsson *et al.*, 1991). Tropical hibiscus grown on the side of a building or near a heat source is more likely to survive the frost. Hardy hibiscus also benefits from warm temperatures for bud growth, and approximately 23 ^{0}C is ideal. Unlike tropical hibiscus, the hardy varieties will survive freezing temperatures, although flowering will reduce. A mulch of 30 cm layer will protect the hibiscus if there is a chance of frost occurrence. A soil temperature of 15 to 20 ^{0}C is most suitable for germination of seeds of *H. cannabinus* and *H. sabdariffa* (Chachalis *et al.*, 2008). Effects of forcing temperatures on phenological and morphological development of *H. moscheutos* 'Disco Belle Mixed' were determined by growing plants under long days in glasshouses set at 15, 18, 21, 24 or 27 ^{0}C. Time from the onset of long-day forcing to visible bud (VB) decreased from 101 to 24 days as average daily temperature (ADT) increased from 15.7 to 25.0 ^{0}C, and time from VB to flower decreased from 54 to 29 days as ADT increased from 19.4 to 27.6 ^{0}C (Schoellhorn and Richardson, 2011).

Both tropical and hardy hibiscus have similar **soil** needs, preferring acidic soil with a pH of 5.8 to 6.5 for which dolomite should be used (Woodson and Raiford, 1986; Neumaier *et al.*, 1987; Wang, 1991) though *H. tiliaceus* grows best in slightly acidic to alkaline soils (pH of 5.0-8.5) (Elevitch and Thompson, 2006). Hibiscus grown on alkaline soils may suffer from micronutrient deficiencies (Ingram and Rabinowitz, 2004). Mixing sphagnum peat moss into garden soil or potting mix increases acidity. Ideally, hibiscus should be planted in well-drained soil amended with organic matter. Heavy clay soil is amended through addition of sufficient organic matter at the time of field preparation, however, in such soils the hibiscus should be cultivated on raised beds to eliminate water build-up. Proper **preparation of the soil** prior to planting will help ensure vigorous healthy plants. Most soils are low in organic matter and hibiscus benefits by the addition of this material. A good thick cover of compost, leaf mould, well decomposed manure or well rotten chicken litter, etc. should thoroughly be mixed in the soil at the time of field preparation along with dolomite (70 g per square metre) and a balanced fertiliser (100 g per square metre) and then the soil should be rested for 2-3 weeks. For **planting,** which should be practiced during rainy season to avoid desiccation, the plants should be taken out from containers one by one and planted in the prepared field, in the pits or in the pots. After planting, the soil should be firmed up around the plant with light pressure. When planting is over, the field may be mulched and then irrigated thoroughly to settle the roots in the soil. The roots and canopies of large trees such as banyans, *neem*, mangoes, pines, palms, bamboos and other trees growing taller than hibiscus offer great competition to hibiscus plants set too close to them and often, regardless of watering and feeding, these plants will not prosper. A plant properly located and **spaced** is more self-sufficient to spread and develop its root system and plant growth. For hedging, standard hedge varieties should be planted in 3-row system, from 1st row to 3rd some one metre distance while plant to plant distance in the row may be kept at 50 cm, however, the plants in the 2nd row should be planted at the same distance but alternate to 1st *cum* 3rd rows, though for commercial uses these may be planted from 60-120 × 60-120 cm depending upon the varieties as there are taller types, mediums and short-statured. For mass effects, the distance may be from 75- 120 cm across. While planting against a house, wall or post, there should be clearance of at least 60 cm.

Hibiscus responds well to **nutrition**. Plant grown in pots or cans should receive more frequent applications than those set out in the open ground. A handful of organic manure such as chicken or cow manure, and one gramme per litre of complete chemical fertilizer such as 16-16-16 NPK every 6 to 8 weeks should be adequate. Less frequent fertilizer application is required when a controlled released fertilizer such as Osmocote is used. Seedlings, rooted cuttings, and grafted plants benefit from light application of chemical fertilizer or some form of organic fertilizer at every stage of transplanting. Plants set out in the ground should be given about 100 g of a complete fertilizer every 4-5 months to ensure continuous good growth. Liberal use of organic fertilizer, especially cow manure is beneficial to the plants which will also improve soil texture and long run fertility. A dose of N 15%, P 2.2% and K 16.6% with irrigation water can be given per month from March to October though in winter the dose may be reduced to its half. Dole and Wilkins (1999) have mentioned that the plants should be fed with 250 ppm N during summer months and only 150-200 ppm during winter, and this nitrogen should be in 80:20 (ammonium:nitrate) in summer while 20:80 during winter. Phosphorus build-up or pH imbalance accounts for some 95% of the cases of slow decline in hibiscus (Gerlich, 2006). Other elements crucial to healthy hibiscus growth are magnesium, iron, boron, sulphur, zinc and molybdenum (Broschat, 2006). Elevitch and Thompson (2006) suggested use of Mg, Fe and Cu along with NPK for balance growth of hibiscus plants. Regular drenches of magnesium sulphate at 0.66% will prevent intereveinal chlorosis of lower leaves (Dole and Wilkins, 1999). Fertilizers that are 'slow-release', 'controlled release', 'sulphur coated', or with nitrogen as IBDU or urea formaldehyde have extended release periods compared to fertilizers that are readily water soluble. The amount of water given to a plant and the intervals between such **waterings** must be determined by experience rather than by arbitrary rules. Irrigation practices suitable to one locality may be entirely unsuited to another because of differences in amount of rainfall, physical nature of the soil, humidity and other environmental factors (Larsen, 1914). Plants grown in areas of high rainfall obviously will require less frequent watering than plants grown in areas of light rainfall or along beach areas predominated by sandy soils of low water-holding capacity. Hibiscus should not get too dry or stay too wet. If dried out for too long, the plant can drop its leaves or even go into shock and die, despite renewed watering. If overwatered, especially in pots, fungal diseases can destroy the root system, or the soil can pack and deprive the roots of needed oxygen. The hibiscus requires a great quanity of moisture, but will not tolerate waterlooged conditions. Where drainage is poor due to poor soil types, less frequent irrigation and that too shallowly should be followed or provision for drainage of excess water should be made. For such conditions an irrigation schedule of once a week may be more than sufficient. In areas with good drainage and relatively dry conditions, irrigating twice a week may be necessary. In summer, when the plants are in active growth and flowering, they require more water. While not all cultivars enjoy the same amount of water or tolerate the same amount of dryness, hibiscus in general likes being well watered, especially in hot weather. Even so, hibiscus does not like to be continually wet, as this promotes fungal diseases and packs the soil so robbing it of vital oxygen. Those grown in pots for indoor decoration, should not be left in saucers full of water for too long. During cooler periods of the year, water more sparingly as then these are not in active growth. Occasionally, the plants will lose all their leaves either due to reaction to stress or the natural process of renewing its foliage for a different season. When plants are without leaves, there is relatively little transpiration and it may be days before they need watering again so regular watering such plants can cause root rot. Water is applied in shallow basins around the shrubs or in furrows where a number of plants are grown in a row (Nakasone, 1980). Neumaier *et al.* (1991) suggested use of Chlormequat (CCC) sprays which will reduce watering by 35%.

Pinching, Disbudding and Height Control

Pruning in hibiscus cannot be specifically determined nor a definite schedule set. Although cultural practices and flower production influence plant growth, the plant in pots usually begins to weaken after 3-5 years of active flowering. Sporadic flowering and decrease in size of the leaves and flowers are typical symptoms of decline (Nakasone, 1980). Under local conditions, pruning may be done during May and June as in India these are almost dry months in summer when loss of flowering will be negligible, and the plants will have sufficient time to grow and be in condition to flower afterwards, *vis-a-vis* in the fall and winter months (Bailey and Bailey, 1946). The degree of pruning depends upon the condition of the shrub. Old shrubs with numerous bare branches require a rather severe pruning. As much as half of the top may be removed together with all dead and unproductive twigs, and small branches usually growing in the interior of the shrub (Jayaweera, 1942). Light pruning may be done in some cases where young plants have grown very rapidly without much lateral shoot growth, thus presenting a leggy appearance. The tall branches are headed back to any desired point to force lateral buds to develop which will help to produce a denser and a balanced shrub (Beers and Howie, 1990). After pruning, organic manure or a complete fertilizer mixture at 100 g per plant should be worked into the soil around the trunk followed by immediate irrigation. Within a few weeks new vegetative shoots will appear and shape the plant in pleasing look. In fact, commercially potted hibiscus is of manageable short height and in good shape which requires soft **pinching** to maximize branching for maximum flowering. Pinching is effected normally after 5-6 weeks of propagation when active growth has begun and then after every 3-4 weeks. Branching can also be maximized through spraying plant growth regulators such as mefluidide at 100-500 ppm to shorten internodal length and to increase the axillary branching and then there would not be any need to pinch or disbud the plants when using growth regulators (Woodson and Raiford, 1986). However, the life of such commercial potted plants is roughly for two flowering years. Spray of 200 ppm CCC to 10-cm pots is done at the time when after pinching the new axillary shoots are 3.75-5.0 cm long and then afterwards every 3-4 weeks but when plants reach the desired height the concentration is increased to 1,000 ppm to encourage uniform flowering through complete cessation of internodal length. Through CCC sprayings the cropping can effectively be scheduled. Spraying in propagating chamber when rooting begins and thereafter a few days before the pinch may shorten the production time by 4-5 weeks (Dole and Wilkins, 1999) and reduces the spider mite population drastically (Osborne and Chase, 1990). Paclobutrazol (Bonzi) spraying at 5 ppm or drenching at 0.5 mg per pot depending on the varieties is also very effective (Shanks, 1972; Criley, 1980; Wilkins, 1986) though ancymidol (A-Rest) and daminozide (B-Nine) have been found marginally effective (Adriansen, 1985). **Mulching** preserves moisture from evaporation, suppresses the growth of weeds, provides nutrients to the plants after getting decomposed into the soil if organic mulch has been used and lowers maintenance though harbours insects (Ashworth and Harrison, 1983). Leaf mould, desalted old seaweed, old straw, bagasse, spent mushroom compost, garden compost, quite old pine bark and sawdust, rice husks, groundnut husks, dried water hyacinth, cocopeat, etc. as organic mulch and black polythene can be used as mulching materials. Weed mats also serve to prevent evaporation yet still allow the passage of air and moisture to the soil (Brown, 1996).

Post Harvest

Potted or container-grown plants are highly sensitive to ethylene, water stress and low light conditions during transportation and marketing (Nell, 1993) which may cause buds and open flowers to abscise due to increased abscissic acid (Swanson *et al.*, 1975), therefore, to overcome this problem 4.0 mM STS spraying 10 days before marketing is recommended. Up to 3 days of transportation, plants can tolerate 10-16 ^{0}C of storage temperatures (Høyer, 1986; Nowak and Rudnicki, 1990). Harvesting for loose flowers commences early in the morning once the calyces have reached an optimum size. The flower is separated from the calyx manually and are sent to the markets in baskets.

Insect-pests, Diseases and Physiological Disorders

Caterpillars of various insects including *Helicoverpa* devour foliage, sometimes stripping a plant of all its leaves and *Hibiscus* without leaves cannot produce new growth and is less likely to produce flowers during the growing season. Treatment with chemical insecticides is generally not warranted and isn't always the best course of action. However, when the infestation is serious, pyrethrins, pyrethroids or acephate insecticides are used when weather is clear though these also together kill predatory and beneficial insects. Only apply the insecticide on days when the weather is clear, calm and the temperature is below 32 ^{0}C. *Neem* oil or insecticidal soap will also work as these are not harmful to predatory insects though will control small larvae if applied thoroughly. **Mealy bugs** (*Maconellicoccus hirsutus*) are important pests as apart from sucking the sap from tender parts of leaves, new shoots and floral bases, they excrete honeydew which attracts ants and serves as a medium for the development of sooty mould. They are soft bodied and scale like insects usually covered with powdery or cottony wax like material. They are around 6 mm long when mature and make cottony nests at twig joinings and under leaves. Application of Imidacloprid as soil drench by the end of the winter, followed by a foliar spray of Dinotefuran or Acephate can be useful (Osborne, 2005). **Aphids** (also known as plant lice) are small, soft bodied insects about the size of a pinhead which attack new growth, causing leaves to curl and blossoms to be malformed. They are green to brownish, with one type being black. Their clustering can be found on the underside of leaves. Ladybugs and aphid lions are natural enemies of aphids. Spraying of 0.1% methyl parathion will control this pest effectively. **Thrips** are tiny, yellow- winged and fast moving insects feeding on the tender parts of the plants such as flowers, new leaves and emerging shoots. They leave silvery patches on leaves and cause buds to drop. **Spider mites** (red spiders) are tiny, tan, red or purple and less than 0.5 mm long but one of the most common pests of hibiscus. They are found by checking the undersides of leaves with a magnifying glass. They suck plant juices causing tiny white spots on the leaves. Kelthane should be used to control this pest. **Scales** have also been found infesting hibiscus which can be controlled through spraying of any systemic insecticide. **Whiteflies** (*Bemisia argentifolia, Trialeurodes vaporariorum*) covered with a waxy powder build up their population quickly and excrete honeydew which supports unsightly black sooty mould, and more so become the nuisance for visitors. These can also be controlled through use of systemic insecticides and by the use of a parasitic fungus, *Beauveria bassiana*. **Leaf miners** are small insects that feed between the upper and lower surfaces of the leaves. They leave distinctive trails on the leaves. **Cutworms** are the immature stage of certain moths. Typically they stay in the soil during the day and feed at night at the base of tender plants. Some feed on buds and leaves. Carbaryl as a dust or spray in the late afternoon is effective. Millipedes, pillbugs and sowbugs thrive in moist soil, attacking roots and sprouting seeds. Soil applications of Diazinon could be suggested.

Sooty mould is a black fungus on the upper surface of leaves, growing in the secretion of aphids, mealybugs, many scales and immature whiteflies and making unsightly appearance on the foliage though not injurious otherwise. To get rid of this problem, these pests will have to be controlled. Oil emulsion or miscible oil sprays at dusk will loosen the mould in the night through deposition of dews and then in the morning through jet water spraying the leaves may be got cleaned. Infected leaves should be removed and burnt and the plants sprayed with a fungicide such as Mancozeb (Dithane M45, Manzion, Manzate) will kill the moulds. Ants are fond of the honeydew excreted by aphids and mealybugs, and they may protect and move these pests around from plant to plant. Chlordane and Baygon are effective controls directed to a broad area around the ant hills as well as in the centre of the hill. Several species of fungi may cause soft **rotting** of roots (*Fusarium* sp. and *Phytophthora* sp.) and sometimes also stems. Infected plants will often wilt as though they are short of water. In general, root rots are associated with over-wet conditions brought about by overwatering or poor drainage. Treatment involves watering less frequently and improvement of soil drainage as well as soil drenches with fungicides such as Terrazole (Terrazole WP) or Fongarid. *Cercospora* and *Xanthomonas* though cause similar leaf spots but in former case it has more distinct yellow halo (Daughtrey and Chase, 1992). *Pseudomonas cichorii* causes irregular-shaped angular leaf spot and *P. syringae* small circular lesions on leaves. Plants infected with **viruses** may lose the vigour and may bear deformed or cupped or mottled leaves. Bacterial, viral and *Cercospora* infected plants should be discarded.

Physiological disorders are the result of certain unfavourable factors in the growing conditions such as too warm, too cold or too shady growing conditions; too acidic or too alkaline soil conditions; over- or under-watering; over-fertilising; trace element deficiency; water stress, ethylene build up and dark conditions in plant storage during transportation; etc. **Bud drop** is one of the most common disorders of hybrid hibiscus though greatly affected due to varieties and growing seasons. The most common causes of bud drop, especially in doubles are either inadequate nutrition or water stress,*vis-a-vis* excess of nitrogen, particularly through foliar application. Hibiscus bushes naturally discard their large and old leaves at the bottom several times a year which a few days before dropping turn bright yellow, but when **leaf yellowing** and **dropping** continue towards the top of the bush, it shows that either the plants have been over-fertilized or there had been water-logged condition around the root zone. The yellowing is usually triggered off by an excessively hot day, cold snap, or by spraying with unsuitable insecticides such as Malathion or Lannate. The constant use of such insecticides could lead to complete defoliation. Plants may also suffer from **cold injury** when temperature reaches down to 11 ^{0}C or even below and this causes cessation of complete growth. When transporting the plants for marketing under continuous darkness for more than three days, water stress or exposure to ethylene, the **abscission of buds and flowers** start occurring (Nell, 1993). **Upper leaf chlorosis** generally occurs due to Fe deficiency and high media pH or insufficient root development, and **interveinal chlorosis of lower leaves** due to Mg deficiency (Dole and Wilkins, 1999).

References

Adkins, S., I. Kamenova, D. Achor and D.J. Lewandowski, 2003. Biological and molecular characterization of a novel tobamovirus with a unique host range. *Plant Dis.*, **87**: 1190-1196.

Adriansen, E. 1985. Kemisk vaekstregulering (Danish). In: *Pottenplanter I-Produktion, Metoder, Midler* (eds Christensen, O.V., A. Klougart, I.S. Pedersen and (K. Wikesjø). GartnerINFO, København, Denmark.

Allard, R.W. 1960. *Principles of Plant Breeding*. John Wiley & Sons, New York, USA.

Akpan, G.A. 1988. Morphology, Cytogenetics and Hybridization of Some West African *Hibiscus* Species. Ph.D. Thesis. University of Calabar, Nigeria.

Akpan, G.A. 1991. Hybridization of some *Hibiscus* species in Nigeria. *Nigerian J. Gen.*, **VIII**:45-47.

Akpan, G.A. and M.G. Hossain, 1996. Morphological studies in intraspecific F_1 hybrids of *Hibiscus sabdariffa* L. (Malvaceae). *Trns. Nigerian Soc. biol. Conserv.*, **4**(1): 19-25.

Anderson, N. and J. Pharis, 2003. Kenaf fiber - A new basket liner! *Minnesota Commerc. Fl. Grs. Bull.*, **52**(3): 7-9.

Ashworth , S. and H. Harrison, 1983. Evaluation of the mulches for the use in home garden. *HortSci.*, **82**(2): 180-182.

Bailey, L.H. 1950. *The Standard Cyclopedia of Horticulture.* Macmillan, New York, USA, p. 50.

Bailey, L.H. and E.Z. Bailey, 1946. *Hortus Second.* The Macmillan Company, New York.

Baker, R. and D.J. Phillips, 1962. Obtaining pathogen-free stock by shoot tip culture. *Phytopath.*, **52**: 1242–1244.

Bates, D.M. 1965. Notes on the cultivated Malvaceae ~ *Hibiscus. Baileya*, **13**: 57–130.

Bean, W.J. 1973. *Trees and Shrubs Hardy in the British Isles.* John Murray, London, UK.

Beckett, K.A. 1987. *Hibiscus*. In: *The RHS Encyclopedia of House Plants Including Greenhouse Plants*, pp. 290-291. Salem Hoise Publishers, Massachsetts, USA.

Beers, L. and J. Howie, 1990. *Growing Hibiscus*. Kangaroo Press, Australia.

Beers, L. and J. Howie, 1992. *Growing Hibiscus* (2nd ed). Revisd by Colorcraft Ltd., Hong Kong.

Borssum-Waalkes, J. van, 1966. Malesian Malvaceae Revised. *Blumea*, **14**: 1-251.

Broschat, T. K. 2006. Effects of phosphorous and phosphoric acids on growth and phosphorus concentrations in container-grown tropical ornamental plants. *Hort Tech.*, **16**: 5-6.

Brown S.H. 1996. Response of hibiscus to organic mulch. *Proc Fla St. hort. Soc.*, **109**: 30-33.

Brown, M.S. and M.Y. Menzel, 1952. Additional evidence of the crossing behavior of *Gossypium gossipioides. Bull. Torrey bot. Club*, No. 79, pp. 285-292.

Carpenter, W.J. 1989. Medium temperature influences the rooting of *Hibiscus rosa-sinensis* L. *J. env. Hort.*, **7**: 79-84.

Carpenter, W.J. and J.A. Cornell, 1992. Auxin application duration and concentration govern rooting of hibiscus stem cuttings. *J. Amer. Soc. hort. Sci.*, **117**: 68-74.

Chachalis, D., N. Korres and E.M. Khah, 2008 . Factors affecting seed germination and emergence of Venice mallow (*Hibiscus trionum*). *Weed Sci.*, **56**(4): 509-515.

Cobley, L.S. 1976. *An Introduction of the Botany of Tropical Crops* (2nd ed. revised by Steel, W.M.), p. 272. Longmans Ltd., London, U.K.

Criley, R.A. 1980. Potted flowering hibiscus. *Flor. Rev.*, **165**(4290): 48-49, 64.

Das, P., P. Ghosh, S. Dube and S.P. Dhua, 1974. Induction of somatic mutation of vegetative propagated ornamentals by gamma radiation. *Technology*, **11**(2,3): 185-188.

Daughtrey, M. and A.R. Chase, 1992. Hibiscus. In: *Ball Field Guide to Diseases of Greenhouse Ornamentals*, pp. 123-125. Ball Publishing, Geneva, Illinois, USA.

Dole, J. M. and H.F. Wilkins, 1999. *Hibiscus*. In: Floriculture Principles and Species, pp. 368-372. Prentice Hall, New Jersey, USA.

Egolf, D.R. 1971. *Hibiscus syriacus* 'Diana,' a new cultivar (Malvaceae). *Baileya*, **17**: 75-78.

Egolf, D.R. 1981. 'Helene' rose of sharon (althea). *HortSci.*, **16**: 226-227.

Elevitch, C. R. and Lex A.J. Thompson, 2006. *Hibiscus tiliaceus* (beach hibiscus). The Traditional Tree Initiative- Species Profile for Pacific Island Agroforestry, p. 13. Permanen Agriculture Resources, Hawaii, USA.

Evans, E. and F.A. Blazitch, 1999. Plant propagation by stem cuttings: Instructions for the home gardener. *Horticulture Information Leaflet 8702*. N. C. State University College of Agriculture & Life Sciences.

Fullaway, D.T. and N. L.H. Krauss, 1945. *Common Insects of Hawaii*. Tongg Publishing Company, Honolulu, USA.

Gerlich, R. 2006. Hibiscus Care ~ General Growing Instructions for *Hibiscus rosa-sinensis*. American Hibiscus Society (www.loyno.edu/~gerlich/NOHS/index.html).

Gerstel, D.V. 1973. *The Families of Flowering Plants* (3rd ed.), p. 22.Oxford University Press, U.K.

Gilman, E. F. 1999. *Hibiscus rosa-sinensis*. Fact Sheet FPS-254. University of Florida (http://edis.ifas.ufl.edu).

Golby, E. 1967. Propagation of *Hibiscus rosa-sinensis*. *The Seed Pord.*, **18**(2): 25-28.

Halma, F.F. 1931. The propagation of citrus by cuttings. *Hilgardia*, **6**: 131-157.

Hamilton D. F. and J.T. Midcap, 2009. Propagation of woody ornamentals by cuttings. University of Florida (http://edis.ifas.ufl.edu).

Hartmann, H.T. and D.E. Kester, 1983. *Plant Propagation: Principles and Practices* (4th ed.). Prentice Hall, Inc., Englewood Cliffs, N. J., USA.

Hartmann, H. T., D. E. Kester, F. T. Davies and R. L. Geneve, 1996. *Plant Propagation, Principles and Practices* (6th ed.). Prentice Hall, Upper Saddle River, New Jersey, USA.

Herath, S.P., T. Suzuki and K. Hattori, 2004. Multiple shoot regeneration from young shoots of kenaf (*Hibiscus cannabinus*). *Plant Cell Tissue Organ Culture*, **77**: 49–53.

Høyer, L. 1986. Silverthiosulfate can to some extent prevent leaf, bud and flower drop in *Hibiscus rosa-sinensis* caused by ethylene and darkness. *Acta Hort.*, No. 181, pp. 147-153.

Huang, J.G., Z.F. Fan, H.F. Li, G.F. Tian and J.S. Hu, 2004. First report of tomato mosaic virus on *Hibiscus rosa-sinensis* in China. *Plant Dis.*, **88:** 683.

Hughes, D. and C. Hughes, 2003. *Hibiscus rosa-sinensis*. www.plantsfacts.com/Family/ malvaceae/Hibiscus rosa-sinensis.shtml.

Ingram, D. L. and L. Rabinowitz, 2004 .Hibiscus in Florida. Horticulture Department, Cooperative Extension Service, Florida, USA (http://edis.ifas.ufl.edu).

Jayaweera, D.M.A. 1942. The production of hibiscus hybrids. *Tropical Agriculturist*, **98**(4): 19-25.

Jenderek, M.M. and A.J. Olney, 2001. *Hibiscus syriacus* plant regeneration from callus. *Intern. Pl. Prop. Soc. Prod.*, **50**: 565–568.

Johnson, C.R. and D.F. Hamilton, 1977. Rooting of *Hibiscus rosa-sinensis* L. cuttings as influenced by light intensity and ethephon. *HortSci.*, **12**: 39-40.

JuanJuan, S. and Z. Donghong, 2001. Chromosome numbers and ploidy of several plants in *Hibiscus* L. *J. Trop. Subtrop. Bot.*, **9**(3): 213-216.

Karlsson, M.G., R.D. Heins, J.O. Gerberick and M.E. Hackmann, 1991. Temperature driven leaf unfolding rate in *Hibiscus rosa-sinensis*. *Sci. Hort.*, **45**: 323-331.

Kasseanis, B. 1957. The use of tissue culture to produce virus-free clones from infected potato varieties. *Ann. Apl. Bio.*, **45**: 422–427.

Kelty, M.M. 1984. Container grown hibiscus: Propagation and production. *Proc.intern. Pl. Prop. Soc.*, **34**: 480-486.

Kholkute, S.D., S. Chatterjee, D.N. Srivastava and K.N. Udupa, 1972. Antifertility effect of alcoholic extract of *Hibiscus rosa-sinensis*. *J. Res. on Indian Med.*, **4**: 72.

Kim, J.K., H. So, M.J. Youn, H.J. Kim, Y. Kim and C. Park, 1996a. *Hibiscus sabdariffa* L. water extract inhibits the adipocyte differentiation through the P13-K and MAPK pathway. *J. Ethnopharmac.*, **114**: 260-267.

Kim, M.S., J.K. Kim, H.J. Kim, S.R. Moon, B.C. Shin and K.W. Park, 1996b. Hibiscus extract inhibits the lipid droplet accumulation and adipose transcription factors expression of 3T3-L1 preadipocytes. *J. Altern. & Complement. Medicine*, **9**: 499-504.

Kimbrough, W.D. 1997. *Hibiscus*. In: *Encyclopedia Americana*, p. 174. Grolier Inc., Danbury, Connecticut, USA.

Krüssmann, G. 1962. *Handbuch der Laubgehölze.* Paul Parey, Berlin, Germany.

Kudoh, H. and D.F. Whigham, 2001. A genetic analysis of hydrologically dispersed seeds of Malvaceae. *Amer. J. Bot.*, **88**(4): 588-593.

Kumar, A. and A. Singh, 2012. Review on *Hibiscus rosa-sinensis. Intern. J. Res. in pharmaceut. & biomed. Sci.*, **3**(2): 534-538.

Kuwanda, H. 1964. The newly artificially raised amphidiploids plant named 'Ai-Fuyo' (*Hibiscus mutamoscheutos*), obtained from the progeny of *H. mutabilis* × *H. moscheutos* (Studies on interspecific and intergeneric hybridization in Malvaceae VII). *Jap. J. Br.*, **14**: 27-32.

Kyung, H.Y. and J.H. Kim, 2001a. Effects of flowering season on the compatibilities of interspecific single, three-way and double crosses among *H. syriacus, H. sinosyriacus* and interspecific hybrids (*H. syriacus* × *H. paramutabilis*). *J. Korean Soc. hort. Sci.*, **42**: 568-573.

Kyung, H.Y. and J.H. Kim, 2001b. Relationship between pistil receptivity and cross compatibility in interspecific single, three-way and double crosses among *Hibiscus syriacus, H. sinosyriacus* and interspecific hybrids (*H. syriacus* × *H. paramutabilis*). *J. Korean Soc. hort. Sci.*, **42**: 561-567.

Kyung, H.Y., S.M. Park and J.H. Kim, 2001a. Effects of bud and old-flower pollination on the interspecific single and three-way cross compatibilities between cultivar group of *Hibiscus syriacus* and *H. sinosyriacus. J. Korean Soc. hort. Sci.*, **42**: 575-580.

Kyung, H.Y., S.M. Park and J.H. Kim, 2001b. Embryo abortion and germination inhibition in interspecific single, three-way and double crosses among *Hibiscus syriacus, H. sinosyriacus* and interspecific hybrids (*H. syriacus* × *H. paramutabilis*). *J. Korean Soc. hort. Sci.*, **42**: 737-742.

Larsen, F. E. 1914. *Propagating Deciduous and Evergreen Shrubs, Trees and Vines with Stem Cuttings*, 7 p. (leaflet). Washington State University Cooperative Extension, US.

Lawton, B. P. 2004. *Hibiscus*. In: *Hardy and Tropical Plants for the Garden*. Timber Press, USA.

Lee, D. W. 2007. *Nature's Palette: The Science of Plant Color*. University of Chicago Press, USA.

Mabberley, D.J. 1987. *A Plant Book*, 130 pp. Cambridge University Press, Cambridge, England.

Menzel, M.Y., K.L. Richmond, C.S. Contolini, and P. Huang, 1986. New intergenomic hybrids among diploid species of *Hibiscus* sect. *Furcaria. Amer. J. Bot.*, **73**:3 04–309.

Mohamed B. B., A.A. Sulaiman and A.A. Dahab, 2012. Roselle (*Hibiscus sabdariffa* L.) in Sudan, Cultivation and Their Uses. *Bulletin of Environment, Pharmacology and Life Sciences*, No. 1, pp. 48-54.

Nakasone, F. D. 1980. *Ornamental Hibiscus Propagation and Culture*, 35 p. College of Tropical Agriculture, Hawaii Agricultural Experiment Station, University of Hawaii.

Neal, M. C. 1948. *In Gardens of Hawaii*. Bernice Publisher. Bishop Mus. Sp. Pub. 40. Honolulu.

Nell, T.A. 1993. *Hibiscus rosa-sinensis*. In: *Flowering Potted Plants, Prlonging Shelf Performance*, pp. 48-50. Ball Publishing, Batavia, Illinois, USA.

Neumaier, E.E., T.M. Blessington and T.A. Price, 1987. Effect of light and fertilizer rate and source on flowering, growth, and quality of hibiscus. *HortSci.*, **22**: 902-904.

Nowak, J. and R.M. Rudnicki, 1990. *Postharvest Handling and Storage of Cut Flowers, Florist Greens, and Potted Plants*. Timber Press, Portland, Oregon, USA.

Osborne, L.S. 2005. *The Pink Hibiscus Mealybug* (leaflet). Mid-Florida Res. and Education Center, University of Florida, Bradenton, Florida, USA.

Osborne, L.S. and A.R. Chase, 1990. Chlormequat chloride growth retardant reduces spider mite infestations of *Hibiscus rosa-sinensis. HortSci.*, **25**: 648-650.

Paek, K., J. Hwang, S. Jong and S. Park, 1989a. *In vitro* pollination, ovule culture of *Hibiscus syriacus* and differentiation of adventitious bud and somatic embryo from callus. *J. Korean Soc. hort. Sci.*, **30**: 145-156.

Paek, K., J. Hwang, S. Jong and S. Park, 1989a. Plant regeneration in tissue cultures initiated from immature embryos of *Hibiscus syriacus*. *J. Korean Soc. hort. Sci.*, **16**: 93-104.

Pal, A.K., K. Bhattacharya, S.N. Kabir and A. Pakrashi, 1985. Flowers of *Hibiscus rosa-sinensis*, a potential source of contragestative agent: II. Possible mode of action with reference to anti-implantation effect of the benzene extract. *Contraception*, **32**(5): 517-29.

Pizzetti, I. and H. Cocker, 1968. *Flowers A Guide For Your Garden*, pp. 590-601. Harry N. Abrams, Inc., Publishers, New York, USA.

Popenoe, J. 2008. Chinese Hibiscus (*Hibiscus rosa-sinensis*). University of Florida Landscape Plant Propagation Information (http://hort.ufl.edu/lppi).

Purseglove, J.W. 1974. Tropical Crop: Dicotyledons (vol. 1 and 2 combined), pp. 364-372.. Longman Group Ltd., London, U.K.

Rajeshwari, R., R.V.C. Reddy, M.N. Maruthi, J. Colvin, S.E. Seal and V. Muniyappa, 2005. Host range, vector relationships and sequence comparison of a begomovirus infecting hibiscus in India. *Ann. Appl. Biol.*, **147**: 15–25.

Reynolds, B.D. and W.J. Blackmon, 1983. Embryogenesis and plantlet regeneration from callus of *Hibiscus acetosella*. *J. Amer. Soc. hort. Sci.*, **108**: 307–310.

Roe Sister, M.J. 1961. A taxonomic study of the indigenous Hawaiian species of the genus *Hibiscus* (Malvaceae). *Pacific Sci.*, **XV**: 3-32.

Runkle, E.S., R.D. Heins, W. Carlson, and A. Cameron. 1996. Manipulating daylength to flower perennials. *Grower Talks*, **60**: 66–70.

Ryan, G.F., E.F. Frolich and T.P. Kinsella, 1958. Some factors influencing rooting of grafted cuttings. *Proc. Amer. Soc. hort. Sci.*, **72**: 454-461.

Ryan, M., I. Warner and J.E. Erwin, 2001. Variation in floral induction requirements of *Hibiscus* sp. *J. Amer. Soc. hort. Sci.*, **126**(3): 262–268.

Schoellhorn, R. and A.A. Richardson, 2011. *Warm Climate Production Guidelines for Hardy Hibiscus. Commercial Floriculture Update*. Extension Bulletin No. ENHFL04-011. University of Florida, USA.

Shanks, J.R. 1972. Chemical control of growth and flowering in hibiscus. *HortSci.*, **7**: 574.

Sharma, A.K. and A. Sharma, 1962. Polyploidy and chromosome evolution of *Hibiscus*. *La Cellule*, **62**: 281-300.

Shim, K.K., K.H. Kim and Y.M. Ha, 1993. Characteristics of triploid cultivars 'Diana' and 'Helene' in *Hibiscus syriacus* L. *J. Korean Soc. hort. Sci.*, **34**: 54-67.

Silow, R.A. 1941. The comparative genetics of *Gossypium anamalum* and the cultivated Asiatic cottons. *J. Genet.*, **42**: 113-127.

Singh, F. and T.N. Khoshoo, 1989. Cytogenetic basis of evolution of garden *Hibiscus*. *The Nucleus*, **32**: 62-67.

Soulange, J. G., N. Boodia, C. Dussooa, R. Gunowa, S. Deensah, S. Facknath and B. Rajkomar, 2009. Vegetative Propagation and Tissue Culture Regeneration of *Hibiscus sabdariffa* L. (Roselle). *World J. agric. Sci.*, **5**(5): 651-661.

Srivatanakul, M., S.H. Park, J.R. Sanders, M.G. Salas and R.H. Smith, 2000. Multiple shoot regeneration of kenaf (*Hibiscus cannabinus* L.) from a shoot apex culture system. *Plant Cell Report*, **19**: 1165–1170.

Stephen, S.G. 1950. The genetics of "Corky". II. Further studies on the genetic basis in relation to the general problem of interspecific isolating mechanisms. *J. Genet.*, **50**: 9-20.

Steinberg, S.L., J.M. Zajicek and M.J. McFarland, 1991. Water relations of hibiscus following pruning or chemical growth regulation. *J. Amer. Soc. hort. Sci.*, **116**: 465-470.

Stoutmeyer, V.T., T.J. Maney and S.B. Pickett, 1935. Root formation in softwood cuttings of apple. *Proc. Amer. Soc. hort. Sci.*, **32**: 343-346.

Swanson, Jr., B.T., H.F. Wilkins, C.J. Weiser and I. Klein, 1975. Endogenous ethylene and abscissic acid relative to phytogerentology. *Pl. Physiol.*, **55**: 370-376.

Tachibana, Y. 1958. An interspecific hybrid of *H. mutabilis* L. and *H. moscheutos* L. (Studies on *Hibiscus* III). *J. hort. Asscn. Japan*, **27**: 201-206.

Thomas, B. and D. Vince-Prue, 1997. *Photoperiodism in Plants* (2nd ed.). Academic Press, San Diego, California, USA.

Van de Laar, H.J. 1997. *Hibiscus syriacus. Dendroflora*, **34**: 42-60.

Van Overbeek, J., S.A. Gordon and L.E. Gregory, 1946. An analysis of the function of the leaf in the process of root formation in cuttings. *Amer. J. Bot.*, **33**: 100-107.

Wang, Y.T. 1991. Growth stage and site of application affect efficacy of uniconazole and GA_3 in hibiscus. *HortSci.*, **26**: 148-150.

West, T.P and J.E. Preece, 2004. Effects of thiadizuron and nutrient salt formulation on micropropagation of hardy hibiscus (*Hibiscus moscheutos* L.). *Acta Hort.*, No. 630, pp. 293–297.

West, T.P. and J.E. Preece, 2006. Use of acephate, benomyl and alginate encapsulation for eliminating culture mites and fungal contamination from *in vitro* cultures of hardy hibiscus (*Hibiscus moscheutos* L.). *In Vitro Cell Devolpment*, **42**: 301–304.

Whalley, D.N. 1974. Ornamentals from hardwood cuttings in heated bins. *UK Grower*, **82**: 77-78.

Wilkins, H.F. 1986. *Hibiscus rosa-sinensis*. In: *Hasndbook of Flowering*, vol. V (ed. Halevy, A.H.), pp. 142-143. CRC Press, Boca Raton, Florida, USA.

Wikesjö,K. 1981. Odling av *Hibiscus rosa-sinensis* (Swedish). Konsulentavdelningens Rapporter, Tradgård 205, Swedish Univeresity of agricultural Science, Alnarp, Sweden.

Willis, J.C. 1973. *A Dictionary of Flowering Plants and Ferns* .Cambridge University Press,UK, pp. 12-14.

Willis, J.C. 1988. *A Dictionary of Flowering Plants and Ferns* (8th ed., revised by Airyshaw, H.K.), 558 pp.Cambridge University Press, U.K. pp. 12-14.

Wilson, F.D. and M.Y. Menzel, 1964. Kenaf (*Hibiscus cannabinus*) and roselle (*Hibiscus sabdariffa*). *Econ. Bot.*, **18**: 80-91.

Winters, H.F. 1970. Our hardy *Hibiscus* as ornamentals. *Econ. Bot.*, **B24**:155-164.

Woodson, W.R. and R.J. Raiford, 1986. Induction of lateral branching in Chinese hibiscus with mefluidide. *HortSci.*, **21**: 71-73.

Yang, L.J., M. Hidaka and H.U.T. Masaki, 1995. *In vitro* plant regeneration from leaf and petiole explants of *Hibiscus syriacus* L. *Plant Tissue Culture Letter*, **12**: 173–177.

Yu, T.Y., D.Y. Yeam and Y. Kim, 1976. A study on the breeding of *Hibiscus syriacus* L. on hybridisation among introduced tetraploids and *H. rosa-sinensis. J. Korean Soc. hort. Sci.*, **17**: 107-112.

Zapata, C., M. Srivatanakul, S.H. Park, B.M. Lee, M.G. Salas and R.H. Smith, 1999. Improvements in shoot apex regeneration of two fiber crops: cotton and kenaf. *Plant Cell Tissue Organ Culture*,**56**:185–191.

15

Hymenocallis

Alka Singh and Sanyat Misra

Family: **Amaryllidaceae**

Common names: Alligator lily (*H. palmeri*, syn.*H. humilis*), Basket flower/Basket lily (*H. narcissiflora*, syn. *Ismene calathina, Pancratium calathinum, P. narcissiflorum*), Caribbean lily (*H. caribbaea*), Cayman islands spider lily/Crysolite lily (*H. latifolia*), Crown beauty, Green pine lily (*H. henryae*), Ismene of Peruvian daffodil (*H. speciosa*), Ismene lily/Peruvian daffodil (*H. calathina, H. narcissifolia*), Sea daffodil, Spider flower, Spiderlily (*H. calathina, H. littoralis*), Summer daffodil

Introduction

Hymenocallis is a bulbous flowering plant. This genus has its origin in South America and Africa. *Hymenocallis* is derived from two Greek words; *hymen* meaning membrane and *kallos* meaning beauty, referring to the membranous beauty of its delicate flowers. *Hymenocallis* bears tender bulbs. *Hymenocallis* is native to warmer regions of the New World, i.e. southern part of North America [Flory, 1976; Smith and Flory, 1990; southern United States, (Florida, Georgia, Indiana, Louisiana, New Carolina, Texas, etc.) and Mexico] to the Andes in South America (Flory, 1976; Smith and Flory, 1990; Bahamas, Bolivia, Brazil, Cayman Islands, Chile, Colombia, Cuba, Equador, Guatemala, Haiti, Peru, Surinam, Venezuela, West Indies, etc.) and one species (*H. senegambica*) to West Africa. The genus has around 40 species, however, *Hymenocallis littoralis* is commercially important and is being cultivated on large scale in western India, specially in Gujarat and Maharashtra and it occupies a premium position in the flower market of Mumbai. There is production of around 2000 lakh bundles (50 flower buds per bundle) of spiderlily per annum from Gujarat alone. This genus is widely distributed in Mexico (Leszezynska *et al.*, 2000). Holland being the major producing country, they grow this crop to the extent of 2-3 hectares every year and the varieties they grow are cv. 'Advance', *H.* × festalis, *H.* × *festalis* 'Zwanenburg' and 'Sulphur Queen' (Langeslag, 1989). The genus has medicinal, ecological and nutritive value besides being potential commercial loose flower crop and garden plant. These bulbous plants are suitable for tropical plains and also for hilly altitudes upto 3000 m. These plants are suitable for growing along swamps, water channels and pools, on the borders, along the paths, in the beds, in pots and for mass effects. Though in India these are being grown for loose flowers but these can effectively be used as cut flowers also.

The generic name *Hymenocallis* is derived from the Greek word *hymen* meaning 'membrane' and *kallos* meaning 'beauty' (beautiful membrane), in reference to the membrane that unites and forms the staminal cup. *Ismene* was though being considered a separate genus in the past which has now been included in *Hymenocallis,* and thus *Ismene calathina* is now known as *Hymenocallis calathina*, further changed to *H. narcissiflora* while even earlier it was named as *Pancratium narcissiflorum*. The flowers of *Hymenocallis* closely resemble the *Pancratium* though flowers do not twist and curl in the former while in the latter the flowers are striped green with outer perianth segments smaller and wider.

Hymenocallis is cultivated for its evergreen leaves (most species) and the fragrant white flowers (only the species *H. amancaes* bears yellow flowers) that are of special attraction because of the membranous corona and the sweet fragrance of the flowers. It is suitable for growing in the field as well as in pots. These look attractive even when not in flowering as in tropical and subtropical climates mostly the leaves remain green. Also as cut flower it is attractive but the flowers do not last long. Certain species are used for loose flower production and *Hymenocallis littoralis* is being grown on commercial scale in Gujarat (Saurashtra region) for production of loose flowers as the flowers are supplied to Mumbai daily packed in gunny bags. Its loose flowers are used for worshippings, garland making, bridal car decorations, *mandaps*, pandal and stage decorations. The white delicate flowers of spider lily have spiritual significance and disseminate the message of 'Alchemy' (Mother, 1995). Its white flowers work as base to blend with other coloured flowers and these flowers have intense sweet fragrance, a possibility for extracting essential oil. Some of the species are though winter bloomers others bloom in spring and summer. Certain species are capable of developing hydrophilous roots to survive in watery beds or standing water hence the plants are highly suitable for growing near water bodies. These are most suitable as plants for border plantings in the greenhouse, along side the boundary walls and water channels, in herbaceous border and along the paths, along side the lawn and also in beds in the gardens but these like sunny situations. *H. littoralis* bulbs possess pancratistatin (20-24 mg/kg fresh weight; Pettit *et al.*, 1995) used for preparing anticancer drug (Backhaus *et al.*, 1992) together with lycorine, narciclasine, 7-deoxynarciclasine littoraline, lignan and 13 other known alkaloids (Lin *et al.*, 1995), and bulbs and leaves of *H. expansa* contain (+)-tazettine, hippeastine and haemanthidine (Antoun *et al.*, 1993).

Botany and Genetics

Hymenocallis littoralis is an important commercial crop. Its plant height is 30-70 cm with narrow to broad, sword shaped leaves. The flowers are fragrant white umbels, each flower with slender recurved petals and elongated stamens emerging from a central cup. The perianth tube is long and there are six segments, which are united at the base by a thin membranous cup of corona and so referred to as 'membranous beauty'. The flowers are borne in cluster (2-12) on around 60 cm long stalk arising from the centre of the leaves. It is rightly called spiderlily owing to the shape of petals, which look like legs of spider. The leaves arise in 'V' shape from the bulbs. *Hymenocallis littoralis* plants, the main commercial species in Western India are around 60-70 cm tall with light green foliage. The flowers are borne in a bunch of 10-12 on 60-90 cm long stalk, which open one by one in 4-5 days. The sword-shaped leaves are broad in size (4-5 cm) and the plant flowers through out the year.

The genetic potentiality of this crop is yet to be fully explored. Smith and Flory (1990) conducted cytological studies on *Hymenocallis henryae* and recorded 2n=38 chromosome numbers. Jee and Vijayavalli (1999) reported the karyotypes of eight taxa of *H. daphe* and recorded 2n=42, 44 and 46; in *H. littoralis, H. occidentalis* and *H. rotata* 2n=46; in *H. speciosa* 2n=54, and in *H. harrisiana* 2n=74, and suggested that polyploidy and aneuploidy significantly contributed to the evolution of chromosomes in this genus, with n=23 representing the basic chromosome number of the genus.

Species and Cultivars

The genus *Hymenocallis* possesses around 40 species. Morphological and anatomical investigation of some of the representatives of the genus *Hymenocallis* have been carried out by Artyushenko (1996). Some other types are although observed in Gujarat but molecular characterization is yet to be carried out. Important species are ***Hymenocallis calathina*** with narrow and 30 cm long leaves, flowers white, highly fragrant and many on the stalk, opening in succession in spring, cup fringed with a green throat, and the umbel lasts for two weeks; ***H. caribeae***, an evergreen plant native to West Indies, flowers white, fragrant, 9-10 cm in size, intricately designed and are borne on top of 40-60 cm tall stalks in summer; ***H. clivorum***, a new xerophytic species (found on rocky slopes in tropical deciduous forests) is unusual though its related species are found in or near standing or running water, it differs from *H. pimana* in having slightly wider leaves (9-16 mm v/s 5-11 mm), a shorter staminal cup (18-21 mm v/s 25-50 mm) and more number of flowers (8-11 v/s less than 8) (Laferriere, 1998[a]); ***H. crassifolia***, an evergreen plant in the south in a line from Picayune to Lucedale, but dormant elsewhere during winter, with leaves 60 cm long and 5 cm wide, flowers white and in clusters of four on a 60 cm long stalk, and flowering in summer; ***H. eucharidifolia***, a native to tropical America, leaves 30 cm long, flowers white, funnel-shaped with green tube, 4 or 5 in a cluster on a tall stalk and flowering in spring; ***H. liriosome***, a marshy species producing large (15-20 cm wide) white flowers with tantalizing fragrance, and does well even in upland soils; ***H. littoralis***, a potent loose flower crop in western India flowering through out the year up to 5-6 years from single planting, though in hilly areas in greenhouses it flowers from July to October and bulbs remain dormant between November and March (Biswas, 2006), species possesses some anti-viral activity (Berghe *et al.*, 1978) and lycorine, an antifeedant activity against desert locust, *Schistocerca gregaria* (Singh and Pant, 1980); ***H. macrostephana***, a South American species with strap-shaped 90-95 cm long leaves, spathe long which bears white flowers in cluster (up to 8) with green tube; ***H. narcissifolia***, a deciduous species native of the Andes of South America, with petals similar to spiderlily though with large frilly cup having orange incurved stamens, flowers fragrant and 2-9 on 60 cm tall stalk; ***H. ovata***, a native of W. Indies, leaves strap-shaped and 25 cm long, flowering in September-October, flowers white, fragrant with greenish tube and on a 30 cm long stalk; ***H. pimana*** where bulbs contain toxic and bitter tasting alkaloids but are being used by the mountain pima of Mexico as a famine food after detoxification by boiling in lye (Laferriere, 1998[b]); ***H. riparia***, a very small form of *Hymenocallis* with narrow leaves, rarely exceeding 30 cm in height and produces seven or more flowers during August-September; ***H. speciosa***, a native of West Indies and contains toxic substances even though cultivated widely in many countries, leaves up to 60 cm long and 10 cm wide, flowers white, highly fragrant; etc.

Though only little work is available on evolution of the varieties, however, some hybrids evolved in various countries are *Hymenocallis* ×'Tropical Giant' (Hayward), a very fine landscape plant with dark green leaves, blooming in July in most areas and covering Gulf coastal areas; *Hymenocallis* ×'Bellum', a hybrid with beautiful cream coloured flowers having golden centres; *Hymenocallis* ×'Mystery' with bulbs as large as oranges, leaves large (60 cm long and 5 cm wide) and similar to crinum, blooms fragrant, up to 20 cm wide and 10 or more per stalk, excellent as cut flowers and stalk 60-90 cm high; *Hymenocallis* ×'Sulphur Queen', flowers appearing in June, fragrant, 4 or more, large, yellow with green star in the throat, lasting over a week as cut flowers; *Hymenocallis* ×'Pax', an improved, long lasting, yellow hybrid; *Hymenocallis* ×'Sofforthiae', a most popular fragrant hybrid producing bright golden yellow beautiful blooms in June-July that are most suitable as cut flowers; *Hymenocallis* ×'Zwanenburg', with identical leaves to *H. calathina*, one of the most beautiful of all hymenocallises, the gracefully postured petals are long lasting as cut flowers or as pot plants; *Hymenocallis* ×'Zeylanicum' produces very beautiful flowers similar to *Hymenocallis* ×'Zwanenburg' with narrow linear foliage about 30 cm long; *Hymenocallis* ×'Carribaea', a cut flower hybrid with flowers lasting up to one week; *Ismene* × *H. festalis* (syn. *Hymenocallis* × *festalis*), a hybrid of the *H. narcissifolia* with *H. longipetala*; etc.

Propagation

Vegetative propagation like offsets, division of clumps, bulbs, twin-scaling, etc. is followed in *Hymenocallis littoralis*. Spiderlily is commercially propagated by bulbs of around 5-8 cm size. Spiderlily can also be grown from fresh seeds and offsets. There are some reports on tissue culture of *Hymenocallis littorallis* (Backhaus *et al.*, 1992). Explants of *Hymenocallis speciosa* have been cultured on solid White's medium, supplemented with NAA and BA at 2 ^{0}C in the dark. The regeneration of bulblets by scales is markedly increased by NAA at 0.01 mg/l and BA at 5 mg/l. Regeneration capacity has been found higher in *H. speciosa* where bulblets were found formed on scale base explants (Yanagawa and Ito, 1988).

Cultural Practices

Though spiderlily grows in a variety of **soils** but fertile loamy or medium black soil which is light in texture is most suitable for its cultivation. It can also be grown in heavy black soil and many species even in muddy soils. The soil pH should be normal, i.e. 6.5-7.5. However, it can tolerate salinity/alkalinity also to some extent. The field should be dug up by 40-60 cm depth, tilled and furrowed.

Hymenocallis littoralis which grows commercially in India, and even other species thrive well in tropical climate, i.e hot and humid conditions which are highly favourable for its growth and flowering. However, it can be grown in dry climates and hilly tracts also. Though it shows profuse flowering in bright sunlight conditions but have been observed to flower even in partial shade.

Hymenocallis littorallis grows well in open and full sunlight conditions, however, in hilly areas it prefers greenhouse conditions. Soil should be weed-free and thoroughly ploughed. Hoeing should also be followed when soil matures after watering. Furrows should be made for bulb **planting** at 90 x 20 cm spacing, i.e. 90 cm distance between two rows and 20 cm between plants in a row (Krushikar, 1988). Thus **plant density** is around 45,000 bulbs per hectare of land. Double row system is also followed in which 60 cm distance is maintained between two rows. Spiderlily plants are perennial and exhibit year round flowering. Hence, once planted it thrives well for about 5-6 years. Planting is done in the month of October-November (Krushikar, 1988). Although it can be planted in any season.

Hoeing and **weeding** should be done in the field of spiderlily at regular intervals for good growth of plants. Herbicide Oxyflouorfen can be effectively used to control weeds. All the dead leaves, flowers and other plant debris lying near the plant should be removed from the field. **Earthing up** should be carried out initially when plants attain some 20 cm height and in perennial cropping every year at the start of the growth to cover the exposed roots. This helps in conserving the moisture for healthy growth of the bulbs and plants, *vis-à-vis* lends support for whole growth and for flowering stalk. Earthing up can be repeated every year in the growing season after rains or irrigation water when soil around the bulbs and plants have washed away and roots have been exposed.

The plants of *Hymenocallis littorallis* do not require disbudding, pinching and pruning for flowering. Although, in the month of May when the price of spider lily goes down in the market, it is wise to **prune** the plants heavily of their leaves and allowed to remain on the soil surface. This technique of leaf pruning also serves as mulch for moisture conservation during hot weather. These leaves decompose and provide additional organic matter to the growing plants. Further, a good healthy flush of spiderlily is obtained in the coming months of June to August when the price of spiderlily is high due to festivals in India. Cutting of older leaves at ground level should be regularly followed in February and August (Patel, 2000). Spiderlily is a good **inter-crop** in the orchards of sapota (Chundawat, 1998) and coconut.

Feeding and Watering

It needs good amount of **watering** for proper growth and flowering. Light irrigation should be given at the time of planting. Gradually irrigation should be increased with sprouting of the bulbs and normal frequency of once in 5-7 days should be maintained. During hot summer months, frequency of irrigation should be increased at 3-4 days intervals and during winter months at 8-9 days interval (Patel *et al.*, 1993). Normally around 25 irrigations per year is required though the frequency largely depends upon prevailing weather conditions (Patel, 2000).

Spiderlily being a perennial crop needs good amount of **feeding**. During land preparation and each year thereafter, 30 tonnes/ha of farmyard manure (compost) should be incorporated in the soil before planting. Koladiya (1995) reported maximum plant height, number of leaves, width of leaves and bulb multiplication with application of 400 kg N/ha. In a study conducted on the nutrition requirement for N and P of spiderlily in Maharashtra, the optimum growth and high number of flower yield were obtained with application of 300 kg N + 250 kg P_2O_5/ha (Ghule *et al.*, 2003). According to Patel (2000), every year the crop need for chemical fertilizers is 300 kg N, 225 kg P and 200 kg K per hectare. Phosphorus and potash should be given as basal dose (Patel, 2000). Nitrogen should be given in four split doses, i.e. June, September, December and March. De Hertogh and Le Nard (1993) advocate 2-3 times application of 14 kg of 12:10:18/ha of NPK every season.

Growth and Flowering

The growth, development and flowering studied so far are only on *Hymenocallis × festalis* (Hanks and Jones, 1987) where there is cent per cent flowering only when bulbs are 13/14 cm in circumference, however, 9/10 cm circumference bulbs show only 25 per cent flowering. Hanks and Jones (1987) further recorded that storing the bulbs at 17 or 20 °C flower usually though 5 °C storage is harmful, and storage duration (4-9 months) has little effect on flowering. The growth of the plants and flowering are optimum around 20-30 °C day temperatures and 15-20 °C night temperatures. However, flowering in *Hymenocallis littoralis* is not much affected till 45 °C temperature as observed in Gujarat. Mori *et al.* (1991) recorded earliest flowering at 15 to 20 °C **temperature** and **long day** conditions of 16 hours in spiderlily. Hanks and Jones (1987) while working with *H. × festalis*, advocated ethephon for delaying flowering from 3-6 days and for reducing of scape length to its half. There is need of further studies in this area. However, *Hymenocallis littorallis* behaves as a day neutral plant in Gujarat conditions. According to Idso *et al.* (2000), atmospheric $\mathbf{CO_2}$ enrichment by 700 ppm enhances the above ground biomass by 48% and underground biomass by 56% in *Hymenocallis littorallis* in the Sonoran Desert environment of central Arizona (USA). Further, mean percentage increases in anticancer and antiviral activities concentrations were 6% for a 1:1 mixture of 7-deoxynarciclasine and 7-deoxy-trans-dihydronarciclasine, 8% for pancratistatin, 8% for trans-dihydronarciclasine, and 28% for narciclasine, with a mean active ingredient percentage concentration increase of 12%. Combined with the 56% increase in bulb biomass, the overall increase in active ingredient content was 75% compared with control bulbs.

Post Harvest

Hymenocallis littorallis plants thrive well continuously for 5-6 years and there is no need of **bulb lifting** every year as observed in Gujarat and Maharashtra. After 5-6 years, a group or cluster of 7-8 bulbs is developed, hence the old crop should be uprooted. The bulbs should be separated, dried and after treatment with a fungicide, *viz.* Carbendazim should be planted independently. This crop does not require to be stored at low temperature for dormancy breaking like other bulbous crops. De Hertogh and Le Nard (1993) advocate lifting of the bulbs before the first frost and storing them immediately at a well ventilated place at 25 °C temperature for drying and then storing at 20 °C after grading.

Flower bud harvesting in spiderlily is done in the early morning between 6 and 9 hours. Stage of harvesting in spiderlily is a mature bud when buds have started turning white just to open the next day. **Grading** of the buds is done as per the bud size (length) in two grades and bundles of 50 buds each is made. The bundles are tied with rubber bands and packed in gunny bags or plastic bags. The spiderlily bags are immediately transported to the markets. Production of spiderlily is around 5-6 lakh bundles per annum per ha. The vase life of spiderlily is very short, it is just one day. There is a great fluctuation in the market rate of spiderlily, ranging from Rs.10-Rs.80 per thousand buds, which varies day-to-day. There is need to work out optimum post harvest handling techniques for spiderlily. The buds of spiderlily can be stored in cold storage for a duration of one week with prestorage pulsing with 250 ppm 8-HQC and 2% sucrose for 45 minutes and tissue paper packaging at 2 °C temperature (Bhatt, 2007). Further, there is need to work out pulsing and preservative solutions to enhance the vase life of spiderlily flowers.

Insect-Pests and Diseases

Though *Hymenocallis littoralis* and its other species are free from most of the insect-pests and diseases but a few attacking these are being elaborated herewith along with their suitable control measures.

Red and white lily caterpillar (*Polytera gloriosae*) is also known as spotted ballworm. There is heavy infestation in the field of spiderlily during monsoon in Gujarat (Patel, 1997). The caterpillar feeds vigorously on the foliage as well as the flower buds causing a heavy damage to the crop. Patel and Patel (1997) studied the biology of this pest in spiderlily and suggested Spark 36 EC (0.045%) spray followed by Chlorpyriphos 20 EC (0.03%) or Alphamethrin 10 EC (0.005%) and Quinalphos 20 A.F. (0.03%) sprays. These sprays can be repeated after 15 days interval if needed. Some species of **aphids** have been observed to damage the crop. These tiny insects suck the cell sap of the foliage and petals of spiderlily. Spray of systemic insecticides like Dimethoate or Metasytox at the rate of 2 ml/l of water can effectively control this pest. Major insect attacking this crop is **large bulb fly** (*Lampetia equestris*) which can be controlled through 2h HWT at 43.5 ^{0}C. **Stem nematode** (*Ditylenchus dipsaci*) infests on this crop whose infestation can be brought under check by using healthy bulbs, soil sterilization and field sanitation.

Stagnospora curtisii causes leaf spot and scorch, and fire. **Cercospora leaf spot** is caused by *Cercospora amarillidis.* Large reddish brown to dark brown spots develop on leaf surface and progress by formation of yellow haloes and thin papery centre. These further increase in size, infecting entire leaf and ultimately result into death of the leaves. This indirectly affects the crop production by decrease in total photosynthesis in the plant. Both these diseases can be effectively controlled by Carbendazim at the rate of 2ml/l of water. **Anthracnose** is caused by *Colletotrichum gloeosporioides* (syn. *Glomerella cingulata*). Infection predominantly develops on leaves in form of small grey-brown spots which spreads and finally results in premature drying. As per a survey in the endemic areas of Maharashtra, the anthracnose disease incidence in spiderlity ranged from 26 to 46%. The disease can be controlled effectively by Bordeaux mixture (0.8 and 1%) or copper oxychloride (0.2 and 0.3%) or Triadimefon (0.05 and 0.1%). **Alternaria blight** is causesd by *Alternaria* sp. The disease infection begins first with browning and drying of the tips and margins of the foliage which further proceeds to the middle portion of the leaf. Though this does not harm the quality of the flower buds directly but decreases the production affecting foliage growth. *Phoma narcissi* is observed in *Hymenocallis narcissifolia* causing bulb rot. This disease can effectively be controlled by bulb dressing with a mixture of thiophanate methyl 70 WP (Topsin M, Rocko), Chlorothalonil 500 SC (Kavach, Deconil) and Carbendazim (Bavistin) with Prochlozol 380 EC (Saniewska, 1997). Garlic extract can also be used effectively to control this disease as its action is comparable to Prochlozol. The plants of *Hymenocallis littoralis* are sometimes infested with **lily rust** caused by *Puccinia legenophorae.* The infection is not yet found heavy. Symptoms of lily rust appear as small orangish to reddish-brown swollen spots on leaves covered with rusty coloured dust. These often increase in size and finally rupture the epidermis. The disease can be controlled by spraying with Areofungin or Cholorthalonil (Kavach).

References

Antoun, M.D., N.T. Mendoza, Y.R. Rios, G.R. Proctor, J.M. Pezzuto and A.D. Kinghorn, 1993. Cytotoxicity of *Hymenocallis expansa* alkaloids. *J. Natural Products,* **56**(8): 1423-1425.

Artyushenko, Z.T. 1996. Morphological and anatomical investigation of representatives of the genus *Hymenocallis* (Amaryllidaceae). *Botanicheskii-Zhurnal,* **81**(4): 78-86.

Backhaus, R.A., G.R III Pettit, D.S. Huang, G.R. Pettit, G. Groszek, J.C. Odgers, J. Ho and A. Meerow, 1992. Biosynthesis of the anti-neoplastic pancratistatin following tissue culture of *Hymenocallis littoralis* (Amaryllidaceae). *Acta Horticulturae,* No. 306, pp. 364-366.

Bhatt, S.T. 2007. Effect of Cold Storage Temperatures, Wrapping Materials and Storage Duration on Post Harvest Life of Spiderlily. M. Sc. (Horticulture) thesis submitted in the Department of Horticulture, Navsari Agricultural University, Navsari (Gujarat).

Berghe D.A., M. Ieven, F. Mertens, A.J. Vlietinck and E. Lammens, 1978. Screening of higher plants for biological activities. II. Antiviral activity. *J. Natural Products,* **41**(5)**:** 463-471.

Biswas, B. 2006. Hymenocallis. In: Advances in Ornamental Horticulture (Vol. 3) : Bulbous Ornamentals and Aquatic Plants (ed. Bhattacharjee, S.K.), pp. 211-217. Pointer Publishers, Jaipur.

Chundawat, B.S. 1998. Sapota. Agrotech Publishing Academy, Udaipur, 88 pp.

De Hertogh, A.A. and M. Le Nard, 1993. *Hymenocallis* (General Chapter on Summer Flowering Bulbs), pp. 759-760. In: The Physiology of Flower Bulbs (eds Hertogh, A. De and M. Le Nard). Elsevier, Amsterdam, Holland.

Flory, W.S. 1976. Distribution, chromosome numbers and types of various species and taxa of *Hymenocallis. Nucleus,* **19**: 204-227.

Ghule, A.D., P.V. Patil and K. T. Kantharaju, 2003. Effect of different levels of nitrogen and phosphorus on growth and flowering of spider lily. *J. Maharashtra agric. Univ.*, **28**(2): 128-130.

Hanks, G.R. and S.K. Jones, 1987. The growth and flowering of *Hymenocallis* × *festalis*. *Scientia Hort.*, **30**: 301-313.

Idso, S.B., B.A. Kimball, G.R. III Pettit, L.C. Garner and R.A. Backhaus, 2000. Effects of atmospheric CO_2 enrichment on the growth and development of *Hymenocallis littoralis* (Amaryllidaceae) and the concentrations of several anti-neoplastic and antiviral constituents of its bulbs. *Amer. J. Bot.*, **87**(6): 769-773.

Jee, G. and B. Vijayavalli, 1999. Karyomorphology of eight taxa of *Hymenocallis* from South India. *Caryologia*, **2**(1-2): 59-64.

Koladiya, B.V. 1995. Effect of Different Spacing and Nitrogen Levels on Growth and Flower Production of Spiderlily (*Hymenocallis littoralis* L.). M.Sc. (Ag.) thesis submitted in the Department of Agriculture, Gujarat Agricultural University, Gujarat.

Krushikar, I. 1988. *Lili ni Amdani Apti Kheti-ek Mulakat* (Gujarati). *Krushi Jivan*, **20**(12): 4-7.

Langeslag, J.J.J. 1989. Teelt en Gebruiksmogelijkheden van Bijgoedgewassen. Tweede Uitgave. Ministerie van Landbouw, Nutuurbeheer en Visserij en Consulentschap Algemene dienst Bloembollenteelt, Lisse, The Netherlands, 273 pp.

Lafferriere, J.E. 1998[a]. A new *Hymenocallis* (Amaryllidaceae) from Sonora, Mexico. *Novon*, **8**(3): 242-243.

Lafferriere, J.E. 1998[b]. Native detoxification of bulbs of mountain pima spider lily (*Hymenocallis pimana*). *Economic Bot.*, **52**(2): 207-208.

Leszezynska, B.H., M.W. Borys, S.A. Espejo, Verdonck (ed.), A. Mathe (ed.), P.D. Relf (ed.), E. Matsuo (ed.), G.D. Groening (ed.) and J. Rammeloo, 2000. Mexican geophytes, biodiversity, conservation and horticultural application. *Acta Hort.*, No. 523, pp. 205-210.

Lin, L.Z., S.F. Hu, H.B. Chai, T. Pengsuparp, J.M. Pezzuto, G.A. Cordell and N. Ruangrungsi, 1995. Lycorine alkaloids from *Hymenocallis littoralis*. *Phytochemistry*, **40**(4): 1295-1298.

Mori, G., H. Imanishi and Y. Sakanishi, 1991. Effect of temperature on flowering of *Hymenocallis speciosa* Salisb. *J. Japanese Soc. hort. Sci.*, **60**(2): 387-393.

Mother, 1995. Flowers and Their Messages. Shri Aurobindo Ashram Press, Pondichery, India.

Patel, K.S. 1997. Biology and Chemical Control of Lily Caterpillar, *Polytela gloriosae* Fabricius on Spiderlily. M.Sc.(Ag.) thesis submitted in the Department of Entomology, Gujarat Agricultural University, Gujarat.

Patel, R.L. 2000. Diagnostic Survey for Cultivation Practices and Studies on Nutrient Management in Spiderlily (*Hymenocallis littoralis* L.). Ph.D Thesis submitted in the Department of Horticulture, Gujarat Agricultural University, Gujarat.

Patel, R.L., R.C. Patel and J.S. Katrodia, 1993. *Lily ni kheti padhati* (Gujarati). *Krushi Go Vidhya*, **5**(6): 38-40.

Patel, K.S. and M.B. Patel, 1997. Biology of lily caterpillar, *Polytela gloriosae* Fabricius (Noctuidae: Lepidoptera) on spiderlily, *Hymenocallis littoralis* L. *J. Applied Hort., Navsari*, **3**(1/2): 115-124.

Pettit, G.R., G.R.III Pettit, R.A. Backaus and F.E. Boettner, 1995. Anti-neoplastic agents, 294. Variations in the formation of pancratistatin and related isocarbostyrils in *Hymenocallis littoralis*. *J. Natural Products*, **58**(1): 37-43.

Saniewska, A. 1997. Control of *Phoma narcissi* on the bulbs of *Hymenocallis narcissiflora* Prog. *Pl. Protection*, **37**(2): 387-389.

Singh, R.P. and N.C. Pant, 1980. Lycorine - a resistance factor in the plants of subfamily Amaryllidaceae against desert locust, *Schistocerca gregaria* F. *Indian J. Entom.*, **42**(3): 460-464.

Smith, G.L. and W.S. Flory, 1990. Studies on *Hymenocallis henryac* (*Amarallidaceae*). *Brittonia*, **42**(3): 212-220.

Yanagawa, T. and I. Ito, 1988. Differences in the capacity for the bulblet regeneration between bulb scale explans excised from different parts of *Hymenocallis* and *Ornithogalum* bulbs. *J. Japanese Soc. hort. Sci.*, **57**(3): 454-461.

16

Ixora

Jayoti Mazumder, Sellam Perinban and R.L. Misra

Family: **Rubiaceae**

Common names: Flame of the woods (*Ixora coccinea*), Ixora, Jungle geranium (*I. javanica*), Cambodia-Kam rontea, Bengali-Rangan, Hindi-Rukmini, Malayalam-Chethi, Tamil-Vedchi, Telugu-Nooru, Vietnam-Dun do

Introduction

Ixora derives from the Portuguese rendering of the Sanskrit 'Ikvana' or 'Israra' meaning 'Ishwara' (the Hindu god Shiva), to whom the flowers are offered and *coccinea* means scarlet. *Ixora* is a tongue-twister name as it is pronounced *icks-sore-ah*. This genus has some 500 species of shrubs and trees from Pan-tropical regions, i.e. India, China, Java, Malayan Archipelago, Moluccas, Madagascar, Guinea, East Indies, etc. The ixoras were used as house plant in England around 1870 and numerous varieties were developed during that period and at the turn of the century. All ixoras produce clusters of star-shaped flowers on stem tip or terminal corymbs varying in colour from red to orange to yellow, pink and white and are used for offering to gods, especially the Shiva and for making garlands apart from its effective use in shrubberies, hedging and for garden landscaping. Often the leaves and stems are used as an ablution for infantile and is sedative. The flowers are used externally to sores, chronic ulcers, scabies and in certain types of dermatitis and the roots in case of cholera, diarrhoea, dysentery, gonorrhoea and fever.

Ixora coccinea Linn, a national flower of Myanmar, is a medium size beautiful landscape shrub of Indian origin which is grown throughout the world in the tropics and sub-tropics for its exquisite flowers. In India though it is found throughout but is very common in west peninsula in scrub jungles. Its loose flowers are used for religious offerings to gods and goddesses, especially the Shiva.

Botany and Classification

The *Ixora* L. is the third largest genus of coffee family. Ixoras are popular ornamental plants grown in the tropics for their beautiful inflorescences which are in different shades of white, yellow, orange, pink and red. In general, the classification and identification of the species and cultivars in this genus have been very difficult. The existence of natural hybrids among the species cannot be ruled out. Taxonomists have tried to distinguish the taxa morphologically, especially using the nature of leaves and floral parts. Alternative approaches for intra-and inter-specific classification have been reported based on geographic origin, morphology, karyotype or isozyme analysis (Kaser and Steiner, 1983). These criteria are either influenced by environmental factors and stage of plant development or reveal only limited variation. It's a finest shrub or tree ensuring for itself a perennial position in the garden. The leaves are simple in opposite pairs and terminal corymbs of tubular, mostly 4-petalled flowers which in a few species are fragrant. There are 15-30 or even more magnificent flowers arranged across the corymb. In *Ixora coccinea*, the large terminal flowers of orange scarlet colours are borne in profusion during summers and rains but in places such as Bangalore the flowering continues throughout the year. The shrub is generally 2.0-3.0 m tall with 5.0-7.5 cm long leaves which are bronzy when young, later turning to a glistening dark green. Although most *Ixora* plants of shrubby nature are much-branched and compact so are ideal for hedges, borders, screens, or as a specimen planting. Ixora fruit is a subglobose berry which ripens to red and black.

Ixora displays colourful, nocturnally fragrant, nectariferous, tubular flowers. The flowers are dichogamous. Therefore, it adopts self pollination (Latha *et al.*, 2012). Ixora flowers are strongly protandrous and it has 'ixorid' secondary pollen presentation mechanism (Nilsson, 1998). In Ixora, the natural fruit set decreases with increasing pollination population (Nayak *et al.*, 2010). Both the fertile and sterile appendages are borne on the receptacle (Pandey, 2006). The cultivated shrub such as *Ixora bandhuca* (67.0%) and *Ixora barbata* show low pollen fertility (Bedi *et al.*, 1981). Principal pollination adoption probably occurs by a noctuid moth (Nilsson, 1998). SEM studies of *I. chinensis* in China revealed that if once the terminal apex reaches the inflorescence-bud-stage, it would flower without abortion. The sepals, petals, stamens, and pistil are well developed thereafter and anthesis achieves in January through March in the following year. The haploid chromosome number of *Ixora* is n=11 (Bedi *et al.*, 1981) and *Ixora rosea* is a triploid with 2n=33 (Sharma and Chatterjee, 1960).

Species and Cultivars

Ixora comprises of about 500 species (Mouly *et al.*, 2009) of shrubs and small trees distributed in tropical and subtropical regions of the world and valued primarily for their colourful flowers though one shrubby species ***I. borbonica*** from Reunion Island growing some 2.0 m with off-white small flowers and bearing 15-25 cm long, lanceolate, leathery-textured and lustrous mossy-green leaves veined carmine-red which is valued for its handsome foliage looking like *Codiaeum*. Phylogenetic analyses of Ixoreae reveal that *Ixora* tribe can be subdivided into three major clades: the Mascarene/neotropical/Malagasy/African clade, the Pacific clade, and the Asian clade. Other conflicting positions for the cultivated species are most likely due to anthropogenic hybridization. Most of cultivated species are native to southern Asia and China but only a few of them are grown as landscape plants (Rajasegar *et al.*, 1997).

In India the genus is represented by 46 species (Husain and Paul, 1991; Sivadasan and Mohanan, 1991; Deb and Rout, 1992; Pradeep, 1997). Among them the popular species grown are ***Ixora acuminata***, a glabrous shrub having various types of elliptic to linear-oblong leaves but floral leaves rounded or obovate and sessile and others petioled, cyme 5-10 cm across and corymbose comprising of densely packed florets which are pure white and fragrant; ***C. amboinica***, a shrub bearing glabrous, ovate-oblong, undulate, acuminate, large but short-petioled leaves, cyme trichotomously divaricately compound and flowers long-lasting showy orange-yellow; ***I. barbata*** from India, a large glabrous shrub bearing thin, stalked, elliptic, and a little acute leaves though upper pair sessile, small and cordate, cyme short-stalked and quite wider (maximally 30 cm) than length, and corolla white being woolly at mouth; ***I. chinensis*** native to Malaysia and China, growing up to 1.5 m tall, very similar to *I. coccinea* but smaller in all its parts and petals with blunt tips, a glabrous shrub growing 60-90 cm bearing obovate or obovate-oblong and sessile to subsessile, glossy and deep green leaves, slender-tubed orange flowers in dense corymbs which appear in summer. This species offers many cultivars, many of which produce yellow, yellow-orange, orange, red, pink and white flowers such as 'Alba' (white), 'Rutilans' (crimson), and 'Singapore Yellow' of 1.5-2.0 m in height with bright to dark green leaves and yellow-orange flowers which continue coming up throughout the year; ***I. coccinea*** L. (syn. *I. bandhuca* Roxb.) from India and East Indies grows to 2.5-3.0 m with new growth reddish, bearing elliptic to oblong with rounded or cordate base, glossy and deep green and up to 19 cm long leaves, and 13 cm wide corymbs with 4 cm long orange-scarlet flowers (cultivars in various shades of red, pink, orange-yellow and yellow) having broadly pointed petal tips coming up in summer. The hybrids being the natural crosses among the species account for majority of ixoras for landscaping. The hybrids 'Nora Grant' grows to 1.5 m with large and light to bright green leaves and clusters of pink flowers which continue appearing throughout the year; 'Mauli' grows to 1.5 m with compact growth bearing smaller leaves and orange-red flowers, and 'Rosea' (pink); ***I. congesta*** from India, an evergreen glabrous tree (only cyme pubescent), leaves stalked, acute or acuminate, elliptic or elliptic-oblong and 15-30 cm long, cyme sessile or so bearing orange-yellow flowers which change to reddish; ***I. duffi***, a larger growing glabrous and tall species native of East Indies with large green (30 cm), linear-oblong to oblong-lanceolate leaves and deep red tinged crimson flowers in large corymb (20 cm across) and its most common cultivar is 'Superking' reaching to 3.0 m in height and bearing orange-red and showy flowers in large clusters which continue coming up almost throughout the year; ***I. fulgens*** from India is an erect shrub with polished branches, leaves linear-oblong to obovate-oblong, acute or acuminate with sunken nerves and petioled, cymes corymbose, large, sessile or short peduncled and short pedicellate orange to scarlet flowers; ***I. javanica*** from South East Asia grows from a large shrub to a small tree (7.5 m) in its homeland, bears 20 cm long, ovate-oblong, leathery-textured with waxy shine rich green leaves and 6 cm or more across cluster with 4 cm long salmon-red flowers appearing during summer; ***I. laxiflora*** from Guinea, a 90-120 cm slender shrub, leaves oblong-lanceolate, acuminate and short-petioled, cymes large, open and trichotomous, and flowers very fragrant and white tinged pink; ***I. lutea*** (syn. *I. coccinea* var. *lutea*) of garden origin, which differs from *I. coccinea* by lax inflorescence and pale-yellow flowers though corolla lobes are larger ovate-rhomboid; ***I. odorata***, a small fragrant shrub from Madagascar bears large (up to 30 cm), thick, ovate to obovate-lanceolate, acute or acuminate leaves, cyme 30 cm or more across with purplish branches, and corolla 10.0-12.5 cm long and white but changing to yellowish-brown and highly fragrant; ***I. parviflora***, an evergreen tree from India, bearing subsessile oblong or elliptic-obtuse leaves 7.5-15.0 cm long, cymes sessile bearing flowers in subglobose clusters, and corolla glabrous and white; ***I. taiwanensis*** 'Petites' seldom reaches over 90 cm with quite compact growth, bears tiny leaves and red, yellow or pink flowers. Among hybrids with unknown parentage are 'Angela Busman' (flowers shrimp-pink), 'Frances Perry' (large trusses of deep yellow flowers), 'Gillettes Yellow' (pale-yellow), 'Helen Dunaway' (deep orange), 'Henry Morat' (fragrant pink), 'Herrera's White' (beautiful white), 'Superking' (deep red and floriferous), etc. *Ixora chinensis* is a spontaneous seedling selection of unknown origin (Zaundam, 1997). 'Frankie Hipp' is a variety derived from 'Nora Grant' through spontaneous mutation in Florida and is being grown there (Fulghum Leigh, 2008). Some other hybrids, varieties or species in **whites** are *Ixora colei* (*I coccinea* × *I. chinensis* 'Alba'), an English hybrid developed around 1870 by E. Cole of Manchester (England) but when once lost there, it was again introduced from Calcutta in 1964 by Swedroe's Nursery, is more cold-hardy and prefers filtered situation having compact and low spreading growth bearing massive pure white flowers blooming intermittently during early summer and late fall and remains in flower for a considerable time; 'Herrera's White', a sport of 'Herrera's Pink', originating at Swedroe's Nursery in Fort Lauderadale (Florida), is a cultivar liking partial shade and grows upright and compact with constant blooming habit, producing pure white flowers of medium size; in **yellows** are 'Frances Perry' ('Gillete's Yellow' × 'Singapore No. 1'), developed by Swedroe in 1963, is compact with broader and dark green leaves and bears best pure yellow flowers. *Ixora javanica flava* was introduced by Swedroe in January 1964 from Bangkok (Thailand), has bushy growth and bears apricot-gold rounded flowers as in *I. chinensis* with which it hybridizes easily and has produced numerous hybrids. 'Singapore No. 1', a chance seedling of *I. aureo-rosea* recorded in 1952 from Singapore with similar flowers to *I. macrothyrsa*, likes partial shade and is prolific bloomer bearing orangish-yellow to pinkish flowers as large as 12.5 cm across; in **Orange and salmon** are 'Angela Busman' ('Superking' × 'Singapore No. 1') was developed by Swedroe in 1963, bearing large and dark green leaves, and the flower colour is shrimp-pink to orange when first open but turns old rose as they age. 'Lois Shore' (a seedling of 'Singapore No. 1') originated at Swedroe Nursery in 1965, prefers partial shade and flower colour is soft orange similar to 'Singapore No. 1'; in **Pink and rose** are *I. fraseri* (probably a cross between *I. ambonensis* and *I. chinensis*), a very beautiful shade-loving shrub but highly susceptible to root-knot nematode, first exhibited in England about 1874 and was introduced in Fort Lauderdale around 1952, bearing large and dark green leaves, flowering in very large clusters with open buff-orange flowers fading to salmon; in **pinks** are 'Henry Morat' (*I. acuminata* × *I. coccinea*) developed in Miami (Florida) in 1943, having compact growth with upright branches bearing large and dark bluish-green leaves and fragrant light pink flowers being produced in great profusion over a long period; 'Herrera's Pink', imported from Trinidad in 1952 and introduced in 1960 to the horticultural trade is an upright shrub loving shade, bearing quite large and light green leaves and the flowers are dark rose similar to *I. rosea; I. hydrangeaeformis* (syn. 'Trinidad Pink'), a fast growing, shade-loving, cold-tolerant and tall but compact shrub most suitable for hedging was imported in 1952 from Trinidad and has since been growing in south Florida for its pink flowers similar to 'Henry Morat'; 'Kelly Gent' (*I. acuminata* × 'Singapore No. 1') was developed by Swedroe in about 1964, is profuse seasonal bloomer with beautiful orange flowers at opening but turning salmon with age; 'Pinksing' (*I. acuminata* × 'Singapore No. 1') was developed by Swedroe around 1960 with slightly fragrant rich pink flowers appearing from spring to summer; *I. rosea* was introduced by Edwin A. Menninger through seeds brought into Florida from Madras (India), likes partial shade, is susceptible to nematode and chlorosis and has compact growth with free branching habit, leaves small and dark green, and the flowers are vivid deep rose appearing throughout the growing season; *I. westii* is a natural hybrid recorded in 1882 and was introduced to Fort Lauderadable by Swedroe in 1964 from Calcutta, flowers opening to pale-rose but turning to bright rose afterwards; in **reds** are *I. chinensis* discovered in South China around 1822, has dwarf and compact growth with good cold hardiness, very variable plant with white, yellow, pink and orange forms, so much used by English hybridizers, mostly free-flowering with orange-red flowers; *I. williamsii* (syn. 'Trinidad Red'), a free-flowering plant is highly amenable to severe trimming in late winter, bears very large and dark green leaves and the large flowers are the deepest of the reds; 'Superking' (a *I. macrothyrsa* cultivar), a large bushy plant introduced to the South Florida trade at about 1950 by Swedroe though was already available there since 1932, probably most spectacular of all the ixoras, bearing large brilliant red flowers being produced throughout the year in Florida, leaves very large, up to 25 cm long, new ones usually pale-green but turn dark green as they mature; and with **variegated leaves** are 'Variegated', appearing to be a mutant cultivar of *I. coccinea* in which the majority of the leaves show patches of non-green (white) sectors (Rajaseger *et al.*, 1999). Dwarf cultivars are also available like *I. compacta* 'Sunkist' growing to a height of only 60 cm with orange flowers; *I. taiwanensis,* a dwarf and compact flowering shrub. A new variety named 'Diora' which is of dwarf hybrid class of milk white colour is under patent.

Propagation

Whistler (2000) in Puerto Rico when grew its **seeds** in commercial potting mixture, he obtained 70% germination. Seeds germinate well in acidic soil (Holtum & Enoch 1991) if sown during autumn (Randhawa and Mukhopadhyay, 1994). Through seeds also these normally come true to the type but it takes long time in attaining the full growth.

It's generally accepted that perennial ornamental plants are multiplied and propagated through vegetative means such as cuttings, layering or grafting (Elgimabi, 2008). Traditionally *Ixora* is propagated from stem cuttings of mature shrubs and is raised in polybags though rooting in such cuttings is very difficult. The **cuttings** should be taken from young and fast growing tips from matured shrubs (Khan *et al.*, 2004) in summer or spring. Since rooting is a problem, therefore IAA and IBA rooting hormones and bottom heating should be employed for efficient rooting. Annual and semi-mature lignified twigs of middle and lower portion of *Ixora chinensis* when were subjected to mixture of various hormonal concentrations and rooting media, it was recorded that sand and clay in half medium was best among the various media used and the mixture of IBA 75 mg^{-1}+ NAA 75 mg^{-1} recorded highest root rate and root number (Sheng-Jian and Jun-Bang, 2011). When cuttings of *I. chinensis* were treated with various concentrations of NAA, IBA and NAA + IBA, the latter at 1,000 ppm recorded best rooting as compared to NAA or IBA treatment alone (Jinsui Lin, 2009). The 2 cm bases of *Ixora chinensis* cuttings taken from October to February were when dipped for 15 seconds in solutions of IAA, IBA or NAA, each at 1,000-6,000 ppm, highest rooting (80-86%) was obtained with cuttings taken in October and treated with IBA at 2,000-6,000 ppm though those taken in January showed poorest response (Singh, 1980). The soilless media cocopeat and teak sawdust produce better root growth and leaf formation (Adetimiria *et al.*, 2008). Taking plant ash as basic culture material and 800 ppm IBA treatment, the *Ixora duffii* cv. 'Super King' cuttings presented the best rooting with highest root number. Summer propagation gives the best performance in number and length of both roots and leaves of the cuttings. Plastic tunnels with water mist also give the best rooting and vegetative growth followed by cuttings planted under plastic tunnels without mist, irrespective of season. The control treatment without plastic cover has been found giving the poorest results (Elgimabi, 2008).

There are some difficulties in growing a good quality flowering plant in various agroclimatic conditions. Foliage chlorosis caused by mineral deficiencies, nematodes and other factors often limit the use of many at the showy varieties in the landscape. In effort to minimize the damage caused by these factors, trials on various rootstocks with **grafting** were conducted. It was discovered as early as 1946 that *Ixora parviflora* rootstock gives resistance to nematode infestation, chlorosis club rot and cold damage, *vis-à-vis* provides great vigour to the scion species and varieties. Cleft grafting can be successfully performed with unrooted or rooted cuttings of *I. parviflora* (Mouly *et al.*, 2009).

Propagation through cuttings is not economical since the collection of stem cuttings leads to arrest of growth and development of the mother plant. Moreover, the market demand for propagules is hardly met with such cuttings. Mass propagation for commercial cultivation requires a simple, economical, rapidly multiplying and highly reproducible protocol employing tissue culture technology. Since sexual propagation is time consuming and labour oriented (Lakshmanan *et al.*, 1997) and cuttings being difficult to root, therefore a more economical, rapid and an elaborate means of propagation in *Ixora* is desired in order to meet the ever-rising market demand of this beautiful plant (George and Sherrington, 1993), and the only alternative left with us is **micropropagation**. The callus culture was initiated from leaf (explants) of *Ixora* taken from expanding young leaves in MS salt medium containing various concentrations of 2, 4-dichlorophenoxy acetic acid (2, 4-D) where it was found that callus initiation was significantly higher in MS medium having 2, 4-D at 3 mg/l. Direct root hair formation occurred from callii induced on 2, 4-D at 1 and 2 mg/l. The best combination which showed excellent multiplication was both, i.e. woody plant medium (WPM) and MS supplemented with gibberellic acid at 2.5 uM (Noreen *et al.*, 2001). Nodal segments were found to be best explants for axillary shoot formation on MS medium with 0.5 mg/l BA + 0.1 mg/l NAA. There was rooting on half strength MS medium having 0.1-0.5 mg/l of IAA, NAA or IBA (Amin *et al.*, 2002). *Ixora singaporensis* when cultured in MS medium with $AdSO_4$ (10 mg/l) gave maximum shoots per explant, and when sub-cultured in B_5 medium with IBA 2.0 mg/l, it gave highest number of roots. The shoot tip explants were cultured in WPM supplemented with 0.5 mg/l BAP and incubated for 10 weeks initially and then sub-cultured onto fresh medium after every 4 weeks. The proliferating clusters were cultured in WPM basal medium fortified with peptone (40 mg/l) and 3% sucrose for elongation of shoots, formation of root initials and further proliferation of axillary shoots. The most suitable medium for rooting was half-WPM enriched with IBA at a concentration of 0.05 mg/l (Khan *et al.*, 2004).

Cultural Practices

Basically, *Ixora* prefers tropical to subtropical conditions. It grows best in the warm climate and flowers continuously throughout the year though in cooler climates of sub-tropical and sub-temperate regions it will flower during spring to summer and summer months, respectively. Alamu (2012) stated that minimum **temperature** for its active growth should be 16 ^{0}C and maximum 27 ^{0}C. In ixora, 33 ^{0}C induced the highest leaf carbon assimilation (LCA) (0.40 mg CO_2/m^2 per second) in the greenhouse at 1,200 h, but there were no apparent differences in midday LCA between plants with root zone temperature (RZT) of 28, 33 and 38 ^{0}C in the growth room. Effects of RZT and environment on the daily fluctuations of gaseous exchange processes raise questions about using measurements at only one time during the day to separate treatment effects. When root temperature of *Ixora coccinea* was increased from 34 to 40 ^{0}C, shoot-root weight ratio increased and shoot sugar -starch ratio decreased. In general, 34 ^{0}C appeared to be the optimum root temperature (Ingram *et al.*, 1986). The requirement of humidity by this plant is also high, and the preferable one is generally above 70%. Plants have only moderate drought-tolerant capacity (Ingram *et al.*, 1986). *Ixora* being a tropical plant, prefers full **sunlight** for its growth and development though at flowering certain species prefer filtered situation to avoid blossom burn. Exposure to greater amount of light could result in compact growth and more flower bud formation. Since almost all the ixoras grow and flower in partially-shaded spots, so it can be grown as a good pot plant for terrace and verandas. These plants like a well-drained, moist and acidic soil that is high in organic matter. While preparing its pits for **planting** it would be better to fill the pits with $^{2}/_{5}{}^{th}$ part of good garden soil, $^{1}/_{5}{}^{th}$ part of coarse sand and $^{2}/_{5}{}^{th}$ part of well decomposed organic matter in the form of compost, cow dung manure or FYM or peat moss. The preferred pH of the **soil** is 5.0-6.5. In fact, the ixora is intolerant to alkaline soil that makes the leaves chlorotic or a sticky yellow-coloured. Use of mulch is required to conserve the moisture. The run-off from concrete may cause the soil to become alkaline. To avoid this problem, *Ixora* should be planted at least a few metres away from concrete structures.

It can be planted in every season except during heavy rains. The land should be fertile with proper provision of watering. The land should be ploughed twice to get good tilth and aeration. Since the plant is maintained at least for 10 years, enough quantity of FYM should be applied. Ridges are formed at 2.5-3.0 m spacing and the plant to plant **distance** is kept 0.9-1.5 m. However, in pots also almost the same mixture is used but the size should be 30 cm and above looking into the plant size at final growth. Grafts are used for planting. The polythene sheet should be removed before planting. However, at initial period of growth some intercrops such as onion, marigold, coriander, bean, cowpea, black gram, etc. can be taken but the weeds will have to be removed regularly and the irrigation should be followed taking into account the main crop, the type of intercrops grown, climatic conditions, soil types and natural rains. In case of drip irrigation, it should be followed on alternate days. Per plant of main crop, 50 g of cake (groundnut, castor, *Pongam*, *neem* or *mahua*) should be applied once in 60 days.

In *Ixora*, suckers and sprouts may grow up to 1 m the first year after which the growth slows down. Individual stems live at least for 15 years but by coppicing and suckering a plant may live indefinitely. Maintenance of ixora in hedges requires frequent **pruning** (Whistler, 2000). Taller growing varieties should be pruned sufficiently after flowering, cutting back by about one-third of the plant. Dwarf varieties don't need anything more than light tip pruning when young to promote bushiness (Ellis *et al.*, 2003).

During the most active growth period, from spring through summer, plants should be **fertilized** once a week though during autumn and winter only once a month. Recommended fertilizer formulation is NPK at 4-8-8 (Keeler and Schoellhornm, 1999). Another option is to alternate an all purpose fertilizer, such as 20-20-20 which can specifically be geared to support bloom production (Gilman, 2003). These are moderate feeders, especially when grown under high light, fertilize once a week using ¼ tsp of fertilizer per gallon of water. Using a balanced fertilizer like a 15-15-15 or a blooming fertilizer like a 7-9-5 of NPK can prove effective (Gilman, 1988). During the winter or when growth has stopped the feeding should be stopped, because new growth needs more nutrients that will not be available at that time. Thus more red spots and/or chlorosis, a yellowing of the leaves due to iron or manganese deficiency (another common ailment with ixora) will appear. Use of slow release nitrogen based fertilizer is a better solution for this. The same is the situation even with K as it is not retained in the root zone of sandy soil. Due to P and K deficiencies, purplish leaf spots occur as is also enhanced due to cold snap (Rahuj, 1999). Chlorosis (yellowing of leaves) indicates a high pH in the root zone (alkaline soil). Treatment with iron and manganese reduces this problem. Slow release fertilizer with a ratio of 8N-4P-12K-4Mg + other micronutrients or in the range of 9-11-11 + iron and micronutrients should satisfy ixora requirements (Szilard Paula, 2003). In older plants, 6 grammes of nitrogen per m^2 and in younger plants 12 g of actual nitrogen per m^2 of root zone for the first two years should be applied. Organic manure like cassava peel, organic compost and tithonia manure increases number of branches, internodal length and plant spread (Alamu, 2012). For regular feeding, a 9-11-11 NPK fertilizer first in February, second in May and the last in August should be applied each time with 112 g per plant (Dickey, 1977). Mulching in this will eventually contribute to the build-up of organic matter in the root zone and improve nutrient availability (Thomas and Gilman, 1991). More organic matter will also reduce nematode population that can also plague older varieties of ixora. Often, the plants are grown in containers too long which leads to a compromised root system causing their girdling and thus reducing the nutrient absorption rate (Gilman, 1988). *Ixora* does not like excess **watering** though moist condition is conducive for its proper growth and waterlogged conditions are injurious as root rot occurs.

Growth and Development

Application of GA_3 at 100 ppm on shoot cuttings produced higher number of leaves, leaf area and dry weight per plant in ixora (Fagge and Manga 2011). Trimtect® (8% paclobutrazol as active ingredient) when was applied to trimmed ixora, it reduced the regrowth by 30-70% over a season, produced very compact plants and the leaves produced were darker green (Idun *et al.*, 2011).

Post Harvest

Ixora flowers remain fresh for a week on the plant and up to even three days it does not dry up when severed fresh from the plant. The flowers can be plucked even two days later of the opening. For a few initial years the yield may be less but in subsequent years it goes on increasing up to 5-6 years. Usually it starts blooming after six months of planting and then flower plucking may start on weekly basis or up to two days of opening and this way one average plant can yield up to 5 kg of flowers per plucking which is sold in the market at the rate of Rs. 50-80/kg. Those having red flowers have good demand in the market. In its flower harvesting the ant is a nuisance as this collects honey from the flowers. *I. chinensis* has a longer vase life which makes it a potential cut flower (Chen and Huang, 1996). A new tool, 1-methylcyclopropene (1-MCP) extends the shelf life and produces quality cut flowers. Its cut flowers treated with various concentrations of sucrose, KH_2PO_4 , 8-HQ and citric acid increased the vase life, the best being 2% sucrose + 150 mg^{-1} KH_2PO_4 + 120 mg^{-1} 8-HQ + 30 mg^{-1} citric acid where the vase life was recorded of six days (Luan-Mei *et al.*, 2010). STS solution spraying at 0.3 to 0.4 mmol/l 10 days before storage prevents or reduces flower falling (Luan-Mei *et al.*, 2010) caused by ethylene. For the antioxidant property it can be harvested and stored in air-tight desiccator at 30 °C (Ratnasooriya *et al.*, 2005).

Insect-pests and Diseases

Anthracnose (*Colletotrichum* spp.) attacks various evergreen and deciduous shrubs and trees, including dwarf ixora. This disease occurs in humid weather and causes black, tan or red spots, yellowing, falling and wilting of leaves, and blackish canker on the stems. Plants should be watered only when needed and that too through sub-irrigation system and overhead irrigation will help dispersal of disease through wet leaves. Infected branches and leaves during autumn and winter should be removed. Once infection has spread, no fungicide will be effective. **Leaf spot** is caused by *Pseudocercospora ixoricola* and many other pathogens. This shows discoloured spots, raised areas or brownish yellow patches on the leaves of infected plants. In severe cases, leaf fall may occur but the long-term damage is rare. The symptoms of leaf spot disease are sometimes confused with signs of a bacterial pathogen or insect infestation. If the ixora is healthy and vigorous, the plant can tolerate leaf spot disease with no lasting damage. The disease can be kept under check by over-watering and sanitation. This disease can be prevented by spraying mancozeb 750 WB at 2-4 kg a.i./ha during spring. **Mushroom root rot** (*Armillaria* spp.) infects vascular tissues and once the infection is noticed, the cambium is already damaged and therefore roots also die. Aerial symptoms include leaf dropping, branch dieback, discoloured stems and cankers. As the fungal disease progresses, mushroom clusters will begin to grow at the base of the plant. Mushroom root rot is fatal problem for plants and occurs in soggy soils with poor drainage. The best way to avoid infection of this fungus is to plant the specimens in well-drained soils. Dead or dying plants should be removed with as much root system as possible and the soil should be replaced or sterilized before replanting. Dithane M-45 at 300 mg/l and Ridomyl MZ 500 mg/l drenching may prevent this disease. The black mess is a fungus called **sooty mould** (*Acetomycetes* fungi) which spreads through sap sucking insects such as scales, mealy bugs and aphids. They prefer the shady north side of the house. After feeding, the insects excrete their waste that drips down onto parts of the plants which attracts the black mess fungus. Products containing Imidacloprid will give long-lasting control. Horticultural oil, usually a 2% mixture of a paraffinic mineral oil may also be effective. This fungus can also be washed away by using light insecticidal oil. Most often, however, unaffected new leaves replace the sooty mould affected leaves but even then to check further spread the insects will have to be controlled. A systemic, soil drench product called Bayer Advanced Garden™ Tree and Shrub Insect Control (contains 1.47% Imidacloprid) is reported to give 4 to 12 months residual effects, depending on the organic matter content of the soil.

Ixoras growing in calcareous sandy soils show an intense reddish leaf spot which is a **physiological disorder**. Symptoms appear as irregular diffuse brownish-red blotches on slightly chlorotic but oldest leaves. The symptoms of necrotic spotting appear sharply and prominently on the old leaves of potassium deficient ixoras and infected shoots are able to retain only a few leaves, however, phosphorus-deficient plants though show no spotting but exhibit uniformly brownish-red older leaves and olive-green younger foliage. Plants deficient in both elements display symptoms similar to those observed on landscape plants (Broschat, 1989). Iron and manganese deficiency causes light green to yellow leaf chlorosis with veins remaining dark green. A slow release fertilizer in the range of 9-11-11 + iron and micronutrients should therefore be applied. These problems can be overcome through application of nitrogenous fertilizer as defined under **cultural practices**, organic matter in the root zone too reduces nematode population plaguing older plants, and soil applications of Sequestrene 138 or foliar applications of Sequestrene 330 to alleviate iron and manganese chlorosis.

There have been found only sucking type of insects on ixoras. **Scale insects** will not be dislodged even when dead, **mealy bugs** are usually very white in colour and **aphids** are often green and pear-shaped. Horticultural oil spray, weekly for five weeks in the afternoon can control these insects. **Nematodes** can be controlled with Nemacur or Furadan at 1 g/m^2 followed by irrigation.

References

Adetimiria, V.O., S.K. Kim and M. Szezeeh, 2008. Factors associated with emeragence of ornamental flowers. *Sr. Lanka J. agric. Sci.*, **144**: 68.

Alamu, L.O. 2012. Enhancing environmental management through a luxuriant vegetative improvement of *Ixora coccinea* by means of organic manuring. *Intern. J. Acad. Res.in Business & Social Sci.*, **2**(6): 252-259.

Amin, M.N., A. Shahrear, S. Sultana, M.R. Alam and M.A.K. Azad, 2002. *In vitro* rapid clonal propagation of ornamental plant ~ *Ixora fulgens* Roxb. *Online J. biol. Sci.,* **2**(7): 485-488.

Bedi, Y.S., S.S. Bir and B.S. Gill, 1981.Cytopanology of woody taxa of family Rubiaceae from north and central India. *Proc. Natl. Sci. Acad.*, **6**: 708-715.

Broschat, T.K. 1989. Potassium deficiency in south Florida ornamentals. *Proc. Fla St. hort. Soc.*, **102**: 106-108.

Chen, L.Y. and M.C. Huang, 1996. Effect of shading and winter heating on growth and flowering of Chinese ixora (*Ixora chinensis*). *J. agric. Asscn. China* (New Series), **174**: 72-82.

Deb, D.B. and R.C. Rout, 1992. Two new species of *Ixora* (Rubiaceae subfam. Ixoroideae) from India and Burma. *Kew Bull.*, No. 47, pp. 295-300.

Dickey, R.D. 1977. Nutritional deficiencies of woody ornamental plants used in Florida landscapes. Co-op. Extn. Bull., p. 791. University of Florida, USA.

Elgimabi, M.E.N.E. 2008. Effect of season of cutting and humidity on propagation of *Ixora chinensis. Adv. Boil. Res.*, **2**(5-6): 108-110.

Ellis, W.O., L. Atuah, P. Kumar and J.A. Bakang, 2003. An overview of the floriculture industry in Ghana. *Ghana J. Hort.*, **3**: 47-56.

Fagge, A.A. and A.A. Manga, 2011. Effect of sowing media and gibberellic acid on the growth and seedling establishment of *Bougainvillea, Ixora coccinea* and *Rosa chinensis*. 2. Root characters. *Bayero J. Pure & Appl. Sci.*, **4**(2): 155-159.

Fulghum Leigh, 2008. Flowering shrubs in Florida ~ 'Frankie' and 'Nora' *Ixora. Daily Post in Floridaplants. com.* United States Patent PP09200.

George, E.F. and P.D. Sherrington, 1993. Plant propagation by tissue culture, 709 p. Exegetics Ltd. Press.

Gilman, E.F. 1988. Tree root spread in relation to branch drip line and harvestable root ball. *HortSci.*, **23**(2): 351-353.

Gilman, E.F. 2003. *Ixora coccinea.* University of Florida Cooperative Extension Fact Sheet FPS-291.

Holtum, R.E. and I. Enoch, 1991. Gardening in the Tropics. *Times Publication* (Singapore), **6**:134-137.

Husain, T. and S.R. Paul, 1991. *Ixora manantoddii*, a new species of *Ixora* L. (Rubiaceae - Pavetteae) from India. *Bull. Du Jardin bot. Belgium*, No. 61, pp. 15-19.

Idun, I.A., P.A. Kuah and H.V. Adzraku, 2011. Rooting and vegetative growth response of difficult to root *Ixora coccinea* and *Ficus benjamina* cv. 'Straight' to different stem cutting types and soilless media. *African J. Pl. Sci.*, **5**(3): 773-780.

Ingram, D.L., C. Ramcharan and T.A. Nell, 1986. Response of container-grown banana, ixora, citrus and dracaena to elevated root temperatures. *HortSci.*, **21**: 254-255.

Jinsui Lin, 2009. Effect of plant growth regulator on rooting and cutting of *Ixora chinensis. Chinese Agr. Sci. Bull.*, No. 685, pp. 105-109.

Kaser, H.R. and A.M. Steiner, 1983. Subspecific classification of *Vicia faba* L. by protein and isozyme patterns. *Fabis Newletters*, **7**: 19-20.

Keeler, G. and R. Schoellhornm, 1999. *Ixora* for South Florida. University of Florida IFAS Extension (ENH 955).

Khan, S., M. Iftikhar and B. Saeed, 2004. An economical and efficient method for mass propagation of *Ixora coccinea. Pakistan J. Bot.*, **36**(4): 751-756.

Lakshmanan, P., C.L. Lee and C.J. Goh, 1997. An efficient *in vitro* method for mass propagation of a woody ornamental *Ixora coccinea. Pl. Cell. Rep.*, **16**: 572-577.

Latha, L.Y., I. Darah and S. sreenivasan, 2012. Pharmacological screening of methanolic extract of *Ixora* species. *Asian Pacific J. Trop. Biomed.*, **2**(2): 149-153.

Luan-Mei, L.U.L., J. Lin and M.X. Zhi, 2010. Effects of different preservative solution on cut flowers of *Ixora chinensis. Acta Hort. Sinica*, **37**: 1351-1356.

Mouly, A., S.G. Razafimandimbison, A. Khodabandeh and B. Bremer, 2009. Phylogeny and classification of the species-rich pantropical showy genus *Ixora* (Rubiaceae-Ixoreae) with indications of geographical monophyletic units and hybrids. *Amer. J. Bot.*, **96**(3): 686-706.

Nayak, R., K. Geetha and Priya Davida, 2010. Pollination limitation and effective breeding system on plant reproduction in forest fragment. *Acta Oecologia*, **36**: 191-196.

Nilsson, L.A. 1998. Deep flowers for long tongues, *Trends in Eco. and Evol.*, **13**: 259-260.

Noreen, R., M.A. Khan, M.J. Jaskani and N. Hussain, 2001. Callogenesis and embryogenesis from leaf discs of *Ixora chinensis. Int. J. Agri. Biol.*, **3**(1): 65-67.

Pandey, A.K. 2006. Strructure development and reproduction in flowering plants. Ph.D. thesis. T.M. Bhagalpur University (Bihar).

Pradeep, A.K. 1997. *Ixora sivarajiana*, a new species of Rubiaceae from India. *Nordic J. Bot.*, **17**: 315-317.

Rahuj, B. 1999. *Ixora. J. hort. Sci. & ornam. Plants*, **2**(2): 78.

Rajasegar, G., H.T.W. Tan, I.M. Turner and P.P. Kumar, 1997. Analysis of genetic diversity among *Ixora* cultivars (Rubiaceae) using random amplified polymorphic DNA. *Ann. Bot.*, **84**: 253-257.

Rajasegar, G., H.T.W. Tan, I.M. Turner, L.G. Saw and P.P. Kumar, 1999. Random amplified polymorphic DNA variation among and within selected *Ixora* (Rubiaceae) populations and mutants. *Ann. Bot.*, **84**:253-257.

Randhawa, G.S. and A. Mukhopadhyay, 1994. Floriculture in India. Allied Publishers Pvt. Ltd., New Delhi.

Ratnasooriya, W.D., G. Galhena, S.S.P. Liyanage and J.A.C. Jayakody, 2005. Antinociceptive action of aqueous extract of the leaves of *Ixora coccinea. Acta Biologica Hungarica*, **56**: 1556.

Sharma, A.K. and T. Chatterjee, 1960. Chromososme studies in ixora. *Genetics*, **31**: 421-447.

Sheng-Jian, M.A. and Y. Jun-Bang, 2011. Effect of hormone mixture and matrix on the rooting of *Ixora chinensis. Heiloingjian Agric. Bull.*, No. 685, pp. 99-105.

Singh, S.P. 1980. Response of varying concentrations of auxins to rooting of *Ixora bandhuca* cuttings during winter under intermittent mist. IV. *Progressive Hort.*, **12**: 21-25.

Sivadasan, M. and N. Mohanan, 1991. *Ixora agasthyamalayana*, a new species of Rubiaceae from India. *Bot. Bull. Academia Sinica*, **32**: 313-316.

Szilard Paula, 2003. Ixoras for year round enjoyment. University of Florida, Cooperative Extension Fact Sheet 5/2/2003.

Thomas, H.Y. and E.F. Gilman, 1991. Fertilization recommendations for trees and shrubs in home and commercial landscapes. University of Florida, Cooperative Extension Circular No. 948.

Whistler, W.A. 2000. Tropical Ornamentals ~ A Guide, 542 p. Timber Press, Portland, Oregon, USA.

Zaundam, D. 1997. Ixora Plant Diora. United States Patent PP09814.

17

Jasminum

Sanyat Misra, R.L. Misra, M. Ganga, P. Mekala, V. Jegadeeswari and K. Padmadevi

Family: **Oleaceae**

Common names: *J. angustifolium* (Wild jasmine, Hindi-*banmallika*), *J. arborescens* (Tree jasmine, Hindi-*muta bela,* Tamil-*naba malliga*), *J. auriculatum* (Hindi- *juhi, jui,* Tamil- *ambur mallige, kadaru mallige, malle, mokggu, mugohee, mullai, oosi mallige, pindari, sunika*, Telugu-*virajagi*), *J. beesianum* (Rosy jasmine), *J. calophyllum* (Tamil-*pandal malli*), *J. dicotomum* (Gold coast jasmine), *J. flexile* (syn. *J. caudatum;* Hindi-*malati,* Tamil-*nitya mallige*), *J. grandiflorum* (Catalonian jasmine, French jasmine, Italian jasmine, Royal jasmine, Spanish jasmine, Sanskrit-*jati,* Hindi-*chameli, safed chambeli,* Tamil-*jathimalli, mallai, pitchi*), *J. humile* (Italian jasmine/Yellow jasmine, Hindi-*peeli chameli,* Tamil-*semmalligai*), *J. mesnyi* (syn. *J. primulinum*; Primrose jasmine), J. *multiflorum* (syn. *J. pubescens; Downy* jasmine, Furry jasmine, Pinwheel jasmine, Star jasmine, Winter jasmine, Hindi-*kunda, kundphool,* Bengali-*kundo,* Tamil & Telugu-*kakada*), *J. nudiflorum* (Winter jasmine), *J. officinale* (Common jasmine/Common white jasmine/Jessamine), *J. revolutum* (Himalayan jasmine), *J. sambac* (Arabian jasmine/Bush jasmine/Tuscan jasmine, Hindi-*belaa, champa, moghra, motia,* Tamil- *adukkumalligai, gundumalli, gumdumalligai, malli, malligai, mogri*, Telugu-*malli*), etc.]

Introduction and Origin

Jasmine (*Jasminum* spp.) is one of the oldest fragrant flowers cultivated by man. Jasmines have been cultivated for various purposes and they have adorned the gardens of central Asia, Afghanistan, Iran, Nepal, India and many other tropical and subtropical countries. India exports jasmine flowers to the neighbouring countries like Sri Lanka, Singapore, Malaysia, the Gulf and to certain western countries like England, Canada, U.S.A., etc. having sizeable Asian and Indian population as they have demand for traditional flowers on their respective festivities, and for perfumes the growing markets are in Japan, UK, USA, USSR, Scandinvia, East Europe, etc. The word jasmine is of Arabic (Persian) derivation, i.e. *yasmyn* but in Latin form (*Jessamine, Jasmin* and *Jasminum*) meaning fragrance. There are some 200-300 species widely distributed in the tropical (warmer) parts of Europe (South Europe, Caucasus), Asia (India, China, Iran, Afghanistan), Australia, Pacific region (North Borneo, Bismarck Archipelago, and Madeira Islands in the EN Atlantic Ocean) and Africa (North Africa, South Africa, Canary Islands, Nile), and North America with only one species. Large number of species are centred around India, China and Malaya (Anon., 1959; Abdul Khader and Kumar, 1995). However, a critical analysis revealed that the number of true *Jasminum* species are only 89 (Veluswamy *et al.* (1975). India may be the principal centre of origin since as many as 40 *Jasminum* species are native to this country. Jasmine is one of the leading traditional flowers of India hence is being cultivated throughout the country for loose flower production and for extraction of jasmine perfumes. This flower has a place in Indian folk songs. Since the time immemorial it is considered as a spiritual flower in our country. Flowers are used as loose flowers for making garlands, bouquets and *veni*, for adorning the hairs of women, at religious and ceremonial functions and also for producing *attars*, perfumed hair oils as it easily blends with floral extracts, and for soap and cosmetic industries. Largest area under jasmine production is in Tamil Nadu (Coimbatore, Madurai, Dindigul, Athoor, Nilakottai, Tirumangalam, Kallupalli and Thiruparankundrum), followed by Karnataka (Bangalore, Bellary, Mysore and Kolar). However, other areas for commercial production are Maharashtra (Pune, Aurangabad and Ahmednagar), Gujarat (Ahmedabad, Anand and Vadodara), Haryana (Ambala, Gurgaon and Faridabad), Punjab (Amritsar, Patiala, Ludhiana and Jalandhar), Delhi, Uttar Pradesh (Kannauj, Jaunpur and Gazipur), West Bengal (Ranaghat, Kolaghat, Panskura and 24-Parganas), Assam (Jorhat, Hoja and Alnugarh) and Rajasthan (Jaipur, Ajmer, Udaipur and Kota). *J. grandiflorum* (Spanish jasmine) yields the world famous jasmine oil. Since the genus consists of shrubs, semi-shrubby and climbing species mostly with fragrant flowers, hence its use is multipurpose such as planting in pots and bowls, on roofs, in the front or backyard, on the railings, on porches, pergolas, arches and boundaries, on wasteland slopes for cascading effect, in the fragrant gardens and elsewhere, near the windows of the bed rooms, in groups and on borders.

Botany, Cytology and Breeding

Jasminum is allied to *Ligustrum* but differs in the compound leaves and twin-carpelled fruits. It is also allied to *Olea*. The species are tender or hardy often with angled stems, either deciduous or mainly tropical evergreen shrubs and climbers. Leaves are pinnate, trifoliate or simple that appear in opposite pairs (sometimes alternate). Flowers are tubular with usually 6 (sometimes 4-9) petal-like lobes (convulate in the bud) spreading flat and are borne in terminal or axillary clusters or occasionally forming panicles and sometimes even solitary on the ends of branchlets, tubes white, yellow or sometimes reddish and fragrant in certain cases. Stamens 2 under corolla tube, filaments very short, ovary 2-loculed with 1-4 erect ovules, and fruit a 2-lobed black berry.

Jasmines are generally diploids with 2n=26 (Bowden, 1940). Sax and Abbe (1932) recorded 2n=26 in *J. fruticans*. Taylor (1945) also recorded x=13 in *jasminum* species. Bowden (1940) recorded 2n=26 in *J. affine, J. floridum, J. humile, J. officinale (J. grandiflorum)* and *J. wallichianum* while 2n=52 in *J. nudiflorum*. Krishnaswamy and Raman (1948) observed 2n=26 in *J. sambac* vars 'Sujimallige', 'Iruvantige' syn. 'Virupakshi', 'Elusullumallige' and *J. rigidum*, 2n=39 in *J. sambac* var. 'Dundumallige' and *J. primulinum*, 2n=52 in *J. angustifolium* and n=13 in *J. auriculatum, J. malabaricum* and *J. flexile*. Alikhan *et al.* (1969; c.f. Muthuswamy and Abdul Khader, 1986) and Murthy and Khanna (1971; c.f. Muthuswamy and Abdul Khader, 1986) recorded naturally occurring higher ploidy forms in *J. primulinum* and *J. sambac* var. 'Gundumalli', each with 2n=39, and in *J. angustifolium* and *J. flexile*, each with 2n=52. Muthuswami and Abdul Khader (1986) mentions that a pink-tinged bud variant of *J. grandiflorum* recorded triploid chromosome number with 2n=39 whereas other four pink-tinged bud variants and the one whitish bud variant recorded 2n=26 (diploidy). *J. sambac* vars 'Kasthurimalli', 'Madanban', 'Rambanam', 'Single Mogra' and 'Sujimalli' recorded 2n=26; 'Iruvatchi'with 2n=26, 39; and 'Double Mogra' with 2n=52. Karmakar and Srivastava (1987a,b) noted meiotic abnormalities such as precocious & tubular cytoplasmic connections and sticky chromosomes in *J. grandiflorum* which seem to be responsible for a very poor seed set. Raman (1955) studied the karyotypes in eight species and four *J. sambac* varieties, *viz.* 'Dundumallige', 'Elusullumallige', 'Iruvantige' and 'Sujimallige' and recorded that the chromosomes in all the species though were almost alike and small but were showing sub-median and sub-terminal constrictions.

Muthuswamy and Abdul Khader (1986) reported that success of seed setting was very low, i.e. 0.5-5% when interspecific reciprocal crossing was attempted involving the species *J. auriculatum, J. calophyllum, J. flexile, J. grandiflorum* and *J. sambac* where the seeds failed to germinate but when an induced tetraploid of *J. auriculatum* was used as a recipient and the *J. angustifolium, J. arborescens, J. calophyllum, J. flexile, J. grandiflorum, J. paniculatum* and *J. rigidum* as donor parents, there was 3-16% seed formation but these also failed to germinate, and the same was the case even with the interspecific hybridization among the wild species. This may be due to inherent strong barrier for crossability and hybrid inviability existing in the species, *vis-à-vis* possibility of endosperm antagonism in operation (Muthusway and Abdul Khader, 1986). They further hypothecated existence of pollen sterility as well as seed sterility due to defective egg mechanism. Muthuswamy and Abdul Khader (1986) stated that when inter-clonal crosses in *J. grandiflorum* were made, there was 0.5-35% success in seed formation and 50-100% germination though there had been differences in different crosses as well as in reciprocals; whereas in *J. auriculatum* crosses involving 'Short Round' × 'Short Point', 'Parimullai' × 'Short Point' and an induced tetraploid × one of its diploids, there was 0-27% fruit set but not even a single progeny was worth retaining.

Veluswamy (1981) got success in *J. auriculatum* (11.4-23.1% polyploids) but not in *J. grandiflorum* when treated the seeds and seedlings with 1% colchicine. Muthuswami and Abdul Khader (1986) reports that when seeds of *J. auriculatum* 'Parimullai' and 'Short Point' clones (diploid) were treated with 0.5% colchicine for 12 or 24 hours or their vegetative buds with 0.1-0.5% , only one tetraploid seedling from 'Parimullai' seeds was obtained which was inferior to existing clone, though in *J. grandiflorum* with 0.5% colchicine seed treatment six uneconomical tetraploids were obtained, however, tetraploids in *J. sambac* were more vigorous but production of bold flowers was poor. Natural triploids have been found better in yield than diploids. Muthuswami and Abdul Khader (1986) stated that in *J. auriculatum*, X-ray treatment of active vegetative buds did not alter any variation whereas seed-treatment with gamma rays beyond 10 kR and EMS 0.1% seed-soaking proved lethal, though under gamma rays one mutant with longer corolla tube was spotted but this was susceptible to gall-mite, and even EMS also produced one mutant but with undesirable traits; however, in the softwood cuttings of *J. grandiflorum* gamma rays beyond 30 kR proved lethal though lower doses produced 10 mutants but all were recorded inferior. Kannan *et al.* (2002) tried ^{60}Co gamma rays at 1.5 and 2.0 kR, EMS at 30 and 45 mM and combined treatments of 1.0 kR + 15 mM, *vis-à-vis* 1.5 kR + 30 mM on *J. sambac* cv. 'Gundumalli' and recorded higher mean values of the economic characters, *viz.* length and width of flower bud, length of corolla tube and diameter of the open flowers

Species and Varieties

India is the native haunt for about one-fifth of the total *Jasminum* species and a number of species along with their clones and varieties developed in this country or elsewhere are described here. Bhattacharjee (1980) furnishes a comprehensive list of 45 valid Indian species and their distribution such as *J. adenophyllum* Wall. (natural distribution), *J. anastomosans* Wall. (syn. *J. strinerve* Roxb., *J. angustifolium* Vahl., Deccan Peninsula, S. Travancore), *J. arborescens* Roxb. (syn. *J. latifolium* Roxb., *J. roxburghianum* Wall., Tropical NW Himalaya, Deccan Peninsula, Chota Nagpur, Odisha), *J. attenuatum* Roxb. (Assam Khasi Hills at 200-1300 m altitude), *J. auriculatum* Vahl. (syn. *J. affine* Wt., *J. ovalifolium* Wt., Bengal, S. Travancore and Western Ghats), *J. azoricum* Backer and Bakh. f. (SW India, especially in Ghats), *J. brevilobum* A. DC. (syn. *J. nilagiricum* Pl., Mtns. of Deccan Peninsula, W. Ghats, Nilgiri Hills, Palani Hills, altitude 1,000-2,000 m), *J. brevipetiolum* Duthie (forests of Pilibhit dist. and Kheri of U.P.), *J. calophyllum* Wall. (syn. *J. courtollense* Wt., South Deccan Peninsula, Nilgiri Hills, W. Ghats, Anamalai & Tinnevelly Hills, up to

Jasminum sambac

Jasminum sambac

Ecotypes of *J. sambac*

Ecotypes of *J. sambac*

J. auriculatum

J. auriculatum var. 'Parimullai'

J. auriculatum var. 'Parimullai'

Co. 1 Mullai and CO. 2 Mullai

Co. 1 Mullai and CO. 2 Mullai

J. grandiflorum

J. beesianum

J. beesianum

J. floridum

J. floridum

J. calophyllum

J. dicotomum

J. fluminense

J. humile

1300 m), *J. caudatum* Wall. (*J. ovatum* Wall., Khasia and Mishmi Hills, up to 1300 m), *J. coarctatum* Roxb. (syn. *J. reticulatum* Wall., Assam, Khasia and Lushi Hills, from 0 to 1000 m altitude), *J. cordifolium* Wall. (syn. *J. erectiflorum* Wt., W. Ghats, Coimbatore, Nilgiri, Tinnevelly, South and Middle Andaman), *J. dispermum* Wall. (Temperate Himalayas, Kashmir, Khasia and Jaintia Hills, altitude (1000-2500m), *J. dumicolum* W.W. Smith (Naga Hills and Manipur), *J. flexile* Vahl. (Deccan Peninsula rain forests, West Coast and W. Ghats up to 1700 m), *J. glandulosum* Wall. (syn. *J. dichotomum* Don., *J. hamiltonii* Gamble, *J. nepalense* Spreng, subtropical Himalaya, Kumaon, Khasia Hills, Aka and Lushai Hills at altitudes of 700-2,000 m), *J. grandiflorum* L. (syn. *J. aureum* Don, subtropical NW Himalayas, Nilgiris, Palani and Tennevelly hills, at 700-1800 m altitude), *J. heterophyllum* Roxb. (Mishmi Hills), *J. humile* L. (syn. *J. bignoniaceum* Wall., *J. chrysanthemum* Roxb., *J. inodorum* Jacq., *J. pubigerum* D. Don, *J. revolutum* Sims, *J. wallichianum* (Ldl), NW Himalayas, Nilgiris, altitudes 700-2000 m), *J. lanceolarium* Roxb. (syn. *J. lanceifolium* Roxb., *J. paniculatum* Roxb., Khasia, Naga and Jaintia Hills, altitude 1000-1700 m), *J. laurifolium* Roxb. (Assam, Khasia, Aka and Lushai Hills), *J. malabaricum* Wt. (syn. *J. glabellum* Wall., W. Coast, W. Ghats, Nilgiris, Mysore, Malabar, altitude 1300 m), *J. mauritianum* Boj. (Long Island, Andaman), *J. multiflorum* Roth. (syn. *Nyctanthes multiflora* Burm. f, common in sub-Himalayan tract, moist forests of W. Ghats, throughout India, altitude up to 1700 m), *J. officinale* Linn. (Kashmir, subtropical NW Himalayas, altitude 700-1700 m), *J. pubescens* Willd. (syn. *J. bracteatum* Roxb., *J. hirsutum* Willd., W. Bengal, Bihar, Chota Nagpur, W. Ghats, Himalayas, at altitudes up to 1700 m), *J. pubigerum* Don.(NW India), *J. rigidum* Zenker (syn. *J. cuspidatum* Rottler, *J. lingustrifolium* Wall., *J. myrtophyllum* Zenker, *J. perrottetianum* DC., *J. triflorum* Heyne, Nilgiris, Deccan Peninsula, W. Ghats, at altitude up to 2000 m), *J. ritchiei* Clarke (W. Peninsula, Bababoodan Hills, Konkan, W. Ghats, Mysore, Nilgiris, Wynad, Shevroy, at altitude up to 1300 m), *J. rottlerianum* Wall. (W. Peninsula from Konkan to Travancore, N. Kanara and W. Ghats, common in rain forests, at altitude 1000-1300 m), *J. roxburghianum* Wall. (Mahabaleshwar Ghat, Bababoodan Hills, North Canara, Deccan, Ramandrug Hills in Bellary, Malabar Ghats), *J. sambac* Ait (syn. *J. fragrans* Salisb., *J. quinqueflorum* Heyne, *J. undulatum* Willd., N. Kanara, Konkan, W. Bengal, and cultivated throughout India), *J. scandens* Vahl. (syn. *J. lactum* Wall., W. Bengal, Assam, Khasia Hills, Sikkim - foot hills), *J.*

sessiliflorum Vahl. (Madras, W. Ghats, Tinnevelly Hills), *J. strictum* Haines (Chota Nagpur), *J. subglandulosum* Kurz. (S. Andaman), *J. subtriplinerve* Bl. (syn. *J. trinerve* Wall., N. Bengal, Sikkim, Assam, Khasia Hills, Kachhar, altitude 0 to 1300 m), *J. syringaefolium* Wall. (Assam), *J. trichotomum* Heyne. (syn. *J. gardnerianum* Wt., up to 700 m, in Madras, Nilgiris, Coimbatore, W. Deccan), *J. undulatum* Ker (syn. *J. amplexicaule* Wall., *J. scandens* Griff, up to 1700 m in Assam, Khasia Hills, Sikkim), and *J. wrightii* Zenker. (Coimbatore in Bolampatti valley). Commercially important *Jasminum* species of Indian origin are *Jasminum auriculatum, J. grandiflorum* and *J. sambac.* In other countries, three species that have been cultivated for hundreds of years for their scent are *J. grandiflorum, J. officinale* and *J. sambac* (Pizzetti and Cocker, 1975). Dioscorides mentions of a Persian oil prepared by soaking 'white flowers of violets' in sesame oil and these 'white flowers of violets' were, in fact, the flowers of *J. officinale.*

However, herewith the commercial species and their varieties are described.

Jasminum auriculatum, a scandent shrub having simple or trifoliate shiny leaves, or with minute lateral leaflets, flowers white, scented, star-shaped where some five variants, *viz.* 'Long Point', 'Long Round', 'Medium Point', 'Short Point' and 'Short Round' based on floral characters have been identified by TNAU, Coimbatore while 'Parimullai' (TNAU, 1971) is a long flowering (9 months of the year) clonal selection from germplasm clone 'Medium Point' having medium round buds with a moderate corolla tube length of 1.25 cm and floral bud yield of some 10 t/ha with a concrete yield of 0.29% and is resistant to gall mite, 'CO. 1 Mullai' (TNAU, 1980), a secondary clonal selection from local type 'Long Round' which has white buds with long corolla tube of 1.5 cm (longer than 'Parimullai') and yields around 8,800 kg/ha of flower buds with a concrete recovery of 0.34 %, 'CO. 2 Mullai' (TNAU, 1988) is also a clonal selection with flowers having long corolla tube (1.7cm), flower yield 11,198 kg/ha and is field-tolerant to phyllody and gall mite, 'Aureum', a spontaneous mutant with yellow variegation on leaves is another variety, and a double-flower type is also commonly grown in the garden, however, Chandra (1982) reported that the varieties 'Madanban', 'Mogra', 'Motia' and 'Palampur' are double-flowered.

J. grandiflorum, a native to India is a large shrub having shiny dark green pinnate leaves with 7-9 leaflets of equal size, flowers large white, often purple-tinged outside and highly scented, axillary or terminal and 5-petalled, its ecotypic variants being 'Bangalore', 'Coimbatore', 'Lucknow', 'Pintype', 'Thimmapuram', 'Thrumtype' and the variety 'Arka Surabhi' ('Pink Pin', IIHR-1993) evolved through clonal selection which yields 10 tonnes/ha of flower buds and 0.35% of concrete up to 12 years, and is tolerant to drought conditions, 'CO. 1 Pitchi' (TNAU, 1980) with pink flower bud, a secondary clonal selection from the germplasm collection having pink-tinged white buds with long corolla tube and is suitable as loose flower and for oil extraction, the average flower yield is about 10.15 tonnes/ha with estimated concrete yield of 29.5 kg/ha, 'CO. 2 Pitchi' (TNAU,1991), a high yielding variety with better floral traits is a mutant with bold and long (4.18 cm) pink buds having high flower yield (11,680 kg/ha) with concrete recovery of around 0.30%, and the six clones collected from various states and tried at TNAU, Coimbatore (Veluswamy, 1980) exhibited the mean flower bud yield from 4,329 kg/ha to 10,144 kg/ha while that of jasmine concrete it ranged from 13.85 kg/ha to 29.42 kg/ha which led to release of the var. 'Co 1 Pitchi', and this species also has one variety named 'Triploid', however, Veluswamy (1981) reported that clones of *J. auriculatum* were though superior in flower bud yield but with low concrete recovery, i.e. 288 mg/100 g whereas the clones of *J. grandiflorum* were observed with comparatively poor yield but with high concrete recovery, i.e. 344 mg/100 g.

J. officinale (syn *J. poeticum* Hort.), a native of Persia and India is a climber and may require support, leaves opposite with terminal one longer and lateral leaflets 2-3 pairs, flowers white in 2-10 terminal leafy clusters, calyx teeth linear, corolla-lobes 4-5 and oblong, varieties are 'Affine' (larger flowers tinged pink), 'Aureo Variegatum' (syn 'Aureum'; leaves blotched cream-yellow).

Jasminum sambac, a native to India is a dwarf shrub with simple, opposite and almost smooth leaves having prominent lateral veins, flowers white and highly scented, borne in clusters of 3 to 12, rarely solitary, with about 2.5 cm diameter; cymes 2 or 3 and chotomous, bract linear and simple or ovate, sometimes petaloid, corolla salver-shaped, yellow or white, flowers hermaphrodite having two bilobed anthers included in the corolla tube, filaments very short, the gynoecium has uni-loculed ovule with simple style and a bilobed stigma, ovary superior and bilocular having axial placentation, and its ecotypes being 'Bela', 'Double Mogra', 'Gundumalli', 'Hazara', 'Iruvatchi',

'Kasthurimalli', 'Khoya', 'Madanban', 'Motia', 'Oosimalli', 'Rai Japanese', 'Ramabanam', 'Single Mogra' and 'Sujimalli' while its variety 'Arka Aradhana' developed by IIHR, Bangalore (2000) is suitable as loose flowers as well as for extraction of essential oil which yields 8 tonnes/ha of flower buds with a concrete yield of 14.95 kg/ha.

Other important species are ***J. arborescens*** which is grown in West Bengal and Uttar Pradesh for its white and highly fragrant flowers; ***J. azoricum***, originally an evergreen species and a native of Madeira growing up to 3.6 metres high with robust stems, leaves shiny green and pinnate, flowers fragrant white, star-shaped and appear in small clusters; ***J. beesianum***, a native of China is a hardy and low climber growing up to about 2.4 m with slender grooved stems, leaves simple, opposite, ovate-lanceolate to lanceolate, sharp pointed and dark dull green, and flowers small, fragrant and pink to deep rose; ***J. calophyllum***, a profuse flowering species generally grown in home gardens, flowers white and highly scented, being produced practically round the year, and are normally free from pest and diseases; ***J. dicotomum***, a vigorous climbing woody vine with simple, large and dark green leaves which are usually opposite but may be single or in whorls of three, the fragrant flowers are small, and are produced in much branched tight terminal clusters at intervals throughout the year, unopened flower buds are tinted red on the outside but pure white inside; ***J. dispermum***, a deciduous climber with opposite leaves, leaflets lanceolate and usually in two pairs, flowers fragrant white or slightly tinged pink and borne in axillary or terminal cymes; ***J. flexile*** (syn. *J. caudatum*), a profuse flowering climbing species grown widely in home gardens for its scented flowers which are produced profusely throughout the year, and plants generally free from pests and diseases; ***J. floridum***, a native to China is an evergreen rambling jasmine with angled and glabrous shoots, leaves alternate, leaflets oval or obovate, flowers in terminal clusters, and corolla yellow; ***J. fluminense***, a strong growing woody vine with opposite and compound leaves, consisting of 3 dark green leaflets, central leaflet being larger with a longer leaf-stalk, and the small fragrant flowers appear in loose clusters; ***J. fruticans*** (syn. *J. luteum, J. syriacum*), a native to South Europe, North Africa and Asia Minor is an evergreen or semi-evergreen shrub with angled and glabrous shoots, leaves alternate with narrow, oblong or linear-obovate leaflets, calyx with slender lobes, corolla yellow and globose, shiny black, pea-size fruits appear in autumn; ***J. gracillimum***, a native of North Borneo is a climbing or rambling plant having hairy, ovate-cordate and stalked leaves, and the fragrant flowers are large and white in colour with nine elliptic-oblong petals; ***J. humile*** (syn. *J. farreri, J. revolutum*), a native of India, Afghanistan and China is a diffuse evergreen shrub with purplish, downy, angled and erect shoots, leaves trifoliate (3-7 leaflets) and ovulate-lanceolate, flowers small, yellow and slightly fragrant and are borne in terminal or axillary clusters of 2 to 4; ***J. multiflorum*** (syn. *J. pubescens, J. hirsutum*), a native to India is a climbing strong growing woody vine climbing to a height of 1.6 m or more, a very ornamental species which bears profusely non-scented, large white sessile flowers in terminal umbellate clusters for a long time with 6-9 lanceolate petals, leaves simple, opposite, acute and ovate-lanceolate, and its var. being 'Arka Arpan' (IIHR), a variety with attractive foliage, most suitable for landscape use, having pink and fragrant flowers with flower yield around 6.1 t/ha being on par with var. 'Kakada'; ***J. nitidum***, a native of Admiralty Islands (Bismarck Archipelago) is a half-twining strong growing woody vine with dark-green, glossy, ovate-lanceolate, opposite and simple leaves, and floral buds tinted pink outside but flowers highly fragrant and white, some 4 cm across, and are produced in a few to many-flowered clusters; ***J. nudiflorum*** (syn. *J. sieboldianum*), a native to China is an almost erect deciduous shrub with 4-angled glabrous branchlets, leaves opposite with 3 small ciliate leaflets, flowers yellow, solitary and subtended by numerous small leaf-like bracts, calyx lobes leafy, corolla-segments obovate and normally wavy; ***J. parkeri***, a native of NW India grows some 30 cm in height with spread of 60 cm, is a hummock-forming hardy evergreen shrub with pinnate leaves having 3-5 ovate leaflets, and solitary yellow flowers coming up from the upper leaf axils; ***J. polyanthum***, a native of China is a half-hardy deciduous or semi-evergreen climbing species growing up to 3 metres in height bearing pinnate leaves with 5-7 leaflets and white to pale-pink flowers appearing in axillary and lateral panicles; ***J. primulinum*** (syn. *J. mesnyi*), a native to China (Yunnan) is a rambling free-flowering evergreren shrub quite similar to *J. nudiflorum* but with larger leaves and flowers; ***J.× stephanense*** (*J. beesianum × J. officinale*), raised in France in 1918, semi-evergreen in mild winters, a twiner, mid-way to parents growing up to 5.4 metres high with simple ovate-lanceolate or most of the leaves pinnate having 3-5 leaflets and flowers fragrant pale-pink appearing in lateral and terminal clusters; etc.

Some other important species not described above are *J. angulare, J. floridum* (syn. *J. subulatum*), *J. gracilimum, J. maingayi, J. multipartitum, J. odoratissimum, J. pubigerum, J. rigidum* (*syn. J. lingustrifolium*), *J. simplifolium* (*syn. J. lucidum, J. gracile*), *J. undulatum,* etc.

Propagation

Jasmines are usually propagated by cuttings and layerings. Seed propagation, though uncommon, is necessary for crop improvement through planned hybridization. Multiplication through sucker, grafting, budding and tissue culture can be done but is uncommon. Interclonal variations in **seed germination** have been reported in *J. auriculatum* and *J. grandiflorum* and also among other species. In general, diploid species of *J. auriculatum* and *J. grandiflorum* have better germination potentiality (75-84 per cent) than tetraploids like *J. flexile* (55 per cent). Higher temperature and humidity hasten germination. Seed dormancy has been noted in *J. auriculatum* and longevity of seeds does not exceed 10 months. Other species reported setting the seeds are *J. arborescens, J. angustifolium, J. calophyllum, J. mesnyi, J. pubescens, J. rigidum, J. suavissinum,* etc. (Bhattacharjee, 2001).

The most popular method for jasmine propagation is simple **layering** made in June-July, i.e. start of the rains to October-November, i.e. before the onset of winter, and layers root properly in 90-120 days when these are separated and kept in shade covered with wet gunny bags or root portions of the layers covered in wet sand or in soil until planting permanently in the prepared pits. Stems of *Jasminum nudiflorum* have been found to develop adventitious roots when the branch comes in contact with ground and these can easily be propagated from layers. This method of multiplication, however, involves more time and restricts the number of plants propagated from one plant. Sreelatha (1987) recommended layering in *J. auriculatum* from June to September (rainy season) for maximum success and with the use of IBA for favourable effects on all aspects of root growth. **Cutting** is by far the easiest method of propagation though several factors like the plant species and varieties, the type of cuttings (softwood, semi-hardwood, hardwood, lateral or terminal), stage of the plant, the rooting media and environmental factors play greater role in root formation of cuttings. The percentage of rooting and number of roots per cutting is highest in *J. auriculatum.* The success of rooting is high with semi-hardwood cuttings of *J. auriculatum,* terminal cuttings of *J. grandiflorum* and semi-hardwood *cum* terminal cuttings of. *J. sambac.* Higher success of rooting can be obtained by taking the cuttings during April-September in the tropical climates though from May to September under sub-tropical and temperate conditions. Success in rooting percentage of jasmine cuttings has been found much higher when growth regulators for rooting were used (Bajpai and Parmar, 1958; Muthuswami and Pappiah, 1976; Bose *et al.*, 1983). *J. auriculatum* var. 'Parimullai' which used to be propagated only through layering as its cuttings were very difficult to root, rooted easily to a tune of 60% when its etiolated cuttings were treated with 1,000 ppm IBA through quick dip method for 10 seconds as soaking even in 200 ppm NAA or IBA proved toxic to the stem tissues (Veeraraghavathatham *et al.*, 1985). Treating cuttings with IBA and placing them in intermittent mist improves rooting. Rooting in *J. grandiflorum* is to a tune of 90% with IAA and NAA at 1,000 ppm in case of soft-wood cuttings and 500 ppm in case of hardwood cuttings. At Tamil Nadu (Coimbatore) conditions, cuttings taken in May and planted in sand provides higher success (Anon., 1977). Sand, vermiculite, moss and mixture of these are some of the commonly used media for root formation in cuttings of jasmine. The role of leaves, stem length, woodiness and rooting media on the rooting of cuttings and their further survival was investigated on four different species of *Jasminum* at IIHR, Bangalore and it was concluded that 15 cm tip-shoot cuttings with four leaves when treated with 4000 ppm IBA and planted in vermiculite under mist give best performance in rooting and survival of rooted cuttings. The exception from above generalization, which was recommended by keeping the other factors constant, was 10 cm cuttings of *J. auriculatum,* basal cuttings of *J. multiflorum* and sand medium for *J. sambac.* Bose *et al.* (1983) stated that in mist chambers jasmine cuttings root much better. In mist chambers, terminal and semi-hardwood cuttings in case of *J. auriculatum* and *J. sambac,* terminal (5-nodal) cuttings in case of *J. grandiflorum,* and terminal cuttings in *J. calophyllum* & *J. rigidum* gave excellent results though response with *J. humile* was poor (Anon., 1974). Treatment of cuttings with growth substances before planting has been found very useful for promoting root formation. The highest rooting (up to 90 per cent) has been reported in *J. grandiflorum* through softwood cuttings treated with IAA 1000 ppm and through hardwood cuttings treated with NAA 500 ppm.

A study on interspecific **grafting** by Nambisan and Krishnan (1980) exhibited approach grafting as most successful with a take of 76 per cent when *J. sambac* was grafted on *J. auriculatum*, 72 per cent when it was reverse and 90 per cent when *J. grandiflorum* was grafted on *J. auriculatum* and the time it took for separation of grafts was from 16-24 weeks, however, the T-budding and side grafting did not succeed. The other work in TNAU, Coimbatore (1980) with *J. grandiflorum* as scion on 11 different species using shield and patch budding and side and cleft grafting revealed that it is only patch budding where *J. grandiflorum* was compatible with *J. sambac* cv. 'Madanban', 'Mathuria' and 'Single Mogra' which took a month for bud take.

Successful **callus proliferation** from culture of buds, petals, anthers, internodes and peduncles and regeneration of plants from stem apices of *J. officinale* cultured on Miller's medium containing glucose, vitamins and growth substances are reported. In order to promote leaf production in shoot tips (0.4–0.8 mm long), a modified half strength MS medium containing 1 mg/l GA_3 and 0.1 mg/l NAA was found best. The leafy shoots, about 2 cm long, were then transferred successively to solid MS medium with 0.5 mg/l NAA and 0.05 mg/l BA for 30 days for root formation, followed by quarter strength MS with little or no NAA to promote root growth. They were then potted in a mixture of equal quantities of peat, vermiculite and compost. Apical and axillary buds selected from plants at different times of the year were cultured on MS medium with four different growth substance formulations. Axillary buds were dormant from June to March and when cultured during this period always formed callus followed by roots, regardless of the medium composition. Outside the dormant period, axillary buds responded to treatments in a similar manner to the apical buds which were active throughout the year. On a medium containing 1 mg NAA and 10 mg/l BA, apical buds produced copious callus which surrounded the meristem and impeded differentiation. Roots were produced later, with 1 mg NAA (or 1.5 mg IAA) + 1 mg kinetin/l, a small callus formed after 20-30 days (or 15-20 days with IAA), meristem differentiation resulted in a green plantlet with numerous roots. With 1-5 mg IAA+10 mg/l BA, callus and multiple shoots developed in 45-60 days, followed much later by roots. Plantlets from various treatments were transferred to MS medium containing 10 mg IBA/l but no cytokinin to complete their development before potting.

Cultivation

Misra *et al.* (1978) have given a full account of all the jasmine aspects including its cultivation. Jasmines start blooming normally from the second year of their planting the cuttings though earlier through layering but normal flower production commences from the third year and continues producing normal flowers up to 10-15 years. These perform best in the mild climates of South India as compared to North India which experiences fierce summers and chilly winters. Some one-and-a- half dozen Indian species and their varieties are cultivated in India.

Jasmines prefer well-drained rich sandy loam to sandy-clay **soils** with a pH range of 6.5-7.5. The ideal condition for its proper cultivation is mild tropical climate. Mild winter, warm summer, ample water supply or moderate rainfall and sunny days provide suitable environment for its proper growing. There are a few species which like cool temperature, i.e 7-13 ^{0}C and certain others survive well even there where snowfall occurs from December to March in India and such species *are J. grandiflorum, J. humile* and *J. officinale.* However, barring *J. humile,* other species described above perform well at 13-18 ^{0}C temperature. Most jasmines adjust to higher temperatures of tropical and sub-tropical regions of India easily. *Jasminum* species are usually grown in the open for commercial flower production. In summer, some species require a cool greenhouse for flowering, though they can survive in the open but in winter they should be placed in the greenhouse with a minimum temperature of 14^0 C.

The soil should be thoroughly prepared and kept free from weeds before **planting**. Pits of 45-90 cm^3 are prepared depending on the type of soil and the species or variety to be planted and the pits are exposed for a week. In Andhra Pradesh, for commercial cultivation, the dimension of the pit is kept at 30 × 30 cm (Girwani *et* al., 2011). The pit is filled with top soil after mixing with about 10-15 kg of well rotten cattle manure or compost and then irrigated to settle down the soil. Two parts well rotten manure, one part fresh garden soil and one part coarse sand along with some 400 g BHC per pit should be used for filling the pits a few days before planting and then the pits should be irrigated to settle down the mixture properly. Planting is normally effected just before the monsoon and also during February-march. **Plant density** is one of the factors determining the ultimate flower production. It varies from place to place and depends upon the species and cultivars and also on the soil and environmental conditions. The spacing recommendations for the three commercial important *Jasminum* species furnished are 1.5 × 1.5 metres for *J. auriculatum* and its varieties, 2.0 × 1.8 m for *J. grandiflorum* and its varieties and 1.25 × 1.25 m for *J. sambac* and its varieties. Pits are filled with well rotten cow dung or farmyard manure @ 10 kg per pit along with dried leaf mould and little quantity of bone-meal and Thimet 10-G to check the menace of termites.

Jasmines require moderate **watering**. Once the pits are thoughly watered when the pit-filling is complete and then immediately after planting. Afterwards, the pits should be watered first after five days of planting and then daily in the evening shallowly till the proper establishment of the plants. In off seasons and during summers twice per month while monthly during winter months. However, during blossoming, these should be watered every 5 days. From flowering to pruning the watering is normally withheld, however, fertilizer application is always followed by watering.

After one week of pruning **manures** and **fertilizers** are added in the pits. Misra *et al.* (1978) suggested a mixture of 15 kg farmyard manure, 300 g ammonium sulphate, 750 g single superphosphate and 200 g muriate of potash and this comes roughly to 60:120:120 g NPK per pit. This dose is repeated in June-July in case of *J. sambac*. NPK @ 120:240:120 g per plant in six split doses, i.e. at bi-monthly intervals in case of *J. auriculatum* and 120:240:240 g per plant in two split doses in case of *J. grandiflorum* should be applied for proper flowering. In case of *J. sambac*, the dose may be halved. Muthuswamy (1975) suggested that in case the nutrients are to be supplied through foliage, nitrogen should be reduced to its half (60 g/plant/year) if applied as foliar sprays at monthly intervals though in case of *J. grandiflorum* foliar spray is not as effective as soil application twice of 60:120:120 g per plant along with 10 kg of FYM applied in two splits, once in January and again in July every year. However, the optimum recommendations of each of N, P and K per plant per year is 240 g at bimonthly intervals for Madurai conditions. At Kalyani (W.B.) conditions, the optimum recommendation for *J. auriculatum* and *J. sambac* is $N_{50}P_{15}K_{15}$ (Anon., 1978). Amburani and Vidhyapriyadharshini (2011) conducted an experiment with mullai (*J. auriculatum*) taking FYM @ 25 t/ha, pressmud @ 25 t/ha and neemcake @ 5 t/ha combined with graded levels of inorganic fertilizers (75, 100 and 125% of RDF) and with two levels of vermiwash (with and without application @ 1:5 dilution) and recorded increased plant height under neemcake @ 5 t/ha combined with 125% RDF (150 : 300 : 300 g/plant) and vermiwash. They recorded significantly highest primary shoot length, secondary shoot number, productive shoot number and plant spread under FYM @ 25 t/ha combined with 100% RDF (120 : 240 : 240 g/plant) + vermiwash @ 1:5 dilution. Asha Raj (1999) stated that to *J. sambac*, highest levels of nutrient application, *viz.* 150 kg each of N, P and K per hectare significantly increased the plant height, branch number, plant spread and yield of flowers though essential oil content was found maximum under 100 kg each of N and P and 150 kg K application per hectare. Nirmalatha (2001) with *J. sambac* recorded better main shoot length, primary & secondary branch numbers and plant spread when used 600 kg per hectare each of N, P_2O_5 and K_2O, number of secondary branches was found maximum under 450 kg per hectare each of K_2O and P_2O_5, and the lowest dose, *viz.* 300 kg each of N & P_2O_5 and 450 kg K_2O increased yield and other attributes of flowers. For commercial cultivation, Girwani *et al.* (2011) reported that in Andhra Pradesh, growers practice 10-12 kg of FYM per plant at the time of planting and then every year, for *J. sambac* 90 g N+120 g P_2O_5+240 g K_2O/plant, for *J. auriculatum* 60 g N+120 g P_2O_5+120 g K_2O/plant and for *J. multiflorum* 60 g N+120 g P_2O_5+120 g K_2O, all applied twice (June and January after pruning) in split doses. They report further that a micronutrient spray of $ZnSO_4$ @2.5 g+$MgSO_4$ @ 5 g/l is also beneficial. These doses are for full size plants but in case of smaller plants the doses may be reduced accordingly. This dose may be repeated on the advent of monsoon.

Before one week of pruning, irrigation is withheld till pruning is complete and afterwards when manure and fertilizer application is also complete. Pruning is an important practice for proper aeration of the plants so that occurrence of insect-pests and diseases is minimized, for usual regulation of healthy growth and flowering and for increasing flower bud yield and that too of better quality. **Pruning** is effected during November-January. In Andhra Pradesh, pruning is done in January-February, i.e. 2-3 months before flowering season by withholding water, and to delay the flowering they also pinch the new shoots (Girwani *et al.*, 2011). Dead, diseased, entire past season's growth of the shoots and the overcrowded twigs and entire leaves are removed, to a height of 40-90 cm from the ground, minimum height being retained in case. of *J. sambac* var.

'Motia' while 45 to 90 cm in other cases. Pruning is carried out in the last week of November in *J. sambac,* in the last week of December in *J. grandiflorum* and in the last week of January in *J. auriculatum* in the tropical areas of India, but in sub-tropical conditions of North India it may be carried out during December-January. Muthuswami and Madhava Rao (1980) reported that **pinching** the shoots to stagger flowering delayed blooming by 14.3 days in *J. auriculatum* and by 17.4 days in case of *J. grandiflorum.*

Pal *et al.* (1980) with *J. auriculatum* double types suggested December and January pruning while with *J. sambac* single types as February for better yield under Kalyani conditions. Subramanian (1977) revealed through a study that early pruning, i.e. in December spanned the flowering duration for eight months than only usual five months in late pruned crop of *J. grandiflorum,* highest yield of flower buds being recorded in late December pruned crop and the lowest when pruned in the end of March. *J. sambac* var. 'Single Mogra' when pruned exhibits peak flowering from March to May and no flowering from October to January at Coimbatore conditions. *J. sambac var.* 'Khoya' when is pruned at 20 cm as against 40 cm in January, the branch number is found reduced with reduced yield of flowers though pruning at 40 cm height results into maximum yield. Following pruning, the cut ends are treated with Bordeaux paste to prevent any disease infection. Chemical defoliants like paraquat dichloride, pentachlorophenol, potassium chloride and sodium chloride are employed on *J. auriculatum.* While paraquat dichloride causes quickest and heaviest defoliation, pentachlorophenol and sodium chloride application shows higher yield than the unpruned plants. Use of defoliants as an alternative to manual pruning is thus effective, however, it has some limitations. Spray of either pentachlorophenol 3000 ppm or potassium iodide 4000 ppm chemical defoliant on *J. grandiflorum* is beneficial over conventional pruning for increasing flower yield.

Weeding and **hoeing** for proper aeration should be carried out regularly. Weeds cause enormous losses to jasmine crop and increase the cost of cultivation. Manual weed control is effective if done properly, but this is very expensive. Chemical weed control is comparatively economical, convenient and efficient in eradicating weeds. Mulching can also be adopted to control the weed population.

Growth and Flowering

In *J. auriculatum,* spraying with ethrel at 1000 ppm records maximum growth retardation, CCC at 1000 ppm induces early flowering while B-Nine produces greater number of laterals (Muthuswami, 1975). Pappiah and Muthuswami (1977) reported that 100 and 500 ppm of ethrel reduced the shoot length in *J. grandiflorum.* CCC 500 ppm spraying on *J. grandiflorum* induced earlier flowering with better yield, larger flowers with longer duration of flowering, and high essential oil content (Pappiah and Muthuswami, 1977). At 0.125-0.50 ml/l ethrel application increased plant height and number of shoots in *J. auriculatum* and *J. sambac* (Pal *et al.,* 1980). Bhattacharjee and Divakar (1984) when sprayed ethrel at 100, 500 and 1000 ppm concentrations on *J. auriculatum, J. grandiflorum, J. calophyllum* and *J. flexile* plants, recorded reduced shoot length in former two but increased shoot length in latter two species though first two doses promoted the formation of axillary shoots, stimulated the basal diameter of the sprayed shoots, appreciably increased the flower bud yield in all the species and delayed the flowering. They observed increase in concrete percentage only in *J. grandiflorum* at 1000 ppm though recovery of oil was found increased in *J. auriculatum, J. flexile* and *J. grandiflorum* in 1000 ppm. Pappiah and Muthuswami (1978) recorded reduction in concrete per cent of *J. auriculatum* in 100 and 500 ppm ethrel but with increased size of buds and flowers having delayed floral initiation.

In *J. grandiflorum,* GA_3 at 25-75 ppm significantly increases the length of primary and secondary shoots whereas MH at 1000-3000 ppm, TIBA at 100-300 ppm, CCC at 500-1500 ppm and ethrel at 100-500 ppm reduce shoot length. The highest number of lateral shoots is observed on plants treated with CCC, followed by TIBA and MH. GA_3 treatment results in the least number of lateral shoots. MH increases the number of internodes and GA_3 increases their length. CCC significantly increases the number of leaves and flowers and induces early flowering whereas ethrel delays flowering. CCC treated plants also flower for longest period. In *J. sambac,* GA_3 has been reported to promote shoot length and delay the flowering while CCC and SADH retard the growth and increase the flowering. In *J. auriculatum* cv. Parimullai, CCC has been reported to increase flower yield through induction of more laterals and by extended flowering period. Application of ethrel at 0.125 and 0.250 ml/l and NAA at 25 mg/l on *J. sambac* promote more laterals resulting in increased flower number.

Muthuswami and Madhava Rao (1980) stated that *J. auriculatum* and *J. grandiflorum* can be regarded as responsive to thermo-periodism quantitatively and classified them as day neutral plants requiring temperature alteration for flowering. Spraying of growth regulants along with pruning round the year at monthly intervals revealed that pruning in January and spraying 1000 ppm Alar shift the peak flowering in *J. grandiflorum* to October while pruning in February and spraying CCC 500 ppm shift the peak flowering to November (Vedamuthu, 1981).

Post Harvest

In Andhra Pradesh, flowers are harvested at tight bud stage (8-10 hours before opening), harvesting being carried out from April to July in *J.* auriculatum, from March to August in *J. sambac* and from September to January in *J. multiflorum* with the flower yield of 5.0-7.5 t/ha in *J. auriculatum,* 5.0-6.25 t/ha in *J.* sambac and 5.0 t/ha in *J. multiflorum* and thus generating an income of Rs.1,50,000 to 1,62,500/ha (Girwani *et al.*, 2011). Flowers in jasmine open in the evening. **Harvesting** is done at different stages depending upon their use. Fully developed unopened tight bud stage (8 to 12 hours before blooming) is ideal for concrete extraction, while fully open flowers are harvested in the early hours of the day. Delay in picking beyond 9.30 a.m. results in gradual reduction of concrete yield. Care in **handling** plays a critical role in post harvest flower quality since even the slightest mechanical damage may lead to discolouration and loss in concrete yield or marketable flower quality.

Shelf life of jasmine flowers, from harvest at tight bud stage to withering varies inherently between species such as *J. multiflorum* 48-60 hours, *J. sambac* 28-30 hours, *J. auriculatum* 28-30 hours and *J. grandiflorum* 24 hours. Post-harvest handling like pre-cooling, packaging, chemical treatment, cold storage, etc. have been found to be beneficial in extending the shelf life of jasmine flowers. Treatment of *J. sambac* floral buds with ethylene absorbants delayed bud opening maximally and newspaper lining along also prolonged shelf life, however, low temperature treatment has been found best to increase the longevity of buds by retaining colour and turgidity (Nirmalatha, 2001). Flowers of *J. sambac* treated with sucrose, boric acid, copper sulphate, aluminium sulphate and silver nitrate have been reported to remain fresh upto 72 hours with minimum loss of fragrance. Water and chemical spraying combined with packaging (300 gauge PE without ventilation) has been reported beneficial in extending the shelf life, retention of white colour and freshness of flowers of *J. multiflorum.* Cool chamber storage of *J. multiflorum* flowers packed in 200 gauge PE bags without ventilation has been reported beneficial in maintaining higher quality and extension of shelf life of flowers. PE bags of 200 gauge without ventilation maintained freshness, retained colour and extended shelf life upto 3.33 days under ambient conditions. Pre-cooling of *J. sambac* flowers at 7.5 ^{0}C and chemical treatment with sodiumthiosulphate has been reported to delay bud opening by 14 and 24 hours, respectively and extended shelf life to 56 and 68 hours, respectively. Harvesting of *J. auriculatum* buds before 8 a.m., packing in 500 gauge PE bags without ventilation and using calcium chloride as desiccant in the package have been reported to delay bud opening and extending the shelf life. In general, corrugated cardboard boxes have been found suitable for **packing** for distant markets while bamboo baskets for local markets. Packaging should be functional, economical and attractive besides being acceptable in markets.

Extraction of Essential Oil

Though *Jasminum grandiflorum* is native to Indo-Myanmar region in the foothills of Himalayas but its potential for perfume production was first exploited in France. This was introduced in Morocco, Egypt, South Africa, Syria, Tunisia, Algeria, United Arab Republic, Sicily, India and China in the recent past on plantation scale. All these countries produce jasmine flowers and extract the essential oil which is supplied to Europe, Russia, Japan and USA. In 1972 at Mettupalayam (T.N.), India, its first oil extraction factory was installed but afterwards over 16 units came into offing in the Tamil Nadu (Encee Aromatic Pvt. Ltd., Main Road, Mettupalayam; Nesstate Flora, 4-Veerapandi, Periyanaicker Palayam, Mettupalayam; Leo Aromatics, Veerapandi, Periyanaicker Palayam, Mettupalayam; Aishswarya Aromatic Pvt. Ltd., Tirpur; Jasmine Concrete factory, Sayarafarms, Sayarpuram, Madapam Taluk; Flowers and fragrance, Palaningoondenpodur, Pennimalai, Coimbatore; Jasmine Concrete Factory, Govt. of Tamil Nadu, Tenkasi, Dist.

Tirunelveli; Fragrance India Pvt. Ltd., P.O. Box 165, Vellore; Venus Herbo Aromatics Pvt. Ltd., Muhavoor Road, Seithur-626121; Jardin Aroma Pvt. Ltd., Indiranagar, Chennai-600 020; Indfrag Pvt. Ltd., 125 km Hosur, Kelamangalam Road, Kundumaranpalli-635 113), Karnataka (Natural Essential Oils Pvt. Ltd., 224, 8th Cross, Gokulam, 3rd Stage, Mysore-570 002; A branch of Encee aromatics, Vill.- Hanoor, Kollegal, near Mysore), Andhra Pradesh (Encee Floral Extracts Pvt. Ltd., Hanumpalli, Ramapura, Penukonda-515 110), Maharashtra (Encee fragrances Pvt. Ltd., Vill.- Dhakepal, Taluk Paithan, Distt.- Aurangabad), *etc.* (Srivastava, 1995).

The fragrance in jasmine flowers is due to a volatile oil present in the epidermal cells of the inner and the outer surface of both petals and sepals, and this fragrance in opening buds starts to develop soon after sunset and ceases within a few hours of sunrise. Jasmine **concrete** obtained from *Jasminum grandiflorum* is a wax-like substance containing the natural flower perfume together with some plant waxes, albumin and colouring matter. The natural perfume is available in very small quantity (0.25%) in jasmine flowers in the form of volatile oil. According to Abdul Khader *et al.* (1984), while working at TNAU, Coimbatore, recorded *J. auriculatum* containing concrete recovery of 0.28-0.36% being maximum, *J. grandiflorum* 0.25-0.32% and *J. sambac* 0.14-0.19% being minimum, with the flower bud yield of 4,733-9,152 kg/ha being maximum, 4,329-10,144 kg/ha and 739-8,129 kg/ha being minimum in their repective species. Though *J. auriculatum* yields maximum of concrete but it is only *J. grandiflorum* which has international market.

The usual and simple method of steam distillation for the extraction of the volatile oil could not be adopted in this case as jasmines do not yield the perfume oil on steam distillation. Hence, the **solvent extraction method** is practiced in which the principle is that the odoriferous substance of the flower is allowed to be absorbed by a highly volatile solvent and then the solvent is evaporated leaving the odoriferous principles. This is the gift of the modern age which was first introduced into India in 1957, and this process is used internationally, *viz.* France (the Grasse region in south France where some 50% of the total of the world's jasmine oil is produced), Italy, the UAR, Morocco, Tunisia, Turkey, Algeria, Belgium and Holland. For extraction of concrete, only fresh and fully opened flowers which had opened by the preceding evening are picked up essentially early in the morning by 9 a.m. as delay in picking beyond 9 a.m. results in gradual reduction of concrete yield. The flowers should be harvested and transported in clean baskets or cloth bags. The harvested flowers should be stored in a cool atmosphere and processed within a maximum period of two hours. Processing of jasmine flowers involves two steps, i.e. (i) treatment of the flowers with solvent, and (ii) evaporation. The fully opened flowers are treated with the **solvent** so that all the aromatic components in the flowers get dissolved in the solvent. In this process, the solvent penetrates the flowers and dissolves the waxes, albumin and colouring matter also, which have to be removed later to get pure perfume oil. In solvent extraction process the most important factor is the quality of solvent used. Commercial food grade hexane with a boiling point of 60-80 ^{0}C has been found to satisfy many of the requirements of a good solvent and hence is used in large scale for the extraction of floral concrete throughout the world. Food grade hexane available in the market is quite impure so it requires to be purified before use. Addition of 5% liquid paraffin in the hexane and its distillation at 70 ^{0}C provides an acceptable grade of hexane (Sambandamurthi and Abdul Khader, 1982). Flowers are treated with purified hexane at 2 litres per kg of flowers for about 30 minutes. The complete extraction of the perfume from the flowers can also be ensured by slowly rotating the container having the flowers and solvent for about 20 minutes in the rotary type of extractor. After the complete extraction of the perfume from the flowers, the solvent is filtered and concentrated by **evaporation** at a constant temperature of 75^0 C. At this temperature, the solvent evaporates leaving the perfume and other plant waxes. The vapour of the solvent is condensed into liquid again for recycling. The concentrated liquid containing perfume, waxes, pigments and the solvent is distilled in a vacuum distillation unit where the complete removal of the solvent takes place, leaving the floral concrete in the form of molten wax. Based on the above principle, TNAU, Coimbatore has designed a pilot plant for extraction of floral concrete. **Flower** and **concrete yield** in jasmines vary considerably according to the species and cultivars, management practices, etc. It is to be noted that concrete from *J. grandiflorum* alone has international market. In a varietal evaluation, cultivar 'CO.1 Pitchi' was found to be the highest yielder producing over 10 tonnes of flowers per hectare with an estimated concrete yield of 29 kg (0.29%) (Muthuswami and Abdul Khader, 1986).

Volatile oil in jasmine is also extracted through **water distillation** but it is not economical as through this method not more than 0.07 per cent oil is extracted. Another method is also sometimes used and this is called **enflurage** process which in India was introduced in 1942. Through this process a good percentage of oil is extracted. Under this process, firstly *til* seeds (*Sesamum indicum*) are soaked in water for dehusking and then dried in the sun and fresh jasmine flowers and dehusked *til* seeds are thinly spread one over the other for 10-12 hours daily and this process is repeated until all the derhusked *til* seeds are fully saturated with jasmine perfume. This process takes about a week or so and thus normally 1 kg of dehusked *til* seeds consume 3 kg of fresh jasmine flowers for complete processing. Now these *til* seeds are distilled and the jasmine vapours are absorbed by sandal wood oil for production of *jasmine attar*. But better than this is the solvent extraction method.

Insect-Pests, Diseases and Physiological Disorders

Budworm (*Hendecasis dublifascialis*) caterpillars destroy the blossoms by boring the buds and infest inside eating the petals during rains, being serious on *J. sambac*. These make the webs with silken threads, fill the excreta in the webbings which make the buds turning pink and premature falling. During peak period of moth emergence from the pupa, light trap should be used and during peak infestation period 5 per cent neem seed kernel extract should be sprayed. During off season the beds should be raked to expose the pupae for predators and 10 per cent carbaryl should be applied around the basin once during April-May and the other during August-September. Monocrotophos at 2 ml/l spraying by evenings when the caterpillars are active is quite effective. **Bud- and shoot-** or **gallery-worm** (*Elasmopalpus jasminophagus*) feeds inside the buds making exit holes at the base, being serious on *J. auriculatum* (Ranjith, 2013). It causes webbing of terminal leaves, shoots and flowers. **Shoot borer** (*Sycophila* sp.) has been found infesting on *J. grandiflorum* where the larvae tunnel inside the shoots making a series of grooves, *vis-à-vis* entry holes and pupating inside (Ranjith, 2013). Damaged shoots should be removed and destroyed, however, spraying with 0.02% Nuvacron will also control this pest. The maggots of **blossom midge** or **gallfly** (*Contarinia maculipennis*) enter the buds from corolla bases and feed in them and have been observed infesting only on *J. sambac* and *J. auriculatum*. Its infestation disfigures the buds through purple discolouration 4-5 days after oviposition (Ranjith, 2013). Control of bud and shootworm as well as blossom midge is through spraying Flufenoxuron 10 DS (Cascade) 0.1% (75-100g a.i./ha) or Novaluron 10 EC (Rimon) 3ml/l i.e. 75-100 g a.i./ha. **Leaf webworm** (*Nausinoe geometralis*) and **leaf roller** (*Glyphodes celsalis*) caterpillars feed on the leaves, especially lower ones during rains and the terminal shoot during dry weathers. The pest is controlled through 2 ml/l spraying of Hostathion during peak period of infestation. The nymphs and adults of the **tingid bug** (*Corylhauma ayyari*), **jasmine bug** (*Antestiopsis cruoiata*) and **green plant hopper** (*Flata oellata*) suck the sap of tender parts, causing leaf yellowing and ultimately their drying and falling. While controlling the worms, the bugs and leaf hoppers will automatically be controlled. **Thrips** (*Isothrips orientalis* syn. *Thrips orientalis*) suck the plant sap from buds and flowers which cause brown streaking on the petals and falling off (Ranjith, 2013). Control measures are the same as in case of worms.

Cyclamen mite (*Steneotarsonemus pallidus*) damages severely to *J. sambac* crop. **Eriophyid mite** (*Aceria jasmini*) is a major pest of *J. auriculatum* and *J. pubescens* (Ranjith, 2013) and there these feed on buds, leaves and tender stems causing velvet-like hairy growth on the leaf surface which hampers plant growth and flower production. Spraying and rubbing of Rogor on the affected parts will control this pest. **Gall mite** (*Eriophyes* spp.), also known as blister mite, rust mite or bud mite is a havoc in South India. This devastates *J. auriculatum* crop during rains and winters by burrowing under the epidermis of the leaves and tender shoots causing whitish bloated galls which later on coalesce, causing drastic reduction in flower yield. TNAU's var. 'Parimullai' is resistant to gall-mite. **Red spider mite** (*Tetranychus urticae*) is severe during summer when the weather is quite dry and hot. This mite feeds on the under surface of the foliage causing them to yellow and finally these drop off. Spraying with monocrotophos 2 ml/l, wettable sulphur 3 g/l, Fenpyroximate 0.1 per cent or Chlorfenapyr at 0.1 per cent is quite effective.

Root-knot **nematodes** (*Meloidogyne incognita*) infest on *J. flexile*, *J. nudiflorum* and *J. sambac* roots making small swelling and enlargement of rootlets which consequently cause yellowing of foliage and in its serious attack even entire plant dies. It can be controlled by raking Carbofuran granules in the soil, followed by watering.

Major diseases affecting jasmine production are **leaf blight** (*Alternaria jasmini* and *Cercospora jasminicola*) which causes brown spots on the leaves irrespective of their age, and these spots coalesce during humid conditions, blight, leaf margins curl inward and finally the leaves fall off. The symptoms are quite conspicuous on the young branches. This problem is very serious on *J. grandiflorum*. Immediately after appearance of the disease, the crop should be sprayed at 'run off' stage with carbendazim 0.1% and repeated once after 7-10 days. Ramakrishnan and Kandaswamy (1980) suggested spraying of benlate 0.4%, bavistin 0.1%, Dithane M-45 0.2% or Bordeaux mixture 1.0% at monthly intervals from May up to pruning though number of sprays may depend on weather conditions and severity of the incidence. Soil application of *Pseudomonas fluorescens* at 25 g/m^2 and its 0.5% foliar spraying at monthly intervals will control this problem effectively. **Leaf spot** (*Curvularia paradisi*) causes brown spots and leaf defoliation which may be controlled through 0.2% Dithane Z-78 fortnightly sprayings. **Wilt or root rot** (*Fusarium solani* and *Sclerotium rolfsii*) is caused as whitish colony with abundance of micro and macro conidia with chlamydospores. Ravikumar *et al.* (2004) succeeded in controlling its infection through drenching twice with benomyl 0.1% + copper oxychloride 0.3% and carbendazim 0.2% + copper oxychloride 0.3% at 15 days' intervals. **Powdery mildew** (*Oidium jasmini*) that spreads through wind-borne conidia, appears as white powdery patches on the upper surface of the leaves which later on coalesce and cover the entire leaf surface. Wettable sulphur at 0.2 per cent or carbendazim at 0.1 per cent spraying will control this disease. **Rust** (*Uromyces hobsoni*) infect the crop during humid weathers of rainy season and the species susceptible to this rust are *J. auriculatum, J. angustifolium, J. communis, J. flexile* and *J. sambac*. *J. auriculatum* is highly susceptible to rust. In its infection both sides of leaves (lower side being more conspicuous) show orange aecial cups and after sometime entire leaf becomes yellow and crinkled, and the disease may cover even the stems causing splitting of the bark and death of the branches. Kandaswamy and Ramakrishnan (1980) suggested sulphur dusting @ 20-25 kg/ha. This can also be controlled through spraying with copper oxychloride or Bordeaux sprayings. **Blossom blight** (*Phoma* sp.), **root rot** (*Clitocybe tabescens*), **crown gall** [*Agrobacterium* (*Erwinia*) *tumefaciens*], etc. also infect jasmine though in India these have yet not been found serious.

Viruses such as chlorotic ring spot (JCSV) which spreads through whitefly (*Bemisia tabaci*) and grafting (Ramakrishna *et al.*, 2004 reported it transmitting by approach grafting from infected to healthy plants of *J. sambac* within 35 days of grafting), but not through the sap and is characterized by development of typical mosaic symptoms consisting of yellow chlorotic spots on leaves. The disease is controlled through Acetamiprid 20 SP at 15-20 g a.i./ha spraying which controls the vector. Jasmine mosaic virus causes stunting and yellowing of entire plant, *vis-à-vis* little leaf, mottling and chlorotic flecking in the form of streaks and rings, more prominently on the older leaves. Disease spreads through grafting and also through insect-vectors. *J. pubescens, J. sambac* and *J. simplicifolium* are highly susceptible to this disease though *J. auriculatum* is relatively tolerant. Control of insect-vectors will prevent spread of the disease. Phyllody (mycoplasma) infection transforms the flowers into malformed and reduced greenish flower-like structure bearing greenish and ovate corolla lobes. Cuttings from such plants should never be taken. The disease can be controlled through the use of 250 ppm of tetracycline hydrochloride. Virus infected plants should be rogued out and burnt.

Except copper deficiency in *J. multiflorum* and *J. nitidum*, no any other **physiological disorder** has been observed in this crop. Copper deficiency can be corrected through application of copper chelates or $CuSO_4$.

References

Abdul Khader, J.B.M., N. Chezhiyan and V. Ponnuswami, 1984. Utilization of jasmine flower for floral concrete extraction. *Proc. Sem. on Utilization of Value Added Agro-Products*. TNAU, Coimbatore.

Abdul Khader, J.B.M. Md. and N. Kumar, 1995. Genetic Resources of Jasmine. In: *Advances in Horticulture*, vol. 12 - *Ornamental Plants* (eds Chadha, K.L. and S.K. Bhattacharjee), pp. 121-132. Malhotra Publishing House, New Delhi.

Amburani, A. and H. Vindhyapriyadharshini, 2011. Response of mullai (*Jasminum auriculatum*) to organic and inorganic nutrients. *J. Ornam. Hort.*, **14**(1-2): 100-103.

Anonymous, 1959. Wealth of India, vol. V. CSIR, New Delhi.

Anon. 1974. *Second Annual Progress Report.* ICAR Scheme for intensification of research on the development of rose and jasmine varieties with high essential oil content. T.N.A.U., Coimbatore.

Anon. 1977. *Fifth Annual Progress Report.* ICAR Scheme for intensification of research on the development of rose and jasmine varieties with high essential oil content. T.N.A.U., Coimbatore.

Anon. 1978. *Seventh Annual Progress Report.* ICAR Scheme for intensification of research on the development of roses and jasmine varieties with high essential oil content. BCKV, Kalyani, Dist. Nadia (W.B.).

Asha Raj, 1999. Nutrient Requirement for Bush Jasmine (*Jasminum sambac* Ait.). M.Sc. Thesis. College of Agriculture, Kerala agricultural University, Vellayani, Kerala.

Bajpai, P.N. and A.S. Parmar, 1958. Effect of some plant growth regulators on the rooting of cuttings of *Jasminum sambac. Sci. Cult.*, **23**: 489-490.

Bhattacharjee, S.K.1980. Native jasmines of India. *Indian Perfumer*, **24**(3): 126-133.

Bhattacharjee, S.K. 2001. Jasmine. In:*Handbook of Horticulture* (ed. Chadha, K,.L.), pp. 569-572. ICAR, New Delhi.

Bhattacharjee, S.K. and N.G. Divakar, 1984. Growth, flowering and essential oil content of different species of *Jasminum* as affected by ethrel. *Indian J. Hort.*, **41**(1-2): 131-136.

Bose, T.K., D.P. Mondal and D.K. Pramanik, 1983. Propagation of ixora, hibiscus and jasmine from cuttings under mist. *Progressive Hort.*, **15**(3-4): 43-50.

Bowden, W.M. 1940. Diploidy, polyploidy and winter hardiness relationships in the flowering plants. *Ann. J. Bot.*, **27**: 357-371.

Chandra, V. 1982. Commercial exploitation of *Jasminum* species in India. In:*Cultivation and Utilization of Aromatic Plants* (eds Atal, C.K. and B.M. Kapur). CSIR, Jammu Tawi.

Girwani, A., Lalitha Kameswari, A.L.N. Prasad, K. Purushotham, Ramesh Kumar, Gunjeet Kumar and Tarak Nath Saha, 2011. *Prospects of Floriculture in Andhra Pradesh*, pp. 16-17. Directorate of Floricultural Research (ICAR), Pusa Campus, New Delhi-110 012.

Kannan, M., V.A. Sathiyamurthy and V. Shankar, 1992. Mutagenetic studies on *Jasminum sambac*. In: *Floriculture Research Trend in India* (eds Misra, R.L. and Sanyat Misra), pp. 209-21. Published by Indian Society of Ornamental Horticulture, Division of Floriculture and Landscaping, IARI, New Delhi-110 012.

Karmakar, P.G. and H.C. Srivastava, 1987a. Meiosis in triploid *Jasminum grandiflorum. Cytologia*, **52**: 775-777.

Karmakar, P.G. and H.C. Srivastava, 1987b. Sterility and meiosis in five species of jasmines. *Curr. Sci.*, **56**: 173-174.

Krishnaswamy, N. and V.S. Raman, 1948. Cytogenetical studies of the Indian jasmines. I. Taxonomy and chromosome numbers of four varieties of *J. sambac* and other species. *J. Indian bot. Soc.*, **27**(1): 77-83.

Misra, R.L., O.P. Chaturvedi and L.P. Tiwari, 1978. Jasmine-a most homely flower of India. *Delhi Garden Mag.*, pp. 1-5. Pub. by Delhi Agri-Hort. Soc., Div. of Floriculture and Landscaping, IARI, New Delhi-110 012.

Muthuswami, S. 1975. *Studies on Some Aspects of Growth and Flowering in J. auriculatum* Vahl. *and J. grandiflorum* L. Ph. D. Thesis. Tamil Nadu Agril. University, Coimbatore.

Muthuswami, S. and C.M. Pappiah, 1976. A note on the effect of growth regulators on rooting of stem cuttings of *Jasminum sambac* Ait. var. Gundumalli. *S. Indian Hort.*, **24**(1): 30-32.

Muthuswami, S. and J.B.M. Abdul Khader, 1986. Jasmine. In: *Ornamental Horticulture in India*, pp. 13-26. ICAR, New Delhi.

Muthuswami, S. and V.N. Madhava Rao, 1980. The concept of relative temperature disparity in flowering of *Jasminum* species. *National Seminar on Producation Technology of Commercial Flower Crops*, pp. 1-14. TNAU, Coimbatore.

Nambisan, K.M.P. and B.M. Krishnan, 1980. National Seminar on Production Technology of Commercial Flower Crops, pp. 19-20. TNAU, Coimbatore.

Nirmalatha, J. 2001. Nutrient Requirements and Post Harvest Studies on Bush Jasmine (*Jasminum sambac* Ait). M.Sc. Thesis. College of Agriculture, Kerala Agricultural University, Vellayani, Kerala.

Pal, P., R.G. Maity and T.K. Bose, 1980.Effects of growth regulators on *Jasminum sambac* var. Khoya and *J. auriculatum* var. Double. *National Seminar on Production Technology for Commercial Flower Crops,* pp. 35-38. Held on August 29-30, at TNAU, Coimbatore.

Pappiah, C.M. and S. Muthuswamy, 1977. Effects of growth regulators on growth and flowering in *Jasminum grandiflorum* L. clone Thimmapuram. *South Indian Hort.*, **25**(2): 68-74.

Pappiah, C.M. and S. Muthuswamy, 1978. Effects of growth regulators on growth, flowering, flower quality and essential oil content of *Jasminum auriculatum* Vahl. var. 'Parimullai.' *South Indian Hort.*, **26**(2): 66-71.

Pizzetti, I. and H. Cocker, 1975. Jasminum**.** In: *Flowers - A Guide for Your Garden* (vol. I), pp. 684-694. Harry N. Abrams Inc., Publisher, New York.

Ramakrishnan, I. and T.K. Kandaswamy, 1980. *National Seminar on Production Technology of Commercial Flower Crops*, pp. 107-108. TNAU, Coimbatore.

Raman, V.S. 1955. Cytogenetics of the Indian jasmines. II. The somatic chromosomes. *Cytologia*, **20**: 29-31.

Ramakrishna, M.V.S., M.S. Kulkarni and B.S. Reddy, 2004. Studies on transmission of jasmine chlorotic spot virus (JCSV) in Karnataka. *J. Ornam. Hort.*, **7**(3-4): 353-362.

Ranjith, A.M. 2013. Jasmine. In: *Identification & Management of Horticultural Crops*, pp. 141-142. New India Publishing Agency, Pitam Pura, New Delhi-110 088.

Ravikumar, M.R., S. R. Patil, S. Jahagirdar and B. Pampanagouda, 2004. Management of jasmine root rot caused by *Fusarium solani* in northern Karnataka. *J. Ornam. Hort.*, **7**(3-4): 349-352.

Sax, K. and E.C. Abbe, 1932. *Chromosome Atlas of Cultivated Plants* (eds Darlington, C.D. and E.K. Janaki Ammal in 1945). George Allen and Unwin Ltd., London. Original paper published in *J.Arnold. Arbor.*, **13**:37.

Sambandamurthi, S. and Md. J.B.M. Khader, 1982. Essential oil of jasmine. *Proc. Natl.Sem. Medicinal and Aromatic Plants.* TNAU, Coimbatore.

Sreelatha, U. 1987. Effect of Growth Regulators on Rooting of Cuttings and Layers in Jasmine (*Jasminum auriculatum* Vahl.). M.Sc. Thesis. Kerala Agricultural University, Vellanikkara, Thrissur (Kerala).

Subramanian, R. 1977. *Studies on the Effect of Pruning on Growth and Development of 'Jathimalli'.* M.Sc. (Ag.) Thesis. Tamil Nadu Agricultural University, Coimbatore.

Srivastava, H.C. 1995. *French Jasmine.* In: *Advances in Horticulture ~ Medicinal and Aromatic Plants,* vol. 11(eds Chadha, K.L. and Rajendra Gupta), pp. 805-823. Malhotra Pub. House, New Delhi

Taylor, H. 1945. Cytotaxonomy and phylogeny of the Oleaceae. *Brittonia*, **5**: 337-367.

Veeraragavathatham, P., V.N. Madhava Rao and K.G. Shanmugavelu. 1985. Effect of etiolation in rooting of stem cuttings of jasmine. *Indian J. Hort.*, **42** (3-4): 287-289.

Veluswamy, P. 1980. Breeding for high essential oil content in jasmine. *Natl. Sem. Prod. Tech. Commerc. Fl. Crops*, pp. 5-6. T.N.A.U., Coimbatore.

Veluswamy, P. 1981. *Genetic Studies in Jasminum auriculatum* Vahl. *and J. grandiflorum* L. Ph. D. Thesis. T.N.A.U., Coimbatore.

Veluswamy, P., T. Thangaraj and S. Muthuswamy, 1975. Study on germination of seeds of some *Jasminum* species and clones. *South Indian Hort.*, **23**:71-72.

18

Lawsonia inermis

Sanyat Misra and R.K. Dubey

Family: **Lythraceae**

Common names: Benjaati (Oriya), Egyptian privet/Reseda (Egypt), Henna (Urdu), Hina (Arabic name of the drug), Hinah (Punjabi), Goranti (Telugu), Goranti/Mailanchi (Kannada), Marithondi/Mullugoranatla/Muruthani (Tamil), Mehndi (Hindi/Gujarati/Marathi), Mailanchi (Malayalam), Mayilanchi (Kannada), Mehedi (Bangla), Mendika/Raktagarbha (Sanskrit), Mignonette (West Indies), Mohuz (Kashmiri)

Introduction

The genus *Lawsonia* was named so for John Lawson who in 1709 published an account of his travels to North Carolina. It is a glabrous and much-branched handsome shrub native to tropical and subtropical regions of North Africa, West and South Asia and northern Australasia in semi-arid zones, bearing abundance of fragrant flowers. Henna's indigenous zone is the tropical savannah and tropical arid zone, in latitudes between 15° and 25° N and S from Africa to the western Pacific rim, up to an elevation of 2,000 metres. This plant is native to Algeria, Cyprus, Egypt, Ethiopia, India, Indonesia, Iran, Iraq, Jordon, Kenya, Kuwait, Lebanon, Libyan Arab Jamahiriya, Malaysia, Morocco, Oman, Philippines, Qatar, Saudi Arabia, Syrian Arab Republic, Tanzania, Tunisia, Turkey, Western Sahara, Yemen, *etc.* It is cultivated as an ornamental plant, especially as the windbreaks and as fences, throughout India, Iran and along the coast of Mediterranean Sea. It has only one species *Lawsonia inermis* where var. *alba* bears white, *rubra* rose and var. *miniata* cinnabar-red flowers. Sometimes it becomes arborescent, i.e. 1.8-7.5 m height. Young branches appear green but afterwards become reddish and quadrangular. Bark of the younger branches is greyish-brown and unarmed and this is known as juvenile phase but after two years the branches become spine-tipped and then plants become senile.

Henna symbolizes fertility. Henna has been used cosmetically and medicinally for over 9,000 years. It is one of the oldest cosmetics in the world. It is cultivated for its leaves although stem bark, roots, flowers and seeds have also been used in traditional medicine. Its use became popular in India because of its cooling effect in the hot Indian summers. Chaudhary *et al.* (2010) stated that all parts of the plants have a medicinal value. Its leaves produce the henna, an orange-yellow-red dye which in Arabian and Asian countries is used by ladies to decorate their nails, fingers, hands, feet in various patterns and also against greying hairs including the beards in males. In Hinduism, Buddhism and Islam religions, the mehndi is certainly used during marriage. The plant is reported to contain carbohydrates, proteins, flavonoids, tannins and phenolic compounds, alkaloids, terpenoids, quinones, coumarins, xanthones and fatty acids. The plant has been reported to have analgesic, hypoglycemic, hepatoprotective, immunostimulant, anti-inflammatory, antibacterial, antimicrobial, antifungal, antiviral, antiparasitic, antitrypanosomal, antidermatophytic, antioxidant, antifertility, tuberculostatic and anticancer properties. It is now considered as a valuable source of unique natural products for development of medicines against various diseases and also for the development of industrial products. Henna leaves, flowers, seeds, stem bark and roots are used in traditional medicine to treat a variety of ailments as rheumatoid arthritis, headache, ulcers, diarrheoa, leprosy, fever, leucorrhoea, diabetes, cardiac disease, and hepatoprotective. It is used for alleviating jaundice, skin diseases, venereal diseases, smallpox and spermatorrhoea. Root is astringent and is said to be a potent medicine for gonorrhoea and herpes, *vis-à-vis* for the treatment of hysteria and nervous disorders (Reddy, 1988) and its pulp may be used against sore eyes and applied to the heads of children for boils. Cambodians drink it as a diuretic decoction and in combination with prepared indigo it becomes a powerful abortifacient. The bark decoction is applied to burns and scalds. It is given internally in case of jaundice, enlargement of the spleen, calculus, as an alternative in leprosy and obstinate skin infections. Aninfusion of the flowers is a valuable application to bruises. Decoction of the flowers is an emmenagogue. Seeds are deodorant, the powered seeds are used for liver disorders and associated problems and when mixed with *desi ghee* (clarified butter), it becomes effective against dysentery.

Its petioles have high contents (0.55–1.5%) lawsone (2-hydroxy-1, 4-naphthoquinone); 1, 4-naphthoquinone; 2-methoxy-3-methyl-1, 4- naphthoquinone; flavonoids, coumarins, and phenolic acids; 5–10% gallic acid and tannin; about 11% sugars, resin, and others. Two xanthones (laxanthone-I and laxanthone-II) and a substituted coumarin named lacoumarin (5-allyloxy-7-hydroxycoumarin) have been isolated from the whole plant and are probably present in the leaves. The aerial parts also contain the triterpenes lawsonin (3-α-*E*-ferulyloxy- urs-11-en-13-β-ol) and lawsonic acid (3-α-*E*-ferulyloxy-lup-20(29)-en-28-oic acid). Lawsone is the major active principle (coloring and pharmacological) in henna. It is not present in the bark, stem, or root of the henna plant. Its concentrations in the leaves vary with climatic conditions as in hot localities its yield is better. Apart from this, henna also contains isoplumbagin lawsoniaside, lalioside, lawsoniaside B, syringinoside, daphneside, daphnorin, agrimonolide 6-*O*-ß-D-glucopyranoside, (+)-syringaresinol *O*-ß-D-glucopyranoside, (+)-piunoresinol di-*O*-ß-D-glucopyranoside, syringaresinol di-*O*-ß-D glucopyranoside, isoscutellarin3ß, hennadiol, (20S)-3ß, 30-dihydroxylupane, lawnermis acid, 3-methyl-nonacosan-1-ol, laxanthones I, II, III and lacoumarin, *etc.* Its essential oil contains -(Z)-2-hexenol, linalool, α-ionone, ß-ionone, α-terpineol, terpinolene, õ-3-carene and Y-terpineol (Makhija, www.pharmatutor.org).

Botany

It is widely cultivated in the tropical and sub-tropical regions of the world. Its branches are more or less spinescent, leaves are dull green to greenish-brown, opposite, sub-sessile, entire, elliptic to broadly lanceolate with entire margin, attenuate into a petiole or a narrow base, acute, 1.5-5.0 cm long and 0.5-2.0 cm wide, and the petioles are short and glabrous. Small tetramerous flowers (1 cm across) appear in terminal panicled cymes which are pyramidal in shape. They are very fragrant, the fragrance being reminiscent of tea rose (*Rosa chinensis*), and are used to extract a perfume, which people use as base for local scents. Calyx tube absent to 5 mm, 4-lobed with a spread of some 7.5 mm; petals 4, ovate and crumbled; stamens normally 8 though sometimes 4 and 12; ovary 4-celled with a long style, stigma is capitate, and the capsules are globose, brown, exserted, opening irregularly in four chambers with many seeds, each being some 3 mm across, smooth, pyramidal, hard and brown.

Cultivation

It produces highest dye content in temperatures between 35-45 ^{0}C. During the onset of precipitation intervals, the plant grows rapidly; putting out new shoots, then growth slows. First cutting of the leaves may be done when the plants are two years old and then these are cut at 15 cm height. The plants produce their best yields during the first 4-8 years after planting but are often left in the field for up to 15-40 years. It prefers a dry to well-drained fertile soil at a sunny position. It is tolerant of poor, stony and sandy soils, though removing a large quantity of nutrients from there. Henna removes a large amount of nutrients from the soil, i.e. a yield of 1,000 kg dry leaves removes 180-190 kg nitrogen, 100-150 kg potassium and 10-30 kg phosphorus. Its preference to soil pH is 5.0-7.0, tolerating from 4.3-8.0. For commercial production these are planted one metre apart. Planting is done normally during rainy season by making pits some 40-50 cm deep and these pits are filled with sandy loam soil mixed with well-rotten farmyard or dung manure at the rate of at least two kilogrammes per pit. When plants have attained 40 cm of height, these are gently removed from the nursery, cut back to 15 cm of the height and then transplanted. After planting the plants are watered thoroughly and then after these should be irrigated every week until these are firmly established. Established plants are highly tolerant to drought. The leaves gradually yellow and fall during prolonged dry or cool intervals. It does not thrive where minimum temperatures are below 11 ^{0}C and the temperatures below 5 ^{0}C will kill these plants. The concentrations of lawsone in the leaves vary with climatic conditions, hot localities yield henna with higher lawsone content than temperate areas. There has been a report indicating lawsone to be a degradation and autoxidation product of primary glycosides called hennosides A, B and C. Yield of the dry leaves may be 2-4 tonnes per hectare under irrigated condition though in the rainfed conditions the yield may be 700-1,500 kg per hectare. However, under intensive cultivation, the leaves may be harvested twice in a year from the second year onwards. It would always be better if the plants are given soil feeding of two kilogrammes of well-rotten farmyard or dung manure per pit in the spring.

Propagation

Its minute **seeds** (7,00,000 seeds/kg) germinate best around 25 ^{0}C but seeds are usually pre-germinated before sowing due to their hard seedcoats. Therefore, they are first steeped into water for 3-7 days by changing the water daily and then these are heaped for a few days by keeping these moist and warm. When seeds start swelling, these are sown in the nursery where the soil should be kept moist daily. After germination when plants reach 40 cm tall, they are lifted from the nursery, cut back to about 15 cm of height and then transplanted. In case of vegetative propagation, softwood or hardwood **cuttings**, both root easily but such cuttings should have 6-8 buds. These are also propagated through layers though is not economic as compared to cuttings. Yakandawala and Adikari (2014) studied the propagabilty of henna by taking three maturity stages (softwood, semi-hardwood and hardwood) of 20 cm long cuttings at least with three nodes, and they recorded best rooting, *vis-à-vis* flowering under softwood cuttings.

Insect-Pests and Diseases

This plant is almost free of insect-pests and diseases. Sometimes, **beetles** of the white grubs (*Holotrichia insularis*) have been recorded in Rajasthan feeding on mehndi plants. These emerge in June after the break of the monsoon, feed at night and defoliate the plants. Chlorpyriphos 0.4% or carbaryl 0.1% spraying during the night will kill the adults. **Whiteflies** and **spider mites** (*Tetranychus urticae*) have also occasionally been found feeding on mehndi plants. A **black root rot** caused by *Corticium koleroga* and a **bacterial leaf spot** caused by *Xanthomonas lawsoniae* have been reported from India. Neither by any insect-pest nor by the disease the loss in this crop is significant.

References

Chaudhary, G., S. Goyal and P. Pooniya, 2010. *Lawsonia inermis* Linnaeus: A. phytopharmacological review. *Intern.J. pharmac. Sci. & Drug Res.*, **2**(2): 91-98.

Makhija, Inder Kumar. *Lawsonia inermis* from traditional use to scientific assessment; PHARMATUTOR-ART-1022.

Reddy, K.R. 1988. Folk medicine from Chittoor district Andhra Pradesh, India used in the treatment of jaundice. *Intern. J. Crude Drug Res.*, **26**: 137-140.

Yadav, S., A. Kumar, Jyotsna Dora and A. Kumar, 2013. Essential perspectives of *Lawsonia inermis*. *Intern. J. Pharmaceut. and Chem. Sci.*, **2**(2): 888-898.

Yakandawala, K. and A.M.L.G. Adikari, 2014. *Lawsonia inermis* (Lythraceae): from the wild in the garden. *J. envoron. profess. Sri Lanka*, **3** (2): 48-59.

19

Nelumbo (Nelumbium)

Sanyat Misra and R.L. Misra

Family: **Nymphaceae** or **Nelumbonaceae**

Common names: American lotus/Duck acorn/Lotus lily/Pondnuts/Water chinquapin/Winkapin/Wonkapin/Yanquapin/Yellow lotus (*Nelumbo lutea,* syn. *N. pentapetala*), Chinese water lily/East Indian lotus/Hindu lotus/Indian lotus/Sacred lotus (*Nelumbo nucifera, N. indica, N. speciosa, Nelumbium speciosum, N. nelumbo*), Magnolia lotus (*Nelumbo nucifera* var. *alba*), Nelumbo, Nelumbium, Sacred lotus of the Nile, *Sanskrit* and *Hindi*: 'Ambuj,' 'Jalaj,' 'Kamal,' 'Neeraj,' 'Padma,' 'Pankaj,' 'Pundareek,' 'Purain,' 'Pushkar,' 'Raajeev,' 'Saroj,' 'Toyaj,' 'Vaarij,' etc., *Assamese*: 'Podum,' *Bengali*: 'Padma,' *Gujarati*: 'Suriya kamal,' *Kashmiri*: 'Pamposh,' *Khasi*: Soh-lapudng, *Kanarese*: 'Kamola' and 'Tarareddade,' *Malayalam*: 'Senthamara' and 'Thamara,' *Marathi*: 'Kamal,' *Mundan*: 'Kombolba,' 'Salukidha' and 'Upalba,' *Oriya*: 'Padam,' *Punjabi*: 'Kanwal' and 'Pamposh,' *Tamil*: 'Ambal' and 'Thamarai,' and *Telugu*: 'Kalang' and 'Erra-tamra'

Introduction and Origin

Banerji (2006), and Sanyat Misra and Misra (2013) have given a comprehensive account of *Nelumbo*. Since 'water garden' is a feature of Floriculture where lotus occupies special place under deep water aquatics and its nature of growing in water bearing beautiful flowers categorizes this plant as most important water plant but one should never forget that lotus has perennial rhizomatous rootstock also which naturally puts it under bulbous ornamentals. Moreover, it is also a matter of pride for the 'specialists of bulbous ornamentals' that it is a rhizomatous plant which likes deep water for its growing. Apart from this, *Nuphar, Nymphaea* and *Victoria amazonica* which also occupy special place in water gardens as deep water aquatics also have rhizomatous rootstocks.

Nelumbo is the Ceylonese (Sri Lankan) vernacular name for *Nelumbo nucifera* but usually it is written as *Nelumbium.* The yellow-flowered, *Nelumbo lutea* is native to the New World, i.e.North America (from New York and S. Ontario to Minnesota, Nebraska, Florida, Texas and Louisiana) and South America (West Indies and Colombia) whereas *N. speciosa* (*N. nucifera*) is native to the Old World i.e. Asiatic, distribution ranging through southern Asia, the Philippines, Indonesia, Cambodia, Myanmar, Korea, Japan, China, Taiwan, India, Thailand, Sri Lanka and northern Australia. It is also found in Hawaii. It is naturalized in the Volga River delta where the river flows to the Caspian Sea. The first fossil record of lotus is available with Pleistocene epoch of Kashmir, and subsequently from the early tertiary period rocks of Assam as had been evident through the impressions of leaves and rhizomes of lotus found in a collection from the Eocene bed near Damalgiri, reinforces the belief that lotus is indigenous to Indian subcontinent.

These are bold plants which beautify the water-stagnated area and whose beauty is superimposed during moonlight-nights due to its shadowing effect on water surface provided the pond is free of weeds and has at least 60 cm depth of water. Its mass effect is very effective when planted in scattered patches. Since its rhizomes wander to a great distance (9-12 metres per year) so the clumps are easily formed within a year of planting. As compared to small ponds, their large pond effect is more pleasing and satisfying, *vis-à-vis* provides a very cooling effect during summer. These can also be planted in large tubs or other containers kept at a well-ventilated and aerated place in the house for indoor enchantment, or lowered in the swimming pools for purifying the water. It does well at a sunny and open location in still water or even in slow moving water. Over and above, it is used as an impressive cut flower where its flowers last for 3-4 days in ordinary tap water, as loose flowers for offering to gods and goddesses, on various other occasions and also for making garlands.

Its rhizomes are white to cream or sometimes orange and are known as *kamal-kakdi* (in U.P., M.P., Bihar, Rajasthan), *pein* or *bhein* (in H.P. and Punjab), etc. which are used for pickling and as vegetables being rich in vitamins, fibres, starch, sugar, proteins, fat and calcium. Rhizomes also yield a type of edible flour or arrowroot which is aromatic and sweet in taste. Rhizome flours offer a remedy against diarrhoea and dysentery in children and its paste being anti-fungal is used against ringworm and cutaneous infections. Its tender leaves, pedicels (petioles), flower stalks (peduncles) and even flowers are also cooked as vegetables. Milky and viscid juice obtained from leaves and flower stalk is effective against diarrhoea. The torus (fruiting body) embedded with edible carpels is also sold in the market as *kamalgatta* because of the edible carpels which are sweet for taking raw or by roasting or boiling. The seeds are also dried and ground for making a typical flour to be eaten during fast with milk and water or for loaf making. The flour is rich in ascorbic acid, iron, calcium, phosphorus, fibre, starch, sucrose, fat and protein. Its plants have the ability to absorb heavy metals from the polluted lakes and ponds and purify the water polluted due to industrial effluents, but to a limited extent. The nelumbine alkaloid obtained from its petiole, pedicel and seed embryo is a cardiac poison. Rural folks use its leaves as dining plates on various social gatherings.

History

There are references of lotus (Sacred lotus of the Nile) from the ancient Egypt *vis-a-vis* frequently depicted as Egyptian murals. Egyptians admired and depicted them some 4,000 years B.C. and had offerings to their dear ones on their death as had been evident during Israelitish captivity of Egypt where in the funeral wreaths 'Egyptian blue lotus' and 'white lotus' were found in the tomb of the Egyptian Pharaoh, Rameses the Great II (13th century B.C.). Both these were, in fact, not lotus but *Nymphaea caerulea*, a sky-blue, narrow petalled tropical waterlily and *Nymphaea lotus,* a white-petalled night-blooming waterlily as Egyptians knew these as the real lotus. In fact, *Nelumbo nucifera* was introduced into Egypt in 523 B.C. at the time of Persian invasion and then it became a most common plant at the banks of Nile due to its seeds being farinaceous which were being ground for making a type of flour and was added to milk and water for making loaves but now it has totally disappeared from Egypt which corroborates the belief that in the first place it was not indigenous to Egypt (Anon., 1989). The lotus has a very old and close association with Indian culture and religion, even for more than 6,000 years. Its references are preserved in history, Indian culture, Indian arts and crafts, paintings and frescoes, sculptures, architecture, cave murals, epics and scriptures, in mythological legends and in ancient Sanskrit literature (with the names of jalaj, kamal, neeraj, padma, pankaj, pundareek, pushkar, rajeev, saroj, vaarij, etc.) *vis-a-vis* Hindu religion as from ancient times it is sacred to Hindus. The flowers are given utmost reverence. It is symbolic to the most esteemed position of the human being, head being held high, pure and undefiled in the sun and his feet rooted in the world of experience. The plant is rooted in utter mud from which clean, pure, erect and unblemished flowers and leaves appear which educate us to be simple, honest, pure, straightforward, genuine and sincere in our day-to-day worldly endeavours. To reach highest level of conciousness (the thousand petalled lotus) through yoga, one adopts lotus posture. It is sacred even to Buddhist and Jainism as in their ancient literature there is mentioning of lotus. Lord Buddha is said to have appeared in this world on the leaves of a lotus and the plant followed the spread of Buddhism as far as China and Japan. It is also considered a symbol of life as in very early civilizations the flower was an emblem of female beauty and fertility. Moreover, it is only Buddhist faith which discovered the 'water garden' in real sense as Britishers did not have the concept and they were making ponds for fishes to feed the people of monasteries, but Buddhists discovered it as in front of the Bodh temples they used to create lively water gardens full of lotus and waterlilies. In the end of 19th century when Sir Latour Marliac of France created a large variation of shape, size and colour in waterlily flowers with differing lengths of pedicels (flower stalk) *vis-à-vis* a large variation of shape and size of the leaves, coupled with invention of cement at the same time at a different place, the real water garden came into a proper form. Dravidians, natives of South India evidently used lotus flowers as their food, during pre-historic period, i.e. before Indus Valley Civilization (3,000-2,000 B.C.) of the 'Mohen-jo-dero' and 'Harappa' period where this plant is no where mentioned. The earliest literary reference of lotus is made by the Aryans in the *Rigveda* (2,000-1,500 B.C.). However, afterwards its mentioning is discovered in poems, plays, *shlokas, kathas, Vedas, Upnishads, Puranas* and in other ancient Sanskrit literatures. In Hindu religion, the lotus is the centre of universe, flowers representing the country and its leaves surrounding cultures and the countries. There is a belief that the flower seated with Brahma (the Creator) in the centre arose from the navel of the Great God, the 'Vishnu'. The Brahma's role was to recreate the universe after the great flood (*pralaya*) which he did by using various parts of lotus. In the Hindu pantheon, the other two brahminical deities, Vishnu (the Protector) and Shiva (the Destroyer) are also associated with lotus. The fourth arm of Lord Vishnu (three arms holding the conch '*shankh*', the wheel '*chakra*' and the club '*gada*') holds a lotus '*padma*', a symbol of the created universe. There is a myth that Lord Brahma when was once sitting at a waste water side, he noticed a solitary lotus leaf floating in the moving waters then he dived in the water in the form of a boar to find the earth below, from where he broke out a chunk which he brought out on the surface of water, and spread it out on that lotus leaf which had been the beginning of our earth. The lotus is also associated with goddess Lakshmi who sits on a fully opened flower holding one lotus flower in her right hand. The epic Mahabharata also mentions that Lakshmi emerged from a lotus which appeared from the Vishnu's forehead. *Pattupatti* classic written during 500-600 A.D. in Tamil also mentions that lotus rose from the navel of god Vishnu. All the gods and goddesses of Hinduism [Brahma, Vishnu, Mahesh (Shiva), Durga, Lakshmi, Saraswati, Parvati, Ganesh, Rama, Agni and Surya], Buddhism (Buddha and Bodhisattva), and Jainism (Vardhman Mahavir and others) are associated with lotus. Almost all the famous monuments and temples have depiction of lotus in one form or the other and various poets, artists, sculptors, kings and Mughal emperors of contemporary and ancient periods were inspired by lotus. Its prominence continues even in the modern times as one 'Lotus Temple' has been created in Delhi by Bahai faith. Still it is a custom to offer 108 lotus flowers while worshipping goddess Durga during 'Durga Puja Festival'. This is why the lotus flower has been bestowed with 'National Flower of India'.

Botany

Nelumbo has long chains of tuberous rhizomes creeping underground and growing to several metres, up to 12 metres. The first young leaves when emerging out of the water are tightly rolled and come out forcefully even piercing its own parts and on opening these are round, flat and surface-floating, followed by waxy and bluish-green adult leaves attached with quite long (60-240 cm) cylindrical leaf stalks in its centre and on which leaves of inverted parasol-shape remain intact usually well above the water level. The diameter of adult leaves is from 45-90 cm with frilled edges, being larger in *Nelumbo nucifera* than *N. lutea*. Sprinkling of water on the leaves appears as mercury beads, and during rains the rain drops merge and collect in the centre but when it becomes heavy, leaves tilt to one side to expel the collected water. The flowers are fragrant, single, quite showy, large, magnificent, velvety white, cream, greenish-white, yellow, sulphur-yellow, pink, rose, rosy-carmine, red, mauve, etc. but the true species bears sulphur-yellow flowers with dark yellow stamens in *Nelumbo lutea* (the American species), and deep rose (on opening) to pale-rose and then creamy-white (with age) with yellow stamens in *N. nucifera* (the Asiatic species), flowers being larger in the Asiatic species and are borne on a strong peduncle which is higher than the leaves. Sepals are 4-5, petals many (usually 16-36; rarely up to 160 as in *N. nucifera* var. 'Krishna', a pink double race collected by NBRI, Lucknow from a pond in Midnapore, W.B.), erect-spreading, stamens many with short filaments, and fruit a large, very attractive, funnel-shaped but flat-topped with perforated receptacle containing many carpels. The funnel-shaped seed pod which is up to 10 cm in diameter looks like the rose of a watering can but with flat and cushioned tops, initially yellow but maturing to a brown woody texture and this is also used in the dry flower arrangements. The surface of these pods contains many (7-15) edible nut-like seeds individually in its sunken cavities. The length of viability of its seeds is enormous as a paleobotanist of Japan collected its seeds from Manchuria estimated to be 1,000 years old and these successfully germinated. Its centuries old seeds collected from various tombs also germinated. Sherriff and Taylor (1948) also state on the viability of its seeds exceeding that of any known species of flowering plant as they mention that seeds estimated to be nearly 400 years old have been found viable. The basic chromosome number is n = 8. Sen and Bhaduri (1976) reported the diploid chromosome number 2n = 16 in *Nelumbo nucifera* vars 'White' and 'Pink'.

Species and Varieties

What details have been given above about *Nelumbo* holds true for the species, *viz., N. lutea* and *N. nucifera* including the colour of the flowers except that the leaves, the flowers, the petioles, the peduncles and the rhizomes in latter species are significantly larger as compared to the former. Bailey (1960) and Beckett (1987) have mentioned that the genus *Nelumbo* comprises only two species, *N. lutea* (*N. pentapetala, Nelumbium lutea*) from USA and *N. nucifera* (*N. speciosa*) from Asia. Many literature sources mention certain other species but those are in fact the various forms of these elemental species. *N. lutea* is though bold and quite beautiful but could not gain that popularity which it deserved. *N. flavescens* bears a strongly scented soft lemon-yellow flower and introduced by Sir Marliac is a smaller form of *N. lutea* and has a red spot on the base of each petal and centre of each leaf, and var. 'Grossherzog Ernst Ludwig' (*N. flavescens* x *N.* 'Osiris') bears salmon-rose flowers. *N. alba* is in fact a form of *N. nucifera* having white flowers, *N. n.* var. *alba grandiflora* (syn. *N. grandiflora*) has white flowers with golden stamens but larger with flowers (more than 30 cm in diameter), *N. n.* var. *alba striata* (syn. 'Empress') produces lovely fragrant and globular flowers where petals are tipped red with red stripes on edges, *N. n.* var. *alba plena* (syn. 'Shiroman') is a Japanese variety with large, globular, fully double and creamy flowers tinged with faint green but become snow white with age, the flower size under optimum conditions is almost 30 cm in diameter and most suitable as cut flower when the cut end is sealed, *N. n.* var. *pygmaea alba* is a dwarf variety with about 15 cm diameter of leaves and 10 cm across flowers of white colour, *N. n.* var. *pygmaea alba plena* is a dwarf and double variety of white colour, *N. n.* var. *pygmaea rosea* is a dwarf variety with flowers of bright rose colour, *N. n.* var. *rosea* produces deep rosy-pink flowers, *N. n.* var. *rosea plena* bears double and deep rosy-pink flowers, *N. n.* var. 'Osiris' produces deep rose flowers, *N. n.* var. *pulchra* bears dark rosy-red flowers, *N. n.* var. *pekinensis rubra* bears rosy-carmine flowers, *N. n.* var. *pekinensis rubra plena* produces very large, double and rosy-carmine flowers, *N. n.* var. *gigantea* bears very large and rose-purple flowers, *N. n.* var. 'Kinshiren' bears flowers of white colour shaded with pink, *N. n.* var. *kermesina* with light rose flowers, vars. 'Angel Wings' produces tulip-shaped large (20-25 cm flower diameter) fragrant and white blooms where inner petals are beautifully rolled, 'Chawan Basu' is a semi-dwarf and free-flowering variety suitable for tub culture where white flowers are edged with pink, 'Charles Thomas'- dwarf pink variety suitable for small tubs and pools, 'Maggie Belle Slocum'- large (25-30 cm across) and mauve flowers with inner petals beautifully rolled and suitable for tub culture, 'Momo Botan'- small plants with 30-45 cm across leaves, flowers fully double, paeony-shaped and carmine-red, and suitable for small pools and tubs, 'Mrs Perry D. Slocum'- most beautiful and vigorous among all the existing varieties with double flowers about 30 cm across, opening rose-pink but changing to creamy-yellow with pink edges and then to complete creamy-yellow, all the stages appealing quite good but the second stage is most attractive, etc. Across the country from Gujarat (tropical to subtropical) to Sikkim (sub-temperate and temperate) and Andaman & Nicobar Islands (absolute tropical), and from Kashmir (temperate, subtemperate and subtropical) to Cape Comorin (absolute tropical), the lotus plants can be found growing in the natural way in plenty in the ponds, lakes, water reservoirs and even in quite slow moving waters predominantly with rose colour though white to purple forms are also available frequently.

These are mostly found growing along with waterlilies. In West Bengal and Orissa, almost every pond has this plant and that too with great variation in flower colours. Most of these races (some 35) of *Nelumbo nucifera* collected from 11 states of India and 25 races collected from the botanical gardens of Australia, Brazil, Japan, Thailand, United Kingdom and United States of America are being maintained at NBRI, Lucknow. There is much scope of improvement in this crop using both the species *vis-à-vis* their existing varieties and the result will be overwhelming when both the species are combined together.

Propagation

Lotus is propagated vegetatively through rhizomes and sexually through seeds. The multiplication through rhizome is very easy as the rhizomes having three nodes with at least one active bud are planted horizontally 7-10 cm deep in natural pond's or lake's soil which is normally most suitable for their growth and multiplication. In case of artificially created pools, the bottom should be filled to a depth of 45-50 cm with 3 parts clayey soil being rich in organic matter and 1 part of FYM, plus 100 g each of *neem* cake and diammonium phosphate per square metre area. The water level at any time should not be less than 30 cm. When flowers start emerging these are pulled up at bud stage to divert this energy towards rhizome formation. By the end of October the rhizomes may be lifted and kept at warmer and well aerated place for starting the buds. These are now ready for marketing. The yield of rhizome is approximately 30-50 quintals per hectare depending upon the variety, climate and the soil conditions.

Through seeds these are grown in February-March in tropical and subtropical conditions but in April-May in subtemperate and temperate conditions. The viability of lotus seeds is extremely high, manifold than any other flowering plant. Scarified (sand paper rubbing at both the ends) and stratified (in H_2SO_4) seeds are sown for better results. Properly scarified seeds germinate within a week at room temperature, and after next 10-20 days the seedlings become ready for transplanting. At 10 days of germination these attain 2-leaf stage. Transplanting is done in the pools having very shallow water but with the growth the level of the water will have to be increased. For planting one hectare of land, 10-12 kg seeds are required for nursery raising.

Cultural Practices

The cultivation of lotus is most simple. Wherever these are to be cultivated, it should be at an open and sunny situation with the pH reaction of the water as well as of the soil not more than 8.5, preferably 5.5 to 7.0. The rhizomatous rootstocks are planted when growth on them is visible. However, if before initiation of growth the rhizomes are dug up they should be kept at a warm place for starting the growth or should immediately be planted in an open pond having warmer water. The best time for planting in the subtropical, subtemperate and temperate areas is April when water temperature has risen substantially. However, in the tropical climates it can be planted throughout the year, preferably when most of the plants from a lotus pool start the new growth. This period normally occurs under tropical conditions after about one month of cessation of flowering of most of the plants. It should not be planted at a situation where water may freeze up to the bottom as their rhizomes rot. Even in 'Dal Lake' of Srinagar there is one lotus garden. Wooler and many other lakes in Kashmir harbour lotus gardens in them. The rhizomes having at least three nodes (at least one node should have started) are set horizontally 7-10 cm below the soil at a shallow bank of the lake or pond in a manner that these may not move from its place of planting so after planting a brick or stone should be laid over the rhizome to keep the rhizome securely in position. If by any reason the planting is delayed, these may be planted timely in pots or other containers from which these plants should be planted in the pond when congenial time comes. The natural soil and deposits in the lake is normally the same which these plants require. In created ponds, the bottom of the pool should be filled with a thick layer (approximately 45 cm deep) of 2 parts fibrous loam mixed with 1 part of well-rotten cow dung or farmyard manure or 3 parts clay soil and 1 part FYM, and per square metre of area some 100 g each of *neem* cake and diammonum phosphate should be mixed. In case if properly decomposed manure is not available and there is sufficient time for planting, the fresh dung and the litters should be stacked in layers with clayey soil in the prepared pit which may be turned over twice or thrice during spring for proper decomposition. For tub planting also 2:1 fibrous loam and well-rotten cow dung manure should be used but during active growth the plants should frequently be fed with liquid fertilizers. Wester (1951) When used 10-6-4 fertilizer containing ammonium sulphate supplied at 25-50 g per cubic feet of soil in case of *Nymphaea zenkeri* var.; 'H.C. Haarstick', recorded superiority in case of leaf area, flower size than the same combination containing nitrate of soda or cow manure. Timely planted pools may start producing flowers from April end and may continue blooming up to September. Seed setting also takes place by September end to October. The flowers last for about four days and then one by one petals start shedding causing protrusion of the yellow torus with flat but spongy top which turn to green after some time and then woody brown at maturity bearing 7-15 seeds per fruit. Rhizomes can be harvested by October end.

Insect-Pests, Diseases and Weeds

Weed is not a problem when lotus has been grown in tubs as the growing weeds are pulled up gently but in ponds since other livestocks like fish, frogs and other animals are there so spray of chemicals may prove injurious to such animals. However, Simazin 3-6 ppm or Monuron 4-12 ppm is safe for lotus *vis-a-vis* livestocks and will also kill the weeds. In ponds, water hyacinths (*Eichhornia crassipes*), water lettuce (*Pistia stratiotes*), azolla, *Typha*, and *Cyperus papyrus* may sometimes create the problem out of which *Typha* and *Cyperus* can easily be removed by pulling them, and to a greater extent even water hyacinths may also be collected and taken out but azolla algae and the blanket weeds can be controlled only by the use of $KMnO_4$ and if there is no fear of livestocks even $CuSO_4$ can also be used.

Lotus leaf spot is caused by *Cercospora* and *Ovularia* species, starting as dark patches on the leaves or drying of the leaf margins. Such leaves should immediately be removed and burnt. At initial attack a mild solution of Bordeaux mixture will also prove effective if sprayed at every 2-3 days. Sometimes even root rot appears due to attack of *Phytophthora* species which is very difficult to control, however, such plants should be uprooted alongwith the adhering soil and destroyed.

Small dark brown beetles (*Galerucella nymphaeae*) which are almost double the size of ladybird, attack the lotus during summer. Though its attack is not so common but once these attack the crop and if these are not checked in time these build up their population (several generations occur in one season) soon and become highly devastating as larvae feed on the leaves. These can be dislodged in the water by jetting the foliage with fresh water where fishes and frogs will swallow them. A solution containing powdered *Bacillus thuringiensis* bacteria when sprayed on such plants parasitize the larvae and kill them. Various aphids such as *Rhopalosiphum* species or greenflies feed on the leaves sucking the sap which hampers the vigour of the plants. Jetting the foliage with fresh water will dislodge them in water to be swallowed by the inhabitants of water. Slugs and snails may also feed on the leaves and flowers which can be trapped by floating lettuce or cabbage leaves in the vicinity for 24 hours and all these pests will start feeding on these leaves and then these are collected and destroyed. Leaf stalk borers are quite damaging for the crop which can be controlled by cutting back the affected parts and destroying them, and then the new growth will appear soon.

References

Anonymous, 1989. *Nelumbo* - Lotus. In: *Collins Guide to Waterlilies and other aquatic plants*, pp. 99-105. Text and photographs by Stapeley Water Gardens. Published by William Collins Sons & Co. Ltd., London.

Bailey, L.H. 1960. *The Standard Cyclopedia of Horticulture* (vol. II), pp. 2117-2118. The Macmillan Co., New York (USA).

Banerji, B.K. 2006. *Nelumbo*. In: *Advances in Ornamental Horticulture* (vol. 3): *Bulbous Ornamentals and Aquatic Plants* (ed. Bhattacharjee, S.K.), pp. 311-321. Pointer Publishers, Jaipur (India).

Beckett, K.A. 1987. *The RHS Encylopedia of House Plants Including Greenhouse Plants*, p. 356. Salem House Publishers, Massachusetts, USA.

Sanyat Misra and R.L. Misra, 2013. *Nelumbo* (*Nelumbium*). In: *Commercial Ornamental Bulb Science*, pp. 306-312. Westville Publishing House, New Delhi.

Sen, S. and P.N. Bhaduri, 1971. Cytotaxonomy of Nymphaceae. *J. Cytol. Gent.*, **6**: 67-89.

Sherriff, G. and G. Taylor, 1948. The sacred lotus- *Nelumbo nucifera*. *J. roy. hort. Soc.*, **73**: 216-217

Wester, H.V. 1951. Comparative growth response of Haarstick water lily to ammonium and nitrate nitrogen in 10-6-4 fertilzer as compared to cow manure. *Proc. Amer. Soc. hort.Sci.*, **58**: 324-328

20

Nerium

B.K.Banerji, R.L. Misra, Atul Batra and Sanyat Misra

Family: **Apocynaceae**

Common names: Oleander, Rose bay, St. Joseph's flower (*Nerium oleander*), Sweet-scented oleander (*N. odorum*)

Introduction and Origin

Nerium is the ancient Greek name for oleander, bestowed by Linnaeus in 1735. Even otherwise *neros* in Greek signifies water due to its being found growing wild in the swampy regions. In western Europe it has been in cultivation for more than 400 years and in China and Japan for certuries. The genus was known to the Greeks. John Gerard in his garden in England had a white and a red variety in 1596. John Parkinson was also growing this plant in his garden through the seeds he had received from John More of Spain in 1629. The ancient city of Volubilis in Morocco took its name from the old Latin name of the flower. Its common name in Italy and France is St. Joseph's flower as it is said that when St. Joseph seeking the hand of Mary went to a temple and placed his staff among other people, it immediately burst into leaf and flower and then the priest thought only St. Joseph worthy of Mary and he married both of them. Oleander is native of North Africa, the Eastern Mediterranean basin and South East Asia, especially Iran, India and Japan. It is so widely cultivated that no precise region of origin, perhaps in southwest Asia has been identified (Pankhurst, 2009). However, Bingtao *et al.* (2009) suggested that oleander is native or naturalized to a broad area from Mauritania, Morocco and Portugal eastward through the Mediterranean region and the Sahara, to the Arabian Peninsula, Southern Asia and as Far East as Yunnan in South China. In the Atlas Mountains they are found growing up to 1,800 metres elevation. These produce brilliantly coloured blooms in the range from white to yellow to blood-red to crimson in large trusses.

Plate-1: Different cultivars of *Nerium* in blooming stage and with their close up views of flowers

Uses

The plant is ideal dust protector and can be seen in the dividers of the highways to clean the atmosphere from the dust apart from providing beauty through its evergreen nature and beautiful flowers emitting fragrance in certain cases. These are most suitable for planting in dust-polluted areas, at railway stations and along the railway track, outside bars and cafés, beaches, neglected terraces, as hedges, in pots and tubs, in the corner of a garden in groups or alone and on the banks of rivers, etc. It is very hardy and versatile plant which if once established does not require further care. Plant is ever-blooming, but flowering is very poor during winter under subtropical conditions. It is most commonly known as oleander, from its superficial resemblance to the unrelated genus *Olea* (olive) which in Latin means oil as *Nerium* when is infected with some serious disease, probably bacterial, though appearing healthy but such plants exude an oil with rancid odour from its basal part of the trunk and the exposed or shallow roots, saturating the soil with this oil in the vicinity and probably due to this reason it may have been named oleander.

Toxicity

Of all the garden plants, it is one of the most poisonous plant in all its parts including nector and its dead wood as these contain numerous highly toxic compounds causing poisoning to animals and people, especially young children (Goetz Rebecca, 1998). The smoke by burning its plants or any of the plant parts is also poisonous. A 10-20 leaves consumed by an adult may prove fatal and even a single leaf could be lethal to an infant or child though if taken in wine with rue they are effective against snake venom. In Libya, the natives make poultices of chopped up leaves to bring abscesses to a head and an ointment is also made for curing scabies. If its leaves are steeped in a pond or water reservoir, it proves lethal for sheeps and goats when they drink it. Despite a lack of any proven benefits, a range of oleander-based treatments are being promoted on the Internet and in some alternative medicine circles, drawing a warning letter from the U.S. Food and Drug Administration (Anonymous, 2010). Oleander in all its sappy parts contains the toxin oleandrin and neriine (cardiac glycosides; Knight, 1999) therefore its ingestion may cause gastrointestinal as well as cardiac effects, whereas its bark contains rosagenin known for its strychnine-like effects which with skin contact causes reactions. The gastrointestinal effects are seen as nausea and vomiting, excess salivation, abdominal pain, diarrhoea (with or without blood), and colic disorders, especially in horses (Trevino Monica, 2009). Cardiac reactions consist of irregular heart rate, sometimes characterized by a racing heart at first and then slowing to below normal or the erratic heart beat. This also affects the central nervous system where drowsiness, tremors or shaking of the muscles, seizures, collapse and even coma may occur which can lead to death. Oleander sap in skin contact causes skin irritations, severe eye inflammation and irritation, and allergy reactions characterized by dermatitis. Florentines at one time called the oleander *ammazzclasina* meaning donkey killer. In 1796, some French soldiers in Corsica died instantly after eating some meat skewered and cooked on a stick of oleander. Sleeping on the pallets made by oleander branches also causes serious sickness and death. According to the Toxic Exposure Surveillance System (TESS) in 2002, there were 847 cases of its poisoning reported in the United States. There are innumerable reported suicidal cases of consuming mashed oleander seeds in southern India. Around 0.23 mg per pound of body weight it is lethal to many animals. Oleander poisoning requires immediate medical attention (Anonymous, 2005). Induced vomiting and gastric lavage are protective measures to reduce absorption of the toxic compounds. Charcoal may also be administered to help absorb any remaining toxins (Watson, 2003). Further medical attention may be required but will depend on the severity of the poisoning and symptoms. Temporary cardiac pacing will be required in many cases (usually for a few days) till the toxin is excreted.

Botany

The plant is an erect, glabrous and evergreen shrub with cylindrical stem attaining the height of up to 6 metres when growing naturally, is fast growing with a good spreading habit, branches being quite smooth, and the linear-lanceolate, simple, entire and leathery leaves that are 6-20 cm long, being produced in opposite pairs or whorls of 3 to 4 or more (Bailey, 1960; Beckett, 1987), and are usually erect and beautifully dark green with a paler central vein though in variegated types the foliage colour is creamish along with various combination of green. Inflorescence terminal, with compound cyme having up to 80 flowers, flowers often fragrant, 3-5 cm across in terminal cymes, calyx 5-lobed and about $^{1}/_{4}{}^{th}$ length of corolla, corolla salver shaped, opening to 5 flat but right-twisted lobes, tube slender-cylindric, throat dilated, limb 2.5 to 5.0 cm in diameter, and the colour of flowers ranging from white to red via cream, yellow and pink, the stamens attached below throat, anthers plumose and are 2-tailed at the base and tapering at the apex into long thread-like appendage, style is 1 but ovaries 2 and the fruit is long narrow pod (follicle) measuring 5 to 23 cm with 2 chambers filled with a large number of twisted and downy seeds. Over 400 varieties are described in colours ranging from white to cream, pink, lilac, red, carmine, purple, yellow, salmon, apricot, flesh, copper and orange, several with variegated leaves and variegation being from white to yellow with usually rose-pink double flowers. Mitotic and meiotic chromosome number in various population of oleander showed that the basic chromosome number in species is 11 (2n=2×=22, Noori-Daloii *et al.*, 1996; El-Shazly, 2002).

Species and Varieties

Nerium has been defined with 2 to 3 species (Beckett, 1987) and over 400 named cultivars in various colours such as white, cream, light yellow, yellow, pink in various shades, red and violet-red as well as doubles, dwarf-compact and variegated ones. Bailey (1960) and Beckett (1985) described two species, *viz. N. odorum* (syn. *N. indicum*; Pizzetti and Cocker, 1975), a stout and erect shrub native to Iran, India and Japan which bears leaves in three's, flowers rosy-pink, some 5 cm across, in clusters of up to 80 and sweetly fragrant, anther's appendages protruding and segments 4-7; and *N. oleander* (syn. *N. lauriforme* or *N. laurifolium*), a native of Mediterranean region and Orient and most tropical and subtropical areas of the Old World bears leaves in pairs or three's, flowers originally single white though varieties with single and double flowers in range of white, cream, various shades of yellow, pink, reddish-orange, apricot, red, ruby-red, dark wine-red, crimson, carmine, reddish-purple and bicoloured, anther appendages seldom protruding, segments 3 to 4-toothed, and its var. *Loddigesii* bears variegated flowers in white, pink and red; while Hay and Beckett (1971) described three species. Beckett (1985) mentioned the varieties of *N. oleander* as 'Album Plenum' (double white), 'Luteum Plenum' (double creamy-yellow), 'Roseum Plenum' (double pink) and 'Variegatum' (leaves margined cream, flowers pink). Bailey (1960) mentioned the varieties of *N. oleander* as *N. album, N. atropurpureum, N. carneum* and *N. roseum*. Pizzetti and Cocker (1975) mentioned varieties as 'Cardinal' (best among all the reds), 'Mrs. Swanson' (double pink) and 'Sister Agnes' (pure white) apart from certain varieties as mentioned above by others. In India, throughout the country except the temperate regions the varieties are available as tall and dwarfs, loose and compact, the variegated ones, and singles and doubles in various colours, *viz.* white, cream, orange-pink, various other light to dark shades of pink, rose, red, carmine and purple-red. This is one of the most homely flowers of the country for worshipping, therefore, in the villages one may find it growing easily near the temples and wells as an easy access to its flowers for offering to gods and goddesses though it is never used in making the garlands due to its toxicity.

Propagation

It can be grown through seeds sown during spring at warm places or in the warm greenhouse at 16-18 ^{0}C temperature but flower colour may be different (Huxley *et al.*,1992). Vegetatively the plant is propagated through hardwood and semi-ripe 10-15 cm long cuttings which root easily if planted in moist sand during summer or at 18 ^{0}C bottom heat and kept at a humid place or under mist. Cuttings root well in a medium containing equal parts of peat and sand by volume. Cuttings inserted in a bottle filled with water and then bottles buried upto the neck at a sunny position will also help it rooting well. Rooted cuttings are planted in the soil medium but under filtered situation. However, through cuttings the plants should be allowed to flower only in the fourth year and not before. The third alternative is ground layering or grafting during summer, especially in case of good varieties where one gets the larger size of plants to flower in the second year, so this way one can save a lot of time in its propagation.

Cultural Practices

Oleanders are hardy from zones 8 to 10 but damage can occur at freezing **temperatures**. It is very suitable for growing in the regions experiencing more than 4.5 ^{0}C temperature, better if temperature is even above 7 ^{0}C. These can survive short period of frost also. However, these have tolerance to cold to some degree as in extremely cold winters in subtropical regions of the country the plants may completely defoliate though death is rare. Damaged plants may require pruning to re-shape the canopy and to remove dead limbs. Full **sun** is always better for its cultivation but is tolerable even to partial shade though its preference is for dry and warm climates. Oleanders are extremely tolerant to a wide range of **soil** types, from heavy clay to well-drained sandy loam, poor, dry and wet soils. Best growth can be obtained when plants are watered in time of drought. The plants can also tolerate relatively high levels of sodium chloride and similar salts in the soil. These offer an excellent alternative in too alkaline (calcareous) soils where even azaleas also can not be grown. In *Nerium,* the flowers are borne on one year old canes which should therefore be well ripened so that before flowering these are in order to set many strong buds. However, after every flowering the undesired branches should be removed to send out new growths. Some gardeners severely **prune** the plants in early to late fall to stimulate new growth though this practice is not necessary. Moreover, this is also grown in several different forms including rounded shrubs and small trees so to maintain these forms the pruning is effected. Removing the dead, damaged and sick branches whenever these occur can improve plant health. Oleanders can withstand severe pruning and many mature, established plants have been rehabilitated through the use of careful pruning. Once properly established these become care-free and do not demand anything afterwards but for their excellent growth and flowering regular **watering** during excessively warm seasons and sporadically during winters should be followed though even without watering these survive as these are highly drought tolerant. Though these are not demanding but regular **fertilization** will make the specimens superb bloomers. While planting, each pit will have to be incorporated with some 4-5 kg of compost or farmyard manure and that would be sufficient for them for one year. Every year before flowering the soil may be incorporated with the same quantum of FYM in the pit. If further need be, some 15-20 grammes of nitrogen per pit may be applied. For planting, the pit size should be 60 × 60 cm and the **distances** apart 3 × 3 m for flowering though for making hedges these should be planted at 25 cm distance.

Insect-Pests and Diseases

Oleander **scale** (*Aspidiotus nerii*) has been found attacking a wide range of plants such as oleander, orchid vine, palm, peach, olive, pear and ivy and its attack speckles the leaves which turn yellow and fall off. Through use of its natural enemies, the horticultural oils and contact insecticides this can be controlled successfully. Scales can also be sponged off. The bright orange larval stage of green-shaded adult polka-dot wasp moth (*Syntomedia epilais*), commonly known as **oleander caterpillar**, feeds on new shoots and foliage and its severe infestation defoliates the entire plant thus making it more susceptible to scale insects and diseases. These can be killed through hand picking and killing them by dipping into soapy water. Spraying with 0.2% Metacid-50 will also kill them. **Aphids** (*Myzus persicae*) are a problem especially in the tropical regions having continuous warmer climate. Bright yellow and black aphids suck the plant sap secreting the honeydew which attracts the sooty mould. Spraying with nicotine decoction, eucalyptus oil, garlic extract, natural sprays and insecticidal soaps controls the aphids. The plant has also been found being infested by **mealy bugs** which can be easily dislodged by the hose.

Oleander is infected by a number of pathogens such as **cotton root rot** (*Phymatotrichopsis omnivora*, syn. *Phymatotrichum omniverum*) which is a very serious fungus of oleander where first symptom appears as leaf wilting and then of the whole plant still attached with the leaves, and this becomes very serious during summer months. The infection has been observed only on mature plants with infection on entire tap roots. Though there is no any consistently effective control but one may get an alternative only in resistant varieties to be planted in the soil infected area. **Sooty canker** is caused by the fungus, *Hendersonula toruloidea*, and as the name implies, it causes canker, wilting and dieback of the branches. Leaves on affected branches are often small and wilt during summer. Brownish and moist areas appear on limbs during its first infection, and afterwards the bark here starts cracking or peeling away revealing brown to black masses of fungal spores and staining of bark and sapwood and then their death afterwards. Sunburnt spot on the limbs become highly susceptible to its infection therefore pruning should be carried out only if necessary and that too in the manner that no part of the branch is exposed directly to sun. **Bacterial gall** is caused by *Pseudomonas syringae* cv. *savastanoi*. Bacterium being systemic enters the plants through wounds or frozen parts and causes gall formation on stems, leaves and flowers generally after spring rains, and its infection becomes severe after a cool wet spring. Pruning the affected parts 20 cm below and disinfection of pruners in 10% common household bleach solution after every cutting will prevent its spread. *Xylella fastidiosa* bacterium causes **oleander leaf scorch** by colonizing the xylem tissues. Its symptoms appear during summer months on one or certain other branches as initial yellowing of leaves, followed by browning and necrosis of the tips and margins. Careful destroying of such branches and treatment with an antibiotic oxytetracycline will keep this disease under check. There is improper or not at all **bud-opening** due to imperfect ripening of the woods, therefore, it would be better to take the flowers only fourth year of its planting.

References

Anonymous, 2005. *Nerium oleander* L. (PIM 366). International Programme on Chemical Safety: INCHEM.

Anonymous, 2010. *Nerium oleander* L. *Germplasm Resources Information Network*. USDA. http://www.ars-grin.gov/cgi-bin/npgs/html/taxon.pl25229 .

Bailey, L.H. 1960. *Nerium*. In: *The Statndard Cyclopedia of Horticulture* (vol. II), pp. 2138-2139. The Macmillan Company, New York, USA.

Beckett, K.A. 1985. *Nerium*. In: *The Concise Encyclopedia of Garden Plants*, pp. 262-263. Orbis Publishing Ltd., London.

Beckett, K.A. 1987. *Nerium*. In: *The RHS Encyclopedia of House Plants including Greenhouse Plants*, p. 360. Salem House Publishers, 462, Boston Street, Topsfield, Massachusetts, USA.

Bingtao, Li, J. Antony, M. Leeuwenberg and D.J. Middleton, 2009. *Nerium oleander* L. *Flora of China*. Harvard University, England.

El-Shazly, M.M. 2002. Observation on oleander (*Nerium oleander* L., Apocynaceae). *Ecosystem in Gija*, Egypt. Proc. 4th intern. Conf. on Urban Pests (eds Susan, C. Jonee, Jing Zhai and W.U.H. Robinson), pp 225-234.

Noori-Daloii, M.R., R. Aliyari and H. Ebrahimzodeh, 1996. Study of chromosomes and soluble proteins of *Rhazya stricta* Decasine and *Nerium Oleander* L. *J. Sci. I.R.*, Iran., **7**(4): 201-216.

Goetz Rebecca, J. 1998. Oleander. *Indiana Plants Poisonous to Livestock and Pets.* Cooperative Extension Service, Purdue University. Oleander Poisoning. *MedlinePlus.* U.S. National Library of Medicine (http://www..nlm.nih.gov.medlineplus/ency/article/002884.htm).

Huxley, A., M. Griffiths and M. Levy, 1992. *The New RHS Dictionary of Gardening.* Macmillan.

Knight, A.P. 1999. Guide to Poisonous Plants ~ Oleander. Colorado State university, USA.

Pakhurst, R. 2009. *Nerium oleander* L. *Flora Europaea.* Royal Botanic Garden, Edinburgh, U.K.

Pizzetti, I. and H. Cocker, 1975. *Nerium.* In: *Flowers ~ A Guide for Your Gardens*, pp. 898-904. Harry N. Abrams. Inc. Publishers, New York.

Trevino Monica, 2009. Dozens of horses poisoned at California farm. CNN, Crime.

Watson, W.A. 2003. *Annual Report of the American Association of Poison Control Centers Toxic Exposure Surveillance System. J. Emergency Med.,* **21**(5): 353-421.

21

Nyctanthes arbor-tristis

Jayoti Majumder, Pragya Ranjan, Babita Singh, R.L. Misra and Sanyat Misra

Family: **Oleaceae** or **Nyctanthaceae**

Common names: Coral jasmine, Night flower, Night flowering jasmine, Night jasmine, Queen of the night, Sad tree, Tree of sadness, Tree of sorrow, Bangla-*Sephalika/Seoli* , Gujarati-Jayaparvathi, Hindi-*Harsingar*, Kannada-Parijatha, Malayalam-*Paarijatakam*, Manipuri-*Singarei*, Marathi-Parijathak, Oriya-Gangasiuli, Sanskrit-*Paarijat,* Tamil-*Pavizhamalli,* Telugu-Pagadamalle, Indonesian/Sundanesse/Javanese-Srigading, Lao/Sino-Tibetan-Salikaa, Malay-Seri gading, Thai-Karanikaa, Vietnamese-Lal tau)

Introduction and Origin

The genus *Nyctanthes,* a native of India though now recorded elsewhere too [Southern Asia, stretching across northern Pakistan and Nepal through northern India to southeast Thailand (Champa Rani *et al.*, 2012)], is represented usually by one member, *N. arbor-tristis* which is a small mythological tree (growing to a height of up to 9-10 m with a life span of 5-20 years) bearing white to cream and highly fragrant flowers, expanding by evening and falling off before sunrise. It is the official flower of the state of West Bengal (India) and for Kanchanaburi Province (Thailand). In Tripura, folk people predict rainfall and weather variation through flower phenology of *Nyctanthes* which help them to plan agroforestry activities and disaster prevention (Acharya, 2011). Two more species are reported which are described further under the column '**Species and Varieties**'. It is often planted near Hindu temples in India and Sri Lanka as well as in Malaysia and Indonesia. As per Hindu mythology, this is a tree from heaven brought to earth by Lord Krishna, and when he found his two wives Satyabhama and Rukmini quarrelling themselves as to where this should be planted, he planted it in the courtyard of Satyabhama in such a way that the tree should shed its flowers even in the courtyard of Rukmini. There is another legend that a princess known as Paarijataka fell in love with the sun but when sun deserted her she committed suicide and from there a tree sprung which blossoms only during nights and sheds the flowers (as the tear-drops) before the dawn, just to hide it from the sight of the lover (sun). Its flowers are much like jasmine in fragrance, colour and shape, and are valued as loose flowers. It is commonly found in rocky ground as the under-growth in down hillsides in dry areas of India in the outer Himalayas, in the tracks of Jammu and Kashmir, Nepal to east of Assam, Bengal, and Tripura extending through the Central region up to Godavari in the south. On whole country basis, flowering usually occurs from July to October (Anon., 1997) though in the northern plains of India from September to November. *Nyctanthes* means 'night-blooming' and *arbor-tristis* means 'tree of sadness' as it blooms only during night hours and sheds its flowers before day-break. Beside the ornamental values, *Nyctanthes* is one of the most versatile medicinal plants.

Medicinal Potential

All the parts of *Nyctanthes* have been used as traditional medicine for household remedies against various human ailments from antiquity (Samal *et al.*, 2007). It was revealed through phytochemical analysis that leaf, fruit and seeds contain phytosterols, phenolics, tannins, flavonoids, glycosides and saponins. The secondary metabolites such as glycosides and alkaloids are the largest groups of chemicals produced by this plant (Kapoor *et al.*, 1971; Dhingra *et al.*, 1976). Champa Rani *et al.* (2012) have listed the chemical constituents and their medicinal properties recorded at various times in the leaves, stems, bark, flowers, floral oil and seeds. The leaves contain d-mannitol, β-amyrin, β-sitosterole, flavonol, glycosides-astragaline, nicotiflorin, oleanolic acid, nyctanthic acid (dye), tannic acid, ascorbic acid, methyl salicylate, carotene, friedeline, lupeol, mannitol, glucose and fructose, iridoid glycosides, benzoic acid, and derivatives of kaempferol; the flowers contain essential oils (α-pinene, p-cymene, 1-hexanol), dye nyctanthin (Puri *et al.*, 1994) allied to crocetin from saffron (*Crocus sativus*) and mannitol, tannin and glucose, carotenoid, glycosides, *viz.*, β-monogentiobioside ester of α-crocetin (or crocin-3), β-monogentiobioside-β-D monoglucoside ester of α-crocetin, β-digentiobioside ester of α-crocetin; the seeds contain arbortristoside A&B, glycerides of linoleic, oleic, lignoceric, stearic, palmitic and myristic acids, nyctanthic acid, and 3-4-secotriterpene acid; the bark contains glycosides and alkaloids; the stem contains glycoside-naringenin-4'-0-β-glucapyranbosyl-α-xylopyranoside and β-sitosterol; and the flower oil contains α-pinene, p-cymene, 1-hexanol methyl heptanone, phenyl acetaldehyde, 1-deconol and anisaldehyde perfume. The essential oil extracted from the flowers is similar to jasmine, hence used as perfume. The seeds contain about 15% of a pale-yellow brown oil nyctanthic acid and beta-sitosterol, and the powdered seeds ameliorate scalp scurf. The bark contains a glycoside and alkaloids, suspected of being poisonous to animals and humans. The flowers contain an abundance of mannitol and the bright orange corolla tubes contain a saffron-yellow colouring matter, once used for dyeing silk, sometimes in conjunction with safflower (*Carthamus longa*) and indigo (*Indigofera* spp.). Locally the dye (nyctanthin) is used for dyeing cotton cloth and as a cheaper substitute for saffron in colouring the robes of Buddhist priests and to maintain its colouring either alum or lime juice is added in the dye bath (file://H:/harsingar/nyctanthes). The earliest record about the use of *Nyctanthes* for prevention of diseases can be traced in '***Rigveda***', perhaps the oldest repository of human knowledge having been written between 4,500 and 1,600 B.C. In '***Atharvaveda***', which is a later production, the use of plants as drug is more varied and it is the '***Ayurveda***' which is considered as '***Upveda***' that the definite properties of drugs and their uses have been given in more details (Agarwal and Pal, 2013). Leaves are used in case of antibacterial, anthelmintic, anti-inflammatory, hepatoprotective, immunopotential, anti-pyretic, antioxidant and antifungal activities. Taking 3 leaves with 5 black pepper seeds cures the females having gynaecological troubles. An anti-inflammatory activity of the leaves has recently been recorded, and an insecticidal effect of an extract from shade-dried leaves has also been reported. Leaves are used to treat the enlargement of spleen. The bitter leaf juice acts as cholagogue, in piles, liver disorders, chronic fever, malarial fever, obstinate sciatica, rheumatism, as a diaphoretic and diuretic, as laxative, in case of loss of appetite, as mild bitter tonic, and as purgative in case of infants, and an extract with honey or sugar mixed with common salt is administered to children for the expulsion of roundworms and threadworms. One teaspoonful of leaf decoction with honey cures fever, malaria and blood dysentery. In treatment of ringworm and certain other skin related problems, the aqueous paste of leaves externally is very effective. Leaf juice is also used in the treatment of chronic fevers in combination with other herbs such as ginger, basil, clove, mint, pippali and vasa. Herbal oil prepared by boiling fresh leaves in mustard oil is effective against skin ailments. Leaf juice with honey is effective for treatment of dry cough. Leaf juice is also used as an antidote to reptile venom. Traditionally, the powdered stem bark is given in rheumatic joint pain, in treatment of malaria and as an expectorant (Badam *et al.*, 1988; Aminuddin *et al.*, 1993). Some 5 grains of the bark eaten with betel leaf and nut promotes expectoration of thick phlegum. Seeds have antibacterial, antifungal, immunomodulatory and antileishmanial activities. In case of piles, fresh paste of crushed seeds externally on piles along with daily one seed with water internally are useful treatment. The decoction of its seeds is used as a hair tonic. Washing hair daily with this tonic helps to get rid of dandruff and lice (Banerjee *et al.*, 2007). Its flowers have diuretic, anti-bilious, antioxidant, anti-inflammatory, sedative and antifilarial activities, and to treat gout the flower decoction used for a week relieves the persons suffering from gout; the bark is anti-microbial; the stem is antipyretic and antioxidant; and the root decoction is used in case of spleen enlargement. In India, Indonesia and Malaysia the flowers are used medicinally to provoke menstruation.

Botany and Cytology

It is a coarse shrub or tree some 9-10 m tall with scabrous (granular or rough) stem bearing 4-angled spreading branches with scabrous, decussately opposite, ovate, cordate, 6.0-12.0 cm × 2.0-6.5 cm, cuneate to subcordate at base, acute or acuminate at apex, short-stalked, above highly scabrous and pubescent beneath, shortly petiolate, entire or dentate leaves. Flowers are few (2-7 in a corymb) in a sessile involucrate cluster (head) which are disposed in dichotomous cymes. Calyx is 6 to 8 mm long, narrowly campanulate, hairy outside, glabrous inside, truncate or obscurely toothed or lobed, and ciliated. The actinomorphic white to creamy flowers have 4-9 contorted lobes, 5-15 mm long and orange tubes, with two nearly sessile and epipetalous stamens (sometimes 3) which alternate with two locules (sometimes 3) of the ovary, and the style is short with 2-lobed stigma. Patel (1960) noted that *Nyctanthes* is heterostyly, as is commonly the case in *Jasminum* species. The regular flowers are arranged on cymose head. The fruit is a cordate to almost orbicular flat and dry chartaceous capsule, about 18 mm across, brown, 2-celled, opening transversely from the apex into two one-seeded compressed parts having 1 erect seed in each.

The somatic chromosome number of *Nyctanthes* has been recorded as 44 (Bolkhorskikh, 1969) and 46 (Kundu and De, 1968). Both these numbers occur in the Oleaceae as *Mendora scabra* also possesses 2n=44 (Taylor, 1945). Kapil and Vani (1966) reported that *Nyctanthes* differs from the Verbenaceae in two important characters, such as (i) the antipodal cells of *Nyctanthes* being small and ephemeral, whereas in Verbenaceae they are large and persistent, and the (ii) following the system of Soueges, the embryogeny of *Nyctanthes* is of the Megarche type III of Second Period, whereas that of Verbenaceae is Megarche type IV of First Period. Devi (1975) found that the embryology of *Nyctanthes arbor-tristis* and *Jasminum* is basically similar. Pollen and seed in *Nyctanthes* are highly sterile (Bentham and Hooker, 1976).

Species and Cultivars

Nyctanthes bearing pretty white flower, diffusing a more exquisite fragrance contains two other species naming *Nyctanthes multiflora* known as *malati chiana* and *Nyctanthes acuminata* known as *bunga malati susun* (Daniel and Sabnis, 1984). *Nyctanthes sambac* and *Nyctanthes undulata* are just the synonyms of *Jasminum sambac* and *J. undulata*, respectively (Kapil and Vani, 1966). About varietal evolution, till now a very little effort has been made, however, it has one variegated variety which is under evaluation (source:Website: *My Garden Frenzy* 2013) though various floral forms with 4-9 lobes having differing lobe-length & width, and differing colouration (various shades of orange) of floral tube exist in the nature. 'Karna Phool' and 'Seeya Shringar' are Indian cultivars.

Propagation

The success rate of **seed** germination of *Nyctanthes arbor-tristis* is poor as hardly there is 10 per cent germination under favourable conditions. In an investigation it was revealed that the phenolic compounds interfere in its germination therefore these compounds should be leached out of the imbibed seeds. The pericarp assisted by the seed coat acts as the store of such inhibitory compounds. The best result is obtained by removing both the coverings. Such effect was due to absence of the barrier and a major percentage of the phenolic inhibitors. Leaching of such interfering substances was restricted by treating seeds prior to germination with solution of antioxidants like polyvinylpyrolidone (PVP) and polyvinylpolypyrolidone (PVPP), resulting in better germination (Bhattacharya *et al.*, 1999). Gibberellic acid 200 ppm induces the seed germination (Champa Rani *et al.*, 2012). Seedlings raised in April are transplanted in May or June. It grows to a height of 2 m by August and flowering starts in September to October the same year (Champa Rani *et al.*, 2012). The required temperature for seed germination is 18 to 22 ^{0}C. It should be periodically watered. The proper drainage facility should be provided. The seeds take 14-20 days for germination.

It can easily be propagated by **vegetative** means, i.e. cutting or layering. It coppices readily and is not browsed by goats or cattles (Champa Rani *et al.*, 2012). Clonal propagation, i.e. cutting or grafting becomes necessary to perpetuate the desirable variations selected from a large plant population. Many selections at NBRI, Lucknow have been propagated by budding (Dey, 1996). For layering the area (ringed) must be cut around the trunk and then repeated approximately 2.5 cm above or below the original cut. In case of the larger branch, there would be greater distance between the cuts.

Attempts have been made for regeneration of night jasmine directly from nodal **explants** and indirectly from cotyledons, leaf, internode and nodal explants excised from one-month-old seedlings, as well as from mature trees of *Nyctanthes arbor-tristis*. Cotyledonary explants obtained from mature green seeds produce maximum percentage of callus induction on 2, 4-D (0.5 mg/l) and on GA_3 (2 mg/l) alone, as well as on medium supplemented with combinations of 2,4-D (0.5 mg/l) + kinetin (0.5 mg/l) (Rout *et al.*, 2007). Internodal and nodal explants respond well to 2,4-D (0.5 mg/l) + kinetin (0.5 mg/l) for callus induction as compared to other combinations of growth regulators (Jahan *et al.*, 2011). The explants excised from one month old seedlings show better response in terms of callus induction and growth as compared to explants obtained from mature trees. Out of the four explants used, cotyledonary and leaf explants have been found to be the best in terms of callus induction, growth and differentiation than the intermodal and nodal explants. However, only the calli obtained from cotyledonary explants showed shoot formation on BAP (1.0 and 2.0 mg/l) + GA_3 (2mg/l) after sub-culturing (Iyer *et al.*, 1998). Direct regeneration of multiple shoots from nodal segments of mature tree and of seedlings is achieved on 1 mg/l BAP alone and 2 mg/l BAP + 2mg/l GA_3, respectively. Available reports also indicate rapid multiplication of *N. arbor-tristis* through *in vitro* axillary shoot proliferation and growth followed by successful *ex vitro* establishment of regenerated plants. An efficient, rapid and reproducible plant regeneration protocol for *N. arbor-trisis* has been developed using cotyledonary node explants excised from 15 days old aseptic seedlings. Two cytokinins, TDZ (1.0 µM) and BA (2.5 µM) added as supplements to MS medium were found to be effective in maximum shoot induction. The shoot multiplication was significantly enhanced by BA and NAA combination. The regenerated shoots when sub-cultured in hormone-free MS medium, showed considerable increase in shoot multiplication and shoot length by the end of 4th sub-culture passage. Shooting was achieved when the basal cut ends of regenerated shoots were dipped in 200 µM IBA for half an hour followed by transplantation in plastic pots containing sterile soilrite. The plantlets with well developed shoots and roots were successfully established in earthen pots containing garden soil with 85% survival rate, in greenhouse (Bhattacharya *et al.*, 1999). An alternate protocol for *in vitro* clonal propagation of *Nyctanthes arbor-tristis* was also developed, using MS basal medium supplemented with 1.0-1.5 mg/l 6-BA, 50 mg/l adenine sulphate (Ads) and 3% sucrose. Inclusion of IAA in the culture medium along with BA + adenine sulphate promoted a higher rate of shoot multiplication. Maximum mean number of micro-shoots per explant (6.65) was achieved on the MS medium supplemented with 1.5 mg/l BA, 50 mg/l Ads and 0.1 mg/l IAA after 4 weeks of culture. The elongated shoots rooted within 13 to 14 days on ½ strength MS medium supplemented either with IBA, IAA or NAA with 2% sucrose. Maximum percentage of rooting was obtained on medium having 0.25 mg/l IBA, 0.1 mg/l IAA and 2% sucrose. Between different media the pH levels and growth regulators tried, the optimum condition for maximum regenerative response was obtained on MS + Kn (2.5 µM) + NAA (0.5 µM) media at 5.8 pH, forming cultures with 23.26 ± 0.89 number of shoots and 6.36 ± 0.80 cm shoot length after 8 weeks of culture. Rapid differentiation of multiple shoots was also observed in 94% of nodal explants of one year old *Nyctanthes arbor-tristis* plants (Jahan *et al.*, 2011). The *in vitro* raised plants could be grown normally in the soil condition (Rout *et al.*, 2007).

Cultural Practices

Flowering in *Nyctanthes* is usually found from July to October when there is transition of weather from summer to winter (Champa Rani *et al.*, 2012) though in northern Indian plains it occurs from September to November. *Nyctanthes* prefers tropical to subtropical condition for growth and flowering. The plant requires full **sunlight** to partial shade (Champa Rani *et al.*, 2012). The plant can be grown up to 1,500 m altitude at the equator, within a wide range of rainfall patterns, from seasonal to non-seasonal. It is a tropical plant and generally grows well from 16 to 23 ^{0}C **temperatures**. The plant can't survive in severe frost. Flowering is initiated by shortening day length of summer and early fall. Subjecting plants to artificial light after darkness from September to December will disrupt flowering preferably 50 per cent or higher. The intensity and diurnal variations of light affects the growth and flavonoid content of plants. In case of *Nyctanthes arbor-tristis*, roots collected in morning from plants growing in intense light conditions could be a good source for flavonoid extraction as the content in each gram of plant material is quite high (Yadav and Kumar 2012). It tolerates moderate *shade* and is often found as undergrowth in dry deciduous forests (Sandhar *et al.*, 2011).

It can survive in varied kind of **soils**, pH ranging from 5.6 to 7.5. Fertile and well-drained soil is necessary for a successful plant growth. For its growing the type of soil is not important but the soil should be free of stones, moisture-retentive, well-drained and rich in organic matter. On the other hand, infertile soil that has good physical properties can be made productive by using organic matter, lime, commercial fertilizer, and other soil improving materials. If the soil sticks together in a ball and does not readily crumble under slight pressure by the thumb and finger, it is too wet for ploughing or working and working at this stage will cause its caking as it dries, making it unsuitable for young plants. For planting, the soil should be dug up at least 45 cm deep. For **soil preparation**, an abundance of organic material should be ploughed into the soil and allowed to compost over the winter. Good loam should be used for the topsoil, e.g. humus, peat moss, well-rotted manure, well-sifted leaf mould or heavy sand. Wood ashes are fine for spring, and lime may be used for loosening the soil. For planting as individual trees, the pits of the size of 60 ×60×60 should be dug up before the monsoon, should be filled with coarse sand, FYM or compost or any other properly decomposed manure and good sandy loam soil, all in equal proportion. The pits should be allowed to weather up to spring and when seedlings have attained the good size, these are transplanted in such pits and watered. Planting is normally carried out during monsoon season so that atmosphere is full of relative humidity and the mixture of the pit is also getting ample watering.

If the soil has already been cultivated and used in previous years, there is little to do other than to plough in with additional **organic material**. The organic material may be in the form of composted manure or any good commercial complete plant food distributed at a rate of 1.5-2.0 kg per square metre. The compost or farmyard manure should settle naturally. *Nyctanthes* is moderate feeder, especially when grown under high light. A balanced 15-15-15 NPK fertilizer or a blooming fertilizer such as 7-9-5 is ideal for its growth. Synthetic fertilizers in excess can burn the roots and leaves or whole plant but organic manure apart from providing nutrients to the plants also improves the soil. Nutrition to the plants is given when these are in their active growth.

Flowers bloom at night and fall from the tree before the morning. After the seed sowing in April, only the healthy seedlings should be allowed to grow. When the plant reaches 10 cm height, **pinching** should be done to induce side shoots. The grown up plant should be **pruned** in November after the flowering is over, and at this time heading back of the flowering shoots is done to induce new shoots for the next year flowering (Dey, 1996).

Height Control

Plant growth regulators also promote growth and flowering in *Nyctanthes*. Sumagic is a triazole PGR that has a higher activity and has also been very effective in height control of perennials like *Nyctanthes*. In Oleaceae it can be applied as foliar spray with 15 ppm Sumagic under growing conditions with excessive foliage (Latimer *et al.*, 2003). *Nyctanthes* is responsive to a tank mix of 5,000 ppm B-Nine for growth control.

Post Harvest

Typically, the flowers are **harvested** at bud stage during early morning. The flower buds are harvested on the basis of colour, as firmness and size are variable depending on the weather. The buds have to be white, as green ones may not emit the characteristic fragrance they are known for. Open flowers are generally not harvested as a larger amount of them is needed to extract oils and they lose their fragrance soon. Leaf and flower extract are stored at 4 ^{0}C (Hussain and Ramteke, 2012).

Insect-Pests and Diseases

Powdery mildew caused by *Oidium* spp. can do some damage to *Nyctanthes* foliage, more seriously in regularly pruned plants as its infection is favoured by the presence of tender and expanding young buds, leaves and other aerial parts. It can be controlled by dusting with sulphur (Pathak *et al.*, 1992). *Nyctanthes* should be carefully watched for **whitefly** and **spider mites** during high temperatures. Their attack can be checked by spraying with 0.2 per cent Rogor.

References

Acharya, S. 2011. Prediction of rainfall variation through flowering phenology of night-flowering jasmine (*Nyctanthes arbor-tristis* L.) in Tripura. *Indian J. Tradit. Knowl.*, **10**: 96-101.

Agrawal, J. and A. Pal, 2013. *Nyctanthes arbor-tristis* Linn—A critical ethnopharmacological review. *J. Ethnopharmac.*, **146**: 645-658.

Aminuddin, S.A., R.D. Girach and S.A. Khan, 1993. Treatment of malaria through herbal drugs from Orissa, India. *Fitoterapia*, **LXIV**: 545-548.

Anonymous, 1997. *The Wealth of India: A Dictionary of Indian Raw Materials and Industrial products*, vol. VII, pp. 69-70. National Institute of Science Communication, CSIR, New Delhi.

Badam, L., R.P. Deolankar, S.R. Rojatkar, B.A. Nagsampgi and U.V. Wagh, 1988. *In vitro* antimalarial activity of medicinal plants of India. *Indian J. Med. Res.*, **87**: 379-383.

Banerjee, A., A. Poddar, S. Ghanta,A. Chakraborty and S. Chattopadhyay, 2007. *Nyctanthes arbor-tristis* Linn—spectrum of its bioactivity potential. *Planta Medica*, **73**: 603-617.

Benham, G. and J.D. Hooker, 1976. Oleaceae. *Genera Plantarum*, **2**:672-680.

Bhattacharya, S., B. Das and T.K. Ghose, 1999. Investigation on seed germination of *Nyctanthes arbor-tristis* (Oleaceae) in relation to the total phenol content. *Science & Techn.*, **27**: 321-327.

Bolkhorskikh, 1969. *Chromosome Number in Flowering Plants.* A Federov, Leningrad, USSR.

Champa Rani, Sunaina Chawla, Manisha Mangal, A.K. Mangal, Subhash Kajla and A.K. Dhawan, 2012. *Nyctanthes arbor-tristris* Linn. (Night Jasmine): A sacred ornamental plant with immense medicinal potentials. *Indian J. Traditional Knowledge,* **11**: 427-435.

Daniel, M. and S.D. Sabnis, 1984. A numerical evaluation of the taxonomical status of *Nyctanthes. Indian bot. Reporter* (Special Volume), pp. 52-57.

Devi, H. M. 1975. Embryology of *Jasminum* and its bearing on compostion of Oleaceae. *Acta bot. Indica.* **3:** 42-61.

Dey, S. C. 1996. Horticulture. In: Fragrant Flowers for Homes and Gardens, Trade and Industry, pp. 33-121. Abvinav Publication. New Delhi.

Dhingra, V.K., T.R. Seshadri and S.K. Mukerjee, 1976. Carotenoid glycosides of *Nyctanthes arbor-tristis* Linn. *Indian J. Chem.*, **14**: 231. file://H:/harsingar/nyctanthes

Hussain, A. and R. Ramteke, 2012. Flower extract of *Nyctanthes arbor-tristis* modulates glutathione level in hydrogen peroxide treated lymphocytes. *Pharmac. Res.*, **4**: 230-233.

Iyer, R. I., V. Mathuram and P.M. Gopinath, 1998. Establishment of callus cultures of *Nyctanthes arbor-tristis* from juvenile explants and detection of secondary metabolites in the callus. *Curr. Sci.* ,**74**: 243-246.

Jahan A. A., M. Anis and I.M. Aref, 2011. Assessment of factors affecting micropropagation and *ex vitro* acclimatization of *Nyctanthes arbor-tristis. Acta Biologica Hungarica*, **42**:45-66.

Kapil, R. N. and R.S. Vani, 1966. *Nyctanthes arbor-tristis* Linn.:embryology and relationships. *Phytomorph.*, **16**: 553-563.

Kapoor, L.D., S.L. Kapoor, S.N. Srivastava, A. Singh and P.C. Sharma, 1971. Survey of Indian plants for saponins, alkaloids and flavonoids, II. *Lloydia*, **34**: 94.

Kundu, B. C. and A. De, 1968. Taxonomic position of the genus *Nyctanthes*. *Bull. Bot. Survey India*,. **10**: 397-408.

Latimer, J., H. Scoggins and V. Groover, 2003. Virginia Tech research reveals the PGRs that work best for crops such as penstemon, verbena, chelone and lantana. Webbsite: www.*Greenhouse product news. Scrophulariaceae and Verbenaceae Response to PGRs.*

Patel, N.K. 1960. Some preliminary observations on the floral structure of Oleaceace. *Curr. Sci.*, **29**: 59.

Pathak, R.K., M. Madhulika and S.N. Sachan, 1992. Powdery mildew and cultural practices. *Indian J. Forestry*, **5**: 73.

Puri, A., R. Saxena, R.P. Saxena, K.C. Saxena, S. Vandita and J.S. Tandon, 1994. Immunostimulant activity of *Nyctanthes arbor-tristis* L. *J. Ethnopharmac.*, **42**: 31-37.

Rout, G. R., A. Mahato and S. K. Senapati, 2007. *In vitro* clonal propagation of *Nyctanthes arbortristis* Linn. A medicinal tree. *Hort Sci.* (Prague), **34**: 84–89.

Samal, D., S. Das and S.P. Basu, 2007. Phytoconstituents and therapeutic potential of *Nyctanthes arbor-tristis* Linn. *Pharmacol. Rev.*, **1**(2): 344-249.

Sandhar, H. K., M. Kaur, B. Kumar and S. Prasher, 2011. An update on *Nyctanthes arbor-tristis* Linn. *Internationale Pharmaceutica Sciencia.* **1**: 77-81.

Taylor, H. 1945. Cytonomy and phylogeny of Oleaceae. *Brittonia*, **5**: 337-367.

Website: My Garden Frenzy : *Nyctanthes arbor-tristis.*

Yadav, S and P. Kumar, 2012. Effect of light intensity on height and production of flavonoids in *Nyctanthes arbor-tristis..J. Pharmacy Res.*, **5**:1-5.

22

Nymphaea

R.L. Misra and Sanyat Misra

Family: **Nymphaceae**

Common names: Blue lotus/Blue lotus of India/ Indian blue lotus (*Nymphaea stellata*), Blue waterlily/ Cape blue waterlily/Cape waterlily (*N. capensis*), Bohemian waterlily (*N. candida*), Cape Cod waterlily (*N. odorata* var. *rosea*), Chinese pygmy waterlily/Dwarf waterlily/Pygmy waterlily/Small white waterlily (*N. pygmaea* 'Alba'), Dauben's waterlily/Blue pygmy waterlily/Madagascar aquarium lily (*N. daubeniana*), Egyptian blue lotus (*N. caerulea*), English white waterlily/Platter dock (*N. alba*), Mill pond lily (*N. odorata minor*), Pond lily, Rice-field waterlily (*N. odorata* var. *gigantea*), Pygmy waterlily (*N. pygmaea*, syn. *N. tetragona*), Royal purple waterlily (*N. capensis* var. *zanzibariensis*), Swedish waterlily (*N. alba* var. *rosea*), Sweet-scented waterlily (*N. odorata*), Waterlily, White lotus/ Egyptian lotus (*N. lotus*), White Nile lotus (*N. dentata*), common *Hindi* names- 'Bera', 'Kuinberi', 'Konkabeli', 'Kumud' and 'Kumudini'; *Bangla*- Rakto-kombal and Shaluk; *Oriya*- Dhabdakain and Rangkain; *Marathi*- Kamal and Raktakamal; *Gujarati*- Lalkamal and Neelophal; *Punjabi*- Chhota kamal; *Murdan*- Pundi salukid; *Khasi*- Soh-lapudng; *Kanarese*- Huvu Malayalee and Nyadule; *Malayalam*-Nacrambal and Periambal; *Tamil*- Alli-tamarai and Vellambai; and *Telugu*- Alitamara and Tella-kalava

Introduction and Origin

A very few bulbous ornamentals survive only in water and one of them is waterlily which is rhizomatous and bears most spectacular blooms in various colours, shapes and sizes and its leaves are also attractive, *vis-à-vis* used as dining plates in social gatherings in rural India. It not only survives in deep water but its survival is in danger if it is not grown in water. Since its blooms are very attractive and the rootstock is rhizomatous hence this has been chosen under rhizomatous bulbous ornamentals being fully conversant that this is most popular water plant. Its flowers are very formal in shape and rather more attractive than lotus. Many are fragrant and there is great variation even in the size of the flowers, *vis-à-vis* leaves.

There are references of lotus (Sacred lotus of the Nile) from the ancient Egypt, *vis-a-vis* frequently depicted as Egyptian murals. Egyptians admired and depicted them some 4,000 years B.C. and had offerings to their dear ones on their death as had been evident during Israelitish captivity of Egypt where in the funeral wreaths 'Egyptian blue lotus' and 'white lotus' were found in the tomb of the Egyptian Pharaoh, Rameses the Great II (13th century B.C.) and in case of the Princess Nzi-Khonsu, the whole body was found covered with several layers with petals of the same flowers (Frances Perry, 1962). Both these were, in fact, not lotus but *Nymphaea caerulea*, a sky-blue, narrow petalled tropical waterlily and *Nymphaea lotus*, a white-petalled night-blooming waterlily, because as per the knowledge of Egyptians these were being considered there then as the real lotus though these were in fact waterlilies. Though Egyptians comparatively had more veneration for *N. caerulea* but Indians have more for *Nymphaea lotus*.

Nymphaea takes its name from *Nympha* which in Greek and Roman mythology means a 'nature-goddess' or from *Nymphe* one of the 'water nymphs', the mythological Greek deity (Beckett, 1987). As per a Greek legend, the name *Nymphaea* derives from *nymphaia*, a flower that arose from a nymph who had died of jealousy. In the 1st century A.D., the Roman historian and naturalist ~ Pliny, associates waterlilies with a nymph who had died from unrequited love for Hercules (Anon., 1989). It inhabits the North and South Temperate and Tropical zones, *viz.*, Europe [Central (Bohemia province of Czech republic), North, and West (Sweden)], North America [Canada (Ontario), Mexico, Tropical, North, East and Central America (Texas, Delaware to Florida, North Idaho, Massachusetts, New Jersey, New York, Cape Cod, etc.)], South America [British Guiana (now renamed as Guyana), Brazil, Caribbean Sea Islands (Cuba, West Indies)], Africa [NE Africa (Egypt), NC Africa, E Africa (Zanzibar), W. Africa (Sierra Leone), WC Africa (Cameroon), West Coast of Africa, South Africa], South and East Asia (India to Java, Philippines, China and Japan), Australia, especially the Cape York Peninsula, and Russia (East Siberia).

N. rubra unripe fruits are used as vegetables, and flowering stalks as vegetables as well as salad. Flowers are astringent and cardiotonic. An infusion prepared from *Nymphaea* fruits and flowers is diaphoretic and is effective for treatment of diarrhoea and an infusion of rhizomes and stem is diuretic, *vis-à-vis* rhizomes are useful for treating diseases of the uninary tract. Rhizomes are considered as demulcent and are effective against dysentery and dyspepsia, and its powdered form is used in treating dyspepsia, diarrhoea and piles. Fresh rhizome decoction is also effective against diarrhoea. Alcoholic extracts of the rhizomes are sedative and spasmodic in action because of 'nymphacine' alkaloid which is available in all parts of the plant except the seeds. Its rhizomes being rich in starch are also used as emergency food after boiling or baking. Apart from starch, it contains 10% crude fibre, 6.4% crude protein, 10.8% ash and an alkaloid 'nymphacine'. Rhizomes are also used in tannin industry. The collection of the seeds for culinary purposes is done before dehiscence of the fruits. The seeds from the fresh fruits are extracted out and dried for various uses. The fresh seeds are eaten raw or boiled, and during fast by flouring and cooking in the form of bread or porridge. Seeds are also parched and puffed and eaten. Seeds are free from alkaloid and contain 47% starch and fatty oils but its excess use may cause toxicity. Its leaves are used as dining plates like to those of lotus.

Botany

Nymphaea is a herbaceous plant having rhizomatous rootstocks, roots appearing from the stem just below each leaf, leaves broadly ovate to rounded, entire or dentate or sinuate, fissi-cordate and often sub-peltate, 5-60 cm across, mostly divided into two lobes from the point of attachment of the petioles, the leaves emerging from the rhizomes in spiral form, floating at water surface irrespective of the rise or fall of water level due to the long and coiling petioles which may be 15 cm to 2.0 m long, and the leaves are of three types: i) a very thin, fragile and submerged ones having short petioles, ii) floating leaves thicker in texture having stomata and palisade cells on the upper side only, and iii) aerial leathery leaves which sometimes bear stomata on the lower surface. Flowers are extra-axillary which arise in their own spiral or in spiral form as members of the leaf spirals, are borne just above the water from white to purple via cream, yellow, orange, pink, rose, red, dark or bright red, mauve and purple, 2.5-30.0 cm across, opening at a particular time each day though this timing may differ from species to species, each flower opening from 2-7 successive days or nights (about an hour later to open and an hour earlier to close on the very first day or night as compared to other subsequent days or nights) and then the flowers through spiral coiling of the peduncle are down in the water (in shallow water this simply bends down) where seeds ripen within 6-10 weeks. Flowers have 4 sepals and many petals, carpels and stamens. Fruit is round, very tight, berry-like with numerous small (similar in appearance but smaller than seeds of *Pennisetum typhoides*) seeds which can be eaten raw. After ripening, the fruits burst and the seeds through the buoyant aril float on the surface of water for many hours to decay the aril so that these may settle down for germination.

Banerji (2006) has reviewed the cytological studies carried out in the crop. Cytological studies in India revealed that *Nymphaea* has a basic chromosome number n = 14, with ploidy levels from 2x = 28 to 16x = 224. Sen and Bidhuri (1971) investigated 16 different species and varieties and recorded 2n = 56 in *N. alba, N. laydekerialea, N. lotus* var. *dentata, N. marliacea rosea, N. mars* and *N. stellata* var. 2, 2n = 70 in *N. rubra* and *N.* 'Sioux', 2n =84 in *N. pubescens, N. marliacea chromatella, N. stellata* var. 1 and *N. t.* var. *helvola*, and 2n = 28 & 56 in *N. caerulea*. This has also been recorded that the hexaploid (2n = 84) cultivars bear small flowers (4-10 cm diameter), and the diploid [2n = 28 (2x)] & triploids [2n = 154 (11x), 224(16x)] larger flowers (12-24 cm). The cytological analysis of 19 taxa including wild indigenous forms of India and cultivated forms of America revealed diploid, triploid, tetraploid and hexaploid levels based on x = 14 (Gupta, 1980). *N. dentalamagnifica* and cv. 'Sunrise' showed presence of polysomaty. Majority of taxa show abnormalities (desynopsis, chromosome association and segregational irregularities) at meiosis which indicate their hybrid origin (Gupta, 1978). Subramanyam (1979, 1980) recorded diploid to triploid (11X = 2n = 154, and 206) and further stated that the probable origin of the triploid *N.* × *dauberiana* (2n = 42) through hybridization was between tetraploid *Nymphaea* × *micrantha* and diploid *N. caerulea*. Subramanyam and Khoshoo (1981) stated that induced hexaploidy restores fertility in a sterile triploid *N.* × *dauberiana* 2n = 3x = 42. Diploids are found in Africa and India and the polyploids from other larger areas adapting diverse climatic conditions and they are self as well as cross-pollinated. The species of single group hybridize readily among themselves, and from the white lily, *N. lotus* var. *dentata* almost all shades of colours, *viz.*, from pure white to dark crimson-red, have been obtained. Similarly, *N. zanzibariensis* also when planted in the same pond with others of the same group it gives rise to many fine varieties. Outside of single groups no genuine hybrids are produced. Between the apocarpous and syncarpous species, a hybrid is not possible.

In between 1,888 to 1,900, Monsieur Joseph Bory Latour Marliac (*b.* 1830 at Granges in Lot-et-Garonne) who after marriage settled down at Temple-sur-Lot, France. He noticed carmine-red sport of *Nymphaea alba* in a lake at Nerike, Sweden whose pollen he used to cross *N. alba,* and *N. odorata* to find the pink forms in these species, and when he used *N. mexicana,* he obtained the first hardy yellow form, *N. marliacea chromatellus* in 1,881 (Frances Perry, 1962). He developed numerous superb varieties by using *N. alba* as one of the parents which was called the first or Marliacea group, and the second group was developed with a hybrid of *N. alba* var. *rubra* and *N. tetragona* (syn. *N. laydekeri* var. *rosea* surely containing the blood of *N. alba* var. *rosea* and *N. mexicana*) and nearly all kinds of both the groups are entirely sterile. The technique employed to evolve Marliacea hybrids still remains a mystery. It was further revealed that the present day garden *Nymphaea* are the result of repeated crossing involving about 12 elemental species. However, between 1,883-1,890, already the dwarfs, giant-flowered, double-forms, red, pinks, and various other colours came into existence.

Classification, Species and Varieties

The genus *Nymphaea* comprises 35-50 species (Beckett, 1987) with cosmopolitan distribution. *N. alba* is found in Kashmir lakes up to 1,800 m elevations. *N. pubescens* (syn. *N. nouchali*) and *N. stellata*, the largest ones are found in warmer parts of India in lakes, ponds and other water bodies (Sanyat Misra and Misra, 2013). *N. tetragona* (syn. *N. pygmaea*), the dwarf ones, are recorded in the Himalayas and in the swamps of Khasi hills at altitudes ranging from 1,200-1,800 m. Some five species occur in India. According to Bailey (1929), there are eight species of the first importance, *viz., N. alba, N. flavo-virens, N. lotus, N. mexicana, N. odorata, N. rubra, N. tuberosa* and *N. zanzibariensis*. The best waterlilies for amateurs are: i) the tender day-blooming ones being *N. daubeniana* (dwarf blue), *N. flavo-virens* (white), Mrs. C.W. Ward (pink), *N. pennsylvania* (light blue) and *N. zanzibariensis* (deep blue), ii) tender night-blooming kinds being *N. dentata* (white), *N. omarana* (magenta), and 'Frank Trelease' (dark crimson), and iii) hardy ones being *N. chromatella* (yellow), *N. gladstoniana* (white), *N. laydekeri rosea* (dwarf pink), 'Wm. Falconer' (dark red), *N. tetragona* (white), 'W.B. Shaw' (pink), and *N. tetragona helvola* (dwarf yellow). For convenience, these may be divided into 'hardy waterlilies' and 'tropical waterlilies', and further both into 'day blooming' and 'night blooming'. Beckett (1987) has given an account of some five species such as *N. coerulea, N. capensis, N. flava* (*N. mexicana*), *N. pygmaea* (*N. tetragona*), and *N. stellata*; and Brickell (1994) has described a few species and many varieties. Looking into the importance, the choicest ones are:

N. alba, a native of Europe (Britain and other parts), North Africa and Asia in large lakes and slow-moving rivers up to 1.2 m deep water but is not suitable for domestic pools due to its robust growth with about 1.8 m length of quite thick and black rhizomes having long fine hairs. It is hardy with leaves 10-30 cm across, rounded, dark green but often red-green beneath and red when young, and flowers day-blooming (7 a.m. to 4 p.m.), white but stamens yellow, faintly fragrant, cup-shaped but later changing to star-shape and 10-13 cm wide, and its varieties and forms are 'Froebeli' with fragrant and deep red flowers, and *N. a. rubra* with rosy-pink flowers; *N. alba* × *N. odorata rosea* with large cup-shaped free-flowering sterile flowers often raised above the water which are white but turning paler with age, and it has three groups, *viz.*, i) flowers almost white flushed rose (*N. marliacea albida* with 12-15 cm across flowers, *N. m. gladstoniana* with fragrant white 25 cm across flowers, var. 'Goliath' and *N. m. lusitana* with large rosy flowers), ii) flowers light pink (*N. marliacea carnea,* flesh-colour, 'Wm. Doogue', 'Mark Hanna', *N. colossea,* the very large forms being 'Leviathan' and *N. m. somptuosa*), and iii) rose coloured flowers (*N. m. rosea, N. nobilissima*, 'James Hudson').

N. caerulea, a hardy species native of Egypt, and North & Central Africa and highly similar to *N. stellata* except that leaves are wavy-margined, its large rootstocks are eaten in parts of Africa, a prolific bloomer and easy to grow but not very showy, used extensively in hybridization programme, leaves ovate, mid-green but purple spotted beneath, lobes overlapping and 30-40 cm across, day-blooming (7.30 a.m.-12.00 a.m.) and held high above the water, flowers star-shaped, 8-15 cm across, pale-blue with yellow stamens but blue anthers, opening for 3 days, bears 14-20 petals per flower, and 'Blue Beauty' is its one of the outstanding varieties.

N. candida (syn. *N. biradiata, N. semiaperta, N. alba minor, N. nitida*), a hardy species native of Central and North Europe, Asia and the Himalayas and is suitable for shallow water, leaves have prominent veins underside, basal lobes of the leaves overlap slightly, this species is similar to *N. pygmaea* 'Alba' but the flowers are larger with small red stigmas while in 'Alba' the stigmas are yellow, flowers white with tinged green sepals, cup-shaped, 8 cm across and are held above the water.

N. capensis, a native of South and East Africa and Malagasy is very similar to *N. stellata* but leaves are up to 40 cm across and the flowers have more petals (20-30), easy to grow, a day bloomer and tropical in nature, the green leaves are large, rounded and wavy-edged with splashes and blotches of purple beneath, the species is very beautiful and fragrant, and the flowers are star-like, stand well when cut uniformly, rich sky-blue with yellow stamens and back of outer anthers blue, opening for 4 days from 7 a.m. to 4 p.m., 15-20 cm across and stand 12-25 cm above water.

N.* × *caroliniana, a day-blooming hardy form believed to be *N. odorata rosea* × *N. tuberosa* [syn. *N. odorata caroliniana* (*carolinensis*), *N. odorata superba, N. tuberosa superba*], has stout rhizomes, leaves 30 cm across, rounded, green above and red beneath, flowers star-shaped with many petals, ivory-white or delicate rosy-pink with yellow stamens, fragrant, and 18-25 cm across.

N.* × *chromatella (syn. *N. marliacea chromatella,* probably a hybrid of *N. mexicana* × *N. alba* or *N. tuberosa*), hardy and free bloomer, leaves orbicular, dark-green marked bronze but coppery with purple streak when young, 8-20 cm across, flowers bright yellow with golden stamens, and 8-15 cm across with numerous incurved petals.

N.* × *daubeniana (a beautiful German hybrid of *N. micrantha* possibly with *N. caerulea*), a day-blooming tropical waterlily, leaves heart-shaped, brownish-green splashed with chocolate markings, not more than 25 cm across, strongly viviparous, prefers shallow water hence most suitable for tub or aquarium planting and tolerates shade to some extent, at a time produces several delicately fragrant and blue blooms touched with green underneath having lemon-yellow stamens, and up to 5 cm across.

N.* × *dentata (possibly a botanical variety of *N. lotus*), a native of Egypt and Sierra Leone is tropical night-blooming species, leaves glabrous, flowers white, 25-30 cm across and petals narrow, and its forms *N. d. grandiflora, N. d. magnifica* and *N. d. superba* (syn. 'Juno') are large and floriferous forms, the flowers of 'Superba' are quite large (22-25 cm across) and fragrant, and are held on hairy flower stalks, regenerating true through seeds.

N.* × *flavo-virens (syn. *N. gracilis* hort. and not Zucc.), a native of Mexico, day-blooming and tropical, propagates easily through rhizomes, leaves suborbicular, angle of lobes rounded, leaves 38-44 cm across, flowers white, fragrant, 15-20 cm across, sepals green, petals 16-20, stamens deep yellow and about 60, its progeny with *N. capensis* var. *zanzibariensis* is similar to *N. zanzibariensis* but has slender and acute petals and sepals, with less number of floral parts, and it has number of varieties and forms, the notable ones being 'Grossherzog Ernst Ludwig' (deep blue), 'Wm. Stone' (bright blue), 'Mrs. C. W. Ward' (pink), *N. gracilis rubra* (red), etc.

N. gigantea, a native of Australia is day-blooming and tropical in nature, leaves round, toothed and wavy margined, mid-green above and tinged pink to purple beneath, up to 60 cm across, lobes often overlapping, flowers star-shaped, sky- to purple-blue, 15-30 cm across, open for 7 days from 9 a.m. to 6 p.m., stamens 350-750 and bright yellow, sepals green, petals 18-50, and its white (*N. alba*) and pink forms (*N. rosea*) occur in Australia.

N.* × *gladstoniana (an hardy and sterile hybrid between *N. alba* and *N. odorata rosea* as described under *N. alba*), leaves round, dark green but bronze when young, 27-46 cm across, wavy and toothed margined, lobes overlapping, flowers fragrant, star-shaped, white with yellow stamens and 12-25 cm across.

N.* × *laydekeri (a day-blooming hardy hybrid between *N. alba rubra* and probably *N. tetragona*, most suitable for tub culture), rootstock erect and unbranched, leaves small to medium green spotted brown, flowers small to medium, pink, rose, red or purplish, and its various garden forms are *N. l. rosea* (pink turning to dark rose, flower size 5.0-7.5 cm across), *N. l. r. prolifera* (many-headed), *N. l. liliacea* (6 cm across, fragrant and soft rose turning to rose and crimson), *N. l. fulgens* (crimson to magenta with fiery red stamens), *N. l. lucida* (red), *N. l. purpurea* or 'Purpurata' (crimson-red flecked white with bright orange stamens), *N. l. rubra* (red), *N. marliacea ignea* (orange-pink), *N. m. flammea* (pink), *N. m. rubra punctata* (red), *N. sanguinea* (red), etc.

N. lotus (syn. *N. edulis, N. thermalis*), a native to the Tropics of the Old World, particularly Egypt and a night-bloomer, rootstocks thick and succulent, leaves deep green above and under surface brownish and hairy, round with slightly ruffled edges and toothed, 30-50 cm across, flowers star-shaped, white but outer broad petals suffused pink, 13-25 cm across, sepals green, petals 19-20, stamens yellow and 96 to103, and blooming during nights and days both, i.e. from 7.30 p.m. to 11 a.m., and this species has several varieties and garden forms.

N. micrantha, a tropical species native of West Coast of Africa, leaves elliptic, lobes spreading and bearing bulbs at the top of petioles for producing new small plantlets which grow from the central leaf vein resembling as the piggy-back plant (*Tolmeia menziesii*), and at this condition it may produce flowers along with its parent which looks extra-interesting, and this condition is known as vivIparity as is also found in *N. daubenyana*, flowers small, bluish-white and 7.5-13.0 cm across.

N. mexicana (syn. *N. flava*), a hardy native of South U.S.A. (Florida) and Mexico, rhizomes erect, tuber-like and 7.5-12.5 cm in diameter, leaves ovate and cup-shaped, dark green blotched brown above and dark crimson-brown with small blackish dots below, margins toothed and wavy, lobes overlapping, flowers 10 cm across and slightly fragrant, standing 10-12 cm above the water, bright yellow with light golden-yellow stamens numbering about 50, petals 23, and flowers opening from 11 a.m. to 4 p.m.

N. odorata, a hardy species native to North America, rootstocks horizontal and easy to remove, some 90 cm long and 2-5 cm thick, and are whitish with black hairs, highly suitable for shallow water and even in marshy land, leaves ovate to round, 15-30 cm across, dark green but purplish-red when young, green, reddish-green to dark red beneath, lobes diverging and may slightly overlap, flowers white, most fragrant among the American species, 7.5-13.0 cm across, opening for 3 days from 6 a.m. to 12 a.m., flowers freely, sepals green tinged reddish-brown exterior and purple inside, petals 23-32, stamens 55-113 and yellow, filaments white and petaloid, and its noteworthy forms and varieties are *N. o. alba* (white), *N. o. caroliniana* (syn. *N. o. superba, N. tuberosa superba,* believed to be hybrid between *N. o. rosea* and *N. tuberosa,* fragrant and rosy-pink), *N. o. gigantea* (pure white), *N. o.* 'Lactea' (opening white tinged apricot but soon turning to pure white), *N. o. minor* (dwarf, pure white, star-shaped and sweetly scented), *N. o. rosea* (pink), *N. o. sulphurea* (yellow), *N. o.* 'Sulphurea Grandiflora' (multi-petalled, sulphur-yellow, standing well above the water, shy bloomer but award winner), *N. o. turicensis* (sweetly scented pale-pink), etc.

N.* × *pennsylvania (a hybrid with *N. caerulea* and *N. zanzibariensis*) with beautiful, large deeper blue flowers, opening from 8 a.m. to 5 p.m., petals 23 and stamens 140.

N. pubescens, though beautiful and frequently used in the hybridization programme but not in cultivation, a hardy species native of India to Java, Philippines, America and Australia, leaves ovate, 25-30 cm long, green above and dull purplish-green beneath, flowers day-blooming, cup-shaped, small to medium (2.5-5.0 cm across), and white with yellow stamens.

N. pygmaea (syn. *N. tetragona* 'Alba'), a hardy species native to Ontario, North Idaho, East Siberia, China and Japan, the smallest *Nymphaea* in cultivation, ideal for aquaria, tubs, sinks and shallow pools, rootstock an erect and black rhizome covered with long hairs, no side-shoots from single crown, grows well from seeds which are quite large, leaves horse-shoe-shape, dark green blotched brown above and reddish beneath, 7.5 to 10.0 cm across, lobes diverging, flowers square at base, white, 3.75-6.25 cm across, free-bloomer and open for 3-4 days from mid-day to 5 p.m., petals 13-17 and stamens yellow and about 40, and its various attractive forms are *N. fennica* (botanically very similar to the parent, but purple-flowered, collected from Finland), *N. tetragona* var. *helvola* (syn. *N. helvola, N. pygmaea helvola,* probably a cross between *N. mexicana* and *N. tetragona*), ideal for planting in small pools and in tubs, a hardy type similar in shape to *N. tetragona,* leaves oval, 7.5-13 cm across and blotched brown all over, blooms yellow with orange stamens, star-shaped, small (5 cm across), floating and open in the afternoon like to that of *N. tetragona, N. himalayensis* (from Himalayas, smaller than the type), *N. orientalis* (collected from Japan, a miniature with coppery-green leaves), *N. pygmaea* 'Rubra', probably a natural hybrid, though it is slightly larger than many of the other *pygmaea* types but it is a true miniature and ideal for very small bowls and tubs, flowers up to 6 cm across are dark pink or rose, and is difficult to propagate, *N. vomerense* (probably a cross of *N. alba* and *N. tetragona,* flowers small and white), *etc.*

N. rubra, a native of India and is night-blooming, foliage bronze, leaves 30-45 cm across, flowers deep purplish to bright red, opening for 3-4 nights from 8 p.m. to 11 a.m., sepals dull purplish-red, 7-nerved, petals 12-20 with rounded apex, and stamens 55 and red which turn brown, and is most suitable for producing fertile hybrids with *N. lotus* and *N. l. dentata,* hence various colours starting from pink to deep purplish-crimson are available in the ponds such as i) delicate pink (*N. smithiana, N. laelia, N. l. colorans*), ii) light pink (*N. dedicatissima, N. deaniana*), iii) intense pink (*N. boucheana, N. indica spira,* 'Reine d'Italie'), iv) rosy-pink (*N. kewensis, N. micheliana*), v) red-pink (*N. startevantii*), vi) tulip shape (*N. rubicunda,* 'Hofgartner Graebner'), vii) magenta (*N. omarana*), viii) flowers wide open but petals narrow (*N. ortgesiano-rubra, N. mariae-legrangei, N. bissetii,* 'Adele', 'Niobe', *N. tulipifera*), ix) rosy carmine ('Diana', *N. indica* 'Brahma', *N. rubra rosea, N. devoniensis, N. arnoldiana,* 'Pres. Girard'), x) carmine (*N. krumbiegelii, N. columbiana, N. diana grandiflora*), xi) dark red (*N. purpurea,* 'Doctor Florenze', 'Geo. Huster'), xii) crimson to deep purplish crimson ('Frank Trelease', 'Rufus J. Lackland', 'James Gurney, Jr.'), etc.

N. stellata, this species is highly similar to *N. caerulea,* a strong and easy growing tropical species native of India and SE Asia, rhizome growth is very fast, leaves irregularly sinuate-toothed, purple beneath, up to 30 cm wide, flowers highly fragrant, sky-blue, star-shaped, 7.5-10.0 cm across and open for 3 days from 8 a.m. to 2 p.m. with 11-14 petals and 33-54 stamens, and its white and pink forms occur in India.

N. tuberosa, a hardy species native to NE United States which is moderate bloomer, ideal for large ponds having 60-120 cm water depth, the rhizomatous rootstocks, producing spontaneously separate tubers 2.5-7.5 cm long which help in getting it multiplied quickly, leaves more veiny above, petioles striped-brown longitudinally, flowers white, 10-23 cm across, opening for 3-4 days from 8 a.m. to 1 p.m., petals concave and seeds quite large (4.4 mm), and its garden forms are *N. t. robusta* and 'Helen Fowler', and the varieties *plena* ('Richardsonii') contains more petals than the original type, *N. odorata maxima* has white flowers whereas 'Rosea' bears pale-pink flowers.

N. zanzibariensis (syn. *N. capensis* var. *zanzibariensis*), a form collected from Zanzibar is tropical in nature and day-blooming (11 a.m. to 5 p.m.), leaves suffused-violet underneath, 20-38 cm across, leaf margins sinuate-dentate, flowers opening 3-5 days, deep blue and 30 cm across, sepals exterior green, margins purple and deep within, petals 18-24, and stamens 136-242 and dark blue while anthers yellow, and its forms *azurea* is light blue, *rosea* is pink, and there are many other forms and varieties such as 'Amethyst', 'L. Dittmann', 'Kobald', 'Countess of Warwick', 'Lord Brooks', etc.

Other noteworthy hardy varieties are: **Red:** 'Attraction' (garnet-red, flecked white, turning deep red with age), 'Charles de Meurville' (plum-coloured streaked white), 'Conqueror' (rose-crimson, outside petals and sepals white, stained and streaked red), 'Ellisiana' (brilliant red), 'Escarboucle' or 'Aflame' (brilliant red, sometimes the first blooms in each season are white, or white and red), 'Froebelii' (deep wine-red, slightly fragrant), 'Gloriosa' (crimson turning to brilliant red, semi-double, fragrant), 'James Bridon' (deep pink to rose-crimson), 'Newton' (vermilion-red, flowers stand well above the water), 'Rembrandt' (maroon-red with a pink flush on outer petals), 'Rene Gerard' (large rose-pink, blotched and striped with deep crimson), 'Sirius' (deepest red), 'Splendida' or 'Splendide' (ruby-red), 'Sultan' (cherry-red, petals tipped and flecked white), 'Vesuve' or 'Vesuvius' (crinkled petals and stamens deep red tipped yellow), 'William Falconer' (deeper red), etc.; **Pink:** 'Amabilis' or 'Pink Marvel' (light pink to salmon), 'American Star' (pale-pink, fading to white at tips, starry flowers with narrow pointed petals), 'Arethusa' (salmon-pink flushed with darker rose-pink), 'Brackleyi Rosea' (rose-pink, fragrant), 'Colossea' (pale flesh-pink), 'Eugene de Land' or 'Eugenia de Land' (one of the *N. odorata* hybrids, scented, semi-double, rich pink), 'Fabiola' (rich pink), 'Firecrest' (pale-pink changing to deep pink, and fiery-red stamens, faintly scented), 'Gloire du Temple sur Lot' (pale blush-pink but turns almost pure white the third day), 'Helen Fowler' (rich pink, well above the water, fragrant and large), 'Lustrous' (soft pink), 'Masaniello' (rose-pink with crimson overlaying at petal bases), 'Mary Patricia' (pink and peach), 'Mme Wilfron Gonnere' (double pink), 'Mrs. C.W. Thomas' (large, semi-double, pink, fragrant), 'Mrs. Richmond' (sweet-scented, pale rose-pink on opening, turning to deep crimson), 'Odalisque' or 'Opalisque' (probably a natural hybrid of *N. tuberosa*, starry blooms, soft pink turning rose), 'Pearl of the Pool' (bright pink), 'Pink Opal' (delicate pink), 'Pink Sensation' (fragrant, silvery-pink), 'Ray Davies' (pink, large and double), 'Rose Arey' (deep clear pink with yellow stamens, being borne 10-15 cm above the water level, fragrant and very beautiful), 'Rose Magnolia' (syn. *N. tuberosa* var. *rosea*, pale-pink), 'Rosy Morn' (large pink), 'Somptuosa' (double pink, deliciously scented), 'William B. Shaw' (an *N. odorata* hybrid, delicate rose-pink, fragrant, and held above the water), etc.; **White:** 'Albatross' (snow white), 'Caroliniana Nivea' (highly scented, large, double white), 'Gonnere' or 'Crystal White' or 'Snowball' (double white), 'Gladstoniana' (semi-double white, 15-30 cm across flowers), 'Hall Miller' (rich cream, flowers held well above the water), 'Hermine' or 'Hermione' (pure white, standing above the water), 'Gonnère' (fully-double white, 15-20 cm across), 'Loose' (white, sweetly scented, flowers held almost 30 cm above the water), 'Nivea' (fragrant, semi-double white 10-15 cm across), 'Virginalis' (large, semi-double, white), 'Virginia' (double, pure white, turning yellow in the centre), etc.; **Yellow, Copper and Changeable ones:** 'Aurora' (pale-yellow changing to copper and deep rose), 'Charlene Strawn' (fragrant yellow), 'Colonel A.J. Welch' (canary-yellow, flowers holding just above the water, reproduces viviparously from spent blossoms), 'Comanche' (apricot-yellow changing to rich coppery-red while still the outer petals remain yellow, flower held above the water), 'Graziella' (yellow, flushed with pink changing to copper-red), 'Indiana' (small flowers, peachy-pink changing to orange-red), 'Moorei' or 'Mooreana' (large pale lemon-yellow), 'Paul Hariot' (yellow, changing to coppery-red, slightly fragrant, flowers held well above the water), 'Solfatare' (pale creamy-yellow flushed rose turning reddish-orange, flowers held above the water), 'Sioux' (pale-yellow on opening, turning to deep orange, and then finally to a coppery-red), 'Sunrise' (deep primrose-yellow, holding well above the water level), etc.

The choicest tropical waterlily varieties are:

Day-blooming

Red: American Beauty, Jack Wood, Rio Rita

Pink: Castaliiflora, Evelyn Randig, General Pershing, Emily Grant Hutchings, Mrs. George C. Sturtevant Hitchcock, Leading Lady, Persian Lilac, Pink Perfection, Pink Platter, Pink Star

White: Isabelle Pring, Janice Wood, Louella G. Uber, Marian Strawn, Mrs. George H. Pring, Ted Uber

Yellow: African Gold, Aviator Pring, Eldorado, Mrs. Edward Whitaker, St. Louis, Trailblazer, Yellow Dazzler

Rosy-yellow: Afterglow, Albert Greenberg, Talisman

Purple: Director George T. Moore, Edward D. Uber, King of the Blues, Midnight, Mrs. Martin E. Randig, Pacific, Panama, Royal Purple

Blue: August Koch, Bagdad, Blue Beauty, Bob Trickett, Dauben, Leopardess, Margaret Mary, Margaret Randig, Pamela

Night-blooming

Red: Devoniensis, Frank Trelease, H.C. Haarstick, Mrs. John A. Wood, Red Flare, Rubra Rosea

Pink: Mrs. George C. Sturtevant

White: Janice Ruth, Juno, Missouri, Sir Galahad, Trudy Slocum, Wood's White Knight

Rosy-yellow: Emily Grant Hutchings, Maroon Beauty

Varieties for Special Purposes

A. Large or Medium Pond Varieties (water depth 45-70 cm)

White: 'Albatross,' 'Caroliniana Nivea,' 'Colossaea,' 'Gladstoniana,' 'Gloire de Temple-sur-Lot,' 'Gonnère,' 'Marliacea Albida,' *Nymphaea alba, N. candida, N. lactea, N. odorata, N. pygmaea alba, N. tuberosa* var. 'Richardsoni,' 'Virginalis,' *etc.*

Yellow: 'Aviator Pring,' 'Colonel A.J. Welch,' *N.* 'Marliacea Chromatella,' 'Moorei,' *N. odorata* 'Sulphurea,' *N. pygmaea* 'Helvola,' 'Sunrise,' *etc.*

Pink: 'Amabilis,' 'Baroness Orczy,' 'Brackleyi Rosea,' 'Colossea,' 'Fabiola,' 'Fire Crest,' 'Formosa,' 'Laydekeri Lilacea,' 'Leviathan,' 'Jean de Lamarshalle,' 'Marguerite Laplace,' 'Marliacea Carnea' ('Morning Glory'), 'Marliacea Rosea,' 'Masaniello,' 'Mme Wilfron Gonnère,' 'Mrs Richmond,' 'Mrs C.W. Thomas,' *N. odorata* 'Turicensis,' *N. odorata* 'W.B. Shaw,' *N. tuberosa rosea*, 'Rose Arey,' *etc.*

Red: 'Arethusa,' 'Atropurpurea,' 'Attraction,' 'Bory de Saint Vincent,' 'Charles de Meurville,' 'Conqueror,' 'Escarboucle,' 'Gloriosa,' 'Froebeli,' 'James Brydon,' 'Marliacea Rubra Punctata,' *N.* 'Laydekeri Purpurata,' 'Newton,' 'Picciola,' *N. pygmaea* 'Rubra,' 'René Gérard,' 'William Falconer,' *etc.*

Blue: 'Blue Beauty,'

Striped: 'Darwin' (fragrant, red with heavy white variegation), 'Eucharis' (rose, heavily flecked and splashed white), 'Graziella' (reddish-yellow flowers becoming paler with age and foliage variegated with purple), 'Indiana' (flowers opening apricot-red, darkening to rich copper-red, and foliage suffused with purple), 'Livingstone' (fragrant, red, striped white), 'Paul Hariot' (flowers opening apricot-yellow, turning orange-pink, and finally almost red, foliage maroon spotted), 'Sioux' (soft yellow suffused red, deepening to reddish-copper), 'Solfaterre' (star-shaped flowers are yellow flushed rose, foliage green mottled with maroon), *etc.*, all suitable for small and medium-sized pools.

B. Small Ponds or Tubs (water depth 30-45 cm)

White: 'Albatross,' *N. caroliniana nivea*, 'Hermine,' 'Lactea,' *etc.*

Yellow: 'Aurora,' 'Comanche,' 'Graziella,' 'Indiana,' 'Paul Hariot,' 'Phocbus,' 'Robinsoniana' ('Robinsoni'), 'Sioux,' 'Solfatare,' *etc.*

Pink: 'Eugenia de Land,' 'Laydekeri Lilacea,' *Nymphaea caroliniana, N. odorata rosea, N. o.* 'Turicensis', *N. o.* 'W.B. Shaw,' 'Somptuosa,' *etc.*

Red: 'Ellisiana,' 'Froebeli,' 'Fulgens,' 'Laydekeri group,' 'Purpurata,' 'Sanguinea,' 'William Falconer,' 'Vesuve,' *etc.*

C. Pygmy Varieties for Rock Garden Pools and Bowls (water depth 7.5-23 cm)

White: *Nymphaea candida, N. odorata minor* (*N. o. pumila*), 'Pygmaea Alba,' *etc.*

Yellow: *N. odorata sulphurea*, 'Pygmaea Helvola,' etc.

Red: 'Maurice Laydeker' (flowers flecked white), *etc.*

Cultural Practices

The rootstock varies from species to species. In case of *N. odorata* and *N. tuberosa*, it extends for several feet, bearing tubers at intervals which can be detached easily and planted. Certain other types have stout and branching rhizomes, extending up to 60 cm or even more in some cases which are black outside and yellow within and run horizontally exposing the upper surface out of the mud but securely held in the mud by their true roots. Tubers or rhizomes are thinned out in April and then planted in a pond having soil depth of 20-30 cm for large plants, 15-20 cm for medium type of plants, 10-15 cm for small types and 7-10 cm for miniatures. While planting in containers, the soil is saturated before planting. In the ponds these are planted in the shallower water than is actually required as waterlilies have the tendency to grow in the deeper water. Main offsets are also detached from the mother plants with a sharp knife after lifting and washing them in April-May. Offsets are planted separately in the small earthen pots which are placed gently in a tank or pond having clear water at a sunny location in a way that water is about 1-2 cm over the pot rims while the mother rhizome is planted at its original place. After these offsets are properly rooted, these are planted at their desired place permanently.

The temperature for tropical waterlilies should be maintained at 18-21^0 C (Hay and Beckett, 1971) while raising them. When growth starts, first a few thin shoots appear which develop arrow-shaped leaves and roots, and then very soon the surface leaves appear. This is the stage when the young plants are separated just above the rhizome and planted as elaborated above or should be potted in a small pot filled with growing medium and then putting these pots in shallow water, and when convenient or when these attain a stage of 3-4 new leaves these should be planted out at their permanent positions.

The plantlets emerging in the leaf centre of viviparous waterlilies (*N. daubenyana, N. micrantha*) should also be detached and planted the same way or the part of such leaves should be touched to the soil mixture filled in a pot or directly to the mud nearby and when the plants start such plants are removed and planted. The variety 'Colonel A.J. Welch' reproduces viviparously from spent blossoms so the plantlets from this may also be treated as those detached from the viviparous leaves.

Species usually breed true through seeds. Certain horticultural varieties and even certain cultivated varieties also breed true through seeds. Where propagation through rhizomes in such varieties and species is cumbersome, these can be raised through seeds. In certain cases the seeds are quite large with excellent germination. Seeds can be sown in pans filled with good garden loam, clayey soil, fibrous loam and leaf mould, all in equal volume. The seeds should be sown thinly and shallowly in March-April and then the pan should be filled with water slightly above the soil level and the pans should be kept at sunny locations but at a temperature range of 18-22^0 C with utmost care of these pans unless the seeds germinate and attain 3-leaf stage. Under open conditions when temperature starts rising above 25^0 C these should be shifted to well-ventilated glasshouses so that the seeds as well as emerging plants may be saved from sun-boiling. After attaining 3-4 leaf stage, the seedlings are planted at their permanent locations as elaborated under viviparous plants and offsets.

The soil of the pond possesses all the conditions including what these plants require, however, the cemented ponds require filling of mixture to a depth of 30 cm for larger varieties and as low as 10 cm for miniatures. Any soil which is proper for growing most of the ornamentals or vegetables will be suitable even for waterlilies with slight modifications. A compost consisting of $^2/_3{}^{rd}$ good soil + $^1/_3{}^{rd}$ well-rotten compost or dung manure and a little quantity of bone meal is suitable for its cultivation. Bone meal is relatively slow to release its nutrients and it encourages the plants to develop strong root system. Once established, these will continue flowering year after year but tropical waterlilies once in a year and the hardy ones once in two-years require to be fed where slow-release fertilizers should be used. Wester (1951) with *Nymphaea zenkeri* var. 'H.C. Haarstick' reported that 300-600 g/m^3 of 10-6-4 NPK fertilizer containing ammonium sulphate supplied to the soil was found quite superior with regard to leaf area, flower size and flower production compared to cow manure or nitrate of soda replacing the ammonium sulphate.

The flowering commences in April and continues till autumn. When growth stops in the autumn, the water level, if possible, should be gradually reduced for facilitating the rhizomes to mature and the old leaves should be cut from the base (not pulled up). After maturing, the rhizomes may be lifted, surface-dried and stored in moist sand at a well ventilated place at 10^0 C or should be left as such in the tank or pond.

Insect-Pests and Diseases

Pests and diseases afflicting waterlilies are difficult to control through chemical means because of the other occupants of the water such as fish and tortoises and to maintain proper ecological balance in the surrounding. Though no serious insect-pests attack waterlilies, *vis-à-vis* fish also control various insects by devouring the larvae but even if the problem is menacing, handpicking of the infected areas is recommended. Where no any other alternative is left except the chemical control, the affected plants should first be taken out of the pond or pool and tubs, and immersed in dilute concentrations of insecticides or fungicides for a few hours and then such plants are thoroughly rinsed and planted back after removing the diseased parts. Various insect-pests which attack waterlilies are furnished in various publications (Frances Perry, 1962; Anon., 1989). **Waterlily aphids** (*Rhopalosiphum nymphaeae*) are troublesome when these colonize the leaves. At initial stage of infestation these should be physically removed. Even through force-spraying of water through fine jet on the leaves will dislodge the aphids in the water and the fish present there will devour on them. '**Waterlily beetle**' (*Galerucella nymphaeae*) lays eggs on the waterside vegetation and also on waterlily leaves from which larvae hatch out within a week which feed voraciously on the waterlily leaves and flowers. Force water jetting for aphids will control even beetles. Its population can also be kept under check through proper sanitation of the surroundings. A solution containing powdered *Bacillus thuringiensis* bacteria spraying will parasitize the larvae. '**Brown China mark moth**' (*Nymphula nymphaeata* and *N. stagnata*) lays cluster of eggs on the underside of leaf edges from which white larvae hatch and infest on the leaves by making holes around the leaf margins and nesting by making a protective dwelling case on the leaves. The spraying with *Bt.* bacterium or removal of the whole aerial part will keep the menace under control. '**Leaf mining midge**' (*Cricotopus ornatus*) adult lays eggs on the leaves. Larvae after hatching start tunnelling the surface tissues of the leaves, the damage being more conspicuous on pygmy waterlilies though the varieties with leathery leaves are more resistant. Removal of infested leaves and use of *Bt.* bacterium spraying will keep this problem under check. '**Caddis flies**' (*Trichoptera* species) lay the eggs during the evening in a form of protective jelly on submerged or other aerial parts. Larvae after hatching spin a web of thread around them for hiding but in the night these come out and feed on the aerial parts. If the goldfish is there in the water body this devours on these larvae. *Bt.* spraying will control this pest also. **Mosquito larvae** attack the leaves and buds of young plant which can be devoured if golden carp, golden rudd, golden orfe and minnows are introduced in the water. The '**common pond snail**' or 'fresh water whelk' (*Limnaea stagnalis*) sometimes becomes a pest of waterlily when alternative food is insufficient. Sometimes in search of algae it reaches the waterlily plants and starts feeding them. These can be trapped and killed by floating cabbage or lettuce leaves for 24 hours in the pool.

Various fungi cause '**leaf spot**' by initially developing the dark patches on the leaves, drying of the leaf margins and turning up. Such leaves should immediately be removed and burnt. Three sprayings at alternate days with mild solutions of Bordeaux mixture will also control this problem. '**Crown rot**' (*Phytophthora* species) causes blackening of the crown and the stem bases and their subsequent rot. Through this infection first leaves start yellowing and when rotting starts a typical smell emanate from the rotted parts. Such plants along with the adjoining soil should be removed and burnt and replaced by another healthy plant. Dalmer (1962) used cheescloth bags containing captan, ferbam, maneb and phaltan at quantities sufficient to saturate the water in saparate tanks and recorded captan and phaltan limiting the spread of the **leaf spot-rot** *Helicoceras nymphaerum* three weeks after application. He reported that captan at one tablespoonful of 50 per cent w.p. per 3 gal. sprayed onto floating leaves at 2.5 gal/9m^2 of tank surface controlled the development of rot and reduced the population of leaf chewing insects, so he recommended saturation of tank water with captan+a captan spray on the leaves as most effective.

'Blanket weed', 'green water' and 'dark or milky water' problems are encountered in waterlily cultivation. Blanket weeds to some extent may be removed through net or by the use of certain algicides which will also together control greening of the water but the latter problem appears soon which can be checked by proper aeration of the water *vis-à-vis* direct exposure to sun. Through balanced planting which may not hinder availability of mineral salts to the algae will provide cleaner water. Dark or milky water can be cured by collecting rotting plant debris from the base of the pond.

References

Anonymous, 1989. *Collins Guide to Waterlilies and Other Aquatic Plants*. William Collins Sons & Co. Ltd., London.

Bailey, L.H. 1929. *The Standard Cyclopedia of Horticulture* (vol. II), pp. 2306-2315. The Macmillan Co., New York.

Banerji, B.K. 2006. *Nymphaea*. In: *Advances in Ornamental Horticulture*, vol. 3 (ed. Bhattacharjee, S.K.), pp. 322-333. Pointer Publishers, Jaipur (India).

Beckett, K.A. 1987. *The RHS Encyclopaedia of House Plants Including Greenhouse Plants*, p. 363. Salem House Publishers, Massachsetts, USA.

Brickell, C. 1994. *The Royal Horticultural Society Gardeners' Encyclopedia of Plants & Flowers*, pp. 551-552. Dorling Kindersley Ltd., London.

Dalmer, J.G. 1962. A method to control leaf spot-rot of water lilies in greenhouse propagating tanks (abstr.) *Phytopath.,* **52**: 165

Frances Perry, 1962. *Water Gardens*, pp. 12-14, 45-61, 96-98, 153-157. Penguin Books Ltd., Middlesex, England.

Gupta, P.P. 1978. Cytology of aquatic ornamentals . II. Cytology of *Nymphaea. Cytol.*, **43**: 477-484.

Gupta, P.P. 1980. Cytogenetics of aquatic ornamentals. VI. Evolutionary trends and relationships in the genus *Nymphaea. Cytol.*, **45**: 307-314.

Hay, R. and Beckett, K.A. 1971. *Nymphaea*. In: *Reader's Digest Encyclopaedia of Garden Plants and Flowers*, pp. 471-474. The Reader's Digest Association Ltd., London.

Sanyat Misra and R.L. Misra, 2013. *Nymphaea*. In: *Commercial Ornamental Bulb Science*, pp. 313-323. Westville Publishing House, New Delhi.

Sen, S. and P.N. Bidhuri, 1971. Cytotaxonomy of Nymphaceae. *J. Cytol. Genet.*, **6**: 67-89.

Subramanyam, G.V. 1979. A trisomy *Nymphaea* hybrid. *Curr. Sci.*, **48**(1): 34-35.

Subramanyam, G.V. 1980. Meiotic behavior of *Nymphaea* × *duberiana* Hort. *Indian J. Hort.*, **37**: 314-315.

Subramanyam, G.V. and T.N. Khoshoo, 1981. Cytogenetics of cultivated nymphaeas. II. Induced hexaploidy and restoration of fertility in a sterile triploid. *The Nuceus*, **24**(182): 56-60.

Wester, H.V. 1951. Comparative growth response of Haarstick water lily to ammonium and nitrate nitrogen in 10-6-4 fertilizer as compared to cow manure. *Proc. Amer. Soc. hort. sci.*, **58**: 324-328.

23

Paeonia

Jayoti Majumder, T.N. Saha, P. Sellam, Sanyat Misra and R.L. Misra

Family: **Paeoniaceae,** formerly **Ranunculaceae**

Common names: Balkan paeony (*P. mascula*), Brown's paeony (*P. brownii*), California paeony (*P. californica*), Chinese paeony/ Shaoyao (*P. lactiflora*), Chinese tree paeony (*P. suffruticosa*), Delavay's tree paeony (*P. delavayi*), European paeony (*P. officinalis*), Fringed paeony (*P. tenuifolia*), Golden paeony (*P. mlokosewitschii*), Greek paeony (*P. parnassica*), Japanese paeony (*P. japonica*), Jishan paeony (*P. jishanensis*, syn. *P. spontanea*), Keserwan paeony (*P. kesrouanensis*), Ludlow's tree paeony (*P. ludlowii*), Majorcan paeony (*P. cambessedesii*), Moutan/tree paeony (*Paeonia suffruticosa*), Osti's paeony (*P. ostii*), Qiu's paeony (*P. qiui*), paeony, peony, piney, Rock's or tree paeony (*P. rockii,* syn. *P. suffruticosa* subsp. *rockii*) and Veitch's paeony (*P. veitchii*)

Introduction and History

Genus *Paeonia* commemorates the mythical Greek physician Paiŏn who treated Greek gods, and in gratitude for his many cures he was changed into a plant by Pluto (Beckett, 1978). Theophrastus noted that the name *Paeonia* was derived from Paeon, a disciple of Aesculapius, who with a poultice prepared with *Paeonia officinalis* healed Hades of a wound inflicted by Hercules (Pizzetti and Cocker, 1968). It is the lonely genus of the family, with 33 species of herbaceous perennials and deciduous shrubs from temperate South Europe, temperate Asia (Himalayas, Yunnan, Tibet, Bhutan) and Pacific Coast areas of NW North America, though sole American species, *Paeonia brownii* is of little interest. It is a favourite flower for centruries. Gardeners throughout Europe, Asia and North America regard paeony as one of the easiest and most rewarding plants to grow. With its long-lasting and colourful blooms, the paeony is also popular as a cut flower used in wedding bouquets and in the large floral arrangements.

Paeonies are one of the oldest cultivated ornamental plants. Its history follows two separate paths, the Western species *Paeonia officinalis* which is valued for its medicinal properties, and the Asiatic species which in American and European gardens are cultivated only for their ornamental potential. John Gilbert Baker in 1884 divided paeonies into herbaceous types - where the disc is never wrapped around or overlaying the base of the carpels, and shrubby or woody types - where the disc wraps round or overlay the base of the carpels, and this division is still valid. Among the herbaceous paeonies, and apart from the hybrids of the European *Paeonia officinalis*, it is only the Chinese *Paeonia lactiflora* hybrids which have been widely cultivated in European and American gardens, however, the remaining species were being grown only by some specialist growers out of curiosity and not for their decorative values as in case of *P. lactiflora* and *P. officinalis* hybrids. Its cultivation dates back nearly 1,400 years ago in China (Dole and Wilkins, 1999) as *Paeonia lactiflora* has been grown in China since the fifth century B.C., where it is known as a symbol of prosperity,wealth, and sensuality (Goody, 1993). The history of cultivation of *P. lactiflora* in China is older than *P. suffruticosa*. At 900 B.C. when in China there was a craze for Moutan (*Paeonia suffruticosa*, syn. *P. moutan*; the tree paeony) in C. & SE China, this period matched even with cultivation of *P. lactiflora* with the main centre being Yangchow in China where its cultivation continued until 1,600 when Chu family gardens had its 60,000 plants with one variety or the other. *Paeonia lactiflora* was introduced into European garden cultivation probably in 1820, and since then more than 3,000 hybrids have been raised (Pizzetti and Cocker, 1968). In about A.D. 750, the Moutan, *P. suffruticosa* was introduced into central and south-east China from the wild from Shensi, Kansu and Szechuan provinces of China. The summer palace of emperor Ming Huang was then planted with 10,000 Moutan bushes of various colours, and Quyang Hsiu (1,007-1,072) recorded and described over 90 cultivars with their grafting method of propagation and other cultivation practices. For centuries the Loyang area was the centre of Moutan cultivation but after 1,700 this was shifted to Ts'aochow and Fahuahsiang near Shanghai. However, this reached to Kew gardens in 1789 through Dr. Duncan of the East India Company but this died soon. Again in 1794, a ship *Triton* reached London with seven live plants of tree paeony on board, one for the King, two for Sir Joseph Banks and four for another person, but during this difficult voyage only five plants survived. Robert Fortune in 1842 introduced about 40 varieties into Europe. The Chinese had developed more than 300 paeony varieties of *P. suffruticosa* with flowers in various shades of white, yellow, pink, amaranth-red, purple, violet-blue, striped, bicoloured and various other shades, and they trained the plants in various shapes, *viz.* fan-shaped, globular, as espaliers, as miniature trees, and so on. In 1903, the Qing Dynasty declared the paeony as the national flower of China, though now the Republic of China on Taiwan designates the plum blossom as the national flower. In Japan, *Paeonia suffruticosa* is called the 'king of flowers' and *Paeonia lactiflora* the 'prime minister of flowers'. Abbéy Delavay discovered in 1882 *P. lutea* from Yunnan, Tibet and Bhutan. This grows about 1.8 m high (Everard and Morley, 1970).

In times of famine the roots of *Paeonia officinalis* have been cooked and eaten, and the Tarters ate *Paeonia lactiflora* roots without any ill-effects, and only because of this, Salisbury named this species as *Paeonia edulis*. In certain parts of Japan, the natives prepare a delicious dish from the flowers of *Paeonia suffruticosa* (Pizzetti and Cocker, 1968). Early records are of the flower's medicinal uses dating back to 200 B.C., although their specific use as a treatment is not known (Rogers, 1995). In Japan, *Paeonia lactiflora* used to be treated as foreign medicine. Its root was used as a treatment for convulsions. However, Halda and Haldova (2004) point out that the flowers are quite toxic and children and pets should not ingest its any part. However, Hippocrates and Galen were so impressed with its healing virtues that they recommended it for relieving insomnia and nightmares, for treating the epilepsy, eclampsia or insanity, *vis-à-vis* as a cure for blisters, and for persisting coughs and catarrh. It is also believed that seeds of *Paeonia officinalis* help ease the discofort of teething. Modern researches have confirmed that the essential drug is contained in the flower petals, which should be gathered when fully expanded and shade-dried, *vis-à-vis* in the seeds and in tuberous roots. Over 262 compounds have been obtained so far from the plants of Paeoniaceae. These include monoterpenoid glucosides, flavonoids, tannins, stilbenoids, triterpenoids and steroids, paeonols, and phenols. Biological activities include antioxidant, antitumor, antipathogenic, immune-system-modulation activities, cardiovascular-system-protective activities and central-nervous-system activities.

Morphological Descriptions and Cytology

Though the paeony generally refers to herbaceous varieties, but there are three general forms. Two major forms of the genus are herbaceous and tree, while the third as 'Itoh' or intersectional hybrids, developed by crossing herbaceous and tree paeonies. These Itoh hybrids take their name from Toichi Itoh who first successfully hybridized them in 1948. Crossing the herbaceous and tree forms of paeonies allows hybridizers to create plants taking the bush form of the more commercially popular herbaceous varieties, but in colours these are only tree varieties. The focus of this study is on herbaceous paeonies, which are commercially the most popular ones in the United States, and have been most extensively studied in Alaska. Thus the term paeony in this work refers to herbaceous paeonies unless otherwise specified.

Paeonies have undergone extensive reticulate evolution comprising of three sections with some 35 diploid (2n =10) and tetraploid species. Diploid species of this section have never been considered to be of hybrid origin. All the tetraploid species are from herbaceous section of *Paeonia*, majority of tetraploids occurring in the Mediterranean region are almost allopolyploids whereas the origin of the putative allotetraploids is not known. Stebbins (1937) stated that crossing between *P. albijora* and *P. triternata mlokosewitschii* has never been possible irrespective of the both being diploids, and when two quite similar species such as *P. anomala* and *P. albijora* are crossed, one faces a very difficult task in getting any success, but the crossing two differently appearing species such as *P. anomala* with *P. triternata mlokosewitschii* or *P. anomala* with *P. tenuifolia*, one very likely gets the success. However, tetraploids if are placed together in the garden, these are likely to cross naturally though such hybrids are partially or fully sterile. He further demonstrated that *P. albiflora* though does not cross with other diploid species but with the species and varieties of the two tetraploid species, *viz. P. officinalis* and *P. tomentosa*, the success is most likely to be obtained. The diploidy status with 2n = 10 had been obtained in *P. albiflora, P. anomala,*'Veitchii', *P. anomala* 'Woodwardii', *P. tenuifolia, P. tenuifolia* × *P. albiflora, P. tenuifolia* × *P. anomala* 'Veitchii', *P. tenuifolia* × *P. anomala* 'Woodwardii', *P. trilernata* 'Mlokosewitschii', *P. trilernata* 'Mlokosewitschii' × *P. anomala* 'Veitchii' and *P. trilernata* 'Mlokosewitschii' × *P. anomala* 'Woodwardii'; 2n =15 in *P. albiflora* × *P. officinalis* & *P. albiflora* × *P. tomentosa*; and 2n = 20 in case of *P. tomentosa*. The F_1 population obtained from the same cross as in case of *P. albijora, P. officinalis* or *P. Suffruticosa*, provides the most striking results with great variation in F_1 population. This is because all these three species have been cultivated for centuries with a large number of their horticultural varieties, the result therefore reflects the heterozygous character of one or both the parents in the progeny. Sometimes, occasional dwarf plants with strap-shaped petals or any other drastic variation is observed in F_1 hybrids of *P. albijora* × *P. tomentosa, P. albijora* × *P. officinalis, P. albijora* × *P. tenuifolia* and *P. delavayi* × *P. suffruticosa*. The F_1 population is quite homozygous in case when the species used in crossing programme are pure with no much horticultural cultivars. The mode of inheritance in case of triploid hybrids is not better understood.

Paeonies, especially the herbaceous types are fully hardy and tuberous rooted. These grow from an underground crown and have either pointed or large, bulky roots. After the blooming season, stem buds—called 'eyes'—form at the base of stems. These buds are the source of new stems in the spring. Tree paeonies, some of which are rhizomatous, grow to eye level on woody stems with few branches. The stems of these peonies stay alive all winter, and they bloom early. They are valued for their bold leaves, the beautiful large flowers, and in some species colourful seed pods. The leaves are alternate and composed of several leaflets of irregular shape and dissimilar size, which may be unlobed or further lobed. Flowers above the leaves are large, showy, globular or bowl-shaped and terminal, with one to three lateral buds, blooming in May and June only in the temperate regions. Flowers open out flat when fully open, and of the herbaceous species are excellent for cutting. Fruit is a cluster of 2-8 pod-like follicles, usually red within. Seed pods of many species mature and ripen during autumn, and these are open wide revealing large glossy blue-black seeds when these are fertile, and oftenly small and red when sterile, which on sowing may take up to three years to germinate. From seeds, herbaceous paeonies should not be disturbed at least for 3-4 years in the nursery rows and may then be set out in permanent positions from September to March for flowering the same year. Tree paeonies, from seeds, take 10-12 years for flowering. First the seeds are sown in a mixture of peat and leafsoil in the seed bed at a shady location, and after germination should be transplanted individually in pots and grown there for 2-3 years and then are planted at permanent positions.

Classification, Species and Varieties

Paeonia lactiflora is long-lived and cold-hardy herbaceous perennial. The present day common modern herbaceous paeony hybrids are largely the result of crosses between *P. lactiflora* and *P. officinalis* (Nau, 1996). All forms of paeonies are perennials, with herbaceous types blooming once per year for months together during spring to summer, while tree and Itoh varieties potentially bloom twice in a season. In herbaceous varieties stems emerge in the spring with renewal buds originating on the crown. In the fall the leaves of the plant senesce and die, and the plant enters dormancy. Herbaceous peonies are very long-lived, potentially producing flowers for 30 years or longer before it becomes necessary to the clustering of roots and then these are replanted separately.

Tree paeony belongs to the section *Moutan* DC and the genus *Paeonia* L. (Stern, 1946; Pan, 1995). In addition to this woody group, there are two herbaceous sections in this genus, sect. *Paeonia* and sect. *Onaepia* (Stern, 1946; Pan, 1995). Bailey and Bailey (1976) stated the division of 33 *Paeonia* species into three sections, four subsections, 16 groups and 12 botanical varieties. The genus *Paeonia* consists of more than 33 species, distributed widely in the temperate region of the world (Pan,1995), out of which the section *Onaepia* has only two species endemic to North America, whereas sect. *Paeonia* comprises about 22 species found in Europe, North-West Africa and Asia, spreading from Portugal and Morocco to Japan (Stern, 1946; Pan, 1995), and the sect. *Moutan,* following a recent classification system of Hong and Pan (1999), contains eight species, and three of them each having two subspecies, distributed in southwestern, central and northern regions of China (Hong and Pan, 1999).

Tree paeonies are not used for cut flowers, though they are popular in gardens. The paeonies are classified according to their flower types (Halda and Haldova, 2004) as under.

Single: Flowers are mostly cup-shaped, each with 1-2 rows of large, mostly lightly ruffled and incurving five or more guard petals with a boss of pollen bearing stamens in the middle. Flowers often stand erect without support. Important varieties under different colours are (i) white – 'Le Jour', 'Krinkled White', 'Hedgemaster', etc., (ii) pink – 'Sea Shell', 'Helen' etc., and (iii) red – 'Arcturus', 'Kickapoo', etc.

Semi-double: These bear flowers similar to singles but have 2-3 rows of petals. No clear distinction is found between the guard petals and many of the petaloides. Petaloides may be in all stages of development, and a few residual pollen bearing stamens may be scattered among the petaloides. Flowers will droop to the ground without support.

Double: Flowers are rounded with usually 1-2 outer rows of large, mostly little ruffled and incurving petals. Other petals are smaller and oftenly more densely arranged with reduction in size towards the centre. These are fully double in which carpels and stamens are fully changed to petaloides and the guard petals are not distinguished from them. However, sometimes a few stamens may be found. Flowers will droop to the ground without support. Important varieties in different colours are (i) white and blush – 'Alma Hansen', 'Avalanche', 'Amelia Olson', 'Festiva Maxima', 'Elsa Sass', etc., (ii) light pink - 'Myrtle Gentry', 'Mrs. Franklin D. Roosevelt', 'Therese', 'Doris Cooper', etc., (iii) medium to dark pink - 'Queen of Hamburg', 'Sarah Bernhardt', 'Mons. Jules Elie', 'Mrs. Livingston Farrand', etc. and (iv) red - 'Karl Rosenfeld', 'Phillipe Rivoire', 'Kansas', 'Highlight', 'Longfellow', 'Felix Crousse', etc.

Japanese: A transition type between the singles and doubles. There are five or more petals and the stamens are partially transformed into short, quite narrow petals which are referred to as staminodes. May require support to hold flowers erect. Important varieties under this category are (i) white and blush – 'Plainsman', 'Isani-Gidui', 'Moon of Nippon', etc., (ii) 'Westerner', 'Tokio', etc., and (iii) 'Charm', 'Mikado', 'Mrs. Wilder Bancroft', etc.

Anemone: Japanese and anemone types are very similar. In the anemones, however, the stamens are fully developed into narrow petals called petaloides without any residual pollen bearing stamens. May require support to hold flowers erect. Varieties are 'Colonel Heneage' (dark rose-crimson), 'Evening World' (blush-pink), 'Globe of Light' (fragrant, rose-pink), 'Kelway's Majestic' (fragrant, cherry rose-pink),

Fern-leaf paeony (*Paeonia tenuifolia*): It has finely divided foliage and produces single, dark red flowers. They are one of the earliest paeonies to bloom in spring. Plants commonly grow 35 to 45 cm tall. The foliage dies back by midsummer. 'Rubra Plena' has double, dark red flowers. Their nature and cultivation are almost the same as for garden paeonies.

Colourwise cut flower varieties

(i) Red: Early - 'Convoy', 'Firebelle', 'Flame' (single), 'Kansas Red Charm', 'Raspberry Ice', 'Red Grace', 'Red Red Rose', 'Richard Carvel', 'Rosedale', etc.; midseason – 'Big Ben Howdy', 'David Harem', 'Felix Crousse', 'Felix Supreme', 'Gerry Mt', 'Karl Rosenfield', 'Lora Dexheimer', 'Midnight Sun' (Japanese), 'Paul M. Wild', 'Red Comet', 'Renato', 'St. Helens', 'Shawnee Chief', etc.; and late – 'Old Faithful', 'Philippe Rivoire', etc.

(ii) Pink: Early - 'Coral Charm', 'Cytherea' (semi-double), 'Edulis Superba', 'Fairy's Petticoat Monsieur', 'Jules Elie', 'June Rose', 'Mister Ed', 'Princess Margaret' (semi-double), etc.; midseason - 'Angel Cheeks', 'Gay Paree', 'Largo' (Japanese), 'Mrs. F.D.R.', 'Mrs. Livingston Farrand', 'Nancy Nicholls', 'President Taft', 'Queen of Sheba', 'Raspberry Sundae', 'Riene Hortense', 'Sea Shell' (single), 'Sweet 16', 'Top Brass', 'Walter Faxon', etc.; and late – 'Hansina Brand Minuet', 'Hermoine Nick Shaylor', 'James Pillow', 'Sarah Bernhardt', 'Vivid Rose', etc.

(iii) White: Early - 'Charlie's White', 'Duchesse de Nemours', 'Festiva Maxima', 'Mme de Verneville', 'Miss America' (semi-double), '69A', etc.; midseason – 'Ann Cousins', 'Baroness Schroeder', 'Bridal Icing', 'Bridal Shower', 'Festiva', 'Florence Nicholls', 'Gardenia', 'Gay Paree', 'Lois Kelsey', 'Marshmallow Puff', 'Mother's Choice', 'Supreme', etc.; and late – 'Elsa Sass', 'Louise Marx' (Japanese), 'Lullaby', 'Mary E. Nicholls', 'Mrs. Frank Beach', etc. 'Alice Harding' (fragrant, double), 'Cheddar Cheese' (double, ivory-white), 'Dresden' (single, ivory-white, tinged blush-rose-pink), 'Duchesse de Nimours' (syn. Mrs. Gwyn Lewis; double, fragrant, white), 'Krinkled White' (single, white), 'Laura Dessert' (fragrant, double; creamy blush-white), 'White Wings' (single, fragrant; white), 'Whitleyi Major' (ivory-white), etc. are other outstanding cultivars.

Pizzetti and Cocker (1968) mention that *Paeonia officinalis* flowers though has a very specific unpleasant odour but the flowers of *Paeonia lactiflora* and its modern, particulary the French hybrids emit an amicable rose fragrance and some of the such varieties are 'Adolphe Rousseau', 'Auton's Pride', 'Baroness Schroeder', Belle Chinoise', 'Big Ben', 'Dr. J.H. Neelsy', 'Eloise', 'Flower Girl', 'Georgina Shayler', 'Kelways Glorious', 'Laura Desert', 'Martin Cahuzac', 'M. Jules Elie', 'Mrs. Franklin D. Roosevelt ', 'Myrtle Gentry', 'Old Siwash', 'Philippe Rivoire', 'Siloam', 'Sistie', 'Walter Faxon', etc. It is mentioned that all the tree paeonies and their hybrids are fragrant but this does not exactly seem so. However, the varieties 'Alice Harding', 'Asahi Minato' (White Port), 'Chromatella', 'La Lorraine', 'L'Esperance', 'Mine d'Or', 'Roman Gold', 'Saigyo Sakura' (Sakura Cherry), 'Sang Lorraine', 'Shikoden' (Palace of Violet Light), 'Souvenir de Maxime Cornu', 'Surprise', etc. are highly fragrant.

There are more than 30 species in **herbaceous type** of paeony such as *Paeonia abchasia, P. anomala, P. bakeri, P. broteri, P. brownii, P. californica, P. cambessedesii, P. caucasica, P. clusii, P. coriacea, P. daurica,P. emodi, P. hirsuta, P. intermedia, P. japonica, P. kesrouanensis, P. lactiflora, P. macrophylla, P. mairei, P. mascula, P. mlokosewitschii,P. obovata, P. officinalis, P. parnassica, P. peregrine, P. rhodia, P. sinjiangensis, P. sterniana, P. steveniana, P. tenuifolia, P. tomentosa, P. veitchii* and *P. wittmanniana*. Among **woody types** there are about 8 species reported so far and they are *Paeonia decomposita, P. delavayi, P. jishanensis* (syn. *P. spontanea*),*P. ludlowii, P. ostii, P. quui, P. rockii* (syn. *P. suffruticosa* subsp. *rockii*) and *P. suffruticosa*. A brief description of a few are given below:

P. cambessedesii: A native of Balearic Islands and Corsica is a half-hardy clump-forming perennial growing 45 cm high and wide, bearing attractive some, 10 cm long and biternate foliage, dark green above with prominent veins and suffused purple-red on stalks and under-surface, leaflets ovate to lanceolate. Flowers are single, deep rose-pink with purple pistil, and some 6.5-10.0 cm across which appear during mid-spring.

P. delavayi: A native of SW China, close to *P. lutea* but with different flower colours, is a fast-growing and clump-forming open deciduous shrub growing up to 2 m tall and 1.2 m wide. Leaves are bipinnate, 15-45 cm long, biternate, upper surface dark green but glaucous beneath, divided into acuminate and oblong-elliptic leaflets, mostly with reddish stalks. Terminal and dark red flowers which appear in May-June are bowl-shaped with slightly pendulous nature, 6.25-10.0 cm across with yellow anthers, single and in clusters of 1-3, with prominent leafy bracts below. Its var. 'Alba' is pure white.

P. emodii: A native of NW India (Kashmir Himalayas), *P. emodii* allied to *P. lactiflora* but with 7.5-13.0 cm wide white flowers usually with one pistil, is a clump-forming perennial growing about 1.2 m tall with spread of about 1 m, bearing dark green foliage with paler underside and topped by tall stems which bear several large, single, fragrant and pure white flowers with golden-yellow anthers appearing in June and July. It can best be grown under sheltering shrubs.

P. lactiflora (syn. *P. albiflora, P. sinensis*): A native of Siberia, Mongolia, China and Tibet, this herbaceous perennial grows to 60-90 cm tall and with 45 cm spread, bearing mid to deep green biternate leaves with elliptic to lanceolate leaflets which are sometimes lobed. Flowers are single, fragrant, pure white but sometimes pink or red with yellow stamens, 7.5-10.0 cm across, and appear in May and June. The true species is seldom seen but its hybrids are tall (0.75-1.05 m), hardy, showy, double, mostly fragrant, 10.0-18.0 cm across and open from late May to July. Its recommended varieties are 'Adolphe Rousseau' (double, purple-red), 'Albatron' (double, white), 'Albert Crousse' (fragrant, double; shining pink flaked carmine in the centre), 'Augustus John' (single, deep cherry rose), 'Aureole' (single, rose-purple), 'Balliol' (single, maroon-red), 'Baroness Schroeder' (fragrant, double; flesh-pink turning white), 'Bowl of Beauty' (purple, filled with pale-yellow staminodes), 'Bower of Roses' (double, rose-crimson), 'Bowl of Beauty' (semi-double, soft pink but stamens golden), 'Canary' (double, yellow and cream), 'Cherry Hill' (double, cherry red), 'Chromatella' (extra double, sulphur-yellow), 'Duchess de Nemours' (double, white), 'Eva' (single, crimson-pink), 'Felix Crousse' (double, crimson), 'Festiva Maxima' (double, white flecked red-purple), 'Gold of Light' (single, pale rose-pink with a boss of yellow petaloids), 'Karl Rosenfeld' (double, wine-red), 'La Cygne' (double, greenish-white), 'Lady Alexandra Duff' (double, fragrant; soft pink ageing white), 'Lovely' (double, carmine), 'Madame Calot' (fragrant, double; pink shading to white), 'Madame Ducel' (double, pale flesh-pink), 'Martin Cahuzac' (double, purple), 'Mons. Jules Elie' (double, silvery lilac-pink), '*P.* × *lemoinei*' (*P. lutea* × *P. suffruticosa*), 'Pink Delight' (single, pale-pink), 'President Roosevelt' (double, dark red), 'Sarah Bernhardt' (fragrant, double; rose-purple, edged pink), 'Solange' (fragrant, double; buff salmon-cream fading to white), 'Shirley Temple' (double, pale-pink ageing white), 'The Bride' (syn. Whitleyi Major; single white with yellow stamens), 'The Moor' (single, deep maroon-crimson), 'Victoria' (single, crimson), and 'White Wings' (single, white with yellow centre).

P. lutea (syn. *P. delavayi* var. *lutea*): A native of SW China and Tibet, allied to *P. delavayi* but yellow flowers cup-shaped and 5.0-8.0 cm wide, is a deciduous and upright shrub growing 1.8 m in height and 1.5 m in spread. Leaves dark green above with prominent veins, glaucous-green below, leathery, 30-40 cm long, highly segmented, and each segment-edges deeply indented. The fragrant flowers are single or semi-double, golden-yellow with crinkled petals, petal-bases marked crimson at the base, 5.0-8.75 cm across and appear in June. Its var. 'Chrometella' (*P. lutea* × *P. suffruticosa*) bears double sulphur-yellow flowers; 'L'Esperance' bears quite large, semi-double and yellow flowers with a carmine blotch at the base of each petal; and var. *ludlowi* from Tibet grows 2.5 m tall and wide, bearing large green leaves divided into acuminate and deeply cut leaflets, and bowl-shaped golden-yellow flowers.

P. mascula (syn. *P. corallina*): A native of South and East Europe is a clump-forming paeony growing 1 m in length and spread, bearing dark red stems with biternate, bright dark green and smooth leaves and the leaflets are narrowly to broadly elliptic and sometimes with 2-3 narrow lobes. Flowers are single, purple- or carmine-red, sometimes white, having bosses of golden-yellow anthers with purple filaments and some 7.0-13.0 cm across which appear during early summer. *P. m* ssp. *arietina* (syn. *P. arietina*) from SE Europe and Turkey grows about 75 cm long with 60 cm spread and is a tuberous form, bearing hairy leaflets with lower ones having 12-15 leaflets, and single reddish-pink flowers some 7.5-12.5 cm across which appear in May and June.

P. mlokosewitschii: A native of Caucasus is a clump-forming herbaceous perennial with 75 cm of length and breadth. Soft blusish-green and biternate leaves divided into numerous obovate and pointed leaflets some 10 cm long, and occasionally edged red-purple, are topped with 10.0-11.25 cm across, single, large and lemon-yellow flowers with numerous golden stamens, and the flowers appear in April and May.

P. obovata: A native of Siberia, China to Japan and Sakhalin grows up to 60 cm long with about 50 cm spread. Leaves grey-green but tinted copper when young, biternate, 15 cm long, and terminal leaflets obovate. Rose-purple to white flowers some 7 cm across open widely in early summer. Its form 'Alba' bears deeply divided dark green foliage and single white flowers some 7.5-10.0 cm across in May or June.

P. officinalis (syn. *P. fulgida*): A native of France and Albania is a clump-forming tuberous paeony with about 60 cm length and breadth and red flowers. It is one of the most grown garden paeonies in its hybrid forms. Its leaves are biternate, dark green above and paler beneath with pinnatipartite leaflets, leaves deeply cut into numerous oblong-lanceolate and some 2.5 cm wide segments though basal foliage is often entire, sometimes hairy below and 15-20 cm long. Single crimson flowers some 7.5-13.0 cm across appear during May and June. This medicinal species is now seldom seen and is represented by its large, beautiful and double hybrids with flower diameter ranging from 12-15 cm. Var. 'Alba Plena' grows 75 cm in length and breadth and bears double white flowers, occasionally spotted pink; 'China Rose' grows 45 cm in length and breadth and bears quite attractive dark green foliage and single dark salmon-rose flowers with orange-yellow anthers; 'Rosea Plena' bears deep pink flowers, and 'Rubra Plena' grows 75 cm in length and breadth, bearing broadly oval leaflets and ruffled pinkish-crimson flowers.

P. peregrina (syn. *P. decora, P. lobata*): A native of Portugal, Spain, Italy, Balkan Peninsula and Romania, very close to *P. officinalis*, is a clump-forming herbaceous tuberous perennial growing 1 m in length and breadth, bearing biternate shining mid-green leaves at upper surface but glaucous and little hairy beneath, leaflets elliptic but some deeply lobed. Bowl-shaped single red flowers some 7.5-12.0 cm across appear in May and June. Its var. 'Sunshine' is beautiful salmon-scarlet.

P. potaninii (syn. *P. delavayi angustiloba*): A native of SW China is a rhizomatous and highly invasive woody tree paeony with deciduous nature, growing to 60-90 cm in height but with indefinite spread. Leaves are deeply divided in irregular fashion into narrowly oval leaflet lobes. Small, single and deep maroon-red flowers which are sometimes white and some 5.0-6.0 cm across are borne in early summer. The bracts below the flowers are inconspicuous. It is very similar to dwarf *P. delavayi* though *P. delavayi* is not invasive. Var. *trollioides* bears yellow flowers and finely divided leaves.

P. × smouthii (syn. *P. laciniata*; *P. lactiflora × P. tenuifolia*): A hybrid of garden origin forms clumps and grows up to 60 cm in height and about 45 cm in spread, bearing lacy dark green leaves and fragrant single crimson flowers some 7.5-10.0 cm across with yellow stamens, flowers appearing in May and June.

P. suffruticosa (syn. *P. arborea, P. moutan*): A tree-type deciduous paeony native of China, Tibet and Bhutan with height and spread of 1.5-2.2 m, bearing glabrous, bipinnate, about 45 cm long, pale to mid-green leaves which are entire when are borne from the base. The stems are erect and much-branched. Large cup-shaped single (sometimes semi-double) white or rose-pink flowers some 15 cm across are borne in May with a magenta-purple to rose-pink blotch at petal bases. Its hybrids bear flowers up to 30 cm wide in the colour range of white, yellow, pink, crimson, carmine, purple and violet oftenly with suffused and striped petals. Its recommended varieties all semi-double or double are 'Duchess of Kent' (pink), 'Godaishu' (semi- or fully double white), 'Hanadaijin' (masses of double purple), 'Hana Kisoi' (double pale-pink), 'Joseph Rock' (syn. Rock's Variety; semi-double white), 'Kamadu Nishiki' (double 20 cm across lilac-pink flowers striped white at the petal edges), 'Kao' (semi-double or double scarlet), 'King George V' (scarlet flecked white), 'Kokkuryu Nishiki' (single purple-red with lighter stripes), 'Mrs William Kelway' (white), 'Montrose' (pale-lilac), 'Reine Elizabeth' (double salmon-pink and flushed copper-red), 'Renkaku' (double white), 'Tamafuyo' (double pink), etc.

P. russi: A native of Sicily, Corsica and Sardinia is a beautiful paeony due to its single reddish-pink flowers some 12.5 cm across where rounded petals overlap each other and with yellow stamens. Leaves are green at the upper surface with prominent veins, glaucous below, entire, 12.5 cm long and pointed at both the ends. It grows to 35-40 cm high. Flower appears in April-May.

P. tenuifolia: A native of SE Europe to Caucasus bears fringed, i.e finely divided fennel-like foliage, quite different from other species. It is a fast clump-forming evergreen herbaceous species which grows about 50 cm in height and spread with wide-spreading roots and basal growths. The cup-shaped dark crimson or reddish-purple flowers having 7.5 cm diameter appear in May-June, bossed in the centre with yellow stamens.

P. veitchii: A Chinese native is a clump-forming herbaceous perennial growing with 75 cm of length and breadth. Bright green leaves are divided into oblong to elliptic leaflets. Cup-shaped single nodding flowers of purple-pink colour appear in early summer.

P. wittmanniana: A native of Caucasus is a clump-forming herbaceous perennial growing up to 1 m in length and breadth. Leaf are shiny dark green above and paler below with segments broadly oval. Large primrose-yellow flowers are single and bossed with yellow anthers on purple-red filaments.

Propagation

Herbaceous and Itoh paeonies are propagated by root division, and sometimes by seed. Tree paeonies can be propagated by grafting, root division, cuttings and from seeds, although root grafting is most common commercially (Coit, 1908). Propagating perennials from **seed** is the fastest and least expensive way to grow large quantities of young plants. However, due to extensive hybridization or cross pollination, all the perennials will not always produce seeds. Many desirable paeonies are infertile as neither they produce fertile pollen nor seeds. Many of the large double paeonies fit this class. However, if these large flower buds are removed, many of the paeonies will often produce small single flowers which can often produce fertile pollen and seeds. Growing new plants from seeds is the only way a paeony breeder can develop new varieties. In the late summer the pods are collected, dried, seeds extracted and sown in September in pots or pans of seed compost (peat + leafsoil) in a cold frame at a shady location. Sometimes the germination can be delayed for 3-5 years. If gertminated timely, it may form a shoot the next spring or summer. For early germination the seed can be subjected to warm-moist treatment. These are grown for 3-4 years in the nursery rows before setting out in permanent positions from September to March. In the beginning of October, 10-15 cm pots should be filled with moist potting soil. Seeds are sown at 2.0-2.5 cm apart in rows apaced 4 cm apart in the nursery bed or individually in the pots, however, for better results the pots can be placed in plastic food storage bags by twist-ting them and stored for three months at 20 ^{0}C storage temperature. During this period the radicles develop which can be examined by checking the soil periodically. When the roots and radicles are developed the pots are required to be placed for 2-4 months until spring at a cold spot having 6 ^{0}C temperature.

The easiest and most satisfactory method of propagation is by **root division**. Division of young roots helps increasing the stock quickly and produces plants that flower more freely. Young roots are straighter, smoother and easier to cut evenly. Two-three year old stock is best though older stock roots are quite large and interwined so while dividing them there is huge wastage and the process is a bit more cumbersome. A few cultivars have a way of growing in separate pieces, each piece with a small crown of its own. This may make division easier, but it prevents development of a good large plant. For taking root divisions, the roots should be lifted and divided only when the plants have gone into dormancy. The stems and the leaves at the ground should be sheared off and the rootstock is taken out gently from the soil without damaging the root system, whole soil from the rootstock should be washed out and then these are divided with a sterilized and sharp knife, each division with 10-15 cm stubs having 2-5 strong buds and a generous portion of fleshy root. Small thread like roots from the stubs should be removed. Such divisions should be placed in a cold frame for the first year. Very small or too frequent root divisions will delay the flowering and will affect the vigour of the plants adversely. If a fleshy piece has broken off near the crown, it will often develop buds and form a crown of its own. When dormant, paeonies withstand considerable exposure and can be shipped long distances safely. It is preferable to remove rooted suckers from paeony plants and replant them.

While herbaceous paeonies can be propagated by division, the woody stems and extensive root systems of tree paeonies make division impractical, therefore, tree paeonies are grafted upon the herbaceous ones. Herbaceous paeonies can also be grafted onto an herbaceous root but there is not often a reason to do this. A divided herbaceous paeony will recover and flower again within two to three years. **Grafting** should be carried out from mid-July to August under temperate conditions of India, when the foliage of the herbaceous plants begins to change colour and the stems of the tree paeony have matured but have not yet completely lignified. Cleaned, undamaged and disease-free tuberous roots of herbaceous paeonies, measuring 15-20 cm long with roughly finger thickness (10-15 mm), can be gathered on a table in a shed onto which the selected tree paeony growths are to be grafted. These stocks should be dipped for 1-2 hours in a 1:10 bleach solution for sterilization. Most cultivars of *P. lactiflora* are suitable, such as 'Krinkled White,' which has long straight roots. The growths of the scion plant should be 2.5-5.0 cm long taken from the tips of the current season's vegetation. A V-shaped clean cut is made on the top of the tuberous rootstock and of the similar size and shape the wedge is prepared at the base of the scion, then both are fitted together and then the grafts are tied securely. When aligned together the thin green cambium layer of the scion must be firmly and securely in contact with the cut surface of the root with no gap whatsoever anywhere on the contact surfaces. To keep both the contact surfaces together, a piece of rubber band under slight tension is wrapped around a few times. Then this whole graft union zone is sealed with grafting tape. The film will protect the graft union from water and from the infection of diseases, *vis-à-vis* will keep the union intact at its place. However, after performing the grafting operation on the table, its tying with raffia and then wrapping around the entire union in a layer of clay will also provide the same results as wrapping with polythene. The graft should be planted out at a shaded place at 45 cm apart where it can be left undisturbed for next two years. While planting, the grafted zone should remain above the soil surface, however, many others recommend planting such grafts 10-15 cm below the soil surface to force the tree paeonies to start producing their own roots, better suckering and for developing a good bushy plant.

Shrubby paeonies may also be increased by **layering**. In this case, the branches are pegged down in March which will take some two years to form their own roots and then these should be separated and planted separately. In case of **cuttings** in the shrubby species, 15-20 cm long hardwood cuttings are taken in October or November and inserted in the already prepared shaded nursery beds or in the cold frame where these are grown at least for two years for proper development and then these are lifted with ball of earth and planted in the already prepared designated beds for flowering. Usually only a small percentage of cuttings root and others dry up. Use of rooting hormones may encourage rooting and rooting percentage which is still a researchable field.

Micropropagation is the most effective option in getting required quantity of planting material of an elite germplasm in limited period of time. Traditional propagation approaches require a considerable amount of propagation material and can only be carried out at a specific time in the vegetative season. There are also some problems with actual production, such as a long breeding cycle, a low vegetative reproduction coefficient, the low seed-producing capacity of some precious cultivars, and the poor development of seed (Li *et al.*, 1984). Other problems include strong variability in seedlings, lack of seed set in double flowers, epicotyl dormancy, high production costs, etc. (Bouza *et al.*, 1994; Harris and Mantell, 1991; Orlikowska *et al.*, 1998). All of these problems seriously limit the industrial and commercial production of tree paeony nursery stock, and thus the ever-increasing demand cannot be satisfied. Also, traditional propagation approaches have limited the preservation of germplasm resources and the development of new varieties (Li, 2007). Consequently, scientists are forced to explore new technologies for the mass production of tree paeony. Studies on tree paeony tissue culture began in sixties when Demoise and Partanen (1969) induced callus from the embryos of tree paeony. Different explant sources have already been tested with variable success (Meyer, 1976; Li *et al.*, 1984; Buchheim and Meyer, 1992; Bouza *et al.*, 1994). Moreover, little information is available on different cultivars. The major obstacles are high contamination rates (Wang and van Staden, 2001), excessive browning of the tissues (Buchheim and Meyer, 1992) and development of callus preventing further bud development (Li *et al.*, 1984). So far, although successes have been reported in clonally propagating tree paeony through axillary bud culture (Li *et al.*, 1984; Bouza et al., 1994), these *in vitro* techniques have not reached the scale of commercial protocols. Propagation via somatic embryogenesis is a valuable alternative for micropropagation by axillary buds (Bhojwani and Razdan, 1996). Somatic embryogenesis in herbaceous paeony has been reported by Lin *et al.* (1987) from callus derived from zygotic embryos, and Roberts and Sunderland (1977) from microspore and anther cultures. However, no valuable information is available for *P. suffruticosa*. Rather *et al.* (2014) with herbaceous paeony var. 'Sara Bernhardt' by taking vegetative shoot tips when the plants were in full bloom, reported that when explants were surface-sterilized with carbendazim (0.01%) +mercuric chloride (0.1%)+ethyl alcohol (70%) yielded maximum aseptic cultures (73.33%) and good explant survival (70%). They recorded highest explant establishment (43.33%) with BAP (1 mg/l) but maximum shoot length (2.36 cm) and leaf number per shoot (7.31) with BAP+GA_3 (0.50+0.10 mg/l) though proliferating cultures (72%) with good shoot multiplication (5.60) and shoot length (1.64 cm) were recorded with BAP+kinetin (0.50+0.50 mg/l). BAP (1.0 mg/l) yielded maximum shoots per explant (6.20) but minimum shoot length (1.22 cm), though rooting and roots per shoot were maximum (53.33% and 2.62, respectively) with IBA (025 mg/l) compared with IAA. *Ex vitro* survival of rooted plantlets under greenhouse was recorded 46.67%.

Cultural Practices

Paeonies prefer fertile clay loam and a well-drained but moist garden **soil** having a pH of 6.5-7.5, at a sunny or filtered location. Good drainage will ensure healthy crop. Paeonies may be grown in most soils provided they are not too wet in the winter. Even on slightly chalky soils, its performance is better as it prefers slightly alkaline conditions. In acidic soils a handful or two of lime at planting should be mixed in the soil in the area of one square metre. A heavier soil is preferable to a light one, but it is always beneficial to mix some organic matter in the form of decayed manure, farmyard manure or compost @ 3-5 kg per square metre of area, which will also improve moisture retention of the soil. Since it is highly perennial crop as once planted the plants are left in the same field for up to 12 years or more or sometimes even up to 30 years so the soil should be quite rich when paeonies are to be planted. For **field preparation**, one deep ploughing should be carried out to facilitate spreading of roots properly. Two more subsequent ploughings are required for making a fine tilth. After every ploughing the field should be planked after taking out the rootstocks of the perennial weeds. After third ploughing the beds of convenient sizes are prepared giving clearance at least of 60 cm between the beds to facilitate essential field operations and making provision for main irrigation channel. There should be sufficient moisture in the soil at the time of bed preparation. Beds should be quite even for smooth and equal distribution of irrigation water.

For **transplanting** of the seedlings or vegetatively propagated plants in the garden, spacing in the field is critical for adequate light and air movement. Care should be taken not to plant the roots too **deep** with the eyes not more than 3.5 cm below the soil surface. The plant should be removed from pot with earth ball and placed suitably at 60 cm apart in the prepared bed during September or October. Paeonies go **dormant** in late August, so planting is carried out at any time between September 1 to late October or sometimes even during November if it is not too cold and the ground has not freezed. Planting in early autumn gives the soil ample time before frost to settle around the newly planted roots. Stock lifted the previous fall and kept in cold storage over winter may be planted in the spring. Such paeonies will be less vigorous the first year because the root system will have a shorter time to develop. In America, traditionally 105 cm × 92 cm (line to line and plant to plant, respectively) **spacing** is normally provided (Howard, 1968) but for economical production the recommended spacing is 90 × 40 cm per plant (Dole and Wilkins, 1999). The **mulching** is required to avoid drying up of the mixture at any level. The plants obtained from seeds may take a minimum of three years after transplanting to come into bloom though in case of tree paeonies, the plants require 10-12 years to attain the stage of flowering. Planting at a location with full sunlight and away from the competitive effects of tree roots is necessary, however, the plants may require a sheltered location during the peak of summer when it is too hot and windy. Since it may take at least three years for first flower harvest in case of herbaceous paeonies, so the place where these are to be planted should have protection against winds and hot sun. At the time of planting, the roots should be moist until they are planted in the ground but in case the roots are drying, these should be soaked in water for several hours before planting. If planting is delayed, repack roots in the container, wetting the packing material thoroughly. They can be kept 1-2 weeks without damage. For **storing the roots** longer or even up to one year without any serious damage, a trench is dug up of suitable dimension where these may be planted 15 cm apart.

Though herbaceous paeonies are quite hardy but first year of their planting these require protection against freezing where organic **mulches** such as straw, dry leaves or any other such materials should be used to cover them. This will not be necessary after the first winter, except in the case of tiny divisions or young seedlings, which should be mulched for several years. Manure or the dead leaves and stems of the paeony should not be used as a mulching material. Removal of mulch on all paeonies is necessary as soon as the shoots break through the ground in spring. Young paeony plants are particularly slow in shoot appearance and may be a month later than established ones.

Removal of lateral flower buds growing in the leaf axils, should be done when axillary buds are barely large enough to handle, i.e. 0.6 cm in diameter and with pedicel length of 1.3-2.5 cm. **Disbudding** enhances the growth of the terminal flower bud because all the plant's resources are diverted towards development of a single flower per stem. Removal of the terminal bud to promote lateral bud development or spray types is done with certain cultivars for particular markets. If quantity is desired and smaller flowers with little stem lengths are acceptable, lateral buds may be allowed to develop. Decker and Weinard (1931) observed little difference in yield of flowers when only three leaves were left on a flowering stem or if half of the leaves was removed. Paeonies grown for cut flowers may not require **staking** in normal weather conditions, as the flowers are cut when in bud. The taller herbaceous species, especially varieties of *O. lactiflora* and those species or varieties bearing double and large flowers may require stakes to support the flowers against breaking off during windy or rainy days or when the flowering stems bend to the soil due to lodging.

Always the soil should be tested for nutrient content and only then any **fertilizer** recommendation should be made. The increased water requirement of cut flowers creates an increased requirement for fertilizers too. Decker and Weinard (1931) recorded yield differences due to different fertilizer rates, though cultivars also showed difference. Too much use of nitrogen reduces the flowering capacity of the plants (Armitage, 1989). The application of fertilizer should coincide with crop needs. Nitrogen at the rate of 450-900 g per year is sufficient for 90 square metre area, half as basal dose and the remaining half in March -April. K_2O and P_2O_5 fertilizers as basal dose per year can also be given at the same rate as to that of nitrogen. Too much nitrogen reduces the flowering capacity. Overhead **watering** is not recommended in case of paeony as this may physically damage the flowers, may cause petal spotting, may splash the soil onto the foliage and promotes the spread of disease. Drip irrigation is recommended because it places water on the ground where it is needed, and not on the flowers or foliage. The amount and frequency of water required will vary with the weather and crop maturity. Irrigation schedule should be based on the soil moisture status around the root zone. Excess watering may invite root rot fungi and insufficient water supply will reduce crop production and quality. While the paeony will withstand dryness to the point of drought without succumbing, a reasonable amount of moisture is essential for the best development. Paeonies should have a liberal supply of water, especially while in bloom.

Growers must control **weeds** in field production of paeonies, as competition with weeds reduces the quantity and quality of floral production. A bed full of weeds also increases the time required to harvest and raises the labour cost. Manual weeding, use of herbicides and mulching are effective in keeping down the weed population. In 1993, Acclaim (fenoxaprop-ethyl) and Dachthal (dimethyl tetrachloroterephthalate) are registered for use on paeonies. Weed barriers or mulches prevent weed growth and restrict soil splashing on foliage and flowers from rain and irrigation. The greatest benefits from these materials occur in the first growing season when the paeony plants are becoming established. If a mulch is used over winter, to avoid diseases, its removal should be done as soon as paeony shoots start appearing. Paeonies develop a strong system of feeder roots near the surface of the soil, so the plant must be cultivated with great care to avoid injury to the roots.

Growth and flowering

In paeonies, the flower bud differentiation initiates after the senescence of the ensuing flowers and the development continues until reaching the state of dormancy during autumn (Byrne and Halevy, 1986). Growth **temperatures** significantly affect the dynamics of bud emergence, stem elongation and flowering in herbaceous paeony. Kamenetsky *et al.* (2003) when exposed the container-grown paeonies to three regimes of day/night temperatures, i.e. 16/5, 22/10 or 28/10 ^{0}C, recorded that at low temperatures the bud emergence was slow but with better flower quality and the first stems were observed only 50 days after exposure to 16/5 ^{0}C. The stems reached 55 cm long after 85 days of growth at these temperatures. When the temperatures of 22/10 ^{0}C were maintained, this led to fast sprouting of the stems after 50 days of growing, at 28/10 ^{0}C bud emergence was fastest, and the maximum number of shoots was observed only after 36 days of growth. Growth temperatures significantly affected the flowering date. Continuous growing at 22/10 ^{0}C led to flowering after 83 days. Under sustained high day/night temperatures of 28/22 ^{0}C, almost all the stems aborted their flower buds, mostly at an early stage of development. The transfer of the plants from 22/10 ^{0}C to 28/22 ^{0}C after 30 days of growing also caused abortion of most flower buds. In contrast, when plants were grown for the first month at 28/10 ^{0}C and then transferred to 22/10 ^{0}C, flowering was not significantly different from those plants grown continuously at 22/10 ^{0}C (Kamenetsky *et al.*, 2003). However, exposure to higher night temperatures of 22 ^{0}C at the first 20 days of growing caused a very high abortion rate of young flower buds.

In **forcing**, the bud abortion is a common problem. Late in the autumn, aboveground shoots of paeony senesce and die. In order to facilitate the growth in the spring, the dormancy of the underground buds should be broken by exposure to **low temperature** (Kamenetsky *et al.*, 2003). Paeonies, especially the herbaceous ones are the crops of the cold climates and require 7 ^{0}C temperature treatment for 3-4 weeks to initiate shoot elongation, *vis-à-vis* further floral development (Dole and Wilkins, 1999). However, Byrne and Halevy (1986) reported that the paeony required 6 ^{0}C treatment for a minimum of 4 weeks to flower, and for a minimum of 6 weeks for maximizing flowering shoots. Six weeks temperature exposure at 1 ^{0}C was found at par to 6 ^{0}C treatment where maximum number of flowering shoots appeared. They further reported that in California when plants were treated with 5 ^{0}C for 4, 6 and 8 weeks, exhibited mean forcing time of 67, 67 and 69 days, respectively at 21-26/17 ^{0}C day/night temperatures. Evans *et al.* (1990) obtained flowering in the var. 'Scarlet O'Hara' in 72, 59, 52 or 29 days after cold-treating at 6 ^{0}C for 6, 8, 10, 14 or 20 weeks, respectively, and then forcing at 18 ^{0}C under natural daylengths starting December 15, whereas 'Krenkeld White' was when subjected to 4 or 6 weeks cold treatment flowered in 72 days. The chilling required for dormancy release was determined by exposing field-grown plants to ambient winter-cold weather in northern Israel (16-26 ^{0}C day and 8-10 ^{0}C night from November to February) and covering the greenhouses with plastic sheets at different dates (Halevy *et al.*, 2002). The accumulated chilling units were calculated according to Fishman *et al.* (1987) and Erez *et al.* (1988). Chilling saturation was reached at 42 chill units for var. 'Sarah Bernhardt' and at 36 chill units for var. 'Duchesse de Nemours'. When container-grown plants were cooled artificially, dormancy release was found to be best under chilling regimes of 2 ^{0}C for 60 days or 6 ^{0}C for 70 days (Kamenetsky *et al.*, 2003). Higher temperatures were less effective. Comparison of the sprouting of plants exposed to 2, 6 or 10 ^{0}C for 60 days clearly indicates that the effect of chilling on dormancy breaking intensified as the chilling temperature was lowered, and that the effect was more pronounced on the number of sprouting stems than on the stem length. Freshly harvested root divisions when prerooted at 16 ^{0}C and then subjected to 6 ^{0}C for 0, 6, 8, 10, 14 and 20 days, Evans *et al.* (1990) recorded flowering in all the treatments except control, however, 10 week's cold treated prerooted plants were found optimum for rapid flowering with more number of flowers.

Application of **GA$_3$** directly to the dormant buds on the crown can partially complement the cold requirement for breaking dormancy and promote bud sprouting and shoot growth (Halevy *et al.*, 2002). GA$_3$ treatment does not promote flower initiation and differentiation in paeonies, and is not effective if applied before flower formation (Halevy *et al.*, 1995). Drenching of GA$_3$ at 118 mg a.i./pot or 2,000 ppm onto potted noncooled roots produced shoots in all the plants but with abortion of the flowers (Evans *et al.*, 1990). As per the usual practice in Israel, GA$_3$ soil-drenching at 100 ppm and then partially field-vernalization of the plants with cold treatment cause shoot elongation and flowering, and then subsequently these are grown under plastic greenhouses for European market sales. This way they make first harvesting in the second year with maximum production in the fifth year. However, photoperiod treatment has not been found effective.

Post Harvest

Harvesting of paeonies should be done at least thrice in a day, i.e. during early morning when the weather is best, by late afternoon and by the evening when plants and flowers are free from dew or moisture, beginning the fourth year after planting. At 25 ^{0}C and above the bud development is hastened. Optimum productivity is reached the fifth year and may continue more than 25 years. The length of the harvest season depends on spring temperatures. Warm spring temperatures shorten paeony harvest, while cool spring temperatures lengthen it. Generally, the harvest season for commercial cut flower cultivars lasts 10–12 days, though the harvest period for each cultivar may be as short as 2–4 days. Late-season cultivars and fully double cultivars often fail to open under sudden hot temperatures. It is difficult to describe the proper stage at which to cut paeony flowers. The experienced growers can determine the harvest stage through 'feel', i.e. squeezing of each bud. Doubles and reds are generally harvested at a softer stage than other types and colours. In fact, the paeonies are harvested at proper bud stage through 'feel' when the buds have attained 2.5-

4.4 cm diameter, green sepals have separated exposing the petal colour and when stalk is at least 30 cm long, and best when it is 37.5-45.0 cm long. These specifications may deviate slightly based on varietal character and vigorosity and whether the side buds were disbudded or not.

Freshly harvested stems should not be placed in water immediately due to being blasted or may open prematurely. However, after 20-30 minutes these should be placed in disinfected containers filled with clean water where these should remain for 1-2 hours at 2 ^{0}C. Rehman *et al.* (2012) studied the effect of various pulsing treatments, *viz.* sucrose 5%, aluminium sulphate 300 ppm, citric acid 1,000 ppm, potassium permanganate 100 ppm, tartaric acid 500 ppm and calcium hypochlorite 100 ppm, where they observed sucrose + aluminium sulphate as the best treatment as in the treatment stems maintained better water relation (water uptake, water loss, water balance, fresh weight change) which consequently brought about best vase life of 9.14 days, followed by 9.12 days in sucrose 5% + citric acid 1,000 ppm, compared to control with 6.02 days.

Insect-Pests and Diseases

Insects are normally of little concern to paeony crop, however, sometimes **flea beetles** (*Phyllotreta* species) become a problem during warm weather. They are particularly active on bright sunny days and in their severe attack the growth of the young plants is checked. Proper garden sanitation will expose the hibernating places of these beetles in the debris. BHC dusting on the plants will kill these beetles. **Thrips** are tiny, dark, elongated insects that feed on leaves and blossoms. They cause distorted and discolored flowers and buds. Spraying plants with an insecticidal soap will control its infestation. If this is not successful, apply Malathion at 0.2% spraying. Gardeners are often concerned about the presence of large numbers of **ants** on their paeonies. The ants are not harmful, they simply feed on the sweet substance secreted by the flower buds. Nematodes may also infest this crop and the best control measure for them is HWT.

Botrytis blight (gray mould) is a fungal disease which infects a wide array of herbaceous annual and perennial plants. Although there are several species of *Botrytis* which can cause blights, the most common one is *Botrytis cinerea*. Botrytis infections are favoured by cool (15.5 ^{0}C), rainy and drizzly weather continuing over several days. One *Botrytis* blight fungus with a strict host preference is *Botrytis paeoniae,* which infects only paeony and becomes very serious during wet seasons. It causes considerable die back of tree and herbaceous paeonies, the latter being killed often completely. To manage the disease all the affected parts should be sheared off and burnt, *vis-a-vis* overwatering is avoided. The plants should be thinned out to allow air and light penetration. Spraying with fungicide such as Bordeaux mixture, or Dithane M-45 at 0.2% sprayings at weekly intervals during the inclement weathers i.e. windy, chilly and humid will keep this disease at bay. Top-dressing of sand around crown often helps to prevent the disease. The causal organism of ***Phytophthora* blight** is *Phytophthora cactorum*. In its infection the stems, leaves, buds and blossoms become brown and leathery. Black cankers form on stems and cause them to fall over. Application of mancozeb to control *Botrytis* blight will also control this disease. ***Sclerotinia* stem rot** and ***Verticillium* wilt** are also problems in paeony cultivation. These should be controlled in the same manner as *Botrytis.* The causal organism for **red spot** disease is *Cladosporium paeoniae*. It creates red spot or measles and small, dark red, circular spots on leaves. Spots coalesce to form blotches that are dark purple on lower leaf surface. Application of mancozeb to control *Botrytis* blight will also help prevent other fungal leaf spots. Leaf spot is also caused by *Gloeosporium* sp., *Alternaria* sp. and *Septoria paeoniae* which will be controlled automatically while controlling *Botrytis.* Sometimes viruses become a nuisance. In this case prevention is better than cure, so whole plants will have to be uprooted and burnt or buried.

References

Armitage, A.M. 1989. *Paeonia.* In: *Herbaceous Perernnial Plants*, pp. 437-443. Varsity Press, Athens, Georgia, USA.

Bailey, L.H. and E.Z. Bailey, 1976. *Paeonia.* In: *Hortus Third : A Concise Dictionary of Plants Cultivated in the United States and Canada*, pp. 810-811. Macmillan Publishing, New York, USA.

Beckett, K.A. 1978. *The Concise Encyclopedia of Garden Plants.* Orbis Publishing Ltd., London.

Bhojwani, S.S. and M.K. Razdan, 1996. *Plant Tissue Culture: Theory and Practice.* Elsevier Science Pub., Holland.

Bouza, L., M. Jacques and E. Miginiac, 1994. Requirements

for *in vitro* rooting of *Paeonia suffruticosa* Andr. cv. 'Mme de Vatry'. *Sci. Hortic.*, **4**: 223-233.

Buchheim, J.A.T. and M.M. Meyer, 1992. Micropropagation of paeony (*Paeonia* spp.). In: *Biotechnology in Agriculture and Forestry* (ed. Bajaj, Y.P.S.), Vol. 20, pp. 269–285. Springer- Verlag, Berlin, Heidelberg.

Byrne, T.G. and A.H. Halevy, 1986. Forcing herbaceous peonies. *J. Amer. Soc. hort. Sci.*, **111**: 379-383.

Coit, J. Eliot, 1908. *Peony. Bulletin* (Cornell University), No. 259, 104 pp.

Decker, S.W. and F.F. Weinard, 1931. Peony fertilizers and cutting. *Proc. Amer. Soc. hort. Sci.*, **28**: 423-426.

Demoise, C.F. and C.R. Partanen, 1969. Effects of subculturing and physical condition of medium on the nuclear behavior of plant tissue culture. *Am. J. Bot.*, **56**: 147-152.

Dole, J.M. and H.F. Wilkins, 1999. *Floriculture Principles and Species*. Prentice-Hall, Inc., Upper Saddle River, New Jersey, USA.

Erez, A., S. Fishman, Z. Gat and G.A. Couvillon, 1988. Evaluation of winter climate for breaking bud rest using the dynamic model. *Acta Hort.*, No. 232, pp. 76-89.

Evans, M.R., N.O. Anderson and H.F. Wilkins, 1990. Temperature and GA_3 effects on emergence and flowering of potted *Paeonia lactiflora*. *HortSci.*, **25**: 923-924.

Everard, B. and B.D. Moraley, 1970. *Wild Flowers of the World*. Peerage Books, London.

Fishman, S., A. Erez and G.A. Couvillon, 1987. The temperature dependence of dormancy breaking in plants: Computer simulation of processes studied under controlled temperatures. *J. Theor. Biol.*, **126**: 309-321.

Goody, J. 1993. *The Culture of Flowers*. Redwood Press Ltd., Melksham, Wiltshire, Great Britain.

Halda, J.J. and J. Haldova, 2004. The Genus *Paeonia*. Timber Press, Portland, USA.

Halevy, A.H., M. Levi, M. Cohen and V. Naor, 2002. Evaluation of methods for flowering advancement of herbaceous peonies. *HortSci.*, **37**: 885-889.

Halevy, A.H., D. Weiss, V. Naor, M. Cohen, M. Levi and D. Skuier, 1995. Introduction of herbaceous peony as commercial cut flower in Israel (in Hebrew). *Dapei Maida*, **5**:58-62.

Harris, A.R. and S.H. Mantell, 1991. Effects of stage: subculture durations on the multiplication rate and rooting capacity of micropropagated shoots of tree paeony (*Paeonia suffruticosa* Andr.). *J. Hort. Sci.*, **66**: 95-102.

Hong, D.-Y. and K.-Y. Pan, 1999. Taxonomical history and revision of *Paeonia* sect. *Moudan* (Paeoniaceae). *Acta Phytotaxonomica Sinica*, **37**: 351–368.

Howard, G.S. 1968. How depth of planting affects the garden peony. *HortSci.*, **3**: 279-280.

Kamenetsky, R., A. Barzilay, A. Erez and A.H. Halevy, 2003. Temperature requirements for floral development of herbaceous peony cv. 'Sarah Bernhardt'. *Scientia Hort.*, **97**: 309-320.

Li, P. 2007. *Studies on the Micropropagation of Tree Peonies*. Ph.D. Thesis. Beijing For. Univ., China.

Li, Y., D. Wu, S. Pan, S. Xu, Z. Wei, Z. Xu and X. Li, 1984. *In vitro* propagation of *Paeonia suffruticosa* (Chinese). *Kexue Tongbao*, **29**: 1675–1678.

Lin, J.J., J.A. Thomas and M.M. Meyer, 1987. Tissue culture and embryoid formation in *Paeonia lactiflora* Pall. *Am. Peony Soc. Bull.*, No. 263, pp. 24–30.

Meyer, M.M. 1976. Culture of *Paeonia* callus by tissue culture techniques. *Am. Peony Soc. Bull.*, No. 218, pp. 27–29.

Nau, J. 1996. *Ball Perennial Manual: Propagation and Production*. Ball Publishing, Batavia, Illinois, USA.

Orlikowska, T., A. Marasek and D. Kucharska, 1998. Regeneration of *Paeonia mlokosewitshii* Lom. and *P. tenuifolia* L. *in vitro* from different explants. *Acta Soc. Bot., Poloniae*, **67**: 223-227.

Pan, K.-Y. 1979. The analysis of distribution pattern in the Paeoniaceae and its formation. *Acta Phytotaxonomica Sinica*, **33**: 340–349.

Pizzetti, I. and H. Cocker, 1968. *Flowers ~ A Guide for your Garden*, vol. 2, pp. 629-641. Harry N. Abrams, Inc., Publishers, New York, USA.

Rather, Z.A., I.T. Nazki, Z.A. Qadri, M.A. Mir, K.M. Bhat and G. Hussain, 2014. *In vitro* propagation of herbaceous peony (*Paeonia lactiflora* Pall.) cv. Sara Bernhardt using shoot tips. *Indian J. Hort.*, **71**(3): 385-389.

Rehmaan, S.I., Z.A. Qadri. M.Q. Sheikh, F.U. Khan and Zahoor Ahmed, 2012. Effect of different pulsing solutions on keppeing quality of cut peony cv. 'Sarah Bernhardt'. *J. Ornam. Hort.*, **15** (1-2): 108-115.

Roberts, M. and N. Sunderland, 1977. Pollen culture of *Paeonia. John Innes Inst. Ann. Rep.*, **68**: 60–61.

Rogers, A. 1995. *Peonies.* Timber Press Inc., Portland, Oregon, USA.

Stebbins, G.L. 1937. Cytogenetic studies in *Paeonia*. II. The cytology of the diploid species and hybrids. Publication of the University of California, Berkley, California, USA, pp. 84-107.

Stern, F.C. 1946. A study of the genus *Paeonia*. A publication of The Royal Horticultural Society, London, 155 pp.

Wang, H. and J. van Staden, 2001. Establishment of *in vitro* cultures of tree peonies. *South African J. Bot.*, **67**: 358–361.

24

Pandanus

Sashikala Beura and P.N. Jagadev

Family: **Pandanaceae**

Common names: Screwpine/Walking palm (*Pandanus*) in English; Blue screw pine (*P. baptisii*); Common screw pine (*P. utilis*); *Gandhapuspa/Suchikapuspa* in Sanskrit; *Kedagi/Kedagai/Muchali* in Tamil; *Keoda/Keya/Ketaki/Kea* in Bengali; *Keora/Keura/Ketgi/Gagandhul* in Hindi; *Keora* in Urdu; *Kewoda* in Gujarati; *Kiya/Kewda* in Odiya; *Leram* in Nicobarese; *Mogali/ Mugali* in Telugu; *Palliyambu kaita* in Malayalam; *Talemara* in Kannada

Introduction

The genus *Pandanus,* a latinized form of a Malayan name, a native of Tropical Africa, Asia, Australasia mostly on sea coasts and islands (Beckett, 1987), popularly known as either screwpine or walking palm is neither a pine nor a palm, but is once thought to be related to the pineapple (Macoboy, 1986). Maneka Gandhi (1989) mentions that the genus, *Pandanus* is a Latinized form of a Malayan word *Pandan,* means stick man. Also it is mentioned that *Pandanus* derives from *Pandang,* a Malayan word signifying conspicuous (Chittenden, 1956; Macoboy, 1986). *Pandanus fascicularis* Lam., is the most common of the 36 species found in India (Deshaprabhu, 1966). *P. fascicularis* commonly known as *kiya* or *kewda,* occurs widely in the coastal districts of Odisha, Andhra Pradesh, Tamil Nadu, Gujarat and to some extent in Badayun, Balia and Gajipur districts of Uttar Pradesh (Dutta *et al.,* 1987). It is estimated that in the Ganjam district of Orissa, there are about 3 to 4 lakh *kewda* shrubs (Deshaprabhu, 1966) *Pandanus* is a native of the old world extending from Africa to Australia (Chittenden, 1956; Mott, 1975). About 600-650 species of *Pandanus* are widely distributed in the moist tropics of the old world and *kewda* is the most widespread species being recorded from Mauritius Islands in the west to Polynesian Islands in the east (Deshaprabhu, 1966).

The common name, screw pine, is due to the saw-edged leaves borne in a 3-rowed phyllotaxy like a screw up the stem that are usually twisted, appearing the leaves to run in well marked spirals towards the tips of the branches which together with the large cone-like fruits give a vague resemblance to pineapple (Mott, 1975; Purseglov, 1975). Generally, the flowers are worn in the hair by the girls to win the love of their lovers. Legends also reveal that Parvati approached Shiva with *kewda* flowers, which attracted Shiva's attention and he got disturbed in his penance. *Kewda* (*Pandanus fascicularis*) is commercially important due to the production of *kewda* perfume from its male spadices (popularly known as flowers). Its commercial exploitation for perfume is centred mainly in Ganjam district of Odisha, accounting for about 90% of India and 50% of total world production (Dutta *et al.*, 1987) and to a lesser extent in a few places of Tamil Nadu, Andhra Pradesh and Uttar Pradesh. *Kewda* attar, *kewda* water and *kewda* oil (*rooh kewda*) are among the important perfumery products due to its rapid increasing demand in India and abroad, especially from Arab countries.

The roots being fibrous are used by the basket-makers for binding. When cut into lengths and beaten out, commonly used as brushes for painting and white washing. The root fibre may be used as a substitute for bristle in brush making. In the Ganjam district of Odisha, the aerial stilt roots are generally collected by low income group people at the end of rainy season (i.e from September onwards) and are processed by cleaning and soaking in water for few hours and then dried under the sun. The recovery of dried fibres is about 15% of the fresh roots by weight. The clean dried fibres are sold to the middlemen @ Rs. 1,000-1,500 per quintal, which they transport to the neighbouring states like Andhra Pradesh, Madhya Pradesh, Tamil Nadu and Karnataka for industrial use. The fibre heaps may also be observed in the local markets. The dried root fibres are used for making mats, ropes and nets for fishing, packing materials for fruits and vegetables, binding the cattle, binding material for fencing and thatching the huts and thick fibres as brush for white washing of the walls of the cemented house. The leaves are pungent, bitter, with flavour and is said to be valuable in leprosy, small pox, scabies, syphilis, leucoderma and diseases of heart and brain, tumours and aphrodisiac (Deshprabhu, 1966). The aerial root is also used medicinally by the Sinhalese. The anthers of the male flowers are given in earache, headache, leucoderma, eruptions and diseases of blood. The juice obtained from the whole inflorescence (spathes removed) is useful in rheumatic arthritis in animals.

Kewda Perfume Industry

Till today in Odisha, the flowers are mostly collected from the shrubs grown wild on natural plantations. Some initiatives may encourage the rural educated youths for establishing the distillation units in the production areas as the cost involvement to establish an indigenous distillery is not very high. Hence, the industry has a great potential for expansion by exploiting the over-loaded natural plantations of *kewda* in the coastal belt. The white spathes of the male inflorescence emit a fragrant smell, which is much valued in India and Maynmar and used for hair decoration of ladies, flavouring catechu and kept in boxes to lend a lasting fragrance to the stored garments. *Kewda* attar, water and oil (*rooh kewda*) are prepared from the spathes by hydro-distillation. *Kewda* attar is one of the most popular perfumes in India, where it has been used since ancient times. It blends well with almost all types of fancy perfumes and is used for scenting clothes, flavouring betel (pan masala) and tobacco, bouquets, lotions, cosmetics, soaps, hair oils, and *agarbatties*. On festive occasions, *kewda* water is used for flavouring varoius foods, sweets, syrups and soft drinks, which are especially popular in North India. In Arab countries, *kewda* water is used during bathing. *Kewda* perfume is obtained by distilling male spadices in water and the *otto* content is 0.03% on weight of spadices. Its commercial exploitation is centred mainly in Ganjam district of Orissa. The perfumes are prepared indigenously by the traditional hydro-distillation method. Althouth this industry is centuries old, it came into prominence during the last four decades, with a sudden spurt in the consumption of *pan* masala and chewing tobacco.

Kewda distillation is carried out in the fire heated stills using equipment of traditional type. Flowers mixed with good quantity of water @ 60 litres/1,000 flowers, are put in a big copper still (*deg*) and directly heated at the bottom, with ordinary wood fire over an open country furnace. The vapours produced are condensed and collected in a receiver copper vessel (*bhabka)* connected with '*deg*' through a bamboo condenser (*chonga*). Attars are manufactured by absorbing flower distillates either over sandalwood oil or liquid paraffin (commercially known as white oil) and some producers also use Mobil oil as base oil. Liquid paraffin and Mobil oil as base are used to manufacture low grade perfumes. Keeping base oil as same in '*bhabka*', a few more distillation batches are run with fresh flowers till the desired concentration of natural oil in the base oil is built up. Several grades of *kewda* attar are prepared depending on number of flowers (1,000 to 10,000) distilled per kg of sandal wood oil and in high grades the number may go up to 1 lac. Cost of attar for 1,000 flowers distilled is worked out to be around Rs.15,000 per kg. Similarly, distillation of 1,000 flowers yields 18 litres of *kewda* water and the grade is known as *khazara*, which costs about Rs.150-450/- per litre depending on quality. *Kewda* oil is not extracted on a large scale due to its high solubility in water; however, about 1 lac flowers are required to produce 1 kg of oil (*rooh kewda*), costing around Rs.2,50,000. Kannauj (U.P.) is considered as the perfume capital of India where 80% of the perfume trade is linked with *pan* masala and chewing tobacco industries. The volume of business in this trade is to the tune of Rs. 500-800 crores, of which *pan* parag alone shares about 50% and rest 50% is shared by the brands like Baba jarda, Prabhat jarda, Gopal jarda, Rajanigandha, Radha, Ashiqui, Kesar, etc. In this industry, of the total perfumes used, *kewda* alone accounts for about 75% by value, hence the importance of *kewda* is amply clear in the perfume trade. The cost of production of *pan* parag is about Rs. 750/- per kg of which Rs. 200/- is spent towards perfumes and *kewda* is the major one, which accounts for about Rs. 150/-. The procurement procedure by the user industry is to call for tenders and the competitive bidder is awarded with the contract for the year. The price of the oil is fixed based on the prevailing price of the flower in Ganjam district and supply is made on monthly basis. The payment is made 60-90 days after delivery of the consignment. Small quantities of *kewda* attar or absolute are also being exported abroad through the exporters from Kannauj, Delhi and Mumbai. For export, the merchant exporters contact the dealers of the importing countries and send the sample along with a quoted price. Once the sample is approved, the price is negotiated and the schedule of the supply is fixed. The merchant exporters procure the consignments from the distillers or traders and export the same acting as middlemen or commission agents. Sometimes the big traders with their own distillation units pool the perfumes from the small distillers and directly contact the importers abroad and execute the orders for export to international market. The traders get maximum margins in the business.

Botany

Kewda is an evergreen monocot in the order Pandanales. Of the three families in the order, only Pandanaceae, an old world family includes woody plants with three genera while other two, *viz.* Typhaceae and Sparganiaceae include monotypic herbaceous marsh-plants with two ranked leaves and creeping stem that are well known in Britain (Rendle, 1979). The members of Pandanaceae are largely confined to the tropics. The largest genus *Pandanus* contains trees, shrubs and climbers and distinctive in the wild as often bear the prop roots which support their erect stems and branches. Morphological evidences suggest that these plants are more primitive than palms although there is no fossil record (Everette, 1969). The genus has a number of useful species yielding minor products used in tropical Africa, Australia, Islands of the Indian Ocean and the Pacific. Thick and strap-shaped leaves oftenly serrated at the margins are usually crowded at or near the stem tips in spiral form hence the vernacular name screwpine. The flowers are dioecious and petalless, the males being carried in the branched spikes and the females in rounded heads that develop into tight cone-like fused clusters of multiple berries. Purseglove (1975) reported its basic chromosome number as 30 and 32.

No breeding work has so far been done on *kewda* except selection of some promising morphotypes (Jagadev *et al.,* 1993). Varietal improvement in *kewda* by breeding poses problems mainly due to dioecious nature and long life cycle of the plants, besides heterozygosity. Lack of promising strains is a major drawback. The wild growing population of *kewda* has a lot of variation which has never been documented and utilized, though this offers a good scope for its improvement by resorting to selection and subsequently propagating them vegetatively.

Species and Varieties

Pandanus is highly polymorphic with numerous forms and varieties, and selected in various countries for specific uses and maintained by vegetative propagation, such as ***P. leram,*** the Nicobar-breadfruit is grown in the Andaman Islands for its edible and starchy fruit that is enjoyed locally and it bears longer leaves; ***P. utilis*** (common screw pine), a native of Malagasy and on the remote Island of Mauritius, 90-180 cm long, stiff, green and slightly glaucous underneath leaves with small red spiny teeth, and the leaves are used for extraction of fibre for making bags; and ***P. veitchii***, a species from Polynesia is most commonly grown in greenhouses and outdoors with short stem (about 2 m tall) and when fully grown branches out freely, bearing arching, spiny and wide creamy-white margined leaves some 60-90 cm long, and here the prop roots and branching stems are particularly well developed attracting attention from many people who might not otherwise notice the stranger from the tropics (Lorus and Milne, 1967). However, these three species have striking foliage effect and especially useful in seaside gardens (Thomas, 1965). An unarmed form of ***P. laevis*** Warb. is cultivated for the fragrant bracts of the inflorescence, while forma *samak,* a group of prickly *Pandanus,* is preferred for the tough leaves suitable for matting; though another form with longitudinal yellow bands on leaves (var. *variegatus)* is grown as ornamental (Deshaprabhu, 1966). In the Pacific Islands, some of the forms such as *P. l. pulposus* are cultivated for their edible fruits (Purseglove, 1975). Burkill (1966) stated that the common wild armed form of the sea-coasts was var. *littoralis* and that flower growers had selected various races for cultivation, which included the prickly var. *samak* and thronless var. *laevis,* grown in India, Malay and Indonesia and are used for making mats. The number of commercially cultivated species in *Pandanus* is very few, although many names occur in horticultural literature. In South India, *Pandanus* is known so far to have six wild or naturalized species, viz., *P.fascicularis* Lam., *P.foetidus* Roxb., *P. furcatus* Roxb., *P. kaida* Kurz., *P. thwaitesii* Mart. and *P. unipapillatus* Dennst. (*P. canaranus* Warb.) (Bhat, 1992). Without flowers and fruits, it is difficult to know what species are actually in cultivation, or how accurate may be the popular descriptions and illustrations. For the same reason it is impossible to construct and accurate botanical key that will be of practical use to the gardener. Some of the good garden kinds are unplaced botanically, particularly the variegated or striped leaf kinds, which are sterile or the fructification insufficiently studied. ***P. baptisii*** (blue screw pine), a native of New Britain is short statured non-flowering species with <40 cm long, deep blue-green and striped (longitudinal creamy-white or yellow) leaves with spineless margins. ***P. odoratissimus***, a native of Sri lanka and also recorded from Malaysia and Philippine Islands may be as escape, is a slow-growing small tree most suitable for growing in containers, bearing 90-150 cm long and linear leaves with slender pendent tips, and the margins and the midrib-backs of leaves are armed with sharp spiny teeth. ***P. sanderi***, a native of Timor (Indonesia) is a slow-growing some 2 m tall plant bearing linear, minutely toothed and some 45-75 cm long leaves with a yellow central stripe on rich green background.

Named varieties are not available in *kewda*. Some growers in Ganjam district of Orissa have, however given names as 'local' and 'mala' (dwarf). Although variations are more spectacular in *kewda*. A detailed study of these variants and their systematic position is still lacking. To study the presence of variability followed by selection of

types in the natural strands of *kewda* plantations under farmer's field condition, a detailed survey was made in the coastal districts of Odisha during 1989 to 1993. Based on the variability present, six distinct morphotypes were selected (Jagadev, *et al.,* 1993a), which exhibit variations in different quantitative and 6 qualitative characters particularly flower bearing, weight, colour and aroma of flower and interestingly, in adaptability with regard to location, land type and soil texture. Deshaprabhu (1966) reported that fertile and well-drained soil was preferred for raising scented types in India. Hence, a comprehensive survey is desirable to identify some promising genotypes from the natural variability existing along the coastal belt which possesses the largest acreage under *kewda* plantations for production of *kewda* flowers.

Propagation

Kewda is exclusively propagated through ground suckers and leaf bearing terminal cuttings. In its growing areas, predominantly the terminal shoot cuttings are used in commercial plantations. However, the seeds are also capable to germinate and give rise to plants, but it is not in practice due to hard mesocarp and seed coat (Bailey, 1950, Deshaprabhu, 1966, Mott, 1975; Maharana and Jagadev, 1993).

In *kewda*, the flowers develop into berries which fuse to form a multiple fruit. The receptacle and bracts are included in the fruit. The **seeds** are deep seated and very difficult to extract, hence seeds are sown along with the fibrous drupes. Each drupe contains 2-9 seeds. Seeds collected from ripe fruits are sown during spring. The seeds usually take about 3 months for germination due to the hard seed coat and impervious fibrous covering. One hour sulphuric acid scarification of the drupes reduces the germination period by about one month which ultimately increases the germination percentage as compared to direct sowing, however, scarification for more than an hour is deterimental (Maharana and Jagadev, 1993). Bailey (1950) reported that seed soaking in tepid water for 24 hours would aid the germination. The seed propagated plants take about 2-3 years more for flowering as compared to vegetatively propagated plants. **Vegetative propagation** through ground suckers and crowns from flowering and non-flowering shoots was found to be more dependable for preserving and multiplying the genetically uniform selected types. Success from leafless old stem cuttings is very low and most of them dry in 5-6 months due to poor development of roots. Experimentally it was observed that plants raised from leafless stem cuttings will die irrespective of their size. Ground suckers were found to be the best planting material followed by the crowns. (Maharana and Jagadev, 1993).

Cultural Practices

Kewda, a perennial evergreen plant is salt tolerant, good soil binder, not browsed by the cattle, once planted can live for more than 40 years, and can successfully be grown on marginal, waste, marsh land and coastal sand. They grow wild beside the rivers or near the sea-shore moist conditions and grow in saline soil. It can also be used along the banks of ponds, canals and rivers to check soil erosion and as fence on the bunds around the crop fields. Soil has a pronounced effect on its plant growth and *kewda* productivity. Well drained sandy-loam soil with adequate supply of moisture is best suited for its vigorous growth and flower production and, thickly distributed profitable plantations are found in such soil. But, in heavy clay soil with inadequate moisture supply, the plants grow vegetatively with poor flowering. Flower production in saline land situation is better than dry land situation. Higher flower production is observed in sandy loam soil as compared to clay loam or sandy soils. *Kewda* is essentially a crop of the tropics and sub-tropics requiring reasonably high rainfall, high temperature and high humidity. It can be grown below 300 m from mean sea level though thrives best at 100 m mean sea level. In practice, the cuttings are collected and scattered on the ground for 2-3 days. Before planting, a few leaves closed to the base of the cuttings are removed to expose the root development sites. The cuttings are planted in trenches of 45 cm depth horizontally or at an angle of 30°. The new leaves appear on the growing points one month after planting. It is usually planted on the field bounds for fencing and subsequently used for collection of flowers. A 30 year old plant can produce a crown of 6-7 m, for which it is difficult to develop a commercial plantation in a close spacing. It has been observed in the natural plantations that plants may be spaced at 8-10 m distance to achieve a full length of plantation life. It is, therefore, suggested to follow a paired-row-method of planting maintaining a spacing of 4 m between plants and rows of a paired-row and 6 m between two paired rows; which provides all the conveniences for plant growth up to 10-12 years of planting and then pruning of a few branches is required.

Exact practice of **manuring** and **fertilization** is hardly followed in *kewda* plantation as no any scientific support is available to the growers. In general practice, each plant is dressed with 1-2 baskets of FYM or organic waste compost before planting. It responds well to application of inorganic fertilizers and an annual application of 60g N, 30g P_2O_5 and 30g K_2O per plant in June (i.e. just before monsoon) is beneficial. The recommended fertilizer dose per hectare is 30:15:15 kg NPK. **Weeds** are not a problem for its plantation because of salinity and littoral type of soil in *kewda* growing areas. Weeds don't grow, but plantations closed to the canal embankment and field bunds compete with some weeds during the initial stages of growth. Once the plant growth peaks up, weed growth is automatically suppressed due to shade and development of extensive root system. Hence, no practice of weeding is generally followed. *Kewda* is though tolerant to drought but for its proper growth and development it requires plenty of **water.** The crown portion dries during summer and favours fungal growth during rainy season. It is, therefore, desirable to provide some water during summer to sustain the vegetative growth. It grows luxuriantly in soils having high water table. Soils deficient with moisture do not allow producing stilt roots and as a result the crown growth is suppressed, and instead favours production of suckers. In the drier inland districts of Odisha, the plants don't produce stems and remain bushy throughout. Flowering is only noticed in the plants with many branches in the areas having high water table, and continuous water supply favours flower production. The places receiving rain for a longer period provide a continuous harvest of flowers. The flowering season only coincides with rainy days and moderately to winter. In *kewda* growing areas some wild melons grow as **intercrops** which are highly compatible with *kewda.* Since it takes more than 5 years to flower, some legumes and melons can be grown as intercrops as soil condition is unsuitable for growing any other crop. Intercropping keeps the soil loose and supplies nitrogen to the crop.

Post Harvest

The flowers are mostly collected from the natural plantations within a radius of 15 km around Chhatrapur, Tulu and Keluapalli areas. Plucking of flowers is though by hand but always with a hooked stick. Collection by cutting down flowering stalk is not recommended as it may damage the floral primordia and arrest further flowering. The growers use to pluck the flowers from the field by engaging their family members of which 70% by male, 25% by female and rest 5% by child members. Collection of flowers is completed by early morning, so that those can be carried in baskets the same day to the distilleries by 9.0 a.m. for distillation as longer storage results in poor oil quality and poor recovery. The collection of flowers from natural plantations can be considered as a family business since most of the family members from children to adults are engaged in collection during early morning hours and the individuals with head-load of flowers even in small quantity start moving to the distilleries. Sometimes middlemen also work for their supply to the distilleries. At the factory site, the collected flowers are counted and recorded in a register from 7.30 a.m. to 10.00 a.m. The flowers delivered to the factory are spread in a cool store house. The height of the heap of flowers sometimes reaches almost to 1.0-1.5 m.

Insect-Pests and Diseases

Kewda plants are hardly attacked by the insect-pests and diseases. However, a few fungal and bacterial diseases are found to reduce the yield of flowers significantly. Some important insect pests and diseases of this plant with their nature of damage and suggested control measures are discussed here. During the dry season, leaves are found to be attacked by bag worm (*Manatha albipes*). Larvae feed on chlorophyll and veins of leaves leaving the mid-rib resulting in perforation of leaves. In case of heavy infestation, the leaves become dry. Extensive damage arrests the vegetative growth resulting in significant reduction in flowering. Spraying of any contact insecticide such as monocrotophos @ 0.05% a.i. is suggested at an inteval of 15 days if the population of insects is noticed on the plants. Some saprophagous beetles and thrips are observed on the inner side of bracts and on the ardroecious spikes of male spadices. Although, the beetles are found feeding on the pollens of the flower, but at times these cut the flower bracts and their excreta also deteriorates the quality of flower. Further research to know as regards to their role as a pollinator or any harmful effect is yet to be worked out. While controlling bag worm will control these pests also, however, a predatory lady bird beetle (*Micraspis discolor*) feeds upon small insects like thrips and aphids.

In summer months, **bacterial bud rot** (*Pseudomonas solanacearum*) of central shoot is sometimes noticed. The disease is characterized by the presence of water soaked, discoloured leaves in the whorl, being more pronounced at the basal portion. Diseased leaves can be pulled out easily from the bunch, and emergence of new leaves is hampered. In advanced stage, the diseased leaf with necrotic lesions show the presence of slimy bacterial oozing, and lead to premature drying of plants. In some localities, plants to a tune of 15 % suffer from this disease. The soil preventive measures are to be taken by mixing with stable bleaching powder @ 12 kg/ha before 15 days of planting or the planting materials should be treated with plantomycin @ 0.2% solution for 15 minutes. During rainy season, the **leaf blight** is caused either by *Botryodiplodia theobromae* (Suryanarayanan and Narasimhan, 1981) or by *Alternaria tenuis* Nees (Purseglove, 1975) and the extent of damage is roughly to a tune of 5-7%. The disease is characterised by extensive yellowing of leaves. The large spots turn deep brown and often cover the entire leaf leading to premature defoliation and significant reduction in flowering. The symptoms are mostly exhibited under high humidity and high temperature. Two sprayings of Dithane M-45 @ 0.4% or Blitox 50 @ 0.2% or Bavistin @ 0.15% at an interval of 10-15 days is suggested. The fruit is rarely eaten by the local folk; however, it can be processed at ripening stage. At this stage, the **central core rotting** of fruit is noticed during rainy season, resulting in its premature dropping. Use of disease-free planting material reduces the incidence of this disease.

References

Bailey, L.H. 1950. Pandanus. In: *Standard Cyclopedia of Horticulture*, vol. 3, pp. 2449-2451. Macmillan Co., New York.

Beckett, K.A. 1987. The RHS Encyclopedia of House Plants Including Greenhouse Plants, pp. 175-176. Salem House Publishers, Topsfield, Massachusetts 01983, USA.

Bhat, K.G. 1992. Studies on screwpines (*Pandanus*) of Karnataka-new records. *Indian J. Forestry*, **15**: 359-360.

Burkill, I.H. 1966. A Dictionary of Economic Products of the Malay Peninsula. Ministry of Agricultural Co-operation. Kuala Lumpur.

Chittenden, F.J. 1956. Pandanaceae. In: *Royal Horticultural Society Dictionary of Gardening* (2nd ed.) vol. 3, pp. 1476-1477. Oxford at Clarendon Press, U.K.

Deshaprabhu, S.B. 1966. Pandanus. The Wealth of India :Raw Materials, vol. 7, pp. 216-221. CSIR, New Delhi.

Dutta, P.K., H.D. Saxena and M.Brahmam, 1987. Kewda perfume industry in India. *Econ. Bot.*, **41**:403-410.

Everette, T.H. 1969. Angiosperms:The Monocotyledons. *The Living Trees of the World*, pp. 66-68. Thames & Hudson Ltd., London

Jagadev, P. N., B. Naik, S.C. Sahoo, S. Beura, and T. Maharana, 1993. Variability, selection and planting material in kewda (*Pandanus fascicularis* Lam.). *Indian J. Genet.*, **53**:310-314.

Jagadev, P.N., S. Beura.and T. Maharana, 1993a. Prospects of aromatic plants in Orissa. *Orissa J. Hort.*, **21**:95-98.

Lorus and Margery Milne, 1967. Cattails, burreeds and screwpines. In: *Living Plants of the World*, pp. 270-272. Chanticler Press, New York.

Macoboy, S. 1986. *Pandanus.* In: *What Tree is That?*, p. 158. Tiger Books Intl. Ltd., London.

Maharana, T. and P.N. Jagadev, 1993. Studies on propagation. In: Final Report of ICAR Project on Productivity and Industrial Use Study in Medicinal and Aromatic Plants of Orissa with Special Reference to *Pandanus* and *Cymbopogon*, pp. 14-16. OUAT, Bhubaneswar.

Maneka Gandhi, 1989. Screwpine. In: *Brahma's Hair : Mythology of Indian Plants*, pp. 121-124. Rupa & Co., New Delhi.

Mott, R.C. 1975. Foliage plants. In: The Complete Book of House Plants, p. 73. Octopus Books Ltd., London.

Purseglove, J.W. 1975. Pandanaceae. In: *Tropical Crops:Monocotyledons* (1st ed.), pp. 511-515. ELBS Longman Group Ltd., London, U.K.

Rendle, A.B. 1979. Pandanales. The Classification of Flowering Plants (Gymnosperms and Monocotyledons) (2nd ed.), vol. 1, pp. 187-197. Vikas Publ. House Pvt. Ltd., Ghaziabad (U.P.).

Suryanarayanan, T.S. and C.L. Narasimham, 1981. A leaf blight of *Pandanus odoratissimus* caused by *Botryodiplodia theobromae. Indian Phytopath.*, **34**: 254-256.

Thomas, A. 1965. Plants with noble foliage and water plants. *Gardening in Hot Countries*, pp. 112-113. Faber & Faber Ltd., 24, Russel Square, London.

25

Polianthes tuberosa

Sanyat Misra and R.L. Misra

Family: **Agavaceae,** formerly **Amaryllidaceae**

Common names: Tuberose**,** Hindi-**Rajanigandha**, and Malayalam-**Nishigandha**

Introduction and Origin

Polianthes takes its name from the Greek *polios* meaning 'white' or 'bright' and *anthos* meaning a 'flower', and contains one of the largest cultivated species of cut flowers, *Polianthes tuberosa,* which was known to Clusius in the 16th century, and the double form was known in the early part of 18th century. It is native to Mexico. At one time, the genus *Polianthes* contained red- and orange-flowered species which have now been placed under *Bravoa*. Tuberose is a sun-loving plant. Flowering occurs throughout the year in tropical climates, from summer to full autumn in the sub-tropical regions and during summer to rainy seasons in the temperate regions of the country. If watering is not the limiting factor and there is protection against hot summer sun and chilly winters in sub-tropical conditions, in the tropical and subtropical regions it flowers throughout the year, peak being from late-rainy season to full autumn, i.e. September to November. Flowering is scarse during hot summers and chilly winters though 'Calcutta Single' is exception. The common name tuberose is not well understood but surely it has nothing to do with rose but seems to be a corruption of the word "tuberous" (tuber-like), referring to the plant's tuberous rootstock. Tuberose is the most homely flower of India. It is being grown in every part of the country irrespective of the climatic conditions prevailing there. Though its flowers are white but they have intense pleasing fragrance and are available in two forms: single and double. Its 1150 g of flowers yield 1 g of oil which is used extensively in perfumery. The intensity of fragrance is more in singles than the doubles. On the temperate region of the country as well as on the lower hills, the doubles are also quite fragrant. It is used as cut flowers, as loose flowers, as border plants, near windows for night fragrance, and for extraction of essential oils. Ali *et al.* (1981) isolated two sapogenins, hecogenin and 9, 11-dehydrohecogenin, and two saponins, becogenin-3-*0*-glucoxyloside and 9, 11-dehydrohecogenin-3-*0*-glucoxylogalactoside from whole plant of *Polianthes ruberosa*.

Botany

Its rootstock is a bulb though has a rhizomatous base with thickened roots, leaves are succulent, basal, very dense, long, narrow and strap-shaped, the inflorescence stalk is strong, erect, bracteate and up to 75-100 cm, flowers white, sweetly fragrant, waxy-textured, produced in a terminal raceme, long, cylindrical and tubular, bent near base and then flaring in funnel-form, and are borne in pairs in the axils of the bracts. Segments 6, unequal but short, stamens 6 and filaments thread-like, short, anthers dorsifixed and erect, style thread-like, stigma sickle-like, ovate and 3-lobed, ovary inferior and 3-chambered, fruit a capsule with black and flat seeds.

There are some 12-15 species under the genus *Polianthes,* nine having only white flowers, but except *Polianthes tuberosa* others seem to be obsolete. Almost all the *Polianthes* species and varieties are fertile except the double form 'The Pearl' under *P. tuberosa* which possesses n = 25 chromosomes and it is hypothesized that double forms are derived from the single form by the loss of the 5 pairs of smallest chromosomes possibly due to spontaneous abnormal meiotic or mitotic divisions. The chromosome numbers in single tuberose is recorded n = 30 which falls under four groups, i.e. 5 large pairs, 4 medium pairs, 7 medium-small pairs and 14 small pairs. Single forms are fertile whereas doubles are sterile, therefore, in a breeding programme double form is normally not used as male parent. There is self-incompatibility whether it is single or double form. Krishnan (2003) through 15 Gy and 20 Gy ^{60}Co gamma rays and 1.0 and 2.0 per cent EMS induced nine vM_2 stable mutants which revealed the differences in banding pattern of esterase, peroxidase and catalase between parents and mutants through isozyme analysis. Rajvir singh *et al.* (2008) gave pre-planting treatment of EMS at 0, 0.25, 0.50 and 0.75 per cent to the bulb tips of tuberose cv. Double and obtained delayed sprouting in all the treatments, in vM_1 increased duration of flowering under 0.25% EMS, and in vM_1 and vM_2 both longest rachis, maximum spike weight and maximum bulb weight with 0.25% EMS, but no any genetical modification was observed.

Classification, Species and Varieties

Except *Polianthes tuberosa,* other 11 species are not of any commercial importance. All the varieties available in the country or elsewhere fall under only one species, *P. tuberosa.* Though there is no any special classification of tuberose but on the basis of the flower form, it is classified as 'Single' or 'Double'. Initially only two varieties were introduced in the country *viz.* 'The Pearl' (double) and 'Mexican Single'. Afterwards, two varieties, 'Swarna Rekha' (double) and 'Rajat Rekha' (single) were developed through ^{60}Co gamma treatment from N.B.R.I., Lucknow. These varieties are only collector's item and do not have any commercial value as both flower seldom. However, there are golden stripes on the leaves of 'Swarna Rekha' and silver stripes on the leaves of 'Rajat Rekha'. Like to that of 'Swarna Rekha', one more variety was developed in Sikkim which also has golden stripes on the leaves but slightly different, and this is known as 'Sikkim Selection'. Its spike is taller than others and grows more than one metre in length. I.I.H.R., Bangalore developed four varieties *viz.* 'Prajwal' (quite sturdy, tall and vigorous and the best among all the singles existing in the country), 'Shringar' (single), 'Suvasini' (double) and 'Vaibhav' (double). One variety has been developed by A.N.G.R.A.U., Hyderabad, probably a selection and has been named as 'Hyderabad Single' which is very similar to 'Shringar'. One variety as 'Phule Rajani' has been developed by M.P.K.V. Pune centre. This single is also quite compact and beautiful. This variety at bud stage is green. Many states in the country grow single and double forms except those described above, which they claim their own but, in fact, these are one, the single one is 'Mexican Single' and the double one 'The Pearl'. All the varieties at the bud stage are pinkish except 'Phule Rajani', 'Mexican Single' (Calcutta Single) and 'Suvasini' which are green at bud stage. Thus in real sense only 11 varieties of tuberose exist in the country. The singles being 'Hyderabad Single', 'Mexican Single'(Calcutta Single), 'Phule Rajani', 'Prajwal', 'Rajat Rekha', 'Shringar' and 'Sikkim Selection' while doubles are 'Suvasini', 'Swarna Rekha', 'The Pearl' (Calcutta Double), and 'Vaibhav'. The Dutch-based 'Ludwig & Co., Holland', in 2013, released coloured tuberoses through genetic engineering, and now this firm is licensee for these five varieties which have high fragrance. Out of these, the boldly coloured single-flowered varieties are 'Cinderella' (lavender-pink) with stalk length of 35 cm, 'Sensation' (pink) with stalk length of 45 cm, 'Super Gold' (deep yellow) with stalk length of 65 cm and 'Yellow Baby' (yellow) with stalk length of 35 cm; and the double-flowered one is 'Double Pink' (pink) with stalk length of 65 cm (www.floraculture.au). Also in Taiwan the work in the direction of conventional breeding is going on and they have achieved success in developing pink, yellow and orange variants for multi-purposes such as bedding, cut flower and pot-growing.

Propagation

Tuberose is propagated vegetatively through bulbs, lateral bulblets forming on the side of the main bulb (also known as offsets), through division of bulbs (bulb segments) and through micropropagation. The most common method for tuberose multiplication is through **bulbs** and **bulblets**. Spindle-shaped bulbs having diameter of 2 cm are flowering size but other smaller ones (bulblets) are only planting stocks. The basal plate of tuberose is so formed that one confuses it with a rhizome and some also call these as tubers though these are bulbs. Though one year planted bulb is not so confusing but the bulbs taken out from perennial cropping give this impression. The formation of lateral buds is so fast that basal plate extension is necessitated to accommodate these bulblets and on lifting when bulbs and bulblets are taken out one by one, a big chunk of the disc (basal plate) is left in the end. In a two-year cropping, a plant of certain varieties produces about 5-10 large (flowering size) bulbs and about or more than 50 bulblets. On planting, in one year, 2-3 large bulbs and 15-30 small bulbs (bulblets) are formed. If cared well, bulblets also grow full flowering size bulbs in one year, if planted singly after separating these from the clump. Since tuberoses are, by and large, grown as perennial crop, i.e. 2-4 years so because of their dependency on the mother bulb as these are sharing the same disc for years, the bulblets get only a small share of nutrients hence their development is very slow. Four-year perennial cropping is not profitable as quality of the spike becomes very poor and so is the bulb and bulblet formation. After planting the large-sized bulbs, first year one gets only 1-3 spikes, but second year 4-6 with slightly poor quality of spikes, while the third year though there is increase in the number of spikes, but quality is obviously poor. Therefore, only two-year cropping should be adopted. It would be better if large bulbs are separated every year for flowering and bulblets for raising the stock. Removal of the old basal-rhizomatous-bulbs from offsets before planting lowers the number of new-forming offsets (Cirrito and De Vita, 1982). If cared well, many of the large bulblets will also provide quality spikes the first year.

Full-sized bulbs and large bulblets may also be **fractionated** into ½-$^1/_8$, each piece having a part of the disc as well as fleshy scales, treated with some effective contact fungicide and planted in sandy loam soil rich in humus. Each segment will provide a new plant. Sharga (1980) selected the tuberose bulbs ranging from 0.5 to 3.0 cm in diameter and divided these into six grades, each grade bulb fractionated longitudinally into two halves but when compared with entire bulb the flowering was considerably delayed, the bulb production was comparatively better though bulblet production was poor, and the segments only from the large diameter bulbs (>2 cm) regenerated well.

Micropropagation may also be adopted for quick multiplication. Bulb segments on MS medium, with addition of growth regulators and vitamins and incubating at 22-25 ^{0}C using continuous fluorescent light of 3,500 lux, may provide some 500-800 plants in one year. Narayanaswamy and Prabhudesai (1979) cultured tuberose explants of bulb scale, bracts and petals on a medium containing 2, 4-D (2 ppm) or kinetin (1 ppm) and the explants differentiated into numerous embryoid-like structures though these embryoids developed only a primary root and were devoid of shoot apical meristem. Bose *et al.* (1987) cultured tuberose cv. 'Single' explants from bulb scales, on MS medium containing 2, 4-D at 0.1 mg/l and obtained good callusing which regenerated into plantlets, the medium containing kinetin at 0.05-0.2 mg/l initiated shoot appearance before the rooting, and when the callus was subcultured on a medium containing NAA at 2 mg/l, this resulted in root and shoot initiation in 15-20 days with good plantlet production. Krishnan (2003) found scale stem section of the bulb the best as the ideal explant source and in somatic embryogenesis the inflorescence segments containing immature flower buds, and he found MS medium supplemented with BAP 6.0 mg l^{-1} + KIN 4.0 mg l^{-1} as the best for high rate of multiple axillary buds though for maximum root initiation MS medium supplemented with IBA 4.0 mg l^{-1} + 0.2% activated charcoal. However, direct organogenesis was obtained from inflorescence segments in MS medium supplemented with NAA 0.2 +BAP 2.0 + KIN 1.0-3.0 mg l^{-1}.

Growers should not be concerned with seed propagation as it is a tedious job where there is only little success and that too with time constraints. Moreover, at flowering seed-raised plants may not come true to the type. 'Hyderabad Single', 'Shringar', 'Mexican Single' (Calcutta Single), 'Phule Rajani', and 'Sikkim Selection' all singles produce seeds which are small, shining black, light and flat. Immediately, when capsules mature, these should be taken out and sown on raised beds having well workable sandy-loam soil rich in organic matter. These beds should lightly be watered daily. Within a month or two, these will germinate.

Cultural Practices

Reasonably fertile sandy-loam **soil** which is light and well-drained but with good water-retention capacity, having pH range of 6.5-7.5, is most suitable for tuberose growing, though tuberoses can successfully be grown up to 8.5 pH provided the field is rich in organic matter. This prefers a sunny situation. Even partial shade hampers the growth. It grows very well from 18-35 ^{0}C temperatures.

Its proper **planting time** in the subtropical regions is February and in temperate regions when danger of frost is over, i.e. March-end to early-April. In tropical regions, it can be planted any time, preferably from October to January. If planted at proper time, these require 20 to 50 days for sprouting, the smaller ones sprouting earlier and the larger ones later as was experienced by Pathak *et al.* (1980) when they planted small (5-10 g), medium (11-15 g) and the large bulbs (19-35 g), though for flower emergence from emergence these took 70 days for small ones, 58 days for medium ones and 45 days for larger ones. In Italy, Farina and Patertniani (1986) recommended tuberose planting in early June for highest flower production when they studied the varieties 'La Perla' (The Pearl, double), 'Fiorentina' (semi-double) and a single-flowered type; however, Cirrito *et al.* (1982) stated best planting time in Italy from 14th to 29th June for bulb enlargement when they planted the bulbs having 6.0-8.0 cm circumference. They are planted shallowly just to cover the tips of the bulb with the soil. For its planting the **soil preparation** should be done as for potato planting. Soil should be well dug up to 30 cm deep thrice, every time followed with planking. The soil should have sufficient moisture at the time of third ploughing and planking so that after planting the field may not require immediate watering. Now the beds of convenient sizes are made after leaving the space for bunds and channels. **Bulb size** for the purpose of cutting should be 8.0-9.0 cm in circumference (Farina and Paterniani, 1986) or at least 2.5 cm diameter (Brundell and Steenstra, 1986), though Curir *et al.* (1984) reported flowering in Italy with 4-7 cm circumference of bulbs provided such bulbs attain 6.0-6.5 cm of circumference at flowering but still the larger bulb size is required for production of commercial flower heads. Brundell and Steenstra (1986) reported increased yield of inflorescences and lateral bulbs by planting large size of bulbs, clumped bulb giving greater response than individual bulbs though bloom quality increases with individual bulbs. Lopes (1971) planted the bulb-transplants weighing 70, 200 or 340 g at population levels of 30, 40 or 50 thousand per hectare and recorded no effect of planting density on flower production, length of floral stalk or plant weight though larger transplant size was associated with greater flower production and plant weight only. However, Sharga (1982) when used bulbs weighing from 3.0 to 49.5 g with single-flowered variety, obtained satisfactory results when the size was 19.4 g and above. Yadav *et al.* (1984) with tuberose var. 'Single' obtained 2.6-3.0 cm bulb diameter with highest yield of spikes and flowers when planted 6 cm deep. For 1-year crop, the proper distance found is 15 × 15 cm and for 2-year crop 17 × 17 cm, planting only one bulb per hill in both the cases (Patil *et al.*, 1980). Patil *et al.* (1987) when planted bulbs of the diameter of 1.5-2.5 or 2.6-3.0 cm at 15 × 20 cm, 20 × 20 cm or 25 × 20 cm spacings for a 3-year cropping, they obtained highest top quality flower yield from smaller bulbs planted at 15 × 20 cm distance. **Planting** is done at 30 cm apart in the rows if it is to be made a two-year crop. However, for a three-year cropping, it requires more distances and that is 40-45 cm apart. After planting, pendimethalin (stomp) pre-emergence **weedicide** at the rate of 3 litre/ha after mixing the chemical in 5,000 litres of water should be applied evenly in the planted area just to wet the surface. This will prevent germination of annual weeds up to 75 days. Then after, only stray weeds are to be removed manually because the main crop covers the in-between area with its own growth of leaves which normally does not permit other weeds to grow. In case, weedicide has not been used, organic or black polyethylene mulching is used. In Italy, Cirrito *et al.* (1980) obtained control of almost all the weeds except *Cyperus longus* with use of Treflan (trifluralin) at 2 kg/ha in tuberose cv. 'Perla', though other herbicides they used were Alipur (chlorbufam), Prevenol (chlorpropham) and Tok E25 (nitrofen). Pre-plant application of atrazine at 3.0 kg/ha caused maximum weed control in tuberose cv.

'Single' and yielded 10.2 tonnes/ha of good quality flowers compared with 5.0-10.0 tonnes in unweeded control (Yadav and Bose, 1987). Panwar *et al.* (2010) encountered weeds in the cultivation of tuberose cv. 'Double', some of them being *Amaranthus* sp., *Bracharia reptens, Corchorus* sp., *Cynodon dactylon, Cyperus rotundus, Euphorbia* sp., *Sesbania aculeata, Solanum nigrum, Sorghum halepense, Trianthema portulacastrum, etc.*, for which they used Atrazine and pendimethalin at 1.0, 1.5, 2.0 kg/ha as pre-emergence, while glyphosate at 1.0, 1.5 & 2.0 kg/ha and XL-71AG at 0.6, 0.9 & 1.2 % 30 days after emergence by dissolving in 625 l water for one hectare, and recorded post-emergence treatment quite injurious to the crop.

In case weedicide is not used, **hoeing** and earthing up should also be carried out as and when required. Hoeing will keep down the weeds and will permit aeration while **earthing** will give good support to the inflorescences to stand erect against winds and rains.

Irrigation and Feeding

First light **irrigation** is followed immediately after, if pendimethalin weedicide has been applied in the field. Whether it is temperate or subtropical region but its growing season passes through spring to autumn hence it requires copious watering but shallowly every time, starting from its sprouting to leaf senescence. In completely tropical region also, the watering frequency is the same. In light soils, it should be watered every 6 days while in heavy soils 8 days. Watering very much depends on the prevailing weather conditions. During rains, it does not require watering. If it is watered well regularly, one can get the flowers throughout the year in the tropical regions, from May-June to October-November in the sub-tropical regions, and from June end to October in the temperate regions. In Egypt, El-Roumi (1981) provided 40 litres water/m^2 at every 7 or 14 days and found better tesults where tuberose 'Pearl' plants were irrigated every week. In Italy, Cirrito *et al.* (1981) when drip-irrigated at 2,000-8,000 m^3/ha/year in a 3-year cropping the tuberose cv. 'La Perla' planted in late April at a density of 36 plants/m^2, recorded best yield of commercial-calibre bulbs, i.e. 30-32/m^2. Erection of polyhouses over the plantings during winter season, i.e. from November to March in India in sub-tropical and temperate regions will make the flowers available regularly for all the 12 months. Even otherwise, the variety 'Mexican Single' which is good only as a loose flower blooms throughout the year under Delhi condition if the crop is not damaged due to frost in winter. If it does not occur, to some extent leaves even of other varieties remain green even during winter otherwise senesce, and new growth starts only by April. Under Delhi condition, there is profuse flowering from September to November.

At the time of preparation of soil, well rotten **compost** or farmyard manure at the rate of 300 quintals/ha should be incorporated in the soil by mixing thoroughly. Being a perennial crop it requires heavy feeding. For tuberose crop, Cirrito (1972) recommended 24 kg each of N and P_2O_5 and 36 kg/1,000 m^2 of K_2O in Italy. In a 3-year trial with tuberose cv. 'La Perla', at Palermo (Italy) on alkaline sandy soils. Cirrito and De Vita (1981) when applied 500 kg/ha of chicken manure as pre-planting and 800 kg/ha ammonium sulphate, 1,000 kg/ha rock phosphate and 180 kg/ha K_2SO_4 as post-planting, they obtained highest production of new lateral bulbs measuring 4-6 cm out of which some 92% were ready for flowering. Cirrito (1975) in Italy stated that tuberose bulbils for getting saleable bulbs should be applied with 20 kg N, 40 kg P_2O_5 and 60 kg K_2O/1000 m^2 at planting. Tuberose bulbs of 12-18 g size or those weighing 19-25 g planted in late April and applied with PK basal dressing though nitrogen in the form of ammonium sulphate at two split-doses, one at 2-3 leaf stage and the other before flowering, resulted into earliest and highest percentage of flowering and with highest number of flowers per spike with those receiving 75 kg N/ha with large bulbs (Mitra *et al.*, 1979). A soil deficient in N and P and high in K, with cv. 'Single', Nanjan *et al.* (1980) recorded highest flower production when applied 200 kg N, 60 kg P_2O_5 and 0 kg/ha K_2O. With cv. 'Single', application of 300 kg/ha N in two split doses, half at planting and the other half 40 days later, and 200 kg/ha P_2O_5 gave highest flower yield and plant growth (Yadav *et al.*, 1985). Bankar and Mukhopadhyay (1985) with tuberose cv. 'Single' when applied 25-35 g N, 60-80 g P_2O_5 and 60-80 g/m^2 K_2O; P_2O_5 improved only floral quality though N had a significant beneficial effect on all the parameters studied including spike durability in the field (22.8 days) which is longest in the highest N rate. Mukhopadhyay and Banker (1985) when applied 20 g/m^2 N in one dose or in 2-4 split doses to tuberose cv. 'Single', they recorded greatest plant height and number of spikes per plant with application of 1/3rd N before planting, 1/3rd after 30 days and remaining 1/3rd at 60 days after planting. By applying 30 g/m^2 N, Muralee Manohar (2000) recorded longest rachis length in tuberose, though 20 g/m^2 N and 15 g/m^2 P_2O_5 application was responsible for increased weight of bulbs and bulblets.

In a 2-year tuberose cv. 'Single' crop, Mukhopadhyay and Bankar (1986) when applied 0-20 g N, and 0-40 g/m^2 each of P_2O_5 and K_2O, they found only N at the highest rate improving plant growth, spike yield, flower quality and production of daughter bulbs but not the main bulb. In a 2-year tuberose crop of cv. 'Double' when N at 0, 5, 10, 15 or 20 g/m^2, half at planting and remaining half as top-dressing at flower emergence; and P_2O_5 and K_2O each at 0, 20 or 40 g/m^2, both as basal dose, N improved vegetative growth, flowering and bulb production, and advanced flowering, and the highest N rate produced highest number of flower spikes (20.09)/m^2, however, the effect of P and K with respect to duration of flowering, spike number and rachis length could be seen only in the second year (Bankar, 1988; Bankar and Mukhopadhyay, 1990), and in the end there was recommendation that per square metre the 20 g each of N, P_2O_5 and K_2O should be applied for best results. In a 2-year pot-culture of tuberose, Sharga and Motial (1983) when applied foliar spray of N as urea at 2.14, 4.28 or 6.42 g/l; orthophosphoric acid at 0.9, 1.8 or 2.6 ml/l; and/or potassium citrate at 2.2 or 4.42 g/l, they obtained highest yield of essential oil from plants receiving medium N rate, highest P and lower K rate when sprayed 50 ml of solution every 15 days starting on June 2. In a sand culture experiment, Jana *et al.* (1974) found increased number of leaves, bulbs, spikes and flowers with high N and P though their deficiencies suppressed flowering whereas low K reduced the number of flowers on the spike. Sharma *et al.* (2008) recommended 200 kg N, 70 kg P_2O_5 and 40 kg K_2O per hectare as optimum for tuberose cultivation under Haryana condition. They found 200 kg/ha nitrogen or 70 kg P_2O_5, followed by 40 kg K_2O increasing plant height, spike length, flower yield & quality and bulb & bulblet yield. Potassic fertilizers should only be incorporated in the soil if soil is deficient in potash. Therefore, it would be better if soil analysis has already been done so that only recommended dose of NPK or any other nutrients may be given. Nitrogen 200 kg and P_2O_5 60 kg/ha will give better yield of flower and bulbs. In potash deficient soils, 50 kg/ha K_2O will benefit the crop. Whole of the potassium and phosphorus and half of nitrogen may be applied in the soil while preparing the land for tuberose cultivation. Half nitrogen may be applied when inflorescence starts emerging. In the standing crop, always after fertilizer application, irrigation should be carried out. Another year also the same quantity of fertilizer may be used if it is a two year crop, and subsequently in the third year also the same quantity if it is three-year crop. Application of *Azospirillum, Azotobacter,* and phosphobacteria alone or in combination, and also when combined with inorganic fertilizers give excellent results in terms of number of spikes per plant, number of flowers per spike, number of bulbs and bulblets and their sizes per plant. Application of granular superphosphate at 60 g/m^2 combined with phosphobacterin increased the number of flowering plants, number of flowers per spike and floral durability (Aikincai, 1960). Chaudhary (2008) used VAM, nitrogen at 50, 100 and 200 kg/ha, phosphorus at 25, 50 and 100 kg/ha, and their combinations in tuberose 'Double' planting, and recorded increased growth and flowering parameters with increase in N and P applications, and the results in VAM was found to be at par to phosphorus at 50 and 100 kg/ha. Tuberose responds well to certain micronutrients. Application of zinc @15 kg/ha or spraying the crop with 0.25% $ZnSO_4$ results in better production of flowers, bulbs and bulblets. $CaCl_2$ and NaCl salinity (EC 1.5-15.0 mmhos/cm) has detrimental effect on bulb sprouting, flowering time, spike length, flower number and flower size, *vis-à-vis* essential oil content, the effect being quite drastic at early stages of growth though later stage tolerates quite high salinity levels, however, the uptake of Na^+, K^+, Ca^{++} and Mg^{++} is also badly affected (Singh *et al.*, 1974). Malini and Khader (1989a, b) when provided 6,000, 8,000 or 10,000 ppm $CaCl_2$ or NaCl to tuberose 'Single' with irrigation water and assessed their effect at 80 and 145 days after planting, they recorded that plants were more tolerant to NaCl than $CaCl_2$ and where both were added even at lower level the growth was utterly poor. Control flowered in 122 days though at highest salinity level it was delayed up to 165 days, and the flower yield in control per plant was found to be 24.7 g though at highest salinity level it as 9.8 g per plant.

Growth, Development and Flowering

The bulbs have vegetative apex protected by a layer of 3-4 sheath leaves, and after a certain growth these change to generative phase and from then it takes hardly 25 days to form a complete flower inside. When such bulbs are planted in field, they initiate flowers in 90 days. Tuberoses do not have true dormancy and if the plants are getting irrigation water regularly these remain green throughout the year and flower though severe winter temperatures force them to senesce but soon after weather improves the bulbs remaining in the field start sprouting. However, for their forcing in greenhouses the dormancy of the bulbs can be broken by 4% thiourea dip for one hour, ethylene chlorhydrin is effective only in case of large bulbs, and carbon tetrachloride dip is partially effective in case of large bulbs but is harmful to small ones (Hakim *et al.*, 1958). With respect to floral bud differentiation in tuberose, Kosugi and Kimura (1961) in Japan when planted the bulbs on April 1, through microscopical examination recorded floral bud initials on June 11 when plants were 19.5 cm high with 19.8 leaves, and when mean minimum temperature was 13.4 ^{0}C, and by June 18 the flower bud primordia had differentiated into two parts, on June 25 the sepals were first observed on the two primordia when mean minimum temperature was 19.3 ^{0}C and then the plant height was 29.2 cm with 23.3 leaves. Under protected cultivation of tuberose cv. 'Pearl' with no heating arrangement, Brundell and Steenstra (1986) in New Zealand obtained 10 days advance flowering with improved flower quality, increased lateral tuber production and 150% increase in the stem yield. They recorded 50-60% overall increased floral-stem yields when planted pre-heat-sprouted bulbs. The growth and differentiation of the apex depends on the supply of soluble carbohydrates from the scales. Storage temperature influences the number of bulbs lifted and the quality of flower spike. Storage at higher temperature (30 ^{0}C) for longer duration, though advances flowering but reduces the quality of the flower and bulb yield. However, storage at 10 ^{0}C for one month is optimum which improves the yield of flowers, but prolonged storage decreases bulb yield. Planting of the fresh bulb will lead only to profuse vegetative growth and little flowering, hence, for sometime the bulbs should be rested to encourage better growth and flower production.

For maximum root growth, soil temperature should be above 20 ^{0}C. For maximum bulb production, atmospheric temperature should be in between 20-30 ^{0}C. If bulb growth does not terminate soon after flowering, new lateral buds will initiate flowers. After senescence of aerial parts due to frost, flower and bulb growth stops. In tuberose, the lateral bulb production is prolific, and it takes 2-3 seasons to attain the flowering size. In perennial cropping after senescence of the foliage (not chopping of the foliage), certain lateral buds mature by attaining proper growth and start flowering later in the season.

Application of GA_3 and chlormequat stimulate the enlargement of laterals as well as the central bud by reducing the number of laterals. Muralee Manohar (2000) when used GA_3 on tuberose, he recorded longest rachis and its highest weight in 50 ppm, and enhanced floret opening, *vis-a-vis* maximum number of floret-opening at a time in 100 ppm; though IAA 25 ppm enhanced the vase life. By spraying the plants of tuberose cv. 'Single' at 4-5 leaf stage and 30 days later with 10-1,000 ppm GA_3, B_9 and ethephon, and 500-2,000 ppm CCC, Jana and Biswas (1979, 1982) recorded maximum number of flowers (35.5) per plant with 2,000 ppm CCC, 1,000 ppm B_9 or 10 ppm GA_3, B_9 being more effective though 10 ppm GA_3 brought about earliest floret opening. Basu and Bose (1980) reported advanced flowering by 9 days and increase in flower number from 35.4 to 44 per plant when treated the plants of tuberose cv. 'Single' with 200 ppm ancymidol at 5-6 leaf stage. Pre-treatment of tuberose bulbs with 300 ppm GA_3 or kinetin, or 10 or 50 ppm ABA showed that GA_3 was not found effective but kinetin and ABA stimulated the sprouting appreciably though both inhibited flowering, and the ABA being very effective for sprouting in case of small bulbs, *vis-à-vis* in stimulating flowering in large bulbs (Pathak *et al.*, 1980). Spraying of tuberose 'Single' plants 40 days after planting and twice more at fortnightly intervals with 25-100 mg/l GA_3 or 500-2,000 mg/l ethephon, Mukhopadhyay and Bankar (1983) recorded reduced plant height, bulb and bulblet production with increased ethephon concentrations though GA_3 increased spike length and number of florets per spike at 75 and 100 mg/l but inhibited bulb production at all the concentrations. Soaking of freshly harvested bulbs of tuberose 'Single' for 24 hours in GA_3 at 10, 100, 250 and 500 ppm and planting in 30-cm pots resulted into delayed sprouting from 8.1 days in control to 17.4 at 500 ppm, reduced plant height from 42 to 34 cm and number of flowering bulbs per plant but increased the rachis length by 4.85 cm at 100 ppm (Mukhopadhyay and Sadhu, 1985). Storage of bulbs (1.5-2.0 cm diameter) at 4-10 ^{0}C for 10-30 days and then soaking in 200 mg/l GA_3 or 2,000 mg/l thiourea for 6 hours improves plant growth, causes early emergence of flower spikes with improved quality and increases the yield of spikes and flowers (Dhua *et al.*, 1987). Choudhary (1987) when used 500, 1,000 and 1,500 ppm MH, 500-1,500 ppm CCC and 50-150 ppm GA_3, he recorded MH decreasing plant height at all the levels, though 500 ppm MH, 100 ppm ethephon, 1,000 ppm CCC or 50 ppm GA_3 gave best results with respect to number of flowers per plant. Planting of dormant bulbs of tuberose after dipping for one hour in solutions of 50-400 ppm GA, and 500-5,000 ppm ethephon and CCC, Ramaswamy *et al.* (1977/1978) recorded 100 ppm GA and 500 ppm CCC advancing flowering for more than a fortnight though flower yield was highest only with 1,000 ppm CCC. Hassan and Agina (1980a, b) through summer (15th March) and autumn (early September) plantings of tuberose cv. 'Double' and treating by spraying and through drenching on 19th April (repeated thrice at monthly intervals) to summer planted one and spraying in early October in autumn planted one with CCC at 1,000, 2,000 or 3,000 ppm, and recorded decreased fresh bulb weight, delayed flowering and slightly increased number of florets with spraying which later on was found significantly decreased after drenching with 3,000 ppm in summer planted crop, though in autumn planted crop the treatments extended the flowering period, and increased the number of floral stalks & the florets with increase in the concentrations and brought about some cold resistance in the plants. Itty (2001) got maximum height reduction in tuberose plants when used paclobutrazol at 100 ppm at 60 days after planting or 200 ppm at 30 days of planting and the latter treatment induced earlier flowering but when applied at flowering it produced longest rachis with larger size of floret. Paclobutrazol at 100 ppm applied at 30 days after transplanting brought size increase of bulblets. Highest plant spread was recorded when bulbs were dipped in CCC at 1,000 ppm, and its 500 ppm recorded more number of small sized bulbs and bulblets. BA 50 ppm treatment given at 60 days after planting produced more tillers and delayed floret opening but longer spikes, and its 50 ppm applied at flowering increased bulb size and weight.

Though tuberose is photo-insensitive, long-day exposure promotes vegetative growth and accelerates flowering with increased spike length while low light intensity and duration promote leaf length.

Post Harvest

Spikes are harvested preferably in the morning when 2-3 florets have opened. After harvesting, immediately their cut ends are placed in a bucket containing water. Spikes should be harvested from the ground level with a sharp knife or should be taken out by giving a gentle push from the sides at ground level from where it breaks from the plant. Picking for loose flowers may also be done in the morning only for 2 hours and only opened and opening buds are picked. One person can harvest some 15 kg of loose flowers in 2 hours duration. The yield of loose flowers may vary from 12,000-30,000 kg/ha/year, depending upon the varieties chosen and the cropping year. Care should be taken that picking should not injure the spikes because other opening flowers from the same spike are to be harvested later. These loose flowers are used for various floral decorations, hair adornments and for making garlands. The var. 'Mexican Single' (Calcutta Single) is meant only for loose flower production. Loose flowers are packed in hessian cloth lined bamboo baskets of various sizes, holding 5-20 kg of flowers and are transported the same day to the designated market where it is sold by weight. If these flowers are marketed the next day, there will be weight loss up to 40%.

After cutting, the flower spikes are also graded (on the basis of flower quality, spike thickness, spike length and rachis length), dressed and bundled in fifties or hundreds, and then wrapped in newspapers, and then finally packed in hessian cloth which may accommodate some 600-1000 spikes, depending on the varieties. The refined way of packing them is to wrap the spikes in tissue paper or corrugated brown paper and then with alkathene papers, and then packed in corrugated boxes, specially prepared for the purpose or in rectangular bamboo baskets lined with hessian cloth. These can easily withstand 48 hours of dry storage during transit.

Cut spike post harvest life depends on the variety, the cropping season, the cultural practices adopted, soil moisture status during whole cropping season, carbohydrate reserves and endogenous growth regulators present, the growing season (status of temperature), photoperiod and light intensity, relative humidity, bulb size and spacing. High nitrogen levels administered to the crop produces tall and succulent spikes which may easily break even without winds. Low light intensity but high temperature and excessive moisture reduce the cut flower life. Attack of pests and diseases also affect flower life adversely. Cut flower life improves when bulbs at planting are treated with daminozide or chlormequat 1,000-5,000 ppm, MH 1,000-2,500 ppm or GA_3 10-1,000 ppm. Panwar *et al.* (2008) when sprayed tuberose cv. Single plants at 4-5 leaf stage with GA_3 (25, 50, 75 and 100 ppm) and placed the spikes in vase solution of 4% sucrose +200 ppm cobalt sulphate, they recorded that increasing concentrations of GA_3 increased the fresh weight and vase life of spike as well as fully opened florets. Benzimidazole or GA_3 (0.01-0.03%) in combination with daminozide, ascorbic acid, glucose and hydroxyquinoline (0.001-0.003%) improve the opening of young green unopened buds or matured half-bloomed flowers and increase the vase life. Calcium nitrate 0.01% + citric acid 250 ppm + sucrose 3% provide maximum vase life (14 days) to tuberose cut flowers. Normal vase life of tuberose cut flower is 7-8 days. Pulsing with 10% sucrose + 250 ppm aluminium sulphate for 12 hours and the cut spikes stored for 5 days, prolongs life of the cut spikes and opening of all the florets are obtained. Khondakar and Mazumdar (1985) when subjected the 30 cm long cut spikes of tuberose in different solutions for up to 96 hours, they found 3% sucrose + 0.03% 8-HQC + 0.01% $AgNO_3$ quite effective. Naidu and Reid (1989) stated that pre-treatment of cut spikes with 20% sugar for 15-20 hours and 1.5% sugar as vase-solution improves display life of stems before or after storage. The response of the sucrose pulsing and the presence of open flowers on the spike increases solution uptake, and as there is very little ethylene evolution by the florets so exogenous ethylene exposure is not injurious to these spikes. Gowda (1990) recorded longest vase life of 12 days when the cut flowers were held in 1% sucrose + 200 ppm aluminium sulphate or 2% sucrose + 400 ppm aluminium sulphate. Itty (2001) recorded maximum days for complete opening of florets when tuberose spikes were pulsed with 250 ppm 8-HQS and its 400 ppm gave maximum opened florets at a time though pulsing with $AgNO_3$ 50 ppm exhibited maximum vase life when held in 0.25 mM $AgNO_3$ solution. Citric acid 250 ppm + sucrose 4%, followed by aluminium sulphate 250 ppm also improved post harvest life of tuberose cut flowers (Vijayalaxmi *et al.*, 2011).

Tuberoses have only white flowers. Hence, for obtaining flowers of various colours, *viz.*, yellow, rose, red, scarlet and blue, one will have to dissolve chemicals like bromocresol green and bromocresol blue for various shades of blue, phenol red for yellow, erythrocin red for rose, ammonium purpurate for rhodomine red and cosin for scarlet colour when pulsed at 0.1% each for five hours as water solution. This process is known as **tinting**. Sambandhamurthi and Appavu (1980) used tuberose var. 'Mexican Single' with stalk lengths of 30, 45 or 60 cm with two basal flowers open and stood in solutions containing bromocresol green, bromophenol blue, eosin, erythrocin red, phenol red or ammonium purpurate, each at 0.025-0.2% for different durations and found colour changes ranging from blue, red, scarlet, rose to yellow.

Lifting and Storage of Bulbs

After flowering is over, the growth of the plant also ceases and in the sub-tropical region of Delhi as well as whole of North India, the winter also sets in November. By 20th of November, it becomes very cold (temperature going below 10 °C, and sometimes even 5 °C, while in January it may go even below 5 °C) which forces the leaves to die down though October-November is the peak flowering time. It is therefore advisable to lift the bulbs (clumps) by December to save the crop from rotting. These, if not lifted, will start sprouting by April provided the field has been watered from March onwards. Even after collapse of the foliage, the field should be cleaned of the weeds. In the temperate climate of India, the clumps should certainly be lifted by November as winter sets in there and continues till mid-March. In tropical regions of the country though all the weathers exist but winter is very mild so lifting may be carried out when peak flowering is over.

After harvesting (lifting), the clumps should be dried in single layer under a shed which is well aerated. These clumps are now separated into bulbs and lateral bulblets, are graded and then again dried. Those bulbs which are spindle-shaped and at least 2 cm (>6/8 cm circumference) or above in diameter are flowering size and those less than 2 cm (<6/8 cm in circumference) are planting stock. The grading may be done on the basis of diameter, circumference or weight. The bulbs having circumference of 8/10, 10/12 and 12-14 cm are the commercial grades. Through good cultural practices, below 2 cm diameter bulbs may also produce flowers of commercial grade. However, <1.5 cm bulbs may not produce commercial spikes. Large bulbs produce more laterals than the laterals themselves.

The yield of bulbs per year is roughly 20 tonnes per hectare, if raised through bulbs having diameter of 2.5-3.0 cm. However, the collective yield in a 3-year cropping period will come to about 50 tonnes.

Flowering may be advanced by storing the bulbs at higher temperature i.e. 30°C for longer duration, but the quality of the flower will be poor and bulb yield will be reduced. Lower temperature (10 °C) for longer duration also hampers the bulb production, but if stored at this temperature only for one month, more flowers with better quality are obtained. Storage temperature of at least 18 °C for 4-6 weeks, or 30 °C for six weeks has been advocated for production of more commercial size bulbs. Initiation and development of the flowers occur after planting of the bulbs. Initiation is affected by physiological stage of the plant (bulb size, number of leaves, soil and air temperature). Earlier planting provides longer harvest duration of spikes, with more number of spikes and better quality blooms. In Delhi, the optimum planting season for tuberose is February when there exists low soil and air temperatures that is why flower harvesting period is prolonged though flowering starts slightly late.

Insect-Pests and Diseases

Tuberose is a very sturdy and easy growing crop which is normally not affected by any serious insect-pest or disease. However, sometimes a few observed are described here under.

Grasshoppers (*Hieroglyphus* spp.) feed on the young leaves and flower buds, especially during rainy season which may be controlled by spraying 0.2% methyl parathion. The adult **weevils** feed in darkness on the leaves making notches on the edges which may be trapped and killed as they are only a few. Their larvae feed on the roots and tunnel into the bulbs whose attack may be prevented by applying Furadan granules @ 3-4 kg/ha at the time of bed making. **Aphids** (*Aphis* spp.) are very small in size and multiply rapidly. They feed on the growing points and floral buds which may be controlled by spraying nicotine solution. **Thrips** (*Taeniothrips* spp.) are minute insects which suck the sap of the leaves, buds and flowers, and act as a carrier agent in causing 'bunchy top' disease where the inflorescence is malformed. While controlling the grasshopper, this will also be controlled. **Red spider mites** suck the sap on foliage causing yellow stripes and streaks, and in severe cases leaves turn yellow, silvery or bronze and finally deformed. Kelthane spraying will prevent its infestation.

Two **nematode**, *Aphelenchoides besseyi* causes greasy streak on foliage and *Meloidogyne* spp. (*M. incognita, M. javanica, M. arenaria, M. acritata*) cause root galling, poor growth of plants with leaf tip burn and yellowing, and suppress spike emergence in their severe infestation to tuberoses. *A. besseyi* was first reported as foliar nematode of tuberose by Holtzmann (1968), whose control was suggested by Trujillo (1968) as HWT, bed fumigation and post-planting drenches with methyl bromide. The treatment advocated for controlling larvae of the weevil will keep these pests also under check.

Stem rot (*Sclerotium rolfsii*) is caused at the soil level. Leaves loose their greenness and whole leaf rots and gets detached from the plant. Due to its infection, round and brown sclerotia are formed on and around the infected leaves. Ultimately the plants become too weak to produce flowers. Mercuric chloride 0.1% and commercial formalin 0.2% have excellent control of the disease. Dusting 20% brassicol has been found quite effective. **Botrytis spots and blight** (*Botrytis elliptica*) is a problem under cool cum moist growing conditions. Initially when noticed should be controlled by spraying with 0.2% maneb weekly and the field humidity should be kept under check. Before planting, the bulbs should also be treated by dipping them for one hour in 0.2% Bavistin or benomyl. ***Alternaria* leaf spot** (*Alternaria polyanthi*) is also seen on tuberose leaves. **Flower bud rot** (*Erwinia carotovora*) sometimes infects the floral buds which may be controlled through mercuric chloride or streptocyclin. Sometimes, **leaf mottling** is caused due to a virus. Homer and Pearson (1988) described a partial purification procedure for a flexuous filamentous virus from leaves of tuberose exhibiting mottle symptoms. Particular problems associated with the purification included particle aggregation and the highly viscous nature of the plant sap. The former difficulty was partially alleviated by addition of 1.0 M urea to the extraction buffer and the latter by treatment of the extract with cellulose. The mean virus particle length from negatively stained leaf dips was 742 nm which is characteristic for potyviruses. In addition, TEM of thin sections of infected leaf tissue demonstrated the presence of several types of inclusion bodies similar in appearance to inclusions previously described in association with potyvirus infections in other plants. Such plants should be destroyed to check further spread.

References

Ailincai, N. 1960. The influence of some fertilizers on the flowering of *Polianthes tuberosa* (Italian). *Lucr. Sti. Inst. Agron. Isasi*, pp. 355-360.

Ali, A.A., S.A. Ross, A.M. Moghazy and M.A. El-Shanawany, 1981. Steroidal saponins from *Polianthes tuberosa. Fitoterapia*, **52**(3): 137-139.

Bankar, G.J. 1988. Nutritional studies in tuberose (*Polianthes tuberosa*) cv. 'Double'. *Progressive Hort.*, **20**(1-2): 49-52.

Bankar, G.J. and A. Mukhopadhyay, 1985. Response of *Polianthes tuberosa* L. cv. 'Single' to high doses of NPK. *South Indian Hort.*, **33**(3): 214-216.

Bankar, G.J. and A. Mukhopadhyay, 1990. Effect of NPK on growth and flowering of tuberose cv. Double. *Indian J. Hort.*, **47**(1): 120-126.

Basu, T.K. and T.K. Bose, 1980. Uses of ancymidol on flowering bulbous plants. *Sci. Cult.*, **46**(9): 323-325.

Bose, T.K., B.K. Jana and S. Moulik, 1987. A note on the micro-propagation of tuberose from scale stem section. *Indian J. Hort.*, **44**(1/2): 100-101.

Brundell, D.J. and D.R. Steenstra, 1986. The effects of protected cultivation, pre-sprouting and lateral tuber removal on tuberose production. *Acta Hort.*, No. 177 (vol. II), pp. 361-367.

Choudhary, M.L. 1987. Effect of growth regulants on growth and flowering of tuberose (*Polianthes tuberosa* L.). *Ann. Agric. Sci., Ain shams Univ.*, **32**(2): 1451-1457.

Chaudhary, S.V.S. 2008. Influence of VAM, nitrogen and phosphorus on growth and flowering of tuberose (*Polianthes tuberosa* Linn.). *Haryana J. hort. Sci.*, **37**(3/4): 289-290.

Cirrito, M. 1972. Preliminary studies on fertilizing to increase the size of tuberose bulbs (*Polianthes tuberosa* L.). *Annali dell'Istituto sperimentale per la Floricoltura*, **3**(1): 1-11.

Cirrito, M. 1975. The effect of manuring and bulbil circumference on the enlargement of bulbs of tuberose (Italian). *Annali dell'Istituto sperimentale per la Floricoltura*, **6**(1): 27-43.

Cirrito, M. and M. De Vita, 1981. A comparison between different types of manuring in the culture of tuberose for rhizome enlargement. *Annali dell'istituto Sperimentale per la Floricoltura*, **12**(1): 41-51.

Cirrito, M. and M. De Vita, 1982. A preliminary trial on the prevention of lateral development of offsets on tuberose (*Polianthes tuberosa*) cultivated for rhizome enlargement (Italian). *Annali dell'istituto Sperimentale per la Floricoltura*, **13**(1): 43-51.

Cirrito, M., M. De Vita and G.V. Zizzo, 1980. Results of triennial experiments on weed control under gladiolus and tuberose (Italian). *Annali dell'Istituto sperimentale per la Floricoltura*, **11**(1): 123-148.

Cirrito, M., M. De Vita and G.V. Zizzo, 1981. Different watering rates and mulching with culture of tuberose for rhizome enlargement (Italian). *Annali dell'Istituto Sperimentale per la Floricoltura*, **12**(1): 53-65.

Cirrito, M., G.V. Zizzo and M. De Vita, 1982. The effects of different planting times of tuberose offsets of circumference 6-8 cm on tuberose rhizome enlargement (Italian). *Annali dell'Istituto Sperimentale per la Floricoltura*, **13**(1): 33-41.

Curir, P., A. Fusconi, S. Sulis and L. Parisi, 1984. Aspects of flower differentiation of the meristematic apex of *Polianthes tuberosa* L. cv. Perla during the bulb enlargement phases (Italian). *Ann. Instituto Sperimentale Floricoltura*, **15**(1): 73-88.

Dhua, R.S., S.K. Ghosh, S.K. Mitra, L.P. Yadav and T.K. Bose, 1987. Effect of bulb size, temperature treatment of bulbs and chemicals on growth and flower production in tuberose (*Polianthes tuberosa* L.). *Acta Hort.*, No. 205, pp. 121-128.

El-Roumi, M.K. 1981. Effect of irrigation frequency on the growth and flowering of tuberose *Polianthes tuberosa* L. *Agricultural Res. Rev.*, **59**(3): 305-312.

Farina, E. and T. Paterniani, 1986. The cultural programming of tuberose for cut flowers. Results of two years of trials in the region of western Liguria (Italian). *Annali dell'Istituto Sperimentale per la Floricoltura*, **17**(1): 49-63.

Gowda, J.V.N. 1990. Effect of sucrose and aluminium sulphate on the post harvest life of tuberose double. *Curr. Res. Univ. agric. Sci.* (Bangalore), **19**(1): 14-16.

Hakim, S., A. El-Gamassy and M.R. Shedid, 1958. A study of the rest period of *Polianthes tuberosa*. *Ann. agric. Sci., Cairo*, **3**(1): 101-113.

Hassan, A.H. and E.A. Agina, 1980a. The effect of chlormequat (CCC) on the growth and flowering of tuberose. I. Autumn planting. *Ann. agric. Sci., Moshtohor*, **12**: 211-221.

Hassan, A.H. and E.A. Agina, 1980b. The effect of chlormequat (CCC) on the growth and flowering of tuberose. II. Summer planting. *Ann. Agric. Sci., Moshtohor*, **13**: 145-152.

Holtzmann, O.V. 1968. A foliar disease of tuberose caused by *Aphelenchoides besseyi. Plant Dis. Reptr*, **52**: 56.

Horner, M.B. and M.N. Pearson, 1988. Purification and electron microscopy studies of a probable potyvirus from *Polianthes tuberosa* L. *J. Phytopath.*, **122**(3): 261-266.

Itty Shiji, K. 2001. Regulation of Flowering and Post Harvest Behaviour of Tuberose (*Polianthes tuberosa* L.). M.Sc. Thesis. Kerala Agricultural University, Vellanikkara, Thrissur (Kerala).

Jana, B.K. and S. Biswas, 1979. Effect of growth substances on growth and flowering of tuberose (*Polianthes tuberosa* L.). *Haryana J. hort. Sci.*, **8**(3/4): 216-219.

Jana, B.K. and S. Biswas, 1982. Effects of growth regulators on growth and flowering of tuberose. *South Indian Hort.*, **30**(2): 163-165.

Jana, B.K., S. Roy and T.K. Bose, 1974. Studies on the nutrition of ornamental plants. III. Effect of nutrition on growth and flowering of dahlia and tuberose. *Indian J. Hort.*, **31**(2): 182-185.

Khondakar, S.R.K. and B.C. Mazumdar, 1985. Studies on prolonging the vase life of tuberose cut flowers. *South Indian Hort.*, **33**(2): 145-147.

Kosugi, K. and Y. Kimura, 1961. On flower bud differentiation and flower bud development in *Polianthes tuberosa* L. *Tech Bull. Fac. Agric. Kagawa*, No. 12, pp. 230-234.

Krishnan, Anu G. 2003. *In Vitro* Multiplication and Genetic Improvement of Tuberose (*Polianthes tuberosa* Linn.). Ph.D. Thesis, Kerala Agricultural University, Vellanikkara, Thrissur (Kerala).

Lopes, L.C. 1971. The effect of transplant size and population on the flowering and growth of *Polianthes tuberosa* (Brazilian). *Revista Ceres*, **18**(99) 399-405.

Malini, G. and M.A. Khader, 1989a. Effect of NaCl and $CaCl_2$ on the growth attributes of tuberose (*Polianthes tuberosa* L.) var. Single. *South Indian Hort.*, **37**(4): 239-241.

Malini, G. and M.A. Khader, 1989b. Effect of different salinity levels on flowering and yield of tuberose (*Polianthes tuberosa* L.) var. 'Single'. *South Indian Hort.*, **37**(6): 357-358.

Mitra, S.N., P.S. Munshi and S. Roy, 1979. Effects of different levels of nitrogen and bulb size on growth and flowering of tuberose (*Polianthes tuberosa* Linn.). *Indian Agric.*, **23**(3): 185-188.

Mukhopadhyay, A. and G.J. Bankar, 1983. Regulation of growth and flowering in *Polianthes tuberosa* L. with gibberellic acid and ethrel spray. *Scientia Hort.*, **19**(1/2): 149-152.

Mukhopadhyay, A. and G.J. Bankar, 1985. Effect of split application of nitrogen on growth and yield of *Polianthes tuberosa* L. cv. 'Single'. *South Indian Hort.*, **33**(1): 60-62.

Mukhopadhyay, A. and G.J. Bankar, 1986. Studies on nutritional requirement of tuberose. *South Indian Hort.*, **34**(3): 167-172.

Mukhopadhyay, A. and M.K. Sadhu, 1985. Studies on the effects of preplanting treatments of bulbs of *Polianthes tuberosa* L. with gibberellic acid. *Progressive Hort.*, **17**(2): 125-128.

Muralee Manohar, 2000. Regulation of Growth and Flowering in Tuberose. M.Sc. Thesis. Kerala Agricultural University, Vellanikkara, Thrissur.

Naidu, S.N. and M.S. Reid, 1989. Postharvest handling of tuberose (*Polianthes tuberosa* L.). *Acta Hort.*, No. 261, pp. 313-317.

Nanjan, K., K.M.P. Nambisan, D. Veeraraghavathatham and B.M. Krishnan, 1980. The effect of nitrogen, phosphorus and potash on the yield of tuberose (*Polianthes tuberosa* L.). In: *Natl Sem. On Prod. Techn. for Commercial Fl. Crops*, Coimbatore (India), pp. 76-78.

Narayanaswamy, S. and V.R. Prabhudesai, 1979. Somatic pseudoembryogeny in tissue cultures of tuberose, *Polianthes tuberosa. Indian J. exp. Biol.*, **17**(9): 873-875.

Panwar, R.D., S.S. Sindhu, J.R. Sharma and R.B. Gupta, 2008. Spraying effect of gibberellic acid on vase-life of tuberose (*Polianthes tuberosa* L.). cv. Single. *Haryana J. hort. Sci.*, **37**(1/2): 87-88.

Panwar, R.D., S.S. Sindhu, J.R. Sharma and R.B. Gupta, 2010. Weed management studies in tuberose (*Polianthes tuberosa* L.) cv. Double. *Haryana J. hort. Sci.*, **39**(3/4): 304-307.

Pathak, S., M.A. Choudhuri and S.K. Chatterjee, 1980. Germination and flowering in different-sized bulbs of tuberose (*Polianthes tuberosa* L.). *Indian J. Pl. Physiol.*, **23**(1): 47-54.

Patil, B.A., D.P. Bhore, J.D. Patil and A.V. Patil, 1980. Effect of planting density on production of flower stalks and flowers in tuberose (*Polianthes tuberosa*). In: *Natl Sem. on Prod. Techn. for Commercial Flower Crops, Coimbatore*, pp. 71-72.

Patil, J.D., B.A. Patil, B.B. Chougule and N.R. Bhat, 1987. Effects of bulb size and spacing on stalk and flower yield in tuberose (*Polianthes tuberosa* L.) cv. Single. *Curr. Res. Reporter, MPAU*, **3**(2): 81-82.

Rajvir Singh, N.R. Godara and R.K. Goyal, 2008. Effect of ethylmethane sulphonate on germination, floral attributes and bulb production in tuberose (*Polianthes tuberosa* L.). *Haryana J. hort. Sci.*, **37**(3/4): 283-284.

Ramaswamy, N., C. Paulraj and P. Chockalingam, 1977/1978. Studies on the influence of growth regulators on flowering and yield of tuberose (*Polianathes tuberosa* L.). *Annamalai Univ. agric. Res. Ann.*, **7/8**: 29-33.

Sambandhamurthi, S. and K. Appavu, 1980. Effect of the chemicals on the colouring of tuberose. In: *Natl Sem. on Production Techn. for Commercial Fl. Crops,* Coimbatore (India), pp. 73-75.

Sharga, A.N. 1980. Maximising tuberose (*Polianthes tuberosa* Linn.) multiplication by using bulb segments. *Progressive Hort.*, **12**(2): 71-77.

Sharga, A.N. 1982. Effect of bulb size on vegetative growth and floral characters of tuberose (*Polianthes tuberosa* Linn.). *Progressive Hort.*, **14**(4): 258-260.

Sharga, A.N. and V.S. Motial, 1983. Studies on the effects of foliar nutrition on essential oil content of tuberose (*Polianthes tuberosa* Linn.). *Indian J. Hort.*, **40**(3/4): 260-265.

Sharma, J.R., R.D. Panwar, R.B. Gupta and Surender Singh, 2008. Nutritional studies in tuberose (*Polianthes tuberosa* L.). *Haryana J. hortic. Sci.*, **37**(1/2): 85-86.

Singh, R.S., V.S. Motial and L.B. Singh, 1974. Studies on salt tolerance of tuberose (*Polianthes tuberosa* Linn.). *Plant sci.*, **6**: 80-88.

Trujillo, E.E. 1968. Diseases of tuberose in Hawaii. *Circ. Univ. Hawaii Coop. Ext. Serv.*, No. 427, p. 13.

Vijayalaxmi, M., A. Manohar Rao, A.S. Padmavatamma and A. Sivashankar, 2011. Influence of different holding solutions on vase life of tuberose. *J. Ornam. Hort.*, **14**(3/4): 106-111.

Yadav, L.P. and T.K. Bose, 1987. Chemical weed control in tuberose and gladiolus. *Acta Hort.*, No. 205, pp. 177-185.

Yadav, L.P., T.K. Bose and R.G. Maiti, 1985. Response of tuberose (*Polianthes tuberosa* L.) to nitrogen and phosphorus fertilization. *Progressive Hort.*, **17**(2): 83-86.

Yadav, L.P., T.K. Bose and R.G. Maiti, 1984. Effect of bulb size and depth of planting on growth and flowering of tuberose (*Polianthes tuberosa* L.). *Progressive Hort.*, **16**(3/4): 209-213.

26

Rosa

(With Special Refrence To Loose Flowers)

R.L. Misra and Sanyat Misra

Family: **Rosaceae**

Common names: Albas (vigorous shrub roses deriving from *R.*× *alba*), Apple rose (*R. pomifera,* syn. *R. villosa*), Austrian briar/Austrian copper (*R. foetida*), Austrian copper (*R. foetida bicolor*), Autumn musk rose (*R. moschata*), Banksian rose/Banks's rose/Lady Bank's rose (*R. banksiae*), Blush Noisette (*R.* × *noisettiana*), Bourbon rose (*R.*× *bourboniana* and descendants of *R.* × *odorata* and *R. damascena*), Burgundy rose (*R. parvifolia,* syn. *R. burgundiaca*), Burnet rose (*R. spinosissima*), Burr/Chestnut rose (*R. microphylla*), Cabbage rose/Provence rose (descendants of *R.* × *centifolia*), Cherokee rose (*R. laevigata*), China/Bengal rose/Monthly rose (*R. chinensis,* syn. *R. indica*), Crimson Chinese rose (*R. semperflorens,* syn.*R. bengalensis*), Damask rose (originating from *Rosa* × *damascena*), Dawson rose (*R. dawsoniana*), Dog rose/Wild briar (*R. canina*), Dwarf polyanthas (progeny of *R. multiflora* crossed with China tea and Noisette roses, have dwarf habits with profusion of blooms), Eglantine/Sweet briar (*R. rubiginosa*), 'Félicité et Perpétue' (*R.* × *felicita*), Field rose (*R. arvensis*), Floribunda roses (derivatives from crosses between the Dwarf Polyantha and early Hybrid tea roses), French rose/Provins rose (*R. gallica,* syn. *R. rubra*), Green rose (*R. viridiflora*), Himalayan musk rose (*R. brunonii*), Hybrid perpetuals (descendants of Portland, Bourbon and China roses and predecessors of Hybrid teas, highly vigorous shrubs but less perpetual than the modern Hybrid teas, very popular in Victorian times), Hybrid Rugosas (some of the best varieties of *R. rugosa* within it or outside with highly scented, sometimes with repeat-flowering), Hybrid musks (perpetual flowering shrubs derived from *R. moschata*), Hybrid tea (successors to the Hybrid perpetuals, and excellent as cut flowers), Gallicas (a largest group of old shrub roses and closest in appearance to the wild *R. gallica*), Macartney rose (*R. bracteata*), Manipur Tea rose (*R. gigantea*), Memorial rose (*R. luciae*), Miniature roses (derivatives of dwarf semi-double *R. chinensis* 'Minima' known as 'Roulettii' introduced into Switzerland by Dr. Roulet, have long flowering season with the beauty and form of the China rose, and 'Roulettii' though very beautiful, has further been improved by crossing with Hybrid tea and Floribunda), Moss rose (*R.*

× *centifolia muscosa*), Noisette rose/Champney rose (*R.* × *noisette*), Penzance briars/ Hybrid sweetbriars (those developed by Lord Penzance at the end of 19th century being fragrant and free-flowering shrubs often prone to mildew and appearing in abundance in small clusters), Portland roses (predecessors of the Bourbons and Hybrid perpetuals showing the influence of *R. chinensis*), Polyantha rose (*R. multiflora,* syn. *R. polyantha*), Pompon rose (*R. pulchella,* syn. *R. dijonensis,* 'De Meaux,' 'Pomponia'), Scotch rose/ Burnet rose (derivatives of *R. spinosissima,* syn. *R. pimpinellifolia*), Sweet briar/Eglantine (*R. rubiginosa*), Rose, Tea rose (*R.* × *odorata*), Threepenny-bit rose (*R. farreri persetosa*), White rose of York (*R.* × *alba*), in Sanskrit it is known as *japa* and in Hindi-*gulab*

Introduction and Origin

Some philologists said that the name rose derives from the Celtic word *rhodd* or *rhudd* meaning red, from which the Greek word *rhodon* was also derived. Varrone asserted that the Latin word *Rosa* derives from the Greek *rhodon.* As per common belief, *Rosa* is ancient Latin name for a rose, which comprises some 100-250 species, depending on the botanical authority but nearer to the first number seems to be more realistic as larger number of botanists have recognized over 100 species. Linnaeus in 1753 had said that "*Rosa* species are very difficult to classify, and those who have seen only a few are better able to distinguish one from another than those who have examined many" and likewise Roy E. Shepherd through his treatise '*The History of the Rose*' published by Macmillan, New York stated that there are an indeterminate number of species of roses less than two hundred. They are native to colder and temperate regions of the Northern Hemisphere from just below the Arctic Circle to North America extending to North Mexico, in north-east Africa to Ethiopia and in Asia to India, China and Japan. The garden roses (*Rosa* × *hybrida,* all non-species roses; Gudin, 2003) are most beautiful creations of nature and are universally acclaimed as 'queen of flowers.' Apart from being admired for its beauty as specimen plant, it is used in the border, as hedges, as climbers for training on walls, pergolas and pillars and various other landscape uses, for loose flowers in worshipping, for making garlands and *rangolies,* as cut flowers and for making bouquets and various other decorations, as preserves, for making rose water and *attar,* and various other pharmaceutical uses. Scented roses like *Rosa* × *damascena, R. bourboniana,* and *R. centifolia* are utilized for extraction of essential oils. Rose water, rose *attar, gulroghan, gulkand,* rose dry petals and rose oil are commercial products for international and national markets. Apart from these the roses are being used for preparation of rose honey, rose petal preserves, rose petal sugar, candied rose petals, rose hip extract, rose hip purée and rose hip jam. **In India, *R.* × *bourboniana, R. indica, R.*× *damascena,* all the cut flower varieties and all the species roses are being used as loose flowers.** They are also grown for their beautiful hips which are also one of the important sources of vitamin C. Rose water is used in eye problems, and petals in nerve complaints or in preparation of tea in relieving brain fatigue. Sometimes it is also grown for its beautiful and shining foliage. Because of the diversified growth habit, exquisite shape, variation in size and form, attractive colour, delightful fragrance and numerous varieties, roses have gained wide acceptability. It is a good plant for garden decoration and landscaping. Roses can fulfil the requirement of trees, shrubs, climbers, hedges and edges in the landscape planning and design of gardening. Rose (*Rosa* × *hybrida*) occupies first position in the international cut flower trade. Conceptually, the rose in India had been used since the time immemorial but only as loose flowers.

With the increasing demand of novel cultivars in the world trade, a large number of cultivars are being bred every year with more and more novel characters. Approximately eight billion rose stems, 80 million potted rose plants and 220 million garden rose plants are sold annually world over. About 31% of all cut flowers traded at the European auctions are cut-roses with a total value of about 858 million EUR (Heinrichs, 2008). In India, 4,330 hectare area, producing 874 million stems valued at Rs.44.0 crores, is under rose cultivation in the states of Delhi, Tamil Nadu, Karnataka, Maharashtra and West Bengal (www.nabard.org).

Botany, Cytology and Breeding

The Genus is divided into 4 sub-genera of which sub-genus *Rosa* includes nearly all the species (Gudin, 2000). The sub-genus *Rosa* is divided into sections of which the actual number is debatable. These are woody plants (shrubs or climbers/ramblers) bearing bristly to hooked-prickly stems; alternate, pinnate and toothed leaves, single or clusters of widely expanded 5-petalled flowers (*R. sericea* is 4-petalled) which are mostly fragrant; and the fruits (hips or heps) are globular to ovoid with hollow receptacle filled with nutlet-like achenes. The rose genome is complex comprising of 5 haploid chromosome sets each made of 7 chromosomes (Nybom *et al.,* 2005). The chromosome number of wild species varies from 2n = 2x=14 to 2n= 8x=56, with most species being diploid or tetraploid (Nybom *et al.*, 2005). In species, the difference between the number of chromosomes is generally linked to the climatic conditions, as northern species as a rule have a higher number of chromosomes when compared to sub-tropical species (Pizzetti and Cocker, 1975). In addition to wild species, there are thousands of varieties in rose. *Rosa* × *hybrida* is not a species in botanical terms, but is a complex artificial species category (Gudin, 2003) that may be used to describe non-species roses and particularly the modern rose cultivars.

Recurrent flowering is one of the key characters crucial for the success of roses as ornamental crops as it leads to superior genotypes flowering throughout the whole growing season. This trait was most probably introgressed from *R. chinensis* in the early 19th century and is inherited as a single recessive gene. The inheritance of double (>5 petals) v/s single (5 petals) flowers was shown to be caused by a single dominant gene (Debener, 1999). Moss phenotype originated from sports of *R. centifolia* in the 17th century and was shown to be inherited as a single dominant gene in crosses between tetraploid moss roses and tetraploid non-moss varieties (De Vries and Dubois, 1984). Several studies have shown that prickles on stems are inherited as single dominant gene (Debener ,1999; Linde *et al.*, 2006) and that they are independently inherited from prickles on petioles (Lal *et al.*, 1982; Crespel *et al.*, 2002). Dwarf phenotypes, probably introgressed from the diploid *R. chinensis minima* (Sims), were used to breed miniature roses (De Vries, 2003). Wylie (1955) had suggested that there is a single dominant gene for glossy leaves, climbing habit and double flowers while polygenes govern characters like vigour, fragrance, thorniness, neck strength, leaf width, floral stalk length and shape and size of buds and flowers. The inheritance of fragrance is controlled by polygenes (Kaicker *et al.,* 1983).

Mr. B. S. Bhattacharjee, Deogarh (Bihar) is called the father of rose **breeding** in India. He released his first variety 'Rama Krishna Deva' in 1941. Rose improvement research was initiated at the Indian Agricultural Research Institute in 1956 by the Floriculture Unit then in the Division of Horticulture. The work started with the organization of a National Rose Collection at the Institute. The breeding of roses started from 1957 keeping in view the fact that the roses bred under temperate climatic conditions may not perform well like those bred under tropical conditions. As a result of this, the first two rose varieties namely 'Rose Sherbet' and 'Delhi Pink Pearl' were evolved by late Dr. B. P. Pal in 1962. The development of new rose varieties at IARI created a great deal of interest in rose cultivation in the country. Conventional hybridization and mutation breeding were employed for evolving new varieties. The main objectives of breeding were to evolve varieties with attractive flower colour and form, fragrance, floriferousness, long stems, disease and pest resistance and suitability for growing under sub-tropical conditions. Novel varieties were evolved by mutation breeding. Swarup *et al.* (1971) reported induced mutations in roses by treating cvs 'Christian Dior', 'Kiss of Fire' and 'Queen Elizabeth' with different doses of gamma rays and released the cvs named 'Pusa Christina'- a pink coloured mutant of cv. 'Christian Dior' and 'Abhisarika'- a pink striped mutant of 'Kiss of Fire'. A mutant 'Madhosh' in IARI was evolved in 1978 from cv. 'Gulzar' by the use of EMS (0.25% for 8 hours) to the budwood.

Steve Connor (2015) stated that scientists (Jean-Louis Magnard and co-workers) in France (University of Lyon in St. Etienne) have identified enzyme in the var. 'Papa Meilland' that is key to fragrance in rose. A study into the chemistry of rose scents has found a new biochemical pathway in the petals of the plants which produce a sweet-smelling fragrance that could be reintroduced into rose varieties that have lost their smell. This particular enzyme playing the key role in producing sweet frangrance in roses has been named RhNUDX1, which is found in the cells of the flower petals and which generates the well-known fragrant substance called monoterpene geraniol, the primary constituent of rose oil.

Classification, Species and Varieties

Genus *Rosa* is endemic to temperate regions of the Northern hemisphere including North America, Europe, Asia and the Middle East, with the greatest diversity of species found in Western China. It is also distributed in the warmer areas such as New Mexico, Iraq, Ethiopia, Bengal and Southern China (Nybom *et al.*, 2005). There are no endemic species in Southern hemisphere. Among the many classification systems, ARS Classification Scheme 2000 is widely used (Cairns, 2003). There are 3 sub-divisions in this system, *viz.* (i) **Species roses** such as *R. canina, R. chinensis, R. foetida, R. gallica, R. gigantea, R. moschata, R. multiflora, R. phoenicea, R. rugosa, R. wichuraiana* and *R. rubra* as has been concluded through various taxonomical and cytological studies (Hurst, 1941; Wylie, 1955; Maia and Vénard, 1976) since these 11 species have exclusively been used in the evolution of modern cultivars. As many as 16 species and four hybrid species were found growing wild in different phytogeographical zones of India (Rathore and Srivastava, 1992). The species endemic to Himalayas are *R. brunonii, R. gigantea, R. involucrata, R. longicuspis, R. leschenaultiana, R. macrocarpa, R. macrophylla, R. moschata, R. sericea,* and R. *webbiana* whereas the species growing wild are *Rosa ecae, R. banksiae, R. beggeriana, R. foetida, R. × bourboniana, R. × centifolia, R. × alba, R. × damascena, R. × bracteata, R. laevigata, R. multiflora, R. indica, R. chinensis* and *R. gallica. Rosa bracteata* and *R. laevigata* perform well under sub-temperate climate and possess excellent vegetative and floral characters, justifying their use as donor parents (Viraraghavan, 1991). Various localized rose species and their variants like *R. × bourboniana* (*Gulkhand* preparation) and 'Kakinada' rose (loose flowers) are potential candidates for Geographical Indicators. The cultivation of *R. × bourboniana* throughout the country has resulted in the evolution of many ecotypes, *viz., Desi, Cheenia, Kaithali* and *Meeruti* (Shrivastava and Chandra, 1985); (ii) **Old garden roses** which have been defined by ARS as those cultivated rose types prior to 1867 and are further divided into 21 types. By the 18th Century, five broad Old European classes had emerged, *viz.* Gallica (*R. gallica*), Alba (*R. alba*), Damask (*R. damascena*), Centifolia (*R. centifolia*) and Moss rose (*R. centifolia muscosa*) (Marriott, 2003). During the later part of the 18th century, the old European roses were crossed with 'China rose'. The China group had distinct features that were lacking in the European roses such as low branching habit, recurrent flowering, crimson colour, new fragrance, distinct flowers with high centre and slender flower buds that unfurled on opening (Higson, 2007). The crossing of the China roses with the Old European roses gave rise to a number of new classes like Portland, Bourbon (Autumn Damask × Parson's Pink China), Noisette (Parson's Pink China × *R. moschata*), Hybrid Perpetual (Portland x Bourbon × Hybrid China) and Tea (China × Bourbon or Noisette) (Marriott, 2003 and Higson, 2007), thus these become as Alba, Ayrshire, Bourbon, Boursalt, Centifolia,, Damask, Hybrid Bracteata, Hybrid China, Hybrid Eglanteria, Hybrid Foetida, Hybrid Gallica, Hybrid Multiflora, Hybrid Perpetual, Hybrid Sempervirens, Hybrid Setigera, Hybrid Spinosissima, Miscellaneous, Moss, Noisette and Tea roses; and (iii) **Modern Roses** (1867+) which in the broader term originated largely from 10 species, *viz. R. canina, R. chinensis, R. foetida, R. gallica, R. gigantea, R. multiflora, R. moschata, R. phoenicia, R. rugosa* and *R. wichuraiana* (Gudin, 2000). The first of the modern roses were the HTs derived from crosses between Hybrid Perpetual and Tea roses (Marriott, 2003) and thus was having the blood of *R.× damascena, R. moschata, R. chinensis, R. gigantea* and *R. gallica*. The first recognized HT was 'La France' discovered in 1867. In the early 1900s' a French breeder Joseph Pernet-Ducher made crosses between Hybrid Perpetuals and *R. foetida* to introduce yellow colour. Most Hybrid Tea rose cultivars are mainly upright mostly with single, well shaped flowers having high spiraling centres at the end of long stems; sturdy often shiny petals; characteristic pointed buds; large glossy leaves; and strong stems. The Hybrid Tea roses have contributed to many of the modern rose classes including Polyantha (*R. multiflora* × Hybrid Tea), Floribunda (Polyantha × Hybrid Tea), Grandiflora (Floribunda × Hybrid Tea) and Miniature (Polyantha × *R. chinensis minima*) (Marriott, 2003) and thus the groups falling under Modern roses are Floribunda, Grandiflora, Hybrid Kordesii, Hybrid Moyesii, Hybrid Musk, Hybrid Rugosa, Hybrid Wichuraiana, Hybrid Tea, Large Flowered Climber, Miniature, Mini-Flora, Polyantha and Shrub roses as per ARS Classification Scheme (Cairns, 2003).

For garden purposes the roses can be classified into two classes; (A) **Modern Garden Roses** into two classes, *viz.* (a) non-climbing, and (b) climbing, and again the both into two groups, *viz.* non-recurrent and recurrent. In non-climbing non-recurrent types are the shrub roses, while the non-climbing recurrent types contain three types, *viz.* (i) repeat-flowering shrubs, (ii) miniatures and

(iii) bush types (large-flowered HTs, cluster-flowered Floribundas, and Polyanthas), whereas (b) climbing class non-recurrent types further contain (i) non-recurrent rambler, (ii) non-recurrent climber, (iii) non-recurrent climbing, and (iv) miniature, while (b) climbing class recurrent types further contain (i) recurrent rambler, (ii) recurrent climber, (iii) recurrent climbing, and (iv) recurrent miniature; (B) **Old Garden Roses** again into two classes, *viz.* (a) non-climbing, e.g. Alba, Bourbon, Boursault, China, Damask, Gallica, Hybrid Perpetual, Moss, Portland, Provence (Centifolia), Sweet Briar and Tea and (b) cimbing, e.g. Ayrshire, Boursault, Climbing Tea, Noisette and Sempervirens; and (C) **Wild Roses** which are classified only in (a) non-climbing, and (b) climbing types (Hellyer. 1982).

Since the beginning of the 19th century rose breeding has flourished, leading to the actual probable number of some 20,000 varieties cultivated in the world (Hareing, 1960). Some of the important varieties listed as **Hybrid Teas** are 'Aalsmeer Gold', 'Abhisarika', 'American Heritage', 'Anuraag', 'Anvil Sparks', 'Arjun', 'Bajazzo', 'Bhim', 'Blue Moon', 'Bonne Nuit', 'Brandy', 'Careless Love', 'Century Two', 'Charles Mallerin', 'Christian Dior', 'Command Performance', 'Crimson Glory', 'Doris Tysterman', 'Dr. B. P. Pal', 'Eiffel Tower', 'Enna Harkness', 'First Prize', 'Frau Karl Druschki', 'Friendship', 'Garden Party', 'Ganga', 'Gladiator', 'Golden Giant', 'Granada', 'Happiness', 'Haseena, 'Inge Horstmann', 'Jadis', 'June Bride', 'Kiss of Fire', 'Lady X', 'Louisiana', McGredy's Sunset', 'Mister Lincoln', 'Montezuma', 'Mridula', 'Mrinalini', 'Nurjehan', 'Oklahoma', 'Papa Mielland', 'Paradise', 'Pasadena', 'Peace', 'Peter Frankenfeld', 'Pusa Ajay', 'Pusa Arun', 'Pusa Shatabdi', 'Pusa Mansij', 'Raja Surendra Singh of Nalagarh', 'Raktagandha', 'Raktima', 'Sonia Meilland', 'Sugandha', 'Super Star', 'Virgo', etc.; as **Floribundas** are 'Africa Star', 'Arunima', 'Banjaran', 'Charisma', 'Charleston', 'Delhi Princess', 'Else Poulsen', 'Flamenco', 'Gene Boerner', 'Iceberg', 'Madhura', 'Mercedes', 'Mohini', 'Neelambari', 'Pink Parfait', 'Prema', 'Pusa Baramasi', 'Pusa Gaurav', 'Pusa Komal', 'Pusa Muskan', 'Sadabahar', 'Suryakiran', 'Sea Pearl', 'Shocking Blue', 'Summer Snow', etc.; as **Polyanthas** are 'Baby Faurax', 'Cameo', 'Chattilon Rose', 'Echo', 'Nartaki', 'Paul Crampel', 'Swati', 'Vatertag', etc., as **Miniatures** are 'Baby Gold', 'Baby Masquerade', 'Blue Mist', 'Chandrika', 'Coralin', 'Cri-Cri', 'Dazzler', 'Deep Velvet', 'Dwarf Queen', 'Honey Comb', 'Joseph Wheatcroft', 'Little Buckaroo', 'Little Flirt', 'Mimi', 'Okington Ruby', 'Perle de Montserrat', 'Perle de Alcanada', 'Pixie Gold', 'Pour Toi', 'Pushkala', 'Red Elf', 'Scarlet Gem', 'Starina', 'Sweet Fairy', 'White Madonna', 'Zwerghoenigin', etc.; as **Climbers** are 'Casino', 'Cocktail', 'Delhi Pink Pearl', 'Delhi White Pearl', 'Golden Showers', 'Marechal Niel', 'Pinata', 'Prosperity', 'Royal Gold', etc.; and the varieties suitable for **Polyhouse** cultivation are 'Aalsmeer Gold', 'Bianca', 'Black Magic', 'Dallas', 'First Red', 'Grand Gala', 'Grand Prix', 'Jacaranda', 'Konfetti', 'Noblesse', 'Ravel', 'Red Velvet', 'Skyline', 'Tineke', 'Virginia', 'Vivaldi', etc. as **long stemmed ones** whereas 'Europa', 'Flirt', 'Frisco', 'Golden Times', 'Kardinal', 'Kiss', 'Lambada', 'Ohio', 'Prophyta', 'Souvenir', 'Tina', 'Vanilla', etc. as **medium stemmed ones**.

In India, it is only *Rosa damascena* which in Kannauj and Haldighati is being cultivated for essential oil extraction. This species is also being used for loose flowers along with 'Bourbon roses' and *R. indica*. There are certain other roses which have local importance for loose flower production such as 'Kakinada' in Andhra Pradesh and 'Titri' in Rajasthan. *R.*× *damascena* is of hybrid origin but with doubtful parentage though is one of the oldest rose in cultivation. It played an important role in developing hybrid garden roses and its progenies by crossing with *R. indica* varieties resulted into Hybrid Perpetuals from among which are the famous 'Frau Karl Druschki' whose double white flowers continue to appear from late spring to autumn, and 'Souvenir du Docteur Jamain' which continues producing intensely fragrant wine-coloured suffused-purple continuous flowers. The original *R.* × *damascena* is a hardy deciduous shrub with large spines and upright growth, bearing 5 oval dentate leaflets and white suffused pink or red long-stalked flowers which are strongly fragrant and in clusters of up to 12. *R.* × *damascena* var. *trigintipetala* is the very important variety being used in Sofia (Bulgaria) for extraction of rose essential oil. *R. indica*, a Chinese species with a variable habit is a large shrub of loose habit, mostly behaving as a semi-climber and flowers for long period but intermittently, i.e. from May to September on the hills though from November to April in the plains. Its leaves are composed of 3-5 oval, pointed and dentate leaflets some 3.75-7.50 cm long, and pinkish brick-red flowers are 5-petalled, solitary, 5 cm in diameter and slightly ruffled. Its monumental varieties are *R. indica chinensis mutabilis* which is a small upright shrub bearing fragrant , single, apricot-pink shaded carmine flowers but changing to crimson with age; *R. indica fragrans* (syn. *R. fragrant*), a continuous and deliciously tea-scented rose is the founder of the modern strain of the HT roses, *vis-à-vis* the parent of the two most famous climbers, 'Maréchal Niel' (yellow) and 'Gloire de Dijon' (yellow-and-orange) and *R. indica* var. *viridiflora* (syn. *R. monstrosa*) where the petals are transformed into leaf-like organs, a curious true green rose which, in fact, was discovered in 1855 in the garden of M. Verdier in Paris who had received it from America.

A few of the fragrant rose varieties are 'Abraham Darby', 'Agnes', 'Alberic Barbier', 'Albertine', 'Alchymist', 'Alec's Red', 'Aloha', 'Angel Face', 'Anurag', 'Apple Jack', 'Apricot Nectar', 'Apsara', 'Arizona', 'Arthur Bell', 'Austins Belle Isis', 'Banana Buff Beauty', 'Belle Amour', 'Belle Storey', 'Blanc Double de Coubert', 'Blush Noisette', 'Bobbie James', 'Break O'Day', 'Buff Beauty', 'Cathedral', 'Celestine', the varieities of *R. × centifolia*, 'Cerise Bouquet', 'Charugandha', 'Cherish', 'Cherry Louis-Philippe', 'China Apricot', 'Chinatown', 'Chrysler Imperial', 'City of Gloucester', 'Clair Matin', 'Claire Rose', 'Command Performance', 'Compassion', 'Coquttes de Blanche', 'Coral Dawn', 'Cornelia', 'Cramoisie Superieur', 'Crimson Glory', 'Cuisse de Nymph', 'Danäe', 'Daphne', 'Delhi White Pearl (clg.)', 'De Meaux', 'Distant Drums', 'Double Delight', 'Dr. M.S. Randhawa', 'Double Fragrance', 'Duchesse de Brabant', 'Electon', 'Etoile de Hollande (clg.)', 'Everest', 'Felicia', 'Fragrant Cloud', 'Fragrant Hour', 'Francesca', 'Friendship', 'Fritz Nobis', 'Frühlingsanfang', 'Gardenia', 'General Jacqueminot', 'Gertrude Jakyll', 'Gloire de Dijon', 'Golden Afternoon', 'Granada', 'Granny Grimmetts', 'Gulzar', 'Henry Hudson', 'Heritage', 'Honeysweet', 'Honorine de Brabant', 'Indian Princess', 'Intrigue', 'Jawahar', 'Jayne Austin', 'Jens Munk', 'Jude the Obscure', 'Jwala', 'Kamla Devi Chattopadhyay', 'Korresis', 'Lady Hillingdon', *R. laevigata*, 'L'Aimant', 'La Reine des Violettes', 'Lavender Crystal', 'Lavender Lassie', 'Love', 'Mme Gregoire Staechelin', 'Madame Hardy', 'Madame Issac Pereire', 'Madhura', 'Maggie', 'Maiden's Blush' (syn. *R. × alba carnea*), 'Marechal Niel', 'Margaret Merrill', 'Marie Pavie', 'Max Graf', 'McCartney Rose', 'Monsieur Tillier', 'Moonlight', 'Mother Teresa', 'Mridula', 'Mrs. John Laing', 'Nastarana', 'New Dawn', 'New Zealand', 'Nishada', 'Nova Zembla', 'Nurjehan', 'Nymphenburg', 'Old Blush', 'Old Pink Moss', 'Papa Meilland', 'Parson's Pink China', 'Penelope', 'Perception', 'Petite de Hollande', 'Perfume Delight', 'Preyasi', 'Prima Ballerina', 'Queen Margarethe', 'Rajhans', 'Ralph's Creeper', 'Ranjana', 'Reine des Violettes', 'Roger Lambelin', '*R. × alba* var. *suaveolens*' (widely cultivated for attar production), '*R. arvensis*', '*R. banksiae* var. *alba*' (single), '*R. banksiae banksiae*', '*R. bracteata*', '*R. brunoni*', '*R. carolina*', '*R. centifolia*', '*R. chinensis* var. *mutabilis*', '*R. chinensis* var. *fragrans*', '*R. cinnamomea*', '*R.× damascena*', '*R. eglanteria*', '*R. gallica*', '*R. gallica complicata*', '*R. gallica conditorum*', '*R. gallica violacea*', '*R. laevigata*' (syn. *R. sinica*), '*R. moschata*', '*R. moschata* var. La Mortola', '*R. mulligani*', '*R. multiflora*', '*R. muscosa purpurea*', '*R. nitida*', '*R.× odorata gigantea* var. Cooper', '*R. phoenicea*', '*R. rubriginosa*' (syn. *R. eglanteria*, foliage deliciously fragrant to that of apple or sweet peas), '*R. rugosa*', '*R. rugosa rubra*', '*R. sempervirens*', '*R. wichuraiana*', 'Rose à Parfum de L'Hay', 'Rose de Resht', 'Rosemary Harkness', 'Roseraie de l'Hay', 'Rose Sherbet', 'Rubra', 'Safrano', 'Sanders White', 'Sarah van Fleet', 'Saratoga', 'Scented Air', 'Sceptred Isle', 'Sheer Bliss', 'Shreyasi', 'Sir C.V. Raman', 'Slater's Crimson', 'Sombreuil', 'Souvenir du Dr. Jamain', 'Souvenir de Mme Boullet', 'Sugandhini', 'Sujata', 'Sundowner', 'Sunsprite', 'Sutter's Gold', 'Sweet Chariot', 'Sweet Surrender', 'Tiffany', 'Trier', 'Tuscany', 'Variegata di Bologna', 'Warwick Castle', 'Wendy Cussons', 'White America', 'White Dawn', 'White Maman Cochet (clg.)', 'Wife of Bath', 'Woburn Abbey', 'Zéphirine Drouhin', etc. and their loose flowers can be used at various social auspicious occasions for emitting highly pleasing fragrance and certain of them may be grown for extraction of essential oils.

Propagation

Rose can be propagated by seed only when developing new varieties as well as species roses as by and large these are self-fertile. Hybrid roses do not appear true to type through seeds so they are usually propagated through vegetative means such as cutting, layering, grafting and budding. Seed propagation is also used to multiply rootstocks of *R. canina*.

Seeds are botanically achenes with a very hard seed coat and these are harvested from mature and ripe hips. In the autumn as soon as the rose hips are swollen and fully ripe, the seeds are extracted, air dried at 10-20 ^{0}C. The hardness of the seed coat can also be softened through acid scarification as well as through stratification, the latter being an ideal and widely adopted method for breaking dormancy. In scarification the hardened seed coat is softened so that it may imbibe water so that seed germination is speeded up but since the rose seeds contain an immature embryo which results into seed dormancy therefore it is only stratification which is effective in this case. The seeds are first placed in an open dish with moist filter paper or into a mix or moist peat and sand in a plastic bag or in the media like sphagnum moss, and wet stratified at 1.6-4.4 ^{0}C for 2-6 months depending on the genotype. The low temperature-treated seeds are sown in raised beds or plugs in the month of October-November. The seed raising medium should be the fine sterilized compost (mixture) comprising of good garden soil, leaf mould and FYM or cow dung manure, all by volume in equal proportion along with a little quantity of some effective fungicide to control damping off pathogens. This mixture is gently firmed up (no hard firming) as hard compact medium does not permit air penetration and too loose a medium is susceptible to creation of large air pockets that facilitate breaking up of water columns essential for capillary rise. Depth of medium should be some 7.5 cm, and the distance among the seeds should be some 5 cm apart. The temperature for germination should be 20-30 ^{0}C, sunlight should be amply given, the place where seeds have been sown or the containers should be kept at a quite airy place and the mixture should be moist (neither wet nor dry) throughout, to facilitate germination. The seeds may take up to one year for germination. After germination, when the seedlings are large enough to handle i.e. formation of first 2-true-leaf stage, these being highly fragile should be pricked out quite gently into individual pots or other containers to avoid competition among the seedlings for light and nutrients. Such pots or containers should be kept in a cold frame until properly established, then for hardening these are taken out during day time, and after full acclimatization these may be kept in the open and potted repeatedly until these are large enough to be planted out directly in the beds outside. These seedlings flower the very first year but the proper colour, shape and size of blooms are visible only in the next year and then only one can assess whether to retain the plant or discard.

Propagation through **layering** (air and ground) is though uncommon but is sometimes practiced in case of climbers and ramblers. **Inarching** is also sometimes followed but is quite uncommon in modern roses. **Stenting** is a quick method of propagation of roses based on grafting a selected cultivar on an unrooted cutting of a suitable rootstock, resulting into a complete plant in 3-4 weeks and such plants are called stentlings. The technique used is whip grafting. A piece of stem with a five leaflet leaf and a single dominant axillary bud is used as a scion. A piece of stem consisting of a single internode without buds or leaves is used as rootstock. Scions and stocks are held together with the help of a peg. The basal end of a rootstock is dipped in a talc with IBA at 0.4%. The rooting medium is a mixture of peat and perlite in 1:1 ratio. The relative humidity and temperatures are maintained at 100% with intermittent misting and 22-27 ^{0}C temperatures, respectively.

Stem cuttings are used for clonal propagation of rootstock, *viz. R. canina, R. gallica*, etc. They are also used for propagation of vigorous cultivars. Climbers, ramblers, miniatures and polyanthas are also propagated through cuttings. Propagation of roses by cuttings is slowly gaining prominence in many countries but it takes longest time of about three years for the new plants to establish, however, miniatures take a little time. IBA 1,000-2,000 ppm may be used for rooting of cuttings. The cuttings of the pencil thickness should be taken during autumn when flowering is over, from the current season's growth and the leaves should be removed. The 2/3rd of the cuttings should be inserted in the prepared bed in slanting position and 1/3rd should be left above. The size of the cuttings should be 20-25 cm but in miniatures 5-10 cm, at least with 3 nodes. Distances apart should be sufficient so that when after rooting these are lifted others may not be disturbed, say some 25 cm apart. Hard cuttings are suitable in those roses that are related quite closely to wild species such as ramblers, as well as miniatures which come up quickly by this method and therefore are propagated commercially. The roses of complex parentage such as bush roses do not root so easily and if rooted also these grow slowly. However, the roses grown on their own roots do not produce suckers. The planted cuttings first develop the callus at the base during autumn and root during spring. If afterwards any flower appears, those are removed to save the energy for developing a strong vegetative growth and when during the following autumn these grow about 25-30 cm in height, these are transplanted to their permanent positions and treated just like the grown up plants. In utterly low winter temperature areas the semi-hardwood cuttings are more successful. In this case during summer but after flowering, the 10-15 cm long cuttings after trimming the soft tips are taken from above a bud where the shoot is beginning to turn woody and are inserted in the medium (peat or peat substitute and sand both in equal volume) as for hardwood cuttings. Water loss from such cuttings is very fast so these should be kept in cold propagating chamber.

Commercially, the roses are propagated only by 'T', inverted 'T' or shield **budding.** The ideal time for budding in North India is from January to March. In case of areas with mild climate, it can be done throughout the year. For budding and grafting the rootstocks are very important. Climate, compatibility, soil, pests and diseases and utility play an important role in the selection of **rootstocks.** *Rosa indica* var. *odorata* is recommended for northern plains. This can withstand dry and wet conditions, *vis-a-vis* salinity. This is also tolerant to powdery mildew and insect-pests. *Rosa bourboniana* is widely used in North India for producing standard roses as it throws vigorous and strong shoots. It has not gained prominence because of susceptibility to powdery mildew. *Rosa multiflora* is widely used as rootstock in South and Eastern India and in the temperate hills, also as it is resistant to nematodes. *Rosa canina* is commonly used as rootstock in Europe. This is propagated by seed and is resistant to drought and alkaline conditions. *R. manetti*, Dr. Huey, *R. laxa* (syn. *R. ceriifolia froebelii*, *R. dumetorum laxa*, popularly known as 'Laxa') and Natal briar are also used as rootstocks in certain other countries. Natal Briar is gaining prominence as a greenhouse rootstock. *R. eglanteria* (syn. *R. rubiginosa*) as rootstock is very thorny, and slow-growing, hence is most suitable for pot-roses and miniatures.

Cultural Practices

Rose production is favoured by an extended growing season with a majority of sunny days. At least 6 hours of **sunshine**, preferably morning sun is quite suitable for its growing. Temperature has a significant effect on quality aspects such as stem length and diameter, bud and leaf size. Four year old plants of rose cv. Dallas were grown at a minimum temperature of 12 ^{0}C with 2,000 lux supplementary lighting (total light period of 16 h/day) from HPS lamps during the period of November-February. Increased flower quality was obtained together with 18 per cent higher cut flower production over control. No further increase in yield was observed when irradiance was raised to 3,000 lux. The ideal **temperature** for rose cultivation is 15-27 ^{0}C. Best commercial quality is obtained when the night temperature is around 16 ^{0}C and over all humidity 60-65%. Rose plants in polyhouse respond well to **CO_2 enrichment** with a decrease in the rate of lower bud abortion, an increase in lateral sprouting and an increase in the fresh weight of flower stems (Mortensen and Moe, 1983)

Prevention of flower bud abortion can be a result of alteration in partitioning of assimilates between upper buds. An additional possibility is that the effect of the CO_2 on the flower bud retention is due to anti-ethylene action of CO_2. The increase in number of flowers and quality may be the result of CO_2 on photosynthetic process, partitioning of assimilates and the effect of CO_2 on ethylene. Increasing CO_2 levels during day time to 800-1,000 ppm are beneficial for rose production due to increased dry matter production.

Roses prefer well-drained **soil** with a pH near 6.5-7.0; however, they are tolerant to a range of 5.5 to 8.0 soil pH. Roses are not salt tolerant, so EC values should be less than 2.0 dsm^{-1}. Sandy loam **soil** is best for successful rose cultivation. These are planted in beds of different shapes. The beds are prepared by incorporation of organic manure and basal dose of phosphorus in the soil. Some 300 quintals of well decomposed FYM, compost or cow dung manure per hectare should be applied in the soil at the time of **field preparation** and at least three ploughings should be carried out every time followed by planking, all sorts of weeds and their rootstocks should be taken out and then beds of convenient size and shapes should be prepared. **Planting** should be done in such a way that the bud union should be 2.5-5.0 cm above the soil level. After planting the plants are to be watered copiously. In case of bare root roses, the roots along with the stem up to the bud union should be kept in water for six hours before planting. In north Indian plains and Eastern India, planting can be done from October to February while in Southern India from October to December. In temperate areas though planting can be done during autumn but it would be better to plant them in March. In areas with mild climate, planting can be done throughout the year. Planting distance for garden roses varies from 30-75 cm between plants and rows depending on the class, variety, soil and climate. In **polyhouses**, rose is planted on raised beds having one metre width, 30 cm height and length as per the structure. Planting in double lines at 30 × 20-30 cm is preferable in polyhouse to accommodate 7-13 plants per square metre. The soil may be sterilized with formaldehyde and thoroughly mixed with FYM, phosphatic and potassic fertilizers.

A **fertilizer** dose of 520 kg N, 868 kg P_2O_5 and 694 kg K_2O per hectare per year is recommended for high density planting (30 × 30 cm) of 'Super Star' under open field conditions (Bhattacharjee and Damke, 1994). Potassium application rates affect the number and quality of flowering stems and reduce diseases. Potassium rates closer to 50 g per square metre per year provide maximum quantity of commercial stems regardless of the cultivar. Heavy dose of organic manure like FYM at the rate of 50-100 tonnes per hectare is recommended in rose. Bhattacharjee (1994) observed that in Delhi condition the soil application of sulphur at 10 kg/ha, magnesium sulphate at 50 kg/ha and calcium sulphate at 50kg/ha significantly improved flower yield and quality in cv. 'Raktagandha'. Drought stress leads to defoliation and sun burn of canes and may contribute to spider mite infestation. However, overwatering or poorly drained soils may lead to root disease or nutritional problems. **Water** stress applied at the stage of leaf primordium formation results in delaying the rose production cycle but does not have any negative effect on the quality of flower shoots or the buds. The stage prior to petal initiation is more sensitive. Water stress at this stage affects the quality of flower buds (reduction in the number of well formed petals and height of the flower bud). Water stress imposed prior to stamen initiation seems to be the most damaging on rose development affecting the quantity (up to 70% reduction of production) and the quality of flowering shoots and buds (reduction on the stem length and fresh weight of the flowering shoot and on the diameter and length of the bud as well as number of well-formed petals in the flower bud). Water stress applied after stamen and carpel formation does not have any negative effect on the number of flowering shoots that reached the marketable stage or on the quality of flowers produced. Generally, a rose plant requires water at 8-10 ml/m^2 a day. During rains and in winters the rose beds are menaced with weeds, the species differing in both the seasons. After pruning thick 20 cm mulching with straw, fallen tree leaves, cocohusk, cocopeat, cocoa shells, bark chips, saw dust, rice husk and chopped bagasse will not allow the weeds to germinate. Spreading of black polythene sheets in between the plants and bunds also do not permit weeds to grow. Rajamani *et al.* (1992) in roses recorded better control of monocot weeds with Glyphosate 1.0 kg a.i/ha and dicot weeds with Oxyfluorfen 0.5 kg a.i/ha.

Whether it is tropical or temperate regions, the open cultivated roses are subjected to **wintering** prior to pruning. About one fortnight before pruning, in India the wintering is normally practiced at the start of October, and all around the plants but slightly away from the base the soil is dug up and turned over and left as such for about one fortnight even without watering. This encourages certain week roots to dry up which ultimately causes drying of weak and diseased shoots which are cut back at the time of pruning which is normally done one fortnight after wintering. After pruning the plants are watered. The practice of wintering followed by pruning *cum* watering encourages vigorous growth of the plants.

Roses require annual **pruning** to speed up the natural process of new, vigorous and disease-free shoots so that new flushes of quality blooms appear in the season. This removal of old growth sometimes acts as a form of dormancy in warmer climates which is otherwise lacking. Pruning shapes the plants properly, removes the old and diseased shoots, invigorates an old and unproductive rose, slows down the plant processes for a short period of time, produces a few larger blooms with long stems or many smaller blooms for good pot display, controls the blooming time in the season and forces the plants for recurrent or remontant flowering. While pruning, a pair of sharp and high quality secateurs should be used and clean and angled cut just above 4th or 5th bud facing outside should be made. Proper time for pruning is when the rose plants go into dormancy or after flowering is over, i.e. normally when there is transition in weather from summer to winter in sub-tropical areas, however, in those areas where winter is always severe better the plants should be pruned in spring but in cases where it is tropical climate all through the year the pruning should be carried out in the cooler months so that the dormancy is induced and the plants may rest for some time. All the newly planted roses are pruned drastically to encourage a vigorous root and shoot system though '**climbers**' are treated otherwise, and here only weak, dead or unwanted shoots are removed. '**Ramblers**' (*R. wichuraiana*) are pruned only to stop them crossing their boundaries. '**Species roses**' require no pruning and the plants are initially shaped through this but in the subsequent years when plants become too dense and produce a few flowers these are thinned out and the plants are again shaped. In case of '**Alba**', '**Provence**', '**Damask**' and '**Moss roses**', the main and side shoots are reduced after flowering. '**Gallica roses**' produce many twiggy shoots so after flowering these are thinned out to permit proper air circulation and pruned to give a proper shape to the plant. Since '**standards**' are made at a rootstock of the height of 1.1 to 1.2 m with large or cluster-flowering bush or from small bush roses budded onto it, these are pruned as per the specification of the budded cultivars. These are normally pruned to 1/3rd in length to shape the standards with balanced heads. '**Weeping standards**' are made on taller rootstocks, usually at the height of 1.5 m so here only those varieties are budded where canes grow in pendent fashion. In this case there is limited scope of pruning, however, dead and any other unwanted cane is removed otherwise only flowered portion after flowering is removed. '**Miniature roses**' require shortening of any overgrown shoot, dead or diseased ones, and twiggy and dense shoots for thinning. Those miniatures which are not doing well despite all the precautions, should be treated like to those of large-flowered miniatures or cluster-flowered bush roses, and leaving only the strong shoots all are removed along with 1/3rd of even the strongest ones. **'Bourbons'**, **'Hybrid Perpetual'** roses and the hybrids derived through *Rosa rugosa* as are generally remontant in nature so are pruned just like large-flowered (HT) bush roses but lightly. **'Shrub'**, **'Old garden roses'**, **'Rugosa'**, **'China'**, and **'Hybrid Musks'** since produce blooms even without pruning for years as these flower on two years or more older shoots so these are also pruned lightly just after flowering without damaging the flowering woods. Only dead, diseased and weak woods are therefore removed so that plants remain in shape and healthy. For improving the bloom quality or to increase the number of blooms these are thinned and pinched. A light pruning each year in winter (October) may be done along with complete removal of a few of the twigs from the base to allow them remaining remontant and to encourage the appearance of new vigorous shoots that may flower the following year and this way all the older shoots one by one will be replaced to perpetuate the flowering. In **'Modern Garden bush roses'** (**Hybrid Teas**) the dead, diseased, thin or twiggy shoots are removed along with those crossing horizontally so that proper air circulation is maintained. Centrally leading shoot is cut back to a height of 20-25 cm, especially in temperate areas though in the tropical and subtropical regions the pruned height should be maintained at 45-60 cm. For producing exhibition blooms, main shoots are removed leaving only 2-3 buds. In HTs, strongly growing only 4 to 5 canes are retained. In case of **'Cluster-flowered bush roses'** (**Floribundas**) also the dead, diseased, intermingled, thin, twiggy and unproductive shoots are cut back like to those of HTs. The side shoots in smaller types are cut back by about 1/3rd and in taller growing types by 2/3rd. Hard pruning reduces the number of blooms so such pruning in 'Floribundas' is recommended only for exhibition blooms otherwise not, however, the height of the pruned plant should be around 30-40 cm in case of smaller types and from 20-30 cm in case of taller types. After the pruning operation, the pruned cuts should be sealed with a proprietary fungicidal paste like Bourdeaux mixture or Blitox paste. A dormant spray of pesticide may be given before the sensitive buds sprout. Fertlizers can be applied after 2-3 weeks.

Bending, an alternative to pruning, originated in Japan which allows harvesters to cut longer stems by cutting back to the knuckle rather than a 5-leaflet leaf. All weak shoots are bent down to fill any area void of foliage and thus attain a desirable leaf area index to optimize photosynthetic potential and facilitate the transport of sugar to the developing shoots. The shoots which arise on dormant buds following knuckle cuts are generally more vigorous and produce high quality and superior grade flowers. This vigour may be attributed to the fact that cane diameter is larger at knuckle and since weak shoots are bent, all remaining shoots are exposed to better light and less competition. Two practical benefits of the arching technique is that the annual cut back is eliminated and roses are grown at working level height. Bending is necessary for keeping enough leaves on the plant which are required for production of carbohydrates. Pegging down technique is also a sort of bending which is although effective but time consuming and is usually practiced on Bourbons and Hybrid Perpetual roses that produce long shoots with flowers only on their tips. These shoots instead of being pruned are bent over gently and pegged into the ground or the tips are pegged in the ground with the help of wires and the side shoots are cut back to 10-15 cm so that many lateral shoots are encouraged for copious flowering in the following season.

Post Harvest

Loose flowers are harvested at fully opened stage early in the morning and are stored at 2-3 ^{0}C temperatures and 90-95% R.H. until marketing but only for a week. These are kept in small portable baskets sprinkled with water to avoid drying till marketing. Such flowers are required for worshipping in the temples, at religious places, for making garlands, for making *rangolies* and for social festivities. However, for extraction of essential oil the just open flowers are harvested early in the morning and in baskets are immediately taken to the place of distillation. For cut flowers, the roses should be harvested at the tight bud stage when one or two petals begin to unfold. The stage of harvest depends on the variety, distance to market, climate and consumer preference. Roses cut too early may develop bent neck. Flowers should be cut in the morning or evening. They should be cut leaving two 5-leaflet leaves on the stem. After cutting, they are immediately placed in a hydrating solution to maintain turgidity. After they are cut, they are cooled or graded. The flowers should be transported to cool rooms. Precooling removes the field heat and improves the post harvest life. The hydrating solution may be acidified with citric acid 300 ppm to improve the uptake of solution. The vase life can be improved by using floral preservatives in vase solution. Aluminum sulphate and citric acid at 300 ppm improve the vase life. The flowers can be wet-stored at 2-3 ^{0}C with a R.H. of 90-95%. After cooling the flowers are shifted to grading room. All the inferior stems and those infested with pests and diseases are removed. The flowers are sorted to different grades manually or automatic graders. Long stemmed varieties are graded from 40 cm onwards with a difference of 10 cm. The short stemmed varieties are graded from 40-65 cm with a difference of 5 cm. The graded stems are made into bundles of 20 each. The buds are wrapped with corrugated paper. The leaves are removed from lower 5 cm portion of the stems and the graded stems are bundled into 20 stems each. These bundles are packed in fiberboard boxes. The stems should be tightly packed to avoid movement during transport.

Insect-Pests and Diseases

There are a wide variety of insects that affect roses. Insect-pests degrade the ornamental value of roses in the garden and decrease productivity in commercial rose ventures. In case of severe infestation, they can even cause death of plants. Vijayan Nair (1989) through a survey in Thiruvananthapuram recorded thrips (*Rhipiphorothrips syriacus, Scirtothrips dorsalis*), scales (*Aonidiella aurantii*), the leaf feeding beetles (*Adoretus* spp.), and the mites (*Tetranychus cinnabarinus, T. neocaledonicus*), out of which *Tetranychus neocaledonicus* was recorded as the most important pest in all the locations, followed by *Adoretus* spp., *Aonidiella aurantii, Rhipiphorothrips syriacus, Scirtothrips dorsalis* and *Tetranychus cinnabarinus*. However, in general the pests attacking roses are described hereunder.

Aphids (*Macrosiphum rosae*) are small, slow-moving, soft-bodied insects which start infestation early in the season rasping the sap by piercing their mouth parts on tender parts such as shoots, buds and flowers and the infested flowers become malformed. These also secrete sweet honeydew which attracts the ants and on which even fungi grow to produce black sooty mould which also degrades the ornamental value of rose. The aphid population is kept under control by natural enemies like lady bird beetles. Excessive use of some of the insecticides increases the risk of aphid outbreak. In home gardens, in case of smaller populations they can be managed by spraying water. Severe infestations can be controlled by spraying insecticides like Monocrotophos, Acephate, Imidacloprid, Malathion and Dimethoate at 0.1 per cent. **Thrips** (*Rhipiphorothrips cruentatus, Scirtothrips dorsalis*) are one of the most important insect-pests of roses whose attack is noticed on new growths appearing after pruning, the first one which appears blackish-brown in the open field and the latter one which appears creamy in polyhouse roses. There are many other species also, but flower thrips and western flower thrips are two of the most common ones. Immature thrips are usually light yellow to lemon coloured and spindle-shaped. Thrips damage the rose flowers and tender leaves mainly by puncturing cells with their needle-like mandibles and sucking the sap so reducing the beauty of the flowers and foliage as these also leave silvery and bleached areas which afterwards turn brown, crinkled, distorted, curled, and burnt margins of the deformed buds and at heavy infestation even flowers fail to open. Low levels of thrips injury tend to be more obvious on white or light-coloured blooms than on darker blooms, but severe injury will damage blooms regardless of colour. Thrips can be effectively controlled by spraying insecticides like Dimethoate, Phophomidon, Metasystox or Chlorpyriphos, each at 0.05% or Fenitrothion at 0.075%. Spinosad is one of the most important insecticides that provides very good control of thrips. **Leafhoppers** (*Edwardsiana rosae*) are small, light- to pale-green hopping insects where nymphs and adults both suck the cell sap from undersides of leaves and tender stems, causing leaves to become spotted or turning yellow. A white or yellow stippling of the leaves is one of the most common symptoms in roses. Leaf hoppers can be effectively controlled by spraying Malathion, Carbaryl, Imidacloprid or Permethrin at 0.2%. **Spider Mites** (*Tetranychus urticae*) are very minute 8-legged insects not visible to the naked eye. They suck the cell sap from underside of the leaves, and their heavy infestation appears as bleached or bronzed, curling and dropping of the leaves and death of the plant. Feeding by individual mites causes localized cell death, resulting in light-coloured stippling. In the initial stage mite infestation is only on the undersides of leaves. Under heavy infestation the mites produce webbing and occur on the tops of leaves and other plant parts. Populations of plant-feeding mites are often kept in check by naturally occurring predatory mites and other predators. Outbreaks of spider mites often occur following insecticide treatments targeted against other pests because these treatments destroy the predatory mites. Foliar application of Carbaryl, Acephate, or Pyrethroid insecticides has a tendency to trigger mite outbreaks. Mites are favoured by hot-dry weather, especially if accompanied by dusty conditions. Keeping plants well watered during periods of drought helps reduce the potential for mite outbreaks. Washing foliage, especially the undersides with a water spray can also help control or prevent mites. Mites can be controlled by spraying miticides like Dicofol, Milbemectin, Malathion, Bifenzate and Dimethoate from 0.5 to 1 ml per litre. Vertimec (Abamectin 0.5 ml/l), Cascade (Flufenoxuron 1 ml/l) and Apollo (Clofentezine 0.6 ml/l) control the mites completely in one fortnight of application. **Scale insects** (*Aulacaspis rosae*) can be easily detected by the reddish encrustations on the shoots. Their bodies are covered with a hard scale-like covering, depending on the species, may be rounding, elliptical, tear-shaped or oyster-shaped and often blending with the bark colour, and these continue damaging the tissues by sucking the plant sap with their thread-like sucking mouth parts. Their attack reduces plant vigour and heavy infestation may cause even death of whole plant as they probe and feed on plant cells. The infested parts can be pruned to prevent further spread. They can be effectively controlled by spraying insecticides like Malathion, Rogor or Dimethoate at 0.2%. **Mealy bugs** (*Pseudococcus* spp.) are comparatively small insects with white cottony filamentous growth on the exterior and these also stick to the plants sucking the sap from the shoots and other tender parts. They can be effectively controlled by spraying Monocrotophos and Dimethoate at 0.2% spraying. **Whiteflies** (*Trialeurodes vaporariorum, Bemisia tabaci*) are very small insects that are covered with a white waxy powder. Immature whiteflies are scale-like immobile insects sticking to one place at the underside of leaves and through their piercing and sucking mouth parts sucking the sap while adults which also are clustered usually at the underside of leaves are seen flying in a crowded form around the infested plants when disturbed. Similar to aphids, they build up their heavy population rapidly, and while sucking the plant sap usually from the underside of the leaves these produce honey dews attracting the ants and sooty moulds and make the flower unfit for marketing. They can be effectively controlled by spraying Imidacloprid, Acephate or Malathion at 0.2% or neem oil (20 ml/l). **Termites** (*Odontotermus obesus*) start destroying the rose plants from underground parts even before the plants are fully established. They are very difficult to control as its queen colonizes more than 60 cm below the soil and laying millions of eggs daily. One of the most difficult things with termites is that by the time they are evident much of the damage has already occurred to the plant. They can be controlled by soil application of Chlorpyriphos at 0.5%, or through Carbofuran granules at 5 g/m^2 followed by watering. The **larvae** (caterpillars) of various moths or butterflies have been found infesting roses through their chewing type of mouth parts and feeding on the leaves, resulting in leaf skeletonization or defoliation. They can be effectively controlled through 0.2% Acephate, Carbaryl, Dichlorvos or Spinosad sprayings and also through use of *Bacillus thuringiensis*. **Chafer beetles** (*Oxycetonia versicolor, Adoretus* spp.) and **ash beetle** (*Myllocerus* spp.) feed on the growing points making irregular holes and punches on the leaves. The grubs feed on the roots. The insecticides like Carbaryl, Acephate or Chlorpyriphos at 0.2% controls these beetles. **Digger wasps** (*Crabro* sp.) damage the plants after pruning by digging into the pruned stem through cut ends. Their digging also facilitates the entry of weak pathogens that cause die back. The pruned ends should be effectively sealed with a proprietary fungicidal paste to prevent the damage. In case of small gardens a few drops of insecticides like Dimethoate may be dropped into the tunnel. **Leaf cutting bees** (*Megachile* spp.) build their nests in hollow stems, reeds or pipes. They stuff the nests with pieces of leaves cut from various plants and feed them to developing larvae. Roses are one of the preferred plants for cutting leaves. Though the damage is minor but adult bees cut semi-circular holes in rose leaves, usually affecting only a few leaves on any one plant. They need not be controlled as their control may be harmful to the bee population. Vijayan Nair (1989) reported 0.05% spraying of Monocrotophos, Dimethoate or Fenithion as quite effective for all the rose pests he recorded in Thiruvananthapuram though in case of *Scirtothrips dorsalis* 0.1% was effective. For integrated insect control, oxymeton methyl 0.05% 30 days after planting, *neem* kernel extract 4% (40 days after planting), methyl parathion 0.05% (50 days after planting), *neem* oil 2% (60 days after planting) and *neem* kernel extract 4% (70 days after planting), followed at 10 day's interval provides significant control of aphids, beetles, caterpillars, thrips, and many other pests of roses under field conditions (Tejaswini *et al.*, 2007).

In roses, most common disease is **die back** (*Diplodia rosarum*) which starts from top to downwards, especially through wounds or pruned ends. In its infection, initially the affected stem-part becomes black but afterwards whole plant dies. The affected part should be cut immediately in slanting position with a sharp knife in one attempt and burnt; *vis-à-vis* cut ends should be sealed with Bordeaux paste or copper oxychloride. Slant cut will avoid water stagnation at the point of cut. Carbendazim 0.2% spraying also controls this disease. **Black spot** (*Diplocarpon rosae*) disease appears as dark brown circular spots with fringed borders which after sometime expand, coalesce, leaves turn brown and start falling. The spread of the disease is checked by burning the plant debris and through spraying with 0.2% Bayleton (Tridumefon), Captan, Chlorothalonil, Fenarimol or Benlate or 0.5% Tilt (Propicenazole). *Rosa laevigata, R. multiflora, R. roxburghii, R. rugosa, R. virginiana* and *R. wichuraiana* are highly resistant to black spot. **Powdery mildew** (*Sphaerotheca pannosa*) appears as powder-greying of entire plant, especially on the abaxial surface of young leaves and floral buds so the flowers do not open. The conidia spreads through wind and the disease is favoured by humid or excessive dry weather conditions and also when days are warm and nights are cool. Sanitation of the plant debris in the surrounding will prevent the spread of the disease. It can be controlled through spraying with 0.05% Karathane followed by 0.2% Sulfex or Topaz (Penconazole 0.05%). **Downy mildew** (*Peronospora sparsa*) is quite prevalent in polyhouse roses where its infection is noticed on young plants with purplish-red to dark irregular spots on leaves during extended period of cool and humid weather in early to mid-spring. Leaf falling is most common and mostly in its severe infection the entire plant defoliates. Downy mildew is controlled through 0.20-0.25% spraying with Chlorothalonil, Fosetyl-aluminium, Ridomil, Cuman-L, Aliette or Aliette 0.25% + mancozeb 0.20%. **Rust** (*Phragmidium mucronatum*) **i**s evident by reddish-orange pustules on broken part of stems, petioles and on leaflets only in the temperate regions during rainy seasons, and these pustules become blackish at later stage. Defoliation of the plant occurs only in its severe infection. This disease is controlled through use of sulphur and Oxycarboxin.

References

Bhattacharjee, S. K. 1994. Influence of secondary nutrients on *Rosa hybrida* cv. 'Raktagandha'. Floriculture-Technology, Trades and Trends (eds Prakash, J. and K.R. Bhandary), pp. 100-102. Oxford & IBH Publishing Co. Ltd., New Delhi.

Bhattacharjee, S.K. and M.M. Damke, 1994. Response of Super Star rose to nitrogen, phosphorus and potash fertilization. *Indian J. Hort.*, **51**(2): 207-213.

Cairns, T. 2003. Horticultural Classification Schemes. In: Encyclopedia of Rose Science, Vol. I (eds Roberts, A.V., T. Debener and S. Gudin), pp. 117-124. Elsevier, Holland.

Crespel, L., M. Chirollet, C.E. Durel, D. Zhang, J. Meynet and S. Gudin, 2002. Mapping of qualitative and quantitative phenotypic traits in *Rosa* using AFLP markers. *Theor. Appl. Genet.*, **105**: 1207–1214.

De Vries, D. P. 2003. Breeding/Selection strategies for pot roses. In: *Encyclopedia of Rose Science* (eds Roberts, A. V., T. Debener and S. Gudin), pp. 41-48. Elsevier Ltd., Oxford, UK.

De Vries D. P. and L.A.M. Dubois, 1984. Inheritance of the recurrent flowering and moss characters in F_1 and F_2 Hybrid Tea × *Rosa centifolia muscosa* (Aiton) Seringe populations. *Gartenbauwissenschaft*. **49**: 97–100.

Debener, Th. and M. Linde, 2009. 'Exploring Complex Ornamental Genomes: The Rose as a Model Plant'. *Critical Reviews in Plant Sciences*, **28**(4): 267- 280.

Debener, T. 1999. Genetic analysis of horticulturally important morphological and physiological characters in diploid roses. *Gartenbauwissenschaft*. **64**: 14-20.

Gudin, S. 2000. Rose: Genetics and Breeding . In: *Plant Breeding Reviews*, **17**: 159-189. John Wiley & Sons, Inc., USA.

Gudin, S. 2003. Breeding Overview. In: Encyclopedia of Rose Science (eds Roberts, A.V., T. Debener and S. Gudin), **I**: 25-33. Elsevier, Oxford, U.K.

Hareing, P.A. 1960. Modern Roses 9. American Rose Society, Shreveport. LA.

Hellyer, A. 1982. The Collingridge Illustrated Encyclopedia of Gardening, pp. 175-176. Collingridge Books, England.

Higson, H. 2007. The History and Legacy of China Rose, Quarryhill Botanical Gardens, Glen Ellen, California (www.quarryhillbg.org/page14.html).

Heinrichs, F. 2008. International Statistics of Flowers and Plants. *AIPH/Union Fleurs*. **56**: 16–90.

Hurst, C.C. 1941. Notes on the origin and evolution of our garden roses. *J. Royal hort. Soc.*, **66**:73-82.

Kaicker, U. S., A.P. Singh and R.S. Malik, 1983. Breeding roses for fragrance. Abst. No. 988 XV *International Genetic Congress* , U.K., p. 554.

Lal, S. D., J.N. Seth, J.P. Yadav and N.S. Danu, 1982. Genetic variability and correlation studies in rose. I. Phenotypic variability, heritability and genetic advance. *Progressive Hort.*, **14**: 234–236.

Linde, M., A. Hattendorf, H. Kaufmann and T. Debener, 2006. Powdery mildew resistance in roses: QTL mapping in different environments using selective genotyping. *Theor. Appl. Genet.*, **113**: 1081-1092.

Maia, N. and P. Vénard, 1976. *Cytotaxonomie du genre Rosa et origine des rosiers cultivés*, pp. 7-20. In: Travaux Sur rosiers de Serre. FNPHP, Antibes.

Marriott, M. 2003. Modern (Post 1800). In: Encyclopedia of Rose Science (eds Roberts, A.V., T. Debener and S. Gudin), **I**, pp. 402-409. Elsevier, Oxford, U.K.

Mortensen, L.M. and R. Moe, 1983. Growth responses of some greenhouse plants to environment. VII. Effect of CO_2 on photosynthesis and growth of roses. *Meldinger Norges Landbrukshøgskole*, **62**(3): 1-11.

Nybom, H., G. Werlemark, D.G. Esselink and B. Vosman, 2005. Sexual preferences linked to rose taxonomy and cytology. *Acta Hort.*, No. 690, pp. 21-27.

Pizzetti, I. and H. Cocker, 1975. *Rosa*. In: Flowers - A Guide for Your Garden, pp. 1126-1172. Harry N. Abrams. Inc., Publishers, New York.

Rajamani, K., S. Thamburaj, T. Thangaraj and S. Murugesan, 1992. Studies on the effect of certain herbicides in rose cv. 'Happiness.' *South Indian Hort.*, **40**(2): 121-122.

Rathore, D. S. and U.C. Srivastava, 1992. Rosa Species . NBPGR. Publ., New Delhi, 113 pp.

Shrivastava, H. P. and V. Chandra, 1985. Roses for perfumery industry-*R. bourboniana*. *Indian Rose Annual*, **4:** 198-213.

Steve Connor, 2015. Roses may soon smell much sweeter. *Sunday Times of India* (New Delhi), July 5, p. 23.

Swarup, V., U.S. Kaicker and H.S. Gill, 1971. Induced mutations in French bean and roses. *Proc. int. Symp. on 'Use of Isotopes and Radiation in Agriculture and Animal Husbandry Research'*, pp. 13-25. Held in IARI, New Delhi from Nov. 30 to Dec. 2, 1970.

Tejaswini, P. Naveen Kumar, R.L. Misra, K. Sujata, Sangama, N. Ramachndran, Jhansi Rani and Suchitra Pushkar, 2007. *Rose* (AICRP on Floriculture Tech. Bull. No. 25). ICAR, New Delhi-110 012.

Vijayan Nair, V. 1989. Biology and Control of Pests of Rose (M.Sc. Thesis). College of Agriculture, Kerala Agricultural University, Vellayani, Kerala.

Viraraghavan, M. S. 1991. Rose breeding for the tropics. *Indian Rose Annual*, **9**: 45-60.

Wylie, A. P. 1955. Master Memorial Lecture on History of Garden Roses II. *J. roy. hort. Soc.*, **89**: 8-24.

27

Rudbeckia

Jayoti Majumder, Sellam Perinban, Sanyat Misra and R.L. Misra

Family: **Asteraceae**

Common names: Black-eyed Susan/Brown-eyed Susan/Yellow daisy (*Rudbeckia hirta,* syn. *R. flava*), Brown betty, Brown daisy, Cone flower/ Conedisk/Conedisk sunflower/Tall coneflower (*R. laciniata*), Gloriosa daisy (*R. hirta* Tetra 'Gloriosa' in particular, and all other perennial *Rudbeckia* hybrids in general), Coneflower/Golden jerusalem/Orange coneflower (*R. fulgida,* syn. *R. aspera, R. scabra*), Poorland daisy, Sweet cone flower (*R. subtomentosa*), Thible flower (*R. bicolor*), Yellow ox-eye daisy

Introduction and Origin

Rudbeckia is a perennial, primarily known for its magnificent mid- to late-summer yellow-orange-petalled, black-centered flowers that rise above the dark green basal foliage. It is a native of mid-west North America. It has been found growing in a wide range of soils and climates in the eastern two-thirds of the United States (east of the rocky mountains) from new foundling to Mexico. *Rudbeckisa* is named by Linnaeus for Olof Rudbeck (1630-1702), a Swedish anatomist, botanist and antiquarian, and his son of the same name (1660-1740), both professors at Uppsala University. *Rudbeckia* is a popular crop and is valued for its ornamental diversity, its colourful golden ray corollas and prominent disc-shaped receptacles that can bloom from May (mid-summer) through October. *Rudbeckia* species generally flourish in full sun to semi-shady situations and well-drained soils and are commonly found in recently disturbed habitats such as roadsides or old fields and thrive during primary and secondary succession. In fact, it is among the most satisfactory of all the herbaceous perennials and annuals because of being completely hardy and resistant to heat and drought, cultivation being quite easy with low maintenance requirements and the crop being quite uncaring to inclement weather conditions even when in bloom. They are suitable for mass planting in borders with other summer bloomers, in informal plantings, and cutting gardens. They are naturals for the meadow or prairie garden, where their showy blooms add welcome colour.

There are reports about allelopathic potential of western coneflower (*Rudbeckia occidentalis*). The roots but not seedheads of *Rudbeckia hirta* can be used as an astringent in a warm infusion as a wash for sores and swellings. The Ojibwa natives of America used it as a poultice for snake bites and to make an infusion for treating colds and worms in children. The plant is diuretic and was used by the Menominee and Potawatomi. Juice from the roots had been used as drops for earaches.

Botany and Genetics

Rudbeckias are mostly perennials but the family also includes some annuals which usually behave as biennials. Plants are tall with branching stems. The foliage is a bit scratchy but planting *Rudbeckia* in the back row hides the foliage and gives advantage only of the blooms. Size can vary from 30 to 90 cm in cultivated ones depending upon soil conditions, the variety and the frequency of watering these receive, though in the wild these grow as long as 45 cm to 2.4 metres. Stems are striate, terete, glabrous, and somewhat glaucous. The leaves are broad lance-shaped generally alternate, entire to undulate, generally non-glaucous, scabrous-pubescent; basal leaves oval or ovate, often abruptly acute at the apex, the blade truncate or broadly cuneate at the base, decurrent 1-3 cm down the petiole, to 24 cm long and 16 cm wide, the principal lateral veins arcuate, the petiole to 36 cm long; and flowers that are daisy-like in form (cone-like or with projected disc). The flowers are much larger than daisies and are arranged around a prominent cone-shaped head. The outer series reflexed, the innermost phyllaries subtending ray flowers. In some species, the petals point out and down (decumbent) as the flowerhead opens. Ray corolla pale yellow, reflexed with age, to 3 cm long and 9 mm wide. Disc flower corolla 4.5 mm long, the limb 3.7 mm long, the teeth erect, 0.8 mm long, and the style branches obtuse and 1.6 mm long. Fruits are mostly dark purple but light purple at the apex and 3.0-4.5 mm long. Pappus is an irregularly toothed or with lacerate crown and 1.0-1.6 mm long.

The literature on the **genetics and breeding** of *Rudbeckia* has been quite scanty. Species in *Rudbeckia* subgenus *Rudbeckia* have a basic chromosome number of x = 19 and polyploids have been reported to exist in *R. hirta, R. fulgida* var. *speciosa* and *R. triloba* (McCrea, 1981; Urbatsch *et al.*, 2000). Further, reported chromosome numbers in *Rudbeckia*, broadly defined, appear to be based on x = 16, 18, or 19. Reported full complements (2n) are 32 (only in *R. amplexicaulis*), 36, 38, 54, 57, 72, 76 and 102 (Urbatsch and Cox, 2006; Palmer *et al.*, 2009). Also, occasional aneuploids have been found in the *R. fulgida* complex, and in the *R. hirta* complex (Palmer *et al.*, 2009). Numbers above 2n = 38 occur mostly in taxa with relatively large or divided lower leaves, including *R. triloba* and *R. laciniata*. Such taxa occur in relatively mesic or subhydric habitats, where denser ground vegetation may have selected plants with more capacity to change their foliar architecture, from basally-concentrated coverage early in the season to taller flowering stems later in the season. Within sect. *Rudbeckia*, the reported numbers are mostly 2n = 38 or 76, but some forms of *R. triloba* have ca. 57. In contrast, numbers for sect. *Macroline* are mostly 2n = 36, but some forms of *R. laciniata* have 54, 72, 102, or perhaps more. Both *R. triloba* and *R. laciniata* may be largely apomictic in their reproduction from seed (Gustafsson, 1947). However, there is little information available on ploidy levels of specific cultivars. Pseudogamy, a form of apomixis, has also been implicated as the sole means of reproduction in triploid *R. triloba* (McCrea, 1981). A greater understanding of reproductive pathways and occurrence of apomixis would provide valuable information for both breeding and commercial propagation of *Rudbeckia*. The objectives of this research were to determine ploidy levels and relative DNA contents of selected species and cultivars, to evaluate self-compatibility and crossability between species, to determine interploid crossability among *R. hirta* cultivars, and to assess reproductive pathways in triploid *R. hirta* to better facilitate future *Rudbeckia* breeding programmes.

Species and Cultivars

Rudbeckia comprises some 25 species of hardy herbaceous perennials, biennials or annuals, all native to North America, concentrating in the southeastern USA. Several taxa are rare to endangered, with small ranges and restricted habitats. The genus has also become widely known in Europe and elsewhere through introduction for horticultural interest. Several species and species complexes contain significant variation that has been hard to describe in traditional taxonomic terms. *Rudbeckia fulgida* Ait. and its allies form a complex of taxa that has been particularly difficult to understand and variously treated (Small, 1933; Urbatsch and Cox, 2006; Weakley, 2012). *Rudbeckia* includes two genetically and morphologically distinct subgenera, *Macrocline* and *Rudbeckia* (Urbatsch *et al.*, 2000). Species in the subgenus *Rudbeckia* include many desirable ornamentals and range from opportunistic annuals to persistent perennials. The annual species *R. hirta* includes cultivars with a diverse range of flower colours and forms, including bicolouration and double inflorescences (greater number of ray florets). Petal colours range from a lemon-yellow and gold exhibited by 'Prairie Sun' to the deep red and mahogany expressed by 'Cherokee Sunset'. However, *R. hirta* is short-lived and susceptible to diseases, including *Rhizoctonia* blight (*Rhizoctonia* sp.) and *Cercospora* leaf spot (*Cercospora* sp.) (Harkess and Lyons, 1994). Other *Rudbeckia* species, including *R. fulgida, R. missouriensis* and *R. subtomentosa* are reliable perennials having disease resistance and superior performance in southern climates (Armitage, 1997). Interspecific hybridization between durable perennial species and showier annual species could lead to development of valuable new cultivars. Out of all the species the most cultivated ones are *R. amplexicaulis, R. bicolor, R. fulgida, R. hirta, R. laciniata, R. maxima, R. newmanii, R. nitida, R. subtomentosa* and *R. triloba* which are being briefly described below after scanning various literature sources (Pizzetti and Cocker, 1968; Everard and Morley, 1970; Hay and Beckett, 1971; Beckett, 1978).

Rudbeckia amplexicaulis: An annual or biennial from Texas, Kansas, Louisiana, Mississippi and Missouri which grows about 0.9 to 1.5 m long. Stems are grey-green, glabrous and much-branched. Its leaves are stalkless and clasp the stems and the flowerheads, are alternate and oval-oblong. Its calyx is formed of 6-8 quite small acuminate-linear scales, central disc maroon, ligules some 1.25 cm long, yellow and the base purple-maroon. It is propagated by seeds.

Rudbeckia bicolor (*syn. R. hirta pulcherrima, R. serotina*): An annual or biennial species from Arkansas, Texas and Georgia with stems growing some 60 cm in height, covered with silky hairs. Leaves mostly stalkless, at base are obovate though others elongated-lanceolate, generally obtuse and some 6.3 cm long and hairy. Flowers 5.0-7.5 cm across, ligules 2.5 cm long, yellow but at base marked with purple-maroon, and central disc dark purple. It is propagated from seeds. Its popular varieties are 'Herbtwalds' (yellow to maroon and bronze, self- , bi- or tri-coloured); 'Sonnenuntergang' (yellow but central disc maroon-black); 'Superba' (petals yellow upside and purplish-maroon below); *etc.*

Rudbeckia fulgida (syn. *R. aspera, R. scabra*): A perennial species from United States (in the northeast from Michigan to Pennsylvania, south to Virginia and Florida, and west to Missouri and Texas) in the dry river beds, light woodlands and near to swamps. It grows to 45 to 55 cm in height with purple stems and leaves. Stems are covererd with fine silky hairs. Leaves variable in shape and size, lower ones long-stalked but above clasping the stems, cordate-lanceolate or linear. Calyx green and hairy. Flowers long-stalked, orange-brown, either solitary or more together and axillary with about 20 ligules some 3.8 cm long, central disc blackish-maroon. This species is suitable for planting at sunny and drier locations. It is propagated through division and seeds. Its varieties *compacta* is very compact and dwarf; *fulgida variegata* bears yellow marked leaves so should be propagated through division; 'Goldsturm' is better than the original species and quite beautiful; *sullivanti* is suitable for growing in moist positions and bears large but irregularly dentate foliage which is sometimes hairy; *variabilis* grows up to 90 cm in height and produces many strong stems bearing yellowish-maroon flowers with dark purple discs; *etc.*

Rudbeckia hirta (syn. *R. flava*): An annual or biennial species is quite invasive and widely diffused throughout North America. It is short and stocky and grows to 45-60 cm in height with erect stems having rough surface and short silky bristles. Leaves are stalkless, entire, hairy, basal spatulate and upper ones elongated or lanceolate. Flowers are long-stalked, solitary, golden-yellow and the central disc initially black but afterwards dark maroon, and petals some 5 cm long and darker at the base. It can be grown at dry locations. It is propagated through seeds. Its var. *vomerensis* has larger flowers with petal colours from golden-yellow to canary-yellow.

Rudbeckia laciniata: A perennial rhizomatous species from Manitoba, Saskatchewan and Quebec to Florida through the New England states to the southern states and as far west as Colorado, Montana, Idaho and Arizona in moist positions. It grows from 1.8 to 2.4 m high and likes damp sites with more water in summer than other species for its growing. Its stems are glabrous with branched upper parts, the leaves are alternate and with 3-5 elongated unequal segments which are pointed, and the flowers are yellow and 10.0-12.5 cm across. The ligules are drooping, entire or with dentate tips. The species is propagated by seeds but its varieties by divisions. It has numerous beautiful varieties such as *foliis variegatus* with irregularly variegated leaves in yellow and white and propagated only by division; 'Golden Glow' which is highly invasive with double and pendulous flowers suitable for mass effect; 'Goldquelle' is suitable for growing in sun or shade and is propagated by division; *hortensia* with double flowers; *humilis* having leaves divided into three rounded segments; *etc.*

Rudbeckia maxima: A perennial species closely allied to *R. nitida* and *R. scabrifolia*, from Central and South USA, is clump-forming and grows from 1.2 to 2.0 m tall, bearing glaucous green, elliptic and some 30 cm long leaves, and the flowerheads have very prominent, greenish-black, cylindrical-conical discs some 4-8 cm long and golden rays florets 4-5 cm long. It is found in the moist pine woods and temperate plains.

Rudbeckia newmanii (syn. *R. fulgida speciosa*): A perennial rhizomatous species found in the moist positions in the midwest from Ohio to Missouri, Pennsylvania to Michigan, and from Arkansas to Alabama, and is a polyploid. Its stems bear silky bristles, slightly marked red and branched in the upper part. The leaves are long-stalked though in the upper part stalkless and stem clasping, generally lanceolate and irregularly dentate. Flowerheads are large yellow with orange-brown to brown-purple bands but central disc maroon-black, and the ligules are greenish below and some 3 cm long. Propagtion is through division and seeds.

Rudbeckia nitida: A perennial species from North America from Georgia to Texas and Florida, is found growing in moist semi-shady positions. Stems are glabrous and grow to 1.8 m in height and branch above. Lanceolate leaves are shining light green, little leathery, rarely divided, dentate, stalkless above but below ovate. Flowers are large, calyx segments green and oval-acuminate, central disc greenish-yellow, ligules yellow, 8-12 and elongated-ovate. Propagation is by division and seeds. Its best variety is 'Herbstsonne' (syn. 'Autumn Sun') which grows up to 2.1 m high though in rich moist soil it may go up to 3.6 m high having single golden-yellow flowers some 10 cm across with green cone. This variety has replaced the original species. Its leaves are mid-green and deeply lobed.

Rudbeckia subtomentosa: A perennial species from Midwestern United States grows up to 1.8 m high. Entire plant is covered with grey-green hairs. Leaves are mid-green, ovate, rarely 3-lobed, some 12.5 cm long, dentate and covered with fine grey hairs. Flowers yellow, some 7.5 cm across, central disc dark brown and scarsely raised, and more button-like than cone-shaped. Ligules yellow but darker at bases.

Rudbeckia triloba: An annual or biennial species from United States (from New York to Michigan, Iowa, Kansas, south to Florida, Louisiana and Oklahoma) in the damp and light woodland soils. It grows 60 cm to 1.5 m high, is much-branched and hairy. Leaves oval-lanceolate, upper ones little dentate, basal ones 3-lobed, broadly dentate and narrowing at the base. Flowerheads dull and profuse, central disc purple-black, ray ligules 8-10, some 1.9 cm long and intense yellow though base sometimes orange to maroon or purple-maroon. It is propagated through seeds.

The **varieties** vary greatly, from dwarf (30 cm) ones such as 'Becky' and 'Toto', to the giant *Rudbeckia maxima*, which can reach 2.7 m tall. *R. fulgida* var. *sullivantii* 'Goldsturm' is long- blooming and virtually pest-free. *R. hirta* 'Cherokee Sunset' grows up to 60 cm and bears semi-double to double flowers in shades of yellow, orange, red, bronze and mahogany though the flowers are short-lived but there is copious seed setting. *R. hirta* 'Indian Summer', an annual, growing up to 0.9-1.2 m, and bearing traditional daisy-like short-lived large yellow flowers which set seeds naturally.

Propagation

Rudbeckia is easy to establish and can be cultivated by spreading seeds in a prepared area. Seeds can be sown outdoors in spring, summer or fall. They can also be started indoors in early spring and moved outdoors when the danger of frost is over, but few people do this since they are so easy to start right in the garden. In fact, many species are used in prairie restorations. The seeds are placed in full sun to light shade, and once the plant is established, it requires moderate watering. Requiring very little maintenance, plants will naturalize easily and reseed themselves after the first season. After sowing, the radicle emergence occurs in 10–14 days. The seeds need to be covered thinly. The media pH should be kept at 6.0–6.2 with EC <1. Light is necessary for germination. A light source of 10–100 fc is necessary if utilizing a chamber. The temperature should be 24–25 ^{0}C until radicle emergence. Moisture for germination should be of wet (4+) level from day 1–14. On day 15, alternate between wet (4+) and moist

(3+); dry back must occur within 18 hours. The humidity should be 100% from day 1-14 or until seed coats are shed. After seed coats have been shed, humidity is lowered to 40%. Providing horizontal airflow to aid in drying down the media through evapo-transpiration, allows better penetration of oxygen to the roots. The plant should be transplanted at 25 cm distance to the main field after 30-45 days of sowing.

Most perennials also produce viable seeds, so they can also be raised through seeds similar to those of annuals. However, for perennials the seeds will have to be raised in the nursery beds and when seedlings are 10-15 cm high, these should be transferred to the nursery rows and grown on until October when they can be planted in the flowering site. Through seeds whether of annuals or perennials, these show variation in growth habit, flowering and flower colours and they do not appear to the type. However, for getting early and by all means true to type plants, these will have to be propagated only through division of clumps, from autumn where flowering is over to spring when danger of frost is over but the temperature is still between 16-20 ^{0}C. Aerial part is headed back and the whole clump of the plant is lifted and divided, each division having a green stem and roots. These are separately planted in the prepared field and watered. If these have been planted during October-November, these are lightly watered at fortnighatly intervals during whole winter. Those planted during spring should be watered each week for proper establishment and sprouting.

Cultural Practices

This crop is a subject of temperate to sub-temperate regions and is not very particular about soil types as the species have been found growing in arid zones and along the dry rivers, in moist or swampy land, in the vicinity of water sources especially the rivers, along the roadsides and fallow lands, in the wood- and prairie-lands, on the mountains and plains even in between the rocks and so on. Certainly, the growth may be invigorated when optimum conditions are met with. Any soil which is well-cultivated, well-drained, moderately rich in organic matter, pH of the soil being 5.5 to 6.5, or even up to 7.5, and the location being open and sunny so that it gets optimum light of 2,500 - 4,500 fc and a temperature range of 18.0-19.5 ^{0}C, and all these will support good stand of the crop. Keever *et al.* (2001) studied the effect of night-interrupted (NI) lighting (beginning late winter, i.e. February 1, February 15, March 1, March 15, and a natural photoperiod) on *Rudbeckia fulgida* cv. 'Goldsturm' produced outdoors in a southern nursery setting of United States, which accelerated flowering by 26-46 days in 1999 and 51-75 days in 2000. Rudbeckias are planted at a distance of 25 to 40 cm (annual to perennial) apart. However, the taller annual species and varieties and all the perennial species and varieties require **staking** at the exposed positions.

After first flowering the plants of *Rudbeckia hirta* and *R. nitida* are cut back hard to encourage emergence of new flowering shoots soon after which they will continue flowering up to full October. Latter species can grow as tall as 3.6 metres if the soil is rich and moist. First dead-heading of faded floweheads and then afterwards cutting back of entire stem from the ground in late October to November, followed by application of liquid feed along with 2 per cent urea in late spring when sprouts have started coming up, the crop stand will be excellent having extended period of blooming and with more yield of flowers. In certain cases in exposed cold gardens, if the crowns are protected during winter with a heap of sand, soil, wellrotten compost, ashes or bracken, these will flower in the second or third years, as in case of *R. hirta* 'Tetra Gloriosa'. Dry soil requires mulching with peat or wellrotten manure early in spring. Throughout the growing period the weeds will have to be taken out so that the crop may not starve of nutrients and light, and may not harbour insect-pests and diseases.

In case of perennial rudbeckias, the soil is kept moist even during off period when the crop is neither in growth nor in flowering so that plants may not dry up. However, during crop season, as that falls during summer to autumn so the crop requires weekly irrigation so that neither crop stand is poor nor flowering is adversely affected. After the flowering is over, watering is restricted just to moisten the soil but when during spring or summer the growth restarts, further it requires watering as usual. Liquid fertilizer of slurry, i.e. one litre of fresh cow dung is diluted with three litres of water and kept in an earthen pitcher for 7-15 days depending upon the weather conditions (if not so hot then for 10-15 days but if the weather has become quite hot, say above 27 ^{0}C, then only for one week). If the dung is fermented, it is further diluted to its four times, then filtered , and then with addition of 2 per cent urea it is sprayed weekly to fortnightly to the plants from the stage when the plants have attained 30 cm height and up to initiation of flowering. The slurry can be applied to the plants after cutting the stems of the first flush to encourage emergence of new shoots for sustained and subsequent flowering. Fertilizers 14-4-14 or 17-5-17 NPK at 50 - 100 ppm nitrogen or calcium nitrate/calcium chloride at 50 ppm once a week from March 15 to September 21 but only after testing the soil. When the potted plants are required to be dwarfed, B-Nine (daminozide) at 1,250–2,500 ppm or Sumagic (uniconazol) may be sprayed.

Post Harvest

The flowers are harvested at bud stage or when just begin to open. To prolong vase life, immerse cut-stem ends in boiling water for about 30 seconds before adding to the floral arrangement. However, the flowerheads are held to the side to keep it out of the steam.

For seed harvesting when the petals start to drop from faded flowers, the cone in the centre of the flower will begin to swell and change the colour to dark brown, and below the cone the flower stem will change colour from green to a lighter brown. At this stage seeds are harvested by removing the cone. The cone can be removed by simply pulling it gently off the stem. The seeds can be stored for one season at room temperature after drying them.

Insect-Pests and Diseases

Tarnished plant bug causes distortion of leaves and flower buds. These can be eliminated by introducing big-eyed bugs or damsel bugs. Insecticidal soaps can also be effective in controlling this pest. Chemicals used are Azatin, Ornazin and Orthene which are effective insecticides that control this pest through spraying at 0.2%. **Lacebug** damage is distinguished from other chewing insects as it leaves behind the tiny black specks of faecal droppings. Shiny black nymphs are found on the underside of leaves. Release of green lacewings in the garden will naturally control the lacebug nymphs as lacewing larvae feed upon them. Spray of plants with horticultural oil is another treatment to control this pest. Washing the plants with insecticidal soap or spray with Malathion at 0.2%, Merit or Orthene will also control this pest. **Thrips** infestation is clearly visible when leaf sheaths are pulled apart, revealing both its adults and larvae. Thrips carry diseases called tospoviruses. Release of predatory mites or pirate bugs will control this pest naturally. Fine spray of horticultural oil will also control this pest. Effective control with Acetamiprid, Avid, Azatin, Ornazin, BotaniGard or Conserve sprayings can also be obtained. These all the pests are of minor importance as their damage does not warrant any significant damage to this crop.

Downy mildew (*Plasmopara halstedii*) appears as dark and blotchy areas on upper leaf surfaces and grayish white fuzzy growth on underside of the leaves. Proper air circulation should be provided around the plants, planting should also be in such a manner that one plant may not touch the other plants and sprinkler system of irrigation should be avoided. Sulphur fungicide should be used on the affected plants. **Rusts** (*Aecidum batesii, Puccinia dioicae, Uromyces perigynius, U. rudbeckiae*) appear in the form of dusty and rusty red spores in blister-like spots on leaves and stems. Some *Carex* species (sedges) act as alternate host of *Uromyces* and *Puccinia*. If grown close to *Rudbeckia*, the crop may be badly affected. Such plants should be uprooted and destroyed, proper aeration in the plantings should be allowed and over-watering should be avoided so that moisture level in the field is lessened.

Bacterial leaf spots (*Pseudomonas* spp. and *Xanthomonas* spp.) create angular, brown to purplish spots primarily on lower leaves but afterwards entire leaf is killed. This is favoured by humid conditions. Sprinkler irrigation should be avoided, and at the end of the season, whole plant debris from the vicinity should be removed and composted or burnt.

Viruses that infect rudbeckias are '*Rudbeckia* mosaic' (RuMV), 'potato yellow dwarf' (PYDV), 'tomato spotted wilt' (TSWV), 'tobacco streak' (ToSV) and 'Bidens mottle' (BiMV). Leaves may be distorted and the plants stunted. Some are spread only by insects while others through infected hands and contaminated tools. 'Aster yellows' is straw-yellowing of whole plant including the colour of the flowers and eventual death. Another type of mycoplasma also affects the plants where leaf-like structures are formed where flower parts are located. In case of 'witches broom' disease there is heavy and abnormal branching, growth is stunted and the leaves are yellowed. To control all the viral or the diseases caused by mycoplasma, the plants should immediately be uprooted and burnt. Leafhoppers should be controlled with insecticides as soon as these appear in the field so that the spread of the disease from infected to healthy plants is arrested.

References

Armitage, A.M. 1997. *Herbaceous Perennial Plants: A Treatise on Their Identification, Culture and Garden Attributes* (2nd ed.). Stipes, Champaign, IL.

Beckett, K.A. 1978. *Rudbeckia*. In: *The Concise Encyclopedia of Garden Plants*, pp. 361-362. Orbis Publishing Ltd., London, Great Britain.

Everard, B. and B.D. Morley, 1970. *Wild Flowers of the World*, p. 165. Peerage Books, London, Great Britain.

Gustafsson, A. 1947. Apomixis in higher plants. Part III. Biotype and species formation. Lunds Universitets Arsscrift N.F. Aud. 2, **43**: 183–370.

Harkess, R.L. and R.E. Lyons, 1994. *Rudbeckia hirta* L.: A versatile North American wildflower. *HortSci.*, **29**: 134, 227.

Hay, R. and K.A. Beckett, 1971. *Rudbeckia*. In: *Reader's Digest Encyclopaedia of Garden Plants and Flowers*, pp. 633-634. The Reader's Digest Association Ltd. 1971, London, Great Britain.

Keever, G.J., R. Kessler, Jr. and J.C. Stephenson, 2001. Accelerated flowering of herbaceous perennials under nursery conditions in the Southern United States. *J. env. Hort.*, **19**(3): 140-144.

McCrea, K.D. 1981. Ultraviolet floral patterning, reproductive isolation and character displacement in the genus *Rudbeckia* (Compositae). Ph. D. Diss., Purdue Univ., West Lafayette, IN, USA.

Palmer, I.E., T.G. Ranney, N.P. Lynch and R.E. Bir, 2009. Crossability, cytogenetics and reproductive pathways in *Rudbeckia* subgenus *Rudbeckia*. *HortSci.*, **44**: 44–48.

Pizzetti, I. and H. Cocker, 1968. *Rudbeckia*. In: *Flowers A Guide for Your Garden*, pp. 1173-1178. Harry N. Abrams, Inc., Publishers, New York, USA.

Small, J.K. 1933. Manual of the Southeastern Flora. Univ. of North Carolina Press, Chapel Hill.

Urbatsch, L.E. and P.B. Cox, 2006. *Rudbeckia*. In: *Flora of North America, North of Mexico* (vol. 21), pp. 44-63. Oxford Univ. Press, New York, USA.

Urbatsch, L.E., B.G. Baldwin and M.J. Donoghue, 2000. Phylogeny of the coneflowers and relatives (Heliantheae: Asteraceae) based on nuclear rDNA internal transcribed spacer (ITS) sequences and chloroplast DNA restriction site data. *Syst. Bot.*, **25**: 539–565.

Weakley, A.S. 2012. *Flora of the Southern and Mid-Atlantic States*. Univ. of North Carolina Herbarium, Chapel Hill. http://www.herbarium.unc.edu/flora.htm

28

Tagetes

Ajit Kumar, Manish Kapoor, Manju Rana, Sanyat Misra, R.L. Misra and V.K. Rao

Family: **Asteraceae,** and **subfamily: Asteroideae**

Common names: African/American marigold/Aztec marigold/Big marigold (*Tagetes erecta*), Bush marigold/ Signet marigold/Striped marigold (*T. tenuifolia* syn. *T. signata*), Climbing marigold (*T. sarmentosa*), Black mint/French marigold/ Southern cone marigold/Stinking Roger/ Wild marigold (*T. patula*), Lemmon's marigold/Mountain marigold (*Tagetes lemmonii*), Marigold, Mexican/ Sweet-scented marigold (*T. lucida*), Marigold, Hindi-*gainda,* Nepali-*shaepatri* meaning 100-leaved flower]

Introduction and Origin

The generic name of *Tagetes* is said to have derived from *Tages*, the boy-god with the wisdom of age who arose from a freshly ploughed field to instruct the Etruscans deity in the art of haruspicy (Pizzetti and Cocker, 1975), though this legend is not certain. The genus was erected by Tournefort. In the language of flowers, *Tagetes* represents divination. The common English name marigold is derived from 'mary's gold', a name exclusively applied to calendula. Marigold was first discovered by Portuguese in Central America, probably Mexico in the 16th century. Neither African nor French marigolds came from Africa or France as both are native to Mexico from where these were brought to Europe by Cortz and were in cultivation in England by 1547, and John Winthrop, Jr. introduced these to USA in 1631from where it spread to the rest of the world. There are some 60 species, all diffused throughout the warmer regions of the New World from New Mexico and Arizona to Argentina (Pizzetti and Cocker, 1975). They introduced these flowers to Europe and India and now these plants are widely cultivated in the sub-continent. Fuchs was the first authority to apply the name of *Tagetes,* which was later adopted by Linnaeus (Kaplan, 1960). Among the Gonds, the aboriginal tribes of Central India, legends connected the origins of marigold with Gonmuli ~ the God of the Gonds. According to Hindu mythology, the marigold existed in India even during Ramayana (400

B.C.) and Mahabharata (8 B.C.) periods as its references are made in these scriptures about its attractive flowers and its medicinal properties. However, in India this plant was introduced by Portuguese. African marigold (*Tagetes erecta*) was first introduced from Mexico into Spain in the early 16th century. It became popular in southern Europe under the name 'Rose of the Indies' and in Italy it was growing in 1532. In 1535, Emperor Charles V reintroduced this plant into Europe under the name 'Flos Africonus' when he had launced an expedition to free Tunis from Moors. African marigold continued to be called 'Flos Africonus' up to 18th century although in Fuch's herbal in 1542 it was correctly mentioned as *Plantas Tagetes indica* (Arora and Pal, 2003). French marigold (*T. patula*) followed similar route from Mexico and was called as *Tanacetum peruvinum* by Parkinson due to the reason of growing it in Peru. Coats discussed that it was believed by James Justice as a Chinese plant and was first sent to Royal Garden, Paris where it was named as French marigold (Arora and Pal, 2003). Major states growing marigold in India are Karnataka, Tamil Nadu, Andhra Pradesh, West Bengal, Maharashtra,Punjab, Haryana, Uttar Pradesh, Rajasthan, Himachal Pradesh, Uttarakhand, etc. (Kameswari *et al.*, 2011).

Marigold, a versatile crop with golden harvest is grown commercially in different parts of the country for loose as well as for cut flower production. It has beautiful flowers with wide adaptability to various soils and climate conditions, and has long duration of flowering which make it one of the most popular loose flower crops with sizeable acreage. Besides, the use of flower for various religious and social functions, *vis-à-vis* garland making, it has an important role in landscaping due to plant's variable height and colours ranging from orange, yellow, gold, cream, apricot, red to black-red. The dwarf type French marigold is most ideal for bedding, sunken garden, rockery, edging, hanging and window boxes, and for container gardening.

Marigolds are well known for their medicinal (Abad *et al.*, 1999), insecticidal (Perich *et al.*, 1995) and nematicidal properties (Govindaiah *et al.*, 1993). Leaf extract is good remedy for earache and the paste of leaf is used externally for the treatment of boils and carbuncles. Flower extract is used for blood purification, bleeding piles and as a remedy for ulcers and eye diseases. Marigold oil is reported to possess repellent properties against ants, flies and mosquitoes and has even anti-fungal properties. In nematode-susceptible crops such as curry plant, okra, tomato, tuberose, etc., marigold can be grown as companion crop. Also, these days farmers are planting marigold in the rows between tomato crop as the former acts as the trap crop against fruit borer which is a prolific pest of the latter.

Carotenoid, a natural pigment mostly composed of esters of xanthophylls (lutein), extracted from the flowers of orange-flowered varieties of marigold, is being used commercially for intensifying the yellow pigmentation of the yolk in poultry eggs (El-Boushy and Raterink, 1992) and broiler skin (Erkek *et al.*, 1993) in many countries including Western Hemisphere and Asian region, by and for various companies who produce feed additives. Lutein, a best source of carotene is extracted from the flower petals of African marigold. Total carotenoids range from 18 to 24.7 g/kg of marigold flower in dry weight. Essential oil is extracted from South American marigold/wild marigold (*Tagetes minuta*). Signet marigold flowers are also edible. The petals from their tiny flowers add bright colour and a spicy tang to tossed salads. The petals are chopped to make a tangy garnish for boiled eggs, steamed vegetables or fish dishes. Besides, marigold flower meal also acts as a source of an emulsifying gum (Medina *et al.*, 1993).

Botany and Cytology

Though basically it is an annual growing up to 120 cm tall but it can be perennialized through soft cuttings and also by keeping its growing atmosphere quite humid (not waterlogged) throughout as can be observed along side water bodies such as hand pumps and country wells by which their branches sprawl and send out roots from the stems and make a huge clump. The plants are stout and plant spread is noticed from 15 cm to 90 cm and sometimes covering an unbelievable space. Its most species have free-branching habit where dark green, finely segmented (sometimes fern-like) and strongly foetid pinnate leaves present a good look. The lower leaves are broad and spatula alternately along the stem. Inflorescence is capitulum in which disc florets are usually actinomorphic and ray florets are always highly zygomorphic. Flowers are scented and hermaphrodite and with variation in colours like yellow and gold to orange, red and mahogany. The diameter of floral head is around 4-6 cm and sometimes exceeding even 15 cm in case of African marigold. Marigold is a sexually propagated crop and the open-pollinated varieties are vulnerable to cross-pollination. There is no seed pod and the seeds are held together in a dense upright bundle. The seeds are long, thin, black woody sticks with a straw-coloured end. There are many seeds on a seedhead. Seeds are crescent to horse shoe shaped with rough exterior. The work on chromosome number of diploid and tetrapolid species of marigold was compiled by Towner (1961). Diploid (2n=24) somatic chromosome numbers have been recorded in *Tagetes erecta* (Eyster, 1941), *T. elongata, T. jaliscensis* (Towner, 1962), *T. lemmonii* and *T. Tenuifolia* (Eyster, 1941) though in *T. lucida* the diploid number was recorded 2n= 22. Tetraploid species with 2n= 4x= 48 are *T. biflora, T. mendocina, T. minuta* (Eyster, 1941), *T. patula* and *T. remotiflora* (Eyster, 1941). Cytological and morphological evidence show that *T. patula* is an allotetraploid species (2n=48) which probably originated by crossing *T. erecta* with *T. tenuifolia* or species closely related to them (Towner, 1961). Towner (1956, 1958) reported that crossing among *T. erecta, T. tenuifolia* (*syn. T. signata*), and/or *T. patula* is difficult as resultant hybrids are almost completely sterile which also suggests that *T. erecta* and *T. tenuifolia* or any other closely related diploid species might have been involved in evolving *T. patula* through amphiploidy. However, F_1 fertile hybrids which are morphologically distinct were obtained through crosses involving *T. jaliscensis* and *T. erecta* (Towner, 1962). Undisturbed segregation occurred in F_2 generation for marker genes like chlorophyll-defect, disc anthocyanin and tubular ray on four pairs of chromosomes. Synthetic allotetraploid *T. erecta-tenuifolia* having 2n=48 chromosomes was produced by Towner (1962) with colchicine-induced chromosome doubling. In this synthetic allotetraploid, evidence of heterogenetic pairing is expected unless such pairing had been reduced or eliminated by differential affinity (Darlington, 1937). Absence of multivalent, bridges or fragments in *T. erecta-tenuifolia* seemed to breed true retaining fertility in advance generations so that if heterogenetic pairing was present also, it was not being transmitted further. Therefore, a few lagging univalents in the synthetic allotetraploid did not result in aneuploid progeny. Studies on the nature and the limit of polyploidy in marigolds were carried out at NBRI, Lucknow (Jalil *et al.*, 1974). Hexaploidy may not be successful here as highest level is tetraploidy with about 50 per cent of the *Tagetes* species. DNA average content can also be used to determine ploidy levels in the genus. Serrato-Cruz *et al.* (2000) determined association between nuclear DNA content and ploidy level in seven populations of *Tagetes* by flow cytometry using *Zea mays* as an internal standard. Chromosome counting performed in cytological preparations stained with acetocarmine on *T. erecta, T. foetidissima,* two ecotypes of *T. tenuifolia* and two unidentified wild populations showed a DNA average count of 2.61±0.23 pg, which corresponds to ploidy level 2x=24. DNA average content in *T. patula* was 5.35±0.13 pg with ploidy level of 4x=48.

Breeding

Sharga *et al.* (1972-1973) and Arora and Singh (1980) mention that a wide array of marigold germplasm was **introduced** at NBRI, Lucknow and PAU, Ludhiana and through their assessment, NBRI recommended African marigold cvs 'Alaska', 'Doubloon', 'Fire Glow', 'Golden Jubilee' and 'Yellow Fluffy' for loose flower production while 'Geraldine', 'Golden Climax Giant', 'Orange Fluffy', 'Orange Mums', 'Sovereign', 'Sun Giants', 'Super Chief Double' and 'Yellow Climax' for cut flower production, though PAU recommended African marigold cvs 'Giant Double African Orange' and 'Climax' & French marigold cvs 'Butter Scotch', 'Red Brocade' and 'Rusty Red' for loose flower production and 'Sussana', 'Tetraploid' and 'Valencia' for bedding purpose. A large number of marigold varieties were also introduced at IARI, New Delhi and as all these varieties were in heterozygous condition, single plant selection were made so the varieties 'Pusa Narangi Gainda' and 'Pusa Basanti Gainda' were released through **pedigree breeding** method. Janakiram and Rao (1992) while studying 11 African marigold lines in Bangalore, recorded positive and significant associations which were attributed to pleiotropic effect of a single gene or to the location of two independent genes controlling these traits in close proximity. Singh and Sen (2000) and Sreekala *et al.* (2002) also recorded high heritability with high genetic advance with regard to flower yield, likewise Kishore and Raghava (2001) recorded highest heritability values for days to flowering, flower diameter and shelf life of flowers, followed by plant spread, number of primary branches and seed yield, along with high heritability and low genetic gain for days to flowering, flower diameter and shelf life. Seeekala *et al.* (2002) while working on carotenoids, recorded higher phenotypic than genotypic variance on total carotenoids which revealed highly significant effects of the environment on various traits, *vis-à-vis* highly significant heritability for total xanthophylls and xanthophyll esters thus indicating the possibility of genetic improvement of these traits through selection. Kennedy (1997) recorded high heritability along with high genetic advance for vigour index and germinating percentage in *T. erecta*, indicating that these traits are amenable to selection. Studies on breeding system in three commonly grown species, *viz. Tagetes erecta, T. patula* and *T. tenuifolia* were carried out at NBRI, Lucknow (Khoshoo, 1979). Of these, two species *T. erecta* and *T. patula* were reported to be self-compatible as well as cross-pollinated, while *T. tenuifolia* was found to be self-incompatible. The ratio of cross and self pollination was observed to be 35:65 in *T. erecta*, whereas, in *T. patula* this ratio was 70:30.

In annuals, there are two mechanisms of pollination, i.e. self-pollination (inbreeding types) and cross-pollination (outbreeding types). Many species of flowers have an intermediate breeding system with varying proportions of selfing and crossing. Marigold is preferentially an outbreeder crop. Various breeding methods for marigold improvement like hybridization, heterosis breeding, male sterility, mutation breeding and biotechnology can be adopted for breeding of new and novel cultivars for qualitative as well as quantitative characters. **Hybridization** can be intervarietal and interspecific for the production of F_1 hybrids as they are uniform in flowering with large blooms and high in yield. Towner (1961, 1962) did interspecific hybridization between *T. erecta* × *T. tenuifolia* and *T. erecta* × *T. jaliscensis*. Red and Gold dwarf hybrids resembling as French marigold but producing large flowers in large number as African marigold were evolved by crossing involving *T. erecta* with *T. patula*. Relichova and Petrova (1988) used a nuclear gene for pollen sterility and recorded statistically significant influence of both the parental lines and their interactions for the traits like plant height, flower number per head per plant and head diameter. In African marigold, Singh (1970), and Singh and Swarup (1971) studied **heterosis** on a diallel set of F_1 hybrids (excluding reciprocals) involving 7 parents. The highest heterosis was observed for number of flowers. The crosses 'Alaska' × 'Hawaii', 'Alaska' × 'Cupid Orange Mum' and 'Katrain Local' × 'Cupid Orange Mum' showed the maximum heterosis percentage for flower size, flower weight, earliness in flowering and flower number, respectively. Reddy *et al.* (1989) studied heterosis on a diallel set of crosses including reciprocals involving 7 parents. The cross combinations 'African Double Yellow' × 'Spun Gold', 'Giant Double African Orange' × 'African Double Yellow' and 'Giant Double African Orange' × 'Coimbatore Local' which had high specific combining ability effects and heterosis (%) for a number of characters were recommended for exploitation of heterosis in commercial cultivation of African marigold. Barad *et al.* (1993) evaluated 15 F_1 hybrids of *Tagetes erecta* raised from one-way crosses made in a diallel fashion involving 6 parental lines where the cross 'Giant Double African Orange' × 'Cracker Jack' was superior for commercial uses. Relichova and Petrova (1988) observed greatest heterosis effect for number of heads per plant and head diameter in African marigold. Similarly, Gupta *et al.* (2001) carried out line × tester analysis involving 3 male sterile lines, *viz.* MS_7, MS_8 and MS_{12}, and 11 diverse pollinators, *viz.* Sel. 7, 8, 14, 19, 21, 22, 27, 28, 29, 31&56 as testers. Significant heterosis was observed in 33 F_1 hybrids along with their parents for all the characters except stalk length and number of seeds, whereas highest heterosis was observed in flower yield which was 310.2% for the winter crop and 535.7% for the summer crop. Sreekala and Raghava (2003) studied heterosis in African marigold for carotenoid content and its correlation with esterase polymorphism and found that the selection of parents with wider variation in their esterase profiles may possibly be exploited for genetic enhancement of carotenoids in *T. erecta*. Mohanty *et al.* (2003) studied heterosis in African marigold cv. 'Bhubaneswar Local Orange', 'Sutton's Double Orange', 'Giant Orange Single', 'Early Yellow Single' and 'Newtech Orange', and a French marigold cv. 'Royal Bengal'. The mean sum of squares due to parents v/s crosses was highly significant for all the characters, indicating the presence of significant heterosis. **Male sterility** is the absence of functional male gamete (pollen) with functional female gamete which has sporadic occurrence in nature probably due to mutation. A monogenic recessive male sterile apetalous mutant was located in one population of *T. erecta*. Rao *et al.* (1993) established 18 male sterile lines at IIHR, Bangalore. Male sterility in marigold is of two types, i.e. apetalous and true doubleness, and for greater reliability apetalous type of male sterility (femina type). The genic male sterility is governed by a single recessive gene, i.e. *msms* (Gupta *et al.*, 1999). This recessive gene is maintained through crossing with heterozygous *(Msms)* maintainer (Singh, 2000). A cross of *msms* with *Msms* gives a ratio of 1:1 for male sterility and male fertility. The male sterile lines can be used for production of F_1 hybrid seeds on commercial scales. Some of the important F_1 hybrids in African types are 'F_1 Gold Coin Series', 'F_1 Climax Series', 'Space Age Series', 'Galore Series', etc., in French marigold are tetraploid hybrids and some interspecific hybrids are 'Red and Cold hybrids', 'Nugget', 'Red Seven Star', etc. The cost of production of hybrid seeds like hand emasculation and pollination can be reduced drastically by using available genic male sterility in African marigold. Rao *et* *al.* (1990) stated that the ability to fully exploit the use of a genic male sterility depends on the ability to clonally propagate pure strains of male sterile plants from selected individuals. Kumar (2002) and Kumar *et al.* (2003, 2004) standardized *in vitro* techniques for culture establishment and multiplication, rooting and hardening of male sterile lines of marigold and found that F_1 hybrid seed plants produced from *in vitro* derived plants were superior to the hybrid seed plants derived from *in vivo* raised male sterile plants. In *Tagetes erecta*, Jacobs *et al.* (1995) treated seeds with EMS, and after screening more than 300 M_2 generation plants, they obtained two **mutants** displaying aberrant thiophene composition.

Regarding **inheritance** of characters in marigold, doubleness has been reported as monogenically dominant over singleness (Singh, 1970; Singh and Swarup, 1973a), gene for doubleness being designated as R while for recessive allele r which governs singleness. Regarding floral colour, they reported that orpiment, tangerine and yellow colours are governed by two independent genes, dominant T governing the tangerine and dominant Y the yellow colours, the factor I inhibiting the action of dominant gene T and dominant gene Y, and the presence of tangerine colour is due to presence of dominant Y, recessive I and recessive t. Production of orpinant colour involves the presence of dominant T, dominant I and dominant Y genes but the genotypes tt II yy and tt ii yy can also produce orpiment colours. Singh and Swarup (1973b) studied the inheritance of the characters such as days for flowering, flower weight and flower numbers in 7 inter-varietal crosses of African marigold where in case of days for flowering they recorded significant additive gene effects in two crosses and significant dominant effects in three crosses, and the mean effects were found greater than additive and dominant effects. Epistatic gene effects were observed quite pronounced than additive and dominant gene effects, and the contribution of dominance × dominance interaction to the inheritance was recorded more. With regard to flower weight, the dominance played a major role than addictiveness as out of 7 crosses, 6 showed significant dominance effects and the magnitude of dominance effects was greater than additive gene effects. The absolute magnitude of the total epistatic gene effects was observed to be larger than mean effects in 6 crosses and the contribution of additive × dominance and dominance × dominance interactions was greater as compared to additive × additive interaction. For number of flowers, dominance effects were significant in 2 crosses, additive gene effects in 3 crosses and epistatic effects in 6 crosses, and dominance × dominance epistatis contributed more towards inheritance of this character.

Plant **transformation** work in marigold was carried out by Mukundan and Hjortso (1990), Buitellar *et al.* (1992) and Croes *et al.* (1994). They transformed roots of *Tagetes* with different strains of *Agrobacterium rhizogenes* to study the effect on thiophene (a medicinal compound) synthesis. In addition to this, marker genes are also being introduced in the *Tagetes* species to identify and for removal of the male fertile plants before flowering in hybrid seed production programme. Mohamed *et al.* (2000) with *T. minuta* drought tolerant somaclone when tested in the greenhouse for growth at 40% soil-field capacity for two months, recorded a significant tolerance compared with other regenerated and control plants. This somaclone had lower water potential, greater accumulated biomass and a higher related growth rate.

Classification, Species and Varieties

Genus *Tagetes* comprises some 50 species of annual and perennial herbaceous plants of which the commonly cultivated species include *Tagetes erecta*, *T. lucida*, *T. minuta*, *T. patula*, *T. sarmentosa* and *T. signata* (syn. *T. tenuifolia*). *T. erecta* and *T. patula* are most commonly grown species for their exquisite blooms and *T. minuta* for its high content of essential oil. In marigold, there are many types and strains available, varying in plant height, growth habit, flower shape and size, mainly in two main groups, African and French. ***Tagetes erecta***, an annual, highly branched, bushy and semi-hardy species of Mexican native is most familiar among growers and can grow as tall as 120 cm, leaves profuse, brilliant green and pinnately divided, leaflets elegantly divided into serrated lanceolate segments, and flower head quite wide (up to 15 cm), pompon-like, double and bears flowers of yellow, orange or bronze colours with many petals (ligules) though now hybrids and forms extend the colour range through various shades of gold, lemon-yellow and reddish-orange. African marigolds are both heat and cold tolerant. The florets of *T. erecta* are rich in orange-yellow carotenoid, lutein and are used as a food colour in the European Union for foods such as pasta, vegetable oil, margarine, mayonnaise, salad dressing, baked goods, confectionery, dairy products, ice cream, yogurt, citrus juice and mustard. However, in the United States, the powders and extracts are only approved as colorants in poultry feed. ***Tagetes lacera***, native to California (USA) is a glabrous annual or sometimes persisting as a short lived perennial with erect, faintly striate stems to 1 m tall, leaves mostly opposite but upper most alternate, 7-15 cm long, leaflets 5-11, lanceolate, acute to subcuneate at base, dark green above, flowers consisting of disc corolla some 7-8 mm long, ray floret 5-8, oblong, 2.0-2.5 mm wide, 8-10 mm long. This species has profuse flowering habit with yellow flowers. ***Tagetes lemmonii*** is an evergreen shrubby plant which grows up to 60-70 cm, stems upright to spreading with ferny, finely divided, slender, opposite, pungent leaves having 2-3 cm long leaflets, and flowers showy and 2-3 cm across. Established plants are highly heat and drought tolerant. ***Tagetes lucida***, a native to Mexico is though a tender perennial species but is cultivated as an annual and grows up to 30-80 cm in height, stems erect, branched, compact and bushy, leaves sessile, entire, evenly dentate, foetid with agreeable fragrance, glabrous, oblong-lanceolate, opposite or alternate and 5-10 cm long, inflorescence corymbose with small (12 mm in diameter), hermaphrodite yellow to orange-yellow terminal flowers produced in dense clusters and these have much more agreeable odour than other species. A yellow dye can also be obtained from the flowers. The true species is used for culinary purposes, the young leaves are substituted for terragon or other aromatic herbs in salad and for flavouring vinegar. It contains oil having anise like odour. Its flowers are suitable for cutting. The most cultivated form in gardens is the var. *floridus*. ***Tagetes minuta*** is a tall (1-2 m) upright marigold, leaves pinnately compound or divided, opposite below and alternate or opposite above, 5-15 cm long, leaflets 11-17, narrowly lanceolate, sharply toothed, 2-4 cm long, inflorescence panicle-like (cymes) consisting of numerous narrowly cylindrical flower heads 7-10 mm long and 3-4 mm in diameter with small yellow to orange-yellow flowers. It is used as a source of essential oil for perfume industry as flavouring agent for food and tobacco products. In South Africa, this species is cultivated in the reclamation of disturbed land. ***T. patula***, an annual native to Mexico, grows up to 75 cm, bushy, branched and spreading, stems often veined in dark violet or reddish-brown, leaves alternate or opposite, pinnately divided, and leaflets serrated, and flowers solitary, terminal, single or double, up to 6.5 cm wide with outer petals wide-spreading, central disc tubular and small. French marigolds appear in the typical shades of yellow and orange but also in a deep mahogany-red similar to the red to black-red, bicolours and tricolours. ***T. sarmentosa*** is probably a form of *T. lacera*, a climber-simulating species growing up to 3.0 metres high and with about 1.0 metre spread, stems loose, open and erect, leaves aromatic, each stem with a solitary bloom, flowering late, scented, single, 3.75 cm wide, flower colour variable but more or less rich golden-yellow marked with reddish-orange, and most suitable for training against trellis. ***T. tenuifolia***, a native of Mexico is dwarf, half hardy annual with a much-branched but compact and bushy, growing from 20 to 75 cm tall, leaves alternate or opposite finely cut, small, divided into 12 oblong-linear, sharply serrated, very narrow delicate segments, imparipinnate to odd pinnate in shape and sessile, flowers numerous, small, terminal, single, some 3.0 cm wide with 5 rays, rotate shape and bright yellow with 4-5 external oval ligules some 12 mm long, and yellowish marked orange or dark purple. Two of its best varieties are 'Lemon Gem' and 'Ursula.' Other important species of the genus are *T. biflora*, *T. elongata*, *T. jaliscensis*, *T. pusilla*, *T. mendocina*, and *T. remotiflora*.

Various popular marigold varieties developed in India under **African marigold** are 'Pusa Narangi Gainda' (Cracker Jack × Golden Jubilee) having vigorous and uniform growth up to 75 cm tall with dark green leaves and carnation-type double flowers in orange colour and of the 7.8 cm diameter with presence of disc florets, taking about 125-136 days for flowering after planting, and flowers used for religious purposes and for carotenoid production with potential fresh flower yield of 350 quintals per hectare; and 'Pusa Basanti Gainda', a cross between Golden Yellow × Sun Giant, most suitable for bedding and for growing in pots, late in flowering, i.e. 135-154 days, a tall, vigorous, uniform with dark green foliage, flowers double and compact, carnation-type, sulphur- yellow colour, with a size of 6.9 cm having invisible disc florets and with 7-10 rows of whorls in its inflorescence. Other important varieties of African marigold are 'Alaska', 'Apricot', 'Arka Bangara' (petaloid sterile flowers), 'Burpee's First White', 'Burpee's Giant Fluffy', 'Chrysanthemum Charm', 'Climax', 'Cracker Jack', 'Crown of Gold', 'Cupid', 'Cupid Yellow', 'Giant Double African Orange', 'Giant Double African Yellow', 'Golden Age', 'Golden Climax', 'Golden Jubilee', 'Guys and Dolls', 'Inca Hybrid', 'Mr. Moonlight', 'Odourless', 'Orange Hawaii', 'Pineapple Crush Improved', 'Pumpkin Crush Improved', 'Snow Bird', 'Spun Gold', 'Spun Yellow', 'Sugar and Spice', 'Sunset Giant', 'Sutton's Double Orange', 'Sutton's Double Yellow', 'Yellow Fluffy', 'Yellow Stone', etc. and of **French marigold** are 'Pusa Arpita' (yellow and medium size) variety recently released from IARI, New Delhi, 'Bolero', 'Bonanza Flame', 'Bonanza Orange', 'Boy Scout', 'Butter Scotch', 'Carmen', 'Dainty Marietta', 'Fiesta', 'Giant Crested', 'Golden Boy', 'Gold Finch', 'Goldie', 'Happy Orange', 'Happy Yellow', 'Harmony', 'Harmony Boy', 'Honeycomb', 'Honey Sophia', 'King Tut', 'Lemon Drop', 'Melody', 'Naughty Marietta', 'Orange Boy', 'Orange Sophia', 'Petite Gold', 'Petite Spray', 'Petite Yellow', 'Queen Sophia', 'Red Brocade', 'Red Cherry', 'Red Glow', 'Red Pygmy', 'Rusty Red', 'Scarlet Sophia', 'Spray Boy', 'Star Dust', 'Tangerine Yellow', 'Valencia', 'Yellow Boy', etc. French marigolds are grouped as (i) **dwarf doubles** growing up to 30 cm and available in various colours of yellow, orange, golden-yellow, reddish-brown, mahagony-red and bicolours; (ii) dwarf double scabious flowers have crested centres and bear golden-yellow and golden-orange with red markings; (iii) dwarf double petite, growing up to 20 cm with large number of flowers in golden-yellow, orange,yellow and maroon colours; (iv) French dwarf single growing up to 35 cm and bearing compact golden-yellow flowers with maroon eye, mahagony-red with yellow eye and yellow with brown eye; (v) dwarf triploid F1 hybrid growing to about 40 cm, blooming earlier with profuse flowering in yellow, golden-yellow, orange and brownish-red colours, and (vi) tetraploid hybrids are dwarf and flower early in brillianat orange, fully double and carnation-type blooms. The F_1 hybrids are dwarf, highly floriferous and free flowering with uniform golden, yellow heads that bear nearly one and a half times more flowers by weight per unit area. In marigold, a F_1 hybrid 'Triploid' (2n=36) has been developed at NBRI, Lucknow by using male sterile African diploid marigold (*T. erecta*, 2n=24) and male fertile French tetraploid marigold (*T. patula*, 2n=48). At IARI, New Delhi, a F_1 hybrid of French marigold obtained from crosses between 'Petite Orange' × 'Rusty Red' and 'Naughty Marrieta' × 'Pygmy' showed increase in weight and number of flowers, respectively than that of parents. In African marigold, F_1 hybrid between 'Hawaii' × 'Cupid Orange' and 'Alaska' × 'Spungold' were superior to parents. Other important commercial F_1 hybrids are 'Apollo', 'Beauty Gold', 'Beauty Orange', 'Beauty Yellow', 'Climax', 'Deep Orange Lady', 'Diamond Jubilee', 'First Lady', 'Golden Climax Improved', 'Golden Jubilee', 'Gold Galore', 'Gold Lady', 'Inca Gold', 'Inca Orange', 'Inca Yellow', 'Moon Shot', 'Orange Jubilee', 'Primrose Lady', 'Pusa Sankar 1', 'Royal Gold', 'Royal Orange', 'Royal Yellow', 'Showboat' (3x, sterile), 'Toreador', 'Tresbien Orange', 'Tresbien Yellow', 'Yellow Climax', 'Yellow Galore', etc. **Interspecific** hybrids which are released involving crosses of *T. erecta* and *T. patula* are 'Red Gold', 'Red Glow', 'Yellow Nugget', 'Burpee's Gold', etc. There are some other interspecific hybrids such as 'Marigold Snowman' (flowers 7.5 cm across, ruffled and white) developed by Burpee Co., USA grows up to 60 cm height; 'Vanilla Improved' (flowers 6-8 cm across, fully double and creamy-white) with compact growth growing up to 40 cm; 'Zenith Series', a triploid between African and French, richly coloured flowers are large as *T. erecta* and heavy bearer as *T. patula*, flowering continuously for longer duration with 7.5-10.0 cm flower diameter; etc.

Propagation

Propagation of marigold is through seeds as well as by cuttings. Usually seed propagation is preferred over cuttings as plants raised from seeds are tall, vigorous and heavy yielder. **Seed propagation** is easiest method of propagation of various cultivars of African marigold (*Tagetes erecta*), French marigold (*Tagetes patula*) and Signet marigold (*Tagetes tenuifolia*). The seeds of these species are easily available and germinate quickly (5-7 days) at temperature ranging from 18 to 30 ^{0}C. Seeds are sown in pots, boxes or in nursery beds depending upon the requirement of the seedlings. For winter season crop, seeds are sown in mid-September and seedlings are transplanted in mid- October. Similarly, sowing for summer and rainy season crops is done in January-February and mid-June, respectively. The transplanting of seedlings is done one month after sowing of the seeds. Marigolds are multiplied through **cuttings** to derive uniform plants for various vegetative and floral attributes of a particular variety. Cuttings of about 6-10 cm length are taken from the apical portion of the shoot which gives out roots adventitiously. Cuttings root best when planted in sand or vermiculite. Root promoting hormones like auxins (IAA & IBA at 500-1,000 ppm) can be used for enhancement of rooting. After rooting the cuttings are planted in the prepared beds.

Micropropagation in marigold is reported from various explants such as leaf segments (Kothari and Chandra, 1984), leaf callus (Kothari and Chandra, 1986), disc florets, capitula (Mohan Ram and Mehta, 1982), cotyledon and hypocotyl (Mohamed *et al.*, 1998, 2000), shoot tips and intermodal stem segments (Kumar *et al.*, 2003). Through micropropagation time required to produce and commercialize a new marigold cultivar is reduced drastically. Qi *et al.* (2011) achieved plant regeneration of *Tagetes patula* from anther explants via adventitious shoot differentiation from callus cultured on MS basal medium containing BA (2.2 μM), NAA (1.82 μM or 2.7 μM) and 3% sucrose, and frequencies of callus induction and shoot regeneration were 100 and 70.5%, respectively. This regeneration system is therefore useful for developing an efficient genetic transformation system.

Cultural Practices

Marigold requires mild climate for luxurious growth and flowering. Many marigold cultivars are **facultative short-day** plants, meaning that plants flower more quickly under short day lengths but will eventually flower regardless of photoperiod. African cultivars display more of a photoperiodic response than French or triploid types. The critical photoperiod for *Tagetes erecta* is between 12.5 and 13 hours. Cultivars of the African types can achieve faster flowering if provided with artificial short days. Photoperiod control is rarely needed or practiced for the French and triploid marigold production. However, marigold can be grown during winter; summer and rainy season indicating that it will eventually flower regardless of photoperiod. Whether grown in 8-h short days or long days effected by night breaks, was found unfavourable for flowering in *Tagetes* though promoted leaf growth (Peat and Summerfield, 1977). The optimum **temperature** range for its proper growth is 18-20 ^{0}C. Night temperatures higher than those normally used in commercial production systems promoted vegetative growth in *Tagetes patula* seedlings; in the short (8 h) daylength regimes with a day temperature of 16 ^{0}C, cool nights (8 ^{0}C) had a detrimental effect, especially when imposed for 6 weeks from pricking-out rather than later; whether warmer nights from pricking-out promoted flowering in *Tagetes* but if applied later (during weeks 7-11) the positive effects were largely on vegetative growth (Summerfield *et al.*, 1977). Temperatures above 35 ^{0}C restrict the growth of the plants, which leads to reduction in size and number of flowers. In severe winter, marigolds are damaged by frost. Marigold is easy growing crop and can be grown in different types of **soils**. African (tall) marigolds can be grown best in rich, well drained, moist soils whereas French (dwarf) marigolds are best cultivated in light soils. However, sandy loam soils having pH 6.0-7.5 and EC less than 1.0 mmhos/cm with good aeration and drainage are considered ideal for marigold. Lower pH prevents iron and manganese speckling, necrotic margins on older leaves and death of the growing tip so such conditions should be avoided. For preparing its **nursery**, the beds are normally raised ones for better drainage and more aeration. Seed rate for raising nursery in one hectare area is 800 g to 1 kg. Seeds should be spread thinly in the nursery beds and covered with fine soil, sand or FYM. Seeds should be sown in lines spaced 6-8 cm and at a depth of 2 cm. Thick sowing of seed results in week seedlings which take longer time to regain growth due to etiolation. After seed sowing, beds should be covered with mulches such as grasses or dried leaves for better germination. The beds should be watered with a fine nozzle can and should be kept moist. When the seed starts germinating, mulch should be removed in the evening hours to avoid sun-burning if exposed in the mornings. To minimize mortality of plants in the nursery, the soil should be drenched with 0.2% Captan. For winter season crop, the seeds are sown in second week of September, for summer crop in second week of February and for rainy season crop, the nursery should be raised in first week of June. In the mean time the beds for transplanting should be prepared by ploughing the field thrice at a depth of 30 cm followed by planking after each ploughing. All the foreign materials such as polythene, woods, grits and stones, *vis-à-vis* rootstocks of the perennial grasses should also be taken out after each ploughing. Farmyard manure at the time of first ploughing should be incorporated in the soil at the rate of 3 kg per square metre for a bumper crop. Marigold seedlings should be **transplanted** in the well prepared field during evening hours for maximum survival of the seedlings. At the time of transplanting, seedlings of optimum growth having 3-4 true leaves should only be taken for raising a good crop. For commercial cultivation, African marigolds should be transplanted at a **spacing** of 40 × 40 cm whereas French marigolds at 30 × 30 cm. After transplanting, seedlings should be lightly pressed around the root zone to remove the air pockets to help them establishing properly. The field should be lightly irrigated immediately after transplanting. Rao and Reddy (2002) reported June planting best for maximum flower and seed yield of African marigold cv. 'Pusa Narangi Gainda' under Hyderabad conditions. It was also observed that irrespective of planting in the month of June, July or August, maximum flower production was recorded only during September and October, which coincided with the short day length confirming that short day lengths are more congenial for flower bud induction and flowering in African marigold. Marigold is often intercropped with other plants. Rotation with marigolds reduces diseases as well as nematode population of other crops.

Pinching is the removal of terminal portion in marigold plants generally after 40 days of marigold transplanting to suppress apical dominance so the plant growth becomes bushy, and thereby precocious lateral shoot formation occurs, *vis-a-vis* more number of flowers are produced. Pinching at 40 days after transplanting causes maximum number of leaves per plant with extended flowering span while plant spread and number of secondary branches per plant are recorded maximum under pinching at 20 days after transplanting (Sharma *et al.*, 2006).

Marigold responds well to **nutrition**. For obtaining good vegetative growth and higher flower production, the soil should be incorporated with wellrotten FYM @ 20-30 tonnes/ha and a basal dose of fertilizers @ 200 kg of nitrogen and 80 kg each of phosphorus and potash/ha. It is advisable to apply half nitrogen along with full dose of phosphorus and potash at the time of field preparation but at last ploughing or preferably on the date of planting and the remaining 50% nitrogen should be top-dressed at the time of pinching. Ghosh and Pal (2010) noticed the response of African marigold cv. 'Siracole' (Table 1) to organic and inorganic nitrogen sources. Sources of nitrogen significantly influenced the vegetative as well as flowering attributes. Application of 50% N as mustard oil cake as basal + 50% N as urea as top-dressing recorded highest value for different parameters, including chlorophyll content in leaf tissues and anthocyanin in petals, but not the plant height. After transplanting, immediately these are **watered** and then thrice every third day for better establishment of the root system and the crop but thereafter it is given only when there is proper requirement and that is 5 to 7 days during summer, 10-12 days during winter but during rains as per adjustment of the weather. Water is best applied by furrow irrigation at timely intervals. Watering frequency depends upon soil type, weather conditions, season of planting, etc. Plants complete vegetative phase in about two months and enter into reproductive phase. Inadequate moisture level in the soil during plant growth and flowering adversely affects flower yield. **Weeds** not only compete with the main crops for space, nutrients, light and water but also harbour various pests and diseases which affect the plant growth and flower production adversely so these should be removed timely either manually or using chemicals. In general, 2-3 manual weedings are required during the entire cropping period to keep the field weed-free and when the plants have grown sufficiently, there being sufficient horizontal growth of the major crop to cover whole space which suppresses the weed growth itself. Chemical weed control has also been tried and found successful. Tenoran 7 kg/ha, Ramrod (propachlor 65%) 7 kg, Amiben-Granulat (chloramben), Kerb 50W 3 kg, Legurame flussig [(carbetamide (70%)] 10 litres and Ronstar (oxidiazon [75%]) 8 litres were tested for their selectivity in *Tagetes erecta* where Tenoran was found giving best results; Kerb 50W and Legurame gave satisfactory weed control; Amiben-Granulat, Kerb 50W and Legurame flussig were well tolerated; Ramrod lacked adequate herbicidal activity but Ronstar caused severe necrosis (Anon., 1974). Nimje and Agrawal (2002) reported that soil solarization during May-June (45 days) has been found to be effective in controlling weeds in marigold (*Tagetes* sp.). Acharya *et al.* (2003) found that pendimethalin @ 1 kg/ha + manual weeding at 30 and 60 days after transplanting reduced mean weed dry weight and also produced the highest flower yield in African marigold cv. 'Pusa Basanti Gainda'.

Marigold is a cross pollinated crop, hence requires an isolation distance of about 1.0 to 1.5 km. However, various commercial grown species, i.e. *T. erecta*, *T. patula* and *T. minuta* lack natural cross pollination and can be grown at a given place in different beds without any isolation distance. The seed crop is of six months duration. The

Table 1: Effect of different sources of N on N, P, K and chlorophyll content in leaf tissue and anthocyanin in petals of African marigold cv. Siracole*

Treatment	Nitrogen (%)	Phosphorus (%)	Potassium (%)	Chlorophyll (mg/g)	Anthocyanin (mg/100g)
N sources					
N1	2.69	0.24	3.76	1.16	0.74
N2	3.23	0.27	3.84	1.25	0.53
N3	3.69	0.28	3.89	1.28	0.56

seed yield varies in different species, being 312 to 375 kg/ ha in African marigold cv. 'Giant Double African Orange' and 1,000 to 1,250 kg/ha in French marigold. For higher vigour and viability, the seeds should be dried to a safe moisture limit (7 ± 0.5%), treated with Thiram @ 2g/kg seed and packed in polyethylene bags. Seeds stored in this manner have higher vigour and viability up to eight months. However, seed can also be stored in cloth bags but the viability is only for 6 months.

Control of Growth and Flowering

Growth regulation in marigold can be achieved by using plant growth regulators and controlling growing environment. African cultivars are more responsive to photoperiod than French or triploid types. Cultivars of the African type can achieve faster flowering if provided with artificial short days. Plant growth regulators play an important role on growth and yield attributes of marigold. Application of CCC at 3,000 ppm recorded a minimum plant height (51.77 cm) and maximum number of branches per plant (74.46) and appearance of first flower bud took highest days (49.02 days) after transplanting (Khandelwal *et al.*, 2003). GA_{4+7} at 25 ppm weekly application to *Tagetes patula* during seedling growth promoted reproductive development whether grown in 8-h short days or long days effected by night-breaks, and the latter treatment proved ineffective though promoted leaf growth (Peat and Summerfield, 1977). GA_3 at 200 ppm registered maximum flower yield per plant (82.62 g), and per hectare (9617.48 kg) as compared to control (59.46 g and 7018.37 kg, respectively) in African marigold (Varma and Arha, 2004). In marigold, various vegetative growth parameters like plant height, number of branches per plant, leaf area per plant, etc. were enhanced with application of either GA_3 @ 150 ppm or NAA @ 150 ppm when given as double spray. These two plant growth regulators also increased various floral attributes like duration of flowering, number of flowers per plant, average flower weight, diameter of flower and yield of flowers when applied @ 200 ppm as double foliar spray (Kanwar *et al.*, 2011).

Post Harvest

For cut flower production newly openend blooms are removed early in the morning or late in the evening when the temperature is low and immediately the cut ends are put in a container filled with palatable water. Later, the lower leaves on the stems are stripped to avoid direct contact of the water as this fouls the solution and then the cut flowers are arranged in the vase with fresh water and preservatives. For loose flower production, marigold flowers are harvested by plucking the flowers when they attain the full size. Plucking of flowers can be easily done through hands as the flower stalk is hollow which breaks easily when twisted between thumb and fingers. For long life of flowers after harvest, field should be irrigated before harvesting of flowers. Harvesting of flowers should be done in the cool hours of the day. Regular plucking of flowers improves the yield of flowers. Plucked flowers should be covered with moist gunny bags to avoid dehydration if stored overnight. Flowers can be stored at 8-10 ^{0}C for 3 days. Fresh flowers are packed in gunny bags or in polythene bags for local markets and in the bamboo baskets for carrying to distant markets. Willoughby *et al.* (2000) developed a mechanical harvester for marigold flowers. The yield of marigold flowers varies considerably depending upon the season, fertility of the soil and variety. Yield of fresh flowers in case of African marigold is 11 to18 tonness or 1.5-2.5 millions (in numbers) whereas in case of French marigold it is 8 to 12 tonness or 6-8 millions/ ha. The shelf life of loose flowers can be extended by soaking them in 0.1% $Al_2(SO_4)_3$ for 2 hours. Irrigating the field before plucking of the flowers also increases flower longevity. Marigolds make excellent cut flowers also for both live and dried arrangements. These cut flowers can last for up to three weeks but with addition of floral preservatives to the vase water. Marigolds are excellent for dried arrangements and floral crafts too. Perfectly blossomed flowers are cut, stem foliage removed and then these are hanged upside down in a warm and dry place for their drying. The flowers will shrink a bit but their colour will remain strong. Wires in case of need may be inserted in the stems for support in an arrangement. For transportation to local market, flowers packed in corrugated fibreboard boxes (CFB) recorded lowest physiological loss in weight (0.33 per cent) and highest xanthophyll recovery (1.45g/kg) in the flowers (5 per cent and 1.15g/kg, respectively in flowers transported in open condition without any packaging material). If packed in polythene bags, followed by cold storage treatment, they last for longer duration (Bhattacharjee, 2001).

Insect-Pests, Diseases and Physiological Disorders

Alternaria leaf spot (*Alternaria tagetica*) appears as minute, purplish brown and circular spots on leaves and stems which enlarge at later stage of infection and girdle the stem. This fungus was first reported from West Bengal (Shome and Mustafee, 1966). Indofil M-45 (mancozeb) (0.25%) significantly controls the disease. **Inflorescence blight** (*Alternaria zinnae*) causes inflorescence blight in marigold. Its infection appears as elongated lesions on the inflorescence, and on the leaves there is formation of large irregular blotches which are light tan to dark brown with zonations. Spraying the plants with 0.2% Dithane M-45 will control this disease. **Flower bud rot** (*Alternaria dianthi*), first reported from West Bengal by Mondal and Chaudhuri (1976), appears as rot of the floral buds and its infection causes older leaves developing few deep brown necrotic spots, and in humid weathers its infection becomes serious. There is formation of dry rotting of young flower buds and the sepals and peduncles turn brown-scorched with necrotic discolouration. Mature buds are less susceptible to this infection but they also fail to open. This disease can be controlled by spraying the plants with 0.2% Dithane M-45 at fortnightly intervals. **Flower bud rot** (*A. alternata*) bcomes serious in humid weathers. Its infection appears as dark brown spots on unopened buds so buds normally fail to open properly. Its control is the same as to that of *A. dianthi*. **Marigold blight** (*Colletotrichum capsici*) has been recorded in India infecting petiole, peduncle and branches. Dithane 0.2% spraying will control this disease also. **Septoria leaf spot** (*Septoria tegeticola*) causes oval to irregular and gray to black spots with tiny dots of fungal fruiting structures first appearing on lower leaves and then spreading upwards. To control this disease overhead irrigation is avoided. Propiconazole, Chlorothalonil, 0.1% Bavistin or 0.2% Dithane M-45 sprays control this malady. **Cercospora leaf spot** (*Cercospora thunbergiae*) produces brown and circular-oblong spots on the leaves of *Tagetes erecta*. It can be controlled by spraying 0.2% Dithane M-45. However, **leaf-spot** or **blight** (*Cercospora tageticola*) on French marigolds causes dark brown circular spots with a prominent grey margin and can be controlled as *C. thunbergiae*. Though an uncommon disease reported from West Bengal by Kar and Mandal (1973) on French marigold, **leaf spot** or **blight** (*C. tagetis*) causes indefinite, irregular and blackish coalescent spots. This can also be controlled through spraying with 0.2% Dithane M-45. *Botrytis, Phytophthora cryptogea, Pythium ultimum, Sclerotium rolfsii* and *Sclerotinia* cause stem cankers, collar rot, wilting and death of the plants at any stage of plant growth (Post, 1949; Forsberg, 1979; Horst, 1990; Daughtrey and Chase, 1992; Sohi, 1992). *Botrytis* infection can be manipulated through reduced humidity level but once its infection has been established, it can be controlled through spraying of Dithane Z-78 0.2% twice or thrice at weekly intervals. *Pythium, Phytophthora* and *Sclerotinia* infection causes lower-leaf yellowing, rotting of basal portion of stems and roots, and seedling damping off. Thiride at 0.2% drenching may control these pathogens. **Damping off** (*Pythium* sp., *Rhizoctonia solani*) appears as brown necrotic spots girdling the radicle which later on extend to plumule. The infection causes pre-emergence mortality of seedlings. Under post-emergence symptoms, lower part of hypocotyls has water-soaked, brown and necrotic rings leading to death of seedlings. When such seedlings are pulled, the root system appears partially or fully decayed. To control this disease proper drainage in the nursery beds should be ensured and nursery beds should be drenched with Brassicol @ 0.3%. **Powdery mildew** (*Oidium* sp. and *Leveillula taurica*) appears as whitish, tiny, superficial spots on leaves, later on spreading to whole of the aerial parts and so whole plant is covered with whitish powder. It is effectively controlled by spraying the plants with 0.5% Karathane (40 E.C.) or dusting of the plants with wettable sulphur at 15 days interval. Seed inoculation with *Pseudomonas fluorescens* strain E6 frequently restricted growth of marigold by 7-13% (Yuen and Schtoth, 1986).

Viruses such as 'cucumber mosaic virus' and 'aster yellows virus' are quite common which cause viral diseases in the marigold plants. Aster yellows is caused by mycoplasma-like organism and its infection causes yellowing or clearing of the veins in leaves, shortening of the internodal regions of the main stem and production of long axillary breaks and this looks like witche's broom, and sometimes forming even phyllody where floral parts convert to greening. Its vectors are leafhoppers. CMV is a serious problem during rainy season and its infection causes streaking or mottling of leaves and stunted growth. Its vector is aphid so Malathion or dimethoate spraying at 2 ml/l at regular intervals will keep this malady under check. Whether some control measures have been applied or not but just to check further spread of the disease, rogueing should be carried out regularly.

Bronze speckling, a specific physiological disorder has been observed in African marigold cvs 'First Lady' and 'Voyager' which has been attributed to iron toxicity (Albano and Miller, 1998). This disorder is characterized by specific symptomology distinguished visually by speckled patterns of chlorosis and necrosis, *vis-a-vis* downward curling and cupping of leaves. In 'First Lady', the older leaf tissues were observed accumulating more total Fe so these showed severe symptoms than younger tissues. Albano and Miller (1996) stated that cultivars vary with regard to iron toxicity, however, leaf tissue iron level should not exceed 500 ppm.

Red spider mites (*Tetranychus* sp.) are spinning creatures, becoming a havoc during dry and hot weather and that too on outdoor plantings. They are red and brown and continue sucking the sap from the leaves and other tender organs. They are active during flowering period. Their infestation causes discolouration of leaves, and plants give dusty appearance. Sulphur at 3 g/l or Kelthane 2 ml/l spraying is effective against them. **Hairy caterpillar** (*Diacrizia obliqua*) is a polyphagus pest and eats away foliage of the plants. Nuvan 50 E.C. or Thiodan 35 E.C. @ 1ml/l water is effective to control this pest. **Leaf hoppers** (*Empoasca fabae* and *Macrosteles fascifrons*) cause considerable damage to the foliage particularly during rainy season and that too more on French marigolds. Leaf rolling and curling at the edges along with wilting of shoots at the tips in its severe infestation are observed. At initial infestation, 2-3 sprayings either of Dimethoate at 0.2%, Acephate at 0.1%, methyl parathion or Quinalphos (0.05%) at fortnightly intervals will control this pest. **Bud borer** (*Helicoverpa armigera*) moth generally lays eggs in the young buds and the hatched larvae bore into the growing flowers and feed on various floral parts and foliage. Collection and destruction of larvae, *vis-à-vis* 0.1% methyl parathion spraying at laying stage will control this pest. **Aphids** (*Aphis gossypii*) are green to bluish-grey or black tiny insects usually occurring in clusters and damage the tender plant parts by sucking the sap and producing honey dew to attract sooty mould. Its infestation deforms the leaves and flowers, turning them yellow, *vis-à-vis* leaf drying. These insects can be knocked down through jet water spraying, and also by spraying with Malathion 0.2%, Monocrotophos 0.05%, Phosphamidon 0.02% or Dimethoate 0.03% at 10-15 days intervals. **Thrips** (*Frankliniella occidentalis*) may sometimes become problematic as this transmits tomato spotted wilt virus. While controlling the aphids the thrips will also be controlled.

References

Abad, M.J., P. Bermejo, S. Sanchez-Palomino, X. Chiriboga and L. Carrasco, 1999. Antiviral activity of some South American medicinal plants. *Phytotherapy Res.*, **13**: 142-146.

Acharya, M.M., S.L. Mundra and L.K. Dashora, 2003. Integrated weed management in African marigold (*Tagetes erecta* L.) under sub-humid southern plains and Aravalli hills of Rajasthan. *Indian J. Weed Sci.*, **35**(3/4): 309-310.

Albano, J.P. and W.B. Miller, 1996. Marigold varieties vary in susceptibility in iron toxicity (abstr.). *HortSci.*, **31**: 674.

Albano, J.P. and W.B. Miller, 1998. Marigold cultivars vary in susceptibility to iron toxicity. *HortSci.*, **33**: 1180-1182.

Anonymous, 1974. *Ann. Report German Pl. Prot. Serv.*, No. 21, pp. 221-223.

Arora, J.S and P. Pal, 2003. Marigold. *In*: Commercial Flowers, vol. 2 (eds Bose, T.K., L.P. Yadav, P. Pal, V.A. Parthasarathy and P. Das). Naya Udyog, Kolkata.

Arora, J.S. and J. Singh, 1980. Performance of marigold cultivars under north Indian conditions. *Nat. Sem. Prod. Tech. Comm. Flower Crops* (abstr.), pp. 81-82. TNAU, Coimbatore.

Barad, A.V., S.C.P. Sachan and M.U. Kakadiya, 1993. Heterosis in African marigold (*Tagetes erecta* L.). *Golden Jubilee Symp. Hort. Res. Changing Scenario* (abstr.), p. 91. IIHR, Bangalore.

Bhattacharjee, S.K. 2001. Post harvest management of flowers. *In*: *Handbook of Horticulture* (ed. Chadha, K.L.), pp. 969-981. Indian Council of Agricultural Research.

Buitellar, R.M., M.T. Cesario and J. Tramper, 1992. Elicitation of thiophene production by hairy roots of *Tagetes patula*. *Enzyme and Microbial Tech.*, **14**(1): 2-7.

Croes, A.F., J.J.M.R. Jacobs, R.R.J. Arroo and G.J. Wullems, 1994. Thiophene biosynthesis in *Tagetes* roots: molecular versus metabolic regulation. *Proc. Workshop Primary and Secondary Metabolism of Plants and Plant Cell Cultures* III. Leiden, the Netherlands, 4-7 April, 1993. *Plant Cell Tissue Organ Cult.*, **38**(2-3): 159-165.

Darlington, C.D. 1937. Recent Advances in Cytology (2nd ed.). Blakiston, Philadelphia, USA.

Daughtrey, M. and A.R. Chase, 1992. Marigold. In: *Ball Field Guide to Diseases of Greenhouse Ornamentals*, pp. 145-147. Ball Publishing, Geneva, Illinois, USA.

El-Boushy, A.R. and R. Raterink, 1992. Egg yolk pigmentation. *World Rev. Animal Production*, **27**: 49-62.

Erkek, R., F. Sevgican, F. Kirkpinar and A.M. Talug, 1993. The effects of marigold extract and paprika meal on broiler pigmentation. *Ege Universitesi Ziraat Fakultesi Dergis*, **30**: 9-16.

Eyster, W.H. 1941. Genetic studies in the genus *Tagetes. Proc. VIII Intern. Genet. Cong.*, Edinburgh 1939. Issued as suppl. Vol. of *J. Genet.*, pp. 117-118 (abstr.).

Forsberg, J.L. 1979. Marigold. In: *Diseases of Ornamental Plants* (Special Publication No. 3), pp. 121-122. College of Agriculture, University of Illinois, Urbana, Illinois, USA.

Horst, R.K. 1990. Marigold. In: *Westcott's Plant Disease Handbook* (5th ed.), pp. 722-723. Van Nostrand Reinhold, New York, USA.

Ghosh, P. and P. Pal, 2010. Response of African marigold cv. Siracole to organic and inorganic nitrogen sources. *Indian J. Hort.*, **67**(3): 372-375.

Govindaiah, S.B. Dandin, T. Philip and R.K. Datta, 1993. Effect of marigold (*Tagetes patula*) intercropping against *Meloidogyne incognita* infecting mulberry. *Indian J. Nematol.*, **21**: 96-99.

Gupta, Y.C., S.P.S. Raghava and R.L. Misra, 1999. Inheritance of male sterile apetalous inflorescence in African marigold. *J. Ornam. Hort.* (New Series), **2**: 65-66.

Gupta, Y.C., S.P.S. Raghava and R.L. Misra, 2001. Heterobeltiosis in African marigold (*Tagetes erecta* L.). *Indian J. Genet.*, **61** (1): 65-68.

Jacobs, J.J., R.R. Arroo, E.A.- de Koning, A. J. Klunder, A.F. Croes, G.J. Wullems and E.A. De-Koning, 1995. Isolation and characterization of mutants of thiophene synthesis in *Tagetes erecta. Plant Physiol.*, **107** (3): 807-814.

Jalil, R., T.N. Khoshoo and M. Pal, 1974. Origin, nature and limit of polyploidy in marigolds. *Curr. Sci.*, **43**(24): 777-779.

Janakiram, T. and T.M. Rao, 1992. Genetic improvement of marigold. *Nat. Sem. Comm. Floriculture in India: Present and Potential* (abstr.), p. 35. IIHR, Bangalore.

Kameswari, P. Lalita, A. Girwani, P. Naveen Kumar, Gunjeet Kumar, Tarak nath Saha and Ramesh Kumar, 2011. *Marigold*, 24 p. DFR Extension Bulletin No. 4. Directorate of Floricultural Research (ICAR), Pusa Campus, New Delhi-110 012.

Kanwar, J., S.K. Khandelwal and S.K. Goel, 2011. Effect of plant growth regulators on growth and yield of African marigold (*Tagetes erecta* Linn.). *J. Ornam. Hort.*, **14**(1&2): 104-108.

Kaplan, L.1960. Historical and ethanobotanical aspects of domestication of *Tagetes. Econ. Bot.*, **11**: 200-202.

Kar, A.K. and M. Mandal, 1973. New *Cercospora* spp. from West Bengal. II. *Indian Phytopath.*, **26**: 685-690.

Kennedy, R.R. 1997. Genetic variability for seed characters in marigold. *Madras agric. J.*, **84**(6): 328-330.

Khandelwal, S.K., N.K. Jain, and P. Singh, 2003. Effect of growth regulators and pinching on growth and yield of African marigold (*T. erecta.* L.). *J. Ornam. Hort.*, **6**(3): 271-273.

Khoshoo, T.N. 1979. Cytogenetics in Relation to Plant Evolution and Improvement. In: Progress in Plant Research - Silver Jubilee Publication, NBRI, 2 (eds. Khoshoo, T.N. and P.K.K. Nair), pp. 1-74.. Today and Tomorrow Printers and Publishers, New Delhi.

Kishore, N. and S.P.S. Raghava, 2001. Variability studies in African marigold. *J. Ornam. Hort.* (New Series), **4**(2): 124-125.

Kothari, S.L. and N. Chandra, 1984. *In vitro* propagation in African marigold. *HortSci.*, **19**: 703-705.

Kothari, S.L. and N. Chandra, 1986. Plant regeneration in callus and suspension cultures of *Tagetes erecta* L. (African marigold). *J. Pl. Physiol.*, **122**: 235-241.

Kumar, A. 2002. *In vitro* production of male sterile lines of African marigold (*Tagetes erecta* L.) for F_1 hybrid seed production. Ph.D. Thesis. IARI, New Delhi.

Kumar, A., S.P.S. Raghava, S.K. Singh and R.L. Misra, 2003. Micropropagation of male sterile marigold plants for F_1 hybrid seed production. *J. Ornam. Hort.* (New Series), **6**(1): 1-6.

Kumar, A., S.P.S. Raghava, S.K. Singh, and R.L. Misra, 2004. Comparison of seed derived with micropropagated male sterile plants of *Tagetes erecta* L. for F_1 hybrid seed production. *J. Hort. Sci. & Biotech.*, **79** (2): 260-266.

Medina, A.L., J.N. BeMiller, J. Janik (ed.) and J.E. Simon, 1993. Marigold flower meal as a source of an emulsifying gum. New Crops. *Proc. IInd natl. Symp.: New Crops, Exploration Research and Commercialization, Indianapolis, Indiana*, pp. 383-393. October 6-9, 1991.

Mohamed, M.A.H., P.J.C. Harris and J. Henderson, 1998. An efficient *in vitro* regeneration protocol for*Tagetes minuta. Plant Cell Tissue Organ Cult.*,**55**: 211-215.

Mohamed, M.A.H., P.J.C. Harris and J. Henderson, 2000. *In vitro* selection and characterization of a drought tolerant clone of *Tagetes minuta*. *Plant Sci.*, **159**: 213-222.

Mohan Ram, H.Y. and G. Mehta, 1982. Regenaration of plantlets from cultured morphactin-induced barren capitula of African marigold (*Tagetes erecta* L.). *Plant Sci. Letter*, **26**: 227-232.

Mohanty, A., C.R. Mohanty and K.C. Mahapatra, 2003. Heterosis studies in African marigold. *J. Ornam. Hort.* (New Series), **6**(1): 55-57.

Mondal, N. and S. Chaudhuri, 1976. Flower bud rot of marigold (*Tagetes erecta* L.) caused by *Alternaria dianthi* Stevens & Hall. in West Bengal. *Curr. Sci.*, **45**: 75.

Mukundan, U. and M.A. Hjortso, 1990. Thiophene accumulation in hairy roots of *Tagetes patula* in response to fungal elicitors. *Biotech. Letters*, **12**(8): 609-614.

Nimje, P.M. and V. Agrawal, 2002. Effect of soil solarisation on weed density and growth of chilli, brinjal and marigold. *Indian J. Weed Sci.*, **34**(3/4): 324-326.

Peat, J.R. and R.J. Summerfield, 1977. Environmental and cultural effects on vegetative growth and flowering of selected "bedding" ornamentals. II. Night-break lighting and gibberellic acid applications. *Sci. Hort.*, **7**(1): 81-89.

Perich, M.J., C. Wells, W. Bertsch and K.E. Tredway, 1995. Isolation of the insecticidal components of *Tagetes minuta* (Compositae) against mosquito larvae and adults. *J. American Mosquito Control Assoc.*, **11**: 307-310.

Pizzetti, I. and H. Cocker, 1975. *Tagetes*. In: *Flowers ~A Guide for Your Garden*, pp. 1268-1274. Harry N. Abrams, Inc., New York.

Post, K. 1949. *Tagetes* (Marigold). In: *Florist Crop Production and Marketing*, pp. 820-821. Orange Judd Publishing, New York, USA.

Qi, Y., Y. Ye, and M. Bao, 2011. Establishment of plant regeneration system from anther culture of *Tagetes patula. African J. Biotech.*, **10**(75), pp. 17332-17338.

Rao, C.C. and K.M. Reddy. 2002. Effect of planting date on African marigold. Floriculture Research Trend in India, pp. 191-195.. *Proc. Nat. Symp. on Indian Floriculture in the New Millennium* 2002. ISOH, New Delhi.

Rao, M.K., K.U. Devi and A. Arundhati, 1990. Applications of genic male sterility in plant breeding. *Plant Br.*, **105**: 1-25.

Rao, T.M., S.S. Negi and S.P.S. Raghava,1993. Male sterility in marigold. *Golden Jubilee Symp. Hort. Res. Changing Scenario* (abstr. 33.60), p. 91. Ashoka Hotel, Bangalore.

Reddy, N.T., S. Muthuswamy, I. Irulappen, and M. Abdul Khader, 1989. Heterosis and combining ability for yield and yield components in African marigold (*Tagetes erecta* L.). *South Indian Hort.*, **32**(1): 51-61.

Relichova, J. and I. Petrova, 1988. Combining ability and heterosis in *T. erecta* L. *Sbornik-UVTIZ, Genetica-a-slechteni*, **24** (1): 75-78.

Serrato-Cruz, M.A., M. Hernandez-Rodriguez, Y. Savidan and N.M. Barcenas-Ortega, 2000. DNA content and ploidy level in *Tagetes* spp. using flow cytometry. *Agrocientia*, **34**(6): 729-734.

Sharga, A.N., V.S. Motial and K.K. Basario, 1972-1973. Performance studies of some marigold varieties. *Progressive Hort.*, **4**(3/4): 71-76.

Sharma, D.P., M. Patel. and N. Gupta. 2006. Influence of nitrogen, phosphorus and pinching on vegetative growth and floral attributes in African marigold *(T. erecta* Linn.). *J. Ornam. Hort.,* **9**(1): 25-28.

Shome, S.K. and T.P. Mustafee, 1966. *Alternaria tagetica* sp. nov. causing blight of marigold (*Tagetes* sp.). *Curr. Sci.*, **35**: 370-371.

Singh, B. 1970. Morphological and genetical studies in marigolds (*Tagetes erecta, T. patula* and *T. tenuifolia*). Ph.D. Thesis, Agra University, Agra.

Singh, B.D. 2000. *Plant Breeding: Principles and Methods.* Kalyani Publishers, New Delhi.

Singh, D. and N.L. Sen, 2000. Genetic variability, heritability and genetic advance in marigold. *J. Ornam. Hort.* (New Series), **3**(2): 75-78.

Singh, B. and V. Swarup, 1971. Heterosis and combining ability in African marigold. *Indian J. Genet.*, **31**: 407-415.

Singh, B. and V. Swarup, 1973a. Inheritance of qualitative characters in African marigold. *Indian J. Genet.,* **33**: 373-377.

Singh, B. and V. Swarup, 1973b. Inheritance of flowering, flower weight and flower number in African marigold. *Indian J. Genet.*, **33**: 172-175.

Sohi, H.S. 1992. *Diseases of Ornamental Plants in India.* ICAR, New Delhi.

Sreekala, C. and S.P.S. Raghava. 2003. Exploitation of heterosis for carotenoid content in African marigold (*Tagetes erecta* L.) and its correlation with esterase polymorphism. *Theoretical and Applied Genetics*, **106** (4): 771-776.

Sreekala, C., S.P.S. Raghava, R.L. M.isra and S.R. Voleti, 2002. Assessment of variability for carotenoids and yield components in African marigold. *J. Ornam. Hort.* (New Series), **5**(2): 5-7.

Summerfield, R.J., E.M. Dawson and J.R. Peat, 1977. Environmental and cultural effects on vegetative growth and flowering of selected "bedding" ornamentals. I. Night temperature. *Sci. Hort.*, **7**(1): 67-79.

Towner, J.W. 1956. Cytogenetic studies in *Tagetes*. Ph.D. Thesis. University of California, Los Angeles, USA.

Towner, J.W. 1958. Amphiploidy in the cultivated marigolds (*Tagetes*). *Proc. Xth int. Genet.Congr.*, **2**: 296. USA.

Towner, J.W. 1961. Cytogenic studies on the origin of *Tagetes patula*. 1. Meiosis and morphology of diploid and allotetraploid *T. erecta* × *T. Tenuifolia*. *American J. Bot.*, **48**: 743-751.

Towner, J.W. 1962. Cytogenetics of *T. juliscensis* × *T. erecta*. *American J. Bot.*, **49**: 164.

Varma, L.R. and Arha, 2004. Regulation of flowering in African marigold (*Tagetes erecta* L.) by the application of GA_3, ethrel and MH. *J.Ornam. Hort.*, **7**(3-4):168-170.

Willoughby, R.A., J.B. Solie, R.W. Whitney, N.O. Maness and M.D. Buser, 2000. *ASAE Annual Intern. Meeting*, Milwaukee, Wisconsin, USA, July 9-12, pp. 1-14.

Yuen, G.Y. and M.N. Schroth, 1986. Interactions of *Pseudomonas fluorescens* strain E6 with ornamental plants and its effect on the composition of root-colonizing microflora. *Phytopath.*, **76**(2): 176-180.

29

Thevetia

R.L. Misra, Atul Batra, B.K.Banerji, A.K.Dwivedi and Sanyat Misra

Family: **Apocynaceae**

Common names: Lucky nut/Oleander/Trumpet flower/Yellow oleander (*Thevetia peruviana*, syn. *T. nereifolia*)

Introduction

Thevetia is named so to honour Andre Thevet (1502-1592), a French monk who travelled to Brazil and Guyana. In 1980, it was proposed to transfer all *Thevetia* species [except *Thevetia ahouai* (L.) DC.] to *Cascabela* and the name *Cascabela thevetia* (L.) Lippold was proposed for *Thevetia peruviana* but this was not accepted. This genus is found wild in entire Tropical America including Mexico (central and southern), Belize (Central America) and West Indies (Peru) including Paraguay. In its native habitat, *Thevetia peruviana* is found in evergreen lowland or riparian forest at 50–200 m altitude. It is quite tolerant to drought and moderately tolerant to salt. *Thevetia* is a genus of 8-10 species of evergreen shrubs or trees. *Thevetia peruviana* is most homely traditional flower in India and is being grown by rural gardeners, especially in North india, to attend to various social functions, family ceremonies and for offering to gods and goddesses. In rural areas, there are gardens planted exclusively with this plant. Almost every garden in rural India also has a few plants of this ornamental. Its flowers are offered, especially to lord Shiva, it is considered so auspicious. Its flowers are also used for making garland in combination with other flowers.Though predominantly it bears yellow flowers but now there are various other forms in orangish-yellow, orange, orange-red, yellowish-white or whte and reddish-white colours. It is most suitable for planting on dividers of wide roads and highways, also used for making hedges, screens, as foundation plantings and as borders. In frost prone areas it is used as container plant. In the winter season plants are brought inside a greenhouse and used

Plate-1: Fig. 1a, 2a and 3a Thevetia Trees of Yellow, White and Orange cultivars; Fig 1b, 2b and 3b close up view of flowers of Yellow, White and Orange cultivars; fig. 1c Fruit of Thevetia Yellow cultivar

as a house plant. It is usually cultivated in home gardens for use in local medicines apart from taking its flowers for offering to gods and goddesses and other ornamental values. People in Ghana and Uganda prepare tool handles and building poles through its wood. Seed oil of *Thevetia peruviana* is used for preparing paint having antifungal, antibacterial and anti-termite properties (Kareru *et al.*, 2010). Leaf extract and seed oil have shown strong antifeedant effects on slugs, insects including termites and nematodes, and are also toxic to *Rhizoctonia solani, Bacillus subtilis* and *Staphylococcus aureus*.

Toxicity and Medicinal Uses

Thevetia peruviana is though toxic even then it is used in the tropics as various folk medicines. Bark or leaf decoction is said to be an effective cure against intermittent fever, as an emetic and for loosening the bowels. In Senegal its macerated leaves and bark are used to cure amenorrhea. In Mali its latex is used to soften corns and calluses. In Côte d'Ivoire and Benin, the leaf sap is used as a nasal drop to revive fainted people and to cure cold, while as eye and nasal drops to cure violent headaches. Luo people in Kenya use the crushed leaf extract in water against cold, and seeds as purgative. Katende *et al.* (1995) while describing its cultivation in Uganda and Benin, reported that an infusion prepared through its roots is used to treat snake bites. Leaf decoction is purgative against intestinal worms, against fever and to treat jaundice, and bark & seeds to kill rats in Ghana. In Cameroon and southern Africa the seeds are used as an arrow and ordeal poison, and in India, Sri Lanka and Ghana, people use its seeds for committing suicide and other crimes. In India, its seed oil is applied externally to cure skin infections. Saxena and Jain (1990) reported its kernel oil as a very effective bactericide. Gupta *et al.*

(2011) found it inhibiting spermatogenesis in rats and suggested its use for developing male contraceptive. Ovine polyclonal anti-digitoxin Fab fragment antibody (DigiTAb; Therapeutic Antibodies Inc.) can be used to treat digoxin poisoning but for many countries the cost is prohibitive (Eddlesston *et al*, 2000).

Any part of *Thevetia peruviana* if swallowed to its lethal dose, due to latex being extremely poisonous, especially to vertebrates, the side effects may appear as irritation, pains and redness of mouth and lips, gums, tongue, nausea, cramping, vomiting, depression, irritability, fast breathing, sweating, stomachache, diarrhoea, headache, confusion, visual disturbances, constricted pupils, bradycardia, gastric and cardiotoxic effects and finally death. The leaves are not eaten by cattle, goats and other livestocks though a few bird species are however known to feed on them without any ill effects and these are Asian cuckoo, red-whiskered nightingale, white-brow nightingale, red-vented nightingale, brahminy myna, common myna and common grey hornbill (Krishnan 1952; Neelakantan 1953; Raj and Sanjeeva, 1959, 1963; Raja Singh *et al.*, 1970; Kannan, 1991). Smoke from burning plant material can also be toxic. Plant sap causes allergic skin reactions. Many cases of intentional and accidental poisoning of humans are known (Shannon *et al.*, 1996). Studies on the toxicity of *Thevetia neriifolia* has been done by Gulati *et al.* (2000). Clinically it has been proved that all forms of cardiac insufficiency can be treated with peruvoside, though with caution that is why its use is barred in Western medicines. Peruvoside is also used in anti-herpes treatment. Thevetin 'A' though less effective but is a glycoside identical to ouabain and is used effectively in case of cardiac decompensation, however, the more potent than thevetin 'A', neriifolin and cerberin, is cerberoside (thevetin 'B'). Leaves yield a range of flavanone and flavonol glycosides, and several of them have shown inhibitory effects against HIV-1 reverse transcriptase and HIV-1 integrase inhibitory, antimicrobial, antipyretic, antifungal, haematological and tissue activities. The phytoconstituents isolated from the latex, more virulent ones from seeds, and other plant parts are cardiac glycoside (cardinolides), thevetin 'A' and 'B', peruvoside (cannogenin thevioside), cerberin (2'-O-acetylneriifolin), neriifolin, thevetoxin, cerberin (2'-O-acetylneriifolin), ruvoside (cannogenol-thevioside), the viridoside, perubosidic acid (perusitin) and terpenoids. All the toxins are similar to digoxin which is extracted from *Digitalis purpurea*. The glycosides, especially thevetin, though act as heart stimulant but in its natural form it is extremely poisonous. All the plant parts including seeds and fruits are also poisonous and if its eight seeds are ingested it proves fatal. Antidotes for treatment include atropine and digoxin antibodies and treatment may include oral administration of activated charcoal (Rajapakse 2009; Roberts *et al.*, 2006; Bandara *et al.*, 2010).

Botany

Thevetia is an evergreen, compact, large upright ornamental shrub or tree with beautiful flowers, closely related to *Nerium* and *Plumeria*. The plant attains the height up to 8 m, though normally it is only around 4 metres, with glabrous branchlets. Young developing stem is cane-like and green in colour turning silver-gray with age and a full-grown stem diameter may be somewhere near to 30 cm. Bark is studded with lenticels, colour is gray to gray-brown with a thin outer skin, and shallow fissure forming with maturity. The plant produces willow-like lush green, glossy covered with waxy coating, about 13-15 cm long and evergreen, linear-lanceolate with long acuminate tips and distinct midribs, arranged spirally, simple and entire, almost sessile leaves which are alternate, 1-nerved or lighly feather-veined and very attractive. Inflorescence is a terminal or seemingly axillary cymes, bracts small and linear. Flowers a few per inflorescence with pleasing fragrance, 8 cm long, continue appearing throughout the year with a lean period from December to February, peak being from May to September, long funnel-shaped, bisexual and are white, bright yellow and orange. Sepals ovate and 5-lobed, corolla funnelform, 5-lobed, spirally twisted, overlapping and broadly oblong, alternating with stamens, stamens inserted near apex of corolla tube, anthers almost sessile, ovary superior and consisting of 2 carpels fused in lower half, style long and slender, pistil head consisting of a 5-ridged basal part and a cone-shaped apex. Flowers are visited by hummingbirds, tailor bird and various species of butterflies. Fruit a deep red-black drupe, obtriangular (as a rugby ball) with a transverse and notched ridging and contains normally a large and sometimes 2-4 nut-like seeds which are obovoid, flattened and their size is 2 × 1.5 cm. The seed contains 57-63% oil which after purification consists mainly of oleic acid (60%), palmitic acid (16%), stearic acid (11%), linolenic acid (7%),and linoleic acid (5%). The oil can be detoxified and could serve as vegetable oil for domestic use. Variation in oil composition of *T. peruviana* has been reported by Ibiyemi *et al.* (2002). The mitotic chromosome count 2n = 20 was the first report on the species *Thevetia ahouai* though later on too the same count was reported on *T. peruviana*.

Species and Varieties

Thevetia comprises 8-10 species and is related to *Cerbera*, which also has spirally arranged leaves and drupe-like fruits, but differs in its fruit consisting of 2 free parts. *Thevetia peruviana* (*T. nereifolia*) is a large yellow-flowered shrub native to Central America. It can grow up to 8 metres tall though normally its height is around 4 metres, with a rounded to upright habit. The 8-cm yellow flowers unfurl from the ends of the new growth during summer to autumn. Flowers after pollination, develop into fleshy green fruits which after maturity turn black and fall to the ground where the fleshy part rots away leaving a 3.5 cm hard nut containing two seeds and these nuts are sometimes strung together and used as body ornament. A selection of ornamental *Thevetia peruviana* with white flowers is traded as cv. 'Alba.' Plants attain more than 3 m height and about the same dimension for spread, stem diameter more than 15 cm, foliage lush green, simple with entire margin, surface glabrous and covered by waxy layers which provide shining to the lamina, length 11-14 cm and width is about 0.6-1.5 cm, midrib very prominent though side veins inconspicuous, lamina surface waxy and smooth, flower bisexual with imbricate aestivation, pentamerous, size 2.5 cm across, tube length 5.5 cm, dorsal view of flower white in case of cv. 'Alba,' whereas in others it is variously coloured such as orange, yellow and intermediates, calyx green, and sepals 5-parted. The other species of importance is *T. nitida* native to West Indies where leaves are oblong-lanceolate, acuminate and margins revolute, 20-25 cm long and some 5.0 cm wide, flowers large, and corolla white but throat yellow.

Propagation

Thevetia peruviana can be propagated by seed or cuttings. Seed can be stored for up to 3 months before sowing and the germination rate is about 80% when sown in early spring. There are about 300 seeds/kg. Cuttings may be taken in late summer. Seeds are sown directly in the prepared nursery beds, each at 25 to 30 cm across and when after growing these attain size of some 15-25 cm, these are transplanted in the prepared beds or pits or in the container and watered immediately. Seeds germinate in 4 to 8 weeks and for attaining plant size of 15-25 cm these may require further 2-3 months. After planting at its permanent place, tree guards are put around the plant for protection from damage. Plants from seedlings start blooming the third year but initially its flowers should be removed to have a good vegetative growth. Through cuttings these are usually multiplied by semi-ripe terminal shoots or by stem sections measuring some 15-20 cm putting two nodes buried in the soil for proper anchoring and root development and the rest above for aerial growth. From cuttings these start blooming from second year. *In-vitro* propagation by direct shoot morphogenesis without any intermediate callus phase is also possible. This method may be used for mass propagation of superior plant material.

Cultural Practices

Thevetia peruviana is drought-tolerant and tolerates most soils though fertile well-drained loamy soils with sufficient leaf mould are most suitable. It accepts rather poor and dry soils as well. It does not have a particular preference for soil pH, however, a soil pH of 5.5 to 7.5 is quite suitable for its cultivation. The plants have no problem even in those soils where pH is as high as 8.5. Plant grows well in full sun or in slightly partial shade and in the soils having average moisture. More shading forces the plants becoming unsightly and lanky, *vis-à-vis* without flowers, and such plants dry out soon. For their planting in the beds for screening and hedging a strip of bed at least 60 cm wide is prepared by digging the soil to a depth of some 45 cm and the soil is fortified with coarse sand and wellrotten cow dung manure and then these are planted in two rows, distance between the rows some 30 cm and from one plant to the other some 25 cm in a manner that plants of second line fit in their row in the centre of two plants in the first row. Immediately after planting, the plants are watered thoroughly to settle down the roots properly. For garden planting the pits of 60 × 60 cm size are prepared and filled with the taken out soil from the pit mixed with some 3 kg coarse sand and 5 kg of farmyard manure, and in these pits the seedlings or plants prepared from cuttings are planted and then watered thoroughly. The distance from one pit to the other should be maintained at 2.5 metres across. The plants are shallow-rooted and should be protected from strong winds. Container plants do best in fertile soils with good drainage. Pruning is one of the most important intercultural practices for healthy growth and for generating profuse flowering. Stem tips of young plants are pinched out (tip pinching) in winter to encourage bushy growth, and established plants are pruned lightly after flowering or shortly before the growing season for providing shape and size to look more attractive after flowering or shortly before the growing season to induce profuse flowering. It is very essential to give a proper shape to the plants growing on road dividers of the highways for smooth traffic flow and for easy movement of pedestrians. The plants meant for hedging and screens are also pruned regularly after the flowering is over or at start of the winter to keep them in shape. In frost- prone areas the container plants in winter are brought inside the house as house plants or in a greenhouse for protection against frost and cold. In areas without a seasonal climate, *Thevetia peruviana* flowers and fruits throughout the year, but with a peak in flowering during the rainy season. The ripe fruits remain on the plant for a long time. The plant coppices well. It shades out neighbouring plants and its litter is allelopathic. In early stage of growth, plant requires periodic watering but once it is established it needs less frequent. Container plants are watered moderately and regularly but when soils dry out between watering. The mature plant need not require watering in normal climate but in summer months if it is watered, the growth and flowering will be optimum. Mulching is required at juvenile stage and in newly transplanted plants to prevent the water loss from the ground during hot summer and to protect the plants from hot winds and sunny weathers during May-June in the subtropical areas of India. In nursery, drip irrigation should be used for a better growth of the plants and to take maximum advantage of the limited water supply. Watering in winter is though important to save the plants from cold injury but only sparingly to avoid the soil being completely dry. Once established properly, these may not require further feeding, though container plants are provided with liquid feeding every month with slurry. Flowers are picked early in the morning and kept at 16-18 ^{0}C temperature to prepare them for sending to the market the same day.

Insect-Pests and Diseases

Thevetia plant enjoys a very good health throughout and no major insect or disease is reported in this plant. Sometimes mealy bugs, aphids and scales attack the plant but injury is mostly insignificant. Caterpillars may also chew on the foliage. Leaf spots may sometimes be encountered so in such cases such leaves are immediately removed and burnt. *Thevetia peruviana* is the preferred host of the parasitic weed *Cuscuta reflexa* which can be kept under check through thorough pruning of the plants and burning of the pruned branches.

References

Bandara, V., S.A. Weinstein, J. White and M. Eddleston, 2010. A review of the natural history, toxicology, diagnosis and clinical management of *Nerium oleander* (common oleander) and *Thevetia peruviana* (yellow oleander) poisoning. *Toxicon*, **56**(3): 273-281.

Eddleston, M., S. Rajapakse, S. Rajakanthan, L. Jayalath, Sjöström and W. Santharaj, 2000. Anti-digoxin Fab fragments in cardiotoxicity induced by ingestion of yellow oleander: a randomised controlled trial. *Lancet*, **355**: 967-972.

Gupta, R., J.B. Kachhawa, R.S. Gupta, A.K. Sharma, M.C. Sharma and M.P. Dobhal, 2011. Phytochemical evaluation and antispermatogenic activity of *Thevetia peruviana* methanol extract in male albino rats. *Hum Fertil.* (*Camb.*) *Mar.*, **14**(1): 53-59.

Gulati, A., S.K. Jain and P.S. Srivastava, 2000. Experimental studies on *Thevetia neriifolia* Juss-a review. *Indian J. Chem.* (Sect. B), *Organic including Medicinal*, **39**(11): 808-812.

Ibiyemi, S.A., V.O. Fadipe, O.O. Akinremi and S.S. Bako, 2002. Variation in oil composition of *Thevetia peruviana* Juss. 'Yellow Oleander' fruit seeds. *J. Appl. Sci. environ. Manag.*, **6**(2): 61-65.

Kannan, R. 1991. "Koels feeding on the yellow oleander.". *Blackbuck*, **7** (2): 48.

Kareru, P.G., J.M. Keriko, G.M. Kenji and A.N. Gachanja, 2010. Anti-termite and antimicrobial properties of paint made from *Thevetia peruviana* (Pers.) Schum.oil extract. *African J. Pharm. Pharmacol.*, **4**(2): 087-089.

Katende, A.B., A. Birnie and B. Tengnäs, 1995. Useful trees and shrubs for Uganda: identification, propagation and management for agricultural and pastoral communities. *TechnicalHandbook* 10. Regional Soil Conservation Unit, Nairobi, Kenya, 710 p.

Krishnan, M. 1952. "Koels (*Eudynamis scolopaceus*) eating the poisonous fruit of the yellow oleander. *J. Bombay Nat. Hist. Soc.*, **50** (4): 943–945.

Neelakantan, K. K. 1953. Common grey hornbill (*Tockus birostris*) eating fruits of the yellow oleander (*Thevetia neriifolia*). *J. Bombay Nat. Hist. Soc.*, **51** (3): 738.

Raj, P.J. and Sanjeeva. 1959. Birds eating poisonous fruit of yellow oleander (*Thevetia neriifolia*). *J. Bombay Nat. Hist. Soc.*, **56** (3): 639.

Raj, P.J. and Sanjeeva, 1963. Additions to the list of birds eating the fruit of yellow oleander (*Thevetia neriifolia*). *J. Bombay Nat. Hist. Soc.*, **60** (2): 457–458.

Rajapakse, S. 2009. Management of yellow oleander poisoning. *Clinical Toxicol.*, **47**(3):206-212.

Raja Singh, G. Simon and V. Irene, 1970. Birds and mammals eating the fruits of yellow oleander (*Thevetia peruviana*). *J. Bombay Nat. Hist. Soc.*, **67**(3): 572–573.

Roberts, D.M., E. Southcott, J.M. Potter, M.S.Roberts, M. Eddleston and N.A. Buckley, 2006. Pharmacokinetics of digoxin cross-reacting substances in patients with acute yellow oleander (*Thevetia peruviana*) poisoning, including the effect of activated charcoal. *Therapeutic Drug Monitoring*, **28**(6): 784-792.

Saxena, V.K. and S.K. Jain, 1990. *Thevetia peruviana* kernel oil: a potential bactericidal agent. *Fitoterapia*, **61**(4): 348–349.

Shannon, D., Langford and J. Boor Paul, 1996. Oleander toxicity: an examination of human and animal toxic exposures. *Toxicology*, **109** (1): 1–13.

Williams, J.K. and J.K. Stutzman, 2008. Chromosome number of *Thevetia ahouai* (Apocynaceae: Rauvolfoidae: Plumerieae) with discussion on the generic boundaries of *Thevetia*. *J. Bot. Res. Inst. Texas*, **2**(1): 489–493.

Part II

General Aspects

30

Biotechnological Interventions in Floriculture

Pushpalatha G.

Introduction

Biotechnology is a field of applied biology that uses genetic, bioprocess engineering and various other technologies on living organisms for the production of desirable bio-products. From decades, the farm scientists have bred the ornamentals or food crops by using biotechnological tools, to produce highest yield that meets the growing population. The plants being sedentary organisms get persistently exposed to different environmental stress (biotic and abiotic) conditions. For instance, bacterial blight of geranium can infect the plants easily, spores survive for years and it is devastating. Similarly, abiotic stresses like high temperature, salt and drought stress can negatively affect flower pollination by decreasing the amount of viable pollen grains, amount of nectar produced by flowers and finally becomes unattractive towards pollinators, as a result the seed set is hampered and flowers senesce. Plants that undergo stress either die or get acclimatized slowly by exhibiting epigenetic inheritance. Flowering plants are not exception from these problems. To overcome such adverse effects of environment, many biotechnological tools have been employed. In the last few years, the biotechnological research system has made remarkable effort in floriculture to develop stress tolerant ornamentals, improving their production, or for incorporation of rare colours which is far from being expedited.

Floricultural crops are ideal for yield improvement using biotechnology. Some techniques of tissue culture like embryo/haploid culture, somaclonal variation and somatic embryogenesis in carnation, chrysanthemum, rose and gerbera (Hutchinson *et al.*, 1992) have successfully been implicated. These techniques are still to get expedited for productive outcome. Further, the techniques of genetic engineering, plant molecular biology, genomics, proteomics and metabolomics possess better potential. Some of the achieved biotechnological aspects in floriculture have been cited and discussed in the chapter.

Plant Tissue Culture and its Role in Floriculture

One of the major branches of biotechnology is the plant tissue culture which includes micropropagation to produce the desired plant species. This technology is the most efficient and commercially viable due to the quality production of plants. It results in large number of disease-free, elite plant production, independent of the season. This is a simple and cost-effective *in vitro* technology. The viability of such raised plants can further be assessed using various gene markers.

A. Totipotency Concept

The concept of totipotency explains the process of regeneration and is a developmental plasticity. This was first demonstrated during 1950s by Steward and Reinert. It is a normal undifferentiated cell-like in meristems. Most of the plants get regenerated from a single cell but all plant cells are not totipotent. The totipotency regeneration expresses in mature cells to re-enter the cell cycle and resume cell division, this process is known as dedifferentiation. This leads to direct organized development like in epidermal cells of *Trifolium* hypocotyls (Maheswaran and Williams, 1985), where somatic embryos (direct embryogenesis) or shoots and roots (direct organogenesis) develop. On the other hand, the intervening callus (indirect organogenesis) may develop. The expression of totipotency depends on 'competence' (ability of cells to get induced along a developmental pathway) and 'determination' (cells get irreversibly committed to a particular pathway). For instance, *Convolvulus* exhibits an initial competence (root or shoot formation) and later the cells become determined.

Various factors like explants, growth promoters and nutrient media are responsible for ultimate success in micropropagation. A report by Simmonds and Nelson (1989) proved the superiority of *Begonia hiemalis* for shoot multiplication. In *Rosa hybrida*, 6-fold multiplication was prominent using axillary bud explants . There was 6-fold of multiplication in *Morus* species with axillary bud explants (Chitra and Padmaja, 2002). In *Chrysanthemum morifolium*, the juvenile explants performed better than mature explants (Lu *et al.*, 1990). Somaclonal variations are more common from floral parts of *Chrysanthemum* (Malaure *et al.*, 1991a,b).

B. In *in- vitro* Differentiation of Plants

Differentiation refers to the development of organized structures, usually from undifferentiated tissue. It also develops from previously specialised cells that would not normally give rise to organized multicellular growth (e.g. pollen grains). In plant tissue culture, undifferentiated tissue is referred to as callus (Fig.1a, 1b) although a callus can contain meristematic nodules. Development of organized structures can follow one of three pathways and it is explained with an example of *Gerbera* here:

(a) **Shoot regeneration (Fig.1c, 1d)**
(b) **Root regeneration (Fig. 1f)**
(c) **Somatic embryogenesis in which there is a bipolar structure (Fig. 1e).**

Repeated division of immature post-meiotic pollen grains (microspores) leads to production of haploid plants, and is another type of embryogenesis. The phenomenon was first discovered in 1966 by Guha and Maheshwari when studying meiosis *in vitro* in *Datura innoxia* anthers. Ploidy was determined in callus of *Gerbera* and there 37 (80.4%) haploid regenerants, seven (15.2%) diploid and two (4.3%) mixoploids with both haploid and diploid cells were found (Miyoshi and Asakura, 1996).

I. Plant growth regulators and differentiation - indirect organogenesis

An understanding of the mechanisms underlying regeneration of whole plants or parts of plants and the influence of the growth regulators, i.e. auxin and cytokinin were well demonstrated by Skoog and Miller. In tobacco stem pith cultures, the regeneration of shoots was reported under high ratio of auxin to cytokinin (Fig.1d). In general, auxins like IAA (indoleacetic acid), NAA (α-naphthaleneacetic acid) and IBA (indolebutyric acid) will stimulate regeneration of roots, and cytokinins such as BAP (6-benzylaminopurine) and kinetin will promote regeneration of shoots or embryos.

The two groups of plant growth regulators play an important role in unlocking totipotent expressions. In response to injury or wound, dedifferentiation and callus formation occur naturally. The process triggers the plant regeneration from somatic cells by involvement of auxin

Fig. 1: *Gerbera* **on MS medium supplemented with, (a) 1mg/l BAP and 2mg/l 2,4-D, (b) 1 mg/l BAP and 2 mg/l NAA, (c) 2 mg/l zeatin and 0.5 mg/l IBA as well as (d) 2 mg/l BAP and 0.5 mg/l NAA, and both showed regeneration of shoots from petiole explants, (e) 0.1mg/l BAP and 2 mg/l NAA as well as (f) 2 mg/l NAA, both showed development of roots from leaf explants.**

Source: Hasbullah *et al.* (2008)

Fig. 2: *Gerbera* ~ (a) plantlets to be transferred to soil, (b) soil transferred plants, (c) soil transferred plantlets after eight weeks, and (d) soil acclimatized plants after six months.

Source: Hasbullah *et al.* (2008)

and cytokinins (Potrykus, 1989). However, additional growth regulators are required for callus growth under *in vitro* conditions. An attempt to obtain long-term callus cultures was far from being elucidated before IAA was chemically characterized in 1934. Auxin is essential for dedifferentiation and 2,4-dichlorophenoxyacetic acid (2,4-D) is used to promote callus, which further enhanced by cytokinin. In tissues with a high endogenous level of auxin, culture of explants on a medium containing cytokinin as the only growth regulator may lead to development of shoots with very little callus. Without dedifferentiation, it is not possible to move to the next stage of totipotent expression, i.e. plant regeneration. Although auxin stimulates initial cell division, continued presence of auxin can inhibit organized outgrowth of the cells. The removal of auxin occurs when it degrades in the medium either by the tissue or *via* chemical reactions like photo-oxidation. Conversely, cytokinins promote outgrowth of shoots but are normally kept at very low concentration during root regeneration.

II. Role of plant growth regulators in direct organogenesis

Direct organogenesis bypasses the callus phase, by forming somatic embryos. Most evidences suggest that the direct embryogenesis originates from cells which are embryogenic in nature. Such pre-embryogenic cells express only in favourable conditions like wounding or application of exogenous growth regulators, for cell division and embryogenesis. They tend to be more responsive than indirect organogenesis. In *Trifolium repens* hypocotyl epidermis, the cytokinin BAP promotes reorientation of cell division that leads to initiation of a promeristemoid. The subepidermal cells develop into shoots in *Abies amabilis*. The haploid embryos were developed from *Brassica napus* through anther culture. Cytokinin suppresses secondary embryoid formation and promotes normal leafy shoots which suggests that it has a role in shoot development and embryogenesis. Likewise, Tran-Thanh van *et al.* (1974) reported the effect of growth regulators on the direction of differentiation in *Nicotiana tabacum*.

III. Somatic Embryogenesis

The capacity of flowering plants to produce embryos is not restricted to the development of the fertilised egg. Embryos can be produced under *in vitro* conditions in cultured plant tissues. This phenomenon was first observed in 1958 by Steward and his team. Somatic embryogenesis is obtained either directly from the epidermis of immature or mature zygotic embryos or indirectly on callus or cell suspensions from the cultured explants. Frequently, the primary embryo is accompanied by direct secondary embryogenesis which becomes difficult in the production of true bipolar somatic embryos (Sangwan *et al.*, 1991). Although, micropropagation is commercialized worldwide, it requires a great deal of labour while, somatic embryogenesis could be employed for large scale propagation with relatively less labour and time. In addition, the somatic embryos are encapsulated to produce artificial seed (Redenbaugh *et al.*, 1986; McKersie and Bowley, 1993). Successful production of *Eucalyptus* species through somatic embryogenesis has been reported (McComb *et al.*, 1989; Watt *et al.*, 1991) including *Lysinema ciliatum* a plant from the family Epacridaceae (Senaratna *et al.*, 2000) and *Swainsona formosa* (Sudhersan and AboEl-Nil, 2002). Formation of embryos followed by their encapsulation to produce artificial seed has enormous commercial potential especially in plantation crops such as *Eucalyptus* and pine. First somatic embryo development and regeneration of the cut rose cultivars 'Domingo' and 'Vickey Brown' was successful from callus of leaf explants on MS media with low concentrations of kinetin and NAA (Wit *et al.*, 1990). Cheong and Pooler (2004) reported various factors affecting the somatic embryogenesis from root explants of *Prunus incisa* cv. February Pink. Root explants in the study were induced on MS medium containing 1.0 μ*M* IBA which produced more somatic embryos than explants induced on medium without IBA. Approximately 50% of the roots induced on medium containing 1.0 μ*M* IBA produced somatic embryos on medium containing 10 μ*M* 2,4-D and 3% sucrose. Efficient *in vitro* flowering through somatic embryogenesis and fruiting was achieved in *Saposhnikovia divaricata* (Qiao *et al.*, 2009). Here somatic embryogenesis was well illustrated (Fig. 3) in *Pelargonium zonale* and *P. peltatum* (Garcia-Sogo *et al.*, 2012, shown and referred in the Fig. 3 b).

Fig. 3: ***In vitro*** **regeneration of transgenic plants via somatic embryogenesis in** ***Pelargonium peltatum*** **and** ***via*** **organogenesis in** ***P. zonale*****. (a)** ***P. peltatum*** **WT plants. (b)** ***P. peltatum*** **inflorescence. (c)** ***P. zonale*** **WT plants. (d)** ***P. zonale*** **inflorescence. (e-f). Production of somatic embryos in a callus of** ***P. peltatum*** **cultivated in selective morphogenesis induction medium (MIM). (g).** ***P. peltatum*** **developing embryos in selective elongation medium (EM). (h).** ***P. peltatum*** **transgenic plantlets in rooting medium (RM). (i-j). Adventitious buds in a callus of** ***P. zonale*** **in selective MIM. (k).** ***P. zonale*** **shoot elongation in selective EM. (l).** ***P. zonale*** **rooted plantlet in RM. (m-n). Detection of transformation events in both** ***Pelargonium*** **spp., GFP green fluorescence is clearly visible in the initial whitish callus with a disorganized growth. (o-p). Chlorophyll shows strong red autofluorescence that could mask the green fluorescence of transformed cells, it becomes increasingly difficult to identify in the subsequent organogenic callus and in the adventitious buds. (q-r). Green fluorescence was observed in regenerated shoots but is masked by the chlorophyll in the young leaves. (s-t). Green fluorescence can be observed in the periphery of young leaves where chlorophyll does not accumulate. (u-v). Green fluorescence is especially evident in the roots, where the chlorophyll is absent. In general, within the same organ, GFP detection varied in different tissues or cell types depending on their chlorophyll content.**

Source: Garcia-Sogo *et al.BMC Plant Biology* 2012 12:156 doi: 10.1186/1471-2229-12-156

IV. *In Vitro* Flowering

The transition from the vegetative to reproductive phase in plants is an important developmental process. It is of practical importance in plant breeding. From the results of previous research, it is evident that the majority of plants use environmental cues to regulate flowering. Environmental variables with regular seasonal patterns such as temperature, photoperiod and irradiance are the key signals in floral induction. These factors are perceived by different plant parts. Strong and diverse interactions between the environmental variables are required for floral induction to occur in many species.

Figure 4: (a) http://users.ugent.be/~pdebergh/tot/tot2_d01.htm, (b) http://www.gmo-safety.eu/database/898.sterile-poplars-prevent-outcrossing-forest-ecosystems.html and (c) http://peakplants.blogspot.in/2012/06/in-vitro-flowering.html

Classical physiological, genetic and grafting experiments, though invaluable in deciphering various aspects of flowering, have failed to unravel the true nature of the flowering stimulus or the mechanisms by which various environmental cues induce flowering (Fig. 4). Novel approaches involving *in vitro* flowering and molecular techniques offer unique opportunities to investigate flowering process from new perspectives especially in species which are difficult to flower or produce flowers only once in several years (Taji *et al.*, 2002; Lakshmanan and Taji, 2003).

V. Embryo Culture and Rescue

Embryo culture is one of the oldest applications of *in vitro* techniques. The main use of embryo culture has been for the rescue of hybrid embryos where incompatibility often leads to embryo abortion. These embryos can often be excised before they abort and cultured to maturity *in vitro*. It can also overcome seed dormancy. Drew *et al.* (1995) reported the application of embryo rescue on crosses between papaya and relative species. The inter-specific hybrids were successfully obtained from different cross combinations between 12 cultivars (8 diploids and 4 tetraploids) of *Begonia semperflorens* and *B. pearcei* by *in vitro* culture of mature and immature embryos on MS medium containing 30 g l^{-1} sucrose, 0.1 mg l^{-1} α-naphthylacetic acid, 0.1 mg l^{-1} 6-benzyladenine, 2.5 g l^{-1} gellan gum, and 10 mg l^{-1} gibberellic acid. Some of the cross combinations resulted in higher plantlet yields up to 5.7/ovary in immature seeds of 12-16 days after pollination, whose hybridity was confirmed through flow cytometric analysis and random amplified polymorphic DNA analysis (Chen and Mii, 2012). Embryo rescue is being promoted in tissue culture to produce virus resistant ornamental varieties. Inter-specific sexual hybridizations combined with embryo rescue techniques have been successful in the development of virus and fungal resistant plants apart from getting novelties in terms of yield and flower colours.

VI. *In Vitro* Pollination and Fertilization

In vitro pollination and fertilization is a technique in which male and female gametes are isolated and introduced to each other under conditions suitable for zygote development. It involves pollen tube penetration of the embryo sac *via* manipulation of maternal tissue, by methods other than the normal *in situ* process. Embryo rescue is most applicable where *in situ* pollination and fertilization are possible. However, in some cases where incompatibility prevents *in situ* pollination or fertilization it may be possible to carry out *in vitro* fertilization. The point of incompatibility between genotypes may be anywhere along the developmental pathway from formation of compatible pollen to embryo growth. A common point of incompatibility is at the stigmatic surface. The incompatibility reaction between the stigma and the pollen grain either prevents pollen germination or it inhibits pollen tube growth down the style to the embryo. In these cases it may be possible to culture the female floral organs (ovary and stigma) by removing the stigmatic surface and then applying the pollen to the remaining surface (Williams and Taji, 1987). In extreme cases it is possible to inject the male nuclear material directly into the egg cell. This may lead to normal development of the embryo and seed within the ovarian tissues or embryo rescue may be necessary.

VII. Micrografting

There is also scope for blending the traditional horticultural practice of grafting with micropropagation. Grafting enables the combination of selected stocks or scions which are not successful on their own roots as well as providing various stress and pest resistance or growth controlling properties. Micrografting may involve either scions or rootstocks taken from *in vitro* stock plant banks utilizing the phenomenon of rejuvenation that occurs *in vitro* to enable adventitious rooting of mature rootstock selections. Micrografting may be carried out aseptically using *in vitro* rootstocks with either *in vitro* or disinfested *ex vitro* scion material, or on micropropagated rootstocks after transfer to the nursery. Kawaguchi and Taji (2002) was successful in micrografting *Swainsona formosa* to a number of rootstocks improving the plant's resistance to a number of soil-borne diseases.

VIII. Protoplast Culture and Fusion

A protoplast is a plant cell from which the cell wall has been removed. After the removal of the rigid cell wall, the cell is surrounded by a thin plasmalemma membrane. Many uses of protoplasts are being explored. The interest in protoplast fusion technique is related to the prospect that wider crosses than are possible by sexual means due to incompatibility, may be achieved with protoplast fusion. The fusion of plant protoplasts is not a particularly new phenomenon. Küster (1909) described the process of random fusion in mechanically isolated protoplasts. However, in 1960 the work of Cocking, and the enzymatic removal of cell wall have made protoplast isolation and fusion a practical reality. Takebe *et al.* (1971) were the first to successfully obtain divisions and regeneration of whole plants in tobacco mesophyll protoplasts. Carlson *et al.* (1972) obtained the first somatic hybrid in *Nicotiana tabacum*. Melchers *et al.* (1978) developed the fusion of protoplasts of potato and tomato and obtained the so called 'pomatoes' and 'topatoes' which were intergeneric hybrids. Successes with somatic hybridization have mainly involved the production of useful and fertile hybrid plants as a result of simple additive combination of the complete genomes of two unrelated species. However, production of cybrids which contain the nuclear and cytoplasmic genome of one parent and only the cytoplasmic genome of the second parent, and their potential use has been described by Gleba and Shlumukov (1990). Use of protoplast technology is in its infancy with Australian native plants. Many laboratories are attempting the use of this technology for direct gene transfer. Teulieres and Boudet (1991) reported some preliminary attempts to isolate protoplasts from different species of *Eucalyptus*.

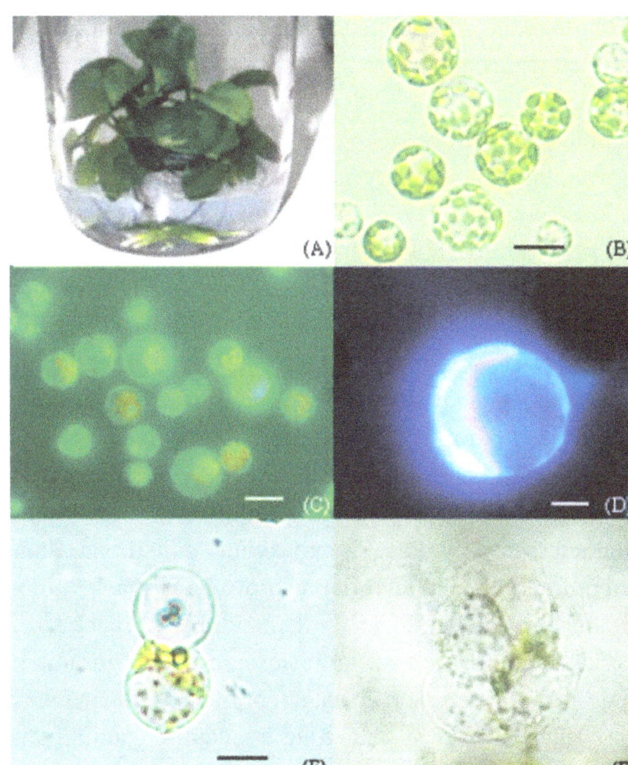

Fig. 5: Isolation, culture and cell division of aquatic flowering plant, *Anubias nana* Engler protoplasts. Six-week-old plantlets suitable for the isolation of leaf plantlets (A), protoplasts after purification with 18% sucrose solution (B), vigorous protoplasts fluoresce a yellow-green colour when stained with FDA (C), protoplast culture in agarose bead, cell wall regenerated within 24 h, detected by calcofluor white staining (D), first cell division after 7-9 days of culture (E), small cell colonies after culturing for 30 days (F).

Source: Pongchawee et al., International journal of of Botany, 2(2), 193 - 200, 2006

Many attempts have been made to transfer cold tolerance from *Lavatera thuringiaca* into *Hibiscus rosa-sinensis* (Vazquez-Thello *et al.*, 1996) but plants were not regenerated in this instance. Subsequently, Horita *et al.* (2003) generated fertile somatic hybrid plants between the Oriental hybrid lily cvs 'Acapulco' and 'Shirotae' and *L. formolongi* cv. 'Hikucho'. Electrofusion-treated protoplasts divided only in the presence of nurse cells. Molecular analysis using CAPS markers and flow cytometry confirmed the somatic hybrid nature of the regenerated plants.

IX. Somaclonal Variation

Somatic hybridization and transformation technologies provide reliable approaches for combining inter-specific and intergeneric traits and targeted modification of genomes. Superimposed on these procedures may occur somaclonal variation, or in the case of protoplasts, so-called protoclonal variation (Davey *et al.*, 2005). Somaclonal/protoclonal variation may involve expression and release of naturally occurring genetic variation as a result of culture and may be regarded as a simple form of genetic manipulation (Kantharajah and Golegaonkar, 2004). Interestingly, this should not be seen in any adverse way by those with concerns relating to genetic manipulation technologies. Thus Mizuhiro *et al.* (2001) reported that protoplast-derived plants of *Primula malacoides* showed alterations in morphological characteristics compared to plants from *in vitro*-germinated seed. Specifically, such changes involved significantly reduced petioles and variation in floral morphology, particularly pedicel and peduncle lengths, inflorescence number, and corolla size. In general, protoplast-derived plants formed compact inflorescences with multiple small flowers. Red pigmentation at the centre of flowers was paler than in controls, while the periphery of some petals had a green pigmentation. Similarly, Jelenic *et al.* (2001) reported that potato plants regenerated from protoplast-derived callus were all morphological and cytological chimeras, with alterations in general appearance, leaf morphology, and tuber characteristics. Wherever the maximum genetic variation is required, somaclonal variation provides a useful adjunct to the more technologically demanding approaches of somatic hybridization and transformation. The attraction of protoclonal variation is that it requires no knowledge of the genetic basis of specific traits, it negates recombinant DNA techniques, it does not require mutagenic agents, specialized apparatus, or containment measures, and it can be exploited through routine culture procedures.

X. Plant Genetic Engineering and Techniques of Plant Transformation

Few decades ago, several research groups attempted to introduce foreign DNA into plant genomes. For example, seeds of a thiamine-requiring mutant of *Arabidopsis thaliana* were incubated in a solution of bacterial DNA. Plants were recovered which could grow without thiamine and it was thought that the bacterial DNA had somehow entered the plant cell and enabled the mutation to be corrected (Kleinhofs and Behki, 1977). At this time, there were no reliable methods for detecting foreign genes within plant genomes so it was difficult to prove whether the bacterial DNA had been stably incorporated. Analysis of the 'corrected' lines indicated the likely presence of bacterial DNA but contamination with wild-type seed could not be ruled out. Direct transformation of intact plants with naked DNA was largely forgotten for several years. However, the approach of these early workers does not seem so outlandish. We now have a greater understanding of what is required to express foreign genes in plants, coupled with several procedures to deliver foreign DNA into plant cells and methods for proving that foreign DNA has been integrated into the host genome. Gene transfer experiments are a routine tool for answering questions relating to most aspects of plant development. In addition, improvement of most major crop species by addition or deletion of specific genes is the subject of worldwide research.

Transferring DNA into plants is the ability to recognize individuals, whether at the cellular or whole-plant level, which have incorporated the foreign DNA (i.e. genetically transformed) against the many individuals. Usually only a small percentage of cells or plants are successfully transformed in any one experiment. In the early 1980s, a new approach was put forward in microbial molecular genetics to enable selection and recovery of transgenic plants. This was the introduction of antibiotic resistance genes which allow transformed cells to grow in the presence of levels of antibiotic which are toxic to wild-type parental cells. The first marker used in this way was a gene encoding neomycin phosphotransferase (NPTII) which when fused to a promoter allowing expression in plant cells, conferred resistance to kanamycin. Alternatives include other antibiotic resistance genes such as hygromycin phosphotransferase, and genes encoding resistance to herbicides such as glyphosate or phosphinothricin. Of course, selectable markers are only one aspect of creating stable transgenic plants. We also require systems for efficient DNA delivery and integration into the plant genome. In addition, there are several approaches to insert foreign genes into plants. These include direct DNA uptake by protoplasts [electroporation, polyethylene glycol (PEG) - induced uptake, micro-injection or sonication], direct DNA delivery by biolistic methods or indirectly *via* vectors (viral or bacterial). The system which has proved most successful is based on the Ti plasmid of *Agrobacterium tumefaciens.* Further discussion about both *Agrobacterium*-mediated transformation and direct gene transfer methods will give some comprehensive ideas.

(i) *Agrobacterium*-mediated gene transfer

Plant transformation is the process of stable introduction of transgene(s) into the genome of the plant. Various methods have been developed to achieve this and many plant species have been transformed successfully. Some transformation methods are based on utilizing *Agrobacterium,* a pathogen of dicotyledonous plants that transfers genes into the plant genome. *Agrobacterium tumefaciens* is a soil-borne, gram-negative bacterium, which has been of great interest both of biotechnological and scientific ramifications. It is the causative agent of crown-gall disease, which is an economically important disease of many plants such as grapes, walnuts, roses and apples. The ability of *Agrobacterium* to cause tumorous tissue growth - crown galls depend on the ability to transfer bacterial genes into the plant genome.

(ii) Crown-gall disease and Ti plasmid

Agrobacterium tumefaciens, the causative agent of crown gall disease (Fig. 5, 6) is found in the rhizosphere, where it normally survives on nutrients released from plant roots. However, if a plant is wounded or damaged, the bacterium can infect the plant at the wound site and causes disease symptoms. The crown gall formation depends on the presence of a plasmid in *A. tumefaciens* known as 'transfer inducing (Ti) plasmid, which possesses a 'transfer DNA (T-DNA)'. T-DNA is transferred from the bacterium into the plant cell, where it becomes integrated into the genome of the host plant. It carries genes that encode proteins involved in both hormone biosynthesis of auxin and cytokinin as well as novel plant metabolites like opines and agropines. Opines and agropines are not normally part of plant metabolism and are very stable chemicals, which provide a carbon and energy source that only *A. tumefaciens* can utilize. Thus, *A. tumefaciens* has therefore developed the ability to genetically transform plant cells to usurp the biosynthetic machinery and produce nutrients.

Fig. 6: (A) A gall caused by *Agrobacterium tumefaciens*, and (B) Wild type (left) and genetically engineered (right) walnut microshoots.

(A) Courtesy of [http://biologi.uio.no/plfys/haa/gen/gmo.htm Halvor Aarnes A gall caused by Agrobacterium tumefaciens (B) http://ucanr.edu/blogs/blogcore/postdetail.cfm?postnum=7343 Wild type (left) and genetically engineered (right) walnut microshoots.

Many of the genes required for DNA transfer reside on a section of the Ti or Ri plasmid known as the virulence (*Vir*) region. *Vir* genes are prokaryotic in structure and so function in bacterial cells, unlike those in T-DNA which have eukaryotic features, as noted above. Wounded plant cells produce phenolic compounds such as acetosyringone (Fig. 6) that gets detected by the protein product of *Vir* genes. The *Vir*A is responsible for the expression of other *Vir* genes. The *Vir* proteins result in T-DNA transfer into the plant cell nucleus, probably in a manner akin to conjugative plasmid transfer between bacteria (Zupan and Zambryski, 1995). Classical experiments in the 1920s, by A.J. Riker, a pioneer in the field, showed that tumour formation occurred most effectively if plant tissues were incubated at 22 ^{0}C but above 28 ^{0}C there was a complete absence of tumour development. It was later reported a probable relationship of T-DNA transfer which has a temperature optimum around 20 ^{0}C (Fullner and Nester,

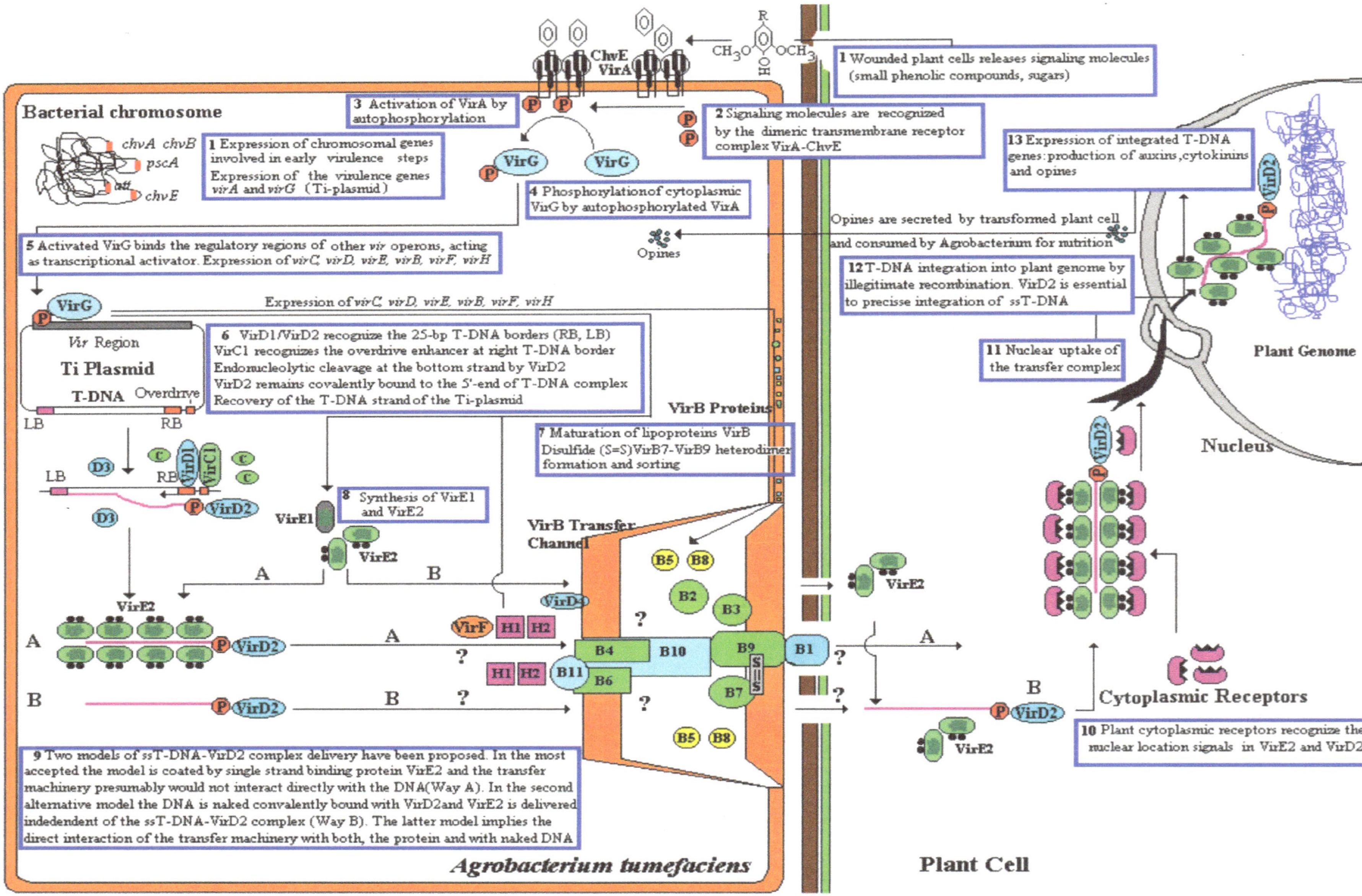

Fig. 7: *Agrobacterium*-mediated gene transfer. The T-DNA transfer is represented according to updated knowledge on this process. The most important events of gene transfer are briefly mentioned in chronological order in the figure.

Source: Image from Electronic Journal of Biotechnology

1996). Thereafter, the method of *Agrobacterium*-mediated transformation was used in improvement of many ornamental plants for their biotic/abiotic stress tolerance and high rate of production. Such instances are explained below.

In Elatior Begonia (*Begonia × hiemalis*), young leaf disc explants were infected with *Agrobacterium tumefaciens* strains AGL0 and LBA4404. Each strain was possessed with a binary vector plasmid, pIG121Hm and β-glucuronidase (GUS) gene with an intron as a reporter gene. Along with these, both neomycin phosphotransferase II and the hygromycin phosphotransferase genes were used as selection markers. Explants were inoculated on modified MS medium supplemented with 1.0 mg/l BA, 0.5 mg/l IAA, 300 mg/l ticarcillin, and either 100 mg/l kanamycin and 5 mg/l hygromycin, or 300 mg/l kanamycin for selection and regeneration. Out of 500 explants infected with AGL0, 16 plantlets were regenerated. Out of 628 explants infected with LBA4404, two plantlets were regenerated four months after inoculation. Around 10 *in vitro* transgenic plants were obtained from AGL0 infected explants, which was confirmed by southern blot analysis of the GUS gene and histochemical assays of GUS activity in plant tissues (Kishimoto *et al.*, 2002). Likewise, a report on chrysanthemum was put forward recently by Shweta *et al.* (2013) to develop a simple and efficient protocol for *Agrobacterium*-mediated transformation, leading to 2.2% transformation frequency rate.

The *pSAG12::ipt* transgenic *Pelargonium* plants showed delayed leaf senescence, increased branching and reduced inter-nodal length as compared to control plants. Leaves and flowers of the *pSAG12::ipt* plants were reduced in size and displayed a more intense colouration. In the transgenic lines carrying the *PsEND1::barnase* construct no pollen grains were observed in the modified anther structures, which developed instead of normal anthers. The locules of sterile anthers collapsed 3–4 days prior to floral anthesis and in most cases, the undeveloped anther tissues underwent necrosis. Thus Garcia *et al.* (2012) reported that the chimaeric construct *pSAG12::ipt* can be useful in *Pelargonium* spp. to delay the senescence process and to modify plant architecture. In addition, the use of engineered male sterile plants would be especially useful to produce environmental-friendly transgenic plants carrying new traits by preventing gene flow between the genetically modified ornamentals and related plant species. These characteristics could be of interest from commercial point of view, both for *Pelargonium* producers and consumers. Several such studies were reported in ornamental flowers. Few more gene transfer methods used and their successful stories are discussed here.

Alstroemeria, like other monocot ornamentals, has been generally recalcitrant to genetic transformation techniques that are routinely applied in dicotyledonous plants. FEC induced from stem tissue of seedling plants was transformed with particle bombardment by pAHC18 that contained the luciferase gene as a reporter gene (Lin *et al.*, 2000). Plants were obtained from 10 independent lines. After one year of maintenance, however, only a few plants were still luciferase-positive. Furthermore, FEC was also used to transform using *Agrobacterium tumefaciens* (van Schaik, 1998), however, no transgenic plants were produced. It was concluded that FEC might be an alternative source for genetic modification and an ideal explant without severe somaclonal variations provided young FEC was used for transformation.

Furthermore, in many monocotyledonous ornamentals, particle bombardment and *Agrobacterium*-mediated transformation have been used for the production of transgenic plants with improved agricultural traits. Particle bombardment has been applied in *Alstroemeria* (Lin *et al.*, 2000), *Gladiolus* (Kamo *et al.*, 1995, 2000, 2005), *Dendrobium* (Kuehnle and Sugii, 1992), lily (Watad *et al.*, 1998), and tulip (Wilmink *et al.*, 1995). *Agrobacterium*-mediated transformation has been applied in *Alstroemeria* (Akutsu *et al.*, 2004a,b), *Anthurium* (Chen and Kuehele, 1996), *Cymbidium* (Yang *et al.*, 1999), *Iris* (Jeknic *et al.*, 1999) and *Phalaenopsis* (Chai *et al.*, 2002; Belarmino and Mii, 2000). However, in spite of several successful reports, *Agrobacterium*-mediated transformation is still cumbersome for quite a number of monocot ornamentals. In the past decades, most of the reports on the transformation of monocot ornamentals used the GUS gene as a reporter gene because of its accuracy, fast and convenient characteristics.

An intergeneric hybrid between *Dendranthema morifolium* (chrysanthemum) var. 'Zhongshanjingui' and *Artemisia vulgaris* (mugwort) 'Variegata' was attempted (Deng *et al.*, 2010). The hybrids showed much higher resistance to chrysanthemum aphid (*Macrosiphoniella sanbourni*) than maternal chrysanthemum by inoculation test. The leaves of the hybrid developed a higher density of trichomes and secretory glands compared to the maternal chrysanthemum. GC-MS analysis revealed that approximately 51% of the essential oil in the hybrid leaves were monoterpenoids and sesquiterpenoids, while the proportion in the chrysanthemum was approximately 37%, and in the mugwort was approximately 90%. It is inferred that higher aphid resistance in the hybrid mainly owed to the leaf micro-morphology and bioactive essential oil content.

Junping *et al.* (2011) reported that the three chrysanthemum cultivars 'Keiun', 'Han6' and 'Jinba' vary markedly in their resistance. Of the three cultivars, the most resistant ('Keiun') produced the longest, highest and densest trichomes, the largest and fullest gland cells, and the most wax on the lower leaf epidermis. Superoxide dismutase (EC 1.15.1.1), peroxidase (EC 1.11.1.7), ascorbate peroxidase (EC 1.11.1.11), polyphenol oxidase activity (EC 1.14.18.1) and phenylalanine ammonia lyase (EC 4.3.1.5) were enhanced by aphid herbivory. In the two more resistant cultivars ('Keiun' and 'Han6'), the activity of superoxide dismutase and ascorbate peroxidase enzymes rapidly increased following infestation, and their levels remained high from 72 to 158 hours after inoculation. It was also reported that the two antioxidant enzymes contribute to aphid resistance of chrysanthemum cultivars. All three cultivars showed quick responses to aphid infestation by increasing polyphenol oxidase and phenylalanine ammonia lyase activities during the early period after inoculation. Activities of these two defense enzymes were higher in the two resistant cultivars 72 h after inoculation, suggesting involvement of these two enzymes in aphid resistance.

The genes (ETR1, EIN2, EIL1 and EIL2) specific for ethylene perception and signal transduction were used to produce transgenic *Petunia*. The ethylene hormone plays an important role in various physiological processes like floral senescence, adventitious root formation, seed germination and scent production (Underwood *et al.*, 2005).

Genetically modified ornamental plants have received much attention because of their worldwide popularity. The most common targets are species such as carnation and rose, and the most valuable traits are novel flower colours and increased vase life for cut flowers. The blue-mauve carnation flowers that are commercially known as 'Moondust' and 'Moonshadow' were generated by an Australian biotechnology company 'Florigene'. Such colours were never possible before in these species and are due to insertion of novel genes for part of the flavonoid biosynthesis pathway. The other slow-senescing 'long life' carnation was reported with a gene down-regulating ethylene biosynthesis (Brian *et al.*, 2010).

XI. Pathogen-Free Plants, and Germplasm Storage

The meristem culture is intensively used worldwide to produce pathogen-free plants. Such approach is required specifically for virus-free plants, as bactericidal and fungicidal agents can be used successfully to eradicate bacteria and fungi, respectively (Bhojwani and Razdan, 1983). Thus, meristem-tip culture is often coupled with thermotherapy and chemotherapy for virus eradication (Kartha, 1981).

Traditionally, pathogen-free seeds, somatic cells, gametic cells and shoot apices can be stored to maintain germplasm to regenerate the whole plant when necessary (Kartha, 1981; Bhojwani and Razdan, 1983). Many *in vitro* approaches have been developed but, cryopreservation (Withers, 1985) is the most widely used method. In this method, cell suspensions, shoot apices, asexual embryos and young plantlets, after a treatment with cryoprotectant, they are frozen and stored at a temperature of liquid nitrogen (-196 ^{0}C).

XII. Genomics and Proteomics in Floriculture

Whole transcriptome analysis is of growing importance in understanding the expression of genetic variants. Analysis of genome-wise differential RNA expression provides researchers with greater insights into its biological pathways and molecular mechanisms that regulate cell fate, development and disease progression. Transcriptomic analysis has been reported in several flowering plants and identified many differentially expressed genes. *Amaranthus hypochondriacus* is a C4 plant, known for its ability to tolerate stressful conditions and produce highly nutritious seeds. A report suggested a transcriptomic analysis in this stress tolerant crop (Delano-Frier *et al.*, 2011). A large scale transcriptomic analysis was performed and generated stem and biotic stress-responsive gene expression profiles. A total of 2,700,168 raw reads were obtained from six 454 pyrosequencing runs, which were assembled into 21,207 high quality sequences (20,408 isotigs + 799 contigs). The average sequence length was 1,064 bp and 930 bp for isotigs and contigs, respectively. Only 5,113 singletons were recovered after quality control. Contigs/isotigs were further incorporated into 15,667 isogroups. All unique sequences were queried against the nr, TAIR, UniRef100, UniRef50 and Amaranthaceae EST databases for annotation. Functional GO annotation was performed with all contigs/isotigs that produced significant hits with the TAIR database. Only 8,260 sequences were found to be homologous when the transcriptomes of *Amaranthus tuberculatus* and *A. hypochondriacus* were compared, most of which were associated with basic house-keeping processes. Digital expression analysis identified 1,971 differentially expressed genes in response to at least one of four stress treatments tested. These included several multiple-stress inducible genes that could represent potential candidates for use in the engineering of stress-resistant plants (Delano-Frier *et al.*, 2011), a trait that has potential biotechnological applications in agriculture. Likewise, a database set of an expressed sequence tag (EST) for carnation cultivar was generated by high-throughput sequencing using 454 pyrosequencing technology. Transcripts involved in various biological pathway of carnation flower, *viz.* chlorophyll, carotenoid metabolism, anthocyanin and ethylene were identified (Tanase *et al.*, 2012), which provides an EST resource for understanding the genetic diversity of carnation. A recent study on microarray-based transcriptomic analysis in the stem of *Petunia hybrida* (line W115) pointed to a possible role of trehalose metabolism and SnRK1 (sucrose non-fermenting 1-related protein kinase) in sugar sensing during this early step of AR formation (Ahkami *et al.*, 2014). The transcripts involved in peroxisomal β-oxidation were up-regulated during different phases of adventitious root formation. In addition to metabolic pathways, the analysis revealed the activation of cell division at the two later phases and in particular the induction of G1-specific genes in the maintenance phase.

To gain a better understanding of the cold signaling pathway and the molecular metabolic reactions involved in the cold response, Wang *et al.* (2014) have reported a genome-wide transcriptional analysis using RNA-sequencing in *Lilium lancifolium*, a very important cold-resistant lily for cold resistance breeding. Approximately 104,703 million clean 90-bp paired-end reads were obtained from three libraries (CK 0 h, cold-treated 2 h and 16 h at 4 ^{0}C), 18,736 unigenes showed similarity to known proteins in the Swiss-Prot protein database and 15,898, 13,705 and 1849 unigenes aligned to existing sequences in the KEGG and COG databases (comprising 25 COG categories) and formed 12 SOM clusters, respectively. Based on qRT-PCR results, a three signal regulation pathways, *viz.* the Ca2+ and ABA independent/dependent pathways were studied. The cold signals to signal transduction genes such as LlICE and LlCDPK and transcription factor genes such as LlDREB1/CBF, LlAP2/EREBP, LlNAC1, LlR2R3-MYB and LlBZIP were expressed highly in bulb. Likewise, LlFAD3, Llβ-amylase, LlP5CS and LlCLS responded to cold and enhanced adaptation processes that involve changes in the expression of transcripts related to cellular osmoprotectants and carbohydrate metabolism during cold stress in *L. lancifolium.* A little is known about the temporal and tissue specific expression pattern of transcription factors (TFs) in the petal senescence of ethylene-insensitive species and its developmental processes. Thus, a study was reported on the flower senescence of the ornamental plant *Gardenia jasminoides*, a non-climacteric flower (Tsanakas *et al.*, 2014). *De novo* assembly of transcriptome sequencing resulted in 102,263 contigs with mean length of 360 nucleotides that generated 57,503 unigenes. These were further clustered into 20,970 clusters and 36,533 singletons. The comparison of the consecutive developmental stages resulted in 180 common, differentially expressed unigenes. A large number of 'simple sequence repeats' (SSRs) were also identified comprising a large number of dinucleotides and trinucleotides. The prevailing families of differentially expressed TFs comprise the AP2/EREBP, WRKY and the bHLH. There are 81 differentially expressed TFs when the symptoms of flower senescence become visible with the most prevailing being the WRKY family with 19 unigenes. No other WRKY TFs had been identified up to now in petal senescence of ethylene insensitive species. The expression of 16 genes that belong to the TF families of WRKY, bHLH and the ethylene sensing pathway was later validated using qRT - PCR. The identification of such transcription factors would facilitate further to investigate their role in gardenia flowers in order to develop tools to delay petal senescence.

The whole-transcriptomic analysis is useful in identifying the genes involved in viroid infection of commercially important crops. Chrysanthemum stunt viroid (CSVd) is one of the main viroid pathogens infecting chrysanthemum species and is responsible for serious economic losses. To identify the genes involved in viroid infection, Jo *et al.* (2014) reported a list of ESTs of chrysanthemums infected by CSVd using a high-throughout next generation sequencing (NGS) method. The cDNA library was prepared from the total RNAs extracted from leaves of chrysanthemum infected with CSVd and was sequenced using the FLX 454 sequencer. In the report, a total of 11,600 ESTs were obtained from 99,750 reads after *de novo* assembly. A BLAST search annotated the functions of the 11,600 chrysanthemum ESTs. In addition, GO terms, enzyme codes, and InterPro domains for each chrysanthemum EST were obtained using the Blast2GO programme with default parameters. Gene ontology enrichment analysis identified 144 significantly enriched GO terms, which were further divided into biological processes (93 GO terms), cellular components (19 GO terms), and molecular functions (32 GO terms). CSVd infection affected the mRNAs involved in various metabolic pathways, stresses, transcription, translation, transport, chloroplasts, plasmodesmata, actins, and microtubules. Comparative analysis revealed 70% of the chrysanthemum ESTs were orthologous to those of five representative plant species. Jo *et al.* (2014) reported complete information on a large number of chrysanthemum genes, which will be valuable for studying plant-viroid interactions.

During 90's, a report on the seed protein content in wild and cultivated forms of *Amaranthus* species was reported to be 13-21% variation. The results of polyacrylamide gel electrophoresis indicated that the seed proteins of *Amaranthus* are heterogeneous and showed 38 different molecular weight bands. Zheleznov *et al.* (1997) confirmed that the electrophoretic analysis of seed storage proteins is promising in analysis of phylogenetic relationship between *Amaranthus* species.

A proteomic analysis of cut chrysanthemum cv. 'Jinba' in adventitious root formation (ARF) demonstrated an association of certain proteins to get an insight into the molecular mechanisms (Liu *et al.*, 2013). It was analysed using 2-dimensional gel electrophoresis and MALDI-TOF/TOF biotechnological approaches.

The report illustrated 19% proteins that were related to carbohydrate and energy metabolism, 16% to photosynthesis, 10% to protein fate, 7% to plant defense, 6% to cell structure, 7% to hormone related, 3% to nitrate metabolism, 3% to lipid metabolism, 3% to ascorbate biosynthesis and 3% to RNA binding and 23% were unknown proteins. Around 20 types of differentially accumulated proteins including ACC oxidase (CmACO) were further analysed at the transcription level in

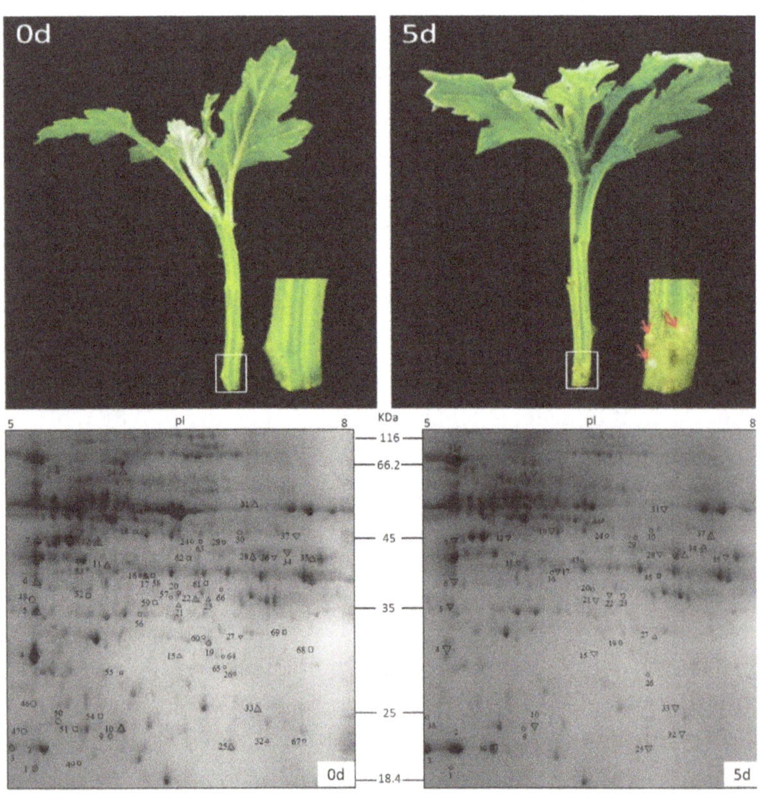

Fig. 8 (a): Morphological and histological properties of adventitious root primodium formation in the stem base of chrysanthemum 'Jinba' cuttings. Anatomy of adventitious root primodium formation morphology of 0 d and 5 d cuttings. The arrow indicated the adventitious root, zoomed image shows the cutting position. 8 (b): Comparative proteomic 2-D maps of soluble proteins extracted from 0 d and 5 d chrysanthemum cutting bases. Each map depicts one representative gel (of three replicates). Proteins were separated in the first-dimension by their pIs on IPG strips pH 5–8 and in the second-dimension by their molecular masses on 12.5% SDS-polyacrylamide gels. Proteins were stained with colloidal Coomassie Blue R-350. A total of 69 protein spots showing differences between the 0 d and 5 d maps are numbered. Squares indicate spots that are only present at 0 d or 5 d. Triangles indicate spots that differed between 0 d and 5 d (up-regulated, up-pointing triangle; down-regulated, down-pointing triangle).

(Source: Liu *et al.*, 2013)

chrysanthemum. Such studies suggested that the proteins identified will be valuable for studying molecular mechanism controlling ARF (Liu *et al.*, 2013).

The proteomic analysis in bundle sheath and mesophyll chloroplasts of *Amaranthus cruentus* var. 'Amaranteca' under salt stress was reported very recently. Differentially accumulated protein spots in photosynthetic tissue showed the Kranz anatomy, and the thylakoid proteins showed some differences at photosystem structure level in *Amaranthus*. Likewise, the presence of ribulose bisphosphate carboxylase/oxygenase (RuBisCO) was confirmed by immune-histochemical staining of chloroplasts (Ramo *et al.*, 2014) and concluded that *Amaranthus cruentus* var. 'Amaranteca' as a C_3-C_4 photosynthetic plant.

QTLs Associated in Floriculture

Various prospecting genes, alleles and QTL's associated with various traits of ornamental plants collectively represent a potential information that can be applied to programmes of their genetic improvement (Bretting and Widrlechner, 1995). In this direction, the use of molecular markers such as 'restriction fragment length polymorphisms' (RFLPs), 'random amplification of polymorphic DNA' (RAPDs), 'amplified fragment length polymorphisms' (AFLPs) and 'microsatellites DNA like simple sequence repeats' (SSRs), 'simple sequence length polymorphisms' (SSLPs), 'short tandem repeats' (STRs), 'simple sequence motifs' (SSMs) and 'sequence target microsatellites' (STMs) for genetic diversity analysis and as a high priority tool in the process of development of ornamental species and cultivars, and the exploitation of species diversity (Dore *et al.*, 2001). In addition, these tools can be utilized in marker assisted breeding, to analyse phenotypic traits of interest and to develop genetic maps. The molecular markers can also be used to analyse parentage of individuals in breeding population with application towards intellectual property rights (Caetano-Anolls *et al.*, 1991).

Markers can be used in mapping of both qualitative and quantitative inherited traits that get affected by more than one gene and influenced by the environment. The genetic diversity in Chilean and Brazilian *Alstroemeria* species assessed by using AFLP and first QTLs in the F_1 population of *Alstroemeria* were mapped and reported by Han *et al.* (2000, 2002). Among them, one QTL for leaf length was mapped on linkage group three of the parents near marker E+ACCT/M+CGCA-1165 with 20% and 14.8% phenotypic variation, two putative QTLs for leaf width on A002-3 and A002-6 and three putative QTLs involved in leaf length/width ratio, accounting for 46.7% of the phenotypic variance in total. Significantly, two main QTLs, *viz.* S+AC/M+ACT-1162 and S+AC/M+AGA-1465 involved in control of flower colour, accounted for up to 60% of phenotypic variance, were also reported (Han *et al.* , 2002). Likewise, Hodges *et al.* (2002) reported five floral-trait-specific QTLs in the F_2 population of *Aquilegia formosa* and *A. pubescens*.

Few other QTLs specific for biotic and abiotic stress tolerance were also reported in several ornamental plants. A single QTL locus for resistance to bacterial wilt (*Burkholderia caryophylli* - Burkholder) was identified in linkage group (LG) B4 among the F_2 population of 90 plants derived from a cross between a highly resistant line (85-11) and a susceptible cultivar (Pretty Favvare) of carnation (*Dianthus caryophyllus*). It explained 63% of phenotypic variance for bacterial wilt resistance (Yagi *et al.*, 2012). Similarly, a study was reported to determine the mode of inheritance for powdery mildew (PM) resistance in UFGE 31-19 (gerbera breeding line), one of its PM-resistant (PM-R) progeny, UFGE 4033 was crossed with PM-susceptible (PM-S) cultivar, 'Sunburst Snow White', and their progeny were evaluated for PM severity. Bulked segregant analysis (BSA) identified 17 molecular markers present in UFGE 4033 and the PM-R bulk but absent in 'Sunburst Snow White' and the PM-S bulk. Eleven of the molecular markers were mapped to one genetic linkage group and two regions on this linkage group together explained 71.1% of the phenotypic (PM severity rating) variance in the segregating population. It was proposed that the two regions, *viz. Rpx1* and *Rpx2* were resistant to *P. xanthii*. The results provided a foundation for use of UFGE 31-19 and progeny UFGE 4033 in gerbera for disease resistance breeding (Song and Deng, 2013) programmes. Feng *et al.* (2014) reported 97% polymorphism when 14 different sequence-related amplified polymorphism (SRAP) primers were used. A large genetic diversity at the interspecies level was recorded. The report also showed that SRAP marker system facilitated germplasm appraisal, evolution and genetic diversity studies in the genus *Dendrobium*.

During past decade, the application of molecular markers have been more widely applied to food and fibre crops than to ornamentals, which was probably reflected the levels of funding availability to the researchers in floriculture, as molecular approaches are costly and difficult to initiate without considerable expenditure. However, we are now able to see an ever-increasing reliance on genetic marker technology in programmes of evaluation and enhancement of floricultural germplasm, as the technology expedited more ubiquitous in research organizations. Traditional breeders of floricultural crops have much to gain by applying these tools to their selection programmes. We can also expect large-scale genetic mapping efforts towards production of floricultural crop genera in the near future.

Floriculture Development Scheme under Government Sectors

A tremendous effort in the world trade on floriculture in the production of cut flowers, traditional and loose flowers, indoor plants, other most commonly used ornamentals in the garden and for landscaping, dry flowers, flower seeds and other planting materials, and tissue-culture plantlets is gaining attention. The developed countries are importing flowers in a large scale to meet their internal demand. For instance, roses are the leading cut flowers usually imported into the United States. In 2012 in USA, the imported value was at $367.3 million roses, $128.9 million fresh mums and $82.5 million fresh carnations. The largest suppliers of imported nursery products were Canada, followed by Netherlands and Colombia. Other importers include the Netherlands, Germany, Switzerland, Denmark, Sweden, Belgium, England, Middle East countries, etc. In this scenario, India's contribution towards floriculture production is not at all encouraging as it counts to $30 lakhs (AGMRC). In this regard, many schemes have been set under the government of India to focus on the promotional and awareness by motivating the farmers to grow traditional as well as non-traditional floral crops for commercial purposes. The government is facilitating the floriculture farmers in terms of infrastructure like greenhouse and other essential requirements. In this regard, an unit at every state/district will serve to provide technical extension and training guidelines. Based on the existing popularity and market demands, the crop selection is being made. The most likely recommended crops include roses, carnation, chrysanthemum, orchids, *Lilium*, gerbera, helichrysum, statice, *etc.* as cut flowers.

In general, the total international trade in floriculture and live plants during the past decade was of the order of US $ 20,000 million. The leading flower producing countries is the Netherlands, and Germany is the biggest importer of flowers. Other leading floriculture countries are Japan, Italy, USA, Spain, Columbia and Israel. India had the largest area under floriculture amongst all the countries, which has 55,000 ha (1992-93). Amongst the states in India, Tamil Nadu has got the largest area under floriculture followed by Karnataka, West Bengal, Andhra Pradesh, Maharashtra, Rajasthan and Delhi. India's export of floriculture products during 1991-92 was estimated at Rs. 14.80 lakhs, which works out to a mere 0.3 percent of the world market for floriculture products. Biotechnology has got a tremendous potential to contribute to sustainable agricultural and horticultural production. India with its substantial area under horticulture can be a major contributor in the global biotechnology scenario. The introduction of various *in vitro* techniques has revolutionized the floriculture sector in India in recent years. The greatest stimulus provided by the tissue culture technology lies in increased speed of clonal multiplication of shoot meristem culture of highly desired strains of plant material besides producing pathogen-free plants. The three major areas of bio-technology, which can help us in crop improvement in general and floriculture in particular are micropropagation systems, including tissue culture, genetic engineering and recombinant DNA technique. The micropropagation industry in India has made rapid strides. From a mere 0.5 million plants in 1987, the production of plants through micro-propagation has gone up to 22.0 million in 1994 and 82.0 million in 2014. The foreign technology in floriculture generally revolves around the greenhouse where plants are grown under fully protected conditions. This is required in view of the externally cold climate in these regions. The greenhouse conditions protect the flowers against precipitation, wind and too much radiation, besides creating a micro-climate suitable for their growing. Globally, the trend in floriculture production has been enormously dynamic during the past four decades. Netherlands increased the area by 70% and West Germany increased its consumption of floriculture products by 150%. Per capita consumption of floriculture products was highest in Switzerland, followed by Holland, Germany, Belgium and Sweden. India has got the major advantage of producing a significant proportion of its flowers during winter, when the demand is very high in western countries. With mild winters in most parts of India, it is possible to produce them at a comparatively lower cost. India has got a number of other advantages like favourable climatic conditions, varying agro-climatic zones, availability of cheap labour, sound research infrastructure, advanced tissue culture facilities, etc. The conventional, biotechnology-based technologies, genomics and proteomics are comparatively well developed in India. It is found that tissue culture is best suited for the development of floriculture in a broad way. Greenhouse technology could be used only in places of high altitude with frost and very cold climate or in case of special varieties commanding a good export market, *vis-à-vis* in the tropical and sub-tropical regions of the country when summer is very hot so under polyhouses the desired crops may be given required temperature for harvesting the crops throughout the year. Data on various financial parameters are available from two sources, *viz.* Biotech Consortium India Ltd. (BCIL) and National Horticulture Board (NHB). The result of these two studies indicate that the adoption of tissue culture technology will be fully beneficial with BC ratio varying from 1.36 to 1.75 in case of BCIL and 1.21 to 1.49 in case of NHB. Further, the investment will yield a return of 566% over 12 years.

In government schemes, a facility of floriculture nursery can be implemented. These nurseries will facilitate in propagating flower seedlings, to supply the materials to the public and interested farmers to cultivate these in large scale. The government sectors also implemented certain schemes to support tissue culture, biotechnology-genomics and proteomics research. The biotechnological research approaches would enhance the production of high yielding flowering plants and resistant to biotic as well as abiotic stresses. Thus, India has to do a lot to expedite the quality-quantity production and improve agribusiness world-wide, through biotechnological approaches.

References

Ahkami, A., U. Scholz, B. Steuernagel, M. Strickert and K.T. Haensch, 2014. Comprehensive transcriptome analysis unravels the existence of crucial genes regulating primary metabolism during adventitious root formation in *Petunia hybrida. PLoS ONE*, **9**(6):e100997. doi:10.1371/journal.pone.0100997.

Akutsu, M., Takuma Ishizaki and Hiroji Sato, 2004a. Transformation of the monocot *Alstroemeria* by *Agrobacterium tumefaciens. Molec. Breed.*, **13**(1): 69-78.

Akutsu, M., Takuma Ishizaki and Hiroji Sato, 2004b. Transformation of the monocotyledonous *Alstroemeria* by *Agrobacterium tumefaciens. Plant Cell Reports*, **22**(8): 561-568.

Belarmino, M.M. and M. Mii, 2000. *Agrobacterium*-mediated genetic transformation of a *Phalaenopsis* orchid. *Plant Cell Rep.*, **19**: 435-442.

Bhojwani, S. S. and M.K. Razdan, 1983. *Plant Tissue Culture - Theory and Practice, Developments in Crop Scienc.* Elsevier, Amsterdam, the Netherlands.

Bretting, P.K. and M.P. Widrlechner, 1995. Genetic markers and plant genetic resource management. *Plant Br. Rev.*, **13**: 11-84.

Brian J. A., P.E. Kriedemann and C.G.N. Turnbull, 2010. *Plants in Action - Adaptation in Nature, Performance in Cultivation.* Australian Society of Plant Scientists, Australia, http://plantsinaction.science.uq.edu.au/edition1.

Caetano-Anolls, G., B.J. Bassam and P.M. Gresshoff, 1991. DNA amplification fingerprinting: a strategy for genome analysis. *Plant Mol. Biol. Rep.*, **9**: 294-307.

Carlson, P.S., H.H. Smith and R.D. Dearing, 1972. Parasexual interspecific plant hybridization. *Proc. Natl. acad. Sci.*, **69**: 2292-2294.

Chai, M.L., C.J. Xu, K.K. Senthil, J.Y. Kim and D.H. Kim, 2002. Stable transformation of protocorm-like bodies in *Phalaenopsis* orchid mediated by *Agrobacterium tumefaciens. Scientia Hort.*, **96**: 213-224.

Chen, F.C. and A.R. Kuehnle, 1996. Obtaining transgenic *Anthurium* through *Agrobacterium*-mediated transformation of etiolated internodes. *J. Amer. Soc. Hort. Sci.*, **121**: 47-51.

Chen Y. M. and M. Mii, 2012. Interspecific hybridization of *Begonia semperflorens* (section Begonia) with *B. pearcei* (section Eupetalum) for introducing yellow flower color. *Plant Biotech.*, **29**: 77–85.

Cheong, E. J. and M.R. Pooler, 2004. Factors affecting somatic embryogenesis in *Prunus incisa* cv. February Pink. *Plant Cell Reports*, **22**(11): 810 - 815.

Chitra, V.D.S. and G. Padmaja, 2002. Seasonal influence on axillary bud sprouting and micropropagation of elite cultivars of mulberry. *Scient. Hort.*, **92**(1): 55-68.

Davey M. R., P. Anthony, J.B. Power and K.C. Lowe, 2005. Plant protoplasts: status and biotechnological perspectives. *Biotechnology Advances*, **23**: 131–171.

Délano-Frier, J. P., H.A. Arnaut, K.C. Castillo, G.C. Arroyo, P.A.C. Arbeláez and L.H. Estrella, 2011. Transcriptomic analysis of grain amaranth (*Amaranthus hypochondriacus*) using 454 pyrosequencing: comparison with *A. tuberculatus*, expression profiling in stems and in response to biotic and abiotic stress. *BMC Genomics*, **12:** 363.

Deng Y., S. Chen, A. Lu, F. Chen, F. Tang, Z. Guan and N. Teng, 2010. Production and characterization of the intergeneric hybrids between *Dendranthema morifolium* and *Artemisia vulgaris* exhibiting enhanced resistance to chrysanthemum aphid (*Macrosiphoniella sanbourni*). *Planta*, **231**(3) 693 - 703.

Dore, C., F. Dosba and C. Baril, 2001. *Proceedings of the Iinternational Symposium on Molecular Markers for Characterizing Genotypes and Identifying Cultivars in Horticulture. Acta Hort.*, No. 546. Held in Leuven, Belgium.

Drew, P.D., G. Franzoso, K.G. Becker, V. Bours, L.M. Carlson, U. Siebenlist and K. Ozato, 1995. NF kappa B and interferon regulatory factor 1 physically interact and synergistically induce major histocompatibility class I gene expression. *J. Interferon Cytokine Res.*, **15**: 1037-1045.

Feng, S.G., J.J. Lu, L. Gao, J.J. Liu and H.Z. Wang, 2014. Molecular phylogeny analysis and species identification of *Dendrobium* (Orchidaceae) in China. *Biochem. Genet.*, **52**(3-4): 127-136.

Fullner, K.J. and E.W. Nester, 1996. Temperature affects the T-DNA transfer machinery of *Agrobacterium tumefaciens. J. Bacterol.*, **176**(6): 1498.

Gleba, Yu and L.R. Shlumukov, 1990. Selection of somatic hybrids. In: *Plant Cell Line Selection: Procedures and Applications*, pp. 257-286. VCH, Weinheim New York Basel Cambridge, USA.

Han, T., M. J. de Jeu, H.J. van Eck and E. Jacobsen, 2000. Genetic diversity of Chilean and Brazilian *Alstroemeria* species assessed by AFLP analysis. *Heredity*, **84**: 564-569.

Han, T., H.J. van Eck, M. J. de Jeu, and E. Jacobsen, 2002. Mapping of quantitative trait loci involved in ornamental traits in *Alstroemeria. HortSci.*, **37**: 585-592.

Hasbullah, N.A., R.M. Taha and A. Awal, 2008. Growth optimization and organogenesis of *Gerbera jamesonii* Bolus *ex.* Hook f. *in vitro. Pakistan J. biol. Sci.*, **11**: 1449-1454.

Hodges, S.A., J.B. Whittall, M. Fulton and J.Y. Yang, 2002. Genetics of floral traits influencing reproductive isolation between *Aquilegia formosa* and *A. pubescens. Amer. Nat.* (159 Suppl.), **3**: S51-60 (doi: 10.1086/338372).

Horita, M., H. Morohoshi and F. Komai, 2003. Production of fertile somatic hybrid plants between Oriental hybrid lily and *Lilium* × *formolongi. Planta*, **217**: 597-601.

Hutchinson, J.F., V. Kaul, G. Maheswaran, J.R. Moran, M.W. Graham and D. Richards, 1992. Genetic improvement of floricultural crops using biotechnology. *Australian J. Bot.*, **40**(6): 765 -787.

Jelenic, S., J. Berljak, P. Drazena and S. Jelaska, 2001. Mixploidy and chimeric structures in somaclones of potato (*Solanum tuberosum* L.) cv. Bintje. *Food Technol. Biotechnol.*, **39**: 13-17.

Jeknic, Z., S.P. Lee, J. Davis, R.C. Ernst and T.H.H. Chen, 1999. Genetic transformation of *Iris germanica* mediated by *Agrobacterium tunefaciens. J. Amer. Soc. Hort. Sci.*, **124**: 575-580.

Jo, Y., K.M. Jo, S.O. Park, K.H. Kim and K.W. Cho, 2014. Transcriptomic landscape of chrysanthemums infected by chrysanthemum stunt viroid. *Plant Omics J.*, **7**(1):1-11. ISSN:1836-3644.

Junping H., F. Chen, S. Chen, L. Guosheng, Y. Deng, W. Fang, L. Zhaolei, G.,Zhiyong and H. Chunyan 2011. Chrysanthemum leaf epidermal surface morphology and antioxidant and defense enzyme activity in response to aphid infestation. *J. Pl. Physiol.*, **168**(7): 687-693.

Kamo, K., A. Blowers, F. Smith, J.V. Eck and R. Lawson, 1995. Stable transformation of *Gladiolus* using suspension cells and callus. The establishment of cell suspension cultures of gladiolus that regenerate plants. *JASHS*, **120**(2): 347-352.

Kamo, K., A. Blowers and D. McElroy, 2000. Effect of the cauliflower mosaic virus 35S, actin, and ubiquitin promoters on *UidA* Expression from a *bar-uidA* fusion gene in transgenic *Gladiolus* plants. *In Vitro Cell. & dev. Biol.-Plant*, **36**(1): 13-20.

Kamo, K., A. Gera, J. Cohen, J. Hammond, A. Blowers, F. Smith and J.V. Eck, 2005. Transgenic *Gladiolus* plants transformed with the bean yellow mosaic virus coat-protein gene in either sense or antisense orientation. *Plant Cell Reports*, **23**(9): 654-663.

Kantharajah, A.S. and P.G. Golegaonkae, 2004. Somatic embryogenesis in eggplant. *Scientia Hort.*, **99**(2): 107-117.

Kartha K. K. 1981. Meristem Culture and Cryopreservation Methods and Applications. In: *Plant Tissue Culture: Methods and Applications in Agriculture* (ed. Thorpe, T. A.), pp. 181-211. Academic Press, New York, USA.

Kawaguchi, M. and A. Taji, 2002. Grafting of Sturt's Desert Pea (*Swainsona formosa*) onto various Leguminosae plants under *in vitro* and *in vivo* conditions. In: *International association for Plant Tissue Culture and Biotechnology: The Importance of Plant Tissue Culture and Biotechnology in Plant Sciences* (eds Taji, A. and R. Williams), pp. 301-309. University of New England, Armidale, Australia.

Kishimoto, S., R. Aida and M. Shibata, 2002. *Agrobacterium tumefaciens*-mediated transformation of Elatior Begonia (*Begonia × hiemalis* Fotsch). *Plant Sci.*, 162(5): 697-703.

Kleinhofs, A. and R. Behki, 1977. Prospects for plant genome modification by nonconventional methods. *Ann. Rev.Genet.*, **11**: 79-101.

Kuehnle, A.R. and N. Sugii, 1992. Transformation of *Dendrobium* orchid using particle bombardment of protocorms. *Plant Cell Reports*, **11**: 484-488.

Küster, E. 1909. Über die Verschmelzung nachter Protoplasten. *Ber. Stsch. Bot. Ges.*, **27**: 589-598.

Lakshmanan, P. and A. Taji, 2003. *In vitro* flowering. In: *Encyclopedia of Plant & Crop Science* (ed. Goodman, R.M.). Marcel Dekker, Inc., New York, USA.

Lin, H.S., C. van der Toorn, K.J.J.M. Raemakers, R.G.F. Visser, M.J. de Leu and E. Jacobsen, 2000. Development of a plant regeneration system based on friable embryogenic callus in the ornamental *Alstroemeria*. *Plant Cell Reports*, **19**(5): 529-534.

Liu R., S. Chen, J. Jiang, L. Zhu, C. Zheng, S. Han, J. Gu, J. Sun, H. Li, H. Wang, A. Song and F. Chen 2013. Proteomic changes in the base of chrysanthemum cuttings during adventitious root formation. *BMC Genomics*, **14**: 919.

Lu, C.Y., G. Nugent and T. Wardley, 1990. Efficient, direct plant regeneration from stem segments of chrysanthemum (*Chrysanthemum morifolium* Ramat cv. Royal Purple). *Plant Cell Rep.*, **8**: 733-736.

Maheswaran, G. and E.G. Williams, 1985. Origin and development of somatic embryos formed directly on immature embryos of *Trifolium repens in vitro*. *Ann. Bot.*, **56**: 619-630.

Malaure, R.G., G. Barclay, J.B. Power and M.R. Davey, 1991a. The production of novel plants from florets of *Chrysanthemum morifolium* using tissue culture. 1. Shoot regeneration from ray florets and somaclonal variations exhibited by the regenerated plants. *J. Pl. Physiol.*, **139**:8-13.

Malaure, R.G., G. Barclay, J.B. Power and M.R. Davey, 1991b. The production of novel plants from florets of *Chrysanthemum morifolium* using tissue culture. 2. Securing natural mutations (sports). *J. Pl. Physiol.*, **139**: 14-18.

McComb, J.A., I.J. Bennett, P.G. van der Moezel and D.T. Bell, 1989. Biotechnogy enhances utilization of Australian woody species for pulp, fuel and land rehabilitation. *Australian J. Biotech.*, **3**: 297-301.

McKersie, B.D. and S.R. Bowley, 1993. Synthetic Seeds in Alfalfa. In: *Synseeds Applications of Synthetc Seeds to Crop Improvement* (ed. Redenbaugh, K.), pp. 231-255. CRC Press, Boca Raton, USA.

Melchers, G., M.D. Sacristán and S.A. Holder, 1978. Somatic hybrid plants of potato and tomato regenerated from fused protoplasts. *Carlsberg Res. Comm.*, **43**: 203-208.

Miyoshi K. and N. Asakura, 1996. Callus induction, regeneration of haploid plants and chromosome doubling in ovule cultures of pot gerbera (*Gerbera jamesonii*). *Plant Cell Reports*, **16**: 1-5.

Mizuhiro, M., K. Ito and M. Mii, 2001. Production and characterization of interspecific somatic hybrids between *Primula malacoides* and *P. obconica*. *Plant Sci.*, **161**: 489-496.

Potrykus, I. 1989. Gene transfer to plants: assessment and perspectives. *Physiologia Plant.*, **79**: 125-134.

Qiao Q., F.W. Xing, Y.P. Xiao and H.F. Chen, 2009. Somatic embryogenesis and *in vitro* flowering in *Saposhnikovia divaricata*. *J. Plant Growth Regul.*, **28**: 81-86.

Ramo, A.J., J.A. Huerta-Ocampo, A. Barrera-Pacheco, Antonio de León-Rodriguez, S. Baginsky, and Ana P. Barba de la Rosa, 2014. Comparative proteomic analysis of amaranth mesophyll and bundle sheath chloroplasts and their adaptation to salt stress. *J. Pl. Physiol.*, **171**(15): 1423-1435.

Redenbaugh, K., B.D. Paasch, J.W. Nichol, M.E. Kossler, P.R. Viss and K.A. Walker, 1986. Somatic seeds: encapsulation of asexual plant embryos. *Biotechnol.*, **4**: 797-801.

Sangwan, R.S., Y. Bourgeois and B.S. Sangwan-Norreel, 1991. Genetic transformation of *Arabidopsis thaliana* zygotic embryos and identification of critical parameters influencing transformation efficiency. *Mol. Gen. Genet.*, **230**: 475-485.

Senaratna, T., D. Touchell, E. Bunn and K. Dixon, 2000. Acetyl salicylic acid (aspirin) and salicylic acid induce multiple stress tolerance in bean and tomato plants. *Plant Gr. Reg.*, **30**: 157-161.

Shweta, S., K. Surinder, G. Minerva and T. Manisha, 2013. *Agrobacterium*-mediated genetic transformation of chrysanthemum (*Dendranthema grandiflora* Tzvelev) with rice chitinase gene for improved resistance against *Septoria obesa*. *Pl. Path. J.*, ISSN 1812-5387, DOI: 10.3923/ppj.2013.

Simmonds, J.A. and S.D. Nelson, 1989. Improved micropropagation of *Begonia* × *hiemalis* by maintaining donor plants in long day conditions. *Hort. Sci.*, **24**: 831-832.

Song, X. and Z. Deng, 2013. Powdery mildews resistance in *Gerbera*: Mode of inheritance, quantitative trait locus identification, and resistance responses. *J. Soc. Hort. Sci.* (JASHS), **138**(6): 470-478.

Sudhersan, C. and M. AboEL-Nil, 2002. Somatic embryogenesis of Sturt's desert pea (*Swainsona formosa*). *Curr. Sci.*, **83**(9): 1074-1076.

Taji, A., P. Kumar and P. Lakshmanan, 2002. *In Vitro Plant Breeding*. CBS Publishers & Distributors Pvt. Ltd., India.

Takebe, I., G. Labib and G. Melchers, 1971. Regeneration of whole plants from isolated mesophyll protoplasts of tobacco. *Naturwissenschaften*, **58**: 318-320.

Tanase, K, C. Nishitani, H. Hirakawa, S. Isobe, S. Tabata, A. Ohmiya and T. Onozaki, 2012. Transcriptome analysis of carnation (*Dianthus caryophyllus* L.) based on next-generation sequencing technology. *BMC Genomics*, **13**: 292.

Teulières, C. and A.M. Boudet, 1991. Isolation of protoplasts from different *Eucalyptus* species and preliminary studies on regeneration. *Plant Cell Tissue Organ Cult.*, **25**: 133-140.

Tran-Thanh van, M., Thi D. Nguyen and C. Averil, 1974. Regulation of organogenesis in small explants of superficial tissue of *Nicotiana tabacum* L. *Planta*, **119**: 149-159.

Tsanakas, G. F., M.E. Manioudaki, A.S. Economou and P. Kalaitzis, 2014. *De novo* transcriptome analysis of petal senescence in *Gardenia jasminoides* Ellis. *BMC Genomics*, **15**: 554.

Underwood, B. A., D.M. Tieman, K. Shibuya, R.J. Dexter, H.M. Loucas, A.J. Simkin, C.A. Sims, A.S. Eric, H.J. Klee and D.G. Clark, 2005. Ethylene-regulated floral volatile synthesis in *Petunia* corollas. *Plant Physiol.*, **138**: 255–266.

Van Schaik, C.E. 1998. Regeneration and transformation of *Alstroemeria*. Thesis. Department of Plant Breeding, Wageningen Agricultural University, Wageningen, The Netherlands.

Vazquez-Thello, A. Jun Li Yang, Makoto Hidaka and Takeshi Uozumi, 1996. Inherited chilling tolerance in somatic hybrids of transgenic *Hibiscus rosa-sinensis* × transgenic *Lavatera thuringiaca* selected by double-antibiotic resistance. *Plant Cell Reports*, **15**(7): 506-511.

Wang, J., Y. Yang, X. Liu, J. Huang, Q. Wang, J. Gu and Y. Lu, 2014. Transcriptome profiling of the cold response and signaling pathways in *Lilium lancifolium*. *BMC Genomics*, **15**: 203.

Watad, A.A., D.J. Yun, T. Matsumoto, X. Niu, Y. Wu, A.K. Kononowicz, R.A. Bressan and P.M. Hasegawa, 1998. Microprojectile bombardment mediated transformation of *Lilium longiflorum*. *Plant Cell Reports*, **17**: 262-267.

Watt, M.P., F. Blakeway, C.F. Cresswell and B. Herman, 1991. Somatic embryogenesis in *Eucalyptus grandis*. *South Afr. Forestry J.*, **157**: 59-65.

Williams, R.R. and A.M. Taji, 1987. Effects of temperature, darkness and gelling agent on long term storage *in vitro* shoot cultures of Australian woody plants. *Plant Cell Tissue and Org. Cult.*, **11**: 151-156.

Wilmink, A., B.C.E. van de Ven and J.J.M. Dons, 1995. Activity of constitutive promoters in various species from the Liliaceae. *Plant Mol. Biol.*, **28**: 949-955.

Wit, J.C., H.E. Esendam, J.J. Honkanen and U. Tuominen, 1990. Somatic embryogenesis and regeneration of flowering plants of rose. *Plant Cell Reports*, **9**: 456–458.

Withers, L.A. 1985. Cryopreservation of Cultured Cells and Meristems. In: *Cell Culture and Somatic Cell Genetics of Plants*, vol. 2 (ed. Vasil, I.K.), pp. 253-316. Academic Press, Inc., Orlando, USA.

Yagi, M., T. Kimara, T. Yamamoto, S. Isobe, S. Tabata and T. Onozaki, 2012. QTL analysis for resistance to bacterial wilt (*Burkholderia caryophylli*) in carnation (*Dianthus caryophyllus*) using an SSR-based genetic linkage map. *Mol. Breed.*, **30**(1): 495-509.

Yang, J., H.J. Lee, D.H. Shin, S.K. Oh, J.H. Seon, K.Y. Paek and K.H. Han, 1999. Genetic transformation of *Cymbidium* orchid by particle bombardment. *Plant Cell Reports*, **18**: 978-984.

Zheleznov, A. V., L.P. Solonenko and N.B. Zheleznova, 1997. Seed proteins of the wild and the cultivated *Amaranthus* species. *Euphytica*, **97**: 177–182.

Zupan, J.R. and P. Zambryski, 1995. Transfer of T-DNA from *Agrobacterium* to the plant cell. *Plant Physiol.*, **107**(4): 1041-1047.

31

Genetically-Engineered Ornamentals

Tanya Thakur and H.S. Grewal

Introduction

Artificial genetic modification through biotechnology is now well recognized for production of novel traits in living organism, especially plants. Biotechnology offers new means of improving ornamental crops by transformation of existing adapted crop species with addition of one or more genes of desirable traits with minimal effects on other desirable characteristics. Transformation systems have currently been developed for over 30 species of herbaceous ornamentals and several woody species. Several groups are working to introduce virus resistance into various ornamentals including chrysanthemums, gladiolus, lilies and various orchids. The traits of interest include plant architecture, flower colour and longevity, and resistance to environmental stresses including drought, pests and diseases (fungal, bacterial and viral). A few genetically modified ornamental crops have been commercialized, notably carnations with altered blue flower colour, and there is future expectation of the introduction of effective virus-resistance into commercial floral crops. These are the modern tools for future generation to tackle the environmental challenges. Broadly, there are two steps in this direction.

i. **Isolation of gene of interest (foreign DNA)**: The gene can be isolated using *restriction endonuclease* enzymes (molecular scissors) to cut DNA into fragments and gel-electrophoresis to separate them out according to length. If the DNA sequence is known, but no copies of the gene are available, it can be artificially synthesized.

ii. **Cloning gene of interest or recombinant DNA**: Cloning is the laboratory process used to create recombinant DNA. There are two widely used methods to direct the replication of any specific DNA sequence chosen by researcher, i.e. (i) molecular cloning which involves replication of DNA within a living cell, and (ii) polymerase chain reaction which replicates and amplifies DNA in the test-tube free of living cells.

In this direction, in all, there are five steps.

- **Choice of host organism and cloning vector** where cloning vector is a DNA molecule that replicates within a living cell. Vectors are generally derived from plasmids or viruses and represent small segments of DNA that contain necessary genetic signals for replication, convenience in inserting foreign DNA, identifying cells that contain recombinant DNA, and expressing the foreign DNA. Bacterium *Escherichia coli* and plasmid vectors are in common use because they are technically sophisticated, versatile, widely available and offer rapid growth of recombinant organisms.

- **Preparation of plasmid** where plasmid is treated with a *restriction endonuclease* to cleave the DNA at the site where foreign DNA will be inserted. The cleaving of vector DNA and foreign DNA is accomplished with the same restriction enzyme (EcoRI). Vector molecules with dephosphorylated cleaved ends are unable to replicate, and replication can only be restored if foreign DNA is integrated into the cleavage site. Plasmid also contains a promoter (initiating transcription) and terminator region (ends transcription) as well as a selectable marker gene (for screening of transformant).

- **Creation of recombinant DNA with DNA *ligase*** where plasmid DNA and foreign DNA are simply mixed together and exposed to an enzyme DNA *ligase* that covalently links the ends together. This joining reaction is often termed **ligation**. The resulting DNA mixture (recombinant DNA) containing randomly joined ends is then ready for introduction into the host organism.

- **Introduction of recombinant DNA (rDNA) into host organism**: Here recombinant DNA (rDNA) is moved back into a living cell, referred to as the host plant. When microorganisms are able to take up and replicate rDNA, the process is termed transformation. As host organism divides, replication and cloning of rDNA occurs

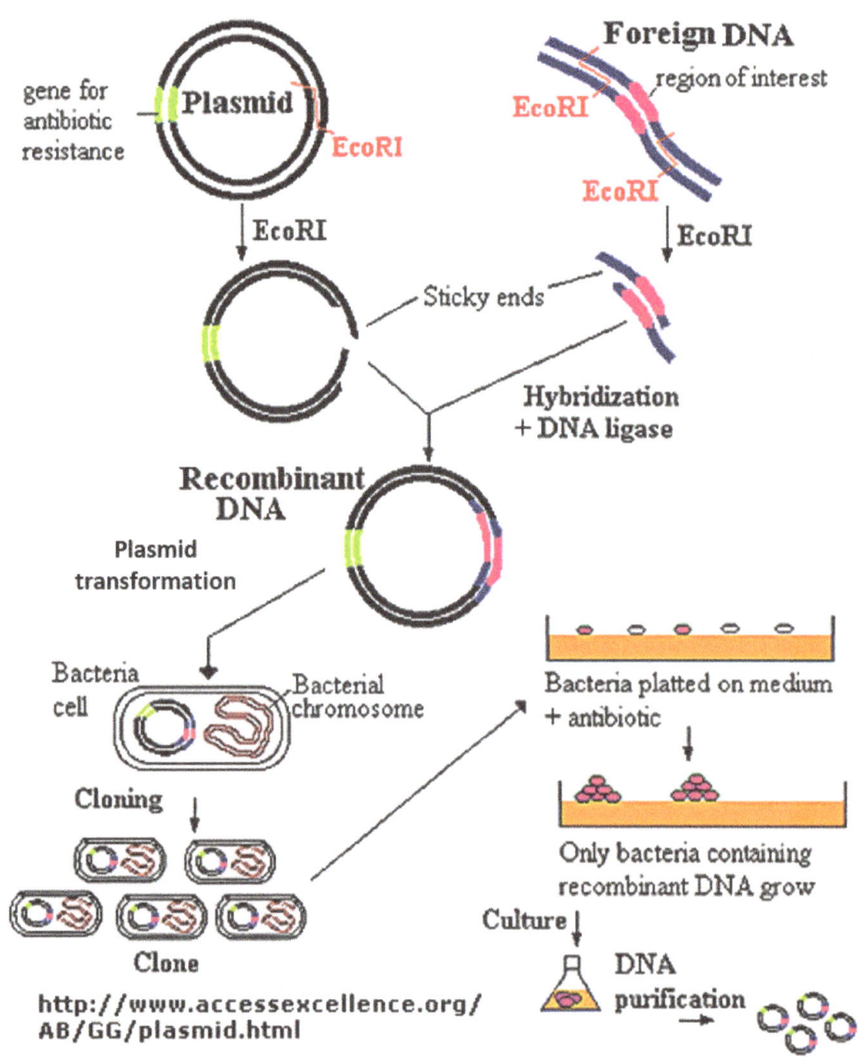

within the host cell. **In transformation system**, the transgenic ornamentals are produced using different methods, including *Agrobacterium*-mediated transformation in chrysanthemums (Courtney-Gutterson *et al.*, 1994), biolistic or particle bombardment in gladiolus (Kamo *et al.*, 1995), and direct electrophoresis of DNA into meristems in orchids (Griesbach., 1994). *Agrobacterium* or biolistic bombardments are most commonly used, and can be combined (Hansen *et al.*, 1997). While Agrobacterium-mediated transformation was initially largely restricted to dicotyledonous species, but various modifications have resulted in successful transformation of monocotyledonous species such as lily. Production of transgenic plants requires a successful transformation event at the cellular event, expression of a marker gene allowing favourable growth of transformed cells and selection of transformants, and successful regeneration from the transformed cells into whole plant (Robinson and Firoozabady, 1993). To date, over 30 species of transformed ornamentals have been reported, including most of the major crops such as chrysanthemums, carnations, orchids and roses (Deroles *et al.*, 2002).

- **Selection of organisms containing rDNA (transformants)**: When bacterial cells are used as host organisms, the marker gene (plasmid) such as GUS or selectable markers such as antibiotic (ampicillin) or herbicide resistance (Deroles *et al.*, 2002) are used to screen transformants which confer resistance to an antibiotic that would otherwise kill the cells. Cells containing the rDNA will survive when exposed to the antibiotic, while those that have failed to take up rDNA will die (http://en.wikipedia.org/wiki/Molecular_cloning).

Genetic Engineering to Improve Traits in Ornamental Crops

1. Biotic Stress Resistance

Many genes have been transferred into flowering ornamentals to give resistance to biotic stresses such as fungal, bacterial, viral and insect attacks. Some of the examples of biotic resistance through gene transfer in ornamentals are:

- **Fungal resistance**: In carnation, osmotin, chitinase and PR-1 genes were transferred to give resistance against *Fusarium oxysporum* f. sp. *dianthi* (Zuker *et al.*, 2001a). Likewise, in chrysanthemum, chitinase gene from rice gave resistance against *Botrytis* (Takatsu *et al.*, 1999); in geranium, onion anti-microbial protein against *Botrytis cinerea* (Bi *et al.*, 1999); and in rose, anti-microbial peptide Ace-AMP I and chitinase gene transfer gave resistance against *Sphaerotheca pannosa* and *Diplocarpon rosae*, respectively (Li *et al.*, 2003; Marchant *et al.*, 1998).

- **Bacterial resistance**: In geranium, cercopin gene gave resistance against *Xanthomonas* (Renou *et al.*, 2000) and onion anti-microbial protein against bacteria (Bi *et al.*, 1999); whereas, in *Oncidium* orchids, ferredoxin-like protein gene transferred from sweet pepper gave resistance against bacterial soft rot caused by *Erwinia carotovora* (Liau *et al.*, 2003).

- **Viral resistance**: Chrysanthemum stunt viroid (CSVd) is a serious pathogen in the chrysanthemum-production industry, as it is easily transmitted mechanically during the course of propagation. Ogawa *et al.* (2005) produced transgenic chrysanthemum plants expressing double-stranded RNA-specific *ribonuclease* gene (pacl) derived from *Schizosaccharomyces pombe* using an *Agrobacterium*-mediated transformation method. The transgenic lines showed a decreased frequency of infection, less accumulation of chrysanthemum stunt viroid and also showed lower infection frequency against tomato spotted wilt virus (TSWV). Thus, indicating that the commercial use of this transgene would limit the spread of viruses to wild populations of *Dendranthema*. This is the first study to demonstrate that a single tolerance gene in transgenic plants can attenuate both viroid and virus diseases.

- **Insect resistance**: Shinoyama *et al.* (2003) introduced a modified *delta-endotoxin* gene, modified *cry1Ab* (*mcbt*) of *Bacillus thuringiensis* into chrysanthemum cultivar 'Shuho-no-

Gene for biotic resistance and their mode of action

Gene for biotic resistance	Resistance against	Mode of action
Bt gene	Insects- Lepidoptera, Coleoptera & Diptera	Pore forming toxin protein released and interfere midgut
ipt gene (isopentenyl transferase)	Insect-tobacco hornworm & peach aphid	Induces cytokinin synthesis which forms insecticidal 2° metabolites
CpTI (cowpea trypsin inhibitor)	Insect- Lepidoptera, Coleoptera, Orthoptera	Interferes or inhibits protein (trypsin) digestion in insect
Serine *proteinase* inhibitor	Insect- whitefly	Interferes or inhibits protein digestion in insect
Chitinase	Fungus- *Rhizoctonia, Fusarium*	Breaks the chitin polysaccharides in cell wall of fungi
Osmotin	Fungus- *Phytophthora*	Increases proline accumulation and decreases oxidative stress
1,3 -β *glucanase* and stilbene *synthase*	Fungus- *Botrytis cinerea*	Breaks glucans in cell wall, synthesizes antioxidant stilbene
Cercopin	Bacterium- *Pseudomonas*	Cell wall lysis and increases electrolyte leakage
Coat protein gene	Virus- TMV	Breaks protein coat (capsid)

chikara' which displays a specific biological activity against lepidopteran insects (*Helicoverpa armigera, Mamestra brassicae* and *Trichoplusia*) . The transgenic plants showed higher feeding inhibition and growth inhibition of the insects compared to non-transformed control plants.

2. Abiotic Stress Resistance

- **Heat stress resistance in chrysanthemum**: Hong *et al.* (2009) reported over-expression of *AtDREB1A* gene from *Arabidopsis* in chrysanthemums inducing heat-stress tolerance. Transgenic with *AtDREB1A* construct and wild type (without transformed gene) plants were exposed to 45 °C as a heat stress treatment for 36 h, followed by a 3-week recovery. Around 70% of the transgenic plants and less than 20% of the wild type plants survived. These results suggest that improvement of heat stress tolerance in transgenic chrysanthemum may be associated with enhanced tolerance of photosynthesis and leaf electrolyte leakage.

3. Flower Colour Modification

Flower colour is, predominantly, due to three types of pigments, *viz.* flavonoids, carotenoids and betalains. Betalains are the least abundant of the three and contribute to ivory, yellow, orange, red and violet colouration in flowers (Forkmann, 1991) of *Bougainvillea* and *Amaranthus.* Carotenoids are lipid-soluble tetra terpenoids and contribute to the yellow hues in flowers such as marigold (lutein) and pansy (violaxanthin). The anthocyanins (class of flavonoids) are the most common of the three types of pigments and contribute to colours from yellow to red to blue. These are water-soluble O-glycosides (Stafford, 1990) and are localized in the vacuoles of petal epidermal cells. The modification of flower colour *via* genetic engineering has generally focused on metabolic engineering of the flavonoid biosynthetic phenylpropanoid pathway (http://en.wikipedia.org/wiki/Anthocyanin).

White colouration develops by down-regulation of an anthocyanin biosynthesis structural gene in flower crops such as carnation (Gutterson, 1995), chrysanthemum (Courtney-Gutterson *et al.*, 1994), gerbera (Elomaa *et al.*, 1993), petunia (van der Krol *et al.*, 1988) and rose (Gutterson, 1995) or by antisense RNA technology in petunia. Zuker *et al.* (2002) unexpectedly found that when they **down-regulated** carnation F3H gene, anthocyanin levels were not only the change observed in the transgenic carnations, but were also more fragrant due to an increase in methylbenzoate, which may be considered a more positive outcome when commercializing these flowers. **Antisense RNA gene** for normal enzyme gene

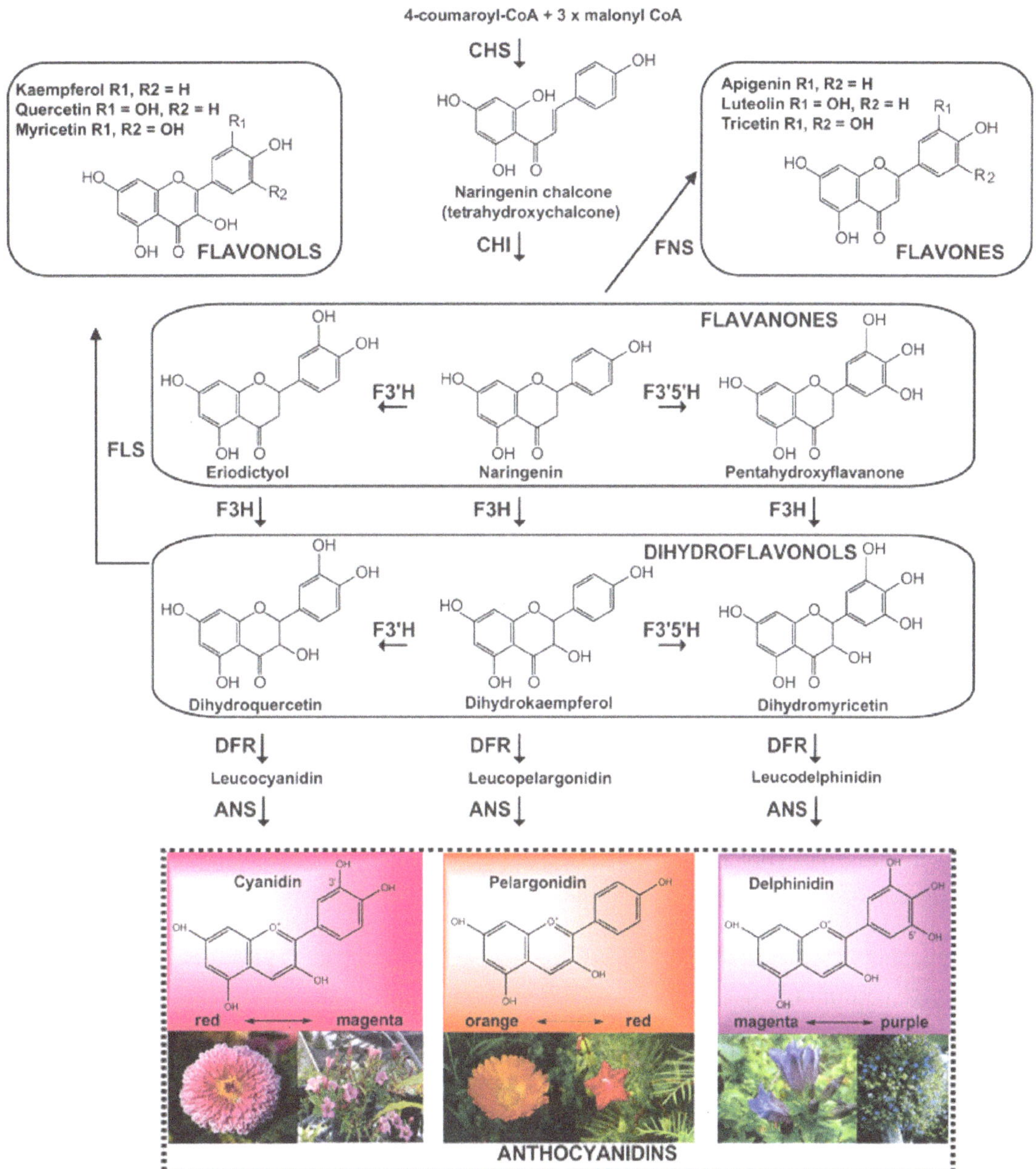

CHS- chalcone synthase
CHI- chalcone isomerase
FLS-flavonol synthase
F3'H- flavanone 3'-hydroxylase
F3'5'H- flavanone 3' 5'-hydroxylase
DFR- dihydroflavonol 4-reductase
ANS- anthocyanidin synthase
FNS- flavone synthase

Anthocyanin biosynthetic pathway

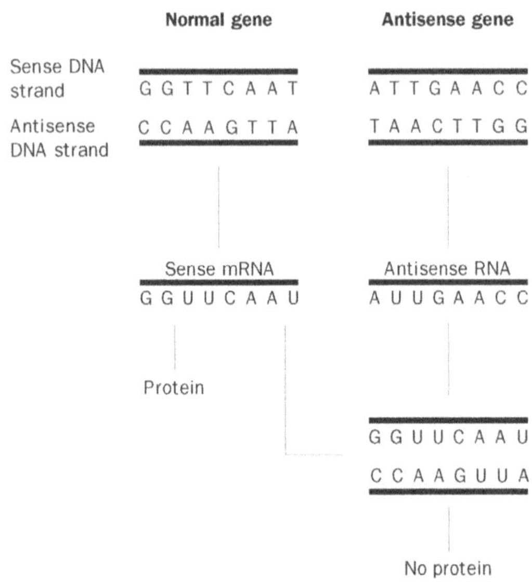

Fig. 1: Outline model for the down regulation of a gene by the use of antisense RNA technology

is inserted into host plant. After transcription of both genes, the sense RNA forms the unstable duplex with antisense RNA, thus inhibiting translation of proteins and no enzyme and pigment formation. White petunias are produced from yellow petunia by antisense RNA technology in which antisense chalcones synthase is inserted to host plant to inhibit sense chalcones synthase activity, which further inhibits the chalcone synthesis, thus inhibiting yellow pigment formation. In *Petunia hybrida*, cyanidin- and delphinidin-derivatives occur, but no pelargonidin-derivatives are produced as pigments, so no orange-red petunia occurs. This is due to the substrate specificity of enzyme dihydroflavonol 4-reductase (DFR) of petunia, which cannot reduce dihydrokaempferol and accumulates dihydrokaempferol, which shows no floral pigmentation. Meyer *et al.* (1987) reported that the transfer of the A_1 gene from *Zea mays* encoding for enzyme dihydroquercetin 4-reductase (DQR) into petunia, reduces dihydrokaempferol and provides the intermediates for pelargonidin biosynthesis, which leads to formation of **orange-red petunias**. Chalcones and aurones contribute to the **yellow colours** as observed in some flowers. Chalcones are present in yellow flowers and enzymes responsible are chalcone synthase, silencing chalcone isomerase and chalcones 2' glucosyltransferase (Ishida *et al.*, 2003; Okuhara *et al.*, 2004) in carnations and 6¢-deoxychalcones in cosmos and dahlia (Davies and Schwinn., 1997). Aurones, the bright yellow flavonoids are found in yellow flowers of *Dahlia*, *Limonium*, morning glory, snapdragon and zinnia. Nakayama *et al.* (2000) reported aurone synthase, more specifically aureusidin synthase (AS), from yellow snapdragon petals. Davies *et al.* (1998) expressed a PKR cDNA from *Medicago sativa* in a white petunia and obtained **pale-yellow** flowers that accumulated the chalcones butein 3-Oglucoside and butein 4-O-glucoside.

Carnation, chrysanthemum and rose make up over 50% of the world cut flower market but lack **true blue colouration** due to absence of delphinidin pigment, and only accumulate pelargonidin and cyanidin derivatives that are not modified with aromatic acyl groups. Thus they have become targets for attempts at engineering the synthesis of delphinidin derivatives to generate blue flowers. Over-expression of a petunia F3'5'H gene under the control of a constitutive promoter in a pelargonidin-producing carnation variety produces petals in which delphinidin derivatives contribute to about 70% of total anthocyanins. However, there was only a slight colour change towards blue. In order to increase the content of delphinidin it was necessary to avoid competition for substrates between two key endogenous enzymes of the anthocyanin pathway (DFR and F3'H) and the introduced enzyme (F3'5'H). To achieve this, white carnation cultivars were selected that were specifically deficient in the DFR gene. Expression of petunia F3'5'H and petunia DFR genes in one such DFR mutant resulted in exclusive accumulation of delphinidin and significant colour change towards blue cv. 'Moondust', the first genetically modified floricultural crop now being sold globally. Expression of a pansy F3'5'H gene and a petunia *DFR-A* gene in transgenic plants was studied which exclusively accumulated delphinidin but at a higher concentration. These flowers had a dark violet colour in cv. 'Moonshadow'. Four **delphinidin-accumulating carnation** varieties developed by **Florigene Australia** (1996) have now been sold for more than 14 years, and these are 'Moonvista', 'Moonshade', 'Moonlite', and 'Moonaqua'. **Florigene Australia** (2004) got success in producing **blue roses**. Expression of a F3'5'H gene from petunia in rose resulted in no or little delphinidin accumulation in the petals of transgenic plants, even though these genes were shown to be functional in petunia and carnation. In contrast, expression of pansy (*Viola* spp.) F3'5'H genes in rose resulted in a significant amount of delphinidin accumulating in petals of the transgenic plants (Brugliera *et al.*, 2004). Rose cultivars that have higher vacuolar pH, large amount of flavonols and no F3'H activity were selected in order to enhance the blue hue of the transgenic petals and to achieve high content of delphinidin. Expression of pansy F3'5'H genes in such cultivars resulted in transgenic lines in which 95% of the anthocyanidins was delphinidin. In order to inhibit the endogenous rose DFR for substrates it was targeted for down-regulation *via* RNAi-mediated silencing .The resulting transgenic plants in which pansy F3'5'H and Dutch iris DFR genes were also expressed, produced flowers that accumulated delphinidin with a concomitant colour change towards blue. This system is called three-gene-packages. Blue flower colour is usually the additive result of various factors such as anthocyanin modification, co-pigment concentration, vacuolar pH, and metal ion type & concentration. Targeting these factors in addition to delphinidin production in transgenic carnation and rose is expected to yield a bluer flower colour (Katsumoto *et al.*, 2007). The Dutch-based breeding company, Ludwig & Co., Holland (2013) is licensee for spectacular ***Polianthes*** series that only recently saw the addition of yellow and pink coloured tuberose varieties with high fragrance.

5. Plant Architecture Modification

Height control in chrysanthemum is a major consideration during its commercial production. Tobacco phytochrome B1 gene was when transferred into plants of 'Iridon' chrysanthemum, the transgenic plants were shorter in stature, had larger branch angles and intense green leaf colouration with higher levels of chlorophyll than wild type plants (Zheng *et al.*, 2001). The commercial application of this biotechnology could provide an economic alternative to the use of chemical growth regulators, thereby reducing production costs. The *rolC* gene was when transferred in carnation cv. 'White Sim', the transgenic carnation exhibited **increased axillary bud break**, three times more flowering stems, generated up to 48% more stem cuttings per mother plant, and the stem cuttings from *rolC* plants exhibited better rooting ability with up to five times higher root dry weight than control (Zuker *et al.*, 2001b).

Boldly coloured tuberose varieties released by Ludwig & Co. (2013)

Cultivars	Yellow Baby	Sensation	Cinderella	Double Pink	Super Gold
Flower	Single	Single	Single	Double	Single
Flower colour	Yellow	Pink	Lavender-pink	Pink	Dark yellow
Bulb size	10 cm	10 cm	10 cm	12 cm	12 cm
Stem length	35 cm	45 cm	35 cm	65 cm	65 cm

(Source: www.floraculture.au)

4. Flower Shape Modification

The ABC model (Coen and Meyerowitz, 1991) and its modified version are known to be applicable to flower shape modification. The ABC model proposed that three functionally different genes, *i.e.* A, B and C specified the four-whorl structure of flowers. Gene A is responsible for sepal development, genes A and B together specify the petals, genes B and C determine the stamens and gene C alone specifies the carpels. Homeotic floral shape gene present in *Antirrhinum* and *Arabidopsis* are deficiencies and agamous, respectively. Constitutive expression of the C gene from *Rosa rugosa* in *Torenia* resulted in a carpeloid structure in place of sepals (Kitahara *et al.*, 2004). Constitutive expression of *Antirrhinum majus* B genes DEF and GLO in transgenic *Torenia* resulted in the conversion of sepals to petals.

6. Vase Life Prolongation

Ethylene is principal hormone affecting flower longevity, which is synthesized from the petals after full opening of the flowers during senescence (Reid and Wu., 1992; Woodson *et al.*, 1992). Sleepiness of carnations (premature wilting of petals before the flowers even open) was known to be the result of ethylene leakage in greenhouse (Crocker., 1913). The biosynthesis of ethylene is essential for senescence by conversion of S-adenosyl methionine (SAM) to 1-aminocyclopropane-1-carboxylic acid (ACC) and the conversion of ACC to ethylene, catalyzed by ACC synthase (ACS) and ACC oxidase (ACO), respectively (Adams and Yang., 1979). **Through genetic engineering**, long vase life can be obtained by antisense ACCS and antisense ACCO technologies, through over-expression of ACC *deaminase* gene which from bacteria into host metabolises ACC before it converts to ethylene, though over-expression of SAM *hydrolase* which converts ACC in other metabolite than into ethylene, and through expression of *isopentenyl transferase* (*ipt*) gene giving increased level of cytokinin to delay senescence. Delayed senescence of **petunia** flower transformed with antisense ACC synthase and oxidase genes. Petunia plants transformed with antisense BoACS1 gene (broccoli ACC synthase) and antisense BoACO1 gene (broccoli ACC oxidase) show reduction in ethylene production, flowers remain fresh for longer but remain still responsive to ethylene. The delayed flower senescence was more pronounced with the antisense BoACO1 than the antisense BoACS1 (Huang *et al.*, 2007). **Antisense ACC oxidase delays carnation petal senescence**. With antisense ACO gene, transgenic carnation cultivars 'Scania' and 'White Sim' exhibit low climacteric ethylene production and delayed petal senescence by increasing vase life to 8-9 days and shows normal senescence by discolouration. However, the ACS and ACO genes can be induced in transgenic flowers by applying exogenous ethylene (Savin *et al.*, 1995).

7. Floral Scent Modification

Floral scent is typically a complex mixture of low molecular weight, mostly lipophilic secondary metabolites emitted by flowers into the atmosphere. The relative abundance and interactions between these compounds give flowers its unique characteristic fragrance. Genetic engineering of floral scent helps to uncover the effect of gene addition or modification on steady-state levels of

Genes responsible for scent production in flowering ornamental crops

Flower crop	Genes responsible (floral volatiles)	Reference
Clarkia breweri	(S)-linalool synthase (LIS) gene	Dudareva *et al.* (1996)
	Isoeugenol-O-methyl transferase (IEMT)	Wang *et al.* (1997)
	Benzyl alcohol acetyl-transferase (BEAT)	Dudareva *et al.* (1998)
	Salicylic acid carboxyl methyl transferase (SAMT)	Ross *et al.* (1999)
	Benzoic acid carboxyl methyl transferase (BAMT)	Dudareva and Pichersky (2000)
	Benzyl alcohol benzyl transferase (BEBT)	Dudareva *et al.* (2000)
Petunia hybrida *P. axillaris*	Benzoic acid/salicylic acid and carbonyl methyl transferase (BSMT)	Negre *et al.* (2003)
	Benzyl alcohol /phenyl ethanol benzyl transferase (BPBT)	Boatright *et al.* (2004)
	Geraniol/citronellol acetyl transferase	Shalit *et al.* (2003)
	Benzyl alcohol /phenyl ethanol benzyl transferase (BPBT)	Boatright *et al,* (2004)
	Orcinol-O-methyl transferase (OOMT)	
Rosa chinensis	Phloroglucinol-O-methyl transferase (POMT)	Lavid *et al.* (2002)
Antirrhinum majus (Snapdragon)	Mycene synthase Ocimene synthase	Dudareva *et al.* (2000)
Stephanotis floribunda (Madagascar Jasmine)	Salicylic acid carboxyl methyl transferase (SAMT)	Pott *et al.* (2002)
Arabidopsis thaliana	(S)-linalool synthase (LIS)	Chen *et al.* (2003)
	Caryophyllene methyl transferase (BSMT)	
Vanda Mimi Palmer	Linalool synthase (LIS), acetyl-CoA acetyltransferase (ACA), 1-deoxy-D-xylulose	
	5-phosphate synthase (DXPS), 3-hydroxy-3-methylglutaryl-coenzyme A reductase (HMGR)	

secondary metabolites and fluxes in the biosynthetic pathways. Verdonk *et al.* (2005) identified ODORANT1 (ODO1) as a key regulator of floral scent biosynthesis for the regulation of volatile benzenoids in ***Petunia hybrida*** cv. 'Mitchell' flowers. Transcript levels of ODO1 increased before the onset of volatile emission and decreased when volatile emission declined. Down regulation of ODO1 in transgenic plants strongly reduced volatile benzenoid levels through decreased synthesis of precursors from the shikimate pathway. *Arabidopsis* production of anthocyanin pigment-1 (PAP1) transcription factor was introduced into ***Rosa hybrida*** cv. 'Pariser Charme and increased levels of phenylpropanoid (colour and scent), as well as 6.5 times higher levels of terpenoid were recorded in PAP1-transgenic rose lines (Zvi *et al.*, 2012). In scentless ***Eustoma grandiflorum***, aroma was introduced through genetic engineering using *Clarkia* BEAT gene. Aranovich *et al.* (2007) transformed lisianthus plants with the *Clarkia breweri* gene coding for benzyl alcohol acetyl transferase (*BEAT*) and catalyzing the synthesis of the volatile compound benzyl acetate to induce a fragrance in their flowers. An external supply of benzyl alcohol induced 5 to 7 times higher production of benzyl acetate in detached flowers and leaves of transgenic lisianthus plants. When fed with additional alcoholic compounds, i.e. hexanol, benzyl alcohol, isoamyl alcohol, phenethyl alcohol and cinnamyl alcohol, assumed to be used as substrates by BEAT, transgenic *in vitro*-grown lisianthus plantlets produced significantly higher levels of acetates.

Hurdles in Commercializing Transgenic Ornamental Crops

There are barriers to commercialization that apply to genetically engineered varieties though not to conventional ones. The cost and difficulty of meeting regulatory requirements are the largest barriers to commercialization. These costs which primarily relate to the costs associated with analysis and risk assessment, are not borne in developing non-GE varieties. The costs incurred seeking regulatory approval for transgenic plants are estimated to be hundreds of thousands of dollars. Genetically modified carnation with improved vase life has been developed, but not commercialized because of the cost of regulation. Colour-modified varieties of the potted plant *Torenia* have also been developed but not been commercialized, even though the plant is male and female sterile and has no capacity for gene flow. Again, the cost of obtaining regulatory approval is the major factor preventing commercialization. Higher costs are required to provide a detailed molecular characterization, including a verified PCR-based identification test unique for each event, for which a fee of 30,000 Euro per event must be paid prior to validation. Promising new non-GE ornamental varieties can be quickly field-tested around the world to evaluate commercial potential, but genetically engineered plants must receive prior regulatory approval. The approval process can take many months or even several years. Because the regulation of ornamental plants at both the national and international levels falls under the same legislation that is applied to major food crops, the regulatory scrutiny applied to ornamentals is generally equivalent to crops with the exception of food safety studies.

Options to Reduce the Cost of Regulation

- As an alternative to PCR-based unique identification, varieties could be protected by plant breeders.
- Legislation could be modified to allow approval based on phenotype and species, or transformation vector and species, rather than requiring a separate regulatory assessment of every transgenic event.
- The amount of molecular characterization required for non-food genetically engineered varieties could be simplified, reducing the cost of analysis.
- In terms of environmental risk assessment legislation, more emphasis could be placed on the lower capacity for gene flow exhibited by plants that are vegetatively propagated or that are completely sterile, such as the *Torenia* colour modification.
- Legislation could emphasize greater consideration of a history of safe use. For example, colour-modified transgenic carnation varieties have been commercially available for more than 14 years and have a proven history of safe use. This history of safe use could be a valid justification for reducing the amount of molecular characterization required (*Chandler, 2013).*

References

Adams, D. O. and S.F. Yang, 1979. Ethylene biosynthesis: Identification of 1- aminocyclopropane-1 carboxylic acid as an intermediate in the conversion of methionine to ethylene. *Proc Natl Acad Sci USA* **76**: 170–174.

Aranovich, D., E. Lewinsohn and M. Zaccai, 2007. Post-harvest enhancement of aroma in transgenic lisianthus (*Eustoma grandiflorum*) using the *Clarkia breweri* benzyl alcohol acetyltransferase (*BEAT*) gene. *Postharvest Bio and Tech.,* **43**: 255–260.

Bi, Y.M., B.P.A. Cammue, P.H. Goodwin, S. Krishna Raj and P.K. Saxena, 1999. Resistance to *Botrytis cinerea* in scented geranium transformed with a gene encoding the antimicrobial protein Ace-AMP1. *Plant Cell Rep.,* **18**: 835-840.

Boatright, J., F. Negre, X. Chen, C.M. Kish, B. Wood, G. Peel, I. Orlova, D. Gang, D. Rhodes and N. Dudareva, 2004. Understanding *in vivo* benzenoid metabolism in petunia petal tissue. *Plant Physiol.,* **135**: 1993-2011.

Brugliera, F., Y. Tanaka and J. Mason, 2004. Flavonoid 3', 5'-hydroxylase gene sequences and uses therefore. Patent Pulication No. US7612257.

Chandler, S. 2013. Genetically engineered ornamental plants: Regulatory hurdles to commercialization. *ISB News Report.*

Chen, F., D. Tholl, J.C. D'Auria, A. Farooq, E. Pichersky and J. Gershenzon, 2003. Biosynthesis and emission of terpenoid volatiles from *Arabidopsis* flowers. *Plant Cell,* **15**: 481-494.

Coen, E. S. and E.M. Meyerowitz, 1991. The war of the whorls: genetic interactions controlling flower development. *Nature,* **353**: 31–37.

Courtney-Gutterson, N., C. Napoli, C. Lemieux, A. Morgan, E. Firoozababy and K.E.P. Robinson, 1994. Modification of flower color in Florist's chrysanthemum: production of a white-flowering variety through molecular genetics. *BioTech.,* **12**: 268-271.

Crocker, W. 1913. The effects of advancing civilization upon plants. *School Sci Math.,* **13**: 277–289.

Courtney-Gutterson, N., C. Napoli, C. Lemieux, A. Morgan, E. Firoozababy and K.E.P. Robinson, 1994. Modification of flower color in Florist's Chrysanthemum: production of a white-flowering variety through molecular genetics. *Bio Technol.,* **12**: 268–271.

Davies, K. M. and K.E. Schwinn, 1997. *Biotechnology of Ornamental Plants.* In: Biotechnology of Ornamental Plants (eds Geneve R.L., J.E.Preece and S.A. Markle), pp. 259-294. CAB International, Wallingford.

Davies, K. M., S.J. Bloor, G.B. Spiller and S.C. Deroles, 1998. Production of yellow colour in flowers: redirection of flavonoid biosynthesis in *Petunia. Plant J.,* **13**: 259–266.

Deroles, S.C.. M.R. Boase, C.E. Lee and T.A. Peters, 2002. Gene transfer to plants. In: Breeding for Ornamentals: Classical and Molecular Approaches (ed. Veinstein, A.), pp. 155-196. Kluwer Academic Publishers, Dordrecht, England.

Dudareva, N., J.C. D'Auria, K.H. Nam, R.A. Raguso and E. Pichersky, 1998. Acetyl-CoA: benzylalcohol acetyltransferase: an enzyme involved in floral scent production in *Clarkia breweri. Plant J.,* **14**: 297-304.

Dudareva, N., D. Martin, C.M. Kish, N. Kolosova, N. Gorenstein, J. Faldt, B. Miller and J. Bohlman, 2000. (E)-b-Ocimene and myrcene synthase genes of floral scent biosynthesis in snapdragon: function and expression of three terpene synthase genes of a new TPS-subfamily. *Plant Cell,* **15**: 1227-1241.

Dudareva, N. and E. Pichersky, 2000. Biochemical and molecular genetic aspects of floral scents. *Plant Physiol.,* **122**: 627-633.

Dudareva, N., L. Cseke, V.M. Blanc and E. Pichersky, 1996. Evolution of floral scent in *Clarkia*: Novel patterns of S-linalool synthase gene expression in the *C. breweri* flower. *Plant Cell,* **8**: 1137-1148.

Elomaa, P., J. Honkanen, R. Puska, P. Seppanen, Y. Helariutta, M. Mehto, M. Kotilainen, L. Nevalainen and T.H. Teeri, 1993. *Agrobacterium*-mediated transfer of antisense chalcone synthase cDNA to *Gerbera hybrida* inhibits flower pigmentation. *BioTech.,* **11**: 508-511.

Forkmann, G. 1991. Flavonoids as flower pigments: the formation of the natural spectrum and its extension by genetic engineering. *Plant Breed.*, **106**: 1-26.

Grieshach, R.J. 1994. An improved method for transforming plants through electrophoresis. *Plant Sci.*, **102**: 81-89.

Gutterson, N. 1995. Anthocyanin biosynthetic genes and their application to flower colour modification through sense suppression. *HortSci.*, **30**: 964-966.

Hansen, G., R.D. Shillito and M.D. Chilton, 1997. T-strand integration in maize protoplasts after co-delivery of a T-DNA substrate and virulence genes. *Proc. Natl. Acad. Sci.* USA, **94**: 11726-11730.

Hong, B., C. Ma, Y. Yang, T. Wang, K.Y. Shinozaki and J. Gao, 2009. Over-expression of *AtDREB1A* in chrysanthemum enhances tolerance to heat stress. *Plant Mol. Biol.*, **70**: 231-240.

Huang, L., U. Lai, S. Yang, M. Chua, C. Kuo, M. Tsai and C. Sunb, 2007. Delayed flower senescence of *Petunia hybrida* plants transformed with antisense broccoli ACC synthase and ACC oxidase genes. *Postharvest Bio and Techn.*, **46**: 47-53.

Ishida, M., J. Ogata, H. Yoshida, Y. Itoh and Y. Ozeki, 2003. Isolation of a cDNA for the chalcones 2¢-glucosyltransferase gene and its expression profile in carnation flowers. *Plant and Cell Physiol.*, **44**: 158.

Kamo, K., A. Blowers, F. Smith and J. van Eck, 1995. Stable transformation of *Gladiolus* by particle gun bombardment of cormels. *Plant Sci.*, **110**: 105-111.

Katsumoto, Y., M. Mizutani, Y. Fukui, F. Brugliera, T. Holton, M. Karan, N. Nakamura, K. Yonekura-Sasakibara, J. Togami, A. Pigeaire, G.Q. Tao, N. Nehra, C.Y. Lu, B. Dyson, S. Tsuda, T. Ashikari, T. Kusumi, J. Mason and Y. Tanaka, 2007. Engineering of the rose flavonoid biosynthesic pathway successfully generated blue-hued flowers accumulating delphinidin. *Plant Cell Physiol.*, **48**: 1589-1600.

Kitahara, K., Y. Hibino, R. Aida and S. Matsumoto, 2004. Ectopic expression of the rose AGAMOUS-like MADS-box genes 'MASAKO C1 and D1' causes similar homeotic transformation of sepal and petal in *Arabidopsis* and sepal in *Torenia. Plant Sci.*, **166**: 1245-1252.

Lavid, N., J.H. Wang, M. Shalit, I. Guterman, E. Bar, T. Beuerle, N. Menda, S. Shafir, D. Zamir and Z. Adam, 2002. O-methyltransferases involved in the biosynthesis of volatile phenolic derivatives in rose petals. *Plant Physiol.*, **129**: 1899-1907.

Li, X., K. Gasic, B. Cammue, W. Brockaert and S.S. Korban, 2003. Transgenic rose lines harboring an antimicrobial protein gene, Ace-AMPl, demonstrate enhanced resistance to powdery mildew (*Sphaerotheca pannosa*). *Planta*, **218**: 226-232.

Liau, C. H., J.C. Lu, V. Prasad, H.H. Hsiao, S.J. You, L.T. Lee, N.S. Yang, H.E. Huang, T.Y. Feng, W.H. Chen and M.T. Chan, 2003. The sweet pepper ferrodoxin~like protein (pflp) conferred resistance against soft rot disease in *Oncidium* orchid. *Transgenic Res.*, **12**: 329-336.

Marchant, R., M.R. Davey, J.A. Lucas, C.J. Lamb, R.A. Dixon and J.B. Power, 1998. Expression of a chitinase transgene in rose (*Rosa hybrida* L.) reduces development of blackspot disease (*Diplocarpon rosae* Wolf). *Molecular Breed.*, **4**: 187-194.

Meyer, P., I. Heidmann, G. Forkmann and I.I. Sacdler, 1987. A new petunia flower colour generated by transformation of a mutant with a maize gene. *Nature*, **330**: 677-678.

Nakayama, T., K. Yonekura-Sakakibara, T. Sato, S. Kikuchi, Y. Fukui, M. Fukuchi-Mizutani, T. Ueda, M. Nakao, Y. Tanaka, T. Kusumi and T. Nishino, 2000. Aureusidin synthase: a polyphenol oxidase homolog responsible for flower coloration. *Sci.*, **290**: 1163-1166.

Negre, F., N. Kolosova, J. Knoll, C.M. Kish and N. Dudareva, 2003. Novel S-adenosyl-L-methionine:salicylic acid carboxyl methyltransferase, an enzyme responsible for biosynthesis of methyl salicylate and methyl benzoate, is not involved in floral scent production in snapdragon flowers. *Archives Biochem. and Biophysics*, **406**: 261-270.

Ogawa, T., T. Toguri, H. Kudoh, M. Okamura, T. Momma, M. Yoshioka, K. Kato, Y. Hagiwara and T. Sano, 2005. Double-stranded RNA-specific ribonuclease confers tolerance against chrysanthemum stunt viroid and tomato spotted wilt virus in transgenic chrysanthemum plants. *Breeding Sci.*, **55**: 49-55.

Okuhara, H., K. Ishiguro, C. Hirose, M. Gao, J. Togami, N. Nakamura, E. Ono, M. Ochiai, Y. Fukui, M. Yanaguchi and Y. Tanaka, 2004. Molecular cloning and functional expression of tetrahydroxychalcone 2¢-glucosyltransferase genes. *Plant Cell Physiol.*, **45**: 133.

Pott, M. B., E. Pichersky and B. Piechulla, 2002. Evening specific oscillations of scent emission, SAMT enzyme activity, and SAMT mRNA in flowers of *Stephanotis floribunda*. *J. Plant Physiol.*, **159**: 925-934.

Reid, M. S. and M.J. Wu, 1992. Ethylene and flower senescence. *Plant Growth Reg.*, **11**: 37–43.

Renou, J. P., M.S. Hanteville, J.P. Narcy, A. Diolez and D. Florack, 2000. Evaluation of the protection against *Xanthomonas* in transgenic *Pelargonium* containing a chimaeric cecropin gene. *Acta Hort.*, No. 508, pp. 323-325.

Robinson, K.L.P. and E. Firoozabady, 1993. Transformation of floriculture crops. *Sci Hort.*, **55**: 83-99.

Ross, J. R., K.H. Nam, J.C. D'Auria and E. Pichersky, 1999. S-adenosyl-Lmethionine: salicylic acid carboxyl methyltransferase, an enzyme involved in floral scent production and plant defense, represents a new class of plant methyltransferases. *Archives Biochem. and Biophysics*, **367**: 9-16.

Savin, K. W., S.C. Baudinette, M.W. Graham, M.Z. Michael, G.D. Nugent, C. Lu, S.F. Chandler and E.C. Cornish, 1995. Antisense ACC oxidase RNA delays carnation petal senescence. *HortSci.*, **30**(5): 970-972.

Shalit, M., I. Guterman, H. Volpin, E. Bar, T. Tamari, N. Menda, Z. Adam, D. Zamir, A. Vainstein and D. Weiss, 2003. Volatile ester formation in roses: identification of an acetyl-CoA: geraniol acetyltransferase in developing rose petals. *Plant Physiol.*, **131**: 1868-1876.

Shinoyama, H., A. Mochizuki, M. Komano, Y. Nomura and T. Nagai, 2003. Insect resistance in transgenic chrysanthemum [*Dendranthema* × *grandiflorum* (Ramat.) Kitamura] by the introduction of a modified δ-endotoxin gene of *Bacillus thuringiensis*. *Breeding Sci.*, **53**: 359-367.

Stafford, H.A. 1990. Flavonoid Metabolism. CRC Press, Boca Raton, Florida, USA.

Takatsu, Y., Y. Nishizawa, T. Hibi and K. Akutsu, 1999. Transgenic chrysanthemum (*Dendranthema grandiflorum* (Ramat.) Kitamura) expressing a rice chitinase gene shows enhanced resistance to grey mold (*Botrytis cinerea*). *Scientia Hort.*, **82**: 113-123.

Van der Krol, A.R., P.E. Lenting, J. Veenstra, I.M. van der Meer, R.E. Koes, A.G.M. Gerats, J.N.M. Mol and A.R. Stuitje, 1988. An antisense chalcone synthase gene in transgenic plants inhibits flower pigmentation, *Nature*, **333**: 866-869.

Verdonk, J. C., M.A. Haring, A.J. Tunen and R.C. Schuurink, 2005. ODORANT1 regulates fragrance biosynthesis in petunia flowers. *The Plant Cell*, **17**: 1612–1624.

Wang, J., N. Dudareva, S. Bhakta, R.A. Raguso and E. Pichersky, 1997. Floral scent production in *Clarkia breweri* (Onagraceae). II. Localization and developmental modulation of the enzyme S-adenosyl-L-methionine: (iso)eugenol O-methyltransferase and phenylpropanoid emission. *Plant Physiol.*, **114**: 213-221.

Woodson, W. R., K.Y. Park, A. Drory, P.B. Larsen and H. Wang, 1992. Expression of ethylene biosynthetic pathway transcripts in senescing carnation flowers. *Plant Physiol.*, **99**: 526–532.

Zheng, Z., Z. Yang, J. Jang and J. Metzger, 2001. Modification of plant architecture in chrysanthemum by ectopic expression of the tobacco phytochrome *b1* gene. *J. Amer. Soc. hort. Sci.*, **126**(1): 19–26.

Zuker, A., E. Shklarman, G. Scovel, H. Ben-Meir, M. Ovadis and I. Neta-Sharir, 2001a. Genetic engineering of agronomic and ornamental traits in carnation. *Plant Sci.*, **560**: 91-94.

Zuker, A., T. Tzfira, G. Scovel, M. Ovadis, E. Shklarman, H. Itzhaki and A. Vainstein, 2001b. Rol-C transgenic carnation with improved horticultural traits: quantitative and qualitative analyses of greenhouse-grown plants. *J. Amer. Soc. hort. Sci.*, **126**(1): 13–18.

Zuker, A., T. Tzfira, H. Ben-Meir, M. Ovadis, E. Shklarman, H. Itzhaki, G. Forkmann, S. Martens, I. Nata-Sharir, D. Weiss and A. Vainstein, 2002. Modification of flower colour and fragrance by antisense suppression of the flavanone 3- hydroxylase gene. *Mol. Breed.*, **9**: 33-41.

Zvi, M. M. B., E. Shklarman, T. Masci, H. Kalev, T. Debener, S. Shafir, M. Ovadis and A. Vainstein, 2012. PAP1 transcription factor enhances production of phenylpropanoid and terpenoid scent compounds in rose flowers. *Newe Phytol.*, **195**: 335-345.

32

Nutraceutical Pigments from Ornamental Crops (Scope, Potential and *In Vitro* Expression)

K.V. Prasad, T. Janakiram, M.R. Legha, Surendra Kumar, B.S. Hada and Prativa Lakhotia

Introduction

Natural pigments including anthocyanins, carotenoids, betalains, shikonin, and other pigmented phytochemicals are conspicuously found in most of the plants. The anthocyanins, carotenoids, and betalains play important ecological functions in the plants (Grotewold, 2006). Among the different group of pigments available in the plants, anthocyanins and carotenoids are more frequently exploited as the source of nutraceuticals to address a number of human ailments. Recent increase in demand for pharmaceutical, nutraceutical and industrial products result from a focus on health and quality of life. The beneficial phytochemicals have experienced exponential growth in popularity as consumer awareness increased about the alternative medicinal properties of these phytochemicals. This increase in demand has resulted in increased production of these nutraceutical pigments from a number of crops. Anthocyanins are water-soluble vacuolar plant pigments which appear in plants in colours ranging from scarlet to blue. Accumulation of anthocyanins is observed in flowers, fruits, leaves and storage tissues of many plants (Schwinn and Davies, 2004; Tanaka *et al.*, 2008). The anthocyanin accumulation is observed throughout entire plant life. In some plants, these anthocyanins accumulate temporarily in a specific developmental stage or transiently by environmental stresses such as wounding, UV irradiation, drought, pathogen attack and nutrient deficiency (Chalker-Scott, 1999; Harborne and Williams, 2000; Davies, 2004; Grotewold, 2006; Shimizu *et al.*, 2010). The interesting biological activities of anthocyanins with several well recognized dietary benefits in human health have prompted intensive research on these compounds and their effects on human health. Besides being present in the human diet at relatively high concentrations, anthocyanins are not only used as the alternative to artificial red dyes that colour food, which is done not only for safety considerations but also as for their medicinal properties (Sato *et al.*, 1996; Gantet and Memelink, 2002; Hou, 2003; Lila, 2004; Stintzing and Carle, 2004). Recently, interest in anthocyanins has been intensified because these pigments are not only effective antioxidants but they also have multifaceted nutraceutical and pharmaceutical activities independent of their antioxidant capacities that produce health benefits such as antitumor, antiproliferative, anticancer activities, suppression of inflammatory responses, protection against age-related decline in cognitive behaviour and neuronal dysfunction in the central nervous system (Joseph *et al.*, 1999; Downham and Collins, 2000; Kong *et al.*, 2003; Tall *et al.*, 2004; Zhang *et al.*, 2005; McDougall *et al.*, 2005; Simões *et al.*, 2009; de Pascual-Teresa *et al.*, 2010; Ram *et al.*, 2011).

The extraction of pigments from the fresh plant tissues still has some problems such as low metabolite yield, seasonal availability, fast deterioration, inconsistent product quality and pigment degradation due to extraction and storage process (Lila, 2004; Ram *et al.*, 2011). Plant tissue culture technologies offer strategies to minimize these problems, allowing the continuous production of secondary metabolites of commercial importance, offer the rare opportunity to tailor the chemical profile of a phytochemical product, and provides enhanced control over the chemical and physical environments to produce a compound of potentially more value for human use (Meyer *et al.*, 2002; Karuppusamy, 2009; Simões *et al.*, 2009). Moreover, the growing demand in today's marketplace for natural, renewable products coupled with the low product yields and supply concerns of isolation from plant has refocused attention on *in vitro* plant materials as potential factories for secondary phytochemical products, and has paved the way for new research exploring secondary product expression *in vitro* (Mulabagal and Tsay, 2004; Karuppusamy, 2009). Therefore, *in vitro* techniques have been used as the alternative for the production of secondary metabolites including anthocyanins. Anthocyanins can be produced by using different biotechnological approaches, such as cell/callus culture techniques (Jedinák *et al.*, 2004) and there have been several reports regarding the production of anthocyanins using plant tissue cultures. It has been reported that anthocyanin production in cell/callus cultures is stimulated by various culture conditions. Exogenous application of jasmonates greatly stimulated the biosynthesis of a wide range of secondary metabolites in cell suspension cultures (Mizukami *et al.*, 1993; Blechert *et al.*, 1995; Curtin *et al.*, 2003) and in intact plants (Aerts *et al.*, 1994; Rudell and Mattheis, 2008). Elicitation has been shown to be the most efficient strategy that directs to the enhancement in anthocyanin production in plant cell cultures (Zhang and Furusaki, 1999). Methyl jasmonate (MeJA) has been successfully used as an elicitor in some plant species for enhancing the production of secondary metabolites in the cell cultures (Fang *et al.*, 1999; Kim *et al.*, 2004; Thanh *et al.*, 2005; Shimizu *et al.*, 2010). Recently, See *et al.* (2011) showed that MeJA stimulated anthocyanin production in cell suspension cultures of *Melastoma malabathricum*. Zhang *et al.* (2002) reported that anthocyanins accumulation was enhanced by MeJA in suspension culture of *Vitis vinifera*. Some reports suggested that jasmonates function as a messenger in the activation of secondary metabolites including anthocyanins and as a signal transducer under environmental stress (Saniewski *et al.*, 2003). Moreover, these regulators induce defence-related gene expression and biosynthesis of secondary metabolites (Pozo *et al.*, 2005; Balbi and Devoto, 2008).

Scope and Potential

Anthocyanins and Associated Flavonoids

The term anthocyanin is derived from the Greek words meaning flower and blue. Anthocyanins are natural pigments belonging to the flavonoid family. They are responsible for the blue, purple, red and orange colour of many fruits, vegetables and flowers. Anthocyanins constitute a major flavonoid group to facilitate the most widespread secondary metabolites found in most of the terrestrial vascular plants. These natural phenolic substances with variable chemical structures are responsible for cynic colours ranging from salmon and pink through red and violet to dark blue of most flowers, fruits, and leaves of angiosperms (Stafford, 1991; Harborne and Williams, 2000). Over 500 structurally distinct anthocyanins have been identified in nature (de Pascual-Teresa and Sanchez-Ballesta, 2008). The primary player in this group includes anthocyanins, flavonols, flavones, flavanones, and flavan-3-ols. Anthocyanins are glycosides and acylglycosides of water-soluble anthocyanidins. Anthocyanidins vary with different hydroxyl or methoxyl substitutions in their basic structure, flavylium (2-phenylbenzopyrilium) (Andersen and Jordheim, 2006; Table 1.1). These pigments in the diverse flavonoid groups are largely responsible for visually striking colours from shades of red to crimson and blue to purple in flowers, fruits, leaves, and storage organs of plants (Zhang and Furusaki, 1999). Over 500 structurally distinct anthocyanins have been identified in nature (Pascual-Teresa and Sanchez-Ballesta, 2008). Anthocyanins play a major role in plant pollination and attracting insects into the trap apparatus of carnivorous plants, protect the plants from harmful effects of UV rays and work against the invasion of hervivores, insects and microbes (Grotewold, 2006). Anthocyanin pigments serve as a UV screen to plants and protect the plant DNA from damage by sunlight (Hennayake *et al.*, 2006). Phytochemicals in this class are frequently referred to as bioflavonoids due to their multifaceted roles in human health and diet (Skibola and Smith, 2000). The interesting biological activities of anthocyanins have impelled intensive research on the physiological properties of these compounds and their effects on human health. Wide daily diet. These anthocyanins are variously associated with sensory and nutritional quality of our plant foods. Their pharmacological potential (Lila, 2004) together with specific biological activities such as being antioxidants (Acquaviva *et al.*, 2003), anti-inflammatory (Wang and Mazza, 2002), anti-viral (Jassim and Naji, 2003), and cardio-protective (Narayana *et al.*, 2001) could elucidate 'why' and 'when' consumed regularly in the human diet, these compounds have pleiotropic health promoting and ailment preventing activities (Gantet and Memelink, 2002). Majority of the works on anthocyanins focus of the antioxidant characteristics of anthocyanin rich diets and the enhancement of cardio-vascular health (Ganet and Memelink, 2002). At the moment, interest in anthocyanin pigments has intensified since these can be used not only as food and beverage additives to attain the attractive natural red colouration (Jackman *et al.*, 1987), but also used as medicine for inhibiting the growth of cancer cells and tumour formation (Bagchi *et al.*, 2004). In addition to above, anthocyanin pigments also prevent the occurrence of coronary heart diseases and fragility of capillary vessels (Zhang and Furusaki, 1999).

Table 1.1: Chemical structures of naturally occurring six common anthocyanidins. The outlying column on the right side shows the basic structure of anthocyanidin through which most of the anthocyanin pigments are derived.

Anthocyanidin	$R^{3'}$	$R^{4'}$	$R^{5'}$	R^3	R^5	R^6	R^7	Basic structure
Cyanidin (Cy)	–OH	–OH	–H	–OH	–OH	–H	–OH	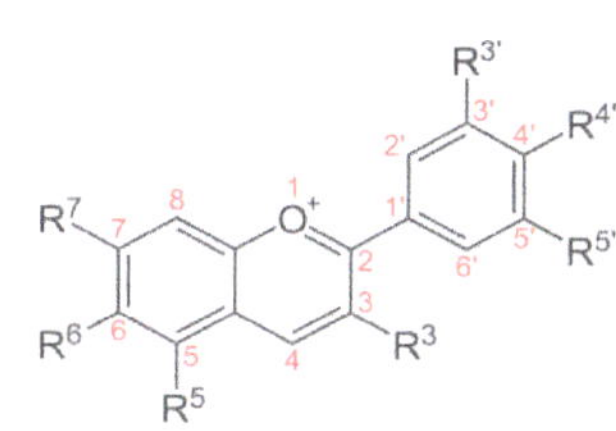
Delphinidin (Dp)	–OH	–OH	–OH	–OH	–OH	–H	–OH	
Pelargonidin (Pg)	–H	–OH	–H	–OH	–OH	–H	–OH	
Malvidin (Mv)	$–OCH_3$	–OH	$–OCH_3$	–OH	–OH	–H	–OH	
Peonidin (Pn)	$–OCH_3$	–OH	–H	–OH	–OH	–H	–OH	
Petunidin (Pt)	–OH	–OH	$–OCH_3$	–OH	–OH	–H	–OH	

occurrence, complex diversity and manifold functions have made anthocyanins a very attractive system for research on a biochemical level. Besides being present in the human diet at relatively high concentrations, anthocyanins are becoming a vital alternative to synthetic dyes because of increased public anxiety over the safety of artificial foodstuff colouring agents (Stintzing and Carle, 2004; Chan *et al.*, 2010). Phytochemicals in this class are often referred as bioflavonoids due to their multifaceted roles in human health and diet (Skibola and Smith, 2000). Anthocyanins are ubiquitous in plant foods and beverages and therefore a significant quantity is consumed in our

Of late, molecular structure of anthocyanins, their biosynthetic pathways including enzymes and genes involved in these pathways and regulatory mechanisms of their expression have been elucidated in detail in high opinion to the development of flower and fruit colours. Expression of the anthocyanin biosynthetic pathway and the buildup of these pigments in tissues are utterly determined in the plant species (Nagira and Ozeki, 2004). Production of anthocyanin pigments *via in vitro* cell cultures has been obtained from several plant species. However, research on regulation of anthocyanins biosynthesis under *in vitro* conditions is still limited, restricting the commercial application of their production by plant cell cultures. Therefore, much effort has been directed toward understanding the regulatory mechanisms of biosynthesis of anthocyanins (Hennayake *et al.*, 2006; Simões *et al.*, 2009). In anthocyanins biosynthetic pathway, two classes of genes have been extensively characterized; the structural genes encoding the enzymes so as to directly participate in the formation of anthocyanins and the regulatory genes that control the transcription of the structural genes (Guo *et al.*, 2008).

Recently a significant finding revealed the role of anthocyanin pigments protecting against **neurological disorders**, especially declines related to aging. *In vivo* evidence demonstrating that rats administered an anthocyanin-rich diet showed significant improvements in protection against age-associated impairments (Hou, 2003). Some of the researchers demonstrated not only protection but also reversal of age-related decrements when rats were fed a diet that comprised of 2% blueberry extract. Anthocyanins were found to confer significant protection from oxidative stress and to be highly bioavailable in endothelial cells, which has direct relevance to neurodegenerative disorders. These polyphenolic compounds enhance red blood cell resistance to oxidative stress. Anthocyanin-rich fruits may be beneficial in reversing the course of neuronal and behavioural aging. Moreover, some of the researchers also concluded that the blueberry extracts have the beneficial effect in Alzheimer's disease (a neurodegenerative illness) in transgenic mice. **Prevention of cardiovascular disease** is possibly the most studied subject and the one for which a great number of epidemiological evidence exists (Zhang and Furusaki, 1999). Anthocyanins are capable of acting on different cells involved in the development of atherosclerosis, one of the leading causes to cardiovascular dysfunction. Vascular endothelial growth factor (VEGF) is a major pro-angiogenic and proatherosclerotic factor but anthocyanins have been shown to prevent the expression of VEGF in vascular smooth muscle cells. Anthocyanin extracts from chokeberry, bilberry and elderberry have shown endothelium-dependent relaxation capacity in porcine coronary arteries. Intake of flavonoids, including tannins and anthocyanins, is associated with reduced incidence of coronary heart disease. A typical daily dietary intake (25–215 mg) without supplementation is sufficient to provide pharmacological benefits. The rich anthocyanin pigment content of red wines, as well as associated flavonoids, are purported to be responsible for the well-publicized correlation between red wine consumption and reduced cardiovascular mortality (the French paradox). Because anthocyanins are potent inhibitors of nitrated tyrosine formation *in vitro*, they may have a role in prevention of atherosclerotic lesions in human coronary arteries (Narayanan *et al.*, 2001). Bioflavonoid mixtures derived from fruit sources (*Rubus occidentalis*, and bilberry) have demonstrated ability in accelerating ethanol metabolism, reducing inflammatory and edematic symptoms, and improving permeability and strength of capillaries (Zhang and Furusaki, 1999). Anthocyanins in red wines and grape juice contributed to the ability of these dietary sources to inhibit heart attacks, by virtue of their antithrombotic potential, and to inhibit platelet function and enhance nitric oxide release. Moreover, anthocyanins have an effect on cholesterol distribution, protecting endothelial cells. Anthocyanins possess anti-inflammatory as well as chemopreventive properties, and therefore are used as **anticancer**, **antitumor** and **antimutagenic** agents. It has been shown that the anthocyanin-containing fraction of red wines is capable of suppressing proliferation of human colon cancer cells and gastric adenocarcinoma (AGS) cells. The proliferation of human hepatocellular liver carcinoma cells is inhibited by raspberry extracts, containing anthocyanins among other flavonoids (Lila, 2004). Some studies reported an inhibitory effect of the anthocyanins on cell proliferation in human vulva carcinoma cells and human colon carcinoma cells. Angiogenesis plays a critical role in the development of cancer. The treatment of human fibroblasts with cyanidin-3-glucoside prevents ochratoxin A-induced DNA damage. Anthocyanins are known to inhibit cycloxygenase enzymes, which can be one marker for the initiation stage of carcinogenesis. Recently, anthocyanins from tart cherries reduced cell growth of human colon cancer cell lines. Anthocyanins and related flavonoids from berries including cranberry, lingonberry, bilberry and blueberry have demonstrated the ability to inhibit not only the initiation stages of chemically induced carcinogenesis but also the later promotion and proliferation stages (Bagchi *et al.*, 2004). Basically anthocyanins may have a beneficial effect on **diabetes** at two levels. On one hand they can interfere with glucose absorption and on the other hand they may have a protective effect on pancreatic cells (Lila, 2004). It has been shown that in rats the simultaneous and daily administration of anthocyanins from red wine together with streptozotocin, a diabetogenic compound, decreases glucose concentration in urine and serum. Additionally, anthocyanins from red wine prevent the generation of free oxygen radicals, and decreases lipid peroxidation. Anthocyanins are also able to stimulate insulin secretion from rodent pancreatic beta-cells. Anthocyanins from *Hibiscus sabdariffa* have shown to protect rat liver from **hepatotoxicity** induced by terbutyl hydroperoxide by lowering the serum levels of alanine and aspartate amino-transferase and reducing oxidative liver damage. In rats in which hepatic injury was induced by the administration of D-galactosamine red potato extracts had a protective effect (Lila, 2004). The protective effect of anthocyanins on **gastric damage** is closely related with the capacity of this group of flavonoids to prevent or ameliorate the inflammatory process. Inflammation is implicated, at least initially, in processes of gastric injury and in this regard, it has been shown that cyanidin protects gastric mucosa from the damage caused by aspirin. Today, the interest in anthocyanin pigments has intensified since these can be used not only as food and beverage additives to obtain attractive natural red colouration (Jackman *et al.*, 1987) but also as medicine for inhibiting growth of cancer cells and tumor formation (Hou, 2003).

The most extensively documented phyto-medicinal role of anthocyanin pigments is in improving **eyesight**, including night vision. Black currant (*Ribes nigrum*) anthocyanins concentrated from juice, when is administered to human subjects, markedly improves visual adaptation to darkness. Anthocyanin pigments have been administered to remedy vision disorders, enhance visual acuity, and to increase capillary resistance. After oral administration of three anthocyanin species to rats, intact pigments can be found in plasma. Similar results have been achieved subsequently in human subjects. Retinopathy and cataracts, serious consequences of diabetes mellitus, can be combated using plant-derived anthocyanin pigments. Anthocyanin extracts from a wide range of plant species including *Ipomoea batatas* and *Brassica oleracea* have demonstrated ability to inhibit AGH (a-glucosidase) enzyme. Consumption of black currant-based extracts or concentrates of anthocyanins has a positive effect on night vision, i.e. dark adaptation.

The effect of anthocyanins on **microbial**, including **viral pathogens** has so far not been studied in depth. Only a few studies exist, however, on the effect of berry extracts on pathogens. Anthocyanins exert significant antimicrobial properties, and in association with other flavonoids has demonstrated quite effective inhibition of aflatoxin biosynthesis (Lila, 2004). Extracts from different berries had an inhibitory effect on gram-negative bacterial growth. Recent studies have demonstrated that blackcurrant concentrates inhibit the growth of different strains of *Staphylococcus aureus* and *Escherichia coli*. Different cyanidin glycosides present in extracts of *Ribes nigrum* have shown antiviral properties towards influenza A and B viruses and herpes-1 virus.

In Vitro Induction of Anthocyanin Pigments

The developments in *in vitro* induction of plant pigments especially anthocyanins, as well as carotenoids are highlighted here with. Conventionally the flowers and leaves are used as sources for the extraction of nutraceutical pigments by solvent extraction method. However, the plant pigments are localized in the flowers and other young vegetative parts thereby restricting their availability only during the cropping season. On the other hand, natural pigments obtained from the fresh plant materials involve the limitations such as low metabolite yield, seasonal availability of raw material, fresh material losses, inconsistent product quality, and pigment degradation caused by storage and the extraction process. Therefore, there is a need to develop a system so as to freely express such nutraceutical pigments at will in controlled conditions so as to harvest them without seasonal boundaries. Plant pigments can be produced by growing plant cells through tissue culture. A considerable amount of research efforts has been directed towards this goal, ranging from strain improvement, medium optimization, and optimization of culture conditions, elicitation and intelligent process strategies. To achieve the commercialization of plant cell suspension technology for pigment production, the major task of related research and development is the achievement of large-scale cultivation of high yielding cell lines, high product level and productivity and the stability of production over extended cultivation time. Different strategies are used to enhance pigment production, which may contribute to scale up and commercialization of plant cell cultures for the production of pigment-based food colorants. The protocols, thus developed would be helpful to upgrade the technology to pilot scale bioreactors to manufacture the pigments in *in vitro* conditions throughout the year in a highly controlled manner. There is a series of distinct advantages to producing plant pigments in *in vitro* culture, rather than *in vivo*. These include the:

- Independent of seasonal boundaries
- *In vitro* culture is a superb model to test elicitation
- Production can be more reliable, simpler, and more predictable
- Interfering compounds that occur in the field-grown plant can be avoided
- *In vitro* cultures can yield a source of defined standard pigments in large volumes
- Compounds produced in *in vitro* cultures can be directly parallel as in the whole plant
- Isolation of pigments can be rapid and efficient, as compared to extraction from complex whole plant.

Earlier attempts were focused on fruit and vegetable crops like grapes, strawberry, blackberry, carrot, radish, etc. though the work on flower crops which exhibit the maximum pigments is negligible, except to some degree in oxalis. The capacity for plant cell, tissue, and organ cultures to produce and accumulate plant pigments has been recognized almost since the inception of *in vitro* technology. With the increasing understanding of the biosynthesis pathway and regulation mechanisms of pigments in the plants, a move has been explored for the development of *in vitro* pigment production technology. The research efforts used to enhance pigment production through *in vitro* technology are reviewed here with.

For **medium optimization**, Hennayake *et al.* (2006) established suspension cell culture lines from *Rosa hybrida* cv. 'Charleston' to understand the influence of *in vitro* environmental conditions on anthocyanin biosynthesis pathway. Anthocyanin production was considerably greater with *Euphorbia millii* (EM) medium than with MS medium at all sugar concentrations. The ratio of NH_4^+ and NO_3^- was found to be one of the important factors in culture media as the nitrogen source, that improved the anthocyanin production by changing primary metabolism of the cells. The expression levels of anthocyanin biosynthesis genes were closely correlated with a temporal build-up of anthocyanins in response to environmental factors. Piovan and Filippini (2007) reported anthocyanin production in *Catharanthus roseus* flowers from both *in vivo* and *in vitro* conditions. These results indicated that in *Catharanthus roseus* cell cultures the accumulation of anthocyanins is dependent upon the factors like phosphate, mineral nitrogen, phenylalanine, trans-cinnamic acid, sucrose and light. Sakurai *et al.* (1997) developed a conditioned medium (CM) from suspension cultures of strawberry (*Fragaria ananassa*) which stimulated anthocyanin accumulation in the suspension-cultured cells of *Rosa hybrida*. Rose cells which were transferred into the CM, expressed anthocyanin formation and accumulation.Sakamoto *et al.* (1994) standardized the optimum culture conditions for anthocyanin production in callus cultures, initiated from stem and leaf tissues of *Aralia cordata*. The highest anthocyanin yield (10.3% dry wt) was obtained on the B5 medium containing 1.0 mg l^{-1} 2,4-D and 0.1 mg l^{-1} kinetin. The highest biomass accumulation was obtained in the callus cultures initiated from stem explants on medium supplemented with 0.90 mM 2,4-D in *Cleome rosea* and reddish-pink regions were observed on callus surface 6–7 months after culture. Anthocyanin production was enhanced by reducing temperature and increasing light irradiation. Pigmented calluses transferred to half-strength MS medium with a 1:4 ratio NH_4^+: NO_3^-, 70 g l^{-1} sucrose and supplementation with 0.90 mM 2,4-D maintained a high biomass accumulation and showed an increase of 150% on anthocyanin production as compared with the initial culture conditions (Simões *et al.*, 2009). Nishiyama and Yamakawa (2004) studied the anthocyanin accumulation in hairy root culture of *Ipomoea batatas* cv. Ayamurasaki. The amount of accumulated anthocyanins in hairy roots increased as a result of basal medium and increased sucrose concentration. The maximum anthocyanins were reported when these cultures were cultured on PRL-4C medium containing 3% sucrose in the continuous light irradiation (25-35 μE m^{-2} s^{-1}).

High **sucrose** (30 g l^{-1}) and nitrogen (100% N of standard MS) concentrations and low temperature (15°C) significantly promoted anthocyanin accumulation in cultured shoots of *Clematis pitcheri*. Reduction of nitrogen compound levels in the medium to 50% and lowering sucrose concentration to 10 g l^{-1} leads to decrease of anthocyanin accumulation in the shoots cultured at 15 and 20 °C (Kawa-Miszczak *et al.*, 2009). Nagira and Ozeki (2004) developed a system in which anthocyanin synthesis can be induced under defined conditions in regenerated torenia (*Torenia fournieri*) shoots. They used different concentration of glucose, fructose, galactose, and maltose for inducing anthocyanin biosynthesis. The anthocyanin content induced by 5% sorbitol together with 3% sucrose was 2-fold greater than that induced by 5% sorbitol and 1.5% sucrose. When the medium contained more than 7% glucose, fructose or galactose, the growth of shoots was strongly suppressed. In fact, sugar acts as a signalling molecule (Solfanelli *et al.*, 2006) whose signal transduction pathways may lead to the activation or inactivation of gene expression in *Arabidopsis thaliana*. Whole-genome transcript profiling reveals that the flavonoid and anthocyanin biosynthetic pathways are strongly up-regulated by sucrose treatment. Yamamoto *et al.* (1989) investigated the optimal conditions for anthocyanin production in suspension cultures of *Euphorbia millii* cell strain obtained by successive cell-aggregate selections. A modified Gamborg's solution containing 5 to 7% sucrose (w/v), a 1:16 ratio of NH^{+}_{4} : NO^{-}_{3}, 30 mM of nitrogen source, 10^{-6} M 2,4-D, 10^{-8} M BA, and low concentration of Fe^{2+} and SO_4^{2-} greatly increased anthocyanin production. The optimum concentration of sugar was 5% and 7% (w/v) for the cell growth and anthocyanin production, respectively. The optimal combination of 2,4-D and BA was found to be 10^{-6} and 10^{-8} M, respectively. Early pigment initiation, enhancement and maximum anthocyanin production from calluses cultured on *Euphorbia millii* (EM) medium supplemented with high concentration of sucrose and under 16/8 h (light/dark) photoperiod regime was reported by Ram *et al.* (2011).

The effect of **plant growth regulators** has been studied by many workers. Hosokawa (1999) reported that the anthocyanin is produced only in L2 cells (sub-epidermis cells) in the flowers of field grown plants while the anthocyanin was produced in both L1 and L2 (epidermal layer cells) cells in the flowers of *in vitro* raised plants on the MS medium containing GA_3 (1 mg l^{-1}) in the tepals of *Hyacinth orientalis*. Makunga *et al.* (1997) studied the anthocyanin accumulation in *in vitro* cultures of *Oxalis reclinata* and reported that the light irradiation induces anthocyanin synthesis in white callus, resulting in a heterogeneous red callus formation. The application of 2,4-D (1.0 mg l^{-1}) inhibited pigment production in the white callus and decreased anthocyanin accumulation in the red callus. Mazukami *et al.* (1988) reported that the callus growth and anthocyanin synthesis were remarkably stimulated by 2,4-D in rosella (*Hibiscus sabdariffa*). The highest anthocyanin yield was observed when 1 µM 2,4-D in combination with 0.1-1 µM kinetin was supplemented to the culture medium. Rau and Forkmann (1989) studied the *in vitro* callus cultures of *Callistephus chinensis* for anthocyanin production. The Gamborg's B5 medium supplemented with 1 mg l^{-1} 2,4-D for callus initiation and slightly modified MS-medium containing 2 mg l^{-1} IAA, 0.2 mg l^{-1} kinetin and 0.1 mg l^{-1} 2,4-D for further sub-cultivation were used. They found that NAA was the most effective auxin for inducing more fresh weight and anthocyanin yield. The over-riding influence of cytokinin source on anthocyanin production *in vitro* was explored using a suspension culture system for *Vaccinium pahalae*. Anthocyanin production reached a maximum after a 16–20 day interval in cultures containing an optimal kinetin concentration, but pigment accumulation peaked at only 12–16 days when BA was used as the sole cytokinin source (Fang *et al.*, 1998). Callus cultures of *Hyoscyamus muticus* cv. HMI-80 I showed anthocyanin pigment accumulation after 15–17 days of inoculation when the cultures were grown on MS medium supplemented with 0.5 mg l^{-1} BAP and 0.5 or 1.0 mg l^{-1} 2,4-D. Cyanidin was found to be the major compound by paper chromatography and spectrophotometric analysis (Basu and Chand, 1996).

The role of **elicitors** are also well established to induce anthocyanin synthesis. Nagira *et al.* (2006) found that the exogenous abscisic acid (ABA) could induce anthocyanin synthesis and chlorophyll senescence in regenerating *Torenia fournieri* shoots on the medium containing a low concentration of sucrose (1.5%). It was revealed that 7% sucrose treatment elevated endogenous ABA levels before the induction of anthocyanin synthesis and chlorophyll degradation. Sudha and Ravishankar (2003) reported the influence of putrescine (Put) on the growth and elicitation of anthocyanin in callus cultures of *Daucus carota* var. Nantes Scarlet by the use of ct-DL-difluoromethylarginine (DFMA), the polyamine (PA) biosynthetic inhibitor and concluded that the addition of Put (0.05 mM) resulted in enhancement of growth and anthocyanin content. The anthocyanin content was found to enhance by 1.68-fold on the 21st day as compared to the untreated controls. The PA inhibitor was found to result in lowering of the growth and the anthocyanin accumulation, which could be partially restored by the addition of Put in combination with this inhibitor. Shimizu *et al.* (2010) reported that anthocyanin accumulation in the root explants of the cultured *Gynura bicolor* plant was more remarkable when 25-50 µM methyl jasmonate (MeJA) was supplemented in the culture medium. However, the response from leaf explants was not encouraging, and the induction was also affected by light irradiation and sucrose concentration in the culture medium. However, they showed that the salicylic acid (SA) and 1-aminocyclopropane-1-carboxylic acid did not induce anthocyanin accumulation in roots. According to their findings, combinations of MeJA and SA did not stimulate the anthocyanin accumulation in the root as found in case of treatment by MeJA alone. Chan *et al.* (2010) reported that the cultures exposed to 10 days continuous darkness showed the lowest pigment content, while the cultures exposed to 10 days continuous irradiance (301-600 lux) showed the highest pigment content in the cells of *Melastoma malabathricum*. The cell cultures incubated at a lower temperature range (20 ± 2 ^{0}C) grew better and had higher pigment content than those grown at 26 and 29 ± 2 ^{0}C. Addition of MeJA was found to have no effect on cell biomass but the presence of higher amount (12.5-50 mg l^{-1}) had caused a reduction in anthocyanin production and accumulation in *Melastoma malabathricum* cell cultures. Cells cultured in the MS medium supplemented with 45 g l^{-1} sucrose without MeJA showed highest pigment content (See *et al.*, 2011). They concluded that MeJA was not necessary but sucrose was needed for the enhancement of cell growth and anthocyanin production in the cell cultures.

Light also has its great influence on anthocyanin induction. Meng *et al.* (2004) suggested that the pigmentation in ray floret (rf) of *Gerbera hybrida* is regulated by light. They reported that when the inflorescence was covered with aluminium foil in *in vivo* condition the pigmentation of the rf was strongly blocked and the gene expression of CHS and DFR inhibited. Similar results were obtained when the detached ray florets were cultured in *in vitro* conditions. Nozue and Yasuda (1985) reported that the anthocyanin synthesis first occurred 24-48 h after exposure to light in the anthocyanoplasts containing anthocyanin in cell suspension culture of *Ipomoea batatas*. They also reported that 2,4-D inhibits anthocyanin synthesis and development of anthocyanoplasts. Vinterhalter *et al.* (2007) studied the anthocyanin production in shoot cultures of *Ceratonia siliqua* at high (45.9 µmol $m^{-2} s^{-1}$) and low (9.2 µmol $m^{-2} s^{-1}$) irradiance levels in relation to carbohydrate nutrition. They reported that the anthocyanins were stimulated by light and carbohydrate nutrition, with sucrose as a choice sugar over fructose and glucose.

Induction of anthocyanin is also influenced by **UV irradiation**. Kakegawa *et al.* (1987) reported that the callus cultures from blue flowers of *Centaurea cyanus* produced maximum anthocyanin under continuous UV irradiation and white light on the seventh day of growth. Maekawa *et al.* (1980) found UV irradiation (below 330 nm) of long duration at low intensity producing considerably more anthocyanins than that of short duration at high intensity in uncoloured detached- and intact-petals of floribunda rose 'Ehigasa'. Irradiation with visible light before or along with UV, markedly stimulated anthocyanin formation as compared to UV irradiation alone. Wellmann *et al.* (1976) reported that continuous irradiation with UV (below 345 nm) light resulted in drastic increase in the activities of enzymes, PAL, CHI and flavanone synthase which stimulated anthocyanin formation in dark-grown cell suspension cultures of *Haplopappus gracilis*.

The Division of Floriculture and Landscaping, IARI, New Delhi initiated the pioneering programme on producing the anthocyanin pigments from rose cv. 'Pusa Ajay' in the tissue proliferated from leaf and petals in test tube by growing them on artificial medium. The expression and the quantity are significant and the pigments can be produced in the laboratory throughout the year without any seasonal barriers. The research paves the way for the large-scale synthesis of rose pigments in a laboratory in ultra pure form for medicinal applications. Division also standardized the protocols for carotenoid expression from the orange-coloured florets of *Calendula officinalis*, one of the richest sources of nutraceutical carotenoid pigments. Four explants, i.e. leaf, sepal, petal and internode of rose cv. 'Pusa Ajay' and two explants, i.e. petal and leaf in calendula were used in the investigation. Pre-culture agitation treatment of different explants in solution having carbendazim (0.1%) + mancozeb-45 (0.1%) + 8-HQC (200 mg/l) for three hours resulted in minimum culture contamination and maximum explant survival in both the species. Surface sterilization of leaf (rose) and petal (calendula) explants with 0.1 per cent mercuric chloride for four minutes gave the maximum number of contamination-free cultures. Best callusing in rose leaf explant was obtained on MS medium supplemented with 4.0 mg/l 2,4-D, and in calendula petal explants, on MS medium supplemented with 2.0 mg/l 2,4-D. Earliest callus induction and highest callusing were obtained in complete darkness in rose leaf explants and calendula petal explants. In case of rose, the maximum response was noticed when the cultures were inoculated on EM (*Euphorbia millii*) medium and incubated at 16/8 (day/night) photoperiod regime. Earliest pigment initiation, intensification, high degree of pigmentation and maximum anthocyanin content were recorded when EM medium was supplemented with 70 g/l sucrose. In case of calendula, maximum culture response was recorded when the cultures were inoculated on MS medium devoid of ammonical nitrogen in both solid as well as liquid media. Whereas, earliest carotenoid initiation and intensification was recorded when solid EM medium was fortified with ABA at 50 µM concentration. However, in case of liquid media, 20 µM ABA concentration was adequate for initiation and intensification of carotenoid pigments. Highest carotenoid content was recorded when the EM solid medium was fortified with 50 µM ABA. Keeping in view the cost economics, 20 µM ABA concentration was found adequate in view of the negligible differences in the quantity of pigments extracted. In case of liquid cultures, MS medium devoid of ammonical nitrogen was found to be ideal for producing maximum amount of carotenoid pigments. Among the other elicitors used, continuous UV-A (320-400 nm) irradiation was also found to be a potential pigment inducer for both anthocyanins and carotenoid pigments (Mohan Ram, 2009). Salient findings of the study are furnished here under.

In rose, the leaf was found to be the most suitable explant which gave the maximum culture initiation, callusing frequency and minimum microbial contamination, though in calendula, highest culture initiation and callusing frequency were obtained through petal explants. In both the plants, treatment of different explants with carbendazim (0.1%) + mancozeb-45 (0.1%) + 8-HQC (200 mg/l) for 3 h resulted in minimum culture contamination and maximum explant survival. In both the cases, surface-sterilization with $HgCl_2$ (0.1%) for 4 minutes followed by washing with sterile distilled water, resulted in maximum contamination-free cultures with maximum survival per cent. In rose, maximum callusing and earliest callus induction were found when 2,4-D was added at 4.0 mg/l in MS medium. In calendula, highest callusing was observed when 2,4-D at 2.0 mg/l was supplemented in MS medium. Earliest callus induction and good callusing were found in the complete darkness in rose leaf explant and calendula petal explant. Supplementation in EM medium of 70 g/l sucrose, rose leaf calli recorded earliest pigment induction, intensification and higher anthocyanin accumulation, *vis-à-vis* high carotenoid induction, initiation and its production in both the crops. Sucrose was more responsive to anthocyanin induction than to carotenoid. With respect to maximum pigmentation frequency and earliest pigment induction, MS medium devoid of ammonical nitrogen resulted in slight improvement in anthocyanin biosynthesis in rose leaf calli. In both the media matrix the calendula petal calli recorded earliest carotenoid initiation, high degree of pigmentation and maximum carotenoid content, however, in solid medium slightly more carotenoid was recorded than the liquid medium.

EM medium supplemented with 20 µM ABA, there was maximum pigment induction in rose leaf calli, and maximum frequency of carotenoid induction in calendula petal calli in both the media matrix. In case of both the plants when 50 µM ABA was added to EM medium, earliest pigment initiation with highest quantity was recorded. Higher pigmentation frequency, earliest anthocyanin induction and maximum quantity were found when rose leaf calli were cultured on EM medium under 16/8 h (day/night) photoperiod though the same was not so significant in case of calendula. Maximum anthocyanin content, early induction and high degree of pigmentation in rose leaf and calendula petal calli were observed under continuous UV-A light irradiation with EM medium.

Betalains

Betalains have been widely used as natural colourants for many centuries, but their attractiveness for use as colourants of foods, drugs or cosmetics has increased recently due to their being reportedly high anti-oxidative, having free radical scavenging activities, and concerns about the use of various synthetic alternatives. The main commercial sources of betalains are powders and concentrates of red beet (*Beta vulgaris*) or cactus pear (*Opuntia ficus-indica*) extracts. Betalains comprise a class of nitrogen-containing plant pigments found in the cell sap of plants. In the ice plant (*Mesembryanthemum crystallinum*), betacyanin accumulation was discussed to screen tissue against damaging UV radiation. Little has been published with respect to the physiological role of betalains in mammals.

Betalains have not been as extensively researched in terms of **health benefits** as the other major plant-derived pigments; however, antiviral and antimicrobial properties are well documented. Recent findings rank beet among the 10 most potent vegetables with respect to their antioxidant capacities. Commercially, betaxanthin pigments have been introduced as a food supplement in order to fortify processed food products with a nutraceutical natural colourant. Betalains reportedly have anti-inflammatory, hepatoprotective activities, and protect low density lipoproteins (LDLs) from oxidation. A striking characteristic of betalain pigments is that they can be naturally absorbed by the body, that means a minimal transformation occurs during metabolism. The antioxidant potential was found to be related to structural features of the respective betalains. Preliminary **cancer-chemopreventive** studies with crude beetroot extract have shown that the protection against skin and lung tumors in animal models is slightly higher than that afforded by anthocyanin pigment, carotenoids. Betanin appears to be a potent anticancer compound. Beetroot extract has demonstrated chemopreventive properties against both lung and skin cancers. Recent reports showed a significant inhibitory effect of beetroot towards skin and lung cancer in mice. Future studies will have to ascertain the alleged promising physiological and pharmacological effects of betalains on humans.

Carotenoids

Carotenoids are the second most abundant pigments in nature, and consist of more than 700 members (Britton, 1998). Carotenoids, a large group of fat-soluble compounds, are responsible for many of the yellow, orange and red colours in plant and animal products (Kirk and Tiliney-Bassett, 1978). Carotenoids make up a diverse class of natural pigments that are of interest for pharmaceutical use, as natural food colourants, nutrient supplements and animal feed. Carotenoids play an important role in plant reproduction, by attracting pollinators and in seed dispersal, and are essential components of human diet (Bartley and Scolnik, 1995). Several carotenoids protect humans by hindering free radical chain reactions caused by ageing, smoking, pollution and other stresses. β-carotene is a precursor of vitamin A, and lutein and zeaxanthin have been implicated in the reduction of age-related macular disease (AMD) (Pratt, 1999; Landrum and Bone, 2004). Dietary carotenoids also provide health benefits based on their antioxidant properties (Botella-Pavia and Rodriguez-Concepcion, 2006). The American Cancer Society states that people who consume greater amounts of carotenoid rich foods have a lower risk of getting cancer. WHO has estimated that more than 100 million children are vitamin A deficient and up to 500,000 of these children become blind each year. Many of these children die within 12 months of going blind. Thus, understanding the regulation of carotenoid accumulation, especially in those crops which are rich in carotenoids, should facilitate improvements in nutritional value with potentially significant health benefits (Howitt and Pogson, 2006). Carotenoids provide protection to vision and eye function, and against macular degeneration and cataracts through reduction in **photo-oxidative damage**. Macular degeneration caused by blue light-mediated free radical damage to the retina, is the leading cause of irreversible blindness in ageing adults. Lutein has been loosely referred to as 'the eye-protective nutrient' and its intake has been shown to be inversely associated with ocular diseases such as AMD and cataracts (Landrum and Bone, 2004). Carotenoids have a photo-protective function in human metabolism, and hence have been used in therapies for photosensitivity diseases. Mechanisms suspected to be involved in the protective effects of these macula carotenoids are filtering blue light, and scavenging reactive intermediates generated in photo-oxidation (Botella-Pavia and Rodriguez-Concepcion, 2006). Carotenoids are credited with biological **promotion of immune system response**. A broad range of studies point to multiple potential mechanisms by which carotenoids may boost the immune system by protecting immune cells from the reactive bactericidal species, increasing natural killer cell activity, and increasing total white blood cells. A series of studies have now demonstrated a correlation between high intakes or high serum concentrations of carotenes and lowered risk of **coronary artery** disease. Inhibition of low-density lipoprotein (LDL) oxidation has been demonstrated, especially with low levels of carotene. Two-week dosage of tomato juice (rich in carotenoids) effectively lowers LDL in male subjects. Another potential mechanism for carotenoids to inhibit cardiovascular disease may be through a feedback mechanism to inhibit HMG-CoA reductase in humans. Carotenoids enhance cell-to-cell communication because they can up-regulate gap junctions. Carotenoids are associated with **inhibition of several types of cancers** including cervical, oesophageal, lung, prostate, colorectal and stomach. In particular, the carotenoid lycopene has been extensively researched for its role in cancer inhibition. Lutein and zeaxanthin intake has been linked to reduced risk of prostate cancer. Both β-carotene and vitamin A supplementation inhibits cell proliferation in carcinogenesis (Pratt, 1999). Inhibitory effect of lycopene on prostate cancer cells has been linked to antioxidant mechanisms. Diets that incorporate diverse mixtures of different carotenoids are recommended for their wide range of activities associated with different individual carotenoid compounds.

Chlorophylls

Chlorophyll is the most important plant pigment and a 'real life force' that nature uses to explode plants into greenery. It is not a single compound, but made up of two classical components—the bluish black chlorophyll 'a' and dark green chlorophvll 'b'. Chlorophyll is photoreceptive, absorbs blue and red spectrum of the visible light (400-700 nm) and reflects the unabsorbed spectrum in the green region. It is the green reflection that makes plant parts look green. The spectacular show of green turning to brilliant shades of yellow and red in autumn before plants shed leaves is nothing but manifestation of pigment change, from one to another. Chlorophyll is present in the membrane of disc-like structures called 'chloroplasts', and though photosynsthesis with its companions works like chemical factory powered by sun to synthesize carbohydrates including simple sugar, and this is the greatest miracle of the nature. Photosynthesis is the most fundamental of all biochemical reactions that helps plants to grow, flower, produce seed and sustain life on earth. Chlorophylls are typically consumed in much higher doses in a diet that incorporates green and leafy vegetables. Chlorophylls in green foods are credited with numerous interventions in **human health maintenance** including reduction of blood pressure, blood sugar control, brain activation, antimutagenic and anticlastogenic effects. In Asian countries, the dietary sources for chlorophyll-rich foods are far more extensive than in Western diets. The dark green vegetables are simultaneously excellent sources of carotenoids, in particular lutein, which contribute with the chlorophylls to the health-protective properties. The anti-mutagenic properties of chlorophylls have been demonstrated in various assays, and clearly, intake of chlorophyll has potential to act as a **chemopreventive** compound in humans. However, much of the research is equivocal. Chlorophyllin has shown both anticancer properties and cancer-promoting activity. Dietary consumption of green vegetables has been associated with the protection against mutagenic and clastogenic activity of genotoxicants. The crude leaf extract or the chlorophyllin proved to reduce clastogenic effects. Research supporting the roles of chlorophyll in cancer chemoprevention indicates that these pigments are capable of deactivating potential carcinogens in the digestive tract. Chlorophyllin significantly reduces the incidence and multiplicity of skin tumors in mice.

Curcumin

Curcumin (turmeric, curry, etc.) is widely consumed, especially in middle-eastern diets, and has a broader range of pharmacological applications. It acts as an anti-inflammatory agent. Low levels of curcumin in the diet of laboratory animals results in significant lowering of blood cholesterol levels. As such, a therapeutic role for turmeric for individuals with diabetes as well as predisposition to coronary heart disease is suggested. Turmeric has demonstrated anti-mutagenic and anti-carcinogenic properties. Curcumin has also been shown to potentiate the beneficial effects of drugs and vitamins.

Conclusion

In recent years several studies have shown that anthocyanins (the water soluble secondary phytochemical) possess a wide range of biological activities (antioxidant, anti-inflammatory, antimicrobial and anti-carcinogenic). In addition, they also display a variety of effects on blood vessels, Alzheimer's disease, cancer, diabetes, cataract and age-related functional decline, platelets and lipoproteins and are able to reduce the risk of coronary heart diseases (Lila, 2004; Stintzing and Carle, 2004; Mazza, 2007; Castañeda-Ovando *et al.*, 2009; de Pascual-Teresa *et al.*, 2010). Today, the interest on anthocyanins production either from fresh plant material or from *in vitro* cultures, has intensified because they can be used not only as a natural food and beverage additives but also as health beneficial nutraceutical compounds (Wang and Jiao, 2000; Collin, 2001; Gantet and Memelink, 2002; Hou, 2003; Lila, 2004, Ram *et al.*, 2011). The pharmacological value of anthocyanins is increasing due to the constant discoveries of their potential role in healthcare and as lead compound for new drug development (Shilpa *et al.*, 2010). Considering the continuous decrease in arable land and increased consideration of environmental problems, production of secondary metabolites by traditional plant cultivation and chemical synthesis may largely be limited in future (Zhao and Verpoorte 2007; Dörnenburg, 2008). Besides this, the pigment obtained from the natural fresh plant materials involve some other limitations such as low metabolite yield, seasonal availability, fresh material loss, inconsistent product quality and pigment degradation during extraction and storage processes (Lila, 2004). As a promising alternative to produce secondary metabolites, plant cell culture technology has many advantages over traditional field cultivation and chemical synthesis, particularly for several natural compounds that are either derived from slow-growing plants or are difficult to synthesise with chemical methods (Dörnenburg, 2008). Large-scale plant cell culture for the production of anthocyanins is an attractive alternative as it offers many advantages, such as controlled supply without seasonal boundaries, a renewable and environment friendly resource, production is more simpler, consistent, predictable and lacks interferences found in whole plants, rapid and easy isolation of active compounds, can be elicited to provoke rapid and prolific accumulation, and a biological factory for the production of high-quality products under strictly controlled conditions (Lila, 2004; Dörnenburg, 2008).

The biosynthesis of anthocyanins in plant tissue culture systems is dependent not only on genetic factors but also on environmental factors such as light, temperature, medium composition, plant growth regulators and nutrition (Zhang *et al.*, 2002; Hennayake *et al.*, 2006). Much efforts have been expended in developing alternative methods to improve anthocyanin biosynthesis under *in vitro* conditions (Simões-Gurgel *et al.*, 2011). Moreover, anthocyanin production through tissue culture technique has been obtained from several plant species such as strawberry (Edahiro *et al.*, 2005), *Torenia fourunieri* (Nagira *et al.*, 2006), *Cleome rosea* (Simões *et al.*, 2009), *Panax sikkimensis* (Mathur *et al.*, 2010), *Bridelia stipularis* (Sreenivas *et al.*, 2011) and *Vitis vinifera* (Saw *et al.*, 2012). However, research on anthocyanin biosynthesis under *in vitro* conditions in ornamental plants which contain good amount of plant pigments, is still limited, thereby restricting the commercial application of plant tissue culture technique for the production of such compounds. Plant tissue culture technique represents a potential renewable source of valuable medicinal compounds, flavours and colourants, which cannot be produced by microbial cells or chemical amalgamation. Evolving commercial importance of secondary metabolites in recent years resulted in a great interest particularly in likelihood to alter the production of bioactive plant metabolites through cell culture technology (Mulabagal and Tsay, 2004). Conventionally, flowers and leaves are used as sources for the extraction of these nutraceutical pigments by solvent extraction technique. However, the plant pigments are localized in flowers and other young vegetative parts thus restricting their availability only during the cropping season, *vis-a-vis* low metabolite yield, fresh material losses, inconsistent product quality and pigment degradation caused by storage and extraction process. In recent years salicylic acid (SA) has been the focus of intensive research due to its function as an endogenous signal mediating local and systemic plant defence responses against pathogens (Vlot *et al.*, 2009; Vicente and Plasencia, 2011). It has also been found that SA, in particular, influences seed germination, seedling establishment, cell growth, responses to abiotic and osmotic stresses (Clarke *et al.*, 2004; Rajou *et al.*, 2006; Vlot *et al.*, 2009; Vicente and Plasencia, 2011), *etc.* It is well established that SA or acetylsalicylic acid can induce resistance to pathogens and the production of some PR proteins in plants, even in the absence of pathogenic organisms (Raskin, 1992). Furthermore, SA treatments were found to enhance the *in vitro* anthocyanin biosynthesis in callus cultures of *Daucus carota*, through involvement of calcium channel (Sudha and Ravishankar, 2003).

Plant pigments can be produced by growing plant cells in tissue culture. Callus culture is the culture of dedifferentiated plant cells induced on media usually containing relatively high auxin concentrations or a combination of auxin and cytokinin under *in vitro* conditions. Callus cultures can be embryogenic or non-embryogenic. Embryogenic calli contain differentiated embryogenically competent cells that can regenerate complete plants through the process called somatic embryogenesis (Zimmerman, 1993). The foremost uses of somatic embryogenesis include clonal propagation of plants, regeneration of haploid or transgenic plants and fundamental study of the process of embryogenesis in plants. Non-embryogenic callus cultures containing more or less homogenous clumps of dedifferentiated cells, are used for secondary metabolite production. Of the tissue culture means, this approach is relatively frequently used for the production of anthocyanins (Jedinák *et al.*, 2004). A substantial amount of research efforts has been directed towards this goal ranging from strain improvement, medium optimization, optimization of culture conditions, elicitation, and intelligent process strategies. To accomplish the commercialization of plant cell culture technology for pigments production the major task of allied research and development is the attainment of large-scale cultivation of high yielding cell lines, high product level and productivity along with stability of production over extended cultivation time.

References

Acquaviva, R., A. Russo, F. Galvano, M.L. Barcellona, G.L. Volti and A. Vanella, 2003. Cyanidin 3-O-b-D-glucoside as DNA cleavage protectors and antioxidants. *Cell Biol. Toxicol.*, **19**: 243–252.

Aerts, R.J., D. Gisi, E. de Carolis, V. de Luca and T.W. Banmann, 1994. Methyl jasmonate vapour increases the developmentally controlled synthesis of alkaloids in *Catharanthus* and *Cinchona* seedlings. *Plant J.*, **5**: 635–643.

Andersen, Ø.M. and M.M. Jordheim, 2006. *The Anthocyanins.* In: *Flavonoids: Chemistry, Biochemistry and Applications* (eds Anderson, Ø.M. and K.R. Markham), pp. 471-530. CRC Press, Boca Raton, Florida, USA.

Bagchi, D., C.K. Sen, M. Bagchi and M. Atalay, 2004. Anti-angiogenic, antioxidant and anti-carcinogenic properties of a novel anthocyanin-rich berry extract formula. *Biochem.*, **69**: 75–80.

Balbi, V. and A. Devoto, 2008. Jasmonate signalling network in *Arabidopsis thaliana*: Crucial regulatory nodes and new physiological scenarios. *New Phytol.*, **177**: 301–318.

Bartley, G.E. and P.A. Scolnik, 1995. Plant carotenoids: Pigments for photo-protection, visual attraction, and human health. *The Plant Cell*, **7**: 1027-1038.

Basu, P. and S. Chand, 1996. Anthocyanin accumulation in *Hyoscyamus muticus* L. tissue cultures. *J. Biotech.*, **52**: 151–159.

Blechert, R.P., W. Brodshelm, S. Hoylder, L. Kammerer, T.M. Kutchan, M.J. Mueller, Z.Q. Xia and M.H. Zenk, 1995. The octadecanoic pathway: signal molecules for the regulation of secondary pathways. *Proc. Natl. Acad. Sci.*, **92**: 4099–4105.

Botella-Pavia, P. and M. Rodriguez-Concepcion, 2006. Carotenoid biotechnology in plants for nutritionally improved foods. *Physiol. Plantarum,* **126**: 369-381.

Britton, G. 1998. Overview of carotenoid biosynthesis. In: *Carotenoids*, pp. 13-147. Birkhauser, Basel, Switzerland.

Chalker-Scott, L. 1999. Environmental significance of anthocyanins in plant stress responses. *Photochem. Photobiol.*, **70**: 1–9.

Chan, L.K., S.S. Koay, P.L. Boey and A. Bhatt, 2010. Effects of abiotic stress on biomass and anthocyanin production in cell cultures of *Melastoma malabathricum. Biol. Res.*, **43**: 127–135.

Clarke, S.M., L.A.J. Mur, J.E. Wood and I.M. Scott, 2004. Salicylic acid dependent signaling promotes basal thermo-tolerance but is not essential for acquired thermo-tolerance in *Arabidopsis thaliana. Plant J.*, **38**: 432–447.

Collin, H.A. 2001. Secondary product formation in plant tissue cultures. *Plant Growth Reg.*, **34**: 119–134.

Curtin, C., W. Zhang and C. Franco, 2003. Manipulationg anthocyanin composition in *Vitis vinifera* suspension cultures by elicitation with jasmonic acid and light irradiation. *Biotech. Lett.*, **25**: 1131–1135.

Davies, K.M. 2004. *An Introduction to Plant Pigments in Biology and Commerce.* In: *Plant Pigments and Their Manipulation*, vol 14 (ed. Davies, K.M.), pp. 1-22. CRC Press, NY, USA.

de Pascual-Teresa, S. and M.T. Sanchez-Ballesta, 2008. Anthocyanins: from plant to health. *Phytochem. Rev.*, **7**: 281–299.

de Pascual-Teresa, S., D.A. Moreno and C. García-Viguera, 2010. Flavanols and anthocyanins in cardiovascular health: a review of current evidence. *Int. J. Mol. Sci.*, **11**: 1679–1703.

Dörnenburg, H. 2008. Plant cell culture technology – harnessing a biological approach for competitive cyclotides production. *Biotech. Lett.*, **30**: 1311–1321.

Downham, A. and P. Collins, 2000. Colouring our food in the last and next millennium. *Int. J. Fd. Sci. Tech.*, **35**: 5–22.

Edahiro, J., M. Nakamura, M. Seki and S. Furusaki, 2005. Enhanced accumulation of anthocyanin in cultured strawberry cells by repetitive feeding of L-Phenylalanine into the medium. *J. Biosci. Bioeng.*, **99**(1): 43–47.

Fang, Y., M.A.L. Smith and M.F. Pépin, 1999. Effects of exogenous methyl jasmonate in elicited anthocyanin-producing cell cultures of ohelo (*Vaccinum pahalae*). *In Vitro Cell. Dev. Biol. - Plant*, **35**: 106-113.

Fang, Y., M.A.L.Smith and M.F. Pepin, 1998. Benzyladenine restores anthocyanin pigmentation in suspension cultures of wild *Vaccinium pahalae. Plant Cell Tiss. Organ Cult.*, **54**: 113-122.

Gantet, P. and J. Memelink, 2002. Transcription factors: tools to engineer the production of pharmacologically active plant metabolites. *Trends Pharmacol. Sci.*, **23**: 563-569.

Grotewold, E. 2006. The genetics and biochemistry of floral pigments. *Ann. Rev. Plant Biol.*, **57**: 761-780.

Guo, J., W. Han and M. Wang, 2008. Ultraviolet and environmental stresses involved in the induction and regulation of anthocyanin biosynthesis - a review. *African J. Biotech.*, **7**: 4966-4972.

Harborne, J.B. and C.A. Williams, 2000. Advances in flavonoid research since 1992. *Phytochem.*, **55**: 481-504.

Hennayake, C.K., S. Takagi, K. Nishimura, M. Kanechi, Y. Uno and N. Inagaki, 2006. Differential expression of anthocyanin biosynthesis genes in suspension culture cells of *Rosa hybrida* cv. 'Charleston'. *Plant Biotech.*, **23**: 379-385.

Hosokawa, K. 1999. Cell layer specific accumulation of anthocyanin in response to gibberellic acid in tepals of *Hyacinthus orientalis. Biosci. Biotech. Biochem.*, **63**(5): 930-931.

Hou, D.X. 2003. Potential mechanisms of cancer chemoprevention by anthocyanins. *Curr. Mol. Med.*, **3**: 149-159

Howitt, C.A. and B.J. Pogson, 2006. Cartenoid accumulation and function in seeds and non-green tissue. *Plant Cell Environ.*, **29**: 435-445.

Jackman, R.L., R.Y. Yada, M.A. Tung and R.A. Speers, 1987. Anthocyanins as food colorants - Review. *J. Food Biochem.*, **11**: 201-274.

Jassim, S.A.A. and M.A. Naji, 2003. Novel antiviral agents: a medicinal plant perspective. *J. App. Microbiol.*, **95**(3): 412-427.

Jedinák, A., J. Faragó, I. Pšenáková and T. Maliar, 2004. Approaches to flavonoid production in plant tissue cultures - Review. *Biologia, Bratislava*, **59**: 697-710.

Joseph, J.A., B. Shukitt-Hale, N.A. Denisnova, D. Bielinski, A. Martin, J.J. McEwan and P.C. Bickford, 1999. Reversals of age-related declines in neuronal signal transduction, cognitive, and motor behavioural deficits with blueberry, spinach, or strawberry dietary supplementation. *J. Neurosci.*, **19**: 8114-8121.

Kakegawa, K., Y. Kaneko, E. Hattori, K. Koike and K. Takeda, 1987. Cell cultures of *Centaurea cyanus* produce malonated anthocyanin in UV light. *Phytochem.*, **26**(8): 2261-2263.

Karuppusamy, S. 2009. A review on trends in production of secondary metabolites from higher plants by *in vitro* tissue, organ and cell cultures. *J. Medicinal Plants Res.*, **3**: 1222-1239.

Kawa-Miszczak, L., E. Wegrzynowicz-Lesiak, E. Gabryszewska and M. Saniewski, 2009. Effect of different sucrose and nitrogen levels in the medium on chlorophyll and anthocyanin content in *Clematis pitcher* shoots cultured *in vitro* at different temperatures. *J. Fruit Orn. Plant Res.*, **17**: 113-121.

Kim, Y.S., E.J. Hahn, H.N. Murthy and K.Y. Paek, 2004. Adventitious root growth and ginsenoside accumulation in *Panax ginseng* cultures as affected by methyl jasmonate. *Biotech. Lett.*, **26**: 1619-1622.

Kirk, J.T. and R.A. Tiliney-Bassett, 1978. Proplastids, etioplasts, amyloplasts, chromoplasts and other plastids. In: *The Plastids, Their Chemistry, Structure, Growth and Inheritance*, pp. 217-239. Elsevier/North Holland Biomedical Press, Amsterdam, Holland.

Kong, J.M., L.S. Chia, N.K. Goh, T.F. Chia and R. Broillard, 2003. Analysis and biological activities of anthocyanins. *Phytochem.*, **64**: 923-933.

Landrum, J.T. and R.A. Bone, 2004. Dietary lutein and zeaxanthin: reducing the risk for macular degeneration. *Agro Food Industry Hi-Tech*, **15**: 22-25.

Lila, M.A. 2004. Anthocyanins and human health: an *in vitro* investigative approach. *J. Biomed. Biotechnol.*, **5**: 306-313.

Maekawa, S., M. Terabun and N. Nakamura, 1980. Effects of UV and visible light on pigmentation of 'Ehigasa' roses. *J. Japan. Soc. hort. Sci.*, **49**(2): 251-259.

Makunga, N.P., J.V. Staden and W.A. Cress, W.A. 1997. The effect of light and 2,4-D on anthocyanin production in *Oxalis reclinata* callus. *Plant Growth Reg.*, **23**: 153-158.

Mathur, A., A.K. Mathur, A. Gangwar, S. Yadav, P. Verma and R.S. Sangwan, 2010. Anthocyanin production in a callus line of *Panax sikkimensis* Ban. *In Vitro Cell. Dev. Biol.-Plant*, **46**: 13-21.

Mazukami, H., K. Tomita, H. Ohashi and N. Hiraoka, 1988. Anthocyanin production in callus cultures of rosella (*Hibiscus sabdariffa* L.). *Plant Cell Rep.*, **7**: 553-556.

Mazza, G.J. 2007. Anthocyanins and heart health. *Ann. Ist. Super Sanità*, **43**: 369-374.

McDougall, G.J., S. Fyffe, P. Dobson and D. Stewart, 2005. Anthocyanins from red wine - their stability under simulated gastrointestinal digestion. *Phytochem.*, **66**: 2540-2548.

Meng, X., T. Xing and X. Wang, 2004. The role of light in the regulation of anthocyanin accumulation in *Gerbera hybrida*. *Plant Growth Reg.*, **44**: 243-250.

Meyer, J.E., M.F. Pépin and M.A.L. Smith, 2002. Anthocyanin production from *Vaccinium pahalae*: limitations of the physical microenvironment. *J. Biotech.*, **93**: 45-57.

Mizukami, H., Y. Tabira and B.E. Ellis, 1993. Methyl jasmonate-induced rosmarinic acid biosynthesis in *Lithospermum erythrorhizon* cell suspension culture. *Plant Cell Rep.*, **12**: 706-709.

Mohan Ram, 2009. *In Vitro* Induction of Nutraceutical Pigments from Rose (*Rosa hybrida* L.) and Calendula (*Calendula officinalis* L.). M.Sc. Thesis. P.G. School, IARI, New Delhi.

Mulabagal, V. and H.-S. Tsay, 2004. Plant cell cultures - an alternative and efficient source for the production of biologically important secondary metabolites. *Int. J. Appl. Sci. Engg.*, **2**: 29-48.

Nagira, Y. and Y. Ozeki, 2004. A system in which anthocyanin synthesis is induced in regenerated torenia shoots. *J. Plant Res.*, **117**: 377-383.

Nagira, Y., K. Ikegami, T. Koshiba and Y. Ozeki, 2006. Effect of ABA upon anthocyanin synthesis in regenerated torenia shoots. *J. Plant Res.*, **119**: 137-144.

Narayana, K.R., M.S. Reddy, M.R. Chaluvadi and D.R. Krishna, 2001. Bioflavonoids classification, pharmacological, biochemical, and therapeutic potential. *Indian J. Pharmacol.*, **33**: 2-16.

Nishiyama, Y. And T. Yamakawa, 2004. Effect of medium composition on the production of anthocyanins by hairy root cultures of *Ipomoea batatas*. *Plant Biotech.*, **21**(5): 411-414.

North, J.J. and P.A. Ndakidemi, 2012. Evaluation of different ratios of auxin and cytokinin for the *in vitro* propagation of *Streptocarpus rexii* Lindl. *Intern. J. Physical Sci.*, **7**: 1083-1087.

Nozue, M. and H. Yasuda, 1985. Occurrence of anthocyanoplasts in cell suspension cultures of sweet potato. *Plant Cell Rep.*, **4**: 252-255.

Piovan, A. and R. Filippini, 2007. Anthocyanins in *Catharanthus roseus in vivo* and *in vitro*: a review. *Phytochem. Rev.*, **6**: 235-242.

Pozo, M.J., L.C. van Loon and C.M. Pieterse, 2005. Jasmonates signals in plant-microbe interactions. *J. Pl. Gr. Reg.*, **23**: 211-222.

Pratt, S. 1999. Dietary prevention of age-related macular degeneration. *J. Amer. Optometric Assoc.*, **70**: 39-47.

Rajou, L., M. Belghazi, R. Huguet, C. Robin and A. Moreau, 2006. Proteomic investigation of the effect of salicylic acid on *Arabidopsis* seed germination and establishment of early defense mechanisms. *Plant Physiol.*, **141**: 910-923.

Ram, M., K.V. Prasad, C. Kaur, S.K. Singh, A. Arora and S. Kumar, 2011. Induction of anthocyanin pigments in callus cultures of *Rosa hybrida* L. in response to sucrose and ammonical nitrogen levels. *Plant Cell Tiss. Org. Cult.*, **104**: 171-179.

Raskin, I. 1992. Salicylate, a new plant hormone? *Plant Physiol.*, **99**: 799-803.

Rau, D. and G. Forkmann, 1989. Anthocyanin synthesis in tissue cultures of *Callistephus chinensis* (China aster). *Plant Cell Rep.*, **5**: 435-438.

Rudell, D.R. and J.P. Mattheis, 2008. Synergism exists between ethylene and methyl jasmonate in artificial light-induced pigment enhancement of 'Fuji' apple fruit peel. *Postharvest Biol. Technol.*, **47**: 136-140.

Sakamoto, K., K. Iida, K. Sawamura, K. Hajiro, Y. Asada, Y. Yoshikawa and T. Furuya, 1994. Anthocyanin production in cultured cells of *Aralia cordata* Thunb. *Plant Cell, Tiss. Org. Cult.*, **36**: 21-26.

Sakurai, M., Y. Ozeki and T. Mori, 1997. Induction of anthocyanin accumulation in rose suspension-cultured cells by conditioned medium of strawberry suspension cultures. *Plant Cell, Tiss. Org. Cult.*, **50**: 211-214.

Saniewski, M., M. Horbowicz, J. Puchalski and J. Ueda, 2003. Methyl jasmonate stimulates the formation and the accumulation of anthocyanin in *Kalanchoe blossfeldiana. Acta Physiologiae Plant.*, **25**: 143-149.

Sato, K., M. Nakayama and J. Shigeta, 1996. Culturing conditions affecting the production of anthocyanins in suspended cell cultures of strawberry. *Plant Sci.*, **113**: 91-98.

Saw, N.M.M.T., H. Riedel, Z. Cai, O. Kütük and I. Smetanska, 2012. Stimulation of anthocyanin synthesis in grape (*Vitis vinifera*) cell cultures by pulsed electric fields and ethephon. *Plant Cell Tiss. Org. Cult.*, **108**: 47-54.

Schwinn, K.E. and K.M. Davies, 2004. *Flavonoids*. In: *Plant Pigments and their Manipulation*, Vol. 14 (ed. Davies, K.M.). CRC Press, NY, USA.

See, K.S., A. Bhatt and C.L. Keng, 2011. Effect of sucrose and methyl jasmonate on biomass and anthocyanin production in cell suspension culture of *Melastoma malabathricum* (Melastomaceae). *Rev. Biol. Trop.*, **59**: 597-606.

Shilpa, K., K. Varun and B.S. Lakshmi, 2010. An alternate method of natural drug production: eliciting secondary metabolite production using plant cell culture. *J. Plant Sci.*, **5**: 222-247.

Shimizu, Y., K. Maeda, M. Kato and K. Shimomura, 2010. Methyl jasmonate induces anthocyanin accumulation in *Gynura bicolor* cultured roots. *In Vitro Cell. Dev. Biol.—Plant*, **46**: 460-465.

Simões, C., C.H.B. Bizarri, L. da Silva Cordeiro, T.C. de Castro, L.C.M. Coutada, A.J.R. da Silva, N. Albarello and E. Mansur, 2009. Anthocyanin production in callus cultures of *Cleome rosea*: modulation by culture conditions and characterization of pigments by means of HPLC-DAD/ESIMS. *Plant Physiol. Biochem.*, **47**: 895-903.

Simões-Gurgel, C., L. da Silva Cordeiro, T.C. de Castro, C.H. Callado, N. Albarello and E. Mansur, 2011. Establishment of anthocyanin-producing cell suspension cultures of *Cleome rosea* Vahl ex DC. (Capparaceae). *Plant Cell Tiss. Org. Cult.*, **106**: 537-545.

Skibola, C. and M. Smith, 2000. Potential health impacts of excessive flavonoid intake. *Free Radic. Biol. Med.*, **29**: 375-383.

Solfanelli, C., A. Poggi, E. Loreti, A. Alpi and P. Perata, 2006. Sucrose-specific induction of the anthocyanin biosynthetic pathway in arabidopsis. *Plant Physiol.*, **140**: 637-646.

Sreenivas, V.K., V.N. Jisha, K.P. Martin and P.V. Madhusoodanan, 2011. *Bridelia stipularis*: a new source for anthocyanin production *in vitro*. *Acta Physiol. Plant.*, **33**: 2051-2056.

Stafford, H.A. 1991. Flavonoid evolution: an enzymic approach. *Plant Physiol.*, **96**: 680-685.

Stintzing, F.C. and R. Carle, 2004. Functional properties of anthocyanins and betalains in plants, food and in human nutrition. *Trends Food Sci. Technol.*, **15**: 19-38.

Sudha, G. and G.A. Ravishankar, 2003. Elicitation of anthocyanin production in callus cultures of *Daucus carota* and the involvement of methyl jasmonate and salicylic acid. *Acta Physiologiae Plant.*, **25**: 249-256.

Tall, J.M., N.P. Seeram, C.S. Zhao, M.G. Nair, R.A. Meyer and S.N. Raja, 2004. Tart cherry anthocyanins suppress inflammation-induced pain behavior in rat. *Behav. Brain Res.*, **153**: 181-192.

Tanaka, Y., N. Sasaki and A. Ohmiya, 2008. Biosynthesis of plant pigments: anthocyanins, betalains and carotenoids. *The Plant J.*, **54**: 733-749.

Thanh, N.T., H.N. Murthy, K.W. Yu, E.J. Hahn and K.Y. Paek, 2005. Methyl jasmonate elicitation enhanced synthesis of ginsenoside by cell suspension cultures of *Panax ginseng* in 5-l balloon type bubble bioreactors. *Appl. Microbiol. Biotech.*, **67**: 197-201.

Vicente, M.R.-S. and J. Plasencia, 2011. Salicylic acid beyond defence: its role in plant growth and development. *J. exp. Bot.*, **62**: 3321-3338.

Vinterhalter, B., S. Ninkovic, B. Kozomara and D. Vinterhalter, 2007. Carbohydrate nutrition and anthocyanin accumulation in light grown and etiolated shoot cultures of carob (*Ceratonia siliqua* L.). *Arch. Biol. Sci.*, **59** (1): 51-56.

Vlot, A.C., D.M.A. Dempsey and D.F. Klessig, 2009. Salicylic acid, a multifaceted hormone to combat disease. *Ann. Rev. Phytopath.*, **47**: 177-206.

Wang, S. and H. Jiao, 2000. Scavenging capacity of berry crops on superoxide radicals, hydrogen peroxide, hydroxyl radicals, and singlet oxygen. *J. Agric. Fd. Chem.*, **48**: 5677-5684.

Wellmann, E., G. Hrazdina and H. Grisebach, 1976. Induction of anthocyanin formation and of enzymes related to its biosynthesis by UV light in cell cultures of *Haplopappus gracilis*. *Phytochem.*, **15**: 913-915.

Yamamoto, Y., Y. Kinoshita, S. Watanabe and Y. Yamada, 1989. Anthocyanin production in suspension cultures of high producing cells of *Euphorbia millii*. *Agric. Biol. Chem.*, **53**: 417-423.

Zhang, W. and S. Furusaki, 1999. Production of anthocyanins by plant cell cultures. *Biotech. Bioproc. Eng.*, **4**: 231-252.

Zhang, W., C. Curtin, M. Kikuchi and C. Franco, 2002. Integration of jasmonic acid and light irradiation for enhancement of anthocyanin biosynthesis in *Vitis vinifera* suspension cultures. *Plant Sci.*, **162**: 459-468.

Zhang, Y., S.K. Vareed and M.G. Nair, 2005. Human tumor cell growth inhibition by nontoxic anthocyanidins, the pigments in fruits and vegetables. *Life Sci.*, **76**: 1465-1472.

Zhao, J. and R. Verpoorte, 2007. Manipulating indole alkaloid production by *Catharanthus roseus* cell cultures in bioreactors: from biochemical processing to metabolic engineering. *Phytochem. Rev.*, **6**: 435-457.

Zimmerman, J.L. 1993. Somatic embryogenesis: a model for early development in higher plants. *Plant Cell*, **5**: 1411-1423.

33

Bio-colours from Ornamentals

Bhavya Bhargava, Y. C. Gupta and Priyanka Sharma

Colours are well known since ancient times for colouring clothes, house paintings, articles and food. These often have different meanings in various cultures. Every individual has its own choice and liking for colour, and nature manifests itself in a wide spectrum of colours. Red depicts the power, control, motivation and love; the blue coveys the message of trust, peace and sincerity; the green signifies the colour of life, growth, health and prosperity; while white conveys the freshness, purity and tranquillity. As far as natural dyes and colours are concerned, India has a virtual monopoly in their production and application.

History

Natural dyes, dyestuff and dyeing are as old as textiles themselves. Man has always been interested in colours. The art of dyeing has a long past, and many of the dyes go back into prehistory. Dyeing and printing were the crafts up to the middle of the 19th century and had a very rich tradition of using natural dyes and colours. There are many references in our literature, and the calico-printing is an example of this. The Caves of Ajanta and Bagh (6th century) are undoubted proof of the natural colours. Numerous specimens of dyeing and printing of ancient periods are still available in museums. It was practised during the Bronze Age in Europe. The earliest written record of the use of natural dyes was found in China around 2,600 BC. Dyeing was known as early as 2,500 BC in the Indus Valley period. This knowledge has been substantiated by findings of coloured garments of clothes and traces of madder dye in the ruins of the Indus Valley Civilization at Mohenjodaro and Harappa (3,500 BC). Natural matter was used to stain hides, decorate shells and feathers, and in cave paintings. Scientists have been able to date the black, white, yellow and reddish pigments made from ochre used by primitive men in cave paintings. In Egypt, mummies have been found wrapped in dyed cloth. Chemical tests of red fabrics found in the tomb of King Tutankhamen in Egypt showed the presence of alizarin, a pigment extracted from madder. In more modern times, Alexander the Great mentioned having found purple robes dating to 541 BC in the royal treasury when he conquered Susa, the Persian capital. Kermes (from the *Kermes* insect) is identified in the Book of Exodus in the Bible, where references are made to scarlet coloured linen. By the 4th century AD, dyes such as woad, madder, weld, Brazilwood, indigo and a dark reddish-purple were known.

Brazil was named after the native woad (*Isatis tinctoria*) there. Henna (*Lawsonia*) was used even before 2,500 BC, while saffron is mentioned in the Bible. The first use of the blue dye, woad by the ancient Britons, may have originated in Palestine, where it was found growing wild. The Phoenicians prepared it until the 7th century, when Arab conquerors destroyed their dyeing installations in the Levant. In the prehistoric times, men used to crush berries to colour mud for their cave paintings. Primitive men used plant dyestuff for colouring animal skin and to their own skin during religious festivals as well as during wars. They believed that the colour would give them magical powers, protect them from evil spirits and help them to achieve victory in war. Dyes might have been discovered accidentally, but their use has become so much a part of man's customs that it is difficult to imagine a modern world without dyes. The art of dyeing spread widely as civilization advanced. Primitive dyeing techniques included sticking plants to fabric or rubbing crushed pigments into cloth. The methods became more sophisticated with time and techniques using natural dyes from crushed fruits, berries and other plant parts, which were boiled into the fabric and which gave light and water fastness (resistance). Some of the well-known ancient dyes include madder, a red dye made from the roots of the *Rubia tinctorum*, blue indigo from the leaves of *Indigofera tinctoria*, yellow from the stigmas of the saffron plant (*Crocus sativus*) and from turmeric (*Curcuma longa*). Until the middle of the 19th century, all the dyes available to man came from natural sources. The range of colour was limited as was the utility of dyes; indigo, tyrian purple, alizarin and turmeric were the prominent vegetable dyes. In 1,856, the first commercially successful synthetic dye mauve was serendipitously discovered by William H. Perkin in England and afterwards a large range of synthetic dyes has been manufactured. After the advent of synthetic dyes, the natural dyes were forgotten and abandoned as part of history for a period of about 150 years. Synthetic dyes and colours dominated natural ones due to their easy manufacturing, low cost, wide range of colours and shades and fastness properties.

Only during the last two decades, when people realized the concern for the environment and health, created an interest in natural dyes. Awareness about the toxicity of synthetic dyes throughout the world resulted in increased usage of natural dyes. Synthetic dyes cause environmental pollution during their production and use. All this happened due to the excessive use of synthetic dyes found in dyeing consumer goods. The global production of synthetic dyes is around 1,000,000 tonnes per annum. This production and its applications release vast amount of waste and unfixed colourant causing health hazards, pollution and disturbance of the eco-system. Several studies have shown that use of synthetic colours in food items causes adverse effects on human beings and animals. Some of the developed countries like Germany and the Netherlands have banned the use of azo-based synthetic dyes for textile dyeing. The European Union too is likely to impose a ban on these toxic dyes. India has also banned the use of 70 odd azo-dyes. Most of the countries have restricted the use of synthetic colours in food applications and their maximum limits have been prescribed. In India, only eight synthetic colours are permitted for food items under Prohibition of Food Adulteration (PFA) Rules.

Plants are the main source of colours and there are more than 300 species which yield colours. India is endowed with rich plant wealth and has a great potential to cope with a major portion of the world demand. The availability of a wide range of climatic conditions is highly conducive for growing a wide variety of plants. The flowers, fruits, seeds, leaves, stem, bark, wood and roots of numerous plants contain appreciable amount of colours and are renewable sources. Similarly, agro-wastes left after extraction of desirable constituent and forest wastes can be utilised for the production of colours.

Natural Colours and Dyes

Natural colour or pigment in the biological system is one that is synthesized and accumulated in or extracted from living cells. In addition, certain colourants such as oxidised phenol, phenolic derivatives and coumarins may be formed by dyeing cells. Natural dyes may be defined as chemicals which are obtained from vegetable and animal sources, without chemical processing. In nature, numerous coloured compounds are obtained but all coloured substances are not dyes. For a chemical to be defined as a dye it must possess a suitable colour which must be fixed in dispersion or solution to a substrate by itself or using a fixing agent, and it must not be fugitive. The applied colour should be fast to sunlight, water washing and to action of mild acid and alkali. The substrate may be textile, timber, leather, hair, fur, food, cosmetics or a pharmaceutical base.

Dyes are not conformed to a particular chemical class of organic compounds. They belong to a wide range of chemical classes of compounds, e.g. tetraterpinoids (carotenoids, xanthophylls), O-heterocyclic compounds (anthocyanins, flavones), tetrapyrroles (chlorophylls), quinones (naphthaquinones, anthraquinones), N-heterocyclic compounds (indigoid), etc. According to the theory of 'Colours and Constitution', chromophores are responsible for colours in compounds. These are unsaturated groups of compounds possessing a multiple bond conjugated system. Some of the important chromophoric groups are nitro, nitroso, carbonyl, azoxy, azo-amino, etc. The depth of colours increases with increase of chromorphic groups. There are other groups called auxochromes which may be acidic or basic and whose presence deepens the colours.

Sources

Natural colourants are obtained from the following categories (Joshi, 2006; Siva, 2007):

Plant-origin: There are some 300 plant species which yield colours such as *Areca katechu, Bauhinia, Bixa, Butea, Crocus, Curcuma* (turmeric), *Indigofera, Lawsonia, Mallotus philippinensis, Prunus* (cherry), *etc.* The colours are extracted from various plant parts such as roots, stems, bark, wood, leaves, flowers and seeds.

Animal-origin: There are various dye-yielding insects apart from the popular ones such as *Coccus cacti* (female cochineal insect whose whole body is dried and edible dye is extracted), *Kermes* spp. (dried bodies of female scale insects are used for extraction of purplish-red dye), *Laccifer lacca* (a South Asian insect, the female of which secretes lac that was used in the past to make shellac), *etc.*

Microorganisms: Certain algae, bacteria, moulds and yeasts also yield colours.

Major Plant Pigments

Pigment Class	Compound Types		Compound Examples	Typical Colours
Porphyrins		Chlorophyll	Chlorophyll a and b	Green
Flavonoids		Anthocyanins	Pelargonidin, Cyanidin, Delphinidin, Peonidin	Red Blue Violet
	Anthoxanthins	Flavonols	Kaempferol, Quercetin, Fisetin, Kaempferide, Morin, Myricetin, Myricitrin, Rutin	Yellow
		Flavones	Apigenin, Biacalein, Chrysin, Diosmetin, Flavone, Luteolin	Yellow
		Isoflavonones	Diadzin, Genistein, Enterodiol, Coumestrol, Biochanin	
		Flavonones	Eriodictyol, Hesperidin Naringin, Naringenin	Colour less co-pigments
		Flavans	Biflavan, Catechin, Epicatechin,	Colour less co-pigments
Carotenoids		Carotenes	Lycopene, -carotene, β-carotene, γ-carotene	Yellow Orange Red
			Lutein, Cryptoxanthin, Zeaxanthin, Neoxanthin, Rhodoxanthin, Violaxanthin, Canthaxanthin, Astaxanthin,	

Pelargonidin, Cyanidin, Delphinidin and Peonidin
Anthocyanins and anthoxanthin
Carotenoids, anthocyanins and flavonols
Anthocyanins and anthoxanthins
Anthocyanins and Carotenoids
Pelargonidin, Cyanidin and carotenoids
Cyanidin and Pelargonidin
Lutein and zeaxanthin
Lutein, zeaxanthin and β-carotene

Anthocyanin and carotenoids
Betacyanin
Cyanidin
Anthocyanin and calcone mix
Cyanidin and delphinidin
Delphinidin

Antirrhinum Pelargonium Portulaca

Anthocyanin Pelargonidin Betacyanin

Bixa orellana *Butea monosperma*

Bixin, norbixin (red) Butrin (yellow and orange)

Chemical Classification

Natural dyes represent a wide range of organic groups having different chemical structures. The commercially prominent dyes can be classified in groups based on their chemical structure (Fig. 1, Table 1).

Classification on Application Basis

Most of the natural dyes have no affinity to substrates. These require pre-treatment with chemicals to create affinity towards fixing their colours. In a broader sense, these can be classified into (i) substantive dyes, and (ii) adjective dyes. Indigo and turmeric dyes are capable of being fixed directly on fibres, hence are called substantive dyes. On the other hand, dyes obtained from most of the plant materials, e.g. madder or logwood, require adequate pre-treatment with suitable metallic salts before dyeing application and these are called adjective dyes. Subsequently, the classification based on application has been extended to their chemical properties and salt-forming capabilities. Presently, dyes have various categories such as **'vat dyes'** which includes indigo dye. Vat has been derived from the wooden vessel (vat) which was used in olden times for the fermentation of dye. Indigo is insoluble in water and hence is reduced to soluble *leuco* form using sodium metabisulphite, hypo or sodium carbonate with alkalis. As such, indigo has no affinity with fibre but after fermentation it becomes a soluble pale-yellow solution which yields blue colour after oxidization with atmospheric air. The strength of the blue colour can be increased by repeated treatments with vat solution followed by oxidation by beating the fibre. For good results, treatment with soap is necessary which facilitates oxidation. **'Direct dyes'** can directly be applied to cellulosic fibre without using any mordant, and no pre-treatment is required. Dyes extracted from turmeric rhizomes, harda fruits and rind of pomegranate, annatto seeds, etc. come under this class. Hot dye bath is used for dyeing applications. **'Mordant dyes'** have no affinity with fibre, however, after mordanting with metallic salts these are capable of being absorbed by fibres, the main reason being for ion of complexes with salts. Alizarin from *Rubia cordifolia* is the first known natural dye, which has no affinity for cotton but after mordanting with alum it is readily absorbed by the fibre. Tannic acid and harda are also used as mordants. **'Acid dyes'** possess either carboxylic or sulphonic acid group(s) in the dye molecule. Crocetin form soon contains two carboxyl groups each at the terminal end and this is the only example of this type of dye. It may be applied directly to silk and wool. Acid groups form electrovalent bonds with amino groups present in the silk and wool. **'Basic dyes'** are mostly the salts of bases with HCl and $ZnC1_2$ which after ionization give coloured cations and form electrovalent bonds with the carboxyl groups of wool and silk. These are applied in mild acidic pH at 60-65^{0}C. Cotton can be dyed with mordanting using tannic acid. Berberine is classified as a natural basic dye. No natural dye can be classified as the true **'disperse dyes'** as these are not transferred directly into the fibre medium but are transmitted through aqueous solution. The highly dispersed particles have little stability in water and serve as

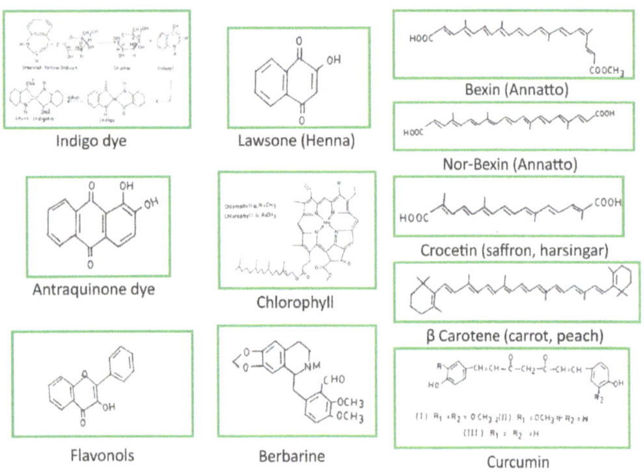

Fig. 1: Chemical structures of prominent natural dyes

Table 1: Classification of prominent dyes based on chemical structure

Chemical Group	Prominent Example	Colour
Indigoids	Indigo, tyrian purple	Blue-pink
Anthraquinones	Madder (Alizarin) Lac Kermes Cochenial	Natural red class dyes
Alpha-napthaquinone	Henna (Lowsone)	Orange
Flavones	Weld (*Reseda luteola*) Wood of pines Dahlia, sunflower, marigold palas, kamala, chrysanthemum, tea, etc.	Natural yellow class of dyes
Anthocyanines	Grape skin extract *Bignonia chica*	Red Orange
Betalains	Beetroot	Red to blue-red
Carotenoids	Annatto (*Bixa orellana*) Carrots Saffron	Yellow-orange Orange Jafran (yellow)
Diferuloyl-methane	Curcumin from turmeric	Yellow
Alkaloids	Berberine	Yellow
Chlorophyll	Leaves of lucerne, nettles, mulberry, green plants, pasture grasses, algae, etc.	Green

reservoirs of saturated dye solution which facilitates the dyeing process. Considering the low relative molecular mass and less solubility, *Lawsonia* (henna), certain flavones and anthraquinones can partially qualify for this class.

Natural Food Colours

We need colour in food because all the senses contribute to the experience of eating. The principal natural colours, most of which in refined form are used as additives, and these are the green pigment chlorophyll, the yellow to red carotenoids, and the flavonoids with their principal subclass the anthocyanins which give flowers and fruits their red to blue pigmentation. There has been much interest in carotenoid, especially ß-carotene (carrots, mango, papaya, etc.) which besides natural orange pigment, is also converted in the body to vitamin A and has antioxidant powers. The natural colours produced by various companies are available in the market as annatto (yellow/Orange), carotene (orange), beetroot (pink/blue/red), chlorophyll (green), beta carotene (yellow/orange), chlorophyllin (green), capsanthin (red/orange), curcumin (yellow), carmine (red), carminic acid (orange/red), lycopene (reddish-orange), lutein (yellow), anthocyanin (red/purple), vegetable carbon (black), etc.

In India, Rule 26 of The Prevention of Food Adulteration Act (PFA), 1954 and The Prevention of Food Adulteration Rules, 1955 &1999, permit certain colours whether isolated from natural sources or produced synthetically in food items, and these are beta-apo-8 carotenal, methylester of beta-apo-8 carotenoic acid, ethylester of beta-apo-8' carotenoic acid, canthaxanthin, chlorophyll, riboflavin (lactoflavin), caramel, annatto, saffron, and curcumin (tumeric). Rule 27 of the PFA prohibits the addition of inorganic matter and pigments in any article of food. The Government of India has restricted the use of synthetic colours in foods due to their adverse effects on human health. At present, Rule 28 of the Prevention of Food Adulteration Act pennants only eight synthetic colours as given in Table 2.

It is mentioned here that reputed companies use permitted colours in safe limits whereas its status at the small or local level is highly unsatisfactory. According to some reports, generally non-permitted colours are being used in the preparation of sweets, confectionery, bakery, ice-cream and other food items. The use of non-permitted colours and excess of permitted colours causes adverse effects on human beings and animals. There is an urgent need to take up the issue at the national level and create awareness in the public for the safe use of edible colours. Natural colours are widely permitted throughout the world but there is no universally accepted definition of colouring matter. Some countries exclude from their list those species which have low flavouring and colouring effect. Turmeric, paprika, sandalwood oil, etc. which have a secondary colouring effect are not classified as colours but declared as ingredients in the normal way. The list of permitted natural food colours in the U.S.A. and the E.U. is bigger than the Indian list. Apart from all 11 sources mentioned in the Indian PFA, it includes vegetable carbon, copper complexes of chlorophyll and chlorophyllins, paprika extracts, lycopene, lutein, beetroot red, anthocyanins, cochineal, carmine, cotton-seed flour, *etc.* Various plant species and their organs yielding colours are furnished in Table 3.

Table 2: Synthetic colours permitted under rule 28 of PFA

No.	Colour	Comman name	Colour index	Chemical class
1.	Red	Ponceau 4 R	16255	Azo
		Carmoisine	14720	Azo
		Erythosine	45430	Xanthen
2.	Yellow	Tartrazine	19140	Pyrazolone
		Sunset yellow FCF	15985	Azo
3.	Blue	Indigo carmine	73015	Indigoid
		Brilliant blue FCF	42090	Triarylmethane
4.	Green	Fast green FCF	42053	Triarylmethane

Table 3: Sources of different natural dyes and colours

Colour	Common name	Botanical name	Parts used
Red dye	Safflower	*Carathamus tinctorius L.*	Flower
	Caesalpinia	*Caesalpinia sappan L.*	Wood
	Madder	*Rubia tinctoria L.*	Wood
	Logwood	*Haematoxylon campechianum L.*	Wood
	Junglee khatti palak	*Rumex dentatus L.*	Wood
	Indian mulberry	*Morinda tinctoria L.*	Wood
	Kamila tree	*Mallotus philippinensis*	Flower
Yellow dye	Golden rod	*Solidago grandis L.*	Flower
	Teak	*Tectona grandis*	Leaf
	Marigold	*Tagetes* sp.	Flower
	Saffron	*Crocus sativus L.*	Flower
	Flame of the forest	*Butea monosperma*	Flower
Blue dye	Indigo	*Indigofera tinctoria L.*	Leaf
	Woad	*Isatis tinctoria L.*	Leaf
	Sunt berry	*Acacia nilotica l.*	Seed pod
	Pivet	*Ligustrum vulgare l.*	Fruit
	Waterlily	*Nymphaea alba L.*	Rhizome
Black dye	Alder	*Alnus glutinosa l.*	Bark
	Rofblamala	*Loranthus pentapetalus*	Leaf
	Custard apple	*Anona reticulata L.*	Fruit
	Harda	*Terminalia chebula*	Fruit
Orange dye	Annato	*Bixa orellena L.*	Seed
	Dahlia	*Dahlia* sp.	Flower
	Lily of the valley	*Convallaria majalis L.*	Leaf
	Nettles	*Urtica dioica L.*	Leaf

Joshi (2006) and Siva (2007) documented the important dye yielding plants with pigments, dye content, habitat and distribution (Table 4).

Role of Pigments in Human Health

1. Anthocyanins

Anthocyanin isolates and anthocyanin-rich mixtures of bioflavonoids may provide protection from DNA cleavage, estrogenic activity, enzyme inhibition, boosting production of cytokinins, anti-inflammatory activity, lipid peroxidation, decreasing capillary permeability and fragility, and membrane strengthening (Mary Ann Lila, 2004).

Prevention of Cardiovascular Disease (CVD)

- Most studied effect, and large number of epidemiological evidence exists
- Red wines and grape juice inhibit heart attacks

Table 4: Important dye yielding plants and their pigments

Plant	Colour obtained	Pigment	Dye content	Habitat and distribution
Acacia catechu (L.f.) Willd.	Brown, black	Catechin, catechutanic acid	The chief constituents of the heartwood vary from 4 to 7% and are distributed throughout the heartwood from the root to the branches.	Occurs throughout India in dry types of mixed forest on a variety of geological formations and soils.
Adhatoda vasica Nees.	Yellow	Adhatodic acid, carotein, lutolin, quercetin	–	Distributed throughout India, up to an attitude of 1300 m; grows on waste land and in a variety of habitats and soil. It is sometimes cultivated as hedge.
Bixa orellena L.	Orange, red	Bixin, norbixin	The dye content is 5–6% by weight of seed. A carotenoid bixin comprises 70–80% in each seed.	The small tree is found to thrive at elevations of 600–900 m; native to tropical America, it has become naturalized in the hotter parts of India.
Butea monosperma (Lam) Taubert.	Yellow or orange	Butrin	–	Commonly found throughout India, except in the arid region. It grows on black cotton soil, even on saline, alkaline and swampy badly drained soils and in barren lands.
Carthamus tinctorious L.	Yellow, red	Carthamin	The chief constituent carthamin ranges from 3 to 6% of the flower.	Cultivated throughout India. It requires fertile, moisture-retentive and well-drained soil.
Curcuma longa L.	Yellow	Curcumin	Percentage of curcumin varies from 5.4 to 8.7.	Turmeric grown generally as an annual crop. It is cultivable from sea level up to 1200 m. It thrives in well-drained, fertile, sandy and clayey, black red soil.
Indigofera tinctoria L.	Blue	Indigotin, Indican	Indigotin content varies according to season and age of the plant. Best grade contains 70–90% in dried leaves.	Distributed commonly in the tropical region.
Lawsonia inermis L.	Orange	Lawsone	The principle colouring matter, lawsone is present in dried leaves at a concentration of 1.0–1.4%.	It is mainly cultivated in Tamil Nadu, Madhya Pradesh and Rajasthan. It can grow on any type of soil from light loam to clay loam, but grows best on heavy soil.
Mallotus philippensis Muell.	Red	Rottlerin	The yield of powder rottlerin is 1.4–3.7% of the weight of the fresh fruits.	Found throughout India; occasionally ascending to 1500 m in the outer Himalayas. Commonly found in Sal and certain shrub and mixed forests.
Morinda citrifolia L.	Yellow, red	Morindone	Roots are dug out when the plants are 3–4 yrs old, dried and sorted for use by the dyeing trade.	A small tree distributed throughout the tropics.
Oldenlandia umbellata L.	Red	Alizarin, Rubicholric acid	–	Prostrate herb distributed in the tropical and subtropical region.
Pterocarpus santalinus L.	Red	Santalin	Red sanders contains 16% of a colouring matter, santalin (santalic acid).	Grows typically on dry, hilly, often rocky ground and is occasionally found growing on precipitous hillside.
Punica granatum L.	Yellow	Petargonidon 3,5,diglucoside	–	Mostly found cultivated in many parts of India, the tree is also common and gregarious in the gravel and boulder deposits of dry ravines and similar places in the outer Himalayas up to about 1800 m.
Rubia cordifolia L.	Red	Purpurin	Purpurin per cent vary from 2.0 to 4.0.	A hardy climber common throughout India, ascending to an altitude of 3750 m.
Semecarpus anacardium L.f.	Black	Bhilawanol	Bhilawanol ranging from 28 to 36% of dry weight of seed.	The tree is common in forests often found occurring with Sal, throughout the hotter parts of India.
Toddalia asiatica (L.) Lam.	Yellow	Toddaline	–	In South India, the plant is common in the Nilgris and Palani hills, and also in the scrubby jungles of Orissa.
Wrightia tinctoria R. Br.	Blue	β-amyrine	Leaves are the source of a blue dye called Mysore pala-indigo and β-amyrine ranges from 3.3–5.0% of dried leaves.	Distributed in Rajasthan, Madhya Pradesh and peninsular India, ascending to an altitude of 1200 m in the hills.

- Protects endothelial cells from pro-inflammatory signalling
- Flavonoids intake is associated with reduced incidence of CVD
- Vascular dysfunction could be ameliorated by blackberry juice
- Dietary intake (25–215 mg/day) is sufficient to provide pharmacological benefits
- Correlation between red wine consumption and reduced CV mortality exists (French paradox)
- Prevents expression of vascular endothelial growth factor (VEGF) in vascular smooth muscle cells
- Bilberry/elderberry extracts show endothelium-dependent relaxation capacity in coronary arteries
- *Rubus occidentalis* and *Sambucus nigra* anthocyanins accelerate the ethanol metabolism, improve permeability and strength of capillaries
- Role in prevention of atherosclerotic (leading cause of CVD) lesions in human coronary arteries

Anticancer, Antitumor and Antimutagenic Activity

- Black raspberry anthocyanins can reduce tumour development
- Anthocyanins have anti-inflammatory and chemopreventive properties
- Red wines suppress proliferation of human colon cancer cells
- Juices (strawberry, blueberry) significantly inhibit mutagenesis
- Cherry's anthocyanins reduce cell growth of human colon cancer
- Anthocyanin inhibits cycloxygenase enzymes which induce carcinogenesis
- Cyanidin-3-glucoside prevents ochratoxin A induced DNA damage in human
- Raspberry extract inhibits proliferation of human hepatocellular liver carcinoma cells
- Delphinidin, cyanidin, pelargonidin, and malvidin inhibit cell proliferation in human vulva and colon carcinoma cells
- Anthocyanins (cranberry, bilberry, blueberry) inhibit the initiation, promotion and proliferation stages of carcinogenesis

Beneficial Effects in Diabetes

- Anthocyanins may have a beneficial effect on diabetes at two levels, i.e. can interfere with glucose absorption, and may have a protective effect on pancreatic cells
- Red wine anthocyanins decrease lipid peroxidation, prevent generation of free radicals, and decrease glucose concentration & serum as has been reported in rats
- It is able to stimulate insulin secretion from pancreatic beta-cells
- Retinopathy and cataracts (serious consequences of diabetes mellitus) can be combated using anthocyanins

Antimicrobial and Antiviral Activity

- Blackcurrant anthocyanin extract inhibits the growth of *Enterococcus faecium, Escherichia coli* and *Staphylococcus aureus*
- Raspberries are the best inhibitors to microbes
- Anthocyanins are quite effective in inhibiting aflatoxin biosynthesis
- Anthocyanins from berries have an inhibitory effect on gram-negative bacterial growth
- Cyanidin glycosides (*Ribes nigrum*) have antiviral properties against influenza A and B viruses and herpes-1 virus

Neuroprotection

- Enhances red blood cell resistance to oxidative stress

- Protects from oxidative stress (related to neuro-degenerative disorder)
- Anthocyanin-rich fruits are effective in reversing the neuronal and behavioural aging
- Blueberry extracts have the beneficial effect on Alzheimer's disease (a neuro-degenerative illness) as has been observed in transgenic mice

Treatment of Vision Disorders

- Most extensively documented phyto-medicinal role
- Enhances visual acuity and increases capillary resistance
- Bilberry extract is effective in prevention of vision disorder and cataract
- Anthocyanin administration for four weeks improves contrast sensitivity
- Consumption of black currant extracts have positive effect on night vision (dark adaptation)

Betalains

These are the water-soluble nitrogen-containing plant pigments found in the cytoplasm of plant tissues of red beets, portulaca, bougainvillea, *etc.*

- Provides insect-repelling signals in cactus thorns, and gives resistance against soil pathogens
- Betacyanins impart the red-violet colouration and provide protection from UV light in ice plants
- Betaxanthins impart visible green fluorescence

Roles in Human Health Maintenance (though not researched extensively)

- Betalains can be naturally absorbed by the body
- Very potent antiradical scavenging activity
- Antiviral and antimicrobial properties are well documented
- Provides increased resistance to oxidation to human LDL
- Introduced as a food supplement (betaxanthin) to fortify the food products

Cancer Chemoprevention

- Betanin appears to be a potent anticancer compound
- Betacyanin-rich beets are most efficient to scavenge free-radical
- Beetroot extract have the chemopreventive properties against lung and skin cancers
- Betalains have slightly higher protection against skin and lung tumours than anthocyanins and carotenoids

Carotenoid Pigments

These are plastid-synthesized and lipid-soluble C_{40} tetraterpenoid pigments and are responsible for red, orange or yellow colouration in the plants. These are found in flowers (*Calendula, Lilium, Narcissus, Rosa, Tagetes, etc.*), fruits (apricot, citrus, papaya, persimmon, pineapple, *etc.*) and vegetables (carrots, pumpkins, squashes, tomatoes, *etc.*).

- >600 carotenoids identified but <50 are abundant in human diet
- Five are important in human health, i.e. β-carotene, α-carotene, lycopene, lutein, and zeaxanthin
- Carotenoids are nutritionally important for eye health, normal cell regeneration, and other health aspects linked to free radicals

Carotenoids and Photo-oxidative Damage

- Provides protection from vision disorders, AMD and cataracts
- Most potent biological quenchers of singlet oxygen
- Its lutein is generally referred to as the 'eye-protective nutrient,' protects the cells from lipid peroxidation, protects from AMD and cataracts, and is used in photosensitive disease therapies

- Lycopene, lutein, and zeaxanthin are excellent antioxidants, protect exposed tissues from light-induced damages, and β-carotene protects cells from lipid peroxidation & membrane damage

Modulation of the Immune System Response

- Credited with biological promotion of immune system response
- Increases total white blood cells
- Increases lymphocyte response to mitogens
- Increases natural killer cells activity in aging subjects
- Protects immune cells from the reactive bactericidal species
- **β-carotene influences immune cells to act more efficiently**
- Many studies point to multiple potential mechanisms by which carotenoids boost the immune system

Inhibition of CVD

- Carotene inhibits the LDL oxidation
- Cholesterol synthesis is inhibited by carotenoids
- Carotenoids inhibit CVD by inhibiting HMG-CoA reductase in humans
- Tomato juice (2-week dose) lowers LDL in male subjects effectively
- High intake of carotenes is correlated with lowered risk of CVD

Cancer Chemoprevention

- Carotenoids inhibit human breast cancer and female lung cancer
- β-carotene inhibits cell proliferation in carcinogenesis
- Inhibits cervical, lung, prostate, colorectal and stomach cancers
- Lycopene-rich foods reduce prostate cancer risk in human
- Lutein and zeaxanthin intake is linked to reduced risk of prostate cancer
- Enhances cell-to-cell communication by up-regulating gap junctions

Chlorophyll Pigments

These are most important plant pigments, a 'real life force', present in the chloroplast membrane and which nature uses to explode plants into greenery. This is the greatest miracle of nature as it produces simple sugar. Its chemical structure closely resembles the human blood. Chlorophyll is consumed in much higher doses (diet having green and leafy vegetables), and its photodynamic properties have the role as therapeutic agents. Chlorophylls are of two types, chlorophyll 'a' (bluish-black) and chlorophyll 'b' (dark green).

Diverse Roles of Chlorophylls in Human Health Maintenance

- Activates the brain
- Controls the blood sugar
- Reduces the blood pressure
- Chlorophylls act as the chemopreventive compound in human
- Reduces the incidence and multiplicity of skin tumours
- Capable of deactivating potential carcinogens in the digestive tract
- Leaf extract of crude Indian spinach reduces clastogenic effects
- Dietary consumption of green vegetables is associated with protection against mutagenic and clastogenic activity

Biosynthesis of Anthocyanins

Anthocyanins are water-soluble pigments that occur in almost all vascular plants. The anthocyanin pigments are responsible for majority of the orange, red, purple and blue colours of flowers. Anthocyanins are derived from a branch of the flavonoid pathway (Fig. 2), for which chalcone synthase (CHS) provides the first committed step by condensing one molecule of *p*-coumaroyl-CoA with three molecules of malonyl-CoA to produce tetrahydroxychalcone. Chalcone provides the precursor for all classes of flavonoids, which include the flavones, flavonols, flavan-diols, flavan 4-ols, proanthocyanidins (condensed tannins), isoflavonoids and anthocyanins. The closure of the C-ring, resulting in the formation of flavanones, is carried out by chalcone isomerase (CHI), an enzyme originally believed to have a structure unique in the plant kingdom. Flavanones (e.g., naringenin) provide a central branch point in the flavonoid pathway and can serve as substrate for enzymes that introduces -OH groups at the 3rd and 5th positions of the B-ring (e.g., F3-H and F3-5-H), or for the hydroxylation of the C-ring by flavanone 3-hydroxylase (F3H), a soluble di-oxygenase. Dihydroflavonol 4-reductase (DFR) provides one entry step to the biosynthesis of anthocyanins, and depending on the plant species, it can utilize as a substrate any one or all three of the possible dihydroflavonols, dihydromyricetin, dihydrokaempferol or dyhydroquercetin, resulting in the formation of the corresponding leucoanthocyanidins, providing structure to the anthocyanin biosynthetic grid. In some plant species, an activity that has sometimes been referred to as flavanone 4-reductase (FNR) reduces naringenin to the corresponding flavan 4-ol.

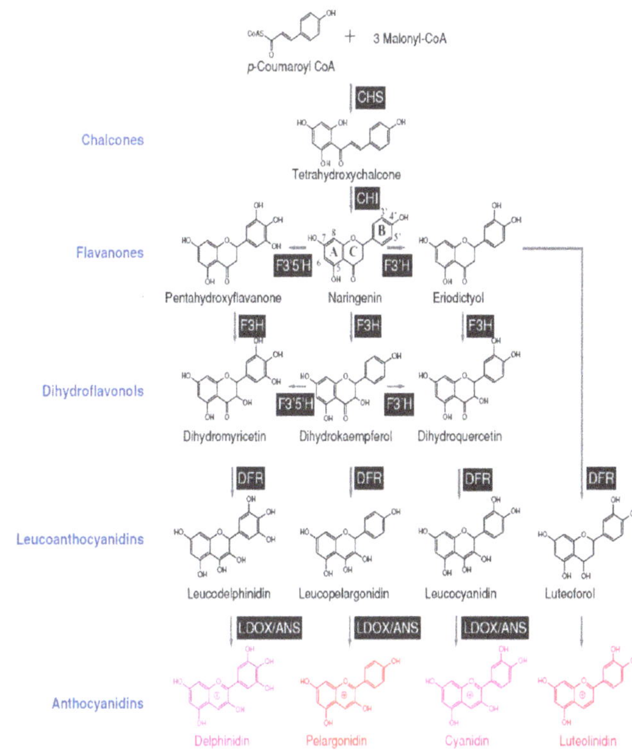

Fig. 2: Schematic representation of the biosynthetic pathway of the most abundant anthocyanin pigments. The names of the compounds are indicated. The enzyme names in black boxes are CHS, chalcone synthase; CHI, chalcone isomerase; F3H, flavanone 3-hydroxylase; F3_H, flavanone 3_hydroxylase; F3_5_H, flavanone 3_5_ hydroxylase; DFR, dihydroflavonol 4-reductase; LDOX/ANS, leucoanthocyanidin dioxygenase/anthocyanidin synthase. The A-, B-, and C-rings with the carbon numbers are indicated in the structure corresponding to the flavanone naringenin.

Biosynthetic Pathway of Some Betalain Pigments

Betalains are water-soluble nitrogen-containing compounds synthesized from tyrosine by the condensation of betalamic acid, a central intermediate in the formation of all betalains, with a derivative of dihydroxyphenylalanine (DOPA). This reaction results in the formation of the red to violet betacyanins, such as those found in red beets or in the flowers of portulaca. The condensation of betalamic acid with an amino acid or amino acid derivative results in the formation of the yellow to orange betaxanthins. Betacyanins and betaxanthins can be further classified into several subclasses, based on the chemical characteristics of the betalamic acid conjugate. The next step in the biosynthesis of betalains involves the formation of an aldimine link between betalamic acid and cyclo-DOPA (to make betanidin) or an amino acid derivative (to make betaxanthin) (Fig. 3).

Biosynthetic Pathway of Some Major Carotenoid Pigments

Carotenoids are plastid-synthesized and localized lipid-soluble C_{40} tetraterpenoids universally distributed in the plant kingdom. In contrast to the dispensable anthocyanins and betalains, carotenoids are essential for plant life, providing important photoprotective functions during photosynthesis and serving as precursors for the

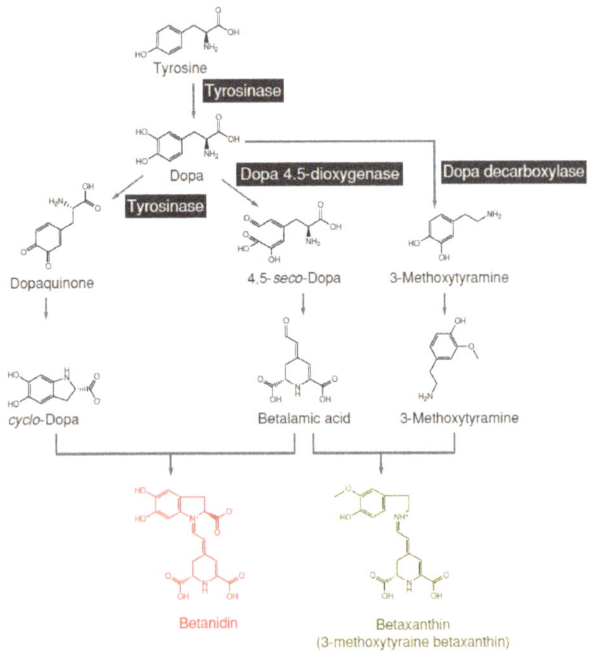

Fig. 3: Schematic representation of the biosynthetic pathway of some betalain pigments. The known enzymes are indicated in black boxes and the compound names are shown.

biosynthesis of the phytohormone abscisic acid (ABA). Carotenoids are also very significant nutraceutical components of the animal diet, serving as precursor for vitamin A biosynthesis and as antioxidant. Carotenoids are responsible for most of the yellow to orange flower colours in ornamentals that include *Calendula, Freesia, Gerbera, Lilium, Narcissus, Rosa, Tagetes, etc.* The first committed step in the carotenoid pathway is catalyzed by phytoene synthase (PSY), resulting in the condensation of two C_{20} geranylgeranyl diphosphate (GGPP) molecules to form phytoene (Fig. 4). Four desaturation reactions, two each catalyzed by the membrane-associated phytoene desaturase (PDS) and ζ-carotene desaturase (ZDS), result in the formation of the pink lycopene from the colourless phytoene. The cyclization of lycopene represents a branch point in the pathway, and two products can be formed depending on the position of the double bond on the cyclohexane ring. On one hand, lycopene β-cyclase, for which there are two forms in tomato, one specific to green tissues (LCY-B) and the other to chromoplasts (CYC-B), first produces γ-carotene containing one β-ring, which is subsequently converted to β-carotene by the same enzyme. On the other hand, lycopene ε-cyclase (LCY-E) produces δ- carotene. The formation of α-carotene, the precursor for lutein, involves formation of a β-ring on δ-carotene by lycopene β-cyclase. The α- and β-carotenes are the precursors for the xanthophylls, which are oxygenated carotenoids generated by β- and ε-ringspecific hydroxylases. β-carotene is converted to zeaxanthin by the carotenoid β-ring hydroxylases (HYD-B).

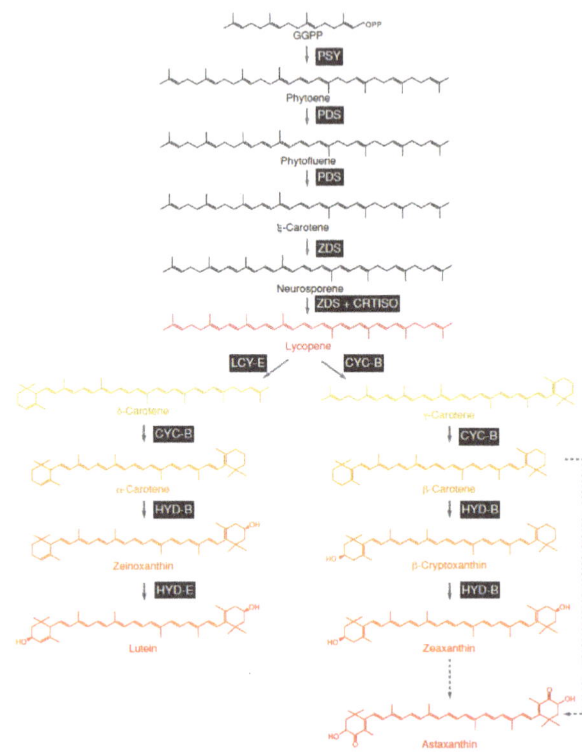

Fig. 4: Schematic representation of the biosynthetic pathway of some major carotenoid pigments. The names of the compounds are indicated. GGPP corresponds to geranylgeranyl diphosphate. The enzyme names, in black boxes, are PSY, phytoene synthase; PDS, phytoene desaturase; ZDS, ζ-carotene desaturase; CRTISO, carotenoid isomerase; CYC-B, chromoplastic form of lycopene Ɛ-cyclase; LCY-E, lycopene ε-cyclase; HYD-B, carotenoid β-ring hydroxylases; HYD-E, carotenoid Ɛ-ring hydroxylase.

Extraction

The vegetable dye's extraction is broadly divided into extraction method and extraction technology.

Methods of Dye Extraction

The extraction method of vegetable dyes basically depends on medium in which the dye is extracted. There are mainly four methods used in natural dye extraction.

i. Aqueous Method

- Boil known amount of dyestuff in 100 ml of soft water at 100 ^{0}C
- Filter the dye solution
- Record the optical density

ii. Alkaline Method

- Prepare 1% alkaline solution with addition of 1g sodium carbonate/sodium hydroxide in 100 ml of water
- Enter the dye material and boil at 100 ^{0}C
- Filter the dye solution
- Record the optical density

iii. Acidic Method

- Prepare 1% of acidic solution by adding 1 ml of HCl in 100 ml of soft water
- Enter the dye material and boil at 100 ^{0}C
- Filter the dye solution
- Record the optical density

iv. Alcoholic Method

- Add 50 ml of alcohol to 50 ml of water
- Enter the dye material and boil
- Filter the dye solution

Extraction Technology

The extraction can be carried out in aqueous, acid or alkaline medium. At present, in India mostly small scale producers/manufacturers are using this method. Even the local dyers are using metallic flax and crude process technology in refined way with the help of blender condenser, distillation plant and drier and crystallization unit with the capacity of 300 tonnes per year. The modern techniques of extraction are carried out with the use of extraction plant, reverse osmosis process and the latest is supercritical fluid extraction method. This method is very common in developed countries.

Solvent Extraction: This technique was developed just before the dawn of 20th century. It has been commercialized in recent years. This technology has been improved to reduce waste generation and eco-effectiveness of extraction methodology. Ultrasonic extraction followed by micro-wave extraction of solid finds extensive use is mainly based on organic solvent extraction. Super Critical Fluid Extraction is a further advancement making significant step over the use of conventional solvent extraction technology. It uses CO_2 as extraction medium. It is non-hazardous and subject to minimum waste generation. This technique is used for the extraction of natural products in food, pharmaceuticals and in chemical industries. Supercritical CO_2 extraction technique is advantageous as it makes possible to work at moderate temperature without effecting the organoleptic qualities and the active ingredients of the extracts are obtained. Moreover, it makes us possible to obtain 100% natural extracts, completely free from extraction solvent residues. At the end of extraction, an expansion phase (achieved by reducing pressure) causes the carbon dioxide to change from the supercritical state to the gaseous state, which enables it to be removed completely from the CO_2 extract obtained.

Microwave Assisted Extraction Technology: The 'microwave assisted extraction technology' is a high-speed method used to selectively target compounds from various raw materials. The technology uses a microwave applicator as the energy source during solvent extraction leading to the following advantages:

- Faster processing
- Better yield
- Improved quality
- Lower energy consumption
- Reduced solvent level, and
- Low capital investment

Continuous Steam Distillation Process: The 'continuous steam distillation process,' as the name implies, is a separation process using steam as a medium but instead of batch type, this process is continuous. The process consists of a totally insulated pneumatic conveying system using super heated steam as a carrier gas.

Mordanting: Alum, chrome, copper sulphate, ferrous sulphate, cream of tartar, stannous chloride and tartaric acid are used as mordants. Three basic methods of mordanting are in vogue on yarns/fabrics. The use of different mordants change the colour of same dyestuff. Thus different colour tones are possible due to mordanting with single mordant or combination of mordants. In **premordanting** method the yarn/fabric is mordanted in the first stage and then dyed in the second stage. By dissolving required amount of suitable mordant in water an aqueous solution is prepared, then yarn/fabric is entered and boiled for 30-45 minutes, then the yarn/fabric is dyed in the prepared dyebath, and then finally washed, rinsed and dried. In **simultaneous mordanting** method the mordant and the dye are applied simultaneously in the same bath, optical density of the extracted dye liquor is recorded, the yarn/fabric is dipped in the extracted dye liquor and boiled for 15 minutes, then required amount of mordant is added to the extracted dye solution, stirred well and boiled for 30-45 minutes, optical density of the dye liquor is recorded and then the fabric is washed, rinsed and dried, and then finally the dyed material is boiled in the mordanting liquid for 30-40 minutes. In **post mordanting** method the fabric is first dyed and then mordanted. Dye solution is prepared, the optical density is recorded, yarn/fabric in the dye solution is dyed, after dyeing again the optical density of the dye liquor is recorded, now aqueous solution is prepared by addition of required amount of suitable mordant, and then dyed material is boiled in the mordanting liquor for 30-45 minutes, and then finally the material is washed, rinsed and dried.

Application Method of Vegetable Dyes in Textile Sector: In **dyeing of wool**, the required amount of dye is added to the dye bath. Sodium hydroxide is added to adjust the pH to 9 so that dye is completely dissolved. It is stirred well to dissolve the dye. Material to the dye bath is added and a temperature of 80 ^{0}C is maintained for 15 minutes. Now acetic acid is added to the dyebath to lower the pH to 4 and then again a temperature of 80 ^{0}C is maintained for 20 minutes. After completion of the dyeing process the soaping of the dyed material is conducted with non-ionic detergent (0.5%) at 60 ^{0}C for 20 minutes, followed by hot washing and then cold washing of the material. For **silk dyeing**, required amount of dye to the dyebath is added, the pH is adjusted to 9 by addition of sodium hydroxide and stirred well so that dye is dissolved completely. Material to the dyebath is added and boiled at 80 ^{0}C for 15 minutes. After the process is complete, acetic acid to the dyebath is added to lower the pH to 5 and at this time also the temperature of 80 ^{0}C for 20 minutes is maintained. Dyed material is now soaped with 0.5% non-ionic detergent at 60 ^{0}C for 20 minutes. Finally the material is hot-washed and then cold-washed. For **cotton dyeing** the required amount of dye is added in the dyebath, stirred well so that dye is completely dissolved and then material is added in the dyebath where 80 ^{0}C termperature for 45 minutes should be maintained. In the end the material is treated with 1% ferrous sulphate solution at room temperature with constant stirring, then soaping of the dyed material is done with 0.5% non-ionic detergent at 60 ^{0}C for 20 minutes. It is possible to dye **synthetic fibres** (nylon, polyester and acrylic fibres) with selected natural dyes. Few dyes have high affinity for synthetic fibres and bright and deep shades can be obtained. The quinon- based dyes exhibit exceptional fastness properties, though dyes based on other chromoforic group may not be so fast. Non-polar dyes such as lawsone and juglone exhibit a behaviour similar to that of dispersed dyes on polyster. Annatto being an acid dye, shows two different behaviours on different fibres. It behaves as a dispersed dye on polyester and as an acid on nylon. Berberine behaves as a typical cationic dye and shows high affinity for acrylic fibre. The process of dyes is endothermic in all cases.

Applications

Indigoid Dyes are blue (C.I. name Natural Blue I) and come mainly from *Indigofera tinctoria* though other sources are *Isatis tinctoria, Indigofera sumatrana, I. erecta* and *I. suffruticosa.* Its leaves are used for extraction of dye. Dye indigo was used prehistorically in India, where *Indigofera* probably originated, hence one of the species

is named as *indicum*. *Indigofera tinctoria*, a leguminous shrub is a 1.2-1.8 m high, found almost throughout India and cultivated in India, China and other countries of the East as a source of indigo. The leaves are the source of the dye, but generally the whole plant is used for extraction of colour, and on per cent dry basis the plant contains 5.11% N, 1.67% K_2O, 0.78% P_2O_5 and 5.35% lime (CaO). The plant contains a maximum dye content of 0.4 per cent. The main colouring matter is indigotin. The colouring matter is present in the form of soluble glucoside known as '*Indican*' which is fermented (hydrolysed) to indoxyl and sugar using alkali and sodium salts. The resultant indoxyl is further oxidised (atmospheric air) to blue indigotin. Blue dye similar to indigo is also obtained from the fleshy leaves of *Isatis tinctoria*. Approximately 20M kilogrammes of dyes are produced annually mainly for blue jeans (Steingruber, 2004). Sales *et al.* (2009) studied the effect of species and sowing date on biomass production and indigo yield from *Isatis* spp. They found early sowing dates giving higher biomass production as compared to spring-sown crop (Table 5). Its dye is extracted through fermentation process. A series of tanks are made in steps at a level where height of the next tank is kept lower than previous one. Raw material as whole plant for fermentation is placed in the uppermost tank with sufficient lime-free water to avoid indigo precipitation. After 10-15 hours when hydrolysis is complete, the yellowish liquor is drained off to another tank where indoxyl is oxidized to indigotin by agitating the liquor through paddling as this way the liquor comes in contact with atmospheric oxygen regularly. Indigotin is precipitated as a blue mass gradually at the bottom of the tank. After a certain period the supernatant blue-mass liquid is removed and taken into another tank where it is heated to prevent further fermentation, filtered and dried into cakes. The quality depends on the quantity of ash present there, however, the rate may range from Rs. 1,000 to 2,000 per kilogramme.

Indigo dye is capable of dyeing both animal and vegetable fibre, although the dyeing procedure varies depending upon the physical structure and chemical property of the fibre. Indigo is a water insoluble dye and cannot be used directly. Firstly, the dye is reduced in alkaline medium to its *leuco* state. The dye is crushed to a smaller size and mixed with water in the presence of

Table 6: Effect of *Isatis* species and sowing dates on biomass production *cum* indigo yield

Seeding date	2002			2003		
	Biomass (t FW/ ha)[a]	Indigo (g/kg FW)	Indigo (kg/ha)	Biomass (t FW/ ha)[a]	Indigo (g/kg FW)	Indigo (kg/ha)[b]
Isatis indigotica						
December	11.4 (4)	0.8	9.5	9.7 (4)	0.7	5.3
January	30.6 (5)	0.8	24.5	8.5 (4)	0.7	5.3
Februray	11.9 (3)	0.6	7.9	9.1 (4)	0.6	5.0
March	14.7 (3)	0.5	7.4	8.0 (3)	0.6	5.0
Isatis tinctoria						
December	75.4 (7)	0.5	54.7	66.4 (5)	0.4	26.2
January	105.7 (7)	0.6	62.8	46.2 (5)	0.7	18.7
Februray	66.3 (6)	0.6	42.6	32.8 (4)	0.5	16.8
March	68.2 (6)	0.6	38.3	33.1 (4)	0.3	10.4
Analysis of variance						
Species (A)	22872.6**	0.008 NS	8327.5**	7679.6**	0.10 NS	991.7**
Species (B)	1095.8**	0.069**	522.2**	401.0**	0.06 NS	66.1**
A x B	152.2 NS	0.014 NS	55.7 NS	351.7**	0.02 NS	61.4**
Error	105.5	0.013	55.5	25.3	0.42	7.9

[A] Data in parenthesis indicate the number of harvests taken during the growing season
[B] For each species and entry, means followed by the same letter are not significantly different according to Tukey's test ($p < 0.05$)
** Significant at $p \leq 0.01$

(Sales *et al.*, 2006)

sodium metabisulphite, hypo, sodium carbonate or alkali. There are different types of vats adopted for dyeing the fibre. Ancient system is **fermentation vat** being applied in most traditional dyeing processes, however, in the recent years the **sodium hyposulphite** and **alkali** are widely used. For cotton dyeing, exclusively **copper salts** are used. In **zinc vat** system, alkaline solution of zinc dust is used as reducing agent. In olden times, **urine** was used as a fermenting and alkaline agent. Potash has also been used for fermentation (**potash vat**). In German vat, soda was used as an alkali. Nowadays, **hydrosulphite vat** is commonly practiced at the commercial level. In dyeing applications, the fibre is dipped in the vat and after squeezing, beaten in the air. It facilitates the oxidation and *leuco* form regenerates back to blue. It is to be mentioned here that weaker vats are always made and concentrated vats are not feasible. Lighter shades can be obtained by few dips and deep shades can be achieved through successive dips. For **dyeing on cotton, wool and silk**, powdered indigo dye 500 g is taken and made into paste with wetting agent, then it is dispersed in 25 litres of hot water where 1.5 kg of sodium bisulphate and 600 g of caustic soda are added and stirred for 20-30 minutes till the solution turns light-yellow, and if it does not appear of this colour some more reducing agent should be added, and the pH is adjusted at 11-12. Yarn in dyebath is dipped for 15-20 minutes and then hanks are taken out and kept in air till their colour becomes blue. This process is repeated until the desired depth of colour is not received. Now these hanks are treated with 2 per cent hydrogen peroxide for 15 minutes and then these are washed with cold water, followed by soap solution thoroughly, and finally with cold water and dried in air. Shade variation is a major problem in indigo dyeing which could be avoided by maintaining a constant 'dye-chemical' equilibrium and standard dyeing profile. If shade is dull, hydrosulphite should be increased. In case of greener shade, lesser quantity of hydrosulphite is recommended. Indigo provides textile dyeing with good light fastness properties, which is highest in wool followed by cotton and silk. The degree of fastness can be improved by treatment with acids, aromatic hydrocarbons and trimethyl ammonium chloride. Rao *et al.* (2008) formulated natural and safe hair colourants. Various combinations of powdered leaves, fruits and flowers of dye yielding and hair-care plants such as henna, indigo, bhringraj, aonla, catechu and centella were evaluated for their colouring and sensitivity reactions. Henna and indigo formulations were found to be suitable natural colourant.

Annatto (*Bixa orellana*) is a tree native to South America known for its reddish orange pigment 'annatto' produced only on the aril portion of its seeds. It is easily multiplied through seeds. It is the most preferred natural food colourant next to saffron, having wide applications in the dairy and cosmetic industries. **Annatto dye** is extracted from seeds, leaves, bark, roots and shoots of *Bixa orellana* which is a small tree growing to 5-10 m in height. It is being abundantly grown in south India. Its prickly reddish-orange pods contain about 50 seeds and one full-grown tree in a good environment can produce up to 200 kg of seeds. The seeds are covered with a reddish aril which is the source of an orange-yellow dye. Traditionally, the crushed seeds are usually soaked in water and then the water is evaporated to make a brightly coloured paste. This paste is then added to soups, cheese and other foods to give it a bright yellow or orange look. Annatto paste is also used as a natural dye for cloth and wool, and is sometimes employed in the paint, varnish, lacquer, cosmetics and soap industries. Annatto seeds contain 40 to 45 per cent cellulose, 3.5 to 5.5 per cent sucrose, 0.3 to 0.9 per cent essential oil, 3 per cent fixed oil, 4.5 to 5.5 per cent pigments, 13 to16 per cent protein, as well as alfa- and beta-carotenoids *cum* other constituents. Annatto oil is extracted from the seeds and is the main source of the pigments bixin and norbixin, which are classified as carotenoids. Major colour constituents are cis-bixin ($C_{25}H_{30}O_4$; oil soluble yellow to orange-yellow) and nor-bixin ($C_{24}H_{28}O_4$; water soluble yellow-orange to orange). Rivera *et al.* (2006) studied genetic improvement in three variants of *Bixa orellana*, i.e. 'Criolla', 'Peruana Verde' and 'Peruana Roja'. Pollen viability at opening was found maximum at day of bud opening, i.e. 86.9% in white and 86.3% in pink variants. Total carotenoids and bixin contents were maximum in 'Peruana Roja' as compared to 'Criolla' and 'Peruana Verde' (Table 7, 8). Maximum amount of annatto pigment Annatto seeds (346 ± 3.8 mg/100 g dry wt.) and maximum root biomass (152 ± 2.5 mg dry wt.) were recorded after 45 and 60 days of growth, respectively, on MS medium containing 0.1 mg/l IBA. Producing annatto pigment from normal root cultures under *in vitro* conditions is a novel approach when compared to the natural annatto pigment that is produced only on the aril portion of seeds. This allows the production of fresh pigment throughout the year. Remainder of the pigment mass in *Bixa orellana* consists of the water soluble yellow dye orellin, methylbixin, β-carotene, cryptoxanthin, lutein and zeaxanthin, the tinctorial strength of which is quite less than that of bixin. Norbixin is the principal colouring matter of water

Table 7: Pollen longevity under different storage conditions for white, pink and purple flower *Bixa orellana* variants

Storage conditions	*In vitro* pollen germination (%)		
	White	Pink	Purple
24 h before bud opening	0.0	0.0	ND
Day of bud opening	86.9 ± 4.3	86.3 ± 5.4	84.9 ± 5.4
24 h at room temperature	38.0 ± 2.5	45.0 ± 3.33	ND
24 h at 4°C	68.5 ± 4.1	73.5 ± 5.3	ND

Data show mean percentage (n = 3 plants per variant with five replicates each) and standard deviation (±).
ND = not determined.

Table 8: Variation in total carotenoids and bixin contents in seeds of three *Bixa orellana* variants

	B. orellana variants (g/100 g dry weight)		
	White flower	Purple flower	Pink flower
	'Criolla'	'Peruana verde'	'Peruana roja'
Total carotenoids	0.589 ± 0.046	0.310 ± 0.021	1.33 ± 0.074
Bixin	0.492 ± 0.052	0.242 ± 0.023	1.10 ± 0.082

Data show mean pigment content (n = 3) and standard deviation (±).

Annatto seeds

soluble products. Bixin is used as a colourant in foods and is believed to have UV protection capabilities and also demonstrates antioxidant and liver protective properties. Annatto preparations are pH sensitive and the sensitivity is of practical significance in low pH foods where the colour may turn pink. It is reported to be most stable at pH 8, with decreased stability in the pH 1 to 8 region. Bixin ($C_{25}H_{30}O_4$) is a carotenoid that is the main colouring component of annatto. It is soluble in fats and oils and the produced colour is found in the fraction of the food. It has a yellow hue, very good oxidation stability, fair light stability, good heat stability but is poor at very high temperatures, i.e. above 125 ^{0}C . The water-soluble annatto is norbixin and is obtained when bixin is saponified and the methyl ethyl group is split off. It is readily soluble in aqueous alkalis with the colouring occurring in the protein and starch fraction of the food. Annatto (norbixin) is used in colouring dairy products such as cheese, cheddar cheese, ice-cream, spreads, *etc.*; flour confectionery as it binds the flour and gives a stable colouring such as biscuits, ice-cream wafers, and snack foods; sugar confectionery with high boiling and in soft drinks as norbixin is acidic and light stable; in powder form in bio-savoury and sweet dry mixes; works as binding constituent to protein-rich food such as fishes; meat products where annatto extracts alone or mixed with carmine are used for colouring chicken and meat dishes; and are also used for dyeing of wool, calico and leather.

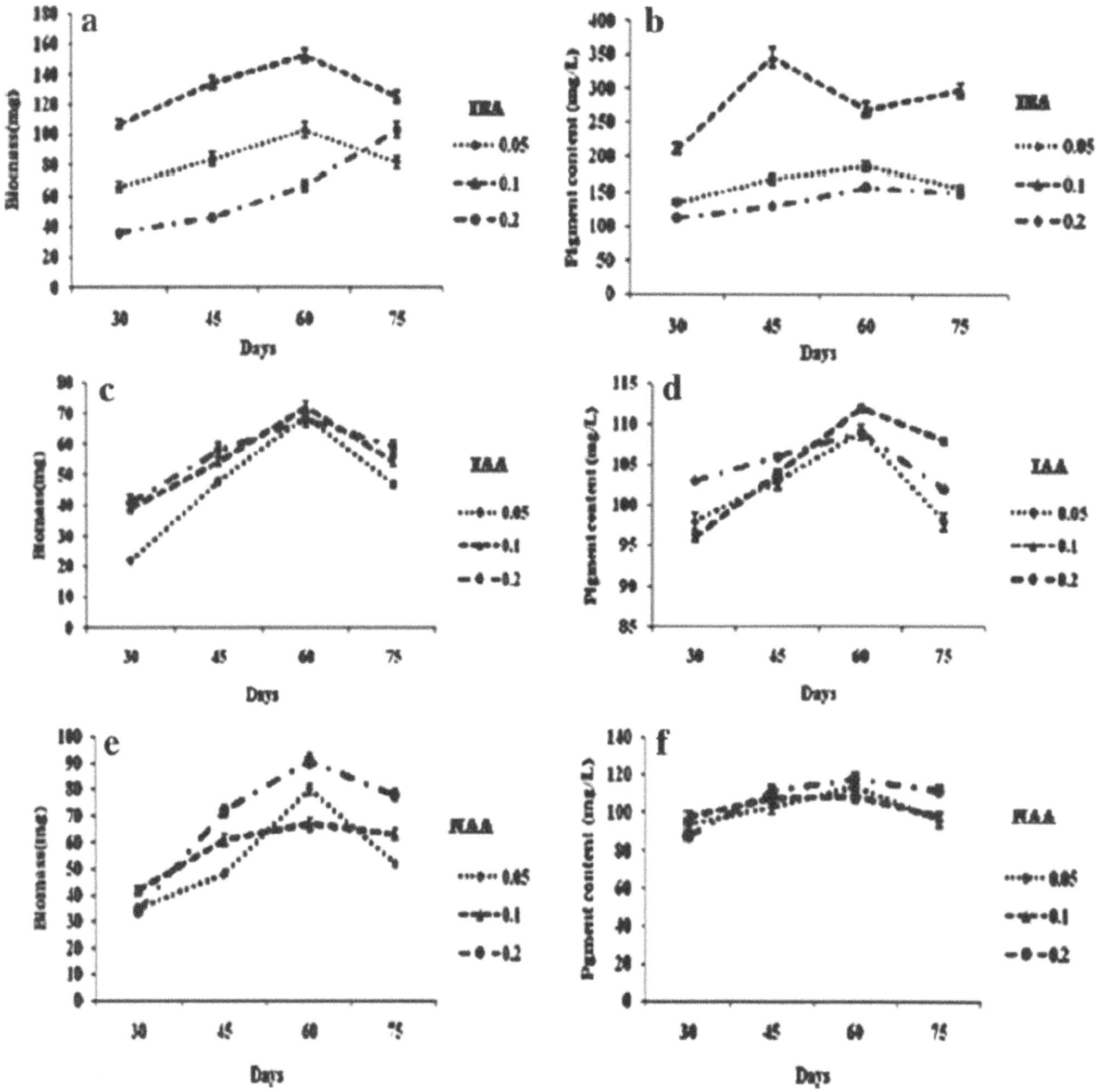

Fig. 5a-f: Production of root biomass (dry weight) and annatto pigment content in *in vitro* normal root cultures of *Bixa orellana* under the influence of auxins; a, c, e biomass of in vitro root cultures under the influence of different concentrations of IBA, IAA and NAA, respectively. b, d, f annatto pigment content in *in vitro* root cultures under the influence of different concentrations of IBA, IAA and NAA, respectively. Values are an average of five replicates (mean ± SE)

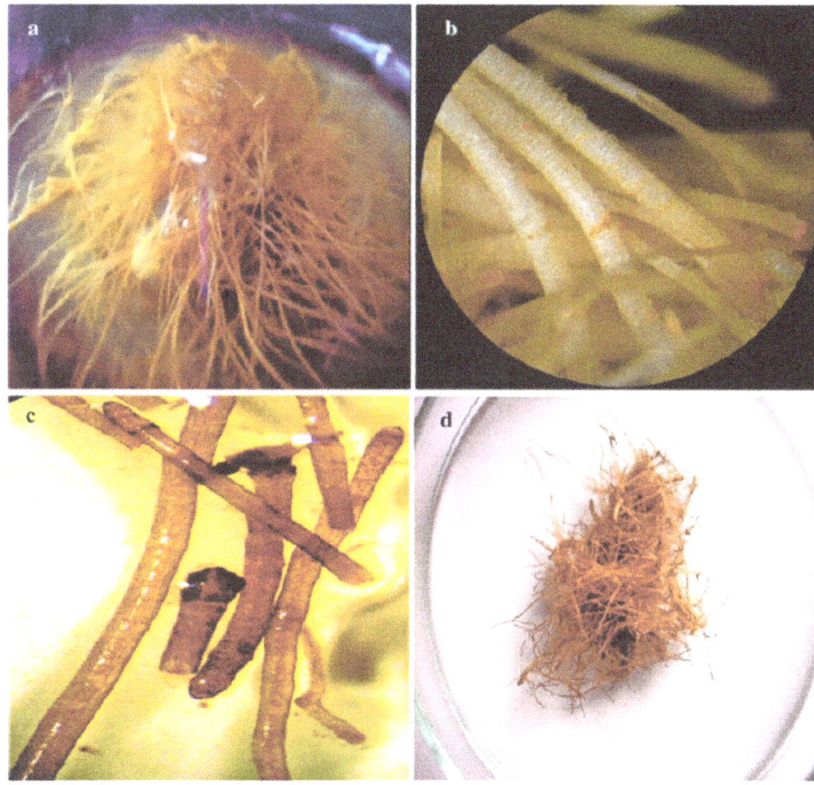

Fig.6a-d: ***In vitro*** **established root cultures of** ***B. orellana*** **on MS broth containing 0.1 mg l^{-1} IBA. b) Microscopical observation of pigment producing areas in** ***in vitro*** **root cultures at 940 magnification. c) Microscopical observation of pigment-producing areas in** ***in vitro*** **root cultures at 920 magnification. d) Harvested and dried in vitro root cultures.**

Henna (*Lawsonia inermis*, commonly known as *mehendi*) also acts as an anti-fungal agent (Bosoglu *et al.*, 1998). Colouring properties are due to lawsone, a burgundy organic compound that has an affinity for bonding with protein. Lawsone is primarily concentrated in the leaves, especially in the leaf petioles. Lawsone content in leaves is negatively correlated with the number of seeds in the fruits (Singh *et al.*, 2005). The dried plant leaves were washed with water to remove the adhering dust, then leaves were dried in an oven at 40 ^{0}C for 24 hours and finely powdered with the help of a grinding machine. One hundred grammes of dried leaf powder sample was taken in saturated Na_2CO_3 solution **(alkalizing process)** and kept for 24 hours at pH 8.5-9.2 at room temperature. This solution after becoming reddish-orange, was percolated with Na_2CO_3 solutions until all of the colour has been removed and then it was filtered again. The reddish orange colour develops due to presence of alkali. Under **acidification process** the alkaline reddish-orange solution is acidified with hydrochloric acid at pH 4-5. After acidification the colour of solution changes from reddish-orange to yellow-orange due to action of the acid. In **liquid-liquid extraction system with chloroform**, the acidified colour solution was taken into separating funnel and shaken with successive portion of chloroform (until 1ml of chloroform extract imparted no colour on shaking with 1ml of sodium carbonate solution). The chloroform extract was dried with anhydrous sodium sulphate and filtered. The chloroform extract produces an orange-yellow colour. This solution is kept away from light until whole of the chloroform is vaporized. Then cauliflower-like clusters of needles with dark mahogany-coloured crystalline mass of about 7.05 g is obtained and is preserved for further investigation. For **degumming of raw silk fibre**, the alkaline soap solution (strength 3.5 g/l) baths are prepared and its pH is adjusted to 10.0-10.5. Degumming is conducted at 95-100 ^{0}C for 1 hour. Now silk fibre is transferred into 2nd bath and heated for 45 minutes maintaining the same temperature. After degumming, the silk fibre is washed well with hot distilled water for 3-4 times. The extracted orange-yellow dye from the leaves is sparingly soluble in cold water. The dyebaths are prepared by adding required amount of dye (0.9 % on the basis of fibre) and alum as an electrolyte (10 % on the basis of the fibre). The pH of dye bath is maintained at 3 by adding acetic acid and sodium acetate. The fibre liquor ratio is maintained at 1:20 during dyeing operation. Before immersing the silk fibre in dye bath, it is soaked well in distilled water and squeezed for even absorption of dye particles. Dyeing is started at 300 ^{0}C, and then the temperature is slowly increased almost to 900 ^{0}C within about 10 minutes, and continued for 60 minutes thereafter with occasional stirring by a glass rod and then allowed for further 30 minutes as the bath cools down. During dyeing, to maintain the level boiling distilled water is added to the dyebaths. After dyeing, the fibres are squeezed over dyebaths. The fibre material is washed with distilled water twice and then dried at room temperature.

Table 8: Effect of dye concentration on dyeing of silk with henna leaf dye Electrolyte concentration (%)

No. of experiments	Dye concentration (%)	Dye exhaustions (%)
1	0.2	97.10
2	0.4	93.00
3	0.6	91.62
4	0.8	88.00
5	1.1	85.70
6	1.3	84.10
7	1.5	81.02

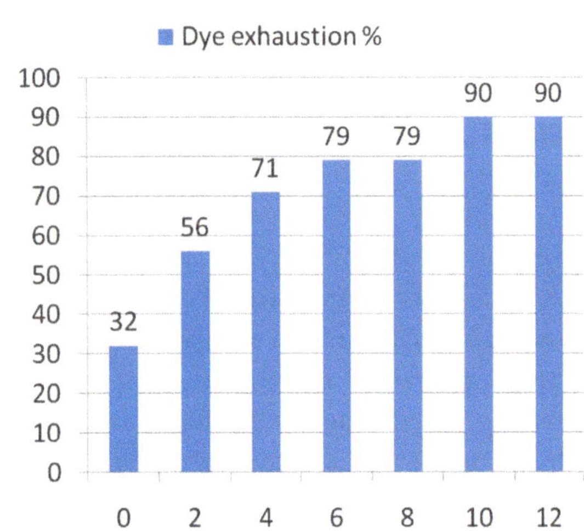

Fig. 7: Effect of electrolyte concentration on dyeing of silk fibre with extracted henna leaf dye

To ascertain the physical properties of dyed fibre such as dye concentration, electrolyte concentration, time, temperature, colour-fastness, etc., the extracted dye was applied on silk fibre and.the results are summarized in Tables 8, 9 and Figs. 7, 8. From Table 8, it is seen that dye absorption by degummed silk fibre decreases with increase of dye concentration in the dyebath. This may be explained that the presence of more dye ions hinder absorption of dye to the fibre, though less ions behave otherwise. It is to be mentioned that with the increase of dye concentration, the absolute quantity diminishes. The size of dye particles in solution usually depends on temperature, concentration of electrolyte and concentration of dyes. The size of dye particles always increases with an increase in dye concentration. It was observed that bright and uniform shades are produced when degummed silk fibre is dyed with 0.8% dye. Above or

Table 9: Effect of time on dyeing of silk fibre with extracted henna leaf dye

No. of expt.	Dyeing time hours	Dye exhaustion %
1	0.5	72
2	1.0	88
3	1.5	88
4	2.0	88
5	2.5	88

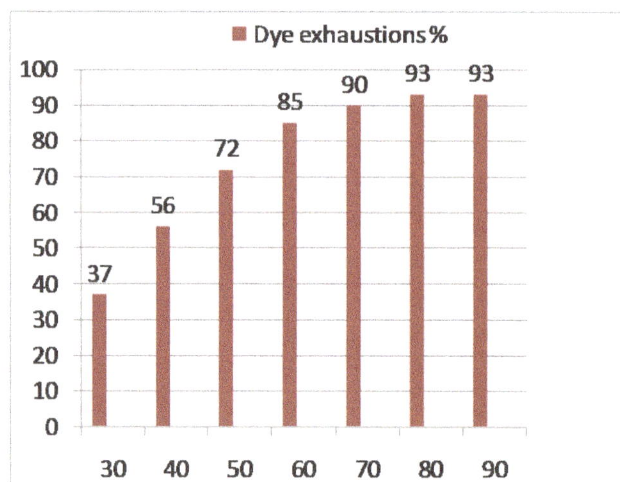

Fig. 8: Effect of temperature (°C)on dyeing of silk fibre with extracted henna leaf dye Dyeing temperature (°C)

below these percentages of dyes, dull and uneven shades are obtained. The effective concentration of dye, e.g. 0.8% corresponds to 88% dye exhaustion from the dyebath. The behaviour of dye absorption in dyeing of degummed silk fibre depends upon the characteristics of dye towards the fibre. From Figs. 7 and 8, it is also observed that the absorption of dye increases with increase of electrolyte concentration in the dyebath up to saturation absorption, and beyond this no change in absorption of dye occurs on further addition of electrolyte. The absorption reaches saturation point when degummed silk fibre is dyed in presence of 10% alum as electrolyte and the corresponding dye exhaustion is 90%. Above or below this percentage of electrolyte, shades become uneven and dull. The behaviour of dye absorption depends upon the characteristics of dyes towards the fibre. Table 9 shows that absorption of dye by degummed silk fibre increases with the increase of dyeing time and it reaches maximum when dyeing time is 1 hour. The absorption of dye from the dyebath at equilibrium dyeing time is 88 %. The absorption remains the same on further increase of dyeing time. This may be explained that the speed of dye diffusion inside the fibre depends on the size of the dye particles and on the state of the fibre, i.e. the smaller the dye particles the greater is the fibre swelling. Fig. 8 shows that absorption of dye by degummed silk fibre increases with increase of dyeing temperature. It seems that dye exists in solution as an aggregate of various sizes, and when the temperature of the solution increases the thermal energy coming from outside impedes aggregation, and therefore increases in the amount of the non-aggregated dye particles in the solution. Hence, the effect of temperature and consequently increase in the kinetic energy of the dye molecules tend to break up large aggregate into smaller units (Chatwal, 1993; Anon., 1965). Purohit *et al.* (2007) prepared natural dyes from different parts of *Delonix regia* flowers and applied on silk fabric. Light fastness study of the dyed fabric showed attractive shades with the dye material (Table 10). Tatsuzawa *et al.* (2010) studied flower pigments from the flowers of *Disa* cultivars. Per cent pelargonidin was found maximum in 'Marlene' as compared to 'San Francisco' and 'Dawn Angel' whereas per cent cyanidin was found maximum in 'Pink Tourmaline' than 'Dawn Angel' and 'Purple Taffy' (Table 11 and Fig. 9). Khalid and Silva (2010) studied the effect of saline irrigation water, flowering stages and their interaction on the flowerhead pigment contents extracted from *Calendula officinalis* and found that saline irrigation water at 0.39 dS/m after 81 days of bud formation gave maximum total flavonoids and total carotenoids, i.e. 25.4% and 24.8% in first season and 26.4% and 24.8% in second season, respectively (Table 12).

Kishimoto and Ohmiya (2009) through HPLC studies on 'Sunny Orange' chrysanthemum reported to have found maximum lutein-5,6-epoxide carotenoid in petals.

Priyanka Kanwar (2010) studied the carotenoid content in different marigold hybrids over two seasons, i.e. summer and rainy, and the hybrid derived from ms_{10} and French Bosnia-4 produced maximum carotenoid content (1084 µg) in rainy season crop though it was minimum (771 µg) in the hybrid ms_7 and Single Petal Red grown in summer season. Corrales *et al.* (2008) extracted

Table 10: Different shades achieved on fabrics with the flower extract of *Delonix regia*

	Parts of the flower Yarn	Solubility	Yarn	Mordant[a]/ developer[b]	Shades
SL-1	Whole flower (part I)	0.5% NaOH	Silk	Turmeric powder[b]	Golden yellow
SL-2	Whole flower (part II)	0.5% NaOH	Silk	10% Alum[a]	Olive green
SL-3	Petal (part I)	0.5% NaOH	Silk	Turmeric powder[b]	Dark tan
SL-4	Petal (part I)	0.5% NaOH	Silk	10% Alum[a]	Olive green
SL-5	Petal (part I)	0.5% NaOH	Silk	–	Saddle brown
SL-6	Petal (part II)	0.5% NaOH	Silk	10% Alum[a]	Olive green
SL-7	Petal + reproductive organ (part I)	0.5% NaOH	Silk	–	Dark brown
SL-8	Petal + reproductive organ (part II)	0.5% NaOH	Silk	Turmeric powder[b]	Brown
SL-9	Calyx	0.2% NaOH	Silk	–	Brown

Table 11 and Fig. 9: Pigments in *Disa* cultivars

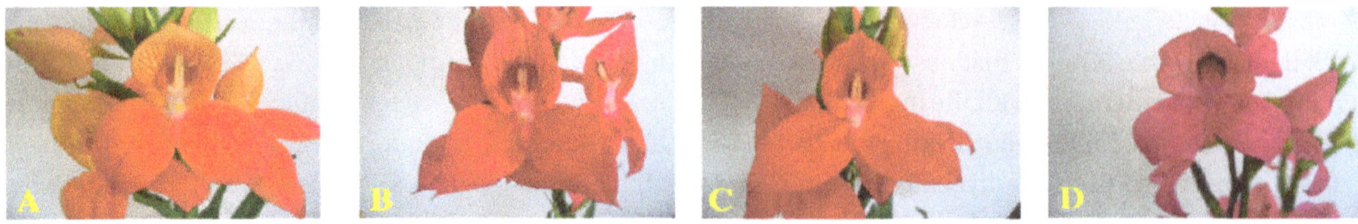

(A) Orange-Red ('Dawn Angel') (B) Red ('Marlene') (C) Red ('San Francisco') (D) Red-Purple ('Purple Taffy')

Flower colours and pigments from the flowers of *Disa* cultivars.

Cultivars	Flower colours[a]	b^*/a^*[b]	Anthocyanidin (%)[c]	
			Pelargonidin	Cyanidin
'Dawn Angel'	Orange-Red 33A	1.07	8	83
'San Francisco'	Red 44A	0.92	19	72
'Marlene'	Red 46C	0.86	20	64
'Purple Taffy'	Red-Purple 67A	0.18	1	82
'Pink Tourmaline'	Red-Purple 66C	0.00	1	90

[a] R.H.S. Colour Chart (The Royal Horticultural Society).

[b] Hue (CIE, 1976).

[c] Percentage of total absorbance of all detected anthocyanidins at 510 nm in HPLC analysis.

CIE: Commision International del' Eclairage $L^*a^*b^*$ system

HPLC: High Performance Liquid Chromatography

Table 12: Effect of saline irrigation water, flowering stages and their interaction on the flowerhead pigment extracted from *Calendula officinalis*

Effect of saline irrigation water, flowering stages and their interactions on the flower heads pigment contents extracted from *Calendula officinalis* L. flower heads.

Treatments		Total flavonoids (mg 100 g^{-1})				Total carotenoids (mg 100 g^{-1})			
Flowering stage	Saline irrigation water (dS m^{-1})	1st season		2nd season		1st season		2nd season	
		Mean	SD	Mean	SD	Mean	SD	Mean	SD
21 DABF	0.39	22.9	±2.0	24.5	±0.5	22.3	±2.0	24.3	±0.3
	1.56	20.8	±1.5	22.3	±0.3	22.1	±1.0	23.9	±0.0
	3.13	19.6	±3.0	17.8	±2.0	22	±2.0	23.5	±0.5
	4.69	14.3	±0.0	15.2	±0.2	17.2	±0.2	18.2	±0.2
	6.25	11.8	±2.0	10.6	±1.0	14.2	±0.2	12.4	±1.0
	7.81	10.8	±0.2	9.8	±1.0	11	±0.0	9.2	±0.2
	9.38	9.6	±0.4	7.3	±0.3	6.3	±0.3	8.2	±1.0
Overall initial flowering		15.7	±5.3	15.4	±6.3	16.4	±6.0	17.1	±6.8
81 DABF	0.39	25.4	±0.4	26.4	±0.0	24.8	±2.0	25.8	±1.0
	1.56	22.9	±2.0	23.4	±0.4	23.9	±1.0	24.9	±1.0
	3.13	20.8	±2.0	18.7	±1.0	23.6	±2.0	24.1	±0.1
	4.69	18.9	±1.0	16.2	±0.2	19.3	±0.3	19.7	±2.0
	6.25	12.6	±0.6	11.5	±0.5	15.3	±0.3	13.5	±0.5
	7.81	11.6	±0.2	10.1	±0.1	12.5	±0.5	9.6	±1.0
	9.38	10	±2.0	8.1	±0.1	6.4	±0.4	8.5	±0.5

Contd...

Treatments		Total flavonoids (mg $100 g^{-1}$)				Total carotenoids (mg $100 g^{-1}$)			
Flowering stage	Saline irrigation water (dS m^{-1})	1st season		2nd season		1st season		2nd season	
		Mean	SD	Mean	SD	Mean	SD	Mean	SD
Overall full flowering		17.5	±5.9	16.3	±6.6	18	±6.7	18	±7.1
111 DABF	0.39	11	±2.0	10.9	±0.0	10.5	±0.5	10.2	±0.2
	1.56	10.9	±1.0	10.3	±0.3	9.9	±1.0	9.8	±0.1
	3.13	9.8	±0.2	9.2	±0.2	9.6	±2.0	9.2	±0.2
	4.69	8.6	±0.2	8.4	±0.4	8.4	±1.0	8.5	±0.5
	6.25	7.4	±0.4	7.6	±1.0	6.8	±0.1	6.9	±1.0
	7.81	6.2	±0.2	5.6	±1.0	4.6	±0.1	4.5	±0.0
	9.38	3.6	±0.0	4.2	±0.0	3.5	±0.5	4.3	±0.3
Overall flowering termination		8.2	±7.7	8	±2.4	7.6	±7.7	7.6	±2.4
Overall salinity	0.39	19.8	±6.8	20.6	±7.3	19.2	±6.9	20.1	±7.5
	1.56	18.2	±5.8	18.7	±6.3	18.6	±6.6	19.5	±7.3
	3.13	16.7	±5.6	15.2	±4.7	18.4	±6.9	18.9	±7.3
	4.69	16.6	±4.5	15.7	±3.7	18.3	±5.0	19	±5.4
	6.25	10.6	±2.6	9.9	±1.9	12.1	±4.0	10.9	±3.2
	7.81	9.5	±2.5	8.5	±2.3	9.4	±3.6	7.8	±2.5
	9.38	7.7	±3.3	6.5	±1.8	5.4	±1.5	7	±2.1
F ratio									
Saline irrigation water		96.6^{***}		511.7^{***}		174.6^{***}		530.0^{***}	
Flowering stage		251.6^{***}		889.5^{***}		455.3^{***}		1283.0^{***}	
Saline irrigation water × flowering stage		6.0^{***}		4.2^{***}		10.9^{***}		44.4^{***}	

DABF: days after bud formation.
$^{*}P \leq 0.05$ according to *F*-values of the two-way analysis of variance (ANOVA-2).
$^{**}P < 0.01$ according to *F*-values of the two-way analysis of variance (ANOVA-2).
$^{***}P < 0.001$ according to *F*-values of the two-way analysis of variance (ANOVA-2).

Table 13 & Fig. 10: Various carotenoids and their percentage as observed in chrysanthemum petals

Peak no. (Fig. 2)	Carotenoids	% of total carotenoids
1	(3*S*,5*S*,6*R*,3'*R*,6'*R*)-5,6-Dihydro-5,6-dihydroxylutein (1)[z]	5.1
2	(9*Z*,13'*Z*)-Lutein-5,6-epoxide[z]	1.8
	(13*Z*,9'*Z*)-Lutein-5,6-epoxide[z]	1.8
3	(9'*Z*,13'*Z*)-Lutein-5,6-epoxide (3)[z]	2.2
4	(9*Z*,13*Z*)-Lutein-5,6-epoxide[z]	2.0
5	(all-*E*)-Lutein-5,6-epoxide (2)	7.7
6	(9*Z*,9'*Z*)-Lutein-5,6-epoxide[z]	2.5
	(9*Z*)-Violaxanthin	2.7
7	(8*S*)-Lutein-5,8-epoxide	5.0
8	(8*R*)-Lutein-5,8-epoxide	1.7
	(9*Z*-8'*R*)-Luteoxanthin	1.8
9	(9'*Z*)-Lutein-5,6-epoxide	16.6
10	(9*Z*)-Lutein-5,6-epoxide	16.9
11	(all-*E*)-Lutein	9.4
12	(9*Z*)-Lutein	11.3
13	(9'*Z*)-Lutein	6.0

[z] Nobel carotenoid as natural products.

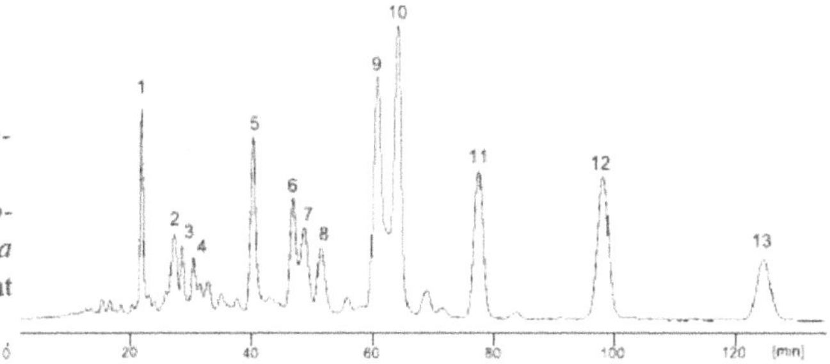

Fig. HPLC separation of carotenoids of an extract of chrysanthemum petals ('Sunny Orange'). Peak numbers as in Table Reprinted from Phytochemistry Vol. 65, Kishimoto et al., Carotenoid composition in petals of chrysanthemum (*Dendranthema grandiflorum* (Ramat.) Kitamura), p. 2781–2787, Copyright (2004), with permission from Elsevier.

anthocyanins from grapes as natural antioxidants or colourants, comparing the various extraction methods {high hydroastatic pressure (HHP), pulsed electric field (PEF) and ultrasonics}. The effect of heat treatment at 70 ^{0}C combined with the effect of different emerging novel technologies such as ultrasonics (35 KHz), HHP (600 MPa) and PEF (3 kV/cm) showed a great feasibility and selectivity for extraction purposes. After 1 hour extraction, the total phenolic content of samples subjected to novel technologies was 50% higher than in the control samples. The application of novel technologies increased the antioxidant activity of the extracts, being 4-fold with PEF, 3-fold with HHP and 2-fold with ultrasonic extraction methods when compared with control. Bechtold *et al.* (2007) studied the pomace extracts from different grape varieties for the contents of extractable anthocyanins, and recorded total anthocyanin concentrations of 24.5-126 mg/dm^3, pigment amount varying with the variety. The extracts were applied in textile dyeing operations using cotton fabric premordanted with tannin as textile substrate. Red/violet shades were obtained that showed good water-fastness; however, limited light-fastness indicated need for improvement to fulfil requirements for textile application. Lutein, a most valued carotenoid pigment with excellent colour value is becoming increasingly popular in the food industry, as chicken feed and in textile colouration. Marigold flowers ranging from yellow to orange-red to chocolate are the rich source for lutein, therefore, Jothi (2008) studied the use of its flower extract as a natural dye. The dye potential of the flower extract was evaluated on pure cotton and silk fabrics under normal dyeing conditions for the dye ability, wash fastness, light fastness, and colour hue. The surface colour was not affected by washing, and the quality of the fabrics was maintained even by washing at 60 ^{0}C for 30 minutes, however, change of some of the colours was noticed after soap-washing. Most of the metal salts exhibited the highest K/S values, due to their ability to form coordination complexes with the dye molecules. These findings reveal that the extract of marigold flowers can be used for colouration of 100% cotton, silk, and wool fabrics. Santos-Bocanegra *et al.* (2004) compared the efficacy of yellow and red xanthophyll extract from *Tagetes erecta* and *Capsicum* sp., with synthetic citranaxanthin, canthaxanthin, and ester of apo-8-carotenoic pigments at different concentrations to different sets of hens to enhance the yellowing intensity in chicken egg-yolk. In each trial, brown laying hens were placed in cages, and after seven days of feeding a sorghum-soybean diet without pigment was administered to one set of hens, while other sets of hens were administered with synthetic pigments, or natural xantophylls, and three combinations of apo-ester and citranaxanthin. After four weeks of feeding with these diets, the colour of 10 egg-yolks per treatment was evaluated every other day or twice a week for 10 times and classified with Roche yolk colour fan (RYCF) or Minolta® refractometer. Results indicated that egg- yolk colour changed significantly according to the pigment added. Hens fed with diets with 7.5 ppm of yellow xanthophylls extracted from *Tagetes* and 4.0 ppm of red xanthophylls from *Capsicum* had egg-yolk classified as colour 11.7 ± 0.1. Capsicum extract levels had linear effect over RYCF. Synthetic carotenoids gave a yolk colour that varied from 13 to 14 in the highest concentration, and from 12 (5%) to 14 (43%) at the lowest concentration (Table 14).

Table 14: Averages of egg yolk colour redness (a), yellowness (b), Roche yolk colour fan (RYCF) number, and percentages of egg yolks with acceptable market colour observed in experiment 1 in hen's-fed diets supplemented with combinations of levels of apo-carotenoic ethyl ester, canthaxanthin, and citranaxanthin, or *Tagetes* and *Capsicum* floral extracts

Treatments	Redness (a)		Yellowness (b)		RYCF		Acceptable yolk color	
							11 or less	12-14
							%	
Control	-7.45^c	± 0.35	34.61^b	± 1.35	3.73^d	± 0.24	-	-
Apo-ester, 2.5 ppm + Canthaxanthin, 1.5 ppm	0.02^a	± 0.35	40.50^a	± 0.94	12.46^a	± 0.17	7.53	92.47
Apo-ester, 2.0 ppm + Citranaxanthin 3.0 ppm	0.63^a	± 0.35	40.26^a	± 0.96	12.61^a	± 0.09	0.05	99.95
Tagetes, 7.5 ppm + *Capsicum*, 2.5 ppm	-2.84^b	± 0.45	42.37^a	± 0.74	10.49^c	± 0.27	65.84	34.16
Tagetes, 7.5 ppm + *Capsicum*, 3.0 ppm	-3.33^b	± 0.44	42.42^a	± 0.90	10.77^{bc}	± 0.11	32.51	67.49
Tagetes, 7.5 ppm + *Capsicum*, 3.5 ppm	-2.98^b	± 0.44	40.63^a	± 0.93	11.17^b	± 0.20	60.00	40.00
Tagetes, 7.5 ppm + *Capsicum*, 4.0 ppm	-2.19^b	± 0.35	42.66^a	± 1.17	11.70^{ab}	± 0.12	12.50	87.50
P-value	<0. 0001		<0. 0001		<0. 0001			
CV %	39.8		7.4		5.2			

^{abcd} Means within a column with different superscript indicate significant differences with $p < 0.05$ (Tukey's test)

Table 15: The antimicrobial activity of natural dyes against test bacteria (inhibition zone, mm)

Test bacteria	Dye			
	Punica granatum	*Rubia tinctorum*	*Mentha piperata*	*Allium cepa*
B. cereus RSKK 863	17.9±0.0	-	10.8±0.6	10.0±0.0
B. megaterium RSKK 5117	12.4±0.1	8.5±0.0	11.8±0.2	13.5±0.0
B. subtilis RSKK 244	17.6±1.7	10.2±0.0	13.5±1.6	9.9±0.0
Shigella sonnei RSKK 877	13.8±0.1	-	-	-
Staphylococcus aureus ATCC 25923	13.7±0.3	9.7±0.0	-	-
P. aeruginosa ATCC 29212	15.8±2.0	-	12.2±1.6	-
P. aeruginosa ATCC 27853	11.6±0.5	-	10.2±0.9	-
Salmonella sp. 21.3	10.5±1.5	-	-	-
Streptococcus epidermis	-	-	10.7±0.6	-

Calis *et al.* (2009) studied the antimicrobial effect of four commercially available dye powder obtained from *Rubia tinctorum, Allium cepa, Punica granatum* and *Mentha* sp. on *Staphylococcus aureus* ATCC 25923, *Shigella sonnei* RSKK 877, *Escherichia coli* ATCC 35218, *Bacillus megaterium* RSKK 5117, *Bacillus subtilis* RSKK 244, *Bacillus cereus* RSKK 863, *Pseudomonas aeruginosa* ATCC 29212, *Streptococcus epidermidis, Salmonella* 21.3 and *P. aeruginosa* ATCC 27853 (Table 15). *P. granatum* dye was most effective against the test bacteria except *E. coli* and *S. epidermidis.* The textile material impregnated with four natural dyes showed maximum inhibition rates of 80, 86 and 52%, respectively against *B. subtilis* of wool samples dyed with *P. granatum, A. cepa* and *R. tinctorum* while maximum inhibition rate of 91% was found against *P. aeruginosa* of wool sample dyed with *R. tinctorum.* Companies in India along with their addresses supplying various dyes are given below.

Companies	Address
Henna Herbs And Things Products manufacturing and exporting, henna hair colour, hair dye, herbal oil, henna hair dyes, natural hair colors, herbal powders, herbal extracts, essential oils, herbal cosmetics, incense ...More	B-86, Sarita Vihar, New Delhi - 110076, India Phone : 91-11-41402999 Mobile : +919873451072 Fax : 91-11-40540770/41402999
Jot Enterpises Products supplying, heena mehndi, black mehndi, herbal mandhi, hair colour, herbal hair colour mehndi	Shop No.-1, Kalish Complex, Opp. Quila Road, Basement, Rohtak - 124001, Haryana, India Phone : 91-1262-247305 Mobile : +919812302221
Min Chem India Products manufacturing and exporting, raw herbs, herbal extracts, phytochemicals, natural essential oils, natural colours, oleoresing, bulk drugs & chemicals	117, Loha Bhavan, P.D' Mello Road, Masjid (East), Mumbai - 400009, Maharashtra, India Phone : 91-22-23485439/32908187 Mobile : +919867308187 Fax : 91-22-23485439/23481619
Shivam Exim Pvt. Ltd. Products manufacturing and exporting, textile printing gum i.e. guar gum and tamarind gum, natural dyes, ready made garments dyed and printed with natural dye.	305, Anand Mangal-3, Opp. Core House, Ellisbridge, Ahmedabad - 380006, Gujarat, India Phone : 91-79-26467686/25430016(R) Mobile : +919824557702 Fax : 91-79-25713186
Shree Krishna Traders Products trading, fruit, grain, vegetable, dyestuffs, organic intermediate, pigments & dyes, edible oil & oil products, sugar, cereals, fruit & vegetables (fresh, processed & dry), pulses	Vishwa Gulab, 28, Dassera Maidan, Ujjain - 456010, Madhya Pradesh, India Phone : 91-734-2530852 Mobile : +919425394066 Fax : 91-734-2530851
Sir Biotech Pvt. Ltd. Products exporting and manufacturing and supplying, all type of natural dyes, punica granatum, indigofera tinctoria, tectona grandis, dipterocarpaceae spp., terminalia chebula, acacia catechu	Ist Floor, Sgm Plaza, 8/226, Arya Nagar, Kanpur - 208012, Uttar Pradesh, India Phone : 91-512-2540884 Mobile : +919336330257 Fax : 91-512-2552794
Victory Dyechem Industries Products manufacturing and exporting, fast colour basis and their salts, solvent dyes, specialty chem and their auxiliaries fast colour, bright orange, brown, discharge binders, ecofriendly plastisol	C-1-B/71, 100 Sheds Area, Chanod, Silvasa Road, Gidc, Vapi - 396195, Gujarat, India Phone : 91-260-2452996 Fax : 91-260-2453058

Drawbacks and Limitations of Natural Dyes

The synthetic dyes and colours dominated the natural dyes since the middle of 19th century because the natural colours possess the following limitations and technical drawbacks:

- **Cost-efficiency of natural dyes:** Natural dyes are many times costlier than synthetic colours due to yield of natural colours in plants being too less and their extraction is labour and fuel (power)-intensive, followed by spray drying or vacuum concentration. On the other hand, raw materials of synthetic dyes are comparatively cheaper (coal tar-based organic compounds) and these can be synthesized in a small space with ease.
- **Availability of raw material and regularity of supply:** Plants yielding dyes grow in a particular season though raw materials for synthetic dyes are available all the year round.
- **Constancy of cost of raw material and possibility of eventual replacement by selected ones:** To maintain regularity in supply of raw material is a major constraint as its price varies a lot, and is affected by location and season, *vis-à-vis* its habitat variation may affect the quality. And there is llittle scope to replace a particular colour from other sources.
- **Colour yield:** This is another bottleneck for natural dyes as colour content is vey less (usually less than 1%) in the concerned plants or their parts as live plants contain more than 80% moisture and the remaining material comprises cellulose, hemicellulose, lignin, glycosides, etc. The yield of dyes extracted from roots, woods and seeds is comparatively quite poor in most of the cases.
- **Complexity of dyeing process and reproducibility of shades:** These factors are the major constraints in the dyeing process. An adequate systematic profile is required to overcome the problem.
- **Limited number of shades:** Natural dyes have limited shades compared to synthetic dyes.
- **Inadequate degree of fastness and fixation properties:** Some dyes are fugitive, with fixation problems. Pre-treatment and post-treatment with mordants are required in natural dyes.
- **Non-standardised extraction and application techniques:** This is another constraint relating to natural dyes. The process of extraction of natural colours and their dyeing applications differ a lot depending upon the particular users, which ultimately causes variation in quality specifications. Lack of proper documentation is the major cause.
- **Limited number of dyes:** Only a limited number of suitable natural dyes are available in the market. Some dyes are very expensive and cannot be exploited for large scale industrial applications.
- **Complexity in blending:** Suitable colours are generally developed by blending of synthetic dyes. In the case of natural dyes, there exists complexity in blending of dyes.
- **Water pollution:** There also exists the problem related to water pollution by heavy metals and organic substances present in plant parts while extracting.

Future Prospects of Natural Dyes

To adopt a more palatial way of life as natural dyes are eco-friendly, people world over prefer these natural products as they like natural foods, herbal medicines, natural curing practices and even biological farming without using synthetic fertiliser and pesticides as in the last one-and-a-half century the excessive use of synthetic chemicals (fertilizers, pesticides, dyes & colours, and derived- products) whose production and application caused human health hazards, environmental (air and water) pollution and disturbed our eco-system. Due to the adverse effect of synthetic dyes, Germany, the Netherlands and many other counties imposed a ban on the use of 70-odd synthetic dyes for textile dyeing. India has also banned the use of 'azo' dyes for colouring textiles and other consumer goods. Similarly, most of the countries have prohibited the use of several synthetic colour agents for colouring food items and permitted a limited number of synthetic colours under specified maximum limit.

Today, a limited range of 'azo' dyes are used in food. All have been extensively tested and permitted in accordance with the rules of the concerned food authorities of different countries. There is also a strong body of environmentalists, scientists, educationists and social organizations for promoting the use of natural colours in food items and also for using natural dyes for textile colouring.

The global trend towards the use of natural products in daily life and awareness for the ecosystem opens new avenues for the increased use of natural dyes and colours because these are eco-friendly, no disposal problem, no chemical reaction in their preparation, non-toxic, safe with no after-effects and sometimes act as health promoters and disease-curers, non-toxic and therefore no-health-hazardous, the wastes can be converted into manures or bio-fertilizers, similar to fixed petroleum reservoirs there is no danger of depletion and according to demand their growing can be increased, and over all there is no toxic pollution. Presently, the international consumption of natural dyes is about 10,000 tonnes per annum, which is about 1 per cent of the total synthetic dye consumption. There is every likelihood that in the coming years the consumption of these colours would be increased many-fold. Plants are the main source of colours and there are 100 active species which yield colours. India is endowed with rich plant wealth and has a great potential to grasp a major portion of the world's demand. The availability of a wide range of climatic conditions is highly conducive for growing wide varieties of plants. The flowers, fruits, seeds, leaves, sterns, bark, woods and roots of numerous plants contain appreciable amount of colours.

To conclude, there is an urgent need for proper collection, documentation, assessment and characterization of colour-yielding plants, supported by research activities to overcome the limitations of natural colours.

References

Anonymous, 1965. *Amer. Fabrics Mag.*, pp. 163-167. Encyclopedia of Textiles. Prentice-Hall, Inc.,USA.

Bechtold, T., A. Mahmud, Amalid and R. Mussak, 2007. Anthocyanin dyes extracted from grape pomace for the purpose of textile dyeing. *J. Sci. Food Agric.*, **87**(14): 2589-2595.

Bosoglu, A., F. Birdane and H. Solmaz, 1998. "The effect of henna (*Folium lawsoniae*) paste in ringworm in calves". *Indian Vet. J.*, **75** (1): 83-84.

Calis, A., G.K. Celik and H. Katicioglu, 2009. Antimicrobial effect of natural dyes on some pathogenic bacteria. *African J.Biotech.*, **8**(2): 291-293.

Chatwal, G.R. 1993. *Synthetic Dyes*, pp. 1-2. Himalaya Publishing House, New Delhi, India.

Corrales, M., S. Toepfl, P. Butz, D. Knorr and B. Tauscher, 2008. Extraction of anthocyanins from grape by-products assisted by ultrasonics, high hydrostatic pressure or pulsed electric fields: A comparison. *Innov. Food Sci. Emerg. Technol.*, **9**(1): 85-91.

Joshi, V.K. 2006. Production and application of biocolour in food: present status and future strategy. In: Symposium on *New Horizons in Fermentation and Food Biotechnology*, p. 14, organised by Punjabi University, Patiala on 21st March, 2006.

Jothi, D. 2008. Extraction of natural dyes from African marigold flower (*Tagetes erecta* L) for textile colouration. *AUTEX Res. J.*, **8** (2): 49-53.

Khalid, K. A. and J.A.T. Silva, 2010. Yield, essential oil and pigment content of *Calendula officinalis* L. flower heads cultivated under salt stress conditions. *Sci. Hort.*, **126**: 297-305.

Kishimoto, S. and A. Ohmiya, 2009. Studies on carotenoid in the petals of compositae plants. *J. Japan. Soc. hort. Sci.*, **78**(3): 263-272.

Mary Ann Lila, 2004. Anthocyanin and human health: An *in vitro* investigative approach. *J. Biomed. and Biotech.*, **5**: 306-313.

Priyanka Kanwar, 2010. "Studies on combining ability and heterosis in French marigold (*Tagetes patula* L.)". Ph.D. Thesis. Dr. Y.S. Parmar University of Horticulture and Forestry, Nauni, Solan (H.P.).

Purohit, A., S. Mallick, A. Nayak, N.B. Das, B. Nanda and S. Sahoo, 2007. Developing multiple natural dyes from flower parts of gulmohur. *Curr. Sci.*, **92** (12): 1682-1684.

Rao, Y. M., Shayeda and P. Sujatha, 2008. Formulation and evaluation of commonly used natural hair colourants. *Natural Prod. Radiance*, **7** (1): 45-48.

Rivera, R., R.M. Escobedo, E. Balam, M. Vera and H. Harries, 2006. Preliminary studies toward genetic improvement of annatto (*Bixa orellana* L.). *Sci. Hort.*, **109**: 165-172.

Sales, E., R. Kanhonou, C. Baixauli, A. Giner, D. Cooke, K. Gilbert, I. Arrillaga, J. Segura and R. Ros, 2009. Sowing date, transplanting, plant density and nitrogen fertilization affect indigo production from *Isatis* species in a mediterranean region of Spain. *Indust. Crops and Prod.*, **23**: 29-39.

Santos-Bocanegra, E., X. Ospina-Osorio and E.O. Oviedo-Rondón, 2004. Evaluation of xanthophylls extracted from *Tagetes erecta* (marigold flower) and *Capsicum* sp. (red pepper paprika) as a pigment for egg-yolks compared with synthetic pigments. *Intern. J. Poultry Sci.*, **3** (11): 685-689.

Singh, M., S.K. Jindal and D. Singh, 2005. "Natural variability, propagation, phenology and reproductive biology of henna". *Henna: Cultivation, Improvement, and Trade*, pp. 13-18. A publication of Central arid Research Institute, Jodhpur.

Siva, R. 2007. Status of natural dyes and dye-yielding plants in India. *Curr. Sci.*, **92**(7): 916-925.

Steingruber, E. 2004. "Indigo and Indigo Colorants". *Ullmann's Encyclopedia of Industrial Chemistry* (ed.Wiley,V.C.H.), Weinheim, Germany. doi: 10.1002/14356007.a14_149.pub2.

Tatsuzawa, F., K. Ichihara, K. Shinoda and K. Miyoshi, 2010. Flower colours and pigments in *Disa* hybrid (Orchidaceae). *South African J. Bot.*, **76**: 49-53.

34

Edible Flowers

Sellam Perinban, Jayoti Majumder and Sanyat Misra

Introduction

Edible flowers have been a part of human diet since ancient times. Consumption of various types of plant food provides excellent health benefits because they are the rich sources of phytochemicals. More than 25,000 phytonutrients are found in plant foods. Some of the most important phytochemicals are carotenoids, ellagic acid, flavonoids, resveratrol, glucosinolates, phytoestrogens, *etc.* These phytochemicals are not only good for human health but for reduction of disease risks as well. The health benefits are attributed to the antioxidant capacity derived from the phenolic compounds present in edible plants. Flower is an important part of plant which contains a great variety of natural antioxidants. The rich colour and fragrance of flowers are contributed by the secondary metabolites such as phenolic acids, flavonoids, anthocyanin, *etc.* This has been used in many culinary dishes like salads, sweets, beverages, *etc.* Edible flower increases the aesthetic appearance of food but they are also mentioned in connection with biologically active substances present in the flowers. The edible flowers from ornamental plants can be included as a new source of nutraceutical due to nutritional and chemoprotective values.

History of Edible Flowers

Flowers as we all know is beautiful to look at but at the same time it would be interesting to know that flowers are not just for eyes and nose. It is also a substance that can be eaten. The history of edible flowers as food component can be traced back to 140 B.C. (first recorded mention about using flower for culinary purpose appears in the literature). Most of the civilizations around the world have incorporated edible flowers as food in their traditional cuisines. Ancient Chinese, Romans, Greeks and Indian herbalists have made detailed recordings about various edible flowers used in authentic food preparations. Realizing the importance of flowers, the early Hindus, Aztecs and Incas have included them in most of their traditional rituals, some of which are still in practice. Dandelion was mentioned as edible flower in the Bible. Capers (*Capparis spinosa*) flower buds have been used as a condiment for over 2,000 years in Europe. Edible flowers such as daylilies and chrysanthemums have been used by the Chinese and Greeks for centuries. Anglo-Norman cuisine of the 13th and 14th centuries utilized a variety of edible flowers such as roses, hawthorn blossoms and elder flowers. Violets, cowslips, pinks, roses and marigolds were used as natural food colourings during the Renaissance. In

England in 1699 John Evelyn wrote *Acetaria: A Discourse of Sallets*, which listed 73 salad herbs with their parts used and the method by which these were prepared to eat. The Portuguese first used the safflower (*Carthamus tinctorius*) in the 1,700's as a substitute for the more expensive saffron. Edible flowers were very popular during the Victorian Era (1837-1890) especially as part of salads. The Victorians added a variety of items to their salads including violets, borage, primroses, lily flowers and nasturtiums. They also candied and pickled these flowers for the usage during winter months. Ancient Egyptisns stored crystallized violets, and English people ground the violets with chicory to prepare different concoctions (Mlcek and Rop, 2011). Historically, flower petals served as chief food garnishes in several salads and sauces. The petals of flowers like carnation, borage, sage, bee balm, nasturtium, violet, daylily and calendula were commonly eaten because of their medicinal properties. Dry flower petals of hibiscus, jasmine, bee balm and rose were used for tea blending. Flower cookery was very popular in ancient Rome, China, India and Middle Eastern Countries. Romans used mallows, violets and roses. The Italians and Spaniards used stuffed squash blossoms, the French used carnation petals in the preparation of classic green liquor while Indians used rose petals, pumpkin flowers and banana flowers in several food preparations.

Sources of Edible Flowers

Based on the source of edible flower it can be from medicinal plants, fruits and vegetables or ornamentals. Many edible herbs are used for culinary and medicinal purposes. They usually taste like a milder version of the herb. Some of the herbs with edible flowers are common lemon verbena (*Aloysia triphylla*), dill (*Anethum graveolens*), chervil (*Anthiscus cerefolium*), borage (*Borago officinalis*), mustard (*Brassica* spp.), pot marigold (*Calendula officinalis*), cilantro (*Coriandrum sativum*), saffron (*Crocus sativus*), globe artichoke (*Cynara scolymus*), fennel (*Foeniculum vulgare*), hyssop (*Hyssopus officinalis*), English lavender (*Lavandula angustifolia*), lovage (*Levisticum officinale*), mint (*Mentha* spp.), basil (*Ocimum* spp.), marjoram (*Origanum marjorana*), oregano (*Origanum vulgare*), rosemary (*Rosmarinus* spp.), sorrel (*Rumex acetosa*), common sage (*Salvia officinalis*), summer savory (*Satureja hortensis*), winter savory (*Satureja montana*), Mexican terragon (*Tagetes lucida*), dandelion (*Taraxacum officinalis*), thyme (*Thymus* spp.), society garlic (*Tulbaghia violacea*), *etc*. Some of the fruits and vegetables with edible flowers are argula (*Eruca vesicaria sativa*), banana (*Musa* spp.), chives (*Allium shoenoprasum*), garlic chives (*Allium tuberosum*), chicory (*Cichorium intybus*), drumstick (*Moringa ovalifolia*), garden peas (*Pisum sativum*), lemon (*Citrus limon*), *mahua* (*Madhuca indica*), mango (*Mangifera indica*), okra (*Abelmoschus esculentus*), orange (*Citrus sinensis*), pineapple guava (*Feijoa sellowiana*), radish (*Raphanus sativus*), runner bean (*Phaseolus coccineus*), squash (*Cucurbita* spp.), strawberry (*Fragaria ananassa*), *etc*.

Lawn daisy (*Bellis perennis*) young leaves and whole small flowers or by separating the petals can be used in salads. The florets of chives (*Allium schoenoprasum*) are sprinkled on salads, chive leaves are used in salads, sandwiches and soups and as a garnish. The leaves of lemon verbena (*Aloysia triphylla*, syn. *Lippia citriodora*) have lemony fragrance. Its leaves infuse a herb tea, finely chopped young leaves for drinks, fruit puddings, confectionery, apple jelly cakes and home-made ice cream. March mallow (*Althaea officinalis*) flowers are tossed on salads. Its young leaves are mixed into salads with oil and vinegar and steamed to serve as vegetables. Its roots are boiled to soften and then fried. Its seeds are either eaten fresh or sprinkled over salads like nuts. *Angelica archangelica* seeds and stems are mixed to flavour certain drinks such as gin, and the leaves are stewed with acidic fruits to reduce sugar requirement. Borage (*Borago officinalis*) flowers are sprinkled over salads and as a garnish; and the young leaves are added to cold drinks for their cucumber flavour and cooling effect. Its chopped leaves are used in salads, yogurt, cheese, pickles and sandwiches. It is also cooked with spinach. *Calendula officinalis* petals are used to add saffron colour with a light tangy flavour to soft cheese, yogurt, butter, milk dishes, cakes, sweet breads, omelettes, fish and venison, and to rice, fish and meat soups. Its leaves are sprinkled in salads and stews. Perennial chamomile (*Chamaemelum nobile*, syn. *Anthemis nobilis* and *Matricaria recutita*) flowers are infused as a tea for general tonic. Good King Henry (*Chenopodium bonus-henricus*) seeds are ground into flour and made into gruel; some 10-15 cm long shoots of pencil thickness are taken, boiled, peeled and eaten as asparagus; young leaves are taken raw in salad, cooked in casseroles, stuffings and soups, and puréed in savoury pies; and spikes are steamed and tossed in butter similar to broccoli. Tansy (*Tanacetun vulgare*, syn. *Chrysanthemum vulgare*) leaves are stewed with rhubarb. Alecost/costmary (*Tanacetum balsamita*, syn. *Chrysanthemum balsamita*) finely chopped leaf in a very limited quantity is used in carrot soup, salads, fruit cakes and in poultry stuffing. *Chrysanthemum coronarium* raw young leaves are used in salads and the yellow flowers are also edible. Chicory (*Cichorium intybus*) seedlings can be eaten fresh; young boiled roots can be used with sauce, and its thick roots are dried in gentle heat, roasted and ground for use as a coffee substitute; and its floral buds are pickled and flowers are used in salads. Clove pink (*Dianthus caryophyllus*) flowers , after removing the bitter white heel of petals, can be used in salads, fruit pies and sandwiches. Its wine-infused petals are nerve tonic. Sunflower (*Helianthus annus*) raw buds are used in salads and its seed kernels are eaten raw or roasted. Curry plant (*Helichrysum angustifolium*) produces evergreen leaves and umbellate flowers. Its leaf sprigs are used in soups, stews, steamed vegetables, rice dishes and pickles for a mild curry flavour, but before serving the sprigs are removed. Sweet rocket (*Hesperis matronalis*) produces beautiful flowers which are used in salads and for decorating the desserts. Hops (*Humulus lupulus*) young side shoots are steamed and used as asparagus. Its blanched young leaves are used in soups and the dried female flowers to flavour the beer and the parboiled male flowers into salads. Hyssop (*Hyssopus officinalis*) leaves and flowers are used in salads. Elecampaane (*Inula helenium*) dried root pieces are eaten or cooked as vegetable. Lavender (*Lavandula angustifolia*, syn. *L. officinalis*) flowers are used to flavour jams and to make lavender vinegar. Bergamot/bee balm (*Monarda didyma*) produces beautiful flowers which can be used in salads. Its leaves are infused or simmered for 10 minutes in an enamel pan for greater flavour as a tea. Evening primrose (*Oenothera biennis*) boiled roots taste like parshnip, are pickled, mixed in salads and are used as appetizer, while its stems and leaves were once popular food with North American people. Cotton thistle (*Onopordum acanthium*) seeds were once used for extraction of cooking oil and for lighting; blanched and peeled young stems are eaten raw with oil and vinegar or steamed and served as asparagus; and the floral discs are cooked and used as globe artichoke. Sweet/knotted marjoram (*Oregano majorana*), French/ pot marjoram (*O. onites*) and winter marjoram (*O. heracleoticum*) have sweet spicy-flavoured pale-green leaves, and oregano (*O. vulgare*) has peppery-flavoured (containing powerful antiseptic thymol) leaves, and with attractive flowers. By laying marjoram stems on barbecue embers, the food emits a faint marjoram flavour. The sweet marjoram leaves are infused as an aromatic tea or for use in salad. Poppy (corn /field poppy-*Papaver rhoeas*; opium poppy-*Papaver somniferum*) seeds are eaten by sprinkling over breads, cakes and biscuits for a pleasant nutty flavour and to curry powder for flavouring and texture. Scented geraniums (*Pelargonium capitatum, P. crispum, P.* × *fragrans, P. graveolens* × *P. tomentosum, P. odoratissimum, P. quercifolium, P. radens*) are variously scented and their flowers are used in salads, and finely chopped or infused in liquid then discarding the leaves and using them to flavour the sauces, custards, jam & jellies, syrups, vinegars, buns, sugar and butter. Summer purslane (*Portulaca oleracea*) leaves in salads provide nutty flavour. Salad burnet (*Poterium sanguisorba*, syn. *Sanguisorba minor*) leaves are used to flavour vinegar, salad dressing, soups, garnishes, vegetables, herb butters and soft cheeses. Cowslip (*Primula veris*) flower is used in wine, pickle and jam, and the leaves in salads and for meat stuffing. Primrose (*Primula vulgaris*) flowers are eaten raw in salads and the leaves are boiled as a vegetable. Rosemary (*Rosmarinus officinalis*) leaves are used to flavour baked potatoes, to make a herb butter for vegetables, and used sparingly to a wide range of meat dishes; while the fresh flowers are used in salads and in the fruit purée. After removing the hairs, the hips of roses are used in syrup, jams, wine and teas; while scented petals after removing the bitter white part are used in salads, apple or cherry pie. These are also used for making preserves, vinegar, syrup, sherbats and rosewater for flavouring sweets and drinks. Garden rue (*Ruta graveolens*) dried and crushed leaves are used as insect-repellent, insecticide and germ-killers. Since taste is bitter so the leaves only in a small amount give unusual muskiness to cream cheese, egg and fish dishes; and the seeds are infused with lovage and mint as a marinage for partridge. Sage (*Salvia officinalis*) flowers are used in salad and infused for a light balsamic tea, while the dipped and fried whole leaves in butter or young leaves in cream for eating with sugar and orange, *vis-a-vis* used for preparing sage vinegar and butter. Soapwort (*Saponaria officinalis*) flowers are tossed on green and fruit salads, and used in brewing industry. Houseleek (*Sempervivum tectorum*) leaves are eaten as salad by Dutch people. Comfrey (*Symphytum officinale*) is a miraculous plant which contains more protein in its leaf structure than any other known member of the vegetable kingdom. Its leaves and roots contain allantoin, a protein that encourages cell division and is a cure from stubborn leg ulcers to broken bones. Though not in abundance, but the leaves contain Ca, K, P, vitamins A,C and B_{12}. Its chopped leaves are used in salads. Cooked as spinach or in fritters, and blanched stems are cooked like asparagus. Variegated nasturtium (*Tropaeolum majus* 'Variegata) leaves and flowers are used in salads and sandwiches. Valerian (*Valeriana officinalis*) roots are added to soups and stews. Mullein (*Verbascum thapsus*) flowers are used to flavour liqueurs. Sweet violet (*Viola odorata*) flowers are eaten raw in salads or made into syrup.

The ornamental plants having edible flowers, *vis-a-vis* medicinal uses are anise hyssop/Korean mint (*Agastache foeniculum, A. rugosa*) which is diaphoretic, acts against cardiac troubles and pectoral problems, and is used for making poultice, *A. rugosa* is used in preparing certain culinary ingredients, and is also used against viral and fungal infections, digestive disorders and nausea; apiary tree (*Robinia pseudoacacia*) is used fror preparing tea, syrup and liqueur, and is effective in joint and stomach pains; balsam (*Impatiens walleriana*) has sweet flavour and is used as garnish, in salads and drinks; begonia (*Begonia tuberhybrida*) is used with citrus, is astringent, ophthalmic, anodyne, antiphlogistic, antispasmodic, antistomachic and is used for preparing poultice; bergamot/bee balm (*Monarda didyma*) is used for preparing tea, jelly and as an ingredient in culinary preparations, *vis-à-vis* is used for toning the skin, and to ease digestive and respiratory ailments; borage (*Borago officinalis*) contains saponins and tannins (pyrrolizidine alkaloids), possesses mild cucumber flavour, is used in salads, desserts, for preparing tea and wine, and is useful in skin treatment, depression, respiratory cough and cold; bougainvillea (*Bougainvillea glabra* and others) is antidiabetic and anti-inflammatory, bougainvillea (*Bougainvillea braziliensis*) is mild in taste and is used as garnish or in salads; butterfly pea (*Clitoria ternatea*) has sweet flavour and is used as garnish or in salad; Californian poppy (*Eschscholtzia californica*) is used as stir fry, tea and in salad; chamomile (*Matricaria recutita*) is anti-inflammatory, anti-allergenic and is used against indigestion; chives (*Allium schoenoprasum*) is used in vinegar and salad, and against chest ailments, bladder and kidney infections; chrysanthemum (*Dendranthema grandiflora*) and oxeye daisy (*Chrysanthemum leucanthemum*) have mild flavour and are used as garnish or in salad, *vis-à-vis* are bitter, purgative, anti-stomachic and is used for preparing expectorant; clover (*Trifolium pratense*) is used as tea, in salad, as in ointment, is anti-cancer, anti-bronchial and anti-arthritis; coral vine/ Mexican creeper (*Antigonon leptopus*) is analgesic, anti-diabetic, anti-inflammatory, anti-thrombin and is lipid peroxidation inhibitory; cosmos (*Cosmos caudatus, C. sulphureus*) has mild aroma, possesses chocolate flavour and is used as garnish or in salad, *C. sulphureus* is antimicrobial, anti-inflammatory, anti-oxidant, anti-genotoxic, and is used against jaundice, intermittent fever and splenomegaly; dahlia (*Dahlia juarezi*) has pungent flavour, is used for skin treatment, against insect bite and is anti-inflammatory; dancing lady (*Fuchsia* spp.) is used for juice extraction, for preparing jelly, in tea, has wound healing properties, and is used against throat infections and in skin treatment; dandelion (*Taraxacum purpurea*) is a source of ß-carotene, and is used as garnish, in salad, for detoxification and against gastroenterological problems; daylily (*Hemerocallis* spp.) is used in salad and tea, *vis-a-vis* as disinfectant, anthelmintic, antiemetic, antispasmodic, diuretic, depurative, febrifuge, laxative, sedative, blood purifier, against cancer, toothache and other pains, and as antidote; dianthus (*Dianthus* spp.) possesses clove and spicy flavour and is antiperiodic, antispasmodic, cardiotonic, diaphoretic, febrifuge, nervine and stimulant; English chamomile (*Anthemis nobilis*) has sweet and apple flavour and is used for preparing tea which should not be taken more than a cup a day but it may be allergic to ragweed sufferers; English daisy (*Bellis perennis*) tastes slightly bitter and is used as garnish or in salads; fruit sage (*Salvia dorissiana*) is used for preparing tea, in salad, for skin and digestive problems and for detoxification; geranium (*Pelargonium* spp.) is highly perfumed and is used for extraction of essential oil; golden rod (*Solidago virgaurea*) for preparing tea and tonic and is diuretic, anti-oxidant, anti-inflammatory, and is used against fungal infection; hollyhock (*Alcea rosea*) tastes slightly bitter and is used for garnishing; lilac (*Syringa vulgaris*) is antiperiodic, febrifuge, vermifuge, as tonic and mouthwash; lavender (*Lavandula angustifolia*) is highly perfumed and has floral flavour; marigolds, *Tagetes erecta* (African) and *T. patula* (French) have strong and bitter aroma and are anthelmintic, carminative, digestive, diuretic, emmenagogue, laxative, ophthalmic, sedative, against skin allergies, anti-stomachic, anaesthetic, febrifuge, hallucinogenic, hypotensive, narcotic and antiviral, though *T. tenuifolia* (Signet) has citrus-like aroma and is used as garnish or in salad and drinks; monstera (*Monstera deliciosa*) can be used for preparing syrup, dessert and salad but the tips of every scale should be discarded due to high accumulation of calcium oxalate crystals, and is used in the treatment of skin problems; nasturtium (*Tropaeolum majus*) has peppery and tangy flavour, is used as garnish, salads and drinks, *vis-a-vis* is antibiotic, antibacterial, antifungal, antiseptic, aperient, depurative, diuretic, emmenagogue, laxative, stimulant and is used as expectorant; pansy (*Viola odorata, V. trtcolor* and *V. × wittrockiana*) has sweet fragrance, is used in candies and salads, *vis-à-vis* the *V. × wittrockiana* is effective against skin ailments and in phytotherapy; pot marigold (*Calendula officinalis*) tastes tangy and peppery and is used as salad, tea, anti-inflammatory, antiphlogistic, antiseptic, antispasmodic, aperient, astringent, diaphoretic, stimulant, cholagogue, emmenagogue, used in homoeopathy, to detoxify, and is used for preparing ointment against dermatitis, vulnerary, warts, etc.; purple cornflower (*Echinacea purpurea*) is used in tea, as salad, as detoxifying agent, in pneumonia, in bronchitis, is anti-fungal, anti-inflammatory and is useful against viral infections; rose (*Rosa* spp.) is highly perfumed and emits sweet to bitter aroma, *vis-a-vis* is astringent, carminative, diuretic, laxative, anti-stomachic, analgesic, antirheumatic, febrifuge, vermifuge, ophthalmic, used as a tonic, used in aromatherapy, bach, salve, and against kidney problems and cancer; rose-of Sharon (*Hibiscus syriacus*) has mild and nutty flavour; safflower (*Carthamus tinctorius*) has bitter flavour and imparts yellow colour to cooked food; snapdragon (*Antirrhinum majus*) tastes bitter and is used in garnishing; tuberose (*Polianthes tuberosa*) is highly perfumed and is used as garnish; tulip (*Tulipa* spp.) is antidote, antipyretic, anti-cancer, depurative, febrifuge, laxative, and as expectorant; *etc.* (Sandra Bastin, 1997; Friedman *et al.*, 2007; Margaret Roberts, 2007; Newman, 2009; Kaisoona *et al.*, 2011).

Edible flowers as Functional foods and Nutraceutical

Functional food and nutraceuticals are gaining popularity due to an increased awareness of the relationship between food and health & lifestyle changes of consumers. There is an extremely wide range of bioactive components that are associated with beneficial effects on human health. Mainly, the phytochemical components which are responsible for the flower colours are good antioxidants and pose good health benefits (Tiwari *et al.*, 2013). Some of the flower colour groups and phytochemicals responsible for their colour are '**red**' for anthocyanin, lycopene, phytoene, phytofluene and vitamin E; '**green**' for chlorophyll, folic acid, glucosinolates, isothiocyanates and indole-3 cardinol; '**green/yellow**' for lutein and zeaxanthin; '**orange**' for α & ß-carotene and ß-cryptoxanthin; '**orange/yellow**' for flavonoids and vitamin C; '**red-purple**' for anthocyanins, ellagic acid and flavonoids; and '**white/green**' for allyl sulphides. The volatile phytochemicals which are responsible for the fragrance of the edible flowers are also contributing to their health benefits. The major volatile phytochemicals present in edible flowers are carbonyl compounds, hydrocarbon esters, pyranones, furanones, volatile sulphur compounds, pyrazines, phenols and terpenes (Jyoti Srivastava and Vankar, 2010).

From the nutritional point of view, the flower can be divided into three major components, which can play a role in human nutrition and health. (Tiwari *et al.*, 2013): (i) **pollen**, whose amounts are very small but is a rich source of proteins, amino acids and carbohydrates, lipids, carotenoids, flavonoids, *etc.* though in taste it is usually not too savorous or distinctive and in some of the cases it may be allergic only to some people when taken fresh with flowers; (ii) **nector**, which is usually a sweetish liquid containing balanced mixture of sugars, amino acids, lipids, phenolic substances, *etc.* and (iii) **petals and other floral parts**, which apart from basic nutrients, possess taste, vitamins, minerals, *etc.* and in the course of flower senescence, the content of saccharose may increase due to increased hydrolysis of fructans.

Beneficial Chemical Components from Edible Flower Colour

Colour is an important organoleptic property of edible flowers. Although this characteristic is determined by many chemical compounds, above all the contents of carotenoids and anthocyanins are very important. In edible flowers, an increased content in the level of anthocyanins is mostly correlated with higher levels of total flavonoids and it can be said that this is one of the factors determining their high antioxidant activity (as compared with most of other plant parts and/or organs). A reduced formation of anthocyanins associated with a simultaneous inhibition of synthesis of total flavonoids is determined genetically by the recessivity of corresponding alleles and causes a light or even white colouration in flowers. There is a common pathway of synthesis of flavonoids and anthocyanins; pale flowers lack enzymes such as flavanone 3-hydroxylase, dihydroflavonol 4-reductase and/or anthocyanidin synthase. Violet and blue colours of flowers are determined by delphinidin-based anthocyanins. Pale cultivars of roses lack delphinidin due to the absence of flavonoid 3,5-hydroxylase. White or creamy flowers show an increased level of leucoanthocyanidins, which is catalysed by enzymes such as chalcone synthase, flavonone 3-b-hydroxylase, and dihydroflavonol 4-reductase (Mlcek and Rop, 2011). *Tropaeolum majus* represents one of the most popular sources of edible flowers. In this species, the colour of flowers is caused, above all, by anthocyanins, on an average, whose concentration is 720 mg/kg of fresh mass and pelargonidin 3-sophoroside constitutes more than 90% of total anthocyanins (Garzon and Wrolstad, 2009), as in chrysanthemums the lutein is the most important carotene. Since the content of this compound is not very high in the majority of staple foods, edible flowers could be one of its valuable sources. This concerns marigold flowers which contain 350-450 mg/kg of fresh mass. As compared with yellow or red cultivars of marigold, the highest content of lutein can be found in flowers of orange varieties (Bhattacharyya *et al.*, 2008). The consumption of marigold flowers shows a significant effect on reduction of the risk of macular degeneration and cataract. It is also reported that *Tagetes erecta* extract exhibits anti-proliferation activity against colorectal adenocarcinoma cells whereas *Antigonon leptopus* acts against the gastric adenocarcinoma and bladder cancer cells. *Bougainvillea glabra* also has anti-proliferation activity (Kaisoona *et al.*, 2012). In daylily flowers, along with lutein, high contents of zeaxanthin and yaniding-3-rutinoside are also found. The colour of the petals in roses is caused by yaniding and pelargonidin, which in combination with carotenes produce attractive red and rosy colouration of flowers. However, delphinine in combination with carotenoids causes an unattractive brownish colour of flowers. Dragon fruit edible flowers (*Hylocereus undatus*) are good source of six different flavonoids (Yan Yi *et al.*, 2012). Flowers of *Robinia pseudoacacia* contains bioactive components of flavonoids, phenolics, ascorbic acid, *etc.* and play a critical role in curing gastritis, sleep disorders, asthma, *etc.* (Hong-fang Ji *et al.*, 2012). *Hibiscus* species are the rich sources of flavonoids and vitamins. The mucilage extracted from hibiscus flowers are used as a substitute for egg. The most important flavonoids found in hibiscus are quercertin and cyanidin and they are also rich in tocopherol. *Hibiscus* flowers are used as medicine for treatment of dysentery, lung and urinary ailments (Puckhaber *et al.*, 2002). The fresh flowers of *Madhuca indica* tree are eaten through various cooked preparations while the dried ones are also cooked for various other preparations apart from preparing country liquor. Juice is extracted from fresh flowers and concentrated as liquid sweetener. Apart from being rich source of sugar and protein, the flowers also contain essential minerals like Ca, P, K and Fe. Riboflavin and niacin vitamin content in this flower is higher than other fruits like apple or mango. It has been used in Ayurvedic medicines for cooling, aphrodisiac, galactagogue, expectorant and carminative values. It is also used for heart disease and piles (Patel and Naik, 2010). *Prunus mume* flowers contain a high concentration of biologically active phenolic compound, chlorogenic acids, which is responsible for its high antioxidant property. These flowers are used for stomachic, alexipharmic, expectorant, and sedative purposes and for the treatment of eye pain and skin disorders (Jiayi Shi *et al.*, 2009). The anti-oxidant activity of *Sophora viciifolia* flowers is contributed by phenolic and flavonoid compounds, particularly luteolin. The flowers of *S. viciifolia* can provide valuable functional ingredients and can be used for the prevention of diseases related to various oxidant by-products of human metabolism (Zhigang Tai *et al.*, 2011). It is also reported that the compounds isolated from the flowers of *Antigonon leptopus* plant have the ability to scavenge reactive oxygen species and reduce oxidative stress *in vivo* which is responsible for its anti-inflammatory character. In addition to its scavenging effect, these compounds may also enhance the endogenous defense systems and have anti-thrombin and anti-diabetic activities (Vanisree *et al.*, 2008).

Active Aromatic Compounds from Edible Ornamental Flowers

The flowers of ornamental plants do not have only antioxidant and ROS scavenging activities but are also significant for anti-inflammatory effects within the human body. Rose flowers are not only the source of anti-inflammatory, anti-bacterial, anti-fungal and anti-viral substances, but emit fragrance whose main components are 2-phenylethanol and terpene linalool (3%), citronellol (7.2%), nerol (4.2%) and geraniol (up to 7%). In American and African alternative (complementary) medicine, the oil from marigold flowers is well known and often used due to its anti-inflammatory properties. Besides aromatic components, the antibacterial and antifungal activities of marigold flowers are caused above all by the flavonoid patuletin. The smell of carnation flowers is caused by a number of compounds, e.g. hexanol, 1-hexanol, 2-hexanol, nonanal, benzylaldehyde, benzylalcohol, benzylbenzoate and caryophyllene. In carnations, the terpene caryophyllene is thought to be the major anti-inflammatory agent. In anise hyssop the major aromatic components are methyl chavicol (60 - 90%), myrcene, limonene and also cineole and globulol. The contents of caryophyllene and anethole may be as much as 6 to 19% (Mlcek and Rop, 2011). Flowers of chrysanthemum, above all, show marked antimicrobial and anti-inflammatory effects due to the content of triterpens arnidiol, 3-monohydroxyterpens, faradiol and heliantriol. As far as these compounds are concerned, they show not only an anti-inflammatory but also a marked inhibiting effect on carcinogenesis in mice (especially in their skin). Chrysanthemum flowers showed an anti tubercular activity, which was attributed to their triterpens (Motohiko Ukiya *et al.,* 2002). In violet flowers, one of the aromatic components is the cyclic terpenoid b-ionone. This substance shows a marked inhibiting effect on the growth of malignant cells; it inhibits, above all, apoptosis of human gastric adenocarcinoma cells. The flowers of garden marigold are also mentioned because of their anti-inflammatory effect, which is caused by benzyl isothiocyanate. This compound is biosynthesized from the glycoside glucotropeolin by splitting with the enzyme myrosin. Two pharmaceutically important bioactive components, *viz.* chamazulene and α-bisabolol are present in oil extracted from chamomile flowers which is an essential oil (Falzari and Menary, 2003).

Quality of Edible Flowers

When defining the suitability of flowers for human nutrition, it is the quality on which demand is based. Besides their whole aesthetic appearance, their suitability for the efficient economic use is also considered. The medicinal soundness of edible flowers is dependent on respecting the limits of the content of toxic agents. Unknown flowers should not be tasted as those may be poisonous for humans such as *Anemone ranunculoides, Colchicum autumnale, Conavallaria majalis, Nerium oleander, Thevetia nereifolia, etc.* (Newman, 2009). When consuming flowers picked up freely in nature, it is always necessary to identify them exactly. Many plants are poisonous and/or in some case certain parts of the plants are edible and others not, hence, it is always necessary to identify the edible flowers properly before using it in any culinary preparation. It is explicitly recommended not to consume edible flowers of ornamental plants originating from non-tested cultivars and/or florist's shops because they could be affected by fertilizers, herbicides and other kinds of pesticides (Lance Gegner, 2004). Besides toxic effects it is also possible that even the flowers, which are sound, clean and seem to be trouble-free, may induce allergic reactions in people who are sensitive to some of their non-defined components. Allergic reactions have often been caused by chrysanthemum flowers; they are manifested, above all, as rash and eczema (Tiwari *et al.*, 2013). Quality criteria involve the capability of individual species to be grown on larger areas and the stability of their yields as well. Apart from this, it is also expected that they will show at least a minimum resistance to diseases and pests. Since flowers are very delicate in nature and their components like pollens and nectars are contributing to their health and sensory characteristics, they cannot be washed or cleaned like other fresh foods and vegetables. Therefore, the edible flower shall be grown pest-free, pesticide-free, organically grown and without any physical contaminants. The resistance to mechanical damages and durability after the harvest represents other quality criteria of individual species and varieties of edible ornamental plants. The flowers are the delicate and sensitive material, being highly prone to microbial decay so after harvest they are immediately placed into perforated plastic bags and/or containers which should protect them against contamination and wilting apart from preventing the vapour condensation inside (Mlcek and Rop, 2011). Harvested flowers are quickly cooled and cleaned, and their shape is partly modified; thereafter they are stored, at the temperature of 1 to 4 ^{0}C for a period of 2 to 14 days (Kelley *et al.*, 2003). Flowers can also be dried, soused in alcohol or sugar, frozen either directly or in form of ice cubes, and added into cocktails, *etc.* The method of sublimation drying is economically and technically rather demanding but very effective as original appearance, colour, shape, and gloss of flowers remain to be preserved . Edible flowers can also be preserved by irradiation techniques which increases the shelf life from 5-12 days (Koike Amanda *et al.*, 2012).

Marketing

As with any crop, it is extremely important to decide on a marketing strategy before you plant. Edible flowers are produced and marketed in much the same way as fresh herbs, although the edible flower market is not as large. Edible flowers are used by chefs as garnishes, in salads and desserts, and for drinks and candy adornments. Market assessment will have to be done before proceeding, concentrating on upscale restaurants in the largest and nearest urban centre. To recognize the unique opportunities that may provide entry into this market, the grower must keep up with food trends. Talking to local chefs/cooks will acquaint you with their needs. Most restaurants/hotels/processing industries demand a consistent supply of any crop, but many edible flowers can be used interchangeably. Since many people are unfamiliar with using edible flowers, it is always a good idea to provide free samples and recipes. Reminding customers that edible flowers mixed in summer salads create unique colours and taste. Often, customers will use these flowers for special events, such as placing of crystallized violets on wedding cakes. It is up to the grower to remind consumers of these special uses. As for pricing, the grower must decide what the market will bear (Kelley *et al.*, 2001, 2002).

Like other fresh horticultural produce, edible flowers are highly perishable and delicate in nature. Edible flowers can complement a cut flower or herb business, providing additional opportunities for value-added products. However, they require a specialized niche market that is not yet developed in India. Flowers intended for human consumption must be grown without pesticides, providing organic growers a production edge. Flowers may be marketed fresh, dried, candied, or in prepackaged salads. Marketing flowers as a package with different colours and varieties of fresh edible flowers in the same container will be appealing to consumers. Value added products that feature edible flowers offer additional marketing opportunities. Simple value addition like making candies, dried flower petals, jelly, extracted waters, *etc.* will increase the consumer acceptance. Minced flowers make a colourful and flavourful addition to herbal butters, cheese spreads, jellies, and jams. Dried flowers could be used in teas or to add flavour to wines. Value-added products, like mesclun mixed with calendula flowers, can generate excitement in the consumer and added income for the grower.

Conclusion

Edible flowers, apart from containing ordinary nutrient components (i.e. proteins, fats, saccharides, and vitamins) like other plant part, they contain many nutraceutical substances which show either protective (chemoprotective) or even curative effects and reduce the risk of the occurrence of various diseases. The most interesting of them are the compounds showing antioxidant effects, e.g. phenolic compounds, carotenoids, etc. Edible flowers may be used not only as decoration or a garnish of various meals and salads, but also for other culinary purposes (baking and the flavouring of sauces, jellies, syrups, vinegar, honey, oils, teas, flower-scented sugars, candied flowers, wine and flavoured liqueurs). Even a small amount of edible flowers may often improve the health condition of consumers. Despite its health benefits, nutritive and chemoprotective properties of several edible flowers are not clearly known or are under study. The composition of flowers for nutraceutical food stuffs is not well documented as in case of medicinal plants and vegetables. Edible flowers are marketed with simple packaging as the sustainable economic business is not fully explored in India. The post harvest handling, packaging, value addition and marketing processes are standardised only for a few very prominent ones such as rose. Consistent production of quality flowers, sustainable supply and quality standards and guidelines for commercial marketing are much needed to boost up edible flower business.

References

Bhattacharyya, S., A. Roychowdhury and S. Ghosh, 2008. Lutein content, fatty acid composition and enzymatic modification of lutein from marigold (*Tagetes patula* L.) flower petals. *J. Indian Chem.. Soc.*, **85**: 942-944.

Denise, D. 1990. The Complete Book of Flowers (2nd edition). North Atlantic books, USA.

Falzari, L.M. and R.C. Menary, 2003. *Chamomile for Oil and Dried Flowers*. BIRDC Publication No. 02/156.

Friedman, H., I. Rot, O. Agami, Y. Vinokur, V. Rodov, and N. Reshnick, 2007. Edible flowers: new crops with potential health benefits. *Acta Hort.*, No. 755, pp. 283-289.

Garzon, G.A. and R.E. Wrolstad, 2009. Major anthocyanins and antioxidant activity of nasturtium flowers (*Tropaeolum majus*). *Food Chem.*, **114**: 44-49.

Hong-fang Ji, Ai-lin Du, Ling-wen Zhang, Chun-yang Xu, Ming-duo Yang and Fang-fang Lil, 2012. Effects of drying methods on antioxidant properties in *Robinia pseudoacacia* L. flowers. *J. Medic. Pl. Res.*, **6**(16): 3233-3239.

Jiayi Shi, Jinyan Gong, Ji'er Liu, Xiaoqin Wu and Ying Zhang, 2009. Antioxidant capacity of extract from edible flowers of *Prunus mume* in China and its active components. *LWT-Food Sci. Tech.*, **42**: 477-482.

Jyoti Srivastava and P.S. Vankar, 2010. *Canna indica* flower: new source of anthocyanins. *Pl. Physiol. Biochem.*, **48**: 1015-1019.

Kaisoona, O., S. Siriamornpuna, N. Weerapreeyakulb and N. Meeso, 2011. Phenolic compounds and antioxidant activities of edible flowers from Thailand. *J. Functional Foods*, **3**: 88-99.

Kaisoona, O., I. Konkzak, S. Siriamornpuna, 2012. Potential health enhancing properties of edible flowers from Thailand. *Food Res. Int.*, **46**: 563-571.

Kelley, K.M., A.C. Cameron, J.A. Biernbaum and K.L. Poff, 2003. Effect of storage temperature on the quality of edible flowers. *Postharvest Biol. Technol.*, **27**: 341-344.

Kelley, K.M., B.K. Behe, J.A. Biembaum and K.L. Poff, 2001. Consumer preference for edible flower color, container size, and price. *HortSci.*, **36**: 801-804

Kelley, K.M., B.K. Behe and J.A. Biernbaum, 2002. Combinations of colours and species of containerized edible flowers: effect on consumer preferences. *HortSci.*, **37**(1): 218-221.

Koike Amanda, C.R., F.T. Rodrigues, Anna Lucia and C.H. Villavicencio, 2012. Shelf life extension of edible flowers by irradiation process. Paper presented to the *16th World Congress of Food Science and Technology*, Brazil.

Lance Gegner, 2004. Edible flowers. www.attra.ncat.org

Margret Roberts, 2007. *Edible and Medicinal Flowers*. Spearhead Publishers, New Africa Books Ltd., South Africa.

Mlcek, J. and O. Rop, 2011. Fresh edible flowers of ornamental plants: A new source of nutraceutical foods. *Trends Food Sci. Technol.*, **22**: 561-569.

Motohiko Ukiya, Toshihiro Akihisaa, Harukuni Takudab, Hiroyuki Suzukia, Teruo Mukainakab, Eiichiro Ichiiishib, Ken Yashukawac, Yoshimasa Kasaharad and Hoyoku Nishinob, 2002. Constituents of Composite Plants III. Anti-tumour promoting effects and cytotoxic activity against human cancer cell lines of triterpenediols and triols from edible chrysanthemum flowers. *Cancer Letters*, **177**: 7-12.

Newman, S.E. and A. Stoven O'Conner, 2009. Edible Flowers. Colorado St. University Extension Fact Sheet No. 7.237 (www.ext.colostate.edu).

Patel, M. and S.N. Naik, 2010. Flowers of *Madhuca indica* J.F. Gmel: present status and future prospective. *Indian J. Natural Products and Resources*, **1**(4): 438-443.

Puckhaber, L.S., R.D. Stipanovic and G.A. Bost, 2002. *Analyses for Flavonoid Aglycones in Fresh and Preserved Hibiscus Flowers*. In: *Trends in New Crops and New Uses* (eds Janick, J. and A. Whipkey). ASHS Press, Alexandria, V.A., USA.

Sandra Bastin, 1997. Edible Flowers (http://www.ca.uky.edu.).

Tiwari, B.K., N.P. Brunton and C. Brennan, 2013. *Handbook of Plant Food Phytochemicals: Sources, Stability and Extraction*. John Wiley & Sons, England.

Vanisree, M., Ruby L. Alexander-Lindo and David L. DeWitt and M.G. Nair, 2008. Functional food components of *Antigonon leptopus* tea. *Food Chem.*, **106**: 487-492.

Yan Yi, Qing-Wen Zhang, Song-Lin Li, Ying Wang, Wen-Cai Ye, Jing Zhao and Yi-Tao Wang, 2012. Simultaneous quantification of major flavonoids in 'Bawanghua', the edible flower of *Hylocereus undatus* using pressurized liquid extraction and high profile liquid chromatography. *Food Chem.*, **135**: 528-533.

35

Nutrition Management in Flower Crops

T. Janakiram, Namita, Sapna Panwar and Ritu Jain

Indian floriculture is characterized by the cultivation of flowers that are basically suited for garland making, worship and decoration. The Indian floriculture industry has been shifted from traditional loose flowers to cut flowers for niche and export markets. India is one of the leading countries in floriculture, where as per NHB database during 2014-2015, india had an area of 2.49 lakh ha producing 4,84,000 tonnes of cut flowers and 16,59,000 tonnes of loose flowers. The major loose flower growing states are Tamil Nadu, Andhra Pradesh, Maharashtra, Punjab, Haryana, West Bengal, Gujarat, Chhattisgarh, Jharkhand, Uttar Pradesh, *etc.* The states like West Bengal, Maharashtra, Andhra Pradesh, Orissa, Karnataka, Gujarat, Uttar Pradesh, etc. have emerged as major cut flower growing states. North Eastern and other Himalayan states in the recent times have become major hubs for growing quality flowers. The *Anthurium* from Mizoram, the carnations from Himachal Pradesh, orchids from Sikkim, and *Gerbera* from Uttarakhand made a significant impact in recent times. Floriculture sector is generating higher income and employment opportunities, promoting greater involvement of women and enhancement of exports. India has now emerged as the second largest grower of flowers in the world. India is exporting flowers and flower products to different countries, *viz.* United States, Germany, the Netherlands, United Kingdom, Japan, United Arab Emirates, Italy, Canada, Belgium, Ethiopia, *etc.*

It is imperative to make a mention that the productivity of most of the flower crops is very low because of improper use of nutrients, unavailability of quality planting materials, lack of adoption of proper planting methods and other agro-techniques. One of the means to improve the productivity is to use adequate amount of fertilizers in balanced proportion which has been given less attention by the flower growers or floriculturists. The available information regarding proper use of fertilizers is very much scattered in flower crops in particular. It is well known fact that balanced nutrition is essential for growth, development and flowering of the crops. All the elements play important role both in the vegetative and reproductive phases and are indispensable for production of foliage and flowers. As nitrogen, phosphorus and potassium are required in large quantities, and hence, affect plant growth more as compared to other mineral nutrients. In addition to these, secondary nutrients such as calcium, sulphur and magnesium are also needed in fairly large quantities than the other essential elements such as iron, manganese, zinc, copper, boron, molybdenum and chlorine. Some of them, *viz.* nitrogen, phosphorous and sulphur are consumed in building up the plant architecture while calcium, potassium and magnesium have both, tissue building and metabolic functions. However, other essential elements such as boron, iron, manganese, copper, zinc and molybdenum also have metabolic functions in the plant life.

In general, the micro nutrients are found naturally in the soil in sufficient quantities, whereas the macro-nutrients are deficient in the soil and needs continuous supply. The availability of these nutrients in soil depends upon the soil pH. Iron, zinc, manganese, aluminum and copper are available at low pH (5.0–7.0), whereas the availability of boron decreases at <5 and >7 pH. However, availability of molybdenum is maximum at pH 6.5. Chelated compounds are used to increase the availability of micronutrients and make them available to the plants. Recommended manures and fertilizers are conventionally applied to the soil either as basal dose or top dressing. In fact, foliar nutrition is an effective method of applying nutrients for better production in crops like orchids and potted foliage plants. The metabolic function of all the elements required for the plant growth and development are described in table 1.

Table 1: Elements, concentration and their role in plant metabolism

Sr. No.	Elements	Quantity in whole plant	Function
	Carbon	45%	All cellular constituents
	Oxygen	43%	All cellular constituents
	Hydrogen	6%	All cellular constituents
	Nitrogen	3-5%	All living matter, amino acids, proteins
	Potassium	1.5-3%	Enzyme system in the change of sugar to starch, citric acid synthesis, in the change of amino acids to proteins, respiration, interaction with iron, enzymes, photosynthesis, buffer
	Calcium	0.1-3.5%	Cell wall, cell permeability, buffer
	Sulphur	0.05-1.5%	All living matter, proteins, nodulations in legumes, allyl oils of mustards, chlorophyll synthesis
	Phosphorous	0.25-0.5%	All living matter, nucleo-proteins, lipids, phosphorylation enzymes
	Magnesium	0.05-0.7%	A part of the chlorophyll molecule, enzyme activator of hexokinase, phosphorylase, carboxylase, dehydrogenase, peptidase, photosynthesis, buffer
	Chlorine	100-300 ppm	With Na and K it helps in maintaining cation-anion balance
	Iron	10-1500 ppm	A part of the porphyrin compounds cytochrome enzyme system, chlorophyll synthesis
	Manganese	5-1500 ppm	Chlorophyll synthesis, stabilisation of H-atoms split from H-OH by hydrogenation in photosynthesis, reduction of nitrates to nitrites, activator of arginase, carboxylases and dehydrogenases
	Zinc	3-150 ppm	Tryptophan synthesis, phosphorylation enzymes, enzymes in chloroplasts
	Copper	2-75 ppm	Enzyme in synthesis of ascorbic acid, activator of polyphenoloxidase, laccase and oxidase
	Boron	2-75 ppm	phosphorylation enzymes, glutamine synthesis, nodulation in legumes
	Molybdenum	Very less	Nodulation in legumes, tannin synthesis, reduction of nitrates to nitrites

Nutritional Management in Ornamentals under Open Field Condition

Rose

Roses prefer farmyard manure (FYM) or compost and leaf mould in addition to concentrated organic manures. The rate of application of organic manure depends upon the source and nature of the soil. The best time of manure application to the established plants is after pruning. Both organic and inorganic fertilization have beneficial effects on growth, development and flower production. The review on nutrition related work during last three decades indicated the significance of developing various technologies for increasing quality flower production. Each of the mineral elements is specific in its functions in plant metabolism. In order to maintain proper health and sustain production of roses, it is essential to apply accurate fertilizer at correct time in appropriate amounts. The nutritional requirement varies with cultivars, soil type and its fertility status. Nitrogenous fertilizers are more important in vegetative growth and phosphates help in production of more and good quality blooms. Although Indian soils are quite rich in potassium, the potash fertilizers in combination with others give good results in some regions (Randhawa and Mukhopadhyay, 2001).

To obtain maximum flower yield in rose cv. 'Super Star', an application of 500 kg N, 800 kg P_2O_5, and 700 kg K_2O along with 10 kg elemental sulphur, 50 kg each of $MgSO_4$ and $CaSO_4$ per hectare per year and foliar spray of $FeSO_4$ (0.1%), $ZnSO_4$ (0.5%) and boron (0.2%) during November and January was found to be the best under Delhi conditions. Balanced feeding with nitrogen and phosphorous imparts vigour to the plant, enhance flower yield and quality. Application of 600 kg nitrogen in 4 split doses and 200 kg phosphorous per hectare in cv. 'Montezuma' proved very effective in Nadia district of West Bengal for improving plant growth, yield and quality of rose flowers (Yadav *et al.*, 1985). The optimum dose of nitrogen and potassium and low amount of phosphorous provides a better vegetative growth with higher yield and anthocyanin pigmentation in rose cv. 'Celebration' (Maharana and Pradhan, 1976). The application of NPK inorganic fertilizers at the rate of 6:12:12 g/plant in Edward rose improved the flower production in South India (Irulappan *et al.*, 1980). The higher flower yield from *Rosa damascena* was obtained with application of 100 kg N and 60 kg P_2O_5 per hectare along with 1% Agromin + 50 ppm NAA (Sreenivas and Gowda, 1999). NPK fertilization in cv. 'Superstar' influenced nutrient content in different seasons of the year (Damke and Bhattacharjee, 1995). It is well known that nearly 60-70 per cent N of total nitrogenous fertilizer is lost during transformation of ammonical form of nitrogen to nitrate form. For increasing N-fertilizer-use efficiency, there is a need to limit the rate of ammonium oxidation. Although a number of nitrification inhibitors are available now days, but all of them have limitations in their usefulness in one or other way. Application of nitrogen along with encapsulated calcium chloride ($CaCl_2$), a nitrogen inhibitor, resulted in better utilization of applied N by the rose plant of cv. 'Raktagandha' and also markedly affected flowering, flower quality and flower yield (Bhattacharjee and Banerjee, 1994).

The secondary nutrients also play an important role for the improvement in production of quality rose flowers. It was observed that the soil application of calcium, magnesium and sulphur appreciably affected the vegetative growth of rose cv. 'Raktagandha' plants under Delhi conditions (Bhattacharjee, 1994). Magnesium increases the stem length, bud size, flower diameter and number of petals per flower. Magnesium and calcium applications increase the longevity of intact flowers. Application of fertilizers in solution form to the soil or hydroponics or in fertigation is practiced for obtaining good exhibition blooms. The advantage of this method is that the nutrient solution reaches the plant root for immediate utilization. Foliar application of nutrients not only rectifies the nutrient deficiency but also increases the flower production. In rose cv. 'Superstar', foliar application of NPK at 200 ppm after pruning (3 weeks) at fortnightly intervals resulted in accelerated plant height, increased number of shoots, enforced basal diameter of shoot, induced early flowering and produced maximum number of flowers with more petals (Bhattacharjee and Rajan, 1995).

Fertilizer recommendation for rose

- Apply a mixture of urea, single superphosphate and potassium sulphate in the ratio of 1:3:2 at 50 g per plant or 10 kg per 100 sq. m. three times i.e. first at pruning, second at the end of December when the first flush is over and third at the end of February when second flush of blooms is over.
- Foliar feeding: Spray 3 g mixture of urea, dihydrogen ammonium phosphate, potassium phosphate and potassium nitrate (2:1:1:1) along with 1.5 g of teepol at 10 days interval until last flush in March.

Gladiolus

Corms of gladiolus are rich in carbohydrates which are sufficient to sustain plant growth for initial few weeks. The cormels, however, require fairly good amount of fertilizers because of limited amount of stored food as they are quite small in size. Gladiolus requires both macro as well as micro nutrients for good growth and flower production. Nitrogen is primarily required to promote vegetative growth. Healthy gladiolus plants should contain 2.5 to 3.0 per cent N on dry weight basis. Nitrogen deficient plants remain weak, show pale-green leaves, and produce shorter spikes with small-sized florets. Application of nitrogen varies with the agro-climatic conditions of the growing locality. Application of excess nitrogen should be avoided as it causes more vegetative growth and lanky spikes. Nitrogen should be applied in two equal split doses, i.e. first when the plant reaches 3rd leaf stage and second when it reaches 6th leaf stage. In addition, as and when the plants show symptoms of nitrogen deficiency, an additional application may be given immediately. Cormels may be given nitrogen in 4-5 applications at about 3 week intervals, first application starting at one month of the crop age. Nitrogen application should be stopped at least six weeks before the harvesting of corms. Chanda *et al.*(2000) reported that nitrogen promotes vegetative growth by increasing more metabolic transport for growth and delays flowering, and potassium influences flowering more than phosphorus. Uday Sharma *et al.* (2008) obtained earlier sprouting of corms, maximum plant height, spike length and number of florets per spike, *vis-à-vis* corm and cormel production with application of recommended doses of NPK along with vermicompost and biofertilizers. In gladiolus, since both the spikes, and corms & cormels are equally important, therefore, early and quick vegetative growth is necessary to obtain long and healthy spikes with more number of buds and to obtain more yield of corms in terms of size and number. Maximum spike length (89.7 cm) was observed in plants treated with 155 g N+77.5 g K in gladiolus cv. 'High Style' (Butt, 2005). Nitrate form of nitrogen is the safest and most suitable source as compared to ammonical form. The use of excessive ammonical nitrogen is detrimental to the crop as it causes high incidence of diseases. Hatibarua and Misra (1999) reported that among the different N sources, calcium ammonium nitrate (CAN) and ammonium sulphate performed better compared with urea. CAN at 30 g N/m^2 was best for increasing plant height, spike length and leaf number, while 40 g/m^2 was best for number of florets per spike.

Phosphorus is an essential nutrient required for the development of good root system. For better growth in general, the leaves of gladiolus should contain 0.3-0.4 per cent phosphorus on dry weight basis. Full amount of phosphorus should be applied as a basal dose. The roots of gladiolus are highly sensitive to high salt concentrations in the soil. Therefore, in heavy soils, the phosphorus application may be delayed till the plant reaches 2-3 leaf stage and develops good root system. Potassium is known to increase resistance to diseases and photosynthetic efficacy of leaves. Potassium deficiency causes reduction in spike length, number of florets and corm size. Healthy plants of gladiolus should contain 3-4 per cent potassium in the leaves on dry weight basis. Potassium may be applied at the time of planting of corms or when the plant reaches 1-2 leaf stage.

Application of NPK at 60:60:60 units significantly improved growth and quality of flowers and corm and cormel yields. A fertilizer combination of N at 50 g/m^2, P at 10 g/m^2 and K at 20 g/m^2 produced more flowers and corms with better quality in cv. 'Tropic Sea (Kumar and Chattopadhyay, 2001). Nitrogen at 30, 60 and 90 g/m^2, P_2O_5 at 15, 30 and 45 g/m^2 and K_2O at 10, 20 and 30 g/m^2 were when applied to gladiolus cv. 'Friendship' administering full dose of phosphorus and potassium and half dose of nitrogen along with 5 kg FYM/m^2 before planting of corms and remaining half dose of nitrogen at 4-leaf stage, Gupta *et al.* (2010) recorded maximum plant height at 60 g nitrogen though no obvious effects were observed with phosphorus and potassium applications. They recorded highest duration of flowering per spike under 20 g/m^2 K_2O, however, spike length, rachis length, number of florets per spike, diameter of florets diameter of corms and number of cormels were found significantly higher under 60 g N, 30 g P_2O_5 and 20 g/m^2 K_2O. Sharma and Singh (2007) reported that application of $N_{40}P_{20}K_{20}$ g/m^2 significantly increased spike quality and its field durability in cv. 'White Friendship' grown under mango orchard in Raipur. A gradual, steady and significant increase in corms produced per plant, corm size and weight including the number of cormels per plant were observed with increasing fertilizer levels, recording highest values at $N_{50}P_{25}K_{25}$ g/m^2. According to Dongardive *et al.* (2007), application of 500:200:200 kg/ha NPK is optimum for growth and flowering of gladiolus cv. 'White Prosperity' in Nagpur. Dongardive *et al.* (2009) obtained maximum corm and cormels with treatment of 500:200:200 NPK kg/ha and in the treatment of vermicompost 8 t/ha+(Azotobacter 5 kg/ha)+PSB 5 kg/ha.

Baskaran *et al.* (2014) tried various bio-fertilizers (phosphorus solubilizing bacteria, *Azospirillum* and *Azotobacter*) and commercial formulations (Annapurna®, Prokissan®, Sumangula®, General liquid®, Chamak®, Twin®, Samaras®, Multinol® and Flower Booster®) and observed *Azotobacter* giving earlier corm sprouting with maximum rachis length, *Azospirillum* giving maximum number of corms per plant, PSB giving maximum plant height and floret diameter, Annapurna inducing earlier flowering with maximum rachis length and number of florets per spike, Sumangala recording increased spike length, and Flower Booster recording maximum weight of corms and cormels with maximum volume and diameter of corms and maximum propagation coefficient. *Bhalla et al.* (2006) with gladiolus cv. 'Red Beauty' recorded that common basal dose + 4% Manchurian mushroom tea (Kombucha) + 6% Panchgavya, common basal dose + 4% Kombucha tea + 4% Panchgavya, and common basal dose + 4% Kombucha tea + 2% Panchgavya to be the three best doses for growth and flowering parameters whereas common basal dose + 4% Kombucha tea + 6% Panchgavya, common basal dose + 6% Panchgavya, and general medium + required fertilizer dose were found best for corm parameters. Coating corms either with rhizobacterin (*Azotobacter chroococcum*; 1010 cfu/g) or cerealin (*Azospirillium brasitense*; 1010 cfu/g) increased the survival percentage. Rizobacterin was more effective in disease control at different irrigation periods than cerealin. Both the biofertilizers increased growth parameters, flower, corm and cormel yields with rhizobacterin being nore superior to cerealin (Hilal *et al.*, 2001). Srivastava and Govil (2007) found better vegetative growth by *Azotobacter* treatment. However, for quality spike production, PSB was found more effective. It was found that treatment of corms with biofertilizers increased the total rhizospheric bacterial population. This indicates that the improvement in various characters of gladiolus is due to the activity of rhizospheric bacteria, which is enhanced by biofertilizer inoculation. Dalve *et al* (2009) suggested use of biofertilizers (*Azotobacter* and *Azospirillum*) with reduced doses of nitrogen significantly influencing the growth, flowering and yield of gladiolus. N 75% + P and K 100%, i.e. 375:200:200 kg NPK ha^{-1} + *Azotobacter* + *Azospirillum* giving the best results, thus saving 25% of nitrogenous fertilizer.

Important micro nutrients for gladiolus are iron, boron, calcium, magnesium, manganese, zinc and copper. Iron deficiency is, however, most prevalent in gladiolus especially in North Western plains of India. Iron deficiency causes interveinal yellowing of new leaves and in severe deficiency, the emerging spikes turn light green to yellow. The deficiency is more pronounced in alkaline soils or in the soil where levels of copper, manganese, zinc or phosphorus are very high. Always apply freshly prepared solution because the ferrous is oxidised to ferric and becomes unavailable to the plant.

Fertilizer recommendation for gladiolus

- N, P, K each @ 200 kg/ha.
- Nitrogen may be applied in 3-4 split applications, i.e. before planting, at 2-3 leaf stage, at spike emergence and after completion of flowering.

Chrysanthemum

The chrysanthemum is known to respond well to fertilizer application. Nitrogen plays an important role during early vegetative stage while potassium during bud appearance stage. The plants need phosphorous during their growth period and can be easily applied as it is released slowly. The recommended dose of NPK fertilization varies with the climatic condition of different regions of our country. Liquid feeding is commonly preferred in greenhouses as the nutrients are readily available to the plants. The toxic effects due to high concentrations coming into indirect contact with roots are less in liquid form. Top dressing is necessary because of depletion of nutrients, chiefly nitrogen and potassium as they are used up by growing plants. Top dressing with 40 g nitrogen and 20 g potash per m^2 is sufficient for better growth and development. This dose can be repeated after one month of planting. In potted plants, the peripheral portion near the rim of the pot is the correct location for top dressing. Nitrogen application at 100, 150 or 200 mg/litre in combination with at least 10 mg sulphur/litre produced plants of commercial quality (Macz *et al.*, 2001).

Fertilizer recommendation for chrysanthemum

- After land preparation, apply 25-30 tonnes per hectare of well decomposed farmyard manure and 125 kg of nitrogen, 400 kg P_2O_5 and 200 kg K_2O/ha as basal dose.
- Top dress the crop with 125 kg nitrogen/ha at the time of first pinching.

Marigold

Information on requirement of macro and micro nutrients is scanty, although marigold responds well to fertilization. Increased number of flowers per plant was obtained with application of 60 kg nitrogen, 90 kg P_2O_5 and 60 kg K_2O per hectare (Anuradha *et al.*, 1990). It was found that nitrogen levels had significant influence on growth, dry matter production, flower yield and floral parameters (Dahiya *et al.*, 1998). Nitrogen fertilization had a marginal effect on the consumptive use of water and water-use efficiency which may be due to higher flower yield at 300 kg N/ha in relation to a very small increase in evaporation (Halepyati *et al.*, 2001). Few attempts have been made for foliar application for higher productivity in marigold. However, in addition to NPK, foliar sprays of 0.2% urea should be done at an interval of 15 days.

Recommended dose for marigold

- 200 kg N and 80 kg P and K per hectare for good vegetative growth and flower yield.
- Full dose application of P and K at the time of land preparation. N in two split doses i.e. 1st at the time of land preparation and 2nd one month after transplanting.

China aster

Application of manures and fertilizers in required quantities is important for proper growth, yield and quality of flowers. The deficiency of nitrogen causes dwarfing of plants and small size flowers. Phosphorus deficiency causes delayed flowering. Nitrogen was the most effective nutrient which significantly increased flower number, diameter and seed yield. Top dressing with 120 kg N, 80 kg P_2O_5 and 120 kg K_2O per hectare at 10-12 leaf stage gave better results than at bud stage or shortly before flowering. The application of recommended dose of NPK and 15 t FYM/ha or 8 t phospho-compost/ha is optimal for high flower yield (Sreenivas and Gowda, 1999). China aster responds well to micronutrients and application of Zn, Cu, B and Mn which improves the quality of flowers. Asters are moderately sensitive to salinity caused by sodium salts which results in decreased plant height and flower size.

Fertilizer recommendation for China aster

- 120 Kg N, 80 kg P_2O_5 and 120 kg K_2O per hectare at 10-12 leaf stage.

Tuberose

Tuberose requires heavy nutrition, therefore fertilizer application is essential for its cultivation. The application of organic matter is essential for promoting growth and more spike production. At the time of land preparation, FYM should be applied at the rate of 20 tonnes per hectare. The application of green manures like *Gliricidia trilobus, Phaseolus trilobous* and *Oxalis corniculata* at the rate of 10 tonnes/ha promoted early spike emergence (De and Dhiman, 2001). In tuberose, nitrogen is more important than phosphorous and potassium for increasing yield and quality of flowers and bulb production. Nitrogen deficiency reduces number of spikes and flowers as well as turning of foliage to pale-green. However, excess nitrogen makes the flower spikes quite tall and soft, further making them vulnerable to wind and pests. Application of 325 kg N and 125 kg each of P_2O_5 and K_2O per hectare resulted in high uptake both at 50% flowering and harvesting stage (Mohanty *et al.*, 1999).

Fertilizer recommendation for tuberose

- FYM @ 3-4 tonnes per hectare.
- 100-250 kg N, 100-150 kg P_2O_5 and 75-100 kg K_2O per hectare.

Jasmine

Cultivation of jasmine is a paying enterprise but the evidences on its nutritional requirement are very meagre. They respond to intensive application of manures, however, heavy manuring encourages vegetative growth and hampers quality and quantity of blooming. A basal dose of 10 kg FYM per plant per year along with inorganic fertilizer dose of 100 g N, 150 g P_2O_5 and 100 g K_2O enhances the flowering of *Jasminum grandiflorum* (Bhattacharjee, 1985). A basal application of 20 kg FYM per plant was given for enhancing the flower yield (Velu, 1988). Application of 10-30 kg of FYM/plant affects the growth performance of different jasmine cultivars. A significant improvement in flower yield and other characters with each increment of N dose up to 100 g/plant/year was recorded in *Jasminum grandiflorum* (Bhattacharjee, 1985). The flower yield in *Jasminum grandiflorum* was found to be enhanced with foliar application of P and K at 120 g per plant (Kumaraguruparan, 1974). Application of nitrogen and potassium in *Jasminum sambac* at 240g/plant/year at bimonthly intervals resulted in higher flower yield. Similar to macro nutrients, micro nutrients supplemented through foliar sprays also increase flower production.

Fertilizer recommendation for jasmine
❖ *Jasminum auriculatum*: 120g N:240 g P :120 g K per plant.
❖ *Jasminum sambac*: 60 g N:60 g P :120 g K per plant.
❖ *Jasminum grandiflorum*: 60g N:60 g P :120 g K per plant.

Annual flowers

The nutritional requirement of **antirrhinum** is relatively low as compared to rose and chrysanthemum (Bose *et al.*, 2003). Jeong and Lee (1992) reported that *A. majus* cv. 'Rocket Mixed' grew best with 25:75 NH_4:NO_3. The nutrient solution containing 160 ppm N, 32 ppm P_2O_5, 240 ppm K_2O, 176 ppm Ca and 64 ppm Mg provide maximum growth and development of the plant grown in sand media. Higher concentration of nutrients caused leaf chlorosis, low dry matter production, shorter stem and reduction in flowers. The application of nitrogen at early stage is beneficial for good vegetative growth. Increased P concentration (3μg/ml) significantly increased plant fresh weight and total number of flowers per spike (Besmer and Koide, 1999). During constant liquid fertilization, about half of the normal concentration of 200 ppm each of N and K is recommended by Sanderson (1975). **Stock** needs fertile soil for its healthy growth and normal flowering. Nitrogen promotes growth of stock plants which is further enhanced by high K (Thomas and Leong, 1980). The application of 30 tonnes of FYM at the time of land preparation and 400 kg nitrogen and 150 kg each of phosphorous and potassium per hectare is sufficient for plant growth. Stocks are subject to potassium deficiency which is seen as burning on the older leaf margins. **Statice** requires balanced fertilizer to meet its demand for nutrients. Field fertilization with FYM at the rate of 50 tonnes per hectare 10-15 days before transplanting is quite beneficial. Application of 100 kg each of N, P and K per hectare is adequate for proper growth and flowering. Half of the N and K and full P should be incorporated in the soil 3-4 days before transplanting and remaining half of the N should be applied one month after first application. Balanced fertilization influences the growth and flowering of ***Zinnia elegans***. Jhon *et al.* (1991) obtained best flowering in terms of number and quality with 150 kg N/ha in 2 split doses, 10 and 30 days after transplanting, in addition to a basal dose of 100 kg/ha each of P and K and 30 tonnes of FYM to *Zinnia elegans* cv. 'Giant Flowered Crimson'. In ***Gaillardia***, for better growth and flowering, liquid feed is given when the plants start bearing flowers. Optimum fertilizer rate for growth and flower yield of *Gaillardia pulchella* cv. 'Yellow Double' was 173 kg N + 103 kg P + 63 kg K/ ha (Singatkar *et al.*, 1995). **Gomphrena** is not a heavy feeder and responds well to a small amount of fertilizer. Well rotten FYM at 10 kg/m^2 should be applied at the time of bed preparation. NPK fertilizer at the rate of 15-20 g/m^2 is applied for better growth and flowering.

Nutritional Management in Ornamentals Under Protected Conditions

Nutrients are usually applied with irrigation water (fertigation) in greenhouse crops. Fertilizer tanks are used for fertigation because of many advantages such as low rate of application, uniform distribution, better nutrient-use-efficiency, easy and better availability to the plants, *etc.*

Rose

Most important aspect in greenhouse cultivation of roses is nutrition. The fertilizer regime is followed only after the soil and water analysis is carried out and quantities are compared with the standards available. It is always desired that the nutrients are applied in smaller quantities more frequently so that plants get regular supply of the necessary nutrients. Commercial grade 100% water soluble or liquid fertilizers are used to enrich the roses that are usually given through irrigation system. During summer months, the nutrient requirements of roses per unit area per day is distributed into three equal parts and is injected into the drip lines when irrigated in the morning, afternoon and evening. However, during rainy and winter seasons, the frequency of irrigation is reduced and the nutrient content is adjusted into one or two irrigations. Rose develops nutritional disorders and deficiencies very quickly and recovers its normal growth very slowly. In order to ensure the maximum availability of all the nutrients, it is advisable to maintain a soil pH between 5.5 to 6.3. When grown under polyhouse conditions it requires heavy dose of nitrogen and potassium, and moderate doses of phosphorous, magnesium, sulphur, manganese, copper and molybdenum. It is very sensitive to iron and boron deficiencies. Generally for cut roses, the basic nutrient concentration is 150-200 ppm. Recommended soil nutrient levels for roses are nitrates 25-75 mg/l, phosphorus 4-6 mg/l, potassium 30-80 mg/l (5% base solution), calcium 80 mg/l (60-70% base solution), magnesium 5 mg/l (10% base saturation) (Bhattacharjee, 1994).

Fertilizer solution for roses

Nutrient	Concentration (mg/l)	Nutrient	Concentration (mg/l)
No_3-N	156	P	46
Ca	132	K	174
S-SO_2	23	Fe	1.5
Mn	1.1	B	0.28
Mo	0.003	Cu	0.15

If the water has a very low EC then the above solution will have an EC of about 1.7 mS/cm

(Source: Singh *et al.*, 2010).

Fertilizer requirement in roses for quality blooms based on foliar nutrient level

Nutrient	Range	Nutrient	Range
Nitrogen	3.5-4.5%	Iron	70-320 ppm
Phosphorous	0.2-0.3%	Zinc	20-80 ppm
Potassium	2.0-2.5%	Copper	5-30 ppm
Calcium	1.0-1.5%	Boron	30-200 ppm
Magnesium	0.28-0.32%	Manganese	35-490 ppm

Gerbera

Gerbera plants need ample amount of macro-nutrients especially phosphorus and potassium for profuse flowering. A standard feeding solution for gerbera contains anions (mmol/l), e.g. NO_3: 8.0-13.0, Cl<4.0; SO_4: 1.3-2.0; HCO_3:0.5 and P: 0.1-0.4, cations (mmol/l), e.g. NH_4<1.0; K_2O 1.0-2.5, Na:0-<4.0; Ca:1.0->4.0; Mg: 0.7-72.0, and micro-elements (micro mol/l), e.g. Fe:12.0-25.0; Mn:1.5-10.0; Zn: 2.0-4.0; B:12.0-30.0; Cu: 0.8-1.0; Mo: 0.5; pH 5.0-6.5; and EC: 0.5-2.0 ms/cm. Application of a combination of NPK (2:1:3) at the rate of 90 mg N, 130 mg P_2O_5 and 400 mg K_2O proves ideal for gerbera. Frequent applications of nutrients at regular intervals are useful for optimum growth and development of gerbera. Application of NPK proved very effective in increasing the number of leaves and suckers and in producing larger number of flower heads of bigger size (Bose and Jana, 1978). The highest number of flowers was obtained by growing gerbera in peat with ammonium nitrate + superphosphate + potassium sulphate at 4 g/l (Gutmane and Grivina, 1975). An excess of N reduced the yield and vase life of gerbera (Moulinier and Montarone, 1978). The optimum level of N as 250 mg/l for the composted pine bark medium was also recorded in gerbera (Strojny *et al.*, 1976). Gerbera cv. 'Lada' grown in loam and peat (1:2) substrate with 4 NPK applications, yielded maximum of 40 flowers per year with nitrogen application in spring, summer, autumn and winter at 40, 20, 10 and 5 mg/100 g of substrate, respectively (Skalska, 1980). In the mixed substrate of peat:soil:sand (3:1:1), NPK fertilizer was applied to determine the optimum time and rate of application in relation to the plant development phases, where N at 0.025 g/kg soil applied at 15-17 days after planting, N: K_2O at 0.1: 0.15 g/kg soil at the state of rapid vegetative growth, N: P_2O_5 at 0.1:0.1/kg soil at bud formation stage and NPK at 0.1: 0.1: 0.23 g/kg soil at the beginning of flowering gave the best results for plant growth and flower production (Mantrova *et al.*, 1982). The most effective dose of fertilizer for *Gerbera jamesonii* hybrid was found to be 60 kg N, 48 kg P_2O and 90 kg K_2O/1,000 m^2/year (Volpi and Farina, 1987). It was observed that number of marketable flowers increased as both N and K rates were increased up to 110 kg/ha (Daufault *et al.*, 1990).

Micronutrients also play an important role in growth and flowering of gerbera. It was suggested that the interveinal chlorosis of gerbera can be corrected by spraying of 0.1-0.2 per cent $MnSO_4.H_2O$ (Voogt, 1976). Further, Fe deficiency could be corrected by application of 1 g chelate per sq. m. grown in a peat substrate having pH range 4.8-5.5, enriched with the fertilizer mixture. The effect of Cu was improved by liming peat and the optimum rates were 60 g $CuSO_4$ and 2 kg $CaCO_3$ per m^3. The application of lime resulted in 46.5 per cent increase in the number of flowers per sq. m. and also improved flower quality. Cultivars 'Hanny' and 'Veromica' exhibited better growth and yield when 60 mg Cu was applied in the form of solution at the rate of 5-6 g $CuSO_4.5H_2O$ per plant at 6-8 weeks interval (Fischer and Forchthammer, 1987).

Carnation

Carnation needs regular supply of nutrients throughout its life cycle. The application of nutrients in small doses but more frequently favours better growth and flower production. Carnation makes a good reserve of nitrogen at the tufting stage which is utilized during flowering. The application of well rotten FYM at 5 kg/m^2 is beneficial for highest number of flower production in carnation cv. 'Marguerite Scarlet'. Application of nitrogen in three split doses, i.e. $^1/_3{}^{rd}$ before transplanting, $^1/_3{}^{rd}$ one month after transplanting and $^1/_3{}^{rd}$ one month after second application increased vegetative growth and production of blue grade flowers as compared to single basal dose or two split doses. Nitrogen and phosphate fertilizers improved the flower yield by increasing the number of branches and shoot growth (Patel and Arora, 1983). In *Dianthus caryophyllus* cv. 'Desio', it was found that increased concentration of NO_3-N decreased the growth related traits of flowers and number of days to flowering (Shim Myung *et al.*, 2001). Application of 1,000 ppm N in two perpetual carnation cultivars, *viz.* 'White Candy' and 'Cargo' showed beneficial effects on plant height, stem length, bud size, flower diameter and flower yield (Verma *et al.*, 2000). Regular soil and tissue tests should be taken to determine existing nutrient levels. In general, the optimal nitrate level in the medium should be 25-40 ppm; phosphorus 5-10 ppm; potassium 25-40 ppm; calcium 125-200 ppm; magnesium 30-40 ppm and boron 30-35 ppm. Tissue nutrient levels are pre-requisite to determine fertilization schedule for quality plant production as are mentioned here in.

Nutrient	Range	Nutrient	Range
Nitrogen	3.2-5.2 %	Magnesium	25-75 %
Phosphorous	0.2-0.3 %	Zinc	25-75 ppm
Potassium	2.5-6.0 %	Copper	10-30 ppm
Calcium	1.0-2.0 %	Boron	30-100 ppm

Chrysanthemum

The crop needs high level of potassium in the medium for good growth and yield. The desirable range of nutrients is 25-50 mg/l - nitrogen, 5-10 mg/l - phosphate, 20-40 mg/l - potassium and 100-150 mg/l calcium. The critical leaf K level is 2.3-2.8%. Boron and copper deficiencies frequently limit the growth of chrysanthemum. Since boron is easily leached from soil, continuous supply through fertigation at low levels is essential to maintain a leaf level of 25-50 ppm. In leaves, copper should be maintained at 8 ppm (Edward, 2007).

Orchids

Orchids receive very small quantity of nutrients under natural conditions. The terrestrial orchids obtain nourishment from the decaying leaves and other vegetative matter. However, epiphytic orchids get nutrients from the first rain and also from the bird-droppings which are washed by rains, from the upper branches downwards. Orchid hybrids, however, require larger amounts of nutrients since growth and flowering rates are faster. The type of nutrients, their quantity and frequency of application depend largely on the type of orchid, potting medium, season of the year, growing conditions, stage of plant development, *etc.* The major nutrients, such as nitrogen is essential for vegetative growth, phosphorus to initiate flowering process and potassium to build up strong roots and for making the plants resistant to biotic stresses. Since most of the cultivated orchids are epiphytic in nature, the media used are highly porous. If nutrients are applied in the media they are prone to leaching and become unavailable to the plants. Hence, foliar feeding is ideal for providing proper nutrition. It is always better to use small quantities of fertilizers at shorter intervals than using large quantities at longer intervals. It helps the plants to grow steadily without any deficiency throughout their growth period but for younger plants, the concentration should be very low. In newly planted orchids, no fertilizer application is necessary till new roots are formed. A fertilizer complex containing nitrogen, phosphorus and potash in equal proportion (e.g. 17:17:17 complex) is ideal for general application. The concentration may be adjusted between 0.2 and 1.0 per cent depending upon the conditions. During vegetative phase a 30:10:10

combination may be used which may be changed to 10:20:20 formulation during blocking stage.

Fermented organic manures such as cow dung, *neem* cake, poultry manure, *etc.* are also used for monopodial orchids. These are diluted 10-15 times with water, filtered and sprayed over the plants. Depending upon the various conditions such as type of orchid, quantity supplied in each application, stage of growth of the plant, season, *etc.*, the frequency of application could vary from thrice a week to once in two weeks under tropical conditions, though once a week is the usual recommendation. As flower buds begin to form interior sheath, a mixture containing more of phosphorus is useful in improving the quality of spike.

It is advised to stop application of inorganic fertilizer about three days prior to harvest of spikes as it may otherwise reduce the shelf life of flowers. An overdose of nitrogenous fertilizers causes delay in flowering and even reduction in bloom quality. Plants should be flushed with plain water every fortnight so that the salt left behind from the fertilizers is not built up to toxic level. Nitrogen starved leaves show yellow colouration, but such symptoms in the new growth can be eliminated by application of nitrogen. Phosphorus deficiency may hamper new growth and plants may remain in vegetative phase. A long period of potash deficiency can result in discolouration of flowers and reduction in the number of flower buds.

Anthurium

It needs adequate amount of nutrients for proper growth and flowering. The application of nitrogen, potassium and calcium markedly improve the yield and quality of flowers. Adequate level of calcium is necessary for obtaining optimum yield and to stabilize spathe colour. The fertilizers are applied to the plants through ground level sprinklers or drip irrigation system. Depending upon nutrient status of the medium, the following combinations of nutrients are supplied through irrigation water by many commercial growers.

Nutrient	Concentration (mg/l)	Nutrient	Concentration (mg/l)
Sulphate	48	Iron	0.80
Phosphate	31	Zinc	0.20
Potash	14	Copper	0.03
Calcium	176	Boron	0.22
Magnesium nitrate	6091	Manganese	0.16

Deficiency of nutrient elements adversely affects the plant growth and development. Potassium deficiency causes leaf necrosis and root tip death. Calcium deficiency results in colour break-down, instability of the middle lamella, cell separation and collapsing of the proximal section of the lobe and spathe. This problem tends to be more serious when substrate pH lowers down to 3-4.

Lilium

Due to high nutrient reserve in bulbs of *Lilium*, fertilization is not required during forcing. Fertilization programme should commence at shoot emergence stage using calcium nitrate and potassium nitrate at a 2:1 ratio on weekly basis and medium should be amended with phosphorous. Constant fertilization throughout the growing period should be done with nitrogen at 200-500 ppm. The tissue nutrient level in *Lilium* is recommended as following.

Nutrient	Range	Nutrient	Range
Nitrogen	2.4-4.0%	Iron	100-250 ppm
Phosphorous	0.1-0.7%	Zinc	30-70 ppm
Potassium	2.0-5.0%	Copper	5-25 ppm
Calcium	0.2-0.4%	Boron	20-25 ppm
Magnesium	0.3-2.0%	Manganese	50-250 ppm

Fertigation Scheduling in Ornamentals under Protected Structures

Crop	Fertigation schedule	Dose (ppm)		
		N	P	K
Rose	Vegetative stage (September-October)	80	50	60
	Flowering & harvesting flush (November-March)	100	60	80
	Flowering and harvesting normal (April-August)	80	50	80
Chrysanthemum	Vegetative stage (September-October)	80	50	60
	Flowering & harvesting flush (November-April)	90	60	80
	Maintenance dose (May-August)	50	30	50
Carnation	Vegetative stage (September-October)	50	30	40
	Flowering & harvesting flush (November-March)	60	40	60
	Maintenance dose (April-August)	40	20	40
Gerbera	Vegetative stage (September-October)	70	50	60
	Flowering & harvesting flush (November-April)	80	60	80
	Maintenance dose (May-August)	40	24	40
Lilium	Vegetative stage (September-October)	60	36	60
	Flowering & harvesting flush (November-March)	80	50	80
	Maintenance dose (April-August)	50	30	50

(Source: Hasan *et al.*, 2010)

Integrated Nutrient Management (INM)

INM maintains soils as storehouses of plant nutrients essential for vegetative growth. Integrated nutrient management (INM) is an approach to soil fertility management that combines organic and mineral methods of soil fertilization with physical and biological measures for soil and water conservation. INM's goal is to integrate the use of all natural and man-made sources of plant nutrients, so that crop productivity increases in an efficient and environmentally benevolent manner, without sacrificing soil productivity for future generations. Balanced application of appropriate fertilizers is a major component of INM. Nutrient conservation in the soil is another critical component of INM. Mulch application, cover crops, intercropping, and biological nitrogen fixation act as physical barriers to wind and water erosion and help to improve soil characteristics and structure. Use of bio-fertilizers is one of the important components of INM due to cost-effectiveness and a renewable source of plant nutrients to supplement the chemical fertilizers for sustainable horticulture.

It was reported that *Azospirillum*-inoculated plants showed early flowering, besides extended crop period and higher flower yield in French marigold (Balasubramanian, 1989). The application of nitrogen at 37.5 kg/ha along with 1,000 ppm ascorbic acid spray, and inoculation with *Azospirillum* resulted in higher flower yield after six months of planting in Edward rose (Preethi, 1990). Phosphobacteria-treated rooted cuttings of *Jasminum sambac* had given higher establishment percentage in the field, as well as increased the plant height and number of branches (Manonmani, 1992). In marigold, plant height was found superior with application of full dose of nitrogen with *Azotobacter* and phosphorous-solubilising bacteria (Chandrikapure *et al.*, 1999). Fungi are also known to play a vital role in releasing phosphates into the rhizosphere from insoluble phosphorous sources. *Penicillium simplicissimum, P. aurantiogriseum* and *P. regulosum* IR-95 solubilise mineral phosphates by secretion of gluconic and citric acids (Reyes *et al.*, 1999). *Aspergillus niger* showed P-solubilisation ability on a variety of carbon sources (Cerazine *et al.*, 1988). *Penicillium balajii* solubilises Ca-P complexes by secretion of oxalic and citric acids (Cunnigham and Kuiack, 1992). The combined application of *Azospirillum* and arbuscular mycorrhizae (AM) even with the reduction of 25% of recommended dose of NPK increased marigold growth (Balasubramanian, 1989).

Plants belonging to Poaceae or Graminae require silica which is not supplied exogenously as a fertilizer. Turf grasses accumulate 4-20% silica in straw and its requirement increases with addition of more nitrogen which makes the plants tender and succulent, rendering a high demand of silica which reverses the nitrogen effect by making sturdy and erect plants. Indian soils exhibit deficiency of zinc with contents much below the critical level of 1.5 ppm (Tiwari and Dwivedi, 1994).

Fertilizer Recommendations from AICRP (Floriculture)

Various trials conducted at coordinated centres under All India Coordinated Research Project (AICRP) on Floriculture revealed that split doses of nutrients are superior to single application at sowing or any other growing stage. Various recommendations thus emerging from the trials in different flower crops at different centres are summarized in Table 2.

Table 2: Fertilizer doses in flower crops recommended by AICRP on floriculture

Crops	Centres	Recommended doses	References
Rose	BCKV, Kalyani	60 kg nitrogen, 20 kg phosphorous and 20 kg potassium	Tejaswini, K. Sujatha, Sangama, N. Ramachandran and B. Jhansi Rani, R. L. Misra, P. Naveen Kumar and Suchitra, 2007. Roses. AICRP on Floriculture (ICAR), IARI, Division of Floriculture and Landscaping, IARI, New Delhi, 60 p.
	MPUAT, Udaipur	Nitrogen 60 kg, phosphorous 40 kg and potassium 20 kg for plant growth, and nitrogen 20 kg, phosphorous 20 kg and potassium 20 kg for higher flower yield.	
	Pune	N:P:K at 60:40:20 g/s.m. is ideal for cv. Gladiator	
Gladiolus	Delhi, Pune, Kalyani and Bengaluru	N 500 kg, P_2O_5 200 kg and K_2O 200 kg/ha. Nitrogen should be applied in two equal doses, first, when the plant reaches 3 leaf stage, and second when it reaches 6 leaf stage.	Arora, J. S., R. L. Misra, Kushal Singh, Premjit Singh and S. K. Bhattacharjee, 2002. Gladiolus. All India Coordinated Research Project on Floriculture (ICAR), Division of Floriculture and Landscaping, IARI, New Delhi, 100 p.
	Ludhiana	400 kg N, 100 kg P_2O_5& 100 kg K_2O/ha in sandy soils. Fe deficiency can be corrected by spraying the plants with 0.2 per cent ferrous sulphate solution twice or thrice at 10-day intervals.	
Chrysanthemum	West Bengal (Kalyani)	30N: 20P: 20 K g/m^2 for optimum growth and flower yield.	Datta, S.K. and S. K. Bhattacharjee, 2001. Chrysanthemum. AICRP (F), ICAR, Division of Floriculture and Landscaping, IARI, New Delhi, 100 p.
	TNAU, Coimbatore & MPKV, Pune	NPK application at 300:200:200 kg/ha along with 10 tonnes of FYM/ha has been recommended for higher flower yield.	
	SKUA S&T-K, Srinagar	N and P_2O_5 20 g each per m^2 has been recommended in cv. Flirt.	
	IIHR, Bengaluru & TNAU, Coimbatore	N 40: $P_2O_5$20: K_2O 20 g/ m^2 has been recommended for cv. Local White and cv. CO1 .	
	Pune,	NPK @ 200: 200: 200 kg/ha.	
	Ludhiana,	NPK @ 400:400:200 kg/ha.	

Contd...

Crops	Centres	Recommended doses	References
China aster	Bengaluru	90 kg N, 60 kg P and 60 kg K/ha with land preparation.	Janakiram, T., M. T. Patil and S.K. Bhattacharjee, 2001. China Aster (*Callistephus chinensis*). AICRP (F), ICAR, Div. of Floriculture and Landscaping, IARI, New Delhi, 16 p.
	Kalyani (W.B.)	400 kg N along with 200 kg P_2O_5/ha for higher flower yield.	
	Pune	300 kg N and 200 kg P_2O_5/ha for flower yield.	
Anthurium	TNAU, Coimbatore	*A. andraeanum* cv. Temptation grown under 75% shade net and sprayed with NPK @ 30:10:10 at 0.2% in combination with GA_3, where 100 ppm producing highest number of flowers and suckers/plant.	Rajeevan, P. K., P. K. Valsalakumari, C. K. Geetha, Leena Ravidas, Vinod Kumar and S. K. Bhattacharjee, 2002. Anthurium. AICRP on Floriculture (ICAR), Division of Floriculture and Landscaping, IARI, New Delhi, 42 p.
Tuberose	BSI, Howrah	200 kg N, 400 kg P_2O_5 and 200 K_2O/ha improving flowering and bulb formation.	Biswas,B., P. Naveen Kumar, S. K. Bhattacharjee, 2002. Tuberose. AICRP on Floriculture (ICAR), Division of Floriculture and Landscaping, IARI, New Delhi, 25 p.
	Pune	150 kg N, 200 kg P_2O_5 and 200 kg K_2O/ha.	
	TNAU, Coimbatore	200 Kg N and 60 kg P_2O_5/ha.	
	Bengaluru	200 kg each of NPK/ha/year .	
Gerbera	SKUA S&T-K, Srinagar	20 g N/m^2 provided increased stalk length, flower diameter and flowers per plant and 10 g P_2O_5/m^2 resulted into more number of flowers with increased vase life.	Mandal, T., Sanyat Misra, R. L. Misra and P. Naveen Kumar, 2007. Gerbera. AICRP on Floriculture (ICAR), Division of Floriculture and Landscaping, IARI, New Delhi, 36 p.
Orchids	BSI, Howrah	Foliar application of fertilizer solution comprising of potassium nitrate (2.63 g), ammonium sulphate (0.44 g), magnesium sulphate (2.04 g), mono-calcium phosphate (1.09 g), calcium sulphate (4.86 g), ferrous sulphate (0.50 g) and manganese sulphate (2 ml of 1% solution) in 4.5 litres of water at fortnightly for improving growth and flowering in *Dendrobium. moschatum* and *Aerides multiflorum.* In *Rhyncostylis gigantea*, foliar spray of 500 ppm each of N, P_2O_5 and K_2O at fortnightly interval was found to be the best.	Rajeevan, P.K., A. Shobhana, J. Bhaskar, S. Swapna and S.K. Bhattacharjee, 2002. Orchids. AICRP on Floriculture (ICAR), Division of Floriculture and Landscaping, IARI, New Delhi, 62 p.
	KAU, Vellanikkara	NPK 10:20:10 at 0.2 per cent weekly twice + BA 100 ppm enhancing number of spikes per year and number of florets per spike in *Dendrobium* cv. Sonia 17.	

Contd...

Crops	Centres	Recommended doses	References
Carnation	Kalyani (W.B.)	400:200:200 kg/ha of NPK for flower production in standard carnation	Gupta, Y.C. and S. K. Bhattacharjee, 2002. Carnation. AICRP on Floriculture (ICAR), Division of Floriculture and Landscaping, IARI, New Delhi, 35 p.
	Ludhiana, Hessaraghatta and Solan	30-40 g N, 20 g P_2O_5 and 10 g of K_2O/m^2. For Ludhiana centre, nitrogen in three equal splits, as basal dose, one and two months after planting proved significantly superior than single or two doses.	

Nutrient Deficiencies and Disorders in Flower Crops

Nutritional disorders are basically physiological disorders in the plants that affect the productivity as well as the quality of flowers. Disturbance in the plant metabolic activities resulting from an excess or deficit of environmental variables like temperature, light, aeration and nutritional imbalances result in disorders. **Deficiency** leads to morphological variation, which results in low productivity of quality flowers. The deficiency symptoms of nutrients can be corrected through foliar feeding even after planting. Nutritional disorders have become widespread with diminishing use of organic manures, imbalanced NPK fertilizer application and extension of horticulture to marginal lands. Therefore, proper application of nutrients/fertilizers is important to prevent the crop from these disorders. Various disorders of flower crops and their symptoms are given in table 3 and below in figures.

Table 3: Nutritional deficiency disorders in commercial flower crops

Crop	Nutritional deficiency	Symptoms	Management
Rose	Iron deficiency	❖ Slow growth and wilting of plants ❖ Interveinal chlorosis in young leaves ❖ Thin leaves and stunted growth ❖ Necrosis at leaf margins ❖ Dull flower colour and small size ❖ Aborted flowering shoot	❖ Soil should be slightly acidic (pH 6 to 6.5) ❖ Use of N fertilizers with higher NH_4/NO_3 ratio ❖ Avoid calcareous soil and bicarbonate rich irrigation water ❖ Spray 0.5 % $FeSO_4$ at pH 4- 5 with surfactant Fe- EDTA chelate at 0.1 %
	Boron deficiency	❖ Malformed flower buds and petals. ❖ Decrease in flower size and number ❖ Die back of stem tip and flowering shoots. ❖ Stunted root growth ❖ Scorched, thicken, cup-shaped and distorted leaves ❖ Shoots become stiff and excessively branched	❖ Foliar spray of 0.2 % boric acid twice at 30 and 45 days after pruning

Contd...

Crop	Nutritional deficiency	Symptoms	Management
	Magnesium deficiency	❖ Large necrotic white areas located symmetrically on both sides of midribs of leaflets between larger veins. ❖ Young leaves become mottled and chlorotic ❖ Immediate drop of injured leaves ❖ Severe root injury	❖ Adding magnesium sulphate at 15-25 g/plant immediately after pruning
Chrysanthemum	Iron deficiency	❖ Interveinal chlorosis in young leaves during early stages which in severe cases become yellowish ❖ Stunted plant growth ❖ The leaves become small and thin	❖ Use of well decomposed FYM ❖ Use of acidic red soil while preparing growth media ❖ Foliar spray of 0.5 % ferrous sulphate at pH 4.5 or Fe-EDTA at 0.1 % twice once at 30 days and another at 45 days after pruning ❖ Maintaining growing media pH at 5.5 to 6.0
	Copper deficiency	❖ Only a few flower buds open completely ❖ Petals curve upwards ❖ In severe cases flower bud initiation is retarded ❖ Terminal leaf base initially chlorotic and even veins are chlorotic in later stages	❖ Copper sulphate application at 10 kg/ha
	Boron deficiency	❖ Loss of turgidity in peripheral flowers before central flowers fully open ❖ The petals are twisted and cup-shaped in severe deficiency	❖ Soil application of boron at 2 kg/ha (15-20 Kg borax/ha) ❖ Foliar spray of 0.2 % boric acid twice once in 3 week after planting and again at 25 % flower bud emergence
Gladiolus	Boron deficiency	❖ Notched marigns of leaves ❖ Transulucent streaks along veins ❖ Break down of petals margins with rolling ❖ Colourless blotches in petals ❖ Roots become brown	❖ Floiar spray of boric acid (0.2 %) twice once in three weeks after planting and again at flower bud initiation stage ❖ Foliar spray of 0.5 % $FeSO_4$ at pH 4-5 a surfactant and 0.2% urea

Contd...

Crop	Nutritional deficiency	Symptoms	Management
	Manganese deficiency	❖ Interveinal chlorosis of young and physiologically mature leaves. The leaves do not turn yellowish in severe conditions as in iron deficiency	❖ Foliar spray of 0.5% $MnSO_4$ or 0.3 % $MnCl_2$ immediately after foliar symptoms appear ❖ Apply NPK fertilizers in band near the cormels to correct the disorder
	Magnesium deficiency	❖ The older or lower leaves are chlorotic with the base of the leaf remaining green while the tip is chlorotic	❖ Applying $MgSO_4$ to soil at 500kg/ha ❖ Foliar spray of 0.5 to 1.0 % $MgSO_4$
Gerbera	Iron deficiency	❖ Interveinal chlorosis in young leaves ❖ Small leaf size ❖ Short flower stalks ❖ The plant dies within a week or 10 days in severe deficiency	❖ Use of Fe-EDTA ❖ Foliar spray of 0.5 % $FeSO_4$ along with 0.5 % urea with suitable surfactant
Carnation	Boron deficiency	❖ Young emerging leaf becomes curved and twisted ❖ Internodes of the upper part of the shoot is curved ❖ The meristem dies as the deficiency becomes severe	❖ Spray 0.2 % boric acid thrice, one at 15 days after planting , second 30 days after planting and third when 25% of the plants have bloomed
	Iron deficiency	❖ Interveinal chlorosis in young emerging leaves ❖ Leaf blades become thin	❖ Correct the pH of the growth medium with gypsum at 0.5 -1.0 t/ha if the pH exceeds 7.5. ❖ Spray of ferrous sulphate at 0.5 % with surfactant at 20 ml/100 l
Tuberose	Calcium deficiency	❖ Spike cracking, but in acute cases buds rot.	❖ Apply 100-120 mg/l calcium sulphate
	Magnesium deficiency	❖ Interveinal chlorosis of older leaves	❖ Foliar spray of 0.5 to 1.0 % $MgSO_4$
	Boron deficiency	❖ Cracking of leaf margin, deformed leaves and stunted inflorescence	❖ Spray 0.2 % boric acid

Conclusion

Fertilizers play an important role in increasing growth and flower yield. Organic manures play a vital role as a component of integrated nutrient management in achieving higher crop yields and soil health for sustainability. Application of appropriate plant nutrients at proper time with suitable method is one of the strategies to reduce various losses of nutrients, which in turn, increases the nutrient use efficiency. Nutrients along with bio-fertilizers are very effective in maintaining soil health and for enhancing yield and quality in flower crops.

Mg deficiency in poinsettia and rose

Symptoms of N and K deficient rose plants

Fe deficiency symtoms in rose and petunia

Gerbera plants deficient in Ca

Molybdenum deficient poinsettia

Calyx splitting in carfnation due to B deficiency

Future Line of Work

There is a dearth of information regarding nutrient management in flower crops. The following strategies need to be prioritized to strengthen this aspect.

- ❖ Need-based and area-specific recommendations need to be developed.
- ❖ Appropriate nutrient management strategies, considering the interactive effects of different nutrients with available soil moisture and other soil attributes are necessary.
- ❖ Identifying critical soil health parameters for soil testing especially for fertilizer management rather than mere analysis of soils for the available nutrient content is also necessary.

- GIS based mapping and their networking to monitor nutrient variations on a dynamic scale for general areas are required to be worked out.
- Development of suitable models and satellite imagery is required to integrate into practical ground situation.

References

Anuradha, K. K., Pampapathy and N. Narayana, 1990. Effect of nitrogen and phosphorous on flowering, yield and quality of marigold. *Indian J. Hort.*, **47**(3): 353-357.

Balasubramanian, J. 1989. Studies on combined effect of *Azospirillum*, VA-mycorrhizae and inorganic fertilizers on growth and performance of French marigold (*Tagetes patula* L.). MSc. (Hort.) Thesis. Tamil Nadu Agricultural University, Coimbatore (India).

Baskaran, V., R.L. Misra, S.K. Singh and K. Abirami, 2014. Response of bio-fertilizers and commercial formulations on growth, yield and corm production of gladiolus. *Indian J. Hort.*, **71**(2): 237-241.

Besmer, Y. L. and R.T. Koide, 1999. Effect of mycorrhizal colonization and phosphorus on ethylene production of snapdragon flowers. *Mycorrhiza*, **9**:161-166.

Bhalla, R., P. Kanwar, S.R. Dhiman and R. Jain, 2006. Effect of biofertilizer and biostimulants on growth and flowering in glagiolus. *J. Ornam. Hort.*, **9**(4): 248-252

Bhattacharjee, S. K. 1994. Influence of secondary nutrients on *Rosa hybrida* cv. Raktagandha. In: Floriculture Technology: Trade and Trends, pp. 100-102 (eds Prakash, J. and K.R. Bhandary). Oxford and IBH publishing Co. Pvt. Ltd., New Delhi (India).

Bhattacharjee, S. K. 1985. The response of *Jasminum grandiflorum* to N, P_2O_5 and K_2O fertilization (2nd trial). *Singapore J. Perf. Ind.*, **13**(2): 102-111.

Bhattacharjee, S. K. and N.K. Banerjee, 1994. Influence of nitrification inhibitor (encapsulated calcium carbide) on *Rosa hybrida* cv. Raktagandha. In: Floriculture Technology: Trade and Trends, pp. 95-99 (eds Prakash, J. and K.R. Bhandary). Oxford and IBH publishing Co. Pvt. Ltd., New Delhi (India).

Bhattacharjee, S. K. and P. Rajan, 1995. Studies on the effect of macronutrient spray on *Rosa hybrida* cv. Superstar. *Indian Rose Annual*, **13**: 82-88.

Bose, T. K. and B.K. Jana, 1978. Effect of different levels of nitrogen, phosphorous and potassium on growth and flowering of gerbera. *Indian J. Hort.*, **31**: 182-185.

Bose, T. K, L.P. Yadav and P. Pal, 2003. Antirrhinum. In: Commercial Flowers, p. 660. Naya Udyog, Kolkata.

Butt, S.J. 2005. Effect of N, P, K on some flower quality and corm yield characteristics of gladiolus. *J. Tekirdag agric. Fac.*, **2**(3): 212-214.

Cerazine, P. C., E. Nahas and D.A. Banzatto, 1988. Soluble phosphate accumulation by *Aspergillus niger* from fluorapatite. *Applied Microbiol. & Biotech.*, **29**: 501-505.

Chanda, S., G. Barma and N. Roychowdhury, 2000. Influence of different levels of nitrogen, phosphorus and potassium on growth and flowering of gladiolus. *Hort. J.*, **13**(1): 64-65.

Chandrikapure, K. R., K.T. Sadawarte, D.M. Panchbhai and B.D. Shelke, 1999. Effect of bioinoculants and graded doses of nitrogen on growth and flower yield of marigold (*Tagetes erecta* L.). *The Orissa J. Hort.*, **27**(2): 31-34.

Cunnigham, J. E. and C. Kuiack, 1992. Production of citric and oxalic acids and solublization of calcium phosphate by *Penicillium bialii*. *Applied environ. Microbiol.*, **58**: 1451-1458.

Dahiya, S. S., N. Singh and S. Singh, 1998. Effect of different doses of nitrogen and phosphorus on growth and flower yield of marigold. *Environ. & Ecol.*, **16**(H): 855-857.

Dalve, P.D., S.V. Mane and R.R. Nimbalkar, 2009. Effect of biofertilizers on growth, flowering and yield of gladiolus. *Asian J. Hort.*, **4**(1): 227-229.

Damke, M. M. and S.K. Bhattacharjee, 1995. Influence of nitrogen, phosphorous and potash fertilization on growth, flowering and soil nutrient content of Super Star roses. *J. Ornam. Hort.*, **3** (1-2): 49-54.

Daufault, R. J., T.L. Phillips and J.W. Kelly, 1990. Nitrogen and phosphorous fertility and plant populations influence field production in gerbera. *HortSci.*, **25**(12): 1599-1602.

De, L. C. and K.R. Dhiman, 2001. Effect of leaf manures, potassium and GA_3 on growth, flowering and longevity of tuberose. *J. Ornam. Hort.* (New Series), **4**(1): 50-52.

Dongardive, S.B., V.J. Golliwar and S.A. Bhongle, 2007. Effect of organic manure and biofertilizers on growth and flowering in gladiolus cv. White Prosperity. *Plant Archives,* **7**(2): 657-658.

Dongardive, S.B., V.J. Golliwar and S.A. Bhongle, 2009. Influence of organic manure and biofertilizers on corm and cormel yield of gladiolus. *Ann.Pl. Physiol.*, **23**(1): 114-116.

Edward, R.E.M. 2007. Fertigation for flower crops. In: Hi-tech Floriculture (eds Reddy, B. S., T. Jankiram, B.S. Kulkarni and R.L. Misra), pp. 50-55. ISOH, New Delhi.

Fischer, P. and N. Forchthammer, 1987. Copper fertilization of gerbera. *Deutscher Gartenbau*, **41**(II): 674-678.

Gupta, R.B., J.R. Sharma and R.D. Panwar, 2010. Growth, flowering and corm production of gladiolus as affected by application of nitrogen, phosphorus and potassium. *Haryana J. hort. Sci.*, **39**(3-4): 286-287.

Gutmane, L. and D. Grivina, 1975. The effect of different kinds and concentrations of fertilizers on the growth and development of gerbera and the nutrient content of the plants. *Izvestiya* AN Latv SSR, **II**: 22-26.

Halepyati, A. S., K. Sujatha and M. Prabhakar, 2001. Effect of irrigation and nitrogen levels on the growth, yield and water use of marigold. *J. Ornam. Hort.* (New Series), **4**(1):30-32.

Hasan, M., Balraj Singh, M. C. Singh, A. K. Singh, S. V. Kaore, Tarunendu, Naved Sabir and B. S. Tomar, 2010. Fertigation Scheduling in Flowers under Protected Cultivation. Fertigation Scheduling for Horticultural Crops, p. 29. CPCT, IARI, New Delhi.

Hatibarua, P. and R.L. Misra, 1999. Effect of nitrogen sources on vegetative and floral characters of gladiolus cv. Dhanvantari. *J. Ornam. Hort.* (New Series), **2**(2): 111-114.

Hilal, A.A., I.S. Elewa, S.E. Hassan and S.A.A. El-Malak, 2001. The effect of fertilization and irrigation on *Fusarium* disease development and yield components of gladiolus. *Egyptian J. Phytopath.*, **29**(2): 97-105.

Irulappan, I., C.M. Pappaiah and S. Muthuswamy, 1980. Effect of farmyard manure and NPK on the flower yield of Edward rose (*Rosa bourboniana* Desp.). National Seminar on Production Technology for commercial crops, held at TNAU, Coimbatore, on Aug. 29-30, pp. 55-56.

Jeong, B. R. and C.W. Lee, 1992. Ammonium and nitrate nutrition of 11 bedding plant species. *Acta Hort.*, No. 319, pp. 505-510.

Jhon, A.Q., T.M. Paul and M.I. Tanbi, 1991. Effect of nitrogen and plant spacing on growth and flower production of *Zinnia elegans* Jacq. *Adv. Pl. Sci.*, **4**: 1-7.

Kumaraguruparan, P. N. 1974. Studies on Foliar Nutrition of Jasmines. MSc (Ag.). Thesis. T.N.A.U, Coimbatore (India).

Kumar, M. and T.K. Chattopadhyay, 2001. Effect of NPK on the yield and quality of gladiolus (*Gladiolus grandiflorus* L.) cv. Tropic Sea. *Environ. and Ecol.*, **19**(4): 868-871.

Macz, O., E.T. Paparozzi and W.W. Stroup, 2001. Effect of nitrogen and sulphur applications on pot chrysanthemum production and post harvest performance. *J. Pl. Nutrit.*, **24**(1) 111-129.

Maharana, T. and R.C. Pradhan, 1976. Effect of NPK and their combinations on growth, flowering and anthocyanin development of hybrid rose cv. Celebrations. *Punjab hort. J.*, **16**(1-2): 77-79.

Manonmani, R. 1992. Effect of Soil Inoculation of *Azospirillum* and Phosphobacteria and Graded Levels of Nitrogen and Phosphorous Biofertilizers on Growth and Yield of *Jasminum sambac* Ait. cv. Gundumalli. M.Sc. (Hort.) Thesis, Tamil Nadu Agricultural University, Coimbatore (India).

Mantrova, E. Z., T. Nikolaeva and V.V. Dvortsova, 1982. Effectiveness of applying fertilizers to gerberas. *Byulletn'Glavnogo Botanicheskogo Sada*, No. 126, pp. 51-57.

Mohanty, B. K., C.R. Sankar and T. Dayanand, 1999. Effect of NP.K and spacing on nutrient content of leaves and uptake in tuberose. *South Indian Hort.*, **47**(1-6): 327-330.

Moulinier, H. and M. Montarone, 1978. The nutrition of gerberas. *Pepinieristes Horticulteurs Maraichers*, **188**: 13-18.

Patel, K. S. and J.S. Arora, 1983. Effect of pinching, source and dose of N on growth and flower production of carnation (*Dianthus caryophyllus* L.) cv. Marguerite White. *Indian J. Hort.*, **41**(1-2): 137-138.

Preethi, T. L. 1990. Studies on the Effect of Nitrogen, *Azospirillum* and Ascorbic Acid on Growth and Flowering of Edward Rose. M.Sc. (Hort.). Thesis, Tamil Nadu Agricultural University, Coimbatore (India).

Randhawa, G. S. and A. Mukhopadhyay, 2001. Rose. In: Floriculture in India, p. 656. Allied Publishers Ltd., New Delhi.

Reyes, I., L. Bernier, R.R. Simard and H. Fema Natoun, 1999. *Microbiology Letter*, **28**: 281-290.

Sanderson, K. C. 1975. A crop to meet the energy crisis-Snapdragon. Part 3: fertilization, watering, temperature, light and photoperiod. *Florists Review*, **156**(17): 59-61.

Sharma, G. and P. Singh, 2007. Response of N, P and K on vegetative growth, flowering and corm production in gladiolus under mango orchard. *J. Ornam. Hort.*, **10**(1): 52-54.

Shim Myung Syun, Kim Sun Hwa and Kim Kisun, 2001. Effects of various composition of nutrient solution on the growth and absorption characteristics of *Dianthus caryophyllus* in closed system. *J. Korean Soc. hort. Sci.*, **42**(6): 631-639.

Singatkar, S. S., R.B. Sawant, S.A. Ranpise and K.N. Wavhal, 1995. Effects of different levels of N, P and K on growth and flower production of gaillardia. *J. Maharashtra agr. Univ.*, **20**(3): 392-394.

Singh, G., K.V. Prasad and T. Janakiram, 2010. Rose: A Production Manual. Westville Publishing House, New Delhi.

Skalska, E. 1980. Gerbera nutrition according to season on the basis of substrate analysis. *Zahradnictvi*, **7**(2): 145-150.

Sreenivas, K. N and J.V.N. Gowda, 1999. Effect of different organic manures on growth and flower yield of China aster [*Callistephus chinensis* (L.) Ness]. *Crop Res.*, **18**: 104-107.

Srivastava, R. and M. Govil, 2007. Influence of biofertilizers on growth and flowering in gladiolus cv. American Beauty. *Acta Hort.*, No. 742, pp. 183-188.

Strojny, Z., M. Saniewski and E. Jesiotr, 1976. Nitrogen and magnesium nutrition of *Gerbera jamesonii* growing on compost pink bark. *Prace Instiytutu Sadownictwa w Skiernewicach*, No. B. 2, pp. 87-95.

Thomas, M. B. and A. Leong, 1980. Combined Proceedings of the *Internnational Plant Propagators Society*, **30**: 156-164.

Tiwari, K. N. and Dwivedi, B. S. 1994. Available zinc status of soils and delineation of the area of zinc deficiency in Uttar Pradesh. *Fertilizer News*, **39**(3): 31-39.

Uday Sharma, S.V.S. Chaudhary and Rajesh Thakur, 2008. Response of gladiolus to integrated plant nutrient management. *Haryana J. hort. Sci.*, **37**(3-4): 285-286.

Velu, G. 1988. Iron requirement of jasmine. *Research and Dev. Reporter*, **5**(1-2): 103-104.

Verma, V. K., O.P. Sehgal and S.R. Dhiman, 2000. Effect of nitrogen and GA_3 on carnation. *J. Ornam Hort.* (New Series), **3**(1): 64.

Volpi, L. and E. Farina, 1987. The nutrition of *Gerbera jamesonii* hybrid interaction relationship of the major plant nutrients. *Colture Protelte*, **16**(3): 100-102.

Voogt, S. J. 1976. Chlorosis of gerbera mainly caused by Mn deficiency. *Vakblad voor de Bloemisterij*, **31**(36): 20-21.

Yadav, L. P., T.K. Bose and T.P. Mukhopadhyay, 1985. Influence of nitrogen and phosphorous on growth and flowering of roses. *Indian Rose Annual*, **4**: 172-178.

36

Flower Seed Production

Manish Kapoor, Ajit Kumar, Ranjan Srivastava, Sanyat Misra, Sunder Pal, Jyoti Bajeli and Mamta Bohra

Introduction

Floriculture is a small, but a significant component of the global economy, estimated to be more than 100 billion US dollars worldwide. The retail floriculture industry has been steadily increasing in size and importance in comparison with overall agricultural income. Great diversity exists in the use of flowers both from a geographical and historic perspective (Kaul and Dadlani, 1995). The incorporation of bedding plants into our gardens can be traced back to the early 19th century when colourful annuals and tender ornamentals were imported to Europe from America, South Africa and Asia. From the onset, domestication of exotic plant species has been a mainstay in the development of garden designs. Seeds are perfect natural packages that facilitate the migration of plant species across the land. Early developments of the flower and seed industries generally involved importing exotic plant materials and improving them through selection in accordance with local climates and prevalent tastes. Starting in the 18th century, mass migration of people throughout the world contributed to the broader distribution of novel ornamental varieties. New trends in desirable flower types and their uses continue to develop, and the flower seed trade evolves to accommodate these trends. The vast diversity of flower crops is an important and unique feature of the flower seed industry and this presents greater technical challenges to it than to other agricultural industries.

The herbaceous ornamental annuals and biennials have varied uses in the garden. Seed often propagates them and a gardener's success depends upon the quality of the seeds sown. Some 1,500 different varieties of flowers are available to amateur gardeners. The major part of the acreage on which they are grown is devoted to producing about 750 of them and the seed values billions of dollars. The acreage of flower seeds thus comprises small plantings of many flower crops that require the best level irrigated lands. The per acre income is two to five times as compared to cereals, but the production costs are correspondingly higher. More than 50 per cent costs are for field labour, which must be of such manual skill that good work approaches a craft status.

Commercial seed production in some selected annuals

Seed production in *Catharanthus roseus*

Seed production in *Tropaeolum majus*

Seed production in *Mesembryanthemum*

Seed production in snapdragon

Seed production in *Calendula*

Seed production in ornamental cabbage

Seed production of Californian poppy (single)

Seed production of Californian poppy (double)

The ornamental industry has changed substantially in the last 20 years. The science of flower seed technology has also advanced rapidly to keep up with new market demands. There is much information scattered in various scientific and trade publications, and more is being kept in the proprietary domain of seed companies. Anyone searching for information on flower seeds find it in short supply and seldom collated into one comprehensive reference. Globalization of seed industry started during and after First World War. The curtailment of European supplies stimulated American growers to expand the production of flower seeds. During World War II, European and Japanese supplies were disturbed and America requested its allies to supply flower seeds. The global effect of World War II resulted in increase in the value of flower seeds. Developed countries like USA, Holland, UK, France and Germany started flower seed production on contract basis in the developing countries of Asia, Africa as the manpower there is quite cheap and existence of a wide range of climatic conditions unlike extreme cold temperature and high labour cost in the developed countries. Hence, various international flower seed companies such as Burpee (USA), Groot (Holland), Royal Sluis and Sluis (Holland), Suttons (England), Tezierand Clause (France), Villmorin (France), *etc.* started seed production on contractual basis in various parts of India, especially Punjab, Haryana, Andhra Pradesh, Maharashtra and Karnataka during last two decades.

Every species of flower grown for seed has its own planting time, culture, problems and harvesting techniques, but there is one basic requirement for good seed production - a mild climate with little rain during the growing and harvesting seasons. Less favourable conditions result in uncertain and usually lower yields and germination percentages, whereas cooler climates are more suitable for higher flower seed production. For the production of row crop flower seeds, the farmer should have the ability, prime acreage, a team of skilled labour and a suitable climate. Greenhouse production is independent of location, having its own tailor-made climate, but it is an entirely different business. The structure of the industry is that the farmer sells to wholesalers who then export the seed to wholesalers abroad who then sell it to retailers there. The wholesalers themselves, however, are major producers of what they sell. They operate their own farms and also contract for bulk production. Many of the flower species represent highly hybridised strains, on which foundation seed stock must be planted each year, not crop seed planted earlier. The wholesaler almost invariably produces this stock on his/her own farm which incorporates research, breeding and trial grounds to keep pace with the rapid changing technology of the industry. Elite strains require planning and supervision to ensure their pedigree and uniform cultural practices. Buyers have specific standards of varietal purity and germination in mind when they buy on an annual contract. The wholesaler has to deliver seeds of such standards which are grown under his/her supervision. A few places in the world have the Mediterranean climate needed to grow a variety of flower crops. Rain is beneficial before flowering though later encourages fungi in the seed head and reduces the yield. Rain on the mature seeds lower germination percentage (Jana, 1998). Nevertheless, most sections of semi-arid climate, even in latitudes where frost is not a problem, may not be suitable for seed farms because high summer temperatures also reduce yield of seed of all but a few annuals. All crops require extensive irrigation during the growth period. A common practice is to withhold the water at the end of the season to encourage rapid and uniform maturation of the seed crop. Flower seed farming has its specific problems and those who can master them tend to be efficient seed farmers year after year. The plants usually grow so slow that weed control becomes a major problem. Chemical weedicides can be used but affect plant growth and seed yield. Thus, most fields need complete hand-weeding every month during the crop season, sometimes combined with thinning in drilled crop. Petunias and many others are usually transplanted manually in the field, because mechanical transplanters can not handle the small, fragile seedlings (Kapoor, 2004).

Mechanical mixing of harvested plants is prevented by alternating plantings of species to be harvested at different times of the year. Special field-labelling techniques are used to identify lots going through the cutting and threshing process to avoid seed mixing. Mixed lots are nearly worthless in an industry where high prices are paid only for pedigreed seed. Many small plots of closely related varieties on the same farm mean that one crop may be unintentionally cross- pollinated by another of the same species rendering the resulting seed virtually useless. Pollination is, however, what makes the wheels running in the seed business. In most species, self and cross-pollinations are accompanied by insects which cannot distinguish between several colours so that such plantings must be isolated from one another. In such cases, strict isolation by as much as 400 metres is necessary. The varietal purity is not solely the concern of the flower-men in the seed industry, but only they set out deliberately to produce mixed varieties. In flower seeds, a mixture of colours of one kind is a saleable or even highly desirable product. Many times this mixture is blended with named colours according to the formula. Seeds of many of the flowers are harvested and threshed by machines used for food grains. Even the fine seed of petunia is handled so. Hand harvesting saves more seed, but to avoid high labour costs mechanical methods are adopted, whenever possible (Kumar and Kaur, 1996). Large scale plantings are harvested occasionally with a combine, but commonly the plants are cut and winnowed until they can dry to moisture content suitable for machine threshing. The diversity of crops leads to several cutting methods, e.g. nasturtium may be winnowed on the ground while the species with seed pods that shatter at a touch must go into large canvas sheets. The stem and leaves of petunia are so sticky that even though the seed shatters out of the capsule, a significant amount of it adheres to the stem and leaves and can be saved if the whole plant is put on sheets to dry. The cutting operation is carried out early in the morning when the morning dew is on the plants to avoid loss of the dry seeds. The drying period varies from 10 to 15 days. Rains at this time are catastrophic. Regardless of the trend towards mechanization, a major part of the harvesting is still manual because the plants have a long flowering period and seed production is continuous. Re-cleaning is basically the same operation as in the rest of the seed industry. Fanning mills, specific gravity separators, disc and roller separators, etc. are used. The difference with flower seed lies in the multiplicity of small lots which necessitate short machine runs and thorough cleaning of the machinery between each run. The high wholesale value necessitates extreme care and elaborate record keeping. Screens, sieves, discs, etc should be chosen judiciously to handle all shapes and sizes of various flower seeds. Relative humidity has an important bearing on the curing of flower seed in the field, it is always better if it is at the lowest. A large part of production is processed to the optimum moisture content for long-term storage and sealed in moisture/vapour-proof containers so that retailers in wet climate can have the vigour, vitality and keeping quality of the seeds as originally harvested.

In India, floriculture has undergone notable transformation in the last three decades as a viable commercial activity. Indian export flori-basket constitutes cut flowers and cut foliage; dried flowers & plant parts; nursery business for multiplication (including micropropagation) and multiplication and supplying of seeds and live ornamental plants including flowering plants, cuttings, buds, bulbs, corms, rhizomes, tubers, etc.; value-added products like garland, bouquet, floral baskets and flower arrangement; plant rental services for supplying ornamental plants for specific periods; extraction of essential oil for cosmetic perfumery and other pharmaceutical uses; etc. Seed production in flowering annuals has a definite place in crop diversification programme owing to better returns. The seasonal flower seed production has climatic advantage in India than the European countries. The most of the European countries are shutting down their seed production programmes due to high cost involved because of low temperature (heating required) and costly labour, as the climatic conditions during winters are unfavourable for seed production, thus resulting in higher cost.

Indian floriculture has transformed from a state of bullock cart transportation to flights through chartered planes in the last 2-3 decades. The green revolution that transformed Indian agriculture from sustenance to a stabilized sustainable reality since the early 60's, has accelerated a high production boom in other related areas too. This green revolution ushers into a glorious 'golden revolution' in horticulture production and floriculture assumes the position of vibrant commercial venture, and flower seed production is an integral part of it. In the present area of globalization and market liberalization compounded by complex problem of deteriorating land and water resources, increasing biotic and abiotic stresses, indication of factor productivity decline, threatened loss of bio-diversity, natural resource bio-degradation, intellectual property rights besides intensifying competition for quality and cost in international floriculture trade, a new paradigm shift is needed to necessitate major push in flower seed production which can help Indian floriculture trade to modulate itself into sustainable system and be globally competitive.

Commercial flower seed production has shown revolutionary growth in terms of number of flowers, numbers of crops, their varieties as well as production. It has been possible as a result of technological advancement in production of flower seed, in terms of infrastructural developments in the area of production, packaging, handling, transportation and marketing of flowers and ornamental plants at international level, *vis-a-vis* human resource development. Many countries are involved in ornamental seed production which is considered as high value and low volume in contrast to high volume and low value seeds of other crops.

Indian Perspective

The annual flowers were introduced in India by Britishers in the end of 18th century. By 1912, the popularity of these flowers increased so much that Suttons and Sons, Royal seed establishment of England opened a branch at Calcutta to meet the increasing demand of seeds of annual flowers. Even before this period, Pestonjee Pocha at Pune had started selling seasonal flower seeds. This company was closely linked with W. Atlee Burpee Seed Company and dominated flower seed production in the country. In the beginning, all seeds were imported but due to interruption in the supplies from overseas, Suttons established farms and started contract farming in Kashmir Valley, Ludhiana, Ranchi, Bangalore, Ootacamund and Kalimpong. In 1965, Dr. Manmohan Attavar established Indo- American Hybrids Seeds Company at Bangalore. This company made a landmark in case of hybrid seed production of petunia and exported it to USA in 1967. Since then, it is exporting even seeds of some other flowers such as antirrhinum, flowering tobacco, *etc.* to USA. The most significant change introduced by the Ministry of Agriculture, Govt. of India was introduction of new policy on seed development in 1988. As per this, Indian farmers can have access to the best planting material in the world. In India, it is estimated that 600-800 ha area is under seed production and out of this, Punjab alone contributes around 40-50% area and earns foreign exchange of around 7.0 crores annually.

Challenges in Flower Seed Industry

Flower seed industry has a great scope of expansion in developing countries of Asia and Africa because of low labour cost and wide range of climatic conditions. In India, beside favourable climate, availability of cheap manpower and focus on crop diversification makes it a strong case for the growth of this industry. Floriculture is a highly fashion-driven industry and new trends in desirable flower type, colour and the flower seed trade evolve to accommodate these trends. Unlike other crops, there is a great diversity in floriculture and hence flower seed industry possesses a real challenge than other agriculture industry as elaborated further.

(i) Continuous quest for new products. Here flower lovers desire novelty in flower crops for different colours and forms. In an attempt to meet the expectation of customers, flower breeders will have to keep abreast with the latest changes in liking of the consumer. Being a quantitative trait, ornamental value is largely subjective and depends greatly on the cultural change in the society.

(ii) Product range. There exists a broad botanical base for flower crops unlike few major field crops, vegetable crops, *etc.* In florist trade, popular flowers such as annual, biennials and perennials are used as bedding plants, pot plants, cut flowers and for interior and exterior decoration. Apart from their aesthetic value, some of the ornamental plants yield products which are eaten as vegetable or otherwise. Various examples like underground stem of lotus (*Nelumbo speciosum*) is used as vegetable and for making pickles in North India. Seeds of aquatic plant *Euryale ferox* when parched or roasted over hot sand gives product called *makhana*, used during fast. Likewise, nasturtium is used as vegetables. Therefore, keeping the broad product range in mind, it becomes imperative to have a very broad knowledge base for the people involved in this highly lucrative industry.

(iii) High seed germination requirement. This is one of the most important challenges of flower seed industry as each seed must germinate in plug trays which are used in the greenhouse production by the 'Plug' growers. In plug plant industry, standard size trays having different number of individual cells called plugs are planted with one seed and sold to end users in retail. These end users then transplant the young plants from plug trays into large containers either manually or by automatic transplanting machine. In order to have high production efficiency, seeds should have a high germination percentage otherwise important resources like greenhouse space, cost of plug trays, nutrients, labours and energy will be wasted. Hence, it becomes imperative for the seed companies to produce seeds of excellent quality with maximum germination percentage for fulfilling the expectation of the growers.

(iv) Serving highly sophisticated customers. Flowers are used for various purposes throughout the year because of the increasing habit of people 'to say it with flowers'. For special occasions when the demand of a particular type of flower is high, seed companies must ensure that they produce the seeds of desired varieties at the required time besides providing guidelines for package of practices on how to induce flowering at the desired period. Use of greenhouses for meeting the specific requirements during specific days can be summarized as programmable production.

(v) Advances in seed technology. In flower seed industry, growers are ready to pay higher price because of high expectations from flower seed performance as compared to field crops or vegetable crops. The additional income from the sale of flower seeds at higher prices also facilitates the flower seed industry to practice seed enhancement technologies which are not found in other crops. For example, seeds of many flower crops are primed for their improved performance or even they are pelleted for greater ease of handling. However, these techniques and equipments are crop-specific and research is needed to standardize them for exploring seed production technologies in various crops.

Major Global Players in Flower Seed Industry

Seed is the most important input which also influences the success of other inputs in crop production. The origin of flower seed industry can be traced back to 1784 when a retail store was opened by David Landreth Company in Philadelphia, USA. This company sold seeds, bulbs and other garden items to flower lovers. Beside retail stores, other three branches of flower seed industry also existed in USA at that time. The first branch was mail-order seed catalogue group which sold flower seeds to consumers. The second branch was packet seed industry, which sold flower seeds procured from growers after making packets through retail stores. The third branch consisted of flower seed companies selling seeds to professional plant or pot plant growers. One of the industrious/religious group, the Shaking Quakers, aka Shakers from Manchester, England started selling herb seeds in 1779. This group shifted to New York in 1784 and started seed production in a networking mode involving 17 independent villages. American seed industry began with Shakers which sold seeds of various herbs in small properly labelled hand-made envelops. The envelops were packed in wooden boxes and transported across towns in horse-drawn wagons and seed boxes were dropped at general stores. Shakers group had an agreement with the store that the stores had not to pay for seed packets until they were sold. The Shakers would collect the money for sold packets and take back unsold packets. The same trend is even followed today in packet seed industry. Basic considerations for flower seed production are assessment of global seed demand and the approximate pricing structure, identification of best areas for quality seed production, determination of export potential based on the existing global demand, and development of basic infrastructure for the production of flower seeds. Major global seed companies are briefly described here under.

George W. Park Seed Company. This company was founded by Mr. George Watt Park (1853-1935). This company is popularly known as Park Seed Company. This seed company was established in 1868 in Libonia, Pennsylvania. It is the oldest and largest family-owned mail-order gardening company in America. George W. Park published his first catalogue in 1868 which consisted of only eight pages and two illustrations. Park also started its magazine, *viz.* Park's Floral Magazine in 1877. This seed company is famous for selling unusual flower species to home gardeners and bedding plant growers. In 1984, millions of tomato seeds produced by Park Seed Company were part of NASA's first commercial payload aboard a space shuttle. The cargo remained in orbit till 1989 and after stay of 5 years in orbit, the seeds were distributed to more than 30 lakhs students for science experiment. This seed company has transformed itself into a strong merchandizing company.

Washington Atlee Burpee and Company. Founder of this company was Washington Atlee Burpee (5th April, 1858-26th November, 1915). It is more commonly known as Burpee Seeds. At the age of 18, W. Atlee Burpee started a mail-order business in Philadelphia, Pennsylvania in 1876. In 1896, free delivery system for delivery of seeds through post offices was extended to rural areas also. The first catalogue published by the company contained 48 pages. The company first sold garden seeds, farm supplies, tools, etc. After the death of its founder in 1915, David Burpee, son of W. Atlee Burpee had run the company. David was interested in war or victory gardens and he prioritized in flowers and initiated several flower breeding programmes. Burpee was the first company in the world to deliver seeds to customers by air in 1911. David Burpee was the first commercial horticulturist to recognize the potential of hybrids. David Burpee and his team produced the first successful hybrid flower in 1934, the double hybrid nasturtium. In 1939, they introduced the first hybrid marigold named as 'Burpee Red' and 'Gold Hybrid' marigold. Burpee and his team also started breeding vegetable crops for home gardeners. During 50's and 60's, Burpee diversified and became pioneer in bedding plant industry especially zinnia, petunia and marigold. W. Atlee Burpee & Co. became the world's largest donor of vegetable seeds, working in close cooperation with the United Nations agency, CARE. In 1991, Burpee Seed Company was acquired by George Ball, Inc., a diversified horticultural family business.

Vaughan's Seed Company. It was established by Vaughan John Charles in Chicago in 1876. In the late 1950s, the company bought the Merion Blue Grass Seed Market and became one of the prominent companies in grass seed market. In 1972, the company also bought the Jacklin Seed Co., Inc. of Dishman, Washington and renamed as Vaughan's - Jacklin Corp. This corporation absorbed many other companies in 1970s. In July 1989, Sandoz Corp. purchased this company and then Novartis Seeds which is world's second largest seed company purchased the Sandoz Corp. This company deals with horticultural plant products, plastic products and grass seed production and marketing. Vaughan Seed Company supplied vegetable seeds and diversified into mail-order catalogue sales and packet seed sales.

Joseph Harris Company. This company was established by Joseph Harris at Rochester, New York. This company sales vegetable and flower seeds to growers all over the world. In 1879, this company published seed catalogue consisting of 44 pages. To promote greater interest in gardening and to encourage more young people to start business, Harris wrote a book "Gardening for Young and Old" in 1883. Mr. Selah Harris, son of the founder started commanding the company from 1892 after death of his father and continued his father's emphasis on careful selection. In 1949, Joseph Harris, son of Mr. Selah Harris became its proprietor and started developing popular new flower and vegetable hybrids and intensified emphasis on disease resistance. This company sells flower and vegetable seeds to bedding plant growers, consumers and market gardeners.

Bodger Seeds Ltd. Company. This seed company was established by Mr. John Bodger in 1990 at California. This seed company played a significant role as a flower seed breeding and production company. It was the first company to introduce first wilt resistance *Aster* in collaboration with University of Wisconsin. Bodger seed introduced first dahlia-flowered *Zinnia.* The company is still managed by Bodger family and has been involved in breeding of flower crops and production of improved flower seeds to home gardener and bedding plant market till 2008. The company was taken over by S and S Seeds Inc., USA in 2009.

George Jacob Ball, Inc. It was established by George Jacob Ball (1874-1949) in Glen Elyn, Illinois, USA. For 13 years he worked with a leading seed company, a rose grower, and a commercial cut flower grower. The founder was able to support his company financially after success of his *Calendula* crop grown by California growers in 1915. This company mainly works with growers and one page seed list was printed and mailed to growers in 1918. George Ball published first issue of Grower Talks in 1937. After death of the founder, his four sons, *viz.* K. George, Victor, Robert and Carl took over Ball Seed Co. This company now possesses 10 other horticultural companies including W. Atlee Burpee & Co., Pan-American Seed Company, Petoseed Co. and a number of other companies and the parent company is called Ball Horticultural Company, and work on to develop improved strains of aster, sweet peas and calendula. This company is breeding, producing, distributing and marketing seeds as well as various other horticultural products throughout the world. In 1995, Anna Caroline Ball became owner and CEO of Ball Horticultural Company.

Henry F. Michell. Henry F. Michell was a Swiss immigrant who started a retail seed business in 1890 at Philadelphia, Pennsylvania. In 1892, his brother Frederick joined him and expanded the business into wholesale market. The company offers bedding plants, greenhouse supplies, structures and equipments. In the 1950s, Frederick's grandsons, Henry F. Michell, III and Frank Michell, Jr., relocated the headquarters to King of Prussia, Pennsylvania.

Sluis and Groot (S and G) Seeds. This company traces its origin to Nanne Jansz Groot who started selling cabbage seeds in the Netherlands in 1818. A cooperative was formed when his grandsons Nanne Sluis, Nanne Groot and Simon Groot joined this company to sell vegetable seeds. This marked the establishment of flower seeds breeding company named as Sluis and Groot (S&G) in 1867. This company started business overseas and had flower seed sale destination in USA and Russia. S&G Company shifted its offices and warehouses to current location at Enkhuizen during 19th century, focusing mainly on uniformity, hybridization and seed quality of flower crops. This company also introduced seed propagated geranium. During 1970s-1990s, new varieties of balsams, begonia, pansy, petunia and verbena flooded the bedding plant market in North America. Sluis and Groot (S&G) Seeds is part of Syngenta, a global leader in agribusiness committed to sustainable agriculture through innovative research and technology.

Sakata Seed Corporation. This seed corporation was established by Takeo Sakata in 1913 with its headquarter in Yokohama, Japan. This company started breeding of flower crops in late 1920s and 1930s and developed many new varieties of carnation, celosia, pansy, petunia, polyanthus, primula and zinnia. In these crops, new varieties were introduced as 'All Double Victorious Mix' in petunia, 'Majestic Giants Mix' in pansy, 'Profusion Cherry' and 'Orange and White' in zinnia. Due to innovative breeding and global recognition, Sakata Seed Corporation is one of the top four flower breeding companies in the world. This Corporation has four business segments, *viz.* **the domestic wholesaling segment** which produces and purchases seeds, bulbs and seedling of vegetable and flower crops and sells products to domestic seed and plant manufacturers and suppliers; **the overseas wholesaling segment** which produces and purchases seeds, bulbs and seedlings of vegetable and flower crops and sells products to seed and plant manufacturers and suppliers located overseas; **the retailing segment** is involved in the purchase of gardening products and sells the products through direct stores, home centres and mail-orders; and **the other's segment** is involved in landscape creation, insurance agency and recruitment services for government and private sectors. This seed corporation has 38 subsidiaries and two associated companies.

Takii Seeds. This seed company was started by Mr. Jisaburo Ohmoriya in 1835. In the beginning, the founder of the company grew vegetables and distributed seeds to farmers. In 1935, Takii Seeds started breeding cabbage and cauliflower varieties. However, in 1978, flower breeding as component was also introduced in celosia, dianthus, pansy, snapdragon, stock, ornamental cabbage, ornamental kale, zinnias, etc. In *Dianthus* 'Snow Fire' and in *Zinnia* 'Red Sun' bred by Takii Seeds got award also.

Colegrave Seeds. This global family operated horticultural seed company was established by Mr. David Colegrave and his wife Nancy of UK in 1962. Initially, founder of this company introduced hybrid seeds of flower crops to bedding plant growers in the country. This company expanded its operation in Australia and France and purchased Waller Flower seeds in Guadalupe, California. It continued its expansion with joint venture and acquisitions in China, France, Germany, Italy and Portugal. In 2001, this seed company was purchased by Ball Horticultural Company owned by Anna Ball and hence now this company is also known as Ball Colegrave. This company offers a comprehensive assortment of annuals, biennials and perennials from seed; seed raised plants and cuttings to the customers and whole ornamental bedding plant industry.

Ernst Benary Seed Growers. It was founded by Ernst Benary in 1843 in Enfurt (former eastern Germany) which was the centre of European ornamental industry. After World War II, Fredrick Ernst Benary re-founded this company in 1946. He started from scratch, travelled all over Europe with his bike to collect seeds for his business. Today Benary is one of the leading breeding companies in the field of bedding ornamentals. This seed company has long been associated with annual rex begonias, tuber and perennial begonias. Benary operated three breeding facilities in Germany, the Netherland and USA and involved in breeding of pansies, rudbackia, sunflower, verbena, zinnias and certain other cut flower crops. Benary stands for premium quality flower seeds and exports them to more than 100 countries worldwide.

Indo-American Hybrid Seeds (IAHS). It is pioneer and innovative Indian seed company established by Dr. Manmohan Attavar in 1965 at Bangalore. This seed company has revolutionized the floriculture industry in India. It made a landmark in case of hybrid seed production of petunia and exported it to US in 1967. This company has 9 regional centres in India with nationwide storage network of dealers and distributors. This company is focussing on research and development (R&D), production and marketing of hybrid seeds in flowers, vegetables, agronomic crops like cotton, rice, sunflower and ornamental plants, biotechnology products and construction of greenhouses. Besides seeds of hybrid petunias, they are also exporting F_1 hybrid seeds of antirrhinum, balsam, geranium, nicotiana and ornamental chillies.

Pan-American Seed Company. This seed company was founded by Mr. Weddle and W. D. Holly in 1946 in Chicago, USA. This company is one of the world's leading breeding and seed production companies on F_1 hybrids and open pollinated flower crops, and markets its products through wholesale and retail distributors around the world to professional greenhouse growers. Much of the world's seed supply for many important bedding plants including grasses, cut flowers and potted plants is made by Pan-American Seeds. This seed company was purchased by George Jacob Ball, Inc. in 1962.

Suttons Seeds. This company was founded by John Sutton (1777-1863) in 1806 and named it as 'House of Sutton' in London, England. In 1832, he was joined by his two sons, Martin Hope and Alfred and the House of Sutton became Suttons and Sons. This seed company is internationally renowned for supply of flower and vegetable seeds, young plants, bulbs, fruit bushes and other horticultural products to amateur gardeners. In 1965, Suttons was the first company to offer seeds in foil packet to amateur gardeners of UK. This company opened its first office outside England in Calcutta, India in 1909. Today Suttons is a part of the international company Vilmorin, the largest packet seed distributors in the world.

Thomson and Morgan Seeds. This company was started in 1855 in a small garden behind a baker's shop in Tavern Street, Ipswich, tended by William Thompson, the baker's son and acquired the name of the 'Baker Botanist'. From the small garden, he moved to a nursery at the edge of Ipswich and then to an even larger one. He began to publish a magazine called 'The English Flower Garden'. In 1855, his first catalogue was published. As one of the most distinguished plantsmen of his day, his skill in raising plants of the Western United States was recognised when in 1876 a volume of Curtis's Botanical Magazine was dedicated to him. Later the Royal Horticultural Society bestowed him with its highest honour, the Victorian Medal of Honour. William Thompson died in 1903, left John Morgan the sole owner of the business. Ten years later in 1913, Joseph Sangster became its partner. He took over the company completely when in 1921 John Morgan died and two years after that Joseph was joined in the company by his son Murray who received horticultural training in Edinburgh as well as having work experience at Slococks, the then well-known nursery of Woking, Surrey. In 1939, Thompson and Morgan became a limited company and when Joseph died in 1952, Murray Sangster took control of the seed company. In 1974, Murray retired and Keith and Bruce Sangster took over the management of what was to become one of the most progressive horticultural companies in the world. In 1994, Keith and Bruce became sole owners of the company. In 1990s Paul Hansord, the current Managing Director, developed an innovative range of young plants which were to be delivered throughout the UK.

Mr Fothergill's Seeds, UK. Mr. Fothergill's Seeds was established in 1978 and began selling seeds to garden shops and centres from the stable block of Lillie

Langtry's former Newmarket home by Jeff Fothergill and Brian Carey. Their sons John Fothergill and David Carey continue to run the business even today. Mr Fothergill's has become one of the most successful seed companies in UK and carries a range that has a wider seed choice than any other company there. The company's growth has come from a commitment to providing the gardener with the best products and the retailer with the best service and Mr. Fothergill's is now one of Europe's largest seed houses. Mr. Fothergill's Ltd. works in partnership with the Royal Horticultural Society, Plant Heritage and East Malling Trust. The company has gained a reputation for innovation, quality and efficiency. Mr. Fothergill's had pioneered many ideas and were the first to display contents on packets; to introduce pictorial wildflower packets; to launch seed collection pack ranges ~ to actively encourage promotion through special packs, give-aways and price offers; and to introduce ranges specifically for children. Mr. Fothergill's still operates from the business started in Suffolk just outside of Newmarket and has factory site at Kentford, incorporates new offices, storage and packing facilities as well a working trial ground to test and evaluate new and existing product lines. Alongside the retail division, it has thriving Export, Mail Order and Promotional divisions also.

Peterpikzonen BV. At the turn of the century, in 1904, Pieter Pik established a company in the West-Frisian town of Andijk in North Holland, specialised in the production and wholesale trade of vegetable seeds. Pieterpikzonen BV remained a family business currently owned and managed by the grandsons Peter and Jac, with its business as a grower, producer and distributor of flower, herb and vegetable seeds. In addition to the bulk supply of seeds, it supplies pictorial seed packets for garden centres, DIY-stores, mail order companies and many other outlets and offers more than 800 different varieties which are distributed through its subsidiary company Tuinplus BV, Tuinboetiek, Buzzy and Hortitops. It has a modern production plant in Heerenveen.

FSH Flecke Saaten Handel Gmbh. It is a privately owned German corporation specializing in the production and wholesale of open-pollinated flower seeds. The company founded in 1982, is located in the town of Wunstorf in northern Germany, close to the city of Hannover. Flower seeds of all types; from aster to zinnia; annuals, biennials and perennials are produced in a wide variety of growing conditions and countries. A 3,000 square meter trial ground is located at the Wunstorf facility, where samples of each seed lot produced by the company the previous season are grown to ensure that flowers are true to type, mixtures are balanced with the proper proportions of colours, and that labelling is correct. Harvested seeds are processed and shipped to Wunstorf for storage and order preparation. Seeds are tested for germination at in-house laboratory of the company. Additional independent laboratories also perform germination and purity tests in order to comply with international trade standards. Consulting for new product selections, special packet-filling and preparation of custom blends are among the additional services FSH can provide to its customers. With expertise in seeds of both cultivated and wild flowers, FSH is a leader in that expanding market.

Seed Den, USA. Dennis Kromer formed The Seed Den, Inc. just recently in 2006. He began his career in the seed industry working for Ferry Morse, moved to Oregon to help in the formation of Daehnfeldt Inc. and most recently started Wild West Seed, before venturing out on his own with The Seed Den, Inc. The Seed Den, Inc. is a wholesale seed supplier that markets to seed distributors, landscape contractors, state highway departments, municipalities, and golf courses.

Hem Genetics. Hem Genetics, a sister company of Hem Zaden BV has been founded in April 1998. With a view to a continued developing consumer trend, away from growing plants from seed to buying plants through supermarkets and garden centres produced by professional growers, Hem Zaden decided to establish Hem Genetics in April 1998. Hem Genetics objective is breeding, development and marketing of both seed- and vegetative-raised ornamental pot- and bedding-plant varieties, aimed at the professional grower market. A number of novel varieties has been recognized, and awarded by the industry associations Fleuroselect, All America Selections and consequently by professional growers worldwide. Hem Genetics breeding put a lot of efforts in its breeding programme developing varieties which require less or no plant growth regulators. These natural dwarf varieties are marked with the "Low Grow Approved" logo. It has succeeded in developing two natural dwarf series of petunia named 'Limbo' and 'Mambo,' one named 'Geranium Nano' and one named 'Salvia Reddy.' Hem Genetics is the only company in the world offering natural dwarf varieties.

Wild West Seeds, USA. It was established in 1998, incorporated in Oregon and was categorized under Field, Garden, and Flower Seeds. The formation of this company brought together three of the industry's well-respected and very experienced seedsmen, *viz.* Ken Hilbert, Ron Hackett and John Wahlert who grew up in a family seed business in Idaho and the owner, Ken Hilbert has been involved in the seed business for over 38 years in several different endeavours.

Goldsmith Seeds. In 1962, Glen Goldsmith founded Goldsmith Seeds in Gilroy, and held the first 'Pack Trial' inviting seed brokers to view seed product in production. Worldwide, Spring 'Pack Trials' are common place in the industry today. In 1966, Goldsmith expanded with a seed production facility in Guatemala. By 1970, the company moved to the current location on Hecker Pass Highway. Goldsmith Seeds received more All American Selection Awards and Fleuroselect, the International Organization for the Ornamental Plants Industry in 2008. In 2008, Goldsmith Seeds was purchased by Syngenta, a global leader in agribusiness committed to sustainable agriculture through innovative research and technology.

Floranova. Established in the UK in 1978, Floranova is now recognised as being one of the world's leading flower seed breeding companies. The company's portfolio also comprises vegetatively propagated products that are licensed to a network of propagators. Floranova is focussed on bringing new, novel, high quality and beautiful products to the global market-place: products that excel for horticultural professionals and gardeners alike. In 2009, a separate corporate 'Vegetalis' entity was formed to specialise in the breeding, production and marketing of vegetable seed for the ornamental sector.

Many seed companies from the USA, UK, Holland, Germany and France have started their flower seed production in India and China which in turn are getting the seeds produced on contractual basis from farmers in different regions having favourable climate conditions. Mostly open-pollinated cultivars have been adopted by the growers.

The Indian Advantage

Most of the flowers are grown under protected conditions such as greenhouses and polyhouses or glasshouses in European and other countries. Due to intense cold and high energy cost, production in these countries is limited during winter months (October to April). Thus they have to depend largely on import to meet their domestic demand as most of the festivals fall during this period when the demand of flowers is at its peak (Kapoor, 2002). As global deficit is predicted, hence, the prospects are good for new entrants like India which can offer reliable and quality products in this trade because of its ideal location to major importers, diverse climatic conditions, cheap labour, low input cost and wide choice of flower species or varieties for different climatic zones. Such a venture will together benefit India more as it will create sound economic future of the country, will generate employment opportunities among rural youths, will earn 20-30 times more foreign exchange as compared to any other agricultural industry and also as it has an annual growth potential of 25 to 30%. In India, about 1200-1500 ha area is under flower seed production. Of the total production, Punjab alone contributes for 45–50% where Sangrur, Ludhiana and Patiala being major seed producing centres. The rest is being contributed by Karnataka (Bengaluru, Raneebenur, Koppal, Itigi and Badami), Andhra Pradesh (Hyderabad and Secundrabad), West Bengal (Kalimpong), Bihar, Jharkhand (Ranchi), Maharashtra, Haryana (Panipat and Sirsa), Himachal Pradesh (Kullu Valley and Solan) and Jammu & Kashmir (Srinagar Valley).

Structure of flower seed industry in India

Export
↓
Contract Farmer
↓
Seed Production Company
↓
Wholesaler abroad
↓
Retailers
↓
Consumers

Most of the seed produced is through contract farming by the contract seed production companies, who themselves are also major producers. The global contract companies provide the contract farmers in India free of cost planting material (seed or nursery), technical know-how, on-site supervision in seed production, seed cleaning, grading and free seed transportation from the farm, cost-to-cost basis agro-chemicals and fertilizers, and over all 100% buy back guarantee. Various seed companies operating in various parts of India are Indo-American Hybrid Seed Company, Indo-Bloom Horti. and Research, Unicorn Seeds, and Exim Seeds all in Karnataka and Andhra Pradesh; Namdhari Seeds, and Horti Flora Seed Farms in Karnataka, Punjab and Himachal Pradesh; Beauscape Farms, Platsman's seeds, Biocarve Seeds, Global Flower Seeds, Biogene Floritechs, and Genesis Seeds in Punjab; Seeds India in West Bengal and Jharkhand; and Suttons Seeds in West Bengal, Jharkhand and Andhra Pradesh.

Depending upon derivation of seeds, different types of flower seeds are sold in the market such as **open pollinated varieties** having a major share and more than 90% seed of open pollinated varieties is produced and exported. The seed of true breeding cultivars is produced by controlled pollination and that of often or cross pollinated species is produced by providing proper isolation distance. Mixture of open pollinated seeds is a blend of different colours of same species or blend of same colour of different species. **F_1 hybrids** are more in demand as they are superior and uniform in plant vigour, earliness, number of flowers and their size. F_1 hybrid seeds are produced as a result of crossing two inbred lines or varieties by using male sterile lines or adopting various field operations like emasculation and pollination. F_1 hybrid seed of marigold and zinnia can be produced using male sterile lines as it saves huge labour costs and other energies which otherwise can be effectively used for some other purposes. However, in pansy and petunia, emasculation and pollination operations are performed for F_1 hybrid seed production. **F_2 strains** result from self fertilization of F_1 hybrids and give wide variation in colour. These types of seeds are most commonly available in pansy, petunia and antirrhinum. **Mixture** refers to the blend of different flower colours in certain proportion to make the garden more colourful.

Types of Annual Flower Crops According to Season

Summer season crops (producing flowers under high temperature) are sown in February end to mid-March in North India, and June to August in South India. These are cosmos, gaillardia, gomphrena, kochia, petunia, portulaca, sunflower, tithonia and zinnia.

Rainy season crops (withstanding heavy rains and high humidity) are sown in June-July in North India and during May-June in South India. Examples are amaranths, balsam, celosia, gaillardia, gomphrena, etc.

Winter season crops (tolerating low temperature) are sown in September-October under North Indian conditions and during February-May in the hills. Most of the winter season annuals seen in our gardens were introduced by Britishers. They are antirrhinum, Californian poppy, calendula, China aster, dianthus, ice plant, larkspur, linaria, pansy, petunia, salvia, stock, verbena, etc.

Flower seed production areas should have good climate, soil and water quality, availability of skilled *vis-à-vis* cheap labour, and should have capability for arranging transportation infrastructures, investment in construction of necessary buildings, and purchase of necessary equipments.

Main flowers for seed production being grown in different zones of India

A. Punjab & Haryana: *Acroclinum, Allysum, Antirrhinum, Bellis, Calendula, Clarkia, Coreopsis, Delphinium, Dianthus, Echium vulgare, Eschscholzia, Gazania, Helichrysum, Lathyrus odoratus, Lupinus, Mesembryanthemum, Matthiola, Nemesia, Phlox, Tagetes, Tropaeolum, Verbena* and *Viola.*

B. J&K and H.P.: *Ageratum, Callistephus chinensis, Celosia, Cleome, Cosmos, Dahlia* (annual), *Helianthus, Portulaca* and *Zinnia.*

C. Jharkhand & West Bengal: *Acroclinum,* butterfly flower, *Centaurea cyanus, Centaurea moschata, Dianthus, Helichrysum, Lathyrus odoratus, Linaria, Nigella, Petunia, Phlox, Papaver, Verbena* and *Viola.*

D. Andhra Pradesh & Karnataka: *Amaranthus, Catharanthus, Celosia argentea, Coleus, Cosmos, Gaillardia, Gomphrena, Helianthus, Impatiens balsamina, Ipomoea alba, Mirabilis jalapa, Salvia, Tagetes, Tithonia, Torenia* and *Zinnia.*

Flower seed production routine includes parental plant culture (stock seed), genetic quality control, pollination management, seed harvesting and extraction, and seed cleaning and conditioning. However, individual plant species or their varieties may have their specific requirements such as soil and climatic conditions, planting time and methods, spacing (plant density), irrigation and nutrition, insect-pests, diseases and physiological disorders, harvesting and threshing techniques which shall be met effectively.

The Flowers Suitable for Seed Production

Abronia umbellata (Sand verbena, Nyctaginaceae). This climbing to trailing seasonal flower is most suitable for the rockery which grows to a height of 45 cm and then trails to 90-150 cm. It takes about 3-4 months from sowing to flowering. The flowers are rosy-lilac and emit at night the sweet fragrance, and resemble verbena.

Achillea filipendulina (Asteraceae). It is a robust perennial plant, growing to a height of 150 cm, with pinnately lobed, toothed and pungent foliage. The bright yellow flowers are densely borne in somewhat convex heads which sometimes measure 12 cm across. The plants need staking.

Achillea millefolium (Milfoil, Asteraceae). The plant is 30-90 cm tall, perennial and bears white flowers in flattish heads.

Acroclinium roseum (Everlasting flower, Asteraceae). The plants are 45-60 cm tall bearing double daisy like flowers of pure white, pink and shades of rose with bright golden yellow and dark coloured centres. The papery flowers are borne singly at the end of long stems. It is useful for cutting and last long when cut and dried. The leaves are narrow and glaucous green.

Adonis aestivalis (Pheasant's eye, Ranunculaceae). Branching annual and perennial rock plants which are 25-30 cm tall and suitable for cultivation in mixed borders, open woodland and wild gardens. The plant has a deep green, finely cut, fern-like foliage and bears small dark crimson buttercup-like flowers of about 2.5 cm across.

Ageratum houstonianum (Floss flower, Asteraceae). The plant height ranges from 15 to 45 cm. The dwarf types are especially suitable for beds, borders or rockeries. The branches are covered with fluffy heads of flowers with colour ranging from pale lavender to deep mauve. Some cultivars bear white flowers.

Agrostemma githago (Corn cockle, Caryophyllaceae). This is a hardy, tall and slender annual growing to a height of 60-90 cm. The pale-lilac large *Viscaria*-like flowers are 5.0-7.5 cm across with radial lines of dark dots in each petal. They are useful for mixed border and pots.

Althaea rosea (Hollyhock, Malvaceae). In temperate countries and in hill stations, this is treated as a perennial, whereas in the plains it behaves as an annual. It is a majestic plant, attaining a height of 1.5 to 2.5 m or more under favourable conditions with rough, large, round and kidney-shaped leaves. The large, single, semi-double or double flowers are borne in the axils of the leaves all along the length of the stem.

Alyssum maritimum (Sweet alyssum, Cruciferae). This forms a low-growing, spreading plant with narrow grey foliage. The plant branches freely and each branch ends in a spike of usually white flowers which are honey-scented. When in flowering, the compact plant looks like a carpet of flowers. They are suited for beds, borders, window boxes, hanging baskets and as the first row in an herbaceous border. *Alyssum saxatile* (yellow alyssum) is other important species for commercial seed production.

***Amaranthus* spp.** (Amaranth, Amaranthaceae). This group comprises some useful garden annuals with bright foliage or blooms or both. It grows to a height of 1.0-1.5 m. It bears tail-like, often with knots at short intervals, drooping flower spikes, white, pale green or crimson in colour. *Amaranthus caudatus* (love-lies-bleeding) and *A. tricolor* (Chinese spinach) are the two important amaranths for seed production.

Ammobium alatum (Winged everlasting, Asteraceae). Here the coarse stems are enlarged through their membranous wings. The small heads of flowers are surrounded by a number of rows of involucral bracts which are papery thus the infloresences become everlasting. The achenes bear a pappus of 2 scales and 2 bristles. Silvery-white floral heads are 2-3 cm wide with central florets yellow.

Anchusa capensis (Cape forget-me-not, Boraginaceae). The plant grows as a biennial or a perennial but can also be grown as an annual in gardens. It grows well in the hills and in plains having a long and cool winter. The plant is compact and grows to a height of 30 to 45 cm. The flowers are deep blue with a white centre and 0.6 to 0.7 cm across.

Angelonia grandiflora (Angelonia, Scrophulariaceae). A very popular, hardy, dwarf perennial which bears terminal racemes of sweet scented purple or white flowers.

Antirrhinum majus (Snapdragon, Scrophulariaceae). This is a branching perennial, 30 to 90 cm tall, with numerous spikes, mostly of crimson and white colour, though in Indian plains it grows as an annual. The colour of the flower varies widely in all the types ranging from white and cream to yellow, shades of pink, scarlet, red, orange, brick red, mauve, rose, maroon and so on.

Aquilegia vulgaris (Columbine, Ranunculaceae). These are pretty perennial plants about 60 cm tall, with handsome fern-like foliage consisting of three leaflets in a leaf. The flowers are double or single and large, 10-12 cm across with a long spur. The flower colour ranges from white to red, yellow, pink, crimson, blue and lavender.

Arctotis grandis (African daisy, Asteraceae). The species is vigorous growing and bushy in habit. The flower stem attains a height of 45-60 cm and thus suitable for cutting. The colour of the flower is white with a lilac centre and the backside of the petals is steel grey.

Aster amellus (Perennial aster, Asteraceae). It grows to a height of 60 cm. The single flowers are small, 2.5-5 cm across, daisy-like, borne in large heads (30-45 cm) of corymbose panicle and are white, lavender, mauve, deep blue, lilac and pink.

Bartonia aurea (Gloden star, Loasaceae). An easy to grow, hardy and showy annual, 45 to 60 cm in height. The flowers are large buttercup-like and glistening orange in colour with a central mass of feathery stamens. It is useful as a bedding plant.

Bellis perennis (English daisy/daisy, Asteraceae). It is a perennial plant but can be best grown as biennial in the hills and as an annual in the plains. It is a dwarf plant, 20-30 cm in height. The leaves are strap-shaped, tapering towards the tip and form into tuft at the base. The flowers are 2.5-5.0 cm across and the flower stalk is about 15 cm long. The flowers are single or double and the petals may be flat or quilled. The flat type is available in white, crimson and pink. The quilled flowers are in a beautiful salmon-pink and crimson in shade.

Berbenia cordifolia (Elephant ear, Saxifragaceae). The plants are 30 cm tall, with large heart-shaped foliage and rose-pink flowers measuring 2 cm across which are borne on medium stalks above the foliage. It is a perennial plant.

Brachycome iberidifolia (Swan river daisy, Asteraceae). It is a beautiful dwarf annual with a slender stem and finely divided leaves. The flowers are 2.5 cm across, borne in profusion and are white, blue, pale-blue, mauve or purple. It is an appropriate plant for rock gardens and herbaceous borders.

Browallia demissa (Amethyst flower, Solanaceae). It is a plant of medium size, rather moderately tall, attaining a height of 30-45 cm and where blue and white flowers resemble *Lobelia*, some 1.0 cm long and are borne in the axils of green, narrow and oval leaves.

Cacalia coccinea (Tassel flower, Asteraceae). A hardy annual 30 to 45 cm tall. The leaves are thin, long, oval and remain attached to the wiry and hairy stem. The small flowers which are 1.5 cm across are borne in clusters on the wiry stems and look like a tassel. Flowers are commonly orange-scarlet or dull red in colour but there is one variety *Cacalia lutea* with golden-yellow flowers.

Calceolaria spp. (Slipperwort, Scrophulariaceae). The plants are remarkable for their brightly coloured flowers and a peculiar shape as that of circular pocket purses. *Calceolaria* includes many species of annuals and perennials. They may be herbaceous or shrubby in habit.

Calendula officinalis (Pot marigold, Asteraceae). One of the most commonly cultivated seasonal flowers. It flowers almost throughout the year. The flowers measure 4-5 cm across and their colour varies from straw to deep orange. The flower may be single or double. Calendula is useful for bedding, cutting, potting and for window boxes.

Callistephus chinensis (China aster, Asteraceae). It is a half-hardy annual which grows some 45 cm in height, bearing mid-green, coarsely-toothed and ovate leaves. Flower heads are daisy-like, solitary and purple though varieties have all the shades such as white, cream, orange, pink, red, maroon and purple. These are chrysanthemum-flowered, giants of California strain, ostrich plume strain, paeony-flowered strain, pompon strain, queen of the market strain and spider-types. All are multiplied through seeds.

Campanula medium (Canterbury bells, Campanulaceae). It is a biennial, bearing bell-shaped flowers of white, blue, pink, mauve or violet colour. The flowers are single or double. The species *Campanula ramosissima* is an annual, 22-45 cm tall. *Campanula pyramidalis*, commonly called 'Chimney bell flower' is another biennial species. *Campanula persifolia* and *Campanula carpatica* are two beautiful perennials.

Catharanthus roseus (syn. *Vinca rosea*, periwinkle, Apocynaceae). The plants are perennial and 30-60 cm tall covered with bright green smooth oblong leaves. The flowers are rosy-purple or white, the latter with or without a very small reddish eye. The plants are free-flowering and bloom round the year. *Vinca minor* (lesser periwinkle) grows to 10 cm high with a spread of up to 120 cm, bearing glossy dark green and elliptic to lanceolate leaves. The flowers are blue and 18-25 mm across. Its varieties have wide range of single or double flowers in the shades of white, wine-red, deep purple, plum-purple, deep blue-purple and blue, and even with variegated leaves.

Celosia argentea (Cockscomb, Amaranthaceae). A very popular and easy to grow annual, 25-50 cm in height but it may grow up to 150-180 cm during rainy season. It has branching growth habit and produces feathery flower spikes in a pyramidal fashion resembling ostrich-plumes, in various colours of silver, yellow, golden-yellow, red and orange. ***Celosia argentea cristata*** (plumed cockscomb) and ***C. plumosa*** are the two important forms for commercial seed production.

Centaurea cyanus (Cornflower, Asteraceae). A hardy upright annual. The plant is slender and 60-90 cm tall. The original blue flower colour sported to a multiple range of colours such as white, pink, maroon and purple. It is useful as cut flower and can be grown in annual and mixed borders.

Centaurea moschata (Sweet sultan, Asteraceae). It bears fragrant flower heads of white, delicate mauve, deep mauve, purple, yellow or deep red on long stems. Plants are about 90-100 cm tall. The flowers are excellent for cutting and look like powder puffs.

Cheiranthus cheiri (Wallflower, Cruciferae). Wallflower is a medium tall shrubby biennial or perennial. There are some early flowering annuals. They are popular for spring bedding and cutting because of their wide colour range and sweet scents.

Chrysanthemum coronarium (Crown daisy/garland daisy, Asteraceae). It is taller and more vigorous plant. The flower shade is yellow and white. *Chrysanthemum carinatum* (tricoloured chrysanthemum/painted daisy) is the best of the annual types. The flower colour is rich crimson but the mixture of colours is available. The plant grows to a height of 60 cm. The large flowers, about 7 cm across show a bright coloured ring at the base of the petals and a dark coloured disc at the centre, thus producing triple ring effect. *C. multicaule* (yellow daisy), *C. paludosum* (creeping daisy) and *C. segetum* (corn marigold) are other annual chrysanthemums for seed production. *Chryanthemum leucanthemum* (moon daisy) and *C. indicum* (Indian chrysanthemum) are though perennials but breed true when grown through seeds and act as annuals.

Clarkia elegans (Clarkia/garland flower, Onagraceae). It is a hardy annual with slender but graceful branches and grows to a height of 60 to 90 cm. The single or double flowers are borne in the axils of the leaves and the petals are beautifully ruffled.

Cleome spinosa (Spider plant, Capparidaceae). It is a tall plant, with sharp spines all over including the leaves. The large clusters of 4-petalled flowers have long prominent stamens and these give the plant its name. The flowers are scented and may be pink, light purple, rose or white in colour.

Clianthus dampieri (Glory pea, Leguminosae). It is one of the most showy winter annuals. The plants are branching and grow to a height of 80 cm. The flowers are quite large, 6-8 cm across and are borne in clusters of 4-6 flowers. The flowers are pendulous and dark crimson with a raised black base which resembles the eyes.

Cobaea scandens (Cup and saucer vine/cathedral bell, Polemoniaceae). It is a perennial climber growing up to 6 m in height, supporting itself through its branched tendrils. Pinnate leaves are mid to dark green with three pairs of leaflets. Each flower comprises the bell-shaped purple corolla on the saucer-like green calyx. The flowers 6.0-7.5 cm long appear freely. It is raised through seeds.

Coleus blumei (Painted nettle, Labiatae). These evergreen plants are though perennial in nature but for thrilling results these are raised through seeds every year though tip cuttings are used to multiply quite promising stocks. It is a foliage plant suitable for pots and in borders.

Out of 150 species, only two are ornamental, *viz. C. blumei* and *C. thyrsoideus. C. blumei* is a shrubby species growing to more than 45 cm in height. The nettle-like leaves, mostly fimbriated and festooned appear in varied patterns of colouraation (white, green, yellow, red, bronze, brown, maroon, purple and mixed). The flowers are small and tubular in blue and white colours. *C. thyrsoideus* grows up to 90 cm in height with spread of up to 60 cm. The leaves are mid-green and roughly cordate. Tubular and bright blue flowers appear in large numbers in terminal branched panicle some 25 cm long.

Collinsia bicolor (Collin's flower, Scrophulariaceae). An easily grown beautiful hardy annual of branching habit. Plants are slender and grow to 40 cm. The flowers are borne in long spikes and are lilac, white and salmon in colour. The plant is suitable as a pot plant and for woodland garden.

Convolvulus tricolor (syn. *C. minor, m*orning glory, Convolvulaceae). It is a hardy annual species with erect and bushy growth, growing up to 40 cm in height. The leaves are dark green and ovate-lanceolate to spathulate. Single rich blue flowers of 3.7 cm width appear profusely from the upper leaf axils, having white to yellow throat colouration.

Coreopsis spp. (Tick seed, Asteraceae). There are two commonly grown species, *Coreopsis tinctoria* and *Coreopsis drummondii.* In *Coreopsis tinctoria* (plain's coreopsis), the flower colour is mainly yellow and crimson-brown. In *Coreopsis tinctoria* var. *marmorata* the flowers are curiously marbled and striped. The plants may be tall or dwarf. *Coreopsis drummondii* has yellow flowers with dark crimson centre. The plants are medium tall. *Coreopsis lanceolata* (lance-leaf-coreopsis) is though perennial but grown as an annual through seeds.

Cosmos bipinnatus (Cosmos, Asteraceae). A delightful annual, mainly grown in the rainy season. The plant is 50-120 cm tall and bears rose, pink, crimson, purple, lavender and white flowers with yellow centre on long stalks. Flowers are large, 10-15 cm across and may be single or double including some tetraploids. *C. sulphureus* (yellow cosmos) is another species for seed production.

Cuphea ignea (Cigar plant, Lythraceae). A dwarf, bushy evergreen parrenial though grown as an annual, having unusually-shaped scarlet flowers with a dark ring at the base resembling a cigar. *Cuphea miniata* is another species with open scarlet flowers having purple bases and white mouth.

Cynoglossum amabile (Chinese forget-me-not, Boraginaceae). A hardy perennial, about 30 cm tall but is grown as an annual. The tiny blue flowers are borne in clusters and are sweet-scented. A white cultivar is available.

Delphinium ajacis (Rocket larkspur, Ranunculaceae). This species includes white, blue, mauve-purple, pink and red shades. The plants grow up to 90 cm in height. It is popular as cut flower and for growing as a background or in front of a tall hedge or a border and in pots. *Delphinium consolida* (imperial larkspur). They are usually tall, 90-150 cm and branching. It is an erect growing hardy annual bearing mid-green leaves cut into several linear segments thus giving a ferny appearance. Racemes are up to 15 cm long bearing dense blue, purple, red, pink or white flowers. *Delphinium hybridum* (perennial delphinium) thrives best in hills and can be grown as annual in northern plains in and around Delhi with long cool winter. It is a robust, 150-180 cm tall plant with flowers in several colours including pure white, deep bright blue and sky-blue, which come true from seed. The other flower colours are purple, lavender, lilac and mauve. *Delphinium grandiflorum* (larkspur) is though perennials but is grown as annual through seeds.

Dianthus barbatus (Sweet William, Caryophyllaceae). This is a biennial plant but an annual strain is available. Some of the early flowering biennials can be grown as annuals. Sweet William is closely related to pink. The plant is about 30-45 cm tall with double or single fragrant flowers borne at the apex of the branches in rounded clusters. The flowers which are white, scarlet, dark red and in different shades of pink are self coloured, edged, mottled or with prominent white eyes.

Dianthus caryophyllus (Carnation, Caryophyllaceae). It is a flowering perennial. There are different types of carnations and each of these can be grown successfully in cooler parts of the plains and at medium to high elevations. It has spray types grown from seeds in the open even in the plains though Sims are grown in polyhouses as cut flowers. The colours range from white to maroon or near black via yellow, pink, red and various other shades, bicolours and mosaic types.

Dianthus chinensis (Chinese pink, Caryophyllaceae). The plants are 30-45 cm tall and bear brightly coloured scented flowers. The flowers are self coloured, exquisitely marked, edged or blotched. Some of them have petals beautifully fringed. There are cultivars in which blooms are borne in upright stems. The flower colour ranges from white to red, scarlet, pink, lilac, violet, maroon-black and mauve.

Digitalis purpurea (Foxglove, Scrophulariaceae). It is a biennial, rarely perennial but sometimes treated as annual. It is grown in the hills. The plant, 60-150 cm tall, is cultivated for the production of digitalin, a medicine. The flowers are bell-shaped, slightly drooping and open at mouth which are sometimes spotted.

Dimorphotheca sinuata (Cape daisy, Asteraceae). The plants are about 30-60 cm tall, branching and bearing daisy like large flowers of various shades such as white, bluish-white, shades of yellow, salmon, rose and reddish-purple. The plants are free-flowering and the flowers remain open during day time only. *Dimorphotheca aurantiaca* (Cape marigold) is other important species for seed production.

Echinacea purpurea (Purple coneflower, Asteraceae). It is a hardy herbaceous perennial and generally grown from seeds though root division is also common for varietal stability. It grows up to 60 cm in height, bearing mid-green, rough, lanceolate and a little toothed leaves. The 10 cm across cone-shaped purple-crimson flowers appear with orange throat. The varieties appear in white, crimson-pink, purple-rose and in various other shades.

Echium plantagineum (Tower of jewels/blue weed, Boraginaceae). It has both tall and dwarf cultivars. The small bell-shaped flowers have a wide range of petal shades in blue, mauve, purple, rose-pink and white. It is useful for mixed borders, rock gardens and for edging.

Eschscholzia californica (Californian poppy, Papaveraceae). In its natural home, the plant is a perennial but is grown in the garden as an annual. The plants are 30-45 cm tall, with blue-green, finely cut foliage and slender stem somewhat resembling a carrot. Some varieties are dwarf, 20-30 cm tall. The flowers are large, about 10 cm in diameter, saucer-shaped, single, semi-double or double and have shades of bright yellow, lemon, orange, crimson, cream and creamy-white. Some are bicoloured, exterior bronze and interior orange.

Euphorbia heterophylla (Annual poinsettia, Euphorbiaceae). It is a hardy annual growing to a height of 60-90 cm. The orange-red flowers are borne in clusters at the end of branches. The leaves surrounding the flowers get attractively coloured during the hot season and resemble *Poinsettia*.

Euphorbia marginata (Snow-on-the-mountain, Euphorbiaceae). It is a bushy annual growing up to 60 cm in height. The bright green leaves are ovate which turn edged and white-veined with maturity. These leaves are used in floral arrangement but since it oozes a milky sap so to stop this the ends of freshly cut stems are dipped in hot water. Small clusters of insignificant white flowers appear during autumn. These are multiplied through seeds.

Felicia bergeriana (Kingfisher daisy, Asteraceae). A dwarf (15-20 cm) annual, bearing numerous small, bright metallic-blue flowers resembling daisy. Plants are useful for rock gardens, for front position in borders and in pots.

Gaillardia pulchella (Blanket flower, Asteraceae). A free-flowering popular annual which blooms in succession for quite some time. The plants are medium and 40-45 cm tall. The flowers are large, 5-7 cm across, single or double and are available in many pretty colours and their combinations such as yellow, lemon, yellow-purple, cream, orange and maroon. *Gaillardia aristata* (blanket flower), a perennial species is grown as an annual through seeds and produces yellow flowers.

Gazania splendens (Treasure flower, Asteraceae). A dwarf, trailing plant with narrow, long, pointed, silvery and hairy leaves. The flowers are in shades of red, pink, orange, yellow and white and are daisy-like in shape. It is grown in rock gardens, in front rows of borders and for edging.

***Gerbera* spp.** (African daisy, Asteraceae). It is a dwarf perennial herbaceous plant. The plants are stemless, some 30-45 cm tall and are hairy throughout. The leaves are 12-20 cm long, 5-7 cm broad, deeply lobed with a fairly long petiole. The flowers are daisy-like, 7-10 cm across but in certain hybrids these may be as large as 15 cm across. Highly improved varieties do not produce seeds but species and little improved varieties do.

Geum coccineum (Rosaceae). A herbaceous perennial suitable for medium and high elevations. The plant flowers freely. The flowering is from summer to autumn. The colour of flower is brick red.

Gilia capitata (Ipomopsis, Polemoniaceae). The plants are 45-60 cm tall with very finely cut foliage, small and lavender-blue flowers. It is suitable for growing in front rows in borders and in rock gardens.

Godetia grandiflora (Satin flower, Onagraceae). It is a dwarf hardy annual with large flowers produced in clusters of white, yellow, orange, rose, crimson, carmine and purple with single or double flowers.

Gomphrena globosa (Globe amaranth/gomphrena, Amaranthaceae). It is a rugged looking annual and does not require any special attention. The plants are bushy in nature and may be tall or dwarf. The flowers are round and clover- or button-like in shades of white, orange, magenta, purple, violet, rose and pink and are everlasting.

Gypsophila elegans (Baby,s breath/chalk plant, Caryophyllaceae). A dwarf, slender, lance-shaped, grey-leaved annual, bearing clusters of small, white or pink flowers. It is grown in beds, borders, rock gardens and for covering open spaces in beds of bulbous plants. Also this is used as cut flower and as filler. *Gypsophila paniculata* is though perennial but is grown as annual through seeds.

Helianthus annuus (Sunflower, Asteraceae). The giant, common yellow sunflower is a coarse looking plant with rough textured leaves. The flowers are 20-30 cm across and are single or double, from yellow to orange and red to black eye. *Helianthus debilis* (beach sunflower) is also important for seed production.

Helichrysum bracteatum (Strawflower, Asteraceae). It is a commonly grown plant, 50-80 cm tall, bearing everlasting flowers of 7-10 cm in diameter. The flowers last long and give a papery feeling when touched. The plant flowers without inhibition.

Heliotropium arborescens (Heliotrope, Boraginaceae). It is a perennial plant but can be grown as an annual. The plants are branching, compact and 30-45 cm tall. The small tubular flowers are borne in long trusses well above the foliage. The plants flower freely and the flower colour ranges from white to blue, purple and various shades of purple. The flowers are vanilla-scented.

Heuchera sanguinea (Crimson bells, Saxifragaceae). A perennial herb, 30-45 cm high, with heart-shaped or roundish, much lobed leaves. The flowers are small, bell-shaped and numerous borne on long stalks. These are suitable for table decoration.

Hunnemannia fumariifolia (Mexican tulip, Papaveraceae). A perennial plant growing to a height of 40-90 cm and bears yellow flower which are 5-7 cm across.

Iberis umbellata (Candytuft, Cruciferae). The plant is 15-30 cm tall and small flowers are borne in broad clusters, the umbels, with a wide range of colours including white, pink, carmine-rose, purple and lilac. *I. amara* (rocket candytuft) is an annual growing to a height of 40 cm bearing mid-green lanceolate leaves and 5 cm long racemes of fragrant flowers in white, pink to red-purple in colour. *Iberis sempervirens* (candytuft) is a hardy evergreen and bushy perennial growing some 20-25 cm high with a spread of 45-60 cm, bearing narrow and oblong leaves, and white flowers in dense and 3.75 cm wide heads.

Impatiens balsamina (Touch-me-not/garden balsam, Balsaminaceae). The plants are 22-60 cm tall, bearing camellia-like double or single flowers with varying colours such as rose, pink, scarlet, red, violet, salmon-pink, purplish-red, lilac, blue, mauve and white.

Ipomoea spp. (Convolvulaceae). It is a genus of some 500 species of tender and half-hardy annuals and perennials, including climbers and shrubs. *Ipomoea quamoclit* (cypress vine, hardy annual) and *I. lobata* (syn. *I. versicolor*, half-hardy perennial), both climbers have been renamed as *Quamoclit pennata* and *Mina lobata*, respectively, though both are grown as tender annuals through seeds. *Q. pennata* grows up to 200 cm, bearing finely divided light green leaves and curved tubular, a little claw-like red and yellow flowers on 3.75 cm long racemes during summer. *M. lobata* grows up to 3.0 metres bearing 3-lobed dark green leaves and some 5.0 cm long crimson flowers on up to 12 on slender spikes which are suitable for cutting. Flowers change to orange, then yellow and finally to white with age. *I. alba* (syn. *Calonyction aculeatum*, moon flower) is a soft-stemmed and evergreen twining climber growing up to 3.6 m, bearing 20 cm long and oval (sometimes 3-lobed) leaves, *vis-à-vis* white tubular, fragrant, 15 cm long flowers opening in night. *I. coccinea* (syn. *Quamoclit coccinea*, red morning glory/ star ipomoea) is a frost-tender annual climber growing up to 3.0 m where leaves are arrow- or heart-shaped, long-pointed and often toothed. Tubular flowers are scarlet with yellow throat and fragrant. *I. horsfalliae* is strong-growing, woody-stemmed, evergreen and twining climber growing up to 3.0 m, bearing 5-7 radiating lobes or leaflets. The clustered flowers some 6 cm long are stalked and deep rose-pink or rose-purple. *I.* × *multifida* (syn. × *sloteri*,.cardinal climber/hearts-and-honey vine) is a frost-tender annual climber growing up to 3.0 m, bearing triangular-oval leaves divided into 7-15 segments. Tubular flowers are crimson with white eyes. *I. nil* (syn. *I. imperialis*, Japanese morning glory) is a half-hardy, soft-stemmed and short-lived perennial climber with hairy stems, and grows up to 4.0 m. Leaves are either cordate or 3-lobed. It bears large and funnel-shaped flowers in variety of colours but with white tubes. *I. purpurea* (syn. *Convolvulus major, C. purpureus*, common morning glory) is a half-hardy, soft-stemmed, short-lived and perennial twiner with hairy stems. It is grown as an annual. Leaves are cordate or 3-lobed and the funnel-shaped flowers are deep to bluish-purple or reddish with white throats, and the sepals are bristly. *I. tricolor* (syn. *I. rubro-caerulea*) is a half-hardy perennial twiner with thin stems, growing up to 2.4 m, though is grown as an annual. The leaves are pale-green and heart-shaped. Flowers are red-purple to blue, 12.5 cm across which open in the morning and fade by the afternoon. All the species described above are grown as annuals through seeds.

Kochia scoparia var. ***trichophila*** (Summer cypress, Chenopodiaceae). It is a dense bushy symmetrical, oval or round annual foliage plant, 60-90 cm tall and 30 cm across. The whole plant is covered with finely cut and narrow pea-green ornamental foliage. On maturity, the plant turns into rich coppery-red and red flowers also appear giving the plant the look of burning bush.

Lathyrus odoratus (Sweet pea, Leguminosae). Sweet pea is valued and widely cultivated for its range of flower colours, elegant form, fragrance and utility as a cut flower. Sweet peas are commonly used as temporary screens to hide unsightly objects or to separate the kitchen garden from the flower garden or alongside a path.

Lavatera rosea (syn. *L. trimestris*, tree mallow, Malvaceae). It is an erect, bushy and very popular hardy annual, growing up to 90 cm in height. It is suitable for cutting and as a border plant. The leaves are pale-green, ovate and lobed. Upper leaf axils produce some 10 cm across rosy flowers freely during rains and autumn. Its varieties even produce white flowers. These are multiplied through direct sowing of seeds.

Limonium sinuatum (Sea lavender/statice, Plumbaginaceae). The species is actually a biennial but is generally grown as an annual. The plants grow to a height of 45 to 60 cm and the stems are hardy and angular. The foliage is grey-green and deeply cut. The flower stalk is winged and the calyx is toothed and straw-like. The flower colours available are white, rose, mauve, lavender and dark blue. *Limonium suworowii* (toad flax) is a half-hardy annual growing up to 30 cm tall and is most suitable for cut flowers. The leaves form a basal rosette, are mid-green, oblanceolate and with wavy margins. Densely packed rose-pink flowers appear in narrow plume-shaped panicles some 45 cm long.

Linaria maroccana (Toad flax, Scrophulariaceae). A commonly grown annual, with small pretty flowers. The plants are 20-35 cm tall, bushy having erect branches. The leaves resemble flax. The mini-antirrhinum like flowers are produced in abundance. The flowers are white, yellow, pink, purple, lavender, rose, blue, mauve and bicoloured having blotches of different colours on the lip.

Linum grandiflorum var. ***rubrum*** (Scarlet flax, Linaceae). The plants are 30-60 cm tall with narrow pointed foliage. The scarlet flowers with dark centres are 2.5-3.5 cm across and are borne in loose panicles. The variety *rubrum* is the best. This showy annual can be used in annual borders or beds for a fine display. *Linum perenne* (perennial flax) is a hardy perennial border plant attaining the height of up to 45 cm. Leaves are grey-green and narrowly lanceolate. Sky-blue flowers some 2.5 cm across appear in abundance during summer. Though the plant lives for 2-3 years only but is usually grown as an annual through seeds.

Lobelia erinus (Lobelia, Campanulaceae). A low growing, 15-30 cm tall perennial, exceedingly beautiful and profusely flowering plant of great garden value. The small flowers are white, different shades of blue, crimson and purple in colour, with yellow or white eyes. The plant may be either compact or trailing in habit. *Lobelia cardinalis* (cardinal flower) is a half-hardy and short-lived border perennial, quite distinct from *L. erinus*. Both are grown as annuals through seeds.

Lupinus spp. (Lupin, Leguminosae). The plants are 45-75 cm tall with foliage deeply-lobed and the long flower spikes growing well above the foliage, bear numerous small pea-like flowers of white, pink and pale blue colours. *Lupinus hartwegii* is an annual species with bushy growth, bearing leaves with 7 or more lobes or leaflets which are hairy. Spikes of more than 30 cm in length appear bearing pink-flushed flowers. *Lupinus luteus* (yellow lupin) bears yellow blooms.

Malcolmia maritima (Virginian stock, Cruciferae). It is a hardy herbaceous annual bearing grey-green, elliptic and blunt leaves. The sweetly-scented flowers appear cross-shaped, some 12 mm across on slender stems. The flowering starts appearing after four weeks of seed-sowing and continues until eight weeks. Flower colours may be white, lilac, rose or red. It is most suitable for sowing in border drifts and for edging.

Malva sylvestris (Common mallow, Malvaceae). Though very similar to *Lavatera* but differs in having 3 bracts below the calyx which are not fused together at the bases. This is an annual and is found in wasteland and along the roadsides throughout the Europe. The purplish-red flowers some 5 cm across appear in trusses in the leaf axils and on cutting the petals drop off soon. The petals are generally bifid. Its leaves are palmately lobed. Decoctions prepared from the leaves soothes upset stomachs.

Matricaria eximia (Feverfew, Asteraceae). The plant is a perennial, 30 cm tall but when raised from seed can be grown as an annual. The flat heads of small double flowers may be ivory-white or yellow in colour.

Matthiola incana (Gilliflower, Cruciferae). It is a genus of annual, biennial and perennial flowers. The small, rosetted and rounded single or fully double, scented flowers are borne on long spikes. There is wide range of flower colours from white to crimson, dark red, pink, rose, mauve, lavender and blue.

Mesembryanthemum criniflorum (Fig marigold/ice plant, Aizoaceae). A handy, spreading, 10 cm tall annual, bearing daisy-like flowers, 2.5 -3.5 cm across, with a wide range of colours including white marked with crimson, pink or buff at the edges of petals or self coloured types in shades of yellow, crimson, apricot, orange, red and white.

Mimulus tigrinus (Monkey flower, Scrophulariaceae). Although they are perennial, they could be conveniently grown as annuals. The plants are 30-45 cm tall and moderately branched. The flowers with two lips are gloxinia-like, open mouthed hence known as monkey flower. Flowers are 2.5-5.0 cm across with many colours including crimson, pink, yellow and other intermediate shades and spotted and blotched generally with brown or maroon.

Mirabilis jalapa (Four O' clock, Nyctaginaceae). A tall, bushy perennial, though generally grown as an annual. The flowers are fragrant, five lobed and 2.5-5.0 cm across with colours of red, yellow or white and often striped. The plants flower in the afternoon or in cloudy weather.

Molucella laevis (Bells of Ireland, Labiatae). Plant grows to 60-90 cm in height. The small white flowers which are inconspicuous are produced in successive whorls in the axils of leaves. This is useful for table decoration.

Monarda didyma (Oswego tea/bee balm/sweet bergamot, Labiatae). It is a perennial border plant, growing up to 90 cm in height, bearing mid-green, hairy and ovate-lanceolate leaves which may be dried and used as a tea. Flowerheads are dense, whorled and up to 7.5 cm across and bear bright scarlet flowers during rains. This has numerous varieties in various shades. Named varieties are propagated through root divisions though others through seeds.

Myosotis alpestris (Forget-me-not, Boraginaceae). Though a perennial but is generally grown as an annual. Both Dwarf and tall types are available. The tiny flowers are borne in clusters and are white, sky-blue, dark blue or pink in colour with a small yellowish eye and fragrant in afternoon.

Nemesia strumosa (Nemesia, Scrophulariaceae). The flowers are white, rich blue, primrose, yellow, orange, scarlet or crimson in colour and are marked with contrasting colours. The large flowered varieties are good for cutting and last as long as 2-3 weeks when kept in low bowls and cut at bud stage. They are good for bedding and borders. The dwarf types can be used with advantage in rock gardens, edgings and window boxes.

Nemophila menziesii insignis (Californian blue-Bell, Hydrophyllaceae). A slender growing and spreading plant, 15-20 cm tall. The plants get covered with numerous small bell shaped and sky blue flowers with a white eye. There are some cultivars having deep purple, pure white and spotted flowers.

Nicotiana alata grandiflora (Jasmine tobacco plant, Solanaceae). It is a perennial species, but is best grown as an annual. The plants are 60-90 cm tall, branching with large, coarse and hairy foliage. The flowers are large tubular or trumpet-shaped and white, blotched with various shades such as violet, lavender, yellow, crimson, pale-green, maroon and so on and are sweet scented.

Nierembergia caerulea (Cup flower, Solanaceae). A dwarf perennial plant, but grown as annual. The small petunia-like or cup-shaped, flattish flowers are violet-blue in colour with a yellow eye and are produced in abundance.

Nigella damascena (Fennel flower, Ranunculaceae). A pretty, upright, 30-45 cm tall plant with finely cut foliage and attractive semi-double white, deep blue or rose coloured flowers which remain surrounded by the leaves.

Oenothera **spp.** (Evening primrose, Onagraceae). It is a biennial some 60-120 cm tall though can be grown as an annual. The flowers are yellow, tubular in shape up to 15 cm across and open in the evening. *Oenothera rosea* (pink evening primrose/sundrops) and *Oenothera speciosa* (white evening primrose) are the two most important species for seed production.

Orthosiphon stamineus (Labiatae). A herbaceous perennial, 30-60 cm tall, bearing pale-lilac-blue flowers, 2.5 cm across and the stamens are much exserted giving the appearance of a cat's whiskers. It is generally grown as a house or greenhouse plant. The plants are grown in semi-shade.

Paeonia officinalis (Paeony, Ranunculaceae). It is a hardy perennial herb grown in beds, borders and shrubberies in the temperate areas for late spring to summer-flowering. These bear bold foliage, showy blooms and in some species the colourful seed pods. *P. officinalis* is clump-forming tuberous perennial with 60 cm height having forms in many colours.

Papaver rhoeas (Corn/common poppy, Papaveraceae). It is about 60 cm tall and bears cup-shaped single or double flowers with frilled silken petals. The flower colour may be white, all shades of pink including salmon, pale-mauve, blue and dark maroon, all with a white base and many have attractive picotee edging. Other poppies for seed production are *Papaver nudicaule* (iceland poppy), *Papaver orientale* (oriental poppy) and *Papaver somniferum* (opium poppy).

Pelargonium spp (Geranium, Geraniaceae). Geraniums are herbaceous perennials, mostly grown as potted plants for their handsome foliage, and flowers borne in clusters. The whole plant on crushing emits a pleasant aroma.

Penstemon spp. (Beard tongue, Scrophulariaceae). It is a semi-hardy to hardy genus of annuals, perennials, sub-shrubs and shrubs most of which are semi-evergreen or evergreen but very often grown as an annual. It is suitable for beds and borders. These bear flowers in various colours such as rose-red, deep wine red, pale-pink, pink and white.

Petunia hybrida (Petunia, Solanaceae). This is of garden origin. The height of the plant may vary between 20-30 cm and 30-45 cm depending upon the type. The flowers are single or double and the petals may be frilled and ruffled in various ways and may be veined or marked beautifully. This includes the varieties of 'Multiflora group' and the 'F_1 varieties', all being weather resistant. The flower colours available are pink and red, pale-blue, violet, yellow, cream, purple and mauve and some cultivars are bicoloured. Petunia is free-flowering plant. 'Grandiflora' is the second group of petunias having plant size similar to *P. hybrida* but with fewer and larger flowers up to 10 cm across and these are less weather resistant. The third group comprises the Nana Compacta (dwarf petunia) where the plants attain 15 cm height with 5 cm across flowers. The 'Pendula' varieties constitute the fourth group of petunias where plants are of trailing habit suitable for covering the ground during summer. These plants grow up to 15 cm height having some 5 cm across flowers, and with a spread of 60-90 cm so are most suitable for planting in hanging baskets and in pots.

Phacelia camanularia (Phacelia, Hydrophyllaceae). A hardy annual, 20-25 cm tall, having attractive grey toothed foliage. It bears clusters of small, bell-shaped, deep gentian-blue flowers with conspicuous anthers which are primrose-coloured.

Phlox decussata (syn. *P. paniculata*, border phlox, Polemoniaceae). It is mostly unbranched and 45-120 cm tall herbaceous perennial plant. The leaves are mid-green oblong-lanceolate, and some 2.5 cm across flowers in white, pink, mauve, violet, purple, crimson and red colours are densely packed in 10-15 cm long panicles. *Phlox drummondii* (annual phlox) is a half-hardy plant with erect stems, growing up to 40 cm in height with light green and lanceolate leaves. These are free-flowering in various shades of white, pink, vermilion, crimson, violet, purple, lavender and red appear in dense heads some 7.5 cm across. Sometimes the flowers are also fragrant. The flowers here are of three types: (i) 'Stellaris (syn. Cuspidata)' which bears star-shaped flowers with pointed petals, (ii) the large-flowered 'Grandiflora' group growing up to 30 cm in height bearing flower colours of cream, rose, red, crimson and violet, and the (iii) 'Nana Compacta' group that grows up to 22.5 cm in height with flowers appearing in various shades. All these are multiplied through seeds and are grown as annuals.

Pimpinella monoica (Lady's lace, Umbelliferae). A tall annual with deeply cut foliage bearing numerous white flowers in large spray on a long stem. The umbels resemble the carrot or the fennel. The plant is grown in the borders and sometime in the shrubbery.

Platycodon grandiflorum (Chinese bellflower, Campanulaceae). It is 30-60 cm tall, bearing bell-shaped flowers of blue or white shade, single or semi-double and about 7 cm across. The shoots die after each flowering but sprout again from the thick roots which survive a number of years underground.

Portulaca grandiflora (Rose moss, Portulacaceae). This is a low growing annual with a trailing habit. The leaves are cylindrical and fleshy. The flower colours are brilliant and available in shades of white, purple, sulphur-yellow, lavender-pink, salmon, scarlet, crimson and orange. The flower may be single or double.

Primula malacoides (Baby primrose, Primulaceae). A herbaceous perennial, 30 cm tall with a rosette of large broad leaves at the base. The flowers, single or double are borne on a long slender stalk in trusses. This plant is good for growing in pots, beds, shrubbery and for cut flowers.

Pyrethrum roseum (Pyrethrum, Asteraceae). The plants are 30-60 cm tall with finely cut leaves and big flowers resembling daisies in colours of white, lilac, rose and red. Sometimes flowers are tipped yellow and the disc is yellow.

Reseda odorata (Mignonette, Resedaceae). The plants are 30-40 cm tall and bear small flowers of bronze, white, red, crimson and yellow shades. The cut flowers last long and emit an agreeable perfume.

Rudbeckia bicolor (Cone flower, Asteraceae). It is a very attractive annual, 30-60 cm tall, bearing large flowers resembling daisies. The yellow flowers are borne on long stalks having a mahogany- crimson central zone and a dark brown central disc.

Saintpaulia ionantha (African violet, Gesneriaceae). African violets are stemless, hairy perennial with ovate or subcordate leaves having many small and a few long hairs on both surfaces. The flowers are of violets shape but larger in size.

Salpiglossis sinuata (Velvet flower, Solanaceae). A tall growing beautiful annual, bearing petunia-like tubular flowers in shades of white, crimson, pink, gold, golden-yellow, scarlet and gold, blue and gold and rich violet in colour. The plant flowers profusely.

Salvia spp. (Sage, Labiatae). There are five species of Salvia which are grown as annuals. ***Salvia coccinea*** grows some 30-60 cm tall and sometimes up to 70 cm. The flower spike is long, erect and bears small crimson and scarlet flowers. *Salvia farinacea* (blue sage) is a perennial species but is grown as annual. The plants grow to a height of 60-90 cm with shinning green leaves, bearing long spikes of lavender-blue flowers. *Salvia horminum* (blue beard/ annual clary) is 30-45 cm tall with small purple blue, pink or white flowers. *Salvia patens* is another perennial species though mostly grown as annual. The plant is 30-75 cm tall bearing long spikes of bright blue flowers. *Salvia splendens* (scarlet sage/red salvias) may be tall or dwarf and bushy. The long terminal spikes with their tubular bright scarlet flowers are borne above foliage. There are cultivars with salmon, pink and violet flowers.

Saponaria calabrica (Soapwort, Caryophyllaceae). It is a dwarf annual bearing sprays of starry pink or white flowers in abundance. It is grown in beds as background, in rock gardens and as edging. The flowers are good for cutting. The flowers are white, pink and red.

Scabiosa atropurpurea (Egyptian rose, Dipsaceae). A colourful branching annual, growing to a height of 60-90 cm. There are dwarf growing varieties. The plants bear handsome globose or rounded heads of flowers borne on stiff stalks and are available in beautiful shades of pink, pale mauve, blue, scarlet, dark maroon and white.

Schizanthus hybridus (Butterfly flower, Solanaceae). The plants are vigorous, branching and grow to 90 cm or more. The colour range of the flower is wide and varying and mostly blotched, veined and spotted exquisitely in contrasting colours thus earning the name of butterfly flower. The small flowers are borne in abundance, measuring 2.5 cm across, resembling orchids.

Senecio cruentus (Cineraria, Asteraceae). It is one of the best shade-loving annuals. The large dark-green heart-shaped leaves are most attractive. Plant grows to a height of 30-60 cm and the daisy like brightly coloured flowers are borne in clusters and may be self coloured or with contrasting centres of white or prominently marked or circled with various shades. *Senecio grandiflora* (florist's cineraria) is also used for seed production.

Silene pendula (Catchfly, Caryophyllaceae). A bushy dwarf annual, bearing single or double flowers in sparse clusters and in colours ranging from white to rose, pink, purple, crimson and salmon.

Solidago canadensis (Golden rod, Asteraceae). *Solidago* comprises a number of perennial species of which *Solidago canadensis* is most commonly grown. The plants are hairy, 90-180 cm tall, stout and with much branched stems. The pretty golden-yellow flowers are feather like and crowded in a long terminal panicle.

Streptocarpus spp. (Cape primrose, Gesneriaceae). These are stemless perennial herbs suitable for medium and high elevations with one leaf or a rosette of several leaves, more generally prostrate and a single leaf may measure from 7.5 cm to 15.0 cm. The showy blooms are borne in erect scapes and the shades include blue, lilac, mauve, pink and white.

Tagetes spp. (Marigold, Asteraceae). Marigolds are one of the easiest annual flowers to cultivate and have wide adaptability to different soil and climatic conditions. The plants, with their attractive flower colours bloom for a considerably long period and the flowers keep remarkably well when cut. *Tagetes erecta* (African marigold), *T. patula* (French marigold) *and T. tenuifolia* (Signet marigold) are important species for seed production.

Thunbergia alata (Black-eyed susan, Acanthaceae). It is an annual twiner growing up to 3 metres in height. The leaves are mid-green and ovate. The flat flowers that appear singly from the leaf axils are some 5 cm across, orange-yellow in colour with dark purple tube and chocolate-brown centre. It is most suitable for growing in pots. These are multiplied through seeds.

Tithonia rotundifolia (Mexican sunflower, Asteraceae). A free-flowering tall growing plant with rough, large, heart-shaped foliage and large single flowers resembling single dahlia. The flower colour is orange with tufted yellow centre.

Torenia fournieri (Wish-borne flower, Scrophulariaceae). This is an erect, much-branched, glabrous annual growing to a height of 30 cm. The flowers appear somewhat like antirrhinum in form and are violet and pale-blue in colour with a blotch of golden-yellow in the throat.

Trachymene caerulea (Lace flower, Umbelliferae). The plants are 45-60 cm tall with finely divided leaves and small lavender blue flowers which are borne in umbrella-like clusters on long stalks.

Tropaeolum majus (Garden nasturtium/Indian cress, Tropaeolaceae). There are four main garden types of nasturtium:(a) 'tall climbing singles' which trail or climb to a height of 180-250 cm, (b) 'dwarf single' (25-30 cm), (c) 'semi-double Gleam' which are semi-climbing in nature and the flowers are scented delicately, and (d) the 'dwarf double'. The nasturtiums have wiry stems, large attractive rounded leaves and spurred flowers, 6-7 cm across. The flower colours include orange, primrose, salmon, orange-red, scarlet and deep mahogany.

Ursinia anethoides (Jewel of the Veldt, Asteraceae). A medium tall shrubby plant with deeply cut foliage and beautiful daisy-like glowing orange flowers with crimson centre, borne on long stalks. The flowers are produced for a long duration and are excellent for cutting, bedding and pots.

Venidium fastuosum (Namaqualand daisy/Cape daisy, Asteraceae). A bushy showy annual, 60–90 cm tall, bearing large some 10 m across deep orange flowers with a black zone and a shining black centre, borne on long stalks. The whole plant is covered with white hairs.

Verbena* × *hybrida (syn. *V. hortensis*, verbena, Verbenaceae). The garden verbenas are compact bushy plants, 20-30 cm tall, with a tendency to trail little. The plants are perennial but treated as annual. The flowers are borne in clusters above the grey-green and toothed foliage. *Verbena bonariensis* is a perennial herbaceous species with branching habit, suitable for border planting which flowers the first season through seeds. It can attain the height of 1.5 m. Its leaves are mid-green and oblong, and the rose-lavender flower heads some 5.0-7.5 cm across appear during summer and autumn.

Viola cornuta (Tufted or horned pansy, Violaceae). The plants are tufted and the flowers are smaller in size. The flowers borne on short stalks are self coloured and include white, blue, violet-blue, yellow, apricot and ruby. *Viola odorata* (sweet violet) is a small herbaceous perennial prized for its sweet-scented flowers. The plants are trailing in habit with round leaves and bear single or double flowers in shades of white, blue and purple. *Viola tricolor* (heartsease) grows up to 15 cm in height with spread of up to 30 cm, bearing mid-green, ovate to lanceolate leaves with rounded teeth. Pansy-like flowers in shades of cream, yellow, dark blue, purple-black or bicolours appear in profusion from winter to spring. *Viola tricolor hortensis* (syn. *V.* × *wittrockiana*, garden pansy) is an all time favourite perennial bushy herb, 15-22 cm tall but treated as an annual in the plains and as biennial in the hills. The flowers are delicately scented especially during the early hours and the size may be 10-12 cm across in giant cultivars.

Zinnia elegans (Zinnia/youth-and-old-age, Asteraceae). It is a popular garden annual and varies in height from very dwarf to tall types. As a result of intensive breeding, various flower forms such as single, semi-double and double varying in size from 2.5–15.0 cm across have been obtained in various colours such as white, cream, primrose, golden-yellow, orange, scarlet, crimson, pink, lilac, purple and chocolate. *Zinnia linearis* (classic zinnia) is a perennial zinnia with neat, bushy erect habit, up to 30 cm tall, producing bright golden-yellow flowers, 3-5 cm across with a dark centre and stripes of lighter shade. These plants flower during summer and the rainy season. *Zinnia angustifolia* (creeping zinnia) and *Z. haageana* (Mexican zinnia) are other species for seed production.

Aspects of Flower Seed Production

Site selection is probably the most important factor in outdoor seed production. The production area should provide the required period of appropriate temperature and moisture conditions for the parent plants to develop and the seeds to ripen fully. There should be a dry period at harvest to allow field drying. The soil type is also an important factor in site selection. Some crops, e.g. pansy, tolerate heavy soils while others, e.g. nasturtium only does well in well-drained fields. One should consider the crop history of the plot as well as crops being grown in the neighboring fields to make sure that insect-pests and diseases (especially seed-borne) can be avoided. Annuals (seasonal flowers) can be grown in a variety of **soils**, but sandy-loam soils with medium fertility and adequate water-holding capacity are ideal for seed production. In heavy and clay soils, the plant-root system does not develop vigorously due to compaction. Sandy and clay soils are not ideal. Water carrying various plant food materials in solution from root to stem may drain out too soon from sandy soils, creating a drought-like condition. It may be held for too long in a clay soil, resulting in poor soil aeration. Thus, in general, the sandy-loam soil is best suited. Such soil is moderately coarse-textured, well-drained and also fertile. Addition of fertilizers and manure will further enhance its capacity to produce a healthy plant. If soil is clayey, addition of grit or sand will make it porous. Plants do not like excess water as it does not allow them to breathe. That is why it is important to make provision for channels to drain out excess water.

The flowers can be cultivated in a wide range of **climatic conditions** – tropical, subtropical and temperate. The subtropical climate with low temperature, almost nearing to freezing point during December–January, sometimes leading to ground frost and in summers with high temperature associated with hot desiccating winds during May–June and hot humid conditions during July–August is most suited for a wide range of flowering annuals (Swarup, 1967). Areas with lots of natural sunlight are best suited for flower seed production, but scorching heat and intense light is also not desirable as it can burn the plants. Shade loving plants should be protected from direct sunlight for quality flower production. Climate is one of the most important factors determining the selection of plants to be grown in a particular season and place. According to climatic requirements, the production of flower seeds is divided into following three climatic zones such as (i) mild areas where calendula, annual chrysanthemum, clarkia, delphinium, godetia, phlox, viola, zinnia, etc. can be grown; (ii) sub-tropical areas where ageratum, antirrhinum, bellis, brachycome, dianthus, dimorphotheca, eschscholzia, iberis, linaria, petunia, portulaca, etc. can be grown; and (iii) tropical areas where ipomoea, salvia, tagetes, zinnia, etc. can be grown.

Nursery bed should preferably be raised approximately 15 cm from the field level and should be more than 1.0-1.5 m wide so that various intercultural operations could be performed without entering the nursery bed. Nursery beds are prepared after mixing equal quantity of garden soil, sand, well decomposed FYM and powdered leaf-mould. Soil may be disinfected with 2% formaldehyde solution and covered with a poly-sheet for at least two days and then polythene should be removed so that the beds are exposed for 5-7 days to get rid of injurious effects of formaldehyde. Some annuals such as sweet peas, balsam and sunflower are sown directly. Other seeds are sown in 1.0-1.5 m wide and 15cm raised nursery beds. The seeds should be sown in lines 4–5 cm apart and then covered with a fine leaf-mould mixture up to a half to 1cm. The water should be applied with a sprinkler, using a fine hose till the establishment of seedlings. The seeds germinate in 3–6 days in summer and in 5–15 days in winter. The seedlings having 3–4 true leaves become ready for transplanting in 20–30 days after sowing in summer and 30–40 days in winter. The herbaceous plants like *Vinca* and many others require pricking. This is the operation in which the seedlings are thinned out from nursery bed and again planted in another bed at a wider spacing before planting in the main field. The seedlings having 3-5 true leaves, i.e. normally 20-30 days in summer and 30-40 days during winter, become ready for transplanting. Watering should be reduced at least a week before transplanting. This helps the plant to bear the transplanting shock.

Main field should be prepared by adding 30-50 tonnes/ha well decomposed FYM one month before transplanting. After proper ploughing, the beds where the seedlings are to be transplanted should be irrigated a few days before transplanting. **Transplanting** should be done in the evening or on cloudy days to facilitate better plant establishment. By making 2-3 cm hole in the bed based on spacing of the crop as in case of tall annuals (coreopsis, gaillardia and helichrysum) 60 × 45 cm^2, medium annuals (petunia, verbena, phlox and California poppy) 45 × 45 cm^2 and dwarf annuals (pansy, ice plant, portulaca and alyssum) 25 × 25 cm^2. **Irrigation** is required immediately after transplanting. Over-watering can harm the plant so water only if 4-5 cm of top soil is dry. The flower crops need light and frequent irrigations. Irrigation should be given at an interval of 4–5 days during summer and 10–12 days in winter. The irrigation should not be restricted at the time of seed ripening.

Insect-pollinated crops are grown in regions having high population of natural pollinators, however, if need be bee hives are also installed. Honey bees are the most common pollinators. Other commercially available insect pollinators include bumble bees, leaf-cutter bees and flies. Climatic conditions influence flowering, pollen production and stigma receptivity. Pollination management for open-pollinated crops begins with selecting production locations naturally conducive to good seed set. Optimum climatic conditions must fit the crop's temperature and light requirements for flowering, pollen production and stigma receptivity. Application of fungicides and insecticides during flowering should be avoided as it hinders stigma and pollen tube growth, kills pollinating insects and thus lowers seed set. In the open field, pollination is done by wind or insects, depending on the specific floral morphology and properties of the crop species. These natural means of pollen transfer are random in nature. Varieties of the same species will cross with each other (Miyajima, 1996). To ensure varietal purity, care has to be taken that different seed crops of the same species are not grown closely together. The **isolation distance** requirements generally range from 50 to 1,000 metres or more, depending on whether the crop is insect- or wind-pollinated. The topography of the production site, as well as the direction of prevalent winds should be considered when determining how far different varieties of the same species should be separated from each other. Some crops like sweet pea have flower structures that allow self-pollination as the flowers develop and mature, so different varieties of these types of crops can be produced as close as 100 metres apart. Seed producers have to keep the required isolation distance in mind when they plan the placement of production fields. Different seed companies operating in the same production area normally exchange information on their production plans prior to planting so that unintended cross-pollination does not occur. The seed crop must be isolated from other nearby field of the same crop and other contaminating crops as per requirement of certification standards. Isolation distances depend upon topography and site, direction of prevailing winds, floral structure, and nearness of the fields of the same crop. Isolation can also be maintained by isolation strips, distant isolation, caging or putting artificial barriers, and time isolation. After the seed crop has been harvested, isolation of seed produce of different varieties is necessary to avoid mechanical contamination.

The crops, their nature of pollination (self or cross) and isolation distance

Mode of pollination	Flower crops	Isolation distance
Self-pollinated	Balsam, Carnation, China pink, Lupin, Sweet williams and Sweet pea.	50 m
Often cross-pollinated	Larkspur, Lobelia, Pansy, Petunia, Phlox, Salvia, Satin flower, Snapdragon, Toadflax and Verbena.	100-250 m
Cross-pollinated	Balsam, Calendula, Candytuft, Clarkia, Cock'scomb, Cornflower, Coreopsis, Cosmos, Gaillardia, Gomphrena, Hollyhock, Marigold, Nasturtium, Portulaca, Strawflower, Sunflower, Sweet alyssum and Zinnia	400-1000 m

Sowing time, isolation distance, spacing, harvest stage and yield of some flowering annuals

Crop	Time of sowing	Isolation distance (m)	Spacing (cm)	Stage of Seed Harvesting	Yield (kg/ha)
Antirrhinum	Sept/Oct	100	45×35	Cut when just about to dry. Spikes mature from lower branch onward	250-350
Calendula	Sept/Oct	400	45×30	Collect head when partially dry because seed shatter when too much dry	400-425
Alyssum	Sept/Oct	400	30×20	Remove when pod just about to dry	250-300
*Celosia**	May/Jun	400	30×30	Collect the heads when dry on plant	450-500
*Coreopsis**	Mar/Apr	1000	60×40	Collect when plants are turning pale	400-500
Dimorphotheca	Sept/Oct	400	45×30	Cut the whole plant when the maximum amount of seed is mature	400-450
*Gaillardia**	Mar/July Sept/ October	400	60×60	Cut the entire plant when maximum amount of seed is mature	600-650
Helichrysum	Sept/Oct	400	60×40	Collect the heads when these become fuzzy	300-350
Impatiens	March/May	400	45×45	Collect individual pod when turning yellow-brown	750-800
Tagetes erecta	Mar/June Aug/Oct	400	60×40	Collect flower head as they dry on plant	500-600
Petunia hybrida	Aug/Sept	100	45×45	Remove seed pod as they begin drying	250-275
Verbena hybrida	Sept/Oct	100	30×30	Remove seed pod as they begin drying	625-650
Phlox	Sept/Oct	100	45×30	Remove as just about to dry	375-400
Viola	Sept/Oct	100	15×15	Collect pod when just about to dry	500-550
*Zinnia**	Jun/July Feb/April	400	45×15	Cut flower head as they dry	400-450

* Sowing times on the hills is Sept-Oct for all the annuals, however, *Celosia*, *Coreopsis*, *Gaillardia* and *Zinnia* should be shown in March-April.

After field preparation, stock seeds are sown directly in the production plots. Some growers raise the young plants in a propagation area before transplanting them into the production fields. The parent plants are usually grown in beds to facilitate irrigation, fertilizer application, and fungicide and insecticide sprays. The plants are checked for genetic uniformity when they begin to flower. Removal of off-types in the population is an intensive activity. This is usually accomplished by a seed company representative possessing detailed knowledge of variety characteristics. A few rounds of **roguing** are often necessary to ensure high genetic purity because not all plants begin to flower at the same time. Timely rouging is extremely important in seed production. The plants which differ from the normal plant population in being weak or sickly or bolters or dissimilar should, therefore be removed at the earliest possible.

To maintain genetic purity, roguing is done at different stages, i.e. at vegetative stage, frequently at flowering, at post-flowering, at pre-harvesting and at harvesting.

As a general rule, **fertilizer** should be supplied in plenty to get high yields of seed, but it varies with crop, soil type and previous crop taken from the same field. Adequate and timely supply of nitrogen is very important for good and healthy growth and increased seed yield (Singh, 2006). In the early stages of growth there should not be too much nitrogen as it causes more vegetative growth, reduced fruiting, and lodging. Along with nitrogen, good supply of phosphorus helps in good root growth, fruiting and seed development and hastens plant maturity, whereas potassium helps to increase disease resistance. General recommendation is to apply 200:80:40 NPK per hectare in improving the seed yield. The nitrogen should be given in two splits, first at 6–8 weeks after transplanting followed by irrigation and the second about four weeks after the first application.

Weed control is another labour-intensive aspect of field production. If left uncontrolled, weeds not only lower seed yield due to competition for nutrients, space, light and water but also harbour pests and diseases. The presence of weed seeds also complicates the seed cleaning process, and noxious weeds if not removed, can make the seed lot unmarketable. There is an increasing use of plastic mulches in seed production fields for weed control and moisture conservation. Good seed yield and quality occur when the environmental conditions are favourable for optimum parent plant growth. Low vigour plants resulting from erratic irrigation and fertilization management are more susceptible to diseases and pests. The plants usually grow so slow that weed control becomes a major problem and most fields require complete hand weeding per month; sometimes it is combined with thinning in drilled crop. Some workers have reported that weeds can be controlled by using chemicals (weedicides) but they have adverse effect on the growth and seed yield of the flower crop (Kaur and Kumar, 2001).

The **virus diseases** are incurable. Only regular checks and removal of affected plant or their parts control their further spread. The **fungal and bacterial diseases** can be controlled to a great extent if they are diagnosed properly. Damping off is most common disease at the seedling stage. Seed dressing with Captan (3g/kg) controls it. Drenching of infected nursery beds with Captan or Thiride (0.2%) will give control of the disease. Stem-rot of many plant tissues caused by fungi and bacteria is another common disease.

Common **insect-pests** of annuals are *Heliothis* sp. and aphids. The larvae of *Heliothis* sp. are polyphagus and attack almost all the plants, most preferred ones being *Antirrhinum, Coreopsis, Cosmos, Dianthus* and *Helichrysum*. The caterpillars can be controlled by spraying of Thiodan 35 EC at 0.2%. The common host plants for aphids are stock, *Nemesia, Calendula* and *Alyssum*. Spraying of Rogor 30 EC or Metasystox 25 EC at 0.1% is effective to control aphids.

Harvesting and Drying

Many popular bedding plant varieties continue to flower and set seed over a long period if the environmental conditions are favourable. Since the seed is harvested only once in the field, determining the proper time to harvest is a critical decision and is based on a compromise between optimum yield and potential seed quality. The onset of unfavourable climatic conditions, such as rainy season or cold weather, also defines when the crop must be harvested. With the exception of a very small production plots and production in countries with low labour-costs, harvesting is done by machines. When the crop is judged ready for harvesting, the plants are cut and placed on canvas tarpaulins to dry in the field. The dried plant materials are threshed by a commercial combine. There are also custom-made harvesting machines for special crops. The field-harvested seed is partially cleaned by scalpers in the open air before being transported to the seed company processing unit. A crop that is harvested too early may germinate well initially, but the seed does not store well. It will also be high in seed moisture content and requires further slow drying. Flower seed drying can be accomplished through sun drying or using a dryer. The seed cannot be threshed from the parent plant with a combine in many cases, because the plant is too green and moist. In those instances, the entire plant is harvested with the seeds and a rotary dryer is used to first dry the plant and subsequently thresh the seeds from the dried plant material. The method of harvesting varies from crop to crop. Most plants continue to flower and set seed till the environmental conditions are favourable. It is critical to determine proper time of harvesting, since harvesting is usually done once in the field. Compromise between optimum yield and seed quality is must. The crop is usually cut and dried in the field. The different methods of harvesting in flower crop are (i) **harvesting the whole plant** as in case of ice plant, linaria, nemophila, ornamental kale, sweet alyssum, etc.; (ii) **harvesting the mature branches** as in case of antirrhinum, clarkia, larkspur, lupins, nemesia, reseda, verbena, etc.; (iii) **harvesting the mature flowers** as in case of chrysanthemum, daisy, gaillardia, gazania, etc.; (iv) **harvesting the mature pods** as in case of pansy, petunia, poppy, etc.; (v) **brooming the field** as in case of coreopsis, garden nasturtium, phlox, etc.

Harvesting of annual flower seeds is the most labour intensive operation than other aspects of seed production. In summer and rainy season, annual seed harvesting is done almost one month after the flowering is completed. In winter season annuals, flowering occurs 90-120 days after transplanting which may last up to 30-60 days or even more resulting in maturity of flowers in various stages. In such cases, it is advisable to harvest individual seed pods/ flower heads/umbels/seed capsule in various flower crops. Harvesting of individual pods is done in balsam, lupin sweet alyssum, heads in gomphrena, marigold, strawflower, sweet Williams, carnation and pinks, capsules in portulaca, etc. In order to save labour costs, instead of harvesting individual seeds/pods/capsules, entire plant is harvested when the maximum number of seeds reach maturity in various crops like African daisy, blanket flower, candytuft, gypsophila, kochia, larkspur, salvia, stock, etc. The harvested seeds are spread over tarpaulin sheet at a well-ventilated place/threshing floor till the complete drying of crop with daily reshuffling of the crop to avoid rotting of the harvested seed crop.

When the crop is properly dried, it is **threshed** either manually or using standard combines. This operation may be repeated few times to get maximum seed yield. During drying process, seeds of some crops like antirrhinum, nasturtium, sweet William, etc. shatter and should be dried and collected on the tarpaulin sheets. Seed **cleaning** is done primarily to remove all debris which range from < 0% to > 80% depending upon crop and harvesting method as well as immature and empty seeds. As flower seeds are harvested over a longer period in staggered manner, environmental conditions change during the production period resulting in heterogeneous nature of seed produce which needs to be cleaned. Seed should be cleaned with the help of blowers for small lot of the seeds. However, for a bigger lot, mechanical cleaning of flower seeds is in practice by various seed companies. If outer coat structures are present on seeds which impede packaging, sowing and water absorption during germination, also need to be cleaned such as removal of hairy layers on the seed coats of anemone and gazania, tail-like structures in marigold and scarification of hard seed coat in case of pelargonium. A complete seed cleaning and grading process generally involves several machines utilizing principles based in different seed characters. **Air screen cleaning** method is of primary importance in seed cleaning and the various processes are involved in seed separation from field debris. Individual flower seeds are separated on the basis of width and thickness. In air screen cleaning method, dust particles are separated by air before the seed passes through a series of screens which are 3-5 in number. The screens are suspended by hanger in such a way that they have horizontal and vertical motion for allowing the raw seed material moment towards screen, and thoroughly stirred. The first screen is a scalping screen which removes large unwanted materials like broken plant parts and dirt clods before they pass through the machine where actual sizing and density separation takes place. The material obtained so is now subjected to second screen which is a grading screen. This screen separates the material into smaller and larger fraction than the screen aperture. The large fraction is the desired whole seed and small fraction comprises of small sized seeds, inert particle and broken seeds. The desired seed is then subjected through air column to remove all light particles. If small seed fraction contains good seed, then screens with small aperture are used to retrieve them. The air screen cleaner is connected to aspiration system for trapping very small sized particle and dust present in the flower seed lot. It is important that all screens should be removed and cleaned carefully in between seed cleaning of various crops for preventing any unwanted seed mixing. Generally, air at high pressure is used to remove seeds from the screen and other unwanted materials present in the machine. The prime seeds collected thus are subjected to further grading.

In order to produce more homogeneous seed, **dimentional seed sizing** is adopted for physical separation by **size and thickness**. The process involves passing of seeds through a set of stacked screens which decrease in aperture size placed horizontally or vertically. The seed travels through the screens and separation takes place by sieving action. There is provision of various types of holes like square, round or slotted holes which affect the sizing process of the seeds producing different results. For example, results of separation through a wire mess screen and a round hole perforated screen will not be the same.

For very small sized seeds, size separation can be done with the help of a sonic siever. The seeds are constantly shaken by up and down action of the sound waves resulting in bouncing of seeds on the screens and sieving through screen layers. The seed which is collected on each screen has almost same size as the size of the aperture of the screen. The seeds keep on passing and retaining on the various screens depending upon the size of the aperture which allow the retaining of seeds or allowing them to pass through and finally they reach the screen with having very small holes to pass again. Another method of seed sizing is according to separation by **seed length**. However, it is important that there must be difference in the length of seeds which are to be separated from each other. This method of seed sizing involves indent cylinder which not only separates seed by length but also separates broken, cracked or weed seeds accurately and quickly. When the seeds go through the cylinder by gravity, the seeds having size identical to that indents of indenting cylinder are fitted there. Shorter seeds are curved upwards out of the seed mass and dropped in a collection trough from which they are collected separately. The seeds which are having more length than the width indent will not be lifted and will remain in the base of the rotating cylinder and are collected in another container. The optimum sized seeds, neither too short nor too long remain in the cylinder and are discharged out through separate spout at the end of cylinder.

Seed grading by density separation is the common commercial practice in flowers. This technique removes not only broken seeds, dirt and inert materials, but also immature, empty and partially filled seeds. It is basically separation by weight and two most commonly used equipments are air column and gravity table. **Air column** works on the principle of terminal velocity in which maximum velocity is attained by a free falling body under a given condition. The seeds are fed at an optimal rate into the column by a vibratory feeder with air flowing through the column. When the seeds fall, the lighter seeds are carried away by the wind to the top of column and are deposited there in collection container. However, heavier seeds fall to the bottom of the column which are collected in another container. Precaution should be taken that not too much seeds should enter the column as it will decrease the accuracy of separation to a greater extent. Therefore, proper knowledge about speed and volume of the seeds entering the column as well as smooth air flow are dispensable for the success of separation through air column. **Gravity table** is another technique for density separation. It is more advantageous than air column as it becomes of higher speed and ensures more volume of seed separation. According to Vaughan *et al.* (1968), seed separation on a gravity separator is based on three principles of stratification such as (i) seeds of same size will be stratified and separated by differences in their specific gravity; (ii) seeds of the same specific gravity will be stratified and separated by differences in their size; and (iii) a mixture of seed differing in both size and specific gravity cannot be stratified and separated effectively. In gravity table, the seed is fed by a vibratory feeder on to a flat, porous deck surface which is made up of woven wire mesh, nylon screen and canvas or cloth material. For maintaining an even distribution of seeds and load of seed on the deck, the speed at which these exit the deck should be the same at which the seeds have been fed on to the deck. There is a fan fitted below the deck for blowing air at a control level of velocity up through the deck surface. The blowing of air is in conjunction with the motion of the deck. Separating the seed into light, medium and heavy layers, the top layer is of light density and bottom layers heavy which can be collected separately in different bins. Factors which influence separation efficiency in gravity table are (i) rate of feeding the seeds into the deck, (ii) velocity of air, and (iii) the pitch and slant of the deck and speed of deck-oscillation. Separation by air gravity table requires knowledge not about only the machine function but also about the principles of separation technique. The technique of separation of seeds by **surface texture** exists since the long time. It is helpful in those seed lots which show no difference in size and minimum variation exists in seed density. Various equipments exist today which can be used for applying the principle of surface texture separation. However, vibratory deck and belt grader are most commonly used equipments. The seed movement in **vibratory deck** takes place across a textured surface of the deck which creates the friction directly related to the surface texture of seeds and the texture of the deck. The amount of friction coming across varies between different types of seeds. For example, a regular shaped, rough coated or broken seeds will come across greater friction than a regular shaped, smooth coated, oval or round shaped seeds. The seeds should be injected in such a manner that the entire surface of deck area is covered which is then vibrated mechanically for continuous seed movement on the deck. The time required for the seeds to cross the deck is directly proportional to the density of vibration and the amount of friction present among seeds and the surface of deck. For instance, broken, rough coated or seeds irregular in shape will have greater friction as the seed edges catch the rough surface of the deck resulting in slow movement of the seeds. However, round or smooth coated seeds experience less friction and hence have free movement across the deck surface. This technique is more successful when there is optimum vibration speed of the deck surface, apparent difference in surface texture, proper speed at which seed is loaded, proper slant and pitch position of deck and the deck covering matching the surface texture of seeds. The use of **belt grader or velvet roll separator** is advantageous over the vibratory deck because more seeds can be processed at a faster speed than the vibratory deck. This machine is designed with belt either vertically inclined, rotating in an upward motion or horizontally inclined moving from left to right. The seed to be graded is fed onto the deck by a vibratory feeder in a single layer. When the belt rotates, seeds with round or smooth surface are rolled on the incline at a higher speed than the set rotation of the belt and can be collected at the base. However, irregular, rough coated or broken seeds will move more slowly due to friction between seeds and surface of the belt and move towards the top of the incline belt and collected there. The success of this technique using belt grader depends upon various factors like proper feed rate of seeds, proper belt speed and inclination and matching of appropriate belt material to the seeds.

There are a few new techniques of seed grading such as **optical sorting technique** where seed is illuminated by light and spectral features of the reflected light are measured. The different seed pigments partly absorb the light in different wavelengths and the light which is reflected from the seed is captured through a detector and analyzed for its spectral information. The spectral features of each seeds are matched with pre-determined values. The seeds with a different brightness or colour from that of the background colour card will lead to change in the signal of the light detector and hence can be separated from the rest of the seeds. **Specific gravity separator** classifies the seed materials according to density or specific gravity. This gravity separator is being used by seed industry for the grading of seed and separation of crop seeds and contaminants. This method is useful in the separation of particles of same size with different density as well as particles of different size with same density. However, this method cannot be employed for separation of a seed mixture consisting of particles of different sizes and density. The various components of gravity separator include an air-blast fan, an air-equalizing chamber, a perforated deck, a feeding device, a variable-speed eccentric, a deck end-raise adjustment, and a deck side-tilt adjustment. The seed mixture which is to be graded is fed at a uniform rate onto the slant of deck and its oscillating motion rolls the seeds over the deck. The air is forced up through the porous deck surface from the equalizing chamber which results in stratification of the material into layers of different densities. The light material floats and heavy material remains attached with the deck. Oscillating motion of deck transfers the material uphill nearly parallel to the discharge on the upper side of deck and collected there, whereas, the light material is carried by air downhill on the lower side of deck and collected there. The discharged material can be further divided into any number of density fraction using movable splitters. For example, for separation of geranium seeds containing seeds of crimson clover, dirt clods, and rocks, by using gravity separator, the geranium seeds and immature clover seeds will be discharged at the lower side of the deck and heavy dirt and rock material at the upper side of the deck, while the good crimson clover seeds can be collected at the middle of deck. **Magnetic separator** exploits advantage of differences in seed coat characteristics. Generally the crop seeds have a smooth coat and contaminants have a rough and sticky surface. The seed lot to be graded is mixed with fine ground iron powder. The powder will stick to the rough seed coat, cracks in seed coat, dirt, chaff or seed with a sticky surface. The treated seed lot is passed over a revolving magnetic drum. The rough textured or other seeds and material which are coated with iron powder either get attracted towards the drum and fall near it or remain adhered to the drum. The materials which remain adhered to the drum are brushed off at the end of each batch through action of brushes or scrappers. However, the smooth seeds which are free of iron powder fall further to the drum in a regular manner and thus separation of good seeds from other materials takes place. The efficiency of this technique depends upon difference in surface characteristics of the component to be separated; greater the difference better will be the separation. Capacity of magnetic separator ranges from 90-900 kg per hour depending upon the type of machine and seed lot to be cleaned. Besides removing particles, magnetic separator also enhances the germination percentage of clean seed lots. As the cracked, immature, misshapen and broken seeds will pick up iron powder and get removed leaving behind the good seeds.

Seed Priming, Pre-germination, Pelleting and Coating

In ornamental plant industry, where plug trays are used extensively for large scale multiplication of plants, predictable and synchronized emergence of seedling have increased. However, the seeds show variable differences in germination due to fungal attack or slow, irregular and too low viability of the seeds under automated conditions. Sometimes, the shape of seed also becomes troublesome in seed sowing. Therefore, to counteract the defective germination and for ease in sowing and synchronized seedling emergence in plug plant industry, various seed treatments like priming, pre-germination, pelleting and coating are done to improve seed characteristics. The concept of **seed priming** is quite old and was practiced for thousands of years in more primitive ways by soaking of seeds in water for some time before they were sown. However, in modern times, seed priming is done to complete the first step of seed germination before they are actually sown. It can be done by partially hydrating the seed to a point where germination begins but no radicle emergence occurs. This technique of seed priming initiates metabolic process within seeds before germination. Seeds are dried usually by exposure to moving air of controlled temperature and relative humidity to the original moisture content after completion of priming treatment. Primed seeds germinate faster than non-primed seeds. This technique offers various advantages like reduced germination duration, improved germination traits, uniform germination as less fungal infection of seedlings, better seedling vigour and crop stand establishment. However, the primed seeds have the disadvantage of reduced longevity if their storage is prolonged. Seed priming can be done using different materials and can be divided into two groups depending upon whether water uptake is uncontrolled, i.e. hydro-priming or controlled osmotic-priming or solid matrix priming (Taylor *et al.*, 1998). In **hydro-priming** method, seeds are either soaked in water or imbibed on moistened blotting papers. The water uptake by seeds is directly proportional to the affinity of tissue for water. In hydro-priming, as the water is not limiting factor, seed may germinate with provision of optimum conditions. Hence, the process of soaking should be interrupted at some specified time to prevent radicle emergence. In **osmotic-priming** the seeds are soaked in osmotic solution of KNO_3, glycerol, manitol and poly-ethylene glycol (PEG), the last one being used most commonly. Unlike hydro-priming, seeds soaked in PEG have restricted water uptake and moisture content rises up to around 40%. The imbibed seeds show reduced metabolic rate than hydro-priming and protrusion of radicle is prevented enabling the continuation of priming treatment for several weeks without radicle emergence. Because of some disadvantages associated with PEG priming, a modification of this technique is membrane priming in which the seeds are separated by a membrane from the PEG solution contained within the walls of double-walled cylinder. Soaking of seeds in salt solution is termed as **halo-priming** and here also partial imbibition of seeds takes place. Efficacy of growth regulator is enhanced when used in combination with priming solution and applied as pre-soaked. For example, addition of gibberellic acid in priming solution exhibited better germination in primula and verbena. **Solid matrix priming** is another method of controlled imbibition of seeds and consists of mixing of seeds with moisture containing matrix like peat, vermiculite, diatomaceous silica and expanded clay. This treatment of seeds with matrix lasts for several days and moisture percentage of seeds and matrix mixture is maintained at such a level that no germination takes place. In this, method, fungicides and growth regulators can also be added. After prolonged treatment, seeds are separated from the matrix and contents are dried.

The technique of **pre-germination** was developed by United Kingdom for commercial application in crops like cabbage, leek and carrot which was further advanced by Syngenta Seeds for impatiens and pansy under the trade names of PreNova and PreMagic. In pre-germination, seeds are allowed to develop until the emergence of radicle, at which time the pre-germination procedure is stopped. Pre-germination differs from priming as limiting amount of water is given to the seeds and the pre-germinated seeds are planted wet using hydro-seeder as compared to primed seeds which can be easily dried and planted dry using conventional planting equipments. Generally, most viable seeds show earliest germination and are eliminated in the process. As most viable seeds selected are at the same stage of development, thereby exhibit almost 100 per cent emergence and high uniformity which especially is important in plug plant production industry. Various **benefits of pre-germinated seeds** include reduced seed rate, uniform emergence of seedlings, higher percentage of usable seedlings, increased yield of the crops and discarding of low-vigour seeds before sowing.

Seed pelleting can be defined as the application of layer of inert material that may obscure the original shape and size of the seed (Butler, 1993). In this technique, flower seeds of any shape and size such as small or large, round or irregular are covered with a coating of inert materials and made into spherical pellets each containing a single seed. The various materials like clay, wood-flour and diatomaceous earth can be used for making pellets. The seeds are placed in a pan or rotating drum containing pelleting material in water at prescribed rate. After completion of pelleting process, pelleted seeds are dried for storage for planting through precision planting equipments. The advantages associated with this technique are reduced seed rate and optimum planting densities thus reducing the cost of thinning operation. Sometimes, multi-pelleting of seeds is also done where sowing of more than one seed per cell is preferred especially in case of lobelia. **Film coating** of those seeds is done which do not require pellleting but require some protection against disease and pest attack. This protection is given in the form of encapsulation of plant protectants by film coating technique. Application of film coating is done by spraying a liquid mixture or powder of a formulated active ingredient, a polymer, a colourant and a filler. In case of a flower seed, batch treaters are used for coating of the seeds. Film coating is advantageous as no health risk is associated with dustless planting using encapsulated seeds; it ensures better seed sowing, improved germination and better crop stand; and it increases visibility of the seed in the soil due to use of colourants. However, sometimes chemicals used in coatings may impede germination, and the use of polymers and plasticizers may be toxic or inhibitory to seed species.

Storage of Flower Seed

The seed produced is an outcome of large investments of time, efforts, labour and money on which there can be no return until the seed is sold. The main purpose of seed storage is to keep the seed in good physical and physiological state from harvesting till seed sowing. The seed must be stored in a structure which ranges from primitive types like clay pots, woven baskets or even pits dug in the ground to the modern storage structures like cold storage, deep freezers, etc. which keep the loss of seed viability and seed vigour to the minimum. On the basis of the duration of storage, seed storage is of three types, *viz.* short term storage, intermediate and long term storage. **Short term storage** of seeds is for a short duration, i.e. 6-8 months (from harvest to the next planting), **intermediate** where carry over seeds are stored for a duration of 12-36 months, and **long term storage** where germplasm, nucleus seed and samples tested in seed testing laboratories for regulatory purpose is done and this storage duration may range from 5-20 years. Stages at which flower seeds are stored can be divided into six such as (i) storage on plants, i.e. from physiological maturity until harvest; (ii) harvesting and processing period; (iii) storage in warehouses; (iv) storage during transit in rail wagons, trucks, railway sheds, etc.; (v) storage in wholesaler's and retailer's stores; and (vi) storage on the user's farm. Principles underlying for effective seed storage of various flower crops should be followed which are storage of only those seeds which are of high quality, i.e. properly cleaned and treated, drying of seeds to a safe moisture level, storage under dry, cool and clean storage conditions, and there should be effective pest control measures.

Factors Affecting Seed Storage

The seeds after proper cleaning are packed in cotton bags and kept in cool and dry places. Before storing, the seed is dried to low seed moisture (sun drying or in seed dryer). The seed is stored at 15°C and RH of less than 30% (Carpenter and Osmark, 1992). For an effective storage of flower seeds, they should be kept dry and temperature should be maintained low. The favourable conditions required for seed storage are just opposite to good seed germination. There are various factors which affect seed storage such as relative humidity which influences the seed moisture content if not stored in moisture-proof containers, e.g. if there is 15% RH the seed will dry down to 6% moisture and will store safely for several years though at 90% RH the seed will dry down to 19% moisture so germination will be very poor after storage for one year. Seed moisture is inversely proportional to the seed storage as well-dried seeds store longer. Seeds become hard if moisture is reduced below 8% and the seeds are known as hard seeds. Hard seeds resist germination under favourable conditions as they absorb water slowly when sown, germinate slowly and produce seedlings over an extended period. The life of seed doubles with every 5 °C decrease in storage temperature (between 0 °C and 50 °C).

Germination of flower seeds stored for variable length of time

S. No.	Crops	Original germination (%)	Storage period (years)	Viability % at end of storage
1.	*Callistephus chinensis*	83.5	2	81.44
2.	*Cosmos bipinnatus*	80.5	5	69.0
3.	*Calendula officinalis*	83.5	9	54.0
4.	*Dianthus caryophyllus*	84.5	10	47.0
5.	*Iberis umbellata*	47.0	1	19
6.	*Delphinium chinensis*	38.5	1	20.5
7.	*Delphinium* hybrid	71.0	1	34.97
8.	*Helichrysum*	72.0	2	58.0
9.	*Kochia*	63.5	1	34.65
10.	*Viola tricolor* var. *hortensis*	81.5	5	45.0
11.	*Petunia hybrida*	81.75	5	73.50
12.	*Phlox drummondii*	48.0	3	26.0
13.	*Lathyrus odoratus*	85.5	5	67.5
14.	*Tagetes erecta*	72.5	5	53.5
15.	*Tagetes patula*	74.0	5	45.0
16.	*Tropaeolum majus*	78.0	10	54.0
17.	*Verbena hybrida*	53.5	3	38.0
18.	*Zinnia grandiflora* var. *robusta*	81.5	10	48.0
19.	*Zinnia* (dwarf)	96.5	10	82.5
20.	*Zinnia* (liliput)	91.0	10	63.5
21.	*Zinnia* (double giant)	96.0	10	56.5

Key Challenges for the Seed Production

1. High seed quality requirement. In response to commercial pressures, significant improvements in flower seed production have been made. A reliable supply of high-vigour seed is now available in a range of high-value bedding plants. However, a great deal of additional work still remains in many annuals and perennials.

2. New crops. Learning to produce a new crop is tedious and time-consuming. Experience with growing existing crops may not carry over to the new products. Details of crop culture, pollination methods, harvest and post harvest handling procedures must be identified. This is difficult because there is usually only one chance of learning per year. As the average market life of a product decreases, mistakes made in early production trials can be costly.

3. Small-scale production. When the total market is divided up by increasing numbers of unique products, the size of production for each product decreases. This causes different problems for field production of flower seeds. The costs of training and general labour have increased. For production of specific crops, it may require different cultural regimes. When grown in the same field, more hand labour will be required to provide appropriate treatments. There are also more training and supervisory costs associated with cultural practices and harvesting of multiple crops as compared with a single crop in the same area of production space. For field production, finding enough isolation to prevent cross-pollination can be a major problem. It is also difficult to find contract growers who are willing to produce crops in small plots.

4. Organic seed. There is an emerging market for certified organic crops, which mandates the use of organically produced seeds. Currently, most commercial flower seeds are produced using conventional agricultural practices as organic certification requires production in fields with limited chemical usage. Most of the synthetic pest and disease control agents are disallowed in organic production. In order to produce organic seed, not only does the seed producer need to go through a 3 to 5 year transition period so that seed fields are free of chemical residues, the producer must also learn an entire new set of crop management techniques, utilizing primarily natural products (McDonald and Kwong, 2005).

5. Skills and techniques. Very different sets of skills are required to address these new flower seed production challenges. Some general background knowledge is available, but much more research is needed to bring about solutions to these issues. It is likely, at least in the short term that these challenges will be met by different specialists rather than one large seed producing company. These issues offer the opportunity for new players to come into the flower seed industry and challenge existing seed producers to evaluate where they can deliver the best value.

Cost Factor

The cost of seed production ranges from Rs.18,750 to Rs.87,500/ha and it generates a net profit of Rs.75,000 to Rs.4,50,000/hectare, depending on the nature of crop, skill of management and seed collection. The per hectare income is 2-10 times as compared to the conventional cereals, the production costs are also correspondingly higher and more than 50% costs are for field labour. However, the labours should be skilled and should be hired from locations where it is damn cheap.

Income per hectare of important flower crops

S. No.	Flower species	Seed yield (kg/ha)	Price (Rs./kg)	Gross income (Rs.)	Net income (Rs/ha)
1.	*Antirrhinum majus*	310	500	1,55,000	1,25,000
2.	*Bellis perennis*	200	1,100	2,20,000	1,95,000
3.	*Coreopsis lanceolata*	500	200	1,00,000	87,500
4.	*Coreopsis tinctoria*	435	225	97,875	87,500
5.	*Cosmos bipinnatus*	625	200	1,25,000	1,06,250
6.	*Cosmos sulphureus*	750	175	1,31,250	1,12,500
7.	*Gaillardia aristata*	625	350	2,18,750	1,93,750
8.	*Gazania splendens*	560	200	1,12,000	93,750
9.	*Helichrysum bracteatum*	310	450	1,39,500	1,15,625
10.	*Mesembryanthemum criniflorum*	625	350	2,18,750	2,00,000
11.	*Papaver nudicaule*	250	500	1,250,00	1,00,000
12.	*Petunia hybrida*	250	1000	2,50,000	2,25,000
13.	*Phlox drummondii*	375	350	1,31,250	1,06,250
14.	*Verbena hybrida*	625	400	2,50,000	2,31,250
15.	*Viola cornuta*	500	1000	5,00,000	4,12,500

References

Butler, R. 1993. Coatings, films and treatments. *Seed World* (October issue), pp. 19-24.

Carpenter, W.J. and E.R. Osmark, 1992. Growth regulators and storage temperature govern germination of flower seeds. *Hort Sci.*, **27**: 1190-1193.

Jana, B.K. 1998. Zinnia. In: Commercial Flowers (eds Bose, T.K. and L.P. Yadav), pp. 837-841. Naya Prokash, Calcutta, India.

Kapoor, M. 2002. Floriculture-A Promising Forex Earner. *The Tribune* (Agriculture Tribune),**121**: 19.

Kapoor, M. 2004. Commercial Production of Flower Seed. *The Tribune* (Agriculture Tribune), **123**: 15.

Kaul, G.L. and N.K. Dadlani, 1995. Prospects of Floriculture in India. Government of India, Department of Agriculture and Co-operation, Ministry of Agriculture. Krishi Bhavan, New Delhi.

Kaur, K. and R. Kumar, 2001. Effect of different weedicides on seed yield in candytuft. Proc. of 4th Punjab Science Congress, held at PAU, Ludhiana.

Kumar, R. and K. Kaur, 1996. Effect of different planting times and method on plant growth and seed yield in *Petunia hybrida. J. Ornam. Hort.*, **4**: 23-26.

McDonald, M.B. and F.Y. Kwong, 2005. Flower Seeds: Biology and Technology. CABI International, London.

Miyajima, D. 1996. Reproductive biology of flowering plants. *Seed Sci. Tech.*, **24**: 465-473.

Singh, A.K. 2006. Flower Crops: Cultivation and Management, 450 p. New India Publishing Agency, New Delhi.

Swarup, V. 1967. Garden Flowers. National Book Trust of India, New Delhi.

Taylor, A.G., P.S. Allen, M.A. Bennett, K.J. Bradford, J.S. Burrisand and M.K. Misra, 1998. Seed enhancements. *Seed Sci. Res.*, **8**: 245-256.

Vaughan, C.E., B.R. Gregg and J.C. Delouche, 1968. Seed processing and handling. Seed Technology Laboratory Handbook No.1. Mississippi State University State College, Mississippi, USA.

37

Light in Regulating Growth and Flowering

Tanya Thakur, H.S. Grewal, Sanyat Misra and R.L. Misra

Introduction

Photoperiod is derived from two words 'photo' meaning light and 'period' meaning a specific length of time. This is how a plant responds with respect to flowering to the relative amount of light. Photoperiodism is phenomenon to regulate the time of flowering in plants, and this involves wavelengths centred around red (660 nm) and far red (720 cm). Red light (660 nm) is most effective in interrupting night length. In fact, photoperiodism is a phytochrome-mediated response, the phytochrome molecule is a light-absorbing pigment which exists in two basic but inter-convertible forms: P_{fr} which absorbs far-red light of wavelength 730 nm and is inactive form and P_r which absorbs red light of wavelength 660 nm and is active form, when red light strikes P_r, this switches to P_{fr} form and similar to this, P_{fr} switches to P_r when far-red light strikes the P_{fr}, during the night or during the dark this P_{fr} slowly (spontaneously) reverts to the P_r form, and under simplest conditions high levels of P_{fr} inhibit flowering in SD (long night) plants and promote flowering in LD (short night) plants, and so surprisingly, using far-red blocking filters encourage flower initiation in SD chrysanthemum plants even under LD photoperiods (McMahon, 1997) .

Photoperiodism is measured in terms of daylength that causes the changes from vegetative to flowering stage. However, this has three factors, (i) **colour**, i.e. quality, the wavelength of the light which is measured in nanometers (nm) with specific wavelengths corresponding to specific colours such as yellow light having a wavelength of approximately 580 nm though generally the lights required for plant growth are in all wavelengths, and the red (700 nm) and blue (470 nm) wavelengths result into greatest growth response; (ii) **intensity** meaning the brightness (quantity/strength) of the light, and (iii) **duration** (**daylength** or **photoperiod**, i.e. the critical period of light) for a particular species or variety; and these all the three factors decide the quality and quantum of light required by the plant. Through photosynthesis, light energy is converted into chemical energy to sustain essential plant growth, and afterwards the light modifies specific physiological responses such as dormancy, seed germination, modified organ formation, flowering and senescence.

Luminous energy, i.e. light visible (colour) to the human eye, is quantified in terms of footcandle (fc) where 1 fc = 10.8 lux, and this emphasizes the green-yellow wavelengths (530 to 580 nm) which the human eyes see the best. However, the photosynthesis is driven by a broader range of wavelengths with a red-blue emphasis. Light meters which measure footcandles are not so punctuated to measure the greenhouse light as there is probability of error up to 45% than what actual photosynthetic energy is there in the radiation (Muckle, 1997), and photosynthetically active radiation (PAR) measures the amount of light energy equally in all wavelengths from 400 to 700 nm without stressing the green-yellow wavelengths most visible to the eye (Dole and Wilkins, 1999). Based on quantum energy, there is a third system of measuring light intensity, where light is explained in terms of tiny particles of energy called photons or quanta and is called photosynthetic photon flux (PPF) and these numbers of photons are measured in moles (mol) or einsteins (E) with 1 mole = 1 einstein = 6.023×10^{23}. This way, in the quantum system, the intensity of light would be measured in the number of photon being transmitted such as $\mu mol \cdot s^{-1}m^{-2}$, which is the preferred unit of measure. Thimijan and Heins (1983) gave approximate conversion factors between fc and $\mu mol \cdot s^{-1}m^{-2}$ as daylight 0.20 = incandescent 0.22 = metal halide 0.15 = cool white fluorescent 0.15 = warm white fluorescent 0.14 = high pressure sodium and mercury deluxe 0.13 = low pressure sodium 0.10, and these given values if are multiplied with those of fc's the $\mu mol \cdot s^{-1}m^{-2}$ for wavelengths ranging from 400 to 700 are obtained. PAR and PPF both, irrespective of human eye, emphasize all wavelengths equally within a specific range such as 400 to 850 nm. Photoperiod affects plant growth in two ways, *viz.* the same intensity of light for shorter duration to the plants provides less total light energy than long photoperiods, and for longer duration it may alter specific physiological activity in many of the species independent of the light intensity, and this is termed as photoperiodism (Dole and Wilkins, 1999).

Plant Growth

Whatever the light energy is required by the plant, if is being received, that is the **light saturation point** for that particular species or cultivar so more than that cannot be used and more than required light energy may even prove injurious causing plant stunting, necrosis or yellowing of foliage or too high of light may appear in the form of sunburning though it very much depends on the species since different species may vary in their light saturation points as cacti and many other succulents have a very high light saturation point while African violets, coleus, and ferns have low requirement. The light intensity at which the species is getting as much light energy during its photosynthesis as it is using during respiration is the **light compensation point**, and plant growing at light compensation point neither accumulates dry weight nor deteriorate, though this point may vary with each plant species. Many of the foliage plants being grown indoors have low light compensation point though outdoor-grown bedding plants have a high light compensation point. By changing the environment of growing, both the points, i.e. light saturation and light compensation, can be changed or altered. To encourage rapid economical growth of the foliage plants for marketing, these can be grown near the saturation point but will have to acclimatize the plants by lowering the light levels before marketing so that respiration rate is also relatively lowered, and so these may be grown in their usual environment after marketing. Under low light, the leaves elongate but become thin with thin cuticle and will have fewer palisade cells with less organized chloroplasts arranged horizontally within the cell than those grown under high light, but in case if the change in environment during acclimatization process is not gradual, and instead is sudden and too great, the plants may deteriorate and die slowly. **Acclimatization** process suits only foliage plants, or others (pot- and cut-flower plants) when they are only in early vegetative phase because afterwards such plants will require maximum light levels to maximize carbohydrate accumulation to have a long postharvest life. Smith (1986) stated that light in far-red wavelengths increases internodal elongation and leaf size, reduces colour of the leaves and flowers, and decreases branching, though plants growing in light high in red wavelengths will be small, well-branched and dark green. Since foliage absorbs red light so the light filtered through leaves will be high in far-red light and will allow its greater proportion to pass through, it is therefore that closer plant spacing decreases the red to far-red ratio resulting into elongated and poorly branched plants (Dole and Wilkins, 1999). However, incandescent lamps are high in far-red wavelengths though fluorescent lamps are high in red wavelengths (Dole and Wilkins, 1999). Manipulation of light quality through filters may control plant height, and the ones with a high red to far-red light ratio will reduce plant elongation and such lights can be obtained by filtering sunlight through a copper sulphate solution (Mortensen and strømme, 1987; McMahon *et al.*, 1991; Rajapakse and Kelly, 1992).

The **greenhouse** construction facilitates efficient utilization of light (natural or supplemental) so it should be constructed in a manner that there is maximum availability of sunlight to the plants inside, and inside it is utter economical to reduce the light levels than to increase it, but this all depends upon the type of glazing materials used. The glass is highly transparent than the polythene, and double-layered polythene has poorest transmission of light energy in the polyhouse. Roof fittings should be as little as possible so that these may hinder very little with light transmission inside, *vis-a-vis* cleaning of glazing material every year during autumn and painting white the internal surfaces of the house will reflect more amount of light in the greenhouse. To allow the plants availing maximum natural light, the plants should be **spaced properly** otherwise the growth of canopy will fill the adjoining space and will not permit penetration of light to its full strength to the level each plant should receive, thereby decreasing the amount of dry weight, *vis-à-vis* plant quality. The required light by each plant can be given by spacing the plants properly as per requirement of each species or varieties, especially during the winter when light level remains very poor, though decreased plant population will cost the grower of its final income dearly. Though expensive but overall income may turn economical by providing **supplemental lighting** for high-value crops such as anthurium, carnations, gerbera, roses, *etc.* and even to low-value crops when these are placed quite close as with seedlings, plugs and cuttings, especially during winter when natural light levels are utterly poor, and this lighting for only a few weeks will increase plant growth and quality over their entire life even when they are not given additional lighting afterwards. The minimum supplemental light requirement may vary from species to species, 300 to 600 fc (60 to 120 $\mu mol \cdot s^{-1} m^{-2}$) is the general recommendation. For providing supplemental lighting to ornamentals during naturally short days to create artificial long days, there are incandescent lamps or HPS lamps for photoperiods, fluorescent lamps (cool white or warm white) and LEDs for germination and in the growth rooms, and high intensity discharge (HID lamps, being of four types such as high pressure mercury, low pressure sodium and high pressure sodium providing supplemental light especially during the nights for greenhouse use, and metal halide for all purpose such as greenhouse and growth rooms) lamp types. HPS lamps are installed to provide a minimum of 10 fc at the top of plants. However, Inc lamps are used most often because they are efficient generators of red light (the LD response is most sensitive to red light) and are inexpensive & easy to install. Use of 60-watt incandescent lamps with reflectors at spacing of 1.5 m apart between lamps and 90 cm above tops of plants is quite effective. There are three possible options of lighting regimes, *viz.* lighting from sunset until later in the night (2200 hours) till critical photoperiod has been met with and this is referred to as **day-continuation (day-extension) lighting**, from 0200 hours until sunrise which is referred to as **pre-dawn lighting**, and lighting in the middle of the night preferably from 2200 hours until 0200 hours which is referred to as **night-interruption (night-break) lighting**. Fewer hours of lighting are needed for night-interruption compared to day-continuation or pre-dawn lighting, so the former is more economical and is used most often for lengthening the daylength. Cyclic lighting is getting popular these days by using incandescent lamps on a timer and setting them to turn on in the middle of the dark period for a specific cycle. Cyclic lighting is only practical with incandescent lamps because the life of HPS lamp depends on the time the bulb is turned on and off. For starting seedling and rooting in cuttings, fully enclosed **growth rooms** can be used efficiently with tiers of fluorescent lights (warm white and cool white types combined) installed 45 to 60 cm apart which will provide sufficient light intensity, *vis-à-vis* heat. However, non-enclosed structures (rooms) will require additional expenditure for installing fans and ventilators for circulating air and for exhausting excess heat. Duration of lighting may vary from species to species though generally it is for 16 hours per day.

During summer, i.e. natural long days, when too high light intensity may prove injurious to the plants by causing growth stunting, leaf yellowing *cum* necrosis, and plant desiccation, it is must to **minimize light energy** by covering the plants with an opaque black plastic or cloth material from 1600 hours to 0800 hours that reduces the light intensity less than 2 fc. A 1.2 m height is recommended between the top of cloth and the bench (bed) top to reduce heat buildup around plants. Several types of materials can be used to cover plants such as **black sateen cloth** which is a jet black woven cloth made from cotton, the **polyolefin sheeting** which is tear-resistant and waterproof, and 4-6 mm **black plastic** which is not tear-resistant. Application of artificial SD treatments during warm weather may result in increased temperatures **causing** delayed flowering in chrysanthemum, kalanchoe, poinsettia and many other plants and this is **called as heat delay**. Use of a blackout system with an aluminum material prevents this delay in flowering. The coloured screens are available in a variety of types (25 to 98% shading) which may be spread on the top but inside of the greenhouse in order to create artificial short days, these are fitted with device (photosensors) to roll these at times, *vis-à-vis* making provision for exhaustion of hot air from the top as well as from the bottom, even if exhaust fans are also required in usual course.

Photoperiodism

Through night interruption lighting (LD) or by black clothing (SD), the daylength, as per specific requirement of a species, can be manipulated conveniently. Extension of light duration during night hours has been found regulating flower initiation and development in many of the species of ornamentals. Apart from flowering, this also influences seed germination, the growth habits of plants, rest in seeds, and the dormancy of underground modified structures of the plants. These responses can be grouped into three categories, *viz.* **short day** (SD) plants where dark period is longer than a critical length as in *Dendranthema* and *Euphorbia pulcherrima*, **long day** (LD) plants where dark period is shorter than a critical length as in *Begonia tuberhybrida*, and **day neutral** (DN) plants where for flowering other stimuli such as cold or total light energy are responsible as geraniums flower more rapidly with increased light energy through supplemental lighting but with same daylength. There are certain species which do not fit into SD, LD or DN categories as there photoperiod responses are more complicated, sometimes LD followed by SD, or SD in tune with cold treatment. Also there are **facultative** (quantitative) SD (*Salvia splendens*) or LD plants where these for flowering require a wide range of daylengths though flowering is hastened under either SD or LD, and increasing the number of LD or SD cycles generally decreases the time to flower though **obligate** (qualitative) SD (*Poinsettia pulcherrima*) or LD plants will not initiate flowering if placed under incorrect photoperiods. All the photoperiodic responses other than flowering such as germination and dormancy may also be termed as obligate or facultative. For photoperiodic treatments of ornamentals for induction of flowering, the total plant size and plant maturity (passing of the juvenile phase) should be taken into account so that effect is conspicuous and size, *vis-à-vis* quality of flower is proper. For quicker results, quality will have to be compromised.

The division of the darker period of night into two or more shorter nights, preferably from 2200 hours to 0200 hours with light intensity of at least 10 fc (2 $\mu mol \cdot s^{-1}m^{-2}$) at plant apex, when the days are short and that too from mid-October to February end under sub-tropical Indian conditions, through supplementary **light interruptions** (night breaks, i.e. NB), is useful in preventing flowering or delaying floral initiation of SD plants, *vis-à-vis* inducing flowering of LD plants though it may differ in species. Days can also be extended by turning on lamps at the end of the day (**day continuation**, i.e. DC) into the night. For this, the strings of incandescent lights with reflectors are used as these produce adequate levels of red light to induce photoperiodic response. Likewise, the **night extensions** can also be achieved by using black cloths placed over the crops, preferably from 1700 to 0800 hours or 1900 to 0700 hours when heat buildup is a problem, in case of SD plants or for inhibiting flowering in LD plants.

Flower Initiation

The events of changes in the morphogenesis and cell differentiation that occur in the apical meristem which leads to production of flower organs is called **floral evocation.** It leads to formation of specialized flower organs like sepals, petals, stamens and carpels. Floral meristem can be differentiated from vegetative meristem in the early stages when these attain the larger size. Transition to reproductive meristem leads to increase in the frequency of cell division in the central zone. The factors which control floral initiation are (i) **photoperiod** which controls the time of flowering through daylength manipulation, (ii) **vernalization** which is a low temperature treatment to control flowering, (iii) **autonomous pathway** which comprises combination of factors involved in RNA processing and epigenetic regulation that down-regulate the floral repressor **flowering locus C** (FLC), and (iv) **GA-dependent short day pathway** which involves severe reduction of endogenous gibberellins as in *Arabidosis* and delays flowering in long days plants and this prevents flowering under long days. This effect of gibberellins on flowering is correlated with expression of LEAFY, a floral meristem identity gene. In 1920, W.W. Garner and H.A. Allard of the U. S. Department of Agriculture, discovered a mutation in tobacco (*Nicotiana tabacum*) variety 'Maryland Mammoth', growing profusely to about 5 m in height but failing to flower in the prevailing conditions of summer when tobacco normally flowers. However, when the plants were grown under the greenhouse during the winter under natural light condition, flowering occurred. The result ultimately led Garner and Allard to test the effect of providing artificially short days by covering plants with black cloth grown during the long days of summer. These artificial short days also caused the plants to flower under the long days of summer. Garner and Allard concluded that the length of the day was the determining factor in flowering and were able to confirm this hypothesis in many different species and conditions. This work laid the foundation for the extensive subsequent research on photoperiodic responses. Photoperiodism in *Xanthium* (cocklebur) is completely dependent on a critical night length of 8.5 hours, but not on day length, therefore, it is called a short-day plant though in real sense it is a long-night plant. Interruption of long night by red light (660 nm) rays is particularly effective to prevent flowering. An intense exposure to far-red light (730 nm) at the start of the night reduces the dark requirement by two hours.

The essential distinction between long-day and short day plants is that flowering in LD plants is promoted only when the day length exceeds certain duration, called the **critical day length**, in every 24-hour cycle, whereas promotion of flowering in SD plants requires a day length that is less than the critical day length. Long-day plants can effectively measure the lengthening days of spring or early summer and delay flowering until the critical day length is reached. Many varieties of wheat (*Triticum aestivum*) behave in this way. SD plants often flower in fall, when the days shorten below the critical day length, as in many varieties of *Chrysanthemum morifolium*. **Long-short-day plants** (LSDPs) flower only after a sequence

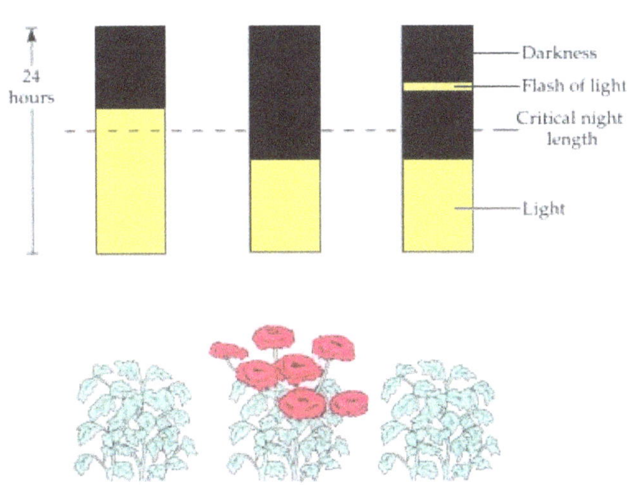

Short-day (long-night) plants

of long days followed by short days. LSDPs, such as *Aloe, Bryophyllum*, *Cestrum nocturnum* (night-blooming jasmine) and *Kalanchoe* flower in the late summer and fall, when the days are shortening. **Short-long-day plants** (SLDPs) flower only after sequence of short days followed by long days. SLDPs, such as *Campanula medium* (Canterbury bells), *Echeveria harmsii*, and *Trifolium repens* (white clover) flower in the early spring in response to lengthening days. **Day neutral plants** (DNPs) flower under any photoperiodic condition and are insensitive to day length. Flowering in DNPs is typically under autonomous regulation, i.e. internal developmental control. Amaranths, balsam, French marigold, geranium, gerbera, rose, stock, sweet William and wax begonia are important day neutral species. The day length of 24 hour cycle measured in hour is called critical day length. It is longest day length for long day plant and shortest day length for short day plant.

Crop	CD (h)
Chrysanthemum	13.5
Henbane	11.5
Morning glory (*Ipomoea*)	15
Carnation	13
Marigold	12.5
Kalanchoe	12
Gladiolus	14.5

Soviet-Armenian plant physiologist Mikhail Chailakhyan in 1937 demonstrated that floral induction can be transmitted through a graft from an induced plant to the non-induced plant to flower through a hypothetical hormone called florigen. Florigen does not exist as an individual substance; rather, florigen's effect could be the result of a particular ratio of other hormones. Later, Anton Lang showed that several long-day plants and biennials could be made to flower by treatment with gibberellin, when grown under a non-flower-inducing photoperiod. This led to the suggestion that florigen may be made up of two classes of flowering hormones: gibberellins and anthesins. It was later postulated that during non-inducing photoperiods, long-day plants produce anthesin, but no gibberellin while short-day plants produce gibberellin but no anthesin. However, these findings did not account for the fact that short-day plants grown under non-inducing conditions (thus producing gibberellin) will not cause flowering of grafted long-day plants that are also under non-inductive conditions (thus producing anthesin).

The photoperiodic stimulus as florigen in both LDPs and SDPs is perceived by the leaves, as treatment of a single leaf of the SDP *Xanthium* with short photoperiods is sufficient to cause the formation of flowers, even when the rest of the plant is exposed to long days. Thus, in response to photoperiod the leaf transmits a signal that regulates the transition to flowering at the shoot apex. The photoperiod-regulated processes that occur in the leaves resulting in the transmission of a floral stimulus to the shoot apex are referred to collectively as photoperiodic induction. Photoperiodic induction can take place in a leaf that has been separated from the plant.

The floral stimulus is transported *via* the phloem once produced, the flowering stimulus appears to be transported to the meristem *via* the phloem, and it appears to biochemical rather than physical in nature. Treatments that block phloem transport, such as girdling or localized heat-killing prevent movement of the floral signal. Because the floral stimulus is translocated along with sugars in the phloem, it is subject to source-sink relations. An induced leaf positioned close to the shoot apex is more likely to cause flowering than an induced leaf at the base of a stem, which normally feeds the roots.

Change in daylength

↓

Photoperiod mechanism in leaf

↓

Florigen hormone

↓

Transported to flower bud

↓

Flowering occurs

Photoperiod Mechanism: Phytochromes

Phytochrome is a homodimer, i.e. two identical protein molecules each conjugated to a light-absorbing molecule. Plants make 5 phytochromes, PhyA, B, C, D, and E that differ in their absorption spectrum, i.e. which wavelengths (e.g., red vs. far-red) they absorb best. At the starting of day, there is high accumulation of P_r in plant

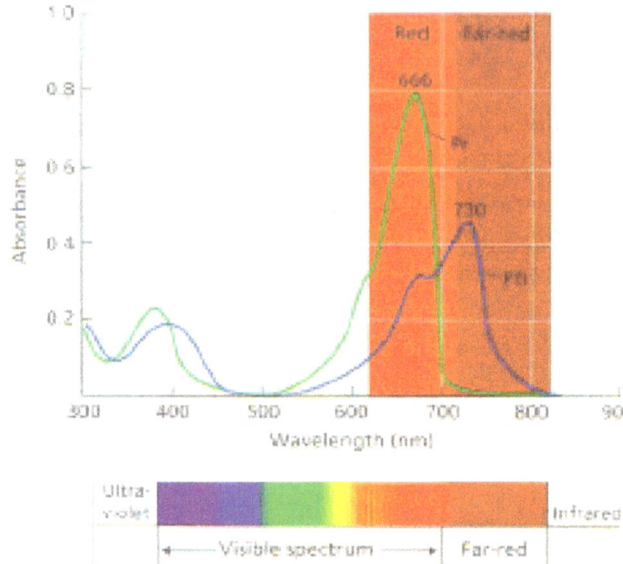

FIGURE 17.3 Absorption spectra of purified oat phytochrome in the Pr (green line) and Pfr (blue line) forms overlap. (After Vierstra and Quail 1983.)

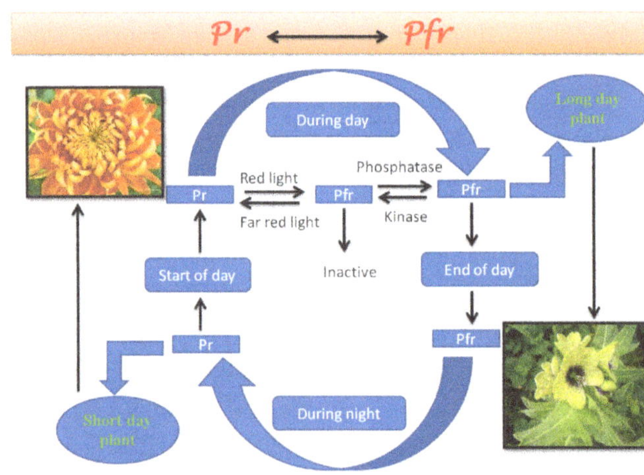

and during day, there is conversion of P_r to P_{fr} and the ratio is in equilibrium at mid-day. At end of the day, there is high accumulation of P_{fr} in plants and during night, there is conversion of P_{fr} to P_r, but still there is 1-2% of P_{fr} present. So this conversion is not 100%. P_r causes flowering in short day plants while P_{fr} causes flowering in long day plants. P_r absorbs red light from sun and gets converted to P_{fr} which activates florigen hormone in long day plants and causes flowering. While P_{fr} absorbs far red light and gets converted (at fast rate or in dark at slow rate) to P_r which activates florigen hormone in short day plants and causes flowering.

For plants with a critical night length, a short flash of light in the middle of the night would make the plant behave as if it had been exposed to a longday. In short

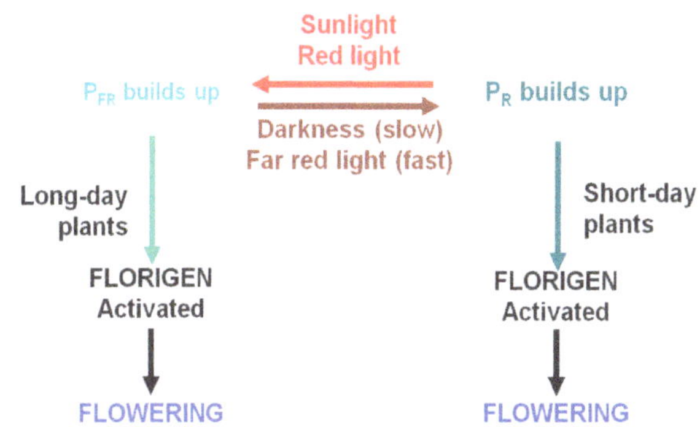

day plants, **night interruption with red light** inhibits flowering but red light followed by far-red light cancels the effect of red light and causes flowering. While in long day plants, night interruption with red light causes flowering

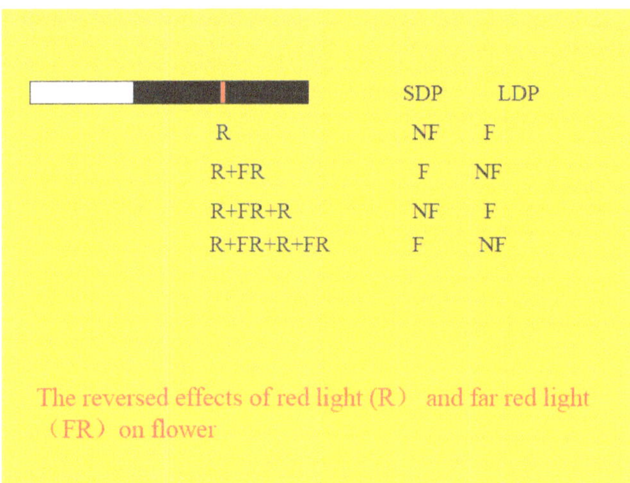

The reversed effects of red light (R) and far red light (FR) on flower

but red light followed by far-red light cancels the effect of red light and inhibits flowering.

Photoperiodically Controlled Ornamental Crops

Aster

Aster is grown as perennial garden flower as well as a cut flower and pot plant with capitulum inflorescence of wide range of ray floret colours, *viz.* blue, purple, violet, pink and white. The life of aster plant lasts for two years, and three flowerings per year can be easily obtained. Long day conditions are beneficial to stem elongation and inhibit flowering, while short days are needed for elongated stem to bloom (Wallerstein *et al.*, 1992). Thus, it is a short day plant. The increase in number of light hours produces increase of stem length, internode length and diameter of capitulum (Gudo and Scordo, 1994). At photoperiod of less than 17.5 hours, lack of flowering and apical bud abortion was observed (Farina *et al.*, 1994).

For timing the flowering in asters, different photoperiodic requirements are needed as per phenological stages of plants. Short days improve plant establishment, i.e. formation of basal leaf rosettes and during this phase photoperiod is usually not regulated. After the formation of the basal leaf rosettes, control of photoperiod starts, i.e. roughly 2-3 weeks after planting. Stem elongation requires long photoperiod, usually longer than 16 light hours (Schwabe, 1985) but it should be stopped when plant attains 40-60 cm of height, and the floral initiation requires short photoperiod, usually less than 10 light hours (Schwabe, 1985). The reaction time from short days to flowering varies from 5 to 10 weeks, and this determines the harvesting date. Flower formation is not affected by daylength. However, a new cycle of scheduling for flowering starts after pruning.

Carnation

Dianthus caryophyllus is a quantitative long day plant with critical photoperiod of 13 hours. Long days promote and short days delay flower initiation, but photoperiod does not affect flower development. Under long days, plants flower sooner and more uniformly with fewer nodes but longer internodes than plants produced under short days. Most cultivars respond to 2-3 week of lighting. Long days can be given in the form of continuous, extension or night break treatment. Long days treatment inhibits lateral branching so stock plants are produced under short days to increase lateral shoot production, better rooting and cutting quality. After pinching, short days are given to promote lateral branching and continued till 70% shoots have 10 visible buds (Healy and Wilkins, 1983). After long days, plants flower in 12-16 weeks depending on temperature, cultivar and light intensity. Long days of 20-30 are required at low light intensity and 7-14 long days at high light intensity. Supplementary lighting during low light intensity can increase growth and development and enhance flowering. A minimum amount of light to sustain carnation growth is 21.5 klux.

Chrysanthemum

Dendranthema × *grandiflorum* (syn. *Chrysanthemum* × *morifolium*) is an herbaceous perennial with composite inflorescence grown as cut flower, pot plant and as a flower for exhibition. The popularity of chrysanthemums has increased with the introduction of the techniques of year round blooming based on photoperiodism. Chrysanthemums are qualitative short day plants, meaning that they flower when the daylength is shorter than the critical daylength and remain vegetative when the daylength is longer than the critical day length. Post (1949) stated that chrysanthemum has two critical photoperiods, daylength of 14.5 hours (August 15) was found necessary for flower initiation, while shorter day length of 13.5 hours (Sept 20) was required for flower development. Chrysanthemum cultivars are classified in terms of response group, based on time required from beginning of short days to 3/4th flower opening. The shorter the response group time, the shorter the critical night length (i.e. longer critical photoperiod) required for floral initiation and *vice-versa*. Long days are provided for vegetative growth and are maintained throughout propagation, early growth, pinching and until laterals are 2.50-3.75 cm long. The night interrupted light can be used during these stages to provide long days throughout the year. Night interruption lighting is usually started at 2200 hours. HID lighting is supplied for 16-24 hours per day to improve vegetative growth during low light period in winters and in propagation areas. Under artificial long day conditions, increased leaf production, longer and broader leaf and increased flowering stems are recorded.

Short days are provided when plant reaches the stem length of 35-50 cm, by covering with opaque blackout material (black satin cloth or polythene) during natural occurring long days of summers (from 21 March to 21 September) or by turning off the lights that provide long days during short days of winters. The blackout material is best applied from beginning of short days until flower bud colour showing from 0500 hours to 0800 hours for minimum of 12-15 hours per day. During summers, the rise in temperature above 18.3 ^{0}C under the black material can cause the delay in flower initiation during early inductive short days called as **heat delay** (Whealy *et al.*, 1987). Therefore, in order to avoid heat buildup, one should wait at least up to 1700 hours to cover the plant. The blackout material must be applied for at least 21-28 consecutive short days for standard cultivars and for 35-42 days for spray cultivars. After 14 consecutive short days the capitulum of inflorescence is completely formed and ray florets begin to initiate. In spray chrysanthemum, when plants attain proper height, they are given 12 short day cycle to initiate inflorescence in entire spray and then an interruption of 10 consecutive long days (night break) and then again returned to short days to complete flower development. These long days interruption elongates the peduncle length and improves spray formation. In standard chrysanthemum, flower diameter can be increased by subjecting the disbudded chrysanthemums to long day conditions after 35th short days, called as **after-lighting**. This after-lighting increases length of ray florets to increase the overall diameter of the inflorescence (Ben-Jaacov and Langhans, 1969).

Dahlia

Dahlia flowers under long day conditions with critical day length of 14 hours, while tuber formation is encouraged under short days. Natural flowering occurs in March-April when days are sufficient to provide long days to prevent tuber formation and encourage flowering. Supplementary lighting as 14 hour extension is better than night interruption to induce off season flowering as night interruption causes early premature flowering but less leaf production (Catlin *et al.*, 2012).

Gypsophila

Gypsophila paniculata is a perennial cut flower, well known for its use as filler in bouquets with white and pink coloured small flowers in panicle. It is an obligatory long day plant with natural flowering in spring and summer and it does not bloom when day length is <12-14 hours. Thus, to force the blooming in winter and spring, supplementary lighting is needed to increase the daylength. It can be grown continuously with 2-3 cycles of flowering per year and can last for 2-3 years. The optimum stem length and flower quality are obtained at 16-18 hour photoperiod. The plant must have 12 nodes to be receptive to long daylengths (Kusey *et al.*, 1981) and needs to receive minimum number of long day cycles to bloom. Long days must be maintained after flower initiation, until the inflorescence expands, to prevent lower yield (Hicklenton *et al.*, 1993). A high light intensity (35 klux) increases number of flowers but minimum light to induce winter flowering is 58 lux. The standard practice to give artificial lighting is when plants have around 20 leaves and an earlier start will produce negative effects like fast growth with reduction in production. This phase occurs 3-5 weeks after transplanting in normal growth conditions. However, plants should receive the long day stimulus before regrowth starts after pruning, so that supplementary lighting must be applied 2-3 weeks earlier. Incandescent lamps are often used for artificial lighting due to their emission in the red and near infrared zone and they emit the light between 550-770 nm. The fluorescent lamps with white cold light are not effective (Shillo and Halevy, 1982), although when they are mixed with incandescent lamps, exhibit an excellent result (Shlomo, 1984). The required light for incandescent lamps varies from 60-100 lux at leaf surface (Shillo and Halevy, 1982) and the structure to support wires and lamps are placed at 1.8-2.0 m height. The distance between lamps is 2.5-3.0 m between the lines and 2.0-2.5 m between the bulbs. Lamps commonly used are 100-150 watt.

Kalanchoe

Kalanchoe blossfeldiana is a succulent that provides consumers with a durable flowering pot plant requiring very little maintenance in the home or elsewhere. Kalanchoes are qualitative short-day plants for flowering with a critical photoperiod of 12 hours; however more rapid floral initiation occurs using a 10-hour photoperiod. Thus, they can be manipulated to flower on a specific finish date of the year much like potmums. Kalanchoe can initiate flowers with as few as two consecutive short days, but as the number of consecutive short days increases, the number of flowers initiated increases exponentially up to 14 to 21 short days. For commercial production, however, at least 42 short days (6 weeks) are recommended for complete induction of flowers and flower buds are usually visible at the end of this period. Returning plants to long days once the short day treatment is completed has the added benefit of reducing elongation of the terminal cyme and producing a more compact plant.

To maintain vegetative stock plants, night-interrupted lighting must be applied every day in the middle of the dark period from mid-September to late-March, i.e. 1-hour in September and March, 2-hours in October and February, 3-hours in November and January, and 4-hours in December. Lighting is provided by spacing 60 watt incandescent bulbs 90 cm above the plants and 1.2 m between bulbs to provide 10-20 fc intensity. Kalanchoe will initiate flower buds naturally from mid-October to mid-March and will flower from late December into January depending on the response group. From March 15 to October 15, black cloth must be pulled over the plants for 14 hours per day for floral initiation.

Kalanchoe cultivars fall into response groups that range from 9 to 13 weeks from the beginning of short days to flower. Within a given response group, flowering is often longer in the winter than in the summer. It is, therefore, important for a grower to keep records of response times for specific cultivars under different environmental conditions. Older plants tend to flower faster than younger plants.

Marigold

All marigolds are quantitative (facultative) short-day plants, but *Tagetes erecta* displays a response more than *Tagetes patula*. The critical photoperiod for *Tagetes erecta* is between 12.5 and 13 hours. If the seeds of cultivars 'Lady', 'Galore', and 'Jubilee' are sown before 15 February, natural short days will accelerate flowering. On the other hand, if seeds are sown after 20 February when usual daylength is longer than the critical daylength, flowering will be delayed and the plants will grow tall. Plants should be grown under artificial short days using black cloth starting the first 2-3 weeks after germination during late February and later. But some cultivars like 'Inca', 'Antigua', 'Discovery' and 'Excel' series are day neutral.

Poinsettia

Poinsettia pulcherrima (Christmas tree) is a popular potted plant with red bracts widely used in Christmas floral display. It is a short day plant in which flower initiation occurs when days become shorter. Under natural conditions, flowers are initiated in early fall, i.e. during late September (23rd to 27th) when day length becomes <12.20 hours (Grueber, 1985) or dark periods become 11.45 hours, but initiation is most rapid at 14.5 hour dark periods at temperatures of 15.5-21.0 ^{0}C. The actual development of flower parts does not occur until 7-10 days later. The development of flower from bud to fully developed stage takes about 8-10 weeks under optimum growing conditions. They are classified into different response groups depending on the time required for a plant to reach maturity following the start of first short days. Most modern day varieties fall in the 8–9 week response group. Two principal conditions for the normal development of poinsettia flower are (i) the day length must continue to shorten for proper flower development, and (ii) temperature must be within the optimum range. If the day length is controlled by lighting and use of black cloth so that poinsettia receives 12 hour of light and 12 hour of darkness, flower bud initiation occurs but flower development is delayed. Therefore, in order to produce poinsettia outside their natural flowering season, it is advisable to reduce day length to 10 hours by covering with black cloth for 14 hour period. The optimum temperature for flower initiation and development is 15.5-21.0 ^{0}C. Following 10 to 14 days of inductive photoperiods, raising the temperature to 9.7-10.0 ^{0}C will favour rapid bract development. About 40-60 inductive cycles are needed for full development of the 22 cyathia and 27 bracts in a normal inflorescence. The natural flowering occurs in late November or early December, and thus the poinsettia becomes the part of Christmas festival. It is necessary to manipulate the flowering of poinsettia as its popularity has created demand in full bloom in early November, so it is necessary to initiate the flowering earlier than under natural day length conditions. Therefore, the use of black cloth to shorten the day length is recommended, usually beginning in early to mid-September and continuing till mid-October when daylength is short, to ensure flower development.

The daylength can be extended by lighting as night-break from September 15th to April 15th to keep the plant vegetative. Incandescent lamps and red light over blue light is more effective in maintaining vegetative stock plants and to prevent flower initiation. Stem elongation is also enhanced by red light. The plants stretch when they are grown close together as light is filtered by leaves before it reaches stem but by providing more space, the unfiltered light reaching stem is less conducive to stem elongation.

Eight-week response cultivars with planting about mid-September will flower in mid-November on natural days (providing no light interference). However, these and the 9-week response cultivars will be too early for Christmas sales unless they are lighted artificially in late September and early October to prevent exposure to the inductive night lengths. Incandescent 60-watt bulbs spaced 1.2 m apart and placed 60 cm over the plants are used from 2200 hours to 0100 hour to provide the artificial **long day conditions**. The 'lights out' date, or start of short days, is determined by counting back from the desired sales date the number of weeks in the response group to which a particular cultivar belongs. The colour of bracts is also created through photoperiodism. It requires 12 hours of darkness at a time for at least 5 days in a row to change the colour.

Sunflower

Sunflowers are facultative short-day plants and when they are planted under natural short days the plants finish short, flower early, and develop many small 'ugly buds' crowding the terminal flower bud. Three weeks of long days (16 hours) will delay flowering by 18 days, nearly double the plant height, double the terminal flower size, and will prevent the formation of 'ugly buds' (Wein, 2009).

Annuals

Calendula officinalis, Centaurea cyanus, Linum grandiflorum, Papaver somniferum, Phlox drummondii, etc. are qualitative long day plants and flower early under 17-h photoperiod, however, flowering is delayed significantly when photoperiod decreases to 11-h (Baloch *et al.*, 2011). *Cosmos* is a facultative short-day plant as it flowers early under short days (constant 9-h photoperiod) and flowering is delayed under long days (9h + 4-h night interruption). *Rudbeckia* is an obligate long-day plant with a critical photoperiod of 13 hours. When the photoperiod is shorter than the critical photoperiod, it remains rosetted. *Salvia* is a long day plant with some dwarf red varieties as day-neutral and some tall varieties as short day plants.

References

Baloch, J., M. Munir, M. Abid and M. Iqbal, 2011. Effects of different photoperiods on flowering time of qualitative long day ornamental annuals. *Pakistan J. Bot.*, **43**(3): 1485-1490.

Ben-Jaacov, J. and R.W. Langhans, 1969. After-lighting of chrysanthemum. *N. Y. St. Flower Grow. Bull.*, No. 285, pp. 1-3.

Catlin, N., B. Krung, R. Lopez and B. Whipker, 2012. Photoperiod critical for avoiding tuber formation in dahlia cuttings. *E-Gro. Alert*, **1**(2): 1-4.

Dole, J.M. and H.F. Wilkins, 1999. Light. In: *Floriculture Principles and Practices*, pp. 36-45. Prentice Hall, New Jersey, USA.

Farina, E., C.D. Guda and E. Scordo, 1994. Flowering and morphogenic response of new aster hybrids to photoperiod. *Physiologia Plant.*, **91**: 312-316.

Grueber, K.L. 1985. Control of Lateral Branching and Reproductive Development in *Euphorbia pulcherrima* Willd. ex Klotzch. Ph.D. Diss., University of Minnesota, St. Paul, USA.

Gudo, C.D. and E. Scordo, 1994. Effect of photoperiod on morphology of aster hybrid. *Colture Prot.*, **23**: 87-92.

Healy, W. and H. Wilkins, 1983. Photoperiod control of lateral branching and flower production in carnation. *Acta Hort.*, No. 141, pp. 151-156.

Hicklenton, P.R., S.M. Newman and L.J. Davies, 1993. Growth and flowering of *Gypsophila paniculata* L. cv. Bristol Fairy and Bridal Veil in relation to temperature and photosynthetic photon flux. *Sci. Hort.*, **53**: 319-331.

Kusey, W.E. Jr., T.C. Weiler, P.A. Hammer, B.K. Harbaugh and G.J. Wilfert, 1981. Seasonal and chemical influences on the flowering of *Gypsophila paniculata* Bristol Fairy selections. *J. Amer. Soc. hort. Sci.*, **106**: 84-88.

McMahon, M.J. 1997.Chrysanthemum cultivars differ in response to photoperiod when grown under far-red absorbing filters. *HortSci.*, **32**: 437 (abstr.).

McMahon, M.J., J.W. Kelly, D.R. Decoteau, R.E. Young and R.K. Pollock, 1991. Growth of *Dendranthema* × *grandiflorum* (Ramat.) Kitamura under various spectral filters. *J. Amer. Soc. hort. Sci.*, **116**: 950-954.

Mortensen, L.M. and E. Strømme, 1987. Effects of light quality on some greenhouse crops. *Sci. Hort.*, **33**: 27-36.

Muckle, E. 1997. Space scientists bring light research down to earth. *Greenh. Business*, **3**(5): 22-23.

Post, K. 1949. Florist Crop Production and Marketing. Orange-Judd, New York, USA.

Rajapakse, N.C. and J.W. Kelly, 1992. Regulation of chrysanthemum growth by spectral filters. *J. Amer. Soc. hort. Sci.*, **117**: 481-485.

Schwabe, W.W. 1985. *Aster novi-belgii*. In: Handbook of Flowering, vol. V (ed. Halevy, A.H.), pp. 29-41. CRC Press, Boca Raton, Florida, USA.

Shillo, R. and A.H. Halevy, 1982. Interaction of photoperiod and temperature in flowering control of *Gypsophila paniculata* L. *Sci. Hort.*, **16**: 385-393.

Shlomo, E. 1984. Flowering in *Gypsophila paniculata*. M.Sc. Thesis, The Herbrew University of Jerusalem, Israel.

Smith, H. 1986. The perception of the light environment. In: *Photomorphogenesis in Plants* (eds Kendrick, R.E. and G.H.M. Kronenberg). Martinus Nijoff, Boston, U.K.

Thimijan, R.W. and R.D. Heins, 1983. Photometric, radiometric, and quantum light units of measure: A review of procedures for interconversion. *HortSci.*, **18**: 818-821.

Wallerstein, I., A. Kadman-Zahavi, H. Yahiel, A. Nissim, R. Stav and S. Michal, 1992. Control by photoperiod and the rhizomatous zone over the production of basal buds and the day length. *Sci. Hort.*, **50**: 209-218.

Whealy, C.A., T.A. Nell, J.E. Barrett and R.A. Larson, 1987. High temperature effects on growth and floral development of chrysanthemum. *J. Amer. Soc. hort. Sci.*, **112**(3): 464-468.

Wein, C. 2009. No more ugly buds! *The Cut Flower Quart.*, **21**(1): 56.

38

Diseases of Ornamentals and their Management

Manas Kumar Bag, Anirban Roy, Pragya Ranjan, R.L. Misra and Sanyat Misra

Growing ornamentals gives us a sense of identity and dignity of earning in our lives. Floriculture has emerged as an important agribusiness, providing employment opportunities and entrepreneurship in both urban and rural areas. In India, earlier times, the floriculture was largely practiced by traditional flower growers. Indian farmers grow flowers on a part of their landholding along with fruits, vegetables, millets, cereals, pulses, oilseed crops, *etc.* By growing flowers, farmers ensure some daily cash-income for their day-to-day needs, which is not possible by cultivating only cereals or coarse grains. Most farmers sell their flowers daily in the local market themselves or sometimes only some growers through middle-men who collect their produce. Time is changing very fast and now, though only a few, but they grow cut flowers under protected houses to ensure best quality and disease-pest-free blooms throughout the year for continuous supply so that the end-users or importers have absolute trust on them. Domestic market of India consumes more than 70 per cent of produce and only a little portion is meant for export. However, if the consignments are found infected with diseases, the value of flowers falls drastically. Diseases are important for the loss they cause to ornamentals, it is important to know the condition under which these infect, the symptoms they cause and the methods by which these should be managed without any loss to the crop. The diseases are caused by fungi, bacteria, viruses and mycoplasma-like organisms as are elaborated below.

Disease Management

Keeping plants healthy is more economical than curing plants after disease incidence. There are several methods to keep plants healthy. Most of diseases develop in the place where hot humid condition prevails. Priorities should be given in maintaining proper drainage and regulation of irrigation to check diseases. The crops may be kept disease-free by procuring or evolving disease-resistant varieties as this is the ideal method to get rid of diseases; soil sterilization to eliminate soil-borne diseases, should be done by exposing the soil through repeated ploughing during hot summer, burning straw or stubles to raise soil temperature, and application of chemicals such as 2% formalin to kill soil germs; eradication of diseases by means of crop rotation, elimination of alternate hosts, removal of infected plant parts, eradication of weeds and wild hosts, and roguing of virus-infected plants; many of the seed-borne pathogens can be managed to a great extent by sowing fungicide-treated seeds; fungicides and antibiotics can be applied as prophylactic sprays to prevent occurrence of any disease, as curative measures to eradicate already established pathogens; and quarantining or legislative measures are other methods in checking the spread of diseases, especially those which are not prevalent in India and this includes compulsory inspection, quarantine and embargo of plant materials or seeds brought from abroad. Best method of managing diseases is to develop the most efficient possible production plans that incorporate as many measures by which plant health can be maximized and opportunities for pest and disease outbreaks minimized — a concept known as integrated pest management, *viz.* prevention, cultural control, scouting, physical control, biological control, and chemical control (Agrios, 1997) as has been briefed here under.

Disease Prevention

Once symptoms of a disease are evident, management can be difficult, costly, or even impossible. Some basic management helps in preventing disease problems from occurring in the first place. The following precautions can reduce the plant disease development and spread.

Exclusion implies that healthy plants or pathogen-free planting materials are kept in an isolated area that excludes plant pathogens. Plants are often grown in a nursery or greenhouse where care is taken to ensure that planting stock and media are pathogen-free. When ordering/purchasing seeds, bulbs, or tubers, or any planting material, watch carefully if they are certified to be pathogen-free.

Avoidance is precaution taken to enable plants to escape pathogens that are present in the growing environment. **Damaged roots** and other plant parts provide entry sites for disease-causing organisms. Make sure the plants are secured during transport and placed gently (not dropped) into planting holes. Once established, avoid damage to trees and shrubs from line trimmers. Such damage interferes with the vascular system of the plant and creates an infection court. Over-pruning weakens trees and shrubs, making them more susceptible to attack by pathogens and pests. No more than 30% of the foliage should be removed at one time. Unless absolutely necessary, do not remove branches that are more than about half the diameter of the trunk. **Manipulating the growing period** to a time when climatic conditions do not favour disease development can help avoid losses. This may be a good option for plants with seasonal versatility and no resistant varieties. Pentas are very susceptible to *Phytophthora* root rot and blight during warm, wet weather. Planting pentas either before or after the rainy season is a good way to avoid losses.

Sanitation practices are cleanliness measures that can reduce unintentional spread of plant pathogens from diseased to healthy plants directly through mechanical transmission or indirectly from pathogen reservoirs. **Mechanical transmission of pathogens** by contaminated hands and tools can readily infect the healthy plants, especially if the causal agent is bacterial or viral, or when present in the vascular system (i.e., vascular wilts). *Fusarium* wilt is a devastating disease that can be transmitted through pruning. Routinely sanitize all items that come into contact with plants, soil, or debris. A 10%-20% solution of household bleach in water makes a good disinfectant for tools. **Pathogen reservoirs should be eliminated**. Any plant showing symptoms of any disease should be kept away from the healthy plants, treat them in an isolated area or destroyed. Plant debris is a potential reservoir for plant pathogens and should be kept away from healthy plants. Workers should disinfest hands, shoes, and tools after handling rogued, diseased, or decaying plant material before resuming regular duties. Weeds also harbour pathogens and spread them so there should be regular weeding of the field and the surrounding areas. Mulch can help control weeds, improve soil structure, and increase water retention in the soil. Occasionally, plant pathogens are spread into healthy plants through irrigation systems. Thus care should be taken when irrigating with surface water.

Cultural Control

Cultural control means employing good horticultural practices to optimize plant health and decrease plant stress, which in turn decreases the likelihood of disease development. The first line of defence in the management of plant disease is to provide the required conditions for **optimal plant growth and development**. The soil type, soil porosity, sunlight exposure, soil pH and microclimate should all be considered before a plant species is chosen to avoid conditions that are not conducive to healthy growth. Planting in improper locations can increase plant stress and the likelihood of pest problems. This may lead to more frequent pesticide applications, which may be ineffective and increase labour costs. Adequate **fertilization** is needed to avoid nutrient deficiencies and keep plants healthy enough to resist and recuperate from diseases. Some diseases can be avoided altogether with careful fertilizer applications. Too much nitrogen encourages excessive growth which makes the plant more vulnerable to plant pathogens. Soils with **poor drainage** become waterlogged if over-irrigated and overwatering is a major catalyst for the development of many root and crown diseases, including those caused by *Phytophthora* and *Fusarium*. In waterlogged conditions, most ornamental plants, especially succulents loose vigour, wilt, and develop leaves that are pale green or yellowish. Ornamental plant cultivars generally are not selected based on their **resistance** to plant pathogens. However, differences in relative susceptibilities of cultivars to some pathogens exist and should be considered when and where appropriate. Many plant pathogens cause disease only on a narrow range of closely related hosts. Continuous cultivation of the same types of plants in an area for long periods of time can result in elevated populations of pathogens. Periodic **rotation of plants** that are susceptible to different pathogens may decrease populations of harmful organisms and reduce the incidence of these problems. Disease outbreaks tend to spread quickly within monoculture systems. When appropriate, use plants of many different families and species, or at least use different cultivars. Plant diversity decreases damage from pathogens that attack specific kinds of plants (Harmon and Bledsoe, 2004; Hong, 2013).

Scouting

Monitoring for plant problems on a routine basis is the cornerstone of any IPM programme and is essential in detecting early pest infestations to prevent economic loss. It is important to correctly ascertain the cause(s) of plant problems so that appropriate measures can be taken to correct them. Before disease management steps are taken, other possibilities, such as insect or mite damage, nutrient or water imbalances, and pesticide damage should be talen into consideration. If the plant's recent history is unknown, ask questions and observe the surroundings to obtain clues that could lead you to a correct diagnosis (Harmon and Bledsoe, 2004).

Signs and Symptoms

Symptoms of plant disease are often similar to damage caused by other stresses. The ability to distinguish between the two is important. One should also be able to recognize characteristics of healthy plants, such as normal speckling, spotting, or variegation that may be confused with disease symptoms. Signs refer to some part of a pathogen and are visible during certain environmental conditions. Signs are typically more difficult to find, but include felty or mouldy fungal sporulation, cobweb-like fungal growth (mycelia), fruiting bodies, i.e. mushrooms (large), pycnidia (tiny), sclerotia, and bacterial ooze.

Thresholds

The presence of disease may not always require treatment. Over time, a level of damage considered economically or aesthetically unacceptable, i.e. economic threshold should be set, as should a damage level at which action must be taken to prevent reaching this point, i.e. action threshold. In most cases, there are no set recommendations for thresholds because various systems differ widely in conditions that affect disease pressure and in plant damage tolerances. In general, the threshold for damage to ornamental plants is very low or zero because of inherent aesthetic considerations.

Physical Control

Physical control of causal organism is obtained through mechanical process applied directly to plants or substrates that reduce or eradicate pathogen populations such as:

Steam sterilization: Steam has been used to disinfest soil to be used for potting and transplanting of high-value crops. An application of aerated steam to maintain a uniform soil temperature of 60 ^{0}C -70 ^{0}C for 30 minutes is sufficient to eliminate most causal organism present in the soil.

Soil solarization: Soil solarization is an option for managing soil-borne pathogens such as *Fusarium, Rhizoctonia, Sclerotium, etc.* In this method transparent polythene sheet is used to trap the radiant energy of the sun, which heats the soil to temperatures sufficient to destroy many plant pathogens without completely eliminating beneficial organisms. Soil solarization is often used when producing high-quality flowers.

Hot water: Seeds, bulbs, corms, rhizomes, tubers and cuttings can be immersed in hot water to kill potential pathogens. Daffodil and Easter lily bulbs are commonly treated with hot water to kill any nematodes that may be present. Accurate time and temperature controls must be in place for the hot water treatment to be successful. The temperature for pathogen inactivation must not exceed the critical damage threshold for the propagative material being disinfested (Agrios, 1997; Hong, 2013).

Biological Control

Biological control is the use of beneficial microorganisms to suppress soil-borne and foliar plant pathogens. Products containing bacteria, such as *Bacillus, Pseudomonas* & *Streptomyces*; and fungi, such as *Gliocladium* **and** *Trichoderma* have reduced a variety of fungal plant pathogens in various experiments, especially when incorporated into transplant media or used as seed treatments. Ongoing research continues to evaluate and develop new biocontrols. The use of organic soil amendments (e.g., compost, sewage sludge, *etc.*) may enhance populations of beneficial microorganisms already present in the soil. Biological control technologies are environmentally friendly IPM tools with great potential, but currently biological controls do not adequately control ornamental plant diseases on their own in most cases.

Chemical Control

Besides IPM strategies, it is sometimes necessary to apply fungicides, bactericides, *etc.* to manage plant diseases. Pesticide application should be need-based rather than routine sprays, because routine spray of pesticides is wasteful and has impact on populations of beneficial organisms. Pesticides may be applied to ornamental plants in a granular form, a liquid foliar spray, or as a drench. Most chemicals used in disease management are applied as sprays aimed at fungal pathogens on leaves and other aboveground plant parts. Pesticides may be applied before planting to soil and planting stock in a drench or dip, respectively. Gaseous soil fumigants are sometimes applied as a preplant treatment in mass plant production systems and kill most organisms in the soil, including fungi, bacteria, nematodes, and weed seed. Fumigants are highly toxic, therefore, extreme safety precautions must be taken to protect workers.

Choosing a Pesticide

Proper diagnosis of disease is essential for determining an appropriate pesticide. Some molecule target specific pathogens, while others are effective against a wider range of organisms. A chemical can be chosen based on the disease group, which describes one or more of the following: the type of symptom exhibited by the plant, e.g. blight, scab, rot, *etc.*; the part of the plant affected, e.g. flower, leaf, root, *etc.*; the type of pathogen causing the disease, e.g. downy mildew, powdery mildew, etc.; or the specific pathogen causing the disease, e.g. *Phytophthora, Stromatinia, etc.* Disease groups describing general symptoms, i.e. leaf spots, blights, scabs, *etc.* may be caused by many different pathogens. It is not likely that every chemical listed for such disease groups will control every possible pathogen for that disease group. Systemic fungicides (s) usually move upward within a plant, though there are a few that also move downward. They may be applied as foliar sprays, seed treatments, root dips, soil drenches, or tree injections. Almost all systemic fungicides disturb only one or a few steps in fungal metabolism. Therefore, resistance to these fungicides generally occurs within a very short period of time if used repeatedly. It is best to use these molecules in combination or rotation with broad-spectrum contact fungicides to stop development of resistance. Eventually use of chemicals must be when environmental conditions are conducive to a disease outbreak. Products with curative activity may be applied after infection but prior to severe disease symptoms. Some chemicals have both protective and curative activity that may depend on the rate of application. Resistance to many contact pesticides is slow to develop or nonexistent because many different metabolic pathways of target organisms are disrupted. Not all pesticides may be applied to all ornamental plants, because of phytotoxicity and the location of the plant (Harmon and Bledsoe, 2004; Hong, 2013).

The virus and bacterial diseases are generally considered to be incurable as they cannot be easily cured with available chemicals. Viruses spread from diseased to healthy plants by insect vectors or through the garden tools. It is advisable that while working with diseased plants the hands or tools should be washed with disinfectant solutions. It is also suggested that while taking cuttings from virus infected chrysanthemum or dahlia with knife or razor blade, virus may spread through these contaminated implements from the diseased to new healthy plants, therefore, it is better snap the cuttings from the mother plants with a gentle downward pull of the hand. This method will avoid spread of the disease. The trouble with virus is that many cultivars are symptomless carriers. In such cases the cultivars will carry the virus but will not show any apparent symptoms. There are some dahlia cultivars which are symptomless carriers of viruses, and they are potential danger to the healthy plants. It is possible to check the spread of the disease by controlling the insect vector. For example, spotted wilt of dahlia is spread by thrips and the recommended spray schedule for thrips can check the spread of virus

Fungal Diseases

Medical practitioner classify human diseases by their effects upon the body, likewise the diseases caused by many fungi on plants are readily recognized because of their peculiar and sometimes distinct symptoms, e.g. spot, blight, rust, downy mildew, powdery mildew, wilt, rot, etc. (Pirone, 1978). Various spots produced by the fungi can be distinguished from other spots upon close examination. Spots of various colours appear on leaves, stems, flowers and fruits which in severe cases coalesce with each other. Blight may first appear as blots but soon after may engulf entire foliage and other tender parts and then over all effect appears as burning of the entire field. Rust diseases in many plants can be identified by their characteristic yellowish, orange to reddish-brown pustules on either surface of leaves. Spore-bearing structures protruding outside the leaf surface can be observed using magnifying glass. Downy mildew is caused by fungi belonging to the family Peronosporaceae, and these fungi produce grey velvety patches on undersides of leaves. Powdery mildew caused by various fungi appears powder-like coating of spores and mycelia on the surface of leaves, stems, flowers and fruits. Wilt disease is caused by fungi, bacteria or viruses that make plants droop and leaves shrivel. Rot disease is decay of the plants or its part and is caused by fungi, bacteria, viruses or nematodes.

Damping off: Damping off is a common disease of seedlings. It is the rotting of seeds (pre-emergence) in the soil and destruction of newly emerged seedlings (post-emergence) by fungi. The entire seed may decay before germination or the seedling may rot just below the soil line. It causes extensive losses to growers through death of seedlings or production of low-quality plants that perform poorly in the field. The conditions favouring 'damping off' are excessive overhead misting and excessive soil moisture, low soil temperatures (below 20 ^{0}C) before germination or high soil temperatures (above 25 ^{0}C after emergence, and overcrowded seedbeds. 'Damping-off' is primarily caused by fungi present in the growing medium, such as *Pythium*, *Rhizoctonia*, *Fusarium*, *Phytophthora*, *Sclerotinia*, *Sclerotium*, *Botrytis*, and many others. Germinating seeds and seedlings, especially weak ones are vulnerable to attack by these fungi during periods of unfavourable growing conditions. Species of *Pythium*, *Sclerotinia* and *Phytophthora* are more likely to cause damping-off in cool and wet soils, whereas species of *Rhizoctonia*, *Fusarium* and *Sclerotium rolfsii* may cause damping-off under warmer and drier conditions, but the predominant fungus causing this disease belongs to the genus *Pythium*. True damping-off may be confused with plant injury caused by excessive fertilization, high soluble salts, drowning in wet soil, desiccation in dry soil, and death of seedlings from excessive heat, cold, fuel fumes or chemical injury. This type of injury often damages the leaves and upper stem before the roots are affected. Typical symptoms of damping-off are rotting stems at or near the soil line and root decay (post-emerging damping off). Affected areas in the seed bed are usually a foot or more in diameter with shriveled brown, collapsed or stunted seedlings. Mouldy fungal growth may be seen on affected plants at the soil line. Germinating seeds can also be attacked by these fungi before they emerge from the soil (pre-emerging damping off), resulting in poor stand. *Rhizoctonia*, *Fusarium*, *Sclerotinia* and *Sclerotium* generally cause post-emergence damping-off by killing the seedling at the soil line. *Pythium* attacks below the soil line, often at root tips as the pre-emergence damping off. The best control of damping-off is to avoid it altogether. Once damping-off has started in a plant bed or seedling flat, it may be difficult to control. The general methods employed to prevent damping-off consist of proper soil preparation and management to provide for good soil drainage, structure, aeration, water-holding capacity and plant nutrition by including proper amounts of fertilizer and lime according to plant requirement and soil test, proper soil treatment with heat or chemicals to reduce the level of fungi causing damping-off, the use of fungicide-treated seed for high germination, proper seeding rates to avoid thick plant stands, poor air movement and low light intensity, proper planting depth and soil temperature to assure rapid seedling emergence, growth, and strict sanitation to avoid re-infesting treated soil with these fungi.

While **root rot** can be caused by several *Pythium* species, the three most commonly encountered ones are *Pythium irregulare, P. aphanidermatum* and *P. ultimum. P. ultimum* and *P. irregulare* are often found in field soil, sand, pond and stream water and their sediments, and dead roots of previous crops. If pond or stream water is used for irrigation, the intake pipe should be well above the bottom so that sediment is not drawn in. If the water supply is suspected of being a source of *Pythium*, it may be necessary to treat the water before use. Slow sand filtration has been shown to be an effective, simple, and inexpensive method for removing *Pythium* from water. Heat, ultraviolet light, ozone and chlorination can be effective but are expensive and require some training for its proper use. *P. irregulare* has been isolated from almost every type of greenhouse crop grown but *P. aphanidermatum* seems to be associated primarily with poinsettia and very few other crops. *Pythium* can be easily introduced into pasteurized soil or soilless mixes by using dirty tools, dirty pots or flats, and by dumping the mixes on benches or potting-shed floors that have not been thoroughly cleaned. When introduced into a soil mix that has been heat-treated for too long or at too high a temperature, *Pythium* can cause severe root rot because it has few competitors to check its activity. *P. aphanidermatum* and *P. irregulare* though do not form the swimming spore stage but pose a threat to crops grown in 'ebb and flow systems' because they form a swimming spore stage that can move in water and also when reservoir becomes fouled with potting mix and plant debris particles that harbour it. This is likely to occur only if irrigation times are long (45 minutes or longer) or if pots are seated in puddles of water because the bench or floor does not drain completely. Covering of ebb and flow system reservoirs should be done to prevent contaminated debris from entering the system. Return water should be passed over a coarse screen to remove potting soil and plant debris in order to help keep *Pythium* out of the reservoir. If *Pythium* infects a cutting bed or if contaminated water is used in propagation, large losses occur. *P. ultimum* is primarily associated with soil and sand. As growers have switched to soilless mixes, this species became less important than when growers used field soil in the potting mix. Almost all plants are susceptible to *Pythium* root rot. Root tips, important in taking up nutrients and water, are attacked and killed first. *Pythium* rots even the cutting bases. In case of *Pythium* infection the plants are stunted, root tips become brown and dead, plants wilt at mid-day but normally recover at nights, plants yellow and die, brown tissue on the outer portion of the root easily pulls off leaving a strand of vascular tissue exposed, and the cells of such roots contain round, microscopic and thick-walled spores. *Pythium* root rot is difficult to control once rot has begun. Every effort should be directed towards preventing the disease before it begins by using heat-pasteurized potting mix (entire pile heated to 82.2 ^{0}C and held at that temperature for 30 minutes). Longer times and higher temperatures will kill beneficial organism in the soil. The treated soil or commercial mixes should be covered and kept away from infected soil. Soil, pots or flats should not be contaminated by placing them on dirty floors or benches or by using dirty tools. Disinfection of the 'ebb and flow reservoirs', flood and drain floors, all the bench surfaces and tools & equipments should be done periodically. Even soil moisture conditions during germination and seedling development should be maintained. Only fungicide-treated seeds should be used or immediately treated after purchasing but one should be careful not to use those fungicides which cause dwarfing or inhibit root formation. It is best to avoid applying the pesticides to seedlings and cuttings. If using a fungicide is unavoidable, it is suggested that only a portion of the crop should be first treated and if no inhibition of germination or rooting is noted, it is safe to use that specific chemical on that variety of crop. In a greenhouse operation which has a history of *Pythium* root rot, application of a fungicide or a biological control agent should be done as early in the cropping cycle as possible. Biological agents should be applied to the potting mix before, during or immediately after transplanting and in the plug trays even before transplanting. No any chemical pesticides should be applied to the potting mix 10 days before or for 10 days after applying the biological control agent. Biological control agents and fungicides may have to be applied more than once in order to maintain adequate protection. Some populations of *Pythium* have resistance to metalaxyl, mefenoxam and/or propamocarb. If these chemicals do not appear protecting the plants, one should switch to another product and contact the plant pathologist specialized in the field. Seeds and cuttings should be started in the pasteurized soil or a soilless mix using only sterile flats or pots in the greenhouse. Most commercially available soilless potting mixes are free of damping-off fungi and do not need to be treated.

Pythium infection causes roots of **African violet** turning brown, leaf yellowing and wilting of plants; death of nursery seedlings in ***Begonia*** as shiny blackened areas develop on the stems and petioles of established plants at and just above the soil line by which plants collapse and die; damping off in ***Callianthus chinensis*** seedling is very serious; common rotting of **Christmas cactus** (*Epiphyllum* spp.) in the nursery; in ***Dendranthema*** it causes brown to blackened stems at soil level, plant stunting and wilting; damping off due to root rot in young seedlings and stunting of older plants in ***Cyclamen***; black leg disease of ***Geranium*** cuttings where brown water-soaked rot occurs at the cutting bases, and this rot progresses up to 6-10 cm above and kills the cuttings rapidly; *P. myriotylum* causes **amaranth** damping off at pre- and post-emergence stages; *Pythium ultimum* var. *ultimum* causes seedling rot of **orchids**, more serious being in community pots in *Cattleya* and *Epidendron*, affected plants turn black and leaves start falling and pseudobulbs also show rotting; *P. ultimum* causes **tulip** rot in the field as becomes evident by gaps and irregular growth of shoots and its infection stunts the plants severely, bulbs develop soft gray spots on the base often with brown borders that releases a distinct odour, flowers wither before maturity, infected roots become fragile and hollow and appear translucent with watery spots (Moore, 1979); *etc.* In case of *Pythium* infection, seeds before sowing should be treated with Captan, carbendazim or Ceresan dry and sown in sterilized soil thinly. Underground organs of the bulbous plants should be planted in well-drained and sterilized soils. Soil sterilization should be carried out either with steam or with formalin before planting. To avoid attack of *Pythium*, field soil or potting mix should be sterilized. Infected plants should not be used for further multiplication. Application of etridiazole, etridiazole + thiophanate methyl, or mefenoxam is suggested to protect healthy plants. Overwatering and water stagnation should always be avoided. Removal of infected plants and spraying with mancozeb 2 g/l can effectively control the disease. Soil drenching with potassium salts of phosphoric acid, etridiazole, mefenoxam, or thiophanate methyl + etridiazole is recommended. In orchids, withholding of watering for a few days and shifting the plants to a less humid part of the greenhouse helps to check the disease. Beneficial effects of fluorescent *Pseudomonas* isolate E113 and *Trichoderma harzianum* against root rot fungi were achieved either by mixing the bacteria through the soil or by dipping the bulbs in bacterial suspensions immediately before planting (Sharma, 2006). ***Amaryllis*** and ***Impatiens*** damping off is caused due to infection of *Phytophthora nicotianae* var. *parasitica* and *Rhizoctonia solani* which can be managed by spraying with Captaf alternate with Bavistin fortnightly in case of *Amaryllis*. In balsam, these fungi infect at pre- and post-emergence as well as to the established seedlings, and the infection becomes very serious under too wet or waterlogged conditions. Both the fungi perpetuate in the soil as saprophytes. To get rid of these pathogens, nursery should be raised in the disease-free soil, planting should not be overcrowded, excessive watering should be avoided and the nursery should be drenched with Captan 0.2%. Benson (1995) recommended use of Metalaxyl and Fosetyl-Al against both the fungi. *Cercospora calenduliae* on ***Calendula*** causes dark brown irregular spots in young seedlings which may coalesce together causing havoc in the seedlings during humid weather, but not in the younger plants. Field should be cleaned of all the plant debris. Bavistin (0.1%), copper oxychloride (0.3%) or zineb (0.2%) weekly sprayings to the seedlings during humid weather is recommended.

Rots: Several pathogens such as *Botrytis, Myrothecium, Phytophthora, Pythium, Rhizoctonia, Sclerotium* and *Thielaviopsis* cause disintegration of living cells of plant in large numbers which results into rotting of the plant parts. Common rot diseases are root rot in African violet caused by *Phytophthora*; a dry rot by *Sclerotium* and rhizome rot by *Botrytis* (Pirone, 1978); bulb rot in *Hyacinthus* and crown rot in *Antirrhinum* by *Sclerotium*; stem rot in *Centaurea* by *Phytophthora, etc.* Rot by *Phytophthora* sp. is caused on ***Calceolaria***, ***Catharanthus***, and ***Kalanchoe*** where its infection causes rotting of the roots and the symptoms appear as dark brown spots of crown and roots at ground level, and disease spreads up the stem causing wilting; *P. cryptogea* causes sudden wilt and leaf browning in ***Gerbera***; *P. parasitica* causes gray-greening of leaves, root rotting and wilt in **Christmas cactus** though in ***Petunia*** it attacks leaves, branches and stems producing water-soaked brownish spots leading to the death of the plants, and the disease is controlled by spraying with 0.3% Difolatan, 0.2% Dithane M-45 or 0.3% copper oxychloride; *P. nicotianae* in ***Poinsettia*** causes soft and wet stem bases, browning and water-soaking of roots, rapid wilt of cuttings but under dry conditions where pith of the lower stem becomes brown and stem develops a gray canker, and in its attack the soil or mix should be drenched with potassium salts of phosphoric acid, mefenoxam, dimethomorph or etridiazole, though in case of ***Cordyline***, leaves at soil levels are water-soaked and have brown, irregular and zonate dead areas which is controlled through dimethomorph spraying; *P. nicotianae* and *P. cryptogea* both cause dark and water-soaked appearance on crowns and roots in **African violet** and ***Sinningia*** and the symptoms spread up to petioles and leaves though these remain firmly attached with the plants even when plant collapses in case of African violet, though in case of ***Sinningia*** both these cause the same symptoms as African violet, young plants are quickly killed and the hypocotyls of mature plants resprout but eventually succumb; *P. cinnamomi, P. cryptogea, P. drechsleri* and *P. megasperma* cause root and stem rotting in ***Cineraria*** at soil level by which plants wilt and die; *Fusarium oxysporum* f. sp. *callistephi, Phytophthora cactorum* and *Pellicularia filamentosa* cause stem rot and wilting in **cornflower**; *Thielaviopsis basicola* causes ***Gerbera*** root rot where roots become dark brown to black, plants yellow and wilt whereas ***Poinsettia*** exhibits longitudinal splits at stem bases, leaf yellowing and falling, roots turning black late in the season and ultimately plants die; *Myrothecium roridum* in ***Gloxinia*** causes easy fragility of stems, darker spots appear on petioles of lower leaves, stems and plant rotting, and for controlling the pathogen iprodione spraying on leaves and stems is very effective together with maintaining the moderate temperature and low humidity, though in case of ***Antirrhinum*** it causes leaf spot and collar rot by producing sunken dark brown and cracked cankers at the collar region, *vis-à-vis* dark brown necrotic spots on leaves and ultimate plant wilting; crown rot of *Antirrhinum* occurs due to attack of *Sclerotium rolfsii* which is most prevalent in sandy-loam soil and for its control Merlo *et al.* (1986) suggested using healthy seeds and sowing in disease-free soil, dipping of seedlings in 0.1% Bavistin for a few hours before transplanting, drenching of field-soil with 0.2% Captan before planting and frequent raking of the soil at the cropping period, though in ***Zinnia*** it causes southern blight which is more serious in sandy-loam soil and makes white wefts of mycelium at stem bases and around the stem numerous reddish-tan to brown sclerotia are produced and the plants ultimately wilt and die at the ground; *Sclerotinia sclerotiorum* causes madwort (foot and stem rot) of ***Alyssum*** where infection appears as brown depressed spots at the collar region which girdles the stem and ultimately the plants wilt and die; *Sclerotium tuliparum* causes ***Hyacinthus*** soft bulb rot initially at the neck but afterwards engulfs whole bulb, *S. bulborum* causes leaf yellowing and falling due to rotting of leaf bases just after flowering and afterwards causes bulb rotting, *Rhizoctonia tuliparum* causes dry rot progressing from the neck of the bulbs downwards, *Embillisia hyacinthi* causes dark to red-brown spots on the bulbs and leaves, *Pythium* sp. makes the roots flaccid and causes decay, *Fusarium oxysporum* f. sp. *gladioli, Penicillium corymbiferum, P. verrucosum, Rhizoctonia solani, Pythium* sp., *etc.* cause root and bulb rot some in the field and others in the storage which to some extent can be controlled through application of Quintozene, however, HWT at 43 ^{0}C for three hours is quite effective in controlling all these hyacinth bulbs and other bulbous plant's diseases. Potting mix or soil should be sterilized, clean pots and tools should be used, infected plants should be discarded immediately, proper sanitation should be maintained and the debris should be destroyed, overhead irrigation should be avoided, 2-3 years crop rotation should be followed and the healthy plants should be protected through metalaxyl, Etridiazole or Etridiazole + thiophante methyl application.

Easter lily bulb storage rot (blue mould) is caused by *Penicillium* spp. and soft rot by *Rhizopus.* Initially bulb scales develop brown spots which are followed by rotting and such bulbs may be soft and afterwards covered with a blue-green mould or white fungal growth. Such bulbs should never be planted. *Fusarium* and *Cylindrocarpon* spp. infections cause dark browning of basal plate by which scales wither. In such cases only sterilized soils should be used for planting and only healthy bulbs should be planted and that too after one hour dipping in 0.2% Captan or thiophanate methyl. Infection of *Sclerotium rolfsii* on ***Narcissus*** causes leaves losing their greenness and formation of round and brown sclerotia, followed by rotting and falling of entire leaves and floral stalks. Later infection makes the plants too weak to produce flowers. To control this malady, mercuric chloride at 0.1%, and 0.2% commercial formalin should be used. Brassicol dusting is also quite effective. *Phialophora cinerescens* causes **carnation** wilt due to attack on vascular tissues which become dark brown and the leaves yellow. It is managed similar to that of *Fusarium* wilt. *Sclerotinia narcissicola* (*Botrytis narcissicola*) causes narcissus smoulder where leaves and flowers are spotted and rot. At lifting, affected bulbs bear masses of black, small and flattend sclerotia (resting bodies) under the outer papery scales and on rotting leaves. Infected bulbs produce distorted yellow shoots with withered dark brown or black leaf tips. Proper sanitation of the field and stores, and avoiding high humidity in the field or in the store will minimize this disease. Within 48 hours of lifting, the bulbs are dipped for 30 minutes in a suspension of 0.2% benomyl, carbendazin or thiophanate methyl and then bulbs are rapidly dried before storage, but in the following season none of these three chemicals are to be applied again. White root rot of ***Narcissus*** is caused by *Rosellinia nectarix* under moist conditions. This root rot disease causes black rotting of the outer bulb scales with white fungal strands in the form of woolly mass near the basal plate. Captan and thiride may control this disease. White mould or *Ramularia* blight of *Narcissus* is caused by *Ramularia vallisumbrosae*. Its infection produces grey or white powder and destroys the foliage and sometimes even stalk. It is controlled by 3-4 sprayings of 4:4:50 bordeaux mixture or tank - mix zineb starting when the leaves are 5-10 cm high. Grey bulb rot of ***Tulipa*** is caused by *Sclerotium tuliparum* which can be controlled by regular spraying with 0.2% Bavistin (benomyl) alternate with 0.2% captan fortnightly to the standing crop from sprouting until leaf senescence. This treatment will prove very effective in keeping almost all the fungal diseases under check. *Verticillium* spp. infects **dahlia** and causes wilting of single twigs or entire plant as it affects vascular tissues so stems become greenish-brown, dark brown or black, therefore dahlias should be planted in sterilized soil or medium. *Verticillium albo-atrum* or *V. dahliae* causes collapse, drying and falling of middle and upper leaves in ***Geranium*** due to infection on vascular system where tissues become brown so the disease is confused with that of bacterial blight. Cuttings should be taken from the culture-indexed plants and the planting should be done in sterilized medium or soil.

Gray mould or *Botrytis* blight: Gray mould or *Botrytis* blight can be a persistent and often severe disease for many of the woody and herbaceous ornameantals in field and in the greenhouses everywhere. It is fast spreading and can grow on many different sources of nutrients, survives well in the greenhouse, and can attack various types of plants. Weak or senescent plants are very prone to its infection, and the fungus can infect the plants at any time in the crop cycle from propagation to maturity (Vann, 2013). *Botrytis* at first appears as a white growth on the plant but very soon darkens to a gray colour. Smoky-gray (dusty) spores form and spread by wind or through water. If humidity is low, infections may be contained within discrete spots, though under high humid conditions it spreads rapidly (Anon., 2013). Release of its spores is spontaneous even during mild winds or under an automated trickle irrigation systems which when turned on, triggers the release of spores. These spores are often found on the outside of seeds. The spores can remain dormant on plant surfaces, sometimes as long as the life of the plant itself. *Botrytis* forms two types of resting structures on or in infected plant tissue, i.e. (i) very dark brown or black multi-celled structures called sclerotia, and (ii) single-celled, thick and dark-walled chlamydospores. *Botrytis* must have some food source before it invades the plant. Nutrients leaking from wounded plant parts or from dying tissues such as old flower petals or through winter injury provide the required nutrients. From this food base, the fungus becomes more aggressive and invades healthy tissues. A dark to light brown rot forms in the diseased tissue. High humidity conditions favour its growth. *Botrytis cinerea* on **African violet** causes small water-soaked lesions on damaged petioles which soon spreads to leaf blades, and flowers fade prematurely. On ***Aster***, it causes brown specks on petals so entire flower becomes brown and dies. Its infection on ***Begonia*** causes tan spots on the leaves and established plants start rotting at the crown. Infected tissues are covered with dusty gray fungal growth. *B. cinerea* on ***Calceolaria*** first causes leaves near the soil surface to rot and then fully covered with gray fungal growth and later on entire plant is engulfed. In its infection, **carnation** petals turn brown and are covered with gray fungal growth. In ***Catharanthus***, its infection rots the leaves near the soil surface and later on engulfs entire plant and flowers and the plants become covered with gray fungal growth. Due to attack of *B. cinerea*, the flowers in **Christmas cactus** rot and under high humid conditions the stems are also affected. Its infection in **chrysanthemum** causes lower petals becoming light brown which afterwards engulfs even the flowers and infected tissues become covered with dusty gray spores. On petals it causes brown water-soaked spots. On ***Cyclamen***, it causes tannin-coloured leaf spots, water-soaked lesions on petals which later on turn tan and entire plant afterwards is covered with gray fungal growth. In its infection, the leaves of **Easter lily** produce oval to circular spots and the flowers brown spots. Stems of established plants rot at the soil line due to *B. cinerea*. Its infection kills the entire plant. ***Geranium*** flowers turn dark and fall prematurely when are infected with *B. cinerea*, and falling of such petals on its leaves, an irregular leaf spot forms.

Cutting stubs have a brown rot that may extend down the stem. Cuttings develop a dark brown rot near the base. Attack of *B. cinerea* on **hollyhock** causes flower rot under humid weather and spoils them. ***Hydrangea*** flower buds due to infection of *B. cinerea* die before opening. However, later infection turns the petals brown and its falling on the leaves causes spot formation there too. *Botrytis cinerea* causes **iris** leaf spots. *Botrytis* flower blight in ***Poinsettia*** is recognized by brown spots formed on flowers, leaves or stem tissues while *Botrytis* stem canker of poinsettia is known by its large, light brown to tan, slightly sunken cankers forming on older stems, especially near large branches or crotches. Defoliation and death of branches occur above cankers that girdle stems. Infection of *Botrytis cinerea* is serious on orchids like *Cattleya, Dendrobium, Phalaenopsis, Oncidium, Vanda, etc.* under cool, damp and still conditions and causes **orchid** flower blight or brown streak. It produces numerous, small, dark brown spots on petals at initial stage, especially on older flowers and sporulates heavily on dead and diseased flowers covering entire plant afterwards. The cropping area should have proper air circulation through proper spacing and excessive humidity, *vis-à-vis* overhead irrigation in the field should be avoided. Yellowing leaves, infected parts and fading flowers should be regularly removed and if possible the smudging should be done in the field during winter. Application of azoxystrobin, chlorothalonil, copper sulphate, copper hydroxide, fenhexamid, fludioxonil, iprodione, mancozeb, manganese + zinc, trifloxystrobin or vinclozolin to the plants when weather is chilly, cloudy and windy will protect the plants against its infection, however, chlorothalonil is applied when the plants are not in flower.

Botrytis gladiolorum infection on **gladiolus** makes pin-head black spots on leaves, stems and flowers which during dry weathers restrict their growth but causes unopening of flowers though during humid weathers entire planting is damaged causing black blight with white fungal growth on dead organs. *Botrytis hyacinthi* causes **hyacinth** fire, destroying the whole foliage but not the bulbs. First water soaked spots appear which turn brown and the leaf tips dark brown similar to that of scorching and during damp conditions a grey mould develops on the affected parts including flowers. These all the blights can be controlled through weekly spraying with Dithane M-45 or Dithane Z-78 at 0.2% during humid, chilly and windy conditions. *Bortrytis* blight in ***Lilium*** is caused by *Botrytis elliptica* which forms reddish-brown spots on the leaves but in Easter lily yellow streaks appear. *Bortrytis cinerea* causes grey mould on the flowers. These may be controlled through the use of copper fungicides. ***Narcissus*** spots and blights are caused by *Botrytis elliptica.* Disease can be controlled by spraying with 0.2% maneb weekly and the field humidity should be kept under check. Before planting, the bulbs should also be treated by dipping them for one hour in 0.2% bavistin or benomyl. **Tulip** fire is caused by *Botrytis tulipae.* The leaves prior to emergence get infected and become stunted, malformed, distorted and fail to unroll. Under moist weather, masses of fungal spores (conidia and sclerotia) are formed on the outer surface of the infected tissues. Leaf spotting is of non-aggressive and aggressive type. Small necrotic areas are generally marked by dark green, water-soaked areas in non-aggressive lesion which do not increase in size and no conidia are produced on them. Aggressive lesions have whitish-grey or brown with water-soaked margins that can enlarge and are able to produce more spores that may result in an epidemic form leading to complete flower loss and a significant reduction in bulb yield. Increase in lesion size depends entirely upon the production of spores and the duration of moisture during infection. Flowers may also be attacked. Fruit capsules are the most susceptible part and sclerotial bodies cover seeds. In the underground infection, bulbs are attacked first. Infection travels from mother to daughter bulb, sclerotia are formed in the outer husk of the affected bulbs. With this, the outermost fleshy scales develop and small round to oval sunken spots with yellow gray border in the centre are formed. The stems produced by these bulbs are shorter and in a few cultivars, a slight streak is apparent on flowers. Infected bulbs should be discarded. Removal of infected shoots just below ground level upon emergence and flower before anthesis is helpful in limiting the disease spread. Bulb dip prior to planting in Benomyl, Thiabendazole or 2% Formalin for 30 minutes reduces the carry over of inoculum and subsequent production of fireheads. Iprodione and Vinclozolin provide significantly higher levels of disease control in the flowering period of the crop. Spraying with Benomyl (0.1%) and Maneb (0.25%) just after leaf development is recommended.

***Fusarium* wilt, Collar rot and *Fusarium* leaf spot**: Wilting is due to deficiency of water in the leaves and stems, may be due to lack of water in soil and the injury to the root system because of infection of pathogens. Fungi like *Fusarium* and *Verticillium* are the most common pathogens causing wilt disease. *Fusarium* wilt is favoured by high temperature and can cause very high percentage of mortality (Fletcher and Martin, 1972). Chief diagnostic symptoms are abnormal growth and stunting of young shoots. Leaves become yellowish and the stems are soft so as to be easily crashed. The entire plant wilts and collapses due to their infection. Plants can be easily pulled out. However, since *Fusarium* attacks the vascular system so mostly the plants become sickle-shaped before yellowing and death. *Fusarium* wilts are very difficult to manage. A single infection is sufficient to introduce the pathogen into the plant and spread to other plants in the same field within a very short period of time. It can survive in the soil for a long time, thus managing through crop rotation, though other cultural practices are not so effective. Once the pathogen has entered the plant, it is not possible to eradicate it by any means. Management based on use of pathogen-free stock, soil free of pathogen and sanitation are much effective in reducing the severity.

Fusarium oxysporum infects ***Aster*** where young seedlings are killed and old plants stunt, water conducting tissues in the plants become brown, leaves become yellow and ultimately the plants wilt generally by bending to one side. Use of either soilless mix or sterilized soil is recommended and while planting the propagating material should also be pathogen-free. *Fusarium oxysporum* f. sp. *dianthi* infects **carnation** where lower leaves yellow and plants wilt to one side. Later infection causes root and stem rot. No symptoms and very little colonization are observed at 14–15 ⁰C though at a temperature of 22 ⁰C the cultivars are clearly differentiated into resistant ones, partially resistant or susceptibile ones. Symptom severity is associated with the extent of colonization. This differentiation is not visible at 26 ⁰C. The pathogen causes disease at 26 ⁰C by a combination of vascular wilting, and rotting of stem base & roots. *Fusarium oxysporum* causes leaf yellowing, discolouration of the vascular tissues and plant wilting in ***Chrysanthemum*** to one side of the plant. *Fusarium oxysporum* f. sp. *cyclaminis* infection on ***Cyclamen*** causes vascular tissues in the tubers turning dark brown to black, heavy yellowing of plants and ultimate death. Only healthy tubers should be planted and that too in the sterilized medium. *F. oxysporum* infection on ***Delphinium*** causes initially minute brown spots on stems which afterwards become darker and enlarge up to 30 cm in the form of stem canker, *vis-à-vis* brown and sunken lesions on petioles, and with the death of canker tissues, the masses of pink spores form in clumps. Vascular tissues are brown and ultimately plants die. Infected plants should be gently uprooted with the adjoining soil and burnt. *Fusarium oxysporum* f. sp. *gladioli* infection on ***Gladiolus*** shows vascular rot of corms so the plants become sickle-shaped, yellow and die. This also causes storage rot of corms by producing sunken and black spots on corm surface. Crop rotation of 3-4 years, soil sterilization, planting of only healthy corms and cormlets, dipping the corms and cormlets before planting and one month after lifting in 0.2% Captan for one hour, and then after sprouting spraying with Bavistin 0.1% alternate with Captan 0.2% fortnightly will keep the crop disease-free. The fungi associated with ***Hyacinthus*** bulb and root rot are *F. oxysporum* f. sp. *gladioli*, *Penicillium corymbiferum* and *P. verrucosum*, some attacking in field while others in storage. HWT for three hours at 43 ⁰C will control all these pathogens. *Fusarium solani* causes foot rot in **cornflower** and the disease becomes apparent when the plants are about to flower. Water-soaked dark brown and depressed lesions appear at the collar region leading to drying of foliage and wilting of plants. This disease can be checked by avoiding waterlogged conditions, use of disease-free seed and by following proper rotation. Infection of *F. oxysporum* f. sp. *lilli* causes bulb rot in Asiatic lilies. Infection starts from the bulb scale bases either during storage or during transit and causes chocolate-brown rot of bulbs. Planting of healthy bulbs in sterilized soil, avoiding over-watering and treating the bulbs for 30 minutes at 30 ⁰C in 0.2% Captan before planting and one month after lifting will keep the crop disease-free. Infection of *Fusarium oxysporium* f. sp. *narcissi* and *F. bulbigenum* causes ***Narcissus*** basal rot by decaying root plate or scale bases in the field as well as in storage when temperature is above 14 ⁰C and the situation aggravates when it rises above 21 ⁰C. Methyl bromide at 50 g/m^2 fumigation in storage or application of Aldicarb 10% a.i. (4g/m^2) twice during growing season controls these pathogens. *Fusarium oxysporum* infection in **tulip** initially causes basal bulb rot, starting with small pale-brown or gray sunken flecks on the outermost fleshy scale bases that become dark brown and enlarge in size afterwards, and engulfs whole bulb gradually. Rotting starts from the bulb bases with exudation, fungus producing white to pink mycelium and spores in warm and damp sites, and such bulbs at affected site shrink and infected tissues appear ring-shaped. Generally, leaves do not wilt but remain upright, often developing a characteristic purple colouration and senesce earlier though flowers produced by such bulbs may blast or deform. Delayed planting and early lifting avoids soil temperature that favours development of disease, and planting at 10 ⁰C soil temperature reduces the infection. Crop rotation of 3-4 years should be adopted. Bulbs should be stored at a well ventilated place at a temperature below 15 °C to avoid build up of ethylene. Bulb dipping immediately after lifting in Bendimidazole is very effective. Bulb dipping for two hours in 0.5% formalin solution also provides excellent results. Spraying of Benlate and Thiophanate at 0.05% gives satisfactory control. Beale and Pitts (1990) when studied antagonistic activities of *Mininedusa polyspora, Streptomyces* sp. and *Trichoderma viride* to *Fusarium oxysporium* f. sp. *narcissi* as pre-planting bulb treatment, recorded 55.72 % and 56.88 % lower disease incidence with first two microorganisms, respectively. Infection of *Fusarium solani* and *Sclerotium rolfsii* is most prevalent on ***Jasminum sambac***. First lower leaves show yellowing then upper ones, and the roots blacken so that gradually whole plant collapses. *Sclerotium* infection symptoms are the same as to those of *Fusarium*, and in addition it forms white mycelia girdling the roots of the wilted plants. Such plants at initial stage of symptoms should be drenched with 1% Bordeaux mixture or Thiride. ***Cordyline*** and ***Dracaena*** are infected by *Fusarium moniliforme* which produces tan to red-brown and oval spots on the tips of young leaves in case of *Cordyline* but close to the base of leaves in case of *Dracaena*, having a bright yellow halo. Overhead watering should be avoided and spraying of chlorothalonil, iprodione, mancozeb, thiophanate methyl or thiophanate methyl + mancozeb is recommended to check its further spread.

On ***Chrysanthemum***, the infection of *Verticillium* sp. causes death of lower leaf margins or the entire leaves and symptoms proceed up to one side of the plant. Only culture-indexed planting materials free of pathogens should be used and planted in the sterilized soil or mix. The soil pH should also be maintained at 6.5-7.0. Use of nitrate rather than ammonical form of nitrogenous fertilizers is recommended. Application of thiophanate methyl as a soil drench is also recommended.

Rhizoctonia*:** The pathogenic soil-borne fungus *Rhizoctonia* is known to cause root rot, stem rot, damping-off and in some cases even the leaf blight, and is favoured by warm (21 °-32 °C) and moderately moist condition (65% soil saturation) and if the plants are injured the higher soil moisture favours the growth. Once the fungus begins to grow in the plant the infected areas decay quickly, causing brown to reddish-brown lesions to form at or just below the soil-line and generally by mid day the plants show wilting. Under favourable conditions, the lesions enlarge and form dry sunken cankers which may eventually girdle the plant. The young hyphae are colourless but turn brown with age. Soil particles often cling to the cankered areas of the plant when removed from the soil because of the coarse, brown mycelium. Leaves touching the soil can be infected. The fungus webs the neighbouring leaves quickly followed by quick rotting under high humid conditions. However, the older plants are quite tolerant. It causes crown rot of **African violet** where infected tissue at soil level show reddening, afterwards entire plant blackens and collapses and the leaves readily detach from such plants. Its infection in ***Begonia rots the stem at soil level where a fine webbing of fungal growth may be seen at the stem and adjoining soil particles. *Rhizoctonia solani* causes collapse of seedlings in ***Impatiens*** coupled with disintegration of the collar region in plants and yellowing and desiccation of older plants; in **carnation** the stems have dry shredded appearance and at soil level these have lesions with brown borders; in ***Dendranthema*** the pathogen causes reddish-brown dead areas girdling the plant at the soil level and such plants on sunny days wilt but recover at night before final collapse; in **ferns** irregularly shaped brown spots form on the foliage close to the crown or sometimes at the top and such spots spread quickly; in **iris** *R. solani* causes leaf yellowing and bulb rot and *R. tuliparum* 'bad-soil disease' and neck & bulb rot which can be controlled while controlling other bulb rots; while in **marigold** it causes brown necrotic spots, girdling the radicle which later on extend to plumule and causes pre-emergence mortality and this problem is controlled by drenching with Brassicol. To get rid of this fungus the infected plants should immediately be destroyed and soil or potting mix should be properly sterilized. Captan can be incorpoprated in the soil before planting to prevent damping off. Iprodione, thiophanate methyl + etridiazole, triflumizole, trifloxystrobin, fludioxonil or copper can be used as a drench or as spraying at the base of the plants. Chlorothalonil treatments should begin at transplant.

Downy mildew (*Peronospora* or *Plasmopara*)**:** Common ornamental hosts include *Aster* spp., *Coreopsis* spp., *Rosa* spp., *Rudbeckia* spp., *Antirrhinum*, *Viburnum* spp. and *Viola* spp. However, most downy mildew fungi have a restricted host range as *Peronospora sparsa* affects only rose and not any other ornamental though *Plasmopara halstedii* infects many members of the daisy family including *Rudbeckia*, but not in other families. Generally, downy mildew is not a major problem for established plants. However, infected plants may show severe symptoms in the spring on young plants. Symptoms of this disease vary by host though there are some similarities. In general, dark angular and often irregular spots develop which are often delimited by veins. Multiple spots may develop on a leaf until the lesions coalesce and such leaves drop afterwards. In certain cases of downy mildews, white or gray fungal growth may be visible on the underside of the leaves, and beneath the discoloured areas on the upper surface. Fungal growth may also be visible on the stems or canes in the period of high humidity. *Peronospora antirrhini* causes ***Antirrhinum*** downy mildew, a very common pathogen on seedlings which on infection produces light brown downy growth on the lower surface of leaves with the corresponding upper surface turning yellow (Paul *et al.*, 1991). Downy mildew on snapdragon is successfully controlled by spraying with Metalaxyl 0.5 g/l water or Manzate 80W 1.5 g/l water. Bordeaux mixture 0.1% or Dithane M-45 0.2% is also quite effective. *P. obducens* causes downy mildew on **balsam** by producing brownish circular to irregular spots, more conspicuous being on lower leaf surface where these are covered with white cottony growth, though spots on the upper surface are initially yellowish and then turning dark brown, starting from the lower leaves and proceeding upwards. Severely affected leaves turn yellow and fall. The pathogens cover whole of the plant afterwards but clarity of infection is seen only on the leaves (Paul *et al.*, 1991), and such plants die even before flowering if there is high humidity coupled with 22-25 ^{0}C temperature. This is controlled by disposal of debris and by spraying with Dithane M-45 or Difolatan 0.2%. *P. tarinosa* infects ***Amaranthus*** and becomes serious when weather is humid and cool. The pathogen grows only on the lower surface of leaves with yellowish patches on upper surface. Such leaves normally drop down and cause severe loss to the crop. Copper oxychloride 0.3% spraying helps in controlling the disease. *Bremia lactucae* on **corn flower**, though not so common, but its infection produces pale greenish or reddish irregular spots on the upper side of the leaves, while the undersides may be covered with a mouldy growth. Affected leaves collapse and the attacked parts die. The infection is severe on young plants in humid weather. This disease should be managed through proper sanitation of the field, by avoiding plant crowding and excessive watering and by use of copper fungicides. In general, the downy mildew is managed by removing the infected stems or canes in late fall or winter. Plants are pruned to increase air circulation and to decrease humidity. During cool and wet weather the disease plants are treated with preventive fungicides such as *Bacillus subtilis*, azoxystrobin, copper, dimethomorph, fosetyl-Al, kresoxim-methyl, mancozeb, phosphite, trifloxystrobin and triadimefon + trifloxystrobin (Armada). To reduce the possibility of the development of fungal resistance to some of the newer systemic fungicides with single or few modes of action, rotate these chemicals with protectants such as mancozeb, and sole use of any fungicide for extended periods of time should be avoided when other reliable products are available, and reference to the label for timing, host plants, and rates should be given (Gould, 2009).

Powdery mildews: Powdery mildew is one of the most common diseases of ornamental plants in the nursery and gardens and the damage can be slight to severe. Greenhouse crops prone to infection include African violet, *Begonia, Dahlia, Gerbera, Hydrangea*, roses, *Verbena, Kalanchoe* and *Poinsettia*. Herbaceous perennials susceptible to powdery mildew include *Aster, Centaurea, Coreopsis, Delphinium, Monarda, Phlox, Rudebeckia* and *Sedum*. Its infection produces white powdery fungus on the upper surface of lower leaves so the leaves may distort, twist and die though in ***Kalanchoe***, the symptoms on leaves is not obvious but leaves develop scab-like spots, become corky and dry. Powdery mildew infection is favoured by high relative humidity at night and 22-27 ^{0}C temperatures. Low humidity favours spore dispersal. The spores are carried by air and germinate on the leaf surface. Water on leaves inhibits spore germination. The fungus grows on the leaf surface but sends fine threads (haustoria) into the cells to obtain nutrients. From the time a spore germinates, new spores require only 48 hours to form. Some powdery mildews are inhibited by free moisture on leaves while others are favoured by wetness on leaf surfaces. Some highly susceptible plants, such as ***Euonymus*** should be replaced with a similar plant that is not susceptible to powdery mildew. On many trees, the disease causes little or no damage. Plants in the landscape which may require fungicide applications to prevent powdery mildew damage include crape myrtle, phlox, rose and zinnia. If damage from powdery mildew is severe and there is a susceptible plant in the landscape, these should be pruned out of diseased portions severely, followed by treatment with sulphur formulations, however, sulphur applications at or above 32 ^{0}C may cause plant injury. Propiconazole and Myclobutanil are quite effective against powdery mildew and many other diseases. Triadimefon is useful on a number of ornamentals though Triforine on roses and certain other ornamentals. Fenarimol is a locally systemic fungicide and is useful in therapeutic control of powdery mildew. In some instances even copper fungicides control powdery mildew (Jones, 1999). When conditions in greenhouses become conducive for 3-6 days for powdery mildew development, in the late afternoon the greenhouses should be ventilated and heated to reduce night humidity, and some anti-transpirant along with fungicide or biological control measures should be applied. Water inhibits spore germination for most powdery mildews, therefore, syringing the leaves at daytime of low humidity greatly inhibits infection and can protect the plant. Syringe only if other leaf diseases are not a problem since other pathogens require water to infect. Syringing may be the best approach at locations where chemical use is not desirable and on crops for which there are no registered fungicides. *Oidium* spp. on **African violet** causes white mealy fungal growth on leaves and flowers so they die and dry. It is controlled by spraying with Myclobutanil, Piperalin or Triadimefon. On ***Antirrhinum***, this powdery mildew becomes serious only during late stages of growth and affects seed production. This can be controlled by spraying with wettable sulphur and 0.2% Karathane. *Oidium* spp. on ***Begonia*** causes white mealy fungal growth on leaves, flowers and stems which can be controlled by applying Dodemorph, Fenarimol, Myclobutanil, Piperalin, thiophante methyl, Triadimefon or Trifloxistrobin. *Oidium* spp. and *Leveillulataurica* spp. cause powdery mildew on **marigold**. In their infection, initially white tiny superficial spots appear on leaves which later on spreads to entire plant in the form of whitish powder. Spraying with 0.5% Karathane (40 EC) or dusting with sulphur powder fortnightly also checks spread of this disease. *Oidium* spp. on ***Petunia*** produces whitish floury patches on leaves which become yellow and dry out and its severe infection kills the plants. Plants should not be kept under shade and should be sprayed with Sulfex 0.2% or Bavistin 0.1%. Infection of *Oidium* spp. on ***Poinsettia*** appears initially as yellow spots on the upper surface of the leaves but afterwards patched white fungal growth is seen on leaves and bracts. Application of thiophanate methyl or kresoxim methyl is recommended as soon as the mildew appears. *Erysiphe polygoni* infection of **sweet alyssum** though uncommon but appears as whitish floury patches on leaves which causes them becoming yellow. Spraying with wettable sulphur 0.2% or Karathane 0.025% controls the spread of disease. ***Zinnia*** powdery mildew caused by *E. cichoracearum* appears as dirty white floury irregular patches on leaves, stems and inflorescenses (Sohi, 1992; Madhu Meeta, 1993) which makes plants to die in its severe attack. The fungus is serious in the humid regions. It is controlled through 0.2% spraying of wettable sulphur or 0.25% Karathane and avoiding shade cultivation. On ***Aster***, **balsam**, ***Chrysanthemum*** and ***Cinerarea***, *Golovinomyces cichoracearum* (syn. *Erysiphe cichoracearum*) infection appears as white felty growth on the leaf surface by which leaves wither and die. The disease is favoured by moderate temperature and humid weather, and appears late in the season. Its spread on *Aster* is checked by applying aliphatic petroleum solvent, Ampelomyces, Fenarimol, fludioxonil mineral oil, kresoxim methyl, mancozeb, Myclobutanil, paraffin oil, Piperalin, Propiconazole, sulphur, thiophanate methyl + mancozeb, Triadimefon, Trifloxystrobin or Triforine, though on balsam, this and other powdery mildew causing pathogens are controlled by spraying with 0.2% wettable sulphur, 0.025% Karathane or 0.1% Bavistin, apart from growing these at a sunny situation with wider spacing. *Erysiphe polygoni* and *Leveillula taurica* and *Oidium* spp. also cause balsam powdery mildew similar to *E. cichoracearum*. *Sphaerotheca fusca* powdery mildew on *Cinerarea* is controlled through use of Triadimefon and through ventilation and heating of glasshouses by evening to maintain low night humidity. *Golovinomyces cichoracearum* infection on **dahlia** causes white mealy and dry fungal growth on the leaf surface, which can be controlled through Fenarimol, mancozeb, Myclobutanil, Piperalin, sulphur, thiophanate methyl + mancozeb, Triadimefon, Triforine and ziram. *Golovinomyces cichoracearum* (*Erysiphe cichoracearum*) also produces powdery mildew on ***Gerbera*** and **hollyhock**. Its attack causes white fungal growth on gerbera but in case of hollyhock white floury patches appear on the leaves, stems, sepals and bracts where leaves turn yellow and dry up. Hollyhock is also infected with *Leveillula taurica*. These are managed through use of Azoxystrobin, kresoxim-methyl and Triadimefon in gerbera but through 0.2% wettable sulphur in case of hollyhock. ***Delphinium*** is attacked by *E. polygoni* which causes dry, white and mealy fungal growth on the upper surface of leaves by which leaves become yellow and die from the base upward. Field should be kept clean of the plant debris and the plants should be treated with Fenarimol, Myclobutanil, Propiconazole, sulphur, Triadimefon, Triforine or ziram when the fungus is first observed. In ***Hydrangea***, *Erysiphe polygoni* causes yellow patches on leaves which become purplish later on. The lower leaf surface forms white and cottony fungal growth. Applicatiopn of Fenarimol, Piperalin, thiophanate methyl or Triadimefon protect the leaves. *Sphaerotheca humili* infection on ***Kalanchoe*** produces tan and scabby spots on the leaves where scab may persist or may be covered with white fungal growth. Mancozeb, thiophanate methyl, thiophante methyl + mancozeb, Triadimefon or Triforine protect the leaves. *S. pannosa* var. *rosae* causes **rose** leaves turning purplish and drop. Karathane 0.1% spraying on the crop controls this problem.

Leaf spots and blight: Spots on foliage and stems are common symptoms of disease, caused by various fungi. Spots are often called 'zonate', marked with concentric zones of different appearance. A blight may be defined as a disease that kills young growing tissues, especially of leaves and twigs (Pirone, 1978). *Sphaceloma poinsettiae* causes ***Poinsettia*** scab by producing small round and light-coloured spots with yellow halos on leaves, particularly along the main vein. Infected stems stretch several inches above the normal crop. Crop canopy should be properly managed with low relative humidity and overhead irrigation should be avoided. Application of azoxystrobin, trifloxistrobin, triflumizole or triadimefon is recommended. ***Narcissus*** leaf scorch is caused by *Stagonospora curtisii* which survives in the dry papery scales of the bulb, and with emergence of leaves, this attacks causing scorched and burnt appearance on the leaf tips but buds and flowers are attacked only in damp weather. Hot water treatment with formalin controls this disease in the first year, and spraying with 0.2% zineb at 10-14 days intervals after removing and burning diseased tips will control it in the second year. ***Narcissus*** fire is caused by *Botryotinia polyblastis* (*Sclerotinia polyblastis*) by causing pale-brown lesions which engulfs entire leaf quickly.Proper sanitation of the field and removal of the infected flowers, *vis-a-vis* fortnightly spraying of the crop with bordeaux mixture (4:4:50) and tank-mix zineb will control this disease.

Amaryllis leaf spots are caused by *Aecidium amaryllidis* and *Cercospora* sp., and these are controlled through spraying with 0.2% Dithane M-45. ***Antirrhinum*** anthracnose (*Colletotrichum fuscum* and *Colletotrichum gloeosporiodes*) is a very serious problem under glasshouse conditions with humid climate or during rainy season. Here plants at all the stages of growth are infected. Spots on older stems are sunken, oblong, pale-yellowish-green to grey with narrow brown borders though on leaves, it produces irregular brownish spots. Several spots may girdle the plant at the base. The fungal acervuli later appear on the spots as minute black dots, which contain a number of long setae (Wiedemann *et al.*, 1988). Diseased plant parts should be removed, debris collected and proper sanitation should be maintained. Overhead irrigation should be avoided. Bavistin 0.1%, boraeaux mixture 1% or zineb 0.2% sprayings are quite effective. *Antirrhinum* blight is caused by (*Phyllosticta antirrhini*). In its infection, cream-coloured or light brown, large circular and zonate spots, sometimes surrounded by halos, appear on the leaves. On the stem, the spots are ashy-gray turning brown. If the spots encircle the stem, the plants show wilting. Pycnidia are epiphyllous, central and dark. Spores are hyaline and 1- celled. Plants of all ages are susceptible. The disease is more serious in summer at temperatures above 16^0 C (Sohi, 1992). Overhead watering should be avoided and field sanitation should be maintained properly. Bavistin (0.1%), bordeaux mixture (1%) or zineb (0.2%) spraying is quite effective to control this disease. ***Aster*** leaf spots are caused by *Alternaria, Ascochyta, Cercospora* and *Septoria*. The spots form first on lower leaves, followed by their wilting afterwards and the disease progresses upward. Overhead irrigation should be avoided, field should be kept clean of all plant debrises and application of mancozeb, iprodione, myclobutanil, fludioxonil, copper sulphate, copper hydroxide, thiophanate methyl + chlorothalonil, or chlorothalonil will check further spread of the pathogen. **Balsam** leaf spots are caused by *Drechslera australiensis*. These produce small dark brown to black and circular to irregular amphigenous spots, 2-3 mm in diameter which on coalescing later on form necrotic lesions. Very often leaf tips and margins also dry. Spraying with 0.2% Dithane M-45 or zineb, and destroying plant debris at the end of the season help in controlling the disease. Similar to *Drechslera, Alternaria tenuis* also causes balsam leaf spots which can be kept under check by using disease-free seeds and spraying the plants with Indofil M-45 (0.2%). *Ascochyta impatientis* leaf spots appear as large pale-brown blotches with paler margins which ultimately results into death of leaves. Pycnidia are black or brown, globose to subglobose, erumpent and ostiolate, and the conidia are 2-celled and hyaline. Its control is similar to *Alternaria* leaf spots. *Cercospora balsaminae* and *C. fukushlana* spots are circular to sub-circular, initially brown but afterwards becoming dingy-grey in the centre with brown margins, and the disease is most prevalent during rainy season. Zineb (0.2%) or Bavistin (0.1%) sprayings controls this pathogen. *Macrophomina phaseoli* causes balsam foot, stem and collar rot. In its infection, dark brown necrotic lesions running parallel to the stem appear from the foot region and proceed upwards resulting in the wilting and eventual death of plants. Affected tissues later turn ochraceous. Numerous dark, hard and irregular sclerotial bodies appear on the diseased tissues. The fungus perpetuates through sclerotia in the soil and has a very wide host range. The disease is serious under stress conditions and kills the plant at any stage of growth including the seedlings (Castillo and Peterson, 1990). To avoid build-up of the inoculum in the soil, diseased plants should be removed and burnt. Proper watering is required to avoid stress conditions. Soil application of Bavistin is also advocated in the nursery. Seed treatment or dipping of seedlings in Bavistin/benomyl solution (0.1%) for a few hours before transplanting gives good control. *Alternaria alternata* causes dark brown, circular or oval spots on both the surfaces of ***Bellis*** leaves. The spots initiate mostly from the leaf margins and result in browning of the apical leaf portion. Indofil M-45 (0.25%) spraying will control this problem. ***Bougainvillea*** leaf spot (*Phyllosticta bougainvilleae*) appears as small to large irregular brownish-grey spots, delimited with brown margins on the leaves. The infected tissues fall off. *Alternaria* leaf spot of bougainvillea (*Alternaria alternata*) is noticed in the form of dark brown circular to irregular spots covering large areas. Those caused by *Colletotrichum gloeosporiodes* and *Colletotrichum dematium* f. sp. *bougainvilleae* are large, irregular and brownish-yellow which start appearing from the leaf margins and then involve most of the leaf area. In all these leaf spots, the infection is also common on dead spines and inflorescence bracts. Most probably the fungus perpetuates on these parts. Affected portions give a blighted look. These diseases are most common during cool humid weather. Dead twigs, leaves and inflorescence should be removed and burnt. Bavistin (0.1%) or zineb (0.2%) spray is recommended. *Alternaria dianthicola* (syn. *A. dianthi*) infection produces small purplish leaf spots with brown centre in **carnation**, *vis-à-vis* leaf yellowing. Chlorothalonil, copper hydroxide, fludioxonil, iprodione, mancozeb, mancozeb + thiophanate methyl or propiconazole sprayings protect the healthy plants. *Septoria obesa* and *S. chrysanthemella* cause grayish-brown circular spots on the leaves of ***Chrysanthemum***. Dithane M-45 or zineb 0.2% will control this problem. *Alternaria sensecionis* on ***Cineraria*** causes dark brown to black leaf spots, engulfing a large portion of leaves. This can be managed by removing and destroying infected parts, maintaining proper sanitation and irrigation and application of iprodione. Infection of *Phyllosticta dracaenae* on *Cordyline* appears as small round to irregular spots with purple borders and yellow halos on older leaves. Overhead irrigation should be avoided and mancozeb should be sprayed to control the disease. *Cercospora* sp. and *Septoria* sp. infect ***Cosmos*** and cause dark brown irregular leaf spots which after coalescing give blighted effect. The disease is quite common during rainy season. Spraying the crop with Dinocap (0.02%) or carbendazim (0.05%) is very effective. *Cosmos* canker or stem blight is caused by *Diaporthe stewarti* on mature plants though quite uncommon. Stems and their branches are infected with brown spots and canker which girdle the stems. Affected stems are weak and fragile. Diseased plants should be removed and burnt. *Cosmos* dieback is caused by *Phoma herbarum* which forms canker on stems and branches, resulting into eventual death of the plants. Bavistin 0.2% spraying will control this pathogen. *Gloeosporium, Phyllosticta* and *Septoria* infection on ***Cyclamen*** produce yellow, brown or gray circular spots on leaves with small dark brown dots within. Proper sanitation, avoiding overhead irrigation and application of thiophanate methyl or chlorothalonil will check spread of the disease. Leaf-underside of ***Geranium*** produces water-soaked spots which enlarge to 6-13 mm in size having concentric rings of dark tissues within due to infection of *Alternaria alternata*. Proper sanitation by removing all the diseased parts, avoiding overhead irrigation and use of azoxystrobin, chlorothalonil, copper hydroxide, fludioxonil, iprodione or mancozeb will protect the plants from further spread of the disease. *Alternaria* spp. causes brown specks on florets and leaves of ***Gerbera***, leaf spots having white centre. Low relative humidity, no overhead irrigation and application of azoxystrobin or fludioxonil will keep this disease at bay. Infection of *Cercospora nebulosa* on **hollyhock** produces minute and light to dark stem blemish or leaf spots which later on grow larger (1.0-3.5 mm in diameter) and circular to sub-circular and the central portion dries up and fall off though on stems these are blackish-gray, oblong and initially solitary but coalescing later on and forming irregular dark brown streaks (Crous and Broun, 1996), and becoming serious during humid weather. *Myrothecium roridum* infection on hollyhock causes circular to irregular and brownish leaf spots. Diseased tissues become dry and brittle with sporodochia. The disease becomes serious during humid weather. *C. altheina* is also reported infecting hollyhock. Controlled irrigation without wetting the foliage, proper sanitation and spraying of Bavistin 0.1% or copper oxychloride 0.35% checks further spread of the disease. *Colletotrichum malvacearum* causes anthracnose or seedling blight, *vis-à-vis* the lower part of stem and roots on **hollyhock**. The disease is very destructive in warm humid weather. Black blotches are formed on veins, leaf blades, petioles and stems. Severely infected leaves curl and dry up. Proper sanitation and burning of the plant debris, avoiding humid conditions and spraying with Bavistin (0.15 %) or zineb (0.25 %) controls this disease. On ***Hydrangea***, *Cercospora* spp. causes tan spots with reddish-brown halloes on leaves. No overhead irrigation and spraying with mancozeb, copper, or chlorothalonil saves the crop from its infection. ***Iris*** leaf spot is caused due to infection of *Heterosporium gracile* (*Didymellina macrospora*, syn. *Mycosphaerella macrospora*). Due to its infection, dark brown or almost black irregular blotches are formed on the leaves, gradually spreading up and down the leaf blades, and its severity causes spots even on flower stems. Ink spot (*Drechslera iridis*) occurs on *I. reticulata* more than Dutch iris. It causes black blotches on the bulb scales and foliage. Before planting, the bulbs should be soaked for one hour in 0.1% benomyl + 0.5% captan, and afterwards fortnightly spraying in the standing crop will prevent both these diseases.

Cercospora jasminicola and *Alternaria jasmini* cause **jasmine** leaf blight, being more serious on ***Jasminum grandiflorum*** though only stray incidence are shown on *J. angustifolium, J. auriculatum, J. communis, J. flexile, J. humile, J. pubescens, J. rigidum* and certain cultivars of *J. sambac*, however, *J. arborescens, J. paniculatum* and certain other varieties of *J. sambac* are almost disease-free. Circular reddish-brown spots some 2-8 mm in diameter are produced on upper leaf surface, spreading more rapidly during rainy season though their appearance is noticed in May-June, peak period being from August to November and their infection causes up to 50% flower yield. Affected leaves become hard & brittle and show inward curling of margins. Their severity causes drying up of entire foliage, vegetative buds and young branches. Monthly sprayings with 0.2% Dithane M-45, 0.1% Bavistin, 0.4% Benlate, 1% Bordeaux mixture or copper oxychloride starting from May until pruning is very effective in controlling the blight. ***Lilium*** black scale is caused by *Colletotrichum lilli* where blackening and rotting of outer scales occur, and on lifting the bulbs show brown and sunken lesions on scales which enlarge in storage and turn black. Lily scale tip darkening and outer scale shrivelling and rotting occur due to infection of *Fusarium* sp. and *Cylindrocarpon* sp. Effectively this malady can be controlled by spraying with 0.1% Bavistin alternate with 0.2% Captan, fortnightly on standing crop. Infection of *Alternaria, Cercospora*, and *Septoria* on **marigold** causes minute brown and circular spots initially on lower leaves that enlarge at later stage leading to premature defoliation and ultimate death of the plants. Inflorescence blight of marigold is caused by *Alternaria zinnae.* Disease is characterized by light tan to dark brown large irregular zonated blotches on the leaves. Flower bud rot of marigold is caused by *Alternaria dianthi.* The disease is known by dry rotting with brown, scratched and necrotic discolouration of sepals and stalk. Dithane M-45 spraying at 0.2% fortnightly starting from first appearance of disease keeps all these diseases under check. *Colletotrichum, Gloeosporium, Phomopsis* and *Phyllosticta* are very common pathogens of cultivated orchids and within a few days of infection sunken spots on leaves appear which later on turn brown. Though these do not kill the plants but leaves give an unsightly look. The disease is favoured by lack of light and warm humid weather. Spraying of Bordeaux mixture (6-6-50 or 4-4-50) will effectively control the disease. Blitox (0.3%) or Difolotan (0.1%) may be sprayed as a remedial mixture. *Cercospora petuniae* causes ***Petunia*** leaf blotch by producing circular to irregular dark brown spots which later on turn grayish, and after coalescing these cause leaf blotches and foliage death. Humid weather favours spread of the pathogen. *Ascophyta petuniae* on petunia causes leaf spots similar to those of *Cercospora petuniae*, and produces 2-celled, oval to elongated hyaline spores in pycnidia. *Phyllosticta petunia* also causes small circular to irregular and salmon to chestnut or grayish petunia leaf spots starting from leaf tips. All these petunia diseases can be managed by proper sanitation by destroying the plant debris and removing the weeds, providing proper air circulation among the plants, avoiding excessive watering and spraying with 0.1% Bavistin or 0.2% Dithane M-45. *Diplocarpon rosae* causes conspicuous circular black spots with fringed margins on either side of **rose** leaves and results into premature leaf falling. All the tools or secateurs used in pruning or cutting should be dipped each time in 70% alcohol or 2% formalin. All the pruned ends should be painted with copper sulphate. Fortnightly, 0.2% Captan or Ferbam should be sprayed on the plants to control the disease. *Alternaria zinniae* leaf spot or blight on ***Zinnia*** appears initially as minute spots which later on enlarge and coalesce to form light to dark brown and large circular to irregular blotches and white patches, sometimes the spots are reddish-brown with grayish centre. Its infection causes spots even on flowers and stems. Leaves dry off in its serious infection.

Maintenance of proper sanitation, destroying all the plant debris, following 2-year crop rotation, spraying the plants with 0.2% maneb, zineb or copper oxychloride and raising the seeds in disease-free areas are effective measures towards minimizing this disease. Zamorshi and Milczarek (1977) suggested use of Thiram (0.3%) and Captan (0.5%) as the best control measure for *Alternaria*. Dithane M-45 was also found to be the most suitable fungicide for seed treatment (Srivastava and Gupta, 1983). Biological control for *Alternaria* is also being tried. Wu and Chou (1995) found that *Bacillus* strain #18 was as effective as spraying Pyrifenose in controlling *Alternaria carthami* on zinnia in both greenhouse and field trials. Complete control was given by Benlate, Topsin-in and Karathane (Powell, 1974). *Cercospora zinniae* leaf spot pathogen of zinnia produces numerous conspicuous, small, circular, white to grayish spots with reddish-brown margins and in severe cases, the leaves turn brownish and dry off. Fruiting is amphiginous but more on the upper surface. The field should be cleared of plant debris regularly, plants should not be ocercrowded, excessive watering should be avoided and trimming of diseased leaves should be carried out regularly. Jindal and Meeta (1991) recommended trimming of diseased leaves and simultaneous application of 0.01% streptocycline + 0.025% carbendazim against *Cercospora* leaf spot and *Xanthomonas* quite effective. Bavistin (0.1%) or Zineb (0.2%) sprays are also quite effective. *Phyllosticta zinniae* also causes zinnia leaf spot of small, reddish-brown, turning grayish afterwards and coalescing to form irregular patches. It also attacks flowers and seeds (Wu and Yang, 1992). Severe infection causes drying of leaves. Spraying the crop with Mancozeb (0.2%), carbendazim (0.05%) or copper oxychloride (0.2%) reduces the disease spread (Wu and Chou, 1995; Yadahalli *et al.*, 1994). *Phomopsis zinniae* causes zinnia twig dieback or blight which results into death of twigs or whole plant. Infected leaves produce dark brown irregular necrotic lesions. Infected tissues turn grayish with numerous dark pinhead-like pycnidia. Proper sanitation, use of disease-free seeds and spraying with zineb 0.2%, copper oxychloride 0.3% or Bavistin 0.1% are very effective in controlling the disease.

Rusts: Symptoms of **white rust** caused by *Albugo* typically include yellow lesions on the upper leaf surface and white pustules on the underside though it is not so common in ornamentals. Glawe *et al.* (2004) from USA and Camele *et al.* (2011) from Italy reported occurrence of *Albugo* on *Lunaria annua*. The pathogen is spread by wind, water and insects. Management includes use of resistant cultivars, proper irrigation practices, crop rotation, sanitation and chemical control. *Albugo bliti* on ***Amaranthus*** causes whitish pustules on the leaves which on maturity turn reddish-brown. Its severe infection causes plants to stunt and distorted. Spraying of 0.3% copper oxychloride reduces the incidence of disease. *A. candida*, though quite uncommon, but its infection on ***Alyssum*** releases white powdery spores on the underside of leaves by which infected leaves turn yellow and dry. Its spread is favoured by humid weather. Since it perpetuates through its zoospores, hence, disposal of plant debris & diseased parts and maintenance of proper sanitation of the field & surrounding areas are quite essential. Sprays of Captafol or Carbendazim (0.2%) are effective to control this disease.

Rusts other than white rust (*Albugo*) affect a wide range of floricultural crops including *Aster,* carnation, *Fuchsia*, *Pelargonium* × *hortorum* (florist's geranium), *Gladiolus, Lilium, m*arigold, *Poinsettia*, snapdragons and *Viola*. Rusts have the potential to negatively impact floriculture production because these pathogens often cannot be detected on infected and symptomless propagation material. Rust fungi are obligate parasites, dependent upon a live host for growth and development and seldom kill plants. However, rust infection reduces plant health and vigour, flower production and aesthetic value. Each type of rust has its own distinctive symptoms and its own specific plant hosts. The disease often first appears as chlorosis on the upper surfaces of leaves but almost all the rust fungi produce powdery masses of spores in pustules, typically on leaf undersides that are yellow, orange, purple, black or brown, with a few exceptions where some produce pustules even on upper leaf surfaces as well. Spores are easily spread on air or with splashing water. Lesions may coalesce resulting in large areas of necrosis, followed by leaf distortion and defoliation (Dicklow, 2013). The rusts can be managed by planting only disease-free plants or cuttings, and the new plantings should be kept in isolation for about three weeks to develop the rusts if present. Infected parts or plants should be removed and burnt. At the end of the growing season, all the plant debris should be cleaned up carefully and destroyed, and the benches and propagating chambers in the greenhouse should be disinfected properly. Plant spacing should be proper to allow proper air circulation. Humidity within the greenhouse should be maintained less than 80% and the air movement should be improved by addition of fans so that moisture condensation on the foliage is avoided. Overhead watering should be avoided, and if necessary, it should be done only in the morning. A rotation of chemicals should be used to check further spread of the rust diseases. *Puccinia antirrhini* is one of the most serious diseases on ***Antirrhinum***. It attacks leaves, young stems and calyces. Brown-coloured pustules develop on the underside of leaves. Infected plants bloom prematurely, bear only small flowers or they may be killed when the infection takes place very early. The disease is favoured by temperature below 21 ^{0}C. It attacks the plants at all stages of growth and is confined to hosts of this family only. Alpha and beta races of the pathogens have been reported (Aitken *et al.*, 1989). Spraying with wettable sulphur (0.25) or Karathane (0.025%) before disease appearance and repeating the spray at periodic intervals will keep this malady under check. ***Aster*** is affected by several rust diseases such as *Coleosporium campanulae, Puccinia asteris, P. campanulae* and other *Puccinia* species. Reddish-orange masses of spores form on the underside of leaves and the heavy infections can cause leaf yellowing and necrosis. *C. campanulae* requires pine as an alternate host, while *Puccinia* species have various sedges and grasses as alternate hosts. The removal of alternate hosts is a good management strategy. On pine alternate host, the blisters of spores form on needles in the spring. **Carnation** rust is caused by *Uromyces dianthi*, and the symptom is characterized by small pustules of powdery brown urediospores. These spores are carried over only on live plants. Small blisters containing rust-red spores form on leaves. Chlorothalonil, mancozeb + thiophanate methyl, Myclobutanil, Propiconazole, Triadimefon, Triforine or ziram should be sprayed to protect the healthy plants. **Chrysanthemum** rusts are caused by *P. chrysanthemi* and *P. horiana*. *P. chrysanthemi* is most common in late summer but is a minor disease in the field, however, uncommon in the greenhose, and is characterized by dirty-brown pustules and yellowish-green spots on upper surface of leaves. Severe infestation may damage large areas of leaves and may lead to defoliation and reduced flower production. Chrysanthemum varieties resistant to rust include 'Achievement', 'Copper Bowl', 'Escapade', 'Helen Castle', ' Mandalay ', 'Matador', 'Miss Atlanta', 'Orange Bowl' and 'Powder Puff' (Dicklow, 2013). Symptoms are white, pinkish or brownish pustules produced on leaf undersides with white, yellow to pale-green lesions on upper leaf surfaces. Chrysanthemum white rust results in leaf distortion, discolouration, defoliation and plant death. White rust is primarily a disease of greenhouse crops and when it occurs outside, the direct sunlight and low humidity kill the spores. ***Fuchsia*** rust is caused by *Pucciniastrum epilobli*. The most serious losses occur during propagation and diseased plants at any stage are not marketable. Affected leaves may be deformed and often defoliation occurs. Weakened plants may occasionally die. The initial source of spores may be from fir or fireweed as alternate hosts, especially if plants are kept outside in summer or from infected stock brought in from other sources. **Hollyhock** rust (*Puccinia malvacearum*) infects leaves, stems, bracts and other plant parts with small raised and dark brown pustules. A dense growth of pustules may destroy large portions of the leaf which later dries out. In humid weather the pustules are ash-grey. The disease is commonly being observed in Himachal Pradesh and Bengaluru, and becomes severe in humid weather. Karathane (0.025 %) or Sulfex (0.2 %) spray is recommended to check this pathogen. The infected seedlings and plant debris should be destroyed. ***Jasminum*** rust (*Uromyces hobsoni*) appears in severe form during monsoon rains on *J. auriculatum*. Other species, *viz. J. angustifolium, J. communis, J. flexile* and *J. sambac* are also susceptible to rust. The leaves show the presence of orange aecial cups on both the sides, but predominantly on the lower surface. Numerous blisters are produced in advanced stages of infection causing yellowing and crinkling of the leaves. The stems and branches are also infected, causing splitting of bark and subsequent death of the branches. The disease can be controlled by dusting sulphur 20-25 kg/ha. Spraying with bordeaux mixture or copper oxychloride is also recommended for its control. ***Geranium*** rust caused by *Puccinia pelaragoni-zonalis* is most serious on florist's geraniums (*Pelargonium* × *hortorum*) though has also been reported on zonal geraniums and seedling geraniums. Ivy geranium (*P. peltatum*), Martha Washington or regal (*P.* × *domesticum*), the scented leaf types and the wild geraniums are resistant. This rust spends its entire life cycle on geranium. Symptoms first appear as small circular yellow spots on the top of leaves opposite the pustules on the flower leaf surface. Chlorotic specks on upper leaf surface appear directly to the opposite pustules of rust-coloured spores on underside of leaf. Spores erupt in concentric rings forming a 'target' spot. The spots on the flower leaves enlarge to blister-like pustules of rust to cinnamon-brown spores which often develop in concentric rings. Discard unwanted geraniums at season's end. Application of Chlorothalonil, Flutolanil, Myclobutanil, Oxycarboxin or Triadimefon will protect the healthy plants from infection. ***Iris*** rust (*Puccinia iridis*) exhibits small and raised orange spots on leaves and stems. Spraying with Plantvax (oxycarboxin) will control this problem to some extent. Immediately after planting, entire foliage of already growing irises should be removed and burnt and then new crop on emergence should be sprayed with oxycarboxin. Unlike most other rust fungi, pustules of ***Poinsettia*** rust caused by *Uromyces euphorbiae* occur on both the leaf surfaces. Spores are cinnamon-brown and occur only on living plants. **Rose** rust caused by *Phragmidium mucronatum* first appears in the spring as bright orange pustules on leaf undersides, leaf stalks and branches only in the temperate regions. In the summer, small raised orange spots appear on the undersides of leaves with small yellow specks on the upper surface beginning on the lower leaves and moving upward. The spots may go unnoticed until the plant begins to exhibit a generally unhealthy appearance and a loss of the lower leaves. Leaves may become dry and twisted before falling off. As the leaves die, the spots darken and produce black teliospores that survive the winter to produce new infections in the spring. This disease may be mild or severe depending on the season and how early the infection began. Pruning and burning of the infected leaves, petals and canes is recommended as soon as the disease is detected. Cultivars vary in their susceptibility to rust diseases. Rust of ***Viola*** caused by *Puccinia violae* is common in the NE region. Symptoms first appear as small pale-green spots on the upper leaf surface. As the fungus develops within the leaf, the corky spots, blisters or pustules containing rusty brown spores develop there.

Bacteria

The general symptom of bacterial diseases is mostly rotting of mass of tissues or sudden wilting and collapsing of the plant. Therefore, the diagnosis of bacterial diseases is not as easy as fungal disease because bacteria in general do not produce any external growth and usually remain confined within the plant tissue. The way of detecting the presence of a particular fungus or bacterium is to isolate them in pure culture in the laboratory from the affected part. *Erwinia chrysanthemi* can cause a mushy, brown, smelly, soft rot or leaf spots. *E. chrysanthemi* and *E. carotovora* survive in plant debris that is not completely decomposed, on or in infected plants, on other greenhouse plants without causing disease, and under some conditions in soil. Both species infect a wide range of plants in the greenhouse. *Pseudomonas cichorii* can cause leaf spots and blights on balsam, chrysanthemum, geranium and many other ornamentals. The spots are generally water-soaked (wet-looking) and dark-brown to black. Depending upon the plant infected, the leaf spots may have a yellow halo (Moorman, 2013). The bacterium *Pseudomonas syringae* is an opportunistic pathogen that can attack a wide variety of woody plants especially when they are damaged by frost or injured in some way. While there are a number of strains or pathovars (pv.) of this organism, pv. *syringae* is most important on woody plants. *Pseudomonas syringae* pv. *syringae* occurs on nursery and landscape plants throughout the world and can cause shoot and flower blights, cankers and diebacks. Common, Persian, Chinese and Japanese lilac leaves develop round to irregularly shaped brown spots with yellow halloes. These spots enlarge and blight entire leaves. Succulent twigs are girdled, killed and turn black red, Norway and Japanese maple leaves can be spotted, their veins blackened and branch tips killed. Other plants known to be susceptible include basswood, saucer magnolia, poplar, dogwood, golden-chain tree, forsythia and some herbaceous plants. In nurseries, even Monterey pine seedlings have been killed by this bacterium. These remain on or in bud and twig tissues, in cankers formed the previous season, and in or on grasses and herbaceous weeds and act as a source for further infection. *Pseudomonas* invades damaged tissue and produces a toxin that kills surrounding cells where the bacteria can then multiply. Also when it is present on plants, it produces a protein, around which ice crystals form. As the ice crystals enlarge, they pierce and severely damage the plant cells. These damaged cells are then colonized by *Pseudomonas*. The bacteria spread by wind, insects, through use of infected budwood and nursery stock or through pruning tools. These are managed by using resistant stocks such as 'Edith Cavell', 'Glory' and 'Pink Elizabeth' common lilac varieties which are less susceptible than many other ones. *Syringa josikaea, S. komarowii, S. microphylla, S. pekinensis* and *S. reflexa* have greater resistance than common lilac *S. vulgaris.* This can be managed through use of healthy stock as planting material, protection from frost, no use of fertilizers during early spring or in autumn which make the plants more succulent, pruning should be carried out in early spring with disinfected pruning shears or tools, sprays of copper or combinations of copper and mancozeb during autumn but not in spring as new tissues may be damaged, however, some of the populations of *Pseudomonas syringae* are resistant to copper while in some plants the application of phosphonate or phosphate chemicals induces a resistance response in plants.

Erwinia chrysanthemi causes bacterial blight of **African violet** and its infection produces dark reddish-brown to black rotting areas on roots and crown and infected petioles develop a greasy appearance. For its management, the planting is recommended in sterilized soil or medium, high temperature and humidity should be avoided, and new plants should be maintained separately to see if any one is having infection. **Carnation** is affected by slow bacterial wilt or stunt due to infection of *Erwinia chrysanthemi*. It causes stunting, curling, twisting and wilting of side shoots, lower and floral leaves which may recover but such plants wilt again slowly taking several weeks. Stunted plants have narrow and yellow-green leaves. Cuttings should be taken only from the culture-indexed plants and planted only in the sterilized soil or medium. On ***Chrysanthemum***, *E. chrysanthemi* causes blight by which cuttings turn dark brown and collapse. Surviving cuttings may be infected but have no symptoms. Established plants wilt during the day when infected and recover at night before final death. Management is similar to carnation. In case of ***Dahlia***, *E. carotovora* causes stem rot so that stems blacken with soft and wet interior which emit foul odour and such plants should be destroyed. *Erwinia carotovora* causes floral bud rot in ***Narcissus*** which can be controlled through streptocyclin or mercuric chloride. *E. carotovora* on ***Cattleya* orchids** shows soft rot and spreads as water-soaked darker lesions from the leaf tips to downward and causes pseudobulbs becoming yellow, soft and pulpy. Streptocycline (0.01%) treatment is effective in its control. *Erwinia* also causes ***Crassula*** basal rot where interior tissues become soft and mushy so stems and branches collapse. ***Cordyline*** stem rot and leaf spots caused by *E. chrysanthemi* and *E. carotovora* pv. *carotovora* make the rotting of cuttings, roots of established plants turning black and dieing, and stems and leaves becoming wet and slimy. Control measures are similar to carnation. Soft rot of ***Cyclamen*** is caused by *E. carotovora* and *E. chrysanthemi* where tubers have wet and soft rot, plants yellow and wilt rapidly. Planting of only healthy tubers and that too in the sterilized medium or soil should be done. These two bacteria also cause stunting of plants, yellowing of leaves, blackened stem bases, emitting of bad odour under wet conditions from rotten parts and ultimate toppling over of entire ***Delphinium*** plant, however, *Paeudomonas syringae* infection kills the seedlings, and forms black and irregularly shaped spots on leaves of adult plants, though on ***Dieffenbachia*** *E. carotovora* and *E. chrysanthemi* form small gray or dark green leaf spots which later on become tan, dark brown or black and irregularly shaped. The bacterial blight of ***Asplenium*** fern caused by *Pseudomonas cichorii or P. gladioli* produces translucent spots all over the leaves that enlarge quickly and turn reddish-brown with purple haloes. **Easter lily twist** occurs probably due to *Pseudomonas* infection where middle stem areas and leaves twist whereas leaves above and below appear normal. *P. cichorii* on ***Geranium*** causes elliptical and 6-13 mm water-soaked leaf spots which later on turn dark brown to black and irregularly shaped, and each spot may sometimes be surrounded with a yellow halo. No overhead watering should be given to keep leaf surface dry. Only culture-indexed cuttings of geranium should be used or grown through seeds. Bacterial leaf spot of ***Gerbera*** caused by *Pseudomonas cichorii* appears as small to large spots that are initially circular but afterwards become irregular and dark brown to black and may also have a concentric ring pattern. *Pseudomonas solanacearum* causes typical ***Cosmos*** wilt. Such plants should be rogued out and 3-4 years rotation should be followed. ***Petunia*** bacterial wilt (*Pseudomonas solanacearum*) is a very serious disease which affects the plants at any stage. Crop rotation, sterilization of soil or media before planting, and use of bleaching powder for treating the soil before sowing are recommended.

Begonia bacterial blight and leaf spot occurs due to attack of *Xanthomonas campestris* pv. *begoniae* where apart from leaf spots it causes systemic infection (Strider, 1975; Harii *et al.*, 1977), especially in Rieger types. Leaves show water-soaked areas surrounded by yellow halos and plants ultimately die. Proper sanitation should be maintained by cleaning and destroying all the plant debris, infected plants and plant parts should be discarded, overhead irrigation should be avoided, and bordeaux mixture, streptomycin and cupric hydroxide (Strider, 1975; Harii et.al., 1977) should be used to combat the disease with moderate success. ***Hedera*** circular dark brown to black bacterial leaf spot with yellow halo is due to infection of *Xanthomonas campestris pv. hederae*. Its infection causes leaf yellowing, death and leaf falling. Potassium salts of phosphoric acid controls this pathogen. Symptoms of *Xanthomonas campestris pv. pelargonii* on ***Geranium*** appear either as well-defined, some 6 mm size sunken spots on leaf underside which ultimately results into spread of bacterium through water-conducting vessels of veins and petioles and causing V-shaped areas from veins to the margins, or lower leaves show marginal wilting while blades and petioles remain turgid, the vascular tissues of the main stem to the side of leaf wilting becomes gray-brown followed by death and falling of lower leaves. *Xanthomonas hyacinthi* on ***Hyacinthus*** causes bulb rot before and after planting, and on cutting bulbs show yellow spots arranged in circle. *Erwinia carotovora* attacks hyacinths and causes soft rot. *Xanthomonas nigromaculans* pv. *zinniae* is the common bacterium infecting ***Zinnia***. Its infection may wither the infected cotyledons or entire seedling within two weeks of seedling emergence. It also causes leaf spots on zinnia (Hernandez and Trujillo, 2000). Strider (1979a) to detect the pathogen in zinnia seed, made dense sowings under warm and humid conditions in the glasshouse and recorded distinct lesions on the cotyledons after 5-14 days. Considerable pre-emergence spread occurred so that a very low percentage of contaminated seed resulted in a high number of infected seedlings, thus facilitating this method of indexing. Bacterial leaf spot is reduced by treating seeds with a 30-minute chlorine bleach soak of 1:2 (bleach: water). Excessive bleach treatment will injure the seeds. After rinsing, the seeds are sown. Several bactericides have been tried for controlling the disease. Earlier, streptomycin was used for eliminating the pathogen. Strider (1979b) reported that streptomycin is effective in limiting the pathogen but germinating seedlings were chlorotic, slightly stunted and not uniform. A combination of trimming of diseased leaves followed by a foliar spray of streptocycline (0.01%) + carbendazim (0.025 %) was found to be the most effective control against *Xanthomonas* as well as *Cercospora* leaf spot (Jindal and Meeta, 1991).

Bacterial fasciation of ***Geranium*** occurs due to infection of *Rhodococcus fascians*. Its infection makes the stems short, thick and fleshy and those stems appearing from the base of the main stem are pale-green though rest of the plant appears healthy. Southern bacterial wilt of ***Geranium*** is caused by *Ralstonia solanacearum* where lower leaves yellow, wilt and fall and the vascular tissues of the affected stems turn brown to black. *Pectobacterium carotovorum* causes bulb decay in ***Lilium***. *Bacillus lilii* also attacks lilies. Since these pathogens enter the bulbs through weak scales or wounds so planting of injured bulbs should be avoided.

Any plant showing bacterial disease symptom should immediately be discarded, in case of bulbous plants, these should be lifted along with adhering soil for destruction, knives should not be used for taking the cuttings or when using these should be thoroughly disinfected regularly, hands should also be washed while taking the cuttings, low humidity in the field or greenhouse should be maintained, overhead irrigation should be avoided, and all the unwanted plants at season-end should be discarded.

Viruses

During the past several decades, the frequency of plant diseases caused by viruses has increased worldwide. Viruses are common pathogens of many ornamental plants. The diseases are many times overlooked since symptoms may be absent at the initial or mild at the later part. Actually no cure is there where plants are already infected. Though in many cases viral infections are just tolerated or ignored unless obvious symptoms arise but detection and identification at early stage may help in managing the plants with available measures. In recent times research in the field of molecular detection (like DNA microarray) system has developed which helps in detection of several viruses infecting ornamentals within five hours of possessing time (Tsai *et al.*, 2012). Affected plants can be discarded to check the spread in healthy plants. Common symptoms are mosaic, distortion, mottling, curling, yellowing, necrosis and spotting. Some of the symptoms produced by viruses resemble physiological disorders. Knowledge about prevention, transmission, maintaining hygiene and proper testing are important for good virus disease management programme.

Viruses are submicroscopic entities capable of causing various diseases. They are a piece of nucleic acid (genetic material) surrounded by a protein coat. Once inside the plant cell, the nucleic acid portion directs the plant cell to produce more virus nucleic acid and virus protein, disrupting the normal activity of the cell. Viruses, in general, cause a disruption of the plant's metabolism which usually results in changes in the pigmentation or shape of the leaves, flowers or fruits, stunting, and/or reduction in the number or size of fruit. By far the most common visible symptom is mosaic or mottle which is a mixture of light and dark green areas on the leaves where uninfected leaves would be uniform in colour (Mathews, 2013). Viruses can multiply only inside a living cell. While some viruses such as 'cucumber mosaic' die quickly if outside a cell or if the cell dies, other viruses such as 'tobacco mosaic' retain their ability to infect for years after the infected plant part dies. Many different viruses can infect plants. Certain crops are well known to be affected by virus diseases including *Dahlia, Gladiolus, Geranium,*

Lilium longiflorum, *Rosa* and *Tulipa*. Depending on which virus is involved, the disease may be spread from infected plants to healthy plants through working tools or human hands mechanically, through various insect-pests such as aphids, leafhoppers, mites, nematodes, thrips, through dodder, and through grafting, budding and cuttings taken from the infected plants. Sometimes, though viruses are present in the plants but are not exhibited so in vegetative propagation of such plants whether cuttings and grafting, the virus infection is exhibited afterwards on the new plants. In fact, the virus particles are found in all parts of the plant except the few cells at the tips of the growing points. It is these few cells that are removed and grown into a healthy plant free of virus by the process called meristem-tip-culturing. Symptoms vary with the virus involved, the species of plant infected and the environmental conditions. In some cases, such as virus disease of geraniums, certain environmental conditions bring out symptoms while other conditions mask or hide. Symptoms associated with virus infections are stunting or reduced growth, mosaic leaf patterning (mixed yellow and green, bronze and green, red and green or yellow colouration), leaf or growing point malformation and distortion even of the floral buds, yellow streaking or spotting on leaves, foliage ring-spots or line patterning, cupping of leaves, floral colour breaking as in tulips, distinct yellowing only on veins, crinkling or curling of leaf margins, *etc.* Some of these symptoms are also caused by high temperature, insect feeding, growth regulators or herbicide treatment, and mineral deficiencies or excesses. Aster yellows which infects many ornamentals including gladiolus and petunia is a though phytoplasmal disease but produces symptoms similar to virus infection such as abnormal branching, yellowing of leaves and formation of leaf-like yellow appendages instead of floral parts.

There is no chemical which cures a virus-infected plant nor there is one that protects the plants from becoming infected. However, in case of virus infection prevention is always better than cure so only virus-free plants should be planted, infected plants should immediately be uprooted and burnt, strict contol of insect, mites, other pests and weeds should be carried out, plant debris should immediately be removed and burnt so that no sheltering of insect-pests remains nor live plant part may get any chance to harbour the viruses, there should be proper field sanitation, any suspiscious plant should be kept in isolation for watching and if possible should be diagnosed through molecular basis, tools used for vegetative propagation should be disinfected every time with chlorine, hydrogen peroxide or ammonium disinfectant for at least 10 minutes and then rinsing thoroughly with tap water, and all the plants, if possible, should be propagated through seeds, however, as certain plants carry viruses even through their seeds so one should be careful. Besides the above precautions, biotechnological *in vitro* culture techniques like meristem culture, thermotherapy, chemotherapy, cryotherapy or a combination of these may be used for their elimination. The plant species, as well as the type of virus determine the choice of a most suitable method (Milošević *et al.*, 2012).

Tomato spotted wilt virus (TSWV) causes ***Amaryllis*** leaves with numerous white or yellow spots. Certain blood red spots may also appear on the leaf edges. Amaryllis mosaic exhibits irregular light and darker green areas on the leaves. Impatiens necrotic spot, tobacco ring spot, broadbean wilt, cucumber mosaic, tobacco necrosis virus, all infect ***Begonia***. Mosaic, malformed leaves, pale yellow ring spots, severe chlorotic mottling of leaves or plant stunting may occur. Fifteen different viruses are known to cause diseases on **carnation**. Leaves may be mottled, have yellow spotting, dead flecks, line or ring spot patterns. Plant may be distorted or have flower colour breaking. Impatiens necrotic spot, streak and mosaic viruses infect ***Cineraria***. Leaves may be mottled, have yellow spotting, dead flecks, line or ring spot patterns. Plant may be distorted, may have flower colour breaking, or may be stunted. Veins may be killed along with petioles and the entire plant. Tobacco mosaic, Cucumber mosaic, Tobacco rattle, Tomato aspermy, Potato virus X, Impatiens necrotic spot infect the ***Cyclamen***. Flowers may be abnormal in shape. Leaves are deformed and streaked with yellow or brown markings or have ring spots in a thumb-print like pattern. Infection of tomato spotted wilt virus produces ring or spotted areas on ***Dahlia*** leaves, which are lighter in colour than the leaf proper. Infection of Cucumber mosaic of dahlia is known by yellow- green banding of the vein and in extreme cases they coalesce and form into mosaic. Impatiens necrotic spot virus causes conspicuous rings, and line patterns may be present or a mosaic of light and dark green may occur on leaves. Cucumber and tobacco mosaic viruses infect ***Delphinium***. Conspicuous rings and line patterns may be present and plants are stunted. Dasheen mosaic virus infects ***Dieffenbachia***. Leaves have a mosaic of light green area where they should be dark green. Plants may be stunted and leaves deformed. Some cultivars die within a month of infection.

Many viruses occur singly or in various combinations in plant tissues of ***Geranium***. Symptoms vary with virus, cultivar, and growing conditions. Symptoms include mottling, ringspotting, leaf cupping, yellowing of veins, leaf malformation, leaf spotting, and loss of zonation. Tobacco rattle, gerbera mosaic, impatiens necrotic spot and cucumber mosaic viruses infect ***Gerbera***. Leaves may be mottled, may have yellow spotting, dead flecks, line or ring spot patterns. Plants may be distorted or have flower colour breaking. Tobacco mosaic, tomato ringspot and impatiens necrotic spot virus infect ***Gloxinia***. Leaves may be downwardly cupped, mottled, or strap shaped. Leaves may have line or ring spot patterns. Shoots may be elongated. ***Hydrangea*** is infected with 'hydrangea ring spot', 'tomato ring spot' and 'impatiens necrotic spot' by which plants show stunting, virescence, distortion and phyllody; yellowing, yellow spotting, dead flecking, line or ring spot patterning, mottling and necrosis of leaves; redness of the leaf edges; and reduced number of flowers and/or colour breaking (Bertaccini *et al.*, 2012). ***Hydrangea* virescence** is a phytoplasmal disease where flowers become green and small, leafy shoots grow from floral parts, plants decline and die. Viruses that attack **hyacinths** are 'hyacinth mosaic virus' (HMV) and 'ornithogalum mosaic virus' (OMV) which cause mosaic and gray leaf mottling, 'tobacco rattle virus' (TRV) causes chrorotic stripes, whereas 'lisser disease' (a mycoplasma) causes plant yellowing. 'Iris mild mosaic virus' (IMMV), 'narcissus latent virus' (NLV), 'iris severe mosaic virus' (ISMV), and 'bean yellow mosaic virus' (BYMV) infect ***Iris***. IMMV, BYMV and ISMV are potyviruses and NLV is carlavirus, all being transmitted through aphids so control of aphids will control these viruses also. The mosaic ***Jasminum*** plants show stunted and yellowish-green appearance with small leaves. Yellowish green to chlorotic flecks of 1-2 mm in diameter and streaks appear irregularly on the leaf, and the streaks form a ring, symptoms being more conspicuous on older leaves. The disease is graft-transmissible on *Jasminum sambac* and *J. simlicifolium*, and *Jasminum pubescens* is highly susceptible though *J. auriculatum* is relatively tolerant. '**Jasmine** phyllody' is suspected to be caused by a mycoplasma. The affected plants produce malformed, highly congested and reduced leaf-like greenish floral-structures (trsnsformation of floral parts into leaf-like structures snd corolla lobes becoming ovate and reduced) instead of fragrant white flowers on panicles. In 'top spotting virus' of ***Kalanchoe*** the plants may stunt without flowering, leaves may be mottled with yellow spotting and dead flecks, *vis-a-vis* line or ringspot patterns. '**Easter lily** curl stripe virus' causes late emergence of sprouts from infected bulbs and soon after emergence the leaves twist with crooked tops and white stripes though mildly affected plants seem normal except for the lower leaves. 'Lily symptomless virus' and 'little mottle virus' cause plant stunting and produce yellow mottle and mosaic on the leaves, and cucumber mosaic virus exhibits mosaic patterns on the foliage and also necrosis on the plants, vector for which is *Myzus persicae, etc.* Affected plants become dwarf and deformed. LSV is transmitted through aphids and causes vein clearing. **Easter lily flecking** is due to infection of CMV in combination with LSV. In their infection, short white lines appear on the leaves, entire plant stunts and flowers become small and sometimes deformed. Lighter striping and mottled pale-yellow discolouration of the foliage and flower stalks occur due to infection of 'cucumber mosaic virus' on ***Narcissus***. 'Narcissi yellow stripe virus' infection on *Narcissus* causes bright yellow stripes or streaks throughout the aerial part from emergence to flowering which reduces yield and quality of flowers and bulbs. Narcissi 'white streak' ('silver leaf') on leaves by which foliage senesces earlier and bulbs also deteriorate rapidly, 'Narcissus latent virus', 'Grand Soleil d'Or and Jonquil mild mosaic' are transmitted through aphids, 'narcissus mosaic' is transmitted mechanically though 'Arabic mosaic', 'raspberry ring spot', ' strawberry latent spot', 'tobacco rattle', 'tobacco ring spot' and 'tomato black ring spot or chocolate spot virus' show large dark brown spots, and all these are transmitted through nematodes.

'Blossom brown necrotic streak' attacks ***Cattleya* orchids** producing brown spots or whole flower break, and yellow streaks on leaves. It is transmitted from one plant to other through cutting knife. ***Cattleya*** and other orchids may have mild 'mosaic flower break' with no distortion or severe colour break with flower distortion and variegation. Leaves mottle and twist sometimes. Transmission is through green-peach aphid (*Myzus persicae*). '***Cymbidium*** mosaic' initially appears as leaf mosaic mottling and then turning to necrotic spots, streaks and rings. In ***Cattleya*** sunken brown to black leaf patterns appear sometimes with rings and more often with black elongated streaks on older leaves. Due to infection of ***Odontoglossum*** ring spot', small necrotic spots and rings develop on the older leaves and the light green to yellow areas on the younger leaves. The leaves may turn yellow and drop in 2-3 months, though yet neither flower symptoms nor any vector have been observed.

Petunia is attacked by 'tobacco mosaic' (TMV), 'cucumber mosaic' (CMV), 'tabacco ring spot', 'aster yellows', 'tomato spotted wilt', 'beet curly top' and 'potato virus X' diseases. In case of TMV infection, the diseased plants show stunted growth with reduction in leaf size. Mottling of leaves with vein-banding symptoms is also common on the lower leaves. TMV is transmitted by sap and *Aphis gossypii.* CMV is transmitted by *Myzus persicae.* ***Zinnia*** is subject to many viral diseases. Curly top, leaf curl, mosaic and spotted with white and yellows are among the most common viral diseases. *Ruga verrucosans* causes curly top and *Letham australiense* causes spotted wilt. Zinnias are badly affected with spotted wilt. At first, the leaves get curled, and after some time the plants get wilted. This disease is very common in the northern plains and heavy rainfall areas, and is transmitted by thrips. However, the hills are comparatively free from this disease. A new virus known as 'tomato aspermy virus' has been identified in zinnia (Kameya *et al,* 1996). It causes chlorotic spots and mosaic symptoms. Sastry *et al* (1973) reported that mosaic disease of zinnia was transmitted by *Aphis gossypii* and *Myzus persicae.* However, 'zinnia mild mottle virus' is transmitted by *Myzus persicae,* but not by *Aphis solanella* (Padmanabhan and Padmanabhan, 1978). Zinnia is a symptom-less carrier of 'watermelon mosaic virus' (Tewari, 1976). It may be of importance in its epidemiology.

Purchasing or planting of only virus-free plants and those which are virus-indexed and further these should be planted in sterilized soil or media. In case of bulbous ornamentals, the bulbs from infected plants should not be saved for further planting. The tools used in dividing the plants should be disinfected and washing of hands should be done thoroughly and frequently when handling plants to prevent mechanical spread. Eliminate weeds within the production area. Use very fine mesh screen to keep thrips out. Greenhouse should be made completely aphid-free since 'lily symptomless virus' and many others are carried by aphids from one plant to the other. Do not use tobacco products when handling plants.Infected plants should be burnt or buried and the insect vectors such as aphids, jassids, leafhoppers, thrips, *vis-à-vis* nematodes should be controlled by spraying certain effective insecticides akin to vector types. The jasmine phyllody may be controlled by spraying with tetracycline hydrochloride (250 ppm). Valuable clonal stocks as in case of orchids should be saved through tissue culture.

Physiological Disorders

Irrigation water colder or warmer than the leaf surface causes **African violet ring spot** due to collapse of cells. Bright yellow or bleached ring patterns are formed on dark green leaves. Therefore, always water should be used at normal temperature and off the foliage. **Dracaena flecking** is due to unknown causes. Here new leaves have small white to yellow spots near the tips. Maintenance of moderate temperature, moisture, and light conditions will combat this problem. Excess of **fluoride** causes toxicity in ***Dracaena*** and long tan to dark brown areas form in the white stripes of cv. 'Warneckii' leaves. Leaf tips are yellow or dead on 'Janet Craig'. Tan dead areas with bright yellow borders form along the leaf margin of 'Massangeana' leaves. A soil pH of 6.0-6.5 should be maintained. Use of fluoride-containing soil amendments including superphosphate and irrigation water containing 1 ppm or above fluoride should be avoided. **Boron deficiency** causes irregularly-shaped small dark brown to black spots at the base of ***Gloxinia*** seedlings, browning spreading up to the leaf blade afterwards, the floral stalks on older plants flower stalks become short, and ultimately such plants die. Boron application can rectify this problem. Combined application of slow-release and soluble fertilizers or only soluble fertilizer with no leaching favours causing plant stunting, leaf yellowing and falling and leaf margin yellowing and burning of ***Poinsettia*** due to over-fertilization. Conductivity readings at or above 1.0 mS for soil or 2.5 mS for soilless mixes (saturated paste extract) are excessive. To get relief from this problem, combined use of slow-release and soluble fertilizers should be avoided, however, the fertilizer water should not stagnate in the planting so there should be proper provision for its leaching. Molybdenum deficiency or acidic pH of potting mixes (4.5 to 5.5 pH) causes leaf yellowing and downward cupping and burning of lower leaf margins in *Poinsettia.* Lime should be applied in the soil to adjust soil pH above 5.5. Use of molybdenum-containing fertilizer, addition of sodium molybdate to soluble fertilizer mix or a complete minor element supplement will solve this problem. Magnesium deficiency in *Poinsettia* causes yellowing between the veins in middle and lower leaves which can be corrected through use of magnesium-containing limestone or fertilizer. Root rot, over-fertilization, low light intensity or lack of moisture causes *Poinsettia* defoliation. To get rid of this problem, even maintenance of moisture level, proper plant spacing, and clean and proper transparency of glasshouse covering should be done, over-fertilization should be avoided and no reason should be permitted which may contribute to root rotting. Restricted root and top growth, and yellowing of lower leaves & burning of margins in *Poinsettia* is favoured by low light and low temperature in late fall, acidic potting mix, and exclusive use of ammonium nitrogen source, and the situation is aggravated due to over-watering during low light and low temperature conditions so such situations should be avoided. Drought conditions cause ***Dieffenbachia*** lower leaf yellowing followed by leaf and entire plant wilting which can be rectified through adjusting even moisture level. **Fern** frond tips and leaflets show brown burning and plants die due to over-fertilization so the pots should have proper provision for leaching of excess fertilizers. Frond lobing of ***Asplenium*** fern occurs due to over-fertilization. Many lobes and indentations in fronds are formed and the tips may be crinkled and dead. Leaching of the pots and reduction in the fertilizer use will rectify this problem. Dought conditions in ***Nephrolepsis*** causes graying of leaves with only a few runners being produced by the plant. The problem can be nullified through even maintenance of soil moisture at all the times. Such conditions also prevail due to root rot or root-knot nematode infestation in boston ferns. Small water-soaked pimples or blisters, known as **edema** form on the underside of lower leaves in ***Geranium***, and this problem is favoured due to cool and cloudy weather. These blisters afterwards become corky-brown, followed by leaf fall in severely affected ones. To do away this problem, the planting should be done in the well-drained mix, the plants should be properly spaced to permit proper air circulation and the watering during cool and cloudy weather should be avoided. **Easter lily leaf scorch** may be due to fluoride toxicity, mineral deficiency induced by root rots and low soil pH. Semicircular dead areas develop along leaf margins and the leaf tips may be engulfed. Use of superphosphate and other fluoride-containing fertilizers should be avoided, soil pH between 6.0-7.0 should be maintained and no condition should be provided for root rotting.

References

Aitken, E.A.B., H.J. Newbury and J.A. Callow, 1989. Races of rust (*Puccinia antirrhini*) of *Antirrhinum majus* and the inheritance of host resistance. *Plant Pathol.*, **38**: 169-175.

Anonymous, 2013. Grey mould. Royal Horticultural Society (http://apps.rhs.org.uk).

Agrios, G. N. 1997. Plant Pathology (4th ed.), 635 p. Academic Press, California, USA.

Beale, R.E. and D. Pitt, 1990. Biological and integrated control of *Fusarium* basal rot of *Narcissus* using *Mininedusa polyspora* and other micro-organisms. *Plant Pathol.*, **39**: 477-488.

Benson, D.M. 1995. Aluminium amendment of potting mixes for control of *Phytophthora* damping off in bedding plants. *HortSci.*, **30**: 1413-1416.

Bertaccini, A., S. Paltrinieri, N. Contaldo, N. Mori, L. Cavicchi and M.G. Bellardi, 2012. Severe diseases induced by viruses and phytoplasmas in *Hydrangea* in Italy. In: *Bioforsk Fokus* (eds Dag-Ragnar Blystad and Carl Spetz), p. 8.

Camele, I., S.M. Mang and G.L. Rana, 2011. First report of *Albugo candida* causing white rust on *Lunaria annua* in Italy. *Plant Disease*, **95**(1): 72.2-72.2.

Castillo, S. and J.L. Peterson, J.L. 1990. Cause and control of crown rot of New Guinea *Impatiens*. *Plant Dis.*, **74**:77-79.

Crous, P.W. and U. Broun, 1996. Cercosporoid fungi from South Africa. *Mycotaxon*, **57**: 233-321.

Dicklow, M.B. 2013. Rust diseases of ornamental crops. Mass Extension Plant Diagnostic Lab, University of Massachusetts (mbdicklo@umext.umass.edu).

Fletcher, J. T. and J.A. Martin, 1972. Spread and control of *Fusarium* wilt of carnations. *Plant Path.*, **21**(4): 182-187.

Glawe, D.A., J.R. Glass and M.L. Putnam, 2004. First report of white rust of *Lunaria annua* caused by *Albugo candida* in North America. *Plant Management Network* (DOI:10.1094/PHP-2004-0317-01-HN).

Gould, A. B. 2009. Downy Mildew in Ornamental Plants. *Plant and Pest advisory*, **15**(10): 1-2.

Harii, J.A., P.O. Larsen and C.C. Powell, Jr. 1977. Bacterial leaf spot and blight of Reiger Elatior begonia: systemic movement of the pathogen, host range, and chemical control trials. *Plant Dis. Reptr*, **61**: 649-653.

Harmon, P.F. and S.D. Bledsoe, 2004. *Professional Disease Management Guide for Ornamental Plants.* A series of the Plant Pathology Department, Florida Cooperative Extension Service, Institute of Food and Agricultural Sciences, University of Florida (http://edis.ifas.ufl.edu).

Hernandez, Y. and G. Trujillo, 2000. *Xanthomonas campestris* pv. *zinniae,* infecting zinnia plants (*Zinnia elegans* Jacq.) in Venezuela'. *Revista de la Facultad Agronomia, Universidad del Zulia*, **17**: 156-163.

Hong, C. 2013. Control of Ornamental Diseases. In: *Home Grounds & Animals*, pp. 4-10. (http://pubs.ext.vt.edu).

Jindal, K.K. and M. Meeta, 1991. Control of bacterial and *Cercospora* leaf spots of zinnia (*Zinnia elegans*). *Indian J. agric. Sci.*, **61**: 156-158.

Jones, R.K., 1999. Powdery mildew of ornamentals and shade trees. *Ornamental Disease Information Note 4*. Plant Pathology Extension, North Carolina State University, USA.

Kameya, M.I., Y. Suzuki, K. Hanada, S. Ito and S. Tanako, 1996. A new mosaic disease of zinnia caused by tomato aspermy virus. *Ann. phytopath. Soc. Japan*, **62**: 175-176.

Madhu Meeta, 1993. Management of powdery mildew of zinnia (*Z. elegans*) with fungicides. *Pl. Dis. Reptr.*, **8**: 67-68.

Mathews, D.M. 2013. Optimizing detection and management of virus diseases of plants. (http://ucanr.org/sites/ucnfa/files/62197.pdf).

Merlo, P.A., M.C. Rollan and B.L. Ronco, 1986. Wilt of *Antirrhimum majus* L. caused by *Verticillium dahliae* Klebahn in the Republic of Argentina. *Rev. Fav. Agron.*, **61-62**: 181-185.

Milošević, S., A. Cingel, S. Jevremović, I. Stanković, A. Bulajić, B. Krstić and A. Subotić, 2012. Virus elimination from ornamental plants using *in vitro* culture Techniques. *Pestic. Phytomed.*, **27**(3): 203–211.

Moore, W.C. 1979. *Diseases of Bulb* (2nd edn., revised by Brunt, A.A.,D. Penn, A.R. Rees and J.S.W. Dickens). Bull. HPD 1, Min. Agr. Fish. and Food, London, 205 pp.

Moorman, G. W. 2013. *Bacterial Diseases of Ornamentals.* State Extension (Penn State University, 4-H. http://extension.psu.edu).

Padmanabhan, C. and G. Padmanabhan, 1978. A leaf curl disease of *Zinnia elegans* Jacq. *Cur. Res.*, **6**: 100-101.

Paul, Y.S., R.C. Sharma and L.N. Bhardwaj, 1991. Additions to fungi of Himachal Pradesh-IV. Peronosporales. *Plant Dis. Res.*, **6**:105.

Powell, C.C. 1974. Chemical control of powdery mildew of zinnia. *Research Summary, Ohio Agricultural Research and Development Center*, **79**: 57-58.

Pirone, P. P. 1978. *Diseases and Pests of Ornamental Plants* (5th edn.), 566 p. John Wiley and Sons, New York, USA.

Sastry, K.S., K.S.M. Sastry and S.J. Singh, 1973. Studies on virus causing mosaic disease on *Zinnia elegans. Indian J. Mycol. Pl. Path.*, **3**: 165-168.

Sharma, P. 2006. Diseases of ornamental plants and their management. In: *Advances in Ornamental Horticulture* (vol. 4) (ed. Bhattacharjee, S.K.), pp. 196-261. Pointer Publishers, Jaipur, India.

Sohi, H.S. 1992. Diseases of Ornamental Plants, pp. 170. ICAR, New Delhi.

Srivastava, R.N. and J.S. Gupta, 1983. Pathogenicity of seed-borne infections of zinnia and their control. *Indian Phytopath.*, **36**: 14-16.

Strider, D.L. 1975. Susceptibility of Reiger Elatior begonia cultivars to bacterial blight caused by *Xanthomonas begoniae. Plant Dis. Reptr*, **59**: 70-73.

Strider, D.L. 1979a. Detection of *Xanthomonas nigromaculas pv. zinniae* in zinnia seed. *Plant Dis. Reptr*, **63**: 869-873.

Strider, D.L. 1979b. Eradication of *Xanthomonas nigromaculans* f. sp. *zinniae in* zinnia seed with sodium hypochlorite. *Plant Dis. Reptr*, **63**: 873-876.

Tewari, J.P. 1976. *Zinnia elegans* Jacq., the latent host of watermelon mosaic virus. *Cur. Sci.*, **45**: 37.

Tsai, C.F., Y.Y. Lin, S.C. Lu and C.A. Chang, 2012. A microarray system for multiplex virus detection in ornamentals. In: *Bioforsk Fokus* (eds Dag-Ragnar Blystad and Carl Spetz), p.39.

Vann, S., 2013. Grey Mold of Greenhouse Ornamentals. Agriculture and Natural Resources. (http://www.uaex.edu).

Wiedemann, G.J., D.O. TeBeest and R.D. Cartwright, 1988. Host specificity of *Colletotrichum gloeosporiodes* f. sp. *aeschynomene* and *C. truncatum* in the Leguminosae. *Phytopath.*, **78**: 986 - 990.

Wu, W.S. and J.K. Chou, 1995. Chemical and biological control of *Alternaria carthami* on *Zinnia*. *Seed Sci. Technol.*, **23**: 193-200.

Wu, W.S. and Y.H. Yang, 1992. *Alternaria* blight, a seed transmitted disease of zinnia in Taiwan. *Plant pathol. Bull.*, No. 1, pp.115-123.

Yadahalli, K.B., K. Srikant, K.H. Anahosur and S. Kulkarni, 1994. *In vitro* and *in vivo* evaluation of fungicides against leaf spot of zinnia caused by *Cercospora zinniae*. *Karnataka J. agric. Sci.*, **7**: 363-365.

Zamorshi, C. and K. Milczarek, 1997. The effect of seed dressing on the germination of seeds of ornamental plants. *Acta Agrobotanica,* **30**: 335-340.

39

Insect-Pests of Ornamentals and their Management

Sanyat Misra, R.L. Misra, Ajit Kumar and Pragya Ranjan

Introduction

Floriculture has emerged as an important agri-business for boosting income of the growers and for providing employment opportunities. Globally, more than 145 countries are involved in commercial cultivation of flower crops. Though the art of growing flowers is not new to India, large scale commercial cultivation, especially the protected cultivation is of recent origin. Enormous genetic diversity, varied agro-climatic conditions and plenty of human resources offer India an unique opportunity for diversification in flower growing. Dry flower industry which contributes a major share (60-70%) to the overall global floricultural trade from India is an example of an effortless venture. Other segments like fillers, potted plants, seeds and planting material, turf grass industry and value added products also contribute a share in the overall growth of the floriculture sector. However, the sector is highly disorganized and no systematic data are available at present. The traditional flower cultivation, comprising of loose flowers mostly for worshipping, garland making and decorations, forms the backbone of Indian floriculture, and this is in the hands of small and marginal farmers. Globalization of Indian economy and subsequent liberalization of 'Seed Act' has paved the way for the advent of protected cultivation in India during early 1990s. Achieving highest targeted goal in floriculture industry demands trained manpower at every stage from production to marketing. Though all the agricultural universities have elaborate floricultural courses at master and doctorate levels but practical know how can be gained only through field experiences by involving in the field for crop cultivation, the seed quality, nursery raising, field preparation, the controlled cultivation techniques, post harvest technology, control of insect-pests and diseases in the field and in polyhouses, as well as at storage levels, and more so about market intelligentia.

The recent decades saw significant gains in the cultivation of commercial ornamentals due to ever-increasing demand for cut flowers both in domestic and export markets. There are several factors which interfere in successful production of flowers both under open and protected conditions. Damage by insect-pests, which results in severe yield losses is one of the important limiting factors. Qualitative loss is far more important than quantitative loss in case of export oriented ornamental crops as slight distortion or discolouration makes flowers unfit for marketing.

The pest scenario in ornamentals grown under protected cultivation is different from open cultivated ornamentals due to various growing conditions. Under protected conditions, the plants are provided with relatively uniform environmental conditions for the optimal growth of the plants and these conditions may also favour quick multiplication of pests. Thus, the incidence of some of the pests may be seen round the year on these crops. Though, the protected cultivation is intended to exclude the pests from the entry inside the polyhouses, practically it may not be possible, as the pests may enter through the plant material or minor damages to the structures or just while opening of the doors.

Successful control of pests depends mainly on understanding of pest, its biology and incidence. Insect-pests are usually managed by adopting a holistic approach, which includes mechanical, physical, cultural, biological and chemical methods. In any case, for prevention of the insect-pests, one should first keep proper hygiene (**sanitation**) of the field so that alternate hosts do not get any place nearby and in the derbris the pest may not get chance to hide, hibernate or pupate. Moreover, propagating material should be healthy and should be taken from only healthy stocks. Optimum growing conditions should be ensured throughout the growing season. There should be constant vigil if any new pest is invading the crop. The diseased plants or its parts should regularly be removed and destroyed or composted. There are certain insects which can be hand-picked and killed or light-trapped. If known, only **resistant cultivars** should be planted. Certain **companion plants** grown in association with a crop can help to reduce pest problem. Deliberate planting of certain host plants may deflect pests from other plants or attract predators as nasturtiums are susceptible to aphids while French marigolds in the vicinity attract hoverflies which predate over the aphids. Following the **crop rotation** every 3-4 year basis helps to prevent soil-borne pests which build up in the soil to the alarming level. **Organic traps** such as greasy yellow strips, the colour to which whiteflies in greenhouses are attracted, should be spread in greenhouses to trap the whiteflies and dispose them off. Inverted flower pots trap the earwigs, and wireworms are trapped in old potatoes and carrots spitted on a stick and buried. Spreading of *Pongamia* or *neem* cakes in the field and spraying their oil repel and kill most of the insects. *Neem* leaf extract spraying also prevents pest infestation. Pyrethrum, derris dust, soft soap and sulphur dust may be used as **organic treatments** in the form of readymade powders or liquid sprays. **Biological control** is applied when a particular pest has become immune to chemicals but this can be accomplished only in greenhouses since outside these are not that much effective. Most such parasites or predators require good daylight and at least 21 ^{0}C temperature for a faster breeding-generation than the pests. Also, when applying biological control measures, the pesticide use should be restricted excepting the insecticidal soaps for all such predators; and pirimicarb against aphids. Predatory mite fly larva of *Aphidoletes aphidimyza* feeds over aphids, *Phytoseiulus persimilis* over red spider mites, *Amblyseius* spp. over thrips, *Cryptolaemus montrouzieri* ladybirds over mealybugs, nematode predator *Heterorhabditis* spp. over vine weevil larvae, parasitic wasp *Encarsia formosa* over whiteflies, *Metaphycus helvolus* over soft scale insects, *Bacillus thuringiensis* bacterium over caterpillars, *etc.* However, the chemical controls should be the last resort. Management measures being followed at present in ornamental crops are mainly confined to use of pesticides. Due to strict quality control norms followed by flower importing countries any minor blemish or harmful pesticide residue on flowers results in rejection of consignment. Accordingly, the management strategies for protected ornamentals should focus on effective, less persistent, non-phytotoxic pesticides along with safer alternatives like botanicals, sticky traps and bio-agents.

General Greenhouse IPM

The primary goal of 'integrated pest management' (IPM) is to optimize pest control in an economically and ecologically sound way. IPM involves the integration of cultural, physical, biological, and chemical practices to grow crops with minimal use of pesticides. Monitoring, sampling, and record-keeping are used to determine when control options are needed to keep pests below an economically damaging threshold. Pest management, not eradication, is the goal of IPM.

Integrated pest management is adaptable to all greenhouse-grown crops and involves specific techniques to manage these as given here with.

- ❖ Exclusion techniques to prevent pests from entering the production area
- ❖ Insect screens to exclude aphids, whiteflies, and thrips from entering through doors and ventilators
- ❖ Monitoring or scouting programme
- ❖ Indicator plants
- ❖ Individual plant inspection
- ❖ Pest identification and life stages
- ❖ Use of yellow or blue sticky cards
- ❖ Record keeping to identify trends and direction for the target pest management programme
- ❖ Cultural practices to prevent problems
- ❖ Sanitation
- ❖ Biological controls by using living organisms to reduce the incidence of pest organisms
- ❖ Use of proper insecticides that interfere with normal insect development or the moulting process. However, insecticides should be used with right procedure and at a right time.

There are several practices that will increase the success of an IPM programme. Some of the simple ways by which the greenhouse crops can be protected from pest attack are keeping the doors closed, keeping people away from the crop area as much as possible, sterilization of growing medium, using resistant varieties, and proper sanitation in the crop area.

Cropwise Insect-Pests and their Management

Quite elaborative account of many of the insect-pests are given by various workers (Brickell, 1992; Dole and Wilkins, 1999; Jhansi Rani, 2001; Parvatha Reddy, 2001; Pratibha Sharma and Bhattacharjee, 2002; Karuppuchamy and Palaniswamy, 2006). On the basis of the plant-parts affected due to infestation of insect and other pests, here with these are being furnished under five broad heads such as:

Problem on entire plant: The problem on whole plant of annuals including balsam; perennials including balsam, sedum and primulas; indoor plants; shrubs including fuchsias; and begonia, cyclamen and many other bulbous plants is caused by **vine weevil larvae** (*Otiorhynchus sulcatus*) which measure some 1 cm long with little curve and brown head but no legs. In its infestation, the growth of plants becomes slow, collapse and die. **Fungus gnats or sciarid flies** (*Bradysia* spp.) infest the plants being grown under cover, *vis-à-vis* seedlings and cuttings fail to grow. Here grayish-brown flies may be seen running or flying over the surface or among the plants. Its white larvae (maggots) some 6 mm long with black heads may be seen easily. Though this feeds on dead roots and leaves but also attacks young roots. This is not a problem on established plants.

Root and tuber problem: This occurs on many of the garden plants due to root-knot nematodes, mainly *Meloidogyne* spp. which exhibits swelling or galling on roots and tubers, thereby plant growth is arrested, leaves may curl and yellow and the plants may be killed. **Chafer larvae** (summer chafer, *Amphimaleon solstitialis*; rose chafer, *Cetania aurata*; cock chafers, *Melolontha melolontha*; and garden chafers, *Phyllopertha horticola*) are 'C'-shaped, some 2 cm long with 3 pairs of legs and a brown head. These attack bulbous plants and young annuals by eating their bases so the young plants die soon. They also infest on leaves, flowers and fruits of various ornamental trees and shrubs. Though damage by adult beetle is seldom but the larvae inhabiting in the soil, seriously damage the roots, tubers, bulbs and stems of many ornamentals including lawn. **Cut worms** (*Agrotis* and *Noctua*) feed on underground parts of annuals and low-growing perennials, where roots may be severed, cavities on tubers may be formed and ultimately plants die. The feeding caterpillars are usually creamy-brown and some 4.5 cm long. **Leatherjackets** are the legless larvae of crane flies (daddy-long legs) some 3.5 cm long with tubular and greyish-brown bodies, e.g. *Tipula* which feed on roots and stems, and in the process the stems are severed from the roots so these die.

Bulb, corm and rhizome problems: Such problems occur when **bulb flies** (*Eumerus* spp. and *Merodon equestris*) infest *Cyrtanthus purpureus, Galanthus, Hippeastrum, Hyacinthoides non-scriptus* (bluebells), *Narcissus, Sprekelia*, etc. The *Merodon* looks like humblebees and lays eggs singly on sound bulbs, larvae are up to 2 cm long and brownish-white and these feed inside the bulbs and fill the bulb centres with muddy excrement. *Eumerus* infests on injured bulbs. Infested bulbs either do not grow or produce only a few grass-like leaves. **Bulb mite** (*Rhizogyphus echinopus*) is small pearly-white soil mite which often invades rotting tissues of gladiolus, hyacinth, narcissus and tulip bulbs as secondary pest. **Bulb scale mite** (*Steneotarsonemus laticeps*) feeds and breeds inside the narcissus bulbs, causing rusted streaks, distortion and stunting on leaves and flowers. The **nematode** (*Ditylenchus dipsaci*) infests *Galanthus, Hyacinthoides non-scriptus, Narcissus, etc.* inside the bulbs as well as foliage which causes bulb rotting, growth distortion and stunting.

Stem, branch and leaf-bud problems: These problems occur as **scale insects** (*Aonidiella aurantii, Aspidiotus dictyospermi* var. *arecae, A. orientalis, A. transparens, Chionaspis* spp., *Iceria aegyptiaca, Iceria formicarum, Iceria purchasae, Lindingaspis rossi,* and *Saissetia hemisphaericum*) feed on the stems of indoor and greenhouse plants, cacti and other succulents, trees, shrubs, roses, *etc.* Their infestation makes the stems sticky and excrete honeydew which entices sooty mould growth. Plants become stunted, show sick growth and sometimes die; and **froghoppers or spittlebugs** (*Philaenus spumarius*) is a creamy sap-sucking insect which secretes a protective froth over themselves. Adults are up to 4 mm long and darker in colour. Its infestation causes little distortion.

Leaf and floral problems: Leaf problems are innumerable in ornamentals. **Slugs** (*Agriolimax reticulatus, Arion, Deroceras, Milax* spp., *etc.*) and **snails** (*Helix aspera, etc.*) feed either after rains or in the night on seedlings, small annuals, bulbous ornamentals, herbaceous perennials, climbers, *etc.* by either completely battering the leaves or making holes in the leaves and stripping in the stems, leaving the silvery slime-trail on the leaves and on the soil surface. Bulbs are also fed with small holes outside. Stems are also damaged underground, leaving the large cavities inside. **Earwigs** (*Forficula auricularia*) are yellowish-brown and some 15 mm long insects with a pair of curved pincers. They are nocturnal though during daytime hide themselves. Both adult and young earwigs feed on young leaves and blooms of annuals, perennials, chrysanthemums, dahlias, clematis, shrubs and many other ornamentals during summer. **Woodlice** (*Armadillium vulgare, Oniscus, Porcellio, etc.*) which are also known as slaters or pill bugs, feed during night on seedlings and other soft growth including flowers and fruits. They are grey to brownish-grey, sometimes with white or yellow markings, hard with segmented bodies and up to 1 cm long. They normally eat on the decaying plant materials or on already damaged plants or their parts. Generally they invade near the shoot tips and cause holes. **Bugs** [Capsid bugs, tarnished plant bug or bishop bug (*Lygus rugulipennis*), the common green capsid (*Lygocoris pabulinus*), and the apple capsid (*Plesiocoris rugicollis*)] are green or brown, 6 mm long and suck plant sap from leaves, shoot tips and flowers of annuals and herbaceous perennials, *Caryopteris, Dahlia, Dendranthema, Fuchsia, Hydrangea*, etc. by making numerous small holes, *vis-a-vis* their toxic saliva kills plant tissues so the leaves are torn. Regarding their damage on the flowers, Capsid bug infestation causes uneven flower development with under-sized ray petals while in *Fuchsia* flowers abort completely. Both young and adult rhododendron bug (*Stephanitis rhododendri*) punctures the lower surface of rhododendron leaves and feeds on the sap so fine light mottling on the upper surface, *vis-à-vis* rusty-brown or chocolate spotting on the lower surface may be seen. **Mealybugs** (*Pseudococcus, Planococcus*) are soft-bodied, wingless, greyish-white and some 5 mm long, often with white and waxy filaments trailing from their bodies. These infest on indoor and greenhouse plants, cacti and other succulents and many other ornamentals. Their infestation causes appearance of a fluffy white substance appearing in leaf and stem axils, they also excrete honeydew to entice sooty mould. Roots are also infested, and there are even root mealybugs (*Rhizoecus* spp.) which feed on roots, especially cacti and other succulents, causing discolouration and wilting. **Aphids** [bean aphid (*Aphis fabae*), beech aphid (*Phyllaphis fagi*), cabbage aphid (*Brevicoryne brassicae*), cherry blackfly (*Myzus cerasi*), cypress aphid (*Cinara cupressi*), *Aphis gossypii,* leaf curling plum aphid (*Brachycaudus helichrysi*), mottled arum aphid (*Aulacorthum circumflexum*), orchid aphid (*Cerataphis orchidearum*), peach-potato aphid (*Myzus persicae*), *Microsiphoniella sanborni,* spruce aphid (*Elatobium abietinum*), tulip bulb aphid (*Dysaphis tulipae*), waterlily aphid (*Rhopalosiphum nymphaeae*), willow stem aphid (*Tuberolachnus saligna*), woolly aphid (*Eriosoma lanigerum*), *etc.*] including those known as blackfly, greenfly or plant lice are up to 5 mm long, winged or wingless, and grey, green, yellow, pink, brown or black and the woolly beech aphid is covered in a fluffy-white wax. While feeding on tender parts of shoots, leaves, buds and flowers, annuals and perennials, bulbous plants, trees, shrubs and climbers, all these lose the plant vigour as their attack is in large number, and these leave sticky excretion to entice sooty mould. Some aphids also infest on roots and the underground parts including the bulbs, *etc.* They also transmit viral diseases. **Whiteflies** (*Aleurodes proletella, Trialeurodes vaporariorum*) are 2 mm long white-winged insects which invade indoor and greenhouse plants. Their nymphs which are scale-like are immobile, flat or oval and whitish-green though adult flies are quite active and suck the cell sap so leaving sticky honeydew to attract sooty mould. These may also transmit viruses. Maggots of **carnation fly** (*Delia brunnescens*) mines the carnation and pink's leaves and stems which causes wilting and death of the infested shoots.**Thrips** (*Taeniothrips simplex, Thrips tabaci, Frankliniella occidentalis, Kakothrips pisivorus, etc.*) nymphs and adults rasp the plant fluid leaving a silver-white discolouration on leaves with tiny black dots on the upper surface of leaves. They are brownish-black insects up to 2 mm long with narrow bodies which are sometimes crossed with pale bands, and survive well in hot and dry conditions. Nymphs resemble the adults except that they are pale-yellow-orange. These infest the annuals and perennials, indoor and greenhouse plants, bulbous plants, *Ligustrum, Lonicera, Syringa* and many other ornamentals. Thrips infesting flowers of *Achimenes, Cyclamen, Gladiolus, Sinningia, Dendranthema, Pelargonium, Primula, Rosa, Saintpaulia, Streptocarpus,* and many other perennials are gladiolus thrips (*Taeniothrips simplex*), garden rose thrips (*Taeniothrips fuscipennis*), polyhouse western flower/rose thrips (*Frankliniella occidentalis*), *etc.* which suck the cell sap by entering and damaging the floral-buds during dry summers. They are up to 2 mm long, narrow-bodied and brown to black though immature nymphs are creamy-yellow. **Leaf miners** [celery leaf miner (*Euleia heracleii*), chrysanthemum leaf miner (*Phytomyza syngenesiae*), beech leaf mining weevil (*Rhynchaenus fagi*), the birch leaf mining sawfly (*Fenusa pusilla*), lilac leaf miner (*Caloptilia syringella*) is a moth caterpillar which is highly abnoxious and damages by burrowing into syringa leaves, *vis-à-vis* leaf-mine blotches on ligustrum and remains active from June onwards, *etc.*] are larvae of various flies, moths and beetles which infest annuals and perennials, chrysanthemum, trees and shrubs including *Ilex* and *Syringa* by mining their leaves and on herbaceous plants even their stems which can be witnessed by the visibility of the white or brown areas (linear, irregular or circular characteristic shape for the particular leaf miner). **Suckers** or **psyllids** [bay sucker (*Triozaalacris*), the box sucker (*Psylla buxi*)] infest annuals and perennials, bulbous plants, trees including *Laurus nobilis* (bay), shrubs including *Buxus* (box) and climbers and their feeding distort and stunt the leaves throughout the summer. Infestation on new shoots of *Buxus* produces cabbage-like clusters of malformed leaves. *Laurus* leaf margins become yellow, thick and curled, though the box leaves are stunted. The nymphs are usually grey-green, small (2 mm long), winged and aphid-like. **Earwigs** (*Forficula auricularia*) is a 1.5 cm long yellowish-brown insect which eats away the petals of annuals and perennials, *Dahlia, Dendranthema,* shrubs and many other ornamentals. **Leafhoppers** [rose leafhopper (*Edwardsiana rosae*), glasshouse leafhopper (*Hauptidia maroccana*), and the rhododendron leafhopper (*Graphocephala fennahi, G. coccinea*), *etc.*] infest on annuals and perennials, pelargoniums, primulas, shrubs, roses, trees, rhododendrons, *etc.*, and their infestation causes pale spotting on upper leaf surface except in rhododendron where *Graphocephala* leafhopper transmits *Pycnostysanus azaleae* fungus from plant to plant so the buds are formed but do not open, turn brown and die but do not fall, and sometimes these develop a silver-grey sheen. These during spring are covered with black fungal bristles. The rose and glasshousse leafhoppers are green or yellow, 2-3 mm long, wider bodied behind the head and then tapering though rhododendron leafhopper is turquoise-green & orange, and some 6 mm long and creamy-white symphs are little active. **Leaf-cutting bees** (*Megachile* spp.) feed on roses and many other shrubs and trees during night by taking out circular to lozenge-shaped uniform cuts from the leaf margins to build their nests. These bees are 1 cm long with ginger-hairs on their abdomen beneath. **Bumble bees** (*Bombus* spp.) sometimes bite the corolla tubes or spurs of antirrhinum and tropaeolum and make small holes. **Millipedes** (*Blaniulus, Brachydesmus, Cylindroiulus* spp., *etc.*) may be black, brown, grey or creamy-white and feed at soil level or underground. The body is hard, segmented with two pairs of legs per segment though centipedes which are beneficial predators bear only one pair of legs. Spotted snake millipede (*Blaniulus guttulatus*) is slender bodied, creamy-white and some 2 cm long bearing a row of red dots along each side. Millipedes have high population in the soil with high organic matter. They infest on the seedlings and other soft growths though the damage is never alarming. **Caterpillars** [brown webber tail moth (*Euproctis chrysorrhoea*), lackey webber moth

(*Malacosoma neustria*), small ermine webber moth (*Yponomeuta* spp.), juniper webber moth (*Dichomeris marginella*), the hawthorn webber moth (*Scythropia crataegella*), *etc.*], first three having hairy and some 5 cm long bodies though caterpillars of the last two species are up to 2 cm long. They feed on annuals, perennials, bulbous plants, shrubs including *Crataegus* and *Cotoneaster*, trees such as *Euonymus, Juniperus, Salix,* and many other garden plants. The caterpillars defoliate the shoots by covering the feeding area with dense greyish-white and silk webbing. Tortrix moth caterpillars (*Cacoecimorpha pronubana*) are dark green with brown heads measuring up to 2 cm long, and feed on leaves either by binding two leaves together, or one leaf folded and then bound in itself or the leaf attached in similar fashion with fruit while dwelling inside and feeding on the inner surface of leaves. When disturbed, these caterpillars wriggle backwards rapidly. Winter moth caterpillars (*Operophtera brumata*) are pale-green and some 2.5 cm long, and feed from bud burst to late spring on leaves, buds, flowers and fruitlets of roses and many other deciduous trees. Cabbage caterpillars [large cabbage white butterfly (*Pieris brassicae*), small cabbage white butterfly (*P. rapae*) and cabbage moth (*Mamestra brassicae*)], where the first one is yellow and black with distinct hairs, the second one pale-green with velvety hairs and the caterpillar of the last one yellowish-green to brown adorned with a few hairs. These feed on nasturtium, the members of Brassicaceae, certain annuals and perennials, *etc.* Their feeding make holes in the foliage from late spring to early autumn and can be easily seen on leaves and flowers. **Beetles** [asparagus beetle (*Creoceris asparagi*), flea beetles (*Phyllotreta* spp.), pollen beetles (*Meligethes* spp.), poplar beetle (*Melosoma populi*) , red lily beetle (*Lilioceris lilii*), viburnum beetle (*Pyrrhalta viburni*), waterlily beetle (*Galerucella nymphaeae*), willow beetle (*Phyllodecta vitellinae*), *etc.*] attack various ornamentals. Adult asparagus beetle is 7 mm long, having reddish thorax and yellow & black wing cases, its larvae are greyish-yellow, and both feed on asparagus stems and leaves from late spring to early autumn and from the feeding place the upper portion dries. Flea beetle is small (2 mm long) with enlarged hind legs for leaping, black to metallic blue and sometimes with a yellow stripe running down each wing case. It feeds on *Cheiranthus, Matthiola, etc.* and its feeding causes small pitting and holes on upper leaf surface which may cause plant death. It overwinters in plant debris. Pollen beetles are small (2 mm) and black, and feed on the pollen of *Narcissus, Lathyrus odoratus, Rosa* and many other garden plants though other parts of the plant are not damaged. Poplar beetle larvae resemble to those of ladybird and feed on the leaf tissues of poplar and skeletonize the foliage leaving only the veins, and the adult beetle eats on leaves making irregular holes. Red lily beetle adult and larvae feed from early spring to mid-autumn on the leaves and flowers of bulbous plants, especially *Fritillaria* and *Lilium, vis-à-vis Polygonatum.* Adult is 8 mm long and bright red though larvae reddish-brown with black head and covered in black excrement. Greyish-brown viburnum beetle larvae are creamy-white 7 mm long with black markings which feed by making holes on viburnum leaves, in early summer by larvae and in late summer by adults. Waterlily brown adult beetle and its fat black larvae feed on the upper surface of waterlily leaves making holes. Such leaves rot and disintegrate. Adults and larvae of willow beetles feed on salix leaves, and their severe infestation causes extensive defoliation. **Vine weevil** (*Otiorhynchus sulcatus*) adult is greyish-black and about 9 mm long with a pair of elbowed antennae and a short snout. It is nocturnal in habit and keeps itself hidden during daytime. It feeds by making notches on the leaf margins generally close to ground from mid-spring to mid-autumn on shrubs, especially azaleas, camellias, *Euonymus,* hydrangeas and many other herbaceous plants. **Sawfly larvae** (*Diprion, Nematus, Pristiphora, etc.*) are generally green, caterpillar-like, some 3 cm long and sometimes with black spots. The larvae of the Solomon's seal sawfly (*Phymatoceraterrima* spp.) are usually greyish-white. These all feed on *Aquilegia, Aruncus dioicus,* conifers, *Geum, Polygonatum* (Solomon's seal), roses, *Salix* (willow) and various other perennials, bulbous plants, trees and shrubs and defoliate the branches or shoots by feeding the leaves bit by bit. **Iris sawfly** (*Rhadinoceraea micans*) caterpillars are bluish-grey and feed on *Iris leavigata* and *I. pseudacorus* leaves during rainy season. **Rose leaf-rolling sawflies** (*Blennocampa pusilla*) infestation causes rose leaflets rolling downwards at the margins to form tight tubes in early summer as black-bodied adult some 3-4 mm long with transparent wings lays eggs in the leaflets and secretes a chemical that causes rolling, and after hatching the pale-green caterpillar-like larvae start feeding inside the rolled leaflets. **Gall midges** [violet gall midge (*Dasineura affinis*), the gleditsia gall midge (*D. gleditchiae*), *Hemerocallis* gall midge (*Contarinia quinquenotata*), *etc.*] are the larvae of midges (flies). The flies are tiny and greyish-brown. They are whitish-orange fly-maggots some 2 mm long that feed within the galled tissues and complete 3-4 generations in one summer. Their infestation on violet (*Viola*) causes thickening of leaves which fail to unfurl though in *Gleditsia* the leaflets swell and fold over to form pod-like galls. In *Hemerocallis*, the fly lays its eggs in the developing buds, which after hatching start feeding between the petals. These maggots are white and some 2 mm long. **Scale insects** [soft scale (*Coccus hesperidum*), mussel scale (*Lepidosaphes ulmi*), horse chestnut scale (*Pulvinaria regalis*), *etc.*] infest the indoor and greenhouse plants, cacti and other succulents, trees and shrubs. They are yellow, brown, dark grey or white, flat or raised, circular, pear-shaped or oval and up to 6 mm long. They feed on lower leaf surface and stems and excrete sticky honeydew which attracts sooty mould. Their infestation slows down the plant growth. **Fasciation** in the form of peculiar flattening of stems occurs on certain ornamentals, especially *Celosia, Daphne, Delphinium, Forsythia, etc.* due to early damage to the growing point caused by mis-handling, frost injury, slug or insect infestation, bacterial infection or genetic malfunction. In roses and certain other ornamentals, **proliferation** occurs where flower stem appears through the centre of the existing flower forming a new flower above the first or several flower buds together surrounded by one set of petals. This occurs because of early damage to the growing point of the bud due to frost or insect attack or due to some virus infection.

Red spider mites (*Tetranychus urticae*) are 8-legged 2-spotted black-marked creatures, yellowish-green though may change to orange-red during autumn, and less than 1 mm long. There are several other species. Their infestation causes pale mottling on the upper leaf surface, makes the leaves becoming dull, yellowish and ultimate premature falling. They also develop a fine silk-webbing on the entire plant. **Bulb scale mites** (*Staneotarsonemus laticeps*) infestation on inside of bulb necks of daffodils and hippeastrum causes stunted growth, leaves becoming curved, and the leaf margins & inflorescence become scarred with saw-toothing. **Michaelmas daisy mite** (*Tarsonemus pallidus*) infests the floral buds and shoot tips of *Aster* which causes appearance of poor quality flowers or flowers may be converted into rosettes of small leaves. **Bryobia mites** (*Bryobia* spp.) are closely related to red spider mites and are troublesome on hedera. They feed on leaves which first develop a light freckling on upper side of leaves which afterwards becomes bronze and wither.

Leaf and bud nematodes, *Aphelenchoides* spp. and certain others infest inside the leaves and buds of annuals and most herbaceous perennials such as anemones, begonias, chrysanthemums, *etc.* and cause brownish-black patches between the larger veins of leaves. *Ditylenchus dipsaci* infests on many of the ornamentals including *Allium* and *Phlox* by feeding on the plant tissues, causing spotting of leaves with rough, small and yellow patches (spickels). *Phlox* becomes stunted with swollen stems and narrow shoot-tip leaves whereas *Allium* stems become soft and swollen.

Though ornamental plants and their insect-pests are enormous but the major ones and that too of major flower crops (open or protected) individually, their symptoms of damage and management options are furnished here with.

Anemone: Woodville (1953) in England found a **wood mouse** (*Apodemus sylvaticus sylvaticus*) damaging and feeding on anemone plantings for which he recommended 1.25-cm mesh wire-netting dug in at the bottom and curved over and down at the top. **Aphids** infest the stems and leaves, making the plants sticky and sooty. **Thrips** also suck the sap from flowers and stems. **Whitefly** (*Trialeurodes vaporariorum*) is a white and 1.5 mm long insect resembling a tiny moth. Their tiny yellow nymphs feed on the underside of leaves and make them chlorotic. Garden **swift moth** (*Hepialus lupulinus*) is effectively controlled by soil drenching of 20% DDT emulsion. Perret (1949) recommended 10% benzene hexachloride for complete control of anemone **flower eating beetle** (*Tropinota squalida*), whereas lead arsenate, calcium arsenate, 1% Dintrophenol, 10% DDT or 10% benzene hexachloride for complete control of another flower eating beetle, *Hoplia aulica*. All other insects can be controlled by spraying with 0.2% Rogor or 0.1% Nuvacron. **Springtails** also infest anemone which can be controlled by Lindane and Aldrin (Jeff, 1956).

Annuals: More or less there are some 500 annuals for landscaping which encounter a large number of insect-pests. The **aphids** are *Aphis fabae, A. gossypii, Aulacorthum circumflexum, Brachycaudus helichrysi, Myzus persicae* and *Microsiphoniella sanborni* which infest on various annuals. These are found clustering on buds, tender leaves and shoots sucking their sap and enticing sooty mould and ants due to secretion of honeydew. Various **thrips** that infest annuals are *Frankliniella intonsa, F. occidentalis, F. tritici, Haplothrips ramakrishnae, Hercirothrips* spp., *Microcephalothrips abdominalis, Thrips hawaiiensis, T.*

nigrophilosus and *T. tabaci*. These are quite tiny insects to hide anywhere on the plant, especially in the leaf axils, buds and flowers. **Leaf miners** are also in plenty to infest on annuals and these are *Liriomyza huidobrensis, L. pusilla, L. trifolii,* and *Phytomyza* (*Chromatomyia*) *sygenesiae*. Their larvae mine the leaves and pupate either in the leaves itself or in the ground, characteristic to their species. **Greenhouse whitefly** (*Trialeurodes vaporariorum*) also feed on various annuals. **Mirid bugs** (*Creontiades pallidifer*) suck plant sap from tender plant parts. ***Bollworm*** (*Helicoverpa armigera*) is polyphagus and feeds on any part of the plant in various ways. **Caterpillar** [*Hydylepta* (*Lamprosema*) *indicata* feeds on tender parts of plants, especially tender leaves and floral buds. **Bihar hairy caterpillar** (*Diacrisia obliqua*) and **tobacco caterpillar** (*Spodoptera litura*) feed on leaves, buds and flowers. The larvae of **stalk borer** (*Papaipema nebris*) and **earwig** (*Forficula auricularia*) also feed on leaves, buds and flowers. The **white grub** (*Holotrichia consanguinea*) feeds on the underground part of stem and roots so plants wilt. An **eriophyid mite** (*Paraphytoptus chrysanthemi*) causes plant phyllody. All these can be controlled by spraying 4% *neem* oil.

Anthurium (*Anthurium* spp.): Anthuriums are highly sensitive to certain pesticides so it is not easy to control the pests effectively. This is invaded by aphids (*Myzus circumflexus*), thrips (*Chaetanaphothrips orchidii*), whitefly (*Aleurotulus anthuricola, Bemisia tabaci*), mites (*Brevipalpus* spp.) and red spider mites (*Tetranychus cinnabarinus*). **Green peach aphid** (*Myzus persicae, M. circumflexus* on *A. scherzerianum*), *Aphis gossypii, Aphis nicotiana, Aulacorthum solani* and *Macrosiphum euphorbiae* suck plant sap from various parts of anthurium, inject into the plants some toxic substances, produce honeydew which attracts ants and sooty mould and transmit viruses. Being viviparous, these build up their population quickly. Aphids may be light green, yellow, pink or red in colour. Proper field sanitation to avoid alternate hosts is a must in its control. Pyrethrum extract gives complete control in greenhouses. Spraying with 0.2% Dicofol, Dimethoate, Malathion or Phosphomidon controls this pest effectively. Field release of parasitic wasp, *Aphidius colemani* 5/m^2 per week will feed on the aphids and will check its menace. **Mealy bugs** are sap-sucking from the lower surface producing honeydew and attracting sooty mould. Drenching the media with systemic insecticide like Dimethoate will control this pest also. Youngs and adult **thrips** (*Chaetanaphorothrips orchidii*, Ranjith, 2013) are light yellow-brown which for sucking the sap bore the cells by feeding on leaves and flowers (spathe and spadix) causing brown stripes on the affected part and mottle them. Its initial infestation is noted somewhere in March and lasts up to October. Their infestation causes white streaking and distortion of spathes. On *Anthurium andraeanum,* it completes its life cycle in 27.4 days. In netted areas, insecticide granules are very effective in checking its menace. Foliar spraying at 10 days intervals with Acephate 0.07%, Cyfluthrin 0.2%, Chlorpyriphos 0.1%, Dimethoate 30 EC at 0.1%, oxydemeton methyl at 0.05% or Phosphamidon 85 WSC at 0.05% controls this pest effectively. Field release of predatory mite, *Amblyseius cucumeris* at 5,00,000/ha is a very effective biological control for this pest. The adults and nymphs of **anthurium whiteflies** (*Crenidorsum aroidephagus, Aleurotulus anthuricola*; Ranjith, 2013), smaller than common glasshouse whiteflies (*Trialeurodes anthuricola, A. vaporariorum*) where adults have white waxy coating, cluster and colonise on the under surface of foliage and petiole sheath making the plants weak with localized necrosis and chlorotic spotting, honeydew deposition and the growth of sooty mould, the nymphs being more dangerous than adults and drain out the plant sap through sucking. Excessive dampness and dark conditions are conducive for its growth. In its severe infestation, there is premature defoliation and sooty mould development. Female lays eggs on the underside of tender leaves and the life cycle of *Aleurotulus anthuricola* lasts for 35.5 days on *Anthurium andraeanum.* Methomyl 0.075% spraying alternating with Triazophos 0.06% at initial stage will prevent build up of its population. **Tomato looper caterpillars**, the **beet armyworms** and certain other lepidopterous insects chew various parts of flowers and leaves which may be controlled through catching and killing or through Lannate 0.1% or Decis at 0.5% sprayings.

Mites [cyclamen mite, *Phytonemus pallidus*; and red spider mites, *Tetranychus urticae, T. telarius* & the privet or false spider mite, *Brevipalpus obovatus*; Ranjith, 2013) are tiny, spider-like, 8-legged, white green (sometimes brown-red) and transparent insects which suck the contents by boring through the plant cells especially on the underside of the leaves or flowers which make the infested part (flowers, i.e. spathe and spadix, and leaves) discoloured to silver-white (red spider mites cause yellow discolouration by feeding on older leaves), being more serious during summer months and later on make the webs. Their adults lay their eggs on the ventral surface of leaves. Spraying of the underside of the leaves and other affected parts with Dicofol 18.5 EC at 0.08%, Vertimec 0.4% or Kelthane 0.08% will control this pest. Field release of predatory mite, *Phytoseiulus persimilis* at 6/m^2 will keep its menace under check. **Scale insects** suck the sap from stems and leaves which can be controlled by wiping the affected parts with methylated spirit soaked cotton-wool swab or spraying with 0.2% Malathion.

Nematodes [*Meloidogyne incognita* (Cadet *et al.*, 1993), *Aphelenchoides fragariae, Pratylenchus infestans, Radopholus citrophilus* (Goo and Sipes, 1997) and *R. similes* (Bala and Hosein, 1996), *etc.*] infest on anthurium. *Meloidogyne* galls the roots and causes brown lesions. *Pratylenchus* attacks only the roots by which leaves yellow and plant may die. *Radopholus* also infests on roots and show dark necrotic lesions which causes stunting and poor growth of the plant with fewer & smaller flowers, and premature leaf yellowing. *Aphelenchoides* feeds in leaves causing necrotic areas between leaf veins. These pests can be put under check by planting comparatively more tolerant species or varieties such as *A. pittieri* and *A. ravenii* instead of *A. andraeanum* cv. 'Midori', by planting healthy stock in media disinfected with steam or methyl bromide, by treating the cuttings for 10 minutes at 50 ^{0}C temperature, by separating planting media from the natural soil of the bed, through proper sanitation, and through precautionary measures. The nematodes can also be controlled through recirculation of water by ozone and activated hydrogen peroxide @ 1 h to 20 g ozone. With its new formulation, only 200 ppm for 24 h is required for their complete elimination (Runia and Amsing, 1996).

Slugs and snails chew over the root tips, leaves, buds as well as flowers, especially during night and batter the leaves which should effectively be controlled through metaldehyde baits.

Antirrhinum: **Aphids** (*Aphis gossypii*) feed on the underside of the leaves and on the tender parts of antirrhinum in gardens and in greenhouses, and the peach aphids (*Myzus persicae*) discolour and distort the leaves similar to virus infection. These are controlled through 0.1-0.2 per cent spraying of Malathion, Lindane, Parathion, or oxydemeton methyl. Lacewing (*Chrysopa carnea*) effectively predates on the aphids if introduced into the greenhouses. **Red spider mite** (*Tetranychus telarius*) is a cumbersome pest during dry weathers. Cyclamen mite (*Steneotarsonemus pallidus*) and a 'red and black stink bug' also attack this crop. These are controlled through the use of Aramite, Dimite, Kelthane or Ovex. **Looper** larvae are controlled through foliar spraying of Trichlorfon or methyl parathion. ***Helicoverpa armigera*** caterpillars are serious pests of antirrhinum which feeds on whole of the plant including the seed-capsules. Its control can be achieved by spraying the plants with 0.3 per cent Thiodan or 0.45 per cent Ripcord. **Root-knot nematode** (*Meloidogyne incognita*) is sometimes found attacking its roots which causes stunting of the plants. Through Carbofuran the nematodes may be kept under check.

Barleria (*Barleria cristata*): **Budworm** (*Trichoptilus* sp.) larvae bore into the buds and feed inside. Its adults are small with plume-like wings. Carbaryl 10 D dusting at 15-20 kg/ha will keep this pest under check. **Leaf folder** (*Syngamia latimarginalis*) larvae feed on the leaves by rolling them. Such leaf rolls should be collected and burnt. Dusting of Carbaryl 10 D or Fenitrothion 5 D at 20 kg/ha will control the leaf folder. **Leaf miners** are of two types, one is a grub of *Trachys* sp. from Coleoptera and the other is a maggot of *Pseudo-napomyza alternanthera* from Diptera. Though in both the cases larval stages mine the leaves and give appearance of scorched blotches, but in former case even adults feed on leaves and flowers by notching them. Spraying with Phosalone 35 EC at 0.2% will control these miners. The nymphs and adults of **tingid bug** (*Habrochila laeta*) feed on leaves causing yellowish spots, and afterwards when infestation is severe, the leaves become brown, shrivel and dry off. By spraying with Dimethoate 30 EC or Malathion 50 EC at 0.1% will give rid of this pest.

Begonia (*Begonia* spp.): Begonia is invaded by aphids (*Aphis gossypii* and *Myzus persicae*), thrips (*Heliothrips haemorrhoidalis* and *Thrips tabaci*), mealy bug (*Planococcus citri*), whitefly (*Trialeurodes vaporariorum*) and black vine weevil (*Otiorhynchus sulcatus*). Nymph and adult **aphids** suck cell sap on the underside of leaves, flowers and other tender parts which causes their yellowing and falling off. **Thrips** cause damage by sucking the sap from the leaves, buds and flowers. The infested leaves show irregular and brownish red streaks on the upper surface, and roughened brownish spots on the under-surface. Aphids and thrips both can be controlled through spraying the plants with Dimethoate 30 EC at 0.1% or Phosphamidon 85 WSC at 0.05%. **Mealybug** nymphs and adults suck the sap and weaken the plants. Its body is elliptical with short bristles or filaments all around the margin and whole body is covered with a waxy mealy substance. Its menace can be checked through spraying with Malathion 50 EC or methyl parathion 50 EC at 0.2%. The nymphs and adults of **whitefly** feed on the underside of leaves by sucking the sap and cause yellowing and drying of leaves.The honeydew secreted by it also attracts sooty mould. Its menace can be checked by spraying with Malathion 50 EC at 0.2%, Phosalone 35 EC or methyl parathion 50 EC at 0.1%. The grubs of **black vine weevil** feed on the smaller roots and tunnel through the longer roots by feeding inside, *vis-à-vis* rarely at the base of the stem so such plants wilt and die soon. Entomogenous fungus, *Metarhizium anisopliae* is very effective in controlling the weevil whether it is greenhouse or field.

Tarsonemid mites (*Polyphagotarsonemus latus* and *Steneotarsonemus pallidus*) also infest begonia by rasping the plant sap which causes streaking and poor growth of the plants. Spraying wettable sulphur 80 WP at 0.3%, Ethion 50 EC at 0.1%, Mitac at 0.05% or Profenofos 50 EC at 0.1% at 15 days interval, followed by 1% *Pongamia* or *neem* oil will give control of these mites

Carnation (*Dianthus caryophyllus*): Misra *et al.* (2006) have given the detailed account of insect-pests and diseases of carnation. Carnation is attacked by eight insect-pests, out of which the most important ones are thrips (*Frankliniella occidentalis, Haplothrips coloratus, H. nigricornis, Thrips flavus, T. hawaiiensis* and *T. tabaci*), bud borer (*Helicoverpa armigera*) and two-spotted spider mite (*Tetranychus urticae*). Both nymph and adult **thrips** suck cell sap from leaves, buds and flowers, causing them to turn yellow and patchy and/or white-streaked, often with black specks and slight crinkling so the flowers become unmarketable (Bhatt, 1993). A severe attack adversely affects the growth and the buds are distorted. Severe thrips attack in greenhouse is often associated with poor growing conditions resulting from under-watering and overheating. Regular watering and maintenance of a cooler, and humid atmosphere can therefore help preventing its infestation. Fortnightly spraying with Acephate at 0.1%, Dimethoate at 0.05%, or Monocrotophos at 0.1% keeps this pest under check. Etofenprox and Imidacloprid are also quite effective. The moth of **bud borer** lays eggs singly on growing buds during September–October and the damage continues up to February–March. Larvae bore into young buds and feed on developing petals making them hallow. Larvae also attack open flowers resulting in dropping of petals and significant yield loss (Singh, 1985). The larvae are hand-trapped and killed. Generally it is controlled by fortnightly spraying with Quinalphos 25 EC at 0.2% or methyl parathion 50 EC at 0.1% but when infestation is severe, the crop is sprayed with Indoxacarb 14.5 SC at 0.1% or Thiocarb 75 WP at 0.1%. Spraying the crop at egg-laying stage with Methomyl 0.05% or Fenvalerate 0.01% is very effective. HaNPV at 250 LE/ha, followed by *neem* formulations at 0.1-0.2% also controls this pest. **Aphids** (*Myzus persicae*) are though of little importance but these suck the sap from the leaves and disfigure the young growth. In severe attack, they leave sticky deposits on the leaves and flower buds. Aphids can be responsible for the transmission of viruses. These can easily be controlled with Malathion or Sevin sprays (0.2%). **Bud Borer** (*Helicoverpa armigera* along with certain others) are mostly a problem of the carnation bud. The eggs are laid in the buds and the larvae eat into the bud and completely damage it. These also feed on other plant parts. These are easily controlled by early application of *Bacillus thuringiensis*. Regular sprays of Sevin or Polyfrin (0.1%) at forthnightly intervals is quite effective for controlling caterpillars.

Two-spotted red spider mite (*Tetranychus urticae*) is a devastating pest on carnation grown under protected conditions, especially during September to November and February to May. Mite colonies are found on lower surface of leaves, covered by dirty webs. Affected leaves get discoloured and dry up. Infestation spreads to buds, flowers and entire plant in severe cases. Attacked plants lose vigour, become stunted and dry up in severe cases. The webs from the plant parts may be dislodged through jet water spraying. Infested plant parts should be removed and burnt. Predatory mite, *Amblyseius tetranychivorus* at the rate of 20 mites per plant will keep the mite population under control. This pest is best controlled with miticide aerosols. Spraying wettable sulphur 80 WP at 0.3%, Triazophos at 0.06%, Profenofos 50 EC at 0.1%, Ethion 50 EC at 0.1% or Mitac at 0.05% at 15 days interval, followed by 1% *Pongamia* or *neem* oil will control the mites, but during summer when infestation is very high, Abamectin 1.9 EC at 0.05%, followed by Flufenoxuron 10 DC at 0.1% or Fenazaquine 10 EC at 0.1% will be very effective

Root-knot nematode (*Meloidogyne* spp.) infest the carnation roots, feed inside and gall them by which plants are stunted, and leaves yellow and fall. Carbofuran 2 kg/ha or *neem* cake at 1 kg/m^2 or *neem* cake 0.5 kg + *Trichoderma viride* 100 g/m^2 is quite effective in checking the menace of nematodes in the crop and field.

China aster (*Callistephus chinensis*): It is attacked by **leaf and flower eating caterpillars** (*Helicoverpa armigera, Phycita* spp.) which are polyphagus in nature and feed on every tender part of the plants, more so on floral buds and flowers including ovaries, stamens and styles. Spraying at egg laying stage with methyl parathion at 0.1% in combination with Diflubenzuron at 0.2% will control this pest effectively. Damage by pest can also be checked by spraying 4% *neem* seed kernel extract and use of 250 LE/ha NaNPV. **Stem borer** (*Platyptilia molopias*) adult moth lays its eggs on the shoot tips, and after hatching the black-headed creamy larvae bore into the stems and lateral branches which causes hollowness and wilting of shoots. Such shoots should be collected and burnt. Carbofuran or Phorate at 1 kg/ha soil application will control this pest. Spraying with Monocrotophos 0.05%, Prefenophos 35 EC or Ethofenprox 10 EC at 0.1% will control this pest. Blue-green **semilooper**, a caterpillar (*Clenoplusia albostriata*) bearing wavy lines, feeds on the leaves and also on the flowers during mid-summer and defoliate the plants. Also sometimes the **cutworms**, **corn earworms** and **cabbage looper** attack this crop (Ranjith, 2013). Spraying with Carbaryl 0.1%, Chlorpyriphos or Quinalphos at 0.05% will control these pests. **Chrysomelid beetle** (*Aulacophora foviecollis*) initially attacks the newly planted crop where adults feed on the leaves and cut the growing shoots, making the plants denuded of leaves and branches, causing serious loss. Their grubs feed on roots and the underground parts of the plants so whole plant dies. Soil application of Phorate at 1kg/ha or Carbofuran at 0.5% a.i. at the rate of 0.5 kg/ha controls the grubs. Fortnightly spraying with Chlorpyriphos, methyl parathion or Quinalphos at 0.05% will save the foliage from adult-infestation. **Aster beetle** or black blister beetle (*Epicauta pennsylvanica*) feeds on foliage and flowers and destroys the whole plant. Weekly spraying of Methoxychlor during the infestation period will control this pest. **Asiatic beetle** (*Autoserica castanea*) feeds on foliage during night and hides itself in the soil at the base of the plant or in the vicinity during day time. Chlordane or Dieldrin soil application will eliminate the pest at its larval stage. Lead arsenate spraying on the plants once or twice during night will kill the feeding adults as well as larvae. Nymphs and adult aphids (*Aphis gossypii*) and **potato aphid** (*Macrosiphum salanifolii*) suck sap from tender parts, especially from lower side of flowers, and petal bases which causes their discolouration, deformation and withering. Numerous **root aphids** (*Macrosiphum* spp., Ranjith, 2013; *Anuraphis maidi-radicis*) are found feeding on the roots which results in yellowing and death of the plants. Chlordane soil application before planting or Lindane spraying on the crop controls this pest. **Leafhoppers** (*Macrosteles fascifrons* and *Orosius albicinctus*) suck foliage sap and cause pale-yellowish tinges on the leaves, blooms becoming off colour with rays being usually yellowish-green, and formation of a large number of adventitious shoots. *Macrosteles fascifrons* is a vector of virus disease caused by *Candidatus phytoplasma asteris*. These are controlled through 0.2% spraying with methyl parathion. **Tarnished plant bug** (*Lygus lineolaris*) is light-brown and variously spotted which feeds on the plants by puncturing the terminal shoots below the flower buds causing the buds or flowers to droop. Spraying with 0.2% Chlorpyriphos or methyl parathion will kill this pest. **Leaf miners** (*Liriomyza compositell*) feed inside the leaves and calyx through tunnelling and in its serious infestation leaves and calyx turn brown and die. Adult fly lays its eggs inside the leaves through puncturing, and after hatching its yellow maggots start feeding inside the leaves by making the mines. Spraying of Monocrotophos, Triazophos or methyl-o-demeton at 0.05% controls this pest. Adult **thrips** (branded greenhouse thrips and western flower thrips) and its nymphs feed on the tender parts of the plant through piercing and in its severe infestation plants deform. These can be controlled by spraying with 0.2% Metasystox. Various species of **whiteflies** (greenhouse whitefly, iris whitefly, silver leaf whitefly and sweet potato whitefly) have been observed attacking this crop (Ranjith, 2013). **Cottony cushion scale** (*Iceria purchasei*) may become a problem in temperate regions (Ranjith, 2013). Sometimes **soft scales** also attack this crop.

Cyclamen mite (*Steneotarsonemus pallidus*) feed inside the folds of young leaves so the plants become stunted with brown-scaring along the stem. Its heavy infestation causes small green rosettes in place of flowers. The infestation of **spider mite** (*Tetranychus telarius*) makes the foliage webby, discoloured and deformed. The scales and mites can be controlled through spraying with Kelthane at 0.1%, Dicofol 18.5 EC at 0.25%, Dimethoate 30 EC at 0.2% or Profenophos 50 EC at 0.1%, followed by repeating it after a fortnight along with *Pongamia* or *neem* oil at 1% will control this pest.

Root-knot nematode (*Meloidogyne incognita*) infest the plants through galling of the roots and **foliage nematode** (*Aphelenchoides ritzema-bosi*) sucking the juice of the plants causing the plants arresting its growth, cupping, yellowing and malformation of the leaves and plant death in its serious infestation. Foliar nematodes cause leaf blight and leaf fall and its serious infestation may kill the plants. Nemacur soil application and Demeton foliar spray will control both types of nematodes.

Chrysanthemum (*Dendranthema grandiflora*): More than 35 pests are known to be the cause of unsalable blossoms and low quality pot plants. The important pests of chrysanthemum are aphids, thrips, leaf miners, whiteflies, gall midge, caterpillars, hairy caterpillars, root cutting grub, mites and nematodes (Waters and Conover, 1969; Hay and Beckett, 1971; Scopes, 1979; Salinger, 1985; Lindquist, 1989; Daughtrey and Chase, 1992). Heinz and Thompson (1997) reported that cultivars vary in their resistance to aphids, leaf miners and thrips. The nymphs and adults of various **aphids** (*Aphis fabae, A. gossypii, Aulacorthum circumflexum, Brachycaudus helichrysi, Macrosiphoniella sanborni, Myzus persicae*) have been found infesting chrysanthemum. These rasp the plant sap from underside of leaves and act as vectors for certain viruses. The wingless (apterous) forms of *Myzus persicae* are yellowish-green or sometimes reddish though others are usually dull reddish and rarely green, measuring about 1.4-2.6 mm. *Microsiphoniella sanborni* is large maroon or dark brown. Newly hatched nymphs of *Brachycaudus helichrysi* are dark green and the adults feeding on bark are chocolate in colour. Adults of *Aphis gossypii* are yellowish green to black in colour, 3 mm long with red eyes while the nymphs are globular and yellowish-green or brown. *Aphis fabae* is a coal-black or very dark olive-green measuring about 1.6–2.6 mm. Aphids on the terminal shoots and on the underside of the leaves feed by sucking the sap. Feeding results in leaf curling, withering of flowers, plant stunting, and severe infestations result in death of entire plant. Severe damage to the top of the plant may reduce the number and quality of flowers. Aphids also excrete honeydew on which sooty mould grows. *Aphidius matricariae* wasps predate on the aphids. Smoking with 98% Deltamethrine at 1 g/m^3 for one hour and Phosphine fumigation for another five hours have been found controlling aphids completely. Spraying *Pongamia* or *neem* oil at 2% is quite effective in controlling their population. Application of *Verticillium lecanii* at 0.3% during evening hours helps in reducing its population. In case of severe incidence, spraying with Malathion, Metasystox, Lindane, Dimethoate 30 EC at 0.2%, oxydemeton methyl 25 EC at 0.2%, Imidacloprid 200 SL at 0.04% or Acephate 75 SP at 0.15% is quite effective. **Thrips** [*Frankliniella intonsa, F. occidentalis, F. tritici, Haplothrips* sp., *Hercinothrips* sp., *Microcephalothrips abdominalis* (Ranjith, 2013), *Thrips hawaiiensis, T. nigrophilosus, T. tabaci*] are slender, white to black and minute insects, whose nymphs and adults continue feeding while hiding themselves in the leaf axils, buds and flowers. They are noticed only when damage in the form of leaf and petal silvering, mottling and distortion of the infested organs caused by them appear. *Frankliniella occidentalis* is of special importance because it transmits the 'tomato spotted wilt virus' and 'impatiens necrotic spot virus' to many floricultural crops. It is replacing many of the other thrips as the predominant thrips-pest in greenhouses throughout the world (Rice Mahr *et al.*, 2001). *Microcephalothrips abdominalis* infests flowers by causing abrasions on petals, due to which the attacked part turns brown. Heavy infestation results in damage to corolla, stamens and developing seed, leading to premature flower drop. Machin and Scopes (1978) recommended 200 g of BHC dissolved in 100 litres of water or 40 g per 100 litres of water Diazinon drenching as the insect pupates in the soil. Thrips hiding in the soil are killed by soil drenching with Chlorpyriphos 20 EC at 0.5%. Spraying Dimethoate 30 EC at 0.2% fortnightly gives satisfactory results. In case of heavy infestation, Ethofenprox 10 EC at 0.1% or Fipronil 5 SC at 0.15% should be sprayed. Two to three times fortnightly spraying with Acephate or cartap hydrochloride at 0.05% is quite effective. **Whiteflies** though have become quite resistant to many of the insecticides but pyrethroids have proved useful. Spreading of bright yellow greasy boards in the beds also attract the whiteflies where these stick and die. **Gall midge** (*Diarthronomyia chrysanthemi*) infests on the stems and causes raised areas while distorts the floral buds. Diazinon is quite effective in controlling this pest. **Leaf folder** (*Hedylepta indicata*) female adults lay pale-white eggs single or in small clusters on the leaf underside. Brown-headed but green larvae hide themselves by folding the leaves and continue feeding to skeletonize the leaves which die afterwards. These also feed on flowers. The larvae may be killed inside the folded leaves or otherwise by pressing with thumb and fingers. Effective insecticides such as Fenvalerate 0.01%, Quinolphos or methyl parathion 0.05% can be sprayed 2-3 times to control this pest. Adults of **bud borer** (*Helicoverpa armigera*) lay single, round and creamy eggs on bracts and petals and larvae feed on buds and flowers, *vis-à-vis* other parts of the plants. Whenever eggs are seen, the plants should be sprayed with 0.01% Fenvalerate or 0.05% methyl parathion. Bacterial spray with *Bacillus thuringiensis* at floral bud stage is very effective in parasitizing these. Spraying of NPV of *Helicoverpa* at 250 LE/ha is also very effective. **Hairy caterpillar** [Diacrisia (*Spilosoma*) *obliquae*] having hairs on its whole body attacks the chrysanthemum plants during rainy season and continues feeding up to onset of winter. These eat up the leaves leaving only papery skeleton which dries up later. Adult lays the eggs in clusters on the underside of the leaves, and after hatching these start feeding gregariously on leaves and skeletonize them though grown up larvae defoliate the plants. At the time of field ploughing in summer the soil should be exposed so that birds prey on their pupae. Wherever the larvae are noticed on the plants, these should be crushed with thumb and fingers. Carbaryl 0.1%, Chlorpyriphos 0.05% or Quinalphos spraying will control the grown up larvae. Brown headed green larvae of the **soybean leaf roller** (*Hydylepta indicata*) eat the buds and flowers by folding the leaves, and the green matter of the leaves becomes completely skeletonized, whereas **semilooper** (*Autographa nigrisegna*) feeds on flowers (Ranjith, 2013). These can be controlled through trapping at earlier stage of their infestation and through 0.2% spraying with Metasystox. **Root cutting white grubs** (*Holotrichia* sp.) are not easy to control, however, at the time of field preparation, soil application of BHC, Thimet 10-G or Solvirex granules will control this pest effectively. **Flower** or **blister beetles** (*Mylabris macilemta, M. phalerata, M. pustulata*) are voracious feeders of petals making the plants completely denuded of flowers, **leaf weevil** (*Alcidodes fabricii*) adults feed on the foliage, whreas adults of the **flea beetle** though also feed on the leaves but forms only small shot-holes (Ranjith, 2013). Since these are nocturnal in habit so their control is a bit difficult, however, trappers in the night may control these pests. The maggots of **leaf miners** such as *Liriomyza huidobrensis, L. pusilla* and *L. trifolii* have been found infesting chrysanthemums through tunneling their leaves and pupating therein (*L. huidobrensis* and *L. trifolii*) or in the ground (*L. huidobrensis* and *L. trifolii*). *Liriomyza trifolii* is a major pest of commercially grown chrysanthemum worldwide (Lindquist, 1983). *Phytomyza syngenesiae* and *Chromatomyia horticola* (Ranjith, 2013) have also been observed feeding on chrysanthemum. The maggots of **stool miner** (*Psila nigricornis*) are larvae of a fly which tunnel into chrysanthemum stools so the production of basal shoots for cutting are inhibited. Their incidence is severe with start of spring throughout summer. Eggs are deposited in leaf tissue and larvae mine leaves by feeding on mesophyll. Severely damaged leaves show extensive mining, and dry up. Affected leaves should be collected and burnt. Yellow sticky traps coated with Pyrazophos 0.5 ml/l, Chlorpyriphos 0.05% or Fenvalerate 0.01% will be very helpful in trapping and killing the adult flies. These can be controlled through use of pyrethroids and organophosphate insecticides. Cypermethrin, Permethrin, Profenofos or Trizophos at 0.06% followed by *neem* or *pongamia* oil at 1% spraying kills the larvae. The larvae can be managed through spraying Abamectin 1.9 EC at 0.04%, but the adults can be controlled by spraying with Dichlorvos 76 EC at 0.1% or Decamethrin 2.8 EC at 0.1%. The green nymphs and adults of **grasshopper** (*Attractomorpha crenulata*) feed on the edges of the leaves and on petals (Ranjith, 2013). These are killed through use of methyl parathion. **Capsid bugs** and **frog-hoppers** also infest chrysanthemums by puncturing the tender stalk below the flower bud as well as the buds and suck the sap, both giving similar symptoms of their attack. BHC or DDT application will control these pests. **Earwigs**, **woodlice**, **wireworms** and **leather jackets** also attack this crop but are of little significance. **Termites** (*Microtermes obesi*) are a serious problem during dry seasons. These feed on the roots causing plant wilting. Chlorpyriphos drenching will control this pest but it is not easy to eradicate its colonies which are many feet below the ground.

Red spider mites (*Tetranychus urticae*) occur in the hot season and feed on the lower surface of leaves and buds, causing white specks in the early stage of feeding but later the buds and leaves turn yellow and fall. Their severe infestation can be noticed through spinning of the webs on the damaged parts. **Broad or yellow mite** (*Polyphagotarsonemus laetus*) is quite minute, measuring about 150 μm long. Its infestation is noticed only when down-curved brittle and distorted leaves with the hairy undersurface are found. Adult females live for about 10 days and produce about five eggs daily. Its life cycle at 20 ^{0}C temperature completes in less than one week (Sunderland *et al.*, 1992). **Spider mite** (*Bryobis* sp.) feed on the upper surface of leaves, starting from spring to rains. Affected leaves become discoloured and fall down. Mites are controlled by spraying with Aldicarb, Cyhexatin (25 g/100 litres), demeton S-methyl (22 g/100 litres), Diazinon (16 g/100 litres), Dicofol 18.5 EC at 0.25%, Dimethoate 30 EC at 0.2%, Ethion at 0.05%, Kelthane, Metasystox, Profenophos 50 EC at 0.1%, Quinomethionate (125 g/100 litres) or Tetradifon. Second spray should be done after one fortnight with *Pongamia* or *neem* oil at 1%.

After 15 days, *pongamia* or *neem* oil spray at 1% should be repeated. In polyhouses, introduction of the predator *Phytoseiulus persimilis* has been found very successful in controlling red spider mite.

Foliar nematodes (*Aphelenchoides ritzemabosi, Belonolaimus longicaudatus, Pratylenchus coffeae, P. penetrans*) cause serious damage during wet season. Field-grown early-flowering plants are normally more seriously affected than the pot-grown greenhouse plants. Foliar nematodes cause triangular browning and black patches in the interveinal regions of the leaves, starting from the lower leaves, which later on turn black and die. For its dispersal, it depends solely on water and enter the leaves through the breathing pores on the underside. *Aphelenchoides ritzemabosi* infestation though starts from the base but on buds and leaves this causes interveinal discolouration and death of leaves though leaves do not fall. HWT of stools for 5 minutes at 46 ^{0}C is quite satisfactory. Spray the aerial part with 0.01% methyl parathion or 0.02% Thionazin to control this pest. The infestation by **lesion nematode** (*Pratylenchus coffeae*) causes heavy root damage and root lesions which ultimately results into plant stunting, premature yellowing, drying of leaves and reduced flower size (Ranjith, 2013). **Root-knot nematodes** (*Meloidogyne incognita*) cause root galling, cupping of the leaves and poor growth. Aldicarb application before planting in the soil has been quite successful venture. Parathion and Mocap (O-ethyl S, S-dipropyl phosphorodithioate) are also reported as quite effective against the nematodes.

Slugs and **snails** feed on the young shoots on chrysanthemum stools. Before planting of the stools, these should be washed thoroughly to free them of their eggs. A proprietary slug bait containing metaldehyde should be placed around the stools in small heaps which will control the slugs within three nights.

Crossandra: Initial stage of caterpillars of **spike borer** (*Helicoverpa armigera*) feed on developing buds and spikes, as well as on leaves and the develoed ones even on the mature spikes, though afterwards they feed on whole plant. Fortnightly spraying of methyl parathion at 0.05%, 2-3 times will kill the borers. Attack of **whitefly** (*Lipaleyrodes* sp.) is seen by the end of rainy season and continues up to autumn. In large numbers these feed on the underside of leaves and their heavy attack causes chlorosis and develops sooty mould. These are controlled by spraying with 0.05% Fenthion or Phosalone two times at fortnightly interval. Nymphs and adults of **crossandra bug** (*Gynencia affinis*) which occur throughout the year but the pest is serious during winter season. These cause brown lesions by rasping the cell sap, and in its severe infestation plants fail to form floral buds. Spraying with Malathion 0.1% or Quinalphos 0.05% will control the bugs. Large number of yellowish nymphs and dark brown adults of **brown** (*Saissetia nigra*) and **white scales** (*Orthezia insignis*) rasp the cell sap from leaf-underside, petioles and stems causing plant stunting and leaf yellowing and falling. *Orthezia insignis* secretes even honey dew which causes attack of sooty mould. **Crossandra midge** (*Contarinia* sp.) is a black-coloured fly which lays its eggs singly in the buds of the inflorescence. The orange maggots after hatching feed on ovaries and pupate therein so the flowers become unmarketable. Soil application of Carbofuran at 1 kg a.i./ha or spraying with Profenofos at 0.05% or Triazophos at 0.06% controls this pest.

Three types of nematodes infest crossandra and these are **lesion nematode** (*Pratylenchus delattrei*), **needle nematode** (*Longidorus africanus*) and **root-knot nematode** (*Meloidogyne incognita*). Their attack causes plant stunting with pinkish to purple and yellowing of leaves, reduction in the size of inflorescence and flowers, and retardation of root growth with spindle-shaped brown to black gall formation. Crossandra should be planted along with marigold and there should be 2 kg/m^2 *neem* cake along with FYM application in the field. *Neem* cake at 1t/ha or Carbofuran 2.5 kg/ha at the time of field preparation will be quite effective.

Croton (*Codiaeum variegatum*): **Mealybug** [*Ferrisia virgata, Planococcus* (*Pseudococcus*) *citri, P. longispinosus, P. lilacinus*) nymphs and adults occur on the lower surface of leaves, base of petioles and at nodes of tender shoots. The body of the insect is covered with a white, wax-like mealy substance. As a result of infestation, plants give withering appearance. *Ferrisia virgata* is abundant in drought conditions. Predatory ladybird beetles (*Cryptolaemus montrouzieri*) and parasitoid (*Leptomastrix dactylopii*) feed on these bugs. Dust of Quinolphos at 1.5% to the affected parts of the plants will kill all the bugs present there. Spraying of the affected patches with Quinalphos 25 EC at 0.2%, Fenitrothion 0.1% or Fenthion 100 EC 0.1% will control these pests.

Scales (*Coccus depressum, Iceria aegyptiaca, Parasaissetia nigra, Parlatoria pergandii, P. proteus* var.

gloveri, Pinnaspis minor, Saissetia coffeae) feed from the lower surface of leaves along the midribs. The affected plants lose their vigour and the leaves dry off. Spraying with Quinalphos 25 EC at 0.2%, Fenitrothion or Fenthion at 0.1% will control these pests. Apart from these pests, crotons are also infested with *Heliothrips haemorrhoidalis* and *Ayyaria chaetophora* thrips which suck the plant sap from the foliage, and the beetle *Sthenias grisator* girdles the stems (Nair, 1975).

Cyclamen: Nymph and adult **thrips** (*Frankliniella occidentalis, Heliothrips haemorrhoidalis*) infest on cyclamen leaves and also the flowers. Their infestation on leaves causes irregular brown-red streaking on the upper side of leaves and rough brownish spots on the undersurface. Predatory mite (*Neoseiulus cucumeris*) weekly release at 200-350/m2 will reduce population of all the thrips drastically (Courcy Williams, 2001). Wilhelm (1997) recommended release of predatory mite (*Amblyseius barkeri*) and predatory insects (*Chrysoperla carnea* and *Orius* sp.) against *Frankniliella occidentalis* in the polyhouse. Spraying the infested crop with Dimethoate 30 EC at 1ml/l, Fenitrothion 50 EC at 3.5ml/l or Malathion 50 EC at 2ml/l will also control cyclamen thrips. Adults and nymphs of **aphids** (*Aphis gossypii*) cause yellowing and distortion of leaves and stunting of plants by rasping the cell sap from tender cyclamen shoots, leaf-undersides and buds and by inviting sooty mould through secretion of honeydew. Spraying with methyl parathion at 2ml/l is quite effective. **Cyclamen mites** (*Steneotarsonemus pallidus*) measuring about 0.25 mm feed on shoot buds and leaves, making them brittle and distorted. **Broad** or **yellow mite** (*Polyphagotarsonemus laetus*) are broader than cyclamen mite, adults yellowish and nymphs creamy-white. Infested leaves curl below and their undersides become shiny. Karuppuchamy *et al.* (1994) recommended introduction of predatory phytoselid mite (*Ambyseius ovalis*) in the polyhouse and Ramesh (1994) recommended spraying of Dicofol 18.5 EC at 2ml/l or Tetradifon 20 WP at 2g/l to control this pest.

Dahlia (*Dahlia variabilis*): Aphids (*Aphis fabae, Brachycaudus helichrysi, Myzodes dianthi, Myzus persicae* and *Neomyzus circumflexus*), thrips (*Heliothrips haemorrhoidalis, Frankliniella tritici* and *Thrips tabaci*), whiteflies, European corn borer (*Pyrausta nubilalis*), capsid bugs, tarnished plant bugs, Japanese beetles (*Popillia japonica*), the carrot beetle (*Ligyrus gibbosus*), mordellid beetle (*Mordellistena ghanii*), the flea beetle (*Nodonta clypealis*), flea hoppers (*Halticus bractatus*), leaf hoppers (*Empoasca fabae*, and *E. solana*), grasshoppers, earwigs, wireworms, red spider and other mites (*Tetranychus urticae*) are the insect-pests which infest dahlia. Several species of **aphids** (*Aphis fabae, Aulacorthum solani, A. circumflexum, Brachycaudus helichrysi, Macrosiphum euphorbiae, Myzus persicae*) invade dahlia. These multiply quickly and their nymphs and adults colonize the soft parts of the plants, suck their juice, weaken the plants, distort the leaves together with their curling and brittleness, and flowers are also distorted, *vis-à-vis* floral buds fail to open. They also transmit various viral diseases from one plant to the other. These can be controlled by spraying Dimethoate 30 EC at 0.1% or Metacid-50 at 0.2%. Olisevič (1959) found good control of aphids (*Myzodes dianthi* and *Neomyzus circumflexus*) through spraying with 0.1-0.2% Mercaptophos and Octamethyl (Schradan) at 15-20 day intervals, first remaining toxic to aphids up to or beyond three months while the second for two months. Various species of **thrips** (*Frankliniella tritici, Heliothrips haemorrhoidalis, Megaleurothrips distalis, Microcephalothrips abdominalis, Thrips hawaiiensis, T. tabaci*) have been recorded invading dahlias. The nymphs and adults rasp the surface of tender parts (flowers and leaves) and feed on exuding juice. Due to their infestation, the lower surface of petals become whitish and wither soon. While controlling the aphid, this will control the thrips also. **Leafhopper** (*Empoasca fabae, Evacanthus* sp.) adults and nymphs suck cell sap from ternder parts of the plants so leaves become discoloured, starting from the margin and spreading towards the mid-vein and plants become stunted. Sood and Kakar (1990) reported that *Evacanthus* sp. causes extensive damage in Himachal Pradesh. Spraying of methyl demeton 25 EC at 0.1% or Phosphamidon 85 WSC at 0.05% controls thrips as well as leaf hoppers. **Budworm** (*Helicoverpa armigera*) larvae feed on flower buds, petals, the reproductive parts of the flower, leaves, pedicels and peduncles. The attacked flower buds remain unopened and infested flowers give ugly appearance. Spraying of Malathion 50 EC at 0.2% will control this pest at bud initiation stage or whenever its incidence on the plant is initially seen. Afterwards also the same treatment can be repeated whenever needed. **European corn borer** larvae feed on the bud ends, floral parts and leaves and cause distortion. The larvae also bore the stems and cause extensive damage. Measures applied for budworm will control this pest also. **Flea beetle** adults and larvae feed on leaves, flowers and seed pods. **Mordellid beetles** bore into the stems and cause extensive damage. Adult **Japanese beetle** feeds on the leaves, flowers and stems but grubs damage the roots and tubers. Commercial preparations of *Bacillus popiliae* (milky spore disease) in soil infects the grubs and kill them. The **carrot beetle** larvae feed on the root and tuberous roots of dahlia. Soil application of Temik 10-G in the soil and spraying of Nuvacron at 0.2% on the aerial parts will control all sorts of beetles. **Earwigs** leave bite marks and make ragged holes in the petals during night as during day time they are hidden within the flowers. Any contact insecticide spraying after setting of night will kill them. **Fleahoppers**, **leafhoppers** and **grasshoppers** feed on the aerial parts of the plants and **wireworms** on roots and tubers so while controlling beetles these insects will automatically be controlled. Neiswander (1955) found 0.065 g of 25% Malathion powder + 0.048 g of 50% DDT powder per litre of water when sprayed at every 10 days starting when plants were 30 cm, found excellent control of potato leafhopper, tarnished plant bug, onion thrips, green peach aphid and common red spider mite (two-spotted spider mite). Widmer (1965) found effective insect control when applied 10% granular Thimet (Phorate) in the well prepared soil at 21 g and 35 g/plant and 10% granular Di-Syston (Disulfoton) at the same rates as Thimet and also at 112 g/plant, followed by immediate irrigation, and found effective control with both the chemicals at 35 g per plant (12 sq. ft.). At 35 g of Dimetilan and Thimet, with Meta-Systox-R for an area of 2 × 6 sq. ft. containing 2 plants, were also recorded very effective.

Both adult and larval stages of **broad mite** (*Hemitarsonemus latus*) which cause curling and distortion of foliage can be controlled by dusting with finely ground sulphur or spraying with water-soluble sulphur at 6 g/litre of water (Anon., 1951) and **red spider mites** (*Tetranychus urticae*) feed on the underside of leaves, causing white specklings and making the webs, and transmit dahlia mosaic virus (Korneeva *et al.*, 1970) which are controlled with Plictran (Cyhexatin) within 11 days and through anise oil which provides acceptable control (Wickizer and Meisch, 1977). Fluvanilate tuber dip or foliar spray will also control the mites. Since resistance of the red spider mites is being developed to organophosphorus compounds, hence, Harris (1971) advocated introduction of the predacious mite (*Phytoseiulus persimilis*) which gave effective control even in the absence of chemical sprays. **Two-spotted spider mite** (*Tetranychus urticae*) feeds on the lower surface of leaves causing white speckling and webbings spun around the buds and leaves. Spraying with methyl demeton 25 EC at 0.2%, wettable sulphur at 0.3% or Karathane at 0.1% will control the mites.

Nematodes like *Meloidogyne hapla* causes root galling, *Trichordoridorus pachydermus* found in the soil has been recorded from Michigan causing poor development of root system with only few feeder roots and leaf notching in *Dahlia pinnata* (Bird and Knobloch, 1976), *Aphelenchoides ritzemabosi* infests leaves and floral buds, and *Rotylenchus robustus* infests whole of the aerial parts, especially the leaves. All these nematodes can be controlled by the use of Furadan @ 4-5 kg when mixed in the soil at the time of bed preparation and by spraying the infested crop with Dimecron or Nuvacron @ 0.1%. Pearman (1959) suggested crop rotation and burning of all the infected plants, *vis-à-vis* soil fumigation with ethylene dibromide, methyl bromide, DD or Namagon. Gill (1981) recommended HWT of infected cuttings or shoots at 50 ^{0}C for 5 minutes or at 44.4 ^{0}C for 30 minutes, and Phorate or Aldicarb at 1.5 kg/ha applied as basal dressing at the time of transplanting.

Delphinium: These are very hardy plants as only **slugs** pose a serious threat though there are certain other insect-pests which may also attack the crop. Slugs check early growth of the plants by feeding on the young shoots and by making holes in the leaves of older plants. They are of four types, *viz.* Grey Field, Black Field, Large Black and Keeled, lattest being more serious during winters which damages the underground plant parts. Contact killers containing alum and the poison metaldehyde baits can control these pests. Cyclamen mite infestation appears as the symptom of the disease since this causes stunting, malformations and blackening of the buds. The larvae of the **moth**, *Tortrix cnephasia* feed on young buds and this can be controlled by spraying the plants with 0.2% Metasystox. This will also control aphids and thrips together. Spraying with 0.1% Keltahne will control the mites. Apart from slugs & snails, and moth larvae, they are infested by **aphids**, **thrips**, **cyclamen mites**, **red spider mites**, and **nematodes**. The nematode, *Pratylenchus pratensis* infests on the larkspur roots and stools which can be worsened by millipedes and by stagnant water. To get rid of this nematode the soil should be disinfected and resistant varieties should be planted.

Gerbera (*Gerbera jamesonii*): **Whiteflies** (*Bemisia tabaci, Trialeurodes vaporariorum*) remain active during moderately hot and humid conditions (February-

September). Small and fragile white adults lay eggs on lower surface of young leaves. Nymphs and adults suck cell sap from lower side of leaves resulting in discolouration, wilting and drying. In case of severe infestation sooty mould develops on honeydew secreted by nymphs. Severely damaged plants produce small and deformed flowers with crooked stalks. Field cleanliness should be maintained, heavily infested leaves should be removed and burnt, and yellow sticky traps should be placed in the field or polyhouse to manage adults. Spraying of Acephate 75 SP at 0.15% alternate with *Pongamia* oil at 1% at 10-15 days intervals will manage the nymphs. Dichlorovos 76 EC at 0.1% spraying will control both nymphs and adults. **Leaf miner** (*Liriomyza trifolii*) infestation is severe during March to June. Adults lay their eggs by puncturing the leaves. Maggots mine the leaves by feeding on mesophyll. Severely infested leaves exhibit extensive mining, become brittle and dry up. Severely mined leaves should be collected and burnt. Yellow sticky traps will also be useful. Maggots can be killed through spraying with Abamectin 1.9 EC at 0.04%. Spraying with Dichlorvos 76 EC or Decamethrin 2.8 EC at 0.1% will manage the adults. **Thrips** (*Thrips palmi*) sucks cell sap from leaves, buds and flowers and damaged parts become brittle and leaves curl up, and the buds and flowers become discoloured and distorted. Spraying Acephate 75 SP at 0.15% or Dimethoate 30 EC at 0.2% with *Pongamia* oil 0.5% 2-3 times at 15 days interval with onset of new flush will control thrips. When infestation is severe, Fipronil 5 SC at 0.15% should be applied. For killing the pupae in the soil, the soil should be drenched with Chlorpyriphos 20 EC at 0.5%. Both of the nymph and adult **aphids** (*Myzus persicae*) colonize on underside of petals from where these spread to entire flower, sucking their sap and making the flowers discoloured, distorted and wilted when attack is severe. Secretion of the honeydews by the aphids attracts sooty mould. *Pongamia* or *neem* oil at 1% spray will provide rid of these pests. Spraying Dimethoate 30 EC or Oxydemeton 25 EC at 0.2% at 10-15 days interval will control this pest. In case of severe incidence, the plants should be sprayed with Imidacloprid 200 SL at 0.05%.

Meloidogyne root-knot **nematode** has been found galling its roots by which plants are stunted, leaves yellow and drop prematurely. Application of *neem* cake at 1 kg/m^2 along with 2.0 kg/ha Carbofuran in the soil gives satisfactory control of the pest.

Gladiolus: Adult **thrips** (*Taeniothrips simplex*, syn. *Thrips simplex*) emerge initially as milky-white but soon turn brown except for the third segment which is light brown. The wings have a light transverse band near the base. A single female lays about 200 eggs. The egg is about 0.3 mm long, opaque white, smooth and bean shaped. Eggs are deposited in the leaf tissue and corms. The incubation period is 5 - 10 days. The two larval stages are light yellow and are usually found beneath the leaves or bracts. The fully developed second instar larva is about the size of the adult. The first pupal stage is distinguished from the second by having forward projecting antennae and short wing pads. The second pupal stage, which is a quiescent period, has the antennae folded over the back and has much longer wing pads. Adults of **honeysuckle thrips** (*Thrips flavus*) are yellow to orange in colour. Legs and forewings are yellow to pale yellow. The hind margin of the pronotum is provided with four long setae. Nymphs are white to pale yellow. Both nymphs and adults feed on the blossoms, as a result they get shriveled and become dark. **Flower thrips** (*Frankliniella* sp.) often invades gladiolus and damages the florets. Certain **other thrips** (*Liothrips* sp., *Thrips hawaiiensis*) also feed on flowers. For deactivation and complete killing of the thrips, the infested corms are stored at 2 ^{0}C for six weeks and then treated with hot-water at 46 ^{0}C. Standing crop infested with thrips should be sprayed with Acephate 75 SP at 0.15%, Dimethoate 30 EC at 0.2% or oxydemeton methyl 25 EC at 0.2% 2-3 times at 10 days interval. Female **cutworm** moth (*Agrotis segetum*) lays eggs near ground level on plant parts. Larvae feed on emerging shoots at night, more serious being up to 3-leaf stage. Afterwards they feed on spikes as well on the corms and grown up blackish larvae cut the plants at soil level. Deep ploughing during hot summer will expose pupae to birds. Spreading of the poison bait consisting of rice bran, jaggery and Chlorpyriphos (10:1:0.5) will kill these pests. In case of severe incidence, spraying of methyl parathion 50 EC or Acephate 75 SP at 0.1% or Quinalphos 25 EC at 0.2% will control this pest. **Gram and pod borer** (*Helicoverpa armigera*) feeds on spikes, flowers, capsules and leaves The adult moths of **leaf caterpillars** (*Spodoptera litura*) lay ash-coloured eggs in clusters on lower side of leaves, and after hatching the early instar larvae feed nearby through scraping and skeletonizing the leaves, especially during night hours. After feeding on the leaves these become shining-green to greenish-brown. After harvesting of the corms the soil should be exposed during summer so that pupae are eaten by birds and other predators. During crop season, the egg masses, early instar larvae and grown ups should be collected and killed. Early instar larvae can be killed by applying 4% *neem* seed kernel. Spraying of Quinalphos 25 EC at 0.2%, Carbaryl 0.2% or Chlorpyriphos 20 EC at 0.25% + Diflubenzuron 25 WP at 0.2% will manage this pest effectively. **Mealy bugs** (*Ferrisia virgata*) though not very common but where dry season exists the nymphs and adults reach to the corms underground and on lifting are carried to the storage where these suck the corm sap causing their shrivelling and drying in severe cases. In the early stage of infestation spraying with *Pongamia* or *neem* oil 2% or *Verticillium lecanii* at 0.3% is very effective. Spraying of methyl parathion 50 EC at 0.1%, Dimethoate 30 EC at 0.2% or Acephate 75 SP at 0.15% fortnightly will control this pest effectively. Carmine adults and green nymphs of the **mite** (*Tetranychus equatoirious*) colonize and infest on gladiolus leaves through sucking their sap when cropping area is not getting proper sunshine and the planting is near the road where dry season also exists. Their infestation web the leaves, cause them to become discoloured, bronzing, weak and hanging. Affected plant parts should be removed and burnt. Fortnightly application of wettable sulphur at 0.3% or spraying with Dicofol at 0.05% 2-3 times will control the mites.

Root-knot nematodes (*Meloidogyne* spp.) infest the gladiolus roots, especially the feeding roots so the roots are galled, the plant growth is arrested and the plants appear sickly. HWT of corms at 57.8 ^{0}C for 30 minutes will kill the nematodes present around the disc, however, the corms before HWT should have been sufficiently cured at a well-ventilated place in shade for about one month after lifting. In standing crop, Carbofuran or Phorate at 1.0 g/m^2 can be used to get rid of this pest.

Hibiscus (*Hibiscus rosa-sinensis*): This crop is favourite for mealybugs (*Maconellicoccus hirsutus*) and whitefly (*Bemisia tabaci, Trialeurodes vaporariorum*). **Mealybug** is a very serious pest whose pinkish nymphs and adults congregate at apical growing points to suck the plant sap which results into crinkling and curling of leaves and its serious infestation causes plants to stunt. Australian ladybird beetle, *Cryptolaemus montrouzeri* is an effective predator of *M. hirsutus*. Spraying with fish oil rosin soap at 2.5% or Dichlorvos 76 EC at 0.1% will control mealy bugs in hibiscus. **Whitefly** nymphs and adults can be found feeding on the undersurface of leaves and their infestation causes yellowing and drying of leaves. Under glasshouse condition, release of *Encarsia formosa* once or twice will control *T. vaporariorum*. Spraying with methyl demeton 25 EC or Parathion 50 EC at 0.1% will control the whiteflies.

Hippeastrum: In India, no any major insect-pest problem has been noticed in hippeastrum plantings but **spider mite**, though very rare, poses a problem while forcing the bulbs which can be controlled by use of sulphur vaporizer. **Aphids**, **mealy bugs** and **thrips** may also attack which may be controlled by spraying with 0.2% Rogor. De Figueiredo and Pereira (1944) found the **caterpillars** of *Xanthopastis timais* moth infesting amaryllis which they controlled by spraying with a mixture of lead arsenate and Paris green. **Bulb scale mite** (*Steneotarsonemus laticeps*) has been found attacking bulbs which can be controlled by exposing the infested bulbs to frost, drenching with Endrin or Endosulfan, dipping bulbs in a Thionazin solution and through HWT (Anon., 1971).

Goodey (1951) found the **nematode** *Rotylenchus* badly infesting the roots of hippeastrums. Johansson and Niedieck (1964) recorded the nematode *Scutellonema brachyurum* infesting amaryllis in Sweden which they controlled through soil treatment with Terracur. Andersson (1971) recorded *Pratylenchus scribneri* infesting rotted hippeastrum roots which he could control through HWT.

Snails also attack the crop seldom which can be controlled through formaldehyde poison baits.

Hollyhock (*Althea rosea*): Nymph and adult **leafhoppers** (*Amrasca devastans*) while feeding on the leaves cause stippling. Methyl demeton 25 EC at 3ml/l or Posphamidon 85 WSC at 0.75 ml/l spraying controls these insects (Anon., 1990). **Caterpillars** of *Helicoverpa armigera* are polyphagus and feed on all the plants parts, starting from hatching from the eggs when these voraciously feed on the leaves, up to the last stage when these feed even on flowers. *Sylepta derogata* larvae feed on the leaves and plants by rolling the leaves and hiding there. Spraying of Quinalphos 25 EC at 2 l/ha by diluting these up to 0.2%. **Spotted ballworm** (*Earias insulana, E. vitella*) larvae damage the shoots, buds, flowers and capsules through feeding and boring these and plugging these holes with their excreta, and therefore, ultimately the buds and flowers dry and drop. These larvae should be hand-picked and killed. Atwal (1986) suggested use of egg parasitoid (*Trichogramma chilonis*) and pupal parasitoid (*Brachymeria nephatidis*). Spraying Malathion 50 EC at 2ml/l will also control this pest. *Chromatomyia* *horticola* **maggots** feed on hollyhock leaves making zigzag trails. *Crocidosema plebeian* larvae also feed on hollyhock leaves and buds. **Tingid bugs** (*Urentis euonymus*) and their nymphs suck cell sap from underside of leaves causing yellowing to browning of leaves, their shrivelling and falling. **Mirid bugs** (*Creontiades pallidifer*) and **red cotton bugs** (*Dysdercus singulatus*) damage the plants by sucking cell sap from tender plant parts and leaves. These are controlled by spraying Malathion 550 EC at 1.25 l/ha (Anon., 1991). **Whitefly** and its nymphs damage the plants by rasping their sap from lower surface of leaves and inviting sooty mould through excreting the honedew, therefore, plant vigour is arrested and plants become deformed. Dimethoate 30 EC, methyl demeton 25 EC or Monocrotophos 36 WSC spraying at 1ml/l of water will control this pest.

Jasmine (*Jasminum* spp.): Jasmine is attacked by various insect-pests. Some of them are important which cause considerable damage to jasmine as elaborated here with. **Jasmine budworm** (*Handecasis duplifascialis*) is a major pest of jasmine, first reported by Hampson in West Africa, India and Sri Lanka in 1896. Later it was recorded in South India and Delhi. The adult is a small white moth with black palpi and three lines of the abdomen. Greenish larva with black head bores into immature jasmine buds and feeds on floral structures and in severe case webbing of buds is also noticed. Dried flowers with frass and characteristic webbing of the adjacent flower buds are symptoms of infestation (Prasad *et al.*, 1983). Amutha (1994) studied the population fluctuation of this pest on five cultivars of jasmine and noticed the prevalence of this insect throughout the year. The damaged buds with larvae should be collected and burnt. Adult moths can be light-trapped and killed. Proper sanitation of the field as well as of bushes should be done with timely and proper pruning. Carbofuran 3G 1 kg a.i./ha should be added in the soil at the time of field preparation. Spraying with NSKE 5%, Malathion 50 EC 0.2%, Profenophos 50 EC at 0.1% or Ethofenprox 10 EC at 0.1% will control this pest. **Gallery worm** (*Elasmopalpus jasminophagus*) is a serious jasmine pest causing webbing of terminal leaves, shoots and flowers. The larva is greenish with red head while thorax and laterals have brown streaks. The larva grows to a length of 2.5 cm and makes silky tunnels for feeding and pupation, *vis-à-vis* for excretion outside the buds. The moth is narrow, elongated and dark grey with pale hind wings. This pest prefers *Jasminum auriculatum* (Amutha, 1994). Its infestation occurs during the early stages, particularly in the vegetative buds and the larva causes numerous holes in the terminal leaves in folded condition. This is controlled by spraying with dimethoate 30 EC at 0.2%. Among both the **leaf webworms**, *Nausinoe geometralis* and *N. neptis*, the former is the major pest of jasmine (Tewari and Mohan, 1981) where moth is brown with hyaline patches on both the wings. Female lays 15-20 eggs singly on the leaves and these hatch in 3-4 days. The larvae attack the lower bushy and shaded portion of the plant in the rainy season and terminal shoots during dry and sunny weather. They skeletonise the leaves by scraping away the parenchymatous tissues. The larvae become fully grown in 12-15 days and pupate beneath, webbings spun over the leaves and the branches for 6-7 days. The pest is common in South India where it completes its life cycle in 22 to 24 days in July and August. The peak infestation is from July to December. These pests are managed by spraying Monocrotophos 36 SL or Quinalphos 25 EC at 0.25%. The greenish larvae of **leaf roller** (*Glyphodes unionalis*) knit together the tender leaves with silk and scrapes their green matter in its early stages and eats bulk of leaf tissues in the later stages. The faecal matter is collected in the webbings. The adult moth is 2.5 cm long and white with a thin brown line along the coastal margin of the fore-wings. Webbings of terminal portions of the twigs and the presence of excreta below the webbings are only due to the attack by *G. unionalis*. Symptoms of *G. celsalis* infestation occurs as leaf folding along the margins and tips in a typical horizontal manner ventrally. Leaf rollers can be controlled by spraying with 0.2% methyl parathion 50 EC.

Tingid bug (*Corythauma ayyari*) and its nymphs feed on the sap of the leaves from their ventral sides especially near the midribs. The infested regions of the leaves turn yellowish which ultimately dry and drop. Spraying with Quinalphos 35 EC at 0.2% will control this pest. **Jasmine bug** (*Antestiopsis cruciata*) at times is found in large numbers feeding on the tender shoots and buds, and prevents flower formation. Nymphs and adults suck cell sap from leaves causing their yellowing and drying. The nymphs are dark brownish-black and round. The adults are dark brown, shield-shaped with orange and white markings on the wings. This can be controlled by spraying with Quinalphos 35 EC at 0.2% (Ramesh, 1994). **Blossom midge** (*Contarinia maculipennis*) was first reported on *Jasminum sambac* in Andhra Pradesh and later recorded on *J. auriculatum* and *J. sambac* in Tamil Nadu. The maggot is seen within the buds at the base of the corolla. The attack by the maggot causes swelling at the base of the bud, plant stunting and ultimate drying. Spraying of Monocrotophos 36 WSC at 0.05 per cent and 0.1 per cent, soil application of Carbofuron 3G at 40 g per bush and foliar application of Cypermethrin at 25 g a.i./ ha, Fentrothion 0.1 per cent and Monocrotophos 0.08 per cent at fortnightly intervals. **Whitefly** (*Aleurotrachelus* sp., *Bemisia giffardi, Dialeurodes kikaldyi*) nymphs and adults feed on the lower surface of leaves and cause little yellowing of the leaves. *Bemisia giffardi* has been found infesting *Jasminum sambac* in Bihar and Tamil Nadu, while *Aleurotrachelus* sp. seriously infests *Jasminum auriculatum* though to a lesser extent *J. grandiflorum* and *J. multiflorum*. Spraying Fenitrothion or Fenthion at 0.1% controls these flies. **Flower thrips** such as *Dendrothrips jasminium, Frankliniella sulphurea, Haplothrips ceylonicus, H. ganglbaneri* and *Thrips orientalis* infest on jasmine crop. The adult of *Thrips orientalis* are black and nymphs are yellowish in colour. They feed and breed in the buds and flowers and rarely on the tender leaves. They occur throughout the year and cause severe leaf damage in Kerala. *Dendrothrips jasminium* and *Frankliniella sulphurea* are also recorded on jasmine in South India. The affected flowers give a decayed appearance and premature falling off. *Haplothrips ceylonicus* and *H. ganglbaneri* have been recorded infesting the flowers. Spraying with Malathion 50 EC at 0.2% or methyl parathion 50 EC at 0.1% will control jasmine thrips.

Jasmine eriophyid mites (*Aceria jasminae, Calacarus jasminae, Phyllocoptes salmaliae*) have been found infesting jasmines in India. Out of these, eriophyid mite, *Aceria jasminae* infests seriously on *J. pubescens* and *J. auriculatum*, producing hairy outgrowths on the leaf surface, tender stems and buds, and plant growth becomes stunted with poor flower production. Its infestation increases from March and reaches to its peak in July (Jose *et al.*, 1999). *Calacarus jasminae* and *Phyllocoptes salmaliae* are reported on *J. sambac* from West Bengal. *Jasminium auriculatum* 'Parimullai' from TNAU is resistant to gall-mite. Mite menace is checked through application of *neem* cake at 250 kg/ha combined with foliar application of 10% NSKE. Fortnightly spray of Phosalone 35 EC at 0.7 kg a.i./ ha or wettable sulphar at 1 kg a.i./ha was found to be effective against *A. jasminae*. Ethion 50 EC at 0.05 per cent and Dimethoate 30 EC at 0.03 per cent reduced the mite population effectively. Diafenthiuron at 0.1 per cent was the most effective chemical, followed by Bromoprophylate at 0.1 per cent or Dicofol 18.5 EC at 0.1 per cent. **Red spider mites** (*Tetranychus cinnabarinus, T. cucurbitae*) while feeding on the undersurface of leaves produce white specks which coalese later on and drastically reduce the photosynthetic activities, as a result the yield is reduced (Karuppuchamy *et al.*, 1986). The latter species often attacks the leaves of *Jasminum sambac* and *J. auriculatum*. Spraying with either Dicofol 18.5 EC at 0.2% or wettable sulphur 35 SP at 0.3% effectively controls *T. cinnabarinus*.

Root-galling (*Meloidogyne incognita*) and burrowing (*Radopholus similis*) **nematodes** infest jasmine crop by which plants are severely stunted, leaves yellow and drop prematurely and the branches dry. Apart from these symptoms, the former nematode causes root-galling though latter one leaf chlorosis and yellow to brown lesions on roots. The infestation of these nematodes can be checked by applying 1t/ha of *neem* cake or 2.5 kg/ha Carbofuran in the soil.

Lawn: **Leatherjackets** are the larvae of crane flies (*Tipula* spp.) and these maggots are tubular, greyish-brown, legless and 3.5 cm long. They feed underground and make the lawn patchy of brown colour. Sometimes maggots are also visible. Patchy areas should be soaked with water and covered with sacks or black plastic sheets for a day so that these pests come out which can be then killed. Carbaryl is also very effective in case of newly hatched larvae. In the lawn grown in sandy soils during beginning of autumn to winter, root feeding larvae of certain beetles infest. These are **chafer larvae** [cock chafer (*Melolontha melolontha*), the garden chafer (*Phyllopertha horticola*) and the welsh chafer (*Hoplia philanthus*) which are plump and 'C'-shaped and about 2 cm long having white bodies, brown heads and three pairs of legs and these feed on roots of lawn grasses. Spraying the lawn with Carbaryl from late spring to early summer will control these. The yellow-brown **ants** (*Lasius flavus*) carry soil particles on the lawn surface in the process of extending their underground nests. After levelling these mounds, watering of the nest site should be done with pirimiphos methyl, HCH, Carbaryl, Permethrin powder or Pyrethrum. In the spring and autumn the worm casts are found on the lawn surface due to **earthworms** (*Allolobophora* spp.). These worm casts should be brushed away, and if necessary, application of Carbaryl is recommended when the soil is warm and damp.

Lilium (*Lilium* spp.): The moth of **leaf caterpillar** (*Brythys crini*) completes its life cycle within six weeks and its larvae feed on leaves. In its severe infestation, the whole plant is defoliated. Another larvae of moth (*Polytela gloriosa*) feed on the green matter of leaves and cause defoliation. The moth is stout with a mosaic pattern of yellow, red and black on the fore wings and black on the hind wings. The smooth larva is pretty looking with black, white and red spots. Round, yellowish eggs are laid in clusters of 15 to 40 on the apical portion of the under surface of leaves. They hatch in 3 to 4 days. The larval period lasts from 16 to 20 days undergoing 5 moults. Pupation takes place in soil for 15 to 20 days. There are two generations per year in North India; the pupae of the second generation hibernate in the soil. They are controlled by spraying with Malathion 50 EC at 0.2%. **Aphids** such as *Dysaphis tulipae, Myzus persicae, Neomyzus circumflexus* and *Rhopalosiphum nymphaease* have been recorded sucking sap from *Lilium* plants. *Dysaphis tulipae* feeds on bulbs which causes stunting and deformed plants. *Myzus persicae* feeds on leaves. *Neomyzus circumflexus* is the crescent-marked lily aphid which sucks the sap from leaves and tender portions of the plants and devitalizes them. *Rhopalosiphum nymphaease* sucks the sap from shoots and floral buds which results in withering of tender shoots and premature falling of buds. While feeding, aphids excrete honeydew on which sooty mould develops, and the affected plants lose their elegance (Butani and Dadlani, 1983). Green peach aphid can be parasitized with predator *Aphelinus* sp. Lacewing fly Ist instar larvae also feed on aphids. To control the bulb infestation of *Dysaphis tulipae* in storage, Carbaryl dusting at 15-20 kg/ha of bulbs before storage is recommended. Dimethoate 30 EC at 0.2% spraying on the crop will control aphids effectively. **Lily bulb mite** (*Rhizoglyphus echinopus, R. robini*) while feeding on the bulbs and roots, enters the bulb scales and burrows into the stems which results into yellowing of plants. In *L. longiflorum*, its infestation causes cessation of growth at 7.5-15.0 cm height, rosetting of leaves, chlorosis and ultimate death. Such infested plants become quite susceptible to pathogens (*Fusarium, Pythium* and *Rhizoctonia*) for secondary infection. For getting rid of this pest, there should be crop rotation of the non-host crops, the field should be properly drained, the bulbs should be treated with hot water for one hour at 44 ^{0}C temperature, and the storage temperature of the bulbs should be 1.6 ^{0}C so that mite feeding can be prevented. Predatory mite, *Hypoaspis aculifer* should be released in the polyhouses which will suppress the mite menace within six weeks (Lesna *et al.*, 2000). **Thrips** (*Liothrips vaneeckei, Thrips simplex*) have preference for lily bulbs and their infestation produces rust-coloured sunken areas at the base of outer scales making them soft so they break off easily. Already infested bulbs produce stunted growth when planted. Bulb dipping in hot water at 43.4 ^{0}C for two hours will control the thrips. Infested bulbs should be stored at 2 ^{0}C for six weeks. Spraying of the infested crop 2-3 times with 0.1% methyl parathion, Acephate at 0.1% at 10 days intervals, or before bud initiation spraying with Malathion 0.1% or Carbaryl 50 WP at 0.2% will control this pest. **Lily beetle** (*Lilioceris lilii*) adults and its larvae are serious pests in certain parts of Canada, U.K. and U.S.A. (Casagrande and Livingston, 1995). These voraciously feed on stems, leaves, flowers and seed pods; and **lily weevil** (*Agaspherops nigra*) grubs feed on the stem underground so these can be controlled by spraying with Carbaryl 50 WP at 0.2%. Smith (1967) observed that **stem borer** (*Papaipema nebris*) bores the lily stem and feeds on stem pith. **Leaf caterpillars** which are black with red-white spots feed on the leaves and its moths (*Polytela gloriosa*) are brown with yellow and red markings. Its larvae can be hand-picked and killed. Another leaf caterpillar, *Spodoptera litura* has also been recorded feeding on lilies (Nair, 1975). After the crop is over, the soil should be exposed so that all the pupae and larvae present in the soil are killed. Spraying of 5% NDKE is also effective.

Lotus (*Nelumbo nucifera*): **Aphids** (*Siphocoryne nymphaeae*) suck the cell sap from shoots and floral buds so immersing for 1-2 days the leaves or other parts of the plants being infested will help drowning the aphids. Jet water spraying will knock the aphids down in the water which will be eaten by fishes. **Thrips** (*Scirtothrips dorsalis*) is a serious pest of lotus that sucks cell sap from the leaves. Spraying the infesting leaves with Acetamiprid, Fipronil, Imidacloprid, Methiocarb or Fenitrothion + Fenpropathrin controls all the larvae and adults within 16 hours of treatment. The caterpillars of **case worm** (*Nymphula* sp.) have tracheal gills and feed on leaves of the lotus. Apart from these, Ayyar (1963) reported **hairy caterpillar** (*Semrya conspersa*) and **tobacco caterpillar** (*Spodoptera* spp.) feeding on lotus. By shaking the plants the insects should be dislodged in the water so that these are eaten by fishes. Generally, insecticides in the water plants are not used.

Marigold (*Tagetes* spp.): **Bud borer's** adult moth (*Helicoverpa armigera, Phycita* sp.) lays its eggs singly on young buds which after hatching start boring into flowers and feed on floral parts. *Phycita* larvae feed on floral bud-heads and flowers. The larvae should be manually collected and killed, *neem* seed kernel extract spray at 4% and spraying of methyl parathion 50 EC at 0.1%, or Fenvalerate 20 EC at 0.05% in combination with Diflubenzuron 25 WP at 0.2% at egg-laying stage on tender foliage will control this pest. HaNPV at 250 LE/ha also reduces the pest damage. **Hairy caterpillar** (*Diacrisia obliqua*) feeds on the foliage and even on other plant parts. This is controlled by spraying with Dichlorvos 76 EC at 1ml/l. **Aphid** (*Aphis craccivora, A. gossypii* and *Myzus persicae*) adults and nymphs mainly infest growing tips and floral parts by sucking the sap and so plants appear sickly. During spring and summer these can be found in large number. These can be controlled by spraying with oxydemeton methyl 25 EC or Dimethoate 30 EC at 0.2%, followed by 1% spray of *neem* or *Pongamia* oil after 10 days. **Tingid bugs** (*Cadmilos retiaris*) feeds on marigold leaves. Its eggs are parasitised by *Trichogramma* spp. *Leptus* mite also parasitizes both the nymphs and adults. **Thrips** have been found infesting on the buds, flowers and young leaves of marigold. These feed on leaves and flowers by sucking of the cell sap so the leaves and flowers get distorted and the petals turn brown and dry. These are controlled by 2-3 sprayings at 10-days intervals with Dimethoate at 0.05% or oxydemeton methyl. **Greenhouse whitefly** (*Trialeurodes vaporariorum*) also feed on tender parts by rasping the sap and through excretion of honeydew entice sooty mould. Release of *Encarsia formosa* and *Chrysoperla carnea* in the greenhouse will ckeck its population. Insecticidal soap is also very effective in controlling this pest. Nymphs and adults of **leafhoppers** (*Empoasca fabae, Macrosteles fascifrons*) by sucking cell sap cause cupping and rolling of leaves and wilting of shoot tips. It is controlled by applying Dimethoate 30 EC or Malathion 50 EC at 2ml/l. **Leaf miner** (*Liriomyza trifolii*) larvae feed on the leaf mesophylls by making serpentine mines so leaves become ugly and yield is also drastically reduced. Parasitic wasp release in the greenhouse reduces its population drastically within two months. **Two-spotted spider mite** (*Tetranychus urticae*) makes its colonies on the lower surface of leaves which are covered by dirty webs. Affected leaves get discoloured and drop. Infested plants get stunted, wilt and dry if damage is severe. To keep away this pest, plants should be jet-water sprayed to dislodge the webs and the infested plant parts should be cut and burnt. Spraying of Dicofol 18.5 EC at 0.1%, followed by 1% spray of *Pongamia* or *neem* oil after 10 days will control this pest.

Oleander (*Nerium* spp.): It is infested with **sphingid caterpillar** (*Deilephila nerii*) which is quite stout, elongated and with a spine at the anal region. Its adult is a robust green moth with yellow markings. The larvae feed voraciously on the leaves and make complete defoliation of the plants (Nair, 1975). The larvae of **leaf butterfly** (*Euploea core*) is blackish-brown and striped with six tentacles, two each at the anterior and posterior ends and a pair in the middle. The adult is black with a series of white spots on wing margins. The larvae can be hand-picked and killed. Spraying with Phosalone 35 EC at 0.1% will control this pest. Nair (1975) stated that stem girdler beetle (*Sthenias grisator*) bores into the stem. Loopers (*Agathia hemithearia, A. lactana*) and woolly bear (*Pericallia ricini*) feed on the foliage of oleander. Nayar *et al.* (1976) reported that scales (*Aspidiotus transparens, Parlatoria calianthiana*) and mealy bugs (*Nipaecoccus vestator*) also feed on oleander.

Orchid (A group with 800-900 genera): It is a large group of herbaceous plants being attacked by a large number of insect-pests such as brown soft scales (*Coccus hesperidum, C. pseudohesperidum, Saissetia coffeae, Lecanium formicarii, Pulvinaria phaiae*), hard scales (*Aspidiotus destructor, A. herderae, A. nerri*), Boisduval scale (*Diaspis boisduvalii*), orchid scale (*Furcaspis biformis*), and certain other scales (*Chrysomphalus dictyospermi, Fiorinia fiorinae, Genaparlatoria pseudoaspidiotus, Hemiberlesia cyanophylli, H. lataniae, Lepidosaphes mackiana, L. noxia, L. machili, Leucaspis cockerelli, Melanaspis aliena, Parlaloria proteus*), mealybugs (*Ferrisia virgata, Nipaecoccus nipae, Planococcus citri, Pseudococcus filamentosus* var. *corymbatus, P. adonidum, P. maritimus*), thrips (*Anaphothrips orchidaceus, A. orchidii, Frankliniella* sp., *Heliothrips haemorrhoidalis, Hercirothrips femoralis, Thrips tabaci*), orchid weevil (*Cholus* sp., *Eucactophagus weissi, Orchidophilus aterrimus*), two-spotted spider mite (*Brevipalpus australis, B. cauliformicus, B. oncidii, Tenuipalpus pacificus, Tetranychus urticae*), slugs (*Deroceras laeve*) and giant African snails (*Deroceras laeve*).

Scales attach themselves to the leaves, petioles and shoots and suck the juice from the cells so plants are stunted and lose vigour. **Armoured scales** are among the most devastating pests of orchids as apart from sucking the sap they also secrete honeydew which entices sooty mould fungus. The **Boisduval scale** is the most common armoured scale infesting many orchid genera, especially *Cattleya* and *Cymbidium* foliage and pseudobulbs. **Soft scales** (*Coccus hesperidum*) infests *Cattleya* and *Dendrobium* orchids, and *Saissetia coffeae* and *Aspidiotus nerri* attack *Cymbidium*; while **orchid scale** attacks *Cattleya; Genaparlatoria pseudoaspidiotus* infests on *Vanda*; *Lepidosaphes hesnoxia* and *L. mackiana* on *Dendrobium*; and *L. machili* on *Cymbidium* orchids. The females of **hard scale** (*Aspidiotus destructor*) while sucking the sap cluster in large numbers on the undersurface of leaves which causes leaves turning yellow and in severe cases their withering and drying up, as a result plant growth is checked and plants appear sickly. Application of Carbofuran 3G at 1 kg a.i./ha will keep **scales** under control. Spraying with Malathion 50 EC at 0.1% or Chlorpyriphos 20 EC at 0.25% + Dimethoate 30 EC at 0.1% or Ethion 50 EC at 0.1% will control these pests effectively.

The adults and nymphs of **mealybugs** cluster around the stems, leaves and their pedicels, and flowers and their peduncles and suck the sap, secrete honeydew and attract ants and sooty mould and in their severe attack the plants lose their vigour. Spray of fish oil rosin soap at 2.5% or Dichlorvos 76 EC 0.1% will control mealybug menace. Two to three times fortnightly spraying with Acephate 0.1%, or methyl parathion 0.04% is also very effective. Nymph and adult **thrips** feed on the foliage and flowers and remain hidden in the flower buds and underside of leaves. Flower buds get disfigured and sometimes even fail to open, and the leaves as well as other plant parts show silvery patches as the thrips puncture the epidermis while feeding on them. In netted areas, insecticide granules are very effective in checking the menace of thrips. Foliar spraying at 10 days interval with Cyfluthrin 0.2%, Chlorpyriphos 0.1%, Dimethoate 30 EC at 0.1%, Monocrotophos at 0.05% or Phosphamidon 85 WSC at 0.05% controls these pests effectively. **Orchid weevil** adults feed on flowers or puncture the flower buds and the grubs bore and feed within the plant tissues. *Orchidophilus aterrimus* is a serious pest of *Dendrobium* orchids. Spraying with methyl parathion 50 EC at 0.1% or Fenvalerate 20EC at 0.05% in combination with Diflubenzuron 25 WP at 0.2% at initial stage of **weevil** infestation will control this pest. Spraying with *neem* seed kernel extract (NSKE) 4% at weekly intervals will also keep this pest at bay. Spraying of HaNPV at 250 LE/ha is also very effective.

Slugs and snails (*Achatina fullica*) are common greenhouse pests where the humidity is high. They feed on the young leaves, roots, flower buds and even open flowers in the night. Slime trails leading to plant injury are indicative of the presence of these pests. Slugs usually hide during the day and become active during night, causing damage by feeding on foliage. Formaldehyde poison baits are very effective in controlling these pests.

Two-spotted red spider mite, *Tetranychus urticae* is a serious pest of orchids during warm and dry seasons. *Brevipalpus australis* and *B. cauliformicus* were also recorded on many orchid genera. *B. oncidii* is a serious pest of *Oncidium* and *Odontoglossum. Tenuipalpus pacificus* is a serious pest of *Phalaenopsis* and certain other orchids. Mites suck the cell sap of leaves, leaving them white-speckled and their heavy infestation causes webbing on plants. Dicofol 18.5 EC at 0.08% spraying will control the mites.

Poinsettia (*Euphorbia pulcherrima*): Both nymph and adult **whiteflies** (*Bemisia argentifolii, B. tabaci* and *Trialeurodes vaporariorum*) rasp the cell sap from the underside of leaves which causes leaf yellowing and drying, *vis-à-vis* invite sooty mould due to secretion of honewdew. Silver-leaf whitefly can be controlled by releasing a coccinellid predator, *Seriangium parcesetosum* (Ellis *et al.*, 2001) in the field while *Encarsia formosa* parasitizes nymphs and reduces their population up to 80% in the greenhouses (McMahon *et al.*, 1994). Myco-insecticide containing *Beauveria bassiana* controls greenhouse *Bemisia argentifolii* (Olsen and Oetting, 1999). *B. tabaci* is also invaded by *Paecilomyces fumosoroseus* fungus (Osborne and Landa, 1984). Sprayng the crop with Malathion 50 EC at 2ml/l, methyl demeton 25 EC at 1ml/l, methyl parathion 50 EC at 1ml/l or Phosalone 35 EC at 2ml/l is also quite effective in controlling the whiteflies.

Rose (*Rosa* spp.): Rose plants are prone to attack of many pests at almost every stage of plant growth and development. **Scale insects** (*Lindingaspis rossi, Aonidiella aurantii, Aspidiotus dictyospermi* var. *arecae, A. orientalis, A. transparens, Chionaspis* spp., *Saissetia hemisphaericum, Iceria purchasi, Iceria aegyptiaca*, and *Iceria formicarum*) have been found infesting on roses. *Lindingaspis rossi* appears as reddish-brown, waxy, scale-like pox marks on the lower portion of the old stems and at times on younger shoots also. The nymphs and adults suck the sap from shoots and stems. The affected portion of the plant dries up. In severe infestation the whole plant may die. *Aonidiella aurantii* is flat, circular and reddish in colour and measures about 2 mm in diameter. Both adults and young ones suck the sap from mature shoots and deplete vigour of the plant (Ramesh, 1994). Their infestation appears in the form of reddish-brown encrustations, and in severe attack the infested parts dry. The infested plant bears a few small flowers (Nandakumar *et al.*, 1988). Greenhouse as well as garden roses are prone to attack by this pest. They infest more severely before and after the monsoon and are often choosy in their preference to certain cultivars of roses. Though no rose plant is immune to this pest (Ramakrishnan, 1984) but some preferred rose cultivars are 'Camelot', 'First Prize', 'Incense', 'Kordes Perfecta' and 'Versailles' among the Hybrid Teas; and 'Ice White', 'Red Pinocchio', 'Zambra' and 'Zorina' in Floribundas. *Aspidiotus dictyospermi, A. orientalis* and *A. transparens* infest shoots and leaves whereas *Chionaspis* sp. infests tender shoots. *Iceria aegyptiaca, I. formicarum* and *I. purchasi* infest twigs of roses in different parts of India (Sohi and Singh, 1995). Selection of scale-free planting material and cutting and burning of infested parts followed by 10% *Pongamia* oil spray, whenever the infestation is noticed, are the key operations to reduce their population. In case of a very severe infestation, application of Carbofuran 3 G at 1 kg a.i./ha after pruning will keep these pests under control. Spraying with Malathion 50 EC at 0.1% or Chlorpyriphos 20 EC at 0.25% + Dimethoate 30 EC at 0.1% or Ethion 50 EC at 0.1% will control these pests effectively.

Several **thrips** (*Lefroyothrips lefroyi, Megalurothrips usitatus, Physothrips andrewsi, Retithrips syriacus, Rhipiphorothrips cruentatus, Scirtothrips dorsalis, Taeniothrips fulvus, T. lefroyi, T. rhopalantennalis, Thrips florum, T. hawaiiensis, T. melanurus* and *T. tabaci*) have been recorded feeding roses in different parts of the country. *Lefroyothrips lefroyi* was recorded infesting rose flowers in Himachal Pradesh (Sood, 1990). *Megalurothrips usitatus* feeds on flowers in many parts of India. *Physothrips andrewsi* also lacerates petals of roses and destroys its beauty. *Retithrips syriacus* attacked leaves show silvery appearance, turn brown, curl up and then fall away. Their feeding also destroys the flowers (Patel, 1988). *Rhipiphorothrips cruentatus* is the most common species attacking roses and other host plants like *Lagerstroemia* sp. and *Punica granatum*. The nymphs are reddish in colour whereas the adults are dark brown or black. The pest breeds throughout the year except in winter when it pupates in the soil. The adult appears in March and lays eggs on the undersurface of leaves in plant tissues. Both nymphs and adults suck cell sap from tender leaves, buds and flowers. Curled leaves with brown marks and deformed buds with burnt margins are the main symptoms of damage. *Scirtothrips dorsalis* infests both under open field and protected conditions. Thrips attack starts on new flush after pruning. Both nymphs and adults suck cell sap from tender leaves, buds and flowers. Curled leaves with brown marks and deformed buds with burnt margins are the main symptoms of damage. Under protected conditions, thrips is a serious problem on roses almost round the year except during November-December and July-August. *Taeniothrips fulvus* has been recorded feeding on leaves and flowers of rose in Himachal Pradesh (Sood, 1990). *Taeniothrips lefroyi, T. hopalantennalis, Thrips florum, T. hawaiiensis, T. melanurus* and *T. tabaci* infest leaves and flowers from March to November and hibernate in the pupal stage from December to March (Butani and Dadlani, 1983). Both nymphs and adult thrips feed by rasping the tissues and sucking the sap which oozes out from the wounds. Affected leaves are deformed with brown or silvery patches or burnt margins. The affected leaves get distorted, wither and drop down. In case of severe infestation, flower buds shed prematurely and show brown patches on the petals. The females lay 60-100 eggs during their life stage. The nymphs hatch in 2-7 days, feed during the first two instars and then pass through two non-feeding pseudo-pupal stages in protected parts of the plant such as leaves or flowers, in leaf litter or in soil. The nymphal and pupal periods are completed in 9-20 and 2-5 days, respectively. The females can reproduce with or without fertilization. Soil should be drenched with Chlorpyriphos 20 EC at 0.5% at 10-15 days interval for killing pupae in the soil. Spraying of Acephate 75 SP at 0.15% or Dimethoate 30 EC at 0.2% with *Pongamia* oil 0.5% 2-3 times at 15 days interval with onset of new flush, and under polyhouse condition, spraying with Fipronil 5 SC 0.15% or Imidacloprid 200 SL 0.04% or Acephate 75 SP at 0.15% will control these pests.

There are various genera and species of **aphids** (*Aphis gossypii, Chaetosiphon tetrapodes, Cinera* sp., *Macrosiphum rosaeformis, M. rosae, M. euphorbiae, Myzus persicae*) which attack roses. Among all the aphids that attack rose, the most common one is *Macrosiphum rosae* which is pear-shaped, soft bodied and light green to dark blackish-green. *Macrosiphum rosae* infests tender shoots (Agarwala, 1983) and *M. euphorbiae* infests shoots and buds of roses in India (Agarwala, 1983; Agarwala and Raichauduri, 1981). In North India, *Macrosiphum rosaeformis* occurs on roses from November to April. The nymphal development of its wingless form is complete in 11-14 days whereas winged forms complete their nymphal development in 14-19 days. *Aphis gossypii* attacks cultivated varieties of roses from September to December with the peak period of activity in October. This aphid multiples well under 20-25 ^{0}C temperatures. The aphid, *Chaetosiphon tetrapodes* sucks sap from tender shoots and *Cinera* sp. infests shoots of roses in Himachal Pradesh (Sood and Kakar, 1990). Aphids of different species are found in clusters on the tender portions of shoots, buds, flowers and leaves. Both nymphs and adults feed by sucking the cell sap. As a result, tender shoots wither, buds fall prematurely and the flowers show malformation and fading. Severe damage to the top of the plant may reduce the number of flowers produced. Aphids also excrete honeydew on which sooty mould grows, and some species even transmit viruses. Aphids feeding on flowers make them unmarketable and their presence is a nuisance on the plants because they leave cast skins stuck to the plant when they moult, which affects the value of the plants and/or flowers. *Neem* or *Pongamia* oil at 0.1% or Dimethoate 30 EC at 0.2% will control the pest when infestation has just begun. However, in severe infestation the sprayings with Imidacloprid 200 SL at 0.04% or cartap hydrochloride 50 SP at 0.1% will control this pest. Spray of *Verticillium lecanii* at 3.0 g/l under high humid conditions is also effective against aphids.

Bud borer (*Helicoverpa armigera*) infests the crop from January to April. Female moth lays cream coloured eggs on young buds. Just-hatched larvae bore into buds by making holes and feed on petals damaging the whole flower (Ramesh, 1994). Collection and destruction of mature larvae reduce borer population. Spraying with methyl parathion 50 EC at 0.1% or Fenvalerate 20EC at 0.05% in combination with Diflubenzuron 25 WP at 0.2% at egg-laying stage on tender foliage will control this pest. Alternatively, spraying with *neem* seed kernel extract (NSKE) 4% at weekly intervals will also keep this pest at bay. Spraying of HaNPV at 250 LE/ha is also very effective. Female adult of **hairy caterpillar** (*Euproctis fracterna*) lays eggs in clusters on the lower side of leaf which on hatching feed gregariously on the surface of leaves and skeletonize them so that such leaves dry up soon whereas grown up black larvae feed on leaves, buds and flowers and defoliate the plants. Their eggs and early instar larvae can be killed by hand-rubbing, grown up larvae can also be trapped and killed and adult moths can be light-trapped and killed. Summer-exposing the soil will give chance to predatory birds to feed over the pupae hiding in the soil. Fortnighly spraying of Quinalphos 0.05% or Acephate 0.1% will control these pests. *Neem* seed kernel extract 4% or *neem* oil 1% spraying will act as antifeedant against early instar larvae. Adults of **tobacco caterpillar** (*Spodoptera litura*) lays eggs in groups on lower surface of leaves. After hatching the black larvae feed gregariously throughout the monsoon during night in the polyhouses on leaves by scraping chlorophyll, growing buds and flowers. The gregarious early instar larvae and subsequent larval stages are collected and killed. Grown up larvae can also be killed by spreading poison baits made of wheat or rice bran, jaggery and chlorpyriphos (10:1:0.5). Spraying with Quinalphos 25 EC at 0.2% at 15 days interval will control this pest at the beginning of incidence but in case of severe attack Indoxicarb 14.5 SC at 0.1% or Thiodicarb 75 WP at 0.1% is very effective. Spraying with SlNPV at 250 LE/ha, followed by *neem* formulations at 0.1-0.2% are also effective. **Chafer beetle** (*Adoretus* spp.) infests roses with onset of monsoon when black to brown beetles emerge from the soil and start feeding on tender shoots, foliage, buds and flowers though its grubs on the roots. Summer ploughing will expose its eggs, grubs, pupae and adults to be predated upon by the birds and other natural enemies. Setting of light traps in the night will help in catching the adults for killing. Spraying with ethanol extract of *neem* kernel at 2.5% will protect the plants from their feeding. Application of Phorate or Carbofuran at 1.0 kg a.i./ha will kill the grubs in the soil. Evening spraying with Chlorpyriphos or Quinalphos at 0.05% to the plants will deter the beetles coming on such plants. The female adult of **ash weevil** (*Myllocerus* spp.) lays its eggs in soil. Its grubs feed on roots and the grey coloured adults on leaves in a regular fashion starting from the margin. Grubs are controlled by application of Carbofuran or Phorate at 1.0 kg a.i./ha in the soil. Adults can be controlled by spraying with 0.1% Carbaryl or 0.05% Malathion. **Whitefly** (*Bemisia tabaci*) remains active during January-June and is problematic in protected cultivation. Small and fragile white adults lay eggs on lower surface of young leaves. Nymphs and adults suck cell sap from lower side of leaves resulting in yellowing and dropping of leaves. In case of severe infestation, sooty mould develops on honeydew secreted by nymphs and significantly affects the growth of the plants resulting in production of small and malformed flowers. Proper field sanitation and use of insect-proof nets help preventing the whiteflies. Removal and burning of heavily infested leaves will control the pest build up. Installation of yellow sticky traps will trap the adult flies. Spraying with *Beauveria bassiana* or *Verticillium lecanii* formulations at 0.2% also checks whitefly population effectively. Spraying with acephate 75 SP at 0.15% at 15 days interval alternating with *Pongamia* oil 1% keeps this insect at bay. In case of severe adult infestation, the crop is sprayed with dichlorvos 76 EC at 0.1%, followed by Lambda cyhalothrin 5 EC at 0.1% or deltamethrin 2.8 EC at 0.1% at 5-7 days interval. **Termites** (*Microtermes obesi*) are highly devastating to the crop especially when soil *vis-a-vis* weather is dry. Initially they feed on roots but afterwards their infestation progresses towards stems by eating the bark as well as a part of the wood. Such plants die soon. Since its queen rests quite deep in the soil so it is very difficult to kill, however, proper irrigation and pre-planting drenching of soil with Chlorpyriphos at 0.05%, or Malathion 0.1% will kill the worker termites in the field and to some extent will destroy their colonies.

Spider mites (*Tetranychus cinnabarinus* and *T. urticae*) are found in large colonies on the undersurface of leaves covered with fine silky webs. The spider mites are 0.5-0.6 mm long. They develop through egg, larva, protonymph, deutonymph and then adult stages. Spider mites can produce a generation in 7-14 days in warm temperatures. In cooler temperatures, generations take more than 40 days and below 12 °C development stops. In greenhouses, adults may go into diapause for several months hibernating in cracks and crevices from October-November through February-March in temperate climates. These mites feed by penetrating the plant tissue with sharp stylets and removing the cell contents. The chloroplasts disappear and the small amount of cellular material coagulates to form an amber mass. In the palisade layers, only the penetrated cells are damaged though adjacent cells show no evidence of injury. Feeding of spider mites results in white specks on leaves which later coalesce and produce white patches resulting in reduced photosynthetic activity (Karuppuchamy, 2001). Ultimately, affected leaves become mottled, turn yellow and fall. Infested flower petals lose their brightness resulting in a direct loss to the grower. Some polyantha and floribunda roses completely lose their leaves prematurely due to the attack of this pest. The damage is more pronounced during warm and dry seasons. In the polyhouse, the prevailing high humidity and temperature is very favourable for the population build up of spider mites. Sanitation, proper irrigation and cutting and burning of severely infested plant parts will keep this problem away. Release of predatory mite, *Amblyseius tetranychivorus* at 20 mites per plant when the population of *T. urticae* is at moderate level will save the crop from mite damage. Spray of *Verticillium lecanii* formulation at 0.5% during evening is also very effective. During export season under protected cultivation, spraying of Abamectin 1.9 EC 0.05%, followed by Flufenoxuron 10 DC 0.1%, Fenazaquine 10 EC at 0.1% or Difenthiuron 50 SC at 0.06%, if necessary, will protect the crop. Spraying with Dicofol 18.5 EC at 0.25%, wettable sulphur 80 WP at 0.3%, Profenofos 50 EC at 0.1% or Ethion 50 EC at 0.1%, followed by *Pongamia*, *neem* or *Jatropha* oil at 0.5% under open cultivation is very effective in controlling the mites.

Lesion nematode (*Pratylenchus* spp.) infestation in roses retards shoot growth, causes leaf chlorosis, destroys feeder roots and causes dark root lesions which ultimately produces poor quality flowers with reduced yield. Rootstock *Rosa indica major* is resistant to this nematode. The roses should be planted with intercropping of marigold. In the soil at the time of land preparation for planting, 2.0 kg/ha Carbofuran is recommended.

Tuberose (*Polianthes*): **Bud borer** (*Helicoverpa armigera*) adult lays eggs singly on developing spikes and on hatching their larvae start feeding the buds and flowers, *vis-à-vis* other plant parts by boring them. This can be controlled by spraying with methyl parathion 0.05%. The ash-coloured adult **weevils** (*Myllocerus* spp.) feed in darkness on the leaves making notches on the edges which may be trapped and killed as they are only a few. The grubs feed on the roots and tunnel into the bulbs whose attack may be prevented by applying Furadan granules @ 3-4 kg/ha at the time of bed making. Fortnightly spraying with Carbaryl 0.2% or Malathion 0.1% at evening will protect the crop from weevil's damage. **Grasshoppers** (*Hieroglyphus* spp.) feed on the young leaves and flower buds, especially during rainy season which may be controlled by spraying 0.2% methyl parathion. **Pentatomid bug** (*Nezar viridula*) lays its eggs in clusters on the leaves. After hatching, the green-coloured nymphs and adults start sucking the sap from the basal portion of the buds and flowers so such buds wither and drop. It should be observed that when eggs are laid these should be killed through hand-pressing. The adults should also be picked and killed. Spraying with Malathion 0.1% will control these pests effectively. Adults and nymphs of\small green **aphids** (*Aphis* spp.) are very small in size and multiply rapidly. They feed on the growing points, spikes and floral buds by sucking their sap from base of floral buds which may be controlled by spraying nicotine solution. Effectively these are controlled by spraying with Dimethoate 0.06% or Monocrotophos 0.05%. **Thrips** are minute insects which suck the sap of the leaves, buds and flowers, and act as a carrier agent in causing 'bunchy top' disease where the inflorescence is malformed. While controlling the grasshopper, this will also be controlled.

Red spider mites suck the sap on foliage causing yellow stripes and streaks, and in severe cases leaves turn yellow, silvery or bronze and finally deformed. Kelthane spraying will prevent its infestation.

Two **nematodes**, *viz. Aphelenchoides besseyi* causes greasy streak on foliage and *Meloidogyne* spp. (*M. incognita, M. javanica, M. arenaria, M. acritata*) cause root galling, poor growth of plants with leaf tip burn and yellowing, and suppress spike emergence in their severe infestation to tuberoses. **Reniform nematode** (*Rotylenchulus reniformis*) also infests tuberose and causes yellowing and drying of leaves with poor plant growth so flower quality becomes poor. Tuberose var. 'Shringar' is resistant while 'Suhasini' is tolerant to root-knot nematodes. *A. besseyi* was first reported as foliar nematode of tuberose in 1968 which can be controlled by HWT, bed fumigation and post-planting drenches with methyl bromide. The treatment advocated for controlling larvae of the weevil will keep these pests also under control. However, soil application of Carbofuran at 2.5 kg/ha or *neem* cake 1t/ha will control their menace.

Zinnia: **Borer** (*Helicoverpa armigera*) larvae feed on leaves and flowers in the season though on other parts when the host population is poor. Larvae should be hand-picked and killed whenever seen. Spraying with Methonyl at 2 ml/l, or HaNPV at 250 LE/ha will control this pest. The adult beetles of the **white grubs** (*Holotrichia consanguinea*) feed during night on the leaves making holes of various shapes and sizes but the grubs feed on rootlets and roots, being more serious during rains. Carbaryl 50 WP at 2 g/l spraying by evening controls the adults (Anon., 1990), however, Thimet 10 G or Phorate application in the soil will control the grubs. **Zinnia flea beetles** (*Longitarsus* spp., *Monolepta khasiensis, M. signata* and *Nisotra* spp.) feed on petals and leaves which can be controlled by spraying Malathion 50 EC at 2ml/l. **Whitefly** (*Bemisia tabaci*) nymphs and adults suck cell sap from leaves and flowers so the flowers are deformed and the leaves turn yellow to brown, curl, shrivel, die and fall. Methyl demeton 25 EC spraying at 3ml/l will control this pest. **Leaf miner** (*Liriomyza compositella*) mines the leaves and stems in zigzag fashion (Udayagiri, 1987), **blossom thrips** (*Microcephalothrips abdominalis*) feeds on flowers and distort them (Sood and Kakar, 1990), **black bean aphid** (*Aphis fabae*), **leafhoppers** (*Graphocephala coccinea, Macrosteles fascifrons*) and long-tailed **mealybugs** (*Pseudococcus adonidum*) suck plant sap and discolour and distort the leaves and flowers. Lady bird beetles (*Cryptolaemus montrouzieri, Scymnus* spp.) naturally feed on mealy bugs. These all the pests can be controlled by spraying Malathion 50 EC at 2 ml/l or Phosphamidon 85 WSC at 0.5ml/l. **Mites** (*Polyphagotarsonemus laetus, Steneotarsonemus pallidus* and *Tetranychus urticae*) damage the crop by rasping the sap from leaves and stems and spin webs on leaves and flowers so whole plant is disfigured and in severe infestation may die. Spraying of Dicofol 18.5 EC at 2 ml/l is quite effective in controlling these mites.

References

Agarwala, B. K. 1983. Parasites and predators of rose infesting aphids (Homoptera: Aphididae) in India. *Entomon*, **8**(1): 35-39.

Agarwala, B. K. and D.N. Raychaudhuri, 1981. Observation on the population trend of rose aphids, and their hymenopterous parasites in Kalimpong, West Bengal (India). *Entomon*, **6**(3): 211-214.

Amutha, S. 1994. Biotechnology and control of three pyralids, *Hendecasis duplifascialis* Hmpsn., *Nausinoe geometralis* (Guen.) and *Margaronia unionalis* Hbn. on *Jasminum* sp. Ph.D. Thesis. University of Madras, Chennai, Tamil Nadu, India.

Andersson, S. 1971. *Pratylenchus scribneri,* a root-damaging nematode in *Hippeastrum* (Swedish). *Växtskyddsnotiser*, **35**(4): 43-47.

Anon. 1951. The broad mite (*Hemitarsonemus latus*) as a pest of dahlias. *Agric. Gaz. N.S.W.*, **62**: 252-253.

Anon. 1971. Bulb Scale Mite. Advisory Leaflet from Min. Agric. Fish. Food, No. 456, 5 pp.

Anon. 1990. Package of Practices for Crops of Punjab, Kharif 1990. PAU, Ludhiana (Punjab).

Anon. 1991. Package of Practices for Crops of Punjab, Rabi 1990-91. PAU, Ludhiana (Punjab).

Atwal, A.S. 1986. Agricultural Pests of India and South-East Asia. Kalyani Publishers, Ludhiana.

Ayyar, T.V.R. 1963. Handbook of Economic Entomolgy for South India. Govt. of Madras, Chennai, India.

Bala, G. and F. Hosein, 1996. Plant parasitic nematodes associated with anthuriums and other tropical ornamentals. *Nematropica*, **26**(1): 9-14.

Bhatt, N. R. 1993. *Carnation*. In: *Commercial Flowers* (eds. Bose, T. K. and L.P. Yadav), pp. 355-416. Naya Prokash, Kolkata, India.

Bird, G.W. and N. Knobloch, 1976. Occurrence of *Trichodorus pachydermus* in Michigan. *Pl. Dis. Reptr*, **60**(1): 76-77.

Brickell, C. 1992. Plant Problems. In: *The Royal Horticultural Society Encyclopedia of Gardening*, pp. 545-572. Dorling Kindersley, London.

Butani, D. K. and N.K. Dadlani, 1983. Pests of ornamental plants in India and their control. *Pesticides*, **17**(3): 18–20.

Cadet, P., E. van den Berg, L. Nema and E. van den Berg, 1993. Parasitic nematodes of flower crops in Martinique. *PHM Rev. Hortic.*, No. 341, pp. 53-58.

Casagrande, R.A. and S. Livingston, 1995. Leaf lily beetle *Lilioceris tilii*. *Quart. Bull. North Amer. Lily Soc.*, **49**(3): 48-50.

Courcy Williams, M.F. De, 2001. Biological control of thrips on ornamental crops: interactions between the predaceous mite, *Neoseiulus* (Acari: Phytoseiidae) and western flower thrips *Frankliniella occidentalis* (Thysanoptera: Thripidae) in cyclamen. *Biocontrol Sci. & Tech.*, **11**(1): 45-55.

Daughtrey, M. and A.R. Chase, 1992. Chrysanthemum. In: *Ball Field Guide to Diseases of Greenhouse Ornamental*, pp. 30-39. Ball Publishing, Geneva, Illinois.

De Figueiredo, E.R., Jr. and H.F. Pereira, 1944. Notas sôbre *Xanthopastis timais* Cram. (Lep. Noct.), praga das Amarylidáceas (*Xanthopastis timais*, a pest of amaryllids). *Arq. Inst. Boil.*, **15**: 289-298.

Dole, J.M. and H.F. Wilkins, 1999. Pest Management. In: Floriculture Principles and Species, pp. 105-122. Prentice Hall, Upper Saddle River, New Jersey, USA.

Ellis, D., R. McAvoy, L.A. Ayyash, M. Flanagon and M. Ciomperlik, 2001. Evaluation of *Serangium parcesetosum* (Coleoptera: Coccinellidae) for biological control of silver leaf whitefly *Bemisia argentifolii* on poinsettia. *Fla Entom.*, **84**(2): 215-221.

Gill, J.S. 1981. Fighting the foliar nematode of chrysanthemum. *Indian Hort.*, **25**(4): 21-22.

Goo, M.Y.C. and B.S. Sipes, 1997. Host preference of *Radopholus citrophilus* from Hawaiian anthurium among selected tropical ornamentals. *HortSci.*, **32**(7): 1237-1238.

Goodey, J.B. 1951. A secondary piliferous layer on the roots of *Hippeastrum*. *Nature*, **167**: 822-823.

Harris, K.M. 1971. A new approach to pest control in the dahlia trial at Wisley. *J. Roy. hort. Soc.*, **96**: 200-206.

Hay, R. and K.A. Beckett, 1971. *Readrer's Digest Encyclopaedia of Garden Plants and Flowers*, pp. 141-149. The Reader's Digest Association Ltd., London.

Heinz, K.M. and S. Thompson, 1997. Use resistant varieties for chrysanthemum pest management. *GrowerTalks*, **60**(13): 82, 84, 86.

Jeff, A.E. 1956. Anemones. *Rep. Rosewarne exp. Hort. Stat.1952-55*, pp. 11-18.

Jhansi Rani, B. 2001. Ornamental Crops: Pest Management. In: *Handbook of Horticulture* (eds Chadha, K.L. and S.K. Bhattacharjee), pp. 825-833. ICAR, New Delhi.

Johansson, E. and H. Niedieck, 1964. *Scutellonema brachyurum*, a nematode pest new to Sweden (Swedish). *Växtskyddsnotiser*, **28**: 94-98.

Jose, F., R.M. Nachiappan and V. Selvarayan, 1999. Seasonal incidence of jasmine eriophyid mite, *Aceria jasmini* (Acari: Eriophyidae) and its influence on the foliar biochemical factors. *J. Acarology*, **14**: 45–46.

Karuppuchamy, P. 2001. Life stages and biology of different groups of mites. In: *Biodiversity and Management of Phytophagous Mites* (eds Rabindra, R. J., P. Karuppuchamy, G. Urnapathy, C. Chinnaiah and V. Balasubramani), pp. 118-123. Tamil Nadu Agricultural University, Coimbatore, India.

Karuppuchamy, P., G. Balasubramanian, P.C. Sundarababu and M. Gopalan, 1994. A potential predator of chilly mite, *Polyphagotarsonemus laetus* (Banks) (Tarsonemidae: Acari). *Madras agric. J.*, **81**: 552-553.

Karuppuchamy, P., N. Dhandapani, P.C. Sundarababu and M. Mohanasundaram, 1986. Efficacy of certain acaricides for the control of carmine spider mite, *Tetranychus cinnabarinus* (Boisduval) on jasmine. *Madras agric. J.*, **73**: 236–239.

Karuppuchamy, P. and S. Palaniswamy, 2006. Insect-Pests of Ornamental Plants and Their Management. In: *Advances in Ornamental Horticulture* (vol. 4): *Ornamental Crop Production Technology* (ed. Bhattacharjee, S.K.), pp. 139-195. Pointer Publishers, Jaipur, India.

Korneeva, I. T. 1971. The development of mosaic virus symptoms in dahlias. *Byulleten' Glavnogo Botanicheskogo Sada*, No. 82, pp. 120-123.

Lindquist, R. K. 1983. New greenhouse pests with particular reference to the leafminer *Liriomyza trifolii*. In: *Proc. 10th International Congress Plant Protection*, pp. 1087-1094. Held at Brighton, England.

Lindquist, R.K. 1989. Insects and related pests. In: *Tips on Growing Potted Chrysanthemums* (eds Tayama, H.K. and T.J. Roll), pp. 40-49. Ohio Coop. Ext. Serv., Columbus, Ohio.

McMahon, R.W., R.K. Lindquist, B.D. Baith, T.L. Makin and M.L. Casey, 1994. Comparison of commercially and locally produced *Encarsia formosa* Gohan on the control of sweet potato whitefly on poinsettias. *Hort. Tech.*, **4**(3): 295-298.

Misra, Sanyat; Pragya; and R.L. Misra, 2006. Carnation. In: *Advances in Ornamental Horticulture* (eds Bhattacharjee, S.K.), pp. 66-80. Pointer Publishers, Jaipur.

Nair, M.R.G.K. 1975. Insects and Mites of Crops in India. ICAR, New Delhi.

Nandakumar, C., P. Reghunath and K.R. Sheela, 1988. On the control of red scale *Aonidiella aurantii* Muskll. on rose. *Entomon*, **13**(3-4): 275–277.

Nayar, K.K., T.N. Ananthakrishnan and B.V. David, 1976. *General and Applied Entomology*. Tata McGraw Hill Pub. Co., New Delhi.

Neiswander, R.B. 1955. Control of insects results in more beautiful dahlias. *Ohio Fm Home Res.*, **40**: 80-81.

Olisevič, G.P. 1959. The use of chemicals with systemic activity for the control of aphids under glass. *Bjull. Glav. Bot. sada*, No. 35, pp. 91-95.

Olson, D.L. and R.D. Oetting, 1999. The efficacy of myco-insecticides of *Beauveria bassiana* against silver leaf whitefly on poinsettia. *J. Agric. & Urban Ent.*, **16**(3): 179-185.

Osborne, L.S. and Z. Landa, 1994. Utilization of entomogenous fungus, *Paecilomyces fumosoroseus* against sweet potato whitefly, *Bemisia tabaci*. Bull. OILB/SROP, **17**(3): 201-206.

Parvatha Reddy, P. 2001. Nematode Management in Horticultural and Plantation Crops. In: *Handbook of Horticulture* (eds Chadha, K.L. and S.K. Bhattacharjee), pp. 888-895. ICAR, New Delhi.

Patel, R. B. 1988. Biology of rose thrips (*Retithrips syriacus* Mayet) and chemical control of pest complex of rose. *Thesis Abstract*, **15**(3): 288–289.

Pearman, J.A. 1959. Diseases of dahlias. *Agric. Gaz. N.S.W.*, **70**: 369-375.

Perret, J.E. 1949. Essais sur deux insectes floricoles. *Rev. hort. Paris*, **121**: 138-13

Prasad, D., K.M. Singh, R.N. Singh and D.N. Methu, 1983. A new predator of new pest of jasmine in Delhi. *Bull. Ent.*, **24**: 140-141.

Pratibha Sharma and S.K. Bhattacharjee, 2002. Plant Protection in Ornamental Crops. AICRP on Floriculture, Tech. Bull. No. 19, pp. 1-116. AICRP(F), ICAR, Division of floriculture and Landscaping, IARI, New Delhi-110012.

Ramakrishnan, T. A. 1984. *The Indian Rose Annual,* **3**: 64–70.

Ramesh, P. 1994. In: *Pests of Floriculture Crops and Their Control.* Kalyani Publishers, New Delhi, India.

Ranjith, A.M. 2013. China aster. In: *Identification & Management of Horticultural Pests,* pp. 131-132. New India Publishing Agency, Pitam Pura, New Delhi-110 088.

Rice Mahr, S. E., R.A. Cloyd, D.L. Mahr and C.S. Sadof, 2001. In: *Biological Control of Insects and Other Pests of Greenhouse Crops.* North Central Regional Publication 581. University of Wisconsin, USA.

Runia, W. Th. and J.J. Amsing, 1996. Disinfestation of nematode-infested recirculation water by ozone and activated hydrogen peroxide. *Intern. Soc. Soilless cult* (*Proc. 9th intern. Congr. Soilless culture*). St. Helier, Jersey, Channel Islands, April 12-19, pp. 381-393.

Salinger, J.P. 1985. *Commercial Flower Growing,* pp. 163-179. Butterworths Horticultural Books, Wellington, New Zealand.

Scopes, N.E.A. 1979. *Pests, Diseases and Nutritional Disorders of chrysanthemums.* National Chrysanthemum society, London.

Singh, S. P. 1985. *Heliothis armigera* (Hub.) as a pest of ornamental plants in Shimla Hills. *Indian J. Plant Prot.,* **II** (1-2): 95.

Smith, F.F. 1967. Controlling Insects of Flowers. *Agric. Ext. Bull. No. 237.* U.S. Deptt. Of Agriculture, USA.

Sohi, A. S. and D. Singh, 1995. Insect-pests of ornamental plants. In: *Advances in Horticulture,* vol. 12 (eds Chadha, K.L. and S.K. Bhattacharjee), 913 pp. Malhotra Publishing House, New Delhi 110005, India.

Sood, A. K. 1990. Insect-Pests of Ornamental Plants in Himachal Pradesh. M.Sc. Thesis. Dr. Y. S. Parmar University of Horticulture & Forestry, Solan. *Abstract,* **16** (1): 29–30.

Sood, A. K. and K.L. Kakar, 1990. Record of insect and non-insect pests of ornamental plants from Himachal Pradesh. *J. Insect Sci.,* **3**(2): 141–145.

Sunderland, K. D., R.J. Chambers, N.L. Helyer and P.I. Sopp, 1992. Integrated pest management of greenhouse crops in Northern Europe. *Hort. Rev.,* **13**: 1–66.

Tewari, G. C. and N.J. Mohan, 1981. Studies on the pests of important ornamental and medicinal plants. *Annual Report.* IIHR,Bangalore (India), pp. 74–75.

Udayagiri, S. 1987. First record of *Liriomyza compositella* Spencer and *L. brassicae* (Riley) (Diptera: Agromyzidae) as a pest of ornamental crops in South India. *Entomon,* **12**(2): 85-88.

Waters, W.E. and C.A. Conover, 1969. *Chrysanthemum Production in Florida* (Bulletin 730). Agric. Exp. Stat., Instt. of Food and Agric. Sci., Univ. of Florida, Gainesville, Florida.

Wickizer, S.L. and M.V. Meisch, 1977. Effects of non-commercial miticides against mites on dahlias and soybeans. *Arknsas Farm Res.,* **26**(6): 13.

Widmer, R.E. 1965. Systemic insect control for dahlias. *Minn. Hort.,* **93**: 26-27.

Wilhelm, C. 1997. Beneficials in the cyclamen crop take over the securing of quality. *TASPO Gartenbaumagazin,* **6**(2): 16.9.

Woodville, H.C. 1953. Damage to anemones by field mice. *Plant Path.,* **2**: 21-24.

Post Harvest Management of Loose Flowers, Cut Flowers and Cut Foliage

Kushal Singh, Pardeep Kaur, R. K. Dubey and Ranjit Singh

Introduction

Cut flowers and greens constitute an important component of floriculture industry the world around. With the development of faster transportation systems, management of growing environment and advancement in postharvest management practices, the volume of trade in cut flowers and foliage has shown considerable increase. International trade of floricultural products ranges from US$40 to 60 billion (Source www.interacen.org). The major importing countries in 2010 included Germany, France, Holland, USA and UK whereas major exporting countries were Holland, Germany, Italy, Belgium and Denmark.

In India, domestic trade of floricultural products has been reported to be around 1,000 crores. According to the estimates, in India the flower crops are grown over an area of about 1,91,000 hactares with the production of 10,31,000 MT loose flowers and about 69,027 lacs cut stems. The value of exports of floricultural products during 2010-11 (up to December) were Rs.4317.28 crores.

The important flowers which dominate the consumer market include rose, chrysanthemum, carnation, orchids, antirrhinum, gerbera, freesia, anthurium and bird of paredise. Apart from these, a wide range of florist greens like ferns, species of asparagus, ruscus, cordyline, *Hedera helix*, dracaena, cyperus species and some wild grasses as well as cut foliage and branches from some woody plants such as eucalyptus, bottlebrush, euonymus, murraya, sago palm, *Cycas circinalis, Livistonia* etc. are also used. Ferns and asparagus remain the most frequently used florist greens due to symmetry of their leaves and mass filling effect. There is also continuous search for new genera and species which are added to the trade every now and then. Flowers like jasmine, marigold, rose, annual chrysanthemum, tuberose and crossandra are used as loose flowers. These flowers are used for offerings in the religious places, and for making garlands, *venis* and *gajras*. Though consumption of flowers and cut foliage is mainly concentrated in the developed countries, developing economies have also witnessed substantial increase in their usage. The global cut flower market is highly competitive, hence the cost of production is a determining factor for export. The cultivation of cut flowers has taken a major shift to many developing countries, especially near the equatorial region where natural climatic conditions are favourable for growing of flowers and human resources are also available at a lower cost. As a consequence, the countries of Central America, Kenya, Ghana, South Africa and Columbia have become major exporters of cut flowers. The trade in cut flowers is quality-based for which utmost care is required to be taken during production as well as post harvest handling practices.

Flowers are highly perishable horticultural commodities. Unlike fruits and vegetables, the flower is composed of heterogeneous parts like petals, sepals, pistils and stamens, each appearing to play a significant role in senescence of petals. Different terminologies such as longevity, shelf life, vase life and keeping quality are used for evaluation of post harvest quality of cut flowers and foliage. The termination of vase life is marked by the wilting of petals, abscission of petals or whole flower or yellowing or blackening of leaves depending upon the species. Similarly, loss of chlorophyll indicates loss of shelf life of green foliage. Factors involved in the decline of vase life of flowers as well as foliage include loss of carbohydrates due to high respiration rates, water deficit conditions induced during post harvest processing chain, sensitivity to ethylene, incidence of pests and diseases, quality of irrigation water and pollutants present in the atmosphere or irrigation water (Woltering and van Doorn, 1988; Nowak and Rudnicki, 1990; Bhattacharjee, 1999; Singh *et al.*, 2001a; Bhattacharjee and De 2006; Singh and Srivastava, 2006). Important aspects of pre- as well as post-harvest management practices of cut flowers and cut foliage are discussed in this chapter.

Pre-Harvest Factors

1. Genetic Factors

Considerable variations occur in the longevity of cut flowers and greens belonging to different species. Varietal differences in longevity have been observed in many species. Some flowers belonging to families Asteraceae and Orchidaceae exhibit longer vase life while flowers like *Iris, Narcissus* and *Tulipa* exhibit shorter vase life. Variations in vase life also occur among the cultivars. The genetic compliment of any flowering species has pronounced effect on the longevity of the flowers. Any gene involved in the natural senescence process, as well as in a response to the stresses that the flower is subjected to by cutting, could affect longevity and flower quality. Besides, metabolic efficiency and structural variations among the species account for inherent variations in vase life. In rose cultivars, the efficiency of the stomatal mechanism which helps the cut stem to maintain high water potential has been shown to account for genetic differences in their longevity. 'Baccara' roses, which have long vase life possess considerable ability to close their stomata upon cutting than the 'Golden Wave' roses which possess short vase life (Mayak *et al.,* 1974). The differences in flower life among rose cultivars have also been ascribed to differential expression of ethylene genes (Müller *et al.*, 2000).

There also occur considerable variations in vase life of cut foliage. The stems of *Ruscus* last for up to two months unlike the other foliage. The species of *Asparagus* also differ in their vase life. The fronds of *A. setaceus* last for about 17 days whereas those of *A. densiflorus* only for 10 days. Similarly, fronds of *Nephrolepis exaltata* 'Bostoniensis' exhibited maximum vase life of 19 days as compared to *N. exaltata* (golden boston fern) with vase life of 10 days (Singh *et al.*, 2004a).

2. Environmental Factors

Light. Endogenous level of carbohydrates exhibits direct relationship with longevity of the flower or the foliage. Brighter light during the period of flower development will produce a cut flower with high carbohydrate level, and hence with longer vase life. Most ornamental plants require nearly 3,200 ft candles light intensity for growth. In general, high light intensities are accompanied by high temperatures, and hence are detrimental to plant growth. The flowers of rose harvested from the lower side of the plant where the leaves were shaded, had shorter shelf life than those cut from the top (Boodley, 1969). Similar effects were observed in chrysanthemums and the effects were totally overcome by the supply of metabolic sugars in the vase water (Halevy and Mayak, 1979). This indicates that carbohydrate status of the cut stem is a critical factor for determining its vase life. Reduced light intensity causes decrease in anthocyanin content and increase in bluing of petals in red cultivars of rose (Halevy and Mayak, 1979). The stems under conditions of low light intensity also grow abnormally tall (the phenomenon referred to as etiolation) and lack sufficient hardening. Low light intensity also causes loss of chlorophyll resulting in leaf abscission and shorter shelf life. Most of the cut green crops require shady or partial sunlight for production of good quality foliage.

Temperature. Temperature interacts with light in its effects on the production of cut flowers or foliage. High temperatures, especially if light is low, can lead to faster utilization of carbohydrate reserves and decreased keeping quality. The temperature requirements of different flower species vary depending upon the climate of their origin. However, moderate temperature conditions (20-25 ^{0}C day/10-16^{0}C night) are optimally suitable for commercial growing of many ornamentals. The flowers such as *Dianthus caryophyllus*, *Freesia*, *Lilium* and *Tulipa* require 20 ^{0}C day and 10 ^{0}C night temperature for proper plant growth. The difference of 10 ^{0}C day and night temperature is optimally required for raising a healthy crop. Ideal temperatures for most of the foliage plants range between 16 ^{0}C and 27 ^{0}C. Temperatures below 10 ^{0}C cause chilling injury especially in tropical foliage plants such as *Aglaonema*, *Aspidistra* and *Maranta*. Technology is now available for modification of temperature conditions to suit a particular crop by using protected structures such as shadenets, UV-stabilized polyethylene, glasshouses and different heating or cooling systems. Many flower crops are hence grown under protected conditions. However, the extent to which the control over modification of environment is to be exercised depends upon the natural climatic conditions prevailing in the region. The temperature requirements of important commercial cut flowers have now been worked out and exploited at a desired time and location. It is, therefore, possible to grow flowers outside their natural habitat by completely or partially modifying the environment. However, modification of the environment considerably adds to the cost of production which is not desirable from commercial viewpoint. As a consequence, in view of rising cost for the production of flowers, countries like Columbia, Kenya and many others in African continent have emerged as major exporters of cut flowers. It is, therefore, essential that before undertaking commercial cultivation of flower crops, mapping of prevailing climatic conditions is of utmost importance. It is imperative to identify regions where natural climatic conditions are optimally suitable for flower cultivation. This not only helps to cut down the production cost but also to increase production of high quality flowers.

Relative humidity. The cut flower crops should be grown under moderate relative humidity (60-80 per cent). High humidity is highly favourable for growth of bacterial and fungal pathogens, especially gray mould whereas low humidity accelerates the transpirational loss of water.

Fertilization and soil. Well-drained sandy loam soils with pH 6.5 are suitable for growing of a large majority of flowers. In greenhouses, however, the crops are also grown on artificial media. Judicious application of nutrients within the optimal range is an important factor for the production of good quality flowers. In plants like rose and gerbera which continue producing flowers over a longer period so plants require continuous supply of nutrients for sustained production of flowers. The nutrients in the greenhouse are therefore provided through automated fertigation systems. The deficiency and excess of nutrients adversely affect vase life. The excess is required to be drained out at regular intervals to prevent high salinity conditions in the root zone. Nitrogen deficiency produces pale-green leaves with low quality flowers. Potassium deficiency also decreases lasting life of cut stems. Micronutrient deficiency symptoms are frequently observed in many floral crops and are required to be addressed immediately. Some gladiolus varieties have been found to exhibit signs of acute deficiency of zinc (Arora *et al.*, 2002).

3. Insect-Pests and Diseases

Flower crops are attacked by a wide range of **pathogens** and **insect-pests** which adversely affect their quality. Seedlings or young plants are damaged by root-rot fungi such as *Pythium* and *Rhizoctonia*. Vascular diseases like fusarial wilt of carnation or gladiolus also cause substantial damage to the flowers and these are very difficult to control. The diseased flowers and leaves also produce more ethylene, lose more water and are also unacceptable in the market. Hence, the crop should be kept free of diseases and pests at all the stages of its growth cycle.

The **pollutants** present in the atmosphere also adversely affect the quality of the flowers and foliage. The ornamentals should not be grown near the industry which releases toxic gases such as nitric oxide, sulphur oxide, *etc.* as these absorb moisture from the air and form weak acids which corrode the plant tissues, especially petals. Growing of ormnamentals should also be avoided near dung heaps, garbage heaps and pits, sewage channels, chimneys, *etc.*

Harvest Factors

The physiological maturity of the bud, time and mode of harvest have considerable bearing on the post harvest keeping quality of cut flowers and foliage. The flower should be harvested at proper harvest maturity, time and using proper techniques.

1. Harvest Stage

Flowers should be cut at a right stage of development. The stages of harvest vary depending upon the species and cultivars. The flowers harvested too immature either do not open or are prone to certain physiological disorders. For example, the roses cut too immature suffer from bent neck. Gerberas harvested at bud stage do not last long due to lack of development of the cavity in the stem which provides an alternate pathway for the uptake of water. Flowers like *Dianthus caryophyllus, Freesia, Gladiolus, Iris, Lilium, Narcissus, Polianthes* (singles), *Rosa* and *Tulipa* can be harvested at bud stage because the buds of these flowers continue to open after harvest when the stems are placed in water. On the other hand, the flowers like *Calendula, Callistephus chinensis, Dahlia, Gaillardia, Gerbera, Gypsophila*, orchid, *Zinnia*, etc. should be cut at fully open stage as immature buds of these flowers fail to open after harvest. Some species such as *Antirrhinum, Delphinium* or *Lupinus* bear many flowers on the same stem. Such stems should be cut when at least one third to half of the flowers on the stem are fully open. Loose flowers such as marigold, rose and annual chrysanthemum are harvested when fully open whereas tuberose and jasmine are harvested at bud stage. The flowers of crossandra are harvested by pulling the corolla out of the calyx. The stage of harvesting the flower also depends upon the season and distance to the market. During summer, the flowers may be harvested at an stage earlier than usual of bud development as high temperatures accelerate opening. Most of the cut greens are harvested when fully mature. The suitable (optimal) stages of harvest for some important flowers and foliage are *Aster chinensis* (fully open), *Antirrhinum majus* (when 1/3rd florets are open), *Anthurium* (when just open), *Calendula officinalis* (fully open stage), *Cattleya* (3-4 days after opening), *Chrysanthemum morifolium* (in standards when outer florets are fully expanded, and in sprays when flowers open but before shedding of pollen), *Cymbidium* (3-4 days after opening), *Dahlia* (fully open), *Delphinium ajacis* (when half florets are open), *Dianthus barbatus* (when half of the florets are open), *D. caryophyllus* (standards-when flowers are half open, sprays-when at least 2 flowers on the stem are open), *Eustoma russellianum* (3-4 floret stage), *Freesia* (when opening of first bud initiates), *Gaillardia cristata* (fully open), *Gerbera jamesonii* (fully open and when outer two rows of heads start shedding the pollen), *Gladiolus* (at 4-5 bottom floret showing colour), *Gloriosa superba* (fully open), *Gypsophila paniculata* (just open), *Hippeastrum* (buds showing colour), *Iris germanica* (bud colour showing stage), *Lathyrus odoratus* (when half of the florets are open), *Liattris spicata* (when half of the florets are open), *Lilium* spp. (buds showing colour), *Limonium sinuatum* (fully open), *Lupinus hartwegii* (when half of florets are open), *Matthiola incana* (When half of florets are open), *Narcissus* (gooseneck stage), *Ornithogalum* (bud colour showing stage), *Phalaenopsis* (3-4 days after opening), *Phlox drummondii* (when half of buds are open), *Polianthes* (single- fully developed unopened buds, double-when 2/3rd florets are open), *Strelitzia reginae* (when fist floret is open), *Tulipa* (bud colour showing stage), *Zantedeschia* (when spathe begins to turn downwards), *Zinnia elegans* (fully open), *etc.*

2. Time of Harvest

Flowers should be harvested either in the morning or evening when temperatures are moderate. High temperatures accelerate the rate of respiration, water loss and ethylene production which shorten flower vase life. In the morning hours, the stems are fully turgid due to low transpirational loss of water during night time. The harvesting in the afternoon has also advantages like high content of endogenous sugars due to photosynthetic activity of the leaves during day time. The buds of tuberose, jasmine and crossandra and loose rose flowers are always harvested early in the morning as these flowers wither very rapidly. The flowering stems with foliage should preferably be harvested in the morning as such stems are highly prone to water loss. In case the stems are harvested in the morning, they should be dried of moisture/dew drops before transportation as the wet flowers are highly prone to infection with pathogens.

3. Mode of Harvest

The stem should always be harvested with smooth cut through use of sharp secateurs or knives. While harvesting, the stem should not get crushed because certain metabolites like sugars, proteins, amino acids, etc. released from the crushed portion of the stem lead to the growth of bacteria at the cut end. The hard stems should be given slanting cut to expose more stem surface for better water absorption. Proper guidelines should be followed for harvesting the flowers. For instance, the stems of chrysanthemum should be harvested at least 10 cm above the ground level. This is because the stem harvested at the ground level is highly lignified and absorbs less water. Similarly, while harvesting roses, sufficient leaf area should be retained on the plant for the production of sufficient photosynthates for sustained production of flowers. Some flowers such as *Dahlia* and *Euphorbia* produce milky latex upon cutting. The latex blocks the stem surface and decreases water uptake by the stem. Such stem should be dipped in alcohol or treated with hot water (85-90 ^{0}C) for few seconds to block the release of latex from the stem. Also there are flowers that require to be pulled up gently rather than being cut as alstroemeria.

Post Harvest Factors

1. Conditioning of Flowers

Conditioning is a process through which the cut stem is made to regain turgidity. The flowers during prolonged storage or transportation lose considerable quantities of water. Such stems are required to be rehydrated. These stems should be placed in warm water containing biocide or water acidified to pH 3-4. Before placing the stem in hydrating solution, the basal 2-3 cm portions of the stem should be given recut preferably under water to remove surface blockages. The stem should then be placed at cooler temperatures (2-4 ^{0}C) to prevent loss of water and to remove heat generated through respiration.

2. Cooling of Flowers and Foliage

High rate of respiration and high temperatures of most greenhouses starve the flowers of their carbohydrate content. It is therefore, necessary to cool the flowers as soon as possible after harvest. Cut foliage, however, require immediate cooling because they are more prone to water loss due to high transpirational surface area. The cooling of the flowers is practiced universally by the commercial growers, exporters as well as wholesalers. Most important way to maintain quality of harvested flowers is to ensure that they are cooled immediately after harvest and that optimum cool temperatures are maintained during distribution. Most flowers should be held at a temperature slightly above the freezing point. Chilling-sensitive flowers (anthurium, bird of paradise, ginger lily, tropical orchids, *etc.*) should be held at temperatures above 10 ^{0}C. Low temperature reduces respiration rate, maintains the carbohydrate content of the flowers, checks undesired opening of the bud, reduces water loss, ethylene production and growth of pathogens and hence helps to keep the flowers fresh for longer duration. Most of the flowers can be cooled slightly above the freezing point, i.e. above 0.5 ^{0}C. Most foliage should be cooled to about 4 ^{0}C.

3. Water Relations

Water is an important component of the flowering bud as well as cut foliage. Cellular turgidity of the petal or leaf is a main determinant of its vase life. Especially, the stems lose considerable quantities of water during storage, transit as well as in vase. The foliage greens or stems with large leaf area like those of rose and chrysanthemum are highly prone to water loss. In general, cut stem continues to exhibit decline in water absorption with time. As water loss exceeds water absorption, the stem develops water deficit conditions leading to wilting of petals as well as leaves (Halevy and Mayak, 1981; Singh and Moore, 1992a; Singh and Arora, 1996; Chawla, 2008). The loss of vase life is accompanied by decline in water potential of the bud (Singh and Moore 1992a; Singh and Arora, 1995; Kumar *et al.*, 2007a). The decrease in water absorption by the stem is caused by microbial growth, mainly bacteria in vase water or on the stem surface (Put, 1986; Zagory and Reid, 1986; Singh *et al.*, 2004b). The cut surface of a flower stem releases proteins, amino acids, sugars, and minerals into the vase water which provides an ideal food for the growth of bacteria, yeasts, and fungi. Microbial growth blocks the xylem tissue at the cut surface of the stem. Microorganisms also decompose cell wall components leading to occlusion of xylem vessels (Singh and Moore, 1992a). Microorganisms also produce ethylene and toxins which accelerate flower senescence. The decrease in water absorption by the stem and stomatal loss of water also leads to the development of air-filled spaces (emboli) within the xylem which causes cavitation and hinders the uptake of water in the stem. Emboli can, however, be removed by recutting 2-3 cm stems under water, acidification of the solution to pH 3 to 4 or by placing the stem neck deep in water. The stems possess the ability to regulate the closing of stomata in response to water stress. The cultivars in which the leaves lack this ability are more prone to water stress. Xylem occlusion also occurs by ethylene produced at the cut end of the stem as in maidenhair fern (Fujino *et al.*, 1981), wounding reaction that involves both peroxidase and catechol oxidase activity (Vaslier and van Doorn, 2003; Loubaud and van Doorn, 2004) and also involves deposition of lignin, suberin and tannins (Tjia and Funnel, 1986; van Doorn *et al.*, 1989). Water deficit conditions are known to cause physiological disorders in the cut stems such as stem break in gerbera and bent neck of rose (Halevy and Mayak 1981; Balestra *et al.*, 2005). Adverse water relations also cause changes in hormonal balance (van Doorn *et al.*, 1995).

Quality of water also considerably affects keeping quality of the flowers and the foliage. Hard water usually contains minerals that make water alkaline. High pH of water reduces water uptake by the stem. Total dissolved solutes (TDS) present in water also affect vase life of the flowers. Total dissolved solutes more than 700 ppm decrease vase life of gladiolus whereas at low concentration of 200 ppm, the vase life is adversely affected in carnation, chrysanthemum and rose (Faernham *et al.*, 1971; Waters, 1968a,b). Tap water also contains minerals and fluoride which are toxic to many flowers. Cut roses are also sensitive to sodium ions whereas water containing fluorides are injurious to flowers such as gladiolus, freesia and gerbera, chrysanthemum and rose. Deionized water, boiled tap water, distilled water or millipore-filtered water are considered appropriate for cut flowers. Acidification of water (pH 3 to 4) also helps to prolong longevity of flowers as low pH decreases bacterial population and increases the rate of water flow in the stem.

4. Use of Floral Preservatives

Use of chemicals for improving vase life of flowers has been in use since long when chemicals like disprin (acetyl salicylic acid), copper sulphate or common salt were used but with little or no success. Because of considerable technological advances made in this field, chemical compounds are now available which can be effectively used to extend vase life of flowers. Commercial formulations are also now available in many countries which can be used by the producer as well as the consumer for increasing vase life. The chemical treatments can be given in the form of pulsing, bud opening solutions or holding solutions and the composition of the formulation varies in different species (Halevy and Mayak, 1981; Nowak and Rudnicki, 1990; Singh *et al.*, 2001a).

(i) Pulsing Solutions

Pulsing refers to short duration treatment of the flowers with chemicals. The effect of such treatments lasts throughout the vase life of the flower. In general, treatments are given at moderate temperatures (20-23 ^{0}C) under well-lighted conditions for 16-20 hours. Pulsing solutions contain sucrose, relatively at higher concentrations. The concentration of sucrose for pulsing varies in different species and depends upon the endogenous levels of carbohydrates in the bud and is very useful for opening of immature buds which contain limited content of soluble sugars. The optimum concentration of sucrose in pulsing solution has been suggested to be 20 per cent for gladiolus and gerbera, 15 per cent for cyclamen, 10 per cent for carnation, bird of paradise and gypsophila and 2 to 5 per cent for rose and chrysanthemum (Halevy and Mayak, 1981; Bhattacharjee and De, 2006). Sucrose is used in combination with a biocide to prevent microbial growth. Considerable variations also occur with regard to sensitivity for applied sucrose among different species. In some species such as chrysanthemum and rose, foliage in particular are most susceptible to damage, *vis-à-vis* petals, by high concentration of sucrose. In general, the flowers raised under protected conditions and those harvested at optimum stage contain sufficient carbohydrates and hence, do not require exogenous supply of sucrose. In such cases, the stems can be given pulsing treatments with biocidal solutions like aluminium sulphate, slow chlorine releasing compounds or quaternary ammonium compounds to prevent the microbial growth or with anti-ethylene compounds to prevent damage caused by ethylene. Buds in flowers like rose exhibit accelerated opening under ambient conditions. Such flowers should be given pulsing treatments under cool temperatures (2-4^{0}C).

(ii) Bud Opening Solutions

The immature flower buds can be made to open in chemical solutions which are referred to as bud opening solutions. The bud opening solutions contain sucrose and biocide and may also contain growth regulators or mineral salts. The bud opening solutions have also been standardised in many flowers such as standard chrysanthemum, carnation, etc. and can be used for opening of the immature buds which otherwise do not open fully in water (Nowak and Rudnicki, 1990; Singh *et. al.*, 2001a). The immature buds of spray chrysanthemum can be made to open in a solution containing 2 per cent sucrose + silver nitrate (25 mg/l) + citric acid (75 mg/l) + 50 mg/l 6-benzyl adenine (Singh *et al.*, 2005a). Similarly, buds of standard chrysanthemum can also be opened in a solution containing sucrose and 8-HQC. The concentration of sucrose required for the opening of buds, however, varies with the cultivar and range from 2 to 30 per cent (Kofranek and Halevy, 1972). Similarly, the immature florets of gladiolus spikes cut at tight bud stage (when one or two florets on the spike show colour) can be made to open with a solution containing sucrose (2 to 4 per cent) and aluminium sulphate (400 ppm) (Singh *et al.*, 2001b).

(iii) Holding Solutions

Holding solutions refer to the chemical solutions in which the cut stem is placed till the termination of its vase life. The composition of the holding solution is more or less the same as that of the bud opening solution. But since the stem is placed in the solution for a longer period, low concentration of sucrose (1 to 2 per cent) is used in such solutions.

Chemicals Used for Improving Vase Life

Sugars. Petal senescence is generally accompanied by hydrolysis of starch and sugars as well as loss of dry weight. The sugar is especially used as sucrose in pulsing , bud opening as well as holding solutions. Mature petals contain reducing sugars as main constituent as absorbed sucrose is readily converted to reducing sugars in the petals. Another disaccharide trehalose also extends vase life of gladiolus (Otsubo and Iwaya-Inoue, 2000). Metabolic sugars like fructose and glucose also improve vase life of flowers (Bhattacharjee, 1998, 1999). Sugars are suggested to improve water balance and osmotic potential of flowers, provide energy for maintaining cellular homeostasis (Kumar *et al.*, 2007b) and also reduce sensitivity to ethylene (Mayak and Dilley, 1986; Paulin, 1986). Vase solution treatment containing sucrose maintained higher activities of antioxidant enzymes, superoxide dismutase (SOD) and glutathione reductase (GR), reduced lipoxygenase (LOX) activity and lipid peroxidation (TBARS) in the petals of the cut spikes of gladiolus (Singh *et al.*, 2007a). Sugars also maintain mitochondrial structure and function and membrane integrity, *vis-a-vis* increase ability of petals to absorb water. Requirement of sugars, however, depends upon the carbohydrate status of the flower bud. The immature flower buds in general contain insufficient soluble sugars. Redistribution of carbohydrates in flower tissue also regulates petal senescence. In daylily, the sugars and other components of the flowers were coordinately exported from the senescing corolla, thereby reducing the energetic cost of a large ephemeral flower (Lay-Yee *et al.*, 1992). In tuberose, substantial demand for carbohydrate in the developing inflorescence causes bud abortion and failure of florets to open unless additional carbohydrates are supplied (Naidu and Reid, 1989).

Biocidal Compounds. These are used to inhibit growth of microorganisms in vase water. Some of the biocides which have been used to extend vase life of flowers are 8-hydroxyquinoline citrate (oxine citrate), silver nitrate, dichloriphen, thiabendazole and organic acids such as citric acid and benzoic acid (Nowak and Rudcinki, 1992; Bhattacharjee and De, 2006). Besides, chlorine is also a very effective biocide but it dissipates very quickly from the solution. Therefore, slow chlorine releasing compounds like DICA (dichloroisocyanurate) and DDMH (1,3,-dichloro-5,5-dimethylhydantoin) are considered more useful. Other sources of chlorine, *viz.* calcium hypochlorite or sodium hypochlorite has also been found effective (Singh *et al.*, 2000, 2004b).

Mineral Salts. Mineral salts especially of potassium, being natural osmolytes are known to improve water balance and longevity of some flowers. Similarly, calcium nitrate, boric acid, nickel chloride, cobalt chloride and zinc ions are also found to improve vase life of some flowers. Cobalt and nickel ions also act as inhibitors of ethylene biosynthesis. Aluminium sulphate forms a constituent part of many preservatives. Aluminium sulphate acidifies the solution, inhibits microbial growth and also maintains water balance in the stem by causing stomatal closure (Nowak and Rudcinki, 1990; Singh *et al.*, 2001; Bhattacharjee and De, 2006).

Plant Growth Regulators

Auxins. Auxins do not appear to play significant role in prolonging vase life of cut flowers. Instead, indole-3-acetic acid (IAA) promotes petal senescence (Singh, 1994) by enhancing ethylene production (Nowak and Rudnicki, 1990). IAA is also known to activate ACC synthase enzyme (Yang, 1980).

Gibberellins. Gibberellic acid is reported to extend vase life of isolated petals of carnation (Garrod and Harris, 1978). Besides, gibberellic acid accelerates opening of immature buds of gladiolus (Rao and Mohan Ram, 1982), reduces ethylene production and delays senescence of cut carnation flowers (Saks *et al.* 1992). Gibberellic acid is also commercially used to prevent post harvest yellowing of leaves in monocotyledonous cut flowers such as *Alstroemeria* and *Lilium* (Serek and Reid, 1997).

Cytokinins. Cytokinins have been reported to delay senescence of various flowers (Halevy and Mayak, 1981). Benzyl adenine inhibits ethylene production from carnation petals (Mor *et al.*, 1985) and also delays increase in ion leakage from the petals of gerbera (Meeteren, 1979). Some rose cultivars with short vase life possess low levels of cytokinins than those with long vase life (Halevy and Mayak, 1981). There occurs inverse relationship between cytokinin content and senescence in some flowers like petunia, roses and carnation (Mayak and Halevy, 1972; van Staden and Dimalla, 1980; Lara *et al.*, 2004). Transgenic plants of petunia overexpressing IPT gene exhibit increased cytokinin content, less ethylene sensitivity and significant delay in flower senescence (Chang *et al.*, 2003). 6-benzyl adenine (BA) in combination with sucrose also facilitates opening of immature buds of chrysanthemum and also significantly delays the yellowing of leaves (Singh *et al.*, 2005a). BA also lowers decline in fresh weight and activities of hydrolytic and proteolytic enzymes in chrysanthemum florets (Elanchezhian and Srivastava, 2001). BA and GA_3 also extended longevity of *Asparagus myrie* and *A. densiflorus* (Singh and Singh, 2002). Thidiauron, a substituted phenyl urea compound also retards post harvest senescence of leaves in cut stems (Ferrante *et al.*, 2002, 2009).

Polyamines. Polyamines which are endogenous growth regulators are also known to possess anti-senescence properties (Pandey *et al.*, 2000). Spermine and spermidine when applied with sucrose increased vase life of gladiolus florets and also retained the activities of superoxide dismutase (SOD), catalase, gluthione reductase and ascorbate peroxidase (Singh *et al.*, 2005b). Spermidine also delayed the senescence of cut carnation flowers by reducing ethylene production, ACC content and the activities of transcript amounts of ACC oxidase and ACC synthase in petals (Lee *et al.*, 1997). Treatment with polyamines does not always increase flower longevity and may even enhance senescence (Downs and Lovell, 1986). No differences in free polyamines were observed in climacteric and non-climacteric flowers of carnation (Downs and Lovell, 1986; Serrano *et al.*, 1991). In the light of these reports, the precise role of polyamines in the regulation of petal senescence needs further investigation.

Growth retardants and inhibitors. Growth retardants such as daminozide and chlormequat chloride have also been reported to extend flower vase life though their mode of action is not fully understood. They are assumed to act through increasing the tolerance to environmental stresses and interacting with native phytohormones (Halevy and Mayak, 1981). Maleic hydrazide is reported to cause significant reduction in abscission of flower buds and increase floral vase life of *Lupinus hartwegii* (Mohan Ram and Rao, 1977).

Nitric oxide (NO). It is a diffusible natural plant growth regulator which plays an important role in hormonal modulation, alleviation of biotic and abiotic stresses and programmed cell death (Setia and Setia, 2006). The role of NO for improving shelf life of cut flowers came to light with the discovery that Viagra, a male fertility drug extended vase life of flowers by retarding the breakdown of cyclic gnanosine monophosphate (cGMP), the production of which is mediated by nitric oxide (Judy, 1999). Nitric oxide donor compound 2,2′-(hydroxynitrosohydrazino)-bisethanamine (DETA/NO) was subsequently found to extend vase life of flowers like chrysanthemum, gerbera, iris, larkspur, lily (Oriental), rose, snapdragon and tulip cut stems (Badiyan *et al.*, 2004). The effect of two NO generating compounds, *viz.* sodium nitroprusside (SNP) and N-tert-butyl-α-phenylnitrone (PNB) on postharvest flower senescence in four advanced breeding lines of *Lupinus havardii* was found to be dependent upon the concentration used and the genotype (Sankhla *et al.*,

2005). NO donors and sucrose, when used in combination, reduced degree of flower senescence. Nitric oxide seems to have commercial potential because solid NO releasing water soluble compounds are available.

Abscisic acid (ABA). ABA is a natural plant growth inhibitor which plays an important role in inducing seed and bud dormancy, maturation of embryo in germinating seeds and closure of stomata during water stress conditions, and in general stimulates senescence of flowers. ABA concentration is found to increase in ageing petals of rose and *Hibiscus* (Kumar *et al.*, 2008a). ABA, however, delays petal senescence in cut roses under well-lighted conditions, by inducing closure of stomata (Kohl and Rundle, 1972) and also reduces leaf damage caused by sucrose in cut 'Kardinal' roses (Markhart and Harper, 1995; Pompodakis and Joyce 2003). The effect of ABA on vase life of cut roses was found to be pH dependent (Pompodakis *et al.*, 2004). ABA increases vase life and reduces vase solution usage of flowers held in low pH (pH 6) solution, indicating induction of stomatal closure. ABA supplied at pH 8 reduces vase life which was associated with development of vase solution microbes at high pH. Pulse treatment with sucrose plus abscisic acid suppressed leaf damage and improved the quality of cut *Eustoma* flowers (Shimizu-Yumoto and Ichimura, 2009).

Ethylene. Ethylene, a simple hydrocarbon gas and plant hormone is present in the atmosphere as a pollutant. Ethylene is generated by senescing and abscising plant organs, mature flowers, ripening fruits and vegetables, diseased, mechanically injured and decaying plant tissues and microorganisms. Ethylene is also produced from coal gas, petroleum gas, exhaust fumes of internal combustion engines, *etc.* (Halevy and Mayak, 1981; Nowak and Rudnicki, 1990; Singh *et al.,* 2001). Concentration of ethylene in the atmosphere varies with the season and the location. Concentration is usually high during autumn and winter as well as at the airports, wholesale distribution warehouses and in the super markets (Morris *et. al.,* 1978; Halevy and Mayak, 1981). Ethylene induces fading, wilting or abscission of petals or whole flowers. It also induces yellowing of leaves and shortens vase life of many flowers considerably (Mor *et. al.*, 1984; Reid, 1989; Nowak and Rudnicki, 1990; Singh, 2000; Singh *et al.*, 2001). In orchids, ethylene causes wilting of sepal tips known as 'dry-sepal' in *Cattleya* and also increases anthocyanin content in labella and gynostemia in *Cymbidium*. The response of ethylene in flower senescence appears to be of physiological in nature as flowers belonging to specific families as Campanulaceae, Caryophyllaceae, Geraniaceae, Labiatae, Malvaceae, Orchidaceae, Polemoniaceae, Ranunculaceae, Rosaceae (only some cultivars of rose are sensitive to ethylene) and Scrophulariaceae are sensitive to ethylene whereas those belonging to Asteraceae, Euphorbiaceae, Iridaceae and Liliaceae do not show sensitivity to ethylene (Woltering and van Doorn, 1988). The list of flowers and cut foliage which show sensitivity to ethylene are *Alstroemeria, Antirrhinum majus, Cattleya, Cymbidium, Delphinium ajacis, Dianthus caryophyllus, Euphorbia pulcherrima, Freesia hybrida, Gypsophila paniculata, Lathyrus odoratus, Lilium* species, *Lupinus hartwegii, Narcissus* spp., *Phalaenopsis* spp., *Phlox drummondii, Petunia hybrida, Rosa hybrida* (some cultivars), *etc.* though insensitive ones to ethylene are *Anthurium andreanum, Chrysanthemum morifolium, Gerbera jamesoni, Gloriosa superba, Gladiolus* cultivars, *Helicornia, Hippeastrum, Iris germanica, Paeonia, Polianthes tuberosa, Protea* hybrids, *Strelitzia reginae, Tulipa* spp., *Zantedeschia* spp., *etc.* Ethylene is synthesised in plant tissues from methionine, a sulphur containing amino acid, through intermediate compounds, SAM (S-adenosyl methionine) and ACC (1-aminocyclopropane-1-carboxylic acid). These steps are catalysed by enzymes ACC synthase and ACC oxidase or EFE (ethylene forming enzyme), respectively (Adams and Yang, 1979).

In ethylene-sensitive flowers, ethylene present in the atmosphere or that produced by the flowers themselves adversely affect their vase life (Halevy and Mayak, 1981; Reid, 1989; Nowak and Rudnicki, 1990; Nell, 1994; Singh *et al.,* 2001). Ethylene-sensitive flowers like those of carnation are reported to exhibit extreme sensitivity to as little as 10 ppb of exogenous ethylene (Woodson, 1991) though variations, however, occur within the species or cultivar, e.g. in roses ethylene accelerates bud opening in cvs Angel, Candia, Capella, Cerisa, Excitement and many others whereas prevents opening of buds in cvs Chantilly Lace, Lovely Girl and St. Louis, though the response varies with the cultivar, however, in some cultivars like Golden Times, Grace and Lady Diana it does not affect opening of buds (Reid *et al.,* 1989). Orchid flowers also exhibit wide range of sensitivity to ethylene as *Vanda* 'Miss Joaquim' is most sensitive, *Cattleya, Cymbidium* and *Paphiopedium* less sensitive, while *Dendrobium* and *Oncidium* are insensitive (Goh *et. al.,* 1985). Exogenously-applied ethylene catalyses the biosynthesis of ethylene from the flowers which is referred to as 'autocatalysis'.

Flower exhibits three phases of ethylene production such as (i) phase of low ethylene production in buds; (ii) increase in ethylene production as the flower matures; and (iii) decline in ethylene production in senescing flowers. Likewise, immature flower buds are less sensitive to exogenously-applied ethylene than the mature flowers. Pollination of flowers stimulates well-programmed sequential production of ethylene from different flower parts but the sequence of ethylene production varies in different types of flowers depending upon their structural organization. In carnation, pollination stimulates ethylene production initially in style and stigma. The stimulus for ethylene production is further transmitted to the ovary, receptacle and petals in chronological order. Style and petals are the main contributors towards total ethylene produced by the flowers (Nichols *et. al.,* 1983; Reid, 1989). In Papilionaceous flower of sweet pea, however, the sheath enclosing the staminal tube contributes largely towards ethylene production, in addition to the style and petals (Singh and Moore, 1994a). Increase in ethylene production is also induced by emasculation, dislodgement of pollinia and auxin (Arditti, 1979). Though initial ethylene production may involve existing RNAs and proteins, subsequent production requires *de novo* synthesis (Arditti, 1979). Pollination-induced ethylene released by the flowers induces 'autocatalytic' ethylene production which initiates physiological and biochemical changes in the petals leading to their senescence. The mechanism through which ethylene production is initiated following pollination is a well-defined and complicated process. High levels of auxin present in the pollen may provide stimulus for enhanced ethylene production by flowers (Burg and Dijkman, 1967) but its movement down the stigma has been reported to be very slow in pollinated flowers (Reid *et. al.,* 1984). High ACC level in the pollen acts as stimulus for ethylene production (Whitehead *et al.,* 1983). Though high ACC level in the pollen is also assumed to cause initial burst in ethylene production, it does not explain sustained production of ethylene from the flowers. *Cosmos* pollen which contains very little ACC also greatly stimulates ethylene production by *Petunia* stigmas (Hoekstra and Weges, 1986). There is thus an evidence to suggest that *de novo* synthesis of ACC is involved in pollination-induced ethylene burst (Hoekstra and Weges, 1986). Flowers of *Cyclamen* offer an interesting example of ethylene-induced senescence in which ethylene or ACC does not induce senescence until the flowers are pollinated (Halevy *et. al.,* 1984). On the other hand in *Petunia,* wounding of stigma and style also mimic pollination effects (Whitehead *et. al.,* 1984). It might thus appear that the nature of stimulus that induces ethylene production in flowers varies with the species. In flowers sensitive to ethylene, vase life can be increased by treatments with inhibitors of ethylene action or synthesis. The flowers insensitive to ethylene appear to have developed an inherent system through which they do not respond to ethylene. The flowers like chrysanthemum do not exhibit climacteric rise in ethylene production (Singh and Moore, 1992 b, 1994b). Exposure of these flowers to ethylene also does not affect vase life (Singh and Moore, 1992 b; Singh and Arora, 1996). In gladiolus which is another ethylene-sensitive flower, floret senescence is also not coordinated by an increase in ethylene biosynthesis though in some cultivars, a moderate climacteric-like peak in ethylene production accompanies floret opening and senescence (Serek *et al.,* 1995a). Senescence in ethylene-insensitive ephemeral flowers of daylily (*Hemerocallis* spp.) is found to be associated with up-regulation of specific genes and at least one of the mechanisms for hydrolysis of corolla proteins (Valpuesta *et al.,* 1995).

Ethylene antagonists. Many of the inhibitors of ethylene action (silver thiosulphate and 1-MCP), and biosynthesis (Ni, Co, aminooxyacetic acid and aminoethoxyvinyl glycine) have been found to improve vase life of flowers (Baker *et al.,* 1977; Brown and Mayak, 1980; Wang and Baker, 1980; Serek *et al.,* 1995a,b). Anionic complex of silver thiosulphate (STS), an inhibitor of ethylene action has been widely used to enhance vase life of flowers (Woltering and van Doorn, 1988; Nowak and Rudnicki, 1990). The use of STS, however, is not recommended due to environmental concerns arising from the presence of silver in silver thiosulphate (Nell, 1992, 1994; Serek *et al.,* 1995b). 1-Methylcyclopropene (1-MCP) which is an irreversible inhibitor of ethylene binding has been recommended to extend shelf life of a range of cut flowers and potted flowering plants (Serek *et al.,* 1995a,b). Being non-toxic to humans, stable at normal conditions and effective at low concentrations, it has tremendous commercial possibilities. 1-MCP is available under the name EthylBloc® and SmartFresh, the former is aimed especially for ornamental crops, and the latter for edible products (Serek *et al.,* 2006). High CO_2 concentration (4 %) also acts a competitive inhibitor of ethylene action.

Methyl jasmonate. Methyl jasmonate, another naturally occurring plant growth regulator has been reported to accelerate ethylene production and senescence in flowers of *Dendrobium* and *Petunia* (Porat *et al.,* 1993).

Refrigerated Storage

Storage is an important operation for orderly marketing of flowers. Therefore, development of suitable storage methods is essential for regulating supply of flowers in the market especially, during periods of overproduction and lean demand.

Various factors such as stage of harvest of the flower, quality of the cut stem, temperature, relative humidity and air circulation in the storage chamber have considerable bearing on the storage behaviour of cut stems (Nowak and Rudnicki, 1990; Singh *et al.*, 2001). The flowers and greens meant for storage must always be cut at an appropriate stage of harvest and should be free from any injury. The stem should also be free from symptoms of diseases and pests as well as deficiency of nutrients. Harvest maturity considerably influences the storage behaviour. The stems harvested with fully open flowers possess shorter vase life after storage than those harvested at bud stage. The flowers such as gladiolus, iris, narcissus, rose and tulip should be stored at bud stage as the buds of these flowers possess the ability to open after storage. Flowers like those of chrysanthemum (standard), dahlia, gerbera and orchids should be harvested when fully open. Relative humidity in the storage chamber should be kept high (90-95 %) to prevent desiccation of the flowers and foliage. Precaution is needed to be taken because conditions of high humidity are also favourable for incidence of gray mould.

The storage chambers should be kept free from ethylene by periodic ventilation and by disinfecting the same at regular intervals. The decaying plant debris and pollinated flowers should be removed from the storage rooms since these are potential sources of ethylene evolution. Ethylene absorbents like brominated activated charcoal or $KMnO_4$ which oxidizes ethylene can also be used to reduce ethylene levels in storage rooms. The fruits and vegetables which release considerable quantities of ethylene should not be stored along with flowers. There are three methods for storage of flowers and cut foliage:

Normal refrigerated storage. In normal refrigerated storage the flowers or foliage are stored either wet or dry. **Wet refrigerated storage** implies storage by placing the stems in containers of waters or in a floral preservative solution. In wet storage, flowers occupy more space in the cold chamber. The bud development and senescence of flowers also occur more rapidly during wet storage than in dry storage (Goszcznska and Rudnicki, 1988). Wet storage is a widely adopted method as it is very useful for day to day handling of flowers (Nowak and Rudnicki, 1990; Mengüc and Zancirkiran, 1996). Stems of asparagus, dahlia, freesia, gerbera and iris are highly suitable for wet storage (Nowak and Rudnicki, 1990). **Dry refrigerated storage** means storing of flowers in water-retentive plastic films which has potential for long duration storage and is more economical than CA-storage or low pressure storage, both of which require specialized and cost-intensive infrastructure. Since flowers stored in polymeric packages take long time to cool, the stems should be pre-cooled before dry storage (modified atmosphere storage, MA) because the composition of CO_2 and O_2 inside the packages experiences changes due to respiratory activity of the cut stem/foliage. The storage response of flowers, however, depends upon the ultimate concentration of CO_2 and O_2 inside the package during storage (Goszczynska and Rudnicki, 1988; Nowak and Rudnicki, 1990; Grover, 2001; Singh *et al.*, 2001a). The high concentration of CO_2 (5-7 %) and low levels of O_2 ($\neq$1%) in the storage packages are considered suitable for better storage of flowers. Since polymeric film sleeves exhibit considerable variations with regard to their permeability to CO_2 and O_2, there is a need to identify the polyfilms which can optimally modify the storage conditions inside the package (Grover, 2001). Flowers like gladiolus and carnation are suitable for storage under dry conditions. Loose flowers of marigold and annual chrysanthemum can be stored at 2-4 ^{0}C for 7-10 days whereas those of rose for 2 to 3 days. For storage, the flowers should be packed in plastic crates or bamboo baskets lined with wax paper, newspaper or polythene sheet and covered with perforated polythene or wet gunny bags. The buds of tuberose and jasmine can be stored in perforated polythene sleeves at about 8 to 10 ^{0}C for about one week. MA storage can offer an alternate method of transport especially surface transport of the cut stems.

Controlled atmosphere (CA) storage. Controlled Atmosphere (CA) Storage refers to storage in gas tight chambers equipped with cooling system in which the concentration of CO_2 and O_2 is strictly monitored. In CA storage, concentration of CO_2 is kept high (5-10 %) and that of O_2 low ($\neq$ 1%). The high CO_2 concentration helps to maintain turgidity of the cut stem by inducing the closure of stomata and also acts as competitive inhibitor of ethylene action. Low levels of O_2 on the other hand decrease destructive oxidative processes and respiratory activity of the flowers and also inhibit the biosynthesis of ethylene. CA storage is not commonly used for storage of flowers and greens because of inconclusive results. The precise conditions for CA storage of different genera and species may also vary which makes difficult the storage of different flowers in the same storage chamber. High CO_2 concentrations also cause toxicity to the flowers especially under low temperature conditions. CA storage, moreover, requires considerable infrastructure and may be beyond the reach of many growers.

Low pressure storage (LPS). It refers to storage under conditions of reduced atmospheric pressure and continuous flow of moisture-laden air. LPS at 40-60 mm Hg has been found to be suitable for storage of flowers (Burg, 1973) but with variable results. Reduced levels of O_2 under LPS conditions also decrease ethylene production. Though LPS is considered suitable for prolonged storage of flowers, the real potential of LPS in flowers or cut foliage has, however, not been realized.

The temperature in the storage chamber is a major determining factor in the storage of flowers. The optimum temperature for storage varies in different species. Though most flowers can be stored at 0.5-1 ^{0}C temperature, those originating in tropical and sub-tropical climates require higher temperatures for storage (Goszcznska and Rudnicki, 1988). The cold chamber should be properly insulated to avoid temperature fluctuations which lead to changes in relative humidity causing condensation of water vapours on the stems. Such humid conditions favourably increase the growth of pathogens especially gray mould.

Flowers during storage are generally kept under dark conditions. Prolonged dark conditions cause yellowing of leaves in many cultivars of *Alstroemeria, Lilium, Dendranthema* and certain other species. The yellowing of leaves in these species can, however, be prevented by the pre-storage treatments with 6-benzyl adenine, thidiazuron or gibberellins though efficacy of the growth regulator varies with the species.

Flowers show decrease in vase life and ability of buds to open following storage (Mengüc and Zencikiram, 1996; Waithaka *et al.*, 2001; Grover *et al.*, 2005, 2006, 2008; Redman *et al.*, 2002). Pre-storage pulsing or post-storage treatments with vase solution have been found to be extremely useful for improving post-storage vase life of flowers (Goszcznska and Rudnicki, 1988; Mengüc and Zencirkiran 1996; Grover *et al.*, 2005, Singh *et al.*, 2007; Singh *et al.*, 2009). Pre-storage application of anti-ethylene compounds to ethylene-sensitive flowers is especially of utmost importance as ethylene build up during storage may accelerate the senescence process.

Grading, Packing and Transportation

The designation of standards for grading of cut flowers is one of the most important but controversial areas in their postharvest handling. Though the standards for cut/loose flowers as well as cut foliage have been suggested by the Economic Commission for Europe (ECE) based at Geneva and the Society for American Florists, there are no uniform international standards for cut flowers or cut foliage. The standards suggested for some important flowers by the Society for American Florists are given in Table 1.

Table 1. Standards for some cut flowers as suggested by the Society of American Florists (Staby *et al.*, 1978)

Gladiolus

Grade	Minimum number of florets	Spike length (cm)
Blue (Fancy)	16	107 and above
Red (Special)	14	97-107
Green (Standard)	12	81-96
Yellow (Utility)	10	81 and below

Carnation (standard)

Minimum flower diameter (mm)	Grades		
	Blue (Fancy)	Red (Standard)	Green (Short)
Tight	50	44	None
Fairly tight	62	56	None
Open	75	60	None
Minimum stem length (cm)	55	43	30

Though grades being followed locally are:

Grade	Stem length (cm)
A	> 45
B	<45 >30
C	<30

Chrysanthemum

Grading of fully open standard chrysanthemum cut flowers for packing depends on stem length & strength and floral colour & diameter as per Society of American Florists (SAF, c.f. Laurie *et al.*, 1968) as mentioned below though in Europe as per EEC, however, pompons are graded into 250-340 g bunches having several stems and the standards of equal size in groups of 10 or 12. Grade 'Yellow' is being omitted as it is almost the grade 'Green' with no specification for floral diameter.

Grade Name	Fancy	Standard	Short
Label colour	Blue	Red	Green
Minimum diameter	14 cm	12 cm	10 cm
Minimum length (flower + stem)	76 cm	76 cm	61 cm

Machin and Scopes (1978) have quoted a metric grade specification for sprays for bulk packing which is given below (stem lengths should not be less than 66 cm for the British markets and those less than 51 cm should be marked short).

Grade	Stems/sleeve	Specification
Gold	(10)	6 flowers or more out and some to come
Silver	(15)	4 or more flowers out and some to come
Bronze	(20)	3 flowers out and some to come
Make-up	(--)	All stems not covered above, filling sleeves to same extent as other grades

Rose

While grading, the stems, leaves and flowers should be blemish- and infestation-free, the leaves should be clean, shining, dark green and healthy with no any discolouration whatsoever or due to micronutrient deficiencies, too open or tight buds, damaged and bullhead types and those not conforming to the varietal specifications are rejected. and as per stem lengths these are graded. Long-stemmed varieties are graded from 40 cm onwards with a difference of 10 cm though smaller ones from 40-65 cm with a difference of 5 cm each. These are graded into 'blue' (the taller ones), 'red' (medium) and 'green' (smaller ones).

The flower should always qualify the quality requirement standards for the particular species. The quality of the flower or foliage is determined by the general appearance such as straightness and strength of the stem, size of the bud, freedom from defects, harvesting stage, uniformity, quality of the foliage, and over all the vase life. The length of the stem which is still a major quality standard for many cut flowers does not always exhibit positive relationship with quality of the flower. Weight of the bunch for a given length, however, reflects the flower quality in a suitable manner.

General Care

Flowers are normally bunched, except for anthuriums, orchids and some other speciality flowers. The number of flowers in the bunch varies according to the market requirement and flower species. In general, the stems are grouped into bundles of 10, 12 and 25 each. The individual bunch/box must contain flowers of the same species or cultivar at same stage of bud development with information regarding the species or cultivar and name of the supplier.

In general, the packing boxes are long, flat and telescopic. The bunches are packed in the boxes in such a way that flower heads face in opposite direction, i.e. towards the width. The second layer is placed opposite to the first layer. In the packing box, the layer of flowers should be covered with soft paper. The flower bunches are wrapped in 3-ply corrugated paper as in rose (Fig. 2) or individual flowers as in gerbera or bunches as in *Lilium* or carnation are enclosed in polythene sleeves (Figs 1, 3 & 4).

The flowers of gerbera are also packed in the boxes provided with special paper inserts for individual stems. The flower heads should be placed few centimetres away from the end of the box to protect the petals from bruising. Flowers such as anthurium and orchids can not be held at low temperatures as these flowers are sensitive to chilling

Fig. 1: Lilium packaging in polythene sleeves

Fig. 2: Carnation packaged in polythene sleeves

Fig. 3: Packaging of cut gerbera in polythene sleeves

Fig. 4: Stem break in gerbera

injury. In these flowers, stem bases are placed in specially designed plastic vials containing water or in moistened absorbent cotton, enclosed in wax paper or polythene foil. After packing, the box is closed with the lid and strapped properly. Mixing of different varieties or stems of different lengths in the same box should be avoided. The stems should be tightly held inside the boxes so that they are not subjected to jerk movements during transport. The boxes should be properly marked with symbols 'fresh flowers' on both sides so that they are handled with care. The dimension of the boxes depends upon the number and length of the stems to be packed and type of the cargo space available so that the boxes fit well into the available space. Most flowers are transported by air but for shorter distances, flowers can also be transported by refrigerated vans.

Before transportation, the flowers are required to be cooled. The simplest method is to pack the flowers in the boxes and cool them by keeping the boxes open in the cold room. This method is, however, time consuming, requires more space in the storage chamber and, hence is not popular.

Forced-air cooling of boxes is the most efficient and practiced method for cooling of flowers. Cool air is sucked or blown through the boxes with closeable flaps. The flowers should be packed carefully in such a way so as to allow the free movement of air through the box. Vacuum cooling is also used to cool the flowers. If the boxes are to remain in a cool environment after cooling, flaps may be left open to allow movement of the cool air in the boxes and to keep the rate of respiration low. In case of boxes meant for transport at ambient temperatures, the flaps should be sealed. Ice can also be placed in the boxes to keep the flowers cool. Care should, however, be taken to ensure that the ice does not touch the flowers and cause freezing injury.

Since loose flowers are grown primarily for the domestic market, there are no grades as well as standard packaging techniques. Most loose flowers are packed in gunny bags or plastic crates and transported under ambient conditions. Gladioli, snapdragons and some other species are often packed in vertical hampers to prevent geotropic bending which reduces their acceptability. Cubic hampers are used for upright storage of daisies and other flowers. A new packaging system, the 'Procona' system, uses plastic container and a cardboard sleeve and allows transport of flowers upright in water. This system is more expensive than traditional boxes but the presence of water is useful to keep flowers fresh, especially when they are not transported under proper temperature conditions.

Marketing

In our country, there is no organized marketing system of flowers or foliage though some efforts have been made in some states to develop Flower Auction Centres. Mostly, the flowers are marketed by growers directly to the retail outlets. One international flower auction centre has been started at Bengaluru by Karnatka Agro-Industry Corporation, though other at Mumbai by Maharashtra Agro-Industries which is yet to be operationalised. Similarly, flower market at Ghaziabad (Uttar Pradesh) is also operational. In the international market, flowers are auctioned in the auction centres. There are seven auction centres, i.e. in VBA Aalsmeer, Naaldwijk/Bleiswijk,

Rijnsburg, Grubbenvorst, Elde, Von Bemmel and Vleuten, the largest being in Aalsmeer. The flowers from different countries of the world are traded at these centres. Nearly 60 per cent flowers are traded in Aalsmeer and Naaldwijk/ Bleiswijk Auction Centres.

Physiological and Biological Basis of Post Harvest Life

Petal senescence is regulated by co-ordinated changes in gene expression and the later stages of senescence share many characteristics of programmed cell death (Rubenstein, 2000; Jones, 2004). During petal senescence, developmental and environmental stimuli enhance the up-regulation of catabolic processes causing breakdown and remobilization of cellular constituents.

The developmental events taking place during senescence also involve physiological changes such as loss of water from the senescing tissue, leakage of ions, transport of metabolites to different tissues, and biochemical changes such as generation of reactive oxygen species (ROS), increase in membrane fluidity and peroxidation, hydrolysis of proteins, nucleic acids, lipids and carbohydrates.

Membrane deterioration during senescence involves changes in membrane fluidity, leakage of ions and loss of differential permeability. There is increase in microviscosity and decrease in phospholipid content of membranes and increase in neutral lipids, mainly due to increased action of phospholipases and acyl hydrolases (Paliyath and Droillard, 1992). Another important event which leads to loss of membrane permeability is oxidation of membrane lipids due to lipoxygenases (Panavas and Rubenstein, 1998; Sylvestre *et al.,* 1989). Due to various oxidative reactions, ROS are known to be involved in death of petal cells. The ROS are produced from hydrogen peroxide, thus the hydrogen peroxide level regulating enzymes show differential expression during senescence (Halliwell and Gutteridge, 1989).

Petal senescence is generally accompanied by increase in respiration and loss of dry matter due to hydrolysis of starch, sugars, proteins and nucleic acids. Respiration of many flowers rises when the flower opens but shows decline with ageing. This may be ascribed to inability of petals to convert sucrose into reducing sugars due to inactivation of invertase. In gladiolus (Kaur *et al.,* 2006) and rose (Yamada *et al.,* 2007; Kumar *et al.* 2008b), invertase activity is found to decrease in senescing petals. Invertase inhibitors have also been reported in carnation (Halaba and Rudnicki, 1986). During the time course of senescence, targeted protein degradation is a critical part. Proteases degrade protein by hydrolysing internal peptide bonds. Among all proteases, cysteine proteases are the most frequent and well-characterized. Cysteine proteases have been shown to be up-regulated and further cloned from petals of *Alstroemeria, Narcissus, Petunia* and *Sandersonia.* Guerrero *et al.* (1998) reported increase of another protease, i.e. thiol in petals of daylily flowers, whose transcript levels are strongly up-regulated during petal senescence. The expression of specific serine proteases also increases during senescence-associated proteolysis in *Gladiolus* flowers (Azeez *et al.,* 2007). RNA and DNA are both degraded during senescence. In petals, DNA and RNA are present in the nucleus, mitochondria, chloroplast (if present) and plastids. Degradation of nucleic acids takes place due to presence of several DNases and RNases. Increase in DNase (endonuclease) activity during petal senescence has been confirmed in *Hemerocallis* (Panavas *et al.,* 1999), *Narcissus* (Hunter *et al.,* 2004) and *Petunia* (Langston *et al.,* 2005). Senescence of petals is also accompanied by continued synthesis of proteins and RNA until very late in senescence which indicates requirement of protein synthesis for onset of senescence (Matile and Winkenbach, 1971; Woodson and Handa, 1987; Woodson, 1991). Petal senescence in some species like gladiolus, iris and narcissus is also inhibited by cycloheximide, an inhibitor of protein synthesis (Jones *et al.,* 1994). This indicates that *de novo* synthesis of proteins is involved in petal senescence.

Most flowers show climacteric rise in ethylene following pollination. The role of ethylene in flower senescence has been studied to a great extent in carnation, geranium, orchids and petunia. Ethylene-induced senescence of carnation flowers is accompanied by changes in mRNA and protein profiles indicating the involvement of ethylene-induced gene expression both at the transcriptional and translational levels. In petunia, out of four ACC oxidase (ACO) genes, gene member ACO1 is expressed specifically in senescing corollas and in other floral organs following ethylene exposure while ACO3 and ACO4 were expressed in developing pistil tissue (Tang *et al.,* 1994). The timing and tissue specificity of the increased expression of ACO transcripts correlate with pollination-induced ethylene production in styles and stigma followed subsequently in corollas leading to senescence (Tang *et al.,* 1996).

Considerable success has already been achieved in genetic transformation of many commercial ornamentals (Robinson and Firoozabady, 1993; Zuker *et al.,* 1995; van-Alvorst *et. al.,* 1996). Antisense expression of ACC synthase and ACC oxidase in transgenic carnations has been successfully achieved to inhibit ethylene production and increase in vase life (Michael *et al.,* 1993; van Alvorst *et al.,* 1995a,b; Veres *et al.,* 2004; YiXun and ManZhu, 2004). Ethylene-induced accumulation of transcripts and their coded proteins which speeded up abscission and reduced useful shelf life of cut flowers of *Delphinium* have been identified (Kuroda *et al* (2004). Florigene, an Australian Biotech company has developed cultivars of transgenic carnation (using agro-bacterium mediated gene transfer system) which exhibit low ethylene production and long vase life (Moody, 1994). Such a system is required to be developed for a diverse range of species to produce flowers with long vase life. The transgenic plants of petunia over expressing IPT gene under the SAG_{12} promoter exhibited significant delays in flower senescence and resulted in the increased cytokinin content and less ethylene sensitivity (Chang *et al.,* 2003). cDNAs encoding enzymes of ethylene biosynthesis in carnation petals have been isolated and cloned (Woodson *et al.,* 1993).

Physiological Disorders Associated with Cut Stems

Cut stems are prone to many physiological disorders which limit their longevity. Some common disorders include bent neck of cut roses, bluing and blackening of petals in red coloured cultivars of rose, toppling of tulips, stem break in gerbera, yellowing and blackening of leaves and geotropic bending of stems and have been discussed as under.

Bent neck of cut roses. The cut rose stems are highly prone to bent neck, a physiological disorder which occurs due to insufficient hardening and water stress conditions in the peduncle region of the flower. Immature flower buds of many cultivars of rose lack proper lignification of the neck region at the time of harvest. The leafy stems of rose are also prone to water stress conditions resulting from transpirational loss of water and/or decrease in water uptake. Under severe water stress conditions, bud being an active sink, draws water from the peduncle which lies immediately below it. This leads to loss of cellular turgidity in the peduncle which bends under the weight of the bud (Fig. 5).

Fig. 5: Bent neck in cut rose

Some cultivars of rose like 'Cara Mia' are highly prone to bent neck. In order to prevent bent neck, the stem should be harvested when the neck region of the flower is properly lignified. The stem should not be allowed to lose water during post harvest handling process. The wilted stems should be placed neck deep in water so that they regain turgidity.

Bluing and blackening of petals in rose. Bluing of petals in roses is another physiological disorder which leads to deterioration of cut red roses. Petal bluing occurs during conditions of carbohydrate starvation due to hydrolysis of proteins and respiratory loss of amino acids which lead to increase in free ammonia in the cells. Increased ammonia levels in the cell increase cytosolic pH which leads to bluing of petals. The rose petals remain red when ammonia levels remain below the equivalency of organic acid levels but turn blue when ammonia levels exceed this level. Bluing of petals has been reported to be due to copigment complex of cyanidin 3,5 diglucoside and quercetin and kaempferol glycosides (Asen *et al.,* 1971). High anthocyanin concentration and copigment phenomena are affected by slight changes in pH. Carbohydrate-based preservatives can be used to prevent bluing of petals. Petal blackening as also occurring in some red cultivars is probably associated with inadequate calcium or boron nutrition.

Stem break of gerbera. Stem break in gerbera is a sudden bending of the stem in many cultivars and is a serious problem affecting vase life of the stem (Fig. 6).

Fig. 6: Cut rose buds wrapped in corrugated paper

The major cause of stem break in gerbera is water imbalance caused by bacterial growth in xylem vessels of the cut stem. The stems of gerbera lose fresh weight which is accompanied by a decline in absorption of water. The petal water potential also shows decrease in these flowers whereas it remains constant in flowers without stem break (Meeteren, 1978). In gerbera, two different pathways for water uptake have been suggested: a direct one through the xylem vessels at the cut surface and an indirect one through the cavity in the stem. Stem break occurs when direct water uptake is inhibited by bacterial activity and the indirect uptake is hampered (Meeteren, 1978). The percentage of stem break varies strongly between summer- and winter-grown flowers. Stem break could be prevented by pretreatment of the stems with sodium hypochlorite, silver nitrate or by addition of silver nitrate or dichlorophen to the vase water, and by handling the stems in such a way that water could enter into the cavity of the stem.

Toppling of tulips. Topple of tulips is a disorder associated with calcium deficiency and high atmospheric humidity. The scape, generally at an upper internode develops glassy or translucent, water-soaked areas shortly before or during flowering. The affected areas shrink forming furrows. The scape collapses and topples under the weight of the flower. Under severe calcium deficiency, the symptoms also appear down on the scape or on the leaves. Spray of dilute calcium nitrate solutions can reduce the losses.

Yellowing and blackening of leaves. Normally, the leaves last longer than the flowers. However, in certain species, leaves show yellowing or blackening before senescence of the flower bud. Since in leafy stems foliage also contribute to the longevity of the stem, yellowing and blackening of leaves mark the end of display life in some flowers. Yellowing of leaves is a major problem in certain cultivars of alstroemeria, chrysanthemum, statice, lilies, *etc.* **Leaf yellowing** occurs due to breakdown of chlorophyll, which is promoted by high temperatures as well as dark conditions during storage. Yellowing of leaves can be delayed by application of gibberellins, benzyl adenine and thidiazuron. The response of growth regulator, however, varies with the species. **Blackening** of leaves is of common occurrence in certain species of *Protea* and results from polymerization of phenols and tannins (Whitehead and Swardt, 1982). It is mainly caused by water stress, high temperature and dark conditions. Sugar-containing vase solutions sometimes help to reduce this disorder. The concentration of sucrose higher than 0.5–1.0 per cent may accelerate blackening since proteas are also sensitive to sugar (Reid *et al.,* 1989).

Geotropic bending. In certain species such as *Antirrhiunum, Delphinium, Gladiolus,* etc., the flowering stems are sensitive to geotropic bending, i.e. the stem exhibits upward bending if placed in horizontal condition for long duration. The bending occurs near the meristematically active tip region. Geotropic bending does not affect vase life but renders the stem unmarketable. Geotropic bending occurs because of more cellular elongation in the lower region of the stem due to downward lateral movement of the auxin. Low temperatures prevent geotropic bending but the stem regains the ability to bend when transferred to ambient conditions. In order to prevent geotropic bending, the stems should always be held vertically during storage or transport.

Postharvest diseases. *Botrytis* blight caused by grey mould fungus (*Botrytis cinerea*) is a major post harvest disease of cut as well as loose flowers (Fig. 7).

Fig. 7: Grey mould in bunched cut roses

This disease develops very rapidly during storage as well as during transport. High moisture conditions during storage are highly favourable for the development of grey mould. The flowers producing lot of nectar may also be more susceptible to grey mould. Adequate precautionary measures should be taken to control this pathogen in the field or greenhouse conditions. Especially for storage, only disease-free flowers or foliage should be used.

Care of Flowers/Cut Foliage at Florist and Consumer Level

By observing the following guidelines, vase life of the flowers/cut foliage can be significantly improved.

- Flowers should always be cut at appropriate stage of development because too immature buds fail to open whereas matured ones deteriorate quickly
- After harvesting the stems should not be placed out of water
- The vases should be thoroughly disinfected with chlorine or other suitable disinfectants
- Water in the vase should be changed at 2-3 day intervals as stale water encourages bacterial growth which interferes with water uptake and decreases vase life
- Basal 2-3 cm portion of the stem should be removed at 2-3 day intervals to expose fresh tissues for improving water uptake by the stem
- Leaves dipping in water/solution should be removed as such leaves rot, impart foul smell and promote growth of the microbes, thereby decreasing vase life of the flower
- Vases should be placed at cool places, and dried & wilted flowers should regularly be removed from the stem as they release high rate of ethylene, thereby causing senescence of other flowers
- Some flowers such as *Narcissus* release some toxic substance and mucilage from the cut surface which decreases vase life of other flowers so use of such flowers should always be avoided with other flowers

References

Adams, D. O. and S.F. Yang, 1979. Ethylene biosynthesis: Identification of 1-aminocylclopropane-1carboxylic acid as an intermediate in the conversion of methionine to ethylene. *Proc. Natl. Acad. Sci.* (USA), **76**: 170-174.

Arditti, J. 1979. Aspects of the physiology of orchids. *Adv. Bot. Res.*, **7**: 421-455.

Arora, J. S., K. Singh and P.J. Singh, 2002. *Gladiolus*. Tech. Bull. No. 14, All India Coordinated Research Project on Floriculture (ICAR), IARI, New Delhi, 100 p.

Asen, S., K.H. Norris and R.N. Stewart, 1971. Effect of pH and concentration of the anthocyanin-flavanol co-pigment complex on the color of Better Times' roses. *J. Amer. Soc. hort. Sci.*, **96**: 770-773.

Azeez, A., A.P. Sane, D. Bhatnagar and P. Nath, 2007. Enhanced expression of serine proteases during floral senescence in *Gladiolus*. *Phytochem.*, **68**: 1352-1357.

Badiyan, D., R.B.H. Wills and M.C. Bowyer, 2004. Use of nitric oxide donor compound to extend the vase life of cut flowers. *HortSci.*, **39**: 1371-1372.

Baker, J. E., C.V. Wang, M. Lieberman and R. Hardenburg, 1977. Delay of senescence in carnation by rhizobitoxine analog and sodium benzoate. *HortSci.*, **12**: 38-39.

Balestra, M. G., R. Agostini, A. Bellincontro, F. Mencarelli and L. Varvaro, 2005. Bacterial populations related to gerbera (*Gerbera jamesonii* L.) stem break. *Phytopath. Medit.*, **44**: 291-299.

Bhattacharjee, S. K. 1998. Handbook of Medicinal Plants, p. 118. Pointer Publishers, Jaipur, India.

Bhattacharjee, S. K. 1999. Post harvest Management of cut flowers, cut foliages and post production management of potted plants. *J. Ornam. Hort.*, **2**: 32-39.

Bhattacharjee, S. K. and L.C. De, 2006. Postharvest handling of cut flowers and loose flowers. In: Amenity Horticulture, Biotechnology and Post harvest Technology (ed. Bhattacharjee, S.K.), pp. 120-169. Pointer Publishers, Jaipur.

Boodley, J. W. 1969. Post harvest life, p. 144-149. In: A Manual on the Culture, Management, Disease, Insects, Economics, and Breeding of Greenhouse Roses (eds Mastalerz, J.W. and R. W. Langhans). Pennsylvania Flower Growers, N.Y. St. Fl. Growers Assoc., Inc.

Brown, R. and S. Mayak, 1980. Aminooxyacetic acid as an inhibitor of ethylene synthesis and senescence in carnation flowers. *Scient. Hort.*, **15**: 277-282.

Burg, S. P. 1973. Ethylene in plant growth. *Proc. Nat. Acad. Sci. United States of America*, **70**: 591-597.

Burg, S. P. and M.J. Dijkman, 1967. Ethylene and auxin participation in pollen induced fading of *Vanda* orchid blossoms. *Plant Physiol.*, **42**: 1648-1650.

Chang, H., L.J. Michelle, G.M. Banowetz and D.G. Clark, 2003. Overproduction of cytokinins in *Petunia* flowers transformed with P_{SAG12}-IPT delays corolla senescence and decreases sensitivity to ethylene. *Plant Physiol.*, **132**: 2174-2183.

Chawla, K. 2008. Effect of GA_3, BAP and refrigerated storage on keeping quality of chrysanthemum (*Chrysanthemum morifolium* Ramat). M. Sc. Thesis. Punjab Agricultural University, Ludhiana, India.

Downs, C. G. and P.H. Lovell, 1986. The effect of spermidine and putresine on the senescence of cut carnations. *Plant Physiol.*, **66**: 679–684.

Elanchezhian, R. and G.C. Srivastava, 2001. Physiological response of chrysanthemum petals during senescecne. *Biol. Plant.*, **44**: 411-415.

Faernham, D. S., C. Barr and A.H. Halevy, 1971. Value of using chemical solutions for conditioning and bud opening carnations. *Flor. Rev.*, **148**: 27-28.

Ferrante, A., D.A. Hunter, W.P. Hackett and M.S. Reid, 2002. Thidiazuron—a potent inhibitor of leaf senescence in *Alstroemeria. Postharvest Biol. Tech.*, **25**: 333–338.

Ferrante, A., A. Mensuali-Sodi and G. Serra, 2009. Effect of thidiazuron and gibberellic acid on leaf yellowing of cut stock flowers. *Central European J. Biol.*, **4**: 461-468.

Fujino, D. W., M.S. Reid and G.E. Vandermolen, 1981. Identification of vascular blockages in rachides of cut maidenhair (*Adiantum raddianum*) fronds. *Scientia Hort.*, **21**: 381-388.

Garrod, J. F. and G.P. Harris, 1978. Effect of gibberellic acid on senescence of isolated petals of carnation. *Ann. Appl. Biol.*, **88**: 309-311.

Goh, C. J., A.H. Halevy, R. Engel and A.M. Kofranek, 1985. Ethylene evolution and sensitivity in cut orchid flowers. *Scientia Hort.*, **26**: 57-67.

Goszczynska, D. M. and R.M. Rudnicki, 1988. Storage of cut flowers. *Hort. Rev.*, **10**: 35-62.

Grover, J. K. 2001. Modified atmosphere packaging and storage of cut flowers in polymeric films. Ph. D. Thesis. Punjab Agricultural University, Ludhiana, India.

Grover, J. S., K. Singh, A.K. Gupta and A. Kumar, 2005. Effect of pre-storage pulsing treatment on storage of gladiolus spikes under modified atmosphere. *Indian J. Pl. Physiol.*, **10**: 248-253.

Grover, J. K., A.K. Gupta, K. Singh, A. Kumar and P. Singh, 2006. Studies on passive modified atmosphere storage of gladiolus spikes. *Adv. Hort. Sci.*, **20**: 175-180.

Grover, J. K., K. Singh and A. Kumar, 2008. Effect of passive hypobaric pressure on storage of gladiolus spikes. *Indian J. Pl. Physiol.*, **13**: 238-242.

Guerrero, C., M. de la Calle, M.S. Reid and V. Valpuesta, 1998. Analysis of the expression of two thiolprotease genes from daylily (*Hemerocallis* spp.) during flower senescence. *Plant Mol. Biol.*, **36**: 656671.

Halaba, J. and R.M. Rudnicki, 1986. The role of enzymes during senescence of cut flowers. *Acta Hort.*, No. 181, pp. 65-74.

Halevy, A. H. and S. Mayak, 1979. Senescence and post harvest physiology of cut flowers. Part I. *Hort. Rev.* (ed. Janick, J.), **1**: 204-236.

Halevy, A. H. and S. Mayak, 1981. Senescence and postharvest physiology of cut flowers, Part 2. *Hort. Rev.*, **3**:59-143.

Halevy, A.H., C.S. Whitehead and A.M. Kofranek, 1984. Does pollination induce corolla abscission of cyclamen flowers by promoting ethylene production. *Plant Physiol.*, **75**: 1090-1093.

Halliwell, B. and J.M.C. Gutteridge, 1989. Free Radicals in Biology and Medicine, pp. 450-499. Clarendon Press, Oxford, UK.

Hoekstra, F. A. and R. Weges, 1986. Lack of control by early pistillate ethylene of the accelerated wilting of *Petunia hybrida* flowers. *Plant Physiol.*, **80**: 403-408.

Hunter, D. A., M. Yi, M. Xu and M.S. Reid, 2004. Role of ethylene in perianth senescence of daffodil (*Narcissus pseudonarcissus* L. 'Dutch Master'). *Postharvest Biol. & Technol.*, **32**: 269-280.

Jones, M. L. 2004. Changes in Gene Expression during Senescence. In: Plant Cell Death Processes (ed. Nooden, L.), pp. 51-72. Elsevier Science, San Diego, CA.

Jones, R.B., M. Serek, C. Kuo and M. Reid, 1994. The effect of protein synthesis inhibition on petal senescence in cut bulb flowers. *J. Amer. Soc. hort. Sci.*, **119**: 1243-1247.

Judy, S. 1999. Viagra makes flowers to stand up straight. *British Medical J.*, **319**: 274.

Kaur, G., P. Singh and U. Parmar, 2006. Changes in sugar metabolism in *Gladiolus tristis* during senescence as affected by ethylene and its antagonists. *J. Plant Sci. Res.*, **22**: 81-89.

Kofranek, A. M. and A.H. Halevy, 1972. Conditions for opening cut chrysanthemum flower buds. *J Amer. Soc. hort. Sci.*, **97**: 578-584.

Kohl, H. C. and D.L. Rundle, 1972. Decreasing water loss of cut roses with abscisic acid. *HortSci.*, **7**: 249.

Kumar, N., K. Dixit, A. Singh and G.C. Srivastava, 2007a. Physiological changes during flower bud opening in rose (*Rosa hybrida* L.). *Indian J. Pl. Physiol.*, **12**: 239-245.

Kumar, N., G.C. Srivastava and K. Dixit, 2007b. Role of superoxide dismutase during petal senescence in rose (*Rosa hybrida* L.). *J. Hort. Sci. & Biotech.*, **82**: 673-678.

Kumar, N., G.C. Srivastava and K. Dixit, 2008a. Senescence in rose (*Rosa hybrida* L.): Role of endogenous antioxidant system. *J. hort. Sci. & Biotech.*, **83**: 125-131.

Kumar, N., G.C. Srivastava and K. Dixit, 2008b. Role of sucrose synthase and invertase during petal senescence in rose (*Rosa hybrida* L.). *J. hort. Sci. & Biotech.*, **83**: 520-524.

Kuroda, S., S. Horose, M. Shiraishi, E. Davies and S. Abe, 2004. Co-expression of an ethylene gene, ERS1, and ethylene signaling regulator gene in *Delphinium* during abscission of florets. *Plant Physiol. Boichem.*, **42**: 745-751.

Langston, B. J., S. Bai and M.L. Jones, 2005. Increases in DNA fragmentation and induction of a senescence specific nuclease are delayed during corolla senescence in ethylene insensitive (etr11) transgenic petunias. *J. exp. Bot.*, **56**:1523.

Lara, M. E. B., M.C.G. Garcia, T. Fatima, R. Ehness, T.K. Lee and R. Proels, 2004. Extracellular invertase is an essential component of cytokinin-mediated delay of senescence. *Plant Cell,* **16**: 1276–1287.

Laurie, A., D.C. Kiplinger and K.C. Nelson, 1968. *Commercial Flower Forcing.* McGraw Hill, New York.

Lay-Yee, M., A.D. Stead and M.S. Reid, 1992. Flower senescence in daylily (*Hemerocallis*). *Physiol. Plant.*, **86**: 308-314.

Lee, M. M., S.H. Lee and K.Y. Park, 1997. Effects of spermine on ethylene biosynthesis in cut carnation (*Dianthus caryophyllus* L.) flowers during senescence. *J. Pl. Physiol.*, **151**: 68-73.

Loubaud, M. and W.G. van Doorn, 2004. Wound-induced and bacteria-induced xylem blockage in roses, *Astilbe*, and *Viburnum*. *Postharvest Biol. & Tech.*, **32**: 281-288.

Machin, B.J. and N. Scopes, 1978. *Chrysanthemums-Year Round Growing.* Blandford Press, U.K.

Markhart, A. H. and M.S. Harper, 1995. Deleterious effects of sucrose in preservative solutions on leaves of cut roses. *HortSci.*, **30**: 1429-1432.

Matile, P. and F. Winkenbach, 1971. Function of lysosomes and lysosomal enzymes in the senescing corolla of the morning glory. *J. expt. Bot.*, **22**: 759-771.

Mayak, S., A.H. Halevy, S. Sagie, A. Bar Yosef and B. Bravdo, 1974. The water balance of cut rose flowers. *Physiol. Plant.*, **32**: 15-22.

Mayak, S. and D. Dilley,1986. Effect of sucrose on response of cut carnation flowers to kinetin, ethylene and abscisic acid. *J. Amer. Soc. hort. Sci.*, **101**: 583–585.

Mayak, S. and A.H. Halevy, 1972. Interrelationships of ethylene and abscisic acid in the miniature potted roses. *J. hort. Sci. & Biotech.*, **75**:12-18.

Meeteren, U. van, 1978. Water relations and keeping quality of cut gerbera flowers. I. The cause of stem break. *Scientia Hort.*, **8**: 65-74.

Meeteren, U. van, 1979. Water relation and keeping quality of cut gerbera flowers. IV. Internal water relations of ageing petal tissue. *Scientia Hort.*, **11**: 83-93.

Mengüc, A. and M. Zencirkiran, 1996. A research on cold storage of *Alstroemeria* cv. Ostara' cut flowers. *Acta Hort.*, No. 429, pp. 591-596.

Michael, M. Z., K.W. Savin, S.C. Baudinette, M.W. Graham, S.F. Chandler, C.Y. Lu, C. Caesar, I. Gautrais, R. Young, G.D. Nugent, K.R. Stevenson, E.L.J. O'Conner, C.S. Cobbett and E.C. Cornish, 1993. Cloning of ethylene biosynthetic genes involved in petal senescence of carnation and petunia, and their anti-sense expression in transgenic plants. In: *Cellular and Molecular Aspects of the Plant Hormone Ethylene* (Proc. intern. symp. on cellular and molecular aspects of biosynthesis and action of the plant hormone ethylene), pp. 298-303, held on August 31-September 4, 1992 at Agen, France.

Mohan Ram, H. Y. and I.V.R. Rao, 1977. Prolongation of vase life of *Lupinus hartwegii* Lindl. by chemical treatments. *Scientia Hort.*, **7**: 377-382.

Moody, H. 1994. Carnations engineered for long life without STS. *Floriculture International* (Dec. issue), pp. 44-45.

Mor, Y., A.H. Halevy, H. Spiegelstein and S. Mayak, 1985. The site of 1-aminocyclopropane-1-carboxylic acid synthesis in senescing carnation petals. *Physiol. Plant.*, **65**: 196-202.

Mor, Y., M.S. Reid and A.M. Kofranek, 1984. Pulse treatments with silver thiosulphate and sucrose improve the vase life of sweet peas. *J. Amer. Soc. hort. Sci.*, **109**: 866-868.

Morris, L. L., A.A. Kader, J.A. Klaustermeyer and C.C. Cheyney, 1978. Avoiding ethylene concentrations in harvested lettuce. *Calif. Agric.*, **32**: 14-15.

Müller, R., S. LindIversen, B.M. Stummann and M. Serek, 2000. Expression of genes for ethylene biosynthetic enzymes and an ethylene receptor in senescing flowers of miniature potted roses. *J. hort. Sci. Biotech.*, **75**:1218.

Naidu, S. N. and M.S. Reid, 1989. Postharvest handling of tuberose (*Polianthes tuberosa* L.). *Acta Hort.*, No. 261, pp. 313-317.

Nell, T. A. 1992. Taking silver safety out of the longevity picture. *Grower Talks* (June issue), pp. 35-38.

Nell, T. A. 1994. Controlling ethylene extends flower life. *Floriculture International* (Dec. issue), pp. 40-42.

Nichols, R., G. Bufler, Y. Mor, D.W. Fujino and M.S. Reid, 1983. Changes in ethylene production and 1-aminocyclopropane-1-carboxylic acid content of pollinated carnation flowers. *J. Pl. Gr. Regul.*, **2**: 1-8.

Nowak, J. and R.M. Rudnicki, 1990. Postharvest Handling and Storage of Cut Flowers, Florist Greens and Potted Plants, p. 209. Chapman and Hall, London.

Otsubo, M. and M. Iwaya-Inoue, 2000. Trehalose delays senescence in cut gladiolus spikes. *HortSci.*, **35**: 1107-1110.

Paliyath, G. and M.J. Droillard, 1992. The mechanisms of membrane deterioration and disassembly during senescence. *Plant Physiol. Biochem.*, **30**:789-812.

Panavas, T. and B. Rubinstein, 1998. Oxidative events during programmed cell death of daylily (*Hemerocallis* hybrid) petals. *Plant Sci.*, **133**:125-138.

Panavas, T., A. Pikula, P.D. Reid, B. Rubinstein and E.L. Walker, 1999. Identification of senescence associated genes from daylily petals. *Plant Mol. Biol.*, **40**:237-248.

Pandey, S., S.A. Ranade, P.K. Nagar and N. Kumar, 2000. Role of polyamines and ethylene as modulators of plant senescence. *J. Biosci.*, **25**: 291-299.

Paulin, A. 1986. Influence of exogenous sugars on the evolution of some senescence parameters. *Acta Hort.*, No. 181, pp. 183-193.

Pompodakis, H. E. and D.C. Joyce, 2003. Abscisic acid analogue effects on the vase life and leaf crisping of cut Baccara roses. *Australian J. expt. Agri.*, **43**: 425-428.

Pompodakis, H. E., D.C. Joyce, L.A. Terry and D.E. Lydakis, 2004. Effects of vase solution pH and abscisic acid on the longevity of cut Baccara roses. *Australian J. hort. Sci. & Biotech.*, **79**: 828-832.

Porat, R., A. Borochov and A.H. Halevy, 1993. Enhancement in *Petunia* and *Dendrobium* flower senescence by jasmonic and methyl ester is *via* the promotion of ethylene production. *Plant Growth Regul.,* **13**: 297-301.

Put, H.M.C. 1986. Investigations into the influence of the microflora from stems of cut flowers on the vase life of *Rosa* cv. Sonia., *Gerbera* cv. Fleur and *Chrysanthemum* cv. Spider. *Acta Hort.*, No. 181, pp. 415-418.

Rao, I.V.R. and H.Y. Mohan Ram, 1982. Specificity of gibberellin and sucrose-promoted flower bud growth in gladiolus. *Ann. Bot.*, **50**: 473-479.

Redman, P.B., J.M. Dole, N.O. Maness and J.A. Anderson, 2002. Postharvest handling of nine specialty cut flower species. *Scientia Hort.*, **92**: 293-303.

Reid, M.S. 1989. The role of ethylene in flower senescence. *Acta Hort.*, No. 261, pp. 157-169.

Reid, M.S., D.W. Fujino, N.E. Hoffman and C.S. Whitehead, 1984. 1-aminocyclopropane-1-carboxylic acid (ACC)-transmitted stimulus in pollinated flowers? *J. Pl. Gr. Regul.*, **3**:189-196.

Reid, M.S., R.Y. Evans and L.L. Dodge, 1989. Ethylene and silver thiosulphate influence opening of cut rose flowers. *J. Amer. Soc. hort. Sci.*, **114**: 436-440.

Robinson, K.E.P. and E. Firoozabady, 1993. Transformation of floriculture crops. *Scientia Hort.*, **55**: 83-90.

Rubinstein, B. 2000. Regulation of cell death in flower petals. *Plant Mol. Biol.*, **44**: 303–318.

Saks, Y., J. van Staden and M.T. Smith, 1992. Effect of gibberellic acid on carnation flower senescence evidence that the delay of carnation flower senescence by gibberellic acid depends upon the stage of flower development. *Plant Gr. Regul.*, **11**: 45-51.

Sankhla, N., W.A. Mackary and T.D. Davis, 2005. Effect of nitric oxide generating compounds in flower senescence in cut recemes of pink flowered *Lupinus havardii* Wats. *Proc. 32nd meeting of Plant Growth Regulation Soc. of America*, p. 11, held on July 24-27, at Newport Beach, California,

Serek M and M.S. Reid, 1997. Use of growth regulators for improving the postharvest quality of ornamentals. *Perishables Handling Quarterly*, **92**: 7-9.

Serek, M., E.C. Sisler and M.S. Reid, 1995a. Effects of 1-MCP on the vase life and ethylene response of cut flowers. *Plant Gr. Regul.,* **16**: 93-97.

Serek, M., E.C. Sisler and M.S. Reid, 1995b. Use of a novel ethylene inhibitor to increase the life of ethylene-sensitive cut flowers and potted plants. In: *Postharvest Physiology, Pathology and Technologies for Horticultural Commodities: Recent Advances* (eds Ait-Oubahou, A. and M. El-Otmani), pp. 460-463. Institut Agronomique & Veterinaire Hassan II, Agadir, Morocco.

Serek, M., E.J. Woltering, E.C. Sisler, S. Frello and S. Srikandarajah, 2006. Controlling ethylene responses in flowers at receptor level. *Biotechnol Adv.*, **24**: 368-381.

Serrano, M., F. Romojaro, J.L. Casas and M. Acosta, 1991. Ethylene and polyamine metabolism in climacteric and non-climacteric carnation flowers. *HortSci.,* **26**: 894-896.

Setia, N. and R.C. Setia, 2006. Nitric oxide (NO) as a regulator and signaling molecule in plants. In: Plant Molecular Physiology: Current Scenario Projections (ed. Trivedi, P.C.), pp. 32-57. Aavishkar Publishers, Jaipur.

Shimizu-Yumoto, H. and K. Ichimura, 2009. Abscisic acid in combination with sucrose, is effective as a treatment to suppress leaf damage and extend foliage vase -life in cut *Eustoma* flowers. *J. hort. Sci. & Biotech.*, **84**: 107-111.

Singh, K. 1994. Effects of spermidine, IAA, ACC and ethylene on petal longevity in carnation (*Dianthus caryophyllus* L.). *Phyton*, **34**: 309-313.

Singh, K. 2000. Studies on ACC and ethylene production by the flowers of sweet pea (*Lathyrus odoratus* L.). *Indian J. Plant Physiol.*, **5**: 180-182.

Singh, K. and J.S. Arora, 1995. Effect of 8-hydroxyquinoline citrate, silver nitrate and Chrysal on vase life of cut chrysanthemum flowers. *J. Ornam. Hort.*, **3**: 32-35.

Singh, K. and J.S. Arora, 1996. Studies on flower senescence in *Chrysanthemum morifolium* Ramat. In: *Current Research in Plant Science* (eds Sharma, T.A., S.S. Saini, M.L. Trivedi and M. Sharma), pp. 181-184. Bishen Singh Mahendra Pal Singh, Dehra Dun.

Singh, K. and K.G. Moore, 1992a. Water relations of cut chrysanthemum flowers. *Adv. hort. Sci.*, **3**: 121-124.

Singh, K. and K.G. Moore, 1992b. Role of ethylene in keeping quality of chrysanthemum flowers. *Adv. hort. Sci.*, **6**: 177-178.

Singh, K. and K.G. Moore, 1994a. Sites of ethylene production from flowers of sweet pea. *Scientia Hort.*, **58**: 351-355.

Singh, K. and K.G. Moore, 1994b. Factors affecting post harvest keeping quality of chrysanthemum flowers. In: *Floriculture Technology, Trends and Trades* (eds Parkash, J. and K.R. Bhandary), pp. 550-565. Oxford & IBH Publishing Co. Pvt. Ltd., New Delhi.

Singh, K. and P. Singh, 2002. Molecular mechanisms of flower senescence. In: R*ole of Biotechnology in Medicinal and Aromatic Plants,* Vol. VII (eds Khan, I.A. and A. Khanam), pp. 173-188. Ukaaz Publication, Hyderabad.

Singh, K., P.J. Singh, J.S. Arora and R.P.S. Mann, 2000. Studies on postharvest management of gladiolus. *J. Ornam. Hort* (New Series), **3:** 107-110.

Singh, K., J.S. Arora and S.K. Bhattacharjee, 2001a. Postharvest Management of Cut Flowers. Tech. Bull. No. 10. All India Coordinated Research Project on Floriculture, IARI., New Delhi, 39 p.

Singh, K., P.J. Singh, J.S. Arora and R.P.S. Mann, 2001b . Effect of vase solutions on keeping quality of different grades of gladiolus. *J. Plant Sci. Res.*, **17**: 23-25.

Singh, K., P.J. Singh and R. Kumar, 2004a. Investigation on post harvest quality of cut flowers as influenced by storage conditions. *Final Report of the ICAR- Funded Ad-hoc Project* from P.A.U., Ludhana, p. 65.

Singh, K., P.J. Singh and R. Kumar, 2004b. Effect of chemicals on keeping quality of cut roses. *Adv. hort. Sci.*, **18**: 161-167.

Singh, K., R. Kumar and M.S. Aulakh, 2005a. Sucrose and cytokinin-induced opening of buds in spray chrysanthemum. *J. Plant Biol.*, **32**: 139-142.

Singh, V.P., D. Kiran and J.S. Arora, 2005b. Alleviation of antioxidant activity in gladiolus flowers during senescence by spermine and spermidine. *J. Ornam. Hort.*, **8**: 167-172.

Singh, V.P. and G.C. Srivastava, 2006. Physiological and biochemical basis of flower senescence. In: Advances in Ornamental Horticulture (ed. Bhattacharjee, S. K.), pp. 105-119. Pointer Publishers, Jaipur.

Singh K., R. Kumar and R. Singh, 2007. Studies on wet-refrigerated storage of gladiolus spikes in relation to pre- and post-harvest treatments. *Asian J. Hort.*, **2**: 218-222.

Singh, K., R. Singh and R. Kumar, 2009. Studies on wet storage of cut rose stems in relation to post storage treatments. *Asian J. Hort.*, **3**: 400-402.

Staby, G.L., J.L. Robertson, D.C. Kipliger and C.A. Conover, 1978. *Chain of Life.* Ohio Florists Association, Ohio State Univ., Columbus.

Sylvestre, I., M.J. Droillard, J.M. Bureau and A. Paulin, 1989. Effects of the ethylene rise on the peroxidation of membrane lipids during the senescence of cut carnations. *Plant Physiol. Biochem.*, **27**: 407-413.

Tang, X., A. Gome, A. Bhatia and W. William, 1994. Pistil-specific and ethylene-regulated expression of 1-aminocyclopropane-1-carboxylate oxidase genes in petunia flowers. *The Plant Cell*, **6**: 1227–1239.

Tang, X. and W.R. Woodson, 1996. Temporal and spatial expression of 1-aminocyclopropane-1-carboxylate oxidase mRNA following pollination of immature and mature *Petunia* flowers. *Plant Physiol.*, **112**: 503-511.

Tjia, B.O. and K.A. Funnell, 1986. Postharvest studies of cut *Zantedeschia* inflorescences. *Acta Hort.*, No. 181, pp. 451-458.

Valpuesta, V., N.E. Lange, C. Guerrero and M.S. Reid, 1995. Up-regulation of a cysteine protease accompanies the ethylene-insensitive senescence of daylily (*Hemerocallis*) flowers. *Plant Mol. Biol.*, **28**: 575-582.

van Doorn, W G., H. Harkema and J.S. Song, 1995. Water relations and senescence of cut iris flowers. *Postthervest Biol. Tech.*, **5**: 345-351.

van Doorn, W. G., E.J. Woltering, M.S. Reid and M.J. Wu, 1989. Reduced sensitivity to ethylene in a group of related carnation cultivars. In: Cellular and Molecular Aspects of the Plant Hormone Ethylene (eds Pech, J.C., A. Latche and C. Balague), pp. 188-194. Kluwer, Dordrecht, The Netherlands.

van Staden, J. and G.G. Dimalla, 1980. The effect of silver thiosulfate preservative on the physiology of cut carnations. II. Influence of endogenous cytokinins. *Z Pflanzenphysiol.*, **99**:19-26.

van-Alvorst, A.C., H. Koehorst, J. De-Jong and H.J.M. Dons, 1996. Transgenic carnation plants obtained by *Agrobacterium tumefaciens*-mediated transformation of leaf explants. *Plant Cell Tissue & Org. Cult.*, **45**: 169-173.

van-Alvorst, A.C., T. Riksen, H. Koehorst and H.J.M. Dons, 1995a. Transgenic carnations obtained by *Agrobacterium tumefaciens*-mediated transformation of leaf explants. *Transgenic Research*, **4**: 105-113.

van-Alvorst, A.C., T. Riksen, H. Koehorst and H.J.M. Dons, 1995b. Shoot regeneration and *Agrobacterium*-mediated transformation of carnation. *Acta Hort.*, No. 420, pp. 92-94.

Vaslier, N. and W.G. van Doorn, 2003. Xylem occlusion in bouvardia flowers: Evidence for a role of peroxidase and catechol oxidase. *Postharvest Biol. & Tech.*, **28**: 231-237.

Veres, A., E. Kiss, E. Toth, A. Toth and L. Heszky, 2004. Effect of an apple-derived antisense ACC-synthase cDNA on the ethylene production and the vase life of carnation (*Dianthus caryophyllus*). *Proc. of the 17th -EUCARPIA-General Congress*, pp 267-271. Tullun, Austria, September 8-11.

Waithaka, K., M.S. Reid and L.L. Dodge, 2001. Cold storage and flower keeping quality of cut tuberose (*Polianthes tuberosa* L.). *J. hort. Sci &. Biotech.*, **76**: 271-275.

Wang, C. Y. and J.E. Baker, 1980. Vase life of cut flowers with rhizobitoxine analogs, sodium benzoate and isopentenyl adenosine. *HortSci.,* **14**: 59-60.

Waters, W. E. 1968a. Relationship of water salinity and fluorides on keeping quality of chrysanthemum and gladiolus cut flowers. *Proc. Amer. Soc. hort. Sci.*, **92**: 633-640.

Waters, W. E. 1968b. influence of well water salinity and fluorides on keeping quality of 'Tropicana' roses. *Proc. Fla. St. hort. Soc.*, **81**: 357-359.

Whitehead, C. S. and G.H. de Swardt, 1982. Extraction and activity of polyphenol oxidase and peroxidase from senescing leaves of *Protea neriifolia. South African J. Bot.*, **1**: 127-130.

Whitehead, C.S., D.W. Fujino and M.S. Reid, 1983. Identification of a ethylene precursor, 1-aminocylclopropane-1-carboxylic acid (ACC) in pollen. *Scientia Hort.*, **21**: 291-297.

Whitehead, C.S., A.H. Halevy and M.S. Reid, 1984. Role of ethylene and ACC in pollination and wound-induced senescence of *Petunia hybrida* L. flowers. *Physiol. Plant.*, **61**: 643-648.

Woltering, E .J. and W.G. van Doorn, 1988. Role of ethylene in senescence of petals- morphological and taxonomical relationships. *J. exp. Bot.*, **39**: 1605-1616.

Woodson, W.R. 1991. Gene expression and flower senescence. In: *Genetics and Breeding of Ornamental Species* (eds Harding, J., F. Singh and J.N.M. Mol), pp. 317-331. Kluwer Academic Publisher, The Nertherlands.

Woodson, W.R., A.S. Brandt, H. Itzhaki, J.M. Maxson, K.Y. Park and H. Wang, 1993. Regulation and function of flower senescence-related genes. *Acta Hort.*, No. 336, pp. 41-46.

Woodson, W. R. and A.K. Handa, 1987. Changes in protein patterns and *in vivo* protein synthesis during pre-senescence and senescence of hibiscus petals. *J. Plant Physiol.*, **128**: 67-75.

Yamada, T., I. Kazuo, K. Motoki, G. Outer and van Doorn, 2007. Gene expression in opening and senescing petals of morning glory (*Ipomoea nil*) flowers. *Plant Cell Reports*, **26**: 823-835.

Yang, S.F. 1980. Regulation of ethylene biosynthesis. *HortSci.*, **15**: 238-243.

YiXun, Y. and B. ManZhu, 2004. Carnation flower vase life prolonged by transformation with antisense ACC oxidase gene. *Acta Hort. Sinica*, **31**: 633-636.

Zagory, D. and M.S. Reid, 1986. Role of vase solution microorganisms in the vase life of cut flowers. *J. Amer. Soc. hort. Sci.*, **111**: 154-158.

Zucker, A., P.L. Chang, K. Cheah, W.R. Woodson, R.A. Bressan, A.A. Watad, P.M. Hasegawa and A. Vainstein, 1995. Transformation of carnation by microprojectile bombardment. *Scientia Hort.*, **64**: 177-185.

41

Flower Drying

R.L. Misra

Though India's export of ornamental planting material is insignificant but out of that dry flowers constitute to a tune 60-70 per cent, and out of the total exports of dry flowers some 90 per cent is being done through a firm in Tuticorin. Its share for export in Europe is estimated to below 1.5 per cent, and below 1 per cent of the world requirement (Singhwi, 1996). The concept of dry flowers came into existence when the developed western countries were snowbound during winter months as there was scarcity of fresh flowers and those available also were being very expensive. People require dry flowers for arrangement during winter, for enjoyment and for celebrating their those festivals which fall during winter months such as 'New Year's Day', 'Christmas' and certain others. Once, the people were confined using only *Contaderia* (silver-cream pampass-grass) and *Physalis alkekengii* (Chinese lantern). As per the international scenario, the top 10 dry flowers are *Helichrysum, Helipterum, Limonium, Nigella, Gypsophila, Delphinium, Amaranthus, Papaver, Carthamus* and *Rosa* (Anon., 1989; Desh Raj, 2006). The leading importing countries for dry flowers are Nigeria, France, United Kingdom, Belgium and Japan, and exporting countries are the Netherlands, Portugal, India, China and Italy for dry flowers though USA, Germany, the Netherlands, France and United Kingdom for dry foliage and its products. India exports dried ornamentals to USA, UK, the Netherlands, Germany, Japan, Singapore and certain other European countries and the major exporting products are lotus pods, camellia, dahlia, bell cups, marigold, jute flowers, wood rose, wild lilies, paper flower and naturally dried plant parts from Himalayan region and this business is confined to about 60 private companies operating from Kolkata (W.B.), Tuticorin (T.N.) and Mumbai close to the sea routes. Through continuous researches, various preservation methods are now developed to preserve the fresh flowers in dried form of many plant species. This way, people are able to preserve the original colours and shape of the flowers and foliage. The biggest advantage of the dry flowers is this, that these last long by remaining fresh for many months, they are eco-friendly and biodegradable, and one gets rid of the tedious job of regular changing the solutions of fresh cut flowers, *vis-à-vis* other operations such as cutting of a portion of stem ends and removal of the faded parts at regular intervals, however, there is greatest disadvantage that neither dry flowers inhale the CO_2 that we emit or that is present in the atmosphere nor that releases O_2 for our use, in turn. Moreover, fresh cut flowers are very expensive though dry flowers are much cheaper and easily affordable. Paul and Shylla (2002) furnished a good account of flower drying.

Various Uses of Dry Flowers

Hanging bunches of dry flowers look attractive in their own right. They are combined with various small flowers, foliage, seed heads and herbs and wired together to form a thick rope which may look quite spectacular when made in the form of a **garland**, either when offered to a person out of emotion or veneration or when hanged in the room at a suitable place, especially over a mirror, over a door, along the beams or around a picture, to celebrate a special event.

Floral baskets are highly attractive containers as these are woven of plant materials and when these are filled with variously coloured dried flowers, aromatic herbs, seed heads, berries and leaves, these become very attractive and fill the atmosphere with a faint plant aroma. However, in case when aroma is quite faint, a few aromatic stems to the basket's rim or handle may be attached for giving arrangement an extra appeal. The stems of the dry material of plants may be fixed at the base of the basket to hold all the materials at place.

A mixture of fragrant and colourful flowers and leaves displayed in bowl is known as '**potpourri**' which in original French means 'a rotten pot,' i.e. a moist mixture of pickled flowers and leaves. Now this 'moist concept' is redundant as it is less attractive though the aroma is long-lasting, however, the dry method is easier, colourful and easily displayable in bowls or potpourri balls. This contains scented or coloured flowers, aromatic leaves, spices and peel and the fixative to preserve the blend. A mixture containing yellow tulip petals for colour; lavender, tansy and mint for scaring flies; and southernwood, mugwort and santolina to scare clothe-moths make a good potpourri (Lesley Bremness, 1988). In pot pourri, spices (allspice and cloves) and citrus peel add extra richness; orris root fixes the scent; and a few drops of essential oil boost the intensity of the fragrance (Lesley Bremness, 1988). She further suggested preparation of a culinary potpourri by blending useful kitchen herbs and spices, such as cinnamon bark and dried leaves, star anise, cloves, bearberry leaf and juniper berries. Important plants used for making potpourri are *Angelica*, bergamot, daisy bush, geranium, lavender, lilac, lilies, hyacinths, marigold, pinks, rose, sage, savoury, sweet cicely, thyme, violets, wall flowers, *etc.* which all are almost fragrant, and others such as driftwood and other items, ferns, lily pods, lotus pods, *etc.*

Dry flower arrangement, similar to that of terrarium, can be made in a transparent plastic or glass container. For this, a thermocol sheet some 2.5 cm thick should be cut as per size of the mouth of the transparent container and pasted centrally inside the container on which the stems of dry flowers are fixed while making arrangements. After the arrangement is complete, at the brim of the container the adhesive is applied and then it is kept in the oven at 45 ^{0}C temperature for about one hour.

Floral crafts with press-dried flowers and foliage, such as 'greeting cards,' 'wall hangings,' 'floral designs,' 'landscape calendars,' *etc.* are prepared by embossing the press-dried plant materials on various backgrounds such as velvet paper; linen, silk or bamboo plates; or wooden boards. Here, first the card is cut to the size, followed by pasting of precut background material all around the card leaving a margin, and then these cards are pressed with a heavy table glass for sometime to settle and dry. Now the plant materials are pasted on the card in desired design with the help of forceps and then again pressed with a table glass for about 30 minutes and then stored at a place away from dust and moisture.

Apart from the ones described above, there are various other uses of dry plant materials such as preparation of **bouquets**, long-lasting **collages**, **floral pictures**, **floral balls**, **pomanders**, festive **decorations** and so on.

Drying and Preservation Methods

Air Drying

In this method the plant materials are hanged upside down (inverted position) or certain which do not bend are positioned erect in the containers but in thin bunches and kept at an average temperature of 10 ^{0}C and at an aerated place having a very poor atmospheric humidity. They are left there till they get fully desiccated. The flowers used for this type of drying should be crisp such as many annuals, *Aconitum, Acroclinum, Allium* including leek flower heads, *Amaranthus, Ammobium, Anaphalis, Astilbe, Banksia, Briza, Bromus, Carthamus, Centaurea, Delphinium, Dryandra, Echinops, Erica* especially the ones having lavender-pink flowers, *Eryngium, Gomphrena, Grevillea, Gypsophila, Helichrysum, Helipterum, Hydrangea, Leucospermum, Liatris, Limonium, Nigella, Paeonia, Protea, Ranunculus*, most varieties of corn and other cereals, seed heads of *Althaea, Cimicifuga, Digitalis,* and *Limonium. Physalis* (Chinese lantern) seed heads, after removal of the leaves, are also dried by hanging and afterwards are arranged well with the preserved leaves of *Fagus. Hydrangea* flower heads are picked by the end of September or when these seem to dry as experienced through touch, the cut stems are stood in the bottles of water and kept at an airy place until the heads become crisp. Sugarcane and pampas-grass plumes are picked only when these are just emerging from their sheaths, they maintain their immature and grass-like natural shape and open indoors into silvery plumes. These may also be tinted in various shades as per requirements. However, large plumes may be broken up into small pieces and arranged with other preserved flowers. Through upside-down hanging, the foliage of *Eucalyptus* turns to a pearly grey-blue colouration Since in this method the petals of the flowers shrink mostly so it is normally not considered an ideal method for broad-petalled flowers.

Embedded Room Drying

In this method the leaves and flowers are **embedded in a desiccator** or drying medium in a metallic, plastic or earthen container at room temperature with ample aeration where these are allowed to dry out completely. In this method the time required for dehydration may vary as per prevailing weather condition. In this method, due to embedding, normally the petals do not shrink and flower colour is maintained better. The desiccant should normally be of the ideal size of 0.02-0.2 mm or 20-200 mesh and heavy so that its weight may keep the plant parts remaining at their position by retaining their original shape. The drying material should neither be very costly as the cost of dried material is almost low, nor should stick to the plant parts during the process of drying. Corn granules, river sand, saw dust, borax, boric acid, alum powder, aluminium sulphate and silica gel (white and self-indicating blue) are the desiccants which are usually used mostly singly and sometimes in combinations for drying such materials. Silica gel is the best among all as it draws moisture from the material quite efficiently. Various types of containers for drying are used looking into the shape and size of the plant material and drying process is effected under sun or solar dryer, in room, in hot air oven or microwave oven. Wherever the materials are dried, there should be equal distribution of heat to all the sides for even drying. Plant parts should also be quite apart from each other while drying. A 2-3 cm layer of desiccant is spread in the bottom of the container and then over this plant parts are arranged and then the desiccant is poured over the material from all the sides until it forms a top layer of about 1.5 cm above the plant material. However, in microwave oven, non-metallic containers such as heat-resistant plastic, glass or China clay is used. Plant parts are arranged properly without having any contact to the sides of the oven.

A version is **sun drying** as here the containers having the embedded plant material are exposed daily to **direct sun** during day time for quick results but by the sunset these are placed under a roof. This practice is repeated daily until the material is perfectly dried. Though this way drying is quick but weather conditions here too play a greater role.

In **solar dryer**, the movable tray in the bottom is fitted to regulate the temperature. Since it is by the use of solar energy so it is very economical though quite effective as the results are at par with hot air oven or microwave oven.

In **vacuum chamber drying**, the embedded plant materials are dehydrated in a thick-walled chamber fitted with a heating device, a vacuum pump and condenser where the materials are dried at low temperature for better quality.

In **hot air oven**, the materials are kept at 45-60 ^{0}C for a few hours to 3 days depending upon the type of plant materials and the moisture content present in them. Combination of embedding + oven drying method with many types of flowers and leaves were tried by Bhutani (1990a,b,c) and Kher and Bhutani (1979a, b) with a little improved method. They held fresh flowers (without stalks) and leaves between the folds of newspaper sheets or blotting paper. Inside the folds, corrugated boards of the size of the sheets are also placed to help escaping the water vapour and then sheets are kept one above the other. The whole stock is placed under a plant-press and the screws are tightened. For specified time these are kept so and then shifted to electrically-operated hot-air oven at specified duration and temperatures. They recommended

35-39 °C oven-drying temperatures for 72 h for pompon dahlias and narcissi, 48 h optimum for *Bougainvillea* and *Viola wittrockiana*, 24 h in case of the leaves of *Adiantum, Grevillea, Nephrolepis, Thuja* and flowers of *Calliandra, Cassia biflora, Haematoxylon, Hibiscus* and *Tagetes*; 40-44 °C for 96 h for gladiolus and very large rose flowers, 72 h for *Tagetes patula* and large to medium rose flowers, 48 h for *Aerva javanica, Callistephus chinensis, Delphinium ajacis, Euphorbia leucocephala, Mina lobata*, rose buds and small flowers, and *Zinnia lineartis*, 24 h for *Delphinium, Euphorbia leucocephala, Galphimia nitida, Lantana camara, L. depressa, L. montevidensis* and *Sambucus canadensis*; and 45-49 °C for 120 h for *Nymphaea*, 96 h for *Tagetes erecta*, 48 h for *Dendranthema* (small-flowered ones), *Dombeya, Gerbera, Gomphrena, Helichrysum, Helipterum roseum, Iberis amara* and *Limonium*, and 24 h for *Ixora* and *Mussaenda*.

In **microwave oven**, the embedded materials in desiccants are dried for 2-5 minutes through electronically produced microwaves at a medium range of 300-700 Hz. in one stroke or in splits, and then the embedded material as such is taken out from the microwave and the desiccators are kept at room temperature for five hours for perfect setting and only then the plant materials from the desiccators are taken out. Split drying is better than one-time drying. Bhutani (1990a) tried microwave drying at NBRI, Lucknow and reported 4 minutes drying optimum for *Nymphaea*; 3 minutes for *Antirrhinum, Bougainvillea* (vars Lady Mary Baring, Partha, Shubhra), *Callistephus chinensis, Chrysanthemum multicaule, C. paludosum, Dahlia* (pompon), *Dianthus caryophyllus, Gerbera, Gladiolus* (yellow cultivar), *Iberis amara, Lagerstroemia indica, Narcissus* and *Scabiosa*; $2^1/_2$ minutes for *Delphinium ajacis, Helichrysum bracteatum, Phlox drummondii, Limonium bounduelli* (yellow type), *L. suworowii* (rosy-pink), *L. sinuatum* (white and lavender) and *Sambucus canadensis*; 2 minutes for *Delonix regia, Helipterum roseum* and *Ixora coccinea;* $1^1/_2$ minute for *Pentas lanceolata* and *Poinciana pulcherrima*; and 1 minute for *Combretum comosum*. Likewise, the other flowers which can be dried through this method are *Anemone, Dendranthema* (small ones), *Dianthus caryophyllus, Iris*, small- flowered orchids, *etc.* The principle of this method is based on liberation of moisture by agitating water molecules in the organic substances with the help of electrically produced microwaves. Though it generates a little heat but drying is very fast. Flowers and foliage embedded in silica gel over non-metallic wares (earthen or glass) are kept in such oven for only a few minutes. After the specified time for the individual flower species is over, the container is taken out for setting so that the moisture of the container gets evaporated and plant material gets fully dried. While taking out the container from the microwave, the container should be little tilted so that desiccant over and around the flowers is removed, and then after setting out period, the dried plant material is lifted by fingers or throw tweezers, cleaned of the desiccant by inverting the plant material and through gentle knocking and brushing. The dried materials should be stored in the glass or plastic jars or better if under desiccators along with crystals of blue silica gel kept in the bottom.

Press Drying

Flowers and foliage are pressed similar to those of herbarium but these become flat and to some extent the foliage also shrink though many of the species maintain their colours and beauty. Such flowers are suitable for making 'greeting cards', 'decorative floral craft items' for commercial use and for home decoration including floral arrangements. In simple method, the flowers or foliage are properly spread on the newspaper sheets and covered, pressed with heavy weight and then kept at an aerated place. In this case sometimes petals show shrinkage. Sprays and individual leaves of *Crocosmia, Fagus* (beech), ferns, iris, whitebeam and many others can be preserved by pressing them under a heavy weight. Such press dried materials retain their colours and brightness which may be stored in a desiccator or at a dry place for further use.

Freeze Drying

This method is very popular in American and European countries though more expensive than even vacuum drying due to necessary equipments but such dried flowers have multi-billion US$ trade due to a very high quality finished product,*vis-a-vis* the simplicity of the process. In this case the floral materials are arranged in a specimen chamber and then frozen onto -35 °C and when any frozen ice crystals on the product is seen that is sublimed or vaporized with the help of heat. After transforming into condensation, the vapour is captured into a separate chamber as ice. By eliminating the water the dry flowers appear with life freshness almost maintaining their original colour.

Water Drying

Baby's breath, cornflower, fever few and hydrangea dry in a better way through this method. While doing this the lower leaves and flowers from the stems are removed and they are stood in about 5 cm deep normal water at room temperature having proper aeration but a very poor humidity. It takes some 6-10 days for proper drying.

Polymer Preservation

Polymer (polyset) method is a chemical pre-treatment some 45 minutes before subjecting the plant material to air drying. This process lessens the drying time, retains the intensity of the original colour of the material, and minimizes the shattering and wrinkling of the petals. Freeze-dried flowers if subjected to this process, the flowers retain their colours for longer duration.

Humectant Preservation

Humectants are hygroscopic chemicals having different moisture absorption isotherm characteristics. These chemicals are from four general groups, i.e. sugar and sugar alcohols (glycerol), polyols (polyethylene glycol), quaternary ammonium compounds (Comfort) and salts (KCl) which are used to maintain the suppleness of dried plant material by attracting water vapour from the atmosphere so that such dried materials become comparatively less brittle than usually dried materials, which helps in checking mechanical damage and shattering to a great extent and such materials become more life-like. Humectants are applied by uptake or immersion, the latter method requiring more volume of solution but here uniformity is maintained, however, before this treatment the waxy plant cuticles are removed by NaOH or organic dip for proper penetration of the humectants. Shorter duration of treatment or lower concentration of humectants will avoid the risk of hygroscopic-sweating of the plant material.

Dehydration through Bleaching and Sulphuring

An aesthetically attractive plant material which due to oxidative browning has become discoloured can be bleached or sulphured for use in various designs through oxidative agents to break down the coloured compounds whereas through reductive bleaching agents to modify into colourless compounds and for this hypochlorite-chlorite peroxide can be used for reducing brittleness, cost and yellowing. Sulphur dioxide is used to bleach coloured, especially green plant materials, and below certain concentration fixes red colours in certain red flowers through acidification of the tissues. Fresh material can be sulphured by injecting SO_2 gas (1-3%) or through burning sublimed sulphur powder (10 g/m^3) into a sealed chamber overnight before ventilating the chamber. These processes are dangerous so should be conducted by trained personnel and with utmost care.

Desiccants (Preservatives)**.** Many flowers with non-succulent stems can have the heads preserved in drying agents, such as **silver sand mixed with borax or silica gel crystals** or powder and these are known as preservatives. After drying, the flowers heads may be provided with false stems made from florists' wire, drinking straws, skewer or any other alike objects. Some 1.25 cm thick layer of preservative over a box is spread over which single leaves, small sprays of berries and flower heads are spread over without touching one another and then more of drying agent is sprinkled over these in a manner that these fall between the petals and then such boxes are placed over a radiator or at an airy place inside until the plant material becomes crisp and dry.

Silica gel is though most expensive but most efficient desiccant which after use can be recycled again and again. The dry crystals are blue but after absorption of moisture these turn pink. These are first ground and then are spread in a layer over which plant material is spread then again the crystals to cover the plant materials and then the boxes are sealed. This process usually takes three days. Single-petalled flowers, without thickened centres are most suitable. Red and orange roses retain their original colours though pink-coloured varieties to some extent lose their colours and white roses turn cream but still remain attractive. *Delphinium* stems, *Fuchsia* sprays, *Gentiana*, *Pelargonium* and *Viola* dry well by this method by preserving their original colours (Hay and Beckett, 1971). *Anemone, Antirrhinum, Cosmos, Freesia, Muscari, etc.* also preserve well with silica crystals.

Borax and alum are lighter in weight than silica crystals so are best suited for delicate flowers such as *Anemone, Cosmos, Delphinium, Ornithogalum, etc.* For use this is mixed with equal part of sand and in this method also drying takes place within 2-3 days. After drying, the desiccants should thoroughly be removed otherwise these may cause spots.

Sand is heavier and requires longer time than other desiccants but is best suited for flowers such as *Dahlia, Paeony, etc.* Before use, the sand should be strained to remove any organic or other foreign materials including ocean salts and then baked in oven to kill micro-organisms.

Foliage sprays can be preserved by standing the stems in a solution of 1 part **glycerine** to 2 parts water to 2 parts glycerine to 1 part of water (v/v), as in warm water, the glycerine mixes well. The cut stem ends are placed in this solution for 3-6 days in case of soft stems and up to 6 weeks for woody stems and together the leaf-tips should be brushed with this solution every few days as solution itself does not reach usually to the leaf tips. When throughout the plant material the colour changes and beads of moisture are seen on the leaves with underside becoming oily, the material is ready. Single leaves are themselves immersed entirely in the solution and when their colour changes these should be taken out, washed in a mild detergent and then placed flat on a newspaper to dry. If colour of the solution is not changed, it can be reused but by addition of a drop of bleach to keep micro-organisms away. The advantage of glycerine over other desiccant is that even after drying the plant material remains supple but normally the original colour is changed which sometimes may be more attractive even. Depending upon the season when the leaves were collected, type of plant species from which these were collected or the type of leaf, the leaves mature, some by changing their colours to cream-white and others to a warm chestnut-brown, tan or green-bronze (Hay and Beckett, 1971). The perfect leaf sprays should preferably be collected from June to early September for a variety of shades. The foliage, berries and *Hydrangea* flowers preserve well by this method. Beech leaves are very easy for such type of preservation. *Choisya ternata* (Mexican orange) leaves dry best by this method by changing to cream colour. The leaves of *Aesculus, Aspidistra, Clematis, Elaeagnus, Eucalyptus, Quercus,* water iris, *etc.* all preserve well through this method. The collection of such materials should be done when weather is dry and the leaves are fully mature. In fact, the glycerine replaces the water content instead of removing it.

Tinting (staining)

To increase the aesthetic value, the fresh, dried or bleached plant materials may be coloured artificially through immersing in such solutions and for this the inflorescences of *Banksia hookeriana, B. prionotes, Polianthes tuberosa,* many of the orchids, *etc.* are most suitable. In dried plant materials, the removal of waxy cuticle through NaOH and addition of surfactants improves the colouration. The plant materials may also be coloured through silver or gold paints.

Potpourri and Its Making

In French, potpourri means 'mixed stew', literally the rotten pot (translation of Spanish *olla podrida*); *pourri,* past participle of *pourrir* 'to rot', a moist mixture of pickled flowers and leaves though now moist potpourri is a thing of past as it does not give attractive look though this way the perfume is long-lasting. Now only dry method is popular for its lasting colour and as it can be kept anywhere in the house (in drawing room, bathroom, dining room, pillows, etc.) without any poor show. In herb-pillows, the potpourri balls are used. In Latin *putris* is for 'rotten'. It is a collection of dried flower petals, leaves, herbs, and spices, sometimes coloured and scented, meant for filling the air with pleasant aroma. The basic ingredients fall into four categories: the flowers for scent or colours or both, aromatic leaves or herbs, spices, and orange (citrus) peels plus a fixative to preserve the blend. Dried red rose petals & orange peels, allspice, cloves and the fixative orris root make a popular blend, and if need be a few drops of essential oil may be used for increasing intensity of the fragrance. Though there are many herbs containing essential oils but dominating subtler scented herbs are avoided to be used for making potpourri.

Rose petals and lavender retain their **perfume** the longest so such flowers should be selected before these fully open and dried. Other flowers are acacia, broom, carnation, elder, freesia, honeysuckle, hyacinth, jasmine, lilac, lily-of-the-valley, lime, Madonna lily, meadowsweet, Mexican orange blossom, mignonette, mock orange, musk mallow, narcissus, nicotiana, orange blossom, stocks, sweet rocket, violets, wallflowers, *etc.* Flowers used for **colours** are from bergamot, borage, calendula, chicory, cornflowers, delphiniums, feverfew, forget-me-not, foxglove, larkspur, lawn daisy, poppy, sage, tansy, tulip petals, viper's bugloss, zinnias and small everlasting flowers along with pussy willow catkins and sweet myrtle buds for extra texture. After the petals in potpourri, the

aromatic leaves which harmonize with scent of the petals are the second largest group, however, those with most powerful scent should be avoided. In this case the leaves should be dried as whole but should be crushed in the blend to release their scent. The plants with aromatic leaves are alecost, balm of Gilead, balsam, poplar buds, basil, bay, bergamot, lady's bedstraw, lemon balm, lemon verbena, melilot, mints, patchouli, scented pelargoniums, rosemary, sage, southernwood, sweetbrier, sweet cicely, sweet marjoram, sweet myrtle, sweet woodruff, terragon, thymes and wild strawberry. Those **spices**, **peel**, **roots** and **wood chips** having strong aroma may be used sparingly at the rate of 15 ml in ground form, mixed with orris root powder, dried and mixed to one litre of flowers and leaves. Selection should be made from among alexanders seed, allspice, aniseed, cardamom, cinnamon, cloves coriander, dill seed, ginger, juniper, nutmeg, star anise, vanilla pods; dried peels of citrus fruits; roots of angelica, cowslip, elecampane, sweet flag, valerian and vetiver; and shreds or rasping of cedarwood, sandalwood or cassia chips. **Fixatives** in powdered form absorb and hold the other scents to last longer, apart from their own, and for this best is orris root with sweet violet scent. Normally it is used at the rate of 60 ml per litre of flowers and leaves. In absence of orris root, gum benzoin which possesses sweet vanilla scent, can be used from 10-15 ml per 1-1.5 litres of petals and leaves. Alternatively, 1-2 crushed tonka beans (*Dipteryx odorata*) which also has vanilla scent can be used. To adjust the desired intensity and depth to a fragrant mixture and to revive an old potpourri which has lost its scent, only a few drops of **essential oil** may be added.

Methods for Making Moist Potpourri

This is in fact prepared from highly fragrant cabbage and damask rose petals which are partly dried until become leathery and then are halved in bulk. The process takes about two days in dry weather. These petals are layered with dry non-iodized sea salt (half coarse and half powdered) in 3:1 (by volume) in a bowl to its 2/3rd of the volume, each layer of petals some 1.25 cm thick. This blend is kept for 10 days in a dry, dark and well-aerated place where it cakes but if froths are coming up, this should be stirred daily and should be kept for another 10 days to settle and cake. These petal-cakes are broken in small pieces and mixed with other ingredients such as 15 ml of crushed hay leaves, 15 ml of crushed orange peel, 100 ml of orris root powder, 30 ml of ground mace, 30 ml of ground allspice, 15 ml of ground cloves, 1 grated nutmeg, 1 crushed cinnamon stick and 225 ml of dried rose buds to 1 litre of already fermented rose petals which is known as **traditional rose and spice mixture** or 1 litre of lavender flowers, 30 ml cloves, 40 g crushed tonka bean, 7 drops of bergamot oil and 3 drops of clary sage oil to every litre of already fermented rose petals which is known as **rondeletia potpourri**.

Methods for Making Dry Potpourri

Dry potpourri is prepared having theme of a fragrance in mind, e.g. citrus or woodland. Paper dried flowers and leaves are mixed with spices already blended with fixative and then one drop of essential oil at a time is sprinkled, if need be, each time followed by stirring till it has started giving desired intensity of perfume. This bowl is now sealed and stored at a dark, warm and dry place for six weeks for curing and then the bowl is unsealed for display. This potpourri is now kept in a clear glass or in a coloured container in harmony with the colour of potpourri between layers of lavender or rose petals.

Cottage garden potpourri contains all the colours of the cottage gardens such as rosebuds, blue larkspur (*Delphinium*), love-in-a-mist (*Nigella*), daisies, lavender, pelargoniums, hops and straw flower (*Helichrysum*). A **soothing potpourri** for a calm atmosphere can be prepared through blending of rose petals and calendula with soothing lavender, meadowsweet, angelica and lemon verbena. **Sweet-smelling** potpourri can be prepared by adding dried yellow petals of tulips for giving colour. Ant and flea repellent potpourri can be prepared by blending in it the pennyroyal; fly repellent potpourri by blending lavender, tansy and mint; and clothe moth repellent potpourri by blending mugwort and santolina. A **culinary** potpourri can be prepared by using kitchen herbs and spices such as bay leaves, berberry leaves, cinnamon, cloves, juniper berries and star anise.

Handling of Dry Flowers

The handling of dry flower materials is comparatively more easier than the fresh flowers which perish soon and where a lot of care is involved for creating congenial environment for their storage and transportation. Dry flower materials are bereft of these problems, albeit these being brittle their packing is problematic as a slight pressure in packing or transportation can damage them.

Therefore, for packing super grade boxes or cartons are used which can bear more pressure during transit. These should also be protected from being moistened at any level. For storing and their transportation, glass desiccators, cartons, tin boxes, and wrapping in wax paper, plastic sheets or herbarium vasculums fitted with cork sheet should be done. Small quantity of silica gel sachets should also be introduced in the packets or cartons for absorbing moisture. The boxes should be stored at dry places having proper labelling over them.

Protection from Insect-Pests and Diseases

Mice, book lice and silver fish are the major pests infesting dry plant materials. Various borers as in case of stored grain-pests also infest these. Solid pest strips of dichlorvos, 0.01% synthetic pyrethroids or ethyl parathion or gaseous phenyl tablets or methyl bromide will control all these pests. If the material is perfectly dry and relative humidity is below 65% moisture level, no pathogen may attack these. The genera of fungi infecting dry materials are *Aspergillus, Penicillium* and *Rhizopus* whose infection, if any, may be negated by burning sulphur or fumigating SO_2 in storage. The plant material before drying should also be treated with 0.2% captan.

References

Anonymous, 1989. Everlasting flowers from Holland ~ a new product line in the ornamental range. *Agri-Holland*, **1**: 1-5.

Bhutani, J.C. 1990a. Capturing nature, a way with flower "everlastings". *Indian Hort.*, **34**(4):15-19.

Bhutani, J.C. 1990b. Dry rose craft. *Rose News*, **9**(11): 8, 11.

Bhutani, J.C. 1990c. Dry rose craft. *Rose News*, **9**(12): 7-11.

Bhutani, J.C. 1992. Export potential of dry flowers and floral craft items. *Proc Natl. Symp. on Export of Fruits, Vegetables and Floriculture Products*, pp. 67-69. National Horticulture Board, New Delhi.

Bhutani, J.C. 1993. Economic potential of dried flowers. *Agril. Marketing*, **36**(1): 43-46.

Desh Raj, 2006. Drying of Attractive Plant Parts and Flowers. In: *Advances in Oernamental Horticulture.* Vol. 5. *Amenity Horticulture, Biotechnology and Post-Harvest Technology* (ed. Bhattacharjee, S.K.), pp. 189-198. Pointer Publishers, Jaipur (India).

Hay, R. and K.A. Beckett, 1971. Dried Flowers. In: *Reader's Digest Encyclopaedia of Garden Plants and Flowers*, p. 238. The Reader's Digest Association Limited, London.

Kher, M.A. and J.C. Bhutani, 1979a. Dehydration of flowers and foliage. *Ext. Bull. EBIS, NBRI, Lucknow*, 20 pp.

Kher, M.A. and J.C. Bhutani, 1979b. *Floriculture in the National Botanical Research Institute. Progress in Plant Research, Silver Jubilee Publication, NBRI, Lucknow*, **2**: 122-136.

Lesley Bremness, 1988. The Complete Book of Herbs, pp. 158-161, 209. Dorling Kindersley Ltd., London, Great Britain.

Paul, D. and Bunty Shylla, 2002. The art of flower drying. *Floriculture Research Trend in India* (eds Misra, R.L. and Sanyat Misra), pp. 41-46. Published by Indian Society of Ornamental Horticulture, Division of Floriculture and Landscaping, I.A.R.I., New Delhi-110 012.

Singhwi, M.R. 1996. Dried Flowers ~An Upcoming Industry. *Flor. Today* (August), pp. 43-44.

42

Flower Arranging and Designing

Alka Singh

Brief History

Knowledge of the floral arts of earlier cultures and civilizations is important so that the flower designer might better understand the derivation of the flower arrangement styles used today. Floral artists of the various periods of continental European art history used lavish quantities of flower types and colours to create a massive display. There is evidence for use of the flowers for decorative purpose during early Egyptian Period (2800-28 B.C.), Roman Period (28 B.C.-A.D.325), the Byzantine Period (A.D.320-600), throughout the Middle Ages (A.D. 476-1400), the Baroque Period (A.D. 1600-1775) as well as during the Victorian period (1820-1901) as described by Berrall (1997) and Hannay Frances (1948). The Italian artist Michelangelo greatly influenced the artistic transition from the classical Renaissance style to the lavish mood of the Baroque period. These arrangements were often created as symmetrical, oval-shaped designs early in the Baroque period, but asymmetrical curves in the shape of a 'S' or a crescent were later adopted. The period of French art history during the short reign of Louis XVI (1774-1793) showed a continued movement to feminity in design styles. This was the result of the queen's direction in setting court styles. Flower arrangements of the Empire era were more compact than those of earlier French periods, with simple lines in a triangular shape and strong colour contrasts. A typical Empire design would be arranged in heavy urn containing an abundance of large, richly coloured flowers (Griner, 2011). The floral art of the American Federal period (1780-1820) took its styles from the Neoclassic and Empire designs that evolved in Europe at that time (Hannay, 1948). The American people wanted to break away from the traditions of England and desired to create their own national art styles. These floral designs were influenced most by the delicate French style, with the individual beauty of the flowers being stressed. In these designs, the height of the flowers was greater than the width of the arrangement. The sleek floral designs gradually gave way to the ornate and stuffy designs characteristic of the Victorian period.

The Oriental style, *i.e. Ikebana* originated in India, emphasized and modified in China and further developed as an art form in Japan. It came to Japan with Buddhism from India *via* China, around 600 A.D. Formal offering of flowers on the Buddhist altar combined with the aristocratic taste for floral decoration and fused into a refined form, which evolved into *ikebana*. It is a disciplined art form, steeped in the philosophy of developing a closeness with nature. From the late Kamakura period to the Muromachi period (late 13th-15th century), large contests of flower arrangement were held in the court on the day of *Tanabata*. History says that it was at the *tanabata-hana-awase* (a floral competition held at Star Festival) that flowers first became the theme of an annual event. The festival dates from the time of Yoshmitsu Ashikaga (1358-1408), who loved flowers and had numerous flowers planted in his garden. People referred to his mansion as 'the palace of flowers'. The art of *ikebana* developed and flourished with the adventment of three leading schools of *Ikebana* in Japan, the Ikenobo school, the Ohara school and the Sogetsu school. The oldest *Ikebana* manuscript, 'Kao irai no Kadensho' dates from 1486. The famous 'Senno Kuden' manuscript, by the founder of the Ikenobo School of *Ikebana*, was written in 1542. Since this time, many styles have appeared, disappeared and developed as referred by Komoda and Pointer (1984). Japan's oldest school of *Ikebana* called ikenobo, having a 1400 year tradition as its heritage was established in the 15th century. The formal teachings of the complex form of *Rikka* were being given at ikenobo. Chabana, the classic *ikebana*, also called Nagarie, meaning 'to throw in', originated in the 16th century, which represented the natural beauty of flowers and did not require any complicated rules in arranging, to that of *Rikka*. *Shoka*, another form of *ikebana* evolved in the beginning of 18th century wherein *Rikka* and *Chabana* interacted with each other. The Ohara School of *Ikebana* dates back to the Meiji Period (1867-1912) that began with the efforts of Unshin Ohara, who although had studied at Ikenobo school, brought novel ways in *ikebana*, making it less rigid and formal with the introduction of moribana. After World War II, *Ikebana* began to be regarded by some as art, with the result that even works of avant-garde *Ikebana* appeared. Wire, metal and stone as well as flowers were used to the extent that it is sometimes difficult to distinguish these works from sculpture. This movement inspired the birth of 'Jiyuka' (free-style) *Ikebana*, which is completely liberated from the ideas and regulations of traditional *Ikebana*. Influence from the artistic movements of the early 20th century led to the development of jiyuka (free-style) arrangement. In the 1930's and then more so in the post-war period, interest in *ikebana* became much more widespread. Led by Teshigahara Sofu, founder in 1927 of the Sogetsu school (Gupta, 2012), zen-eibana or avant-garde *ikebana* introduced all kinds of new materials, such as plastic, plaster and steel. There are about 3,000 *Ikebana* schools in Japan and thousands more are flourishing around the world. The Ikenobo school alone has some 60,000 teachers worldwide and is practised by about 15 million people in Japan, mostly young women. India has a tradition of garlands and floral ornaments, since time immemorial as depicted in our old scriptures and manuscripts, Ramayana and Mahabharata believed to have been compiled in 500 B.C. Beautiful description of floral ornaments adorned by women is found in the famous ancient writings, *viz.* '*Mrachhkatikam*' by Sudraka (100 B.C.) and in '*Abhigyan Shakuntalam*' by Kalidasa (57 B.C.). The traditional as well as modern day flower arrangement styles reflect the developmental and creative changes being followed throughout different historical era.

Flowers truely represent freshness, colour, fragrance and beauty, focusing on aesthetic and positive side of life. Flowers in the form of garlands or bouquets denote freshness, acceptance, a positive outlook and a newly blessed beginning of any important social function. Flowers have hidden message to cheerful life, the reason they have been an integral part of all important personal, religious or social ceremonies and rituals. As civilization progressed, society changed, so did the art forms and the use of flowers. From being used as offerings to Gods and Goddesses, to a mark of honour with laurel wreaths, to a form of adornment, flowers have evolved over time to what we today know as floral design or flower arrangement. Fresh flower arrangements are careful designing, augmenting the tender qualities of flowers in a more beautiful manner. Thus, the fresh flower arrangements with their beauty, charm and fragrance are hidden inspiration for a peaceful mind, motivating force behind creative work performance.

Influence of Flowers on Human Psychology

Flowers display strong effect on immediate emotional behavioral expression and display positive emotion in human beings. Flowers depict human emotions beautifully besides enhancing and boosting creative ability. Offering flowers to someone has been associated with welcome and positive approach. Since flowers enthuse positive emotion in view of happiness in humans wherein its long-term expression leads to a prolonged involvement related to the feelings of social safety and would therefore be associated with social gathering (Ekman and Davidson, 1994) and family rearing (Mastekaasa, 1992). Various sensory elements of flowers like colour, fragrance, texture, softness, freshness, *etc.* in all, directly influence human psyche and awaken his/her senses, making him/her sensitive and aware of the existing wonderful nature. A recent behavioral study conducted in USA, at Department of Psychology, Rutgers University, and at La Salle University, Philadelphia, shows that persons who received flowers demonstrated increased social interaction and happiness (Havliland Jones *et al.*, 2005). Sight and fragrance of flowers combine to influence brain chemistry and stimulate more positive feelings, as reported by Dr. Jeannette Haviland-Jones, professor of psychology and director of the 'Human Emotions Lab' at Rutgers University, New Jersey. Researchers who have assessed the impact of nature and plants on human health have suggested that people-plant/flowers interactions provide physiological stress reduction. Flowers have a long-term positive effect on the mind frame. Flowers induce happy emotions that contribute in developing friendly environment for facilitating better social and personal connectivity within family, friends, as well business relations (Singh and Dhaduk, 2012). Thus, a view to a beautiful flower arrangement is refreshing and energizes human force to work with more optimistic approach (Singh, 2013).

Flower Arrangement

Flower arrangement is the artistic display of flowers in a right manner and proportion so that the same may provide maximum pleasure and satisfaction, and create a quite peaceful, pleasant and congenial atmosphere at the place. It is an art of designing and arranging flowers aesthetically in a vase or bowl retaining their freshness for a longer time. It is organization of design, form and colours towards creating an ambience using flowers, foliage and other floral accessories. It is a throbbing and live creative art, that invokes aesthetic faculties of the mind. Flower arrangement endows with a magic touch to the cut flowers and grants value addition to more than 50% to the cut flowers (Singh and Dhaduk, 2011). The art of flower arrangement is a beautiful and creative hobby, *vis-a-vis* a lucrative business. Fusion of creativity and science of post harvest handling is needed for developing the art of flower arrangement. Flower Arrangement is an important and integral part of life on almost all the occasions and events whether it be wedding, birthday, usual parties on various occasions, religious and other social functions, *vis-à-vis* beautifying homes and offices. There are basically three styles of arranging flowers, *viz.* Oriental (western style), Ikebana (eastern style) and contemporary free-style which are generally composed of cut flowers, foliage, twigs, stems, along with certain accessories (Hiller, 1974; Packer, 1998). Besides, in Indian tradition or to say in whole of Asia, loose flowers are used for designing floral arrangements in the form of *rangoli,* garlands, flower ornaments and hair decoration.

Material for Flower Arranging

1. Containers

In floral design, the container is a foundation or base of the floral composition. A properly selected container should be functional and in harmony and unity with the arrangement (Aaronson Marion, 1983). Properties of container in flower arranging have been well defined by Blacklock (2006) and Ascher (1974). The container must be capable and deep enough for holding an adequate volume of water and to support the weight of the flowers. In case the container is unable to hold water, another container, which can hold water can be placed inside. Glass, metal, plastic, porcelain or glazed pottery container all have unique effect on flower designs. Metallic finishes require protection from the tarnishing that accompanies the use of floral preservatives in the container water. The shape and size of a container often suggests the type of design to be created within it. A large, heavy container requires a taller arrangement to balance the mass of the container. A horizontal dinner table arrangement is most attractive when constructed in a low and flat rectangular container. The lines of the container are repeated or accentuated to provide unity and harmony within a design. The colour pattern found in the container appears best when it either repeats those found in colour scheme of the design or when neutral colours are present in the container. The most utilitarian colours used in floral containers are green, white, gray, brown, tan or black. Green-coloured floral vases are often selected, since nearly all flower stems and foliage repeat this colour. Brown containers appear best when this colour is duplicated in dried flowers, wood or autumn-coloured leaves. Metallic containers having a finish of silver, brass, bronze and gold leaf usually suggest a formal design. The containers with different shapes are termed differently, *viz.* (i) 'Vase', where height of the container is greater than its width,(ii) 'Bowl', where width of the container is greater than the height and it is often used for table arrangements, (iii) 'Pedestal', where the container is elevated on a base that may be short- or long-stemmed and is useful when a tall arrangement is needed yet the mass of a vase container is not desired, and 'S' curve and triangular types look perfect in pedestals, (iv) 'Basket' (made of metal, wood or bamboo), which are used for both vertical and horizontal but usually in informal designs including the 'round', and (v) 'Novelty' which are limitless in possibilities. Care should be taken to assure that they do not overpower the design. These are perfect specially for free styles.

2. Support Material

(i) **Needlepoint holders** ('pin holders' or 'frogs'): Needlepoint holders are customarily used for creation of modernistic and stylized designs by an amateur as well as by an *ikebana* artist as these are popular stem-supporting aids. These are available in various shapes and sizes (round, oval or square) to suit any particular arranging requirements and are made of coated iron, brass or copper. They may be placed in any location at the base of the container and anchored for temporary or permanent use.

(ii) **Chicken wire** (floral netting): The stem support aid used for large floral bouquets usually consists of chicken wire with 2.5-cm mesh and a filler material. The filler material may be of chipped floral foam, vermiculite or the wire may be used to cover a solid block of foam.

(iii) **Floral foam**: Various floral foam products are popular stem-supporting aids for both florist and home flower arrangement (Blacklock, 2006; Westland,1992; Ascher, 1974). These floral foams are manufactured from resins in several shapes, colours and for various design purposes. They may be purchased in either rectangular bricks to be cut to fit individual containers, or in cylindrical blocks for uses in specialized containers. These products are extremely porous, so they provide water to the inserted flower stems used in the arrangement.

3. Tools Used

Various tools required in flower arranging include (a) '**knife**', a short blade made of quality steel is used to give sharp cut at the stem ends as well as floral foam, (b) '**secateurs'**, having short blades with serrated edges are used for cutting both herbaceous and woody plant material, (c) '**pruning shears**', having two cutting blades of quality steel, used for cutting woody materials, and (d) '**ribbon shears**', being similar to ordinary scissors in appearance. Besides, different wires are also needed based on the floral design, such as (a) '**florist wire**', varying in length and thickness (gauge), the heaviest being No. 18 looking into the weight of the flower and foliage to support them, (b) '**chenille stems**', which are like pipe cleaners with extra flocking, and are used both for decorative purposes and to provide water to corsage flowers, (c) '**spool wire**', a heavy gauge wire is needed to wire arrangements such as wreaths and door swags, (d) '**twistems**' are short pieces of wire enclosed in paper for filling a number of roles where support of repair is needed, (e) '**wooden picks**' are generally green wood picks with a length of thin wire attached, which can be fastened to stems or other materials and inserted into a holding substance, are widely used with nonliving arrangements, and (f) '**floral tape**' having adhesive material on one side and is available in assorted colours to blend into arrangements, similar to a waxed crepe paper it is also used in making corsage or boutonnieres, and as has stretching quality, it is used to wrap flower stems when wire is needed for their support.

4. Foliage and Flowers Used in Designing

Line type of design is thin, vertical and tapering and is made by using various flowers such as amaryllis, bells of Ireland, *Dendrobium*, gladiolus, golden rod, heliconia, iris, tuberose, *Vanda*, etc. and branches of bottle brush, *Murraya, Dieffenbachia*, etc. and foliage of areca palm, asparagus, cycas, fishtail palm, pandanus, sansevieria, snapdragon, etc. For mass designing in round shape at the centre of arrangement the flowers and foliage required are anthurium, China aster, dahlia, *Lilium,* fruits of apple, orange, pomegranate, *etc.* and foliage such as money plant, philodendron, syngonium, *etc.* For uncommon and unusual shape the flowers used are anthurium, bird-of-paradise, heliconia, orchids, *etc.* and the fruits of cashew, grapes, banana, *etc.* Fillers are used in case of thin, spreading and small arrangements with the help of candytuft, daisy, larkspur, golden rod, gypsophila, lady's lace, larkspur, marguerite daisy, statice, aglaonema, asparagus, bottle brush, casuarina, eucalyptus, ferns, silver oak, *Murraya, Thuja, etc.*

Classification of Flower Arrangement

Basically three styles of arranging flowers have been evolved, *viz.* **western style**, **eastern style** (*Ikebana*) and **contemporary style** (free style) which are generally composed of cut flowers, foliage, twigs and stems with some accessories as described by Singh and Dhaduk (2006). Besides, loose flowers are also used for designing flower arrangements in form of *rangoli,* garlands, floral ornaments and hair decoration. Formal offering of flowers on the Buddhist altar combined with the aristocratic taste for floral decoration and fused into a refined form, evolved into *ikebana*. This art greatly influenced all aspects of life, culture and the arts. *Ikebana* is a disciplined art form in which the arrangement is a living thing in which nature and humanity are brought together. Steeped in the philosophy of developing a closeness with nature, it is the art of association of ideas and impressions of nature expressed through flowers. Contemporary styles of flower arrangement neither follow geometrical forms of western type flower arrangement nor the strict rules of *ikebana*, and has a large scope of displaying aesthetic sense and playing with and widening one's creative ability. There may be more or less number of flowers and foliage. The arranger can use sculpture, statue, candle or small fountain or even pot plants according to one's imagination and creativity. There are arrangements designed for festivals, celebrations, inaugural functions and even funerals depicting cheerfulness, emotions, changing seasons and melancholy.

1. Western Style

The emphasis is on the use of 'many flowers as a mass' in the oriental (western) style of flower arrangement. This style follows geometrical forms and consists of more number of flowers and depicts a complete and full effect (McDaniel, 1996). Balance, depth and rhythm are very important to be maintained. It displays mass effect hence repetition and continuity are very important to exhibit appealing and aesthetic look. The basic principles involved in western style of flower designs are as under. By visualizing the plan about the shape and size of the arrangement, it is the **design** that decides the structural pattern whether it is a line, mass or a combination of both, depending upon the materials available, location and occasion. Such arrangements, irrespective of whether are symmetrical or asymmetrical, should have both mechanical, i.e. by all means the arrangement being commensurate to the container and visual **balance**. A well-designed arrangement should appear to be stable and self supporting. Balance is lost when the arrangement appears to be divided into several parts of areas and that shows gravitational instability. Dark colours give a visual effect of weight and so are used low in the arrangement though lighter colours are more appropriate for outer and upper portions. The **scale** or proportion of a design is dictated by the size of arrangement and its relationship to its surroundings. Proportion in an arrangement is accomplished by scaling flowers towards the focal point. The flowers are graded in size for placement of increasingly larger flowers in the arrangement until the focal point is reached. Gradation of line, texture, colour and flower form in graceful **rhythmic** form is characteristic of these designs. Motion is created most easily in a design having a curved line. Here the viewer is attracted gently from the top of the arrangement, along the curving elements of line or texture of colour through to its focal point, the centre of interest. Rhythm may be achieved in several ways through opposition, repetition, radiation, or transition. **Harmony** in flower arrangement indicates a pleasant composition of flower shapes, colours, textures and size. There is little or no abstraction and the plant materials are used in a naturalistic way. **Unity** is best achieved by repeating related flower types, colours and textures throughout the design. This principle involves appropriate use of flowers for the occasion and blending of flower colours in the floral arrangement. The central portion of the floral arrangement from where the flowers and foliage of the arrangement appear to be emerging is known as the **focal point (emphasis, *vista* effect** or **accent)** of the arrangement. The most striking contrast in a design is created by use of large and brightly coloured flowers at the centre of interest with less vibrant tones or tints on the outer perimeters of the design. Flowers with unique shapes also attract attention. The **broad types of designs** in western style of flower arrangement include various traditional designs, corsage, boutonnieres and wreaths.

Symmetrically or asymmetrically 'triangular', 'round' ('spherical' or 'globular'), 'oval', 'Hogarth curve' ('S curve'), 'L shape' and 'crescent' are the standard patterns of traditional western style (Larmie, 1980; Griner, 2011; Singh and Kumar, 2007), and their techniques are described here with. The symmetrical **triangular** design is more traditional with line mass effect and this has an added advantage of displaying it in large sizes on any occasion. This is arranged in three basic lines of the triangle in the container, mass flowers are arranged in the geometrical design, the gaps are filled by proper arranging and fixing the fillers, and a complete and perfect look is given by covering the base and sides of arrangement with green foliage. The **round** arrangements are designed on dining or central tables and around the candles or candle stands, with lower heights, to be viewed from all the directions. This arrangement looks very attaractive and has traditional mass effect. The steps involved in this design are arrangement of fillers of required length horizontally as per specification of the shape, arrangement of mass flowers evenly and the whole arrangement should radiate like the sun. The typical '**Hogarth curve**' ('**S curve**') arrangement is an elegant and exclusive design named after the English painter William Hogarth (1697-1764), which is perfect for home display and as a welcome for special guests, and it has an added elegance when designed in pedestal. The 'S' shape basic curved lines are prepared as per requirement by wiring two twigs of flower or foliage, then mass flowers are arranged within the specified design and then finally the gaps in the design are properly arranged by fixing the fillers. '**L shape**' arrangement is a form having similarity with asymmetrical triangle and basically a traditional line arrangement. Here arrangement in the container is made in two basic lines forming 'L' shape first, then mass flowers are arranged in geometrical form, and then finally the gaps are filled by proper arrangement of fillers. The **crescent**-shaped arrangement looks delicate and perfect for evening programmes whether at home or elsewhere in a general get-together, and the beauty is enhanced when it is designed in pedestal and bows. This arrangement denotes innocence and romance in life with cooling effects and for relaxing. Crescent can be prepared by wiring the flowers or foliage twigs. These are designed in a crescent-shaped curves by using two floral stems, then mass flowers are fitted in between to maintain the crescent shape, and then finally the gaps are designed by proper arranging and fixing of the fillers in between.

Traditional western flower arranging design (style) typically refers to all floral designs originating worldwide during middle of 20th century having influence of primarily Asian, European and Early Americans. Culmination of two main trends one from Asia with the Chinese emphasis on naturalism to the highly stylized linear designs of the Japanese and the influence of the massed bouquet art of Europe, which typically included all styles evolved in different periods (Hannay Frances, 1984; Berrall, 1997). Thus the traditional style of flower arrangement combined attention to both line and mass, creating many variations of the line, line mass, and mass arrangements (Packer, 1998; Singh, 2013). These classic designs are based on geometric shapes including ovals, crescents, spirals, triangles, the Hogarth curve, zigzags or other geometric forms, having one natural centre of interest somewhere near the central axis where all lines converge, with no evident crossed lines (Larmie, 1980). Traditional designs were initially arranged in different types of containers like bowl, vase, pedestal or novelty shaped flower pots using different shapes of pinholder for holding flowers. Later, floral foams were being used to make lovely designs in western style of flower arrangement.

In a **traditional line design** (arrangement), the line predominates and is strongly influenced by Asian design. Line designs are simple and clean resulting into a sharp open silhouette. Inspired by the forms in nature, this design depends upon normal patterns of plant growth and existing curves. A set pattern is usually required, e.g. triangle, vertical, horizontal, crescent, zigzag, etc. Various forms of line designs include inverted 'T' arrangement, 'L' pattern, vertical, crescent and Hogarth curve (McDanial, 1996) while that of mass designs include round, triangle, oval or fan shaped. Typical **traditional line mass design** is a blending of European and Japanese design wherein linear qualities of a line design with the additional mass of plant material provides gradation and transition for balance and rhythm. Thus a line mass arrangement is similar to a line arrangement, but there is more massing of the material, creating a greater depth, the third dimension that is practically not much emphasized in line designs. Further, additional material mainly foliage is used at the rim of the container to enhance and strengthen the line and also to emphasize centre of interest. It has an open silhouette, a pleasing outline and an organized placement of flowers and foliage resulting in a predetermined form. A **traditional mass arrangement** is a bold and bountiful highly symmetrical and well balanced arrangement with a range of flowers in many forms, sizes, colours and textures typically arranged in a round, oval, triangular or fan-shaped design (Aaronson Marian, 1983). It has a closed silhouette, more solid than being void. It is directly influenced from European design. Though the arrangement is typically compact but flowers should never be crowded. Focal point is emphasised in the centre using bolder flower forms with brighter hues.

Corsages and **boutonnieres** are small floral decoration designed to be worn and are very popular in western countries (Griner, 2011). Corsage is a small floral arrangement designed beautifully to be attached to the dress or worn on wrist. Corsages are also sold extensively during the Christmas, Easter, and Mother's Day holidays in western countries. One of the more important uses for corsages in Europe is for weddings. Florist's corsage is so formed so that it resembles the arrangement in containers. The bow and trim become the integral part of the design and satin leaves may provide the foliage. The corsage must be constructed securely enough to retain its original design with minimum use of wires and floral tape to avoid large masses of stems at the base. Flowers should be positioned for best effect and firmness in construction. Flower colours should be coordinated with the dress, the complexion and hair colour of the wearer, and the lighting available (day or evening). **Floral wreaths** are formed in a circular garland, usually woven of flowers and foliage, that traditionally indicates honour or celebration. Wreaths are especially very popular on funerals as a symbol of honouring the deceased. Besides, they are also mostly used on occasions such as special celebrations and to welcome the guests as welcome wreaths when dinners are hosted. Wreaths are constructed on the wire frames by tying the greenery with a continuous strands of wire.

2. The Eastern Style ~ the *Ikebana*

Over the seven centuries of its evolution, ***Ikebana*** has developed many different styles of arrangement with different philosophies behind each. The first teachers and students were priests and members of the nobility. *Ikebana* symbolizes living flower. In Japanese *ike* means 'to set in' or 'to live', and *bana* means 'flower'. *Ikebana* is an expression of liveliness and has spiritual significance, depicting life and freshness (Komoda and Pointer, 1984). *Ikebana* finds its basis in the beauty and the meaning the humans have for flowers is a perceived beauty, which originates from the essential bonds of man and nature. Japanese *ikebana* is a creative art, which brings indoors the charm and beauty of landscapes, the seashore or lakeside. It recreates nature on a reduced scale through the arrangement of all types of plant materials gathered from nature, i.e. from gardens, rivers and valleys. Though it is always based on principles of asymmetry yet the arrangement looks quite balanced, where only a few flowers are used to form basic lines of *ikebana* for giving filling effect, the arrangement depicts spaceous and creative view, it signifies life being associated on the theme of nature, and it appears persisting and continuous. There are three basic lines in *Ikebana*, *viz.* **shin** which forms the most important basic line, symbolizes 'heaven' and expresses 'truth', and it equals the length × width of the container × 1.5-3.0 times; **soe** forms the second important line and symbolizes man and expresses 'body of the human being' and its size equals to $^2/_3$-$^3/_4$ of *shin*; and **hikae** (*tai*) is the third important line symbolizing 'earth' which expresses support and denotes base and environment, and its size equals to $^1/_3$-$^1/_2$ of *shin*. So the three basic lines of *ikebana* are 'earth', 'man' and 'heaven'. The shape of *ikebana* container varies from informal to curved, double or partitioned types, tall, shallow or broad, rectangular, round or oval and are available in varied forms of porcelain and lacquer ware lined with copper and glass. Besides, baskets of metal and wood, and mostly traditional bamboos are used as *ikebana* container. A single, hollow bamboo container is cut to have double and triple openings for the purpose of incorporating different floral designs to form one complete composition. 'Ikonobo', meaning arranging flowers of the hermitage along a pool, was the first school of *Ikebana* which started in early 621 A.D. by Buddhist monk Semmu.

Rikka (standing flowers), *Nagiere* (*Chabana*; meaning thrown flowers),*Shoka, Moribana* (heaped up flowers),*Morimono* (triangular arrangement of fruits, vegetables and flowers), *Jiyuka* (*Jiyubana*; freestyle) and *Zeneibana* (avant-garden flower arranging) are the important styles of *ikebana*. The ***rikka*** style developed in Ikenobo, the oldest school of ikebana, is the Buddhist expression of the beauty of nature, with nine branches representing hills, waterfalls, valleys and so on arranged in a formalised way. *Rikka* is the basic and original style of *Ikebana* arrangement that uses tall vase and highlights vertical lines. In *rikka*, the emphasis is on elaborate technique, large-scale symbolism and fixed style where the vases with a narrow opening or tall, jar-shaped containers are used, with the stems being bundled tightly together at the mouth. Crosspieces are used to fasten the stems to the vase. Hidden within the principles of this style is surprisingly fertile ground for variation and adaptation to contemporary environments. A *rikka* arrangement consists of nine parts (*yakueda* or *dogu*) in all and these are *shin, shoshun, soe, uke, mikoshi, hikae, nagashi, do* and *maeoki*. Sometimes not all of the nine *dogu* appear in the arrangement and at other times certain of the *dogu* are exaggerated, depending upon the purpose of the work or the characteristics of the materials. Each of these nine have a specific role and are arranged at a specific angle. ***Chabana*** is done in a tall vase with a few materials, having distinct three main lines or sometimes just two lines. The container is called *hanarie* meaning 'to put in flowers'. The distinctive feature of the *chabana* arrangement, from a technical viewpoint is that flowers in it are not made to stand erect by artificial means, but are allowed to rest in the vase naturally and so the vase used should be tall and narrow-mouthed. ***Shoka*** is a style that is less formal than *rikka*, but at the same time, not as free-flowing as *chabana*. It combines the dignity of *rikka* with the simplicity of *chabana*. Basic *shoka* pattern is upright, the way most plants grow in their natural state. *Shoka* consists of five lines in all with three major and two supporting ones. The basic lines of *shoka* are *shin* (heaven or truth), *soe* (man or support), *tai* or *hikae* (reserve or earth), *do* (trunk or base) (half of the shin in the centre) and *sugat naoshi* (improving line). ***Moribana***, which means 'piled up flowers', is entirely different from the former 'standing' arrangement. *Moribana* is done in flat containers that allows for spreading out sideways. Changing from the former *Ikebana* emphasis on line, where the stems were kept in one clean vertical line at the water's edge, the concept of an arrangement of wider space and depth was born (Care, 1961). There are basic three lines in *moribana*, *viz. shin, soe* and *tai* (*hikae*). Besides, some *jushi* (fillers) are further arranged to enhance the overall look of *moribana*. There are different types of *moribana* based on the angle of the primary (*shin*) line, viz. basic upright, slanting and hanging, *kenzan* (pin holder) with many sharp points into which flowers are inserted and placed as off-centre in the container and never in the centre. In the *moribana* arrangement all the materials used must be fixed firmly in the pin-holder. **Under other types in *ikebana*** are ***morimono*** developed later in the 19th and 20th century with a simpler form. The basic principles remained the same although fruits and vegetables were also used in this style. ***Zeneika*** (*zeneibana*) style was also developed later in the 20th century which allowed the use of sculpture and statues. ***Jiyuka*** (free-style *ikebana*) is the most recent to emerge from Ikenobo's long tradition. As a more personal expression it is suited to contemporary environments and tastes. Free style is sometimes broadly divided into a naturalistic style and a more abstract style. Both styles use plant materials in new ways, yet respect the beauty and essential qualities of each material. Flowing from the arranger's inventiveness in using materials to convey an effect or mood, free style's possibilities are unlimited.

3. Contemporary or Free-Style Designs

The contemporary (free) style of flower arrangement neither follows rules of geometrical shape of western style nor does it follow strict rules of eastern style. The aim of the free style is the beauty and expression. However, it is also based on some principles to maintain aesthetic look (Aaronson Marian, 1984). This style should be practiced after learning the western style of flower arrangement. This style is highly creative and expresses the aesthetic sense of the arranger. The basic principles include **balance**, i.e. where the shape of the free style may not be geometrical yet a perfect balance between the flowers arranged, container and the base or placement of the flower arrangement should be clearly visible. It emphasizes on the use of materials in a floral arrangement which gives stability to the arrangement. The overall arrangement should look in **scale** and **proportion** but free flowing with boundation of particular shape. All the floral material used in the arrangement should be in **rhythm** and should complement each other, *vis-a-vis* container. Whole of the arrangement should be in **harmony** and **unity** to please the viewer's eye. Inspite of various flower materials being used, the arrangement should express unity and oneness. There are no standard patterns of free style. There is wide scope for the use of flower material for the arrangement. Besides, other material like drift- wood, stones, rock, a live plant in pot, statues, candle, candle stand, *etc.* can also be incorporated. Also, there is no fixed rules for the container type, form and plant material and stem supporting material. The free style can imitate natural landscapes with modern outlook in a stylish way. Few contemporary styles have been described by Miller (1981), Healey (1986) and Curtis (2004) as mentioned here with.

Abstract designs are bold in colour and form, innovative and creative in which unusual or reformed plant material is arranged in a non-naturalistic way to create new and exciting shape. Plant material is utilized as pure line, form and texture to create new images. **Assemblage** design is influenced by artists Picasso and Braque's early prototypical assemblages which is a 3-dimensional creative, abstract design combining 'disparate found objects' and plant material into a unified whole. The arranger creatively relates the items through his/her unique elements of line, form, colour and texture. **Biedermeier** is a design in which pavé techniques are used to create concentric rings of flowers arranged in a compact rounded or slightly conical shape. The contrast of colour, form and texture creates interest in the design. Beside flowers, berries, leaves, nuts, small vegetables or fruits may be used in the concentric rings. Inspired from the floral cascades of the Art Nouveau period, the design is characterized by a downward, loose-trailing flow of materials, often heavy with foliage representing a cascading waterfall, hence it is named **cascade** (waterfall) design. Plant material flows from the centre of the design out and over the container edges. Long and pliable flowers and foliage such as orchid sprays, *Clematis*, *Asparagus plumosus*, *A. sprengeri*, grasses, vines, ivies, string smilax, yarn, conifers, twigs, metallic thread, small fragments of mirror, feathers, *etc.* can be incorporated to give the appearance of reflected sunlight off the cascading water. This design combines the geometric form or forms to create a single unit and has a foundation similar to a strong and dominating architectural structure. Developed in the middle of 20th century, **creative floral design** refers to designs in which creativity, imagination, and originality are basic. The arranger creates new concepts and forms by combining features from different styles and periods. Bold in form, size and use of colour, having more than one focal point but strong clarity of design with clear space being reflected are the main characteristic features of creative designs. It is a design in two or more containers of the same shape, but different in size or colour, reflecting an **echo** of the bold one. **Floor designs** are typically very tall, bold, dominant and staged directly on the floor. Proportion is the most important element of design. **Framed spatial** is a smallest design organised as a single unit suspended within a frame. It is a light airy design with an unconventional use of plant material combined into a single, unified design, which is then staged within a volume of space defined by a frame. **Free form** design is a closed silhouette in which outline is free and flowing as an outer periphery of an amoeba and not a geometric shape, with no intersecting or straight lines. Plant material is manipulated by bending or twisting. Negative space is very important, with good asymmetrical balance. The arrangements which do not exceed 12.5 cm in any direction is called the **miniature arrangements**. They can follow any design or style but should be artistic with proper scale and proportion. **Mille fleurs** in French means 'a thousand flowers', and the design is rounded or fan-shaped using different kinds of flowers and colours, arranged in a light, airy fashion in a contemporary vase. New millennium or **modern European arrangements** are distinguished by roundish shape and compactness with an emphasis on texture. Colours may be monochromatic or contrasting. Flowers arranged in natural, casual style, projects that the flowers are casually stuffed in a vase and this is **modern French arrangement**. **Naturalistic designs** are based on nature and capture the essence of a natural landscape scene. The plant material should be of high quality and the entire life cycle of a single flower (often a bulb flower) is represented. The design includes the stem, foliage, buds, blooms and roots in varying stages of development and emphasizes the beauty of the flower. **Panel** is a creative design which includes a panel or panels, plant material and other components. The panels can either be self-supporting or be supported by the other components in the design. **Parallel designs** comprise of three or more units of strong, parallel lines with open (negative) spaces between each unit. These may be horizontal, vertical, oblique or a combination of these having varying heights but in dominating form. **Pavé design** is placing of individual materials closely together to create a flat and jewel-like pattern. This tight clustering style emphasizes contrast, colour and texture. Based on mythology, rebirth and a round symmetrical form are the basis of **phoenix** design. Base is composed of plant material giving traditional and compact look. Overall, it is a radial design with a very little space between the materials. Form, scale and proportion are important elements of this design. A creative design featuring the dominance of the **cylindrical** (tubular, columnar or pillar) form where plant material may be densely packed within the tube and soaring upward, but still echoing the cylinder shape. It is a creative design containing material which reflects an image or light back to the viewer, hence it is called as **reflective design**. These reflective components are integral to the design and must appear throughout the design. **Synergistic**

is the creative design consisting of three or more units, which have a greater impact when viewed together than they would on their own. Although, all the units share the same colours, textures and forms, creating balance, rhythm and harmony, but each unit should not appear as a complete design in itself and therefore there should be some connection among all the units. An **underwater** design must have some or all of its components, including plant material visible under water. All materials in and out of water should appear as a unified design. A clear glass circular container provides the best effect. A properly designed **all-foliage arrangement** can be as attractive and artistic even without using any flower. A drift-wood laden with lichen will make an arrangement very natural. An all-foliage arrangement looks elegant when properly selected foliage of varied colours, shape and forms are used. Tall leaves such as palms, cycas, sansevieria, lilies, pandanus can be combined with dwarf colourful foliage such as alternanthera, chlorophytum, coleus, dracaena, anthurium, calathea, maranta, rex begonia, etc. Colourful leaves of crotons and *Magnolia grandiflora* are also quite suitable for such arrangements. For a cascade arrangement the foliage such as *Pilea muscosa* (*P. microphylla*), *Zebrina pendula* and *Hedera helix* are very useful. A figure or statue of a bird or animal or a natural-looking rock placed at the base of an arrangement can add artistic and natural look to an all foliage arrangement.

Loose Flower Arrangements

The loose flower-arrangements in the form of garlands and ornaments have been the tradition of India since the time immemorial. Basically the loose flowers were used as flower offerings to gods and goddesses and for decoration in temples. The classic poet Kalidasa, in his famous compilation *Meghaduta*, written during the rule of Chandragupta II (A.D.500), mentions that the women of Alkapuri in monsoon decorated their hairs with *kadamba* flowers, wore *sirisha* flowers (*Albizzia lebbek*) in the ears and carried pink lotuses in their hands. Possibly floral **garland** is one of the oldest and traditional method of using flower for decoration. The use of flower garlands have been mentioned in our oldest manuscripts like Ramayana, Mahabharata and Vedas. Even today, flower garlanding is an important custom in marriages and as a welcome gesture to dignitaries and special guests in Indian culture among all religions and castes. The devotees offer garlands of red *Hibiscus rosa-sinensis* to goddess Kali and the same is worn by the sage who worship this particular Goddess. Sweet scented flowers are more preferred than others for making garland. The most sought after flowers are *Jaminum auriculatum, J. grandiflorum, J. officinale, J. pubescens, J. sambac* and scented *Rosa damascena* and other roses. Other flowers such as *Asclepias, Barleria, Callistephus chinensis*, chrysanthemum (annual and perennial both), *Centaurea moschata, Crossandra, Dahlia, Dianthus, Eranthemum, Ervatamia, Gaillardia, Gardenia lucida, Gomphrena, Helianthus, Hymenocallis, Ixora, Michelia champaca, Mimusops hexandra, Nelumbo, Nerium, Nyctanthes, Nymphaea, Paeonia, Pandanus, Polianthes, Rudbeckia, Scabiosa, Tagetes erecta, T. patula, Thevetia*, etc are widely used as loose flowers and for making garland. Often a flower of lotus or a bunch of lotus is used as the locket of the garland or in between the garland of other flowers. Garlands are either made of only one type of flower or a combination of different flowers. The flowers are held together with the help of a cotton, nylon or a silk thread, inserted in the flowers with the help of a needle. For heavy garlands fine wire stings are used. To make the garlands more gorgeous, in between the flowers threads of jari or silk ribbons are also inserted. Decorative foliage of some ornamental plants such as ferns are used in the garland to improve its look. The central focal point of garland is generally of bright coloured flowers to emphasize its look.

Important **floral ornaments** like bangles and floral crowns were worn by women since old time. Even today, in certain Indian dances, the dancer, especially the female dancers, also wear floral bangles along with garlands. In Maharashtra, the expecting mothers, in a ceremony called 'Shrimanth' are honoured like flower queen with floral ornaments like crown, bangles, necklace, earrings, etc. In West Bengal, a floral crown is also made to be used in ceremonies by the central character, especially during *Annaprasana*, a ceremony when the newborn takes its first cereal food, generally cooked rice, at the age of 6-9 months wearing a floral princess in a drama. The flower most used for this purpose is tuberose. At first a crown shaped strings may be used to make the structure and the flowers may be sewn inside the wires. In the countryside, generally, the frame is made of large leathery leaves, such as jackfruit or *kadamba*, on which the flowers are sewn with the help of cotton thread and needle. Ear-rings and *bajubandhs* made of flowers are also used in ceremonial dances. Heroes and heroines in the drama of various figures such as Rama and Sita wear various floral ornaments. The tribal women even today, decorate their ears with flowers as earrings and also put on flower ornaments during festivals and celebrations.

The temptation of women to **adorn hairs** with flowers is an age-old custom. This is the tradition followed from ancient times till today with modern styles. Here again, the fragrant flowers get preference over others. However, in certain cities of other countries such as Hawaii and islands of Tahiti, many large flowering perennials such as chrysanthemum, *Cattleya* orchids, carnation, *etc.* are used for hair decoration. Though it is not very common in North India, the tribal women still adorn their hairs with flowers. In whole of South India, especially Andhra Pradesh, Karnataka, Kerala and Tamil Nadu, the women adorn their hairs with jasmine flowers. In our country, *Veni, gajra* and *broach* are the variations available in flower arrangements for hair styles specially for specific occasions (Singh and Dhaduk, 2006). These are very well accepted in todays modern age and preferred by the women. These special kind of flower arrangements are widely used by the *Bharat Natyam* and *Kuchipudi* dance recitals for hair ornaments (Manivannan, 2012). The ***veni*** is single or multiple rows or chain of flowers bunched into one unit or as loose multiple flower strings, which is applied on long open hairs or plaits (*choti*). These flower chains are nothing but mini- garlands prepared by piercing flowers with needle and thread. Generally, fragrant white flowers like jasmine and tuberose are used. In South India, *Barleria* and *Crossandra veni* is very popular. ***Gajra*** is wider, circular and firm form of *veni. Gajra* making requires proper skill of flower weaving with needle and thread. It needs proper arrangement and combination of other flowers to give a decorative form. The *gajra*s are made out of a single fragrant species of flowers such as jasmine, *Michelia champaca* and tuberose, however, now crossandra and barleria or mixture of flowers and decorative foliage have got place in this venture. Mostly white coloured fragrant flowers are prefered and to add design and appearance rose flowers or petals are also woven along. Several decorative materials such as *jari* or silk strings or threads may be used to make the *gajra* more gorgeous. ***Broach*** is a hair decoration flower arrangement similar to *veni*. It is rather a combination of *veni* and *gajra*. It is prepared using thin wire by spirally and tightly turning around the flower bud pedicel. The merit of broach over *veni* and *gajra* is that it can be made into any shape as desired and after preparing also it can be turned to give any shape. It is sold at a very high price, ranging from Rs. 100 to Rs. 2,000, depending upon size and design. Broach provides most stylish and modern look to the flower decoration. These have good demand during marriage ceremonies. For *broach* making buds of *Ervatamia, Jasminum arborescencens, J. auriculatum, J. grandiflorum, J. officinale, J. sambac* and single *Polianthes tuberosa* are used. Petals of rose are also woven to provide excellent designer look.

The custom of ***rangoli*** has a religious connotation and is done during *puja* (prayer offerings) in front of temples or in some social functions such as marriage, and at various festivals in front of the gate of the house. A colourful 2-dimensional display in an attractive design, geometrical or otherwise, is referred to as *rangoli. Rangoli* drawing with the help of dry colour or a coloured paste on the floor, is a common practice especially among the Hindus in India. *Rangoli* when done with flowers add life and grace to the beautiful display. For making *rangoli*, a selected design is drawn on the ground, followed by arranging of whole flowers or petals and foliage with proper colour combination according to the design (Singh, 2013). As per the designer's choice and creativity, a wide range of flowers and petals can be used for *rangoli* making. Flowers of varied colours such as bougainvillea bracts, jasmine, gaillardia, marigold, amaranthus, cock'scomb, rose, chrysanthemum, aster, plumeria, *etc.* give wonderful effect to the floral *rangoli*. Beside flowers, fruits like jamun, ber, fig, sapota, grapes also look wonderful when used in proper design. Even vegetables like cauliflower, okra, peas, beans, *etc.* can be used in a proper way. Various patterns are used for making *rangolis*, in design that may be traditional and geometrical, floral, sceneries depicting nature or rural life, thematic designs for specific occasions or festive seasons. Specific fragrant *rangoli* and floating water *rangoli* can also be created. Fragrant *rangoli* also has air-freshening effects. It is created by using fragrant flowers such as jasmine, roses, frangipani, champa, gardenia, *etc.* Even few drops of citronella oil can be added, this will keep away the mosquitoes. Water *rangoli* looks very graceful, elegant and soothing to the eyes and at the same time is very simple and less time consuming. Flowers are made to float on the water surface in a desired pattern. Floating candles and *diyas* when combined, add more charm to the floating floral *rangoli* especially during Deepawali. Small coloured stones and shells can be placed at the bottom of the large sized urn or bowl that is selected as the container for holding water. Rose, asters and chrysanthemums give a wonderful effect to the water floral *rangoli*. Jasmine, gerbera, rose, frangipani, gaillardia, *etc.* can also be used.

References

Aaronson Marian, 1983. *The Art of Flower Arranging*. Grower Books, London, UK.

Aaronson Marian, 1984. *Flowers in the Modern Manner*. Grower Books, London, UK.

Ascher, A. A. 1974. *The Complete Flower Arranger*. Simon and Schuster, New York, USA.

Berrall, J.S. 1997. *A History of Flower Arrangement*. The Saint Austin Press, London, UK.

Blacklock, J. 2006. *The Judith Blacklock Encyclopedia of Flower Design*. The Flower Press Ltd, Bournemouth, UK.

Care, R. E. 1961. Japanese Floral Art: Symbolism, Cult and Practice. D. Van Nostrand Co., Princeton, New Jersey, USA.

Curtis, C.D. 2004. *Contemporary Floral Design*. The National Association of Flower Arrangement Societies, London, UK.

Ekman P and R. Davidson, 1994. *The Nature of Emotion*. Oxford University Press, N. Y., USA.

Griner, C. 2011. *The History of Floral Design*. In: *Floriculture: Designing & Merchandising* (3rd ed.). Delmar Thomson Learning, Albany, USA.

Gupta, V. 2012. Ikebana: The Concept. In Compendium 'National Workshop on Floral Craft: The Art and Technique for Value Addition in Flowers', pp. 131-132, organized at NAU, Navsari, 12-13 April, 2012.

Hannay Frances, J. 1948. Period Flower Arrangements. National Council Books, Inc. Philadelphia, PA, USA.

Havliland Jones, J., H. Rosario, P. Wilson and T. McGuire, 2005. An Environmental Approach to Positive Emotion: Flowers. *Evolutionary Psychology*, **3**: 104-132

Healey, D. 1986. *The New Art of Flower Design*. Villard Books, New York, USA.

Hiller, F.B. 1974. *Basic Guide to Flower Arranging*. McGraw-Hill Book Co., New York, USA.

Komoda, S. and H. Pointer, 1984. Ikebana-Spirit and Technique. Blandford Press Ltd, Poole, NewYork, USA.

Larmie, W. E. 1980. *Flower Arranging Basics to Advanced Design*. Prentice Hall, New Jersey, USA.

Mastekaasa, A. 1992. Marriage and psychological well-being: Some evidence on selection into marriage. *J. Marriage and Family*, **54**(4): 901-911.

McDaniel, G. L. 1996. *Floral Design and Arrangement*. Prentice Hall Company, Verginia, USA.

Manivannan, K. 2012. Traditional Floral Crafts of South India. In Compendium, 'National Workshop on Floral Craft- The Art and Technique for Value Addition in Flowers', pp. 136-138, organized at NAU, Navsari, on April 12-13, 2012.

Miller, M.S. 1981. *New Dimensions in Floral Design*. Wegferd Publications, North Bend, OR, USA.

Packer, J. 1998. *The Complete Guide to Flower Arrangement*. D.K.Pub., Inc, New York Press, USA.

Singh, A. 2013. Fresh Flower Arrangement: A Hidden Inspiration to Creative Work. *Floriculture Today*, October 2013, pp. 27-32. Media Today, New Delhi.

Singh A and B.K. Dhaduk, 2006. The Textbook of Bonsai and Flower Arrangement. A publication by ASPEE College of Horticulture and Forestry, NAU, Navsari.

Singh A and B.K. Dhaduk, 2011. Floral Craft: A Profitable Business. In Souvenir: 'National Conference on Recent Trends and Future Prospects in Floriculture', organized by ISOH, at SVP University of Agriculture and technology, Meerut on 5-8 March, 2011.

Singh A and B.K. Dhaduk, 2012. Impact of Flowers on Human Life: A Journey Revealed. In: Compendium, 'National Workshop on Floral Craft: The Art and Technique for Value Addition in Flowers', pp. 1-5, organized at NAU, Navsari, on 12-13 April, 2012.

Singh, A. and P. Kumar, 2007. Flower Arrangement: The Art and Technique. *Floriculture Today*, **11**(10): 22-27. Media Today, New Delhi.

Westland, P. 1992. The Complete Flower Arranger. Ultimate Editions. Annes Publishing Ltd., U.K.

43

Mechanization of Microbial Composting Technology for Commercial Production to Address Depleting Soil Health

Indra Mani, Livleen Shukla and Satish Lande

Introduction

The food grain production of India has been estimated to touch the landmark of 264.38 million tonnes in 2013-14 (3rd Advance Estimates of Crop Production Released, 2014) through proper and efficient utilization of the resources. Consequently, the amount of agro-waste generated in India has been estimated to be about 500 million tonnes annually (Ministry of New and Renewable Energy, 2014). Out of this, 330 million tonnes of wastes are obtained from crops like wheat, maize, rice and sugarcane remains. The recycling of these wastes is both ecological necessity and economic compulsion. Globally, biomass is ranked fourth as an energy resource, providing approximately 14 per cent of the world's energy needs and can efficiently be used as an alternative for fossil fuels. Currently, approximately 50 per cent of the world's biomass is being used by human for food, construction and fuel.

The conventional and traditional method of burning agricultural residues for waste disposal not only causes air pollution but also results in loss of nutrients present in the residues. A current study showed loss of nutrient due to burning in sugarcane, rice and wheat residues to be 0.84, 0.45 and 0.14 million tonnes, respectively with a total of 1.43 million tonnes per year (Niveta Jain *et al.*, 2014). However, the use of agricultural wastes as soil amendments on farmland instead of burning is an attractive alternative because it allows some cost recovery, improves physical, chemical and biological properties of soil and recycles the nutrients into the soil (Edwards and Fletcher, 1988). The agro-waste can be efficiently converted into more fusible form in order to utilize it for further improvement in crop production.

Soil is a dynamic living system and if it is not manured to feed soil biota, soil health gets deteriorated which is essential for cycling of plant nutrients. Organic farming is successful on farms where organic biomass is generated in and around farms and whatever rural organic materials become available should be mobilized for this purpose. Therefore, recycling of crop residues through composting is a highly desirable option for the farmers. Often there is desire and frequently economic necessity to integrate the use of wastes in agriculture to supplement nutrient requirement of crops.

Composting is a microbial process in which biological oxidation of organic matter takes place giving rise to a stable product (de Bertoldi *et al.*, 1983). It is an environmentally sound way to reduce organic wastes and produce either organic fertilizers or soil conditioners. It is in fact a fully sustainable practice, since it aims at both conservation of the environment, human safety and economically convenient production. Besides, the use of compost for plant nutrition in organic farming systems, it is also considered the cornerstone of nutrient resources for conserving the soil fertility.

Enriched compost is attained through management of the composting process by addition of microorganisms capable of tolerating and/or decomposing inhibitory compounds, and/or mixing with other substances that stimulate biological decomposition. Studies show that rock phosphate enriched compost prepared by mixing low grade Indian rock phosphates with rice straw inoculated with the ascomycetes, *Aspergillus awamori* is a sustainable system for P supply to plants (Biswas and Narayanasamy, 2006). Use of *Trichoderma harzianum, T. viride* and non-pathogenic *Fusarium oxysporum* biocontrol strains increases the disease suppressiveness of compost against disease causing fungal pathogens such as *Phytophthora nicotianae* in tomato, *Pythium ultimum* in cucumber, and *Rhizoctonia solani* in bean. The role of microorganisms during composting and future prospects for improving microbial composting has been reviewed (Singh and Nain, 2014).

Nitrogen is essential for the microbial metabolism to oxidize lignocellulose. However, an increase in nitrogen content decreases the lignin breakdown and increases cellulose degradation. Therefore the nitrogen content is important for the decomposition process as high nitrogen content increases the lag phase of composting. Use of chemical nitrogen sources such as urea can be replaced with nitrogen sources obtained from plant and animal origins such as soybean trash, farmyard manure (FYM) and poultry waste.

A consortium of four fungal inoculants, *viz. Aspergillus awamori, A. nidulans, Phanerochaete chrysosporium* and *Trichoderma viride* have been showing promising results on transformation of agricultural residues to compost (Sunita Gaind *et al.*, 2009; Sunita Gaind and Lata Nain, 2010).

Methods for Aerobic Microbial Composting

Various constraints based upon time, labour, land, investment and economy have led to development of several composting methodologies. Introduction of effective microorganisms and mechanization of the process through forced aeration and turning have brought a drastic change in the mode of composting processes. On farm, composting is carried in pit-heaps or windrows and depending upon the principle obtained, the transformation process is followed. Water standoff in the composting pits causes decrease in oxygen availability and perturbs the microbial activity; therefore areas low in drainage system or high rainfall, heap method is followed. Pits used are of the dimension 1.5 m depth, 1.5-2.0 m width and of suitable length. Heaps have a 2 m wide base and 1.5 m height with 0.5 m tapering at the top. In case of windrow composting the composting material is placed in rows and turned periodically using loaders equipped in front of a tractor.

i) **Indore Method.** Decomposition takes place aerobically. Five centimetres thin animal manure (4.5 kg dung, 3.5 kg urine-earth and 4.5 kg of inoculum taken from a 15 day-old composting pit) is applied between thick 15-cm layers of composting feedstock to reach a height of 1.5 m. Turning is carried at approximately six week interval for a total composting time of 4-6 months.

ii) **Bangalore Method.** Decomposition takes place aerobically for first few days (8-10 days) followed by a long anaerobic processes (6-8 months). Generally carried in pits having alternate layers of organic residues and night soil, the pit is completely packed and plastered with mud to prevent loss of moisture and breeding of flies. The materials are allowed to remain in pits without turning or watering for three months.

iii) **Efficient Microorganisms (EM) Method.** This method is based upon aerobic degradation of agriculture residues by microorganisms. Several EM are available, however, the discussed EM pertains to the one developed at Indian Agricultural Research Institute, New Delhi which is known as Pusa Composting Inoculant (PCI) having four fungal inoculants, *viz. Aspergillus* *nidulans* ITCC 2011, *Trichoderma viride* ITCC 2211, *Phanerochaete chrysosporium* NCIM 1073 and *Aspergillus awamori* F18. The PCI is applied to the composting substrate through an accelerator solution containing one packet, i.e. 300 g of inoculant, 5 kg cowdung in 100 l of water applied to 1,000 kg of substrates. Generally carried in pits, the residues are given a turning in 30 days interval for a total composting time of 3-4 months.

Factors Affecting Composting

The most important factors which regulate the composting process are as follows:

A. **Microorganisms and Microbial Inoculants.** Most of the organic wastes and crop residues carry indigenous populations of bacteria, Actinomycetes and other fungi, which contribute significantly in the composting process. The fungi and acid producing bacteria appear during mesophilic stage (Chang Yung and Hudson, 1967), and as the temperature increases above 40 ^{0}C these organisms are replaced by thermophilic bacteria, Actinomycetes and other fungi. Spore-forming bacteria develop at still higher temperature, and finally mesophilic bacteria and fungi reappear as the temperature falls. Many aerobic bacteria are initially present and multiply but after increase in temperature their number decreases. *Bacillus* spp. is involved in degradation of proteins, peptones and blood meal. Actinomycetes degrade starch very actively and also water-soluble fractions of raw materials. Thermophilic bacteria attack proteins, lipids and hemicelluloses.

B. **Carbon/Nitrogen Ratio.** It is the most important factor in composting which is influenced by the proportion of carbonaceous and nitrogenous components present in the residues to be composted. Microorganisms require carbon for growth and nitrogen for protein synthesis. A C/N ratio 30 to 40 is optimal for efficient composting and a ratio of 25 to 40 has been found satisfactory. If the organic material is low in nitrogen, i.e. wide C/N ratio, biological activity decreases and several cycles of microbial activities may be required to complete the process of composting. Thus, the finished compost takes longer time for production. On the other hand if C/N ratio of organic materials is less than 30:1, the proportion of nitrogen is in the excess of requirement of microorganism, and in such conditions although the bioprocess goes on, the unassailable nitrogen is lost by volatilization as ammonia or by denitrification. Low C/N ratio under unfavourable condition causes increasing loss of ammonia. The time of composting of high C/N ratio materials can be reduced by adding a mineral nitrogen sources or a nitrogen source such as activated sewage sludge, legumes residues, grass clipping, green leaves and sugar industry wastes.

C. **Shredding of Materials.** The bioprocess can be accelerated if raw materials are shredded into smaller parts as then it becomes more susceptible to microbial action due to greater exposure of the surface area (Gaur, 1987). The most desirable particle size for composting may be in the range of 3-5 cm although large sizes can be composted satisfactorily.

D. **Mixing of Organic Materials.** The C/N ratio and moisture percentages are two important parameters to be considered when mixing together different kinds of material for composting. There may not be any need of blending if the C/N ratio of the material is between 25 and 50 although 30 to 40 would be a better range to aim for substances poor in nitrogen such as sawdust, coir pith, coir dust, paper, cereal crop residue and stalks, stubbles, *etc.* which can be mixed with comparatively nitrogen rich materials such as leguminous wastes, vegetable and fruit wastes, water hyacinth, biogas slurry, *etc.* to obtain a near optimum ratio around 30 to 40. Similarly, the wet and dry materials can be mixed together to obtain a desirable percentage of moisture and to avoid anaerobic conditions. Soil at the rate of 5 to 10 per cent can be added to moist material to reduce the moisture content and absorb ammonia while composting materials of low C/N ratios. Dry soil may be added if sufficient dry organic wastes are not available. Soil may be added to high C/N ratio organic material to buffer acid conditions and to act as diluents for retarding anaerobic process. It can also be used for improving the appearance of the finished compost by giving it a more granular structure.

E. **Moisture and Aeration.** Moisture content and degree of aeration are important factors in composting. If the amount of moisture in the compost pile is below 40 per cent (w/w), decomposition will be aerobic but slow, as degree of aeration is an important factor in composting. Aerobic decomposition will occur at any moisture content between 30 and 100 per cent if adequate turning is provided but higher moisture content should be avoided. The optimum moisture level for aerobic composting is 50-60 per cent; however, a range of 40-80 per cent will be quite satisfactory depending on the nature of material to be composted. Gaur *et al.* (1980) maintained moisture content of 80-85 per cent for aerobic composting of fibrous materials including straw. For anaerobic composting moisture level is not important.

F. **Temperature.** With the multiplication of the microorganism in the composting mass, the heat of exothermic reactions is retained due to large mass. When the temperature rises above 40 °C, the mesophilic microorganisms are replaced by thermophilic microorganisms. High temperature is essential for destruction of pathogenic organisms and weed seeds. The thermophilic condition generally occurs within 2 to 7 days of the start of the operations. The temperature in the middle of the pile goes up to 55-70 °C after which it gradually cools to ambient temperature. Decomposition is fastest in the thermophilic range, the optimum temperature for organic matter decomposition into carbon dioxide and water has been found to be 60 °C (Wiley and Pierce, 1955), whereas Wiley (1957) and Schulz (1952) reported the maximum temperature of 71 °C to be the best but should not be allowed to exceed this for long as decomposition will be slowed down due to thermal kill of microorganism. Only a few thermophilic organisms carry on decomposition actively above 70 °C. While composting, high C/N ratio materials such as wheat and rice straw, sorghum stalks and *jamun* leaves maintaining C/N ratios of 48 to 50. Gaur *et al.* (1980) observed that temperature of compost beyond 52 °C is not so useful.

G. **pH.** Decomposition will occur satisfactorily in the neutral pH because most of microorganisms grow best under neutral pH conditions. Under aerobic conditions, initially a neutral raw material will experience a decrease in pH. The drop is due to the activity of acid forming bacteria. It is followed by an increase in the pH due to proteolytic metabolism and liberation of ammonia from amino acids and account for subsequent rise in pH. Most materials decomposing aerobically will stay within pH range that is conducive for microbial growth thus eliminating any need for pH control.

Maturity of Compost

The maturity of compost is essential for achieving the objectives of composting. This can be adjudged by studying several indices like volatile fatty acids (VFA), cation exchange capacity (CEC), nitrate – N ratio, reducing sugars, C/N ratio, germination test, and Dewar self heating test (Brinton *et al.*, 1995) for humic substances. Compost prepared from good quality organic material, in general, contains 40 kg nutrients/tonne (1.5 % N + 1.0% P_2O_5 + 1.5% K_2O). The final product should be tested for presence of heavy metals or phototoxic substances which are equally important. Humus content in the compost can be an important criterion to judge the maturity of compost.

Standards for Good Quality Compost

pH	6.5-7.5
C : N	20 : 1
Available P	0.5%
Available K	1.0%
Humus	6 to 8 %
Moisture content	15.0 % to 25.0 %
Colour	Black or brown
Odour	Earthy smell
Conductivity	4.0 dS/m

Various methods to evaluate maturity of compost

Physical	Chemical	Biological
Particle size	pH	Microbes
Colour	EC	Annelids
Moisture	C/N ratio	Worms
Porosity	S	Collembola
Bulk density	Starch-iodine test	Thermopiles
Water holding capacity	Heavy metals	Pathogens microbes
Temperature	Pesticide residues	Verticillium test
Total solids	Protein	Plant growth response test
Volatile solids	CEC	
Total ash content	Humification index	

Mechanization of Compost Preparation

Mechanization of composting process is necessary as it improves efficiency, quality of produce, timeliness of operation along with removal of drudgery and saving of labours. The pile method of composting involves the turning of piles at frequent intervals by employing the compost-turner-*cum*-mixer.

About 100-150 manhours are required for preparing compost by pit (10 m × 1 m × 1 m) method and requires continuous monitoring. The pile method of composting involves laying of materials in the form of piles and subsequent mixing with the help of compost turner-*cum*-mixer. The compost turner-*cum*-mixer consisting of mixing rotor, hydraulic system to operate the rotor, water tank to store water and culture, and side tank for weight balancing. The machine employed for turning helps to increase the capacity, timeliness, precision and uniformity of mixing and creating favourable conditions for microbial growth. The turner-*cum*-mixer reduces the energy consumption per tonne of the material handled and is used for mass production of compost. The compost turner-*cum*-mixer is having provision of adding moisture during operation that adds extra feather to its working ability.

Indian Agricultural Research Institute has successfully developed **PUSA Compost Turner-*cum*-Mixer**, **PUSA Compost Loader**, and **PUSA Automatic Compost Sieving Machine** as elaborated here under for mechanization of the entire composting process starting from handling to sieving at mass scale.

PUSA Compost Turner-*cum*-Mixer

This is a trailing- and offset-type machine used for thoroughly mixing the cowdung, farm residues and biomass for manure preparation. The main components of the equipment are a mixing rotor, a hydraulic system to operate the rotor, a water tank to store water and culture, and the machine also has a side tank for weight balancing. The power source is a 75 hp (52.22 kW) tractor with a facility of adjustment of fixing power take off (PTO) shaft with the shaft of the compost turner-*cum*-mixer. The power is transmitted to both turning rotor and ground wheels by means of a gear box with a speed reduction ratio of 2:1. The frame (280.8 cm × 6 cm × 164.5 cm) is made of rectangular iron sheet steel. The rotor is made from a seamless pipe having a diameter of 400 mm and a thickness of 9 mm, and is fitted with projections made of high carbon steel blades over its body surface. A total of 42 blades arranged in a helical path facilitate thorough mixing and turning of the material. The design takes care of the dynamics of compost handling in an efficient manner. The rotor mounted on a frame makes it a complete mixing unit. Rubber pads are provided for partially covering the mixing unit to check the spillage of compost while mixing. To facilitate safe transport of the equipment, both off-and on-the road, the rotor unit is hinged at one end to be lifted vertically with the help of a hydraulic system. A water pump is provided to facilitate sprinkling of water over the compost material. As this machine is an offset type, a side tank full of dead weight of concrete helps counter balancing the weight of the rotor unit particularly while lifting and lowering.

Specifications of the PUSA compost turner-*cum*-mixer

Components	Dimensions
Mixing rotor diameter (mm)	400
Length and shaft diameter of rotor (mm)	2670, 63
Length of rotor shaft (mm)	2870
Hydraulic system with base plate, hoses (5 numbers) and direction control valve (5 numbers)	1000 mm stroke, 63 mm cylinder diameter
No. of blades over rotor surface	42 (numbers)
Length and width of blade (mm)	260, 80
Dimension of water tank (mm)	1220 × 1350 × 1250
Side tank (mm)	1250 × 730 × 760

The machine can operate in extremely rough conditions and it has been provided with a proper safety system. A wire mesh structure has been mounted on the cabin to protect the driver from blown off pebbles and stones. The turning rotor thoroughly mixes the top windrow material with the interior one. The machine is provided with a water tank connected to a sprinkling system for spraying water to maintain the necessary moisture content during the turning operation. For effective composting, the machine supports straight-shaped blades at a rotor speed of 300 rpm with an operating speed of 2.5 km/h, and keeps the 1.0 m pile height.

Comparison with pit method

Composting method	Pit	Windrow
Cost of preparation of one tonne (Rs/tonne)	3000	300
Saving in cost (Rs/tonne)		2700
Percentage saving in cost		90.4 %
Preparation time (days)	120	45

Advantages

1) Compost preparation by manual method is labour-intensive. Also, handling cowdung manually is full of drudgery. The use of machines overcomes the labour problem and hastens the compost preparation process while imparting dignity to labour.

2) The technology facilitates compost making in 45-60 days, whereas in traditional pit method it takes almost 120-140 days, depending upon the ingredients and seasonal variations.

3) The payback period of the technology is only one year estimated at an annual utility of 722 hours. It offers a cost saving of about 90 per cent over manual pit method if the technology is utilized at its full capacity, i.e. composting of 1.0 lakh tonnes raw material per annum.

4) On an average, three turnings per windrow are required so that the machine can handle a gross material of 3.0 lakh tonnes per annum. The energy consumption and economics are favourable for commercialization of the technology for mass production of compost.

5) Experiments confirmed that the use of the machine led to the breakdown of larger molecules into simpler ones, which in turn, resulted in the decrease in C:N ratio from the initial 54:1 to the final 21:1. Similarly, electrical conductivity increased from the initial 1421 μS/m to the final 1531μS/m at a slightly alkaline medium. The final compost consisted of 0.58% N, 1.2% K and 1.2% P. All the values of the final compost showed conformity with the BIS standards.

PUSA Compost Loader

The design of a tractor front-mounted loader was modified for handling cowdung, biomass and ready compost, and was custom-built. It is used for lifting and carrying of compost material up to a height of 3 metres. The bucket fitted with the machine has a capacity of 600 kg. The machine can handle 12 tonnes of compost per hour. It saves time and cost and takes only 0.5 machine-hours for loading a truck whereas 5 manhours are required in manual operation. The loader is under extensive use at the Biomass Utilization Centre of the Institute. Its estimated payback period is 14 months at an annual utility of 1,000 hours.

Specifications of PUSA compost loader

Components	Dimensions
Bucket size	180 cm × 50 cm × 60 cm
Bucket volume	0.5 m3
Hydraulic lift	3 m
Power source	55 hp (41.25 kW) tractor

Advantages

1) The machine requires less labour and hastens the compost handling process.

2) It helps in shaping the windrows and reshaping after every turning.

3) It facilitates easy and efficient loading of the compost into truck or tractor for transporting to the end-users.

4) It helps in feeding of ready compost into the hopper of the automatic sieving machine for rapid compost sieving in the desired grade.

PUSA Automatic Compost Sieving Machine

The machine is suitable for sieving and separating the finer grade material from the coarse grade one. It is powered by three phase electric motors (2 hp, 3 hp and 3 hp). The capacity of machine is 5.0 tonnes per day. The machine is provided with an automatic belt feeding mechanism, a compost separator unit and hopper for safe feeding, and a belt conveyor based unit for conveying the finished quality compost. Separation in different sizes is done for value addition. The finely sieved compost is very useful for plants in pots and kitchen garden. If packed properly, it can fetch a price of Rs 6.0 per kg. As compost sieving is generally not in practice, vis-à-vis fine and quality compost is not available in the market, to a small extent, low capacity machines are in use which require a lot of labour. The machine is economical, efficient and the estimated payback period of the machine is 30 months at an annual utility of 1,000 hours.

Specifications of Pusa automatic compost sieving machine

Components	Dimensions
Feeding table	92 cm × 62 cm × 45 cm
Rubber conveyer belt	40 cm × 3 ply × 5 m
Rotor	3 m × 1 m diameter
Process conveyer belt	40 cm × 3 ply × 3.5 m
Gear box (3 numbers)	40 : 1, 40 : 1 and 30 : 1
Electric motors, 3 phase (3 numbers)	2 hp, 3 hp and 3 hp

Advantages

1) It helps in separating all unwanted, large size material from compost. The machine is particularly useful if the raw material is domestic waste.

2) Unhygienic and un-degradable materials like polythene, metal pieces, broken glass pieces, stones, *etc.* can be separated effectively.

3) Separation in different sizes helps in value addition. Finer grades are used in flower pots.

4) Can convert waste to the best and adds value to the product.

5) Owing to its high capacity, a large amount of compost can be sieved.

A. *bajra* seed-based inoculant B. Liquid inoculant

PUSA compost inoculants

Windrow making

Application of inoculants

PUSA Compost Turner-cum-Mixer

PUSA Compost Loader

PUSA Automatic Compost Sieving Machine

PUSA Compost Sieving Plant

The package of technology is available for sale under Revolving Fund Scheme (RFS) of the Division of Agricultural Engineering, IARI, New Delhi.

Benefits of compost

- Compost has a definite short term and long term advantages when applied to soil to raise crops.
- Compost is a storehouse of energy or the growth and multiplication of microorganisms and their activities in soil which is vital in plant nutrient cycling of different elements.
- Compost supplies both macro and micro nutrients such as nitrogen, phosphorus, potassium, calcium, sulphur, iron, manganese, zinc, boron for plant growth.
- Compost provides growth-promoting substances, which stimulate early plant growth and maturity of crops.
- Compost acts as a soil conditioner because it improves the physical condition of soil such as texture, the tilth, water holding capacity and cation exchange capacity.
- Compost has been found to be very effective material for erosion control.
- The end-product of composting is safe for human and animals because of all pathogens including *Bacillus*, *Anthrax*, *Salmonella enteridis* and several other pathogens are eliminated due to high temperatures (55-65 ^{0}C) attained in the compost piles.
- Compost increases the biological activity of soil and produces crops with less plant diseases.
- Antibiotic substances produced in composting process control plant and animal pathogens.
- Pathogenic bacteria that are associated with sewage sludge and municipal refuses are destroyed when added to compost heaps

Troubleshooting

Making compost is really quite easy, but having too much of a certain material or letting the compost get too wet or too dry can cause problems.

Problems	Possible causes	Solution
Damp and warm only in the middle of the pile.	Pile could be too small, or cold weather might have slowed down the composting	Composting only in piles, one should be sure that pile is at least 90 cm high and wide. With a bin, the pile doesn't need to be so large.
Nothing is happening	1. Not enough nitrogen 2. Not enough oxygen 3. Not enough moisture 4. Cold weather? 5. Compost is finished	1. Make sure you have enough nitrogen- rich sources like manure, grass clippings or food scraps. 2. Mix up the pile so that it can breathe. 3. Mix up the pile and water it with the hose so that there is some moisture in the pile. A completely dry pile doesn't compost. 4. Wait for spring, cover the pile, or use a bin.
Pile doesn't seem to be heating up at all	Poor aeration or lack of moisture.	Avoid thick layers of just one material. Too much of something like leaves, paper or grass clippings don't break down well. Break up the layers and mix up the pile so that there is a good mix of materials. Shred any big material that isn't breaking down well.

Problems	Possible causes	Solution
Matted leaves or grass clippings aren't decomposing.	Not enough oxygen or the pile is too wet, or compacted.	Mix up the pile so that it gets some aeration and can breathe. Add coarse dry materials like straw, hay or leaves to soak up excess moisture. If smell is too bad, add dry materials on top and wait until it dries out a bit before you mix the pile.
Stinks like ancid butter, vinegar or rotten eggs.	Not enough carbon.	Add brown materials like leaves, straw, hay, shredded newspaper, *etc.*
Odour like ammonia.	Inappropriate materials (like meat, oil, bones), or the food-like material is too close to the surface of the pile.	Bury kitchen scraps near the centre of the pile. Don't add inappropriate materials to compost. Switch to a rodent-proof closed bin.
Attracts rodents, flies, or other animals.	This is normal composting, and part of the natural process.	Not a problem.
Attracts insects, millipedes, slugs, *etc.* and has fire ant problems.	Pile could be too dry, not hot enough, or has kitchen scraps too close to the surface.	Make sure your pile has a good mix of materials to heat up, and keep it enough moist

References

Biswas, D. R. and G. Narayanasamy, 2006. Rock phosphate enriched compost: an approach to improve low-grade Indian rock phosphate. *Bioresource Techn.*, **97**(18): 2243-2251.

Brinton, W. F., E. Evans, M.L. Droffner and R.B. Brinton, 1995. A standardized Dewar test for evaluation of compost self-heating. *Biocycle*, **36**: 1-16.

Chang Yung and H. J. Hudson, 1967. The fungi of wheat straw compost. I. Ecological studies. *Trans. British mycol. Soc.*, **50**(4): 649-666.

de Bertldi, M. de, G. Vallini and A. Pera, 1983. The biology of composting: a review. *Waste Manag. & Res.*, **1**(1): 157-176.

Edwards, C.A. and K. E. Fletcher, 1988. Interactions between earthworms and microorganisms in organic-matter breakdown. *Agric., Ecosys. & Env.*, **24**(1): 235-247.

Gaur, A.C. 1987. Recycling of organic wastes by improved techniques of composting and other methods. *Resources and Conservation.* **13**: 157-174.

Gaur, A.C., R.S. Mathur and K.V. Sadasivam, 1980. Effect of organic materials and phosphate-dissolving culture on the yield of wheat and greengram. *Indian J. Agron.* **25**: 501-503

Niveta Jain, Arti Bhatia and Himanshu Pathak, 2014. Emission of air pollutants from crop residue burning in India. *Aerosol and Air Qual. Res.*, **14**(1): 422-430.

Schultz, K. 1952. Aerobic decomposition of organic waste materials (continuous thermophilic composting). Final Report Project RG-4180 to ICAR.

Singh, S. and L. Nain, 2014. Microorganisms in the conversion of agricultural wastes to compost. *Proc. Indian Nat. Sci. Acad.*, **80**:(2, Spl. Sec.):473-481.

Sunita Gaind, Lata Nain and V. B. Patel, 2009. Quality evaluation of co-composted wheat straw, poultry droppings and oil seed cakes. *Biodegradation*, **20**(3): 307-317.

Sunita Gaind and Lata Nain, 2010. Exploration of composted cereal waste and poultry manure for soil restoration. *Bioresource Techn.*, **101**(9): 2996-3003.

Wiley, J.S. 1957. Report on high rate composting studies. *Proc.12th Indus. Wastes Conf.* (Purdue Univ. *Engg. Bull.*), **17**: 596-603

Wiley, J. and G.A. Pierce, 1955. Preliminary study of high rate composting. *Proc. Am. Soc. Civil Eng.*, (*J. Sanit. Engr. Div.*), **81**(paper No. 846), 27 p.

COLOUR PLATES

Plate 1: (a) Calotropis plant, (b) Mature Stem, (c) Young stem with foliage, (d) Buds, (e) Flowers, (f) Fruit (Page 16)

***Barleria cristata* (Page 24)**

***Fusarium* and Phytophthora wilt in crossandra (Page 42)**

C. guineensis

C. nilotica

C. flava

C. pungens

C.p. cv. 'Orange'

'Delhi'

'Lutea Yellow'

'Sebaculis Red'

C. undulaefolia

'Arka Ambara'

'Arka Kanaka'

(Page 37)

Single-flowered

Anemone

Star shaped

Decorative

Collarette

Peony

(Page 46)

Semi-double

Pompon

Cactus

Plate-1: Close up view of flower heads of different types of dahlias (Page 46)

Eranthemum nervosum

E. nigrum

E. pulchellum

E. roseum

Pseuderanthemum

(Page 102)

Plate-1
Cultivars of tabernaemontana

a. Tree
b. Shrub
c. Variegated shrub
d. Dwarf variegated
e. Shrub with multipetalled flower

Plate-2
Close up view of flowers of different cultivars of tabernaemontana

a. Tree
b. Shrub
c. Variegated shrub
d. Dwarf variegated
e. Shrub with multipetalled flower

(Page 106)

(Page 112)

(Page 124)

(Page 124)

(Page 124)

Jasminum sambac

Jasminum sambac

Ecotypes of *J. sambac*

Ecotypes of *J. sambac*

J. auriculatum

J. auriculatum var. *'Parimullai'*

(Page 184)

J. auriculatum var. 'Parimullai'

Co. 1 Mullai and CO. 2 Mullai

Co. 1 Mullai and CO. 2 Mullai

J. grandiflorum

J. beesianum

J. beesianum

(Page 184)

J. floridum

J. floridum

J. calophyllum

J. dicotomum

J. fluminense

J. humile

(Page 189)

(Page 210)

Plate-1: Different cultivars of *Nerium* in blooming stage and with their close up views of flowers (Page 210)

(Page 254)

(Page 256)

(Page 258)

Plate-1: Fig. 1a, 2a and 3a Thevetia Trees of Yellow, White and Orange cultivars; Fig 1b, 2b and 3b close up view of flowers of Yellow, White and Orange cultivars; fig. 1c Fruit of Thevetia Yellow cultivar (Page: 318)

Fig. 1: *Gerbera* **on MS medium supplemented with, (a) 1mg/l BAP and 2mg/l 2,4-D, (b) 1 mg/l BAP and 2 mg/l NAA, (c) 2 mg/l zeatin and 0.5 mg/l IBA as well as (d) 2 mg/l BAP and 0.5 mg/l NAA, and both showed regeneration of shoots from petiole explants, (e) 0.1mg/l BAP and 2 mg/l NAA as well as (f) 2 mg/l NAA, both showed development of roots from leaf explants.**

Fig. 2: ***Gerbera*** **~ (a) plantlets to be transferred to soil, (b) soil transferred plants, (c) soil transferred plantlets after eight weeks, and (d) soil acclimatized plants after six months.**

(Page 321)

Fig. 3: *In vitro* **regeneration of transgenic plants via somatic embryogenesis in** *Pelargonium peltatum* **and** *via* **organogenesis in** *P. zonale***. (a)** *P. peltatum* **WT plants. (b)** *P. peltatum* **inflorescence. (c)** *P. zonale* **WT plants. (d)** *P. zonale* **inflorescence. (e-f). Production of somatic embryos in a callus of** *P. peltatum* **cultivated in selective morphogenesis induction medium (MIM). (g).** *P. peltatum* **developing embryos in selective elongation medium (EM). (h).** *P. peltatum* **transgenic plantlets in rooting medium (RM). (i-j). Adventitious buds in a callus of** *P. zonale* **in selective MIM. (k).** *P. zonale* **shoot elongation in selective EM. (l).** *P. zonale* **rooted plantlet in RM. (m-n). Detection of transformation events in both** *Pelargonium* **spp., GFP green fluorescence is clearly visible in the initial whitish callus with a disorganized growth. (o-p). Chlorophyll shows strong red autofluorescence that could mask the green fluorescence of transformed cells, it becomes increasingly difficult to identify in the subsequent organogenic callus and in the adventitious buds. (q-r). Green fluorescence was observed in regenerated shoots but is masked by the chlorophyll in the young leaves. (s-t). Green fluorescence can be observed in the periphery of young leaves where chlorophyll does not accumulate. (u-v). Green fluorescence is especially evident in the roots, where the chlorophyll is absent. In general, within the same organ, GFP detection varied in different tissues or cell types depending on their chlorophyll content. (Page: 323)**

Figure 4: (a) http://users.ugent.be/~pdebergh/tot/tot2_d01.htm, (b) http://www.gmo-safety.eu/database/898.sterile-poplars-prevent-outcrossing-forest-ecosystems.html and (c) http://peakplants.blogspot.in/2012/06/in-vitro-flowering.html (Page 324)

Fig. 6: (A) A gall caused by *Agrobacterium tumefaciens*, and (B) Wild type (left) and genetically engineered (right) walnut microshoots. (Page 327)

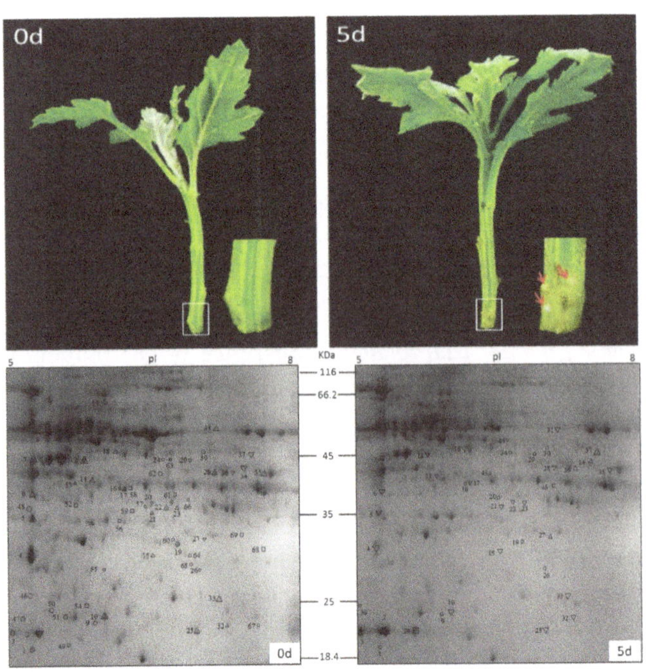

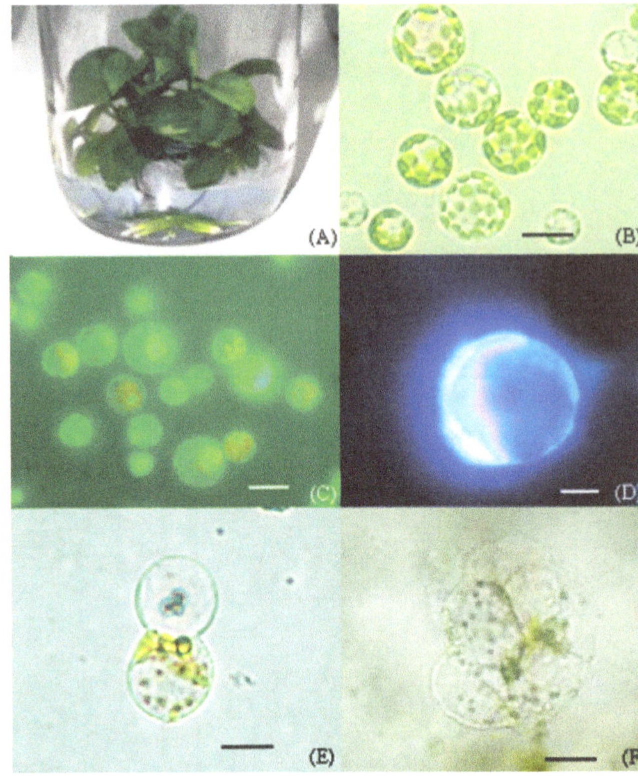

Fig. 5: Isolation, culture and cell division of aquatic flowering plant, *Anubias nana* Engler protoplasts. Six-week-old plantlets suitable for the isolation of leaf plantlets (A), protoplasts after purification with 18% sucrose solution (B), vigorous protoplasts fluoresce a yellow-green colour when stained with FDA (C), protoplast culture in agarose bead, cell wall regenerated within 24 h, detected by calcofluor white staining (D), first cell division after 7-9 days of culture (E), small cell colonies after culturing for 30 days (F). (Page: 325)

Fig. 8 (a): Morphological and histological properties of adventitious root primodium formation in the stem base of chrysanthemum 'Jinba' cuttings. Anatomy of adventitious root primodium formation morphology of 0 d and 5 d cuttings. The arrow indicated the adventitious root, zoomed image shows the cutting position. 8 (b): Comparative proteomic 2-D maps of soluble proteins extracted from 0 d and 5 d chrysanthemum cutting bases. Each map depicts one representative gel (of three replicates). Proteins were separated in the first-dimension by their pIs on IPG strips pH 5-8 and in the second-dimension by their molecular masses on 12.5% SDS-polyacrylamide gels. Proteins were stained with colloidal Coomassie Blue R-350. A total of 69 protein spots showing differences between the 0 d and 5 d maps are numbered. Squares indicate spots that are only present at 0 d or 5 d. Triangles indicate spots that differed between 0 d and 5 d (up-regulated, up-pointing triangle; down-regulated, down-pointing triangle). (Page: 332)

(Page 373)

(Page 373)

(Page 373)

(Page 384)

(Page 387)

Fig. 6a-d: *In vitro* **established root cultures of** *B. orellana* **on MS broth containing 0.1 mg l^{-1} IBA. b) Microscopical observation of pigment producing areas in** *in vitro* **root cultures at 940 magnification. c) Microscopical observation of pigment-producing areas in** *in vitro* **root cultures at 920 magnification. d) Harvested and dried in vitro root cultures. (Page 389)**

(A) Orange-Red ('Dawn Angel')

(B) Red ('Marlene')

(C) Red ('San Francisco')

(D) Red-Purple ('Purple Taffy')

(Page 389)

Mg deficiency in poinsettia and rose

Symptoms of N and K deficient rose plants

Fe deficiency symtoms in rose and petunia

Gerbera plants deficient in Ca

Molybdenum deficient poinsettia

Calyx splitting in carfnation due to B deficiency

(Page 425)

Seed production in *Catharanthus roseus*

Seed production in *Tropaeolum majus*

Seed production in *Mesembryanthemum*

Seed production in snapdragon

Seed production in *Calendula*

Seed production in ornamental cabbage

(Page 430)

Seed production of Californian poppy (single)

Seed production of Californian poppy (double)

(Page 431)

Fig. 1: Lilium packaging in polythene sleeves

Fig. 2: Carnation packaged in polythene sleeves

Fig. 3: Packaging of cut gerbera in polythene sleeves

Fig. 4: Stem break in gerbera

(Page 548)

Fig. 5: Bent neck in cut rose

(Page 550)

Fig. 6: Cut rose buds wrapped in corrugated paper

(Page 551)

Fig. 7: Grey mould in bunched cut roses

(Page 551)

Glossary

Abaxial. Of the surface or part of a lateral organ facing away from the axis and towards the plant base.

Acaulescent. Stem of a plant being absent or appears to be absent, being very short or subterranean.

Achene. A one-seeded dry and indehiscent small fruit (nut) with a tight thin pericarp as in certain members of Ranunculaceae.

Acropetalous. Leaves or flowers that grow in order from the base of a plant or stem towards the apex.

Actinomorphic. Regular flowers with radial symmetry, developing equally on all sides.

Acute. Having a short sharp point.

Acuminate. Where the tips or the base of leaves and perianth segments taper gradually to a point.

Adaxial. Surface or a part of a lateral organ facing towards the axis and apex.

Adnate. Being attached by its whole length to the face of an organ.

Adpressed (appressed). Hairs (indumentum), leaves, etc. lying flat or close to the stem or leaves to which these are attached.

Adventitious. Originating from an uncommon place as viviparously produced plantlets, the roots from the stem or leaf axil than the radicles, or the buds originating along the stem than at leaf axils.

Alternate. Arrangement of branches on a plant or leaves on the stem alternatively to each other at different heights.

Amplexicaul. Leaf sessile at the base as in case of *Tradescantia.*

Androecium. The stamen or stamens (the male component) of a flower as a whole.

Angular. The laterally projecting angles as in case of longitudinal ridging and angular stems.

Annual. A plant completing its entire life cycle (seed to seed i.e. germination, flowering and death) within one season or in a year.

Anther. The pollen-bearing organ of the stamen.

Appendix. The long and narrowed development of the spadix in Araceae.

Apiculate. Of leaf tips which terminate abruptly in a short, often firm and sometimes sharp point.

Aquatic. A plant growing naturally in water entirely or submerged.

Arborescent. Tree-like.

Arboretum. A man-made garden of planted trees as a museum or collection.

Areole. A small pin-cushion like sunken area on the body of cacti functioning as a point for throwing out spines, producing flowers and working as a generating organ.

Aril. Fleshy layer covering the seed of a plant.

Arillate. Fleshy appendage of the hylum or funiculus enveloping the testa of seed entirely or partially.

Aristate. Bearing britles on awns, often appearing as if bearded.

Articulate. Jointed.

Asymmetrical. Informal (irregular) or lop-sided, usually of leaves which have one half larger than the other.

Attenuate. Gradually long and tapering, as in case of a leaf.

Auriculate. Eared as in case of leaves of some plant species in which leaf has two basal lobes.

Awl-shaped. Appearing an awl, a short and broad needle-like tool usually flattened on one face.

Axillary. Arising from an axil (a junction point of stem and leaf).

Baccate. Berry-like.

Back bulb. A pseudobulb in case of an orchid, produced at the back of a flowering stem and one which has borne a flower in the previous season.

Basipetalous. In case of inflorescence where flowers open from apex to base

Beard. Referring to a tuft or zone of hairs on or within a flower as in bearded irises, the tuft or line of hairs on the falls.

Bending. A method used for increasing the yield of flowers and/or fruits through arching or bending of prospective young branches.

Berry. A fleshy and pulpy many seeded indehiscent fruit with a soft covering.

Bidentate. The apex or margins with two teeth, the teeth may also be toothed.

Biennial. The plants that grow and die (seed to seed) within two seasons, producing leaves only in the first season or year, flowers and seed the next.

Bilabiate. Two-lipped.

Bipinnate. Compound leaves with both the primary and secondary divisions pinnated.

Bipinnatisect. Of leaves, bracts or stipules which are twice or doubly pinnatisect, i.e.with leaf lobes which are again lobed.

Bisexual. Flowers having both the sexes, i.e. stamens and pistils.

Blade. The extended portion of a leaf.

Bloom. A flower or a thin layer of white waxy powder on certain plants on their stems, leaves, flowers and fruits.

Bract. A modified leaf usually associated with an inflorescence, in the axils of which flowers arise. Some bracts are scale-like and insignificant, others are large and coloured as in *Bougainvillea* and *Euphorbia pulcherrima*.

Bracteole. A little bract.

Break bud. A flower bud which eventually appears at the end of the solitary main stem if left to grow in natural manner, but before sending out side growths. Normally such buds shrivel and do not develop into flowers.

Bud. A shoot in embryonic form containing within it a miniature stem, leaves and sometimes flowers. Also it is applied to a flower before it opens.

Bulb. A modified usually subterranean bud (growing point) consisting of a short thickened and solid stem surrounded by one or many tightly overlapping fleshy scales (modified leaves) attached to the basal plate (stem) enclosing the growing point.

Bulbil. Diminutive bulbs being formed in the axils of the leaves or in the inflorescences.

Bulblet. Small bulbs being formed by parent bulb underground directly around it.

Burr. Fruits or seeds with hooked or barbed hairs or bristles which cling to animal fur or feathers and are thus dispersed well away from the parent plant.

Caducous. Abscising or falling very early.

Callus (pl. calli)**.** In orchid a ridge-, bump- or club-shape thickening in the labellum.

Calyx. The external whorl of floral organs composed of sepals, mostly protect the petals but sometimes become petaloid.

Campanulate. A broad tube of corolla extending in a flared limb or lobes forming bell-shaped structure.

Candelabra. Flowers of an inflorescence arranged in tiers up the stem.

Capillary. Hair-like.

Capitate. A flower head composed of small flowers arranged in dense cluster as in Asteraceae, or terminating in a knob or in somewhat spherical tip.

Capitulum. A flower head that looks like a large single flower but consists of numerous sessile or subsessile small flowers clustered together on a disc.

Capsule. A fruit becoming dry after ripening and opening through one or more vertical valves.

Carnivorous (insectivorous)**.** A group of plants which catch insects and other small animals such as *Darlingtonia, Dionaea, Nepenthes, Pinguicula* and *Sarracenia.*

Carpel. The floral organ (female sporophyll or a simple pistil) bearing ovules.

Cataphyll. A reduced or little developed leaf, bract-like and often sheathing emerging shoots.

Catkin. Slender inflorescence of small and unisexual, usually wind-pollinated flowers.

Caudate. The apex of a leaf or a perianth-segment tapering gradually to form a long tail-like appendage.

Caudex. A trunk of a tree that bears leaves only at its apex as in palm and tree ferns, or the swollen stem base of certain perennial plants from where new growths appear.

Cauliflorus. Flowers or inflorescences being produced directly from the trunks and larger branches of the trees.

Cauline. Arising from the stem.

Cephalium. The area of stem in cacti, often at or near the apex from which the flowers emerge year after year from a dense mass of woolly hairs or bristles as in case of *Espostoa* and *Melocactus.*

Channelled (candiculate)**.** A narrow leaf or leaflet with up-turned margins and forming a channel or gutter-shape.

Ciliate. Fringed with hairs.

Cladode. A green stem or branch that has taken over the major part of a plant's photosynthetic function.

Clavate. Thickening to the apex from a tapered base. Club or basketball bat shape.

Cleistogamous. Self-pollination occurring in the closed flower.

Climber. A plant not having strong stem to grow erect and requires support to grow straight in terms of twining stems, tendrils, hooked thorns, clinging roots and the leaves with reflexed stalks.

Column. The central part of an orchid flower which combines ovary, stigma and pollinia.

Compound. A division into two or more subsidiary orders.

Connate. Joined together as in case of leaf bases or sometimes flowers.

Conservatory. A structure composed partly or entirely of glass attached to the house and within which a large number of plants are grown.

Contractile. These are roots contracting the plants deeper into the soil akin to their natural habitat.

Cordate. The heart-shaped leaves and leafy stipules which have rounded base and a deep basal sinus or notch where the petiole is inserted.

Coriaceous. Leathery and tough but smooth in texture.

Corm. An annual solid and swollen part of a stem which is bulb-like underground structure having growing point at the apex and the basal plate below which forms the roots, and all around the body it has rings as scars of the previous season's leaves. Every ring has a prominent node just opposite the buds of its adjoining rings and many invisible buds. Every part of the corm has regenerating capacity. Entire corm is covered with shredded tunic as dry leaf bases.

Cormel (Cormlets). A small corm developing from and around the new developing (daughter) corm which normally has a very tough and intact tunic. It is as senile as the corm and can develop the flowers similar to a corm under optimum conditions.

Cormule. Similar and smaller than corm but normally larger than the cormels, with shredded tunic, developing 1-4 laterally or terminally at the first nodes but on the cost of spike. No further spike development in case it is terminal.

Corolla. The collective name for petals and if there are two whorls the inner one.

Corona. An extra cup-like appendage or crown as a development of the perianth (*Narcissus*), of the staminal circle (*Asclepias*), or located between corolla and the stamens (*Passiflora*).

Corymb. An indeterminate flat-topped or convex inflorescence where outer flowers open first.

Costa-palmate. The leaves being halfway between palmate and pinnate as in case of palms.

Cotyledon. The primary or seed leaf appearing immediately after germination.

Creeper. A plant not having strong trunk for growing erect so spreads horizontally prostrate over the ground or on a wall or any other object, usually throwing out clinging aerial roots or adhesive pads.

Crenate. Margins of leaf blades and other flattened organs with blunt and rounded teeth.

Crest. A ridge formed on the falls as in *Iris*, which is most commonly orange or yellow in colour.

Crisped. Of hairs, leaves or petal margins being wavy or waved.

Cristate. Crested or less commonly 'crest-like' or ruffled.

Crown. A corona, or the branching pattern, the foliage and the shape (spreading, conical, dome, etc.) of a tree or a shrub, or a part of herbaceous plant where the roots (or underground structures) and the aerial parts meet, or a plane (i.e. the point on rhizome or root tubers) having strong terminal bud for propagation.

Crownshaft. The sheathing frond bases wrapped around one another to form a smooth, usually green cylindrical shape with an apparent continuation of the trunk as in case of *Roystonia regia*.

Cultivar. A cultivated variety that has originated and persisted under cultivation.

Cuneate. Wedge-shaped or inversely triangular.

Cuspidate. Apices terminating abruptly in a sharp inflexible point.

Cutting. An unrooted part of stem, leaf or root of a plant taken for producing similar plant.

Cyme. More or less flat-topped and determinate inflorescence being formed of main and secondary stem where the central or terminal flower opens first.

Dealbate. Covered with a white powder.

Deccusate. At right angles to one another.

Deciduous. Of leaves, not persistence or evergreen, and generally fall in the autumn.

Decumbent. Having prostrate stems with tips erect.

Dehiscent. The fruit, pollen or spore which spontaneously open when ripe.

Deliquescent. Appearing to melt or become fluid as in petals of *Commelina*.

Dentate. Margins of leaf blades and other flattened organs having shallow and equilateral triangular teeth.

Denticulate. Having many very small teeth.

Determinate. Inflorescences as in cymes that terminate in a bud where terminal or central flower opens first.

Dichotomous. A method of branching in which each axis bifurcates at the tip.

Dicotyledon. A plant having two primary seed-leaves when germination takes place.

Didynamous. Containing two pairs of stamens, one pair shorter than the other.

Digitate. A compound leaf palmately arranged in which the leaflets arise from the same point at the apex of the petiole

Dioecious. Indicating male and female flowers on separate plants.

Diploid. Plants having two matching sets of chromosomes.

Diplont. An organism whose cells (other than reproductive) have a diploid number of chromosomes in their nuclei.

Diplontic selection In case of mutation in plants when once the exhibited mutation does not appear again in (**extra somatic selection**). the successive generation(s) as mutated (abnormal) tissues are not allowed to develop further due to suppression by the surrounding normal tissues, and which results in the loss of the mutation, it is known as *diplontic selection*.

Disc (Disk). The central part of the flower head (capitulum) of a composite plant made up of many tubular flowers, or a fleshy (or raised) development of the torus within the calyx or within the corolla and stamens or composed of coalesced nectarines or staminodes and surrounding the pistil, or the basal plate (condensed stem) of a bulb or a corm.

Discoid. Leaves with round fleshy blades and thickened margins, or the capitula of Asteraceae formed entirely of disc florets.

Dissected. Divided into numerous segments.

Distichous. Arrangement of leaves on stems or its branches in two opposite ranks.

Dormancy. The state of cessation of growth of a plant during adverse environmental conditions, usually during winter.

Dorsal. The back of an organ or the surface turned away from the axis i.e. abaxial.

Drupe. A fruit consisting of a more or less succulent pericarp which encloses a single many-celled stone.

Ellipsoid. Midway between oblong and ovate with equally narrowed or rounded ends and widest across the middle but 3-dimensional.

Elliptic. Midway between oblong-ovate but with equally narrowed or rounded ends and widest across the middle, i.e oval but narrowing towards each of the rounded ends.

Emarginate. Apex shallowly notched with acute indentation, i.e. petals or leaves having a notched tip.

Embryo. It is part of the seed after fertilization but before germination and this develops into a new plant, i.e. the rudimentary plant within the seed.

Endemic. The natural haunts being confined only to a particular region and not elsewhere.

Endosperm. The tissue surrounding the embryo in a plant seed to nourish the embryo while germinating.

Ensiform. Sword-shaped and with an acute point, especially the foliage.

Entire. The leaf margins not toothed.

Epicalyx. A ring of bracts just below the calyx or sepals and resembling them.

Epigynous. The sepals, petals and stamens growing on top of the ovary or appearing so.

Epipetalous. Borne upon the petals of a flower.

Epiphyte. Plants that grow habitually in the wild on the tree trunk or its branches, poles, telephone wires, or rock, *etc.* but are not parasitic..

Equitant. The conduplicate leaves overlapping inside each other in two ranks in a strongly compressed fan as in *Iris* or *Gladiolus* and in other members of Iridaceae.

Evergreen. A plant that remains in leafy condition through all the weathers.

Exotic. An introduced plant not native to that area.

Exserted. Projecting or extending beyond the organs surrounding it as stamens coming out of the corolla.

Eye. The centre of a flower, particularly when of a contrasting colour.

F_1, F_2. Symbols used to indicate first or second generations from a deliberate cross.

Falcate. Curved and tapering to a point like a sickle.

Falls. Outer parts of the flowers which seem to fall downwards or outwards, often bearded, as in *Iris*.

Family. A plant or a group of plants having certain common botanical characteristics, especially in the flowers, and contain a number of genera.

Farina. The mealy or powdery coating on stems, leaves and sometimes on flowers of certain plants.

Farinaceous (Farinose). Having a mealy or granular texture.

Fasciated. The malformation with flattened stems looking as if many stems have joined together as in *Amaranthus* and *Celosia*.

Fascicle. A cluster of roots, stems, leaves, racemes or flowers though arising always independent but appear as if coming out from one common point.

Fastigiate. Of erect habit with all or most stems steeply ascending, creating an overall shape taller than wide.

Filament. The thread-like stalk of a stamen.

Filiform. Thread-like long and very slender leaves, branches, etc. and rounded in cross-section.

Fimbriate. Bordered with a fringe of slender processes usually derived from the lamina rather than attached as hairs.

First crown bud. The first bud which appears at the end of a lateral growth from the main stem of the plant.

Flabellate. A fan like structure.

Flore-pleno. Double flowered, usually with stamens and sometimes with carpels transformed to petals or petaloids.

Floret. A small individual flower, generally one of a group or a cluster.

Flower. A part of the plant containing reproductive organs. It may either be unisexual or bisexual.

Flowerhead (capitulum). A dense cluster of flowers of more or less regular size.

Foliaceous. Leaf-like in appearance or structure.

Form. A variant of a wild species not so distinctive to be called a subspecies or a variety.

Frond. An usually large, much divided or compound leaf which appears consisting of many leaves such as in case of ferns and palms.

Fruit. The seed-bearing organ of a plant.

Frutescent. Shrubby or becoming so.

Fruticose. Though without a single main trunk, but bearing stems or branches similar to shrubs.

Fugacious. Withering soon.

Fusiform. Spindle-shaped.

Funnel-form. Flowers shaped like the inverted cone or funnel where petals may either be fused or overlapping and usually joined into a short tube at the base.

Fusiform. Swollen in the middle and tapering at both the ends i.e. spindle-shaped.

Gamopetalous. Petals united by their margins to form a tubelike corolla.

Gamosepalous. Having sepals united by their margins.

Genotype. The underlying genetic make up of an individual plant or clone.

Genus. The next upper rank above species in botanical classification.

Geophyte. A plant growing with stem or bulb (bulbs, corms, rhizomes and tubers) below ground.

Gesneriad. A plant belonging to African violet (Gesneriaceae) family. The most popular gesneriads are *Columnea, Gloxinia, Saintpaulia* and *Streptocarpus*.

Glabrous. Smooth, without hairs.

Gland. Usually the base of tepals secrete normally a sweet substance to attract insects for pollination.

Glaucous. Coated with a fine bloom mostly grey or blue-green in colour being formed by the waxy coating of the foliage, which may easily be rubbed off.

Globose (Globular). Spherical or near spherical in shape.

Glochid. Tiny barbed spine or bristle growing from areole in cactus.

Grex. A name applied to a group of individual plants, sometimes difficult to separate individually, as they differ slightly or much from each other but with common origin. Usually applied to a batch of hybrid plants of the same parentage, particularly orchids.

Gynoecium. Female organ of a flower, the carpels collectively.

Gynophore. Stalk of a pistil which raises it above the receptacle.

Gynostegium. A covering of any kind of gynoecium, a perianth.

Habitat. The district or geographical area where a plant is found in the wild.

Haft. The narrow basal portion of the 'falls' as in *Iris.*

Halophyte. A plant tolerant or adapted to saline soil.

Half-hardy. A plant requiring a minimum temperature of 10-13 ^{0}C for healthy growth.

Hardy. A plant withstanding prolong exposure to temperature at or below 10 ^{0}C.

Hastate (Sagittate). Arrow-shaped, in triangular fashion, both downward lateral points and one upward.

Heel. A strip of bark and wood remaining at the base of side shoot cutting pulled of a main shoot.

Helophyte. A plant growing seasonally or permanently in mud.

Herbaceous: A plant with no persistent woody stem above ground though it may be evergreen or deciduous.

Hermaphrodite. Flowers having in them both types of sex organs, *viz.* stamens and pistils.

Heterogamous. In Asteraceae where two types of flowers (ray and disc) are borne, or where flowers bear abnormally arranged sexual organs, or the functions being transferred from one sex to another.

Heteromorphic. Where plants assume different forms at different stages of the life cycle, which is opposite to polymorphic where different forms occur at the same stage of plant development.

Heterophyllous. Where a plant bears two or more forms of leaves on the same plant at the same or at different times.

Heterostylous. Species where flowers differ in the presence or number of styles, some being unisexual while some others hermaphrodite.

Hirsute. Covered with long, coarse, spreading hairs.

Hispid. Covered with bristle-like hairs.

Hoary. Leaves and stems being covered with very short, usually branched or star-shaped white hairs.

Homogamous. Either the flowers are hermaphrodite or of the same sex.

Hyaline. The margins of leaves or bracts being transparent or translucent.

Hybrid. The offspring of crosses involving two plants of the same variety or race, or two different varieties, species or even genera.

Hydrophyte. The plants growing in water.

Hypogynous. All or any of the floral organs situated below the ovary.

Hysteranthous. Where leaves develop after the flowers.

Imbricate. Leaves and bracts overlapping in a regular pattern, sometimes even encircling the axis.

Imparipinnate. Pinnate leaf with an odd terminal leaflet.

Imperfect. When flowers become unisexual due to certain parts though present but are undeveloped.

Incised. Having the leaves or petals deeply cut at the margins, generally in an irregular manner.

Incomplete. Where a flower lacks one or more of the four organs.

Incurved. Inwards or upwards curving nature of leaves and petals.

Indehiscent. Fruit or seed pod and spores splitting self for dispersal of seeds or spores.

Indeterminate. An inflorescence not terminating by a single flower but bears many flowers where opening starts from the bottom.

Indumentum. A covering of hairs, scales or scurfs, used in the sense of hairs as in *Anigozanthos.*

Indusium. The membranous covering over the sporangia of ferns which gives each sorus its particular shape.

Inflated. Bladdery, frequently used to describe a type of fruit or seed vessel.

Inflorescence. It refers to the spike of individual flowers on a stem or a branch. This may be a corymb, cyme, panicle, raceme or spike.

Internode. The area of a stem between two nodes.

Introrse. Facing inward or towards the centre.

Involucre. A single, a pair or a ring (whorl or whorls) of small bracts or leaves subtending a flower or inflorescence.

Involute. Rolling inward towards the uppermost side.

Irregular. Asymmetrical or zygomorphic.

Juvenile. Sexually immature phase of a plant's life.

Keel. A prominent ridge running longitudinally down the centre of the under-surface of a leaf, petiole, bract, petal or sepal as a longitudinal ridge on the foliage of *Kniphofia*.

Keiki. An offshoot or offset of certain plants as in orchids.

Labellum. The lowest of three petals in the orchid flower commonly known as lip. It is modified in a variety of ways to aid insect-pollination.

Labiate. The calyx or corolla possessing a lip or lips.

Labium. One of the lip-like divisions of the calyx or corolla.

Laciniate. An irregular but finely cut as if slashed.

Laevigate. Appearing smoothly polished.

Lamina. Leaf blade, the extended part of a leaf.

Lanate. Covered with thick woolly hairs.

Lanceolate. Lance-shaped tapering towards the apex, three to six times as long as broad, broadest point before the middle.

Lateral. Shoots arising on the sides of main or leading stems.

Leggy. Abnormally tall and spindly growth.

Ligule. The strap-shaped petals or leaves.

Linear. Slender and elongated with margins parallel or nearly so.

Lip. A staminode or petal differentiated from the others, or in Labiatae, one of the two distinct corolla divisions, the upper often hooded and the lower often forming a flattened platform for landing of pollinators. In orchids, it is labellum

Lithophyte. A plant growing on rocks or stony soil by deriving nourishments from the atmosphere.

Littoral. Growing on the sea-shore.

Lobe. Divided usually into rounded segments and separated by adjacent lobes with sinuses as in petals and leaves.

Loculicidal. Longitudinal and dorsal splitting through a capsule wall into the capsule.

Long day plant. A plant which requires light for a longer period than it would normally receive from daylight in order to induce flowering, such as *Saintpaulia*.

Lorate. Strap-shaped.

Lyrate. Pinnatifid with a large rounded terminal lobe.

Membranous. Thin, soft, more or less translucent.

Mericarp. A one-seeded carpel where one of a pair splits apart at maturity from a syncarpous or schizocarpous ovary.

Meristem. Tissues yet not differenatiated but capable of developing into organs.

Mesophyte. The vast majority of plants which are midway to hydrophytes and xerophytes.

Midrib. The thicker central vein of a leaf.

Monadelphous. Stamens fused by their filaments into a single bundle.

Moniliform. A cylindrical or terete organ with regular constrictions to appear as string of beads.

Monocarpic. Fruiting or flowering only once after several years of growth and then dying as in case of bromeliads and sempervivums. Technically annuals and biennials are monocarpic, but horticulturally the term is used for plants which live more than two years before flowering as in case of some *Agave* species.

Monocotyledon. Having only one cotyledon or seed leaf at germination.

Monoecious. Male and female flowers being borne on the same plant, with one flower of one sex.

Monopetalous. Only with one petal.

Monopodial. A stem or rhizome growing continuously year after year from the apical or terminal bud, with little or no secondary branching.

Monotypic. It describes a taxon, in particular a genus, family, order, class or division that has only one member at its next lower level.

Mucronate. An apex terminating suddenly with an abrupt spine or spur developed from the midrib.

Mutation. A sudden change in the genetic make-up of a plant, leading to a new feature which may be inherited.

Natural break. Sending out of side growths from the main stem after appearance of the break bud at the end of the main stem which has yet not been removed.

Natural 1st crown. It is, in fact, 1st crown bud but is termed so when the plant has made a natural break - as distinct from a 1st crown bud which appears on a plant that has been stopped.

Nectary. Often a projected or depressed gland which secretes or sometimes absorbs the nectar.

Netted. Net-veined or reticulate.

Node. The part of stem where the leaf is joined and a lateral shoot grows out.

Obconical. Of a leaf, roughly conical but with the stalk at the narrow end.

Obcordate. Like to that of cordate but with the sinus at the apex rather than the base.

Oblanceolate. Similar to lanceolate but with the broadest part above the middle and tapering to the base rather than the apex.

Oblong. Linear and 2-3 times as long as broad with almost parallel sides terminating obtusely at both ends.

Obovate. Ovate but broadest at half past so narrow at the base.

Obovoid. As ovoid but broadest before the half.

Obtuse. Apex or base terminating gradually in a blunt or rounded end.

Odd-pinnate. A pinnate leaf with odd number of leaflets, though also implying that there is a single terminal leaflet.

Offset. A lateral growth at base of the plant used for propagation.

Oleifeous. Those plants which have leaves and seeds producing oil.

Opposite. Of leaves and flowers being borne on different sides of axis with bases on the same level.

Orbicular. Circular or nearly so.

Ovary. Usually the swollen base of a pistil containing one or more ovules, which after fertilization becomes the seeds.

Ovate. Egg-shaped, 1.5-2.0 times as long as broad, broadest before half past and rounded at both ends.

Over-potting. Repotting a plant into a pot which is too large to allow successful establishment.

Ovoid. Egg-shaped ovate but 3-dimentional.

Ovule. One or more minute roundish structures occurring within the chamber of the ovary, and each such structure contains an egg cell which after fertilization develops into a seed.

Palmate. A leaf with several lobes attached to the stalk like the fingers of a hand.

Palmitifid. Palmate with the sinuses reaching about half-way down, i.e. rounded leaves lobed in a palmate fashion to half their length.

Palmatisect. Rounded leaves palmately lobed almost to the base.

Panicle. Inflorescence in the form of a branched or compound raceme.

Papillae (papillose). Small, usually soft nipple-like protuberances on a leaf, stem or fruit.

Pappus. A tuft or whorls of delicate bristles or scales on the place of calyx as in certain Asteraceous plants.

Paripinnate. Pinnate leaf with an even number of leaflets.

Parthenocarpic. Fruiting without fertilization.

Pathenogenesis. Development of seed without fertilization.

Peat. Partially decomposed sphagnum moss or sedge used in making composts. It is valuable for its pronounced air and water-holding capacity and its freedom from weeds and disease organisms.

Pedate. A palmatisect leaf with atleast the two basal lobes again lobed.

Pedicel. The stalk supporting an individual flower or fruit.

Peduncle. The stalk of entire inflorescence.

Peltate. A leaf whose stalk instead of being attached at the edge is attached inside its margin normally at the centre below as in *Nelumbium.*

Pendent. Hanging downwards.

Pentamerous. Organs in a group of five or multiples thereof.

Perennate. The survival from year to year by overwintering.

Perennial. Any plant that lives from several to many years, but in the strict sense applied to non-woody species which produce new stems annually from at or near ground level.

Perfoliate. A leaf whose base clasps the stem as in *Papaver somniferum* or paired leaves which fuse around the stem.

Perianth. The calyx and corolla together.

Perigynous. The flowers in which the perianth and the stamens are basally united and borne on the margins of a cup-shaped rim, itself being borne on the receptacle of a superior ovary, neither above nor below it.

Persistent. An organ neither withering nor falling.

Petal. A part of flower which forms the corolla.

Petaloid. Various organs of a flower adapted to give the appearance of a petal.

Petiole. The stalk of a leaf.

Photosynthesis. The food-making process which occurs usually in the leaf. This process requires water, air and adequate light.

Phyllocad. A flattened stem or branch to function as a leaf.

Phylloclade. Another name for cladode.

Phyllode. An expanded petiole functioning as a leaf blade.

Picotee. Of petals with a narrow margin having a contrasting colour.

Pilose. Covered with long ascending hairs.

Pinching. The act of removing the tip of a growth, either of the main stem or/and of a side growth.

Pinna (pl. pinnae). One of the primary divisions of a pinnate leaf (leaflet).

Pinnate. In compound leaves the arrangement of pinnae (leaflets) in two rows along the rachis. In bipinnate leaves the pinnae are divided into rachillae aand petiolules bearing pinnules (leaflets). Applied also to feather-like arranged veins.

Pinnatifid. Deeply lobed to about half-way down or more with the lobes pinnately arranged.

Pinnatisect. A leaf divided pinnately to the midrib or almost so, but not rounded off into leaflets.

Pinnule. The secondary or more complex division of a fern or palm frond.

Pips. Corms, rhizomes or tubers obtained by ripening-off seedling plants a few months after flowering as in case of anemone and convallaria.

Pistil. Female seed-bearing organ made up of ovary, style and stigma, i.e. gynoecium.

Pitcher plants. Species of carnivorous plants in which the trapping mechanism is of pitcher-like form, with a pool of digestive liquid in the bottom, such as *Cephalotus, Heliamphora, Darlingtonia, Napenthes* and *Sarracenia.* .

Plicate. Pleated or folded lengthwise, mainly of leaves.

Plumose. Feather-like with long fine hairs which further have fine secondary hairs.

Pollen. The male spores which are developed in the pollen sacs or loculi of anthers.

Pollinia. A group of waxy pollen grains found in orchids and in the members of Asclepiadaceae.

Polygamous. A species bearing both unisexual and bisexual flowers on the same or on different plants.

Pot-bound. A plant growing in a pot which is too small to allow proper leaf and stem growth.

Potting on. The repotting of a plant into a proper-sized larger container which will allow continued root development.

Pricking out. The moving of the seedlings from nursery, tray or pot in which they were sown to other receptacles where they can be spaced out individually.

Prickle. A sharp-pointed broad-based outgrowth of a stem, leaf or other organ as in roses, as opposed to a thorn or spine.

Priming. Prior to sowing, it is the preparation of seeds to enhance the partial water inhibitions to help germination and estabilishment as in case of petunia. Polyethelene glycol 600 is quite effective substance used to activate processes leading to germinations, and then seeds are stored dry. Such seeds germinte quite even and more quickly after sowing.

Procumbent. Loosely trailing on ground without rooting.

Prophyll. The bract-like leaves subtending the entire plant or inflorescence in monocots as in *Gladiolus*.

Prostrate. Lying flat on the ground.

Protandrous. In a flower where mature anthers release the pollen before stigma receptivity, but when it continues even after the stigma receptivity it is known as incomplete protandry which enables self-pollination.

Prothallus. A tiny leaf-like body which develops from the spore of a fern, on which are borne the sex cells or gametes which upon fusion give rise to a new fern plant.

Pseudobulb. The fleshy bulb-like thickened aerial stem found in many epiphytic orchids.

Pseudostem. The aerial stem composed of only overlapping sheaths and stalks of basal leaves.

Pubescent. Hairy. When an organ is covered with silky or downy hairs.

Punctate. Dotted with minute, translucent impressions, pits or dark spots.

Pyramidal. Conical in shape but with more angular sides.

Raceme. An indeterminate and unbranched inflorescence composed of pedicellate flowers. Here each flower is borne on an individual stalk along a central stem.

Rachis. The axis of a compound leaf or of an inflorescence.

Radical. The leaves rising directly from the roots.

Ray floret. A small flower with a tubular corolla having expanded and flattened limb in a strap-like ligule (blade), normally occupying the peripheral rings in a radiate capitulum as in Asteraceae.

Receptacle. The elongated and either flat, concave or convex end of the stem from which the perigynous zone (floral parts) originates.

Reniform. Kidney-shaped.

Repent (repens). Growing flat on the ground, usually of stems that root at intervals.

Reticulate. Netted like a fish-net. Network of veins, ribs and colouring. Reticulate tunic is composed of a lattice of fibres in certain bulbs and corms.

Retrorse. A minor organ attached to a larger part such as prickle on a stem, a barb on a leaf or a callus on a lip, and so on, and turned, curved or bent backwards or downwards, away from the apex.

Revolute. Margins rolling towards the dorsal surface.

Rhizome. A specialized slender or swollen, simple or branched subterranean (sometimes close to soil surface, but exposed in epiphytes) and normally creeping stem which produces roots, stems, leaves and inflorescences along its length and at its apex.

Rhomboid. In case of leaves and tepals being angularly oval, the base and apex with acute angles and both the sides having obtuse angles.

Rootstock. The base of a stem from which the roots emerge but in graftage the rootstock refers to the locally adapted plants on whose rooted stems the desired variety is grafted or budded.

Rosette. Leaves clustered together in circular pattern.

Rosulate. Bearing a rosette.

Rugose. Wrinkled, puckered or corrugated appearance of leaves.

Runner. A slender, prostrate stem bearing plantlets at the well-spaced nodes.

Sagittate. Spear- or arrow-shaped leaves where triangular basal lobes point towards the stalk or downward.

Salverform. Where corolla has long and slim tube with an abruptly expanded flattened limb.

Scabrous (scabrid). Rough, usually with short and hard hairs.

Scale leaves. Specialized leaves which are fleshy and scale-like and cover the buds or make the bulbs.

Scandent. Climbing or scrambling stems.

Scape. A leafless erect stalk bearing a terminal inflorescence or flower.

Schizocarp. A dry fruit splitting at maturity into two or more one-seeded carpels which remain closed.

Second crown bud. Removal (stopping) of the tips of the lateral growths (first breaks) from the main stem, forces appearance of further side growths (second breaks) in the leaf axils of these laterals, and these laterals grow longer terminating into a bud at the end, and this is termed as second crown bud.

Secund. Where all the parts of the leaves or flowers are borne along one side of the axis or appear so due to twisting of the stalk.

Segment. One of the parts of a compound leaf or the flowers where sepals and petals are similar.

Sepal. One of the separate parts of the calyx.

Sericeus. Silky-hairy.

Serrate. Leaf blades and other flattened organs having sharp saw-toothed margins curving forward.

Sessile. Without a stalk as in case of leaves amd flowers.

Short day plant. A plant which requires light for a shorter period than it would normally receive from daylight in order to induce flowering, e.g. *Dendranthema* and *Poinsettia*.

Shrub. A woody plant with a framework of branches and little or no central system.

Simple. One piece leaf, neither composed nor divided.

Sinuate. With undulated or wavy margins.

Sinus. The space between two lobes or divisions.

Solitary. Used in case of flowers when they are borne one to a stem of leaf axil.

Sorus. A group of sporangia usually of rounded or oval shape on the back of a fern frond. Each sorus may or may not be covered with an indusium.

Spadix. A spike of flowers on a thick and fleshy axis, often subtended by spathe as in Araceae.

Spathe. A modified, often papery bract or leaf enclosing the inflorescence.

Spathulate (Spatulate). An oblong and spatula-shaped organ which is narrow and tapering at base but rounded at apex.

Spicate. Borne in a spike-like inflorescence.

Spike. An indeterminate and unbranched inflorescence bearing sessile flowers.

Spikelet. A very small, usually compact spike, generally used for the basic unit of a grass inflorescence.

Spine. A hard and sharp pointed structure formed by a modified twig.

Sporangium. An asexually formed spore produced by ferns and certain fungi.

Spore. Minute reproductive bodies formed of one or a few cells together, which give rise to new individuals, either directly as in fungi or indirectly as in ferns.

Sport. A plant which shows a marked and inheritable change from its parent; a mutation.

Stalk. The petiole, peduncle, filament, stipe (stalk of the pistil), etc. which provide stem-like support to their respective organs.

Stamen. Anther and filament constituting the male pollen-bearing part of the flower.

Staminate. A male flower bearing stamens and no functional pistils.

Staminode. A stamen which is sterile and often enlarged and petal-like.

Standard. The perianth segments of inner whorl, often erect as in *Iris*.

Stellate. Star-shaped.

Stigma. The part of the pistil which is receptive when pollen over it is deposited.

Stipe. An alternative name for the leaf stalks of ferns. Also used for the stems of mushrooms and toadstools.

Stipule. A bract-like or leafy appendage, often in pairs, at the base of a petiole, shedding soon.

Stolon. An underground trailing stem giving rise to plantlets at its nodes and apex or cormlets and bulblets at its simple or branched apices.

Stool. It is the root of an old plant with a portion of the old stem and its surrounding young shoots.

Stopping. The act of removing the tip of a growth, either of the main stem and/or the side one(s).

Stove plant. A plant which requires warm greenhouse conditions in winter.

Strain. A selection of a variety, cultivar or species which is raised from seed.

Striate. Fine parallel lines or stripes on the surface.

Style. The area of ovary between stigma and ovary.

Sub-shrub. A plant which at the adult stage has woody stems near the base and a framework of soft green growth.

Succulent. Fleshy, juicy as in cacti and many other succulents.

Sucker. A rooted shoot developing from a underground stem or the plant's root.

Sympodial. In growth of a plant where terminal bud terminates into an inflorescence or dies, and successive growth starts in the ensuing season through secondary axes from lateral buds.

Synanthous. Where leaves appear with the flowers.

Syncarpous. Where two or more carpels fuse in a compound pistil.

Synflorescence. A compound inflorescence which possesses a terminal inflorescence and lateral inflorescences.

Syngenecious. Stamens united into a cylindrical form by the anthers.

Taking a bud (securing a bud). The act of removing all leaf axil shoots and unwanted buds except the bud which is to develop into a flower.

Tender. A plant which requires a minimum temperature of 15.6 ^{0}C. Occasional short exposure to temperatures below this level may be tolerated.

Tendril. A slender twining organ that enables a plant to climb.

Tepal. An unit of undifferentiated perianth which can not be distinguished as a petal or a sepal.

Terete. Slender and cylindrical, circular in cross-section.

Terminal bud. It is a bud which develops surrounded by other flower buds, and this marks end of further vegetative growth.

Ternate. Usually of leaves, in whorls or clusters of three.

Terrarium. A partly or entirely closed glass container used to house a collection of indoor plants.

Terrestrial. Growing on the ground.

Tessellated. Chequered or marked with a grid of small squares.

Tetramerous. With parts in fours or group in fours.

Throat. The opening between the tube and the limb of the tubular part of calyx, corolla or the perianth.

Thyrse. A panicle which is broadest in the middle and tapers to base and apex.

To break. The branching or sending out of side growths.

Tomentose. A downy covering of tiny, woolly and stiff hairs on leaves and other plant parts.

Topiary. Art of clipping and training woody plants to form geometric shapes or intricate patterns.

Toothed. Dentated. Bearing small teeth on the edges.

Torus. A receptacle of a flower, the region in which floral parts are inserted. An elongated torus is a gynophore, the stalk of a pistil which raises it above the pistil.

Transpiration. The loss of water through the pores of the leaf.

Tree. A woody plant with a distinct woody trunk.

Trifoliate. With three leaves or leaflets.

Trimerous. Having parts in threes.

Trumpet. When the tepals (perianth segments) are fused or appearing to be fused into the flared bell of a trumpet as in *Hymenocallis, Narcissus,* etc.

Truncate. Ending abruptly or bluntly.

Tuber. A generally subterranean fleshy and swollen stem with eyes over the surface as in tuberous *Begonia,* or fleshy and swollen roots having buds (eyes) only on the crown, the part which is attached with the aerial stem as in *Dahlia.*

Tubercle. A small wart or knob-like projection on a stem, leaf, fruit, *etc.*

Tuberous. Producing tubers or resembling a tuber.

Tunic. The papery or netted covering of corms derived from widely expanded leaf bases.

Tunicate. Enclosure of a tunic, the papery covering of a bulb or a corm.

Umbel. Umbrella-type inflorescence, normally flat-topped and corymb-like where all the flowered pedicels arise from the same point at the apex of the main axis.

Uncinate. Hooked as in the spines of certain cacti.

Undulate. The wavy margins of leaves, bracts, petals, *etc.*

Unilateral. Arranged only at one side.

Unisexual. Having single-sexed flowers, though both may be on the same plant.

Urceolate. Of flowers, urn-shaped calyx or corolla with an inflated tube narrowed in the mouth but slightly broadening again at the tip.

Vaginate. Enclosed by a sheath.

Variegated. Mottled, streaked, edged or striped with colours of leaves mostly white to yellow other than the normal green.

Velamen. A multilayered corky covering of dead cells on some aerial roots as in some orchids, which enable them to absorb moisture from the atmosphere.

Ventral. The inner (adaxial) of the upper side of a leaf or other surface that faces towards the stem.

Virescens. Appearance of green pigmentation on the tissues not ordinarily green.

Vernal. Appearing in spring.

Verrucose. Covered with small wart-like outgrowths, usually on stems.

Verticil. An alternative name for whorl.

Verticulate. Arranged in a whorl or in a series of whorls.

Vesicle. A small bladder or blister-like organ.

Villous. Covered somewhat shaggily with long soft hairs.

Viscid. Sticky, glutinous.

Viviparous. The buds or bulbs which sprout and become plantlets or the seeds which germinate, while still attached to the parent plant.

Whorl. A circular arrangement of foliage or flowers round an axis, at a node, as in case of leaves of some lilies.

Woolly. For leaves and stems, covered with long soft hairs that are usually white or grey.

Xerophyte. A plant which is able to live under very dry conditions.

Zygomorphic. Capable of being divided into two equal parts in one plane (only along a vertical axis), i.e. bilaterally symmetrical.

Index

A

D

E

F

G

H

R

S

T

U

V

W

Z

9 789359 191027

Printed at Repro India Ltd.